The Art of the Musculoskeletal Physical Exam

John G. Lane • Alberto Gobbi
João Espregueira-Mendes
Camila Cohen Kaleka • Nobuo Adachi
Editors

The Art of the Musculoskeletal Physical Exam

ISAKOS
International Society of Arthroscopy,
Knee Surgery and Orthopaedic Sports Medicine

Springer

Editors
John G. Lane
Musculoskeletal and Joint Research Foundation
San Diego, CA, USA

Alberto Gobbi
O.A.S.I. Bioresearch Foundation Gobbi NPO
Milan, Italy

João Espregueira-Mendes
Clínica Espregueira - FIFA Medical Centre of Excellence
Porto, Portugal

Camila Cohen Kaleka
Cohen Institute
Sao Paulo, São Paulo, Brazil

Nobuo Adachi
Department of Orthopaedic Surgery
Hiroshima University
Hiroshima, Japan

ISBN 978-3-031-24403-2 ISBN 978-3-031-24404-9 (eBook)
https://doi.org/10.1007/978-3-031-24404-9

This Springer imprint is published by the registered company Springer Nature Switzerland AG
The registered company address is: Gewerbestrasse 11, 6330 Cham, Switzerland

Preface

Physical examination techniques have been the basis for medical care for hundreds of years and have been refined over time. Evaluation of the patient and determination of a specific diagnosis is critical in determining the patient's condition and need for treatment. We are fortunate to have developed technologies which facilitate the diagnostic process. One may be tempted to rely exclusively on diagnostic studies such as diagnostic imaging, laboratory testing, and/or invasive monitoring to reach a conclusion. The increased use of telemedicine has bolstered reliance on the results of studies instead of physical examination. This reliance on technology would suggest that physical examination of the patient is superfluous.

As technology is not infallible, the goal of this publication is to educate the reader regarding the importance of a well-performed examination during which the practitioner physically examines the patient and evaluates various characteristics of the musculoskeletal system to reach a diagnosis. Many times, there is a discrepancy between physical examination findings and diagnostic technology results such as from an MRI scan. Having the ability to critically assess the injured body part to determine if the physical findings are consistent with the MRI results is critical to determine if treatment is necessary. Therefore, we have an obligation to our patients to be able to create a differential diagnosis based on the physical examination.

In this book, we are fortunate to have experts from around the world share their experience in evaluating musculoskeletal injuries and conditions. We hope it will improve the reader's ability to physically examine the patient in order to render a diagnosis.

San Diego, CA, USA — John G. Lane
Milan, Italy — Alberto Gobbi
Porto, Portugal — João Espregueira-Mendes
São Paulo, Brazil — Camila Cohen Kaleka
Hiroshima, Japan — Nobuo Adachi

The original version of the book was inadvertently published without including the List of Reviewers in front matter. The book has been updated

Contents

List of Reviewers

Part I - Shoulder
John Lane, MD
Musculoskeletal and Joint Research Foundation
San Diego, CA
United States

Part II – Elbow (note update to Part II)
Luigi A Pederzini, MD
Sassuolo Hospital
Sassuolo
Italy

Part III - Wrist and Hand
Dori Cage, MD
Scripps Mercy Hospital
San Diego, CA
United States

Part IV - Hip/Pelvis
Ehud Rath, Prof.
Tel Aviv Medical Center
School of medicine Tel Aviv University
Israel

Per Hölmich, DMSc Prof.
Sports Orthopedic Research Center - Copenhagen (SORC-C)
Department of Orthopedic Surgery
Copenhagen University Hospital
Amager-Hvidovre
Denmark

Part V - Knee
Alberto Gobbi, MD, FRACS
O.A.S.I. Bioresearch Foundation Gobbi NPO
Milan
Italy

João Espregueira-Mendes, MD, PhD
Clínica Espregueira
FIFA Medical Centre of Excellence
Porto
Portugal

Part VI - Ankle

Gian Luigi Canata, MD
Centre of Sports Traumatology
Koelliker Hospital
Torino
Italy

Part VII - Foot and Toes

Gian Luigi Canata, MD
Centre of Sports Traumatology
Koelliker Hospital
Torino
Italy

Part I

Shoulder Reviewer John Lane

1 Shoulder Anatomy

Kevin Taniguchi, John G. Lane, and Anshuman Singh

1.1 Shoulder Anatomy

1.1.1 Osseous

The primary articulation of the shoulder is the glenohumeral joint which is "ball-and-socket" shape with the concave glenoid fossa of the scapula articulating with the slightly ovoid head of the humerus. The glenoid is shallow with a large radius of curvature permitting for a wide arc of motion as it articulates with the humerus [1]. In order to maintain this relationship, however, it requires stabilization by surrounding bones, ligaments, and muscles.

The scapula is a flat, triangular shaped structure that serves as a skeletal strut for the shoulder joint as well as an attachment site for the various soft tissue structures that stabilize the shoulder. The glenoid is located at its lateral aspect and articulates with the humeral head. The acromion is a hook-like structure projecting off the posterolateral border of the scapula. Anteriorly, the acromion articulates with the clavicle, a broad S-shaped bone connecting the scapula to the sternum, together forming the acromioclavicular joint, allowing scapular rotation while holding the shoulder out to length [2]. The clavicle articulates with the sternum, or breastbone which is the only connection between the shoulder and the axial skeleton. The coracoid is a bony projection off the anterior-lateral aspect of the scapula and serves as an attachment site of several ligaments: the coracoclavicular ligaments, coracoacromial ligaments, and coracohumeral ligaments. These, along with the acromioclavicular ligaments, form the superior shoulder suspensory complex, a ring of bone and soft tissue that is an important biomechanical structure stabilizing the shoulder joint [3] (Fig. 1.1).

1.1.2 Muscles

The deltoid muscle forms the superior-lateral contour of the shoulder. There are three sets of fibers that form the heads of the deltoid; anterior, intermediate, and posterior. These originate on the anterior aspect of the clavicle, acromion, and scapular spine, respectively [4]. The orientation of these fibers allow for the various functions of the deltoid. The anterior fibers assist in forward flexion of the arm and medial rotation. The intermediate fibers allow for abduction of the arm away from the body in the frontal plane, and the posterior fibers assist in extending the humerus [5]. There are three muscles that originate on the cora-

K. Taniguchi
US Navy-Balboa Medical Center,
San Diego, CA, USA

J. G. Lane · A. Singh (✉)
University of California at San Diego,
San Diego, CA, USA
e-mail: jglane@san.rr.com

J. G. Lane et al. (eds.), *The Art of the Musculoskeletal Physical Exam*,
https://doi.org/10.1007/978-3-031-24404-9_1

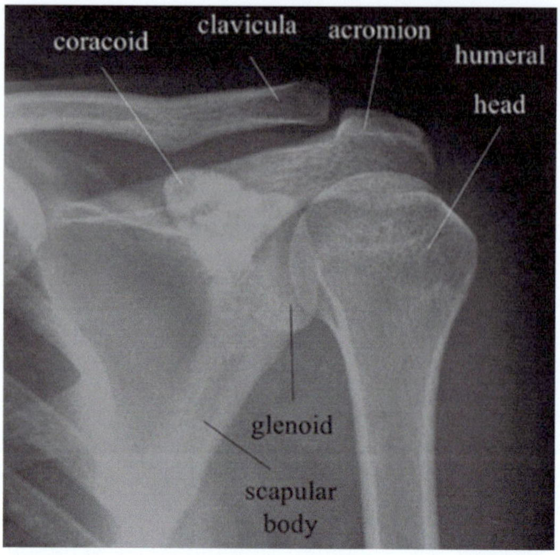

Fig. 1.1 The shoulder has four articulations: the scapulothoracic, sternoclavicular, glenohumeral, and acromioclavicular. The latter two are visible on this anteroposterior radiograph. Bozkurt, M., & Acar, H. I. (Eds.). (2017). *Clinical anatomy of the shoulder: An atlas* (1st ed.). Springer International Publishing

coid process of the scapula; the pectoralis minor, coracobrachialis, and short head of the biceps. The pectoralis minor helps to depress and internally rotate the scapula while elevating the ribs aiding with inspiration during breathing; the coracobrachialis and short head of biceps act to flex the arm [6]. The long head of the biceps is a secondary flexor of the arm, its primary function is supination of the forearm. It originates at the supraglenoid tubercle on the scapula and its tendon runs intra-articularly within the glenohumeral joint as it moves distally towards its insertion on the radius of the forearm. The triceps brachii runs opposite the flexors on the posterior aspect of the arm and is an antagonist to the biceps, coracobrachialis, and brachialis muscles, acting to extend the arm at the elbow. It is made up of three heads; the long, medial, and lateral. The long head originates at the infraglenoid tubercle, and since it spans the shoulder joint, contributes to extension and adduction of the shoulder. The medial and lateral heads originate on the posterior humeral shaft, distal to the long head. The three heads converge to a single tendon attaching at the posterior aspect of the olecranon process of the elbow [7].

The rotator cuff is a key structure in regard to stability and function of the shoulder joint. It is comprised of four muscles: the supraspinatus, infraspinatus, subscapularis, and teres minor. These muscles function to provide rotatory movement at the shoulder joint and maintain the humeral head centered within the glenoid. The supraspinatus originates on the posterior aspect of the scapula and acts to abduct the shoulder. It has its highest mechanical advantage during the first 15° of motion, with the deltoid contributing more to abduction at greater degrees of abduction. The infraspinatus is a thick triangular shaped muscle that is located on the posterior aspect of the scapula and separated from the supraspinatus by the bony spine of the scapula. The infraspinatus acts to externally rotate the shoulder when the arm is at the side. The teres minor is the major external rotator with the arm abducted. The supraspinatus, infraspinatus, and teres minor all attach on the greater tuberosity, a bony prominence on the posterior lateral aspect of the proximal humerus. The subscapularis runs along the anterior aspect of the shoulder acting as secondary restraint to anterior translation of the humeral head in addition to performing shoulder adduction and internal rotation [8]. It attaches to the lesser tuberosity of the humerus, located anteromedial to the greater tuberosity (Fig. 1.2).

1.1.3 Ligaments

While the rotator cuff musculature provides dynamic stability with movement of the shoulder, the ligamentous structures act as static stabilizers. These glenohumeral ligaments are capsular thickenings of the shoulder joint that are check reins to excessive rotational or translational movement of the humeral head within the glenoid. They are described as discrete bands, the superior glenohumeral ligament (SGHL), middle glenohumeral ligament (MGHL), and inferior glenohumeral ligament (IGHL) complexes. The IGHL has both an anterior band (aIGHL) and posterior band (pIGHL).

Unlike most ligaments in the body which impart force through the entire arc of motion, the

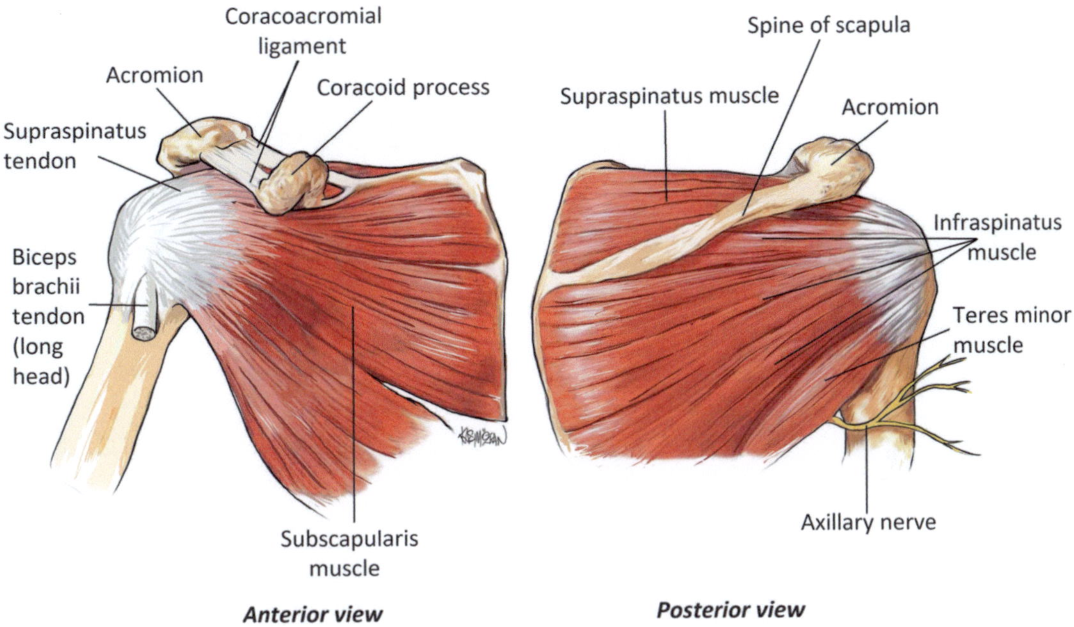

Fig. 1.2 The four rotator cuff muscles are critical to shoulder stability and motion. Yılmaz S., Vayısoğlu T., Çolak M.A. (2020) Shoulder Anatomy. In: Huri G., Familiari F., Moon Y.L., Doral M.N., Marcheggiani Muccioli G.M. (eds) Shoulder Arthroplasty. Springer, Cham. https://doi.org/10.1007/978-3-030-19285-3_1

glenohumeral ligaments act variably depending on the specific position of the arm. The SGHL provides restraint to inferior translation when the arm is at the side, the MGHL resists anterior and posterior translation at the midrange of abduction. The IGHL is the most important contributor to stability overall as it acts during the most common position of dislocation, when the shoulder is abducted 45–90°. The aIGHL is important when the arm is externally rotated, and the pIGHL in internal rotation [9, 10]. The coracohumeral ligament (CHL) supplements the function of the SGHL, running from the base of the coracoid process and attaching to the superior aspect of the shoulder capsule. The CHL, MGHL, and SGHL along with the long head of the biceps tendon travel within the rotator interval, which is bordered by the tendons of the supraspinatus and infraspinatus [11, 12].

The glenoid labrum is a fibrocartilaginous structure that serves as an anchor for the glenohumeral ligaments in addition to deepening the socket of the glenoid to enhance stability of the joint. It also serves as an anchor point for the long head of the biceps tendon at its most superior position. Injuries to the labrum are common and may manifest as shoulder pain, instability, or both depending on their location and severity [13] (Fig. 1.3).

1.1.4 Nerves

The brachial plexus is made up of a series of nerves that convey sensory and motor function to the upper extremity. They are organized as nerve roots branching off the spinal cord at the C5-T1 levels. These nerve roots initially begin at the neck and are subdivided into trunks, divisions, cords, and branches as they move distally down the arm. The dorsal scapular nerve arises proximally from the C5 nerve root to provide motor function to the rhomboid muscles and levator scapulae, which medialize and elevate the scapula, respectively. The suprascapular nerve arises from the upper trunk formed by the C5 and C6 nerve roots and innervates two muscles of the rotator cuff; the supraspinatus and infraspinatus,

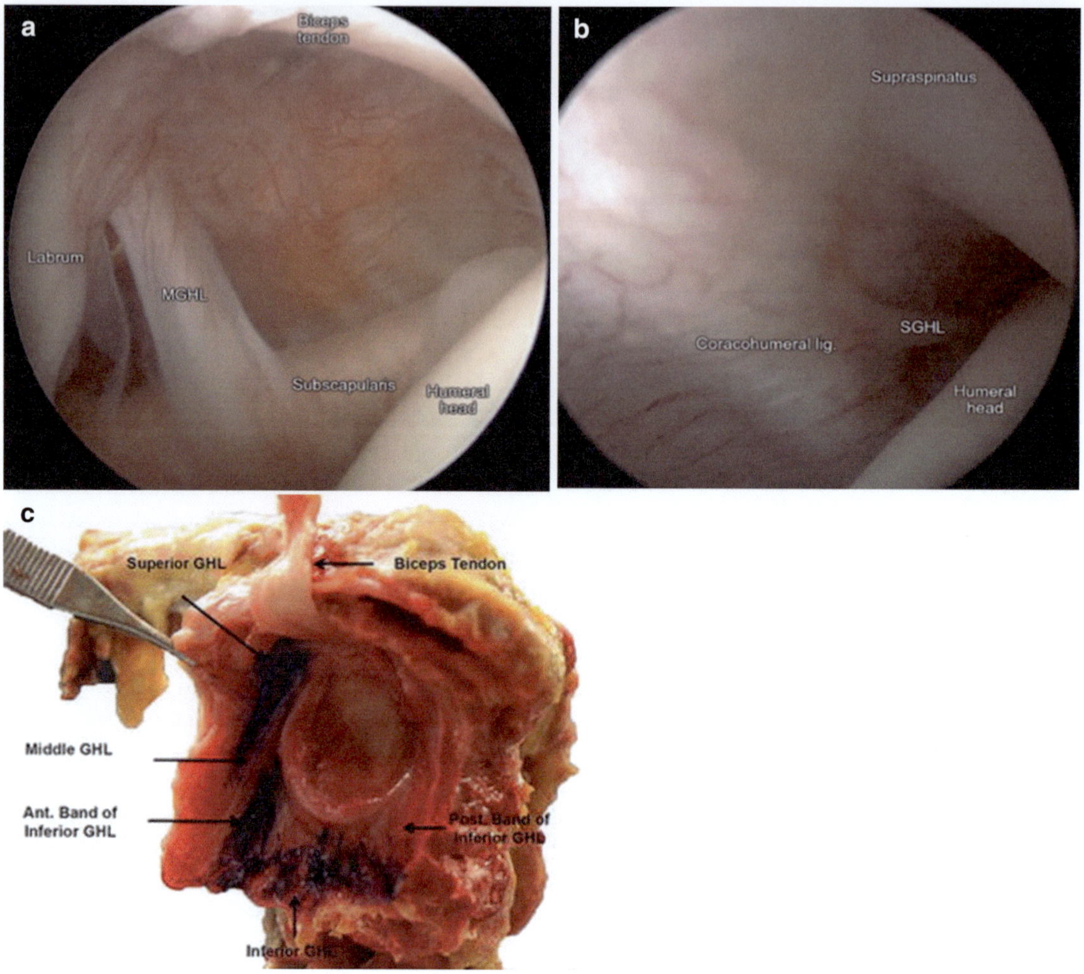

Fig. 1.3 The glenohumeral ligaments are thickenings of the shoulder capsule that serve to as static stabilizers of the shoulder at the end range of motion. (**a** and **b**) Arthroscopic images. (**c**) Anatomic dissection. Apostolakos J. et al. (2015) Glenoid Labrum. In: Bain G., Itoi E., Di Giacomo G., Sugaya H. (eds) Normal and Pathological Anatomy of the Shoulder. Springer, Berlin, Heidelberg. https://doi.org/10.1007/978-3-662-45719-1_9

as well as providing sensory innervation to the glenohumeral joint capsule. Cysts, or abnormal fluid pockets can form at either the suprascapular notch or spinoglenoid notch; fossae about the scapular neck, which can compress the suprascapular nerve causing both supraspinatus and infraspinatus dysfunction if found at the former, or isolated infraspinatus dysfunction if at the latter [14, 15]. The upper and lower subscapular nerves branch off of the posterior cord of the brachial plexus and innervate the subscapularis muscle; the lower subscapular nerve additionally supplies motor function to the teres major muscle. The axillary nerve is a large terminal branch of the posterior cord that innervates the deltoid and the teres minor. Its course has been well described as it wraps from posterior to anterior approximately 5 cm distal to the lateral edge of the acromion. It travels through the quadrangular space along with the posterior humeral circumflex artery, this anatomic space is bordered by the humerus laterally, the long head of the triceps medially, teres minor superiorly, and the teres major inferiorly [16]. It gives off a posterior branch to innervate the teres minor and shoulder joint capsule and an anterior branch to innervate the deltoid muscle. The poste-

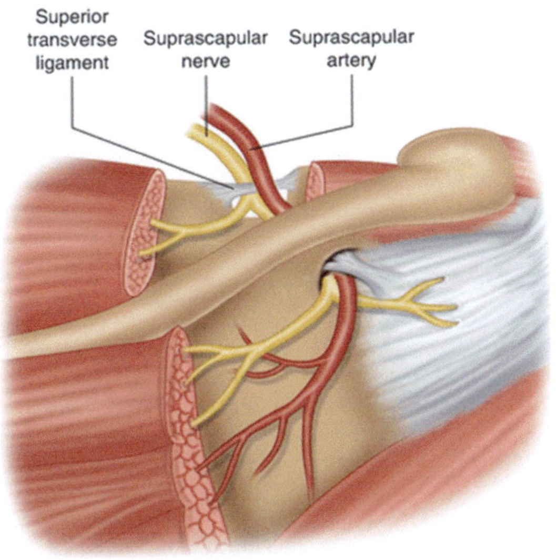

Fig. 1.4 The suprascapular nerve innervates the supraspinatus, infraspinatus and sends branches to the posterior glenohumeral capsule. Martinez, M., Doulatram, G.R. (2018). Suprascapular Nerve Blocks and Neurolysis. In: Manchikanti, L., Kaye, A., Falco, F., Hirsch, J. (eds) Essentials of Interventional Techniques in Managing Chronic Pain. Springer, Cham. https://doi.org/10.1007/978-3-319-60361-2_28

rior branch penetrates the fascia of the deltoid muscle before terminating as the upper lateral cutaneous nerve of the arm which provides sensory innervation to the skin overlying the upper arm [17]. The main terminal branch of the lateral cord is the musculocutaneous nerve which runs in the upper arm, gives off motor innervation to the coracobrachialis prior to piercing its deep surface approximately 6 cm distal to the coracoid process. The musculocutaneous nerve then travels in the anterior arm between the biceps brachii and brachialis, innervating both these muscles. After it exits the interval between these two muscles, it terminates as the lateral antebrachial cutaneous nerve which provides sensation to the lateral forearm [18] (Fig. 1.4).

1.1.5 Vascular

The subclavian artery and its branches provide the blood supply to the shoulder joint. There are several relevant named arterial branches of the axillary artery that are clinically important. The suprascapular artery branches off the proximal aspect of the subclavian artery, it then travels over the superior border of the scapula, in most cases over top of the transverse scapular ligament with the suprascapular nerve running underneath the ligament. The suprascapular artery then supplies the supraspinatus and infraspinatus muscles [19]. After passing the first rib, the subclavian artery changes names to the axillary artery. The axillary artery gives off the anterior and posterior circumflex arteries which supply the deltoid, biceps, coracobrachialis, teres minor, teres major, and triceps muscles. They are however, most important in providing blood supply to the humeral head. The posterior humeral circumflex artery travels with the axillary nerve through the quadrangular space prior to winding around the neck of the humerus to anastomose with the anterior humeral circumflex artery. Based on most recent anatomic studies, the posterior circumflex artery provides the majority of the blood supply to the humeral head [20].

References

1. Bakhsh W, Nicandri G. Anatomy and physical examination of the shoulder. Sports Med Arthrosc Rev. 2018;26(3):e10–22.
2. Willimon SC, Gaskill TR, Millett PJ. Acromioclavicular joint injuries: anatomy, diagnosis, and treatment. Phys Sportsmed. 2011;39(1):116–22.
3. DeFranco MJ, Patterson BM. The floating shoulder. J Am Acad Orthop Surg. 2006;14(8):499–509.
4. Brown JM, Wickham JB, McAndrew DJ, Huang XF. Muscles within muscles: coordination of 19 muscle segments within three shoulder muscles during isometric motor tasks. J Electromyogr Kinesiol. 2007;17(1):57–73.
5. Gagey O, Hue E. Mechanics of the deltoid muscle. Clin Orthop Relat Res. 2000;375:250–7.
6. Rios CG, Arciero RA, Mazzocca AD. Anatomy of the clavicle and coracoid process for reconstruction of the coracoclavicular ligaments. Am J Sports Med. 2007;35(5):811–7.
7. Casadei K, Kiel J, Freidl M. Triceps tendon injuries. Curr Sports Med Rep. 2020;19(9):367–72.
8. Clark JM, Harryman DT 2nd. Tendons, ligaments, and capsule of the rotator cuff. Gross and microscopic anatomy. J Bone Joint Surg Am. 1992;74(5):713–25.

9. O'Brien SJ, Neves MC, Arnoczky SP, Rozbruck SR, Dicarlo EF, Warren RF, Wickiewicz TL. The anatomy and histology of the inferior glenohumeral ligament complex of the shoulder. Am J Sports Med. 1990;18(5):449–56.
10. O'Connell PW, Nuber GW, Mileski RA, Lautenschlager E. The contribution of the glenohumeral ligaments to anterior stability of the shoulder joint. Am J Sports Med. 1990;18(6):579–84.
11. Jost B, Koch PP, Gerber CH. Anatomy and functional aspects of the rotator interval. J Shoulder Elb Surg. 2000;9:336–41.
12. Hunt SA, Kwon YW, Zuckerman JD. The rotator interval: anatomy, pathology, and strategies for treatment. J Am Acad Orthop Surg. 2007;15(4):218–27.
13. Burkart AC, Debski RE. Anatomy and function of the glenohumeral ligaments in anterior shoulder instability. Clin Orthopaed Related Res. 2002;400:32–9.
14. Leschinger T, Hackl M, Buess E, et al. The risk of suprascapular and axillary nerve injury in reverse total shoulder arthroplasty: an anatomic study. Injury. 2017;48:2042–9.
15. Westerheide KJ, Dopirak RM, Karzel RP, et al. Suprascapular nerve palsy secondary to spinoglenoid cysts: results of arthroscopic treatment. Arthroscopy. 2006;22:721–7.
16. Rea P. Chapter 3: Neck. In: Rea P, editor. Essential clinically applied anatomy of the peripheral nervous system in the head and neck. Academic Press; 2016. p. 131–83.
17. Gurushantappa PK, Kuppasad S. Anatomy of axillary nerve and its clinical importance: a cadaveric study. J Clin Diagn Res. 2015;9(3):AC13–7. https://doi.org/10.7860/JCDR/2015/12349.5680.
18. Flatow EL, Bigliani LU, April EW. An anatomic study of the musculocutaneous nerve and its relationship to the coracoid process. Clin Orthop Relat Res. 1989;244:166–71.
19. Singh R. Variations in the origin and course of the suprascapular artery: case report and literature review. J Vasc Bras. 2018;17(1):61–5. https://doi.org/10.1590/1677-5449.008117.
20. Hettrich CM, Boraiah S, Dyke JP, Neviaser A, Helfet DL, Lorich DG. Quantitative assessment of the vascularity of the proximal part of the humerus. J Bone Joint Surg Am. 2010;92(4):943–8.

2 Biomechanics of the Glenohumeral, Acromioclavicular, and Sternoclavicular Joints

Nahum Rosenberg

"It is the fate of detailed 'practical' descriptions to wear the desultory look of curves mapped out with points: each is a series of related but disjoined minutiae—the 'static snapshots' which the mind demands before it can proceed to the direction of a complex, uninstinctive act."

Arnold K. Henry [1]

2.1 General Principles of Shoulder Biomechanics

Normal shoulder mechanics reflect the product of compromise between the necessity of maximal range of active movement of the upper limb with the essential need for joint stability. This is achieved by the dynamics of muscles vs. the constraint of bone and the extent of elasticity of tendinous structures. Notably, the "soft" musculotendinous structures, which are responsible for the shoulder's active movement, also contribute to shoulder stability under the central and local nervous reflex control of different muscle groups that act in unison to provide stability of the basically unstable glenohumeral joint [2].

The overall range of shoulder movement reaches almost 30% of imaginary spherical "volume" when a head humerus is considered the center of rotation [3]. This extensive range of movements occurs primarily through the glenohumeral and scapulothoracic joints with a contribution ratio of 2:1 of movement range, respectively [4]. Two additional joints of the shoulder complex (acromioclavicular and sternoclavicular) have a relatively low contribution to shoulder movement range; their main role is to annex the upper limb to the axial skeleton; therefore, they have stiffer characteristics.

Overall, the torque generated by the shoulder is maximal in the isometric mode [2] and is dependent on muscle size and functional integrity. The maximal isometric strength around the shoulder is higher in the dominant upper extremity (Fig. 2.1) and is found to be higher in men than in women. The dominancy difference also exists in the scapular motion, when a higher range of scapular rotation exists in the dominant limb [5]. The gender difference is not clear enough since even when isometric torque buildup is normalized to a lean body mass, women reach lower maximal torque during isometric force

N. Rosenberg (✉)
Specialists Center, National Insurance Institute, Haifa, Israel

Sheltagen Medical Ltd, Rosenberg, Atlit, Israel

J. G. Lane et al. (eds.), *The Art of the Musculoskeletal Physical Exam*,
https://doi.org/10.1007/978-3-031-24404-9_2

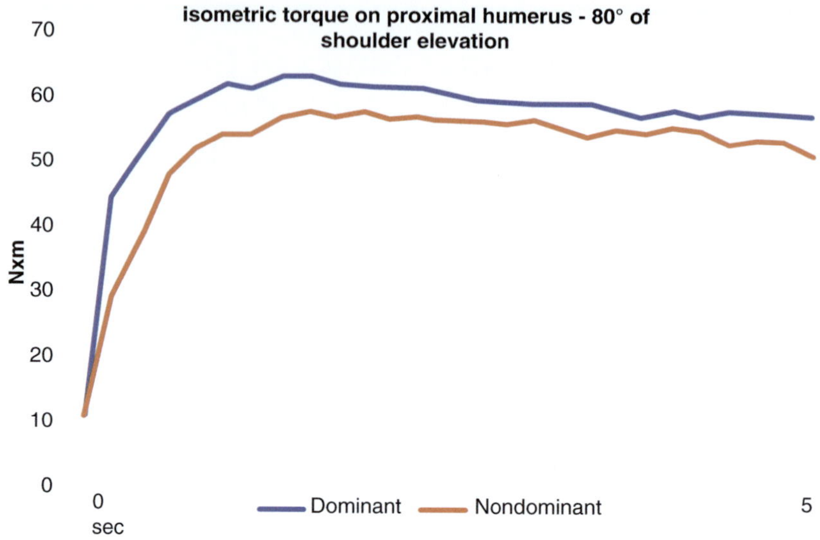

Fig. 2.1 Example of a profile of isometric force buildup in both shoulders of a 46-year-old man with normal shoulders. Higher magnitude torque is generated in the dominant limb. Measurements were made using a dynamometer (1200 readings/s, resolution 0.04 N; Myometer; Atlantech Medical, Nottingham, UK)

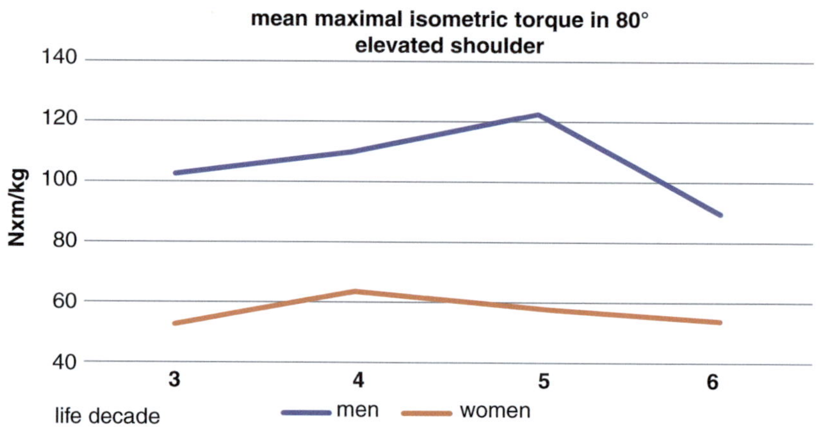

Fig. 2.2 Maximal shoulder torque (normalized to lean body mass) according to age decade. The maximal values in men are in the fifth decade of life and in women in the fourth decade (data from 500 healthy individuals) [4]

generation [6, 7]. The gender difference also exists in the age when maximal shoulder isometric force magnitude is noted in a normal population, i.e., in the fifth age decade in men vs. the fourth age decade in women (Fig. 2.2) [6], indicating that muscle mass is not the only determining factor for the generation of isometric torque around the shoulder axis.

2.2 Glenohumeral Joint

The glenohumeral (GHJ) joint is the main axis of shoulder movement. Its mobility and stability interaction determine most of the shoulder movement. The glenoid concavity depth provides 50% of joint stability, and 20% of stability is contributed by the cartilaginous labrum, which increases the glenoid depth by 50% [8]. An additional 30% GHJ stability is generated by the dynamic effect of the rotator cuff (RC) muscles, the long head of biceps (LHB) that cause compression of the humeral head into the central glenoid [8]. GHJ conformity is also a stabilizing factor accompanied by the joint's capsule components that control humeral head translation during the joint's passive movements [9]. The maximal reaction force on the glenoid can be as high as 90% of body weight at 90° GHJ abduction [4].

The maximal torque on the GHJ is generated mostly by the external group of large muscles, i.e., *latissimus dorsi*, *serratus anterior*, *pectoralis*

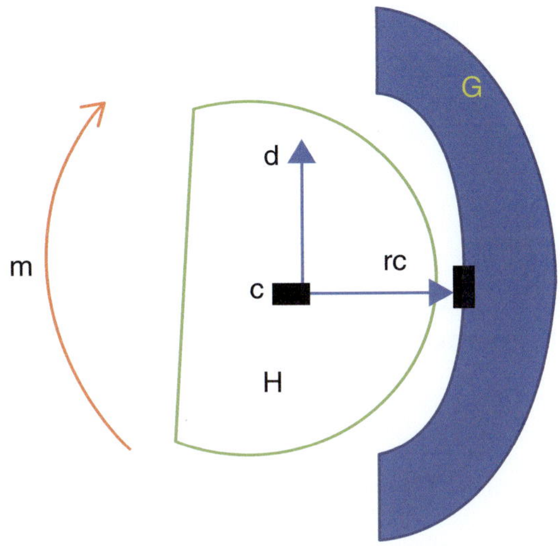

Fig. 2.3 A simplistic representation of force vectors on the humeral head during movement (frontal plane). *H* humeral head; *G* glenoid; *c* center of rotation of the humeral head; *m* rotational movement direction of the humeral head around the c axis; *rc* force vector generated by rotator cuff muscles; *d* force vector generated by external shoulder muscles, mainly by the *deltoid* muscle. *d* represents the translational force on the humeral head that is stabilized dynamically by *rc* into the glenoid fossa; the result is stable rotational movement (m) around the center of rotation c

major, and especially by parts of the *deltoid* muscle [10]. Subsequently, this group of muscles generates dynamic shearing forces on the GHJ. In order to stabilize this shearing torque dynamically, the RC muscles generate maximal force on the humeral head, which is directed to the center of the glenoid fossa, especially in the midrange of shoulder movement (Fig. 2.3) [11]. Since the humeral head is retroverted by 30° in the three-dimensional orientation of the GHJ, the glenoid, which is retroverted on average by 7.4°, provides the main posterior support [4, 10]. The anterior restriction of humeral head translation is based on the static stabilizers that are comprised of the distinct capsular components. The superior and middle glenohumeral ligaments stabilize the GHJ anteriorly and inferiorly when the joint is in adduction. The same effect, but to a lesser extent, is provided by the coracohumeral ligament, and the inferior glenohumeral ligament stabilizes the GHJ when the arm is in abduction and external rotation [4]. These static restrictors prevent a significant magnitude of humeral head translation in the humeral fossa [11], i.e., at shoulder elevation the vertical translation is restricted up to 0.35 mm, while during shoulder flexion the anterior translation is up to 3.8 mm and at extension the posterior translation is up to 4.9 mm [4]. Therefore, the humeral head normally does not translate more than 5 mm in the glenoid fossa. But, since the GHJ is inherently unstable, i.e., the humeral head surface is four times larger than the glenoid fossa surface, dynamic fine-tuning stabilization is essential in addition to static stabilization. This is provided by the RC muscles that finetune the humeral head stabilization by generating a force vector directed to the center of the glenoid fossa (Fig. 2.3).

An additional stabilizing factor of GHJ, which is neither static nor dynamic, is a negative intra-articular pressure that is especially important for inferiorly directed humeral head stabilization at shoulder adduction and can be compromised when the rotator interval is damaged.

Thus, since the anterior shear force on the glenoid is maximal at 60° of shoulder abduction and reaches 40% of body weight [4], the most important anterior shoulder static stabilizer, as mentioned above, is the inferior glenohumeral ligament.

2.3 Acromioclavicular Joint

The acromioclavicular joint (ACJ) is also basically non-congruent. In most individuals (about 50%), it has an "overriding" configuration when the distal clavicle overrides the acromion, and the less common configurations are either parallel articular surfaces in 27% of the population, or the acromion overrides the distal clavicle, i.e., "under-riding" the ACJ [12]. Naturally, all these osseous configurations of ACJ are unstable, and ACJ stability is primarily based on static support of the articular capsule and adjacent ligaments, as well as on some dynamic stabilization by the *deltoid* and *trapezius* muscles anteriorly and posteriorly, respectively. Since these stabilizing structures are not solid, some degree of axial rotation (up to 8° of 45° of maximal clavicle axial rotation) in the ACJ is possible. The acromioclavicular ligament can constrain joint displacement in the range of 50% anteriorly and 90% posteriorly [13]. In the vertical direction, ACJ can withstand up to 635 N [14] when most of the stabilizing resistance is provided by the acromioclavicular ligaments. The coracoclavicular ligament also plays an important stabilizing role when the conoid component acts as an anterior and superior constraint, and the trapezoid component mainly resists axial compression on the ACJ [12]. ACJ is primarily stabilized by the coracoclavicular ligament in the vertical plane and by the acromioclavicular ligaments in the horizontal plane [15]. Due to the high grade of stability, ACJ contributes less to shoulder range movement, and rather behaves as a stable axial fulcrum for combined GH and scapular movement since its semistiff structure is crucial for the mechanically reliable suspension of the upper limb [16].

2.4 Sternoclavicular Joint

The sternoclavicular joint (SCJ) is also unstable in its osseous configuration. This joint is the only articulation of the shoulder to the axial skeleton. Its stability is mainly due to solid restriction caused by the "saddle shaped" articulation of the diarthrodial components (medial clavicle articulation with upper sternum and superior surface of the first rib). The congruity of the SCJ is partly achieved by an intra-articular disc that provides some medial stability of the joint [12]. Additional stabilization is provided by the sternoclavicular ligament. Still, some range of movement exists in the horizontal and frontal planes and around the clavicular axis. The movement does not exceed 50° in all planes [17]. This movement occurs at a shoulder elevation of up to 90° with a 2:5 ratio of SCJ movement to shoulder elevation [18]. The main static posterior stabilization of the SCJ is provided by the posterior joint capsule and is controlled by the dynamic stabilization of the *trapezius* muscle.

The joints at both ends of the clavicle (ACJ and SCJ) "compromise" to some extent their stability primarily to allow the rotational movement of the clavicle along its long axis with high stability in other planes of potential movements, aiming at providing reliable anchorage for effective scapulothoracic rhythm that is controlled and interrelated by glenohumeral and scapulothoracic movements essential for optimal shoulder mobility.

References

1. Henry AK. Extensile exposure. 3rd ed. Edinburgh, London, New York: Churchill Livingstone; 1995. p. 308.
2. Morrey BF, Itoi E, An KN. Biomechanics of the shoulder. In: Rockwood CA, Matsen FA, editors. The shoulder. 2nd ed. Philadelphia, London, Toronto, Montreal, Sydney, Tokyo: W.B. Saunders Company; 1998. p. 233–76.
3. Harryman DT II, Lazarus MD, Rozencwaig R. Chapter 20: The stiff shoulder. In: Rockwood CA, Matsen FA, editors. The shoulder. 2nd ed. Philadelphia, London, Toronto, Montreal, Sydney, Tokyo: W.B. Saunders Company; 1998. p. 1067–8.
4. Halder AM, Itoi E, An KN. Anatomy and biomechanics of the shoulder. Orthop Clin North Am. 2000;31(2):159–76.
5. Matsuki K, Matsuki KO, Mu S, Yamaguch SI, Ochiai N, Sasho T, Sugaya H, Toyone T, Wada Y, Takahashi K, Banks SA. *In vivo* 3-dimensional analysis of scapular kinematics: comparison of dominant and nondominant shoulders. Shoulder Elbow Surg. 2011;20(4):659–65.
6. Schezar A, Berkovitch Y, Haddad M, Soudry M, Rosenberg N. Normal isometric strength of rotator cuff muscles in adults. Bone Joint Res. 2013;2:214–9.
7. Rosenberg N. Rotator cuff isometric strength across the life span in normal population and in patients with rotator cuff pathology. Chapter 2. In: Imhoff A, Savoie F, editors. Rotator cuff across the life span. Berlin: Springer Nature; 2019. p. 11–7.
8. Lippitt S, Matsen F. Mechanisms of glenohumeral joint stability. Clin Orthop Rel Res. 1993 Jun;291:20–8.
9. Karduna AR, Williams GR, Iannotti JP, Williams JL. Kinematics of the glenohumeral joint: influences of muscle forces, ligamentous constraints, and articular geometry. J Orthop Res. 1996;14:986–93.
10. Lugo R, Kung P, Ma CB. Shoulder biomechanics. Eur J Radiol. 2008;68(1):16–24.
11. Bigliani LU, Kelkar R, Flatow EL, Pollock RG, Mow VC, Glenohumeral stability. Biomechanical properties of passive and active stabilizers. Clin Orthop Relat Res. 1996;330:13–30.
12. Renfree KJ, Wright TW. Anatomy and biomechanics of the acromioclavicular and sternoclavicular joints. Clin Sports Med. 2003;22(2):219–37.

13. Stine IA, Vangsness CT. Analysis of the capsule and ligament insertions about the acromioclavicular joint: a cadaveric study. J Arthroscop Relat Surg. 2009;25:968–74.
14. Abat F, Sarasquete J, Natera LG, Calvo A, Pérez-España M, Zurita N, Ferrer J, del Real JC, Paz-Jimenez E, Forriol F. Biomechanical analysis of acromioclavicular joint dislocation repair using coracoclavicular suspension devices in two different configurations. J Orthop Traumatol. 2015;16(3):215–9.
15. Saccomanno MF, Ieso DE, C, Milano G. Acromioclavicular joint instability: anatomy, biomechanics and evaluation. Joints. 2014;2(2):87–92.
16. Abbott LC, Lucas DB. The function of the clavicle. Ann Surg. 1954;146:583–99.
17. Dhawan R, Singh RA, Tins B, Hay SM. Sternoclavicular joint. Shoulder Elbow. 2018;10(4):296–305.
18. Rockwood CA, Williams GR, Yong DC. Disorders of the acromioclavicular joint. In: Rockwood CA, Matsen FA, editors. The Shoulder. 2nd ed. Philadelphia, London, Toronto, Montreal, Sydney, Tokyo: W.B. Saunders Company; 1998. p. 483–553.

3 Evaluation of the Range of Motion of the Glenohumeral Joint

Aaron Martinez-Ulloa, Maria Valencia, and Emilio Calvo

3.1 Anatomy

The shoulder complex is made up of three main joints: acromioclavicular, scapulothoracic, and glenohumeral joint; and four less contributing joints: sternoclavicular, costomanubrial, costosternal, and costovertebral [1].

The glenohumeral joint is formed by the glenoid cavity and the humeral head, stablishing a spheroid joint. By having the greatest range of motion, covering 65% of a sphere, it lacks the stability other joints such as the hip has. Its stability is provided mainly by active muscle control, with a minor role of the glenohumeral capsule, labrum, and ligaments [2]. The following stabilizers have been defined: [3].

- Dynamic stabilizers: biceps tendon, supraspinatus, infraspinatus, teres minor, and subscapularis muscles.
- Static stabilizers: glenoid anatomy, labrum, glenohumeral ligaments, capsule, and articular vacuum.

The musculature of the shoulder can be divided into deep and superficial layers. The deep one, consisting mainly of the rotator cuff which, enhanced by the biceps, stabilizes the humeral head within the glenoid giving an effective fulcrum for the superficial layers to create a stable motion. Superficial layers, being part of it deltoid, trapezius and pectoralis mayor; act as power and controller of the arm motion once the joint has been stabilized [1].

Glenohumeral ligaments are disposed in a multi-directional fashion, providing stability to the joint. Coracohumeral and superior glenohumeral ligament (GHL) act as inferior stabilizers in adduction and external rotation. Middle GHL restrains external rotation and provides anterosuperior stability. Finally, the inferior GHL complex is the most important stabilizer against anteroinferior translation [4].

Each of the restraints and stabilizers of the shoulder joint portraits a particular function. We should be able to examine every single structure individually to identify the issue and localize the possible sources of pain.

When joints are examined passively higher range of motion is observed, whereas in active positioning muscles tend to limit humeral head translation, specially rotational range, keeping the humeral head centered in the glenoid (Karduna et al., 1996).

A. Martinez-Ulloa · M. Valencia · E. Calvo (✉)
Shoulder and Elbow Reconstructive Surgery Unit, Department of Orthopedic Surgery and Traumatology, Fundacion Jimenez Diaz, Universidad Autonoma, Madrid, Spain
e-mail: maria.valencia@quironsalud.es; aaron.martinez@fjd.es; ECalvo@fjd.es

J. G. Lane et al. (eds.), *The Art of the Musculoskeletal Physical Exam*, https://doi.org/10.1007/978-3-031-24404-9_3

3.2 Evaluation of the Shoulder Joint

Shoulder pain makes up for a great number of visits in our daily practice, some articles estimating its annual incidence close to 15/1000 in a GP consultation [5]; but that does not mean we can consider it an easy pathology. The wide spectrum of possible etiologies makes it hard to achieve a proper diagnosis without the use of complementary test. That is one step we can avoid in many cases, especially for soft tissue disorders, by performing a meticulous and thorough physical examination.

We must take into account that pain around the shoulder girdle is poorly localized, where the patient refers its origin is rarely the location of the lesion. The typical pattern for shoulder intrinsic diseases is one that radiates down the arm and stops at the elbow. Whereas ache of cervical origin radiates from the base of the ear towards the scapula. The two branches in which we can sort out shoulder pain are rotator cuff disorders and instability due to osteoarthritis. Some particular occupations and sport activities have a higher risk of suffering shoulder disorders. Other complaints that may arrive at our daily practice are loss of motion, stiffness, neurologic dysfunction, or scapular dyskinesia. Here, a proper clinical evaluation will help us differentiate issues of the shoulder itself or referred symptoms with its origin on the neck. The scapulothoracic and acromioclavicular joints should always be specifically examined.

Examination can be performed in a seating position when the patient has trouble standing up, although it is preferred to evaluate and assess the joint in a standing position. It should always be performed bilaterally from the front and from the back in order to detect any muscular atrophy or scapular dyskinesia.

It is also paramount to examine both active and passive range of motion. When joints are examined passively, higher range of motion is observed, whereas in active positioning muscles tend to limit humeral head translation, especially rotational range, keeping the humeral head centered in the glenoid. By testing joint motion passively, we can stablish the maximum allowance of the soft tissues (ligaments and capsule), somehow ignoring the possible restraint of muscles. On the other hand, active range of movement can be limited not only by soft tissue disorders but by pain which should be addressed. If we have limitation in both active and passive we most likely be looking at shoulder stiffness, the primary reason being adhesive capsulitis or primary osteoarthritis [6].

The angle obtained during evaluation of ROM can be measured using traditional methods, such as employing a goniometer, using angle paintings on the wall and asking the patient to step in front of it. However, more advanced systems that use laser and high-resolution cameras to track the motion performed by the patient have recently been advocated. It has been found that using these new options, we can achieve, higher accuracy than compared to the goniometer or anatomical references such as in the Constant classification [7].

Different scoring systems to evaluate shoulder function are available in the literature, being the Oxford shoulder score and the Constant score the ones being used by most surgeons [8]. Most of them include range of motion as one of the most relevant sections of the test. The Oxford shoulder score is patient assessed, answering 12 questions regarding the ability to perform daily living activities and pain derived from doing them [9]. The Constant score is considered the gold standard in Europe, assessing pain reported by the patient, activities of daily living, ROM, and strength determined by examiner, with a maximum of 100 points. It was designed in 1987 based on four parameters to assess shoulder performance [1]. These parameters are pain (0–15 points), activities of daily living (0–20), range of motion (0–40), and power (0–25) making up to a sum of a hundred points. Several measurements are taken in each, such as forward and lateral elevation, external/internal rotation, or degree of activity level.

Once we have calculated the score we can sort patients as very good function (86–100 points), good (71–85), fair (56–70), and poor (<56). The minimum our patients should be able to do by

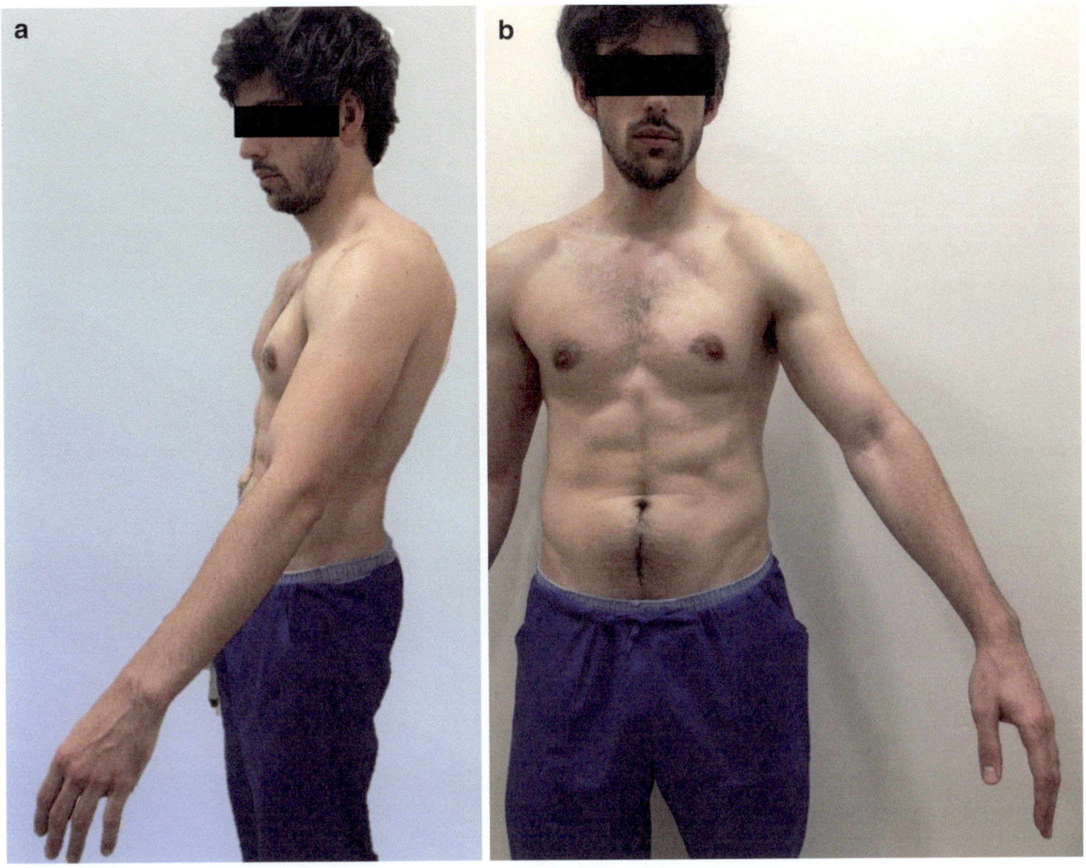

Fig. 3.1 (**a** and **b**) Shoulder resting position

themselves are the daily living tasks such as perineal care task, comb their hair, washing their axilla, and eat with a spoon [10]. If these goals are not met, surgical treatment should be an option.

3.2.1 Resting Position

Posture in which the soft tissue surrounding the glenohumeral joint are in its most relaxed state and the capsule achieve its greatest laxity. The articular surface is minimally congruent. Owing to all these circumstances, the joint allows maximum mobility in passive manipulation.

We should be looking for 40° of abduction or 45% of the maximum abduction, neutral rotation and 30° of anteversion, within the plane of the scapula [11]. It should be noted that this position varies depending on factors such as hyper or hypomobility, and thus the degrees of abduction should be adapted to the maximum range of motion, looking for 45% of the maximal abduction range (Fig. 3.1).

3.2.2 Abduction

Motion of the humerus when elevated in the coronal plane towards the head with a straight arm extended laterally. Normal range goes from 0° to 180° [12]. Since the normal anteversion of the shoulder joint is 30° [3], abduction can be tested maintaining the anatomical forward angulation in order to keep posterior and anterior soft tissues relaxed. Examination looking for an absolutely straight lateral abduction increases the stress on anterior structures, thus giving abnormal results (Fig. 3.2).

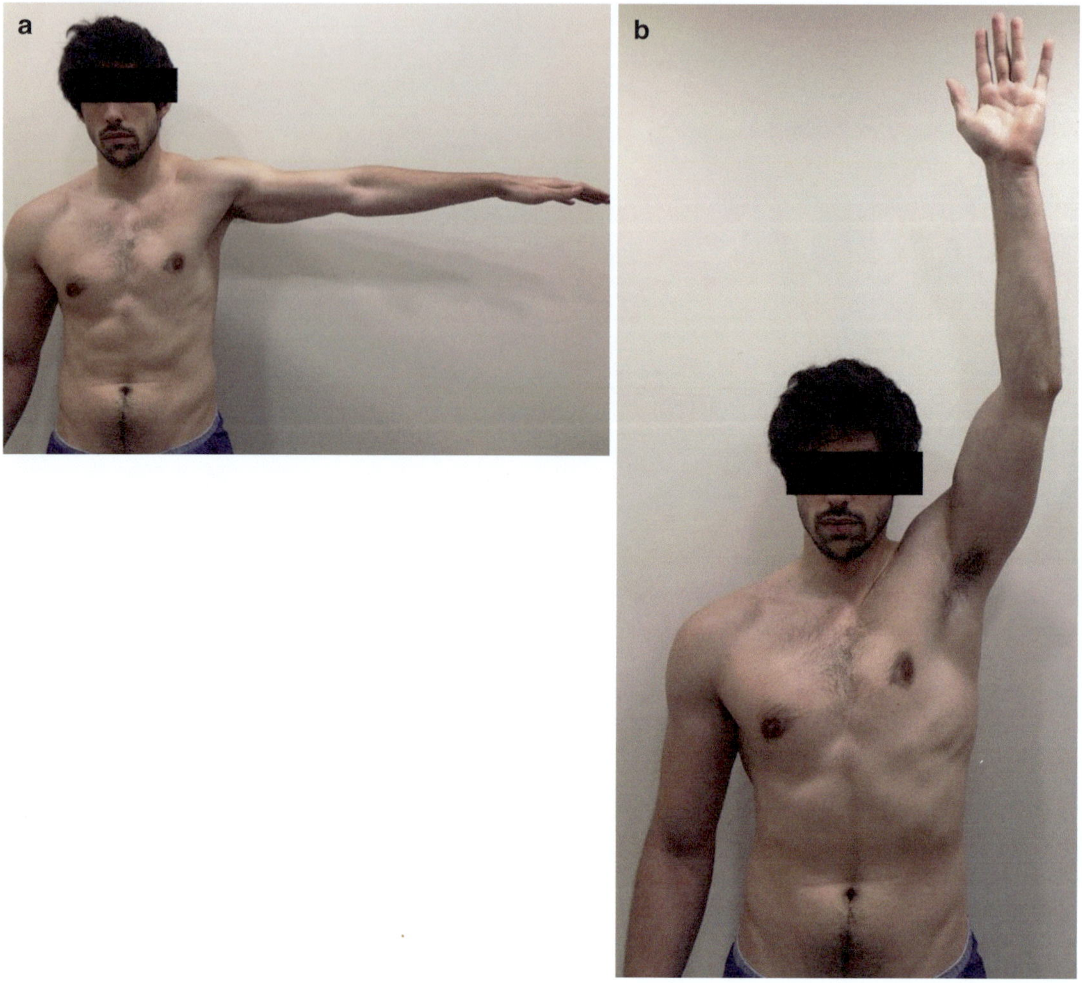

Fig. 3.2 (**a** and **b**) Shoulder abduction to 90 and 180°

The muscles taking part in this action are mainly the supraspinatus, which initiates the first 15° of abduction; and deltoid that takes it from 15 to 90°. Coracobrachialis, biceps brachialis, teres major, pectoralis major, serratus anterior, and trapezius act as secondary abductors.

Abduction can be limited or painful in disorders such as subacromial impingement (pain between 60 and 120°), supraspinatus tears, deltoid disorders, or osteoarthritis of the glenohumeral joint.

3.2.3 Adduction

We can explore resting or horizontal adduction since this position is obtained when the arm is in direct contact to the body (resting adduction) or flexed to 90° in the sagittal plane and rotated internally towards the contralateral shoulder (horizontal adduction). Shoulder strength is maximal performing this action (Fig. 3.3) [2].

Adduction can be measured taking the linear distance (cm) between the lateral humeral epicondyle and contralateral acromion process. The mean standard deviation for this measure is 25 cm [13].

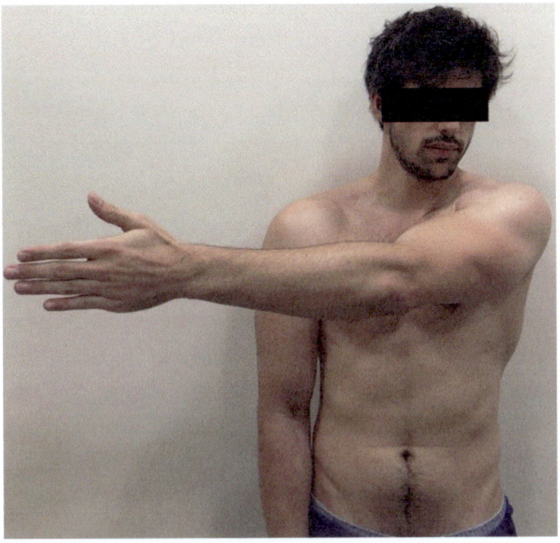

Fig. 3.3 Resting adduction

Powering this action, we have pectoralis major, latissimus dorsi, teres major, teres minor, subscapularis, and coracobrachialis. Should disorders in any of them appear, adduction can be affected.

3.2.4 Forward Flexion

Elevation of the humerus to the front side of the body in the sagittal plane. Normal range from 0° to 180° [14]. It is not specified whether it should be done with the elbow extended or not, but it is normally done in maximum elbow extension.

Muscles responsible for this motion are deltoid (anterior fibers), pectoralis major, and coracobrachialis. Flexion can be affected in deltoid disorders (e.g., axillary nerve palsy), OA or in pathology affecting the pectoralis major or coracobrachialis (Fig. 3.4).

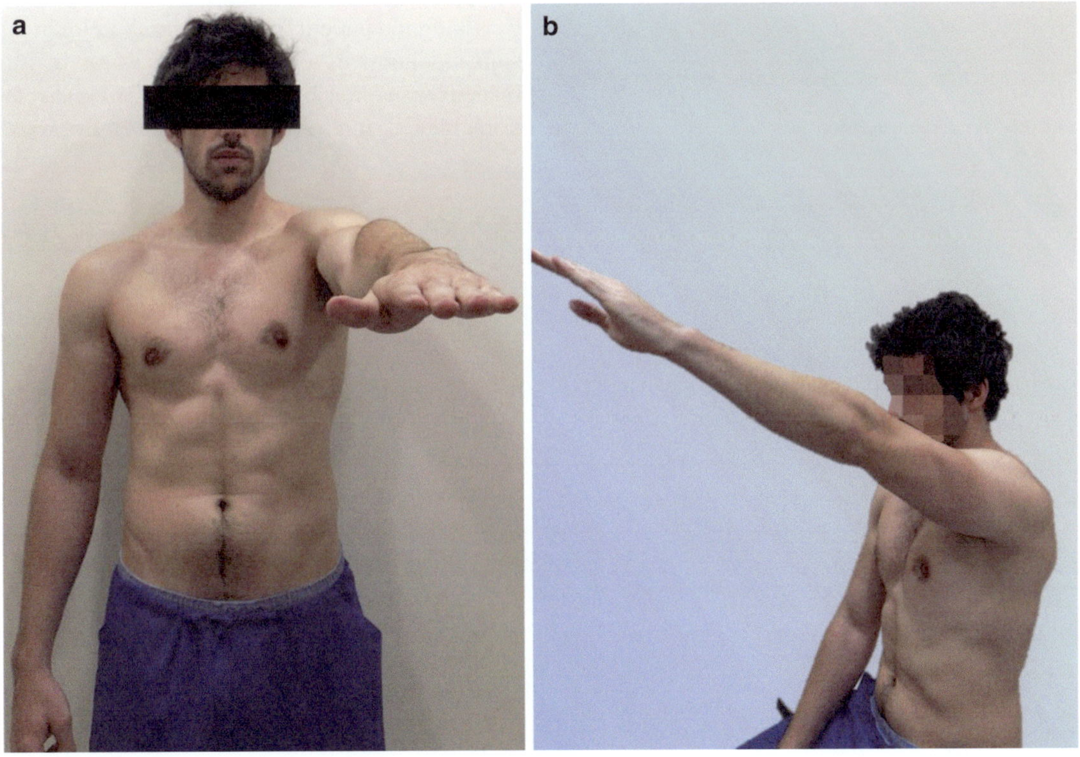

Fig. 3.4 (**a** and **b**) Forward flexion to 90 and 120°

3.2.5 Extension

It refers to the position of the shoulder in which the arm rotates to the back side of the body (Fig. 3.5). The maximum range of motion in this plane is with the humerus slightly extended to the resting position angle or in 90° with the floor in the coronal plane. The maximum angle is around 60° of extension (AAOS, 1965). Involved in this action are deltoid muscle (posterior fibers), teres minor, teres major, the long head of triceps brachialis and latissimus dorsi. Paralysis of any of the muscles may affect the extension in different degrees depending on the muscle involved.

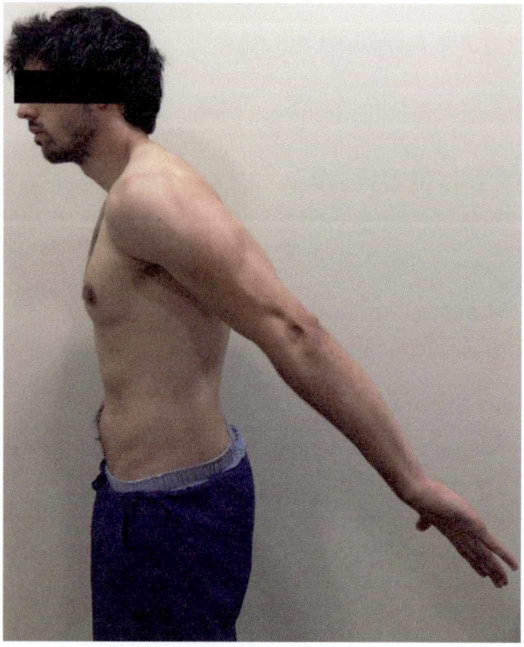

Fig. 3.5 Shoulder extension

3.2.6 Internal Rotation

Motion described as rotation of the humeral head around its center counter-clockwise, with the tuberosities moving forward and inward towards the pectoralis. It can be examined in abduction, resting position or flexion no matter what position the elbow, forearm, or hand are in. Pectoralis major, anterior fibers of deltoid, subscapularis, teres major, latissimus dorsi, and biceps brachialis are all involved.

Typically, we would explore internal rotation by asking the patient to try and reach as up in the spinal column as possible (Fig. 3.6a); in this maneuver, we reach the maximum internal rotation although it also depends on the elbow, wrist,

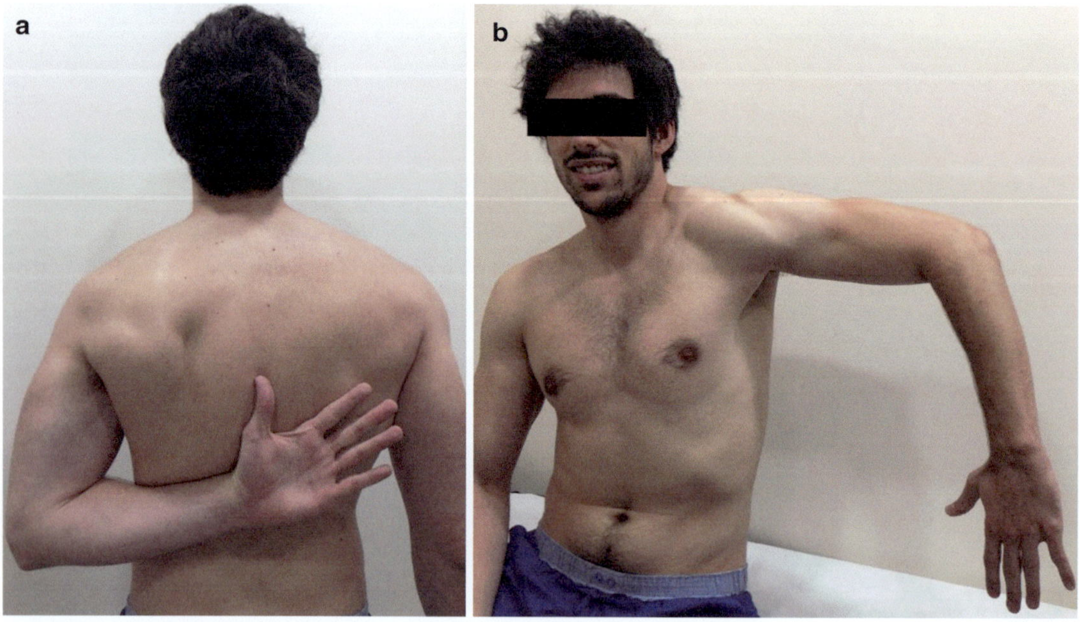

Fig. 3.6 (**a** and **b**) Shoulder internal rotation

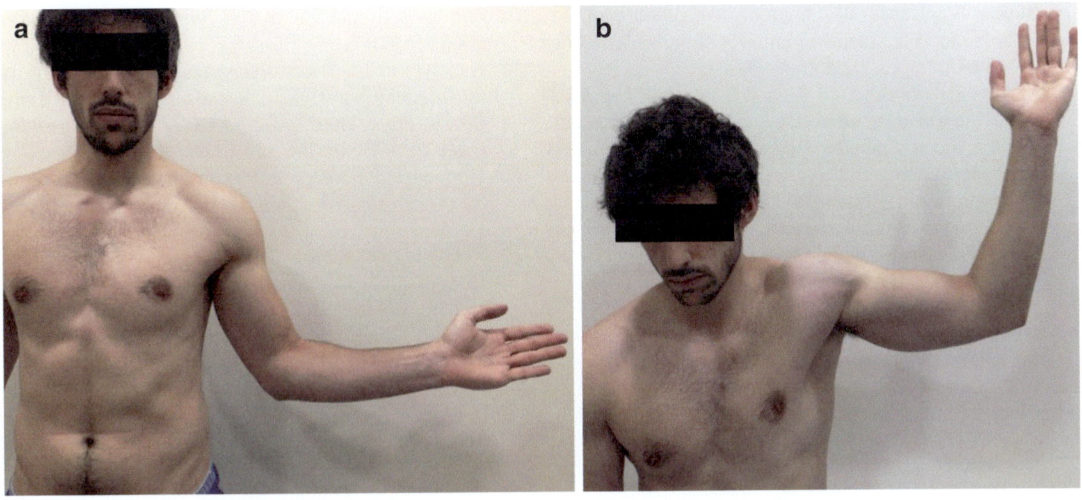

Fig. 3.7 (**a** and **b**) Shoulder external rotation

and hand joint's ROM. When exploring it with the arm at the side, adduction is necessary to move the arm backwards in order to reach the spinal column up to the highest level the patient can reach. Typically, the levels according to Constant are: lateral thigh, buttock, lumbosacral junction, waist (L3), T12 vertebra, and interscapular (T7).

On the other hand, internal rotation done with the arm in 90° of abduction, in which we would record it in degrees (Fig. 3.6b). Normally, the range we could achieve is of 50–60°, slightly greater (around 4°) in the non-dominant arm [13].

Deficit in internal rotation may appear typically in throwers, such as in the glenohumeral internal rotation deficit (GIRD), defined by a loss of >20° in internal rotation compared to the contralateral shoulder, due to labral tears or posterior capsular and rotator cuff tightness. It is related to the overhead throwing motion that is seen in many sports such as tennis, baseball, javelin throwers, handball, softball, or volleyball [15].

3.2.7 External Rotation

External rotation can be described as the opposite from internal rotation, humeral head is rotated clockwise around its axis, moving the humeral head tuberosities towards the back of the body (Fig. 3.7a). The position of the arm changes the maximum ROM available, in neutral position and the elbow 90° flexed we could achieve around 75° of external rotation, whereas in 90° of arm abduction up to 98° of internal rotation (Fig. 3.7b) [13]. Producing this action, we have infraspinatus, teres minor, and long head of triceps brachialis. Performing this action, we find the least powerful movement of the shoulder [2]. Disorders of the rotator cuff are the main reason for active external rotation deficit.

Throwing athletes exhibit a pattern of increased glenohumeral external rotation in their dominant shoulder, partially due to the typical increased humeral retrotorsion, higher laxity of the anterior capsule, and posterior capsulotendinous tightness. This would allow the thrower to achieve the late cocking phase of throwing with less stress to the anterior stabilizers [16].

References

1. Constant CR. Historical background, anatomy and shoulder function. Baillières Clin Rheumatol. 1989;3:429–35.
2. Veeger HEJ, van der Helm FCT. Shoulder function: the perfect compromise between mobility and stability. J Biomech. 2007;40:2119–29.
3. Thompson MMS. Miller's review of orthopaedics. 8th ed. Elsevier; 2019.

4. Burkart AC, Debski RE. Anatomy and function of the glenohumeral ligaments in anterior shoulder instability. Clin Orthop. 2002;400:32–9.
5. Urwin M, Symmons D, Allison T, et al. Estimating the burden of musculoskeletal disorders in the community: the comparative prevalence of symptoms at different anatomical sites, and the relation to social deprivation. Ann Rheum Dis. 1998;57:649–55.
6. Karduna AR, Williams GR, Iannotti JP, et al. Kinematics of the glenohumeral joint: influences of muscle forces, ligamentous constraints, and articular geometry. J Orthop Res. 1996;14:986–93.
7. Wilson JD, Khan-Perez J, Marley D, et al. Can shoulder range of movement be measured accurately using the Microsoft Kinect sensor plus Medical Interactive Recovery Assistant (MIRA) software? J Shoulder Elb Surg. 2017;26:e382–9.
8. Varghese M, Lamb J, Rambani R, et al. The use of shoulder scoring systems and outcome measures in the UK. Ann R Coll Surg Engl. 2014;96:590–2.
9. Booker S. Use of scoring systems for assessing and reporting the outcome results from shoulder surgery and arthroplasty. World J Orthop. 2015;6:244.
10. Magermans DJ, Chadwick EKJ, Veeger HEJ, et al. Requirements for upper extremity motions during activities of daily living. Clin Biomech. 2005;20:591–9.
11. Hsu AT, et al. Determining the resting position of the glenohumeral joint: a cadaver study. J Orthop Sports Phys Ther. 2002;32(12):605–12. https://doi.org/10.2519/jospt.2002.32.12.605.
12. Hoppenfeld S. Physical examination of the spine & extremities. Pearson Education. Prentice Hall; 1976.
13. Kenneth L, Cameron GLV. Clinical descriptive measures of shoulder range of motion for a healthy, young and physically active cohort. Sports Med Arthrosc Rehabil Ther Technol. 2012;4(1):33. https://doi.org/10.1186/1758-2555-4-33.
14. Perry J. Anatomy and biomechanics of the shoulder in throwing, swimming, gymnastics, and tennis. Clin Sports Med. 1983;2(2):247–70.
15. Rose MB, Noonan T. Glenohumeral internal rotation deficit in throwing athletes: current perspectives. Open Access J Sports Med. 2018;9:69–78.
16. Greenberg EM, Fernandez-Fernandez A, Lawrence JTR, et al. The development of humeral retrotorsion and its relationship to throwing sports. Sports Health Multidiscip Approach. 2015;7:489–96.

4 Examination of Range of Motion Scapulothoracic, Acromioclavicular, and Scapulothoracic Joints

Giovanni Di Giacomo, W. Ben Kibler, Francesco Franceschi, and Aaron Sciascia

4.1 Scapulothoracic Joints

Keeping in mind that the shoulder is a ring of a very complex kinetic chain, it is evident that it is important to assess a patient's posture and core stability so as to tailor the right sequence of goals to be obtained over the course of functional shoulder movement.

As Kibler explains [1, 2], the various body segments play specific roles in the kinetic chain activation sequence. The muscles, the joints of the hips, pelvis, and spine ("the core") are centrally located and can perform many of the dynamic stabilizing functions that the body requires if the distal segments are to perform their specific tasks. Thus "core stability" provides proximal stability for distal limb mobility and function.

G. Di Giacomo
Concordia Hospital, Rome, Italy

W. B. Kibler
Shoulder Center of Kentucky, Lexington Clinic, Lexington, KY, USA

F. Franceschi (✉)
UniCamillus-Saint Camillus International University of Health Sciences, Rome, Italy

Department of Orthopaedic and Trauma Surgery, San Pietro Fatebenefratelli Hospital, Rome, Italy
e-mail: f.franceschi@unicampus.it

A. Sciascia
Eastern Kentucky University, Exercise and Sport Science, Richmond, KY, USA
e-mail: aaron.sciascia@eku.edu

When assessing core stability and strength, it is important to evaluate the muscles working in an eccentric, load-absorbing function, the body segments in a closed-chain situation, and the resultant movements in the three planes of trunk motion. In the standing balance test, the patient is asked to stand on one leg and is given no further verbal cue. A positive test result, known as the Trendelenburg sign, is when the hip drops on the unsupported side. This indicates inability to control the posture and suggests proximal core weakness.

The scapula is anatomically and biomechanically intimately involved with shoulder function.

A primary role of the scapula is that it is integral to the glenohumeral articulation, which kinematically is a ball-and-socket configuration.

The second role of the scapula is to provide motion along the thoracic wall, retract (externally rotate) and laterally protract (internally rotate) around the thoracic cage, to maintain a normal position in relation to the humerus.

The third role that the scapula plays in shoulder function is elevation of the acromion, to clear the acromion from the moving rotator cuff to decrease impingement and coracoacromial arch compression [3, 4].

Although rotator cuff fatigue may cause superior humeral head migration to trigger subacromial impingement in this position [5], lower

J. G. Lane et al. (eds.), *The Art of the Musculoskeletal Physical Exam*,
https://doi.org/10.1007/978-3-031-24404-9_4

trapezius and serratus anterior muscle fatigue also may contribute to impingement by decreasing acromial elevation [6].

Finally, the scapula is a link in proximal-to-distal sequencing of velocity, energy, and forces of shoulder function [5, 7, 8].

For most activities, sequencing begins at the ground, and individual body segments (links) are coordinated by muscle activation and body position to generate, summate, and transfer force through these segments to the terminal link. This sequence is termed the kinetic chain [5, 7, 9].

The scapula has to be considered pivotal in transferring large forces and high energy from the legs, back, and trunk to the delivery point, the arm, and the hand [7, 10], thereby allowing more force to be generated in activities such as throwing than could be done by the arm musculature alone. The scapula, serving as a link, also stabilizes the arm to more effectively absorb loads that may be generated through the long lever of the extended or elevated arm [11].

When, for different reasons, there is a breakage of the "kinetic chain," a clinical picture termed scapular dyskinesis can subjectively and objectively be described.

Scapular dyskinesis [12] is defined as observable alterations in the position of the scapula and the patterns of scapular motion in relation to the thoracic cage.

More commonly, the scapular stabilizing muscles (1) are directly injured from direct-blow trauma; (2) have microtrauma-induced strain in the muscles leading to muscle weakness; (3) become fatigued from repetitive tensile use; or (4) are inhibited by painful conditions around the shoulder. Muscle inhibition or weakness is quite common in glenohumeral pathology, whether from instability, labral pathology, or arthrosis [5, 6, 13, 14].

The serratus anterior and the lower trapezius muscles are the most susceptible to the effect of the inhibition, and they are more frequently involved in early phases of shoulder pathology [6, 15, 16].

Scapular evaluation, as previously described, includes distant contributions to normal scapular function and dyskinesis.

The evaluation of the scapula itself should be done mainly from the posterior aspect.

Abnormalities of winging, elevation, or rotation may first be examined in the resting position.

Pure serratus anterior muscle weakness resulting from nerve palsy will create a prominent superior medial border and depressed acromion, whereas pure trapezius muscle weakness resulting from nerve palsy will create a protracted inferior border and elevated acromion [17].

Motion and position should be examined in both the elevating and lowering phases of motion. Muscle weakness and mild scapular dyskinesis are more common in the lowering phase of arm movement.

An effective test for evaluating scapular muscle strength is an isometric pinch of the scapulas in retraction. Scapular muscle weakness may be present as a burning pain in less than 15 s, whereas the scapula normally may be held in this position for 15–20 s without burning pain or muscle weakness. Wall push-ups are effective for evaluating serratus anterior muscle strength.

The scapular assistance test evaluates scapular and acromial involvement in subacromial impingement.

In a patient with impingement symptoms with forward elevation or abduction, assistance for scapular elevation is provided by manually stabilizing the scapula and rotating the inferior border of the scapula as the arm moves. This procedure simulates the force-couple activity of the serratus anterior and lower trapezius muscles. Elimination or modification of the impingement symptoms indicates that these muscles should be a major focus in rehabilitation.

The scapular retraction test involves manually stabilizing the scapula in a retracted position on the thorax. This position confers a stable base of origin for the rotator cuff and often will improve tested rotator cuff strength. (That is, the apparent strength generated by isolated rotator cuff strength testing often improves by retesting in the scapula-retracted position.) The scapular retraction test also frequently demonstrates scapular and glenoid involvement in internal impingement lesions [18]. The positive posterior labral findings on modified Jobe relocation testing will be decreased with scapular retraction and removal of the glenoid from the excessively protracted impingement position.

4.2 Specific Strength Test for Scapulothoracic Muscles

4.2.1 Serratus Anterior

This muscle abducts the scapula, rotates the inferior angle laterally and the glenoid cavity cranially, and holds the medial border of the scapula firmly against the rib cage.

The patient is in a supine position: Abduction of the scapula, projecting the upper extremity anteriorly (upward from the table). *Movement of the scapula must be observed and the inferior angle palpated to ensure that the scapula is abducting.*

The examiner should press against the subject's fist, transmitting the pressure downward through the extremity to the scapula in the direction of the adducting scapula.

When the serratus is weak, the scapula tilts forward at the coracoid process and the inferior angle moves posteriorly and in the direction of the medial rotation. Because some type of substitution can occur, we advise you to perform the test in the sitting position. In this position, the test emphasizes the upward rotation action of the serratus in the abducted position.

4.2.2 Lower Trapezius Test

The muscle with upper and middle fibers adducts the scapula, especially with the medial fibers. It rotates the scapula so the glenoid cavity faces cranially; there is also the pressure of the scapula.

To test the lower trapezius the patient is prone, the arm is placed diagonally overhead, in line with lower fibers of the trapezius, the examiner presses the forearm in a downward direction toward the table. When the muscle is weak, the scapula rides upward and tilts forward with depression of the coracoid process.

4.2.3 Latissimus Dorsi

This muscle medially rotates, adducts, and extends the shoulder joint. The patient is prone in adduction and extension in medially rotated position, the examiner presses the forearm in the direction of abduction and slight flexion of the arm. Weakness interferes with activities that involve the adduction of the arm toward the body or of the body toward the arm.

4.2.4 Rhomboid Test

This muscle adducts and elevates the scapula and rotates it so the glenoid cavity faces caudally. The patient is prone, the position of the scapula is obtained by placing the shoulder in 90° abduction and in medial rotation to move the scapula into the test position. The examiner presses against the forearm in the downward direction toward the table. With the patient in the same position, but the arm in lateral rotation it is possible to test the middle fibers of the trapezius.

4.3 Acromioclavicular Joint

The accurate and complete exam of the acromioclavicular (AC) joint is based on in depth understanding of the functional anatomy of the AC joint. Static descriptions of the AC joint anatomy fail to provide the context for the most effective understanding of AC joint injuries. This requires a more functional description, relating the anatomy to how it facilitates, guides, and optimizes the three-dimensional mechanics of the clavicle, scapula, AC joint, and arm to create motions and forces to accomplish tasks.

Efficient upper limb mechanics requires coupled motions of the clavicle and acromion, with the AC joint acting as a stable articulation. The S-shaped clavicle acts as a (1) strut, maintaining length and stiffness [19, 20]; (2) crank handle, allowing large amounts of distal rotational arcs of motion for short amounts of proximal rotation [21–23]; and (3) the only bony attachment to the axial skeleton. The clavicle has minimal muscular attachments with so much of the clavicular long axis rotation, anterior/posterior motion, and elevation/depression occurring through the influence of scapular motion.

The AC joint is a relatively stiff structure, with strong posterior, superior, and anterior ligament components that are thicker on their acromial insertions than their clavicular insertions. Individual AC joint motions average 5° of acromial elevation and 8° of acromial rotation [24, 25]. A three-dimensional kinematic analysis of the AC joint demonstrated that the scapula rotated 35° on an axis (termed the "screw axis") that passed through the insertions of the AC and coracoclavicular (CC) ligaments, and that with abduction, the lateral clavicle translated 3.5 mm in the anterior/posterior direction and 1 mm in the superior direction [26]. This stiffness creates a strong link that allows rotational and elevation motions produced by the scapula or clavicle to be efficiently transmitted to the other bone of the articulation [27, 28]. Interruptions of the normal integrity of the AC and CC ligaments change the normal linkage between the scapula and the clavicle and can result in dyskinetic motion patterns during limb movement. In addition, indirect AC joint stability and stiffness is maintained by the CC ligaments.

In summary, intact AC joint anatomy is the basis for optimal arm and shoulder mechanics. It creates the most efficient screw axis and allows efficient scapulohumeral rhythm (SHR), the coupled sequenced motion of the scapula and humerus in all phases of arm motion.

Pathology involving the AC joint will create altered motions at the joint and in the surrounding structures and will affect the roles the AC joint plays in maintaining efficient SHR. The physical exam should be organized to identify the altered motions and functions. It should be able to identify not only the pathoanatomy (injury to the bones, ligaments, and joint) but also the effect of the pathoanatomy on the normal mechanics, creating pathomechanics.

The exam comprises visualization, palpation, provocative maneuvers, observation of motion, and corrective maneuvers that may alter the clinical symptoms. Table 4.1 summarizes the possible clinical and examination findings that may be seen.

Visualization should be accomplished by direct evaluation of the symptomatic joint and comparison, if possible, to the asymptomatic contralateral joint. The most common visualized alterations may include: (1) prominence of the distal clavicle due to a superior bone spur from arthritis; (2) an apparent superior position relative to the inferiorly and medially displaced acromion in low-grade or high-grade AC separations (Fig. 4.1); or (3) altered posture of the scapula and arm into protraction due to muscle weakness or imbalance. Any swelling or bruising, which

Table 4.1 Typical exam findings for acromioclavicular joint pathology

Pathology	Inspection	Palpation	Provocative maneuver	Observation	Corrective maneuver
Arthritis	Bone spur	Pain	(+)active compression (+)cross body adduction	Decreased arm motion, pain during flexion	(−)
Excessive distal clavicle excision	Joint deficit	Pain, bony deficit	(+)cross body adduction (+)anterior/posterior laxity	Pain to motion	Decreased with SRT
Low-grade AC separation	Clavicle prominence	Pain	(+)anterior/posterior laxity	Scapular protraction	Decreased with SRT
High-grade AC separation	Clavicle prominence	Pain	(+)anterior/posterior and inferior/superior laxity	Scapular protraction, decreased arm elevation	Decreased with SRT and joint reduction
Distal clavicle fracture	Swelling	Pain over bone	(+)cross body adduction	Decreased arm motion	(−)

(+) = positive; (−) = negative; AC = acromioclavicular; SRT = scapular retraction tests could indicate localized trauma, should be noted. Frequently, there will be minimal visual alterations.

Palpation often reveals the location of the symptomatic pathology. The clavicle should be palpated along its entire length from the sternoclavicular joint to the AC joint. Point tenderness along the bone, especially near the AC joint, will suggest bony injury. The acromion and its extension into the scapular spine as well as the coracoid process and CC ligament area can also be palpated. Finally, direct palpation of the AC joint, making sure that the palpation pressure is localized to the joint and not to the bone, can elicit pain that can confirm the joint is the actual source of some or all of the clinical symptoms. Palpating the AC joint directly can also be helpful in identifying clinically important AC joint pathology that may be associated with other shoulder pathology such as rotator cuff disease or impingement and will need to be addressed as part of the comprehensive treatment.

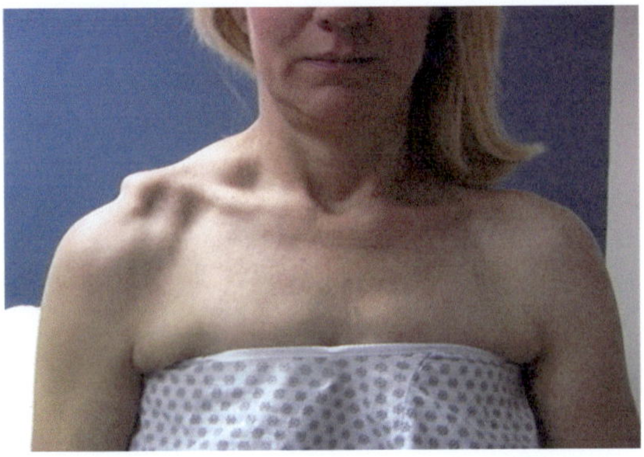

Fig. 4.1 Example of lost integrity of AC joint giving visual appearance of inferiorly position acromion relative to clavicle

Provocative maneuvers may be utilized to reproduce the clinical symptoms and provide information about the anatomic structures that may be involved. The clavicle and acromion may be grasped and mobilized (this maneuver is sometimes labeled the AC shear test [29, 30]) in several directions to load, unload, and shift the loads, and to place stress on the ligaments. Anterior/posterior testing assesses the integrity of the AC ligaments while inferior/superior testing assesses the integrity of the CC ligaments [19, 31]. Compression of the acromion into the clavicle may simulate load bearing and increase or reproduce the pain of an arthritic joint. Several types of horizontal adduction maneuvers of the arm in relation to the body have been described and advocated to produce the position of impingement of the acromion on the clavicle that occurs with increased motion of the arthritic (active compression test), slightly lax joint (cross body adduction test) or with compression against bone spurs (Paxinos test) [32–35] (Figs. 4.2, 4.3, and 4.4). These involve manual or active motion of the extended arm across the body, with a positive test being localized pain and/or crepitus at the joint and reproduction of the clinical symptoms. Ligamentous injury and resulting laxity can be identified by maneuvers that assess joint transla-

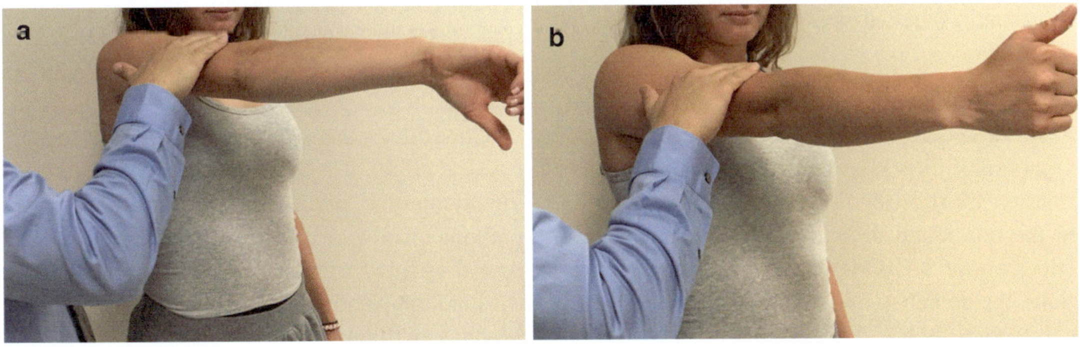

Fig. 4.2 Active compression test. Resistance is applied inferiorly to a patient's arm positioned in forward elevation, 10° of horizontal adduction, and internal rotation (**a**). The test is then repeated with the arm in external rotation (**b**). A positive test is determined as pain with resistance in both rotational positions

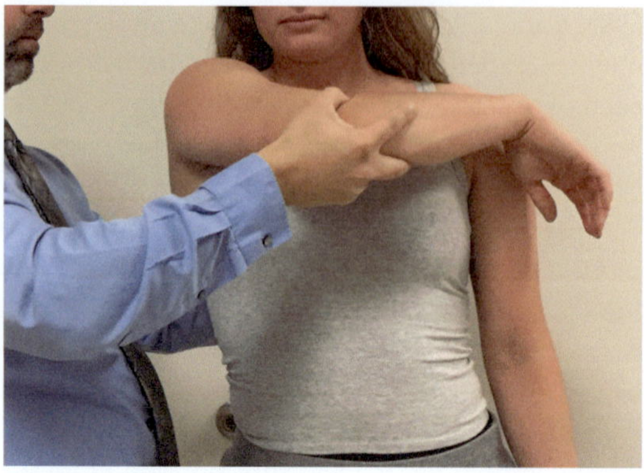

Fig. 4.3 Cross body adduction test. The arm is passively moved across the patient's body to determine if pain occurs at the AC joint

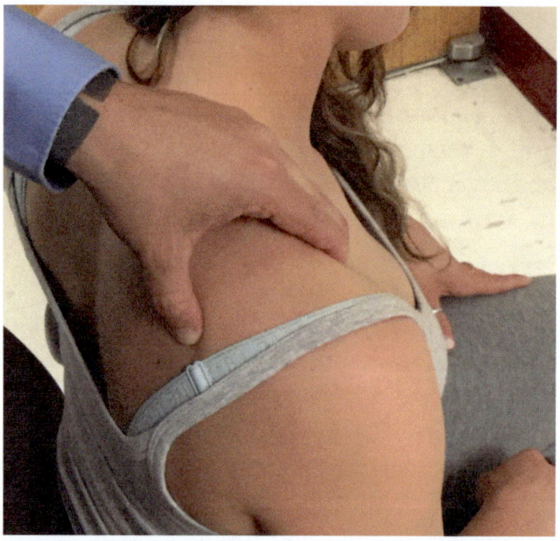

Fig. 4.4 Paxinos test. While grasping the posterior acromion and the middle aspect of the clavicle, pressure is applied to the acromion anterosuperiorly and to the clavicle inferiorly. Pain with pressure indicates a positive test

tion under an imposed load or motion. In low-grade (Rockwood/ISAKOS 1, 2, 3A) AC separations [36–38], with mainly AC ligament injuries, most of the demonstrated laxity will be in the horizontal direction, while in high-grade (Rockwood/ISAKOS 3B, 4, and 5) AC separations, the demonstrated laxity will be in the vertical as well as horizontal directions [36–38]. In addition, in the high-grade injuries, the cross body adduction maneuver will pull the acromion inferior and medial to the clavicle, resulting in

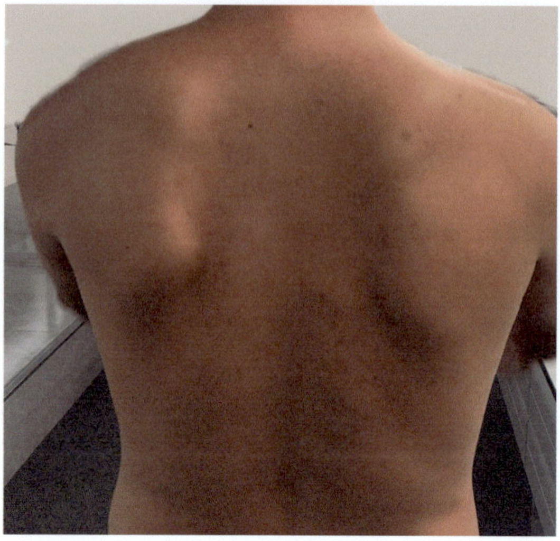

Fig. 4.5 Example of altered scapular positioning due to muscle weakness

the characteristic posture of clavicle prominence, even though the pathomechanical motion and resulting symptoms are due to abnormal acromial motion. Patients with excessive distal clavicle excision following arthroscopic or open surgery may demonstrate pain upon horizontal adduction and horizontal translation maneuvers, due to the bone shortening and ligament injuries.

The effect of AC joint pathology on SHR can be assessed by observation of the motions of the arm, scapula, and clavicle [39]. Limitation of arm motion in flexion and abduction may be seen in AC joint arthritis. In high-grade AC separations or type 2 distal clavicle fractures, the resulting pathomechanics may include excessive scapular protraction, which can be observed an asymmetrical scapular position at rest or medial border prominence upon arm motion in elevation and/or descent [40] (Fig. 4.5).

Corrective maneuvers can be helpful to indicate the effect of modifying the pathoanatomy or the symptoms and noting the change in the pathomechanics or function. This can give clues regarding how the alterations result in clinical dysfunction and suggest treatment options. Manual realignment and reduction of the AC joint, conferring joint stability, may decrease the joint symptoms and may improve the dynamic deficits in normal SHR. This can be accomplished directly by mobilizing the acromion and clavicle, or indirectly by mobilizing the scapula in relation to the stabilized clavicle in the Scapula Retraction Test (SRT) [41, 42]. In this test, the scapula is manually positioned in retraction and held in this position as arm motion or strength is measured or joint stability is tested. Change in symptoms of joint impingement, joint instability, arm impingement, arm strength, or SHR are positive results.

In summary, physical examination of the AC joint can result in specific information, enabling the accurate diagnosis of AC joint pathology and the effect on SHR mechanics, and provide guidelines for treatment.

References

1. Sciascia A, Kibler WB. Alterations in the kinetic chain are found in a high percentage of patients. Conducting the "nonshoulder" shoulder examination. J Musculoskelet Med. 2011;28:61–2, 157.
2. Kibler WB, Press J, Sciascia A. The role of core stability in athletic function. Sports Med. 2006;36:189–98.
3. Ludewig PM, Cook TM. Alterations in shoulder kinematics and associated muscle activity in people with symptoms of shoulder impingement. Phys Ther. 2000;80:276–91.
4. Lukasiewicz AC, McClure P, Michener L, Pratt N, Sennett B. Comparison of 3-dimensional scapular position and orientation between subjects with and without shoulder impingement. J Orthop Sports Phys Ther. 1999;29:574–86.
5. Fleisig GS, Barrentine SW, Escamilla RF, Andrews JR. Biomechanics of overhand throwing with implications for injuries. Sports Med. 1996;21:421–37.
6. McQuade KJ, Dawson J, Smidt GL. Scapulothoracic muscle fatigue associated with alterations in scapulohumeral rhythm kinematics during maximum resistive shoulder elevation. J Orthop Sports Phys Ther. 1998;28:74–80.
7. Kibler WB. Biomechanical analysis of the shoulder during tennis activities. Clin Sports Med. 1995;14:79–85.
8. Kennedy K. Rehabilitation of the unstable shoulder. Op Tech Sports Med. 1993;1:311–24.
9. Elliott BC, Marshall R, Noffal G. Contributions of upper limb segment rotations during the power serve in tennis. J Appl Biomech. 1995;11:433–42.
10. Kraemer WJ, Triplett NT, Fry AC. An in-depth sports medicine profile of women college tennis players. J Sports Rehabil. 1995;4:79–88.
11. Happee R, Van der Helm FC. The control of shoulder muscles during goal directed movements: an inverse dynamic analysis. J Biomech. 1995;28:1179–91.
12. Warner JJ, Micheli LJ, Arslanian LE, Kennedy J, Kennedy R. Scapulothoracic motion in normal shoulders and shoulders with glenohumeral instability and impingement syndrome: a study using Moire topographic analysis. Clin Orthop. 1992;285:191–9.
13. Moseley JB Jr, Jobe FW, Pink M, Perry J, Tibone J. EMG analysis of the scapular muscles during a shoulder rehabilitation program. Am J Sports Med. 1992;20:128–34.
14. Glousman R, Jobe F, Tibone J, Moynes D, Antonelli D, Perry J. Dynamic electromyographic analysis of the throwing shoulder with glenohumeral instability. J Bone Joint Surg Am. 1988;70:220–6.
15. Pink MM, Perry J. Biomechanics of the shoulder. In: Jobe FW, Pink MM, Glousman RE, Kvitne RS, Zemel NP, editors. Operative techniques in upper extremity sports injuries. St. Louis, MO: Mosby-Year Book; 1996. p. 109–23.
16. McClure PW, Michener LA, Sennett BJ, Karduna AR. Direct 3-dimensional measurement of scapular kinematics during dynamic movements in vivo. J Shoulder Elb Surg. 2001;10:269–77.
17. Kuhn JE, Plancher KD, Hawkins RJ. Scapular winging. J Am Acad Orthop Surg. 1995;3:319–25.
18. Kibler WB, McMullen J, Uhl T. Shoulder rehabilitation strategies, guidelines, and practices. Op Tech Sports Med. 2000;8:258–67.
19. Oki S, Matsumura N, Iwamoto W, et al. The function of the acromioclavicular and coracoclavicular ligaments in shoulder motion: a whole-cadaver study. Am J Sports Med. 2012;40:2612–26.
20. Oki S, Matsumura N, Iwamoto W, et al. Acromioclavicular joint ligamentous system contributing to clavicular strut function: a cadaveric study. J Shoulder Elb Surg. 2013;22(10):1433–9.
21. Matsumura N, Ikegami H, Nakamichi N, et al. Effect of shortening deformity of the clavicle on scapular kinematics: a cadaveric study. Am J Sports Med. 2010;38(5):1000–6.
22. Hillen RJ, Burger BJ, Poll RG, van Dijk CN, Veeger DH. The effect of experimental shortening of the clavicle on shoulder kinematics. Clin Biomech. 2012;27(8):777–81.
23. Beitzel K, Sablan N, Chowaniec DM, et al. Sequential resection of the distal clavicle and its effects on horizontal acromioclavicular joint translation. Am J Sports Med. 2012;40:681–5.
24. Ludewig PM, Behrens SA, Meyer SM, Spoden SM, Wilson LA. Three-dimensional clavicular motion during arm elevation: reliability and descriptive data. J Orthop Sports Phys Ther. 2004;34(3):140–9.
25. Ludewig PM, Phadke V, Braman JP, Hassett DR, Cieminski CJ, LaPrade RF. Motion of the shoulder complex during multiplanar humeral elevation. J Bone Joint Surg Am. 2009;91A(2):378–89.
26. Sahara W, Sugamoto K, Murai M, Yoshikawa H. Three-dimensional clavicular and acromioclavicular rotations during arm abduction using vertically open MRI. J Orthop Res. 2007;25:1243–9.

27. Lawrence RL, Braman JP, LaPrade RF, Ludewig PM. Comparison of 3-dimensional shoulder complex kinematics in individuals with and without shoulder pain, part 1: sternoclavicular, acromioclavicular, and scapulothoracic joints. J Orthop Sports Phys Ther. 2014;44:636–45.
28. Ludewig PM, Lawrence RL. Mechanics of the scapula in shoulder function and dysfunction. In: Kibler WB, Sciascia AD, editors. Disorders of the scapula and their role in shoulder injury: a clinical guide to evaluation and management. Switzerland: Springer; 2017. p. 7–24.
29. Magee DJ. Orthopedic physical assessment, vol. 5. St Louis: Saunders Elsevier; 2008.
30. Examination of orthopedic and athletic injuries. Vol 3. Philadephia: F.A. Davis; 2010.
31. Fukuda K, Craig EV, An KN, Cofield RH, Chao EY. Biomechanical study of the ligamentous system of the acromioclavicular joint. J Bone Joint Surg Am. 1986;68:434–40.
32. O'Brien SJ, Pagnani MJ, Fealy S, McGlynn SR, Wilson JB. The active compression test: a new and effective test for diagnosing labral tears and acromioclavicular joint abnormality. Am J Sports Med. 1998;26(5):610–3.
33. Chronopoulos E, Kim TK, Park HB, Ashenbrenner D, McFarland EG. Diagnostic value of physical tests for isolated chronic acromioclavicular lesions. Am J Sports Med. 2004;32(3):655–61.
34. Walton J, Mahajan S, Paxinos A, et al. Diagnostic values of tests for acromioclavicular joint pain. J Bone Joint Surg Am. 2004;86A(4):807–12.
35. McLaughlin HL. On the frozen shoulder. Bull Hosp Joint Dis. 1951;12(2):383–93.
36. Rockwood CA Jr. Injuries to the acromioclavicular joint. In: Rockwood Jr CA, Green DP, editors. Fractures in adults. 2nd ed. Philadelphia: Lippincott; 1984. p. 860–910.
37. Bak K, Mazzocca A, Beitzel K, et al. Copenhagen consensus on acromioclavicular disorders. In: Arce G, Bak K, Shea KP, et al., editors. Shoulder concepts 2013: consensus and concerns—proceedings of the ISAKOS upper extremity committees 2009–2013. Heidelberg: Springer; 2013. p. 51–67.
38. Beitzel K, Mazzocca AD, Bak K, et al. ISAKOS upper extremity committee consensus statement on the need for diversification of the Rockwood classification for acromioclavicular joint injuries. Arthroscopy. 2014;30:271–8.
39. Kibler WB, Ludewig PM, McClure PW, Michener LA, Bak K, Sciascia AD. Clinical implications of scapular dyskinesis in shoulder injury: the 2013 consensus statement from the "scapula summit". Br J Sports Med. 2013;47:877–85.
40. Gumina S, Carbone S, Postacchini F. Scapular dyskinesis and SICK scapula syndrome in patients with chronic type III acromioclavicular dislocation. Arthroscopy. 2009;25(1):40–5.
41. Kibler WB. The role of the scapula in athletic function. Am J Sports Med. 1998;26:325–37.
42. Kibler WB, Sciascia AD, Dome DC. Evaluation of apparent and absolute supraspinatus strength in patients with shoulder injury using the scapular retraction test. Am J Sports Med. 2006;34(10):1643–7.

5 Clinical Tests for Evaluation of Motor Function of the Shoulder

Nedal Alkhatib, Catherine M. Coady, and Ivan Wong

5.1 Introduction

Functional movements of the shoulder constitute synergistic interaction of muscle forces, ligament strains, and bony articulation. Injury to any of these components may interrupt this interaction and disturb the normal function of the shoulder [1]. The shoulder examination involves testing for each one of these components. In this chapter, we will review the clinical tests for shoulder function, and we will focus on the muscular components and their related disorders. The functional shoulder examination involves basic joint examination (inspection, palpation, range of motion, and strength testing) as well as other specialized clinical tests. The diagnostic accuracy of specialized tests is increased by using multiple tests to confirm a suspected pathology since using only one test may result in a lower sensitivity or specificity to rule in or rule out the suspected pathology [2].

N. Alkhatib · C. M. Coady · I. Wong (✉)
Division of Orthopaedic Surgery,
Department of Surgery, Faculty of Medicine,
Dalhousie University, Halifax, NS, Canada
e-mail: nd447030@dal.ca; cathy.coady@dal.ca; iw@drivanwong.com

J. G. Lane et al. (eds.), *The Art of the Musculoskeletal Physical Exam*,
https://doi.org/10.1007/978-3-031-24404-9_5

5.2 History

An essential part of shoulder motor function evaluation is taking a complete history. Pain, weakness, instability, and stiffness are the main complaints of the shoulder in clinical practice. Shoulder instability and stiffness will be covered in other chapters of this book. Analyzing the relationship between pain and shoulder motor function is required to localize the site of pathology. Pain caused by a chronic rotator cuff tear is diffuse, insidious in onset, associated with overhead activities, and is often worse while sleeping. Although shoulder weakness is frequently a result of pain inhibition, rotator cuff pathology and a nerve injury are other important causes, thus, a detailed neurological examination of the patient with shoulder weakness is essential [3].

5.3 Clinical Examination

5.3.1 Inspection

Examination for shoulder motor function starts with inspection of both shoulders, looking for any abnormal posture, skin lesions, muscle atrophy, bony prominences, and shoulder asymmetry. Exposure for shoulder examination is from head to waist for males, and for females, garments are used to cover the breasts. Functional disabilities can often be recognized by observing the patient undressing [3].

Front View: The examiner should look at the head position and level of the shoulders. It is usual for the dominant shoulder to be slightly lower than the non-dominant shoulder; however it is important to assess unleveled shoulders since this may be as a result of Sprengel's deformity or other spinal abnormality such as scoliosis or kyphosis. Other abnormalities that can be seen from the front, include Popeye's sign in the case of biceps tendon rupture, loss of axillary fold with pectoralis major muscle tear, and wasting of deltoid muscle either due to axillary nerve injury or in case of rotator cuff pathology.

Side View: The examiner should inspect for shoulder contour and muscle wasting.

Back View: Again, the examiner should look at posture, shoulder level, spinal deformity, as well as assess for scapula position, and abnormality. The examiner should inspect for supraspinatus and infraspinatus muscle scalloping in the case of rotator cuff pathology or suprascapular nerve compression. Asking the patient to put his hands over his hips helps in making the scalloping over the infraspinatus more obvious [3, 4].

5.3.2 Palpation

Palpation is important to check for any tenderness, swelling, or deformity. During the motor examination for the shoulder, it is valuable to palpate the rotator cuff tendons. The supraspinatus tendon is palpated under the anterolateral border of the acromion with the shoulder in adduction internal rotation and slight extension (forearm against stomach position). The infraspinatus and teres minor tendons are palpated just below the posterolateral border of the acromion with the shoulder in 90° of flexion, 20° of external rotation, and 10° of adduction (i.e., thinkers' position). The subscapularis tendon is palpated below the clavicle, lateral to coracoid, and medial to bicipital groove with shoulder beside the body in neutral position [4, 5].

5.3.3 Range of Motion

Examination for the shoulder's range of motion is essential before testing for muscle strength. Any limitation of shoulder movement may affect the position of the muscle being tested or give a false impression of muscle weakness. For example, limitation of internal shoulder rotation makes lift-off test inappropriate for testing subscapularis muscle function. It is recommended for the examiner to start with evaluating the shoulder's active range of motion. If there is a limitation for any of shoulder movements, gentle passive shoulder movements can be checked. A comprehensive evaluation of the shoulder's range of motion will be covered in another book chapter [3].

5.3.4 Strength Testing

The manual muscle test grading system is the most clinically relevant method to assess for muscle power (Table 5.1). However, the reliability of this method is limited, especially for subtle muscle weakness and muscle imbalance [4]. Using a hand-held dynamometer can help to increase the accuracy of isometric muscle strength testing. Jain et al. introduce a modified protocol for muscle strength testing using a dynamometer for more accurate examination. Muscle strength testing should be performed in both arms for comparison [6].

5.3.4.1 Abduction

This movement mainly measures the power of the supraspinatus muscle. To assess, the exam-

Table 5.1 Grading of muscle strength

Grade	Ability to move
0	No muscle contraction
1	Muscle contraction only
2	Full range of motion only with gravity eliminated
3	Full range of motion against gravity only
4	Full range of motion against moderate resistance
5	Full range of motion against full resistance

iner asks the patient to elevate their shoulder to 90° of abduction and 45° of forward flexion. Keeping their elbow in full extension with the palm facing down, then the examiner places the dynamometer at the lateral epicondyle of the humerus.

5.3.4.2 External Rotation

This movement mainly measures the power of the infraspinatus muscle. To assess, the examiner asks the patient to put their arm at their side with the elbow in 90° of flexion, and arm in neutral rotation with the thumb facing up. Then, the examiner places the dynamometer at the dorsal aspect of the forearm just proximal to the ulnar styloid. It is essential to stabilize the elbow during the examination.

5.3.4.3 Internal Rotation

This movement measures the power of the subscapularis muscle. To assess the strength, the examiner asks the patient to elevate their shoulder to 90° of forward flexion with the elbow in 90° of flexion, then the examiner places the dynamometer at the volar aspect of the patient's hand. The other hand of the examiner should check the patient's olecranon process to make sure the patient is applying internal rotation and not adduction force.

5.3.5 Special Tests

Based on the clinical history and examination, special tests are generally done to confirm the clinical findings. In this chapter, we will cover special tests for shoulder motor function which include tests for the rotator cuff muscles and other shoulder girdle muscles.

5.3.5.1 Rotator Cuff Muscles

Numerous tests were introduced for the examination of rotator cuff muscles; it is not practical to use all these tests during shoulder examination [6]. We will present the most clinically useful tests to detect rotator cuff muscles tear based on their sensitivity and specificity.

Tests for the Supraspinatus Muscle

The supraspinatus muscle is an abductor of the shoulder. In clinical practice, patients with a supraspinatus tear can still do shoulder abduction because of the effect of the deltoid muscle. In order to test supraspinatus muscle, we need to decrease the effect of the deltoid during the examination [3].

Empty Can Test/Jobe Test/Supraspinatus Test (Sensitivity 53–89% and Specificity 65–82%, Fig. 5.1a and b) [6]

With the patient in a standing position, the examiner needs to ask the patient to elevate their shoulder to 90° of abduction and to fully internally rotate their shoulder with their thumb pointing downward, which aims to decrease the effect of deltoid muscle. Then, the examiner forward flexes the shoulder to 30° in the coronal plane. In this position, the examiner applies downward pressure at the distal forearm and asks the patient to maintain this position. Muscle strength can be tested in this position, and it would be a positive test if the patient is unable to maintain his position either due to muscle pain or weakness. Kelly et al. [11] modify this test to 45° of external rotation **full can test** in order to decrease shoulder pain, but both tests have the same accuracy [12, 13].

Drop Arm Test (Sensitivity 10–73% and Specificity 77–98%) [6]

The examiner elevates the patient's shoulder to 90° of abduction, and then asks the patient to slowly lower their arm to their side. If the patient is unable to do that due to severe pain or if their arm suddenly drops, the test is considered positive which suggests a supraspinatus tendon tear [14].

Tests for the Infraspinatus Muscle

Strength

The infraspinatus muscle is the most powerful external rotator of the shoulder in 0° of abduction. To test muscle power, the patient is asked to put their arm at their side with the elbow in 90° of flexion, and the arm in slight internal rotation,

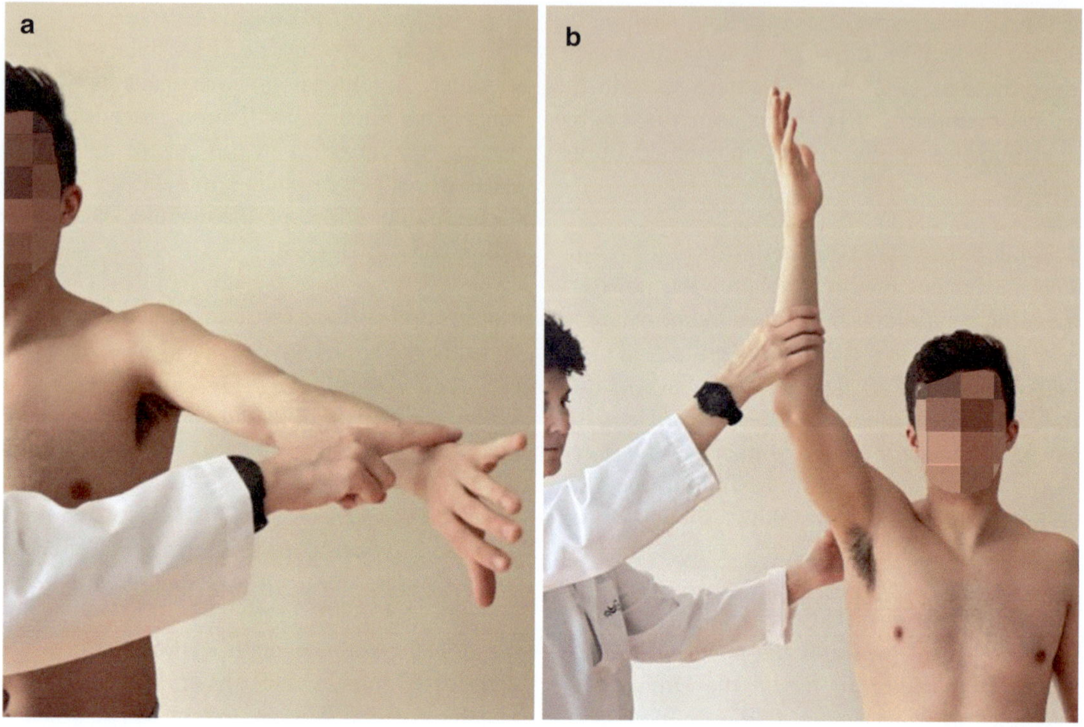

Fig. 5.1 (**a**) Empty can test. (**b**) Jobe test

then the examiner can test for infraspinatus muscle power by resisting external rotation of the patient's arm. During this test, it is important for the examiner to stabilize the patient's elbow for a proper evaluation of muscle power [7].

External Rotation Lag Sign (Sensitivity 46–98%, Specificity 72–98%) [6]

Using the same position for infraspinatus power testing, with the patient's arm at their side and their elbow in 90° of flexion, the examiner should hold the shoulder in 20° of abduction in the scapular plane and in maximum external rotation. The patient is then asked to keep this position while the examiner stabilizes the elbow (Fig. 5.2). If the patient is unable to hold their arm in this position and their arm lags or drops to a neutral position, the external rotation lag sign is considered positive due to infraspinatus muscle tear [8].

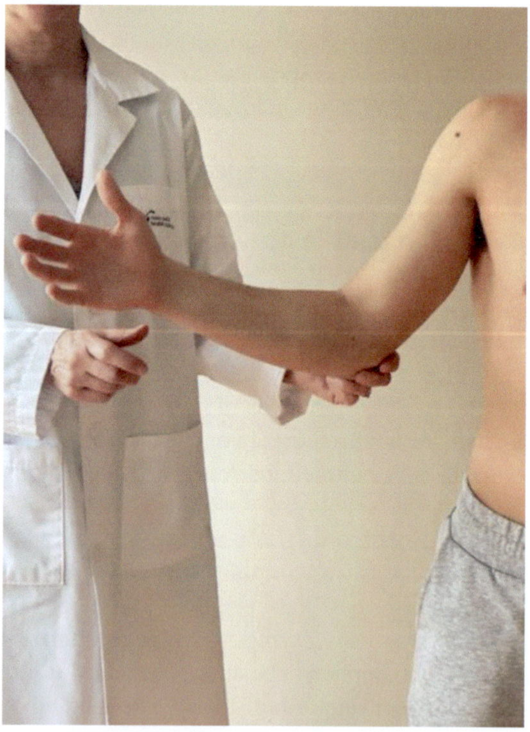

Fig. 5.2 External rotation lag sign

Drop Sign

The same test can be done with the shoulder in 90° of abduction in the scapular plane and the elbow in 90° of flexion. The examiner holds the arm of the patient in maximum external rotation and asks them to keep this position (Fig. 5.3). The test is considered positive if any lag or arm drop occurs. With this position, the posteroinferior cuff muscles (infraspinatus and teres minor) are being tested [8].

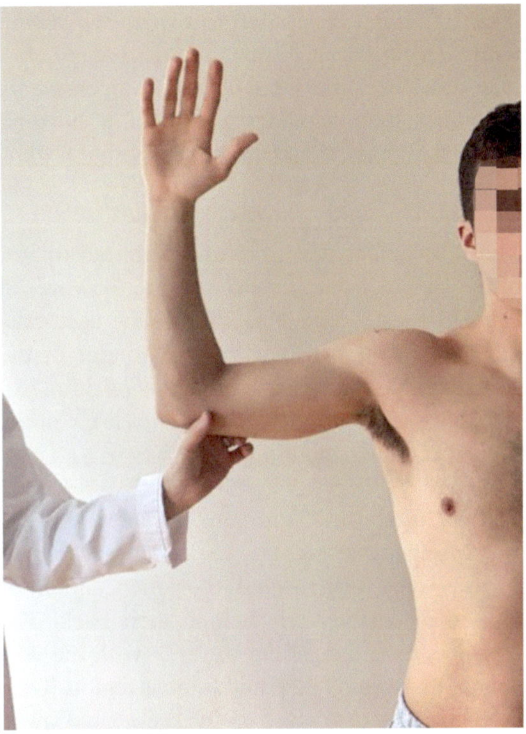

Fig. 5.3 Drop sign

Tests for the Teres Minor Muscle

The teres minor is an external rotator of the shoulder primarily at 90° of abduction [9].

Hornblower's Sign (Sensitivity 100% and Specificity 93%) [6]

This test is designed to examine teres minor strength, with the patient standing or sitting with their arms at their side. The examiner needs to hold the patient's arm in 90° of abduction and elbow in 90° of flexion, and then ask the patient to externally rotate their shoulder against resistance while supporting their elbow (Fig. 5.4a and b). If the patient is unable to do that because of pain or weakness, then the test is considered positive due to a teres minor tear. The test can also be

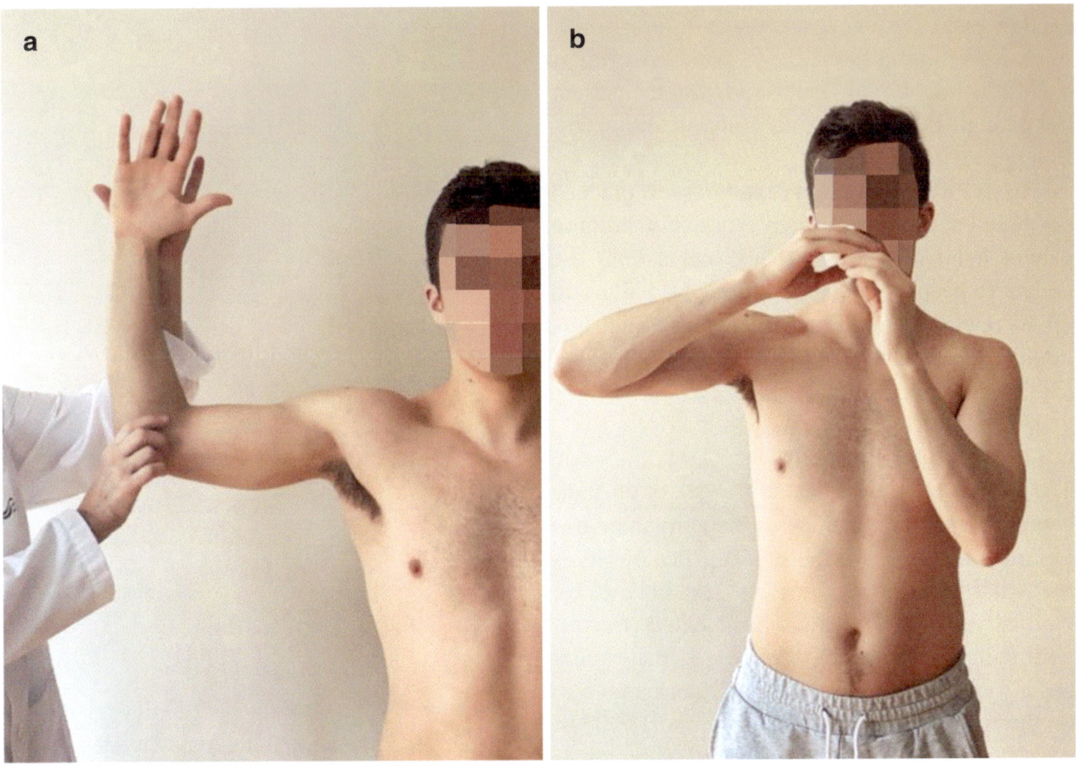

Fig. 5.4 (**a**) Teres Minor strength test. (**b**) Hornblower's sign

done by asking the patient to reach their mouth with their arm at their side, and the patient with teres minor weakness will usually compensate by abducting their shoulder and flexion of their elbow, which is called Hornblower's sign [10].

Tests for the Subscapularis Muscle

The subscapularis is a powerful internal rotator of the shoulder with the arm at side; however, it cannot be tested in this position as the pectoralis major muscle is another internal rotator of the shoulder that will be working in this position [3]. The following tests are used to decrease the effect of the pectoralis major muscle.

Gerber Lift-Off Test (Sensitivity 17–100% and Specificity 60–98%) [6]

In order to perform this test, the patient should be able to reach their arm behind their back. The examiner asks the patient to reach with the dorsum of their hand to the mid lumbar region. Then the patient is instructed to lift their hand away from their back. Inability to do that is considered a positive lift-off sign with a subscapularis muscle tear. If the patient can do the test, then the examiner can test for subscapularis muscle power by resisting lifting the patient's hand off of their back. It is important to make sure that the patient has enough shoulder internal rotation and extension before doing this test.

Passive lift-off lag sign is another test that can be done in this position by passively holding patient hand off of their back and asking the patient to keep this position (Fig. 5.5). Failure to do so is considered as a positive sign and this implies subscapularis muscle injury [15].

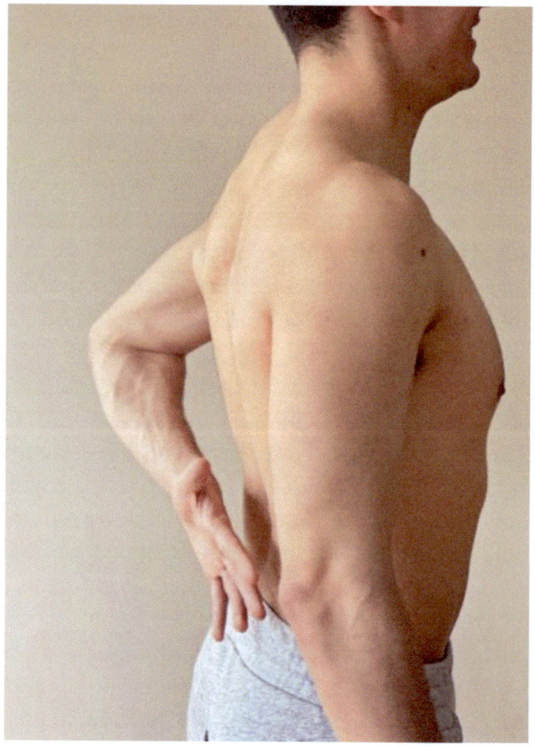

Fig. 5.5 Lift-off test

Belly Press Test/Napoleon Test (Sensitivity 40–43% and Specificity 93–98%) [6]

The examiner asks the patient to press on their abdomen with their hand while keeping their arm in a maximum internal rotation and then observes the wrist position during this test (Fig. 5.6). Normally, the patient should be able to do the test with their wrist straight; however, if the patient has a subscapularis tear, they will not be able to do the test without flexing their wrist. If 90° of wrist flexion is noticed, then the Napoleon test is considered positive with upper and lower sub-

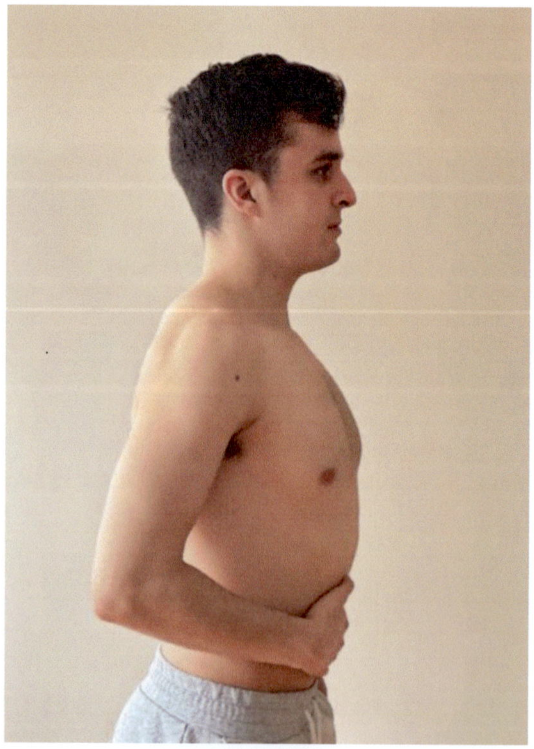

Fig. 5.6 Positive Napoleon test

scapularis tears. If wrist flexion is between 30 and 60°, the test is considered intermediate with a lesser degree of a subscapularis tear [15].

Bear Hug Test (Sensitivity 60% and Specificity 92%) [16]
The examiner asks the patients to put the palm of their hand of the examined shoulder on their other shoulder and to keep their fingers extended and their elbow anterior to their body. Next, the examiner tries to remove the patients' hand from their shoulder by applying an external rotation force perpendicular to the forearm. The test is considered positive if the patients are unable to keep their hand on their shoulder or showed 20% weakness compared to the opposite side.

5.3.5.2 Testing for Other Muscles Around Shoulder Girdle

Trapezius Muscle
In a seated position, the patient is asked to put both of their hands above their head. If the examiner tries to push patients' elbows anteriorly by asking the patient to try to maintain this position, then they can notice the contraction of all parts of the trapezius muscle. An alternative, easier way to examine the muscle is to ask the patient to shrug their shoulder against resistance [5].

Rhomboid Muscle
The examiner asks the patient to put their hands against their hips and to push their elbows posteriorly against resistance [3].

Serratus Anterior
The examiner asks the patient to push against the wall with both arms straight and their palms downward, then the examiner observes for any medial winging of the scapula (Fig. 5.7) [3].

Latissimus Dorsi
With the abducted arm in the scapular plane, the examiner asks the patient to move their arm in a backward and downward direction against resistance as if they are climbing a ladder, then the examiner feels for the muscle contraction (Fig. 5.8) [17].

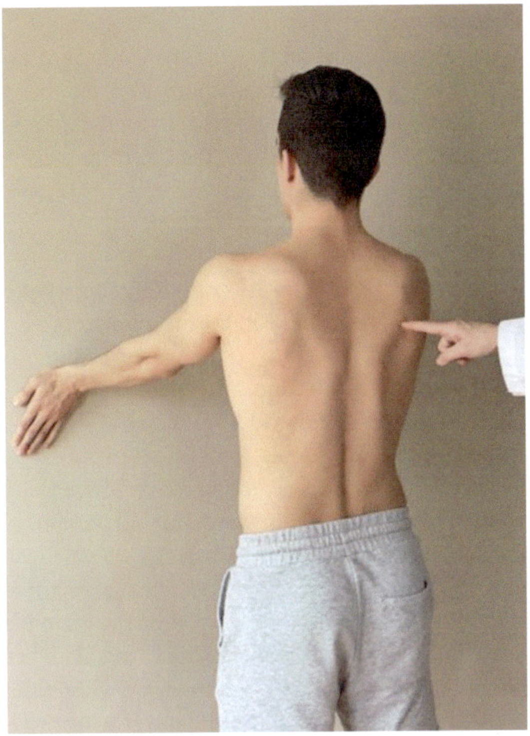

Fig. 5.7 Serratus anterior

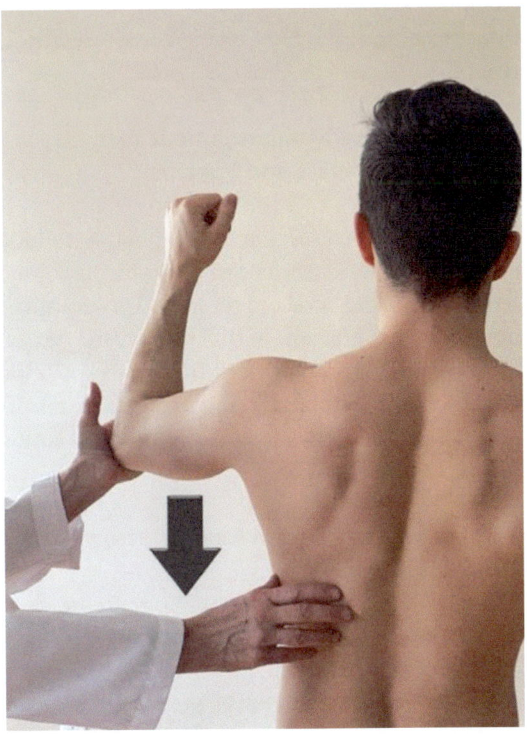

Fig. 5.8 Latissimus dorsi

Deltoid

Deltoid muscle can be tested with the arm at the side or in an abducted position. The examiner asks the patient to abduct, flex, or extend their arm against resistance to test for the lateral, anterior, and posterior deltoid muscle, respectively [3].

5.3.5.3 Impingement Tests

Impingement tests aim to compress the rotator cuff muscles under the coracoacromial arch.

Neer Impingement Sign and Test (Sensitivity 68–89% and Specificity 49–98%) [6]

This test compresses the greater tuberosity against the anteroinferior border of acromion.

The patient's arm is passively elevated in scapular plane with the arm in full extension and internal rotation and at the same time the patient's scapula is stabilized with the other hand. **Neer sign** is positive if there is any pain with shoulder abduction. The **Neer test**, although it is not routinely done nowadays, is performed by subacromial injection of local anesthetic and then the examiner repeats the previous test. If the pain disappears, then the Neer test is considered positive [18].

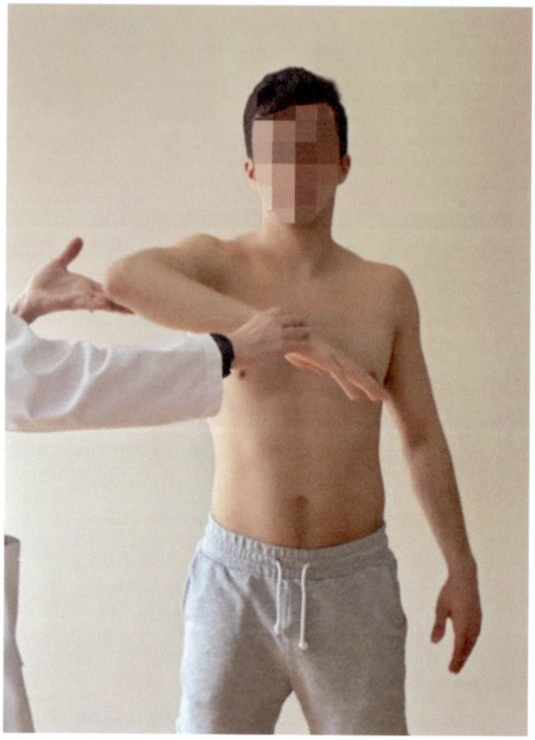

Fig. 5.9 Hawkins-Kennedy impingement test

Hawkins-Kennedy Impingement Test (Sensitivity 72–92% and Specificity 44–78%) [6]

This test compresses the supraspinatus tendon against the anterior surface of the coracoacromial ligament and the coracoid process. The examiner holds the patient's arm in 90° of forward flexion and 10° of adduction and from this position applies internal rotation (Fig. 5.9). The test is positive if there is any pain with internal rotation of the shoulder [19].

5.3.5.4 Biceps Tendon Examination

Speed's Test (Sensitivity 32%, Specificity 75%) [21]

With the elbow in full extension and forearm supination, the examiner asks the patient to resist anterior shoulder forward flexion, and if there is any tenderness along the bicipital groove, the test is positive (Fig. 5.10) [20].

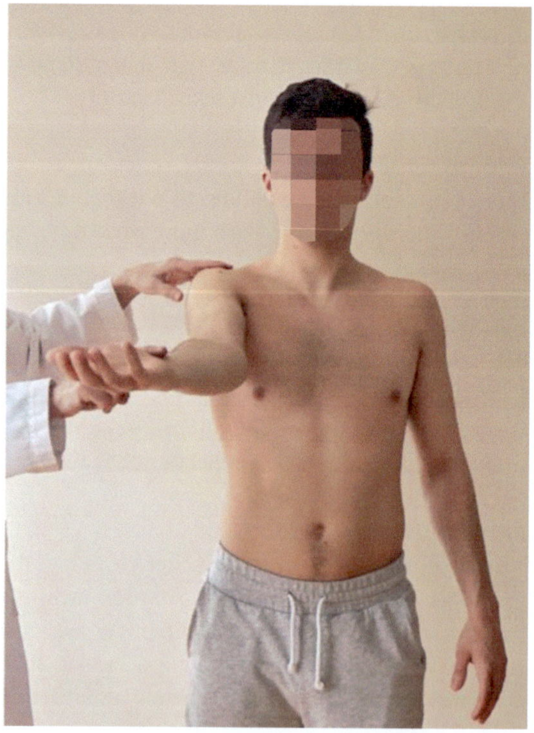

Fig. 5.10 Speed test

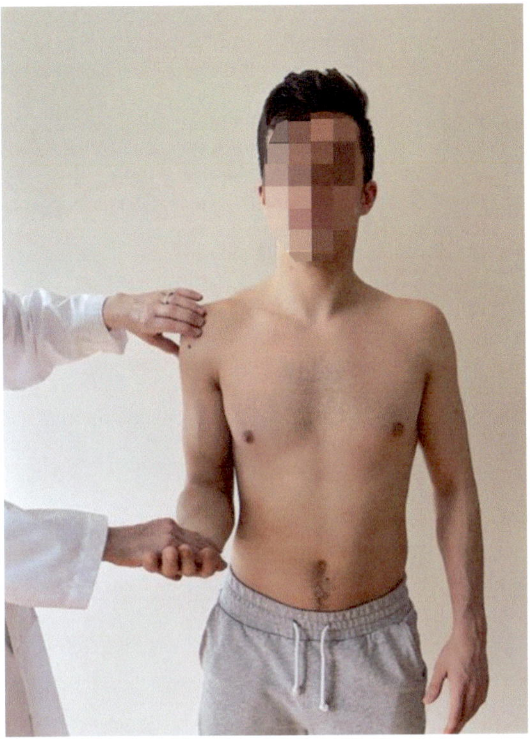

Fig. 5.11 Yergason's test

Yergason's Test (Sensitivity 43%, Specificity 79%) [21]

With the elbow in 90° of flexion beside the patient, the examiner firmly grasps the patient's hand in handshake position and asks him/her to try to supinate his/her forearm against resistance (Fig. 5.11). The test is considered positive if there is any pain or tenderness along the bicipital groove due to presumptive biceps tendinitis. There is a modification for the test to detect biceps tendon dislocation, by resisting external forearm rotation concurrently with supination and palpating for the popping of the tendon out from the groove [22].

5.3.6 Neurovascular Examination

One of the essential components for motor function evaluation is to perform a neurovascular examination of the upper extremity. Sensory and motor evaluation of the nerves (axillary, median, ulnar, radial, and musculocutaneous), as well as vascular examination for brachial, radial, and ulnar pulses, are crucial to exclude other causes of motor function impairment, such as neuropathy and vascular entrapment. This part will be covered later in the book.

5.4 Conclusion

Evaluation of the motor function of the shoulder, especially the rotator cuff muscles, is an important part of clinical evaluations of the patients presenting with shoulder pain or weakness. The clinical tests should be interpreted along with patient presentation and imaging studies. The sensitivity and specificity of these clinical tests vary among the literature, and it is recommended to use a combination of clinical tests to increase the diagnostic accuracy [2].

Acknowledgments Sara Sparavalo and Jie Ma for assistance in writing, language editing, proofreading, formatting, and the preparation of the submission.

References

1. Terry GC, Chopp TM. Functional anatomy of the shoulder. J Athl Train. 2000;35(3):248–55.
2. O'Kane JW, Toresdahl BG. The evidenced-based shoulder evaluation. Curr Sports Med Rep. 2014;13:307–13.
3. Examination techniques in orthopaedics, 2nd ed. Nick Harris and Fazal Ali. Published by Cambridge University Press.
4. Ellenbecker TS. Clinical examination of the shoulder. St. Louis, MO: Elsevier Saunders; 2004.
5. Dhatt SS, Prabhakar S. Handbook of clinical examination in orthopedics: an illustrated guide. Gateway East, Singapore: Springer; 2019.
6. Jain NB, Wilcox RB 3rd, Katz JN, Higgins LD. Clinical examination of the rotator cuff. PM R. 2013;5(1):45–56.
7. Ryan G, Johnston H, Moreside J. Infraspinatus isolation during external rotation exercise at varying degrees of abduction. J Sport Rehabil. 2018;27(4):334–9.
8. Hertel R, Ballmer FT, Lambert SM, Gerber C. Lag signs in the diagnosis of rotator cuff rupture. J Shoulder Elb Surg. 1996;5(4):307–13.
9. Kurokawa D, Sano H, Nagamoto H, Omi R, Shinozaki N, Watanuki S, et al. Muscle activity pattern of the shoulder external rotators differs in adduction and

abduction: an analysis using positron emission tomography. J Shoulder Elb Surg. 2014;23(5):658–64.
10. Walch G, Boulahia A, Calderone S, Robinson AH. The dropping and Hornblower's signs in evaluating rotator cuff tears. J Bone Joint Surg (Br). 1998;80:624–8.
11. Kelly BT, Kadrmas WR, Speer KP. The manual muscle examination for rotator cuff strength: an electromyographic investigation. Am J Sports Med. 1996;24:581–8.
12. Jobe FW, Jobe CM. Painful athletic injuries of the shoulder. Clin Orthop Relat Res. 1983;173:117–24.
13. Itoi E, Kido T, Sano A, Masakazu U, Sato K. Which is the more useful, the "full can test" or the "empty can test", in detecting the torn supraspinatus tendon? Am J Sports Med. 1999;27(1):65–8.
14. The painful shoulder: part I. Clinical evaluation. Woodward TW, Best TMAm Fam Physician. 2000; 61(10):3079–88.
15. Gerber c, Krushell RJ. Isolated ruptures of the tendon of the subscapularis muscle. J Bone Joint Surg (Br). 1991;73:389–94.
16. Barth JR, Burkhart SS, De Beer JF. The bear-hug test: a new and sensitive test for diagnosing a subscapularis tear. Arthroscopy. 2006;22(10): 1076–84.
17. Arcand MA, Reider B. Shoulder and upper arm. In: Reider B, editor. The orthopedic physical examination. Philadelphia: W. B. Salmders; 1999.
18. Neer CS 2nd. Impingement lesions. Clin Orthop Relat Res. 1983;173:70–7.
19. Hawkins RJ, Kennedy JC. Impingement syndrome in athletes. Am J Sports Med. 1980;8(3):151–8.
20. Hegedus EJ, Goode A, Campbell S, et al. Physical examination tests of the shoulder: a systematic review with meta-analysis of individual tests. Br J Sports Med. 2008;42:80–92.
21. Holtby R, Razmjou H. Accuracy of the Speed's and Yergason's tests in detecting biceps pathology and SLAP lesions: comparison with arthroscopic findings. Arthroscopy. 2004;20(3):231–6.
22. Yergason RM. Supination sign. J Bone Joint Surg Am. 1931;13:160.

6 Evaluation of the Stability and Function of the Sternoclavicular and Acromioclavicular Joint

Daniel P. Berthold, Lukas N. Muench, Sebastian Siebenlist, Andreas B. Imhoff, and Augustus D. Mazzocca

6.1 Acromioclavicular Joint

6.1.1 Introduction

Acromioclavicular joint (ACJ) injuries account for 11–12% of all shoulder injuries in the overall population, with the highest prevalence in 20- to 30-year-old male patients in high contact sports [1–3]. Indications for surgery remain highly variable among surgeons and are based on evaluation of clinical examination and patients' concern regarding cosmetic deformity of the shoulder [4]. Thus, a detailed physical examination completed with accurate diagnostic imaging is needed in order to decide between a non-operative care or an interventional procedure. Rockwood described the worldwide best accepted radiographic classification system based on six injury types (Rockwood I–VI) [5]. Non-operative treatment is recommended for patients with type I or II ACJ injury according to the Rockwood classification [6] while high-grade instabilities such as type IV to VI may be addressed surgically. However, it remains under debate if patients with type III injuries benefit from either treatment, with non-operative or surgical treatment showing good, comparable outcomes [7–11]. Thus, in 2014, the Upper Extremity Committee of the International

D. P. Berthold
Department of Orthopaedic Surgery, University of Connecticut, Farmington, CT, USA

Department of Orthopaedic Sports Medicine, Technical University of Munich, Munich, Germany

Musculoskeletal University Centre Munich (MUM), Munich, Deutschland, Germany

L. N. Muench
Department of Orthopaedic Surgery, University of Connecticut, Farmington, CT, USA

Department of Orthopaedic Sports Medicine, Technical University of Munich, Munich, Germany

S. Siebenlist · A. B. Imhoff
Department of Orthopaedic Sports Medicine, Technical University of Munich, Munich, Germany
e-mail: sebastian.siebenlist@tum.de; imhoff@tum.de

A. D. Mazzocca (✉)
Department of Orthopaedic Surgery, University of Connecticut, Farmington, CT, USA
e-mail: mazzocca@uchc.edu

J. G. Lane et al. (eds.), *The Art of the Musculoskeletal Physical Exam*,
https://doi.org/10.1007/978-3-031-24404-9_6

Table 6.1 Diagnostic tests for ACJ and SC Joint instabilities

Diagnostic test	Description	Positive test
O'Brien-Test	1. Patient's arm in 90° of forward flexion with the examiner positioned behind the patient 2. Passive adduction into 10–15° of adduction in complete internal rotation 3. Applying force toward the floor and the patient attempts to resist the downward force	Pain at the ACJ
Hawkins-Kennedy Test	1. Patient's arm in 90° of forward flexion with the elbow flexed to 90° and fixed scapula 2. Internal rotation of the glenohumeral joint	Pain when arm is brought into internal rotation
Cross-body Test	1. Patient's arm is passively abducted in 90° (Fig. 6.5) 2. Cross arm across the patient's chest with arm in internal rotation	Pain when patient's arm is moved across the chest
Paxinos sign [15]	1. Examiner's hand is placed on the affected shoulder with the thumb under the posterolateral aspect of the acromion 2. Index and long fingers rest superior to the clavicle 3. Examiner exerts a force superiorly and inferiorly on the clavicle simultaneously	Pain upon compression of ACJ
Shoulder Shrug-Sign Test	1. Patient is asked to lift the arm to 90° of abduction without elevating the scapula	Inability to lift the arm to 90° of abduction without elevating the whole scapula
Push-Down Test	2. Examiner applies a posteriorly directed force on the medial clavicle	Pain at the SC joint
SC Joint Provocation Test	1. Examiner tries to dislocate clavicular head anteriorly by extending the abducted arm (in 90°) with the elbow in 90° of flexion	Pain at the SC joint, aggravated by any movement of the upper extremity

Society of Arthroscopy, Knee Surgery and Orthopaedic Sports Medicine (ISAKOS) suggested adding Rockwood IIIA and IIIB-type injuries to the historical Rockwood classification with IIIB injuries being classified as unstable ACJ with therapy-resistant scapular dysfunction and persistent horizontal instability [4]. In contrast, type IIIA injuries have been defined as stable without overriding of the clavicle on the cross-body adduction view and without significant scapular dysfunction. However, a definitive consensus regarding non-operative versus operative treatment in patients with type III ACJ injuries is still lacking [7, 8, 10, 11]. Some authors advocate that surgical intervention may be indicated in type III injuries in young, active, highly demanding athletes or manual laborers as well as chronically symptomatic patients [8, 12–14]. Non-operative treatment was shown to restore shoulder kinematics without functional impairment even in elite-level, overhead-throwing athletes [8, 12–14]. Additionally, non-operative management may lead to quicker recovery, including return to sports or work [7].

Thus, clinical examination should focus on detecting pain, vertical and/or horizontal (and rotational) instability and scapulothoracic dysfunction.

Several diagnostic tests have been described in the literature with varying sensitivity (14–79%) and specificity (26–92%) (Table 6.1) being reported [15–18]. However, specificity can be raised to 95–97% by combining two or more diagnostic tests [16, 18].

6.1.2 Clinical Evaluation

Patients should be examined in a standing or sitting position with a downward load being placed on both arms putting an inferior directed stress on the ACJ, making any deformities more visible (Figs. 6.1, 6.2, and 6.3) [4]. Pain, swelling, and point tenderness at the ACJ can be the first indicator for traumatic pathology of the ACJ

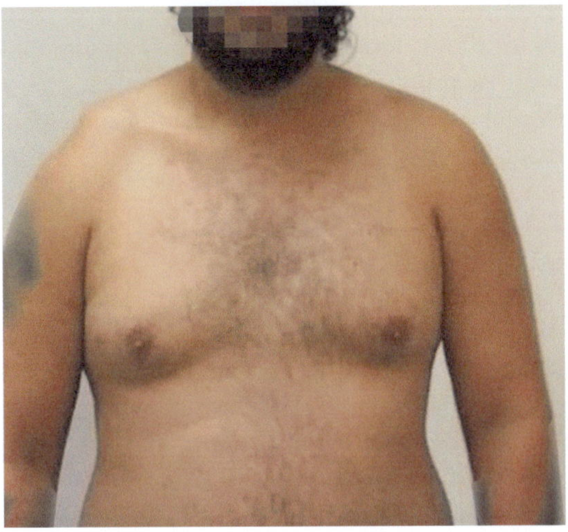

Fig. 6.1 Patients should be examined in standing or sitting position with both arms being pulled downward with stress on the ACJ, making any deformities more visible

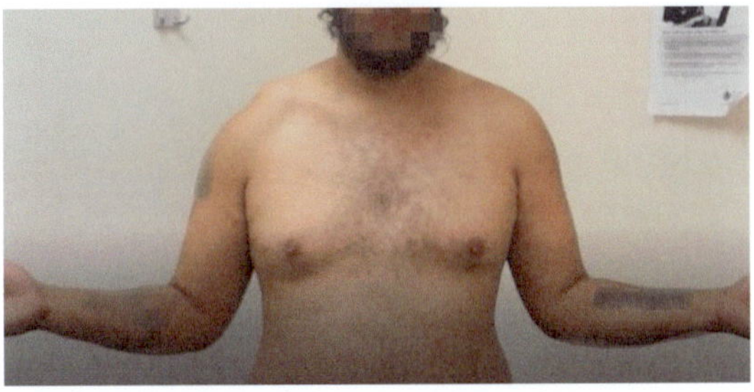

Fig. 6.2 Excessive external rotation may help make any deformities at the ACJ more visible

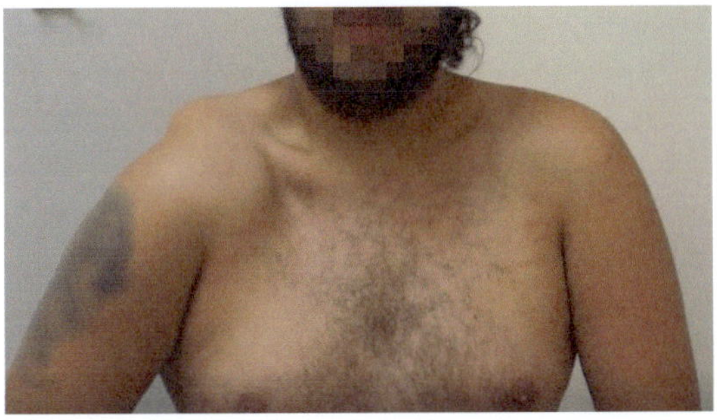

Fig. 6.3 Positive shoulder shrug sign

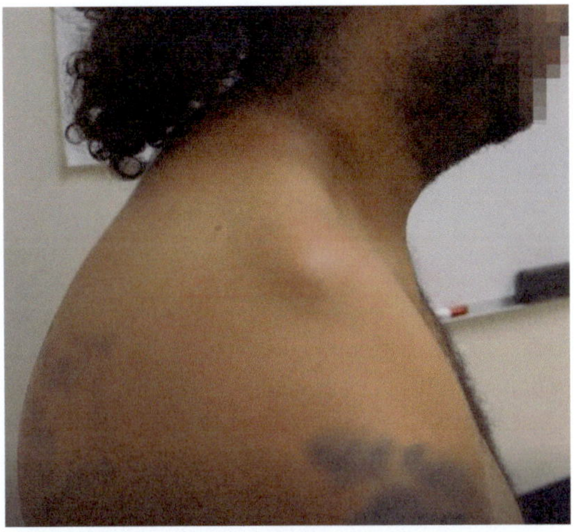

Fig. 6.4 Swelling and a prominent lateral clavicle can be the first indicator for traumatic pathology of the ACJ

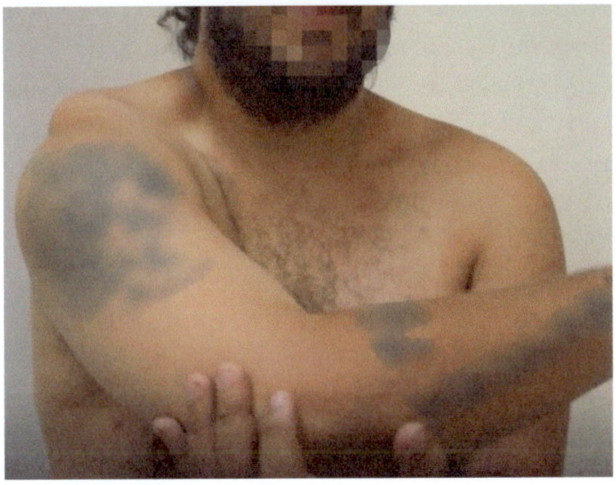

Fig. 6.5 The cross-body test

(Fig. 6.4). As traumatic ACJ dislocations require excessive forces to the shoulder and mostly occur while by falling on an adducted outstretched hand or elbow, intra-articular injuries are found in up to 20% of ACJ Types III–V with superior labral anterior posterior lesions (SLAP) being reported by Tischer and Imhoff as high as 14% in this population [19, 20]. Thus, a thorough clinical evaluation of the shoulder including an examination for neural and vascular injuries should be performed. To this, focus should be on testing vertical and horizontal instability (Fig. 6.5). As slight persistent horizontal instability was shown to have no significant effect on clinical outcomes, any dysfunction to the scapulothoracic joint has to be taken into account [21].

As the integrity of the acromioclavicular (AC) and coracoclavicular (CC) ligaments ensures a physiological motion of the scapula, the scapulohumeral rhythm is highly reliable on proper function of the ACJ [22–24] In high-grade ACJ (IIIB–VI) instabilities, the scapula lacks an anterior strut, and this may result in scapular malfunction such as scapular asymmetry, excessive scapular protraction and internal rotation, and/or anterior tilt [4]. Clinically, this was shown by Gumina et al., whereas almost 70% of the patients (with chronic type III injuries) presented with scapular dyskinesia, and almost 60% of those had a SICK scapula syndrome (scapular malposition, inferior medial scapular winging, coracoid tenderness, and scapular dyskinesis). Of interest, the SICK syndrome was associated with inferior shoulder function [25]. However, in the acute setting, scapular dyskinesia may be able to be assessed due to severe pain which may lead to false-negative results. Thus, any malfunction to the scapula should be assessed at a minimum of 10 days to 3 weeks following injury [4]. Additionally, local infiltration with anaesthetics is a valuable diagnostic tool as relieving pain in the ACJ allows accurate detection of scapular dyskinesia. If patients suffer from disruption of both AC and CC ligaments (Type III), physical therapy may help restore range of motion (ROM) of the glenohumeral (GH) joint as well as improve scapulothoracic rhythm. In case of therapy-resistant scapular malfunction, stabilization of the ACJ is indicated [4, 17, 26].

6.1.3 Imaging of the Acromioclavicular Joint

6.1.3.1 Radiographic Imaging

In current literature, (bilateral) Zanca or panoramic view, (dynamic) modified axillary or Alexander views, stress imaging or specific imaging such as the stryker view may be used for accurate diagnosis of ACJ dislocations [27, 28]. However, a gold standard has still not been proposed [28]. The modified Rockwood classification (Table 6.1) is widely recognized and used for (radiographic) classification of ACJ dislocations. In acute ACJ injuries, radiographs are the preferred diagnostic method due high availability, low costs, and examiner-independence [28].

6.1.3.2 Assessment of Vertical Instability

Current literature shows high inter- and intra-observer reliability for diagnosing vertical instability using bilateral panoramic view with measurement of the CC-distance according to the classification of Rockwood (Figs. 6.6, 6.7, 6.8, and 6.9) [4, 28–30]. Bilateral views allow direct visualization of both CC-distance and

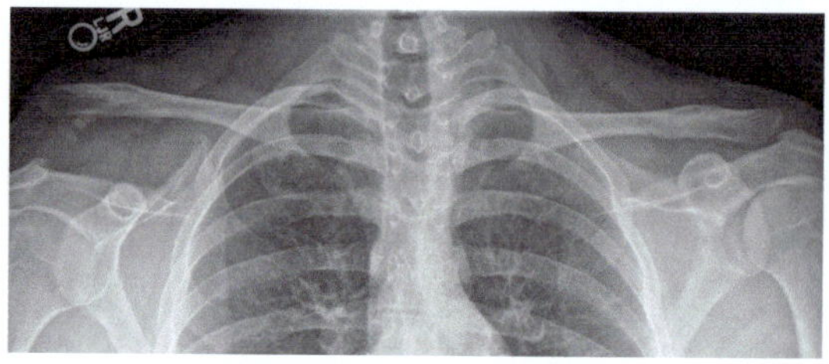

Fig. 6.6 Preoperative panoramic bilateral view, allowing for direct visualization of both CC-distances and direct comparison in both in pre- and postoperative examinations

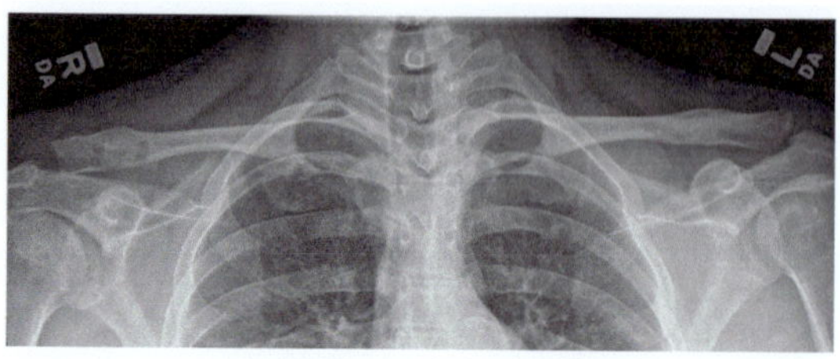

Fig. 6.7 Postoperative panoramic bilateral view using the anatomic coracoclavicular reconstruction technique according to Mazzocca and additional distal clavicle excision (due to ossifications at the lateral clavicle)

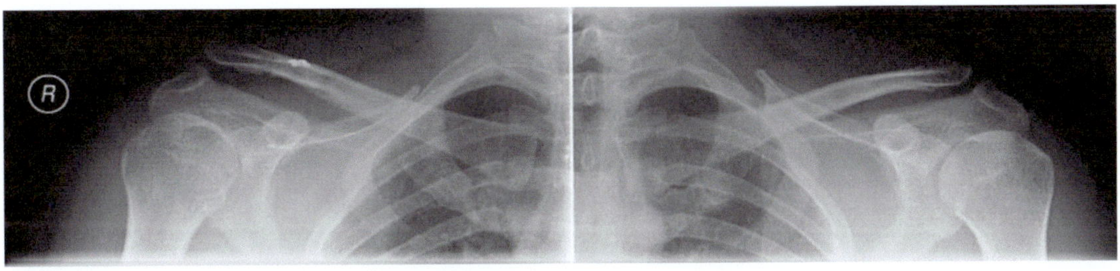

Fig. 6.8 Recurrent ACJ Instability (Rockwood IIIB, right side) after failed AC/CC-reconstruction using an arthroscopic-assisted high-tensile suture and Endobutton technique

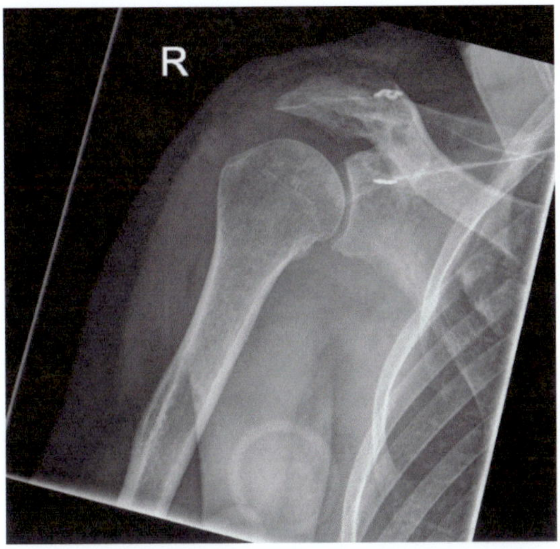

Fig. 6.9 Postoperative plain a.p. radiograph displaying correct CC-stabilization and Endobutton position

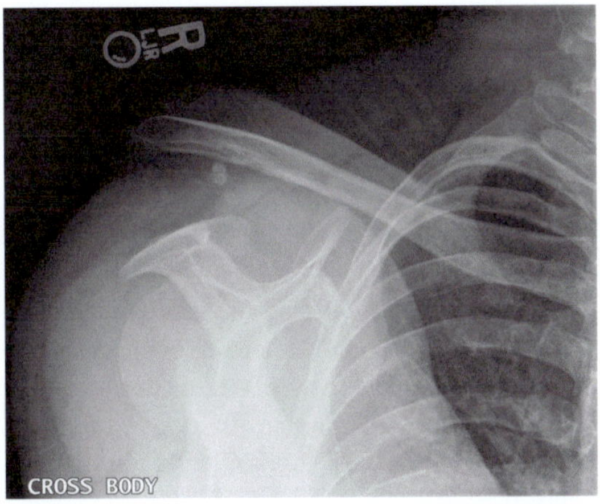

Fig. 6.10 Preoperative modified y-view (cross-body) for detection of dynamic horizontal instability. Note: heterotopic ossification around the chronic insufficient trapezoideum ligament

direct comparison in both in pre- and postoperative examinations.

6.1.3.3 Assessment of Horizontal Instability

Accurate assessment of horizontal instability is shown to be more demanding in clinical practice than vertical displacement with heterogeneous inter- and intra-observer reliability being reported [28]. This is mainly due to the fact that horizontal instability is a dynamic pathology contrary to static radiographic imaging. Therefore, bilateral cross-body adduction view, as proposed by Alexander in 1949, can detect dynamic overriding clavicle (Fig. 6.10) in high-grade ACJ (IIIB) instabilities [31]. However, axillary views (Fig. 6.11) may be more accurate for diagnosis of static horizontal ACJ instability (Rockwood IV).

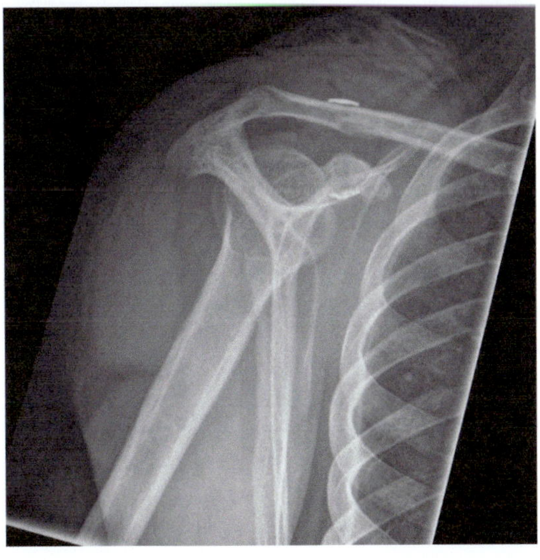

Fig. 6.11 Postoperative axillary view after reconstruction of the AC-joint using the ACCR technique according to Mazzocca

6.1.4 Magnetic Tomography Imaging

Magnetic tomography imaging (MRI) can be a useful tool for detecting intra-articular injuries such as lesions to the long head of the biceps or SLAP-lesions [19, 20]. In the acute setting, MRI may not add additional significant information about vertical and horizontal instability, mostly due to the fact that the examination is performed in supine position contrary to radiographs [28]. However, if necessary, MRI can help differentiate between AC-, CC-, and/or fascia sprain or rupture.

6.1.5 Computer Tomography

Computer tomography (CT) plays a minor role in acute ACJ dislocations [28]. However, concomitant bony injuries such as coracoid or clavicle fractures can be reliable detected. In chronic situations, CT might help surgeons to detect tunnel position and/or widening, insufficient fixations in order to guarantee optimal re-stabilization (Figs. 6.12, 6.13, 6.14, and 6.15) [32–34].

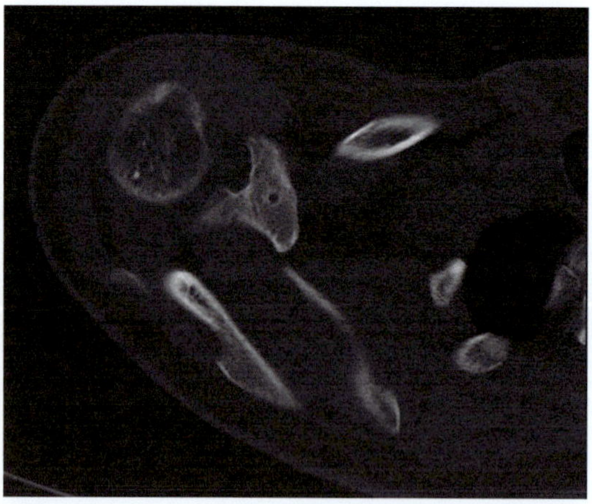

Fig. 6.12 Preoperative CT showing correct placement of the coracoid bone tunnel after recurrent ACJ instability. For this case, the coracoid bone tunnel was re-used during revision surgery

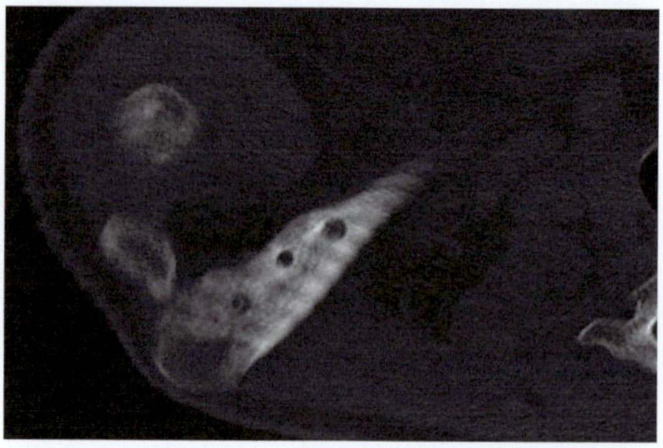

Fig. 6.13 Preoperative CT showing previous clavicular bone tunnel position and location after recurrent ACJ instability. Prior ACJ reconstruction consisted in a hybrid-technique using an allograft (similar to the ACCR technique) looped around the coracoid with an additional high-tensile suture/Endobutton reconstructing technique as an additional healing support

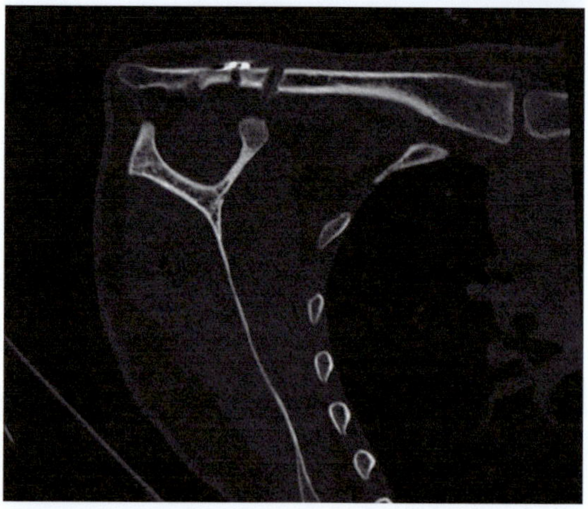

Fig. 6.14 Preoperative CT previous showing clavicular bone tunnel position, location, and tunnel width after recurrent ACJ instability

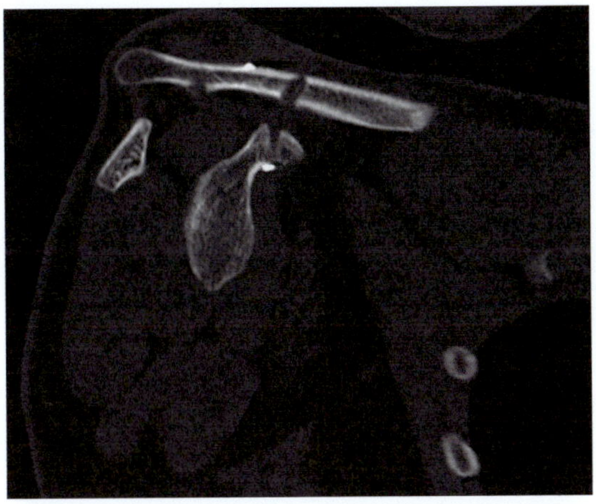

Fig. 6.15 Preoperative CT showing clavicular and coracoid bone tunnel position and tunnel width after recurrent ACJ instability

6.2 Sternoclavicular Joint

Injuries to the sternoclavicular (SC) joint account for as little as 3% of all shoulder injuries and often require high energy trauma to disrupt the SC ligaments [35, 36]. Additionally, close proximity to cardiopulmonary structures may put these structures at risk, especially when facing acute or chronic posterior SC joint dislocations. Unfortunately, injuries to the SC joint are frequently missed, as the pathomechanism of the injury repeatedly leads to more severe, dramatic injuries [37, 38].

Acute dislocations treated non-operatively often result in painful degenerative osteoarthritis, as patients may consult an orthopaedic surgeon years after sustaining the injury [39–41]. Accurate diagnosis of acute SC joint dislocations or fractures is key, with acute posterior dislocations being classified as an emergency, as mediastinal compression, pneumothorax, or complete shock can occur [40, 42]. There have been several reconstruction techniques described in current literature, with biological (auto- or allografts) or suture augmentation may be noted as the current gold standard in treating acute or chronic SC joint dislocations [43–45].

6.2.1 Clinical Evaluation

In the acute setting, dysphagia, cough or hoarseness, venous congestion or a feeling of choking are the first symptoms of acute mediastinal compression [40]. Acute SC joint or clavicle pain and/or deformity or observable step-off deformity are often noted in patients with SC joint injuries. A detailed examination of the rib cage, clavicle, ACJ, and shoulder girdle is important to detect any concomitant injuries such as the "floating clavicle", ACJ instabilities, or fractures [40, 46, 47]. Severe anterior chest pain can be triggered by dynamic arm movement in patients with acute SC joint injuries. In contrast, patients with symptomatic SC joint osteoarthritis often complain about swelling, pain, and local crepitus especially during dynamic arm movement. In these patients, palpable osteophytic prominences may be present [40]. The physician can reproduce these symptoms by the cross-body test, the push-down test, resisted arm abduction, or by dynamic provocation tests. (Table 6.1, Fig. 6.16) [48].

In cases of mild sprains or subluxations, patients may complain about instability. In the case of severe instability, the involved shoulder may be protracted, or patients may hold the involved arm across the chest in an adducted position to support their injured arm. Spontaneous anterior subluxation may also been found in young patients with demonstrated ligamentous laxity.

Atraumatic conditions such as osteoarthritis or rheumatoid arthritis (RA) may also occur [49]. In contrast to traumatic events, RA commonly affects multiple joints or presents bilaterally at the SC joint. Osteophyte prominence, crepitus, and fixed subluxation may also occur [49]. If patients present with unilateral SC joint symptoms, seronegative spondyloarthropathies (e.g.

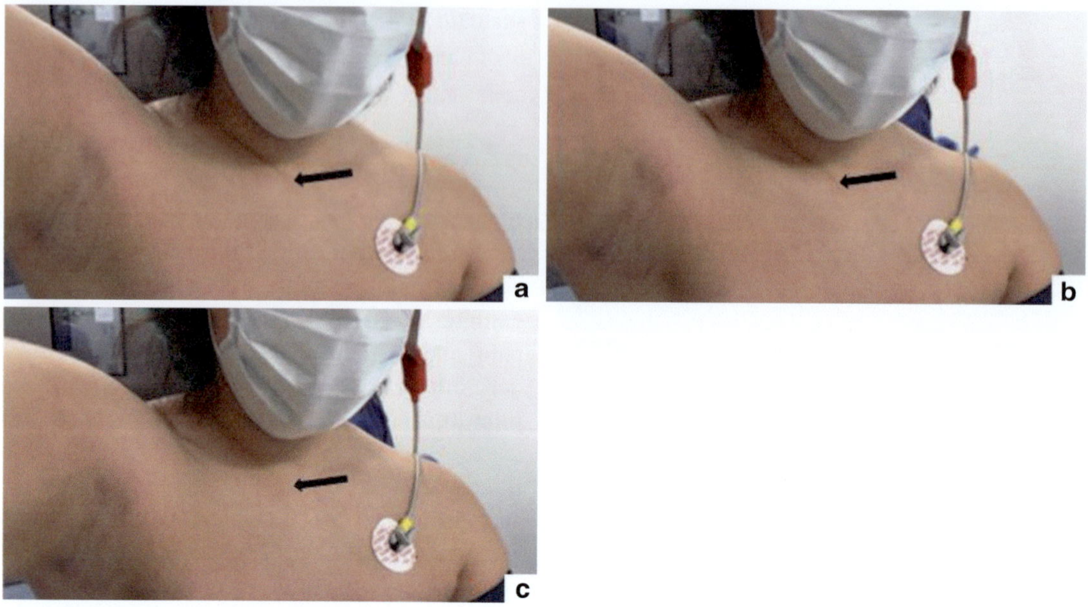

Fig. 6.16 Dynamic examination of a chronic posterior SC joint dislocation. (**a**) The examiner holds the arm in maximum abduction and high external rotation; (**b**) anterior translation of the chronic posterior dislocated SC joint (black arrow) through dynamic arm extension; (**c**) bringing the arm back to maximum abduction and flexion leads to the chronic posterior dislocation

psoriasis, Reiter's syndrome, and Bechterew's disease) have to be excluded.

Septic conditions may be surrounded with fever, chills, night sweating, pain, and swelling (around the SC joint) and may be more common in patients with chronic alcoholism, drug misuse, HIV, osteomyelitis, or immunocompromised patients [49]. Aseptic osteonecrosis such as the Friedrich's disease often include swelling, crepitus, infections, or loss of active shoulder motion [49]. In conclusion, laboratory studies may be helpful to rule out certain inflammatory or infectious diseases and should be considered.

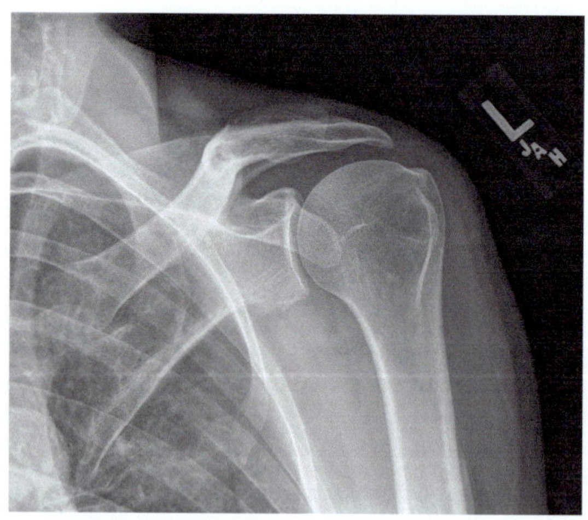

Fig. 6.17 Preoperative a.p. view of the glenohumeral joint including the SC joint

6.3 Imaging of the Sternoclavicular Joint

6.3.1 Radiographic Imaging

Radiographic imaging of SC injuries includes standard AP radiographs (Figs. 6.17, 6.18, and 6.19). However, radiographic imaging of SC injuries remains demanding especially in clinical daily practice because overlapping structures such as ribs and vertebrae can hide the view on the SC joint. In 1990, Wirth and Rockwood described the "serendipity" view, an oblique view

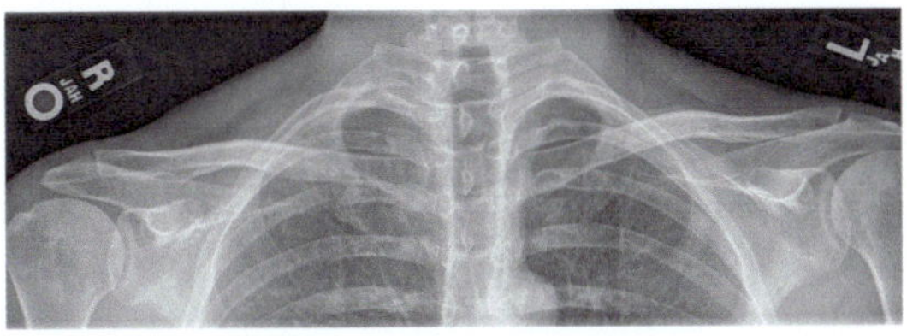

Fig. 6.18 Preoperative bilateral panoramic view to evaluate the SC joint

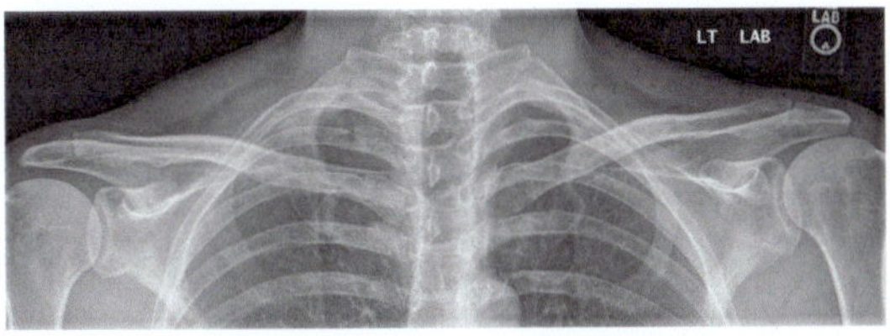

Fig. 6.19 Postoperative bilateral panoramic view to evaluate the SC joint. Those radiographs are highly demanding in clinical practice because overlapping structures such as ribs and vertebrae can hide the view on the SC joint

of the SC joint to visualize both clavicles and help visualize dislocation and osteoarthritis changes to the SC joint [40, 50].

6.3.2 Magnetic Tomography Imaging

MRI plays a minor role in acute SC joint disruption as multiplanar CT scans are favoured in the acute trauma setting due to speed, availability, and ability to distinguish between soft-tissue and bony injuries (Fig. 6.20). However, if patients suffer from symptomatic posterior dislocation, MRI can help visualize concomitant injuries such as vascular lesions [37, 40, 48]. In patients with atraumatic SC joint injuries, MRI may be helpful as an adjunctive method for diagnosing sternoclavicular hyperostosis, avascular necrosis, infections, rheumatoid arthritis, or osteoarthritis [48].

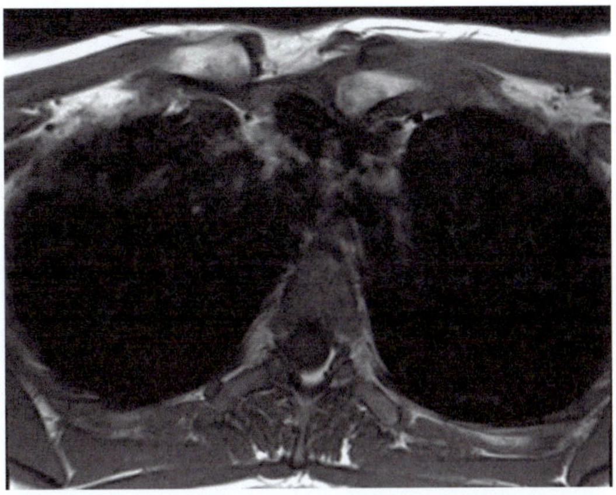

Fig. 6.20 Preoperative MRI of the SC joint showing chronic posterior dislocation. Note the proximity to (neuro) cardiovascular structures

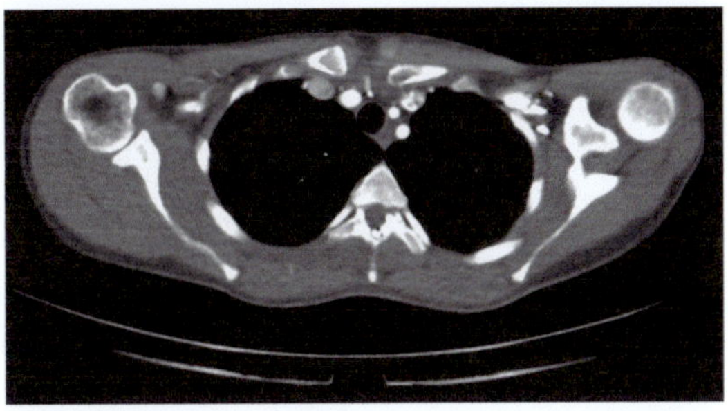

Fig. 6.21 Axial preoperative CT of the SC joint showing chronic posterior dislocation. Note the close proximity to (neuro) cardiovascular structures

6.3.3 Computer Tomography

In acute SC joint dislocations, CT is needed to exclude life-threatening injuries. Further, CT can identify fractures of the clavicle and associated joints, determine the degree of dislocation in comparison to the contralateral, non-injured side (Figs. 6.21 and 6.22) [48, 51]. In case of acute or chronic ST joint infections, CT scans will reveal bony erosions and retrosternal gas or abscess formation [48]. A CT angiogram is

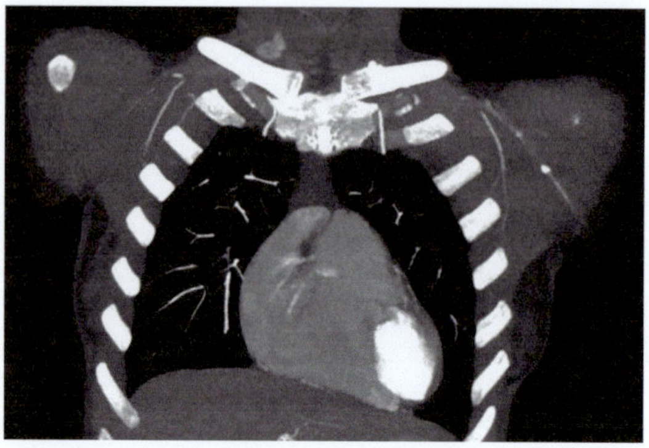

Fig. 6.22 Coronary preoperative CT of the SC joint showing chronic posterior dislocation

indicated before any surgical intervention to evaluate the proximity of vascular and nerval structures.

Additionally, modern 3D-reconstructions may help better visualizing complex dislocations or may help in the preoperative setting (Figs. 6.23 and 6.24).

Fig. 6.23 Preoperative 3D-Reconstruction of the thoracic cage and the shoulder girdle to visualize any dislocation of the SC Joint (in this case: posterior dislocation)

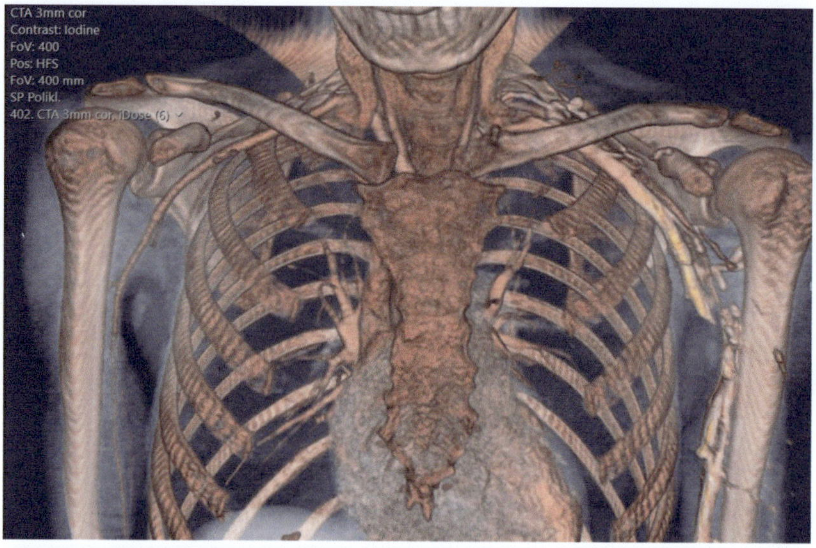

Fig. 6.24 Preoperative axial 3D-Reconstruction of the thoracic cage and the shoulder girdle to visualize the posterior dislocation of the SC Joint

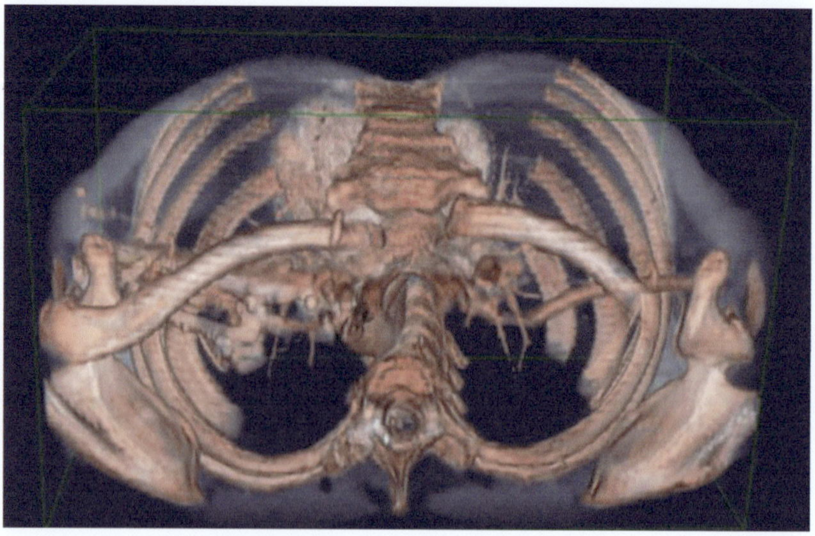

References

1. Braun S, Martetschläger F, Imhoff AB. Acromioclavicular joint injuries and reconstruction. Sports injuries: prevention, diagnosis, treatment and rehabilitation. 2015:1–12.
2. Fraser-Moodie J, Shortt N, Robinson C. Injuries to the acromioclavicular joint. J Bone Joint Surg Br. 2008;90(6):697–707.
3. Skjaker SA, Enger M, Engebretsen L, Brox JI, Bøe B. Young men in sports are at highest risk of acromioclavicular joint injuries: a prospective cohort study. Knee Surg Sports Traumatol Arthrosc. 2020:1–7.
4. Beitzel K, Mazzocca AD, Bak K, et al. ISAKOS upper extremity committee consensus statement on the need for diversification of the Rockwood classification for acromioclavicular joint injuries. Arthroscopy. 2014;30(2):271–8.
5. Rockwood C. Disorders of the acromioclavicular joint. In: Rockwood CA, Matsen FA, editors. The shoulder, vol. 1. Pennsylvania: WB Saunders Company; 1998.
6. Rockwood CA. Injuries to the acromioclavicular joint: subluxations and dislocations about the shoulder. In: Rockwood CA, Green DP, editors. Fractures in adults. Philadelphia: Lippincott, J. B.; 1984. p. 860–910.
7. Beitzel K, Cote MP, Apostolakos J, et al. Current concepts in the treatment of acromioclavicular joint dislocations. Arthroscopy. 2013;29(2):387–97.
8. Frank RM, Cotter EJ, Leroux TS, Romeo AA. Acromioclavicular joint injuries: evidence-based treatment. J Am Acad Orthop Surg. 2019.
9. Mazzocca AD, Conway JE, Johnson S, et al. The anatomic coracoclavicular ligament reconstruction. Oper Tech Sports Med. 2004;12(1):56–61.
10. Millett PJ, Horan MP, Warth RJ. Two-year outcomes after primary anatomic coracoclavicular ligament reconstruction. Arthroscopy. 2015;31(10):1962–73.
11. Moatshe G, Kruckeberg BM, Chahla J, et al. Acromioclavicular and coracoclavicular ligament reconstruction for acromioclavicular joint instability: a systematic review of clinical and radiographic outcomes. Arthroscopy. 2018;34(6):1979–1995 e1978.
12. Gowd AK, Liu JN, Cabarcas BC, et al. Current concepts in the operative management of acromioclavicular dislocations: a systematic review and meta-analysis of operative techniques. Am J Sports Med. 2019;47(11):2745–58.
13. Korsten K, Gunning AC, Leenen LP. Operative or conservative treatment in patients with Rockwood type III acromioclavicular dislocation: a systematic review and update of current literature. Int Orthop. 2014;38(4):831–8.
14. Trainer G, Arciero RA, Mazzocca AD. Practical management of grade III acromioclavicular separations. Clin J Sport Med. 2008;18(2):162–6.
15. Walton J, Mahajan S, Paxinos A, et al. Diagnostic values of tests for acromioclavicular joint pain. J Bone Joint Surg Am. 2004;86(4):807–12.
16. Chronopoulos E, Kim TK, Park HB, Ashenbrenner D, McFarland EG. Diagnostic value of physical tests for isolated chronic acromioclavicular lesions. Am J Sports Med. 2004;32(3):655–61.
17. Kibler WB, Sciascia AD, Morris BJ, Dome DC. Treatment of symptomatic acromioclavicular joint instability by a docking technique: clinical indications, surgical technique, and outcomes. Arthroscopy. 2017;33(4):696–708.e692.
18. Krill MK, Rosas S, Kwon K, Dakkak A, Nwachukwu BU, McCormick F. A concise evidence-based physical examination for diagnosis of acromioclavicular joint pathology: a systematic review. Phys Sportsmed. 2018;46(1):98–104.
19. Ruiz Ibán MA, Moreno Romero MS, Diaz Heredia J, Ruiz Díaz R, Muriel A, López-Alcalde J. The prevalence of intraarticular associated lesions after acute acromioclavicular joint injuries is 20%. A systematic review and meta-analysis. Knee Surg Sports Traumatol Arthrosc. 2020.
20. Tischer T, Salzmann GM, El-Azab H, Vogt S, Imhoff AB. Incidence of associated injuries with acute acromioclavicular joint dislocations types III through V. Am J Sports Med. 2009;37(1):136–9.
21. Scheibel M, Dröschel S, Gerhardt C, Kraus N. Arthroscopically assisted stabilization of acute high-grade acromioclavicular joint separations. Am J Sports Med. 2011;39(7):1507–16.
22. Izadpanah K, Weitzel E, Honal M, et al. In vivo analysis of coracoclavicular ligament kinematics during shoulder abduction. Am J Sports Med. 2012;40(1):185–92.
23. Ludewig PM, Phadke V, Braman JP, Hassett DR, Cieminski CJ, LaPrade RF. Motion of the shoulder complex during multiplanar humeral elevation. J Bone Joint Surg Am. 2009;91(2):378.
24. Seo Y-J, Yoo Y-S, Noh K-C, et al. Dynamic function of coracoclavicular ligament at different shoulder abduction angles: a study using a 3-dimensional finite element model. Arthroscopy. 2012;28(6):778–87.
25. Gumina S, Carbone S, Postacchini F. Scapular dyskinesis and SICK scapula syndrome in patients with chronic type III acromioclavicular dislocation. Arthroscopy. 2009;25(1):40–5.
26. Muench LN, Kia C, Jerliu A, et al. Functional and radiographic outcomes after anatomic coracoclavicular ligament reconstruction for type III/V acromioclavicular joint injuries. Orthop J Sports Med. 2019;7(11):2325967119884539.
27. Berthold D, Dyrna F, Imhoff A, Martetschlaeger F. Innovations for treatment of acromioclavicular joint instability. Arthroskopie. 2019;32(1):11–4.
28. Pogorzelski J, Beitzel K, Ranuccio F, et al. The acutely injured acromioclavicular joint–which imaging modalities should be used for accurate diagnosis? A systematic review. BMC Musculoskelet Disord. 2017;18(1):515.
29. Gastaud O, Raynier JL, Duparc F, et al. Reliability of radiographic measurements for acromioclavicu-

lar joint separations. Orthop Traumatol Surg Res. 2015;101(8 Suppl):S291–5.
30. Schneider MM, Balke M, Koenen P, et al. Inter- and intraobserver reliability of the Rockwood classification in acute acromioclavicular joint dislocations. Knee Surg Sports Traumatol Arthrosc. 2016;24(7):2192–6.
31. Alexander O. Dislocation of the acromioclavicular joint. Radiography. 1949;15(179):260, illust-260, illust.
32. Berthold D, Muench L, Dyrna F, Mazzocca A, Beitzel K, Voss A. Komplikationsmanagement in der Versorgung von Verletzungen des Akromioklavikulargelenks. Arthroskopie. 2020. https://doi.org/10.1007/s00142-00020-00361-00147.
33. Berthold DP, Muench LN, Dyrna F, et al. Radiographic alterations in clavicular bone tunnel width following anatomic coracoclavicular ligament reconstruction (ACCR) for chronic acromioclavicular joint injuries. Knee Surg Sports Traumatol Arthrosc. 2020. https://doi.org/10.1007/s00167-020-05980-z.
34. Dyrna F, Berthold DP, Feucht MJ, et al. The importance of biomechanical properties in revision acromioclavicular joint stabilization: a scoping review. Knee Surg Sports Traumatol Arthrosc. 2019;27(12):3844–55.
35. Groh GI, Wirth MA. Management of traumatic sternoclavicular joint injuries. J Am Acad Orthop Surg. 2011;19(1):1–7.
36. Panzica M, Zeichen J, Hankemeier S, Gaulke R, Krettek C, Jagodzinski M. Long-term outcome after joint reconstruction or medial resection arthroplasty for anterior SCJ instability. Arch Orthop Trauma Surg. 2010;130(5):657–65.
37. Chaudhry FA, Killampalli VV, Chowdhry M, Holland P, Knebel RW. Posterior dislocation of the sternoclavicular joint in a young rugby player. Acta Orthop Traumatol Turc. 2011;45(5):376–8.
38. Perron AD. Chest pain in athletes. Clin Sports Med. 2003;22(1):37–50.
39. Clark RL, Milgram JW, Yawn DH. Fatal aortic perforation and cardiac tamponade due to a Kirschner wire migrating from the right sternoclavicular joint. South Med J. 1974;67(3):316–8.
40. Martetschlager F, Warth RJ, Millett PJ. Instability and degenerative arthritis of the sternoclavicular joint: a current concepts review. Am J Sports Med. 2014;42(4):999–1007.
41. Yang J, Al-Etani H, Letts M. Diagnosis and treatment of posterior sternoclavicular joint dislocations in children. Am J Orthop (Belle Mead NJ). 1996;25(8):565–9.
42. Bicos J, Nicholson GP. Treatment and results of sternoclavicular joint injuries. Clin Sports Med. 2003;22(2):359–70.
43. Martetschläger F, Braun S, Lorenz S, Lenich A, Imhoff AB. Novel technique for sternoclavicular joint reconstruction using a gracilis tendon autograft. Knee Surg Sports Traumatol Arthrosc. 2016;24(7):2225–30.
44. Martetschläger F, Imhoff AB. [Surgical stabilization of acute/chronic sternoclavicular instability with autologous gracilis tendon graft]. Oper Orthop Traumatol. 2014;26(3):218–27.
45. Martetschläger F, Reifenschneider F, Fischer N, et al. Sternoclavicular joint reconstruction fracture risk is reduced with straight drill tunnels and optimized with tendon graft suture augmentation. Orthop J Sports Med. 2019;7(4): 2325967119838265.
46. Eni-Olotu DO, Hobbs NJ. Floating clavicle—simultaneous dislocation of both ends of the clavicle. Injury. 1997;28(4):319–20.
47. Sanders JO, Lyons FA, Rockwood CA Jr. Management of dislocations of both ends of the clavicle. J Bone Joint Surg Am. 1990;72(3):399–402.
48. Iannotti JP, Williams GR. Disorders of the shoulder: diagnosis & management, vol. 1. Lippincott Williams & Wilkins; 2007.
49. Higginbotham TO, Kuhn JE. Atraumatic disorders of the sternoclavicular joint. J Am Acad Orthop Surg. 2005;13(2):138–45.
50. Matsen FA, Rockwood CA, Wirth M, Lippitt S. The shoulder. Saunders; 1990.
51. van Holsbeeck M, van Melkebeke J, Dequeker J, Pennes DR. Radiographic findings of spontaneous subluxation of the sternoclavicular joint. Clin Rheumatol. 1992;11(3):376–81.

7 Evaluation of the Stability and Function of the Glenohumeral Joint

Gregory W. Hall, Anthony Kasch, John G. Lane, and Anshuman Singh

7.1 Introduction

Glenohumeral instability is a condition that primarily affects the young athletic population. The average age in these patients is 24 years of age and males are affected 82% of the time [1]. The shoulder is most unstable when forced anteriorly and inferiorly due to a lack of direct and indirect soft tissue constraint.

For the glenohumeral joint to function properly, coordination of static and dynamic anatomic factors is necessary. The primary static feature is the capsuloligamentous complex which consists of the glenohumeral ligaments. These ligaments provide static restraint to abnormal translation of the glenohumeral joint at the end range of motion. The primary dynamic feature is concavity compression of the glenohumeral joint from the rotator cuff musculature. To understand the interplay of static and dynamic anatomic structures that work together to stabilize the glenohumeral joint, we must first define each individual component.

G. W. Hall
US Navy-Balboa Medical Center, San Diego, CA, USA

A. Kasch
Sanford Health, Bismarck, ND, USA

J. G. Lane · A. Singh (✉)
University of California at San Diego, San Diego, CA, USA
e-mail: jglane@san.rr.com

Table 7.1 Static stabilizers of the glenohumeral joint

Static Stabilizers
1. Osteology
(a) Glenoid
(b) Humeral Head
(c) Coracoid
(d) Acromion
2. Cartilaginous (labrum)
(a) Anatomy, function, composition
(b) Variants
3. Ligamentous
(a) Superior Glenohumeral Ligaments
(b) Coracohumeral ligament
(c) Middle Glenohumeral Ligament
(d) Inferior Glenohumeral Ligament
i. Anterior band
ii. Inferior band
(e) Rotator Interval
4. Pressure (negative intra-articular)
(a) Concavity-compression theory

The static stabilizers of the Glenohumeral joint are listed in Table 7.1.

7.1.1 The Glenoid and Humeral Head

The glenohumeral joint is an articulation between the spherical head of the humerus and the glenoid fossa of the scapula. The glenoid is a pear-shaped superolateral extension of the scapula and has a concave surface that typically sits in 5° of cephalad tilt. In relation to the scapular body, the glenoid is most commonly 5° retroverted but can

J. G. Lane et al. (eds.), *The Art of the Musculoskeletal Physical Exam*,
https://doi.org/10.1007/978-3-031-24404-9_7

range from 10° of anteversion to 7° of retroversion. In contrast, the humeral head is inclined about 130° from the humeral shaft and sits in about 20° of retroversion in relation to the transepicondylar axis of the humerus. The humeral head is three to four times larger than the glenoid fossa which predisposes to joint incongruity and therefore glenohumeral joint instability [2].

Any alteration of normal anatomy of the glenohumeral joint can compromise stability. Whether through congenital defects, traumatic injury or chronic subtle instability, these anatomic changes will impact function of the shoulder. Most commonly encountered are glenoid bone loss and Hill Sachs lesions of the humeral head. Both of these lesions decrease the functional contact zone of the glenohumeral joint which leads to instability.

7.1.2 The Coracoid

The coracoid is an anterosuperior extension of the scapula which serves as the attachment site for the coracobrachialis, pectoralis minor and short head of the biceps. It also serves as the origin of the coracohumeral ligament. While the coracoid provides no direct bony stability to the joint, its location and numerous soft tissue attachments help to provide indirect prevention of humeral head escape from the glenoid as well as contributes to the capsuloligamentous stability via the coracohumeral ligament.

7.1.3 The Labrum

The glenohumeral joint's depth and congruence is augmented by a fibrocartilaginous ring called the labrum [3]. The labrum attaches to the osteochondral rim of the glenoid fossa as well as the periosteum of the glenoid neck. Normal, firm attachments of the labrum increase the depth of the glenoid and provide improved congruence with the large size of the humeral head. In fact, the labrum provides about 50% of the overall depth of the fossa [4]. Superiorly, the labrum is more loosely associated with the glenoid rim and has increased mobility. Here, the labrum is intimately associated with the insertion of the long head of the bicep tendon as it inserts 5 mm medial to the glenoid rim on the supraglenoid tubercle. Inferiorly, the labrum is more round and has a firm, continuous attachment to the glenoid. Attachment sites for the middle and inferior glenohumeral ligaments are located in this inferior glenoid region [5].

Further variation in the attachment of the labrum to the glenoid exists and has been noted in asymptomatic patients 13.5% of the time [6]. These variants include an anterosuperior sublabral foramen, a cord-like MGHL with an associated absent anterosuperior labrum (Buford complex) or without.

Minor injury or separation of the labrum may cause pain. Complete separation of the labrum from the anterior glenoid negates its contribution to joint congruity and depth. Further, separation of the labrum from the glenoid rim may disrupt the attachment of the glenohumeral ligaments negating their contribution to glenohumeral stability. Speer et al. demonstrated that in simulated Bankart lesions in a cadaveric model increased anterior and inferior translation occurred in all degrees in forward elevation [7]. Further, increased posterior translation was noted at 90° of elevation.

7.1.4 Capsuloligamentous Complex

The capsuloligamentous complex of the shoulder consists of the joint capsule itself combined with distinct areas of thickening that have been identified as the coracohumeral ligament and glenohumeral ligaments: the superior, middle, anterior-inferior, and posterior-inferior glenohumeral ligaments. The tension of each of these is not isometric and depends on the position of the humerus in relation to the glenoid. This predisposes the glenohumeral joint to varying risk of instability based on the position of the arm in space at time of injury. While primarily responsible for stability during the end range of motion, studies show the additional contribution of glenohumeral ligament tension to stability during midrange of motion as well [8].

7.2 Coracohumeral Ligament

As described by Burkart, the coracohumeral ligament (CHL) originates from the proximal, dorsolateral aspect of the coracoid [9]. It fans out and consolidates into two distinct bands as it inserts on to the humerus. The main band inserts on to the greater tuberosity of the humerus and the smaller band on to the lesser tuberosity. Transition into the superior glenohumeral ligament exists inferiorly and aids in restraint for posterior translation of the glenohumeral joint in a position of adduction, flexion, and internal rotation. The coracohumeral ligament also limits inferior translation of the glenohumeral joint when positioned in adduction and external rotation.

7.3 Superior Glenohumeral Ligament

The superior glenohumeral ligament (SGHL) originates just anterior to the biceps tendon within the supraglenoid tubercle. It then inserts on the medial ridge of the intertubercular groove of the humerus at the proximal extent of the lesser tuberosity. It works in conjunction with the coracohumeral ligament to stabilize the glenohumeral joint when the arm is placed in a position of adduction and neutral rotation while also preventing inferior translation of the humerus [10].

7.4 Middle Glenohumeral Ligament

The middle glenohumeral ligament (MGHL) also originates nearest to the supraglenoid tubercle but has an additional direct origin from the anterosuperior labrum. The ligament extends distally to insert 2 cm medial to the subscapularis insertion on the humerus. Here, portions of the MGHL do blend with the subscapularis tendon insertion. The MGHL primarily resists anterior and posterior translation of the humerus in a mid-range position of 45° of abduction and 45° of external rotation.

7.5 Inferior Glenohumeral Ligament

The Inferior Glenohumeral Ligament (IGHL) splits into anterior and posterior limbs to form a hammock-like support for inferior stability of the glenohumeral joint. The IGHL originates through direct attachments to the anterior- and posterior-inferior labrum as well as indirect attachments to the glenoid neck. The bands then insert in a V-like coalescence along the inferior humeral head just distal to the articular cartilage border. The central portion of the "V" forms the axillary pouch. The anterior band of the IGHL performs anteroinferior restraint of the glenohumeral joint in 90° of abduction and max external rotation. The posterior band of the IGHL prevents posterior subluxation restraint in 90° of forward flexion and internal rotation [11]. *An anatomic dissection of the Glenohumeral Ligaments are demonstrated in* Fig. 7.1 (Table 7.2).

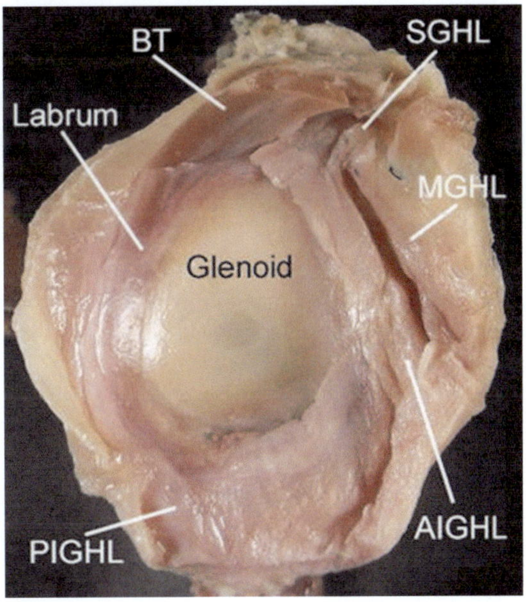

Fig. 7.1 En Face view of the glenoid with capsuloligamentous complex. *BT* long head biceps tendon; *SGHL* superior glenohumeral ligament; *MGHL* middle glenohumeral ligament; *AIGHL* anterior inferior glenohumeral ligament; *PIGHL* posterior inferior glenohumeral ligament

Table 7.2 Dynamic stabilizers of the glenohumeral joint

Dynamic Stabilizers
1. Dynamic
(a) Musculature
i. Rotator cuff
ii. Deltoid
iii. Periscapular
(b) Tendon
i. Conjoined tendon
ii. Long head biceps tendon
(c) Pressure (negative intra-articular)
i. Concavity-compression theory

7.5.1 Rotator Cuff

The rotator cuff is a unit of four individual muscles that coordinate to provide motion, strength, and stability to the glenohumeral joint. The supraspinatus, infraspinatus, subscapularis, and teres minor aid the stability of the joint during mid-range of motion when the glenohumeral ligaments are not at full tension. The main stabilizer of the glenohumeral joint during lifting is the supraspinatus [12]. The stability that the rotator cuff provides is via concavity compression [13]. This concept works by the rotator cuff's compression of the humeral head into the glenoid fossa which provides inherent stability and resistance to shear forces. Concavity compression is possible due to the depth of the glenoid fossa, for which the presence of the fibrocartilaginous labrum maximizes this. The depth of the articulation combined with the strength of compression form the total resistance to instability of the glenohumeral joint [14, 15].

7.5.2 Deltoid

The deltoid's vector of pull and mass (approximately 20% of shoulder musculature) provides additional compression of the glenohumeral joint and resistance to shear force. There are three distinct heads of the deltoid muscle. Anatomic studies have shown that the middle and posterior heads provide the most contribution to dynamic resistance to anterior instability during movement of the arm in the scapular plane [16].

7.5.3 Periscapular Muscles

While indirect in their effect on glenohumeral stability, the scapula and its numerous soft tissue attachments provide indirect stabilization through coordinated movement and positioning of the arm in space. The periscapular muscle group is composed of the serratus anterior, levator scapulae, rhomboids, pectoralis minor, and the trapezius muscle. Given that the glenoid is an extension of the scapula, changes in scapular position will affect the relative inclination and version of the glenoid as it articulates with the humerus. If the scapula is not well controlled and positioned, then the force transfer is felt directly on the static and dynamic stabilizers of the glenohumeral joint which increases their risk of failure. In addition to stabilizing the shoulder for proper glenohumeral function, the periscapular muscles also have a contribution to total force applied to the concavity compression of the joint for dynamic stability.

7.5.4 Long Head Biceps Tendon

The long head of the biceps tendon originates from the superior labrum and supraglenoid tubercle in the glenohumeral joint. It then passes obliquely over the humeral head and enters the bicipital groove before descending in the arm. Numerous cadaveric and biomechanical studies have shown that the long head of the biceps aids in stability of the glenohumeral joint based on load experienced during motion [17]. How this correlates to clinical stability is still debated [18].

7.5.5 Testing

The physical exam of the glenohumeral joint must be preceded by thorough examination of the cervical spine and scapulothoracic region. The patient's entire shoulder must be appropriately exposed to assess for muscular atrophy, scapular winging or other extra-articular causes of shoulder pathology. We recommend having the patient's shirt removed and providing an appropriate gown that protects privacy but also allows for complete examination of the shoulder and arm. The examiner should perform their standard neurovascular assessment of the bilateral upper extremities to rule out neurologic or vascular compromise to the arm.

Once the above is completed, the focused examination of the glenohumeral joint can begin. Full visual inspection should be performed with close attention paid to any atrophy or weakness of the deltoid or rotator cuff musculature. Weakness or injury here can limit the effectiveness of the stability provided from concavity compression. Documentation of the range of motion of the glenohumeral joint should be made while ensuring no scapular motion is included. By having the shoulder completely exposed, the examiner can note when glenohumeral motion stops and when scapular motion begins. The scapula must be stabilized during all exam maneuvers. This can be done with the examiner's opposite hand or the patient can be laid supine to allow the exam table to assist with stabilization.

Full visualization of the scapula also allows for assessment of scapular winging. Medial scapular winging is secondary to weakness of the serratus anterior while lateral winging is a result of weakness of the trapezius muscle (in. spinal accessory nerve). Winging prohibits normal mechanics of the shoulder and predisposes to glenohumeral instability.

Another predisposing factor for shoulder instability includes generalized ligamentous laxity. Beighton's criteria is a widely accepted tool to use for assessment. The examiner will note hyperlaxity at the elbow and knee, thumb, and finger MCP joints as well as forward flexion of the trunk. A score greater than 3 should alert the examiner that there may be an element of ligamentous laxity contributing to a patient's instability. This can have implications on the decision of operative or non-operative management [19]. *The Beighton Criteria scoring are listed in* Table 7.3.

Table 7.3 Beighton criteria

	Left	Right	
Hyperextension of Elbow >10°	1 point	1 point	
Hyperextension of Knee	1 point	1 point	
Touching of thumb to anterior aspect of forearm	1 point	1 point	
Hyperextension of fifth MCP joint	1 point	1 point	
Ability to place palms flat on ground with knees straight			1 point
Total			
Maximum score			**9 points**

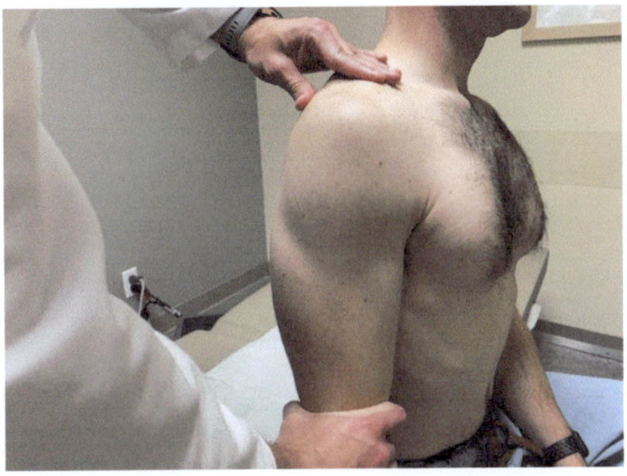

Fig. 7.2 Example of provider pulling distal traction on humerus to elucidate a sulcus sign

There is significant clinical overlap between generalized ligamentous laxity and multidirectional instability. Further assessment of these conditions can be made by examining for a glenohumeral sulcus sign on both shoulders. Originally described by Neer and Foster, the sulcus sign is elucidated by the examiner pulling inferiorly on the arm while the patient is seated and relaxed. If a sulcus >2 cm is noted, then the patient may have a component of multidirectional instability and/or ligamentous laxity [20]. *A photograph of the sulcus sign is shown in* Fig. 7.2.

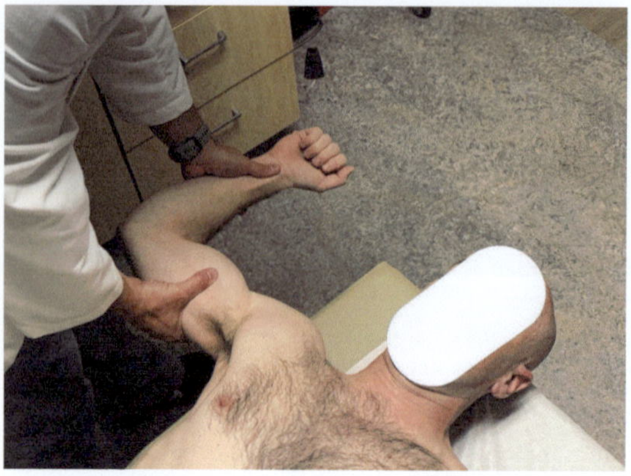

Fig. 7.3 Example of the anterior apprehension test with patient supine

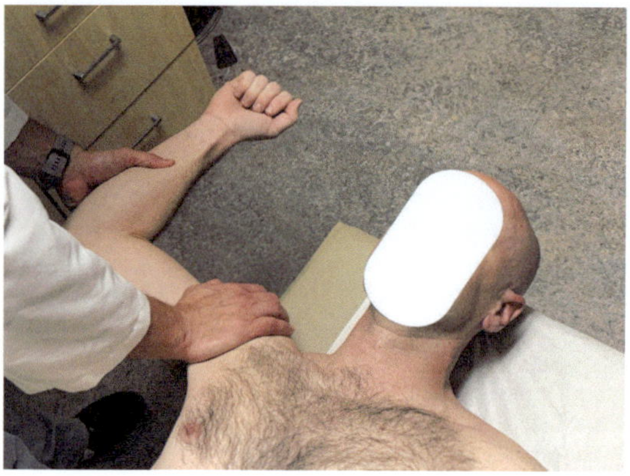

Fig. 7.4 Example of the Jobe relocation test with patient supine

The anterior apprehension test, often performed supine, is evaluated by placing the patient's arm in 90° of abduction and elbow in 90° of flexion, then placing an anteriorly directed force on the posterior humeral head while performing maximum external rotation and observing for signs of pain or apprehension by the patient [21]. *The supine apprehension test is demonstrated in* Fig. 7.3.

If the apprehension test is positive, then a Jobe relocation test should often be performed. The examiner places a posteriorly directed force on the anterior aspect of the humeral head and glenohumeral joint. If this relieves the feeling of instability, then the test is considered positive. If indicated, the examiner may also consider performing the surprise test at this time. Quickly remove the posteriorly directed force over the humeral head and note whether the patient's feelings of apprehension are again present [22]. *The supine Jobe Relocation test is shown in* Fig. 7.4.

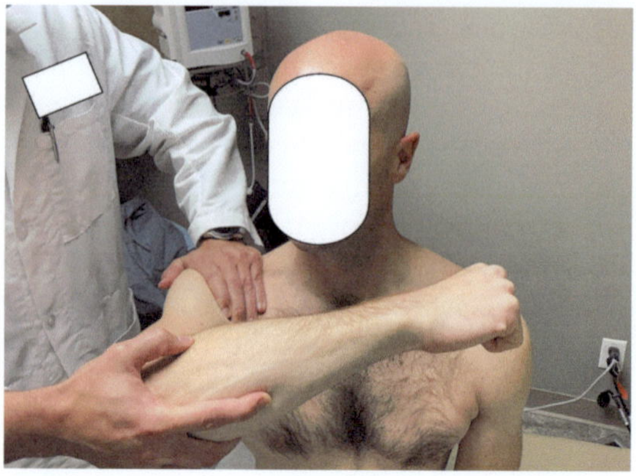

Fig. 7.5 Example of the Jerk test with patient seated

Anterior and posterior translation can be assessed using the load-shift exam. This is accomplished by placing the patient's completely relaxed arm in a position of roughly 80–120° of abduction, 0–20° forward flexion, and 0–30° of external rotation. The patient's arm is then stabilized under the examiner's axilla and the patient's scapula is stabilized with the examiner's other hand. Gently slide the humeral head anteriorly or posteriorly to assess the amount of translation in relation to the glenoid rim. The load and shift test is similar except that it is often performed with the patient's arm hanging at their side. The examiner stabilizes the scapula with one hand and uses their other hand to translate the humeral head anteriorly or posteriorly, again noting the amount of translation in relation to the glenoid rim.

The jerk test is used to evaluate for posterior instability. The test is performed by placing the patient's arm in 90° of abduction and internal rotation. Apply an axial load while adducting the patient's arm across their body and observing for posterior translation, subluxation, or reproduction of pain. *The Posterior Jerk test is demonstrated in* Fig. 7.5.

The Kim test is a similar test used to evaluate for posterior, inferior instability. The patient's arm is placed in 90° of abduction in the scapular plane. Hold the upper forearm and apply an axial load while also forward flexing the arm 45° to create a posterior, inferiorly directed force. Any pain in the posterior shoulder is considered positive for possible posterior, inferior labral pathology. Subluxations or frank dislocations can also be assessed during this exam [23].

References

1. Magnuson JA, Wolf BR, Cronin KJ, Jacobs CA, Ortiz SF, Bishop JY, Baumgarten KM, MOON Shoulder Group, Hettrich CM. Sex-related differences in patients undergoing surgery for shoulder instability: a Multicenter Orthopaedic Outcomes Network (MOON) Shoulder Instability cohort study. J Shoulder Elb Surg. 2019;28(6):1013–21. https://doi.org/10.1016/j.jse.2019.02.020. Epub 2019 Apr 16.
2. Iltoi E. "On-track" and "off-track" shoulder lesions. EFORT Open Rev. 2017;2:343–51.
3. Bankart AB. The pathology and treatment of recurrent dislocation of the shoulder-joint. Br J Surg. 1938;26(101):23–9.
4. Lippitt S, Matsen F. Mechanisms of glenohumeral joint stability. Clin Orthop Relat Res. 1993;291: 20–8.
5. Cooper DE, Arnoczky SP, O'Brien SJ, Warren RF, DiCarlo E, Allen AA. Anatomy, histology, and vascularity of the glenoid labrum. An anatomical study. J Bone Joint Surg Am. 1992;74(1):46–52.
6. Williams MM, Snyder SJ, Buford D. The Buford complex—the "cord-like" middle glenohumeral ligament and absent anterosuperior labrum complex: a normal anatomic capsulolabral variant. Arthroscopy. 1994;10:241–7.
7. Speer KP, Deng X, Borrero S, Torzilli PA, Altchek DA, Warren RF. Biomechanical evaluation of a simulated Bankart lesion. J Bone Joint Surg Am. 1994;76(12):1819–26.
8. Felli L, Biglieni L, Fiore M, et al. Functional study of glenohumeral ligaments. J Orthop Sci. 2012;17:634–7. https://doi.org/10.1007/s00776-012-0261-5.
9. Burkart A, Debski R. Anatomy and function of the glenohumeral ligaments in anterior shoulder instability. Clin Orthop Relat Res. 2002;400:32–9.
10. Turkel SJ, Panio MW, Marshall JL, Girgis FG. Stabilizing mechanisms preventing anterior dislocation of the glenohumeral joint. J Bone Joint Surg Am. 1981;63(8):1208–17.
11. O'Brien SJ, Neves MC, Arnoczky SP, Rozbruck SR, Dicarlo EF, Warren RF, Schwartz R, Wickiewicz TL. The anatomy and histology of the inferior glenohumeral ligament complex of the shoulder. Am J Sports Med. 1990;18(5):449–56.
12. Blache Y, Begon M, Michaud B, et al. Muscle function in glenohumeral stability during lifting task. PLoS One. 2017;12(12):e0189406.
13. Goetti P, et al. Shoulder biomechanics in normal and selected pathological conditions. EFORT Open Rev. 2020;5(8):508–18. https://doi.org/10.1302/2058-5241.5.200006.
14. Lippitt S, Matsen F. Mechanisms of glenohumeral joint stability. Clin Orthop Relat Res. 1993;(291): 20–28.
15. Matsen FA III, Chebli C, Lippitt S. Principles for the evaluation and management of shoulder instability. J Bone Joint Surg Am. 2006;88(3):647–59.
16. Lee SB, An KN. Dynamic glenohumeral stability provided by three heads of the deltoid muscle. Clin Orthop Relat Res. 2002;(400):40–7. https://doi.org/10.1097/00003086-200207000-00006.
17. Elser F, Braun S, Dewing CB, Erik Giphart J, Millett PJ. Anatomy, function, injuries, and treatment of the long head of the biceps brachii tendon. Arthroscopy. 2011;27(4):581–92. https://doi.org/10.1016/j.arthro.2010.10.014.
18. Levine WN, Flatow EL. The pathophysiology of shoulder instability. Am J Sports Med. 2000;28(6):910–7.
19. Beighton P, Solomon L, Soskolne CL. Articular mobility in an African population. Ann Rheum Dis. 1973;32:413–8. https://doi.org/10.1136/ard.32.5.413.
20. Neer CS, Foster CR. Inferior capsular shift for involuntary and multidirectional instability of the shoulder. J Bone Joint Surg Am. 1980;62(6):897–908.
21. Lo IK, Nonweiler B, Woolfrey M, Litchfield R, Kirkley A. An evaluation of the apprehension, relocation, and surprise tests for anterior shoulder instability. Am J Sports Med. 2004;32(2):301–7.
22. Jobe FW, Jobe CM. Painful athletic injuries of the shoulder. Clin Orthop Relat Res. 1983;173: 117–24.
23. Kim SH, Park JS, Jeong WK, Shin SK. The Kim test: a novel test for posteroinferior labral lesion of the shoulder—a comparison to the jerk test. Am J Ports Med. 2005;33:1188–92.

8 Evaluation of the Stability and Function of the Scapulothoracic Joint

Maximilian Hinz, Daniel P. Berthold, Lukas N. Muench, and Knut Beitzel

8.1 Biomechanics

The scapula is fixed anteriorly to the thoracic cage through the acromioclavicular (AC) joint, the clavicle, and the sternoclavicular (SC) joint. Additionally, its posterior connection to the rib-cage is ensured by the trapezius, rhomboids, and serratus anterior muscles [1].

Simplified, the biomechanical properties of the scapula are comparable to those of a triangle (Fig. 8.1) consisting of a stable medial side (thoracic platform) and two mobile sides anterior (clavicular boom) and posterior (scapular body and peri-scapular muscles). The scapula's position is defined by the degree of elevation in the SC joint, length of the clavicle, and tension of the peri-scapular muscles. Thus, the anterior stabilizers are hinged, however, are noted to be static; they elevate the scapula and maintain its lateral position and create the anterior part of the aforementioned triangle. The scapula and the posterior dynamic stabilizers (or peri-scapular muscles) combine into the posterior side and mobilize the scapula along the "scapular track". These two sides are stabilized by the medial side—the thorax. This interaction allows the scapula to accurately position the rotator cuff to move the humeral head and orient the glenoid anterolaterally, thus enabling a functional plane of the shoulder [2].

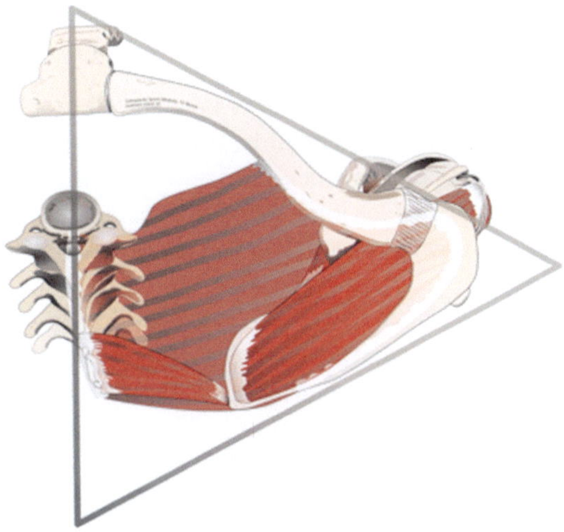

Fig. 8.1 Displaying the scapula's triangle-like build. The triangle's stable medial side is created by the thorax. The length of the clavicle defines the static anterior side of the triangle. The posterior side however, comprising the scapular body and peri-scapular muscles, moves the scapular dynamically along its "scapular track". Picture adapted from [2]

In general, physiological movement of the scapula is achieved by combining three components of motion:

M. Hinz · D. P. Berthold · L. N. Muench
Department of Sports Orthopaedics,
Technical University of Munich, Munich, Germany

K. Beitzel (✉)
Department of Sports Orthopaedics,
Technical University of Munich, Munich, Germany

Department of Shoulder Surgery, ATOS Clinic,
Cologne, Germany
e-mail: knut.beitzel@atos.de

J. G. Lane et al. (eds.), *The Art of the Musculoskeletal Physical Exam*,
https://doi.org/10.1007/978-3-031-24404-9_8

1. upward/downward rotation around a horizontal axis perpendicular to the plane of the scapula
2. internal/external rotation around a vertical axis along the medial border of the scapula
3. anterior/posterior tilt around a horizontal axis along the scapular spine.

Of importance, the clavicle as its connection to the axial skeleton enables two translations:

1. upward/downward rotation (also known as a shrug)
2. retraction/protraction around the rounded thorax [3–5]

Elevation of the humerus is achieved by clavicular elevation, retraction, and posterior axial rotation at the sternoclavicular joint, while the acromioclavicular joint allows for scapular internal and upward rotation as well as posterior tilting [6].

Proper shoulder girdle function requires a complex and coordinated force development and force regulation, which is achieved by the surrounding muscles as well as ligamentous tension of the AC and SC joint [7, 8].

As such, the synergistic humeral and scapular movements are essential to ensure a coordinated scapulohumeral rhythm [9]. Further, during active glenohumeral abduction, the ratio of humeral and scapular movement is 2:1. Of interest, the 2:1 ratio between glenohumeral abduction and scapular rotation may often be observed in the throwing athlete with any disruption resulting in functional impairment [10]. Thus, during the first 30–50° of shoulder abduction, the scapula moves laterally. With increasing active abduction, the scapula rotates approximately 65° (in an arc) until full active shoulder abduction is achieved [10, 11].

The aforementioned, complex interactions between the glenohumeral and scapulothoracic joints highlight that appropriate and correct assessment of the scapulothoracic joint is an important part of the clinical examination. Thus, it seems evident that any intra- or extra-articular shoulder disorders, including rotator cuff tears as well as glenohumeral or acromioclavicular instability, may cause an abnormal scapulothoracic rhythm, often requiring subsequent treatment [12–14].

These abnormal scapular positions (asymmetry or increased topography) and motions have been collectively termed "scapular dyskinesis" [13]. Extensive research has been done on the causative factors for scapular dyskinesis utilizing Moire topography analysis [15], skin electrode monitors [16, 17], and bone pins [18].

8.2 Clinical Examination

One of the most important and common classifications of scapular dyskinesis has been described by Kibler et al. [19] in 2003. Kibler and colleagues classified scapular dyskinesis into three different types (in patients in a resting standing position):

- Type I: prominence of the inferior medial scapular border
- Type II: prominence of the entire medial scapular border
- Type III: superior translation of the entire scapula and prominence of the superior medial border of the scapular

During clinical examination the scapulothoracic rhythm is best assessed from posterior by observing the scapula, while the patient actively raises and lowers the arm.

As such, by carefully observing the descending movement of the arm, any muscle weakness and slight scapular dyskinesis can be observed [10]. Patients may also be asked to retract, protract, and shrug the shoulder so that any—even slight—asymmetry becomes evident.

A common pathology encountered by shoulder surgeons is injury to the AC joint, in which scapular dyskinesis may already be evident 10 days after the trauma, when initial symptoms have subsided [14]. In this setting, scapular dyskinesis develops due to the clavicle's inhibited function regarding the physiological motion of the clavicle as an anterior strut through the AC and coracoclavicular (CC) ligaments [6, 20, 21]. Scapular dyskinesis may develop in (type II) AC joint injuries due to disruption of the AC ligaments and concomitant loss of the clavicular strut. In addition, disruption of the AC and CC ligaments in AC joint injuries (type III or higher) may also cause scapular dyskinesis [4]. The extent to which scapular dyskinesis persists influences the need for surgical intervention in patients presenting with AC joint injury, consequently being a substantial part in recently proposed treatment algorithms [4, 14].

In particular, patients with acute or chronic AC-joint injuries present with scapular dyskinesis that is caused by a protracted and internally rotated scapula. This may result in intractable glenohumeral and lateral shoulder pain as well as limitations in functional range of motion. Chronic functional deficits and persistent pain, observed by any concomitant scapular dyskinesis, may indicate the need for surgical intervention [4, 14].

8.2.1 Scapular Assistance Test

During the scapular assistance test, the examiner pushes upward and laterally on the inferior medial border of the scapula, while active shoulder abduction is performed (Fig. 8.2). As such, the examiner simulates the force of the serratus anterior muscle in the elevation force couple. The test is deemed positive if pain relief is observed [10].

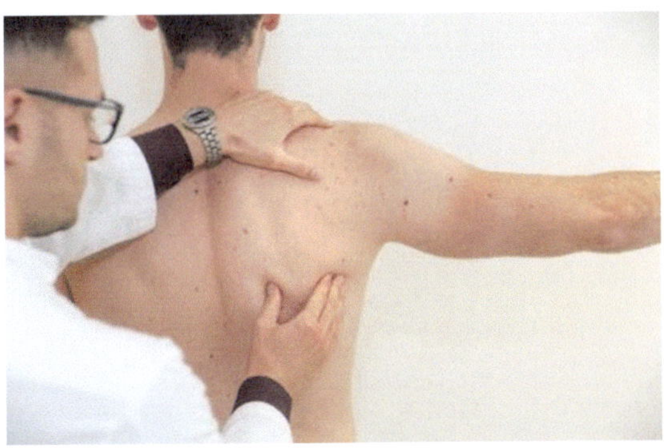

Fig. 8.2 Scapular assistance test

8.2.2 Scapular Retraction Test

A highly valuable tool for detecting full-thickness rotator cuff tears is by following the more traditional Jobe test up with the scapular retraction test [22]. For Jobe test, (Fig. 8.3): the patient is asked to perform internal shoulder rotation, flexion to 90°, and abduction that is collinear to the scapular axis, with then trying to resist downward pressure applied by the examiner. Subsequently, the test is repeated with additional manual stabilization of the patient's medial scapular border by retracting the scapula with the volar aspect of the hand and placing the forearm against it for support (Fig. 8.4) [23].

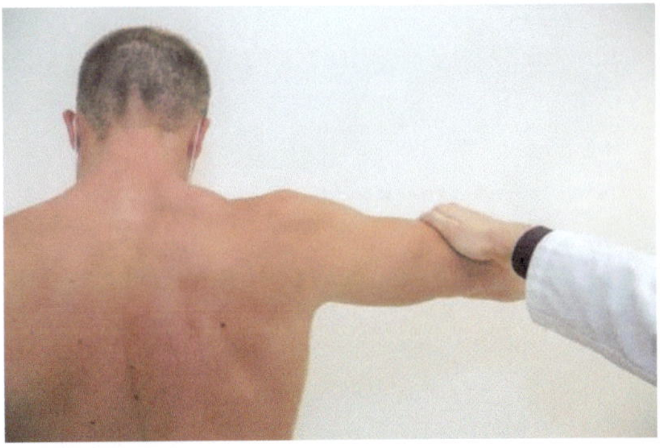

Fig. 8.3 Jobe test

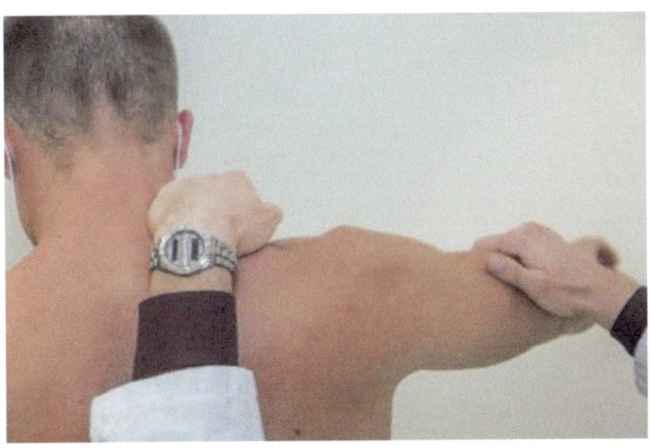

Fig. 8.4 Scapular retraction test

If the stabilization of the scapula leads to a restoration of strength, meaning the patient resisted the examiner's downward pressure during the Jobe test, the scapular retraction test is positive. A positive scapular retraction test indicates an intact rotator cuff.

In contrast, continued weakness or arm dropping is defined as a negative scapular retraction test, indicating a full-thickness rotator cuff tear [23].

8.2.3 Lateral Scapular Slide Test

First proposed by Kibler et al. in 1998, the lateral scapular slide test may be best used during clinical follow-ups, in order to quantify the degree of scapular dyskinesis. The lateral scapular slide test usually evaluates the exact position of the scapula on both the affected and non-affected side. Thus, the distance of the inferomedial border of the scapula to the spine can be measured with both arms hanging on the side; both arms resting on the hips, and with both arms in 90° abduction and maximal glenohumeral internal rotation (Fig. 8.5). Differences of more than 1.5 cm in any position between the affected and non-affected side are considered pathological [10].

Another modality to quantify the degree of scapular dyskinesis has been proposed by Park et al.. The authors reported that a 3-dimensional wing computer tomographic analysis can be an efficient tool with a good interrater reliability for evaluating scapular dyskinesis with a significant correlation to the physical exam [24].

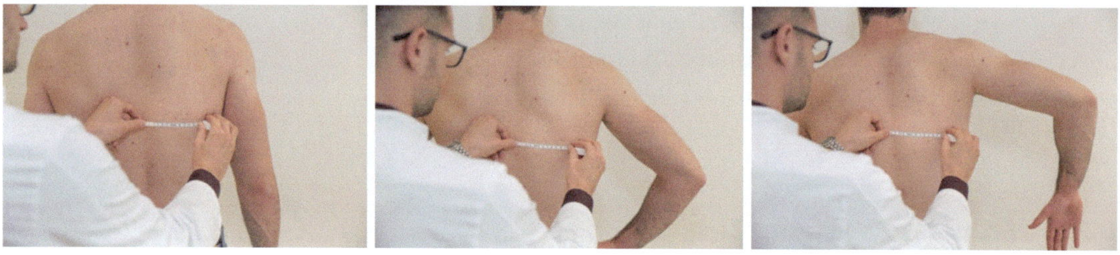

Fig. 8.5 Lateral scapular slide test

8.2.4 Scapular Winging

Various pathologies of the upper extremity have been identified in current literature as causes for scapular winging, with thoracic longus nerve palsy, resulting in concomitant serratus anterior impairment, being the most common one.

The evaluation starts with the arm in the resting position. In patients with thoracic longus nerve palsy, a medialized scapula in the resting position is characteristic. Due to the lack of activating the serratus anterior muscle, patients may have difficulty with shoulder abduction above 120°; concurrently, the degree of winging increases as well [25].

The patient may also be asked to actively perform shoulder flexion. During this manoeuvre, any scapular movement in relation to the thoracic wall can be observed. As such, winging may be present with resisted motion (e.g. when performing wall push-ups) (Fig. 8.6). Thus, scapular winging may occur in static, dynamic, or resisted motions [25].

Furthermore, other pathologies—for example, accessory nerve palsy—have also been identified to be a risk for causing what is referred to as "pseudowinging" [26]. Patients with "pseudowinging" due to accessory nerve palsy—and a concomitant trapezius muscle impairment—present with a depressed shoulder, laterally translated scapula, and a laterally rotated inferior angle. Trapezius wasting, inability to shrug the shoulder, and difficulty to actively perform shoulder flexion and abduction may confirm the diagnosis [25].

In patients with scapular winging, a thorough examination of the adjacent structures including the glenohumeral joint, acromioclavicular joint, and cervical spine should be performed to differentiate scapular winging or "pseudowinging" from other pathologies.

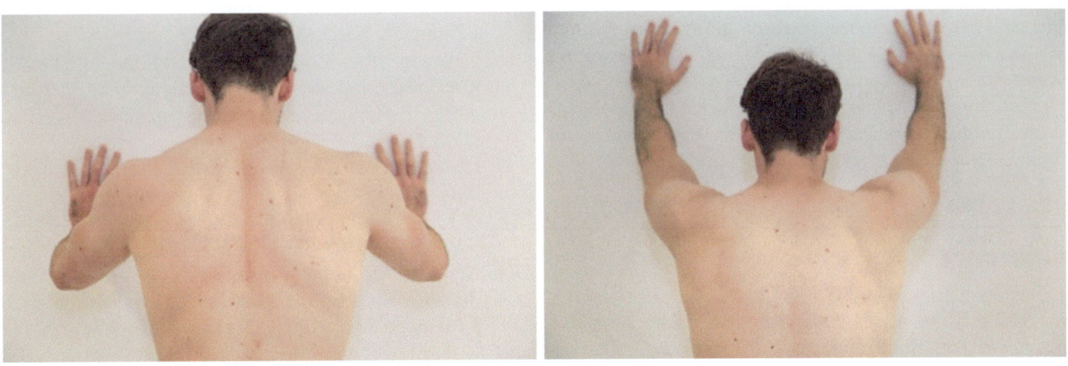

Fig. 8.6 Wall push-up

8.3 Summary

The scapulothoracic joint is stabilized medially by the thorax and both stabilized and statically moved anteriorly by the clavicle and manoeuvred on the so-called scapular track by the periscapular muscles. Together, these structures that influence the position of the scapula also enable the shoulder joint to perform physiologically.

Any form of unphysiological scapula setting or movement, collectively termed scapular dyskinesis, may be caused by various shoulder pathologies, such as rotator cuff tears, shoulder instability, and injuries to the AC joint.

Thoroughly observing the movement of the scapula by applying the aforementioned clinical tests may help the examiner to differentiate between the pathologies that cause an unphysiological position or movement of the scapula and guide their treatment.

Conflict of Interest The authors Hinz M., Berthold D.P., and Muench L.N. declare that they have no conflicts of interest. Beitzel K. reports research grants from Arthrex Inc., is a consultant for Arthrex Inc. and receives royalties from Arthrex Inc.

Funding: No funding has been provided from agencies in the public, commercial, or not-for-profit sectors to complete this manuscript.

References

1. Roche SJ, et al. Scapular dyskinesis: the surgeon's perspective. Shoulder Elbow. 2015;7(4):289–97.
2. Bain GI, Phadnis J, Sonnabend DH. The functional shoulder. In: Normal and pathological anatomy of the shoulder. Springer; 2015. p. 403–14.
3. McClure PW, et al. Direct 3-dimensional measurement of scapular kinematics during dynamic movements in vivo. J Shoulder Elb Surg. 2001;10(3):269–77.
4. Kibler WB, Sciascia A. Current concepts: scapular dyskinesis. Br J Sports Med. 2010;44(5):300–5.
5. Kibler BW, Sciascia A, Wilkes T. Scapular dyskinesis and its relation to shoulder injury. J Am Acad Orthop Surg. 2012;20(6):364–72.
6. Ludewig PM, et al. Motion of the shoulder complex during multiplanar humeral elevation. J Bone Joint Surg Am. 2009;91(2):378.
7. Happee R, Van der Helm F. The control of shoulder muscles during goal directed movements, an inverse dynamic analysis. J Biomech. 1995;28(10):1179–91.
8. Weiser WM, et al. Effects of simulated scapular protraction on anterior glenohumeral stability. Am J Sports Med. 1999;27(6):801–5.
9. Flores-Hernandez C, et al. Scapulothoracic rhythm affects glenohumeral joint force. JSES Open Access. 2019;3(2):77–82.
10. Ben Kibler W. The role of the scapula in athletic shoulder function. Am J Sports Med. 1998;26(2):325–37.
11. Nk P, Walker PS. Normal and abnormal motion of the shoulder. J Bone Joint Surg A. 1976;58:I95.
12. Paletta GA Jr, et al. Shoulder kinematics with two-plane x-ray evaluation in patients with anterior instability or rotator cuff tearing. J Shoulder Elb Surg. 1997;6(6):516–27.
13. Warner JJ, et al. Scapulothoracic motion in normal shoulders and shoulders with glenohumeral instability and impingement syndrome: a study using Moire topographic analysis. Clin Orthop Relat Res. 1992;285:191–9.
14. Beitzel K, et al. ISAKOS upper extremity committee consensus statement on the need for diversification of the Rockwood classification for acromioclavicular joint injuries. Arthroscopy. 2014;30(2):271–8.
15. Gomes PF, et al. Measurement of scapular kinematics with the moiré fringe projection technique. J Biomech. 2010;43(6):1215–9.
16. Ludewig PM, Cook TM. Alterations in shoulder kinematics and associated muscle activity in people with symptoms of shoulder impingement. Phys Ther. 2000;80(3):276–91.
17. Ludewig PM, Cook TM, Nawoczenski DA. Three-dimensional scapular orientation and muscle activity at selected positions of humeral elevation. J Orthop Sports Phys Ther. 1996;24(2):57–65.
18. Karduna AR, et al. Dynamic measurements of three-dimensional scapular kinematics: a validation study. J Biomech Eng. 2001;123(2):184–90.
19. Kibler BW, McMullen J. Scapular dyskinesis and its relation to shoulder pain. J Am Acad Orthop Surg. 2003;11(2):142–51.
20. Seo Y-J, et al. Dynamic function of coracoclavicular ligament at different shoulder abduction angles: a study using a 3-dimensional finite element model. Arthroscopy. 2012;28(6):778–87.
21. Izadpanah K, et al. In vivo analysis of coracoclavicular ligament kinematics during shoulder abduction. Am J Sports Med. 2012;40(1):185–92.
22. Kibler WB, Sciascia A, Dome D. Evaluation of apparent and absolute supraspinatus strength in patients with shoulder injury using the scapular retraction test. Am J Sports Med. 2006;34(10):1643–7.
23. Khazzam M, et al. Diagnostic Accuracy of the Scapular Retraction Test in Assessing the Status of the Rotator Cuff. Orthop J Sports Med. 2018;6(10):2325967118799308.

24. Park J-Y, et al. How to assess scapular dyskinesis precisely: 3-dimensional wing computer tomography-a new diagnostic modality. J Shoulder Elb Surg. 2013;22(8):1084–91.
25. Kuhn JE, Plancher KD, Hawkins RJ. Scapular winging. J Am Acad Orthop Surg. 1995;3(6):319–25.
26. Duralde X. Evaluation and treatment of the winged scapula. J South Orthop Assoc. 1995;4(1):38.

9 Nerve Compressions Around the Shoulder

Daniel Adolfo Slullitel, Glasberg Ernesto, Escalante Mateo and Vega Francisco

The complex anatomy of the shoulder, coupled with its delicate biomechanics, must be understood from an overall perspective.

Its proximity to the brachial plexus and terminal branches make the understanding of this anatomical region even more important.

When the physician is faced with a shoulder pathology, he must not only consider the structures that shape it, but also the surrounding environment, since referred pain often simulates a pathology of the nerves that innervate the shoulder, and the same occurs the other way round.

A diverse situation exists as a result of different variations in size, shape, and location of the anatomical components of this joint, and this is seen among people, between men and women and even within the same person in the contralateral side [1].

The symptomatology that causes nerve involvement that run through the shoulder region, generate symptoms that patients report as pain, burning sensation, paresthesia, or loss of strength.

Regarding nerves, it is crucial to know their pathways, variants, the function that they perform, the symptoms they produce when affected by traumatic, congenital, sports, occupational or degenerative conditions, so that the territory where the pathology of a peripheral nerve is presented can be accurately identified on physical examination, worst case scenario on diagnosis for shoulder problems is probably faced when there is subtle involvement, and this is also ideal time for treatment, before its insufficiency sets the neuromuscular motor unit onto an irreversible path.

Nerve compressions include those which are common and of typical presentation, others which are less common and also those which are common but of an atypical presentation.

Additional studies to physical examination provide a great deal of information, and sometimes transient suppression tests (blockages, etc.) give us a quick diagnosis.

Determining whether pathology affects a muscle or tendon insertion versus nerve involvement requires clinical, imaging and surgical diagnostic tools, which imply expertise and multidisciplinary work.

The focus of this chapter will be on four nerves around the shoulder, for which we will try

D. A. Slullitel (✉)
IJS Instituto Jaime Slullitel, Sanatorio de la Mujer, Rosario, Argentina

UAI Universidad Abierta Interamericana, Rosario, Argentina

G. Ernesto
UAI Universidad Abierta Interamericana, Rosario, Argentina

Sanatorio de la Mujer, Rosario, Argentina

E. Mateo · V. Francisco
Sanatorio de la Mujer, Rosario, Argentina

J. G. Lane et al. (eds.), *The Art of the Musculoskeletal Physical Exam*,
https://doi.org/10.1007/978-3-031-24404-9_9

to describe their anatomy, their symptoms when dysfunction occurs, physical examination, additional diagnostic methods, and possible treatments.

9.1 Suprascapular Nerve

9.1.1 Introduction

Suprascapular nerve involvement was described by Frenchman and Andre Thomas in 1936 and later updated by Kopell and Thompson in 1959 [2]. As it is a mixed nerve, when affected, it causes pain and different degrees of involvement of supra and infraspinatus muscles [3].

Suprascapular nerve compression is a rare disorder and one of its predisposing factors is sports practice, especially in overhead athletes. However, it can affect people who perform different working activities, and in some cases, without apparent cause. Additional electrical and imaging studies usually show loss of function, muscle atrophy in advanced cases and edema in the early stages. One third of the patients are asymptomatic [4].

Two types of presentations are possible. One is related to nerve compression at the suprascapular notch. This form is the most frequent and patient reports pain affecting mostly the posterior aspect of the shoulder. During physical examination, we see atrophy of the supra and infraspinatus muscles and positive signs which are typical when studying these affected muscles, in particular, difficulty to raise the arm overhead. For distal compression (spine-glenoid notch), only the infraspinatus muscle is affected, accompanied by pain that is often vague and difficult to localize [5].

Correct shoulder function requires the work of all its components, including bones, ligaments, muscles, and nerves. A physiological function to take into account is proprioception and, because shoulder joint has a large range of motion and must provide stability for distal joints for proper functioning, this reflexive mechanism must be intact. The shoulder operates by means of dynamic stabilization and, for this purpose, the joint and its components are rich in mechanoreceptors. With injury, receptors sensitivity decreases and this originates proprioceptive problems. Proprioceptive sensitivity loss favors injuries recurrencies [6, 7].

9.1.2 Anatomy

Identifying paralysis or paresis of innervated muscles by the suprascapular nerve, as well as sensory disorders caused when this nerve is affected, requires a detailed knowledge of its anatomy.

Suprascapular nerve may emerge from upper primary trunk although it may also arise directly from C5 root with some fibers emerging from C4 and runs laterally and dorsally towards scapular notch [8] (Photo 9.1).

From its starting point, it runs laterally to the cervical triangle and behind the clavicle passing beneath the trapezius. The suprascapular notch is the most common site of nerve compression and, there, suprascapular artery and vein pass over the transverse ligament [9] (Photo 9.2). After passing through the suprascapular notch, it lies lateral to supraspinous fossa, with giving its motor branches towards the supraspinatus muscle and receiving sensory fibers from coraco-humeral and coraco-acromial ligaments, subacromial

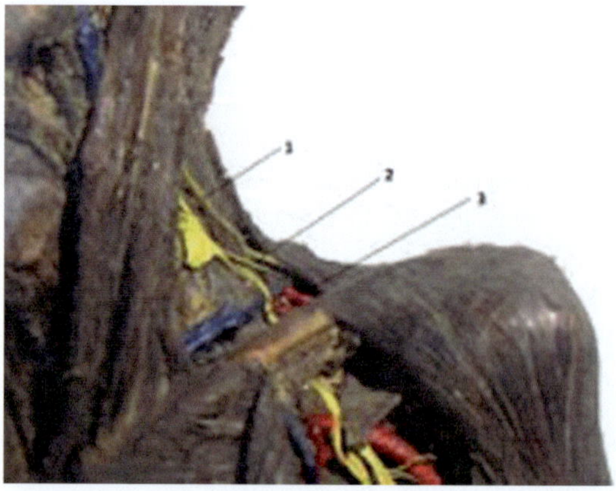

Photo 9.1 Supraclavicular fossa anatomical preparation. (1) Suprascapular nerve. (2) Spinal nerve. (3) Lateral fascicle. Courtesy of Morphological Science Museum National University of Rosario

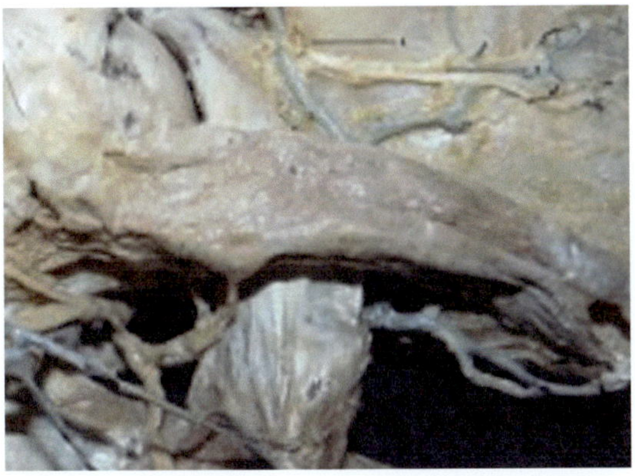

Photo 9.2 Anatomical preparation of the posterior aspect of the shoulder. (1) Spinoglenoid notch. (2) Infraspinatus branch suprascapular nerve. Courtesy of Dr. Ezequiel Zaidemberg

bursa, acromioclavicular joint, and posterior part of the glenohumeral joint, then passing beneath supraspinatus muscle and reaching spinoglenoid notch, from where it passes beneath spinoglenoid ligament (also called transverse inferior scapular ligament) [8–10].

Presence of the suprascapular notch is a consistent finding in human beings, and nerve entrapment at this foramen is rare. Variations in size and shape could predispose to compression.

9.1.3 Compression Sites and Causes

Compression sites are places where scapula and ligament morphology predispose this condition statically, combining with a dynamic component, which is a patient's certain specific activity. The suprascapular notch is the most common site for nerve pathology (Photo 9.3).

There are two important details that make this region a foramen: one is that it is mostly formed by bone and second is that the bone thickness at this level is 2/3 mm. Other aspects to emphasize are that notch diameter does not necessarily predicts a greater foramen since ligament thickness can determine a decrease in volume and also that, being the shoulder a very dynamic region, nerve movement inside this foramen predisposes it to suffering.

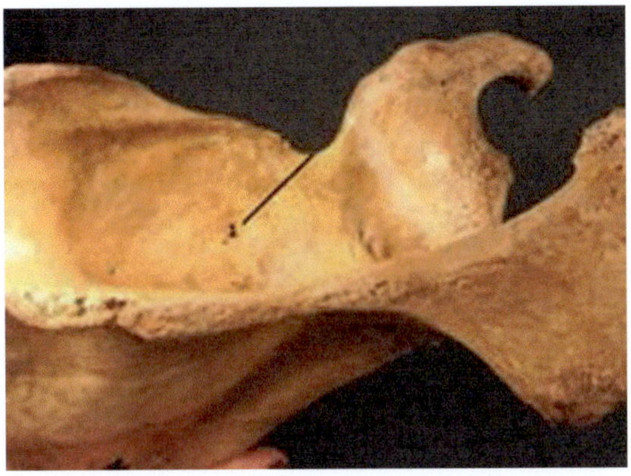

Photo 9.3 Posterior aspect of the scapular bone. (1) Suprascapular notch. Courtesy of Museum of Morphological sciences—National University of Rosario

Anatomical variations have been described at the level of suprascapular notch. Rengachary's morphotypes are six, describing a space that can be wide or narrow and either deep or flat at the same time; it can also be closed by the transverse ligament and its own structure or it can be completely calcified (Photo 9.3). All these variants often cause a harmful sling effect on the nerve [11–13].

The previous causes that compress the nerve are mostly related to dynamic conditions, but rarely the causes could be related to a tumor like lipomas [14] or intraosseous ganglion cyst [15]. Another uncommon factor could be the presence of a paralabral cyst (supraglenoid or spinoglenoid) in the context of a patient with a history of glenohumeral instability [14].

Spinoglenoid notch is a bone depression in the lateral third of the spine, and it is located between scapular spine base and glenoid cavity; it connects supraspinous fossa with infraspinous fossa and serves as passage area of the suprascapular nerve and vessels [16, 17]. It is considered the second most frequent site of compression. Spinoglenoid ligament, also called transverse inferior scapular ligament (the structure covering this notch) is present in 14–100% of people and connects scapular spine to glenoid neck at the posterior capsule. This capsular insertion becomes tense during arm adduction [18–21].

Other less frequent causes of disease at this level of this nerve are: Nerve involvement after an anesthetic block is possible but rare. Repetitive internal limb can generate a dynamic compression in overhead athletes without a clear static (anatomical) component [5].

9.1.4 Physical Examination

The maneuvers we perform during physical examination have a certain degree of specificity and sensitivity.

Most of the times they are supportive and, in some cases, confirm diagnosis. The lesion is generally confirmed by the sum of two or more positive maneuvers in addition to imaging and anamnesis.

During physical examination, we will emphasize individual muscular assessment, since findings seen (supra, infraspinatus insufficiency, or both) are very useful for lesion localization. In cases where infraspinatus is only muscle affected, we suspect that there is spinoglenoid notch compression and, when both are involved, the compression is higher, at the suprascapular notch. Isolated supraspinatus insufficiency suggests certain pathology affecting this muscle. Depending on the site of compression and the exit of its sensitive branches towards the capsule and ligaments, pain is variable and in some cases it may not be present.

9.2 Supraspinatus

9.2.1 Jobe Sign (Empty Can Test)

Examiner stands in front of the patient and places patient's arms at 90° of abduction, 30° of anterior flexion and internal rotation, with thumb downwards; then he/she pushes his/her arm downwards while the patient tries to maintain the initial position (as if emptying a can) [22, 23] (Photo 9.4).

It can be considered positive when it causes pain or weakness. Combination of both responses achieved a sensitivity of 89% and a specificity of 50% for detecting MRI-confirmed supraspinatus tendon ruptures [21]. Kim et al. [24] also calculated the usefulness of both responses. In this study, it showed a sensitivity of 94% and a specificity of 46% when pain occurrence was considered positive; 78% and 71%, respectively, when weakness was considered positive; and 99% and 43% with the combination of both.

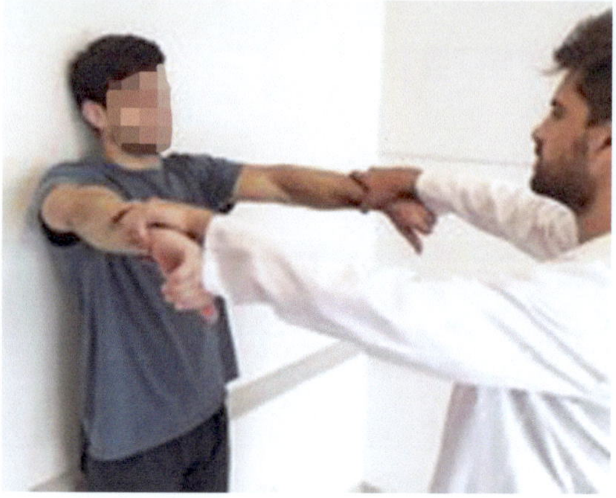

Photo 9.4 Jobe sign

Full can test consists of assessing patient's ability to resist downward pressure on his/her arms at 90° of abduction in the plane of the scapula and at 45° of external rotation. External rotation changes the status of muscle contraction and weakness is considered as positive. Its sensitivity and specificity are similar to those for Jobe's maneuver when pain exacerbation is considered as positive [24, 25].

9.2.2 Other Tests

There are other signs to which we pay attention in the physical examination and are useful to differentiate pathology of the rotator cuff versus suprascapular nerve entrapment.

Codman's point, the site where the supraspinatus tendon inserts into the greater Humerus tubercle, should be evaluated. For this purpose, the patient slightly extends his/her arm and brings it to internal rotation, with his/her hand close to the dorsum of the lumbar spine. Rotator cuff pathology that causes pain without insufficiency is another differential diagnosis [24, 26–28].

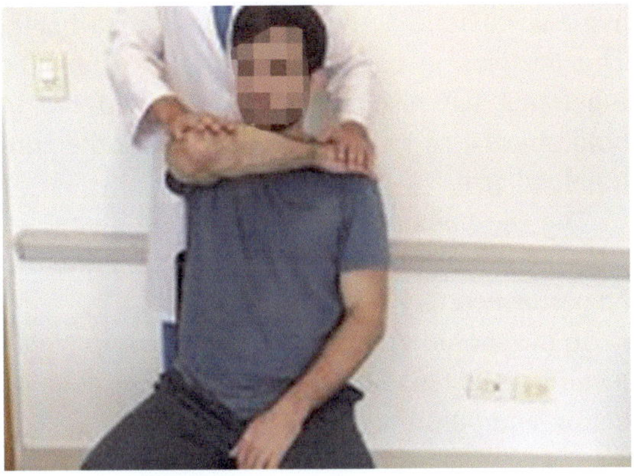

Photo 9.5 Yocum sign

Because of this, the physical examination must go further and must be completed with other maneuvers that evaluate the pathology of the subacromial space elements [27–29].

9.2.3 Yocum Sign (Subacromial Space)

In this maneuver, the patient places the hand of the studied side on the contralateral shoulder and actively raises his/her elbow against resistance without elevating his/her shoulder. It causes pain when there is anterointernal conflict. This maneuver has a sensitivity of 79% and a specificity of 40% as compared to MRI [27, 28] (Photo 9.5).

9.3 Infraspinatus

The inspection of this muscle basically consists of evaluating external rotation. In pure compressions at the spinoglenoid notch level, this is the muscle to be evaluated [26, 30].

9.3.1 Patte Sign

The patient has his/her arm at 90° of abduction flexing his/her elbow at 90°, and he/she attempts to perform an external rotation against the resistance exerted by the examiner. This maneuver has been shown to have a high sensitivity and low specificity for the diagnosis of infraspinatus inflammatory condition [28, 30] (Photo 9.6).

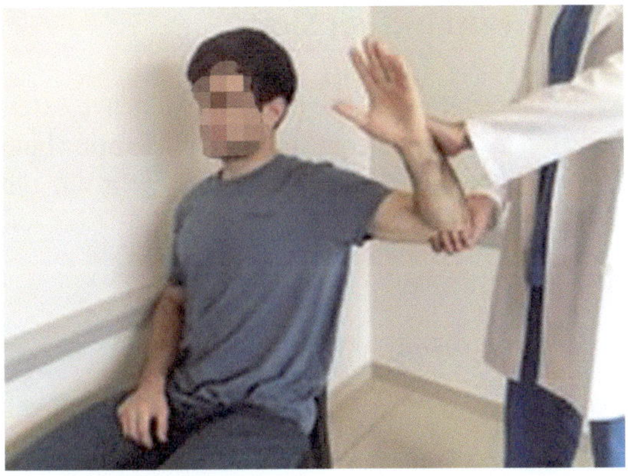

Photo 9.6 Patte sign

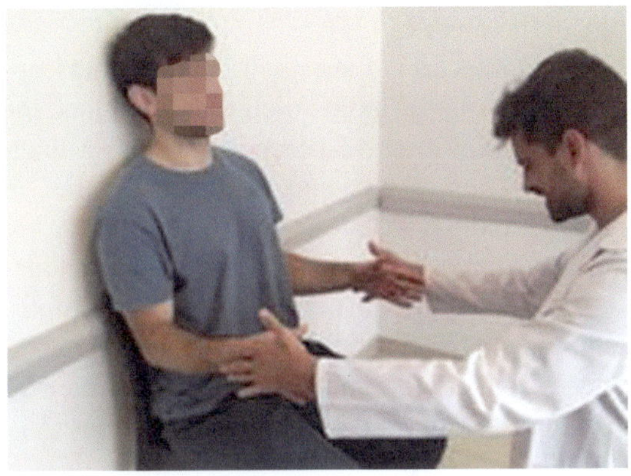

Photo 9.7 Infraspinatus test

9.3.2 Infraspinatus Test (External Rotation Against Resistance)

When seated with his/her arm close to the body, with his/her elbow flexed at 90° and his/her forearm in neutral rotation, patient is asked to perform external rotation against resistance. This maneuver, in which pain occurrence is considered as a positive result, has a sensitivity between 42 and 98% and a specificity between 54 and 98% [30]. If weakness is considered positive, its sensitivity improves [25] (Photo 9.7).

9.3.3 Hornblower Sign

When we want to test the posibility to active externally rotate the arm the physician asked to bring both hands to her mouth and is unable to do it without abducting the affected arm [26].

9.3.4 Complementary Studies

Electrical and imaging studies are an essential tool in the diagnosis.

Simple X-ray AP and lateral view, a CT and an MRI are used to find anomalies around the shoulder, fracture sequelae, tumors, and mostly atrophy of the muscles [31].

Local injection of anesthetic in the suprascapular or the spinoglenoid notch trying to block the nerve is mini-invasive procedure to get a diagnose by suppression, and therefore a relief in shoulder pain.

Electromyography (EMG) and nerve conduction velocity (NCV) studies are the very important and often confirm the diagnosis. Increased latency and fibrillations indicate denervation of the supraspinatus and infraspinatus muscles [32].

9.3.5 Treatment

The first line is a non-operative treatment, usually a rehabilitation process guided by a therapist.

Mostly, suprascapular nerve compressions require surgical treatment, which can be performed openly or arthroscopically when involvement is at the level of the suprascapular notch [10, 33].

The release or sectioning of the ligament on its two locations is sufficient, followed by a rehabilitation process [33].

9.4 Musculocutaneous Nerve

9.4.1 Introduction

Flexion of the elbow contributes with the mobility of the shoulder to carry out most of the hygienic-dietetic activities of the human being. The muscles in charge of it must not only be functional but also provide sufficient force to hold objects, carry out loads, and assist the contralateral arm.

The musculocutaneous nerve is in charge of carrying motor function so that this function can be performed.

In the shoulder, this nerve has a path in which it can be affected and its anatomy will be detailed below, and possible causes of injury to it.

9.4.2 Anatomy

Musculocutaneous nerve originates from lateral fascicle of brachial plexus together with anterolateral branch of the median nerve and its fibers emerge mainly from the anterior C5 and C6 nerve roots with input from roots coming from C7 [34]. It is responsible for motor innervation of coracobrachialis, biceps brachii, and brachialis, while at the sensory level it corresponds to the anterior and posterolateral region of the forearm [35].

Variations have been described regarding its origin and pathway. Although it generally rises directly from lateral fascicle of the brachial plexus, this nerve can originate from lateral and posterior fascicles of the median nerve [36]. It runs through the fibers of the coracobrachialis muscle from posteriorly to anteriorly and it is known as perforating Casserius nerve [36]. The site through which this nerve travels is usually at a distance of 5–8 cm from the coracoid process, but it has been described at a distance of 3 cm [37, 38].

At shoulder girdle, it passes through coracoid process medial side, where at a varying 1–3 cm distance it separates from its base. Here it emerges from lateral fascicle and, along its oblique pathway, it travels towards the coracobrachial muscle which is pierced. Cardoso et al. [39] established a triangular zone where the nerve is located 2.38–4.30 cm distally and 1.03–3.80 cm medially to the lower part of the coracoid process based on cadaveric measurements. Macchi et al. [40] determined that, in cadaveric studies, mean distance from coracoids tip to nerve origin was 2.9 cm ± 0.5 cm and inlet and outlet points through coracobrachial muscle was at 7.7 ± 2.5 cm, motor branch origin for biceps brachii was found to be at 16.9 ± 0.7 from the inferior edge of the coracoid, approximately at 10 mm from the axillary artery [41] (Photo 9.8).

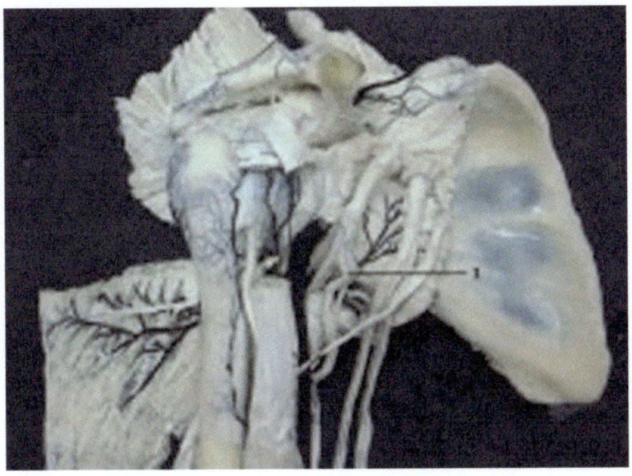

Photo 9.8 Anatomical preparation of the posterior face of the shoulder. (1) Musculocutaneous nerve. Courtesy of Dr. Ezequiel Zaidemberg

9.4.3 Physical Examination

Musculocutaneous nerve is a motor and sensory nerve, innervates the biceps muscle (it innervates biceps short head), brachialis, and coracobrachialis muscle [42].

9.4.4 Motor Testing

Evaluation of this nerve by muscle examination should be comparative. On inspection, atrophy of anterior compartment muscles should be addressed. Loss of forearm flexion strength on the arm should be evaluated both with supinated and pronated forearms. In cases of musculocutaneous nerve injury, the patient will be able to perform elbow flexion with forearm supination, due to supinator longus muscle action, but there will be a noticeable decrease in strength compared to the contralateral side [43].

By placing forearm in prone position, supinator longus muscle action is suppressed and strict anterior brachial action can be observed (Photo 9.9).

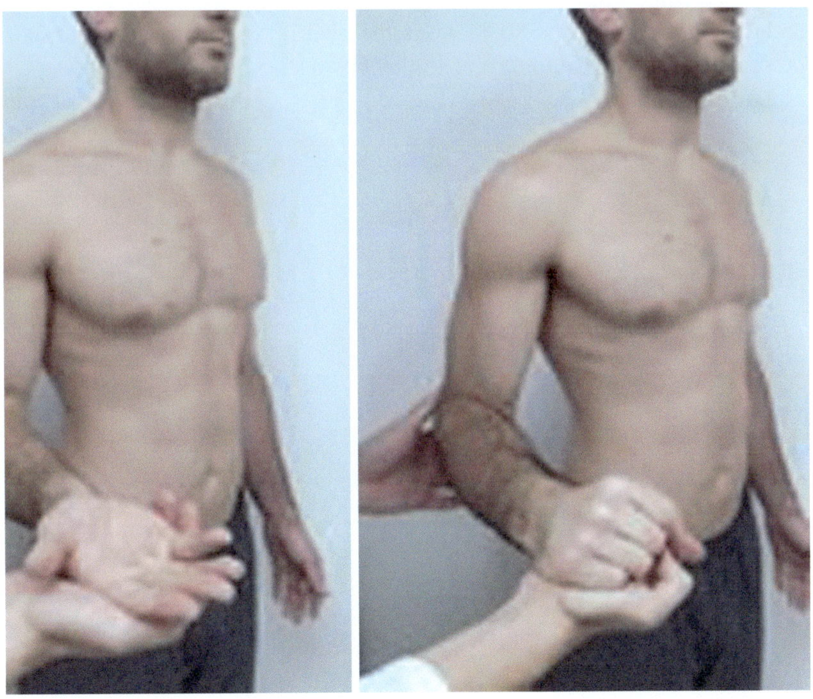

Photo 9.9 Motor testing: evaluation of musculocutaneous nerve

Biceps muscle is fundamentally a supinator and secondary forearm flexor. By means of Yergason's sign, its long head can be isolated when pathology exists. It can be identified early in patients who have supinator longus overload caused by compensation for the insufficiency affecting the other elbow flexors.

9.4.5 Speed Sign (Long Head of the Biceps)

The patient is positioned in front of the examiner with the palm of his/her hand facing upwards, the elbow is extended, and the patient is stimulated to perform anterior flexion of the shoulder in external rotation against a resistance exerted by the examiner. Sensitivity is between 40 and 80%, and specificity is between 35 and 97% [29] (Photo 9.10).

9.4.6 Yergason Sign (Long Head of the Biceps)

The patient is asked to perform supination against resistance while keeping his/her shoulder locked and the elbow is flexed at 80°. Pain in the bicipital region indicates pathology of the biceps tendon or its sheath [44].

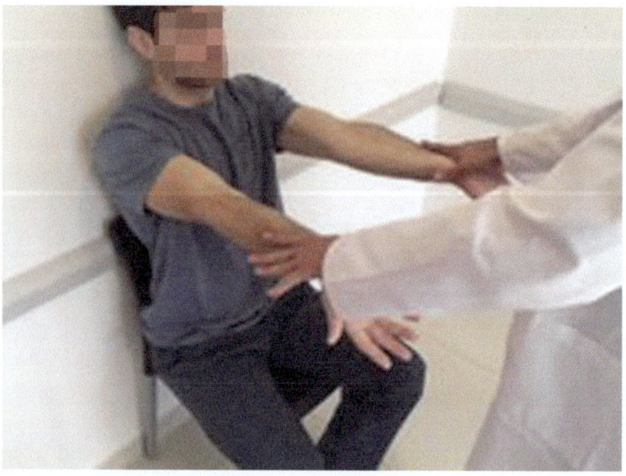

Photo 9.10 Speed sign

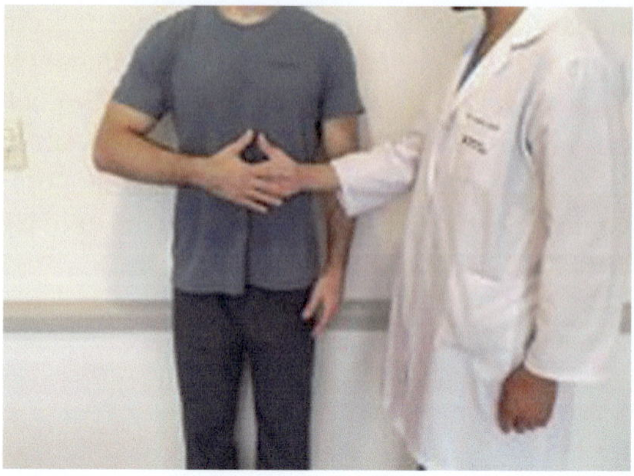

Photo 9.11 Belly press test

9.4.7 Sensitivity Testing

Regarding its sensory evaluation, thermoalgesic and touch sensitivity of the radial side and external part of the forearm anterior aspect should be evaluated [43] (Photo 9.11).

9.4.8 Pathology and Diagnosis

This nerve may be affected due to excessive and prolonged traction, also due to extreme positioning during surgical field setup, open direct injuries or during surgical procedures where retractors or other instruments are placed proximally or medially to the coracoid process. Sport activities that repetitively overload the arm, such as throwing, weight lifting, carrying heavy objects, or sudden events that bring the arm into extension, such as pushing or during a fight, have been described as mechanisms of injury.

Distally, nerve injuries are more related to muscular conditions (hypertrophy, abrupt contraction) leading to mechanical or ischemic injuries [43].

Nerve injuries in shoulder instability surgeries can occur in up to 8.2% [44].

In anterior instability procedures, especially Bristow-Latarjet procedure, coracoid process is transferred to the anteroinferior portion of the glenoid along with the coracobrachialis and to the short head of the biceps to obtain a dynamic reinforcement of the anteroinferior portion of the glenoid capsule when the arm is in abduction and in external rotation [45]. Nerve complications associated with this procedure are estimated to be around 1.2–1.8% and musculocutaneous nerve injuries in particular are estimated to be around 0.6–0.8% of all the complications related to this procedure [46].

When performing these procedures, the position of the shoulder is important in order to protect the musculocutaneous nerve. In relation to the coracoid process, it was found that there is a shorter distance when the arm is in abduction, abduction-internal rotation, and abduction-external rotation with the consequent risk of injury during these procedures. The distance is greater when the arm is in neutral position or at 45° of abduction [47].

The diagnosis is based on the physical examination, imaging, and electric studies.

9.4.9 Treatment

There is limited literature about the treatment of persistent musculocutaneous nerve injury probably because most of musculocutaneous nerve lesions resolve spontaneously within 4 months. Where nerve injury persists, an option can be nerve grafting and that depends on the time of evolution the condition of the stump state and the nerve defect. For injuries where the primary has failed or when the size of the defect is large nerve transfers might be a good option with a good outcome [48].

9.5 Subscapular Nerve

9.5.1 Anatomy

The upper portion of the subscapularis muscle is innervated by the superior subscapular nerve, and the lower portion by its namesake. These nerves most often arise from posterior cord of the brachial plexus.

Sager and Cols [49] demonstrated that superior subscapular nerve originated from the posterior cord of the brachial plexus in 20 cadaveric shoulders, while the inferior subscapular nerve originated in the posterior cord in 17 shoulders and from the axillary nerve in 3 of the 20 shoulders studied. And that the superior subscapular nerve penetrated the subscapularis proximal to the inferior subscapular nerve.

Another study by Kasper and Cols [50] showed considerable variability in the origin of the inferior subscapular nerve. Specifically, the nerve arose from the axillary nerve in 5 of 20 samples (25%), and they also found that the nerve came from the thoracodorsal nerve in two cases, thus demonstrating the variability in the origin of this nerve.

From its origin, it descends and lateralizes superficially on the muscular belly of the subscapularis until it penetrates it.

The superior subscapular nerve and the inferior subscapular nerve insert into the muscle belly 38.5 ± 9.7 mm and 31.9 ± 9.3 mm, respectively, medial to the myotendinous junction with the arm in neutral rotation.

The subscapularis functions as an internal rotator and passive stabilizer against anterior dislocation through inferior fibers that depress the humeral head. By means of this last function, it resists the gliding of the deltoid and helps the elevation, in addition to the glenohumeral compression [50].

9.5.2 Physical Examination

To evaluate subscapularis muscle, internal rotation of the shoulder should be evaluated, which will be painful with active mobility. Passive mobility in external rotation will be slightly increased and vaguely painful in its last degrees.

9.5.2.1 Belly Press Test

Patient is positioned with the elbows flexed and the hands in the anterior part of the abdomen and is asked to compress the abdomen, while examiner exerts a force contrary to that of the patient from the elbows [51].

9.5.2.2 Gerber Test

Patient with the shoulder fully extended, internally rotated and the elbow flexed, so that the back of the hand contacts patients back, and attempts to separate the hand from the back against resistance at the mercy of an internal rotation of the shoulder [51].

9.5.2.3 Bear Hug Test

The patient places the palm of the hand on the opposite shoulder, with the elbow anterior to the body, and maintains the internal rotational force in this position while the evaluator attempts to rotate the patient's arm outward [51].

9.5.3 Pathology and Diagnosis

Due to its deep course, injuries to the nerve are rare. The most frequent, although rare, are iatrogenic lesions in arthroscopic surgeries through posterior portals on the inferior subscapular nerve or in capsular releases due to excess temperature during arthroscopy [52, 53]. The diagnosis consists of performing MRI, ultrasound, and electric studies.

9.5.4 Treatment

Because most of the injuries to this nerve are deep and close to the motor nerve unit, the treatment consists of tendon transfer to activate internal rotation or as a stabilizer of the glenohumeral joint.

9.6 Axillary Nerve (Circumflex Nerve)

9.6.1 Introduction

The deltoid muscle is part of the motor apparatus of the shoulder, playing a key role in elevating the arm through its three muscle bellies in collaboration with the rotator cuff muscles.

Its paralysis not only weakens the shoulder but also causes atrophy in its region, which is very easy to objectify on inspection.

To complete the motor function of the shoulder, the teres minor muscle collaborates in external rotation and with this being able to bring the arm towards the face and head to perform different activities.

9.6.2 Anatomy

The axillary nerve is mixed motor and sensory, whose fibers originate from the C5 and C6 roots (posterior fascicle of the brachial plexus) [10]. It originates in axillary fossa and provides motor branches to deltoid and teres minor muscles, as well as sensory branches to joint capsule, and over the shoulder skin in its lower deltoid region [47].

Its origin is below pectoralis minor lower border and above pectoralis major lower border. It is related anteriorly to axillary artery and posteriorly to the subscapularis. It exits the axilla below the subscapularis muscle where it is reached by posterior humeral circumflex artery and veins, and together cross the axillary space (constituted by the anterior border of the subscapularis and the teres minor superiorly, the upper border of the teres major inferiorly, medially to the border of the long head of the triceps and laterally to the surgical neck of the humerus). It crosses this space in contact with the inferior glenohumeral capsule to sit on the lower border of the teres minor. From there it goes around the surgical neck where it will provide the terminal branches for the deltoid muscle.

On its pathway through the quadrangular space, it provides two collateral motor branches and a lateral superior cutaneous brachial branch.

9.6.2.1 Branches

1. Anterior (motor) branch: innervates the anterior and medial portions of the deltoid.
2. Posterior (motor): provides the motor branches for the lesser teres muscle and provides branches for the posterior deltoid.
3. Superior lateral brachial cutaneous branch: surrounds the deltoid inferiorly and from behind, penetrates its fascia, and ends in the skin of the shoulder and lateral aspect of the arm. It provides small branches for the glenohumeral joint.

The average distance from the anteromedial tip of the coracoid process and from the acromion posterolateral aspect to the axillary nerve is 3.56 cm (+/−0.51) and 7.4 cm (+/−0.99) [47]. Its relation to the humeral head and the lower aspect of the glenohumeral joint does not undergo major changes during external rotation, but abduction at more than 45° decreases the distance of the axillary nerve and the joint and should be avoided in anterior approaches [49].

9.6.3 Injuries

Axillary nerve injuries can be divided into two main groups:

Traumatic and iatrogenic. Within the traumatic injuries occurring in sports, axillary nerve injury is reported as 10% of peripheral nerve injuries and as the most frequently injured nerve at the shoulder level [54]. It should be diagnosed as early as possible since delayed diagnosis and treatment mean a worse prognosis considering repair will not be possible. Its most frequent injury is caused by anterior shoulder dislocation or humeral fracture [55, 56], as well as by direct trauma [57] or compression in the quadrangular space [58].

Iatrogenic lesions on the axillary nerve are the most frequent lesions of the peripheral nerve in shoulder surgeries [52, 53]. Given its proximity to the subscapularis, it is at risk in any surgery involving the anteroinferior aspect of the shoulder as well as during the subscapularis muscle split. It can also be damaged in capsular releases due to excess temperature during arthroscopy. The risk of injury is higher in intramuscular injections of the deltoid or in intra bursal/intra-articular infiltrations due to the lack of knowledge of its anatomical pathway [59].

9.6.4 Physical Examination, Tests, and Diagnosis

Axillary nerve assessment includes both sensory assessment and motor assessment of deltoid and teres minor function.

At the beginning of the physical examination, it is crucial to assess the tonicity of three muscular bellies of the deltoid and its tropism (the contralateral comparison may be used); there may be hypertrophy of any of the three portions and atrophy of the others.

Shoulder abduction can be performed completely without deltoid muscle activity given the ability of the supraspinatus and biceps brachii to replace it. The tropism and functionality of the three muscle bellies of the deltoid muscle should be evaluated since they can be innervated by different branches of the axillary nerve (the anterior portion of the deltoid is always innervated by the anterior branch, the middle portion, 56% by the anterior branch and 44% by the posterior branch; and the posterior portion, 92% by the posterior branch) [47].

9.6.5 Motor Testing

The evaluation of the *anterior belly* of the deltoid is performed in dorsal position with the shoulder at 90° of abduction, the elbow flexed at 90° and in internal rotation. In such position, the patient is asked to elevate the arm and the contraction of the clavicular portion of the deltoid is palpated. The *middle belly* is evaluated with the patient seated with his/her shoulder at 90° of abduction, his/her elbow flexed at 90° and in internal rotation. In such position, the patient is asked to try to raise his/her arm laterally and the contraction of the acromial portion of the deltoid is palpated. Finally, the *posterior belly* should be evaluated in prone position, with the anterior aspect of the elbow resting on the edge of the table at 90° and the forearm hanging free. In such position, the patient is asked to raise his/her arm to separate it from the table in order to palpate the spinal portion of the deltoid [45] (Photos 9.12, 9.13, and 9.14).

Akimbo's test [50] demonstrates the presence of isolated weakness of the deltoid. This is performed by placing both hands on the iliac crest (with internal rotation), pronating the patient's forearms, flexing his/her elbows, and performing coronal abduction of the shoulder. This test is considered positive when the patient is not able to perform it (Photo 9.15).

The motor function of the *teres minor* is evaluated in the same position as the posterior belly of

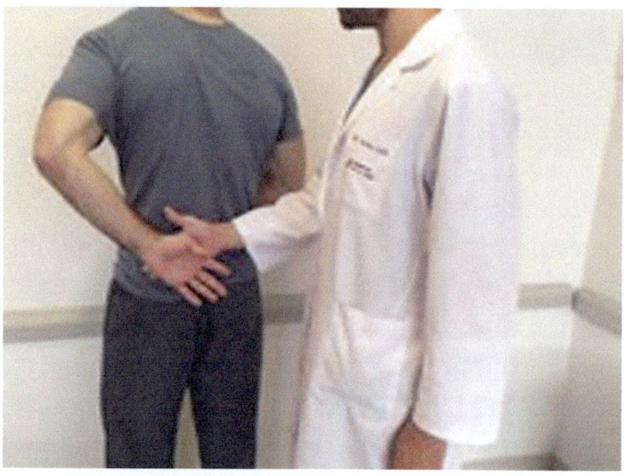

Photo 9.12 Gerber test

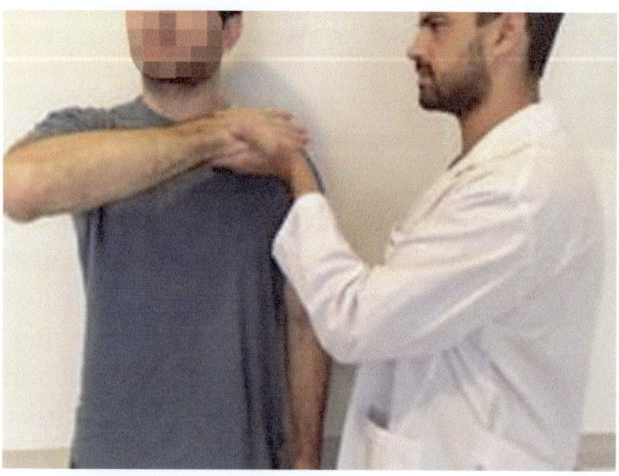

Photo 9.13 Bear Hug test

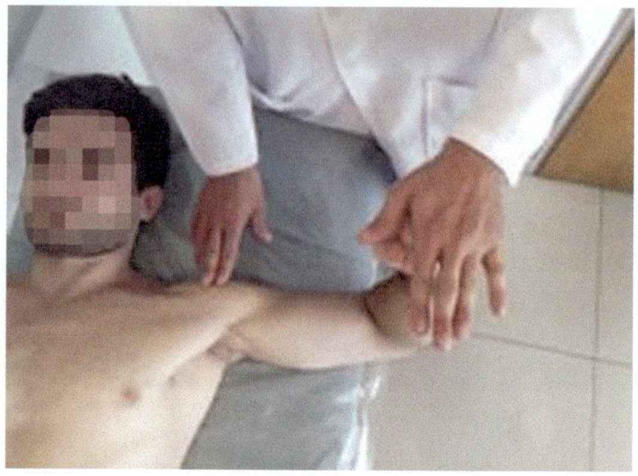

Photo 9.14 Motor Testing: evaluation of the anterior belly of the deltoid

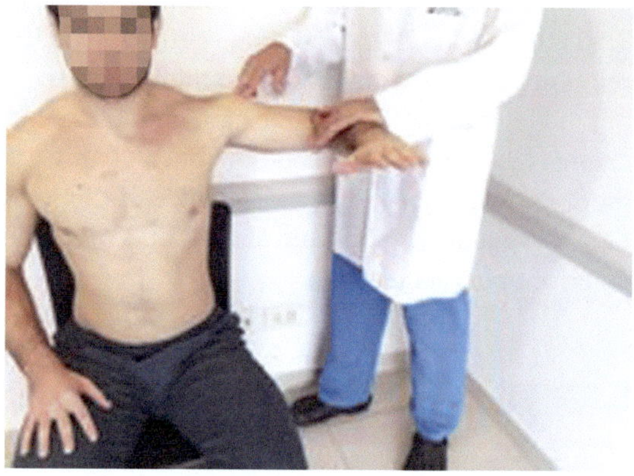

Photo 9.15 Motor Testing: evaluation of the middle belly of the deltoid

the deltoid and the patient should be asked to externally rotate his/her shoulder and the muscle belly should be palpated (external rotator along with the infraspinatus). *Sensitivity* should be evaluated in the superolateral area of the shoulder, a region innervated by the sensory branches of the axillary nerve.

Occasionally, nerve injury may present with a lesion of the artery that accompanies it, creating a large hematoma which may suggest humeral fracture [51].

The diagnosis can be confirmed with an electromyography showing the denervation, whose changes take 7–10 days to appear (Photos 9.16 and 9.17).

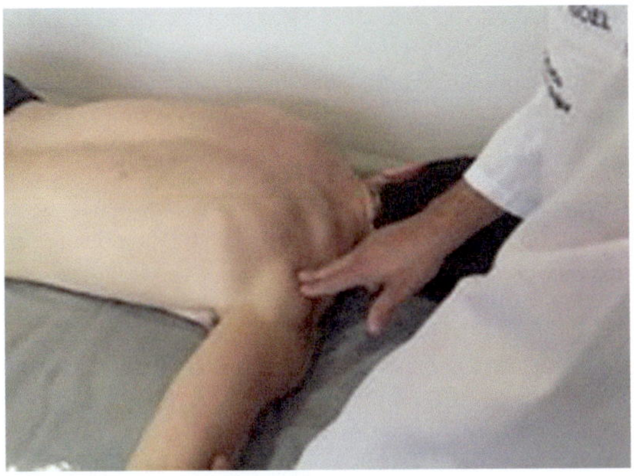

Photo 9.16 Motor Testing: evaluation of the posterior belly of the deltoid

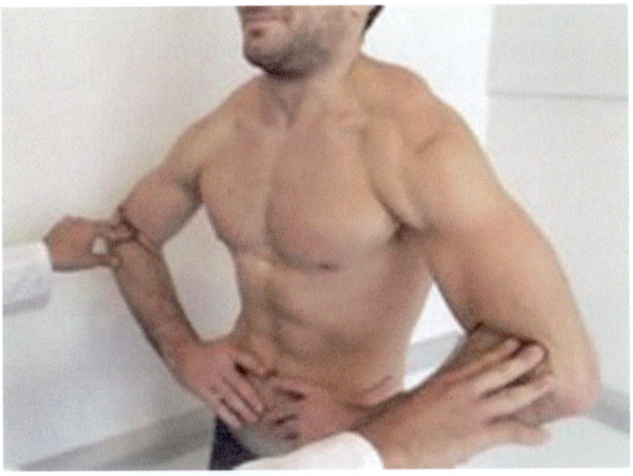

Photo 9.17 Akimbo's test

9.6.6 Treatment

For traumatic injuries or for those produced after the reduction of a shoulder dislocation, a transient neuropraxia may occur, which reverts spontaneously. If an acute and traumatic rupture is identified (it should be suspected based on the energy of the trauma, open wound, and association of vascular lesion), a repair would be the appropriate indication.

When the diagnosis comes late and there is no recovery within 4–6 months, the discussion is focused on nerve transfer, when the motor branch of the triceps (radial nerve) can be used, or a graft to replace the defect caused by retraction. Both have shown good results and no significant differences between them [60].

References

1. Kadi R, Milants A, Shahabpour M. Shoulder anatomy and normal variants. J Belg Soc Radiol. 2017;101(Suppl 2):3. https://doi.org/10.5334/jbr-btr.1467.
2. Pecina M. Who really first described and explained the suprascapular nerve entrapment syndrome? J Bone Joint Surg Am. 2001;83(8):1273–4.
3. Labetowicz P, Synder M, Wojciechowski M, Orczyk K, Jezierski H, Topol M, et al. Protective and predisposing morphological factors in suprascapular nerve entrapment syndrome: a fundamental review based on recent observations. Biomed Res Int. 2017;2017:4659761. https://doi.org/10.1155/2017/4659761.
4. Gosk J, Urban M, Rutowski R. Entrapment of the suprascapular nerve: anatomy, etiology, diagnosis, treatment. Ortop Traumatol Rehabil. 2007;9(1):68–74.
5. Kostretzis L, Theodoroudis I, Boutsiadis A, Papadakis N, Papadopoulos P. Suprascapular nerve pathology: a review of the literature. Open Orthop J. 2017;11:140–53. https://doi.org/10.2174/1874325001711010140.
6. Bozzi F, Alabau-Rodriguez S, Barrera-Ochoa S, Ateschrang A, Schreiner AJ, Monllau JC, Perelli S. Suprascapular neuropathy around the shoulder: a current concept review. J Clin Med. 2020;9(8):2331 https://doi.org/10.3390/jcm9082331.
7. Ager AL, Roy JS, Roos M, Belley AF, Cools A, Hébert LJ. Shoulder proprioception: how is it measured and is it reliable? A systematic review. J Hand Ther. 2017;30(2):221–31. https://doi.org/10.1016/j.jht.2017.05.003.
8. Domingo Ruano Gil Manuel Llusá Pérez, Àlex Merí Vived. Manual y Atlas Fotográfico de Anatomía del Aparato Locomotor. Ed Panamericana. pp 176–77.
9. Yang HJ, Gil YC, Jin JD, Ahn SV, Lee HY. Topographical anatomy of the suprascapular nerve and vessels at the suprascapular notch. Clin Anat. 2012;25(3):359–65. https://doi.org/10.1002/ca.21248. Epub 2011 Aug 18.
10. Momaya AM, Kwapisz A, Choate WS, Kissenberth MJ, Tolan SJ, Lonergan KT, Hawkins RJ, Tokish JM. Clinical outcomes of suprascapular nerve decompression: a systematic review. J Shoulder Elb Surg. 2018;27(1):172–80. https://doi.org/10.1016/j.jse.2017.09.025.
11. Rengachary SS, Burr D, Lucas S, Hassanein KM, Mohn MO, Matzke H. Suprascapular nerve entrapment neuropathy: a clinical, anatomical and comparative study. Part 2: anatomical study. Neurosurgery. 1979;5:447–51.
12. Zehetgruber H, Noske H, Lang T, Wurnig C. Suprascapular nerve entrapment. A meta-analysis. Int Orthop. 2002;26:339–43.
13. Cummins CA, Messer TM, Nuber GW. Suprascapular nerve entrapment. J Bone Joint Surg Am. 2000;82:415–24.
14. Hazrati Y, Miller S, Moore S, Hausman M, Flatow E. Suprascapular nerve entrapment secondary to a

lipoma. Clin Orthop Relat Res. 2003;411(411):124–8. https://doi.org/10.1097/01.blo.0000063791.32430.59.
15. Yi JW, Cho NS, Rhee YG. Intraosseous ganglion of the glenoid causing suprascapular nerve entrapment syndrome: a case report. J Shoulder Elb Surg. 2009;18(3):e25–7. https://doi.org/10.1016/j.jse.2008.10.014.
16. Demaio M, Drez D, Mullins R. The inferior transverse scapular ligament as a possible cause of entrapment neuropathy of the nerve to the infraspinatus. A brief note. J Bone Joint Surg Am. 1991;73:1061–3.
17. Dermirhan M, Imhoff AB, Debski RE, Patel PR, Fu FH, Woo SL-Y. The spinoglenoid ligament and its relationship to the suprascapular nerve. J Shoulder Elb Surg. 1998;7:238–41.
18. Ide J, Maeda S, Takagi K. Does the inferior transverse scapular ligament cause distal suprascapular nerve entrapment?. An anatomic and morphologic study. J Shoulder Elb Surg. 2003;12:253–5.
19. Dermirkan AF, Sargon MF, Erkula G, Kiter E. The spinoglenoid ligament: an anatomic study. Clin Anat. 2003;16:511–3.
20. Plancher KD, Peterson RK, Johnston JC, Luke TA. The spinoglenoid ligament. Anatomy, morphology and histological findings. J Bone Joint Surg Am. 2005;87:361–5.
21. Greiner A, Golser K, Wambacher M, Kralinger F, Sperner G. The course of the suprascapular nerve in the supraspinatus fossa and its vulnerability in muscle avancement. J Shoulder Elb Surg. 2003;12:256–9.
22. Park HB, Yokota A, Gill HS, El Rassi G, McFarland EG. Diagnostic accuracy of clinical tests for the different degrees of subacromial impingement syndrome. J Bone Joint Surg Am. 2005;87:1446–55.
23. Itoi E, Kido T, Sano A, Urayama M, Sato K. Which is more useful, the "full can test" or the "empty can test," in detecting the torn supraspinatus tendon? Am J Sports Med. 1999;27:65–8.
24. Kim E, Jeong HJ, Lee KW, Song JS. Interpreting positive signs of the supraspinatus test in screening for torn rotator cuff. Acta Med Okayama. 2006;60:223–8.
25. Itoi E, Minagawa H, Yamamoto N, Seki N, Abe H. Are pain location and physical examinations useful in locating a tear site of the rotator cuff? Am J Sports Med. 2006;34:256–64.
26. Walch G, Boulahia A, Calderone S, Robinson AHN. The 'dropping' and Hornblower's' signs in evaluation of rotator-cuff tears. J Bone Joint Surg Br. 1998;80:624–8.
27. Codman EA (Ernest Amory), 1869–1940. The shoulder: rupture of the supraspinatus tendon and other lesions in or about the subacromial bursa/ E.A. Codman; foreword by Anthony F. DePalma. Brooklyn: G. Miller. p c1934.
28. Yocum LA. Assessing the shoulder. History, physical examination, differential diagnosis, and special tests used. Clin Sports Med. 1983;2:281–9.
29. Leroux JL, Thomas E, Bonnel F, Blotman F. Diagnostic value of clinical tests for shoulder impingement syndrome. Rev Rhum Engl Ed. 1995;62:423–8.
30. Beaudreuil J, Nizard R, Thomas T, Peyre M, Liotard JP, Boileau P, et al. Contribution of clinical tests to the diagnosis of rotator cuff disease: a systematic literature review. Joint Bone Spine. 2009;76: 15–9.
31. Mellado JM, Calmet J, Olona M, et al. MR assessment of the repaired rotator cuff: prevalence, size, location, and clinical relevance of tendon rerupture. Eur Radiol. 2006;16(10):2186–96. https://doi.org/10.1007/s00330-006-0147-z.
32. Post M, Mayer J. Suprascapular nerve entrapment, diagnosis and treatment. Clin Orthop. 1987;223:419–22.
33. Lafosse L, Tomasi A, Corbett S, Baier G, Willems K, Gobezie R. Arthroscopic release of suprascapular nerve entrapment at the suprascapular notch: technique and preliminary results. Arthroscopy. 2007;23:34–42.
34. Sperber GH. Clinically oriented anatomy. J Anat. 2006;208(3):393. https://doi.org/10.1111/j.1469-7580.2006.00537.x.
35. Carlotto, Jorge Roberto Marcante; Ambros, Luciana Estacia; Dal Vesco, Juarez Antônio y Reichert, Paulo Roberto. Origen y Trayecto Anómalos del Nervio Musculocutáneo. Int J Morphol 2009;27(2), pp.507–508.
36. Latarjet M, Liard A. Anatomía humana. 4th ed. Buenos Aires, Argentina: Editorial Panamericana; 2011. p. 637–8.
37. Flatow EL, Bigliani LU, April EW. An anatomic study of the musculocutaneous nerve and its relationship to the coracoid process. Clin Orthop Relat Res. 1989;244:166–71.
38. Ozturk A, Bayraktar B, Taskara N, Kale AC, Kutlu C, Cecen A. Morphometric study of the nerves entering into the coracobrachialis muscle. Surg Radiol Anat. 2005;27:308–11.
39. Rebouças F, Filho RB, Filardis C, Pereira RR, Cardoso AA. Anatomical study of the musculocutaneous nerve in relation to the coracoid process. Revista Brasileira de Ortopedia (English Edition). 2010;45(4):400–3.
40. Macchi V, Tiengo C, Porzionato A, Parenti A, Stecco C, Bassetto F, De Caro R. Musculocutaneous nerve: histotopographic study and clinical implications. Clin Anat. 2007;20(4):400–6.
41. Bloc S, Mercadal L, Garnier T, Huynh D, Komly B, Leclerc P, Dhonneur G. Shoulder position influences the location of the musculocutaneous nerve in the axillary fossa. J Clin Anesth. 2016;33:250–3.
42. Desai SS, Varacallo M. Anatomy, shoulder and upper limb, musculocutaneous nerve. In: StatPearls. Treasure Island, FL: StatPearls Publishing; 2020.
43. Pećina M, Bojanić I. Musculocutaneous nerve entrapment in the upper arm. Int Orthop. 1993;17(4):232–4.
44. Boardman ND, CoWeld RH. Neurologic complications of shoulder surgery. Clin Orthop. 1999;368:44–50.
45. Yergason RM. Supination sign. J Bone Joint Surg. 1931;13:160. International Orthopaedics, 17(4), 232–234

46. Griesser MJ, et al. Complications and re-operations after Bristow-Latarjet Shoulder Stabilization: a systematic review. J Shoulder Elb Surg. 2013;22(2):286–92. https://doi.org/10.1016/j.jse.2012.09.009.
47. Apaydin N, Bozkurt M, Sen T, Loukas M, Tubbs RS, Ugurlu M, Elhan A. Effects of the adducted or abducted position of the arm on the course of the musculocutaneous nerve during anterior approaches to the shoulder. Surg Radiol Anat. 2008;30(4)
48. Ferris S, Reid I. Contemporary nerve reconstruction for iatrogenic musculocutaneous nerve injury after shoulder stabilization surgery. J Shoulder Elb Surg. 2020;29(9):e341–4. https://doi.org/10.1016/j.jse.2020.03.025. Epub 2020 Jun 9.
49. Sager B, Gates S, Collett G, Chhabra A, Khazzam M. Innervation of the subscapularis: an anatomic study. JSES Open Access. 2019;3(2):65–9.
50. James C. Kasper; John M. Itamura; James E. Tibone; Scott L. Levin; Milan V. Stevanovic (2008). Human cadaveric study of subscapularis muscle innervation and guidelines to prevent denervation, J Shoulder Elbow Surg 17 (4), 0–662.
51. Valerius K, Frank A, Kloster B, Hirsch M, Hamilton C, Lafont E. The book of muscles. 1st ed. Barcelona, Spain: Ars Medica; 2007. p. 52–3.
52. Duralde XA. Neurologic injuries in the athlete's shoulder. J Athl Train. 2000;35:316–28.
53. Visser CP, Coene LN, Brand R, Tavy DL. Nerve lesions in proximal humeral fractures. J Shoulder Elb Surg. 2001;10:421–7.
54. Gurushantappa PK, Kuppasad S. Anatomy of axillary nerve and its clinical importance: a cadaveric study. J Clin Diagn Res. 2015;9(3):AC13-7.
55. Simone JP, Streubel PN, Sanchez-Sotelo J, Steinmann SP, Adams JE. Change in the distance from the axillary nerve to the glenohumeral joint with shoulder external rotation or abduction position. Hand. 2017;12(4):395–400.
56. Fujihara Y, Doi K, Dodakundi C, Hattori Y, Sakamoto S, Takagi T. Simple clinical test to detect deltoid muscle dysfunction causing weakness of abduction—"Akimbo" Test. J Reconstr Microsurg. 2012;2012(28):375–80.
57. Patel J, Turner M, Birch R, et al. Rupture of the axillary (circumflex) nerve and artery in a champion jockey. Br J Sports Med. 2001;35:361–2.
58. Krivickas LS, Wilbourn AJ. Peripheral nerve injuries in athletes: a case series of over 200 injuries. Semin Neurol. 2000;20:225–32.
59. Berry H, Bril V. Axillary nerve palsy following blunt trauma to the shoulder region: a clinical and electrophysiological review. J Neurol Neurosurg Psychiatry. 1982;45:1027–32.
60. Perlmutter GS, Leffert RD, Zarins B. Direct injury to the axillary nerve in athletes playing contact sports. Am J Sports Med. 1997;25:65–8.
61. Murrell GA, Walton JR. Diagnosis of rotator cuff tears. Lancet. 2001;357:769–70.

Evaluation of the Stiff Shoulder

10

Stephen C. Weber, Prashant Meshram, Guillermo Arce, and Edward McFarland

10.1 Introduction

Shoulder stiffness, presenting with or without pain, is a common expressing complaint in the routine orthopedic practice. Despite this, the actual definition of shoulder stiffness remains elusive. Most clinicians would agree that shoulder stiffness equates with loss of passive motion and does not describe the patient who cannot move it, or is unable to move the shoulder because of pain [1]. While shoulder pain can have a variety of sources outside the shoulder joint itself [1], true shoulder stiffness is invariably caused by intrinsic shoulder pathology. Codman was one of the first to explore the subject of atraumatic shoulder stiffness, coining the term "frozen shoulder" [2] to describe the condition later called "adhesive capsulitis" by Neviaser [3]. While Zuckerman [4] surveyed the American Shoulder and Elbow Society members to determine a consensus definition for this syndrome, others have used the Delphi consensus method [5]. The ISAKOS shoulder committee published a detailed guideline describing the terminology recommended for stiff shoulders [6]. This group stated that the term "stiff shoulder" should be used to describe the patient who presents with a restricted range of motion (ROM). The etiology of the stiff shoulder can be due to primary or secondary causes" [6]. Diercks et al. note a specific definition for the actual amount of stiffness required: "A range of motion of less than 100° in forward flexion, less than 10° in external rotation, and less than L5 level in internal rotation is indicative" [5].

Among the causes of stiff shoulder, common are idiopathic (adhesive capsulitis or frozen shoulder), rotator cuff pathology, glenohumeral arthritis, post-trauma, and post-surgery. While numerous causes for the stiff shoulder can be identified, the history, physical examination, and basic laboratory and radiographic studies can be used to diagnose and direct treatment of these patients accurately.

10.2 Patient History

Several authors have noted the importance of taking a thorough history to evaluate the patient presenting with a stiff shoulder [1, 5, 6]. This should include information regarding the initial onset

S. C. Weber (✉) · P. Meshram · E. McFarland
Division of Shoulder Surgery, Department of Orthopedic Surgery, The Johns Hopkins University, Baltimore, MD, USA

G. Arce
Department of Orthopaedic Surgery, Instituto Argentino de Diagnóstico y Tratamiento, Buenos Aires, Argentina

J. G. Lane et al. (eds.), *The Art of the Musculoskeletal Physical Exam*,
https://doi.org/10.1007/978-3-031-24404-9_10

and development of the stiffness and/or pain. Any history of trauma is essential in the characterization of the problem, as well as the onset being gradual or abrupt in nature. The history identifying any significant past medical history (e.g., diabetes, hypothyroidism, malignancy, neurological disease, cardiovascular disease, cerebrovascular accident, hyperlipidemia, and Dupuytren's contracture) and previous injuries should be noted. Accompanying symptoms of weakness and paresthesia are important as they indicate neurological etiology [3]. Night pain is common to many shoulder conditions consistent with the stiff shoulder and signifies substantial impairment in the patient's daily life. Prior treatment received is equally important, such as analgesics, physical therapy, steroid injection, or surgery, and the presence or absence of any benefit should be discussed. This history will help to direct the subsequent examination and evaluation.

10.3 Physical Examination

As with all orthopedic conditions, physical examination of the stiff shoulder requires observation of the shoulder, palpation, ROM, strength, and specialized examination tests. It must be emphasized that visual inspection should be performed from the front, back, and side of the patient, unclothed to the waist. The findings during shoulder examination should be compared to the opposite side [7]. Like the orthopedic examination of any joint, the joints proximal and distal to the involved one should be examined [8]. The resting posture of the shoulder and the pattern of any wasting that is present are important examination findings. Wasting points typically towards the chronic nature of the underlying disease and involved structures. The inspection will also reveal scars from prior injury or surgery. Any abnormalities in the form of weakness or sensory loss should trigger a complete neurologic examination. It is critical to rule out referred pain from the neck as a source of shoulder pain.

Passive and active ROM must be assessed, as this will differentiate the patients with a primary motor problem from those with a primary articular problem [6]. The ROM examination should always be compared to the normal shoulder. The four ROMs usually assessed are flexion, abduction, internal rotation, and external rotation. While these are generally evaluated with the patient sitting upright, flexion, abduction, and external rotation can be evaluated supine especially if pain is suspected to limit the examination.

Differentiation between pain-induced reduction in ROM and true "capsular" pathology can be difficult [5]. In idiopathic frozen shoulder, the loss of passive motion tends to be global. In contracture secondary to other causes such as trauma or spasticity, the movements may be restricted in one plane yet relatively preserved in another [6]. Documenting the unrestricted planes will help prevent an inappropriate release of tissues at the time of any surgical intervention. For some diagnoses, the assessment of internal and external rotation with the arm at 90° of abduction can also be helpful [1]. For example, glenohumeral internal rotation deficit (GIRD) can be common in throwing athletes. Fixing the scapula with one hand while performing active and passive motion can allow the examiner to eliminate scapulothoracic substitution in measuring motion [8]. Observing active motion allows assessment of the deltoid and peri-scapular muscle control. The presence of spasticity can be elicited during the evaluation of the passive motion. The strength of all rotator cuff muscles is assessed and the presence of any lag sign is noted [6]. Specialized tests for rotator cuff muscles like painful arc sign, Jobe's test, drop arm sign, belly press, and bear hug test should be performed [9].

10.4 Diagnostic Investigations

10.4.1 Laboratory Studies

There are no specific laboratory tests for the frozen shoulder [10]. Laboratory studies can elucidate secondary causes of shoulder stiffness. Measures of glucose metabolism, such as fasting blood sugar can detect diabetes. Serum hemoglobin A1C test can reveal poor glycemic control over a period of last 3 months. Elevated triglycerides can similarly be associated with shoulder stiffness [10]. White blood cell count and inflammatory markers such as C reactive protein and erythrocyte sedimentation rate can provide a clue to subtle infection. While numerous other markers of the disease have been described, these are not generally clinically useful. Serum levels of ICAM-1, TIMP1, TIMP-2, and TGF-β1 were significantly higher, but MMP-1 and MMP-2 levels were significantly lower in frozen shoulder [10, 11].

10.4.2 Radiographs

Radiographs in both true anteroposterior (Grashey) and axillary views are essential for optimizing investigation of shoulder problems in general. Several potential diagnoses that lead to shoulder stiffness should have normal radiographs, including idiopathic adhesive capsulitis and rotator cuff tear. Two other common causes of stiff shoulders, osteoarthritis and calcific tendonitis, are easily diagnosed with plain radiographs. The physicians should be mindful of getting radiographs in patients with stiff shoulders as the literature continues to demonstrate cases followed conservatively with a late diagnosis of primary or metastatic malignancy [12]. Radiographs are especially of use in patients with post-surgical stiffness, where hardware placement often gives valuable clues to treatment.

10.4.3 Arthrogram

Although currently of limited use, Neviaser et al. have described the classic findings on arthrogram of loss of the inferior pouch and limited injection volume available when performing the arthrogram [3]. These findings are pathognomonic for adhesive capsulitis although they are also consistent with advanced glenohumeral arthritis which should be ruled out.

10.4.4 Magnetic Resonance Imaging (MRI) and Ultrasound Findings

MRI is rarely required to diagnose adhesive capsulitis [3]. Rather, it is used to exclude other etiologies of stiffness to assist the physician in the correct diagnosis and treatment. While not required, MRI findings in stiff shoulders in general, and idiopathic adhesive capsulitis in particular, have been well established [13–20]. The MRI findings that suggest adhesive capsulitis include soft tissue thickening in the rotator interval, which may encase the coracohumeral and superior glenohumeral ligaments, and soft tissue thickening adjacent to the bicipital groove. The typical MRI findings in patients suffering adhesive capsulitis depend on the stage of the disease. Peri-articular swelling and capsule inflammation are found at the initial period of the freezing stage (Fig. 10.1). Also, fibrosis of the rotator interval is

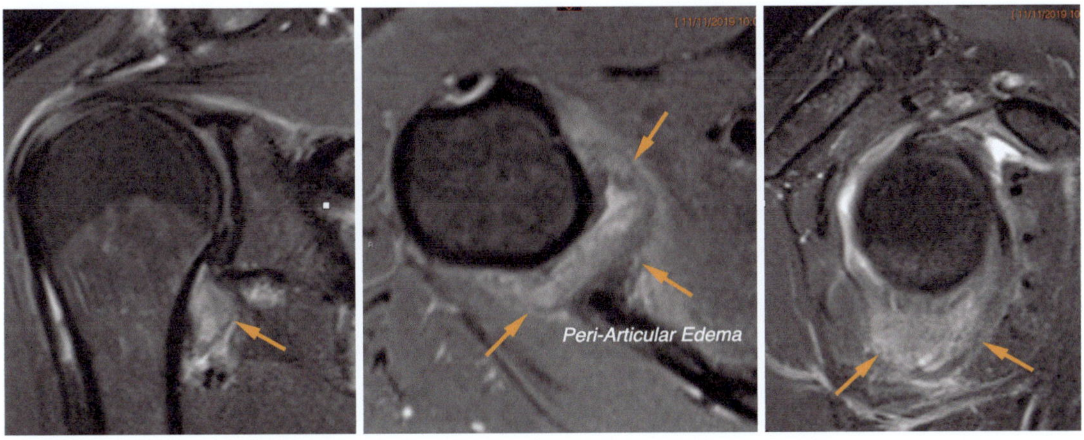

Fig. 10.1 Right shoulder MRI. Adhesive Capsulitis. DP-FS slices. Coronal, axial, and sagittal oblique cuts. The orange arrows indicate the ligaments swelling and capsular inflammation of the initial stage of adhesive capsulitis. (Courtesy F Idoate, Spain)

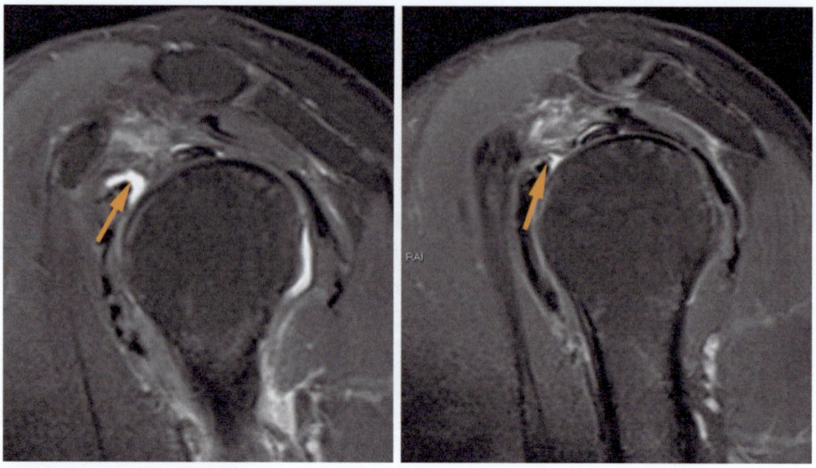

Fig. 10.2 Right shoulder MRI. Adhesive Capsulitis. Sagittal oblique DP-FS cuts. The orange arrows signpost the rotator interval tissue fibrosis between the coracoid tip and the bicipital groove. The coracohumeral ligament, which arises from the coracoid base to the proximal humerus, is also involved. (Courtesy F Idoate, Spain)

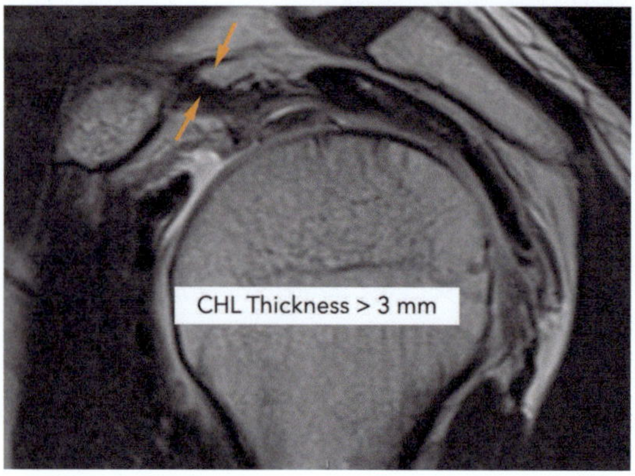

Fig. 10.3 Right shoulder MRI. Adhesive Capsulitis. Sagittal oblique T1 cut. The orange arrows indicate the thickened coracohumeral ligament. It runs from the coracoid to the bicipital groove, and when it has more than 3 mm wide strongly suggests a frozen shoulder syndrome. (Courtesy F Idoate, Spain)

a common finding at the sagittal fast spin-echo cuts demonstrating the origin of pain and the beginning of the motion loss (Fig. 10.2). After a few weeks of this freezing period, the coracohumeral ligament, a key component of the motion restriction, is shown thicker than usual. According to Homsi et al. [21], if the coracohumeral ligament is thicker than 3 mm at the sagittal oblique T1 cut, the diagnosis of frozen shoulder syndrome is ensured (Fig. 10.3) [21]. The range of motion is remarkably reduced in the final frozen stage by a thick and rigid inferior glenohumeral ligament (IGHL) and capsule. The average width of the capsule at the IGHL level is 2 mm. In severe adhesive capsulitis cases, the capsule thickness can reach between 4 and 7 mm wide (Fig. 10.4), [22–25]. It is important to note that the MRI in the patient with clinically established frozen shoulder can confuse, rather than assist, as Loeffler et al. has noted the high number of false-positive MRI scans in the presence of adhesive capsulitis [26]. Given a consistent clinical picture of the etiology of a stiff shoulder, MRI findings, in general, should not direct treatment. However, they need to be carefully ruled in or out during arthroscopic evaluation. Rotator cuff pathology in the presence of a clinical frozen shoulder is uncommon although the radiologic reading of partial rotator cuff tear is more so [19]. In the authors' experience, this finding is often false positive at arthroscopic evaluation secondary to cuff edema from the synovitis of frozen shoulder.

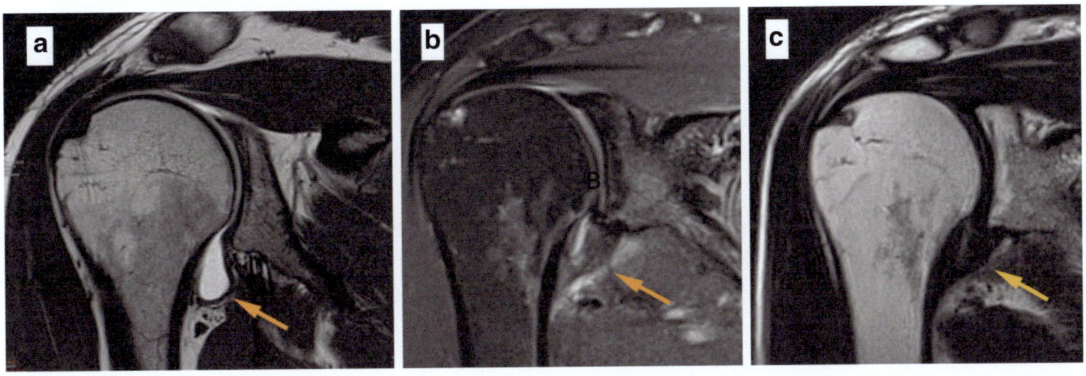

Fig. 10.4 Right shoulder MRI. Coronal slices. (**a**) Normal inferior glenohumeral ligament (IGHL) and capsule width of approximately 2 mm. (**b** and **c**) The orange arrows signpost the stiffened capsule and the thickened IGHL between 4 and 7 mm wide. (Courtesy F Idoate, Spain)

10.5 Etiology of Stiff Shoulder

As well described by Armstrong [1] the stiff shoulder can be broadly classified based on etiology into atraumatic and traumatic etiology. Other authors have divided joint stiffness into intrinsic (joint) and extrinsic (outside joint) [4, 27]. While this classification has some advantages, the primary cause of the stiff shoulder is alterations to the capsule. Bain et al. [6] further divided the causes of stiffness into intracapsular, capsular, and extracapsular (Table 10.1). The anatomic etiology of shoulder stiffness remains a point of discussion, the clinical classification into traumatic and atraumatic remains useful.

Bain et al. note that "The extra-articular causes are outside of the joint and include rotator cuff tendon and muscle. 'Other causes' are entirely separate from the shoulder. The systemic causes will affect a specific anatomical structure, e.g. diabetes causes capsulitis" [6].

In summary, the stiff shoulder can have multiple etiologies. While the diagnosis can be challenging, this can generally be accomplished with a thoughtful history and physical examination, given the etiologies described above. Plain radiographs are required. Increasingly, the pathology creating the stiff shoulder can be further evaluated with ultrasound and MRI imaging.

Table 10.1 Patho-anatomical classification of shoulder stiffness (Reproduced from Bain et al. 2015 [6])

Intra-articular (bearings)	Capsular (constraints)	Extra-articular (motor, cable, levers)	Neurological (control, electrics, sensors)
Articular surface	*Labrum and ligaments*	*Muscles*	*Central*
Osteochondral defect	Deficient	Myopathy	Behavioral, dyskinesia
Degeneration	Tear	Fatty infiltration	Dystonia
Subchondral	*Capsule*	*Tendons and bursa*	*UMN and LMN*
Dysplasia, fracture	Patulous capsule	Tear, calcification	Spasticity
AVN, degeneration	Capsulitis, contracture	degeneration, bursitis	Flaccid paralysis
Synovium	*Congruity*	*Other (external to shoulder)*	*Sensory and autonomic*
Inflammatory	Subluxation	Fracture, malignancy, HO	Charcot joint
Crystalopathy	Dislocation	Skin contracture	Chronic regional pain

AVN avascular necrosis; *HO* heterotopic ossification; *UMN* upper motor neuron; *LMN* lower motor neuron

Acknowledgments

Disclaimer: None.

References

1. Armstrong A. Diagnosis and clinical assessment of a stiff shoulder. Shoulder Elbow. 2015;7(2):128–34.
2. Codman EA. The shoulder. New York: G Miller & Company; 1934.
3. Neviaser AS, Neviaser RJ. Adhesive capsulitis of the shoulder. J Am Acad Orthop Surg. 2011;19(9): 536–42.
4. Zuckerman JD, Rokito A. Frozen shoulder: a consensus definition. J Shoulder Elb Surg. 2011;20(2):322–5.
5. Diercks RLLT. Clinical symptoms and physical examinations. In: Itoi EAG, Bain GI, Diercks RL, Guttmann D, Imhoff AB, Mazzocca AD, Sugaya H, Yoo Y-S, editors. Shoulder stiffness current concepts and concerns. New York: Springer; 2015.
6. Bain GICH. The pathogenesis and classification of shoulder stiffness. In: Itoi EAG, Bain GI, Diercks RL, Guttmann D, Imhoff AB, Mazzocca AD, Sugaya H, Yoo Y-S, editors. Shoulder stiffness current concepts and concerns. New York: Springer; 2015.
7. McFarland EG. Examination of the shoulder: the complete guide. 1st ed. New York: Thieme; 2006.
8. McFarland EG, Kibler WB, Murrell GAC, Rojas J. Examination of the shoulder for beginners and experts: an update. Instr Course Lect. 2020;69:255–72.
9. Park HB, Yokota A, Gill HS, El Rassi G, McFarland EG. Diagnostic accuracy of clinical tests for the different degrees of subacromial impingement syndrome. J Bone Joint Surg Am. 2005;87(7):1446–55.
10. Itoi EHY. Pathophysiology of frozen shoulders: histology and laboratory tests. In: Itoi EAG, Bain GI, Diercks RL, Guttmann D, Imhoff AB, Mazzocca AD, Sugaya H, Yoo Y-S, editors. Shoulder stiffness current concepts and concerns. New York: Springer; 2015.
11. Lubis AM, Lubis VK. Matrix metalloproteinase, tissue inhibitor of metalloproteinase and transforming growth factor-beta 1 in frozen shoulder, and their changes as response to intensive stretching and supervised neglect exercise. J Orthop Sci. 2013;18(4):519–27.
12. Quan GM, Carr D, Schlicht S, Powell G, Choong PF. Lessons learnt from the painful shoulder; a case series of malignant shoulder girdle tumours misdiagnosed as frozen shoulder. Int Semin Surg Oncol. 2005;2(1):2.
13. Emig EW, Schweitzer ME, Karasick D, Lubowitz J. Adhesive capsulitis of the shoulder: MR diagnosis. AJR Am J Roentgenol. 1995;164(6):1457–9.
14. Sugaya H. Imaging of stiff shoulders. In: Itoi E, Arce G, Bain GI, Diercks RL, Guttmann D, Imhoff AB, Mazzocca AD, Sugaya H, Yoo Y-S, editors. Shoulder stiffness current concepts and concerns. New York: Springer; 2015.
15. Lee MH, Ahn JM, Muhle C, Kim SH, Park JS, Kim SH, et al. Adhesive capsulitis of the shoulder: diagnosis using magnetic resonance arthrography, with arthroscopic findings as the standard. J Comput Assist Tomogr. 2003;27(6):901–6.
16. Lefevre-Colau MM, Drapé JL, Fayad F, Rannou F, Diche T, Minvielle F, et al. Magnetic resonance imaging of shoulders with idiopathic adhesive capsulitis: reliability of measures. Eur Radiol. 2005;15(12):2415–22.
17. Sofka CM, Ciavarra GA, Hannafin JA, Cordasco FA, Potter HG. Magnetic resonance imaging of adhesive capsulitis: correlation with clinical staging. HSS J. 2008;4(2):164–9.
18. Tamai K, Yamato M. Abnormal synovium in the frozen shoulder: a preliminary report with dynamic magnetic resonance imaging. J Shoulder Elb Surg. 1997;6(6):534–43.
19. Ueda Y, Sugaya H, Takahashi N, Matsuki K, Kawai N, Tokai M, et al. Rotator cuff lesions in patients with

stiff shoulders: a prospective analysis of 379 shoulders. J Bone Joint Surg Am. 2015;97(15):1233–7.

20. Jung J-Y, Jee W-H, Chun HJ, Kim Y-S, Chung YG, Kim J-M. Adhesive capsulitis of the shoulder: evaluation with MR arthrography. Eur Radiol. 2006;16(4):791–6.
21. Homsi C, Bordalo-Rodrigues M, da Silva JJ, Stump XM. Ultrasound in adhesive capsulitis of the shoulder: is assessment of the coracohumeral ligament a valuable diagnostic tool? Skelet Radiol. 2006;35(9):673–8.
22. Suh CH, Yun SJ, Jin W, Lee SH, Park SY, Park JS, et al. Systematic review and meta-analysis of magnetic resonance imaging features for diagnosis of adhesive capsulitis of the shoulder. Eur Radiol. 2019;29(2):566–77.
23. Sernik RA, Vidal Leão R, Luis Bizetto E, Sanford Damasceno R, Horvat N, Guido CG. Thickening of the axillary recess capsule on ultrasound correlates with magnetic resonance imaging signs of adhesive capsulitis. Ultrasound. 2019;27(3):183–90.
24. Mengiardi B, Pfirrmann CW, Gerber C, Hodler J, Zanetti M. Frozen shoulder: MR arthrographic findings. Radiology. 2004;233(2):486–92.
25. Lee SY, Park J, Song SW. Correlation of MR arthrographic findings and range of shoulder motions in patients with frozen shoulder. AJR Am J Roentgenol. 2012;198(1):173–9.
26. Loeffler BJ, Brown SL, D'Alessandro DF, Fleischli JE, Connor PM. Incidence of false positive rotator cuff pathology in MRIs of patients with adhesive capsulitis. Orthopedics. 2011; 34(5):362.
27. Cuomo F, Holloway GB. Diagnosis and management of the stiff shoulder. In: Williams GR, Iannotti JP, editors. Disorders of the shoulder—diagnosis and management. 2nd ed. Philadelphia: Lippincott Williams & Wilkins; 2007. p. 541–60.

11 Evaluation of the Thrower's Shoulder

Kyle R. Sochacki and Michael T. Freehill

11.1 Phases of Throwing

The overhead throwing motion is classically divided into six different phases: wind-up, early cocking, late cocking, acceleration, deceleration, and follow-through (Fig. 11.1) [1–3]. Although variations can exist, the entire motion has an approximate duration of 2 s with 75% of that time encompassing the wind-up, early cocking, and late cocking phases of throwing [4, 5].

During the first phase of throwing (wind-up), the lead leg is elevated to its maximum height and the center of gravity is shifted over the back leg in preparation for energy transfer in subsequent phases [1]. The shoulder is simultaneously placed in slight abduction and internal rotation. During this phase, there is a relatively low risk of injury to the shoulder as minimal stress is placed on the upper extremity in this position [1, 6–8].

The early cocking phase begins once the lead leg reaches its maximum point (load) and ends when it contacts the pitching mound. The arm moves into abduction and external rotation due to the activation of the deltoid, supraspinatus, and infraspinatus during the later portion of this phase [1].

K. R. Sochacki · M. T. Freehill (✉)
Sports Medicine and Shoulder Surgery, Department of Orthopaedic Surgery, Stanford University, Stanford, CA, USA
e-mail: freehill@stanford.edu

Late cocking occurs between lead foot contact and the point of shoulder maximum external rotation. The scapula retracts providing a stable glenoid surface for humeral head compression while the humerus undergoes additional abduction and external rotation. Activation of the infraspinatus and teres minor in the early portion of this phase leads to increasing humeral external rotation. However, as the shoulder approaches maximum external rotation, the subscapularis, pectoralis major, and latissimus dorsi eccentrically contract to decelerate and stabilize the glenohumeral joint [2, 9].

The time between maximum shoulder external rotation and ball release comprises the acceleration phase of throwing. The scapula protracts as the humerus undergoes horizontal abduction and internal rotation. The triceps is active during the early portion of this phase while maximum contraction of the subscapularis, pectoralis major, and latissimus dorsi occurs later to produce humeral internal rotation [1, 10].

The deceleration phase begins with ball release and ends with maximum humeral internal rotation and elbow extension. This is often considered the most violent and dangerous phase of the throwing motion due to the excessive distraction and shear forces on the glenohumeral joint. Any energy not imparted through the ball is dissipated through the shoulder, resulting in large eccentric loads by the posterior shoulder external rotators (infraspinatus and teres minor) to

J. G. Lane et al. (eds.), *The Art of the Musculoskeletal Physical Exam*,
https://doi.org/10.1007/978-3-031-24404-9_11

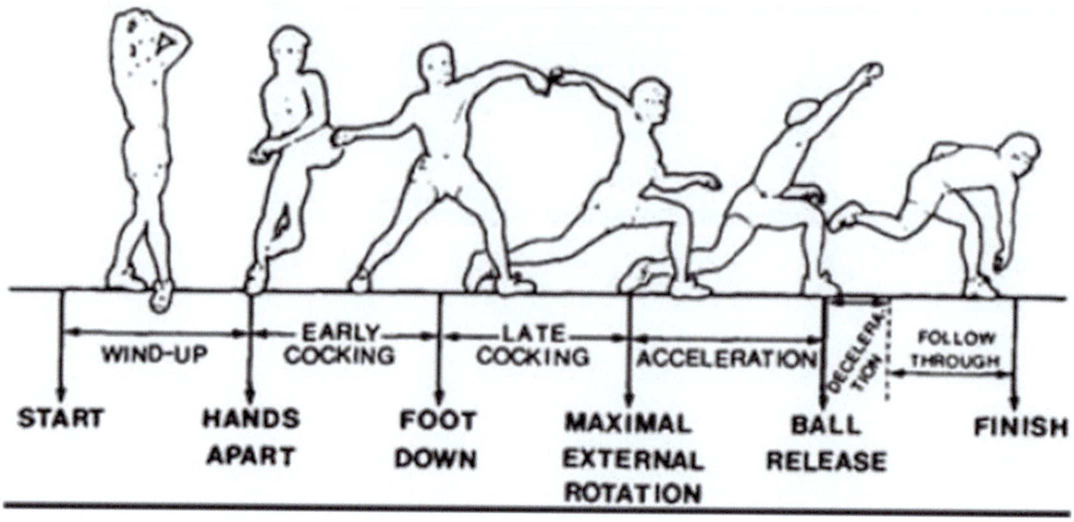

Fig. 11.1 Phases of throwing. Reprinted with permission from Bakshi, Neil MD; Freehill, Michael T. MD The Overhead Athletes Shoulder, Sports Medicine and Arthroscopy Review: September 2018 - Volume 26 - Issue 3 - p 88–94, Wolters Kluwer Health, Inc.

decelerate the arm and limit excessive anterior humeral translation in relation to the glenoid [1, 2]. The forces required to decelerate the arm have been estimated to reach up to 1200 Newtons (N) as the muscles dispel the kinetic energy generated in the earlier phases [2].

The final phase of throwing is the follow-through. The phase proceeds with the body moving forward with the arm until motion is ceased. Arm deceleration continues during this phase as the deltoid and rotator cuff muscles continue to eccentrically contract, and the serratus anterior, trapezius, and rhomboids work eccentrically to decelerate the scapula [11]. At the end of this phase, muscle activity returns to resting levels and the forces on the glenohumeral joint drop drastically.

11.2 Adaptations to the Throwing Shoulder

The stresses placed on the shoulder during throwing often lead to the development of specific adaptations over time. Specifically, the extremes of shoulder motion during the late cocking, early acceleration, and the deceleration phases of throwing can lead alterations in anatomy and kinematics that often begins in early childhood. These soft tissue and bony adaptations that occur are necessary to throw at a high level but can progress to pathologic deficits if not properly managed.

The most common physiologic change to a thrower's shoulder is seen in range of motion. Generally, there is an increase in external rotation and decrease in internal rotation between the dominant (throwing side) and non-dominant side [2]. This reduced internal rotation is caused by posteroinferior capsular hypertrophy and contracture. Despite lack of agreement on a threshold, a comparative loss of internal rotation greater than 20° has been generally termed glenohumeral internal rotation deficit (GIRD) [12–14].

The pathologic consequences of this decreased shoulder internal rotation has been a topic of recent debate with some authors arguing that a total arc of motion or external rotation deficit poses a more significant risk for injury [14–18]. Professional pitchers have been reported to have a mean shoulder external rotation of 141° with the arm at 90° of abduction [19]. Despite this significantly increased external rotation, the total arc of motion was found to be within 7–9° of their non-throwing side [19, 20].

Bony adaptations of the proximal humerus have also been postulated to allow for increased glenohumeral external rotation. Previous studies have demonstrated that the dominant arm of throwers exhibits 10–17° of increased humeral retroversion compared to their contralateral side [21, 22]. It is thought that the repetitive throwing motion inhibits the normal derotation of the proximal humerus that occurs physiologically during growth [23].

In addition to posterior capsule tightness, the shoulder also exhibits anterior capsular laxity secondary to repetitive external rotation during the later phases of throwing [24–28]. This microinstability allows for increased humeral head translation and rotation during the throwing motion. As such, microinstability is likely a necessary adaptation with 61% of professional pitchers demonstrating a positive sulcus sign in their throwing shoulder regardless of symptoms [24, 25, 29].

11.3 Pathophysiology

11.3.1 Glenohumeral Internal Rotation Deficit (GIRD)

As previously stated, the pathologic consequences of glenohumeral internal rotation deficit (GIRD) is debated. Posteroinferior capsular hypertrophy and contracture leads to an exaggerated posterosuperior shift of the humeral head compared to those patients without GIRD [26]. This results in increased external rotation, decreased internal rotation, decreased horizontal adduction, and decreased flexion that likely contribute to the increased risk of shoulder injury in patients with GIRD [17, 30, 31] As such, GIRD has been associated with superior labral from anterior to posterior (SLAP) tears, partial articular-sided rotator cuff tears (PASTA), internal impingement, and external impingement [12, 14, 15, 26, 32, 33].

11.3.2 Internal Impingement

Walch et al. was the first to describe internal impingement as the pathologic contact between the posterosuperior glenoid labrum and articular-sided rotator cuff (posterior supraspinatus and anterior infraspinatus) and greater tuberosity [34]. This has been shown to lead to partial articular-sided rotator cuff tears (PASTA) and SLAP tears [12, 13, 17, 26, 35–38]. Microinstability, shear stresses, and muscle fatigue with imbalance have all been proposed as theories regarding the pathophysiology [39–42]. However, no clear etiology has been established.

11.3.3 Scapular Dyskinesis

Burkhart et al. described a constellation of findings known as the "SICK" scapula (**S**capular malposition, **I**nferior medial border prominence, **C**oracoid pain, and dys**K**inesis) and theorized this to be the primary contributor to the development of SLAP and PASTA tears [12, 25]. The scapula moves into a more protracted position with resultant contracture of the pectoralis minor tendon causing inferior medial border prominence and coracoid pain. The alteration of this static scapular position leads to scapula dyskinesis during arm movement and throwing due to the overuse and weakness of the scapular stabilizing and posterior rotator cuff muscles [25, 43, 44]. As a result, scapula dyskinesis has been implicated in increased scapular anterior tilt, internal impingement, decreased rotator cuff strength, and anterior capsular strain [2, 28, 36, 45–52].

11.4 History

A detailed history is critical to narrowing the differential diagnosis in the overhead athlete. This includes the location of pain, duration and frequency of symptoms, trauma or inciting event, neurologic symptoms, exacerbating or mitigating factors, prior shoulder injuries, and previous shoulder treatments or surgeries. There is additional information specific to pitchers that may be pertinent such as type of pitcher (starter or reliever), phase(s) of throwing when symptoms occur, alterations in mechanics or training regimen, number of innings pitched, number of games, rest over the past year without throwing, loss of velocity, loss of control, and types of pitches where pain occurs.

The athlete will often complain of a "dead arm" with loss of velocity and/or control due to pain, feelings of stiffness, or fatigue [12, 43]. Anterosuperior and posterosuperior shoulder discomfort is typically felt during the late cocking or early acceleration phases of throwing when the arm begins to move forward. Throwers with labral tears may also exhibit varying degrees of mechanical symptoms such as clicking, catching, locking, or shoulder instability depending on the type and size of the lesion.

It is also important to question the athlete about symptoms that may be related to cervical spine or thoracic outlet syndrome. The lumbar spine, core, and hip also play a major role in the throwing motion through a highly coordinated musculoskeletal sequence that transfers energy from the lower body to the throwing arm. As such, history of injury to these areas should be considered as it can increase the stress on the shoulder due to changes in throwing mechanics.

Fig. 11.2 A 12-year-old baseball player demonstrates an inability to balance his trunk during a single leg squat. The pelvic tilt observed is secondary to weak core and hip external rotators

11.5 Physical Examination

A comprehensive physical examination of the athlete including routine and specific examination maneuvers is important to differentiate between the varied pathologies. Examination of the throwing athlete should always focus on the entire kinetic chain including the lower extremities and trunk. The upper and lower extremities should be assessed for alignment, asymmetries, strength imbalances, and range of motion deficits or limitations of any joint. Core muscle control and strength can be assessed with resisted abdominal crunches, single leg stance to assess for Trendelenburg, and observing control while having the athlete perform a single leg squat. Compensatory movements such as pelvic tilt or rotation may indicate imbalances or weakness (Fig. 11.2). A supine examination of the hips should also be performed to assess for range of motion and signs of femoroacetabular impingement.

A focused shoulder examination can then be performed after the remainder of the kinetic chain was thoroughly evaluated. The shoulder examination should always be performed with the shirt off allowing inspection of the entire shoulder complex for signs of bruising, muscular atrophy, swelling, discoloration of fingers or nails, and scapular abnormalities such as winging [12, 25, 43]. The resting position of both scapula with the arms at the sides should be examined for asymmetry of tilt, rotation, and elevation or depression. Injured throwers often demonstrate loss of external rotation control, elevation, and posterior tilt of the scapula that manifests and medial scapula winging [51]. As previously discussed, "SICK" scapular syndrome is characterized by **S**capular malposition, **I**nferior medial border prominence, medial-sided **C**oracoid tenderness, and scapular dys**K**inesis (Fig. 11.3). Repeated forward flexion can aid in diagnosing scapular dyskinesis via fatigue or accentuation of abnormal symmetry between sides. Additionally, the scapular retraction test can be performed by manually placing the scapula in the retracted position or by assisting the scapula by placing pressure on it for support from posterior. Improvements of pain and increasing motion with this maneuver are diagnostic for scapula dyskinesis [53].

After visual inspection, thorough palpation should be performed of the humeral head, greater tuberosity, glenohumeral joint line (posterior), bicipital groove, coracoid, acromioclavicular (AC) joint, and scapula. Active and passive shoulder range of motion of both arms should be performed in the standing and supine position. Comparison of internal and external rotation of the shoulder at 90° of abduction between shoulders is critical (Fig. 11.4). Care should be taken to observe and measure without allowing scapulothoracic contribution. This should also include total arc of motion measurement in both shoulders. Patients with GIRD can present with significantly decreased internal rotation (greater than 20°) compared to the contralateral shoulder [12–14].

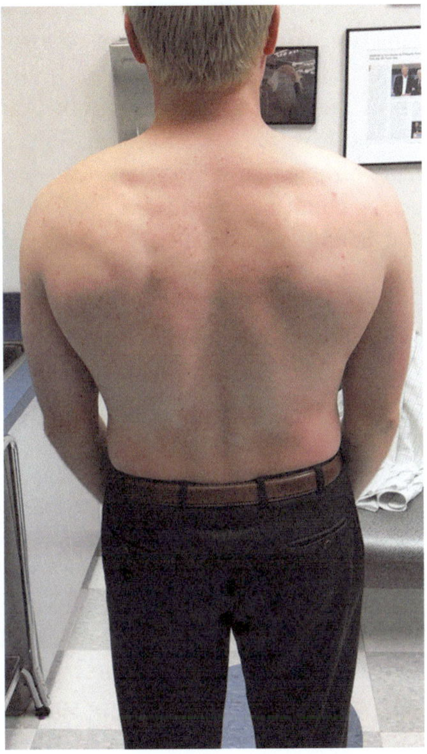

Fig. 11.3 Posterior view of scapulae in a right-handed thrower. Many throwers often exhibit scapula asymmetry. The scapula of the throwing arm (right) is slightly inferior with a more prominent inferomedial border

Comprehensive major muscle and rotator cuff strength testing should be evaluated after range of motion and compared to the contralateral arm. This should include deltoid, teres minor, supraspinatus, infraspinatus, and subscapularis testing. Testing can be done with a dynamometer to increase reliability of measurement. Subacromial external impingement testing should also be completed using the Hawkins and Neer tests.

More provocative testing for overhead athletes should be performed next in the physical examination. The O'Brien active compression test is the most commonly conducted examination maneuver to assess for SLAP tears [54]. The shoulder is positioned in 90° of forward flexion, 20° of horizontal adduction, and maximum internal rotation (thumb down). The examiner then applies a downward force while the athlete resists. The extremity is then externally rotated with the palm facing upward (thumb up) and the maneuver is repeated. Reproduction of pain during internal rotation with decreased pain during

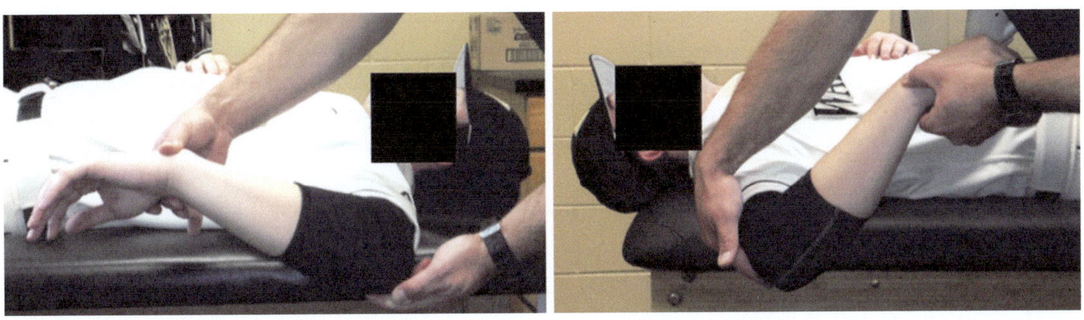

Fig. 11.4 Examination of glenohumeral internal rotation at 90° of abduction in the non-throwing (left) and throwing (right) arms

external rotation is considered a positive test [54]. This test has good sensitivity and specificity in the general population but is significantly decreased in athletes with 78% sensitivity and 11% specificity [55]. SLAP pathology can also be assessed with the Mayo Shear test (dynamic labral shear test) as it reproduces the "peel-back" seen while throwing and has 78% sensitivity [56]. The Mayo Shear test is performed with the patient laying supine on an examination table with table supporting the scapula and the humerus free. The elbow is flexed to 90° while the shoulder is externally rotated by gravity. The shoulder is then passively elevated to full overhead elevation maintaining 90° of elbow flexion from neutral to maximal abduction. A positive test is defined as deep shoulder pain or clicking between 90 and 120° of abduction [56]. The "3-pack" exam of O'Brien's active compression test, throwing test, and tenderness with bicipital tunnel palpation has also demonstrated excellent sensitivity for diagnosing injuries to the biceps and biceps labral complex [57]. The senior author stresses the performance of multiple tests to aid in the diagnosis of SLAP tears in an overhead athlete.

Although less common in a thrower, the shoulder should also be assessed for instability through testing of the anterior and posterior labrum as well as sulcus sign. The anterior labrum can be assessed by bringing the patient's shoulder into a position of 90° of abduction and 90° of external rotation while in the supine position. A positive finding is the subjective feeling of apprehension from impending shoulder subluxation or dislocation. The relocation test can then be performed by maintaining the arm in the position of apprehension and applying a posteriorly directed force on the humerus to stabilize the shoulder [58]. Resolution of the apprehension would constitute a positive test. These physical examination maneuvers have been shown to have a sensitivity and specificity of 90% and 85%, respectively, for the diagnosis of anterior glenoid labral tears with instability [59].

The posterior labrum is assessed with the Jerk, Kim, and posterior Modified Labral tests. The Kim test is performed by placing the arm at 90° of abduction, followed by flexing the shoulder to 45° of forward flexion while simultaneously applying axial load on the elbow and a posteroinferior force on the upper humerus evaluating the posteroinferior aspect of the labrum. The Jerk test can be performed in either the seated or standing position with arm in 90° of abduction, internal rotation, and elbow flexion. The examiner then applies an axial force along the axis of the humerus and adducts the arm to a forward flexed position. This assesses chiefly the midportion of the posterior labrum. Both examinations are positive for posterior labral injury with reproduction pain, while the Jerk test may also produce a clunk as the posteriorly subluxated humeral head reduces. The Kim and Jerk tests have excellent specificity (94% and 98%, respectively) and good sensitivity (80% and 73%, respectively) individually, but when combined, their sensitivity for diagnosing posterior labral injury increases to 97% [60].

The sulcus sign is performed with the patient upright with their arm resting at their side. The examiner then stabilizes the shoulder and applies inferiorly directed force on the elbow. The amount of inferior translation is then graded to indicate multi-directional instability or rotator cuff interval deficiency. The sulcus is graded as Grade 1, Grade 2, or Grade 3 for 1 centimeter (cm), 2 cm, and 3 cm inferior translation, respectively.

Additionally, in throwers, a comprehensive neurovascular examination should be performed routinely as thoracic outlet syndrome can be overlooked in this population. Inspection includes digits and fingernails for petechiae. Examination maneuvers include the Roos, Wright, and Adson tests. The Wright test is performed by placing the arm in hyperabduction and external rotation with the head turned in the opposite direction [61]. The Adson test describes bringing the arm into extension, turning the head toward the affected side, and taking a deep breath. Both the Adson and Wright tests are considered positive with a decrease in radial pulse in the affected extremity [61]. In order to perform the Roos test, the patient places both arms in 90° of abduction with the elbows flexed to 90°. The hands are then opened and closed for 3 min with a positive test causing significant discomfort recreating their symptoms and often preventing them from completing the test [62]. Individually, these examinations lack specificity for thoracic outlet syndrome, but when the Adson and Roos tests are combined, the specificity increases to 82% [63].

11.6 Summary

A comprehensive physical examination of the throwing athlete should be performed focusing on the entire kinetic chain including the shoulder, lower extremities, and trunk. It should be performed in a systematic and reproducible manner as to avoid missing subtle injuries. Additionally, all physical examination findings should consider the adaptations to the thrower's shoulder and be compared to the contralateral arm.

References

1. Dillman CJ, Fleisig GS, Andrews JR. Biomechanics of pitching with emphasis upon shoulder kinematics. J Orthop Sports Phys Ther. 1993;18(2):402–8.
2. Fleisig GS, Andrews JR, Dillman CJ, Escamilla RF. Kinetics of baseball pitching with implications about injury mechanisms. Am J Sports Med. 1995;23(2):233–9.
3. Pappas AM, Zawacki RM, Sullivan TJ. Biomechanics of baseball pitching. A preliminary report. Am J Sports Med. 1985;13(4):216–22.
4. Ellenbecker TS, Roetert EP, Bailie DS, Davies GJ, Brown SW. Glenohumeral joint total rotation range of motion in elite tennis players and baseball pitchers. Med Sci Sports Exerc. 2002;34(12):2052–6.
5. Johnson L. Patterns of shoulder flexibility among college baseball players. J Athl Train. 1992;27(1):44–9.
6. Glousman RE, Barron J, Jobe FW, Perry J, Pink M. An electromyographic analysis of the elbow in normal and injured pitchers with medial collateral ligament insufficiency. Am J Sports Med. 1992;20(3):311–7.
7. Jobe FW, Moynes DR, Tibone JE, Perry J. An EMG analysis of the shoulder in pitching. A second report. Am J Sports Med. 1984;12(3):218–20.
8. Jobe FW, Tibone JE, Perry J, Moynes D. An EMG analysis of the shoulder in throwing and pitching. A preliminary report. Am J Sports Med. 1983;11(1):3–5.
9. Glousman R, Jobe F, Tibone J, Moynes D, Antonelli D, Perry J. Dynamic electromyographic analysis of the throwing shoulder with glenohumeral instability. J Bone Joint Surg Am. 1988;70(2):220–6.
10. Kibler WB. The role of the scapula in athletic shoulder function. Am J Sports Med. 1998;26(2):325–37.
11. Digiovine NM, Jobe FW, Pink M, Perry J. An electromyographic analysis of the upper extremity in pitching. J Shoulder Elb Surg. 1992;1(1):15–25.
12. Burkhart SS, Morgan CD, Kibler WB. The disabled throwing shoulder: spectrum of pathology Part I: pathoanatomy and biomechanics. Arthroscopy. 2003;19(4):404–20.
13. Ruotolo C, Price E, Panchal A. Loss of total arc of motion in collegiate baseball players. J Shoulder Elb Surg. 2006;15(1):67–71.
14. Wilk KE, Macrina LC, Fleisig GS, Porterfield R, Simpson CD 2nd, Harker P, et al. Correlation of glenohumeral internal rotation deficit and total rotational motion to shoulder injuries in professional baseball pitchers. Am J Sports Med. 2011;39(2):329–35.
15. Amin NH, Ryan J, Fening SD, Soloff L, Schickendantz MS, Jones M. The relationship between glenohumeral internal rotational deficits, total range of motion, and shoulder strength in professional baseball pitchers. J Am Acad Orthop Surg. 2015;23(12):789–96.
16. Clarsen B, Bahr R, Andersson SH, Munk R, Myklebust G. Reduced glenohumeral rotation, external rotation weakness and scapular dyskinesis are risk factors for shoulder injuries among elite male handball players: a prospective cohort study. Br J Sports Med. 2014;48(17):1327–33.
17. Keller RA, De Giacomo AF, Neumann JA, Limpisvasti O, Tibone JE. Glenohumeral internal rotation deficit and risk of upper extremity injury in overhead athletes: a meta-analysis and systematic review. Sports Health. 2018;10(2):125–32.
18. Wilk KE, Macrina LC, Fleisig GS, Aune KT, Porterfield RA, Harker P, et al. Deficits in glenohumeral passive range of motion increase risk of shoulder injury in professional baseball pitchers: a prospective study. Am J Sports Med. 2015;43(10):2379–85.
19. Brown LP, Niehues SL, Harrah A, Yavorsky P, Hirshman HP. Upper extremity range of motion and isokinetic strength of the internal and external shoulder rotators in major league baseball players. Am J Sports Med. 1988;16(6):577–85.
20. Wilk KE, Meister K, Andrews JR. Current concepts in the rehabilitation of the overhead throwing athlete. Am J Sports Med. 2002;30(1):136–51.
21. Chant CB, Litchfield R, Griffin S, Thain LM. Humeral head retroversion in competitive baseball players and its relationship to glenohumeral rotation range of motion. J Orthop Sports Phys Ther. 2007;37(9):514–20.
22. Reagan KM, Meister K, Horodyski MB, Werner DW, Carruthers C, Wilk K. Humeral retroversion and its relationship to glenohumeral rotation in the shoulder of college baseball players. Am J Sports Med. 2002;30(3):354–60.
23. Yamamoto N, Itoi E, Minagawa H, Urayama M, Saito H, Seki N, et al. Why is the humeral retroversion of throwing athletes greater in dominant shoulders than

in nondominant shoulders? J Shoulder Elb Surg. 2006;15(5):571–5.
24. Bigliani LU, Codd TP, Connor PM, Levine WN, Littlefield MA, Hershon SJ. Shoulder motion and laxity in the professional baseball player. Am J Sports Med. 1997;25(5):609–13.
25. Burkhart SS, Morgan CD, Kibler WB. The disabled throwing shoulder: spectrum of pathology Part III: the SICK scapula, scapular dyskinesis, the kinetic chain, and rehabilitation. Arthroscopy. 2003;19(6):641–61.
26. Grossman MG, Tibone JE, McGarry MH, Schneider DJ, Veneziani S, Lee TQ. A cadaveric model of the throwing shoulder: a possible etiology of superior labrum anterior-to-posterior lesions. J Bone Joint Surg Am. 2005;87(4):824–31.
27. Mihata T, Lee Y, McGarry MH, Abe M, Lee TQ. Excessive humeral external rotation results in increased shoulder laxity. Am J Sports Med. 2004;32(5):1278–85.
28. Weiser WM, Lee TQ, McMaster WC, McMahon PJ. Effects of simulated scapular protraction on anterior glenohumeral stability. Am J Sports Med. 1999;27(6):801–5.
29. Borsa PA, Wilk KE, Jacobson JA, Scibek JS, Dover GC, Reinold MM, et al. Correlation of range of motion and glenohumeral translation in professional baseball pitchers. Am J Sports Med. 2005;33(9):1392–9.
30. Howell SM, Galinat BJ. The glenoid-labral socket. A constrained articular surface. Clin Orthop Relat Res. 1989;243:122–5.
31. Terry GC, Hammon D, France P, Norwood LA. The stabilizing function of passive shoulder restraints. Am J Sports Med. 1991;19(1):26–34.
32. Laudner KG, Myers JB, Pasquale MR, Bradley JP, Lephart SM. Scapular dysfunction in throwers with pathologic internal impingement. J Orthop Sports Phys Ther. 2006;36(7):485–94.
33. Tyler TF, Nicholas SJ, Roy T, Gleim GW. Quantification of posterior capsule tightness and motion loss in patients with shoulder impingement. Am J Sports Med. 2000;28(5):668–73.
34. Walch G, Boileau P, Noel E, Donell ST. Impingement of the deep surface of the supraspinatus tendon on the posterosuperior glenoid rim: an arthroscopic study. J Shoulder Elb Surg. 1992;1(5):238–45.
35. Halbrecht JL, Tirman P, Atkin D. Internal impingement of the shoulder: comparison of findings between the throwing and nonthrowing shoulders of college baseball players. Arthroscopy. 1999;15(3):253–8.
36. Jazrawi LM, McCluskey GM 3rd, Andrews JR. Superior labral anterior and posterior lesions and internal impingement in the overhead athlete. Instr Course Lect. 2003;52:43–63.
37. Myers JB, Laudner KG, Pasquale MR, Bradley JP, Lephart SM. Glenohumeral range of motion deficits and posterior shoulder tightness in throwers with pathologic internal impingement. Am J Sports Med. 2006;34(3):385–91.
38. Valadie AL 3rd, Jobe CM, Pink MM, Ekman EF, Jobe FW. Anatomy of provocative tests for impingement syndrome of the shoulder. J Shoulder Elb Surg. 2000;9(1):36–46.
39. Burkhart SS, Morgan CD. The peel-back mechanism: its role in producing and extending posterior type II SLAP lesions and its effect on SLAP repair rehabilitation. Arthroscopy. 1998;14(6):637–40.
40. Drakos MC, Rudzki JR, Allen AA, Potter HG, Altchek DW. Internal impingement of the shoulder in the overhead athlete. J Bone Joint Surg Am. 2009;91(11):2719–28.
41. Jobe FW, Pink M. Classification and treatment of shoulder dysfunction in the overhead athlete. J Orthop Sports Phys Ther. 1993;18(2):427–32.
42. Paley KJ, Jobe FW, Pink MM, Kvitne RS, ElAttrache NS. Arthroscopic findings in the overhand throwing athlete: evidence for posterior internal impingement of the rotator cuff. Arthroscopy. 2000;16(1):35–40.
43. Kibler WB, Thomas SJ. Pathomechanics of the throwing shoulder. Sports Med Arthrosc Rev. 2012;20(1):22–9.
44. Burkhart SS, Morgan CD, Kibler WB. The disabled throwing shoulder: spectrum of pathology. Part II: evaluation and treatment of SLAP lesions in throwers. Arthroscopy. 2003;19(5):531–9.
45. Fleisig GS, Barrentine SW, Zheng N, Escamilla RF, Andrews JR. Kinematic and kinetic comparison of baseball pitching among various levels of development. J Biomech. 1999;32(12):1371–5.
46. Ben Kibler W, Sciascia AD, Hester P, Dome D, Jacobs C. Clinical utility of traditional and new tests in the diagnosis of biceps tendon injuries and superior labrum anterior and posterior lesions in the shoulder. Am J Sports Med. 2009;37(9):1840–7.
47. Castagna A, Mouhsine E, Conti M, Vinci E, Borroni M, Giardella A, et al. Chondral print on humeral head: an indirect sign of long head biceps tendon instability. Knee Surg Sports Traumatol Arthrosc. 2007;15(5):645–8.
48. Ide J, Maeda S, Takagi K. Arthroscopic transtendon repair of partial-thickness articular-side tears of the rotator cuff: anatomical and clinical study. Am J Sports Med. 2005;33(11):1672–9.
49. Ludewig PM, Cook TM. Alterations in shoulder kinematics and associated muscle activity in people with symptoms of shoulder impingement. Phys Ther. 2000;80(3):276–91.
50. McClure PW, Michener LA, Sennett BJ, Karduna AR. Direct 3-dimensional measurement of scapular kinematics during dynamic movements in vivo. J Shoulder Elb Surg. 2001;10(3):269–77.
51. Myers JB, Laudner KG, Pasquale MR, Bradley JP, Lephart SM. Scapular position and orientation in throwing athletes. Am J Sports Med. 2005;33(2):263–71.
52. Tate AR, McClure PW, Kareha S, Irwin D. Effect of the Scapula Reposition Test on shoulder impingement symptoms and elevation strength in overhead athletes. J Orthop Sports Phys Ther. 2008;38(1):4–11.
53. Kibler WB, Sciascia A. Current concepts: scapular dyskinesis. Br J Sports Med. 2010;44(5):300–5.

54. O'Brien SJ, Pagnani MJ, Fealy S, McGlynn SR, Wilson JB. The active compression test: a new and effective test for diagnosing labral tears and acromioclavicular joint abnormality. Am J Sports Med. 1998;26(5):610–3.
55. Myers TH, Zemanovic JR, Andrews JR. The resisted supination external rotation test: a new test for the diagnosis of superior labral anterior posterior lesions. Am J Sports Med. 2005;33(9):1315–20.
56. Sodha S, Srikumaran U, Choi K, Borade AU, McFarland EG. Clinical assessment of the dynamic labral shear test for superior labrum anterior and posterior lesions. Am J Sports Med. 2017;45(4):775–81.
57. Taylor SA, Newman AM, Dawson C, Gallagher KA, Bowers A, Nguyen J, et al. The "3-pack" examination is critical for comprehensive evaluation of the biceps-labrum complex and the bicipital tunnel: a prospective study. Arthroscopy. 2017;33(1):28–38.
58. Jobe FW, Kvitne RS, Giangarra CE. Shoulder pain in the overhand or throwing athlete. The relationship of anterior instability and rotator cuff impingement. Orthop Rev. 1989;18(9):963–75.
59. Liu SH, Henry MH, Nuccion S, Shapiro MS, Dorey F. Diagnosis of glenoid labral tears. A comparison between magnetic resonance imaging and clinical examinations. Am J Sports Med. 1996;24(2):149–54.
60. Kim SH, Park JS, Jeong WK, Shin SK. The Kim test: a novel test for posteroinferior labral lesion of the shoulder—a comparison to the jerk test. Am J Sports Med. 2005;33(8):1188–92.
61. Brantigan CO, Roos DB. Diagnosing thoracic outlet syndrome. Hand Clin. 2004;20(1):27–36.
62. Atasoy E. A hand surgeon's further experience with thoracic outlet compression syndrome. J Hand Surg Am. 2010;35(9):1528–38.
63. Braun RM, Rechnic M, Shah KN. Pulse oximetry measurements in the evaluation of patients with possible thoracic outlet syndrome. J Hand Surg Am. 2012;37(12):2564–9.

Part II

Elbow Reviewer Dr Pederizini

12 Anatomy

Nadine Ott and Kilian Wegmann

12.1 Osseous Anatomy

The elbow involves three joints: the radiohumeral, ulnohumeral, and proximal radioulnar joint (Fig. 12.1). The centre of rotation of the elbow runs from the inferior edge of the medial epicondyle to the small tubercle of the lateral epicondyle. The condyles show a 30° anterior flexion in relation to the humeral axis, a 6–8° valgus tilt and a 5° internal rotation in relation to the epicondylar axis. The lateral condyle, the capitellum, is shaped spherically and just covered by cartilage [1]. At the base of the coronoid, the ligament is attached to the sublime tubercle [2–4]. The ulnar collateral complex and the flexor compartment start from the medial condyle, and the ligament inserts to the sublime tubercle. The lateral ligament and the extensor compartment originate from the lateral epicondyle and the ligament is attached to the supinator crest. The radial neck has a 15° angle with the long axis and the greater sigmoid notch angulates 30° with the ulna shaft. While performing osteosynthesis of the proximal ulna, the dorsal, varus, and torsion angulation should be considered [5]. The transverse portion is covered by a smaller area of cartilage, and it divides the sigmoid notch into an anterior part and the olecranon. The bare area of articular cartilage is comprised between the coronoid and the olecranon articular surface [6]. Elbow ossification during the childhood occurs at the six elbow ossification centreds. The order of appearance is capitellum, radial head, internal epicondyle, trochlea, olecranon, and external epicondyle (CRITOE).

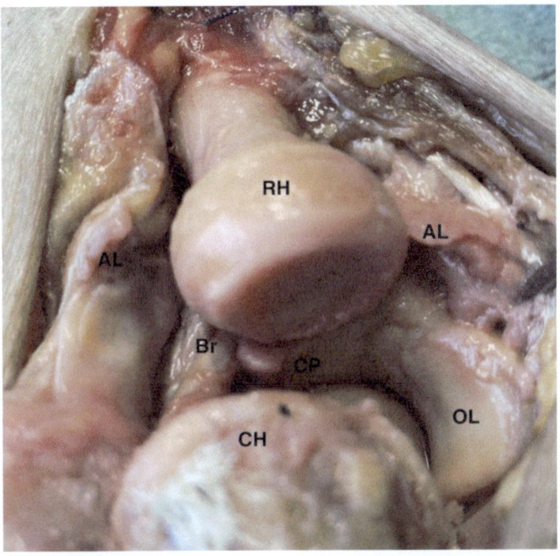

Fig. 12.1 Osseous anatomy. Osseous anatomy of the elbow with the radiohumeral, ulnohumeral, and proximal radioulnar joint; capitellum (CH) with the radial head (RH), coronoid (CP), anular ligament (AL), olecranon (OL), brachialis muscle inserts to the proximal ulna behind the MCL complex

N. Ott · K. Wegmann (✉)
Center for Orthopedic and Trauma Surgery, University Medical Center of Cologne, Cologne, Germany
e-mail: Nadine.ott@uk-koeln.de; kilian.wegmann@uk-koeln.de

J. G. Lane et al. (eds.), *The Art of the Musculoskeletal Physical Exam*,
https://doi.org/10.1007/978-3-031-24404-9_12

12.2 Capsular Ligamentous Complex

The joint capsule surrounds the elbow joint, the olecranon, coronoid, and the radial fossa. Its thickness on the medial and lateral aspects forms the collateral ligaments. The mean capacity of the capsule is approximately 20 mL in 80° flexion of the elbow [7, 8]. In case of a capsular pathology patients, prefer this position [9]. In extension, the anterior capsule stabilizes the elbow joint against varus and valgus stress.

The chorda Oblique has a relevant influence on longitudinal stability. It traverses from the lateral base of the coronoid to the proximal radius and inserts distal on the neck of the radius [10].

12.2.1 Ulnar Collateral Complex

The medial collateral ligament complex is made of three different elements: the anterior, posterior, and transverse ligament. The anterior ligament consists of parallel fibres running from its origin on the antero-inferior aspect of the medial epicondyle and inserts onto the medial coronoid process, sublime tubercle [11]. It can be separated into anterior, posterior, and central fibres which have different functions (Fig. 12.2). The anterior ligament stabilizes against valgus stress [12]. Its anterior fibres are the primary restraint up to 90° flexion, while the posterior fibres are more important between 60° and maximum of flexion [13, 14]. In case of a fracture of the coronoid process, valgus instability should be considered [15]. Fracture type 2.2 (according to the O'Driscoll classification) have a high risk to cause valgus instability [4, 16]. The posterior ligament inserts onto the medial margin of the semilunar notch of the olecranon. It stabilizes the elbow pronation and represents the secondary restraint to the valgus stress when the anterior ligament is ruptured. The transverse ligament connects the medial coronoid to the olecranon processes [17].

12.2.2 Lateral Collateral Ligament Complex

The lateral collateral ligament complex contains four bundles: lateral ulnar collateral ligament (LUCL), accessory lateral ulnar ligament, radial collateral ligament (RCL), and annular ligament (AL) (Fig. 12.3). The AL encircles the radial head

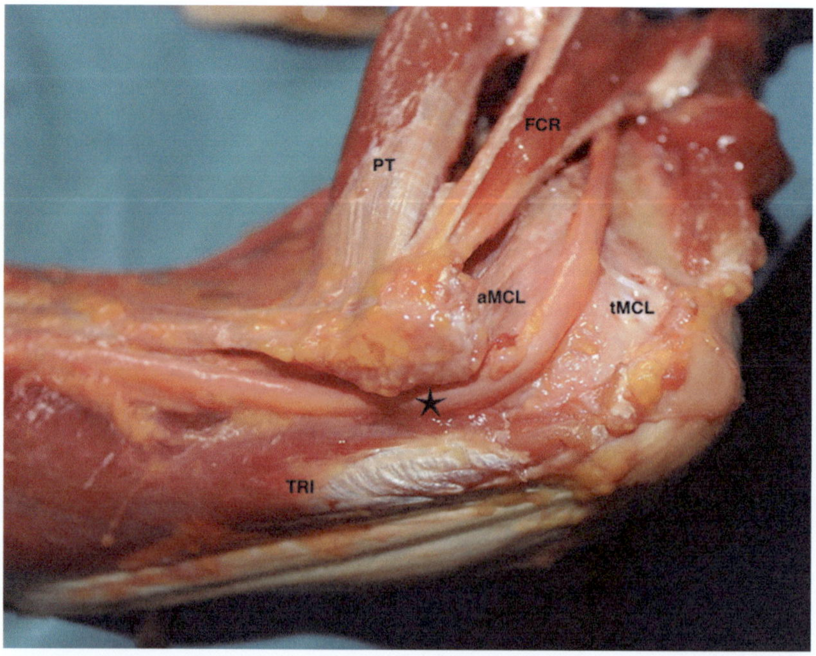

Fig. 12.2 The medial collateral ligament complex has three separate components: anterior (aMCL), posterior (pMCL), and transverse ligament (pMCL); the relation between the MCL complex and the ulnar nerve (star); *TR* triceps muscle, *PT* pronator teres muscle, *FCR* flexor carp radialis muscle

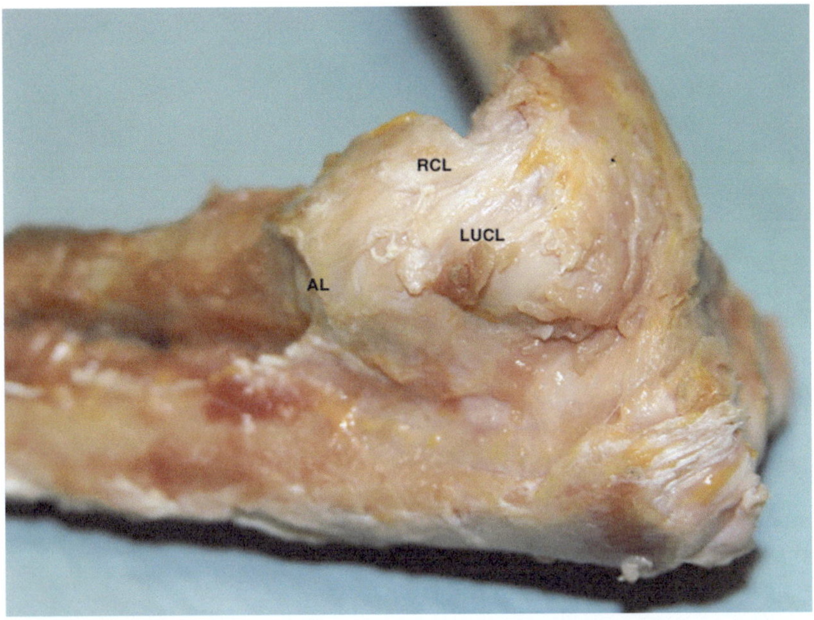

Fig. 12.3 Lateral ligament complex, with the lateral ulnar collateral ligament (LUCL), accessory lateral ulnar ligament, radial collateral ligament (RCL), and annular ligament (AL)

and stabilizes the proximal radioulnar joint [10, 18]. The RCL originates at the lateral epicondyle and extends to the AL, and together stabilizes against varus stress [19]. The anterior and the posterior parts of the RCL are tight in extension and flexion, while the middle part is tight in extension and flexion. The accessory lateral ulnar ligament inserts onto the supinator crest of the ulna and the inferior margin of the AL [11, 20]. The LUCL attachment on the supinator crest of the ulna can vary: with a bilobed insertion (type I), a coalesced with AL (type II) and with a conjoined broad base (type III). The most common way is type III (50%).

12.3 Muscle Layer

12.3.1 Triceps Brachii Muscle

The triceps brachii muscle covers the entire posterior part of the arm and inserts with his three heads (long, lateral, and medial) on the tip of the olecranon. The subtendinous olecranon bursa protects the tendon from the tip of the olecranon. The radial nerve provides muscular branches for every three heads of the triceps. The long head arises from the infraglenoid tuberosity of the scapula, the lateral head origins proximally to the spiral groove of the humerus and the medial head has an extensive origin distally to the spiral groove [21]. The lateral head is the strongest head of the triceps. The triceps is the extensor of the elbow, the long head has also an effect on the shoulder [22]. At the lateral side, the triceps continues with the anconeus muscle and the antebrachial fascia and inserts at the ulna. Snapping of the medial head can lead to triceps tendon and ulnar nerve pathology. The posterior approach through the triceps muscle can be challenging [23, 24].

12.3.2 Anconeus Muscle

The anconeus is triangle shaped, it originates from the posterior lateral epicondyle and inserts into the proximal ulna. The anconeus provides an important landmark for the lateral approach to the elbow. It has a joint stabilizing function [25]. In case of radiohumeral instability or loss of the radial head, it has been used as a local interposition flap between the radial neck and capitellum [26].

12.3.3 Biceps Brachii Muscle

The biceps brachii muscle occupies the entire anterior part of the distal arm and inserts with the two heads onto the bicipital tuberosity. The short head is inserted more distally, whereas the long head is inserted more eccentrically and medially. The long head is attached to the supraglenoid tubercle and the short head to the coracoid process [27]. It is innervated by a branch of the musculocutaneous nerve. The biceps muscle acts on three joints: the glenohumeral, ulnohumeral, and proximal radioulnar joints. Functional independence of the two heads is caused by the different moment arms. The long head has a greater moment arm in supination, while the short head is greater in pronation and neutral position. So, an isolated rupture of each can be explained. The biceps muscle continues to the fascia antebrachia with the lacertus fibrosus; the lacertus fibrosus serves as a stabilizer of the distal biceps tendon (Fig. 12.4). The lacertus is tensed as the forearm flexors contract, subsequently causing a medial pull on the tendon and increase its force. An intact lacertus with a distal biceps tendon tear can decrease the functional deficit.

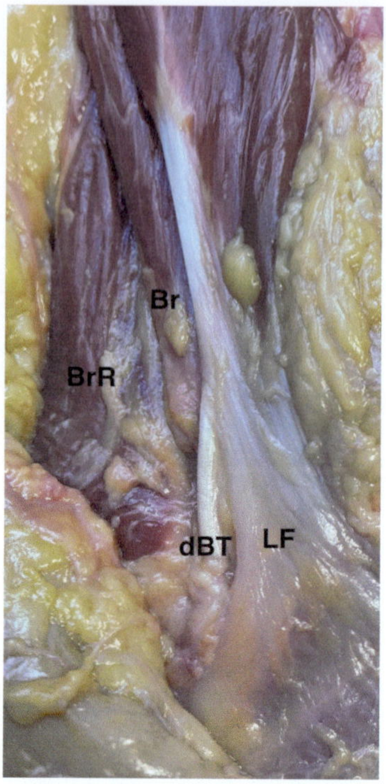

Fig. 12.4 Muscle layer. The biceps muscle continues to the fascia antebrachia with the lacertus fibrosus (LT), the lacertus fibrosus serves as a stabilizer of the distal biceps tendon (dBT); brachialis muscle (Br) is the strongest flexor, it works like an "anterior anconeus" and stabilize with the true anconeus the elbow against rotatory instability; the brachioradialis muscle (BrR) originates from the intermuscular septum and the lateral aspect of the distal humerus and inserts into the distal styloid process of the radius

12.3.4 Brachioradialis

The brachioradialis muscle has the greatest mechanical advantage of any elbow flexor. It originates from the intermuscular septum and the lateral aspect of the distal humerus and inserts into the distal styloid process of the radius. Besides the flexion, it pronates the supinated forearm. It is innervated by a branch of the radial nerve and is the leading muscle of the radial nerve.

12.3.5 Brachialis

The brachialis muscle has two heads, the superficial and the deep head. The superficial head has a long proximal origin with a cordlike tendon that attaches to the anterior proximal humerus, while the deep head takes origin from the anterior distal humerus and inserts obliquely into the proximal ulnar [27]. It works like an "anterior anconeus" and stabilizes with the true anconeus the elbow against rotatory instability. The brachialis muscle is the strongest flexor [28]. The deep head is more involved at the beginning of the elbow flexion, while the superficial head provides greater power once the elbow is flexed. It is innervated by the musculocutaneous nerve, which runs on its distal surface.

12.3.6 Extensor Muscles

The common extensor origin (CEO) of the extensor group of the forearm is the lateral condyle. The extensor group includes the extensor carpi radialis longus and brevis, extensor digitorum communis, extensor digiti minimi, and extensor carpi ulnaris. The extensor radialis longus originates from the supracondylar ridge just below the origin of the brachioradialis muscle. It is an extensor of the wrist, but it can act as a flexor of the elbow. The extensor carpi radialis brevis originates below the extensor carpi radialis longus, while the origin of the extensor digitorum beyond it. The supinator muscle has a complex origin on the lateral epicondyle, lateral collateral ligament complex and ulna. At the proximal border of the superficial part of the supinator could represent an entrapment of the radial nerve, ramus profundus. The lateral approach to the elbow, called Kocher approach, is realized between the extensor carpi ulnaris muscle and the anconeus muscle [29].

12.4 Neuroanatomy

The three major nerves cross the elbow inside fibrous tunnels. The most common sides of nervous entrapment are Osborne's fascia for the ulnar nerve, the lacertus fibrosus or the pronator teres muscle for the median nerve, and the Arcade of Frohse for the radial nerve.

12.4.1 Median Nerve

The median nerve is formed by the lateral (C5–7) and the medial cords (C8–T1) of the brachial plexus. It lies in the cubital fossa, anteriorly to brachioradialis and medially to the brachial artery and the biceps tendon. The median nerve passes below the lacertus fibrosus, between the two heads of the pronator teres and the flexor digitorum superficialis. The joint of the elbow is innervated by sensory branches of the median nerve. Until it passes below the lacertus fibrosus, the median nerve is enveloped by a fascia of the biceps and brachialis muscles. In extension, the median nerve lies 4–7 mm ventral onto the trochlea, while in flexion the distance between the trochlea and the nerve becomes increasingly wider to 12–18 mm [30]. Knowledge of this distance is important for elbow arthroscopy [31]. In case of filling the joint before the arthroscopy, the distance to the osseous structures and the nerves increase. However, the joint capsule is very close the nerves. This should be considered in case of an anterior capsulotomy.

12.4.2 Radial Nerve

The radial nerve is a branch of the posterior cord of the plexus brachialis (C5/T1). Before entering the spiral groove, it provides branches to the medial and the long head of the triceps muscle. It follows the spiral groove of the humeral shaft and enters into the anterior compartment, 13 cm proximal of the lateral epicondyle after perforating the lateral intermuscular septum (Fig. 12.5). It lies between the brachialis and brachioradialis muscle. At the level of the elbow, it passes through the radial tunnel and lies on the joint capsule. It bifurcates into the PIN and

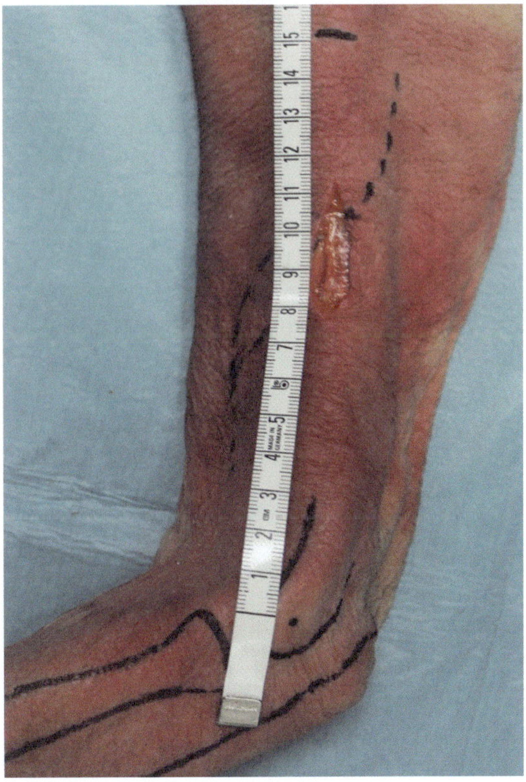

Fig. 12.5 The radial nerve follows the spiral groove of the humeral shaft and enters the anterior compartment 13 cm proximal of the lateral epicondyle after perforating the lateral intermuscular septum. It lies between the brachialis and brachioradialis muscle

the superficial branch [32]. The PIN courses between the two heads of the supinator, supplying the extensor carpi ulnaris and the extensor of all digits [33]. The anterolateral portal for arthroscopy could be a risk for the radial nerve [31]. The rotation of the forearm can influence the close position to the radiohumeral joint. A supinated forearm can increase the distance between the radial nerve and the proximal radius to 3.6 cm [34].

12.4.3 Ulnar Nerve

The ulnar nerve is a branch of the medial cord of the plexus brachialis (C8–T1). The ulnar nerve passes about 10 cm proximally to the medial epicondyle on the arcade of Struthers, a fibrosus tissue that spans from the triceps fascia to the medial intermuscular septum, the cubital tunnel, and the arcade of Osborne. The cubital tunnel is a fibro-osseous tunnel, which is bordered by the cubital retinaculum, a groove in the medial epicondyle and the medial collateral ligament. The cubital tunnel is a cause of common entrapment of the ulnar nerve. The anconeus epitrochlearis muscle is another cause of entrapment of the ulnar nerve, that is present in approximately 2% of cases. After the cubital tunnel, the ulnar nerve provides motor branches to the two heads of the flexor carpi ulnaris. In case of a neurolysis and anterior transposition, these branches should be protected [35].

References

1. Koebke J. Funktionelle Anatomie und Biomechanik des Ellenbogengelenkes. In: Stahl C, Koebke J, et al., editors. Klinische Arthrologie. Landsberg: Ecomed; 1998. p. 11.
2. Cage DJ, Abrams RA, Callahan JJ, et al. Soft tissue attachments of the ulnar coronoid process. An anatomic study with radiographic correlation. Clin Orthop Relat Res. 1995;320:154–8.
3. Morrey BF. An KN Stability of the elbow: osseous constraints. J Shoulder Elb Surg. 2005;14:174S–8S.
4. O'Driscoll SW, Jaloszynski R, Morrey BF, et al. Origin of the medial ulnar collateral ligament. J Hand Surg Am. 1992;17(1):164–8.
5. Windisch G, Clement H, Grechenig W, et al. The anatomy of the proximal ulna. J Shoulder Elb Surg. 2007;16(5):661–6.
6. Eckstein F, Merz B, Müller-Berbl M et al. Morphomechanics of the humero-ulnar joint: II. Concave incongruity determines the distribution of load and subchondral mineralization. Anat Rec 1995b; 243: 327–335.
7. Gallay SH, Richards RR, O'Driscoll SW. Intraarticular capacity and compliance of stiff and normal elbows. Arthroscopy. 1993;9(1):9–13.
8. O'Driscoll SW, Morrey BF, An KN. Intraarticular pressure and capacity of the elbow. Arthroscopy. 1990;6(2):100–3.
9. Burkhart KJ, Hollinger B, Wegmann K, et al. Luxationen und Bandverletzungen am Ellenbogen und Unterarm. Orthop Unfallchir up2date. 2012;7:435–62.
10. Morrey BF. An KN Articular and ligamentous contributions to the stability of the elbow joint. Am J Sports Med. 1983;11:315–9.
11. O'Driscoll SW, Horii E, Morrey B, et al. Anatomy of the ulnar part of the lateral collateral ligament of the elbow. Clin Anat. 1992a;5:296–303.
12. Dugas JR, Ostrander RV, Cain EL, et al. Anatomy of the anterior bundle of the ulnar collateral ligament. J Shoulder Elb Surg. 2007;16(5):657–60.
13. Callaway GH, Field LD, Deng XH, et al. Biomechanical evaluation of the medial collateral ligament of the elbow. J Bone Joint Surg Am. 1997;79(8):1223–31.
14. Miyake J, Moritomo H, Masatomi T, et al. In vivo and 3-dimensional functional anatomy of the anterior bundle of the medial collateral ligament of the elbow. J Shoulder Elb Surg. 2012;21(8):1006–12.
15. Ring D, Jupiter JB. Fracture-dislocation of the elbow. J Bone Joint Surg Am. 1998;80(4):566–80.
16. Farrow LD, Mahoney AJ, Stefancin JJ, et al. Quantitative analysis of the medial ulnar collateral ligament ulnar footprint and its relationship to the ulnar sublime tubercle. Am J Sports Med. 2011;39(9):1936–41.
17. Fuss FK. The ulnar collateral ligament of the human elbow joint. Anatomy, function and biomechanics. J Anat. 1991;175:203–12.
18. Bozkurt M, Acar HI, Apaydin N, et al. The annular ligament: an anatomical study. Am J Sports Med. 2005;33(1):114–8.
19. Morrey BF. An KN Functional anatomy of the elbow ligaments. Clin Orthop. 1985;201:84.
20. Olsen BS, Vaesel MT, Sojbjerg JO, et al. Lateral collateral ligament of the elbow joint: anatomy and kinematics. J Shoulder Elb Surg. 1996;5:103–12.
21. Keener JD, Chafik D, Kim HM, et al. Insertional anatomy of the triceps brachii tendon. J Shoulder Elb Surg. 2010;19(3):399–405.
22. An KN, Hui FC, Morrey B, Linscheid RL, et al. Muscle across the elbow joint: a biomechanical analysis. J Biomech. 1981;14:659.

23. Bryan RS. Morrey BF Extensive posterior exposure of the elbow. A triceps-sparing approach. Clin Orthop Relat Res. 1982;166:188–92.
24. Morrey BF, Sanchez-Sotelo J. Approaches for elbow arthroplasty: how to handle the triceps. J Shoulder Elb Surg. 2011;20(2 Suppl):S90–6.
25. Basmajian JV. Griffin WR Jr Function of anconeus muscle. An electromyographic study. J Bone Joint Surg Am. 1972;54:1712–4.
26. Hwang K, Han JY, Chung IH. Topographical anatomy of the anconeus muscle for use as a free flap. J Reconstr Microsurg. 2004;20(8):631–6.
27. Mazzocca AD, Cohen M, Berkson E, et al. The anatomy of the bicipital tuberosity and distal biceps tendon. J Shoulder Elb Surg. 2007;16(1):122–7.
28. Leonello DT, Galley IJ, Bain GI, Carter CD. Brachialis muscle anatomy; a study in cadavers. J Bone Joint Surg Am. 2007;89:1293–7.
29. Nimura A, Fujishiro H, Wakabayashi Y, et al. Joint capsule attachment to the extensor carpi radialis brevis origin: an anatomical study with possible implications regarding the etiology of lateral epicondylitis. J Hand Surg Am. 2014;39(2):219–25.
30. Gunther SF, DiPasquale D, Martin R. The internal anatomy of the median nerve in the region of the elbow. J Hand Surg Am. 1992;17(4):648–56.
31. Omid R, Hamid N, Keener JD, et al. Relation of the radial nerve to the anterior capsule of the elbow: anatomy with correlation to arthroscopy. Arthroscopy. 2012;28(12):1800–4.
32. Abrams RA, Brown RA, Botte MJ. The superficial branch of the radial nerve: an anatomic study with surgical implications. J Hand Surg Am. 1992;17(6):1037–41.
33. Spinner M. The arcade of Frohse and its relationship to posterior interosseous nerve paralysis. J Bone Joint Surg. 1968;50B:809–12.
34. Diliberti T, Botte MJ, Abrams RA. Anatomical considerations regarding the posterior interosseous nerve during posterolateral approaches to the proximal part of the radius. J Bone Joint Surg Am. 2000;82(6):809–13.
35. Contreras MG, Warner MA, Charboneau WJ, et al. Anatomy of the ulnar nerve at the elbow: potential relationship of acute ulnar neuropathy to gender differences. Clin Anat. 1998;11(6):372–8.

Biomechanics of the Elbow

13

Carina Cohen, Guilherme Augusto Stirma, Gyoguevara Patriota, and Benno Ejnisman

13.1 Kinematics

The normal elbow range of motion (ROM) goes from 0° to 140° for flexion/extension and 80°/90° for pronation/supination, but the functional ROM for most daily activities is 30° to 130° for flexion/extension and 50°/50° for pronation/supination. Daily activities considered are to bring the hand to the face, to drink, eat, dress, perform hygiene with elbow flexion/pronation and throw and push with elbow extension/supination [1]. There is a valgus alignment called carrying angle, which measures 10 to 15° in men and 15° to 20° in women [2].

Variation of flexion and extension axis throughout ROM is described as the screw displacement axis (SDA). The flexion axis is oriented at 3° to 5° of internal rotation to the medial and lateral epicondyles plane and 4° to 8° valgus to the humerus long axis. The axis of rotation planned on the surface of the condyles lean to be smaller on the medial side than on the lateral side, varying from 5.67° to 17.23° (mean, 11.02°) in the axial plane, 7.80° to 19.4° (mean, 11.95°) in the coronal plane. The center of rotation changes during the elbow's movement, recent literature has presented that the elbow joint does not function as a simple hinge joint since its axis translates and rotate. The lateral condyle demonstrated a counterclockwise circular pattern (Fig. 13.1), the axis of rotation changes lightly ulnar and volar in supination, radial and dorsal in pronation and the Radius, as well, moves proximally with pronation and distally with supination [3, 4].

While the congruency of the elbow articular surfaces is perfect, the compression and weight-bearing loads are not equally distributed. Elbow is stilted by the position of the forearm,: in elbow

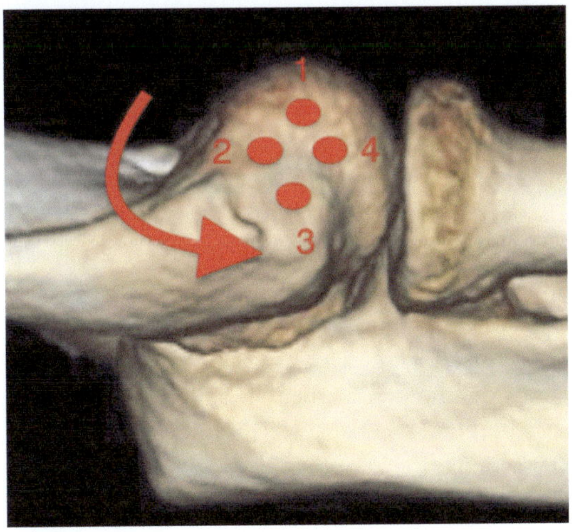

Fig. 13.1 Axis of rotation on the lateral condyle with counterclockwise pattern. (1) 30°, (2) 60°, (3) 90°, (4) 135°

C. Cohen (✉) · G. A. Stirma · G. Patriota · B. Ejnisman
Department of Orthopedics and Traumatology, Sports Traumatology Center (CETE), Federal University of São Paulo (UNIFESP), São Paulo, Brazil

J. G. Lane et al. (eds.), *The Art of the Musculoskeletal Physical Exam*,
https://doi.org/10.1007/978-3-031-24404-9_13

extension the Radial Head has no pressure against Humeral Capitulum, supination decreases the contact, for as much as pronation increases [2]. The contact area of the ulna against the trochlea is on the medial facet of the trochlear notch for any elbow motion [3].

13.2 Elbow Stability

Elbow stability is provided by static stabilizers and muscle dynamic stabilizers. The constraints are the ulnohumeral articulation, anterior and posterior bundle of Medial Collateral Ligament (MCL), Lateral Collateral Ligament LCL complex, radiocapitellar articulation, common flexor tendons, common extensor tendons, and the joint capsule [5]. The elbow becomes progressively unstable when each of the bony structures are removed, especially the radial head and coronoid process [6]. Coronoid Fractures Type III with more than 50% of the coronoid compromised increase results in varus-valgus laxity, even with normal collateral ligaments [7]. The coronoid is extremely important in posterolateral stability in association with the radial head. With 30% of the coronoid height loss and without radial head, the ulnohumeral joint dislocates, and the stability can just be restored with replacement of the radial head [4]. On the other hand, fractures of less than 80% of the olecranon can be removed without compromising elbow stability [8].

The anterior capsule supplies 70% of the soft tissue restraint to distraction, while the medial collateral ligament has this function at 90° of flexion. Varus stress is tested in extension equally by the articulation, lateral collateral ligament, and capsule (Fig. 13.2), while valgus is equally divided among the medial collateral ligament, articulation, and capsule (Fig. 13.3). In flexion, the articulation supplies 75% of the varus stability and in valgus the medial collateral ligament supplies 54% of stability [9].

Forearm rotation is important for elbow stability. With passive flexion, the MCL tear is more stable in supination, while the LCL absence is more stable in pronation. Literature has also

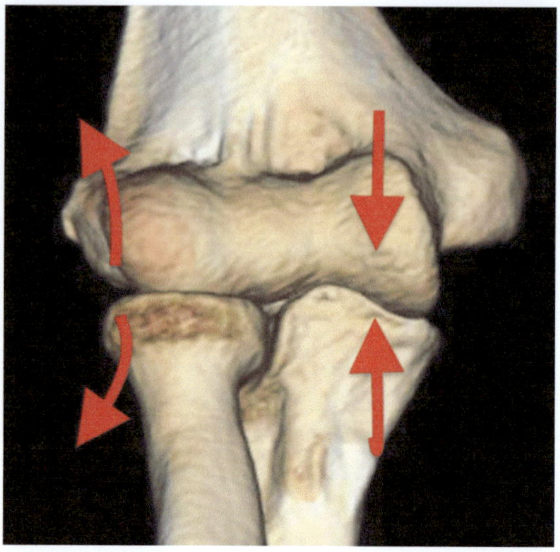

Fig. 13.2 Varus stress (The osseous stability provides the majority of joint stability; while the LCL complex holds out 10% of varus stress)

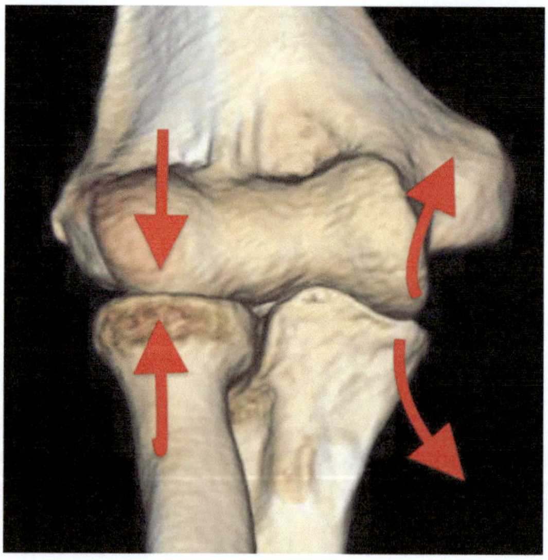

Fig. 13.3 Valgus stress (The MCL is a restraint to valgus stress and the radial head provides a secondary restraint)

shown that the elbow is more stable in supination in coronoid fractures with more than 50% of the coronoid [7].

The anterior bundle of the medial collateral ligament (A-MCL) is the main static stabilizer of the elbow in valgus stress, and this bundle is taut in 0 to 85°, while the posterior band is taut from 55° to 140°. Morrey et al. studied the muscle contributions to dynamic valgus stability of the elbow simulating, in a cadaver model, the contraction of the biceps, brachialis, and triceps. They suggest that the elbow joint forces created by muscles subscribe to valgus stability in a deficient MCL. The flexor carpi ulnaris (FCU) is proposed as a contributor to dynamic valgus stability because the FCU position is in line with the medial ulnar collateral ligament and fine wire electromyographic studies demonstrate that pitchers' players with symptomatic valgus instability have reduced flexor–pronator muscle action [1].

13.3 Joint Forces

The compressive forces at the elbow are significant that some doctors have affirmed that it is "wrong to think of the elbow as non-weightbearing" [10]. Loads are distributed about 43% across at the ulnohumeral joint and 57% at radiocapitellar joint. Force transfer at the radiocapitellar joint is greatest between 0 and 30° of flexion and in pronation position. When the elbow is extended, the higher part of force across the ulnohumeral joint is intent at the coronoid; while when the elbow is flexed, is on the Olecranon (Fig. 13.4) [4, 11].

The compression forces on the elbow have been determined when maximal isometric elbow flexion and extension happens. The largest forces occur during isometric elbow flexion in near full extension, but compressive forces from isometric extension contraction are reduced in more extended positions and largest in flexed positions.

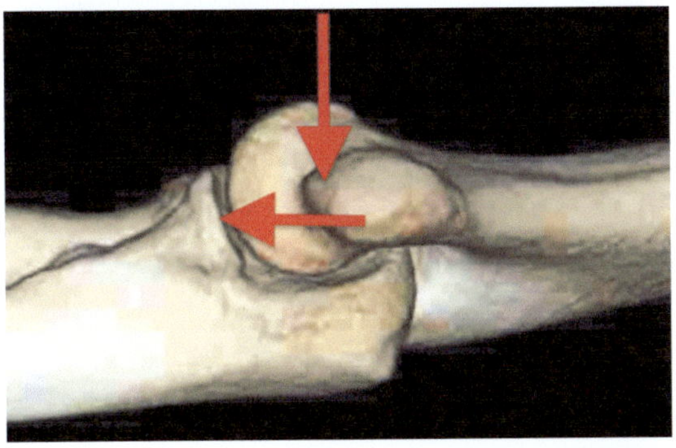

Fig. 13.4 In an extended joint position, forces are driven more anteriorly

The applied muscle action force is assumed to be perpendicular to the forearm. The muscle and joint reaction forces change lightly with the shift of the muscle action. Nonetheless, the orientation of the resultant joint force is responsive to changes in the muscle force line. The guidance of the joint force goes from the central part of the trochlea toward to the rim with the direction of muscle action. This is particularly for joint forces in the trochlea where the forearm axis changes with the elbow joint flexion angle [12]. Considerations about forearm rotation or elbow pronation and supination equilibrium, demonstrate that pronator teres had rotation moment counter the biceps tension. The pronator teres force must be relaxed to maintain the equilibrium, rather than sum to the humeroradial joint tension [10, 13].

References

1. Lockard M. Clinical Biomechanics of the Elbow. J Hand Ther. 2006;19(2):72–81. https://linkinghub.elsevier.com/retrieve/pii/S0894113006000445.
2. Martin S, Sanchez E. Anatomy and Biomechanics of the Elbow Joint. Semin Musculoskelet Radiol. 2013;17(05):429–36. https://linkinghub.elsevier.com/retrieve/pii/S0030589807001228.
3. Goto A, Moritomo H, Murase T, Oka K, Sugamoto K, Arimura T, et al. In vivo elbow biomechanical analysis during flexion: three-dimensional motion analysis using magnetic resonance imaging. J Shoulder Elb Surg. 2004;13(4):441–7.
4. Bryce CD, Armstrong AD. Anatomy and biomechanics of the elbow. Orthop Clin North Am. 2008;39(2):141–54. https://linkinghub.elsevier.com/retrieve/pii/S0030589807001228.
5. Karbach LE, Elfar J. Elbow instability: anatomy, biomechanics, diagnostic maneuvers, and testing. J Hand Surg Am. 2017;42(2):118–26. https://linkinghub.elsevier.com/retrieve/pii/S0031938416312148.
6. Ring D, Jupiter JB, Zilberfarb J. Posterior dislocation of the elbow with fractures of the radial head and coronoid. J Bone Jt Surg Am. 2002;84(4):547–51. http://journals.lww.com/00004623-200204000-00006.
7. Beingessner DM, Dunning CE, Stacpoole RA, Johnson JA, King GJW. The effect of coronoid fractures on elbow kinematics and stability. Clin Biomech. 2007;22(2):183–90.

8. Mckeever FM. Fracture of the olecranon process of the ulna. J Am Med Assoc. 1947;135(1):1. http://jama.jamanetwork.com/article.aspx?doi=10.1001/jama.1947.02890010003001.
9. Morrey BF, Sanchez-Sotelo J, Morrey ME. The elbow and its disorders. Elsevier Health Sciences. 2017;5(1):38–9.
10. Amis AA, Dowson D, Wright V. Elbow joint force predictions for some strenuous isometric actions. J Biomech. 1980;13(9):765–75.
11. Wake H, Hashizume H, Nishida K, Inoue H, Nagayama N. Biomechanical analysis of the mechanism of elbow fracture-dislocations by compression force. J Orthop Sci. 2004;9(1):44–50.
12. Morrey BF, An KN, Stormont TJ. Force transmission through the radial head. J Bone Joint Surg Am. 1988;70(2):250–6.
13. Hems TEJ. The effect of elbow position on the range of supination and pronation of the forearm, Shaaban et al., JHSE, 33E: 3–8. J Hand Surg. 2009;34(1):138–9. http://journals.sagepub.com/doi/10.1177/1753193408098900.

14 Evaluation of Range of Motion

Carina Cohen, Gyoguevara Patriota,
Guilherme Stirma, and Benno Ejnisman

14.1 Introduction

The elbow is a complex hinge joint that comprises three bony articulations responsible for its stability. Physical examination is very important and begins with the inspection of the affected elbow and comparison with the contralateral side. The examiner should observe the resting position of the elbow. In patients with effusion, the elbow is often held in 70–80° of flexion, a position accommodating the greatest capsular volume [1].

14.2 Carrying Angle

The examiner should also assess the carrying angle of the elbow. The carrying angle is a clinical measurement of the angle formed by the forearm (specifically an ulna) and the arm (humerus) with the elbow extended in the coronal plane [2, 3]. In full extension, a normal valgus carrying angle is approximately 11° in men and 13° in women (Fig. 14.1).

The changes in the carrying angle can be seen in sequelae of elbow fractures (supracondylar fractures in children—varus) and in throwing athletes (may indicate an adaptation to repetitive valgus stress) [4].

C. Cohen (✉) · G. Patriota · G. Stirma · B. Ejnisman
Department of Orthopedics and Traumatology, Sports Traumatology Center (CETE), Federal University of São Paulo (UNIFESP), São Paulo, Brazil

J. G. Lane et al. (eds.), *The Art of the Musculoskeletal Physical Exam*,
https://doi.org/10.1007/978-3-031-24404-9_14

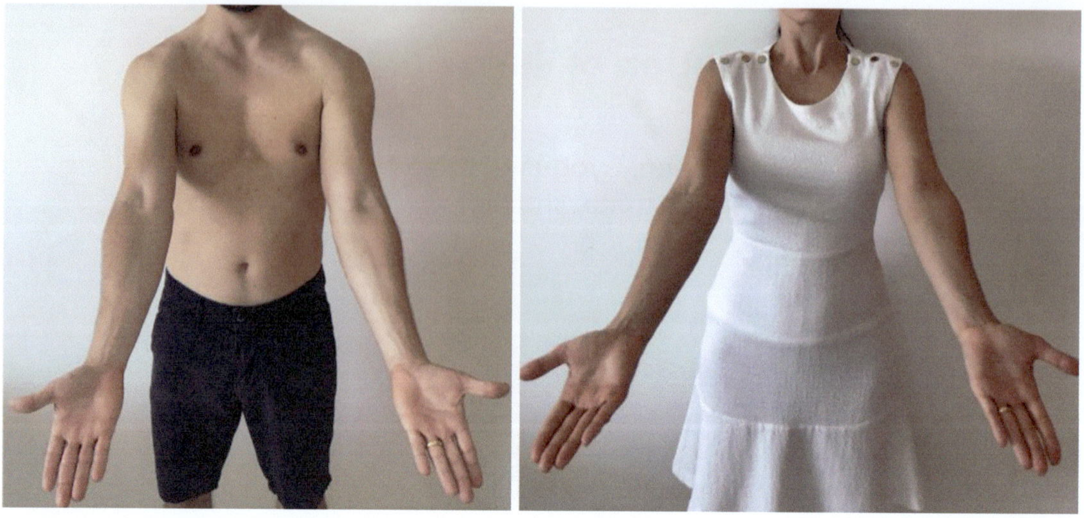

Fig. 14.1 The carrying angle in man and woman

14.3 Motion

The loss of full extension is the first motion alteration caused by the majority of pathologies and the last movement to be regained. Normally, the arc of flexion–extension, although variable, ranges from about 0° to 140° plus or minus 10° [5] (Fig. 14.2). This range exceeds the range normally required for daily living activities, which ranges from 30° to 130°.

Pronation–supination range can change more than the flexion–extension one. Acceptable degree of pronation and supination are 75° and 85°, respectively, whereas functional motion is about 50° in each direction [6]. The examiner should record both active and passive movements. The arm is at the side and the elbow flexed 90° during the assessment of the arc of forearm rotation to avoid abduction of the shoulder that could occur when patients tend to accommodate for loss of pronation (Fig. 14.3).

The measurement of the range of motion with the elbow in extension is inadequate because the values could result overestimated for the compensation of the shoulders (Fig. 14.4).

Chapleau et al. found an association between body mass index (BMI), age, hyperlaxity, arm and forearm circumferences, as well as elbow range of motion (ROM), in healthy adults. Among these factors, BMI and forearm circumferences seem to have a greater effect on ROM. Women had more flexion than men. No clear association was found between the laterality (or hand dominance) and elbow ROM [7].

In patients with extension or flexion contracture, the examiner should evaluate the solid or soft end points as well as the pain or crepitus elicited during the arc of the movement. The examiner should make a careful assessment of any compromised arc of motion from the shoulder to the wrist. Often, the disability will arise from a combination of factors, but it should be stressed that a full ROM at the elbow is not essential for performance of the activities of daily living as described previously. Because the loss of extension up to a certain degree only shortens the lever arm of the upper extremity, flexion contractures of less than 45° may have little practical significance although patients may be concerned about the cosmetic appearance [5, 6].

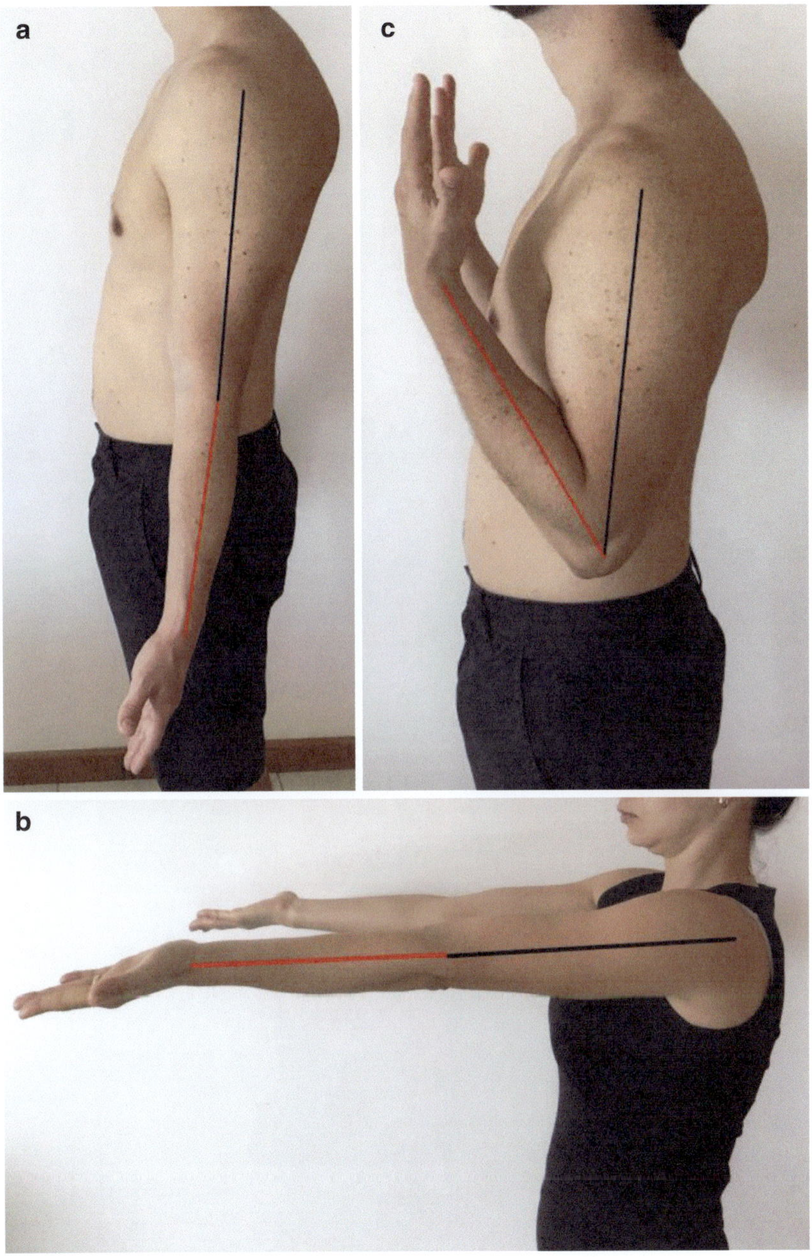

Fig. 14.2 The arc of flexion–extension of the elbow. (**a** and **b**) Extension; (**c**) Flexion

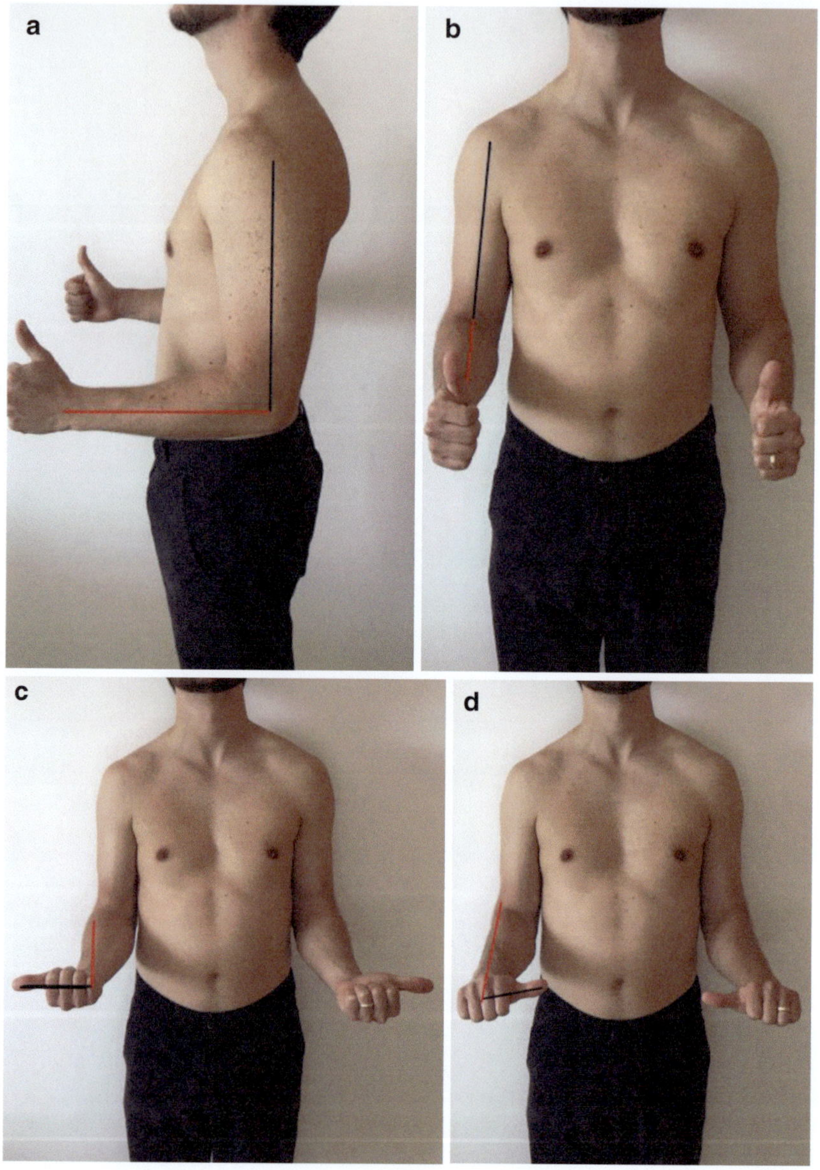

Fig. 14.3 The arc of pronation–supination of the elbow: exam done with elbow flexed 90° (**a** and **b**). (**c**) Maximum supination; (**d**) Maximum pronation

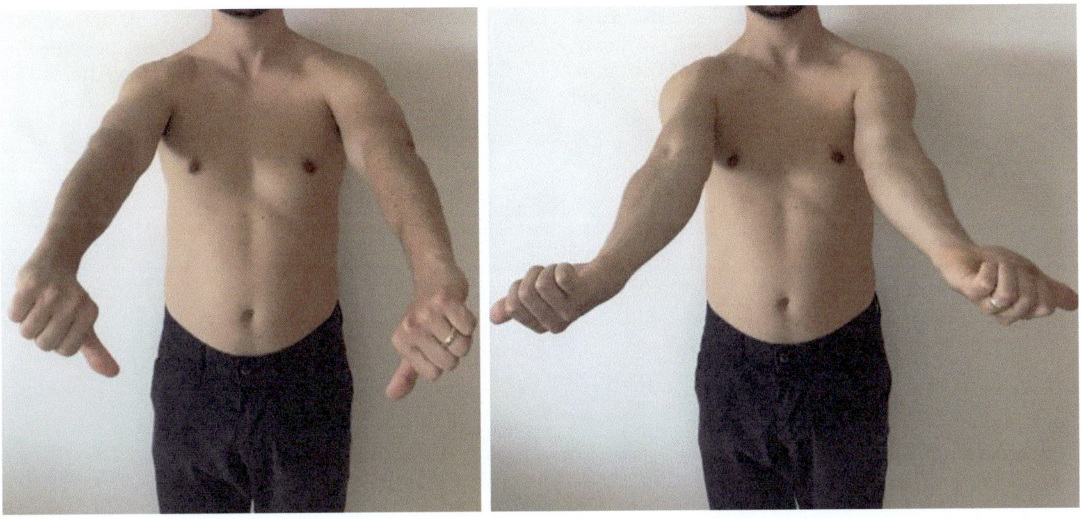

Fig. 14.4 The arc of pronation–supination of the elbow: Incorrect measurement with total extension of the elbow

14.4 Range of Motion in Daily Activities

Latz et al. described elbow's range of motion while driving a car on different road types. When driving a car with a left-sided steering wheel and a manual transmission on the right side, mean pronation of the right elbow is significantly higher than that of the left elbow. Their results suggest that movement restrictions in pronation could possibly affect driving capability earlier than restricted supination [8].

14.5 Limitation of Motion (LOM)

LOM of the elbow joint can derive from different factors, including acute or chronic trauma, burn scar contracture, heterotopic ossification, coma, postoperative scarring spasticity. Intra-articular lesions causing LOM are adhesion, loose bodies, osteochondritis dissecans of capitellum, chondromalacia of the radial head, synovitis, and osteophytes on the olecranon or coronoid process [9].

Function of the elbow is essential for the activities of daily living. The outcomes of treatment are favorable when the etiology is considered in the decision-making process. Nonoperative management should be attempted in all cases of the stiffness of the elbow, except when the stiffness is due to heterotopic ossification or intrinsic causes. Operative intervention should be considered after nonoperative treatment has failed and should be performed within the first year to 19 months after the injury (cases with postoperative elbow stiffness).

References

1. Morrey BF. Applied anatomy and biomechanics of the elbow joint. Instr Course Lect. 1986;35:59–68.
2. Atkinson WB, Elfman H. The carrying angle of the human arm as a secondary symptom character. Anatomy Rec. 1945;91:49.
3. Beals RK. The normal carrying angle of the elbow. Clin Orthop Relat Res. 1976;119:194.
4. Wright RW, Steger-May K, Wasserlauf BL, O'Neal ME, Weinberg BW, Paletta GA. Elbow range of motion in professional baseball pitchers. Am J Sports Med. 2006;34:190–3. https://doi.org/10.1177/0363546505279921.
5. Boone DC, Azen SP. Normal range of motion of joints in male subjects. J Bone Joint Surg. 1979;61A:756–9.
6. Morrey BF, Chao EY. Passive motion of the elbow joint: a biomechanical study. J Bone Joint Surg. 1979;61A:63.
7. Chapleau J, Canet F, Petit Y, Sandman E, Laflamme GY, Rouleau DM. Demographic and anthropometric factors affecting elbow range of motion in healthy adults. J Shoulder Elb Surg. 2013;22:88–93.

8. Latz D, Schiffner E, Schneppendahl J, Hilsmann F, Seiler LF, Jungbluth P, Kaufmann RA, Windolf J, Gehrmann SV. Doctor, when can I drive? The range of elbow motion while driving a car. J Shoulder Elb Surg. 2019;28:1139–45.
9. Nowicki KD, Shall LM. Arthroscopic release of a posttraumatic flexion contracture in the elbow: a case report and review of the literature. Arthroscopy. 1992;8:544–7.

15 Evaluation of Triceps Tendon

Andrea Celli, Nicoletta Fabio, Duca Vito, and Luigi Adriano Pederzini

15.1 Introduction

Triceps tendon injuries are probably among the rarest tendon injuries in the human body. Evidence shows male predominance with wide age variance at occurrence. The most common rupture site is at the tendon's insertion into the olecranon and more rarely at the myotendinous junction or intramuscularly. Triceps tendon ruptures may be due to four major causes: traumatic lesion, spontaneous rupture, overuse injuries, following total elbow arthroplasty. Understanding the anatomy of the triceps tendon is key to good clinical assessment following post-traumatic triceps injury, and it helps to improve the outcome following surgical repair. Elbow extension against gravity or resistance may be difficult or impossible when the triceps distal tendon is ruptured or avulsed from the olecranon insertion.

Complete ruptures of all three tendon insertion heads (long, lateral and medial heads) generally require surgical treatment. Partial lesions are functionally well tolerated in patients with low functional demand.

A. Celli (✉)
Shoulder and Elbow Unit, Department of Orthopaedic and Traumatology Surgery, Hesperia Hospital (Modena), Modena, Italy

N. Fabio · D. Vito · L. A. Pederzini
Department of Orthopaedic, Traumatology and Arthroscopic Surgeries, Nuovo Ospedale di Sassuolo (Modena), Modena, Italy

J. G. Lane et al. (eds.), *The Art of the Musculoskeletal Physical Exam*,
https://doi.org/10.1007/978-3-031-24404-9_15

15.2 Anatomy

Triceps brachii (Fig. 15.1):

15.2.1 Origin

Three heads:

- The long head arises from the infraglenoid tubercle of the scapula.
- The lateral head has a linear attachment from the upper margin of the radial grove of the humerus.
- The medial head originates below the lateral margin of the radial groove that contains the radial nerve. Its insertion covers the entire rear surface of the lower part of the humerus.

15.2.2 Insertion

In the distal third of the posterior aspect of the arm, the lateral head joins with the long head from the superficial tendinous part of the insertion on the posterior surface of the olecranon. The medial head (deep part of the triceps) inserts through muscular and tendinous fibres directly onto the olecranon. There are anatomical differences between the lateral and medial side of the superficial tendon and between the superficial

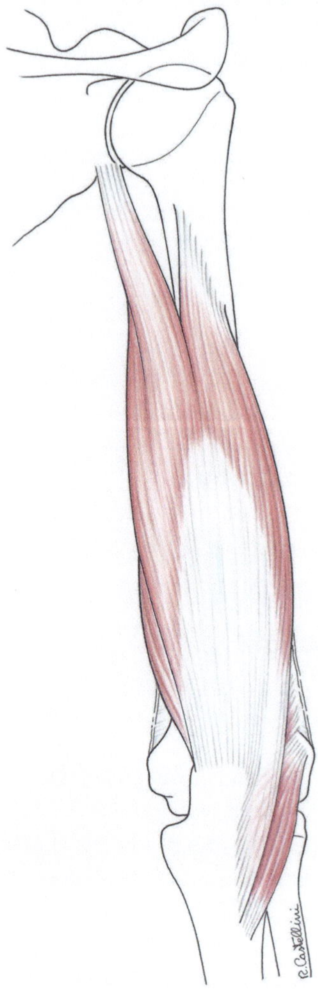

Fig. 15.1 Triceps brachii anatomy. Distal triceps tendon insertion (superficial and deep components): the lateral part is more expansive and relatively thinner in continuity with anconeus muscle and fascia. The medial part is thicker than the lateral aspect like a proper triceps tendon and inserts directly onto olecranon

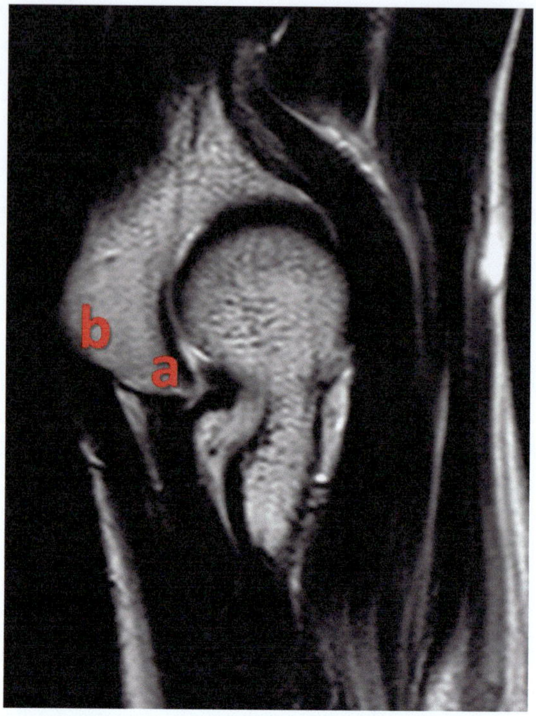

Fig. 15.2 Magnetic resonance imaging describes a bipartite insertion between the deep (**a**) and superficial (**b**) components of the triceps tendon into the olecranon

and deep tendon anatomy. As it approaches its insertion area, the superficial tendon forms two components:

- A lateral part which is more expansive and relatively thinner in continuity with the anconeus muscle and fascia.
- A medial part which is thicker than the lateral aspect like a proper triceps tendon inserting directly onto the olecranon.

In some cases, a well-defined interval is located between the lateral triceps expansion and the medial triceps tendon just proximal to the olecranon. This interval is located along the crest of the ulna and is commonly known as "triceps decussation".

The deep tendon is medially and laterally covered by a thin layer of muscle fibres.

Centrally and laterally, the deep medial head of the triceps shows a broad and flat tendon, while medially it is a narrow-thickened tendon [1]. Medially, the superficial and deep tendons are confluent forming the proper triceps tendon inserting into the medial olecranon footprint.

Magnetic resonance imaging (MRI) often describes a bipartite insertion between the deep and superficial tendon of the triceps into the olecranon [2] (Fig. 15.2). However, all three heads of the triceps contribute to forming the dome-shaped triceps olecranon footprint [3] (Fig. 15.3).

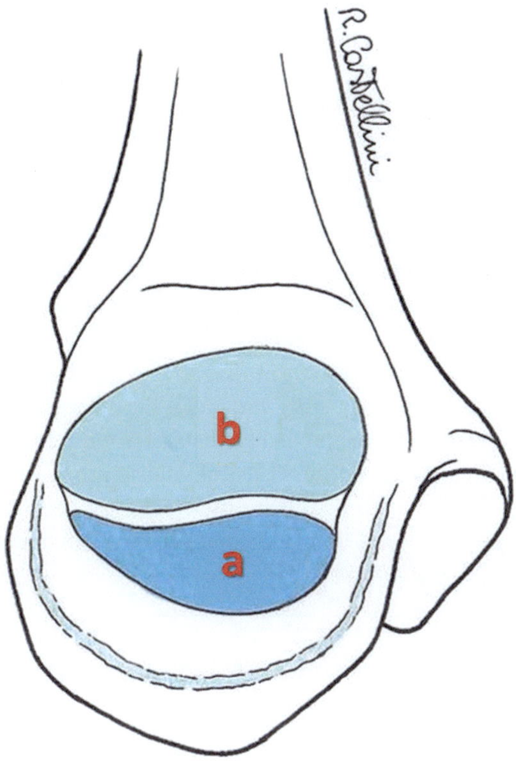

Fig. 15.3 Mean width (medial to lateral distance) (**a**) of the medial head footprint is about 16 mm and the mean thickness 4 mm. Mean width of the common superficial tendon footprint is 19 mm and the mean thickness is 8 mm (**b**)

15.2.2.1 Olecranon Footprint

Mean medial to lateral insertion area width is about 20 mm and mean proximal to distal length is about 13 mm [1]. Mean distance from the olecranon tip to the most proximal aspect of the medial head insertion site ranges between 14.8 mm and 16 mm [1]. Mean medial head footprint width (medial to lateral distance) is 16 mm and mean thickness is 4 mm. Mean common superficial tendon footprint width is 19 mm and mean thickness is 8 mm [1].

From surgical experience when triceps turndown approach is performed and the triceps superficial tendon is detached to create an inverted flap tendon with the base attached to the olecranon, exposing the medial head tendon insertion will show:

- Medially, a distinct thickened tendon in continuity with the remaining central and lateral flat medial head tendon.
- Medial head tendon at olecranon insertion level is difficult to separate from superficial tendon.
- Medial head tendon insertion is covered by a thin layer of muscle. This allows the surgeon to separate the superficial tendon fibres from the deep tendon.

Understanding the anatomy of the triceps tendon is key to good outcomes following post-traumatic triceps repair or triceps reinsertion after deep surgical exposure.

15.3 Pathogenesis

In healthy tendons, traumatic lesion is the most common triceps distal tendon rupture mechanism. In most cases, tendon ruptures more frequently occur with intrinsic pathological disorders such as overuse injuries or previous surgical treatment decreasing tendon tensile strength.

Triceps tendon rupture surgical treatment success depends on a thorough understanding of:

- Status of tendon tissue (with or without pathological disorders)
- Type of rupture (partial or total)
- Duration of symptoms (acute or chronic)
- Associated lesions (radial head, capitellum, medial collateral lesion)

A simplified classification categorizes triceps tendon ruptures in four groups:

- Traumatic lesion
- Spontaneous rupture
- Overuse injuries
- Following total elbow arthroplasty

15.3.1 Traumatic Lesions

Acute triceps tendon ruptures most commonly occur when a patient falls and a forced load is applied to the contracted triceps on an outstretched hand with flexed elbow. Alternatively, the injury mechanism can be a direct trauma to the posterior aspect of the arm. The injury can be more severe when the elbow flexed and, at the same time, the triceps is contracted. Traumatic tears can occur at several different anatomical regions, but they are extremely rare at musculotendinous junction just like those that occur within the muscle belly. In general, tears are most commonly observed at tendon insertion in the form of olecranon avulsions.

Traumatic distal tendon lesions can be partial or total and they are often isolated. Associated lesions were described at radial head [4], medial collateral ligament (MCL), and capitellum [5].

Partial traumatic tendo-osseous avulsions with one small fragment proximal to the olecranon (Flake Sign) are often associated with radial head fractures. "Flake" fractures *with* tendon avulsions from the olecranon can also be associated with injuries to the radial head and the medial collateral ligament [5]. Partial ruptures often involve the medial head of the triceps tendon along with radial head fractures and MCL lesions. These associated lesions suggest that valgus forces applied to the elbow on an outstretched arm can cause rupture of the MCL and at the same time the deep medial head inserted with a narrow-thickened tendon in the medial aspect of the olecranon footprint. Radial head fractures or coronal fractures of the capitellum can be a consequence of direct axial forces being transmitted on the lateral elbow compartment when the lateral elbow is in a valgus and a semi-flexed position while the forearm is pronated. Disruption of the MCL increases triceps medial head eccentric contraction which may induce avulsion "flake" fractures from the olecranon leading to the radial head or capitellum fractures. Triceps avulsions, radial head fractures, and MCL ruptures were jointly reported by Yoon as triad injuries [5].

Based on our knowledge and surgical experience, we point out the following:

- Triceps medial head ruptures or avulsions should be considered when, after trauma, the patient presents triceps insertion pain associated with radial head fracture or MCL insufficiency.
- When a valgus stress is applied to the elbow, the medially distinct thickened tendon of the medial head inserted on the postero-medial olecranon footprint acts as an active stabilizer in association with the MCL and the radial head.
- We advise caution in diagnosing and treating elbow lesions when the radial head fracture is associated with MCL attenuation with swelling and tenderness or ecchymosis around the medial region of the elbow. That is because diagnosis of triceps medial head rupture could be overlooked.

15.3.2 Spontaneous Ruptures

Triceps injury can occur spontaneously due to attrition if tendon integrity is compromised. Several risk factors have been studied for triceps muscle and tendon pathological disorders. These include rheumatoid arthritis, chronic renal failure, endocrine disorders, metabolic bone diseases, as well as steroid use. Patients with metabolic bone diseases such as chronic renal failure usually present elevated levels of parathyroid hormone which could stimulate olecranon bone absorption, thus predisposing to tendon avulsion. Use of local or systemic corticosteroid or anabolic steroids may also predispose to tendon rupture by decreasing tendon tensile strength. It is widely accepted that biosynthesis of collagen is inhibited by glucocorticoids [6]. Tendon rupture is a common manifestation of rheumatoid arthritis in the hand, sometimes occurring in triceps tendon after surgical repair. Tenosynovial pathological tissue proliferation in the tendon combined with low local blood supply and mechanical attrition result in tendon rupture. Alternatively, this could result after surgical reinsertion predisposing to attenuation and rupture of the triceps reattachment.

15.3.3 Overuse Injuries

Differently from acute traumatic lesion injuries, sport-related injuries are most frequently ascribed to repetitive motion that results in pain and the inability to participate in sport activities. Cumulative submaximal loading of the tissue is referred to as "overuse injuries". Complete rupture often occurs through abnormal tendon with intrinsic pathological disorders.

A classification of progressive Achilles tendon disorders [7] can be useful to understand the structural manifestations of triceps tendon overuse injuries:

1. Peritendinitis
2. Tendinosis with or without peritendinitis
3. Partial rupture
4. Total rupture

Triceps tendinosis during sport activities is not infrequent. In cases of chronic tendon pain, the pathological lesion is typical of a degenerative process with areas of marked degeneration and lack of local vascular supply. This characteristic pattern, especially if associated to a repetitive use of local corticosteroids, predisposes to triceps tendon rupture, by possibly decreasing tendon strength. Potentially worsening evolutions of tendinosis disorders begin with partial ruptures but can lead to complete ruptures. Surgical treatment is recommended in chronic tendinosis when clinical and MRI assessment detect partial rupture of the tendon.

15.3.4 Following Total Elbow Arthroplasty

Triceps reattachment failure can be seen following surgical treatment in which triceps takedown was performed. At Mayo Clinic, with over 887 total elbow arthroplasties, this complication was observed in 16 elbows—about 2% of all procedures [8].

When total elbow arthroplasty is performed, surgical triceps takedown or slide is usually well tolerated but some patients have an unsuccessful outcome with postsurgical attenuation or ruptures of the triceps reattachment.

Predisposing factors include inflammatory arthropathy with poor tissue quality. Triceps weakening is frequent, and it is a well-recognized problem following total elbow arthroplasty [8].

15.4 Natural History

Triceps tendon injuries are probably among the rarest tendon injuries in the body. Evidence shows male predominance (3:1, male to female) with a wide age variance at occurrence [9]. The most common rupture site is at tendon insertion into the olecranon and more rarely at the myotendinous junction or intramuscularly. Anzel et al. reviewed more than 1000 tendon injuries, they reported an approximate 0.8% triceps injuries, half of which were associated with open lacerations in the posterior aspect of the upper extremity [3]. Mair et al. described 21 cases of triceps tendon ruptures over a 6-year period with contact athletes (e.g. professional football players) [10]. Kibuule et al. reported that triceps injuries can also occur in adolescents with incompletely fused physis and avulsions of the olecranon apophysis [4].

Several risk factors have been reported for triceps tendon injuries. The main injury mechanism is falling on an outstretched hand. Other conditions that can increase risk of triceps ruptures include sports activities (such as contact activities or body building), use of anabolic steroid drugs or local steroid injection, olecranon bursitis, or secondary to surgical exposure such as following total elbow replacement or release for stiff elbow [8, 11].

The lesions can also occur following direct penetrating trauma in the posterior aspect of the elbow. It is not rare to observe associated injuries, such as radial head fractures or complex fractures of the distal humerus.

15.5 Patient History and Physical Findings

In patients with suspected triceps lesions, the physician should gather a precise history:

- age, dominant arm
- presence of pathological disorders such as overuse injuries
- repetitive use of local corticosteroids
- previous surgeries such as total elbow arthroplasty

Tendon lesions causes can be:

- Traumatic: acute events.
- Pathological: chronic tendinitis in patients with cumulative submaximal loading.
- Iatrogenic: following surgery especially in elderly patient with surgery history of violation of the extensor mechanism.

Acute triceps tendon ruptures are more common, but chronic lesions have also been reported [12]. The patient most commonly describes a sudden pain in the posterior aspect of the elbow after a history of direct blow or falling on an outstretched hand. The usual injury mechanism is a forceful sudden flexion of the extended elbow. Laceration and open injuries with or without elbow fracture dislocation can also cause distal triceps rupture. Post-traumatic presenting signs and symptoms are correlated to:

- lesion types (tendinous, tendon avulsion, or within the muscle belly)
- degree (partial or total)
- timing (acute or chronic)

In general, triceps lesions are characterized by:

- Spontaneous and evocate pain on palpation at the indicated site of lesion. There will usually be tenderness and a palpable defect in the tendon that can be seen proximal to the olecranon.
- Swelling, ecchymosis, and body habitus frequently do not allow the tendon defect to be palpable in its acute stage. Once swelling subsides, most patients demonstrate a palpable gap in the tendon.
- Active extension is typically diminished or absent depending on whether a complete or partial tear is there.

The medial tendon insertion head of the deep triceps tendon can be palpable proximally to the olecranon insertion along the line of the crest of the ulna. Partial medial head ruptures are frequently undiagnosed. Diagnosis is difficult, but care should be taken when the radiograph assesses the presence of flake signs.

Fragment avulsions from the olecranon are also associated with radial head fractures which is frequently reported along with medial collateral ligament (MCL) concomitant injury.

However, when radial head fractures are present and patients feel pain and swelling in the medial compartment of the elbow, MCL, and triceps tendon integrity should always be assessed.

Subjects performing repetitive forceful extension activities (athletes or manual workers) may manifest triceps tendinosis.

Pathological tendon disorders consist in chronic posterior elbow discomfort at active extension, but in an irregular way concomitantly acute tendinitis may occur. Unlike olecranon bursitis or symptomatic olecranon spur can be seen.

Triceps tendon rupture is uncommon without a predisposing factor. Repetitive local steroid injection may be implicated in chronic tendon rupture.

The physician has to analyse chronic tendon ruptures contributing factors for such as pre-existing triceps tendinopathy and steroid intake.

However, surgical treatment should consider triceps tendon degeneration when the surgical reconstructive procedure is performed.

Clinical findings of triceps insufficiency following total elbow arthroplasty include [8, 11]:

- change in the posterior contour of the elbow with visual and palpable prominence of the implant
- presence of olecranon bursitis
- triceps muscle atrophy

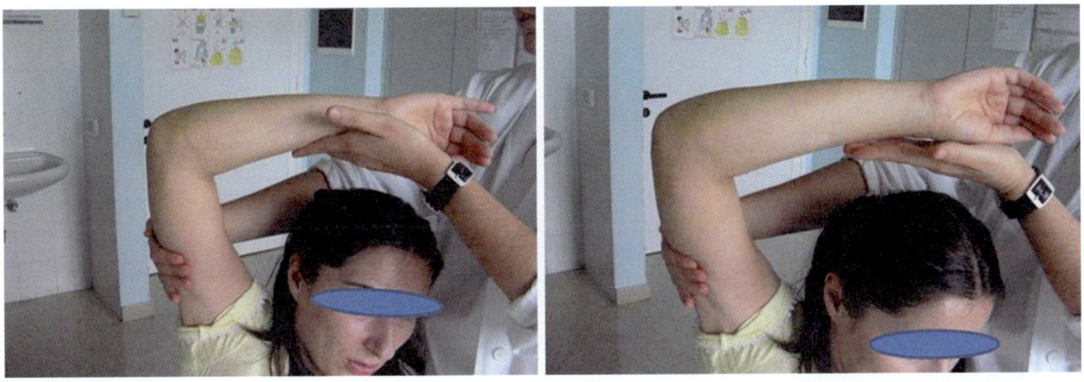

Fig. 15.4 Tendon tear is manifested by loss of extension strength also against gravity and the inability to extend the elbow. Triceps extension tests can also be performed observing the ability of the patient to extend the elbow over his/her head, against gravity

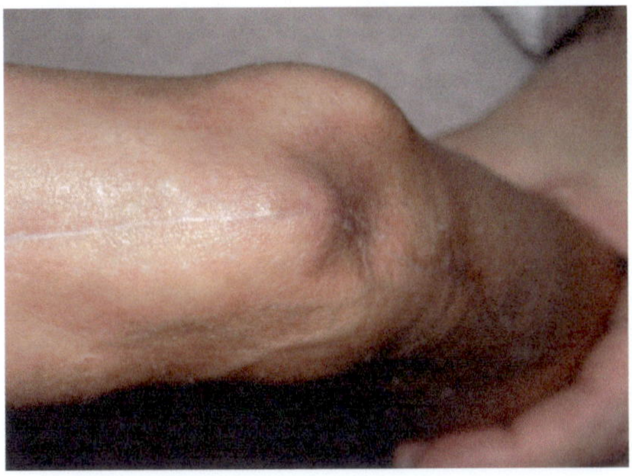

Fig. 15.5 Clinical findings of triceps insufficiency include change in the posterior contour of the elbow with visual and palpable bone prominence

- proximal retraction or lateral subluxation of the extensor mechanism

A universal physical finding with triceps ruptures is the inability to extend the arm against gravity (Fig. 15.4).

In acute injuries extension strength can be difficult to evaluate when traumatic pain is present. In case of questionable diagnoses, re-examination can be performed several days after the injury.

A tear can be seen (Fig. 15.5) and palpated in the tendon during attempts at resisted extension (Fig. 15.6).

Distinguishing between partial or complete ruptures can be a diagnostic challenge. The test to evaluate triceps function against gravity and resistance is performed by the physician with the patient in the prone position and 90° flexed elbow with the upper arm supported by a table and the forearm hanging free (Fig. 15.7).

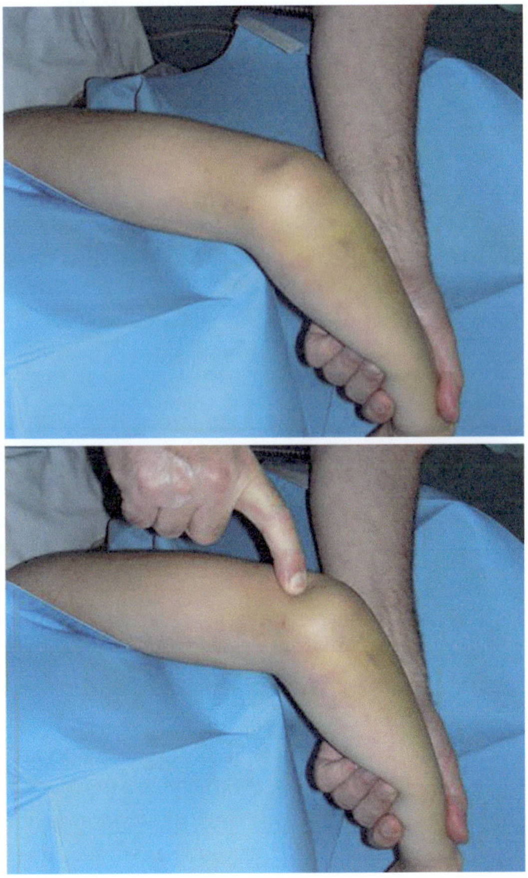

Fig. 15.6 A tear can often be palpated in the triceps tendon

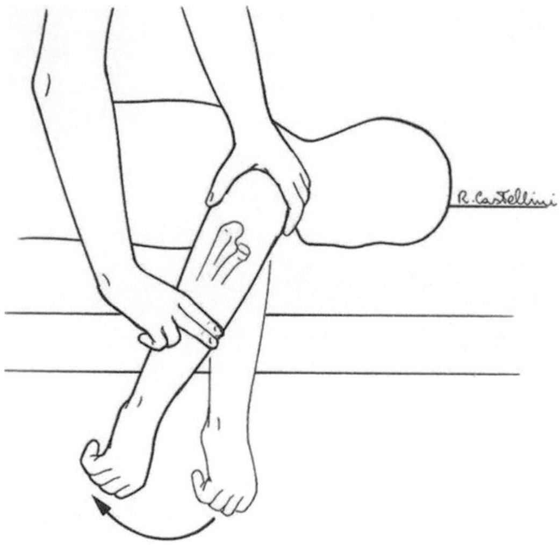

Fig. 15.7 The test to evaluate the triceps function against gravity and resistance is performed by the physician with the patient in the prone position and 90° flexed elbow with the upper arm supported on a table and the forearm hanging free

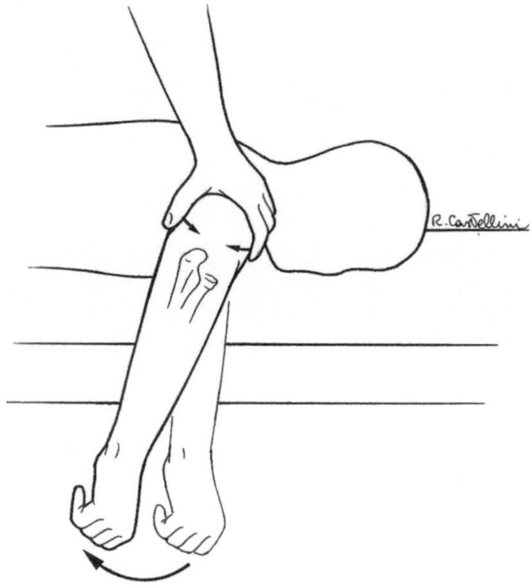

Fig. 15.8 Viegas's test. In prone position, the patient lets their relaxed forearm hang over the table while the physician squeezes the triceps muscle belly. When the tendon lesion is partial, this should produce a slight elbow extension. Conversely, no motion will occur in case of complete rupture

A partial tendon lesion is manifested by weakness and the ability to actively extend the elbow against the gravity, but not against resistance. This finding is likely secondary to an intact lateral expansion or a compensating anconeus muscle.

Total tendon tear is manifested by loss of extension strength also against gravity and the inability to extend the elbow. Triceps extension tests can also be performed observing the patient's ability to extend the elbow over their head, against gravity.

Viegas [13] described a provocative test similar to Thompson's test (used to help in diagnosing Achilles tendon rupture) that can be employed in triceps tendon rupture diagnoses. In prone position, the patient lets their relaxed forearm hang over the table while the physician squeezes the triceps muscle belly. When the tendon lesion is partial, this should produce a slight elbow extension. Conversely, no motion will occur in case of complete rupture (Fig. 15.8).

Celli et al. [14] defined a test called "fall down triceps test" for triceps insertion tendon ruptures. This test assesses the inability of the patients to keep their forearm in maximum extension against gravity. While standing with their shoulder at 90° abduction and internal rotation, the patient's forearm is kept in full passive extension by the examiner placed behind them. Upon dropping the forearm, if the triceps tendon presents a complete rupture, the patient will not be able to maintain the initial position and their elbow will drop down to 90° flexion. Conversely, in case of partial rupture the patient's forearm will only slightly drop down given the patient's effort to maintain limited elbow extension (Fig. 15.9).

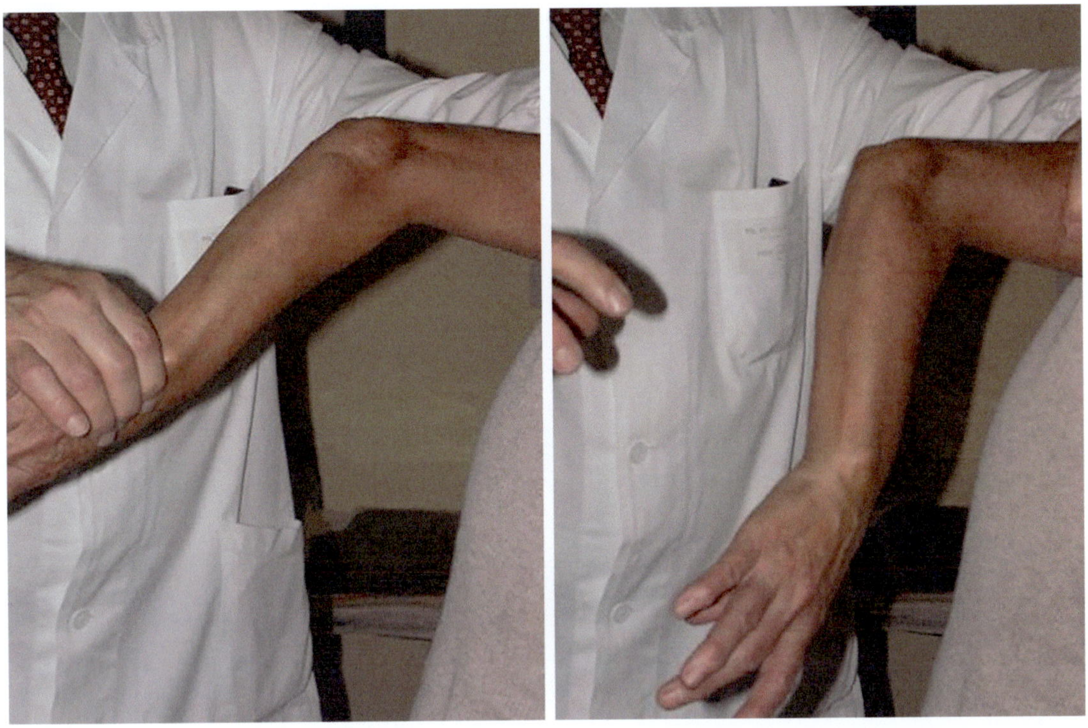

Fig. 15.9 Fall down triceps test. Upon dropping the forearm, if the triceps tendon presents a completer rupture, the patient will not be able to maintain the initial position and their elbow will drop down to 90° flexion. Conversely, in case of partial rupture the patient's forearm will only slightly drop down given the patient's effort to maintain limited elbow extension

15.6 Imaging and Other Diagnostic Studies

Imaging studies help in identifying lesion level (olecranon insertion, myotendinous junction, or intramuscular), distinguishing between partial and complete tear, estimating the amount of the tendon retraction, and excluding associated osseous injuries.

Lateral elbow X-rays are useful in confirming the diagnosis when a small extra-articular avulsion fracture of the olecranon (flake sign) is present. The bony fragment is usually small and easy to ignore but its presence is pathognomonic of distal triceps avulsion (Fig. 15.10).

X-rays (AP and lateral) and computer tomography (CT) are also helpful in aiding diagnosis of injuries associated with triceps rupture, such as ipsilateral radial head and capitellum fractures.

Ultra-sonography may be used although it provides limited anatomical details. Nonetheless, it may be useful immediately after the injury [15], in case of doubtful diagnosis.

Magnetic resonance imaging is the best technique to assess tendon lesions [16] because it provides more details to distinguish among (Fig. 15.11):

- side of the injury
- partial and complete lesions
- degree of tendon retraction
- muscular atrophy

MRIs on the sagittal planes demonstrated the integrity of the triceps tendon: partial tears most commonly occur distal at olecranon insertion and become visible in the form of a small fluid-filled gap within the ruptured distal triceps tendon.

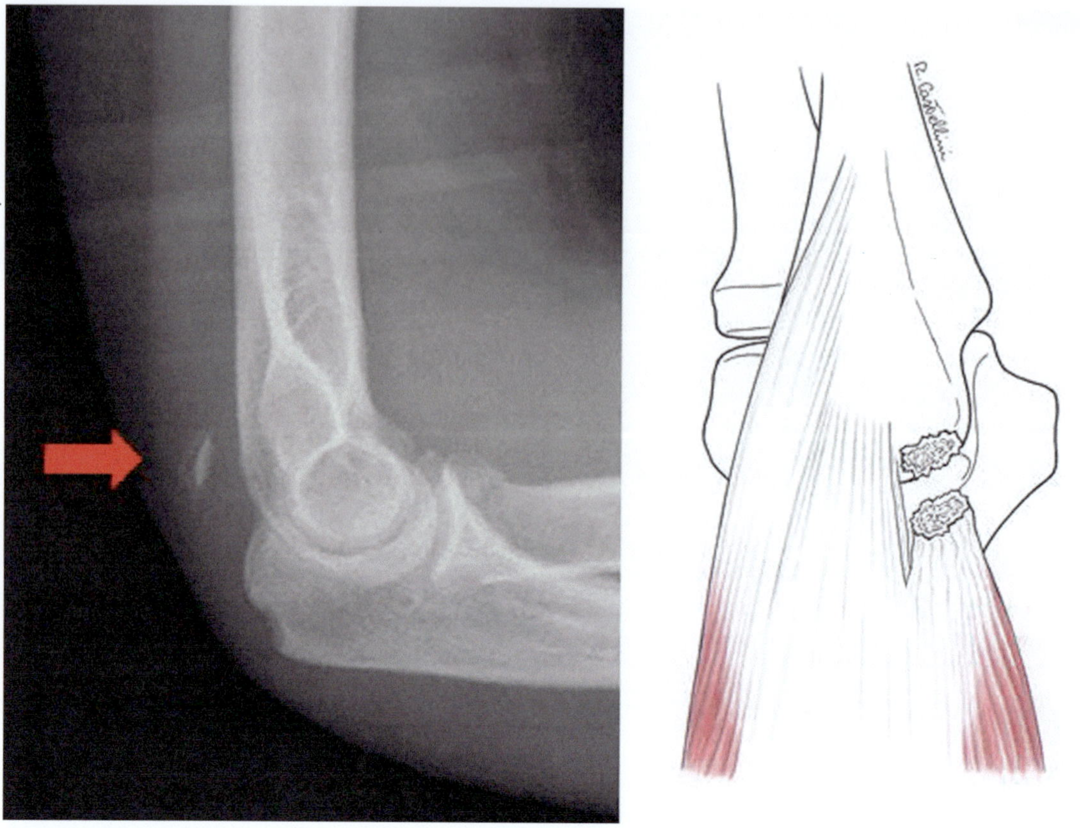

Fig. 15.10 Lateral elbow X-rays are useful in confirming the diagnosis if a small extra-articular avulsion fracture of the olecranon (flake sign) is present

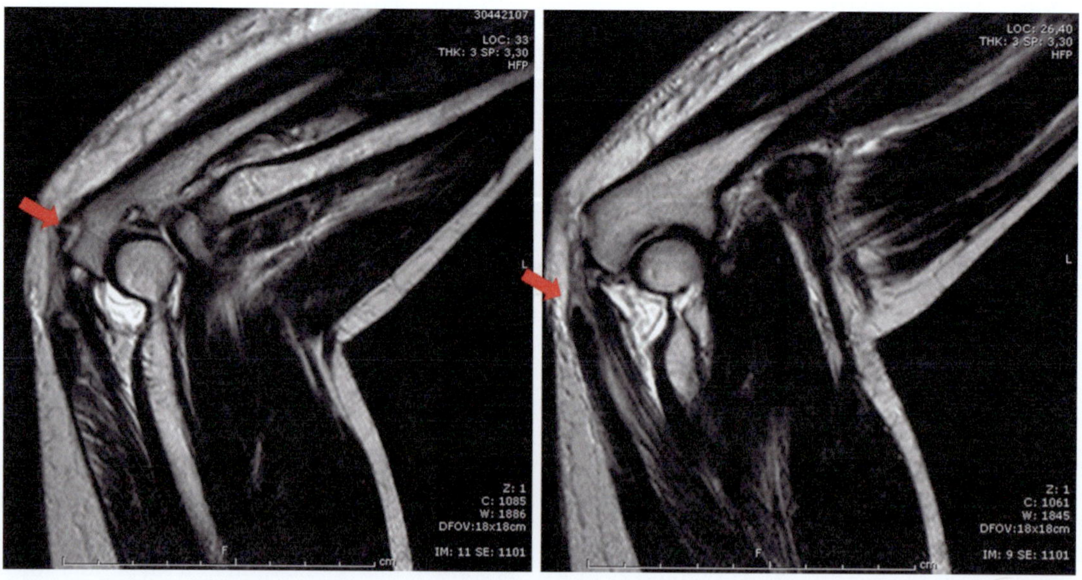

Fig. 15.11 Magnetic resonance imaging provides more details on the side and entity of the lesions, tendon retraction, and muscle atrophy of the superficial and deep heads. Partial tears occur at olecranon insertion and become visible as gap sometimes with a small piece of bone

Complete rupture is characterized by a large fluid-filled gap between the retracted triceps tendon stump and the olecranon process.

In overuse degenerative tendinosis, MRI is characterized by thickening and signal alteration of the distal tendon fibres.

15.7 Differential Diagnosis

- Weakness for neurological radial nerve problems (compression or lesion) due to different anatomical innervation of the triceps muscle. The long head receives the innervation high in the arm, the lateral head at the upper level of the spiral groove, and the medial head receives the innervation longitudinally down the medial and lateral sides of the muscle. All of them are possible points of denervation as a consequence of traumatic or iatrogenic lesions.
- C7 isolated nerve root lesion.
- Triceps tendinitis is an uncommon condition that is rarely reported in the literature by Nirschl as posterior tennis elbow. It involves pain at the triceps insertion which is thought to be due to angio-fibroblastic changes in the tendon [17].
- Rupture can occur at several different levels, but in general it occurs at olecranon insertion. Ruptures at the intramuscular site or at myotendinous level are less frequent.

15.8 Nonoperative Management

Conservative management plays a role in partial triceps injuries involving the muscle belly or musculotendinous junction, or in the partial ruptures of the distal tendon insertion when there is no significant extension power loss against gravity and no resistance according to the patient's age and lifestyle [18].

15.9 Surgical Management

Several surgical techniques and different approaches offer variable options for the surgical management of acute complete or chronic tears. Different surgical methods can be used to restore the extensor mechanism.

The choice depends on tissue quality, on the entity of the muscle retraction, and on the time elapsed from the lesion (chronic lesion more than 6 weeks old).

The surgical options [18] are:

- Direct olecranon repair
- Augmentation with auto or allograft
- Anconeus rotation flap
- Achilles tendon allograft with or without calcaneus bone

References

1. Keener JD, Chafik D, Kim M, Galatz LM, Yamaguchi K. Insertional anatomy of the triceps brachii tendon. J Shoulder Elb Surg. 2010;19:399–405.
2. Belentani C, Pastore D, Wangwinyuvirat M, Dirim B, Trudell DI, Haghighi P, et al. Triceps brachii tendon: anatomic-MR imaging. Study in cadavers with histological correlation. Skelet Radiol. 2009;38:171–5.
3. Anzel SH, Convey KW, Weiner AD, Lipscomb PR. Disruption of muscles and tendons: an analysis of 1014 cases. Surgery. 1959;45:406–14.
4. Kibuule LK, Fehringer EV. Distal triceps tendon rupture and repair in an otherwise healthy pediatric patient: a case report and review of the literature. J Shoulder Elb Surg. 2007;16(3):e1–3.
5. Yoon MY, Koris MJ, Ortiz JA, Papandrea RF. Triceps avulsion, radial head fracture, and medial collateral ligament rupture about the elbow: a report of 4 cases. J Shoulder Elb Surg. 2012;21:12–7.
6. Huxley AF, Niedergerke R. Structural changes in muscle during contraction: interference microscopy of living muscle fibers. Nature. 1954;173:971–3.
7. Smart GW, Taunton JE, Clement DB. Achilles tendon disorders in runners. A review. Med Sc Sports Exerc. 1980;12:231–43.
8. Celli A, Arash A, Adams RA, Morrey BF. Triceps tendon insufficiency following total elbow arthroplasty. J Bone Joint Surg Am. 2005;87:1957–4964.
9. Yeh PC, Dodds SD, Smart LR, et al. Distal triceps rupture. J Am Acad Orthop Surg. 2010;18(1):31–40.
10. Mair SD, Isbell WM, Gill TJ, et al. Triceps tendon ruptures in professional football players. Am J Sports Med. 2004;32(2):431–4.
11. Celli A, Morrey BF. Triceps insufficiency following total elbow arthroplasty. In: Morrey BF, Sanchez Sotelo J, editors. The elbow and its disorders. 4th ed; 2009. p. 873–9.
12. Inhofe PD, Manein MS. Late presentation of triceps rupture: a case report and review of literature. Am J Orthop. 1996;11:290–2.

13. Viegas SF. Avulsion of the triceps tendon. Orthop Rev. 1990;19(6):533–6.
14. Celli A. Chapter 38. Triceps tendon ruptures. In: Wiesel SW, editor. Operative techniques in orthopaedic surgery, Shoulder and elbow part 7, vol. 3. Wolters Kluwer; 2015. p. 3917–33.
15. Kaempffe FA, Lerner RM. Ultrasound diagnosis of tricep tendon rupture: a report of 2 cases. Clin Orthop Relat Res. 1996;332:138–42.
16. Gaines ST, Durbin RA, Marsalka DS. The use of magnetic resonance imaging in the Diagnosis of triceps tendon ruptures. Contemp Orthop. 1990;20:607–11.
17. Nirschl RP. Prevention and treatment of elbow and shoulder injuries in the tennis player. Clin Sports Med. 1988;7:289–308.
18. Morrey BF. Rupture of the triceps tendon. In: Morrey BF, Sanchez-Sotelo J, editors. The elbow and its disorders. Saunders Elsevier; 2009. p. 536–46.

16 Clinical Evaluation of the Distal Biceps Tendon

Deepak N. Bhatia and Gregory I. Bain

16.1 Introduction

Distal biceps pathology is a common cause of pain and weakness in athletes and may result in a Popeye deformity. The spectrum of pathology ranges from tendinosis and low-grade partial tears, and in severe cases may lead to high-grade tears or complete ruptures. The tendon insertion is difficult to examine and palpate precisely due to the deep anatomic location of the bicipital tuberosity; the resultant vague localization of symptoms often results in delayed diagnosis and progression of tears.

A thorough understanding of the extent and course of the DBT is necessary for a systematic clinical evaluation of the distal biceps tendon (DBT) region, and clinical findings may be supplemented with in-clinic sonography for accurate diagnosis.

D. N. Bhatia (✉)
SportsDocs, Mumbai, India

G. I. Bain
Department of Orthopaedic and Trauma Surgery, Flinders University, Bedford Park, SA, Australia

J. G. Lane et al. (eds.), *The Art of the Musculoskeletal Physical Exam*, https://doi.org/10.1007/978-3-031-24404-9_16

16.2 Anatomy

The biceps brachii muscle has two distinct muscle bellies that continue as the long and short components of the DBT. The DBT begins in the distal arm at the musculotendinous junction of the biceps brachii and can be palpated approximately 3 cm proximal to the anterior elbow crease (AEC). The tendon traverses the AEC and courses across the layers of the cubital fossa for a mean of 9 cm (7–12 cm) to insert into the bicipital tuberosity (BT) of the proximal radius [1]. The BT can be located approximately 3–4 cm distal to the AEC; the tuberosity lies in an anterior orientation in full supination and rotates posteriorly in full pronation. The lacertus fibrosus is an aponeurotic band of fascia that is attached to the medial aspect of the DBT; the band courses ulnarly and merges with the superficial fascia that encircles the flexors of the forearm. Contraction of the forearm flexors tenses the BA, pulling the biceps tendon distally and medially. An intact aponeurosis prevents retraction of the ruptured DBT; however, it is often torn or elongated in retracted DBT tears.

16.3 Clinical Evaluation

Patients with non-traumatic DBT pathology (tendinosis, partial tears) may present with forearm and arm pain with overuse and exertion, and occasionally may complain of weakness and fatigue in certain work-related activities. Traumatic ruptures present with sudden severe pain and bruising after forceful eccentric contraction-related activity and a change is the contour of the biceps is noted.

Clinical evaluation of the DBT should involve an assessment of the arm, and the cubital fossa and proximal forearm. The upper limb is inspected to detect any bruising and the biceps contour is compared to the unaffected arm to identify a "Popeye" deformity. Complete rupture of the DBT results in a proximal shift of the muscle belly; the resultant Popeye deformity can be differentiated from a distally shifted biceps seen in a proximal long head rupture (Fig. 16.1). The biceps muscle belly and the DBT are then palpated from proximal to distal, and tautness of the DBT is assessed with isometric contraction. The bicipital tuberosity is palpated for tenderness from the volar and dorsal aspects of the proximal forearm; in a muscular individual, volar palpation is difficult, and dorsal palpation is performed approximately 4 cm distal to the radiocapitellar joint in a fully pronated forearm. Thereafter, elbow flexion and supination strength tests are performed and compared with the unaffected side. Partial or complete ruptures usually present with weakness in supination, and this weakness is best appreciated in the terminal range of motion. Flexion weakness is often mild and strength testing using a dynamometer is useful to quantify the weakness.

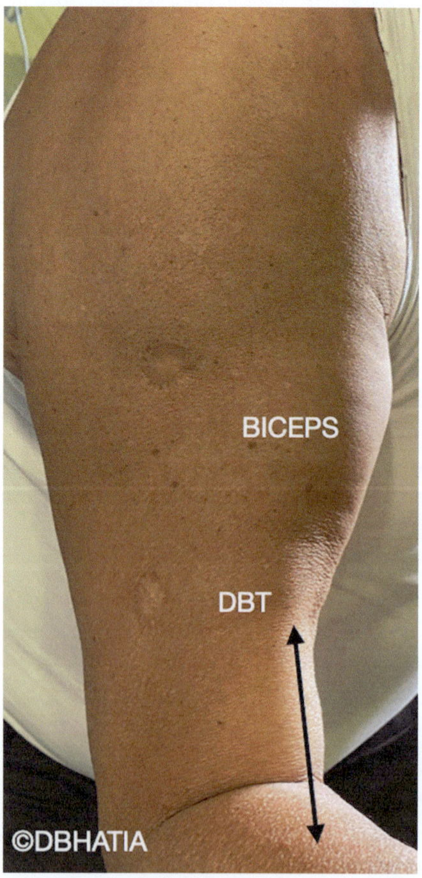

Fig. 16.1 A complete rupture of the distal biceps tendon (DBT) is shown. The tendon has retracted (arrows) into the mid-arm resulting in a "Popeye" deformity

Special tests have been described for assessment of structural integrity of DBT. These tests are useful in complete ruptures and can also be used in the post-surgical period to evaluate healing and repair integrity.

(a) **Biceps crease interval (BCI)**

The Popeye deformity may sometimes be subtle and may not be apparent in early stages of injury. The BCI is useful to quantify the proximal retraction "when visible or palpable alterations in biceps contour and proximal tendon migration are absent or equivocal" [2]. BCI is the distance between the AEC and the cusp of the distal descent of the biceps muscle, as described by ElMaraghy et al. (Fig. 16.2). BCI is measured on both

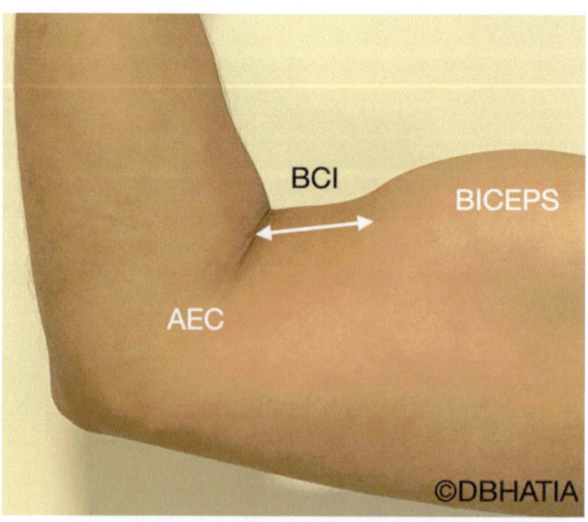

Fig. 16.2 The biceps crease interval (BCI) is the distance between the anterior elbow crease (AEC) and the cusp of the distal descent of the biceps muscle

arms and is expressed as the biceps crease ratio (BCR = affected BCI/unaffected BCI). A BCI > 6.0 cm or a BCR > 1.2 is diagnostic of complete DBT rupture (sensitivity 96%, diagnostic accuracy 93%).

(b) **The Bicipital aponeurosis (BA) flex test**

The BA flex test is performed for assessment of integrity of the lacertus fibrosus in complete DBT ruptures. In a normal arm, the bicipital aponeurosis can be palpated as follows: the wrist is flexed and forearm is supinated to tension the distal attachment of BA. The elbow is flexed to 75° and isometrically contracted to tension the proximal attachment of BA. The BA can then be palpated as it diverges distally in an ulnar direction from the DBT, and the result of the BA flex test is documented as intact or absent (Fig. 16.3). ElMaraghy and Devereaux found that BA remained intact in 59% of complete DBT ruptures and resulted in lesser tendon retraction [3]. The test was found to have 100% sensitivity and 90% specificity, with overall diagnostic accuracy of 94%.

(c) **Hook test**

The hook test was first described by O'Driscoll et al. as a reliable diagnostic test for DBT ruptures [4]. The test can also be used to assess the integrity of the lacertus fibrosus in a complete DBT rupture. The test is performed with the elbow in 90° flexion and forearm supinated. The examiner passes a finger under the lateral aspect of DBT in an attempt to hook it. In an intact tendon, the finger passes under the taut DBT and the taut tendon can be palpated (Fig. 16.4). In an avulsed and retracted tendon, the taut DBT is absent, and the hook test is abnormal. If the tendon is intact and a painful response is elicited (painful hook test), then the test is suggestive of a partial DBT tear. Phadnis and Bain have described four grades of interpretation of the hook test (Normal: Taut, unyielding, and symmetric; A1: Taut, yielding, and asymmetric; A2: Lax and asymmetric; A3: absent cord) [5]. O'Driscoll et al. described the test to be 100% specific and sensitive with a positive and negative predictive value of 100%. In another study, Luokkala et al. described the sensitivity of

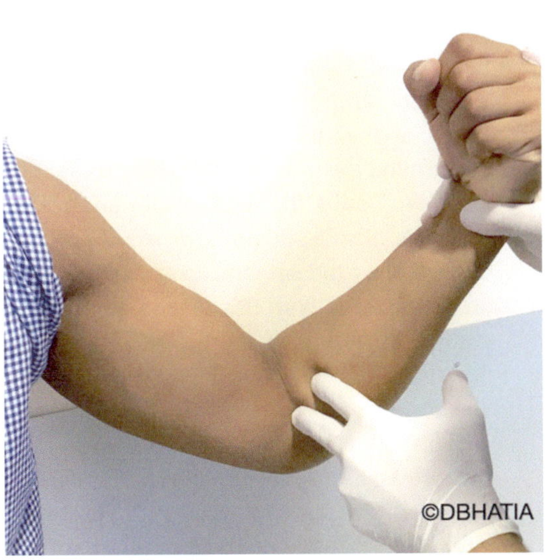

Fig. 16.3 The bicipital aponeurosis can be palpated as it diverges distally in an ulnar direction from the distal biceps

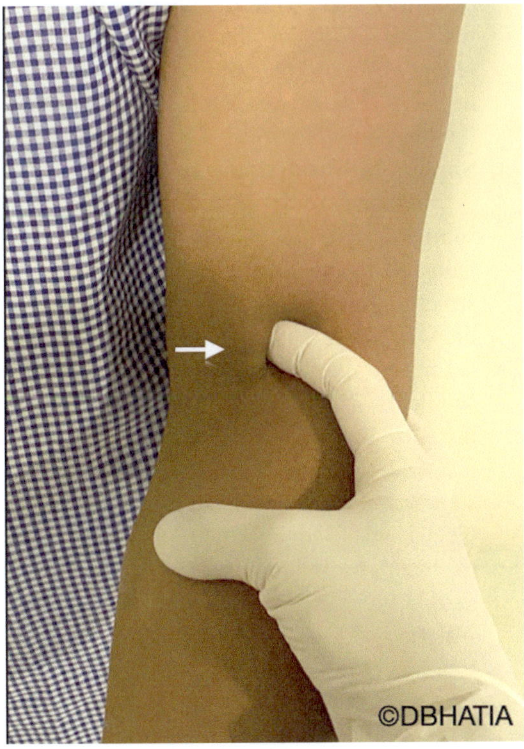

Fig. 16.4 The hook test is demonstrated. The examiner passes a finger under the lateral aspect of distal biceps tendon (arrow) in an attempt to "hook" it

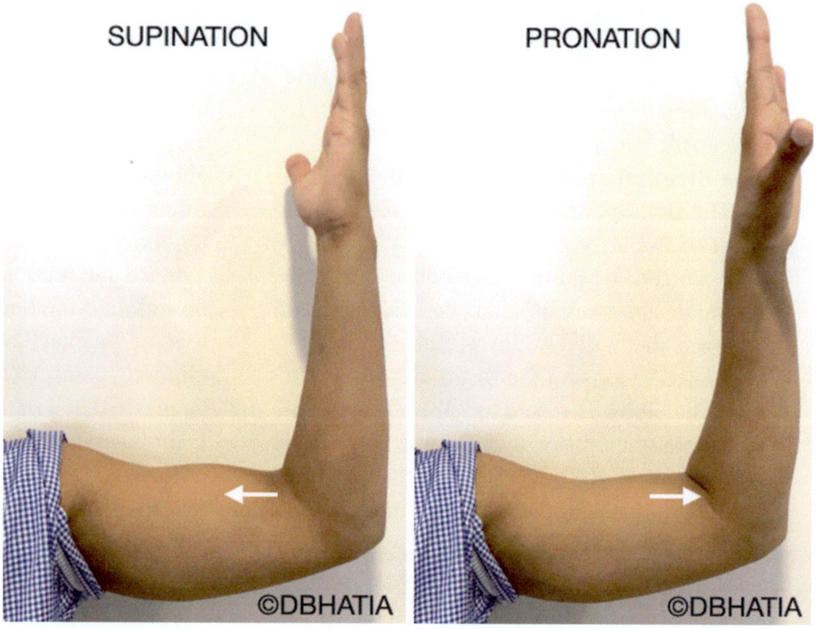

Fig. 16.5 The biceps supination-pronation test is shown. In an intact DBT, the biceps moves proximally with supination and distally with pronation (arrows)

the hook test as 78% (all tears), 83% (complete tears), 45% (intact lacertus fibrosus), and 30% (partial tears) [6]. In delayed cases (8 weeks or more), a positive hook test was suggestive of the need for allograft reconstruction (75% probability), while in a negative test, the probability of reconstruction was only 20%.

(d) **Supination-Pronation test**

Metzman and Tivener described the supination-pronation test as a reliable and pain-free test that would help to evaluate the structural integrity of the DBT and to isolate the DBT from the lacertus fibrosus [7]. They described the test as follows: with both shoulders abducted to 90° and the elbows flexed to approximately 60–70°, the patient is asked to actively perform pronation and supination. In an intact DBT, a change in the biceps muscle contour in the arm is observed; the biceps moves proximally with supination and distally with pronation (Fig. 16.5). Absence of biceps migration with dynamic rotation is suggestive of a complete DBT rupture. The contour of the biceps may be difficult to observe in some patients; in such cases, the movement of the biceps can be palpated during dynamic rotations. The supination-pronation test may be useful in an acute scenario where palpation of the DBT (Hook test) and biceps muscle (squeeze test) cannot be performed due to pain.

(e) **Biceps Squeeze test**

The biceps squeeze test described by Ruland et al. is analogous with the Thompson test has been used in the diagnosis of Achilles tendon rupture [8]. The test is performed as follows: The forearm is placed in a comfortable position in the patients lap in 60–80° flexion and slight pronation. The biceps muscle in the arm is squeezed firmly with both hands at the distal myotendinous junction and around the muscle belly (Fig. 16.6). This "squeeze" pulls the muscle into an anterior bow and results in forearm supination. As described by the authors, an absence of forearm supination with this maneuver indicates complete rupture of the biceps brachii tendon or muscle belly.

(f) **TILT sign**

This sign was described by Shim and Strauch as a sensitive (100%) test for diagnosis of partial tears of DBT [9]. The test is based on the "anatomical rotation of the radial tuberosity in full pronation to allow palpation from the dorsal side of the fore-

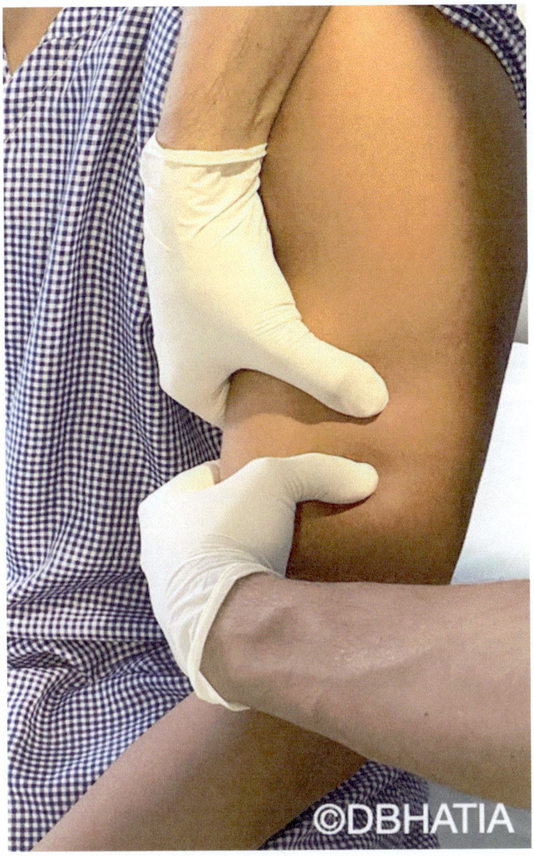

Fig. 16.6 The biceps squeeze test pulls the muscle into an anterior bow and results in forearm supination

arm." The test is performed as follows: The elbow is flexed to 90° and the forearm is rotated to full supination. The dorsal forearm is palpated at the area over the radial tuberosity. Next the forearm is rotated into full pronation and the tuberosity area is palpated in a similar way. Tenderness over the radial (or lateral) aspect of the tuberosity (TILT sign) in full pronation, but not in supination is suggestive of a partial DBT tear (Fig. 16.7).

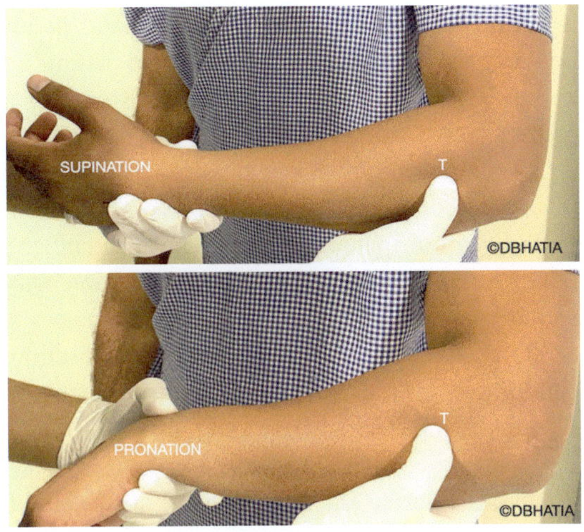

Fig. 16.7 The TILT sign is positive if tenderness can be felt over the radial aspect of the tuberosity (T) in full pronation, but not in supination

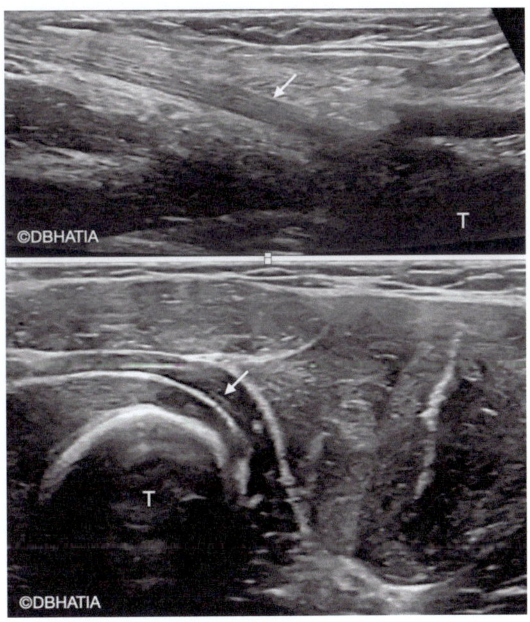

Fig. 16.8 Sonographic evaluation of the distal biceps tendon (arrows) is demonstrated using the probe placed on the volar (top image) and dorsal (bottom image) aspects of the arm and forearm. [T: bicipital tuberosity]

16.3.1 Sonography

In-clinic ultrasonography (USG) is an excellent method of assessing the DBT integrity and can be a useful adjunct to clinical examination (Fig. 16.8). USG is advantageous as it is inexpensive and permits dynamic evaluation and comparison with the contralateral elbow. The discontinuity of the tendon can be demonstrated and the retraction can be quantified in acute and chronic complete DBT tears. Partial tears and bursitis can be assessed and sonography-guided injections can be used for diagnostic and therapeutic purposes. Sonography is also useful to assess tendon integrity in the postoperative period.

In summary, clinical evaluation of the distal biceps tendon (DBT) should involve assessment of the region spanning the lower arm and the upper forearm. Special tests for tendon integrity should be used in combination to evaluate the DBT and the bicipital aponeurosis. In-clinic sonography is an excellent tool to supplement the diagnosis of partial and complete tears.

Conflict of Interest The author certifies that he has no commercial associations (e.g., consultancies, stock ownership, equity interest, patent/licensing arrangements) that might pose a conflict of interest in connection with the submitted article.

References

1. Bhatia DN, Kandhari V, DasGupta B. Cadaveric study of insertional anatomy of distal biceps tendon and its relationship to the dynamic proximal Radioulnar space. J Hand Surg Am. 2017;42:e15–23.
2. ElMaraghy A, Devereaux M, Tsoi K. The biceps crease interval for diagnosing complete distal biceps tendon ruptures. Clin Orthop Relat Res. 2008;466(9): 2255–62. https://doi.org/10.1007/s11999-008-0334-0. Epub 2008 Jun 13.
3. ElMaraghy A, Devereaux M. The "bicipital aponeurosis flex test": evaluating the integrity of the bicipital aponeurosis and its implications for treatment of distal biceps tendon ruptures. J Shoulder Elb Surg 2013;22(7):908–914. doi: https://doi.org/10.1016/j.jse.2013.02.005.
4. O'Driscoll SW, Goncalves LB, Dietz P. The hook test for distal biceps tendon avulsion. Am J Sports Med. 2007;35(11):1865–9. https://doi.org/10.1177/0363546507305016. Epub 2007 Aug 8.
5. Kruger N, Phadnis J, Bhatia D, et al. Acute distal biceps tendon ruptures: anatomy, pathology and management—state of the art. J ISAKOS. 2020;5: 304–13.
6. Luokkala T, Siddharthan SK, Karjalainen TV, Watts AC. Distal biceps hook test—sensitivity in acute and chronic tears and ability to predict the need for graft reconstruction. Shoulder Elbow. 2020;12(4):294–8. https://doi.org/10.1177/1758573219847146. Epub 2019 May 15.
7. Metzman LS, Tivener KA. The Supination-Pronation Test for Distal Biceps Tendon Rupture. Am J Orthop (Belle Mead NJ). 2015;44(10):E361–4.
8. Ruland RT, Dunbar RP, Bowen JD. The biceps squeeze test for diagnosis of distal biceps tendon ruptures. Clin Orthop Relat Res. 2005;437:128–31. https://doi.org/10.1097/01.blo.0000167668.18444.f5.
9. Shim SS, Strauch RJ. A novel clinical test for partial tears of the distal biceps brachii tendon: the TILT sign. Clin Anat. 2018;31(2):301–3. https://doi.org/10.1002/ca.23038. Epub 2018 Jan 9.

17 Evaluation of Elbow Instability with Clinical Testing

Yoav Rosenthal and Mark I. Loebenberg

17.1 Introduction

A comprehensive understanding of the physical examination of the unstable elbow is essential for the clinician to effectively diagnose pathology conditions affecting the elbow joint.

Elbow instability may result from an acute injury due to a specific mechanism, or due to the chronic elbow tension overload, common in overhead throwing athletes. Various physical examination tests, that may improve the physician's ability to accurately diagnose these special elbow conditions, have been well described.

This chapter will provide a thorough review of the different elbow physical examination tests related to elbow instability and will include their original descriptions, as well as the evidence supporting their use.

17.2 Anatomy and Biomechanics

In order to perform a proper physical examination, the physician must first understand basic anatomy and biomechanics of the elbow.

The elbow joint complex possesses two degrees of freedom, thus allowing two types of motion: flexion-extension, occurring at the humeroulnar and humeroradial joints and forearm pronation-supination, or simply, forearm rotation, occurring at the radiocapitellar and proximal radioulnar joints. The carrying angle is defined as the angle between the axis of the humerus and the ulna in full extension and varies between 11 and 16° of valgus [1]. This angle changes linearly during flexion, moving from valgus to varus, as the elbow moves from extension to flexion [2].

The elbow joint constraints to instability may be classified as primary and secondary, or as static and dynamic. The three primary constraints include the ulnohumeral articulation, the medial ulnar collateral ligament (MUCL), and the lateral ulnar collateral ligament (LUCL). The secondary constraints include the radial head, the common flexor and extensor mass origins, and the joint capsule [3]. The static constraints are the bony morphology of the ulna-humeral-radius articulation, the MUCL, the LUCL, and the anterior capsule. The dynamic constraints include the surrounding muscular attachments and their fascial bands [4].

17.2.1 Posterolateral Rotatory Instability

The most common elbow instability pattern was described by O'Driscoll in 1991 [5], and was termed posterolateral rotatory instability (PLRI).

Y. Rosenthal · M. I. Loebenberg (✉)
Shoulder and Elbow Division, Department of Orthopedic Surgery, Rabin Medical Center, Tel Aviv University, Tel Aviv-Yafo, Israel

J. G. Lane et al. (eds.), *The Art of the Musculoskeletal Physical Exam*,
https://doi.org/10.1007/978-3-031-24404-9_17

In this mechanism, a fall on an outstretched hand with the shoulder abducted an axial force is exerted on the elbow as it flexes. As the body internally rotates on the hand and approaches the ground, external rotation and valgus moments are applied to the elbow. This injury progresses, from the lateral to medial side in three stages (Circle of Horii): (1) LUCL disruption; (2) the other lateral ligamentous structures, as well as the anterior and posterior capsules, are disrupted; (3) The MUCL is disrupted either partially (posterior band only) or completely (posterior and anterior bands). LUCL insufficiency is mostly a sequalae of elbow dislocation, especially in young patients. Despite both the LUCL and MUCL are disrupted during an acute elbow dislocation, the chronic, residual insufficiency, usually arises due to improper healing of the LUCL, which culminates in recurrent PLRI [5, 6].

17.2.2 Valgus Instability

Most cases of chronic valgus instability occur as a result of repetitive trauma to the medial collateral ligament (MCL) and occurs mostly in overhead throwing athletes, but may also occur in gymnasts, martial arts fighters, football players, and other contact sports athletes. Extreme valgus stress loads during throwing leads to repetitive microtrauma to the MCL and results in stretching and attenuation of the MCL [7, 8]. As the MCL becomes insufficient, the radial head and the flexor-pronator mass become the primary stabilizers against valgus instability.

Even though injuries to the MCL are usually attritional and although it is widely thought that the medial ligaments heal well after simple dislocations [9], it was shown that up to 58% of patients with history of posterolateral dislocation of the elbow, may suffer from valgus laxity [10].

17.3 History

Patients suffering from PLRI may present with a highly variable history. The patients may have a history of one or more elbow dislocations. Patients often learn maneuvers to reduce the joint and therefore may not require formal reductions [11]. They may complain of clear elbow instability, or more subtly, pain and discomfort, without being aware of gross instability. Other symptoms include giving-way, catching, slippage, snapping, or clicking with flexion and extension, or typically, more prominently during axial loading of the joint, while supination and slight flexion of the elbow [4]. Occasionally, patients may complain of elbow subluxation, occurring with elbow extension, or even with pushing on the armrests, rising from a chair [3, 12].

Patients with an acute injury to their medial ulnar collateral ligament typically present with a history of sudden onset of pain after throwing, with or without an associated popping sensation that occurred during throwing and are unable to continue throwing. Patients with chronic injury to their MCL typically complain of a more gradual-onset of medial elbow pain during the late-cocking or acceleration phase of throwing, affecting their pitching accuracy or velocity. Patients may also complain of ulnar nerve symptoms due to local inflammation of the ligamentous complex, which causes local irritation of the ulnar nerve within the cubital fossa [13–15].

17.4 Physical Examination

17.4.1 Inspection

Clinical examination is typically initiated with inspection of the affected elbow, compared with the contralateral side. Ecchymosis may be identified at the medial side, as a sign of MCL rupture (Fig. 17.1) and may suggest elbow dislocation [16]. If joint effusion is present, the elbow is usually held in the resting position of 70–80° of flexion, to accommodate the increased intracapsular volume [2, 7], and there may be loss of the normal subtle concavity of the lateral soft spot (the triangular region bordered by the tip of the olecranon, the radial head, and lateral epicondyle). The carrying angle is evaluated in full extension and should be compared to the contralateral side.

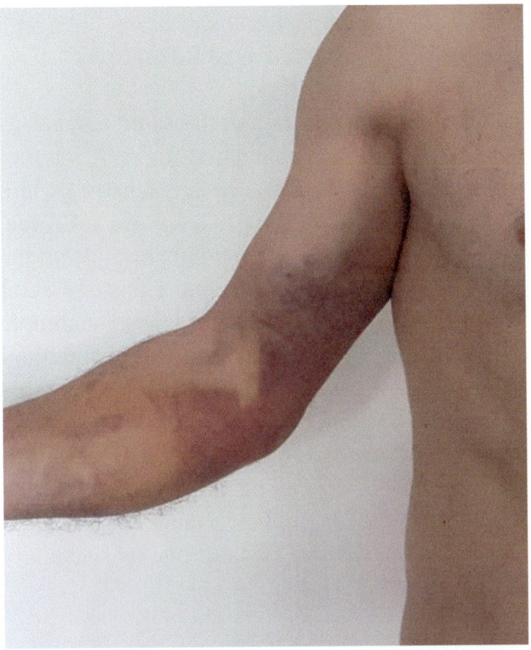

Fig. 17.1 Inspection. Ecchymosis identified at the medial side of the elbow, may be a sign of MCL rupture and possible prior elbow dislocation

17.4.2 Palpation

Fullness to palpation located at the soft spot is indicative of joint effusion or hemarthrosis.

The radial head may be palpated with passive forearm rotation. Focal radial head tenderness may be indicative of radial head fracture. Tenderness at the lateral epicondyle, or rarely at the supinator crest, may be suggestive of LUCL avulsion. Other structures to be palpated on the lateral side are the capitellum and the extensor mass origin [17].

In order to appreciate MCL integrity, it is recommended to flex the elbow to 50–70° of flexion. Subsequently, the flexor pronator mass will move anterior to the MCL, exposing it for direct palpation. If MCL injury is suspected, it should be palpated through its course, from the inferior aspect of the medial epicondyle to the sublime tubercle, to determine the location of the injury [2, 14]. Tenderness to palpation over the MCL is 94% sensitive but only 22% specific for an MCL abnormality [18] since point tenderness over the MCL may vary with the amount of inflammation and edema present. The absence of increased pain with wrist flexion, combined with pain localized slightly posterior to the common flexor origin, differentiates MCL injury from flexor-pronator mass injury [13].

17.4.3 Range of Motion

The normal range the elbow's arc is from −1.5° of extension (hyperextension) to 150° of flexion with a functional range of 30–130°. The normal range of forearm rotation arc averages from 68° of pronation to 74° of supination [19].

The unstable elbow may be congruent in 90° of flexion, however, during elbow extension, the joint may lose its congruency, as may be evident in dynamic fluoroscopy or ultrasound by widening of the ulnohumeral joint space (>4 mm) [20].

Pain at the end point of flexion and extension, localized to the medial side of the olecranon is usually caused by the early degenerative changes consistent with chronic valgus extension overload and posterolateral impingement, typically found in throwing athletes [21] (see 'moving valgus stress test').

17.5 Specific Tests

17.5.1 Physical Examination for the Evaluation of Posterolateral Rotatory Instability

For most elbow instability physical examinations, patient relaxation is critical and is accomplished by two measures: First, by securely holding the patient's arm between the examiner's upper arm and chest, the patient feels that the arm will not be allowed to fall or slip (Fig. 17.2). Second, the patient must be convinced that the examiner is not about to make any sudden or painful maneuvers [22]. In order to calm the patient, adequate time is dedicated to explain to the patient the course of the examination and what he or she is about to feel, and while holding the patient's wrist and forearm firmly and

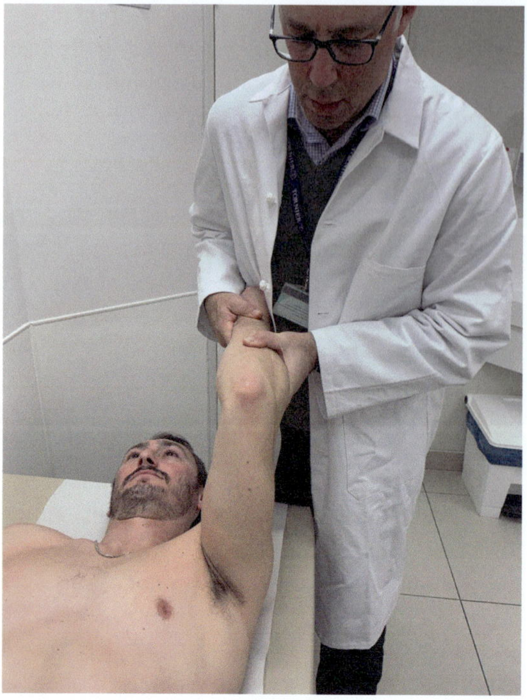

Fig. 17.2 Patients trust and confidence is gained by securely holding the patient's arm between the examiner's upper arm and chest

securely, various gentle, slow repetitive movements are performed in order to provide a sense of security to the patient.

17.5.1.1 Lateral Pivot-Shift Apprehension (Posterolateral Rotatory Apprehension)

The patient is placed in a supine position with the affected extremity overhead, the examiner, standing at the head of the bed, securely grasps the patient's wrist and elbow. Then the forearm is supinated, and mild valgus stress combined with axial loading of the elbow is applied through the wrist during elbow flexion. An apprehensive response with reproduction of the patient's symptoms and a sense of instability is considered as a positive test. Reproducing the actual subluxation and the clunk the occurs with reduction can be achieved under general anesthesia or after lidocaine injection into the joint [12, 23].

17.5.1.2 Lateral Pivot-Shift (Posterolateral Rotatory Instability)

The patient is positioned in the same manner as the lateral pivot-shift apprehension test. Starting from a position of full elbow extension, while applying a valgus and axial compression force, subluxation of the radius and ulna away from the humerus will occur, producing a posterolateral prominence over the radial head and a dimple between the radial head and the capitellum (Fig. 17.3a). As the examiner flexes the patient's elbow to 40° or more, sudden simultaneous reduction of the ulna and radius on the humerus occurs, producing a palpable visible clunk (Fig. 17.3b) [2, 12, 23]. Performing this examination under anesthesia will improve its sensitivity to up to 100% [24].

17.5.1.3 Posterolateral Rotatory Drawer Test

As the two previous described examinations, the patient is positioned supine with the affected extremity overhead. The patient's elbow is supinated and flexed to 40°, while an anterior-to-posterior force is applied to the ulna and radius, causing posterior translation of the forearm, pivoting around the intact medial soft tissue, and a dimple may appear (Fig. 17.4). With joint reduction, the patient may feel the ulna tapping against the humerus, producing a typical clunk [23]. If the patient is under anesthesia, this test can be performed adding a valgus stress to the elbow, causing even more translation.

17.5.1.4 Chair Push-Up Test (Chair Sign)

The patient is seated with the elbows flexed at 90°, the forearm supinated, and arms abducted to greater than the width of the shoulders (Fig. 17.5a). Then, the patient is requested to execute an active sitting push-up, using exclusively upper body forces only (Fig. 17.5b). The test is considered positive if it reproduces the symptoms and causes apprehension as the patient pushes against the armrests. Regan and Lapner reported a sensitivity of 87.5% of this test [25].

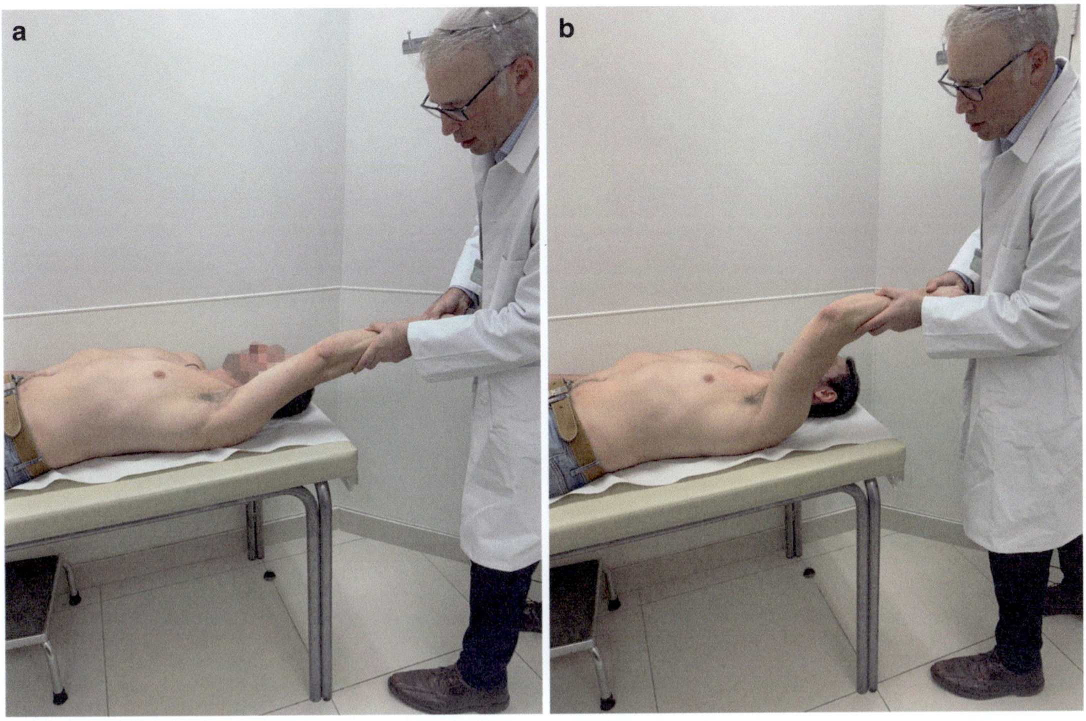

Fig. 17.3 (**a** and **b**) Lateral pivot-shift test. The patient is placed supine with the affected extremity overhead, while the examiner, standing at the head of the bed, securely grasps the patient's wrist and elbow. Then, the forearm is supinated and mild valgus stress combined with axial loading of the elbow (**a**) is applied through the wrist during elbow flexion. At first, subluxation of the radius and ulna away from the humerus occurs. As the elbow is flexed to 40° or more, sudden reduction of the ulna and radius on the humerus will occur (**b**)

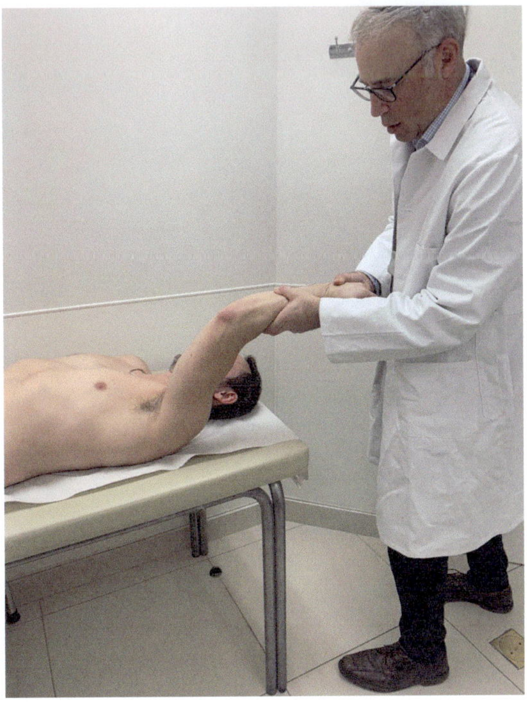

Fig. 17.4 Posterolateral rotatory drawer test. The patient's elbow is supinated and flexed to 40°, while an anterior-to-posterior force is applied to the ulna and radius, causing posterior translation of the forearm

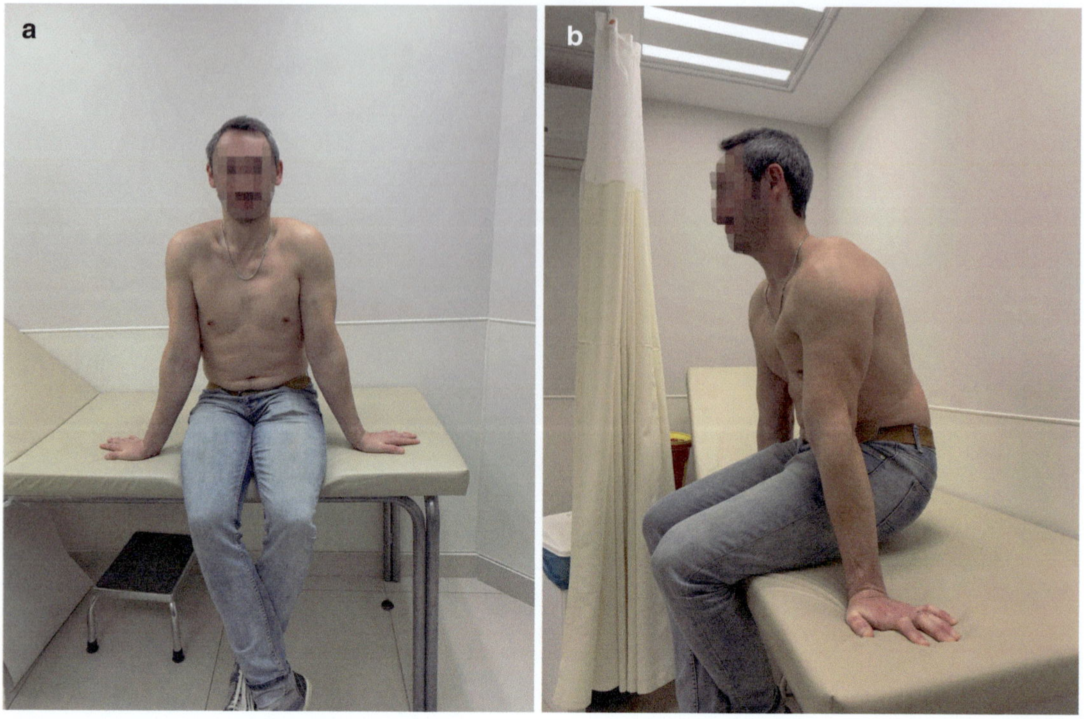

Fig. 17.5 (**a** and **b**) Chair push-up test. The patient is seated on the bed or chair with the elbows flexed at 90°, the forearm supinated, and arms abducted to greater than the width of the shoulders (**a**) Then, the patient is requested to execute an active sitting push-up, using exclusively upper body forces only (**b**)

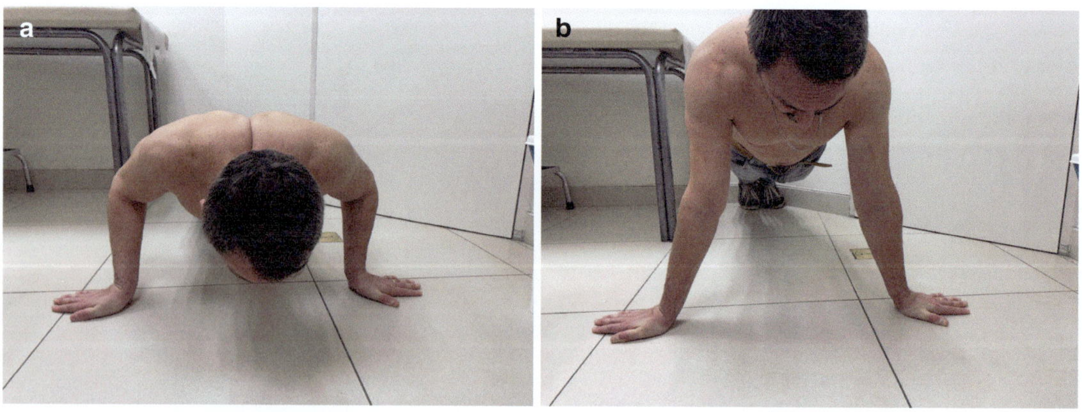

Fig. 17.6 (**a** and **b**) Active floor push-up test. The patient is positioned prone on the floor in "push-up position" with the elbows flexed at 90°, the forearms supinated and the arms abducted to greater than the shoulder width (**a**). Then, the patient is asked to carry out an active floor push-up (**b**)

17.5.1.5 Active Floor Push-Up Test (Push-Up Sign)

The patient is positioned prone on the floor in "push-up position" with the elbows flexed at 90°, the forearms supinated and the arms abducted to greater than the shoulder width (Fig. 17.6a). Then, the patient is asked to carry out an active floor push-up (Fig. 17.6b). This test is considered positive if apprehension, along with guarding, occurs at full elbow extension. The push-up test

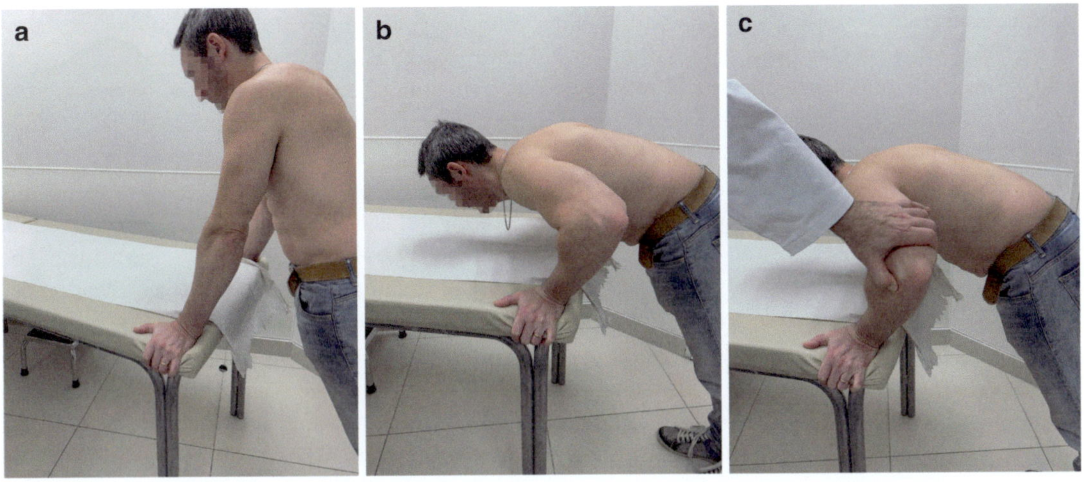

Fig. 17.7 (**a**–**c**) Table-top relocation test. The patient is positioned in front of a table with the hand of the affected elbow is placed over the lateral edge of the table (**a**). For the starting position, the patient is initially asked to perform a push-up maneuver with the elbow pointing laterally to maintain forearm supination. Then, the patient is requested to push down through the hand onto the edge of the table, bringing the chest toward the table (**b**). Then, the push down maneuver is repeated with the examiner's thumb supporting the radial head, preventing posterior subluxation (**c**)

was reported to be 87.5% sensitive, and combined with the chair push-up test, both demonstrated a sensitivity of 100% [25].

17.5.1.6 Table-Top Relocation Test

The patient is positioned in front of a table. The hand of the affected elbow is placed over the lateral edge of the table (Fig. 17.7a). The test consists of three parts: for the starting position, the patient is initially asked to perform a push-up maneuver with the elbow pointing laterally to maintain forearm supination. Then, the patient is requested to push down through the hand onto the edge of the table, bringing the chest toward the table (Fig. 17.7b). The test is considered positive if the maneuver reproduces the pain and apprehension at approximately 40° of flexion. Then, similar to the Jobe relocation test of the shoulder in which the examiner supports the patient's arm to prevent anterior subluxation of the shoulder, the later maneuver is repeated with the examiner's thumb pushing gently over the radial head, supporting it, and preventing posterior subluxation (Fig. 17.7c). Finally, similar to the surprise test, performed subsequently to the Jobe relocation test of the shoulder, removal of the examiner's thumb from the weight-bearing, partially flexed elbow, reproduces the pain and apprehension again. Arvind and Hargreaves, who described this test, reported a sensitivity of 100% [26].

17.5.2 Physical Examination for the Evaluation of Valgus Instability

As opposed to PLRI testing, where the patient is usually positioned supine, for the valgus instability tests, the patients may be placed in various positions: standing or seated upright in a chair, lying prone, or supine, as long as the examiner can adequately stabilize the elbow.

17.5.2.1 Valgus Stress Test

Valgus stress testing is performed to evaluate injury to the anterior bundle of the MCL (aMCL).

The patient is seated and his wrist is secured between the examiner's forearm and trunk.

The affected elbow is pronated and flexed to 20–30° to unlock the olecranon from its fossa, and a valgus force is applied (Fig. 17.8). Valgus laxity is manifested by an increased medial joint space opening (at least 2 mm), compared with the contralateral side. In some cases, as a partial tear of the MCL, only medial-sided pain will may be evident, without joint-space opening [13, 14, 17]. The valgus stress test manifested by laxity was found to be 100% specific but only 18.8% sensitive, whereas the test manifested by pain only was reported to have a sensitivity of 50% and a specificity of 64.7% [24].

17.5.2.2 Moving Valgus Stress Test

The patient is seated upright and the shoulder is abducted 90°. The test is started with the elbow maximally flexed (Fig. 17.9a). A valgus torque is applied to the elbow until the shoulder is maximally externally rotated. While the valgus torque is maintained, the elbow is quickly extended to 30° (Fig. 17.9b). Reproduction of the medial elbow pain at the MCL, with maximal pain experienced between 120° and 70° of extension (the position of late-cocking and early acceleration, respectively) will define this test as positive. The angle of maximal pain is defined as 'shear angle' and is averaged to occur at 90°. The 'shear range' is referred to the range of motion that causes pain while the elbow is being extend with valgus stress. This arc usually occurs at 70–120° of elbow extension. A positive moving valgus stress test is confirmed when the same pain response, but usually to a lesser extent, is elicited by reversing the motion, starting at elbow extension and then flexing it, while maintaining a valgus moment. The moving stress test was reported to have a sensitivity of 100% and specificity of 75% [21].

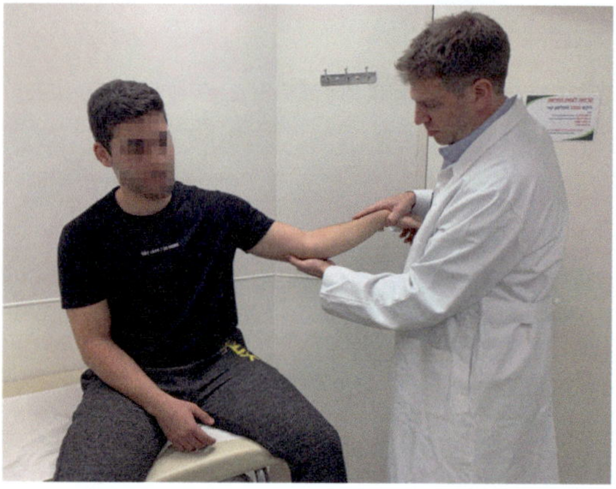

Fig. 17.8 The valgus stress test. The affected elbow is pronated and flexed to 20–30° to unlock the olecranon from its fossa, and a valgus force is applied

17.5.2.3 The Milking Maneuver

By generating a valgus stress on the elbow, the milking maneuver tests the posterior band of the aMCL. This exam can be performed solely by the patient or by the examining physician. Originally, the milking maneuver was described as follows: the patient is asked to flex the affected elbow beyond 90°. The contralateral hand is

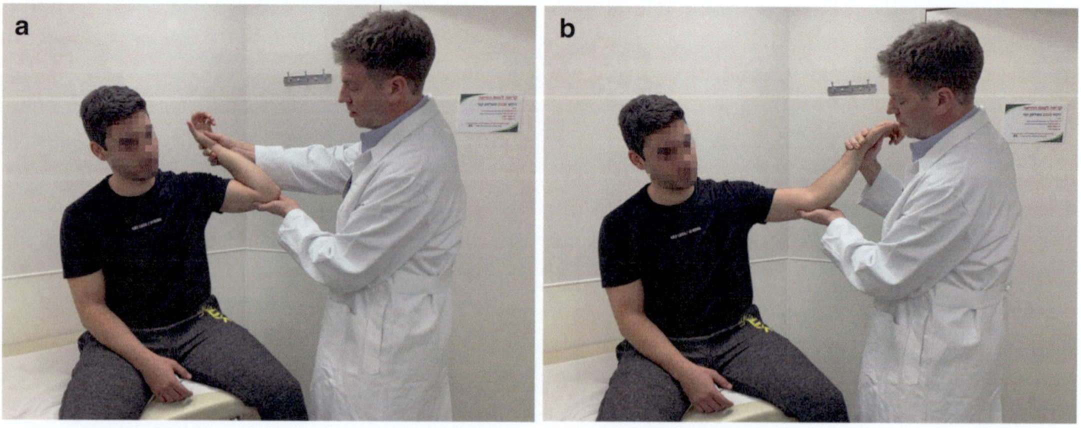

Fig. 17.9 (**a** and **b**) Moving valgus stress test. The patient is seated with the shoulder is abducted to 90°. At first, the elbow is maximally flexed (**a**). Then, a valgus torque is applied to the elbow and the elbow is quickly extended to 30° (**b**)

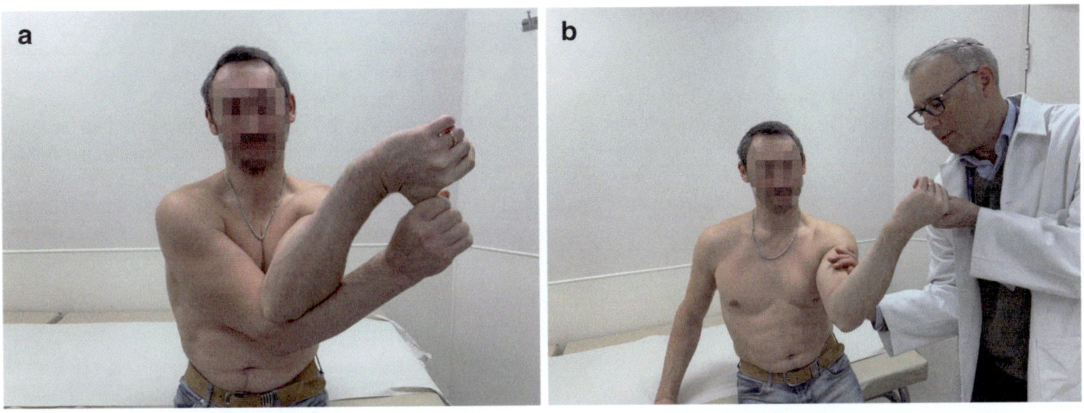

Fig. 17.10 (**a** and **b**) The Milking Maneuver. The patient was asked to flex the affected elbow beyond 90°. The contralateral hand is placed under the affected elbow to grasp the thumb of the affected extremity, thereby exerting a valgus stress on the MCL (**a**). The same maneuver can be generated by the examining physician, (**b**)

placed under the affected elbow to grasp the thumb of the affected extremity, thereby exerting a valgus stress on the MCL (Fig. 17.10a). A positive test elicits pain, apprehension, and instability. This test's sensitivity was reported to be 87.5% [27]. An alternative method is performed by the physician pulling on the patient's thumb, with the patient's forearm supinated, arm extended, and elbow flexed beyond 90° (Fig. 17.10b) [13].

17.6 Summary

Vulnerable to unique acute injury mechanism or chronic, repetitive, stress overload, the elbow is prone to instability injuries. In order to establish the correct diagnosis of type of elbow instability, the physician must have a profound understanding of the elbow's anatomy and biomechanics and master the special physical examination maneuvers described in this chapter. As reported in several studies, in some cases, the physical examination may be more sensitive than advanced imaging modalities, emphasizing even more their importance. This chapter provides the clinician with a detailed review of the various elbow instability physical examination maneuvers.

References

1. Steinberg BD, Plancher KD. Clinical anatomy of the wrist and elbow. Clin Sports Med. 1995;14:299–313.
2. Smith MV, Lamplot JD, Wright RW, Brophy RH. Comprehensive review of the elbow physical examination. J Am Acad Orthop Surg. 2018;26:678–87.
3. O'Driscoll SW, Jupiter JB, King GJ, Hotchkiss RN, Morrey BF. The unstable elbow. Instr Course Lect. 2001;50:89–102.
4. Singleton SB, Conway JE. PLRI: posterolateral rotatory instability of the elbow. Clin Sports Med. 2004,23(629–42):ix–x.
5. O'Driscoll SW, Bell DF, Morrey BF. Posterolateral rotatory instability of the elbow. J Bone Joint Surg Am. 1991;73:440–6.
6. O'Driscoll SW, Morrey BF, Korinek S, An KN. Elbow subluxation and dislocation. A spectrum of instability. Clin Orthop Relat Res. 1992:186–97.
7. Willemot L, Hendrikx FR, Byrne AM, van Riet RP. Valgus instability of the elbow: acute and chronic form. Obere Extrem. 2018;13:173–9.
8. Ramos N, Limpisvasti O. UCL injury in the non-throwing athlete. Curr Rev Musculoskelet Med. 2019;12:527–33.
9. Murthi AM, Keener JD, Armstrong AD, Getz CL. The recurrent unstable elbow: diagnosis and treatment. J Bone Joint Surg Am. 2010;92:1794–804.
10. Eygendaal D, Verdegaal SH, Obermann WR, van Vugt AB, Pöll RG, Rozing PM. Posterolateral dislocation of the elbow joint. Relationship to medial instability. J Bone Joint Surg Am. 2000;82:555–60.

11. Mehta JA, Bain GI. Posterolateral rotatory instability of the elbow. J Am Acad Orthop Surg. 2004;12:405–15.
12. O'Driscoll SW. Classification and evaluation of recurrent instability of the elbow. Clin Orthop Relat Res. 2000:34–43.
13. Chen FS, Rokito AS, Jobe FW. Medial elbow problems in the overhead-throwing athlete. J Am Acad Orthop Surg. 2001;9:99–113.
14. Bruce JR, Andrews JR. Ulnar collateral ligament injuries in the throwing athlete. J Am Acad Orthop Surg. 2014;22:315–25.
15. Hyman J, Breazeale NM, Altchek DW. Valgus instability of the elbow in athletes. Clin Sports Med. 2001;20:25–45. viii
16. Hausman MR, Lang P. Examination of the elbow: current concepts. J Hand Surg Am. 2014;39:2534–41.
17. Cain EL Jr, Dugas JR. History and examination of the thrower's elbow. Clin Sports Med. 2004;23:553–66. viii
18. Timmerman LA, Schwartz ML, Andrews JR. Preoperative evaluation of the ulnar collateral ligament by magnetic resonance imaging and computed tomography arthrography. Evaluation in 25 baseball players with surgical confirmation. Am J Sports Med. 1994;22:26–31. discussion 2
19. Soucie JM, Wang C, Forsyth A, et al. Range of motion measurements: reference values and a database for comparison studies. Haemophilia. 2011;17:500–7.
20. Camp CL, Smith J, O'Driscoll SW. Posterolateral rotatory instability of the elbow: Part II. Supplementary examination and dynamic imaging techniques. Arthrosc Tech. 2017;6:e407–e11.
21. O'Driscoll SW, Lawton RL, Smith AM. The "moving valgus stress test" for medial collateral ligament tears of the elbow. Am J Sports Med. 2005;33:231–9.
22. Camp CL, Smith J, O'Driscoll SW. Posterolateral rotatory instability of the elbow: Part I. Mechanism of injury and the posterolateral rotatory drawer test. Arthrosc Tech. 2017;6:e401–e5.
23. O'Driscoll SW. Elbow instability. Acta Orthop Belg. 1999;65:404–15.
24. Zwerus EL, Somford MP, Maissan F, Heisen J, Eygendaal D, van den Bekerom MP. Physical examination of the elbow, what is the evidence? A systematic literature review. Br J Sports Med. 2018;52:1253–60.
25. Regan W, Lapner PC. Prospective evaluation of two diagnostic apprehension signs for posterolateral instability of the elbow. J Shoulder Elb Surg. 2006;15:344–6.
26. Arvind CH, Hargreaves DG. Tabletop relocation test: a new clinical test for posterolateral rotatory instability of the elbow. J Shoulder Elb Surg. 2006;15:707–8.
27. Veltri DM, O'Brien SJ, Field LD, Deutsch A, Altchek DW, Potter HG. The milking manuever-a new test to evaluate the MCL of the elbow in the throwing athlete. J Shoulder Elb Surg. 1995;4:S10.

18 Neurologic Evaluation of the Elbow and Forearm

José Carlos Garcia Jr, Rafael José Zamith Gadioli, and Leandro Sossai Altoé

The assessment of nerves at the elbow and forearm have become more and more important because the current better understanding of neurologic pathologies associated to these nerves. Indeed, some of these pathologies sometimes can't be assessed even by electromyographic studies or image exams; therefore, the meticulous clinical exam will present an utmost importance on detecting these entrapment syndromes.

On this chapter, our aim is to give a clinical picture of all conditions associated to nerve entrapment syndromes at the elbow and forearm.

18.1 Lateral Cutaneous Nerve of Forearm

Originating from the musculocutaneous nerve, the lateral cutaneous nerve of the forearm innervates the anterolateral region of the forearm until the thenar eminence. Its compression site is in the distal third of the arm, between the biceps and the fascia of the brachial muscle, which tenses in the forearm extension and pronation [1].

Compression of the lateral cutaneous nerve of the forearm is a rare and badly understood condition. It presents a poor clinical picture, with burning pain in the anterolateral aspect of the forearm with worsening in passive pronation and hyperextension of the elbow [2]. In full pronation forced supination of the forearm with elbow flexion can also reproduce the symptoms. In chronic cases, the patient reports vague discomfort in the forearm that can intensify and worsen with pronation supination activities with the elbow extended [3].

On physical examination, an area of hypoesthesia on the anterolateral surface of the forearm can be identified by applying a light touch to the skin with a blind spot.

Thinking about differential diagnoses, the lateral epicondylitis and radial tunnel syndrome are highlighted.

For diagnostic confirmation, the use of electroneuromyography has an action potential with prolonged latency or decreased amplitude; however, a negative result will not predict patient does not present this nerve entrapment. Imaging tests have little or no value.

18.2 Radial Nerve

The radial nerve and its major branches, the posterior interosseous nerve and the superficial radial nerve, are vulnerable to compression forces from the level of the lateral head of the triceps through the region of the elbow, proximal forearm, and even into the distal forearm [4].

J. C. Garcia Jr (✉) · R. J. Z. Gadioli · L. S. Altoé
NAEON Institute, Sao Paulo, Brazil
e-mail: josecarlos@especialistadoombro.com.br

J. G. Lane et al. (eds.), *The Art of the Musculoskeletal Physical Exam*,
https://doi.org/10.1007/978-3-031-24404-9_18

Depending on which branch of the nerve is involved at the elbow, either motor and sensory (posterior interosseous nerve) or just sensory (superficial radial nerve) symptoms can occur. One needs to be aware that when sensory symptoms like tingling or numbness are present the superficial radial nerve or a radial nerve before the division of this nerve are highly suspected. In occasions where pain on the anterolateral aspect of the elbow is present, a posterior interosseous nerve impairment can be suspected, once it is not a sensitive nerve neither tingling nor numbness is related to this nerve [5].

Rarely, motor and sensory involvement can be due to a process in the proximal forearm affecting both branches rather than the radial nerve.

The radial nerve has its origin in the posterior fascicle of the brachial plexus and innervates the triceps, brachioradialis, anconeus, all extensors, supinator, and abductor pollicis longus. In the distal region of the arm, it crosses anterior to 10 cm proximal to the lateral epicondyle. At the level of radiocapitellar articulation, it divides into superficial and deep branches, passing the brachioradial fascia deep, innervating the muscles brachioradialis and extensoris carpi radialis longus. The motor branch of the extensoris carpi radialis brevis originates from the superficial branch of the radial nerve in 58% of the population. In the elbow, the deep branch, crosses between the two heads of the supinator muscle, where it is called the posterior interosseous nerve (PIN) and innervates that muscle [4]. The proximal edge of the supinator muscle forms the fibrous Fröhse's arcade. The superficial radial nerve continues below the brachioradialis muscle until it emerges at the distal third of the forearm to the subcutaneous. Dorsoradial forearm tenderness until the hand can make one suspect of radial-sensitive branch compression. The recurrent vessels of the radial artery cross superficial and deep to the branches of the radial nerve at the elbow, these structures can entrap the nerve, as well as fibrous adhesions of the anterior capsule.

The PIN goes in the dorsoradial direction in the proximal forearm where 6–8 cm distal from the elbow it emerges from the supinator muscle and releases its terminal branches to the extensor digitorum communis, extensor digiti minimi, extensor carpi ulnaris, abductor policis brevis and longus, long and short thumb extensor, extensor indicis proprius, and extensor digiti minimi [6].

Indeed, the compression sites are as follows: spiral groove of the humerus between the intermuscle septum of the triceps, and in the forearm: Fröhse's arch (most common compression site), fibrous band between the brachioradialis and brachial muscles, Henry's vascular complex (Arteria Recurrens Radialis), tendon margin of the extensor radialis carpi brevis, and distal edge of the supinator muscle.

18.3 Posterior Interosseous Nerve

In cases of high bifurcation, the deep branch may be compressed in the lateral intermuscular septum, with local sensitivity and weakness of the wrist and finger extensors. In these cases, the superficial radial nerve is anterior to the intermuscular septum.

Classical clinical presentation of posterior interosseous nerve (PIN) paralysis is typically motor. Due to segmental innervation of the supinator, the proximal or distal location of the compression may determine an occasionally positive electromyographic study. A supine myofibrillation suggests a more proximal compression of the Fröhse arcade [7].

If only the PIN is compressed, there will be a deficit in the extension of the fingers and thumb with radial deviation of the wrist. Because the branches for the long and short extensor carpi radialis brevis and longus of the wrist originate more proximally. In case of partial paralysis or compression of the medial branch, weakness of the extensor carpi ulnaris, extensor digiti minimi, and extensor digitorum will occur, which may lead to an attitude called the "pseudoulnar" claw. In case of compression it occurs in the lateral branch, the weakness will be of the abductor pollicis longus, long and short extensor of the thumb and extensor proper of the fingers [4]. There may be vague pain on the back of the forearm, but without sensory changes. Atrophy may be present in chronic cases.

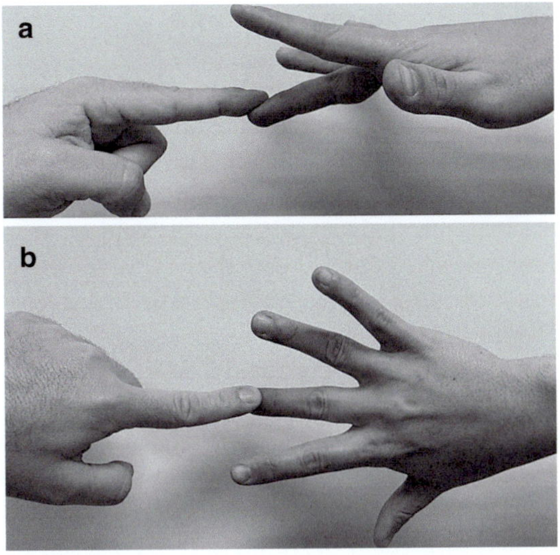

Fig. 18.1 Maudsley test, (**a**) lateral view, (**b**) dorsa view

The Maudsley test (Fig. 18.1) identifies pain at the origin of the extensor carpi radialis brevis during movement of extension against resistance of the third finger with the extended member (differential diagnosis with lateral epicondylitis is needed) [8].

#Local tenderness along the posterior interosseous nerve, and its branches need to be compared with the contralateral side, this region is regularly not very sensitive to pressure. Tenderness is particularly present on the main compression site, most commonly the Fröhse arcade on the anterolateral aspect of the elbow, located 5–10 cm distal and anterior to the lateral epicondyle.

#Supination of the forearm against resistance can reproduce pain in cases of PIN compression in the Fröhse's arcade. This test needs to be done in two stages:

1. Elbow flexion: where pain can be reported in the proximal lateral forearm.
2. Elbow extension: pain can be equal or inferior.

When one tests the elbow in flexion, the supinator muscle is more requested, because biceps is not under full tension. When elbow is extended biceps is under full tension assuming more control over the supination movement. Therefore, if pain is higher in forced supination with elbow's flexion the PIN compression is highly suggested, in opposite side when pain is higher during forced supination and elbow extension distal biceps partial lesions, tendinopathies or biciptoradial bursitis are suggested. It is also not common that biceps pathologies cause tenderness on anterolateral aspect of the elbow when palped. Biceps will also cause high tenderness when patient in full pronation and extended elbow isometrically try to move to flexion and supination against resistance. Indeed, distal biceps tendinopathies are differential diagnosis of PIN in the Fröhse's arcade.

Among the complementary exams that may assist in the diagnosis, there are few findings that define NIP compression. Dynamic electroneuromyography may show changes with denervation of the muscles innervated by PIN, but false negatives will not exclude the PIN syndrome.

18.4 Radial Tunnel Syndrome

The existence of the radial tunnel syndrome (RTS) is sometimes questionable, since the only symptom and complaint of the patient is constant pain, as soon as this condition is not associated with weakness of the extensor muscles and the electroneuromyography is negative.

The RTS, or also called the "tough tennis elbow," presents a characteristic pain clinic located on the radial nerve, ±5 cm lateral and distal to the epicondyle. The pain occurs during supination in a very similar manner to the PIN syndrome [8]. Indeed, RTS is a slight PIN syndrome that predominantly presents sensitive symptoms and fatigue. Tests for RTS are the same of PIN syndrome [9].

18.5 Wartenberg Syndrome

It is the compression of the sensory branch of the radial nerve, described by Wartenberg in 1926. This condition presents as clinical picture a middle or distal third of the forearm tenderness and paresthesia, between the brachioradialis and extensor carpi radialis longus (ECRL). The nerve arises in the proximal forearm at the bifurcation of the radial nerve, goes deeply to the brachioradial in the forearm, where there are 9 cm of the radial styloid emerging between the brachioradialis and the ECRL.

It has differential diagnosis with De Quervain's tenosynovitis.

Compression can be caused extrinsically due to the use of a watch, bracelets, and rubber bands or intrinsically by tumors, traumas, fibrous bands, and anomalous muscles.

Symptoms are such as paresthesia on the dorsoradial aspect of the hand, with worsening with passive wrist flexion with ulnar deviation and wrist in pronation (for 1 min), or active forearm supination and wrist extension against resistance (30 s). The Tinel sign can be positive in the region proximal to the radial styloid (9 cm). If positive over the radial styloid, its compression is just of the dorsal branches. The Finkelstein test can also be positive for performing nerve traction [10].

The use of complementary exams is also restricted in this case, with electroneuromyography of low value. Local anesthetic block can be performed with lidocaine, which can lead to temporary improvement of the condition.

18.6 Median Nerve

The median nerve can be compressed within the elbow region by the supracondylar process, by the Struthers' ligament, in the lacertus fibrosus, in rounded pronator muscle deep head or in the flexor arch. It can even be compressed by vascular malformations, anomalous muscles, and synovial and bursal strains. Distal humerus fractures and elbow dislocation also can cause lesion of the median nerve.

The passage of the nerve through the elbow region has close relationship with these anatomic structures, bringing the possibility of being affected by one or more of these structures. It can present motor and sensitive symptoms combinations, showing symptoms in the elbow, hand, and forearm region [11].

18.7 Pronator Syndrome

This syndrome consists in the compression of the median nerve in the elbow region and also in the proximal part of the forearm. The symptoms are always vague such as: pain, tingling, numbness, tiredness or fatigue, forearm discomfort with proximal irradiation. Laboral or sportive activities with pronosupination repetitive movements can trigger the symptoms.

Generally, the symptoms are developed insidiously, but occasionally a specific event or sudden onset of pain in the forearm are related to the bigger susceptibility of the muscle stress.

It is estimated that ±5% of the symptoms related to median nerve are directly caused by the pronator teres syndrome, some studies suggest even higher percentages [12]. Misunderstood, it is often misdiagnosed as an atypical carpal tunnel syndrome. However, unlike carpal tunnel syndrome, its symptomatology is more related to physical activities, night tingling isn't an important complain and Phalen sign is not present.

The pronator syndrome is characterized by the median nerve compression at the elbow and proximal forearm. Four potential compression sites include the supracondylar process and ligament of Stüthers, the lacertus fibrous, the rounded pronator, and the superficial flexor digitorum superficialis ach.

The compression region most proximal and less usual is the humerus supracondylar process. This bone process, existent in approximately 1–3% of the population, stems from the anteromedial aspect of the distal humerus, proximal 5 cm to the medial epicondyle. The Struthers' ligament is the fibrous band that can arise from the supracondylar process and attach to the medial epicondyle, forming a fibro-bone tunnel through which the median nerve crosses. The entrapment of the median nerve inside this tunnel is also called supracondylar process syndrome [13].

The lacertus fibrosus arises from the distal bicipital tendon and inserts into the antebrachial fascia, crossing the flexor pronator muscle group. The thickened lacertus can produce median nerve compression.

The pronator syndrome compression most frequent region is between the superficial heads (humeral) and deep (ulnar) of the rounded pronator, located from 2 to 4 cm distal to the medial epicondyle. It can be caused by muscle hypertrophy, fibrous adhesions, or other teres pronator anomalies [14].

The physical exam should be always comparative to the contralateral side. The hand's palm symptoms are more related to higher compression sites out of the carpal tunnel once the sensitive branch to this hand's region raises before the carpal tunnel.

Special care must be taken because the forearm pronation can cause tighting of lacertus fibrosus at the forearm, entrapping the median nerve. The extension of the elbow can increase compression on the humeral head of the pronator teres muscle, once it is a biarticular muscle.

The complete understanding related to the main compression site can be a difficult task.

During the physical examination, one needs to be aware in order to keep a suitable position avoiding median nerve compression on the carpal tunnel.

Symptoms can be reproduced by the following tests:

#Middle finger proximal interphalangeal flexion against resistance (Fig. 18.2). The authors use to do this test with the patient's forearm on a table. This test will tension flexor digitorum superficialis arch, another compression site.

Pronation against resistance (Fig. 18.2) for 30 s can reproduce the symptoms. The authors use to do this test with the patient's forearm on a table, with the elbow flexed, making lacertus fibrous relaxed, the same test can be done also with elbow extension to improve tension of the humeral head of the teres pronator.

#The elbow flexion against resistance with supine forearm may trigger symptoms due to pressure from lacertus fibrosus (Fig. 18.3). One needs to be aware because if wrist is hyperextended the flexopronator mass can be overstretched, compressing the teres pronator site.

Direct digital compression by the examiner on the pronator proximal region, approximately 5 cm distal to the antecubital fossa (elbow pit) while making resistance to active pronation. This manouver can reproduce the neurologic symptoms of median nerve entrapment.

The weakness of innervated median muscles is uncommon, but it is indicated the comparison between the two hands strength. The pollicis longus flexor and the digitorum profundus index finger flexor are probably those which will present more evident weakness.

Tinel signal can be present on the compression sites.

It is important to differentiate the simple compression from the double crush syndrome of the

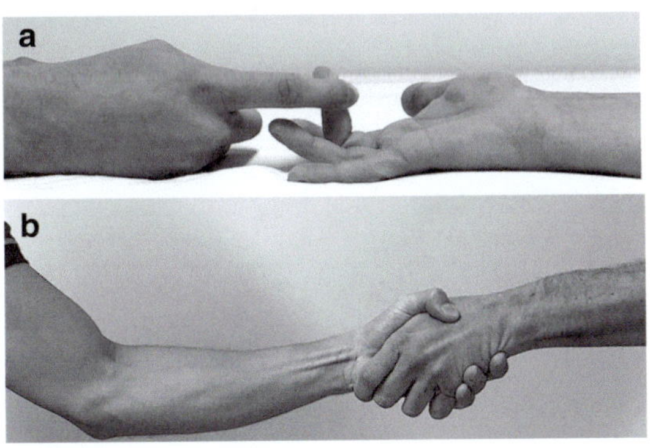

Fig. 18.2 (**a**) Middle finger proximal interphalangeal flexion against resistance, (**b**) pronation against resistance

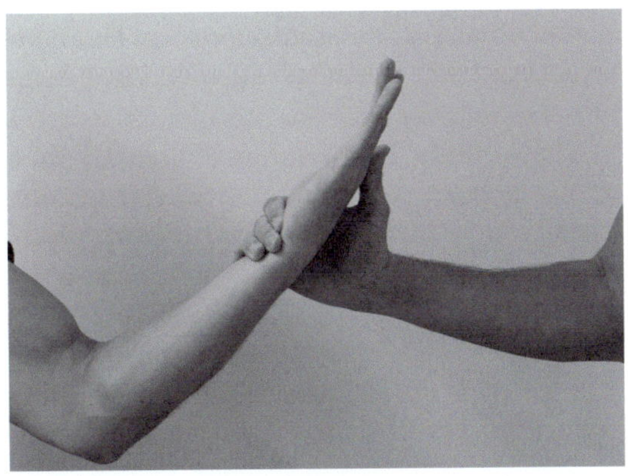

Fig. 18.3 Flexion against resistance, the examiner leaves the carpal tunnel free during this test

median nerve, thus cervical and wrist examination are also necessary.

As a dynamic compression, electroneuromyographic tests in can sometimes show not significant alterations, however, when present will need to be considered [15].

Ultrasound can allow dynamic, real-time visualization of nerves and also provide the means to precisely locate anatomical site of nerve compression. In addition, ultrasound can provide the examiner with more information about any underlying condition like ganglion cysts, lipoma, formation of neuroma, or epineural hematoma [16].

The pronator syndrome can also lead to forearm pronator flexor muscle atrophy, including the rounded pronator, the radial carpal flexor, long palmar, and superficial flexor of the fingers.

Nerve sheath tumors such as schwannoma and neurofibroma can also be related to nerve compression symptoms.

18.8 Anterior Interosseous Nerve

The Anterior Interosseous Nerve (AIN) can be compressed in anatomical sites such as: the supracondylar process at the distal humerus, within the Strüthers' ligament of the supracondylar process, by the aponeurosis of the biceps (lacertus fibrosus), between the superficial and deep heads of the teres pronator muscle (most common site), arcade of the superficial flexor digitorum superficialis, Gantzer muscle and at the Martin–Gruber anastomosis site, which is an anastomosis region that interconnects the motor branches of the median and ulnar nerves, present in 7–23% of the population.

The Strüthers' ligament is present in 0.6–2.7% of the population and is a fibrous band that extends from the anteromedial aspect of the humerus to the medial epicondyle. Compression by the Strüthers' ligament is responsible for 0.5% of all cases. This ligament can originate in the supracondylar process, when present in 1% of the population [4].

The AIN is the terminal motor branch of the median nerve, being responsible for innervating the second and third flexor digitorum profundus tendons, flexor pollicis longus, and pronator quadratus.

The compression of the true AIN has only motor deficit. There will be no sensitivity change in the forearm. The complaint will be of ill-defined pain in the forearm and a report of weakness of the thumb and forefinger [17].

In the physical examination, the main test to be performed is the "ok" sign test, where the patient is unable to perform the OK sign (touching the tip of the thumb to the point of the indicator), performing flexion of the distal phalanx of the thumb and the index finger. The impairment of the flexor digitorum profundus tendon and the second flexor digitorum profundus tendon will impede the patient to flex these tendons. A typical incapacity of these tendons' flexion during the ok is named as the Kiloh–Nevin sign (Fig. 18.4) [18].

The pronator quadratus muscle can be examined separately by pronation against resistance with the elbow in flexion, in order to less tension on the humeral head of the teres pronator. Its examination needs to be comparative to the contralateral side. Other tests can be performed to help identify the most accurate location of the compression. Lacertus fibrosus can be assessed

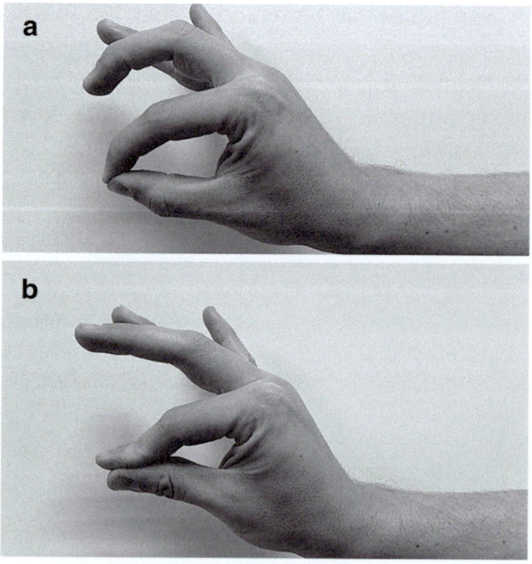

Fig. 18.4 Kiloh–Nevin test: (**a**) anterior interosseous nerve intact. (**b**) Impaired or lesioned anterior interosseous nerve

with supine resistance flexion. The pronator is round, with the pronated counter-resistance in flexion. The flexor arch, with the flexion against resistance of the intermediate phalanx of the annular finger.

Electromyographic studies are essential, which should show normal for sensitive conduction of the median nerve and changes to the quadratum pronator muscle, long flexor pollicis, and flexor digitorum profundus tendon and the second flexor digitorum profundus tendon, with tapered waves, fibrillations, and driving latency. MRI may show muscles with signs of edema and denervation.

18.9 Ulnar Nerve

The cubital tunnel syndrome is the second most common compressive syndrome of the upper limb, second only to the carpal tunnel syndrome.

The ulnar nerve is localized medially, it raises from the Strüthers ach on the medial aspect of the arm. At the elbow, it passes at posterior to the medial epicondyle at the ulnar groove, between the medial epicondyle of the humerus and the olecranon, through the cubital tunnel [19].

In the cubital tunnel, the nerve is permanently subjected to compressive effects each time the elbow is flexed.

The most proximal compression locations are the Arcade of Strüthers and the medial edge of the intermuscular septum. Less frequent compression locations include Osborne's fascia, a fibrous band that connects the proximal edge of the flexor carpi ulnaris muscle to the medial epicondyle and aponeurosis of the flexor pronator muscles [20].

The ulnar nerve is not a fixed structure and needs to move freely both longitudinally and medially during elbow movement.

Two major age ranges are described for cubital tunnel syndrome. The first is between 20 and 30 years old, predominantly secondary to the trauma. The second is developed between 50 and 60 years old, associated with degenerative disease.

Several factors can increase pressure on the ulnar nerve at the elbow. Postural addiction with flexion when sleeping, the hypermobility of the nerve at the cubital tunnel promoting subluxation in the medial epicondyle during the flexion. Other etiological factors of compression are muscle anomalies, such as epitrochlear anconeus, tumors, ganglia, cubit valgus deformity, sequelae of elbow fractures and dislocations, elbow arthrosis, thickening of the Arcade of Struthers, as well as sports [21].

Chronic compression may be secondary to mass lesions, including bursae, ganglia, synovitis, bruise, osteophytes, calcifications, and ectopic ossifications. In athletes, lateral displacement of the ulna secondary to chronic medial collateral ligament laxity or lesion, and cubitus valgus deformity can also cause cubital tunnel syndrome [19].

The clinical picture is characterized by intermittent numbness and tingling in the ulnar nerve autogenous area, being related to the shoulder and elbow position increasing the pressure in the nerve and its symptomatology [22].

Patients can describe a difficulty in fine motor tasks, such as buttoning. Crossing fingers can be difficult because of interosseous weakness. Hypotrophy or atrophy of the intrinsic muscles and adductor pollicis, with ulnar claw, weakness of other extrinsic muscles innervated by the ulnar can be also present.

Pain is not a frequent complaint, there is no change in forearm sensibility. However, less sensibility, tingling, and numbness in medial part of the annular finger and all the little finger are the frequent symptoms [6].

Tinel can be positive and helpful in delimitating the main compression area.

The elbow flexion for 30 s may also reproduce sensitive symptoms in the patient. It happens because the ulnar nerve passes posterior to the elbow's rotation center. Pressures within the cubital tunnel can increase until to sevenfold during elbow flexion.

Collateral medial ligament tests are also important because lesions or attenuation of this ligament can also stretch the nerve when a valgus force is done on the elbow. Valgus stress test will be done with 30° elbow flexion. It is strongly suggested to make this test in supination and pronation because in supination a false positive can take place when the patient presents posterolateral rotatory instability.

Ulnar nerve dislocation can be visible, usually during the elbow flexion. When one has difficulties on visualization one needs to full extends the patient's elbow, feel the medial epicodyle, using second and third fingers, and then the patient flexes his elbow. It is possible to feel the nerve dislocation using this maneuver. Snapping triceps can also be felt or visualized with this same maneuver.

X-ray examen can be useful in the presence of bone lesions or deformities. The electroneuromyography is usually very helpful; however, for some cases of dynamic or light compression it may not be altered. Ultrasound dynamic exams can make possible to determine ulnar dislocations and nerve narrowing. Unlike radial and median nerve neuropathies, ulnar neuropathy usually presents with increased ulnar nerve signal on T2-weighted examination. This change in the signal may be better appreciated in the ulnar nerve because of its larger size. Other pathological processes related to cubital tunnel neuropathy are identified by MRI imaging, including osteoarthritis, synovitis, valgus deformity, anomalous muscles, and tumors [19].

References

1. Bassett FH 3rd, Nunley JA. Compression of the musculocutaneous nerve at the elbow. J Bone Jt Surg Am. 1982;64(7):1050–2.
2. Naam NH, Massoud HA. Painful entrapment of the lateral antebrachial cutaneous nerve at the elbow. J Hand Surg Am. 2004;29:1148–53.
3. Felsenthal G, Mondell DL, Reischer MA, Mack RH. Forearm pain secondary to compression syndrome of the lateral cutaneous nerve of the forearm. Arch Phys Med Rehabil. 1984;65:139–41.
4. Morrey BF, Sotelo JS, Morrey ME. The elbow and its disorders. 5th ed. Philadelphia: Elsevier; 2018.
5. Dang AC, Rodner CM. Unusual compression neuropathies of the forearm, part I: radial nerve. J Hand Surg Am. 2009;34(10):1906–14.
6. Kumar SD, Bourke G. Nerve compression syndromes at the elbow. Orthop Trauma. 2016;30(4):355–62.
7. Nicolle FV, Woolhouse FM. Nerve compression syndromes of the upper limb. J Trauma. 1965;5:313–8.
8. Roles NC, Maudsley RH. Radial tunnel syndrome: resistant tennis elbow as a nerve entrapment. J Bone Jt Surg Br. 1972;54(3):499–508.
9. Moradi A, Ebrahimzadeh MH, Jupiter JB. Radial tunnel syndrome, diagnostic and treatment dilemma. Arch Bone Jt Surg. 2015;3(3):156–62.
10. Lanzetta M, Foucher G. Entrapment of the superficial branch of the radial nerve (Wartenberg's syndrome). A report of 52 cases. Int Orthop. 1993;17:342–5.
11. Dang AC, Rodner CM. Unusual compression neuropathies of the forearm, part II: median nerve. J Hand Surg Am. 2009;34(10):1915–20.
12. Asheghan M, Hollisaz M, Aghdam AS, Khatibiaghda A. The prevalence of pronator teres among patients with carpal tunnel syndrome: cross sectional study. Int J Biomed Sci. 2016;12(3):89–94.
13. Pardini AG, Freitas AD. Cirurgia da mão—Lesões não traumáticas. 2nd ed. Rio de Janeiro: Med Book; 2008.
14. Strohl AB, Zelouf DS. Ulnar tunnel syndrome, radial tunnel syndrome, anterior interosseous nerve syndrome, and pronator syndrome. J Am Acad Orthop Surg. 2017;25(1):e1–e10.
15. Freedman M, Helber G, Pothast J, Shahwan TG, Simon J, Sher L. Electrodiagnostic evaluation of compressive nerve injuries of the upper extremities. Orthop Clin N Am. 2012;43(4):409–16.
16. Babaei-Ghazani A, Roomizadeh P, Nouri E, Raeisi G, Yousefi N, Asilian-mahabadi M, et al. Ultrasonographic reference values for the median nerve at the level of pronator teres muscle. Surg Radiol Anat. 2018;40(9):1019–24.
17. Akhondi H, Varacallo M. Anterior interosseous syndrome. Treasure Island (FL): StatPearls Publishing; 2020.
18. Kiloh LG, Nevin S. Isolated neuritis of the anterior interosseous nerve. Br Med J. 1952;1(4763):850–1.
19. Bordalo-Rodrigues M, Rosenberg ZS. MR imaging of entrapment neuropathies at the elbow. Magn Reson Imaging Clin N Am. 2004;12(2):247–63.
20. Doughty C, Bowley M. Entrapment neuropathies of the upper extremity. Med Clin N Am. 2019;103(2):357–70.
21. Dy CJ, Mackinnon SE. Ulnar neuropathy: evaluation and management. Curr Rev Musculoskelet Med. 2016;9(2):178–84.
22. Wolfe S, Scott W, Pederson W, Kozin SH, Cohen M. Green's operative hand surgery. 7th ed. Philadelphia: Elsevier; 2016.

19 Evaluation of Common Tendinopathies of the Elbow

Alessandro Marinelli, Catello Buondonno, Ahmad Al Zoubi, and Enrico Guerra

19.1 Introduction

Elbow tendinopathies represent a common cause of pain and disability, mainly in manual workers or athletes in their 35–55 years. Based on the location, elbow tendinopathies can be classified as lateral (affecting the common extensor origin), medial (affecting the flexor-pronator muscles origin), anterior (affecting the biceps tendon insertion), and posterior (affecting the triceps tendon insertion). Tendinopathies include *traumatic forms*, usually with acute onset, presenting some degree of tendon tears (from partial to complete lesion) and *degenerative forms,* usually with chronic onset, where the tendon is generally continuous, but shows more or less severe degrees of degeneration. Usually, both acute or traumatic forms can be diagnosed through the patient medical history and the execution of specific provocative examination manoeuvres, without a systematic need of imaging.

19.1.1 Medical History

In elbow tendinopathies, it is important to take an accurate medical history that needs to investigate

- *Patient features*: age, working, sporting and recreational activities, dominant side, compensation claims.
- *Past medical history*: rheumatic diseases, metabolic disorders, drugs use (anabolic hormones, steroids, fluoroquinolones).
- *Characteristic of the pain*: onset (traumatic or insidious), mechanism of injury (if present), type of pain (sharp or dull), unilateral or bilateral elbow involvement, time lasting, pain severity, clinical impairments, activities triggering pain, pain relief modalities and response to medical therapy, evolution of the symptoms over time (Table 19.1).

19.1.2 Clinical Examination

With clinical examination we need to evaluate

- Local conditions: swelling, bruises, anatomical shape modification, precise pain localization and its irradiation, presence of specific trigger points, elbow range of motion and strength. The comparison with the contralateral limb is always recommended.

A. Marinelli (✉) · C. Buondonno · A. Al Zoubi · E. Guerra
Shoulder and Elbow Unit, IRCCS, Istituto Ortopedico Rizzoli, Bologna, Italy
e-mail: alessandro.marinelli@ior.it

J. G. Lane et al. (eds.), *The Art of the Musculoskeletal Physical Exam*,
https://doi.org/10.1007/978-3-031-24404-9_19

Table 19.1 Five useful questions to investigate the diagnosis of elbow tendinopathies

How?	How did the pain start? How was the injury mechanism?
Where?	Where does it hurt? Can you point to the exact spot that hurts?
When?	When did the pain begin? Does it come and go or is it constant?
How much?	How much is the pain? Does it prevent you from participating in your normal activities?
What?	What activities make your pain worse? What medications or treatments make it better?

- Elbow stability, range of motion, strength testing, and neurovascular function
- Conditions in adjacent joints: to detect causes of irradiated pain (like cervical radiculopathy) or predisposing conditions causing elbow tendinopathy (like shoulder stiffness).
- Specific evocative tests: these tests are a keypoint for elbow tendinopathies diagnosis. They have been developed to evoke pain in the specific muscle-tendon unit under investigation, through its specific active contraction or with its selective passive stretching. The comparison with the contralateral limb is always useful.

The stronger is the agreement among medical history, physical examination, and evocative tests, more reliable is the diagnosis of elbow tendinopathy.

Even if in most cases imaging, as plain X-ray, ultrasonography, and magnetic resonance imaging, is not usually required for the diagnosis, it is often requested especially in atypical or persistent cases for diagnosis confirmation, to rule out concomitant causes of elbow pain, to evaluate the severity of the lesion, and to follow up its evolution over time.

Table 19.2 Etiologies of the pain caused by common elbow tendinopathies, with their possible differential diagnosis

Pain location	Common tendinopathies	Differential diagnosis
Lateral	Lateral epicondylitis (tennis elbow)	• Cervical radiculopathy • Radiocapitellar arthrosis • Radial tunnel syndrome • Posterolateral rotatory instability • Synovial plica • OCD capitulum humeri • Panner disease
Medial	Medial epicondylitis (golfer's elbow)	• Medial instability • Cubital tunnel syndrome • Snapping triceps syndrome • Ulnar collateral ligament tear
Posterior	Triceps tendon injury	• Olecranon bursitis
Anterior	Distal biceps tendon injury	• Bicipitoradial bursitis

Elbow tendinopathies are the most common causes of elbow pain; however, we should not forget other possible etiologies (Table 19.2).

19.2 Lateral Epicondylitis (Tennis Elbow)

Lateral epicondylitis (LE) is the most common cause of lateral elbow pain. It consists of a symptomatic tendinosis of the short carpal radial extensor (ERBC) and of the aponeurosis of the common finger extensor at the level of the lateral epicondyle of the elbow.

19.2.1 Clinical Presentation

The typical patient is a middle age person subjected to repetitive movements, hand-arm vibration and awkward postures, as manual workers or recreational athletes. There is an equal gender distribution and the dominant extremity is more frequently affected.

Patients with lateral epicondylitis present pain at or around the bony prominence of the lateral epicondyle that often radiates down to the forearm in line with the common extensor muscle mass, especially during activities involving forearm supination and wrist extension.

The pain can vary in each patient from an intermittent and mild ache to a constant, severe, and sharp pain, causing a disturbance in sleep and limiting the grip strength and sometimes the last degrees of the elbow extension during daily activities.

Usually, lateral epicondylitis starts with insidious onset and gradual progression of the pain.

19.2.2 Physical Examination

It is important to perform a physical examination of the entire upper extremities, beginning from a cervical spine evaluation, moving to the shoulder, then elbow, wrist and hand, followed by comparison with the unaffected, contralateral extremity.

Examination typically reveals localized soreness over the common extensor origin 0* palpation of the lateral epicondyle, especially just anterior and distal to the lateral epicondyle—at the origin of the extensor carpi radialis brevis (ECRB)—elicits pain that often radiates along with the extensor muscle mass.

Sometimes skin hypopigmentation and soft tissue atrophy can be evident on the lateral epicondyle if multiple cortisone injections have been previously performed.

Other diagnoses should be considered if the patient is younger than 25 years or older than 65 years of age, or the onset of the pain is clearly due to an acute traumatic event, or pain is referred at the soft spot level or more distally along the forearm instead of at the lateral epicondyle, or, if during forearm rotation, a crepitus can be detected.

19.2.3 Specific Examination Manoeuvres

To evaluate lateral epicondylitis many specific physical examination manoeuvres have been described. The tests are better performed with the patient comfortably seated with both arms exposed.

- **Cozen's test**: The patient is positioned with the arm forward, the elbow fully extended, the wrist extended, the forearm pronated. The examiner resists to the dorsal flexion of the wrist. This test stresses the whole of the common extensor origin [1]. If the patient holds the wrist in radial deviation, the extensor carpi radialis brevis and longus are selectively activated and the test is even more accurate (Fig. 19.1).

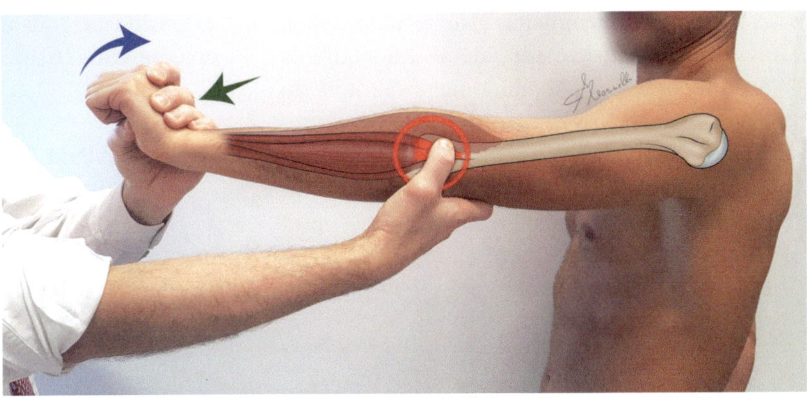

Fig. 19.1 Cozen's test: the resisted wrist extension with radial deviation and full pronation can be considered one of the best tests to confirm the diagnosis of LE. The pain is typically exacerbated by gentle pressure over the lateral epicondyle

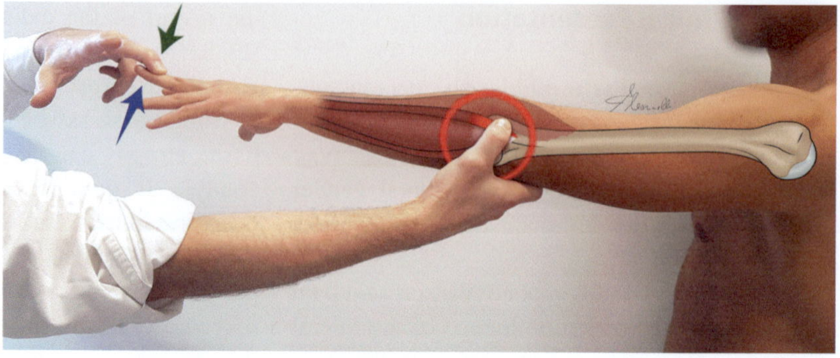

Fig. 19.2 Maudsley's test: the resisted middle finger extension causes pain at the lateral epicondyle in case of LE. The pain is exacerbated by a gentle local pressure over the lateral epicondyle. In case of radial tunnel syndrome, the pain is typically located some centimetres more distal

- **Maudsley's test** (or **resisted middle finger extension test**): The patient is positioned with the arm forward, the elbow extended, the wrist in neutral position, the forearm pronated. The examiner resists to the dorsal flexion of the middle finger: the test elicits pain on the ECRB tendon (Fig. 19.2) and it is particularly painful if a radial tunnel syndrome is associated [2].
- **Mill's test:** The patient is positioned with the arm along the body, the elbow extended with the forearm in full pronation. The examiner flexes the patient's wrist. The onset of pain over the lateral epicondyle suggests ECRB tendinosis [3] (Fig. 19.3).
- **Polk's test**: The patient is asked to grab an object (about 2.5 kg) with the elbow flexed to 100° and forearm pronated [4]. The test is called Laptop test if the raised object is a notebook computer (Fig. 19.4).
- **Chair pick up test**: The patient is asked to lift a chair, placed in front of him/her, with the forearm pronated and a partially extended elbow, using the first three fingers [5] (Fig. 19.5).
- A loss of grip strength has also been described as a diagnostic test for lateral epicondylitis and the use of a dynamometer permits to quantify the relative impairment:
- **Grip strength test** (*sens. 80%; spec. 85%*): The patient is asked to squeeze the dynamometer as strong as possible. Maximal grip strength can be reduced to almost 50% if the test is performed with the elbow in full extension; however, just a reduction in strength of approximately 8% between flexion and extension is considered indicative of lateral epicondylitis [6].

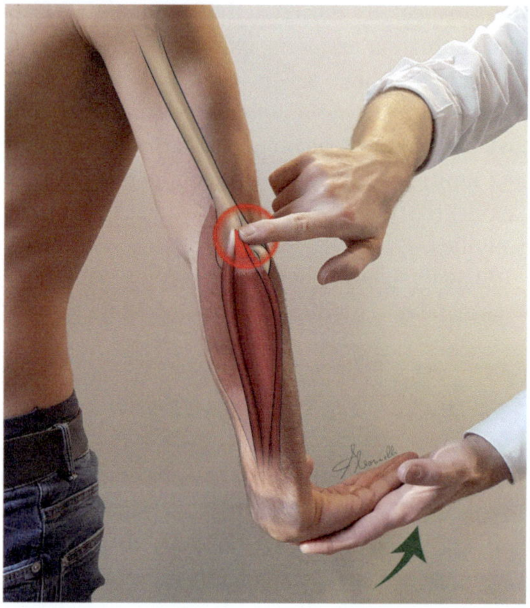

Fig. 19.3 Mill's test: in case of LE, a passive wrist flexion movement causes pain at the lateral epicondyle. The pain is exacerbated by a gentle local pressure

19.2.4 Possible Associated Symptoms

Lateral epicondylitis is often associated with other clinical disorders, like radial tunnel syn-

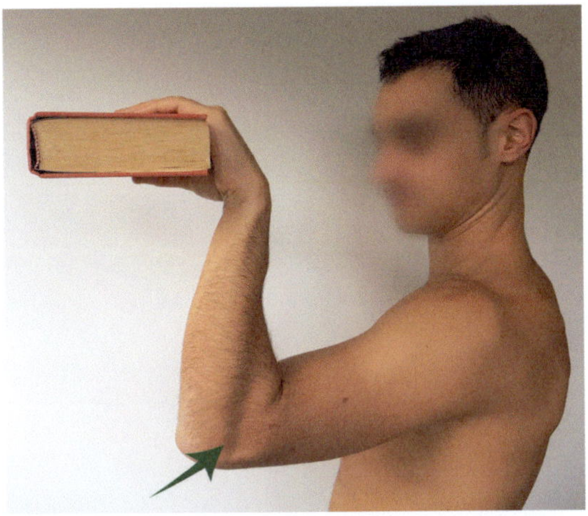

Fig. 19.4 Polk's test for LE: grasping a relatively heavy object (2–3 kg) with the elbow flexed and the forearm pronated, the extensors of the wrist are stressed. In case of L.E. this test causes pain at the lateral epicondyle

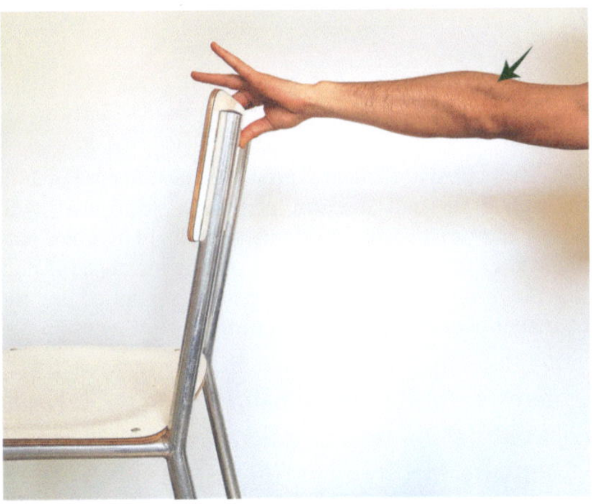

Fig. 19.5 Chair pick up test: in case of LE, lifting the back of a chair with a three-finger pinch (thumb, index, and long fingers) and the elbow fully extended elicits pain at the lateral epicondyle

drome (entrapment of the posterior interosseous nerve) or, cervical radiculopathy, homolateral shoulder stiffness or scapular dyskinesis, which affecting elbow kinematics, can cause elbow over-use.

19.3 Medial Epicondylitis (Golfer's Elbow)

Medial epicondylitis (ME) is a degenerative tendinopathy of the flexor-pronator muscles origin at the level of the medial epicondyle.

19.3.1 Clinical Presentation

Medial epicondylitis usually affects middle aged athletes or workers involved in repetitive wrist flexion and forearm pronation activities.

Patients with medial epicondylitis typically present a subtle onset of pain at the medial aspect of the elbow that often radiates down to the forearm, especially during activities involving forearm pronation and wrist flexion. The pain varies in each patient from a mild and intermittent ache to constant and severe sharp pain.

19.3.2 Physical Examination

Examination typically reveals localized tenderness at the origin of the flexor-pronator mass on the medial epicondyle, exacerbated by resisted wrist flexion performed with the elbow extended and forearm supinated.

19.3.3 Specific Examination Manoeuvres

Several tests have been described to elicit pain in case of medial epicondylitis. These tests are performed with the patient comfortably seated with both arms exposed.

- **Reverse Mill's test**: The patient is positioned with the arm forward, the elbow extended, and the forearm supinated. The examiner passively moves the wrist in dorsal flexion (Fig. 19.6).

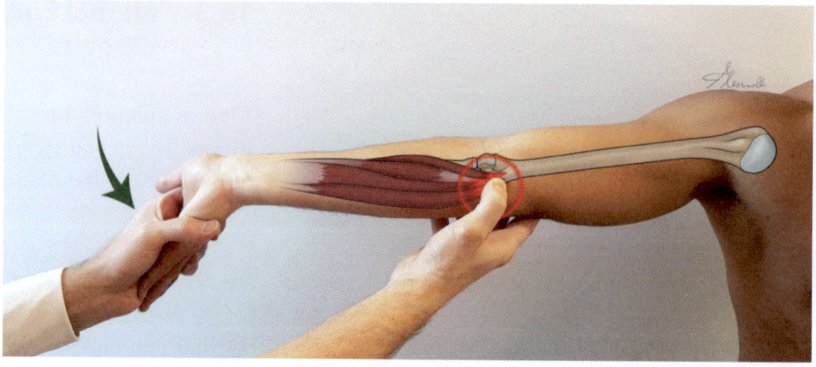

Fig. 19.6 Reverse Mill's test: in case of ME the passive wrist extension, performed with elbow extended and forearm supinated, causes pain at the medial epicondyle, exacerbated by gentle local pressure

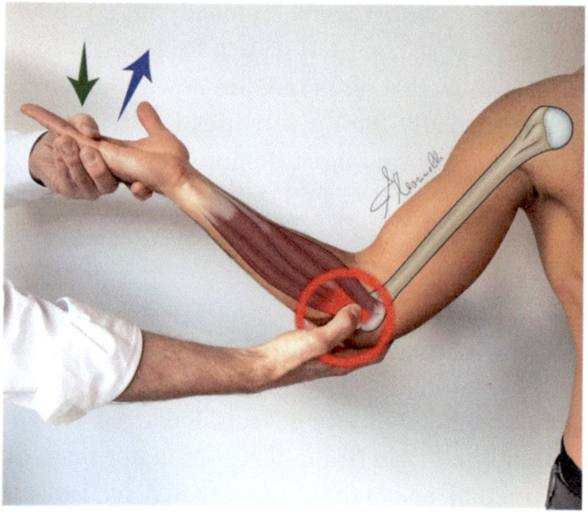

Fig. 19.7 Resisted wrist flexion test: in case of ME the active resisted wrist flexion performed with elbow and wrist partially extended, elicits pain at the medial epicondyle, exacerbated by gentle local pressure

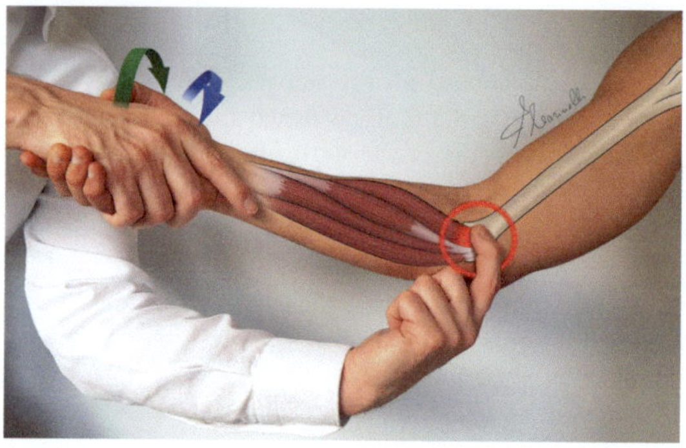

Fig. 19.8 Resisted forearm pronation test: in case of ME, the active resisted forearm pronation performed with elbow and wrist partially extended elicits pain at the pronator teres tendon insertion. The pain is increased if during the test a gentle local pressure is applied by the examiner over the tendon's insertion

- **Resisted wrist flexion test:** The patient is positioned with the arm forward, the elbow and wrist extended and forearm supinated. The patient is asked to actively flex the wrist, against the resistance of the examiner who, in the meanwhile, palpate with his/her thumb the insertion of the patient's flexors mass (Fig. 19.7).
- **Resisted forearm pronation**: The patient is positioned with elbow in 90° of flexion. The examiner hand grasps the patient's hand in a handshake position, while the index finger of the opposite hand rests over the medial part of the tendon insertion on the medial epicondyle. The patient is asked to actively pronate the forearm while the examiner holds resistance, maintaining the hand in neutral position. If this test, that selectively activates the pronator teres muscle, is more painful than the resisted wrist flexion test, it indicates a greater pronator teres involvement (Fig. 19.8).
- **Polk's test**: The patient is asked to grab an object (about 2.5 kg), in front of him/her, with a flexed elbow and forearm supination [4] (Fig. 19.9).
- **Cheek test:** The patient is asked to press his own cheeks with their fingers, keeping shoulders abducted. The pain at the medial epicondyle is caused by the contraction of the flexor-pronator mass [7] (Fig. 19.10).

19.3.4 Possible Associated Symptoms

Medial epicondylitis symptoms can be associated to lateral epicondylitis or to cubital tunnel

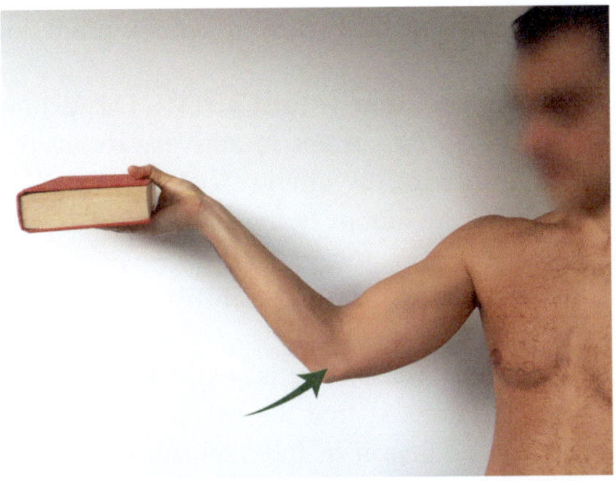

Fig. 19.9 Polk's test for ME: grasping a relatively heavy object (2–3 kg) with the elbow flexed and the forearm supinated puts under stress the wrist flexors muscles, causing pain at the medial epicondyle in case of ME

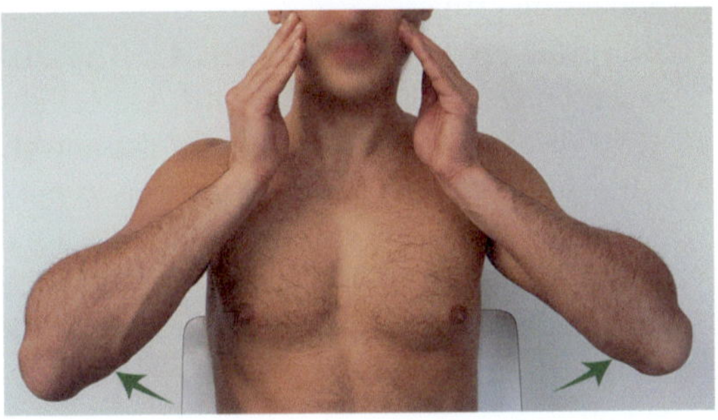

Fig. 19.10 Pressing own cheeks, while keeping the shoulder partially abducted, causes contraction of the flexor-pronator mass and elicits pain at the medial epicondyle

syndrome. Tinel's sign and a full neurological examination, including sensory and motor assessment, permit to rule out ulnar nerve neuropathy. Ulnar collateral stability should also be assessed.

19.4 Distal Biceps Tendinopathy

Injuries to the distal biceps tendon are relatively common and are usually due to traumatic or micro-traumatic lesions, more or less severe, of the tendon at its bone insertion.

The most common lesions are traumatic and acute complete tendon tears, with or without a lacertus fibrosus rupture. Less frequently, the tendon presents a partial thickness that can be insertional or intrasubstance.

19.4.1 Clinical Presentation

Patients with complete distal biceps tear are typically muscular and middle-aged men, between 35 and 55 years, reporting an uncontrolled eccentric load that has led to a forced elbow extension while the bicep was actively contracting. Patients usually report a sudden, painful "pop" at the time of injury followed by the development of a dull ache. The diagnosis of complete distal biceps tendon tear can usually be established only based on patient history and physical examination. However, an intact lacertus fibrosus can make the proximal migration of the muscle belly less evident, an excellent strength conservation in very muscular patients can make difficult the perception of flexion weakness, the absence of significant pain or visible hematoma can hide the severity of the tendon lesion. On the other hand, great attention should be given to avoid missed or delayed diagnoses because late surgery makes reinsertion more difficult, with a higher complication rate.

19.4.2 Physical Examination

A patient with complete tendon rupture typically presents a "Popeye" deformity that is a visible flattening of the distal muscle contour of the arm due to the proximal retraction of the biceps muscle belly. When not easily noticeable, the crease-to-biceps distance between the elbow flexion crease and the round biceps muscle belly can be compared with the opposite arm to confirm the diagnosis.

The presence of ecchymosis in the distal arm and proximal forearm suggests an acute and complete injury although sometimes it does not appear until days after the insult. In partial ruptures and tears, usually ecchymosis may never develop due to confinement of the hematoma by an intact bicipital aponeurosis.

ROM and strength should be assessed compared to the contralateral extremity. Typically, the patient complains of pain in the affected arm at the antecubital fossa during full extension and supination.

Reduction of strength and pain are typically noticed with resisted elbow flexion and even more with forearm supination.

19.4.3 Specific Examination Manoeuvres

- **Hook test:** The patient is asked to look at the palm of his/her hand on the affected side keeping the forearm in active supination, with the shoulder elevated and the elbow flexed at 90°. An intact distal bicep tendon allows the examiner to hook his/her finger around the lateral side of the distal biceps tendon of the patient. If the bicep is torn, the examiner cannot hook his/her finger around any anterior structures [8] (Fig. 19.11). The absence of the tendon compared to the contralateral arm is a reported to have 80/100% of sensibility and 100% of specificity for complete biceps tendon tear.
- However, care must be taken during direct palpation of the tendon in the antecubital fossa as an intact bicipital aponeurosis may be misleading. The presence of a cord-like structure cannot exclude a partial tendon tear.
- **Passive forearm rotation test:** The elbow of the patient is flexed at 90°, relaxed on the patient's side. The examiner with one hand rotates the forearm, while with the other hand palpates the biceps muscle: the absence of proximal excursion during supination and distal excursion during pronation is a positive test for total distal biceps rupture, with a reported sensibility of 95% and specificity of 100% [9].
- **Active forearm rotation test:** The examiner asks the patient to actively rotate the forearm with the elbow flexed at 60°–70°. Lack of the biceps belly migration is a positive test for total distal biceps rupture, with a reported sensibility of 100% (Fig. 19.12).
- **Lag Sign:** The patient is seated with the arm over a table. At first the examiner holds the patient's forearm in full supination and asks the patient to hold this position. Then the examiner releases the forearm and assesses if some degrees of pronation occur. If this happens, it means that the pronation forces exceed the supination forces, as a result of biceps tendon lesion [10].
- **Bicipital crease interval (BCI):** The distance between the antecubital fossa (elbow at 90° of flexion) and the start of the muscle belly is calculated (normal value is 6 cm). A BCI greater than 6 cm indicates a total distal biceps rupture [11]. It has been reported that this test presents a 90% of sensibility and from 50 to 100% of specificity.

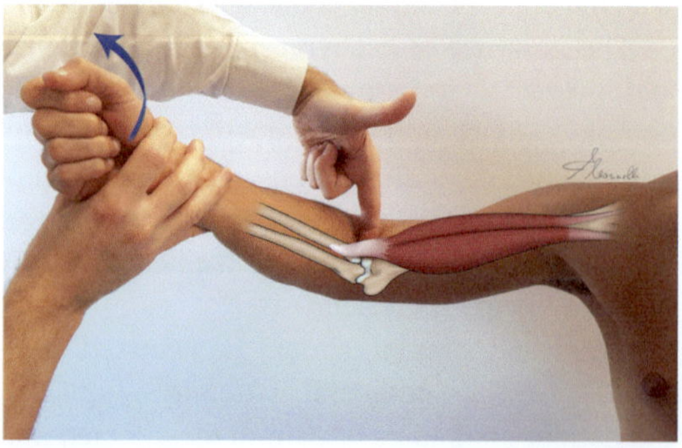

Fig. 19.11 Hook test: the index of the examiner is used to hook the distal insertion of the biceps tendon of the patient, while the patient actively supinates the forearm. If the distal biceps tendon of the patient is intact, a strong cord-like structure can be easily palpated

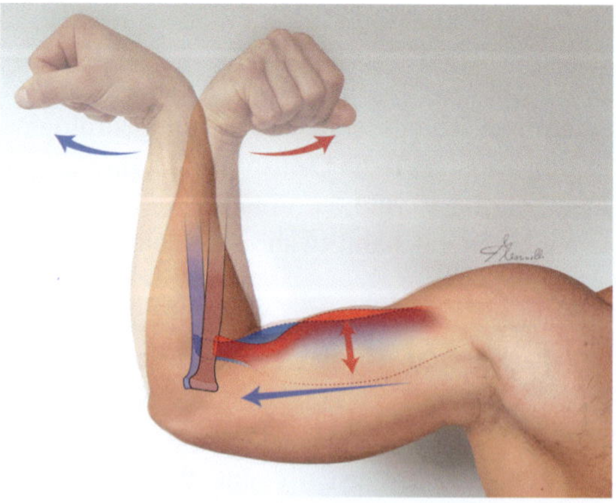

Fig. 19.12 Active forearm rotation test: if the distal biceps tendon is intact, keeping the elbow flexed and rotating the forearm, the arm changes its contour. During the forearm rotation, due to the relative movement of the radial tuberosity, the biceps muscle belly gets longer and stretched with forearm pronation and gets shorter and dumpy in supination

- **Bicipital crease ratio (BCR)** The ratio between the BCI in both arms is calculated. The normal value from 1 to 1.2. A BCR value greater than 1.2 indicates a total distal biceps rupture. This test presents a 96% of sensibility and 80% of specificity.
- **The biceps squeeze test** is similar to the Thompson test for Achilles tendon ruptures. The patient is seated and the forearm rests on a table, slightly pronated and elbow flexed to 60°–80°. The clinician squeezes the distal part of the belly biceps brachii. The lack of forearm supination suggests complete rupture [12]. This test, even if it has been reported to present a good sensitivity, in our experience is not very reliable.
- **Bicipital Aponeurosis Flex Test:** the bicipital aponeurosis, if present, can be felt on the medial side of the elbow while the patient flexes the elbow at around 75°, flexes the wrist, supinates the forearm, and with the hand closed into a fist isometrically contracts the biceps [13].

19.5 Triceps Tendinopathies

Triceps tendinopathy, that generally occurrs in male muscular athletes or manual workers, represents the least common type of elbow tendinopathies.

Triceps tendinopathies are a spectrum of lesion, going from chronic tendinosis to acute tendon rupture.

Triceps tendinosis is an enthesopathy, secondary to overuse caused by repetitive and resisted elbow extension activities, causing degenerative process at the tendon to bone insertion.

Patients with symptomatic tendinosis present recurrent or persistent pain at the olecranon insertion, that increases with resisted elbow extension.

Triceps tendon rupture is usually acute and traumatic, caused by a sudden eccentric load applied to a contracting triceps muscle. Traumatic tendon rupture can be partial or complete.

19.5.1 Clinical Presentation

Patients typically report a sudden, painful "pop" at the time of injury followed by the development of a dull ache and swelling at the posterior elbow.

19.5.2 Physical Examination

On physical examination, the posterior elbow is examined for signs of direct trauma, swelling, ecchymosis, and visible defects. Elbow range of motion and elbow extension strength is compared to the contralateral limb. Tenderness at the olecranon insertion as well as increased pain with resisted elbow extension can be present, with different degrees of severity, in both acute and chronic triceps tendinopathies. In case of complete tendon rupture, a palpable defect may be present but soft tissue swelling and body habitus may limit the clinician's ability to accurately identify the defect; a significant reduction in extension strength is normally present, but the inability to actively keep the forearm in extension against gravity is seen in a minority of cases. In fact, the presence of an intact lateral expansion of the tendon may allow active extension (albeit weak), even in complete ruptures. This may lead to misdiagnosis, or delayed diagnosis of the rupture.

19.5.3 Specific Examination Manoeuvres

- **Fall down triceps test**: The test can be performed in supine position, the shoulder elevated, and the elbow partially extended. The examiner pushes the forearm to flex the elbow, while the patient tries to maintain the elbow extended (Fig. 19.13). In case of tendon lesion, the patient shows weakness extension. However, not all complete triceps tendon tears result in total loss of active elbow extension: an intact lateral expansion

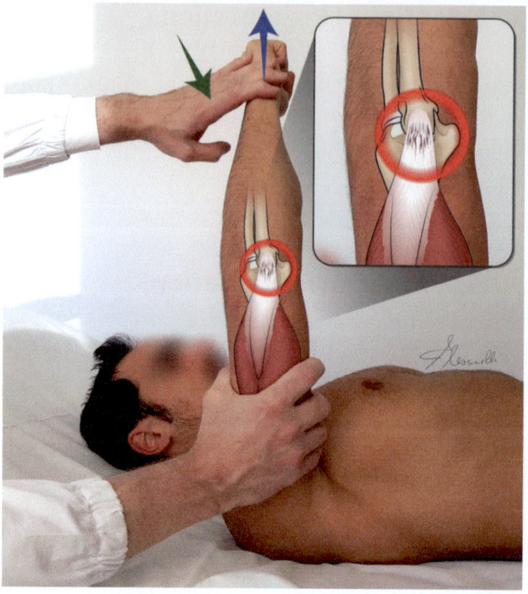

Fig. 19.13 Fall down triceps test: the examiner pushes the forearm to flex the elbow, while the patient tries to maintain the elbow extended: the inability to actively keep the forearm in extension against gravity is a sign of complete rupture

or compensating anconeus may still provide some degree of active elbow extension, albeit the test is painful and the strength weaker.

- **Triceps squeeze test**: This test it is the adaptation of the Thompson test for the triceps tendon. It can be performed on a prone patient with the shoulder abducted and the elbow flexed 90° over the edge of the examination table. In a patient with complete tear of the triceps tendon, squeezing the triceps muscle belly is not able to produce an extension [14] (Fig. 19.14).

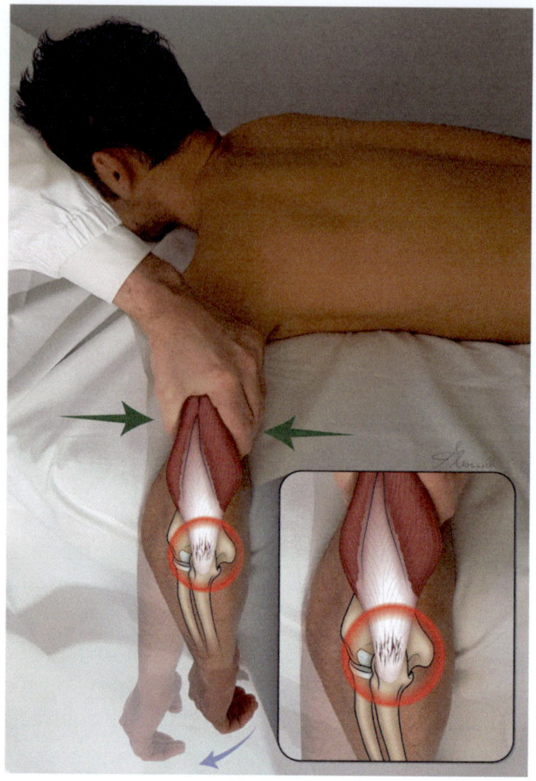

Fig. 19.14 Triceps squeeze test: the triceps muscle belly is squeezed in a patient in prone position over the edge of the examination table, with the shoulder abducted and the elbow flexed 90°. If the triceps tendon is intact, the test will produce some degrees of elbow extension. In case of a complete tear, triceps squeezing is not able to produce elbow extension

References

1. Cozen L. The painful elbow. Ind Med Surg. 1962;31:369–71.
2. Roles NC, Maudsley RH. Radial tunnel syndrome: resistant tennis elbow as a nerve entrapment. J Bone Jt Surg Br. 1972;54(3):499–508.

3. Mills GP. Treatment of "tennis elbow". Br Med J. 1928;1(3496):12–3.
4. Polkinghorn BS. A novel method for assessing elbow pain resulting from epicondylitis. J Chiropr Med. 2002;1(3):117–21.
5. Gardner RC. Tennis elbow: diagnosis, pathology and treatment. Nine severe cases treated by a new reconstructive operation. Clin Orthop Relat Res. 1970;72:248–53.
6. Zwerus EL, Somford MP, Maissan F, Heisen J, Eygendaal D, van den Bekerom MP. Physical examination of the elbow, what is the evidence? A systematic literature review. Br J Sports Med. 2018;52(19):1253–60.
7. O'Driscoll SW. Personal communication. Amsterdam: Elsevier; 2011.
8. O'Driscoll SW, Goncalves LB, Dietz P. The hook test for distal biceps tendon avulsion. Am J Sports Med. 2007;35(11):1865–9.
9. Harding WG 3rd. A new clinical test for avulsion of the insertion of the biceps tendon. Orthopedics. 2005;28(1):27–9.
10. Hertel R. Personal communication. Amsterdam: Elsevier; 2007.
11. ElMaraghy A, Devereaux M, Tsoi K. The biceps crease interval for diagnosing complete distal biceps tendon ruptures. Clin Orthop Relat Res. 2008;466(9):2255–62.
12. Ruland RT, Dunbar RP, Bowen JD. The biceps squeeze test for diagnosis of distal biceps tendon ruptures. Clin Orthop Relat Res. 2005;437:128–31.
13. ElMaraghy A, Devereaux M. The "bicipital aponeurosis flex test": evaluating the integrity of the bicipital aponeurosis and its implications for treatment of distal biceps tendon ruptures. J Shoulder Elb Surg. 2013;22(7):908–14.
14. Viegas SF. Avulsion of the triceps tendon. Orthop Rev. 1990;19(6):533–6.

20 Evaluation of Sports-Related Elbow Instability

Cheli Andrea Filippo, Andrea Celli, and Luigi Adriano Pederzini

20.1 Introduction

The clinical assessment of the elbow in athletes can be extremely challenging.

An adequate evaluation requires knowledge of the anatomy and biomechanics.

The best approach is a step-by-step one: (1) history, (2) inspection, (3) palpation, (4) passive motion, (5) active motion, (6) active motion against resistance, (7) neurologic examination, and (8) lidocaine test.

These eight steps will allow a clinical diagnosis to be made in most cases of athletes with elbow instability: all maneuvers should be performed on both upper extremities for comparison; ipsilateral shoulder and wrist examination should not be forgotten [1].

Overhead and throwing athletes subject the elbow to significant loads, which can lead to multiple degrees of instability. Elbow injuries in the throwing athlete are most of the time the result of the high valgus and extension forces acting on the elbow during the throwing motion. These forces result in tensile stress on medial structures, compression stress on lateral structures and shear stress posteromedially [2].

The height and weight of the athlete should be noted; a higher incidence of ulnar collateral ligament (UCL) injuries has been found in taller and heavier baseball players. This may be because heavier players may throw harder and taller player has longer extremities, resulting in a longer fulcrum and therefore greater forces on the UCL [3].

Instability in elbows can occur on a frontal plane (valgus–varus), on a longitudinal one (posterolateral–posteromedial), or both.

The center of the elbow varus–valgus axis is along the center of the trochlea, which is located medial to the midline. Therefore, valgus elbow stress with compromised medial structures produces less medial gapping than varus stress when the lateral structures are injured [4].

Here is a list of the most common instability tests, that will be described in detail later in this chapter:

20.2 Instability Tests

Medial instability

- Valgus stress test
- Milking maneuver
- Moving valgus test
- Posteromedial pivot shift

C. A. Filippo (✉) · L. A. Pederzini
Orthopaedic Unit, New Civil Sassuolo Hospital, Modena, Italy

A. Celli
Orthopaedic Unit, Hesperia Hospital, Modena, Italy

J. G. Lane et al. (eds.), *The Art of the Musculoskeletal Physical Exam*,
https://doi.org/10.1007/978-3-031-24404-9_20

Lateral instability

- Varus stress test
- Posterolateral pivot shift
- Posterolateral drawer test
- Table-top relocation test
- Stand-up test/chair push-up test
- Push-up test

20.3 Medial Instability

Valgus stability is provided by the osseous anatomy of the olecranon and the humerus, the dynamic muscle forces, and the medial ulnar collateral ligament (M-UCL) complex. The interlocking bony anatomy of the olecranon with the olecranon fossa provides stability from 0 to 20° of flexion. If the elbow is bent more than 20°, the M-UCL is more important, and the majority of stress is placed on its anterior bundle [1].

Lesions of the UCL can be evaluated with the valgus stress test, the moving valgus stress test, and the milking maneuver.

Palpation of the ulnar collateral ligament (UCL) is performed with the elbow in approximately 50°–70° of flexion, allowing the anteposition of the overlying medial muscle mass with respect to the fibers of the UCL. The ligament should be palpated along its entire course, beginning at its origin from the inferior aspect of the medial epicondyle and progressing distally to its insertion onto the sublime tubercle of the proximal medial ulna, searching for tender points that could indicate partial injury. Tenderness over the UCL has an 81–94% sensitivity but only a 22% specificity for UCL tears [5, 6]. When there is an acute UCL injury, ecchymosis may develop along the medial elbow and proximal forearm, reducing even more specificity [4].

The diagnosis of medial epicondylitis must be considered, although the presence of medial epicondylitis does not rule out M-UCL injury because they may coexist. Patients with flexor-pronator epicondylitis have pain with resisted wrist flexion when the elbow is fully extended and localize their pain just anterior and distal to the common flexor muscle origin. In contrast, patients with a M-UCL injury typically have point tenderness about 2 cm distal to their medial epicondyle [4]. M-UCL injury may also be differentiated from medial epicondylitis by performing the ***valgus stress test*** with the wrist in passive flexion and pronation to eliminate tension on the flexor-pronator mass [7]. Medial pain with this modification likely indicates UCL injury rather than medial epicondylitis or muscle injury [2].

The elbow should be assessed for a fixed flexion and valgus deformity, sometimes seen in older baseball pitchers, which can predispose athletes to ulnar neuritis [8, 9].

When the UCL is injured, the ipsilateral shoulder appears to have increased external rotation on physical examination [10].

A thorough neurovascular examination should follow, with attention to the ulnar nerve distribution. Examination of the shoulder and scapula is critical because altered mechanics can alter throwing mechanics through the kinetic chain. A glenohumeral internal rotation deficit is associated with valgus instability in throwers [11].

The neck should be a potential source of referred pain and needs to be considered.

"Surgery thoughts" should request the evaluation for presence or absence of a palmaris longus (PL) tendon: ask the patient to touch the ipsilateral thumb and small finger and flex the wrist to detect the tendon. It is a potential autograft source for UCL reconstruction but is absent in 1 extremity in 3% and absent in both extremities in 2.5% of white people in North America [12].

The most critical component of the physical examination when UCL injury is suspected is assessment of the UCL functional integrity through various tests subjecting the elbow to a valgus force and assessing for medial joint space opening, the quality of an eventual end point, and medial-sided pain [4].

Valgus stress test is performed to evaluate injury to the anterior bundle of the anterior oblique ligament (AOL) of the M-UCL. Although cadaveric studies have suggested 70°–90° elbow flexion as the optimal position to isolate the contribution of the UCL to valgus stability, it is difficult to control humeral rotation and apply valgus stress at that angle. Therefore, testing is best performed at 20°–30° elbow flexion with the forearm pronated. Valgus stress test may be performed with the patient seated upright, supine, or prone. Norwood and colleagues described valgus stress test of the elbow with the forearm supinated at 15°–20° elbow flexion to unlock the olecranon from the olecranon fossa [13].

It is now recognized that forearm pronation prevents subtle posterolateral instability (pseudo-valgus instability) from mimicking medial laxity [14].

To perform the valgus stress test on the right elbow in the seated or supine position, the examiner stabilizes the humerus with the left hand just above the humeral condyles and applies a valgus moment with the right hand while grasping the patient's pronated forearm (Fig. 20.1). In the prone position, the examiner stabilizes the humerus in 90 shoulder abduction with the right hand above the humeral condyles, flexes the elbow 20°–30°, and applies valgus stress with the left hand on the patient's pronated forearm.

Traditionally, the humerus is stabilized, the forearm pronated and the elbow is subjected to a valgus stress at about 20–30 elbow flexion. In a positive test, there is no firm end point and the articular surfaces of the ulna and medial humeral condyle are felt to move apart and the forearm swings out laterally [13]. About half of patients with a torn M-UCL are painful at the test [5]. This test is 66% sensitive and 60% specific for detecting abnormalities of the anterior band of the anterior oblique ligament (AOL) [6].

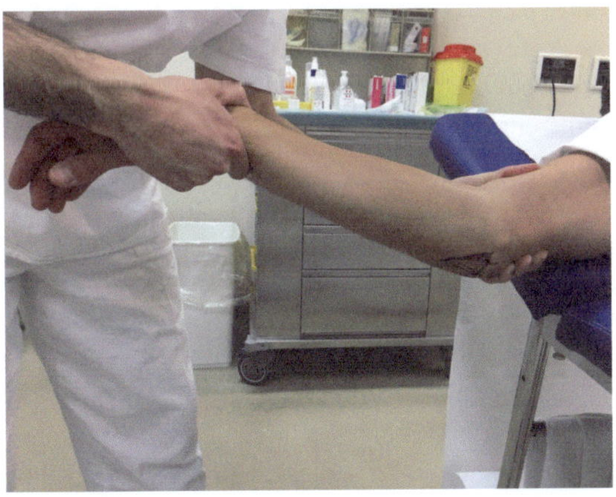

Fig. 20.1 The ***valgus stress test*** is performed with the humerus externally rotated, the elbow flexed to 20°–30° and the forearm pronated

Comparison with the opposite elbow is mandatory.

Detecting significant instability is often very subtle in the throwing athlete because of the relatively small degree of medial opening on examination, even in case of significant ligament injury. The literature shows that a complete sectioning of the anterior bundle of the M-UCL only increases medial opening by 1–2 mm [15, 16]. This fact highlights the importance of comparing findings with the contralateral side. Accordingly, even the most experienced elbow surgeons are only able to detect preoperative valgus laxity on physical examination in between 26 [17] and 82% [5] of patients with operative M-UCL tear.

For pain as an outcome, the test showed 65% sensitivity and 50% specificity. However, by using laxity as outcome, the test had a disappointing sensitivity but perfect specificity of 100% [18].

Veltri and colleagues described the ***"milking maneuver"*** to evaluate valgus stability [19]. While the valgus stress tests the anterior band of the AOL, the milking maneuver tests the posterior band of the AOL. Patients with a UCL injury may report a feeling of apprehension, instability, and medial joint pain [20].

With the patient seated, the examiner grasps the thrower's thumb with the arm in the cocked position (90° shoulder abduction and maximum external rotation, removing shoulder external rotation as a confounding variable, and 90° elbow flexion, with forearm supinated) and applies valgus stress by pulling down on the thumb with one hand. The examiner's hand that is being used to hold the elbow is also used to palpate the medial joint line to feel for joint space opening and an end point (Fig. 20.2). This position is felt to be similar to pulling down on the teats when milking a cow, hence the name [1, 4].

In literature the ***milking manoeuvre*** was described, but no evidence on diagnostic accuracy has been found [21, 22].

The "moving valgus stress test" (Fig. 20.3) is a variation and is performed with the examiner beginning in the position of the standard milking

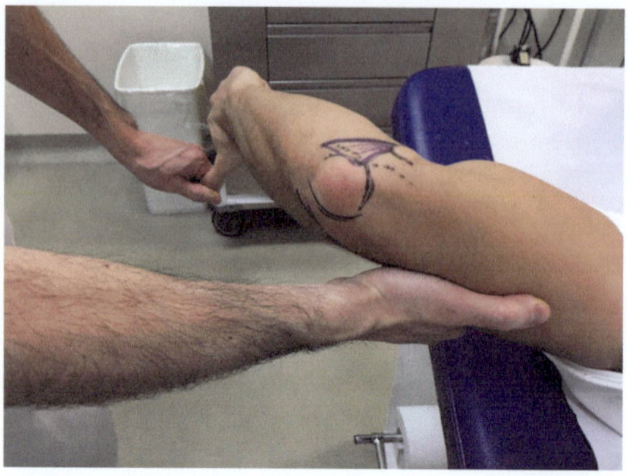

Fig. 20.2 The ***milking maneuver*** and the ***moving valgus stress test*** can be performed with the patient supine and the arm can be supported against applied valgus by the examiner hand

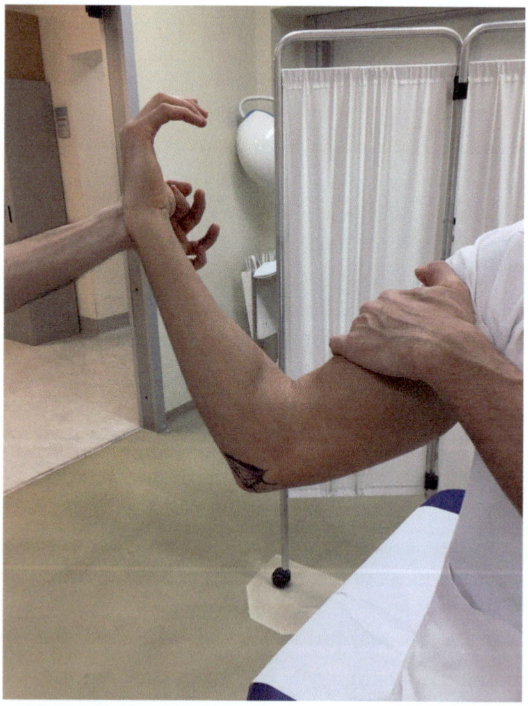

Fig. 20.3 The ***milking maneuver*** and the ***moving valgus stress test*** can also be performed with the patient in the sitting position; and the arm can be supported against applied valgus by the patient hand, in this way the examiners second hand is free to palpate the medial side of the elbow

maneuver and extending the elbow from 120° flexion to 20°–30° flexion while applying valgus stress (pulling the thumb toward the floor).

This alternative way of testing the M-UCL, as described by O'Driscoll and colleagues [23] is performed with the patient in an upright position and the shoulder abducted 90°. Starting with the elbow maximally flexed, a modest valgus torque is applied to the elbow until the shoulder reaches its limit of external rotation. While a constant valgus torque is maintained, the elbow is quickly extended to about 20°–30°. The test is positive when the pain generated by the maneuver reproduces exactly the medial elbow pain at the M-UCL that the patient has with activities. Second, although the patient may experience pain throughout the range, the pain should be maximal between the position of late cocking (120°) and early acceleration (70°) as the elbow is extended (Fig. 20.3).

With the ***moving valgus stress test*** sensitivity for an M-UCL insufficiency is reported at 75% and specificity 100% [18].

There is 100% sensitivity and 75% specificity using arthroscopic valgus stress testing and surgical exploration of the M-UCL as the gold standard [23].

20.4 Lateral Instability

Varus stability is provided by the osseous anatomy of the olecranon and the humerus, the dynamic muscle forces and the lateral collateral ligament (LCL) complex. Varus instability is less common, because a direct impact on the medial side causing varus of the elbow is difficult as the body protects the medial side most of the time [1].

Although lateral instability is uncommon in the throwing athlete, other athletes may present with chronic elbow instability after a traumatic elbow dislocation. Several tests have been described to evaluate the LCL complex [14, 24].

During the ***varus stress test***, the patient's arm is stabilized with one of the examiners hands at the medial distal humerus, and the other hand is placed above the patient's lateral distal radius with the elbow flexed around 20°–30°. An adduction or varus force is applied at the distal forearm by the examiner to test the radial collateral ligament. Varus stress is better applied with the humerus in full internal rotation (Fig. 20.4).

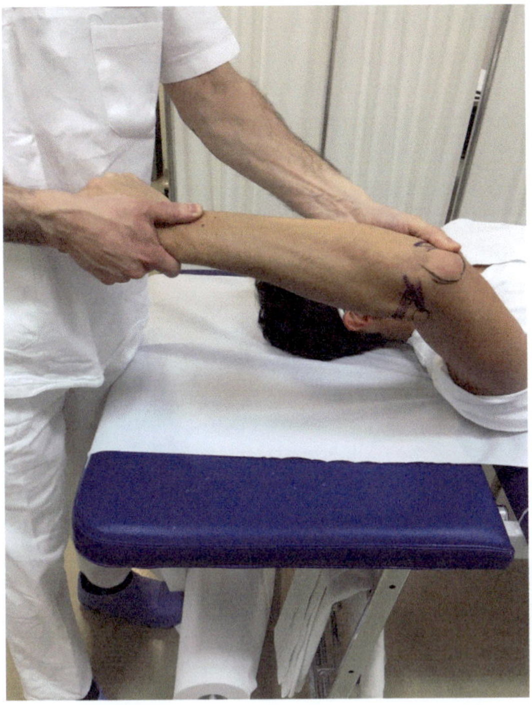

Fig. 20.4 ***Lateral pivot-shift stress test*** by O'Driscoll

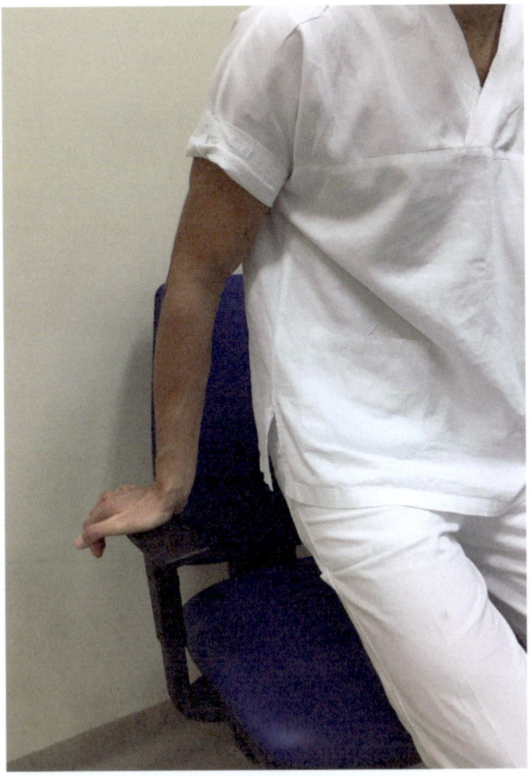

Fig. 20.5 The ***chair push-up test***; the test is considered positive if there is a reluctance to extend the elbow fully as the patient raises his body up from a chair using exclusively upper extremity force with forearm supinated

20.5 Rotatory Instability

A combination of instability on both frontal and longitudinal axis can occur. Posterolateral instability is more common than posteromedial instability.

The most common test for **posterolateral instability** is the ***lateral pivot-shift test*** described by O'Driscoll et al. [14]. During this test, the patient is supine with the affected limb overhead. With the forearm supinated, valgus and axial loading is applied, and the elbow is flexed from full extension (Fig. 20.5).

In posterolateral rotatory instability, as the elbow is flexed, the radial head dislocates: this appears as an osseous prominence posterolaterally. With flexion beyond 40°, the radial head suddenly reduces with a palpable and visible clunk. The test may also be done starting with the elbow flexed and then extending, reversing the sequence.

Although the lateral pivot shift test and posterolateral rotatory drawer test are sensitive for detecting lateral instability, they often require general anesthesia to reproduce the radial head subluxation and subsequent reduction.

In addition to passive tests, there are also active tests to evaluate posterolateral instability. The first is the ***table top relocation test***, an active apprehension sign [25]. The upper extremities are positioned on a table with the elbow at 90° flexion, forearms supinated and arms abducted. The test is considered positive if apprehension occurs as the affected elbow is fully extended from a flexed position together with voluntary and involuntary guarding (Fig. 20.5).

A second active apprehension sign is the ***stand up/chair push up test*** [25]: the patient is seated with the elbows flexed 90°, the forearms supinated and the arms abducted. The test is considered positive if the patient is reluctant to fully extend the elbow while raising his body up from a chair using exclusively upper extremity force as a result of apprehension or elbow dislocation.

A third test is ***the push up test***: the patient lays with chest on the floor, elbows flexed at 90°, shoulders abducted, forearm supinated and then performs a push-up. The test is positive with apprehension or radial head subluxation, but this test is much more uncomfortable for the patient than the other two.

Regan et al. described that these tests are more sensitive than the pivot-shift test in the awake patient and may be easily performed in the clinic environment [25].

Specifically, the ***table-top relocation test***, ***stand-up/chair push-up test*** and ***push-up test*** show similar capacities for a positive finding with sensitivity from 88 to 100%.

The ***pivot shift test*** shows sensitivity of only 38% in the awake patient, but 100% sensitivity if performed under anaesthesia [25, 26].

In **posteromedial instability**, theoretically, a subluxation can be obtained applying varus and axial loading with the forearm pronated (***medial pivot-shift test***). This occurs only in the case of a coronoid fracture or deficit. This test cannot be performed in an awake patient [1].

The ***valgus extension overload test*** is performed to detect the presence of a posteromedial olecranon osteophyte or olecranon fossa overgrowth. The examiner stabilizes the humerus with one hand, and with the opposite hand, pronates the forearm and applies a valgus force while quickly maximally extending the elbow. A positive test causes posteromedial pain as the olecranon tip osteophyte engages into the olecranon fossa [1, 2].

20.6 Micro-instability

Recalcitrant lateral elbow pain is supposed to be related to minor lateral instability of the elbow. Recently two new clinical tests have been described to evaluate this condition:

- ***Supination and Antero- Lateral pain Test (SALT).***
- ***Posterior Elbow Pain by Palpation-Extension of the Radiocapitellar joint (PEPPER).***

In the SALT test we suppose that the examiner's thumb, while gliding along the anterolateral surface of the radial head, can selectively compress the anterior capsule and the synovial tissue lying underneath it. In case of synovial hypertrophy and inflammation, the supination movement pushes this synovial tissue in the sigmoid notch. Compression of the inflamed synovial tissue causes pain.

In the PEPPER test, the examiner's thumb is placed on the surface of the radial head with the elbow in 90° flexion. With extension of the radiocapitellar joint, pressure on the thumb and, indirectly, on the radial head is increased. Compression of a chondropathic radial head might be the main source of pain when performing this test.

SALT has a high sensitivity but a low specificity and is accurate in detecting the presence of intra-articular synovitis. PEPPER test is sensible, specific and accurate in the detection of radial head chondropathy.

Positive findings may be indicative of a minor instability of the lateral elbow in most cases of recalcitrant lateral elbow pain, with at least one test positive in 90% of patients affected by this condition [27, 28].

References

1. Van Tongel A. Physical examination of the elbow. In: Elbow and sport. Milan: Esska Book; 2016.
2. Lyle Cain E Jr, Dugas JR. History and examination of the thrower's elbow. Clin Sports Med. 2004;23(4):553–66.
3. Han KJ, Kim YK, Lim SK, et al. The effect of physical characteristics and field position on the shoulder and elbow injuries of 490 baseball players: confirmation of diagnosis by magnetic resonance imaging. Clin J Sport Med. 2009;19(4):271–6.
4. Hariri S, Safran MR. Ulnar collateral ligament injury in the overhead athlete. Clin Sports Med. 2010;29(4):619–44.
5. Thompson WH, Jobe FW, Yocum LA, et al. Ulnar collateral ligament reconstruction in athletes: muscle-splitting approach without transposition of the ulnar nerve. J Shoulder Elb Surg. 2001;10(2):152–7.
6. Timmerman LA, Schwartz ML, Andrews JR. Preoperative evaluation of the ulnar collateral ligament by magnetic resonance imaging and computed tomography arthrography. Evaluation in 25 baseball players with surgical confirmation. Am J Sports Med. 1994;22(1):26–31.

7. Leach RE, Miller JK. Lateral and medial epicondylitis of the elbow. Clin Sports Med. 1987;6(2):259–72.
8. Bennett GE. Shoulder and elbow lesions distinctive of baseball players. Ann Surg. 1947;126(1):107–10.
9. King J, Brelsford HJ, Tullos HS. Analysis of the pitching arm of the professional baseball pitcher. Clin Orthop Relat Res. 1969;67:116–23.
10. Mihata T, Safran MR, McGarry MH, et al. Elbow valgus laxity may result in an overestimation of apparent shoulder external rotation during physical examination. Am J Sports Med. 2008;36(5):978–82.
11. Dines JS, Frank JB, Akerman M, et al. Glenohumeral internal rotation deficits in baseball players with ulnar collateral ligament insufficiency. Am J Sports Med. 2009;37(3):566–70.
12. Troha F, Baibak GJ, Kelleher JC. Frequency of the palmaris longus tendon in North American Caucasians. Ann Plast Surg. 1990;25(6):477–8.
13. Norwood LA, Shook JA, Andrews JR. Acute medial elbow ruptures. Am J Sports Med. 1981;9:16–9.
14. O'Driscoll SW, Bell DF, Morrey BF. Posterolateral rotatory instability of the elbow. J Bone Jt Surg. 1991;73A:440–6.
15. Field LD, Altchek DW. Evaluation of the arthroscopic valgus instability test of the elbow. Am J Sports Med. 1996;24:177–81.
16. Field LD, Callaway GH, O'Brien SJ, et al. Arthroscopic assessment of the medial collateral ligament complex of the elbow. Am J Sports Med. 1995;23:396–400.
17. Azar FM, Andrews JR, Wilk KE, et al. Operative treatment of ulnar collateral ligament injuries of the elbow in athletes. Am J Sports Med. 2000;28(1):16–23.
18. Zwerus EL, Somford MP, Maissan F, Heisen J, Eygendaal D, Van Den Bekerom MP. Physical examination of the elbow: what is the evidence? A systematic literature review. Br J Sports Med. 2017;52(19):1253 60.
19. Veltri DM, O'Brien SJ, Field LD, et al. The milking maneuver: a new test to evaluate the MCL of the elbow in the throwing athlete. In: Programs and abstracts of the 10th open meeting of the American Shoulder and Elbow Surgeons. Rosemont: American Academy of Orthopaedic Surgeons; 1994.
20. Chen FS, Rokito AS, Jobe FW. Medial elbow problems in the overhead-throwing athlete. J Am Acad Orthop Surg. 2001;9(2):99–113.
21. Miller MD, Thompson SR. DeLee & Drez's orthopaedic sports medicine. 4th ed. Philadelphia: Saunders Elsevier Inc; 2014. p. 721–9.
22. Chen FS, Diaz VA, Loebenberg M, et al. Shoulder and elbow injuries in the skeletally immature athlete. J Am Acad Orthop Surg. 2005;13:172–85.
23. O'Driscoll SW, Lawton RL, Smith AM. The "moving valgus stress test" for medial collateral ligament tears of the elbow. Am J Sports Med. 2005;33(2): 231–9.
24. Olsen BS, Sojbjerg JO, Nielsen KK, et al. Posterolateral elbow joint instability: the basic kinematics. J Shoulder Elb Surg. 1998;7:19–29.
25. Regan W, Lapner PC. Prospective evaluation of two diagnostic apprehension signs for posterolateral instability of the elbow. J Shoulder Elb Surg. 2006;463(15):344–6.
26. Arvind CH, Hargreaves DG. Tabletop relocation test: a new clinical test for posterolateral rotatory instability of the elbow. J Shoulder Elb Surg. 2006;15:707–8.
27. Arrigoni P, Cucchi D, Menon A, Randelli P. It's time to change perspective! New diagnostic tools for lateral elbow pain. Musculoskelet Surg. 2017;101:175–9.
28. Arrigoni P, Cucchi D, D'Ambrosi R, et al. Intra-articular findings in symptomatic minor instability of the lateral elbow (SMILE). Knee Surg Sports Traumatol Arthrosc. 2017;25:2255–63. https://doi.org/10.1007/s00167 017 4530 x.

21 Compartment Syndrome in the Upper Limb

William N. Yetter and Benjamin R. Graves

21.1 Introduction

Compartment syndromes occur when pressures within a fascial compartment overcome the intravascular pressures required to perfuse the soft tissues contained within the compartment [1]. Compartment syndromes fall into two main categories: ***acute compartment syndrome*** **(ACS)** and ***exertional compartment syndrome*** **(ECS)**. ACS is a surgical emergency and must be addressed rapidly to minimize tissue death. In the upper limb, a missed diagnosis or delayed treatment can produce devastating consequences such as Volkmann contracture of the forearm and loss of limb. ECS can be *acute* (AECS) or *chronic* (CECS) and is far less common than ACS. ECS does not typically present as a surgical emergency, rather, patients with ECS will describe progressive pain during physical activity that resolves after cessation of the activity. With ECS, increased intracompartmental pressure (ICP) occurs during sustained activity and produces pain; however, the etiology and pathophysiology of ECS in the upper limb remain poorly understood [1].

W. N. Yetter · B. R. Graves (✉)
Department of Orthopaedic Surgery, Wake Forest University School of Medicine, Winston Salem, NC, USA
e-mail: bgraves@wakehealth.edu

21.2 Anatomy

Separated by the elbow, the upper limb is divided into the brachium proximally, and the forearm distally. The ***brachium*** is divided into two fascial compartments: the **anterior compartment** and the **posterior compartment** (Table 21.1). The muscles of the anterior compartment are all innervated by the musculocutaneous nerve, except for the brachialis, which has duel innervation, receiving radial nerve innervation to its lateral portion. The arterial blood supply for the anterior compartment is from the brachial artery. The median, ulnar, radial, and medial and lateral antebrachial cutaneous nerves all course distally within the anterior compartment, which places these structures at risk for ischemia during a compartment syndrome [2]. Both muscles of the posterior compartment are innervated by the radial nerve, and their blood supply comes from the profunda brachii artery.

Compared to the brachium, the ***forearm*** is more anatomically complex, and has three fascial compartments: the **volar** (flexor), **dorsal** (extensor), and **lateral** (mobile wad) compartments (Table 21.2). Additionally, the volar and dorsal compartments are each subdivided into ***superficial*** and ***deep*** sub-compartments. The interosseous membrane that spans between the radius and ulna bones separates the volar and dorsal compartments of the forearm. Distally,

J. G. Lane et al. (eds.), *The Art of the Musculoskeletal Physical Exam*,
https://doi.org/10.1007/978-3-031-24404-9_21

Table 21.1 Anatomy of the compartments of the brachium

Brachial compartments	Muscles	Innervation	Arterial supply
Anterior compartment	Coracobrachialis	Musculocutaneous	Brachial artery
	Biceps brachii		
	Brachialis	Musculocutaneous and radial	
Posterior compartment	Triceps brachii	Radial	Profunda brachii
	Anconeus		

Table 21.2 Anatomy of the compartments of the forearm

Forearm compartments		Muscles	Innervation	Arterial supply
Volar (flexor) compartment	Superficial	Pronator teres	Median	Ulnar and radial arteries
		Palmaris longus		
		Flexor digitorum superficialis		
		Flexor carpi radialis		
		Flexor carpi ulnaris	Ulnar	
	Deep	Flexor digitorum profundus	Median and ulnar	
		Flexor pollicis longus	Median	
		Pronator quadratus		
Dorsal (extensor) compartment	Superficial	Extensor digitorum communis	PIN	Radial artery branches and posterior interosseous artery
		Extensor carpi ulnaris		
		Extensor digiti minimi		
	Deep	Abductor pollicis longus		
		Extensor pollicis brevis		
		Extensor pollicis longus		
		Extensor indicis		
		Supinator		
Lateral (mobile wad) compartment		Brachioradialis	Radial	
		Extensor carpi radialis longus	Radial	
		Extensor carpi radialis brevis	PIN	

the carpal tunnel is continuous with the volar compartment and is frequently affected in ACS of the forearm.

21.3 Pathophysiology

The most common etiology associated with development of ACS in adults is high energy trauma with long bone fractures. ACS is commonly associated with supracondylar humerus fractures (SCHF), both-bone radius and ulna forearm fractures (BBFF), and with floating elbow injuries (combination of SCHF and BBFF) in children [3]. Fractures are the most common cause of ACS; however, acute compartment syndrome can result from a variety of injuries to the upper limb (Table 21.3). The deep volar compartment is the most commonly affected, while the mobile wad is generally the least affected.

ACS occurs when there is elevated pressure within a fibro-osseous space which ultimately results in decreased tissue perfusion [4]. After an initial insult to the forearm and/or brachium, swelling ensues and ICP rises. This leads to an early collapse of thin-walled venules, which results in decreased venous outflow. Arterioles will remain patent longer due to their muscular-walled composition, which allows them to withstand greater pressures. As ICP continues to rise, the imbalance of inflow and outflow within the compartment increases. As venous outflow obstruction progresses and ICP rises, ultimately, arterial inflow will become obstructed and soft tissue ischemia ensues.

Table 21.3 Causes of compartment syndromes in the upper limb

Type	Causes
Constrictive	Tight dressings, casts, or splints Tourniquet Patient "found down"/ prolonged compression Amniotic band syndrome (in utero)
Intracompartmental fluid extravasation	Penetrating trauma Crush injury High pressure injection injury Fracture hematoma Vascular injury Regional anesthesia Bleeding disorders Reperfusion injury Extravasation of IV infiltrate
Inoculation	Spider bite Snakebite Needle injection (intravenous drug abuse)
Other	Infection Burns Suction injury

As ICP increases, if left unchecked, it will eventually become greater than capillary pressure. At the capillary level, microcirculation ceases when ICP is equal to or greater than diastolic blood pressure (DBP). When this threshold has been reached, the microcirculation providing perfusion to the muscle, nerve, and vascular tissue is decreased, which ultimately causes cell death [5]. The pressure gradient between DBP and ICP is represented as ΔP (termed "delta P"), and the equation $\Delta P = \text{DBP} - \text{ICP}$ is used to determine the pressure gradient and thus perfusion capability of a compartment. A canine study demonstrated that when ICP rose to within 20 mmHg of DBP, or when $\Delta P = 20$ mmHg, muscle necrosis resulted [5]. For this reason, a threshold of $\Delta P \leq 30$ mmHg is used frequently as a cutoff to help guide the decision-making process of when to perform surgery while monitoring a potential compartment syndrome. Because of the difficultly and inconsistency of ICP monitoring, and the vasomotor instability that is common during trauma scenarios, some institutions use ICP pressure monitoring devices with an ICP >30–40 mmHg as a threshold for fasciotomy. Whether using ΔP or ICP to guide treatment, one fact is undisputed and is represented well by the commonly used phrase ***"time is muscle,"*** as muscle necrosis can occur within 4 h in the setting of sustained elevated compartment pressures [6].

There are several theories regarding the pathophysiology of chronic exertional compartment syndrome (CECS). In some patients, ICP rises as muscles swell during activity, which can produce pain, weakness, and sensory paresthesias. The main difference between CECS and ACS is that, with CECS, cessation of the activity that produces the symptoms typically leads to a resolution of the symptoms. CECS of the forearm has been associated with motorcycle sports, and the majority of reported cases involve bilateral forearms. Cessation of the activity will reverse the pathophysiology of CECS so loss of limb and function are uncommon; however, it is still possible to produce irreversible damage if an individual continues the activity despite pain [1].

Acute exertional compartment syndrome (AECS) has also been described but is not well understood. AECS should be suspected when a patient presents with prolonged pain out of proportion to examination after strenuous activity. The majority of research on AECS focuses on the lower extremity. A case series that looked at AECS in the lower extremity found that patients with longer than 24 h of symptoms prior to fasciotomies had substantial muscle necrosis and functional deficit after surgery [7]. One unusual case report describes a weightlifter with bilateral supraspinatus AECS which required surgical release 1 day after a strenuous upper body workout [8]; however, the reported incidence of AECS in the upper extremity is rare [9].

21.4 Physical Examination

The clinical examination for ACS is classically taught using the "6 P's" mnemonic. Pain with passive flexion or extension of the hand, wrist, or elbow, pressure, pallor, pulselessness, paralysis, and paresthesias. Although easy to memorize as a medical student, this classic teaching aid may result in missed or delayed treatment of ACS as it provides no insight into the timing and relative importance of the individual presenting signs. Pulselessness, for example, is typically a ***late sign*** of advanced ACS and occurs after irreversible damage to soft tissues has already taken place. Additionally, pallor, paralysis, and paresthesias are nonspecific signs that may have no relation to elevated intracompartmental pressure.

When examining a patient with potential ACS, the surgeon needs to know the earliest signs of impending trouble so that fasciotomy can be performed before irreversible damage has occurred. The most important early findings on physical exam are: **(1) Pain with passive stretch of the affected compartment, (2) Pain out of proportion to exam, and (3) Pressure with palpation of the affected compartment that worsens over time.**

It is quite helpful to have multiple time points of physical examination so that pressures and a patient's pain responses can be trended as worsening or improving. Additionally, it is important to note that these physical exam findings can vary with different stages of presentation. Pain generally peaks around 2–6 h of ischemia and appears to improve as nerve function deteriorates [10]. This means a "burned-out" compartment syndrome that has already produced muscle and nerve death will typically present as less painful than an early compartment syndrome. It is also important to consider distracting injuries when assessing pain.

Young children are not typically able to accurately report pain, so ACS will likely present differently in the pediatric population. Due to the unreliability of the "6 P's" when examining children, the "3 A's" should be used instead. These are: **(1) increasing need for Analgesia, (2) presence of Anxiety, and (3) progressive Agitation (crying)** [3] (Table 21.4). The physician should maintain a high index of suspicion for ACS in children with these symptoms.

Table 21.4 The "3 P's" and "3 A's"

Adults: "3 P's"	Children: "3 A's"
Pain with passive stretch	Analgesia requirement increasing
Pain out of proportion	Anxiety
Pressure with palpation	Agitation

In contrast to physical examination findings with ACS, the physical examination for a patient with CECS may be normal in a clinic setting. To elicit symptoms, patients with CECS require a dynamic examination that takes place during the physical activity that produces the symptoms. CECS symptoms typically evolve and worsen during prolonged activity and may persist for some time after cessation of the activity [11].

21.5 Clinical Decision-Making

ACS is a surgical emergency and the potential consequences of delayed or missed treatment can be devastating. In essence, "time is muscle," and early release will give the best chance for preservation of viable soft tissue and functional return after recovery. In the case of ACS, solving the "who" and "when" in the equation can be far more challenging than the "how" of surgical compartment release. A physician's ability to accurately examine a patient for ACS, and a well-honed clinical threshold for surgical release, are the two most important factors when determining how to proceed with the treatment for ACS.

Frequent and clear communication between the surgeon, medical trainees, and nursing staff involved in the patient's care is critical and must be prioritized for optimal outcomes. Serial examination by the same provider or team is useful when monitoring progression of symptoms. Compartment pressure monitoring is helpful when treating obtunded patients who are not able to participate in an examination, or for objective information when the clinical examination is equivocal. However, if there is a high clinical suspicion for ACS, fasciotomy should not be delayed.

CECS is generally not a surgical emergency. Patients will report exertional pain that prevents them from continuing activities, and cessation of the activity will produce a resolution of symptoms. With more chronic presentations of CECS, patients may notice a decrease in endurance and performance over time [1]. Elective fasciotomy of the affected compartment(s) may be performed for persistent pain that limits activity.

21.6 Technique: ICP measurement

There are several methods and techniques for measuring and monitoring intracompartmental pressure. When indicated, we prefer the Stryker Intra-Compartmental Pressure Monitor System (Stryker Instruments; Kalamazoo, MI USA) (Fig. 21.1). The Pressure Monitor is assembled with three disposable items that are then used to obtain ICP. ICP should be taken within 5 cm of a fracture or site of injury as ICP is typically highest in this location. When monitoring a forearm for compartment syndrome, it is not always clear which compartment is affected. In this scenario, measurements taken from the volar compartment are most commonly affected due to a higher muscle volume than the dorsal compartment or mobile wad and typically report the highest pressure. The effects of gravity on pooling blood can falsely alter ICP [12], so the forearm should be at the level of the heart and motion should be limited while obtaining measurements. In some intensive care situations, it may be beneficial to monitor ICP continuously. Multiple techniques utilizing arterial transducers have been shown to be effective for this purpose [4, 13].

In addition to a careful and accurate history and physical examination, intracompartmental pressure measurements obtained during provocative maneuvers is the gold standard for diagnosis of ECS. To establish a baseline, a patient's ICP is first measured at rest (<10 mmHg is a normal resting ICP for the forearm). Serial measurements are then taken after the patient performs the activity that causes their symptoms. Meeting one or more of the following criteria is diagnostic for ECS: **(1) Pre-exercise pressure ≥10 mmHg, (2) ICP reaches ≥20 mmHg at 1 or 5 min after exercise, (3) ICP takes longer than 15 min to return to <10 mmHg**. When measuring forearm compartments, the superficial volar compartment as well as the dorsal compartment should be measured [1]. Although some authors suggest releasing each compartment that meets 2 or more of the above criteria, intracompartmental pressure measurements during and after dynamic examination in the clinic setting may be technically challenging and inaccurate. For this reason, some authors recommend relying on history and physical examination alone, foregoing provocative testing, for the diagnosis of ECS [11].

21.6.1 Imaging

Several studies have evaluated the utility of magnetic resonance imaging (MRI) as a diagnostic

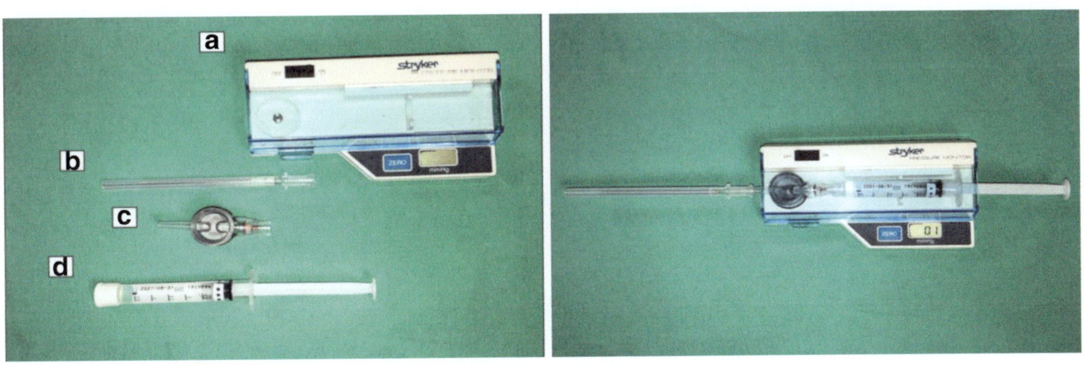

Fig. 21.1 *Left*: Intracompartmental pressure monitoring device components (**a**) device housing/digital display, (**b**) side-ported, non-coring needle, (**c**) diaphragm chamber, (**d**) syringe; *Right:* Assembled device. (Stryker Instruments; Kalamazoo, MI USA)

tool for CECS [11, 14]. Tominaga et al. performed pre- and post-exertional forearm MRI before and after fasciotomies. To initiate forearm ECS, patients used a hand gripping device for 10 min or until onset of symptoms prior to MRI. Before fasciotomy, increased muscle signal on T2-weighted MRI was present. This signal resolved on repeat MRI after fasciotomies were performed. MRI has been shown to have utility as a non-invasive method of evaluating for increased ICP, this technology is not always available and may be cost-prohibitive for many patients.

21.7 Operative Technique: Compartment Fasciotomy in the Upper Limb

21.7.1 Volar Forearm and Anterior Brachium

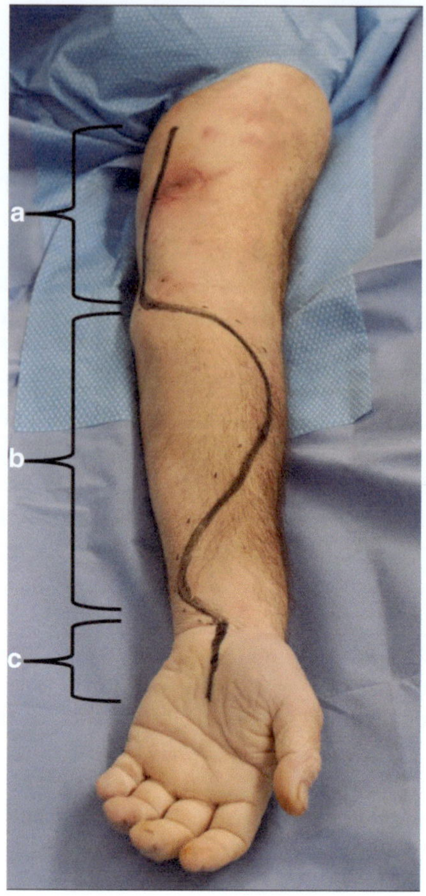

Fig. 21.2 Fasciotomy incisions drawn for the (**a**) anterior brachium, (**b**) volar forearm, and (**c**) carpal tunnel. *(Copyright Benjamin Graves, MD)*

In a cadaver study, Ronel et al. reviewed the anatomy and evaluated techniques for forearm compartment release to determine the safest approach. They showed that an ulnar approach to the deep volar compartment, between the flexor carpi ulnaris (FCU) and flexor digitorum superficialis (FDS), is the safest approach.

To perform an anterior fasciotomy, a volar skin incision is made radial to FCU at the wrist and extended to the medial epicondyle. When fasciotomy of the anterior compartment of the brachium is indicated, the forearm incision can be extended across the elbow between brachioradialis and biceps brachii proximally to connect with a standard deltopectoral incision Additionally, because the volar compartment is continuous with the carpal tunnel, the median nerve is at high risk for ischemia during ACS of the volar forearm compartment and carpal tunnel release should be performed (Figs. 21.2 and 21.3).

When crossing the elbow during forearm and brachial fasciotomies, it is important to release the lacertus fibrosus as this can act as a point of compression for the brachial artery and median nerve as they cross into the forearm (Fig. 21.4).

21.7.2 Dorsal Forearm, Mobile Wad, and Posterior Brachium

A midline approach between the extensor digitorum communis (EDC) and the extensor carpi radialis brevis (ECRB) is safe for fasciotomy of the dorsal compartment. The dorsal incision extends from the lateral epicondyle to the distal

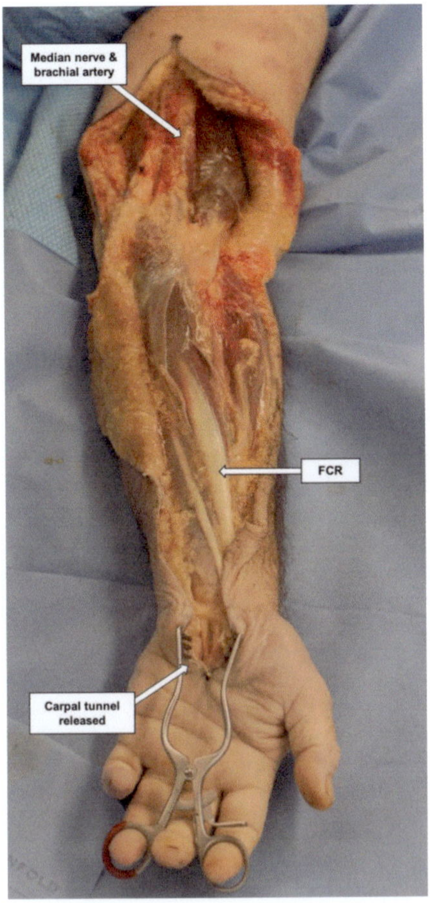

Fig. 21.3 Fasciotomy of the anterior brachium, volar forearm, and carpal tunnel. *(Copyright Benjamin Graves, MD)*

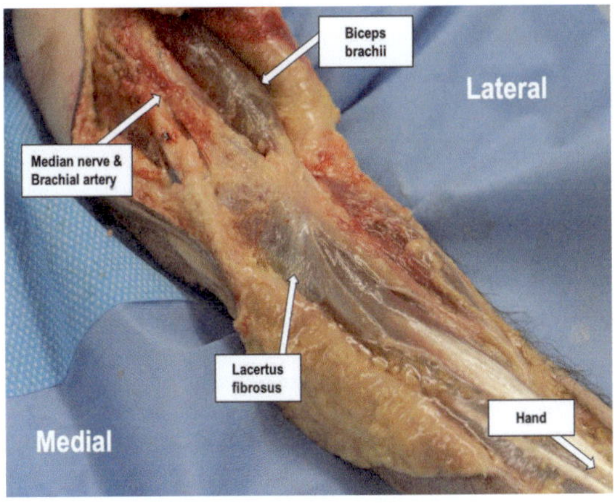

Fig. 21.4 The *lacertus fibrosus* crosses from the biceps brachii musculotendinous junction laterally to the medial flexor pronator muscle mass. This thick band of tissue can act as a point of compression for the anterior neurovascular structures that cross the elbow and should be released. *(Copyright Benjamin Graves, MD)*

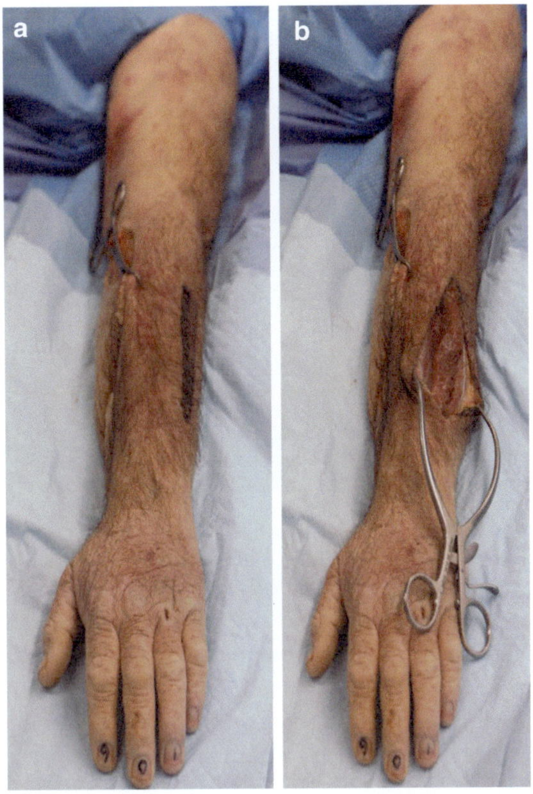

Fig. 21.5 (**a**) Incision drawn for dorsal forearm compartment releases, and (**b**) after fasciotomy of the mobile wad and dorsal compartments. *(Copyright Benjamin Graves, MD)*

radioulnar joint. This will allow access to both the mobile wad and the extensor fascial compartments [15] (Fig. 21.5).

When fasciotomy of the posterior compartment of the brachium is indicated, a midline skin incision made over the posterior arm is safe and provides access to the fascia of the triceps brachii. The anconeus is typically indirectly decompressed following release of the triceps fascia (Fig. 21.6).

After skin incision, a 15-blade knife is used to perforate the fascial sleeve. A Metzenbaum scissor is then used to run this fascial rent proximally and distally, until there are no points of muscle tenting, and the muscle is free to expand. Ischemic muscle is debrided until only healthy muscle remains.

Operative technique for compartment release in CECS is similar to above; however, fascial release is targeted to the affected compartment

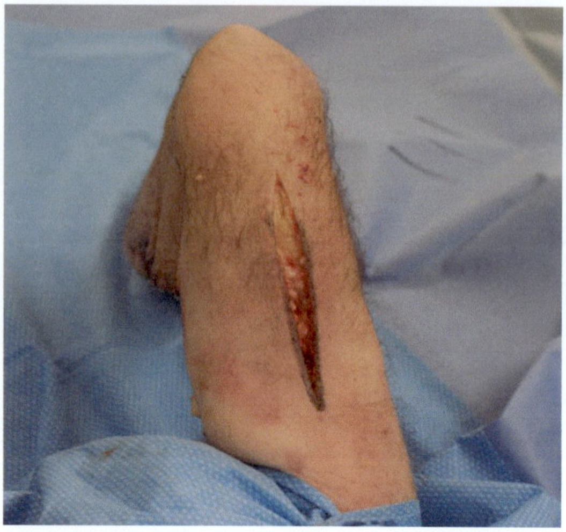

Fig. 21.6 Posterior view of a left arm, with the patient's head to the right, after fasciotomy of the posterior brachial compartment. *(Copyright Benjamin Graves, MD)*

and is typically performed in an elective setting with smaller incisions than are required for ACS. Preoperative evaluation helps determine which compartment will benefit from release [1].

21.8 Post-Operative Management: When to Close

Following fasciotomy for ACS, in the setting of muscle necrosis, patients should be taken back to the operating room for serial debridement until no necrotic tissue is found. Patients often require several trips to the operating room prior to final closure. After swelling abates, if primary closure is not achievable, split-thickness skin grafting is performed to cover healthy muscle and fat.

In the absence of muscle necrosis, the skin is taught after compartment fascial release all incisions should be left open following decompression. However, it is often the case that the skin can be approximated after fasciotomy without undue tension. When this is the case, the skin may be closed over the released fascial compartment. The authors prefer closure in this setting with 2–0 or #0 nylon suture utilizing a vertical mattress configuration. A moist dressing can be applied if the fasciotomy wound is left open, or alternatively, a negative pressure vacuum dressing can be applied. A negative-pressure vacuum dressing will aid in minimizing retraction of the wound edges and maintaining egress of edema fluid from the wound, which will aid in infection control and, ultimately, wound closure. Negative-pressure wound vacuum dressings can also be applied to a fasciotomy incision after the incision has been closed. This so-called incisional vac will help granulation tissue form, will help control wound edema leakage, and will help minimize the requirement for dressing changes.

References

1. García-Mata S. Chronic exertional compartment syndrome of the forearm in adolescents. J Pediatr Orthop. 2013;33(8):832–7.
2. Leversedge FJ, Moore TJ, Peterson BC, Seiler JG. Compartment syndrome of the upper extremity. J Hand Surg Am. 2011;36(3):544–59.
3. Hosseinzadeh P, Hayes CB. Compartment syndrome in children. Orthop Clin N Am. 2016;47(3):579–87.
4. Prasarn ML, Ouellette EA. Acute compartment syndrome of the upper extremity. J Am Acad Orthop Surg. 2011;19(1):49–58.
5. Matava MJ, Whitesides TE, Seiler JG, Hewan-Lowe K, Hutton WC. Determination of the compartment pressure threshold of muscle ischemia in a canine model. J Trauma. 1994;37(1):50–8.
6. Hargens AR, Schmidt DA, Evans KL, Gonsalves MR, Cologne JB, Garfin SR, et al. Quantitation of skeletal-muscle necrosis in a model compartment syndrome. J Bone Jt Surg Ser A. 1981;63(4):631–6.
7. Livingston KS, Meehan WP, Hresko MT, Matheney TH, Shore BJ. Acute exertional compartment syndrome in young athletes. Pediatr Emerg Care. 2018;34(2):76–80.
8. Bunting L, Briggs B. An unusual complication of weightlifting: a case report. Ann Emerg Med. 2014;63(3):357–60.
9. Papachristos IV, Giannoudis PV. Unusual presentation of compartment syndrome. In: Mauffrey C, Hak DJ, Martin III MP, editors. Compartment syndrome: a guide to diagnosis and management. Cham: Springer; 2019. p. 145–60.
10. Stevanovic MV, Sharpe F. Compartment syndrome and Volkmann ischemic contracture. In: Wolfe S, Pederson W, Kozin SH, Cohen M, editors. Green's operative hand surgery. 7th ed. Amsterdam: Elsevier Inc.; 2016. p. 1763–87.
11. Liu B, Barrazueta G, Ruchelsman DE. Chronic exertional compartment syndrome in athletes. J Hand Surg Am. 2017;42(11):917–23.

12. Lipschitz AH, Lifchez SD. Measurement of compartment pressures in the hand and forearm. J Hand Surg Am. 2010;35(11):1893–4.
13. Duckworth AD, Court-Brown CM, McQueen MM. Pressure measurement: surrogate of ischaemia. In: Mauffrey C, Hak DJ, Martin III MP, editors. Compartment syndrome: a guide to diagnosis and management. Cham: Springer; 2019. p. 35–49.
14. Tominaga A, Shimada K, Temporin K, Noguchi R. Post-exertional MRI is useful as a tool for diagnosis and treatment evaluation for chronic exertional compartment syndrome of forearms. J Hand Surg Asian Pac. 2019;24(3):311–6.
15. Ronel DN, Mtui E, Nolan WB. Forearm compartment syndrome: anatomical analysis of surgical approaches to the deep space. Plast Reconstr Surg. 2004;114(3):697–705.

22 Evaluation of Pediatric Elbow Conditions

Andrea Celli, Nicoletta Fabio, Duca Vito, and Luigi Adriano Pederzini

22.1 Ossification Centers

Development and growth of the human skeleton can be divided into three stages: childhood, adolescence, young adulthood.

Childhood terminates with the appearance of all secondary centers of ossification. Adolescence terminates with the fusion of the secondary ossification centers to their respective long bones.

Young adulthood terminates with the formation of new bone growth and with the attainment of final adult skeletal form [1].

Skeletal growth at the elbow occurs following the six secondary ossification centers that appear at characteristic times in male and female development [1]. At birth, the distal humerus appears as a single cartilaginous epiphysis encompassing both condyles and epicondyles with only one physis. This epiphysis differentiates into two epiphyses (the capitellum and trochlea) during the individual's first decade, and two apophyses (the medial and lateral epicondyles) and the radial head and olecranon epiphyses have been developed.

The acronym "CRITOE" is often used as a mnemonic device to list the order of appearance of the ossification centers. That is:

C, capitellum
R, radial head
I, internal (medial) epicondyle
T, trochlea
O, olecranon
E, external (lateral) epicondyle

Appearance of the capitellum at age 1–2 is followed by the radial head at approximately age 4–5. The medial epicondyle develops at age 5–7, the trochlea at age 8–9, the olecranon at age 8–10, and finally, the lateral epicondyle at around 11 (for girls) and 12 (for boys) (Fig. 22.1). Fusion of these secondary ossification centers also occurs in a sequential, age-dependent order. Trochlea, capitellum, and lateral epicondyle fuse before physeal closure [2–4].

Fusion of the ossification centers is typically delayed in male subjects compared to female subjects, and it approximately proceeds as follows: the trochlea, capitellum, and olecranon close at 14 years old; the medial epicondyle at 15; the radial head and lateral epicondyle at around 16.

A. Celli (✉)
Hesperia Hospital, Modena, Italy

Department of Orthopaedic and Traumatology Surgery, Shoulder and Elbow Unit, Modena, Italy

N. Fabio · D. Vito · L. A. Pederzini
Nuovo Ospedale di Sassuolo, Modena, Italy

Department of Orthopaedic, Traumatology and Arthroscopic Surgeries, Modena, Italy

J. G. Lane et al. (eds.), *The Art of the Musculoskeletal Physical Exam*,
https://doi.org/10.1007/978-3-031-24404-9_22

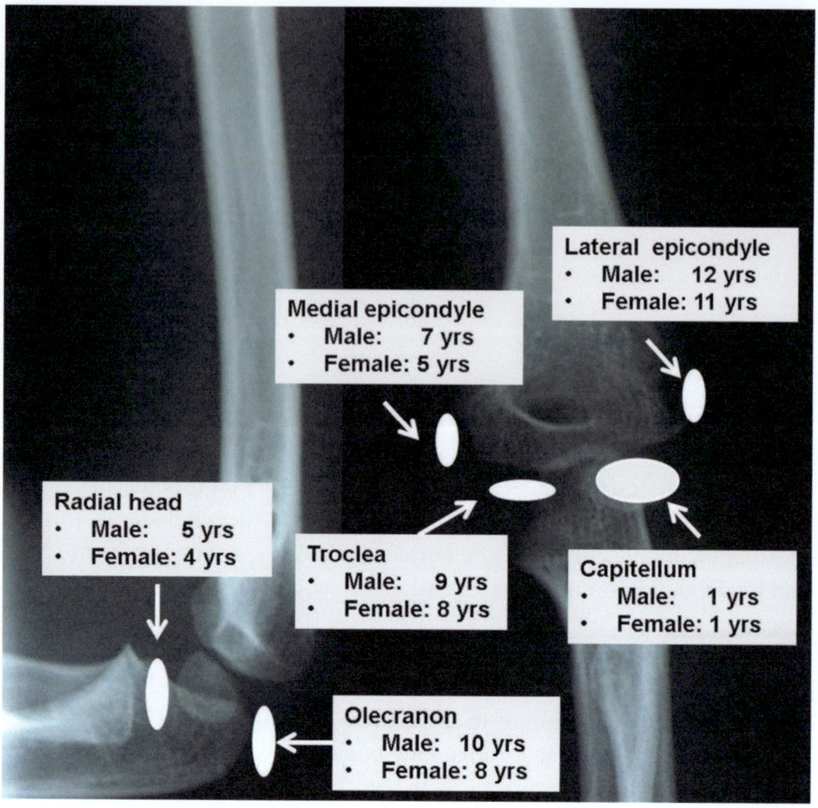

Fig. 22.1 Ossification centers

22.2 The History

22.2.1 Acute Injury

Acute fractures: History is usually indicative of a high-energy trauma or a fall onto an outstretched hand with extended elbow. In these cases, it is important to consider complex fracture-dislocations or associated neurovascular injuries, which may make overall management more complicate.

Age of the patient and degree of skeletal maturity also influence the type of fracture pattern sustained.

22.2.2 Chronic Injury

In chronic overuse injuries with patients in their early childhood whose secondary ossification centers have yet to appear, reported pain is likely due to repetitive injuries to the ossification center and apophysis—most frequently on the medial epicondyle. In adolescents, however, the pain is more likely due to muscle attachments repetitively pulling on the medial epicondyle which eventually results in medial epicondyle avulsion or ulnar collateral ligament injuries.

22.3 Physical Examination

Physical examinations should follow different criteria: inspection, palpation, and "Look, Feel and Move" movement. The examination is completed by specific provocation tests.

Inspection is key when examining younger patients. Skin should be checked for any ecchymosis, abrasions, tissue loss, or swelling indicating an acute injury. Alignment of the upper limb axis—also known as the carrying angle—should be assessed and compared to the contralateral side (Fig. 22.2). Variations in the carrying angle may also be secondary to previous trauma.

Previous humeral fractures can result in cubitus varus (or "gunstock" deformity) or varus malunion (Fig. 22.3). Genetic syndromes associated with congenital elbow abnormalities—for example, congenital radial head dislocations—can also lead to changes to the carrying angle.

Clinical palpations involve the evaluations of the bony landmarks including the distal *humeral*, the medial and lateral epicondyles, the olecranon process, the radial head and capitellum, and the radio-capitellar joint space. Assessment of the lateral soft spot between the capitellum, the radial head, and the olecranon may exhibit a possible joint intra-articular hematoma.

Normal range of motion is defined comparing the motion on the opposite side. Arc of flexion-extension and supination-pronation should be actively and passively evaluated (Fig. 22.4). Stability and neurological deficit examinations

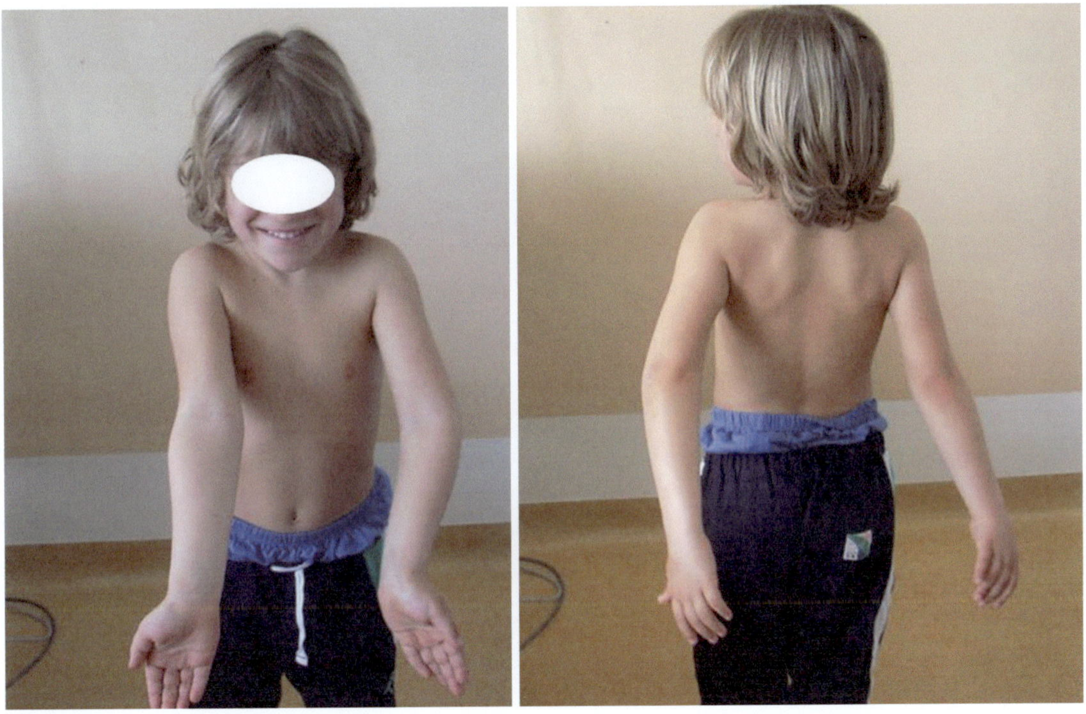

Fig. 22.2 The alignment of the upper limb axis should be assessed and compared to the contralateral side

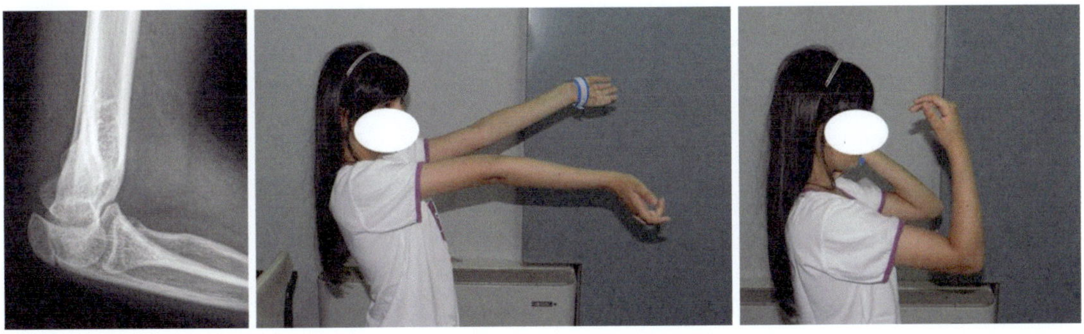

Fig. 22.3 Variations in alignment may also be secondary to previous trauma, resulting in a cubitus varus (or "gunstock" deformity) or varus malunion

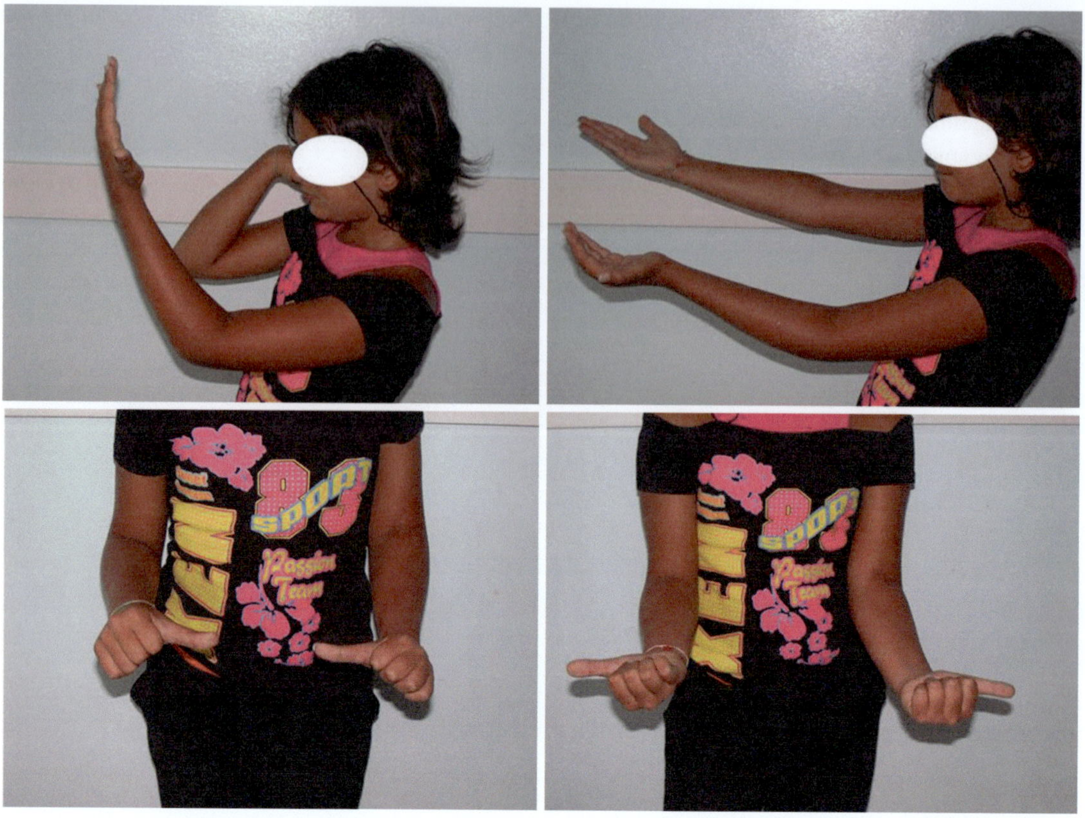

Fig. 22.4 Flexion-extension and supination-pronation should be evaluated actively and passively comparing them to the contralateral side

should be performed and included in the movement evaluation *time as for the adult.*

22.3.1 Imaging Assessment

Radiographic evaluations of immature elbows require knowledge of the normal sequence of appearance of the secondary ossification centers (Fig. 22.1) . Thus, they need to be performed also on the contralateral side to correctly distinguish between pathological conditions and normal conditions on the lateral view (Fig. 22.5). The anterior humeral line along the anterior cortex of the distal humerus normally intersects the middle third of the capitellum. Anomalies suggest underlying a supracondylar fracture (Fig. 22.6). Conversely, the radio-capitellar line through the

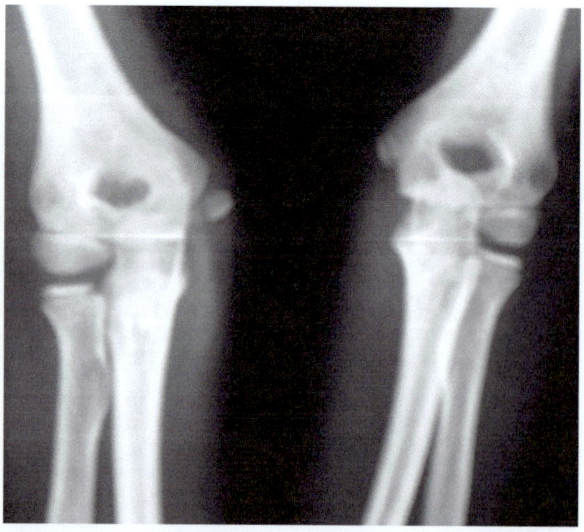

Fig. 22.5 Radiographic evaluations of the immature elbow need to be performed also on the contralateral side in order to correctly distinguish pathology from normal condition

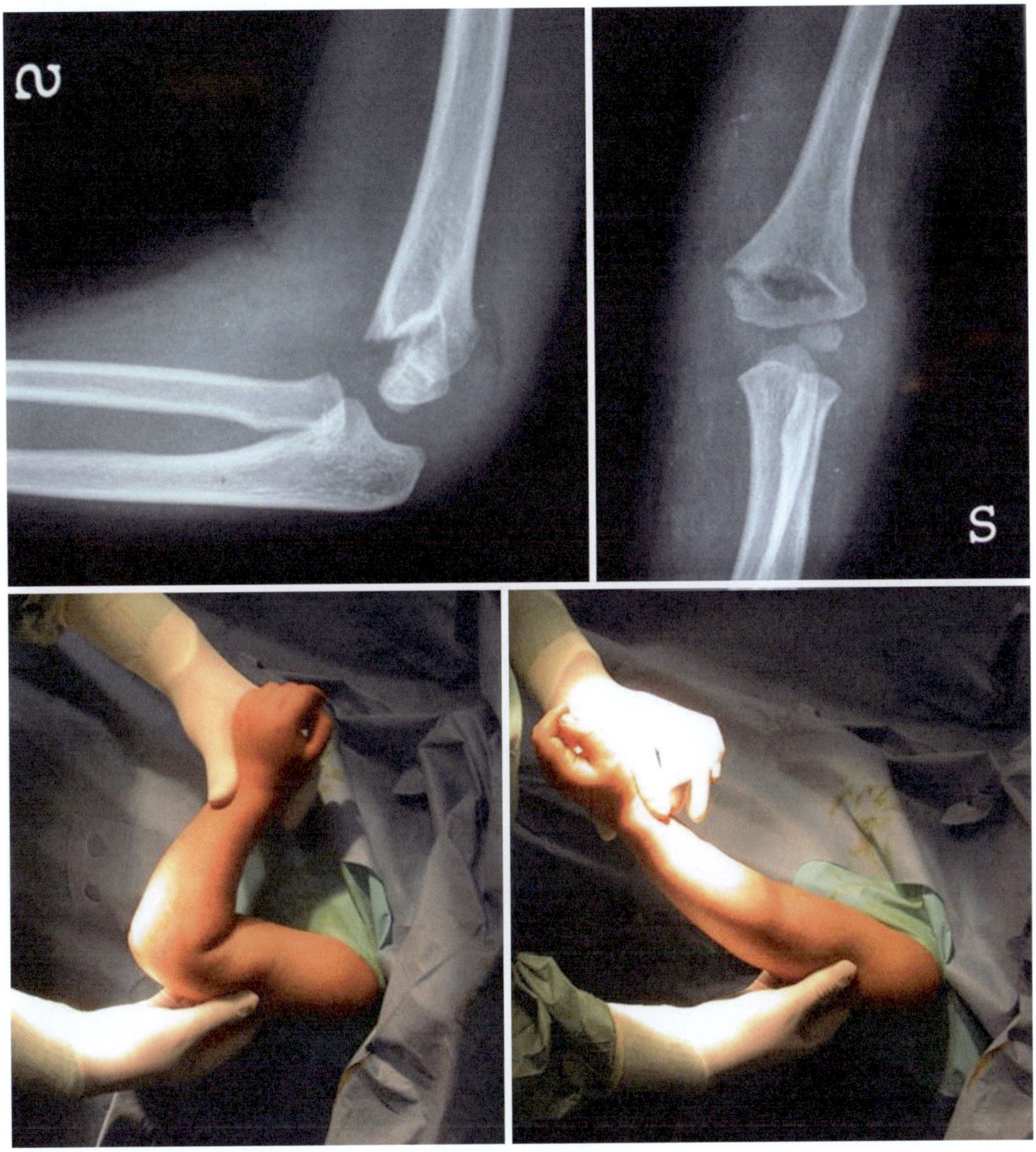

Fig. 22.6 The anterior humeral line along the anterior cortex of the distal humerus normally intersects the middle third of the capitellum, changes on it suggest underlying supracondylar fracture

longitudinal axis of the proximal radius should pass through the center of capitellum. Anomalies are indicative of a radial head dislocation (Fig. 22.7). Posterior and anterior fat pads should not be normally visible. When, instead, they are, joint hematoma may be present. It is also important to evaluate congruency of the joint and presence of displaced medial and lateral epicondyles (Fig. 22.8). A combination of US, MRI, and CT evaluations can be used to better define the pathologies and for surgical management purposes (Fig. 22.9). Dynamic fluoroscopy evaluation is useful to assess elbow stability during provocative tests and during arc of motions.

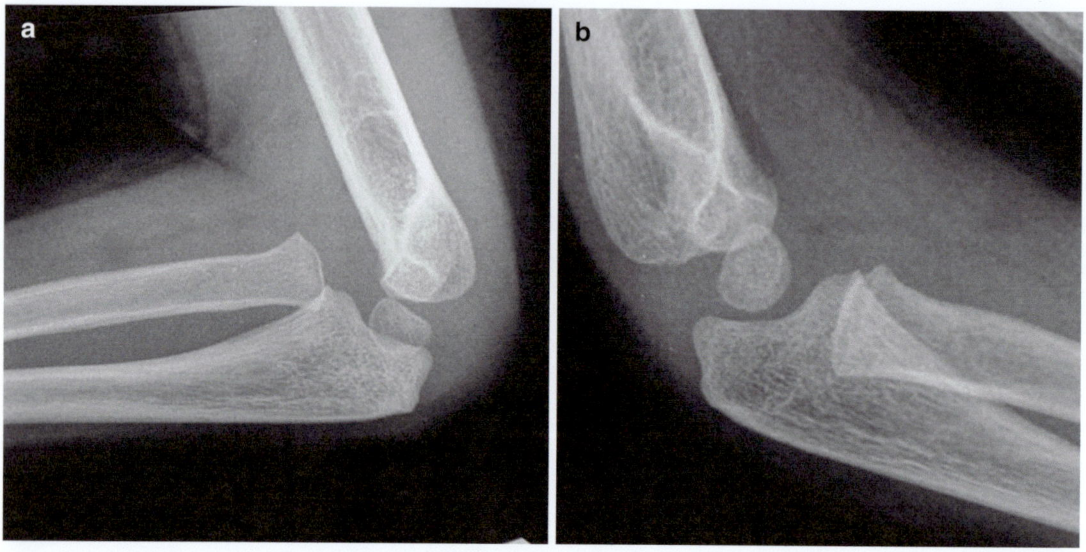

Fig. 22.7 The radio-capitellar line through the longitudinal axis of the proximal radius should pass through the center of capitellum (**b**) changes on it suggests radial head dislocation (**a**)

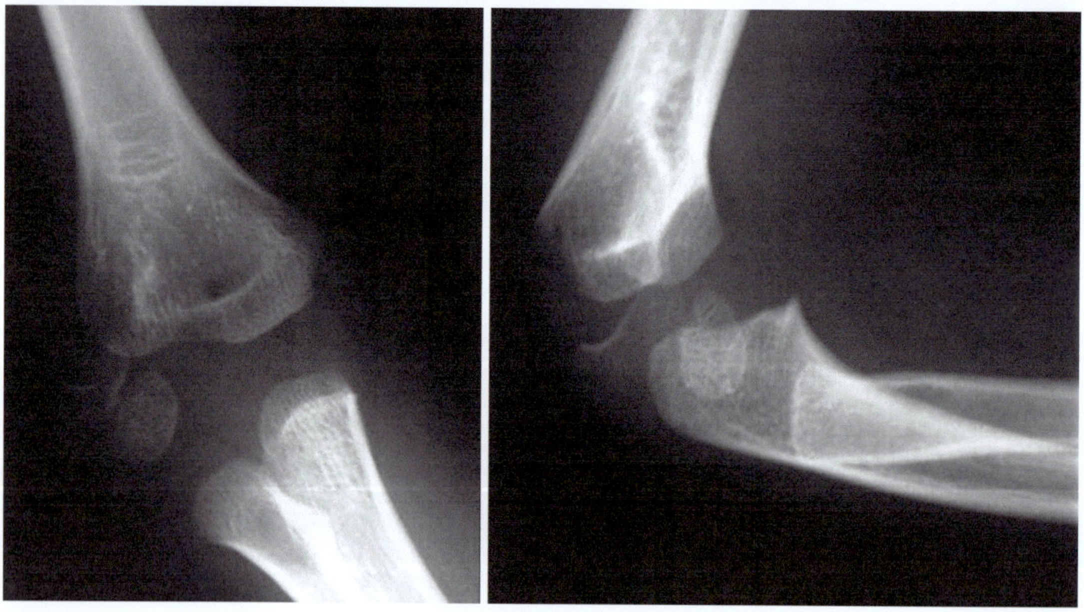

Fig. 22.8 Congruency of the joint and presence of the displaced medial and lateral epicondyles are also important to be evaluated

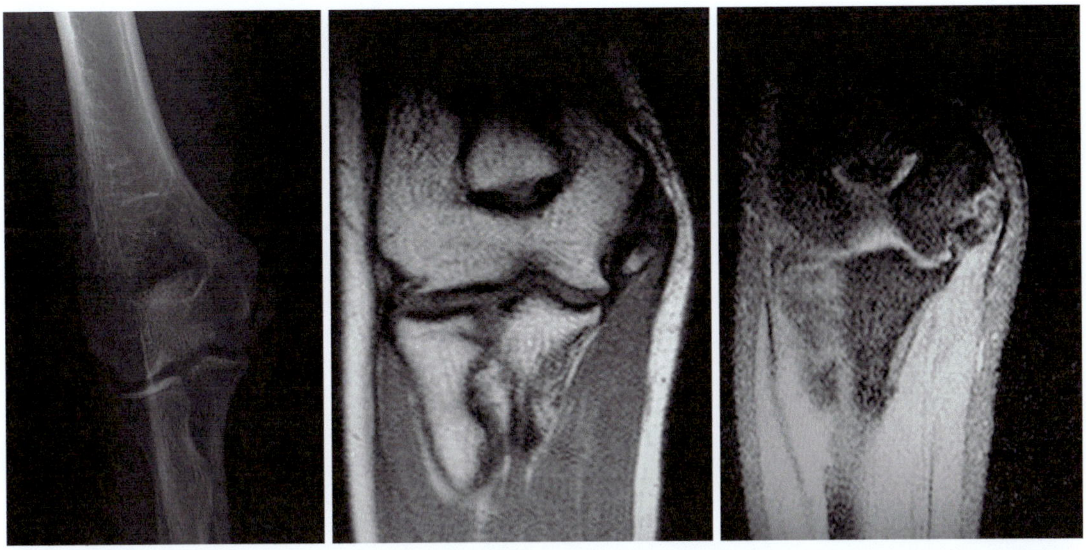

Fig. 22.9 MRI evaluations can be used in association to better define the pathologies and for surgical management purposes

22.4 Pathological Conditions

22.4.1 Supracondylar Fracture of the Distal humerus

In children, supracondylar fractures are the most common type of elbow fractures observed [5, 6]. Hyperextension force during trauma causes posterior angulation or displacement of the distal humeral fragment relative to the humeral shaft. Gartland classification is useful for classification and management [6]. Gartland type I is a nondisplaced or minimally displaced fracture. It is managed with simple immobilization. Gartland type II is a displaced fracture with intact posterior cortex. It is managed with closed reduction and, depending on the degree of displacement, with percutaneous pinning depending (Fig. 22.10).

Gartland type III corresponds to a completely displaced fracture with complete loss of the posterior cortical bone. It can be managed with closed reduction and percutaneous pinning. Open reduction is indicated if closed reduction fails or in case of concomitant involvement of the neurovascular structures [7, 8] (Fig. 22.11).

22.4.2 Lateral Condylar Fracture

Lateral condylar fractures are the second most common elbow fractures in children [9]. Patients usually report lateral elbow pain after falling onto an outstretched arm with their forearm in supination which causes a varus force on the elbow. The fracture originates at the lateral metaphysis of the distal humerus and extends obliquely toward the elbow joint. Milch classification is used [10]:

- Milch type I fracture begins at the lateral metaphysis of the distal humerus and extends

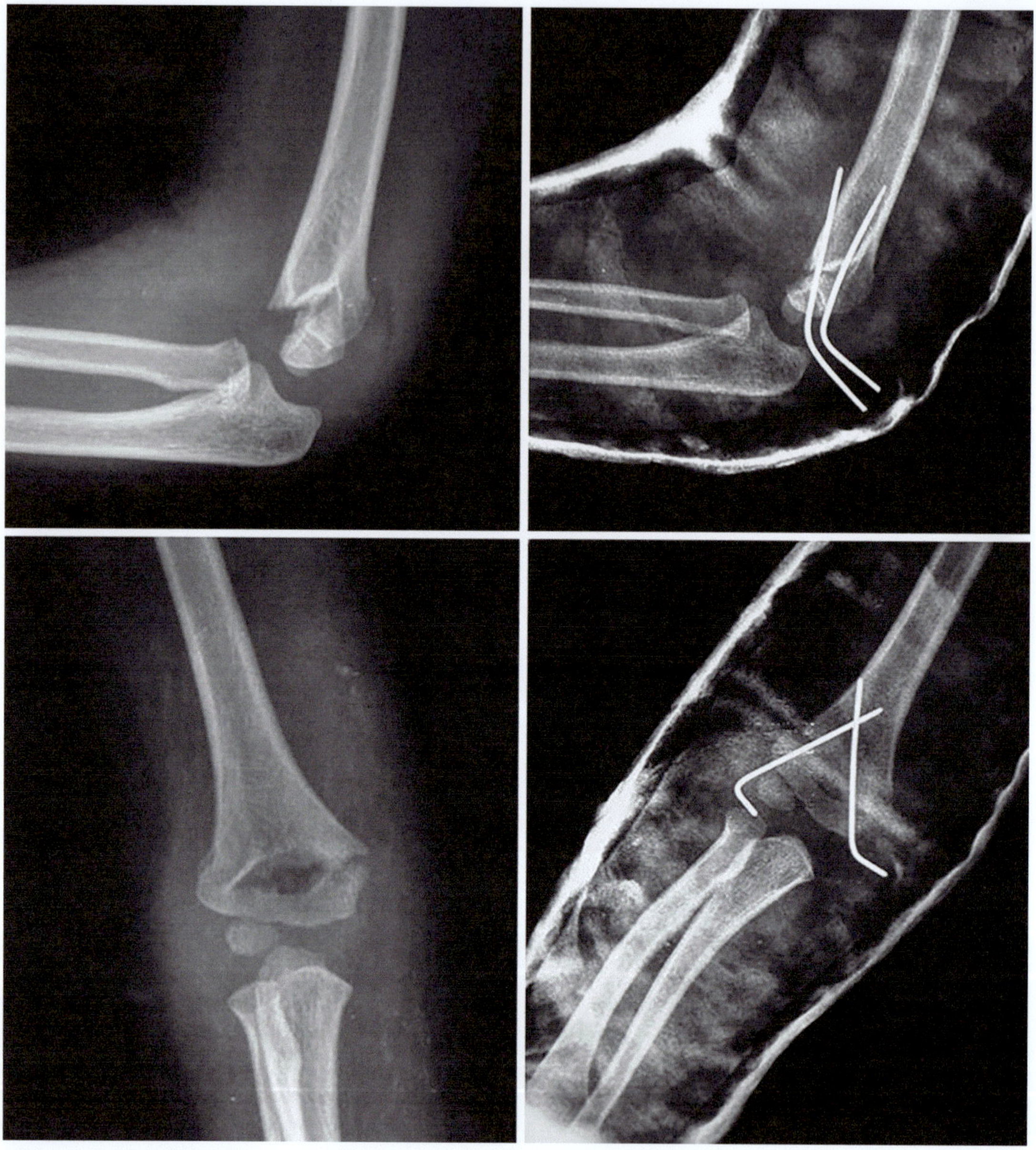

Fig. 22.10 Closed reduction or closed reduction with percutaneous pinning

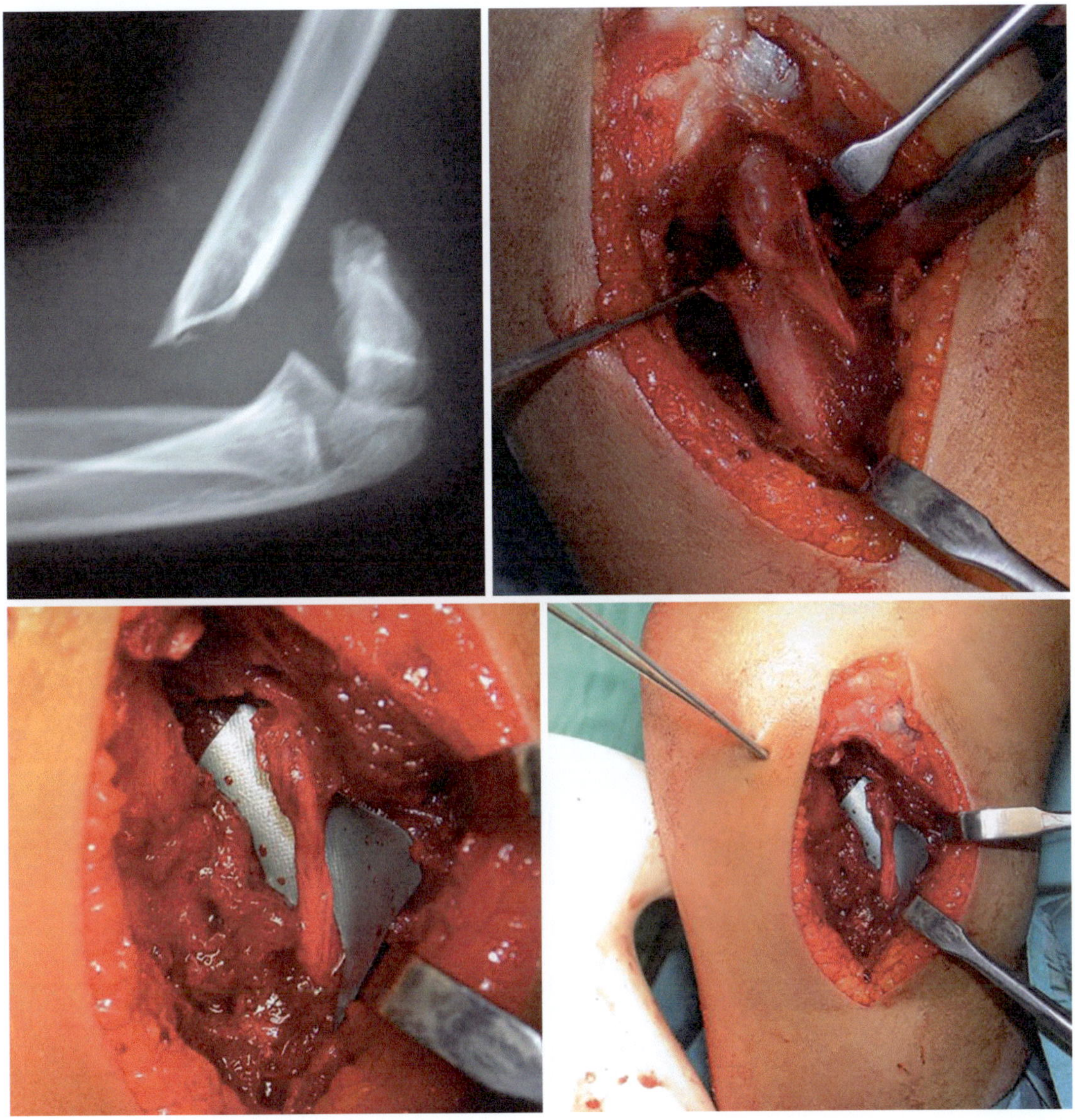

Fig. 22.11 The open reduction is indicated if closed reduction fails or if there are concomitant involvement of the neurovascular structures (radial nerve in the case)

over the capitellar ossification center. The trochlear ridge remains intact. It is considered as a stable fracture (Fig. 22.12).

- Milch type II fracture also begins at the lateral metaphysis of the distal humerus, but it extends over the capitellar-trochlear groove and it involves the trochlea. it is considered as an unstable fracture (Fig. 22.13).

22.4.3 Medial Epicondyle Fractures

It is the most common fractures in juvenile or adolescent throwing athletes [11–15]. Injury mechanism involves medial traction caused by acute valgus stress, associated with a massive flexor-pronator mass muscle contraction which can produce medial epicondylar detachment.

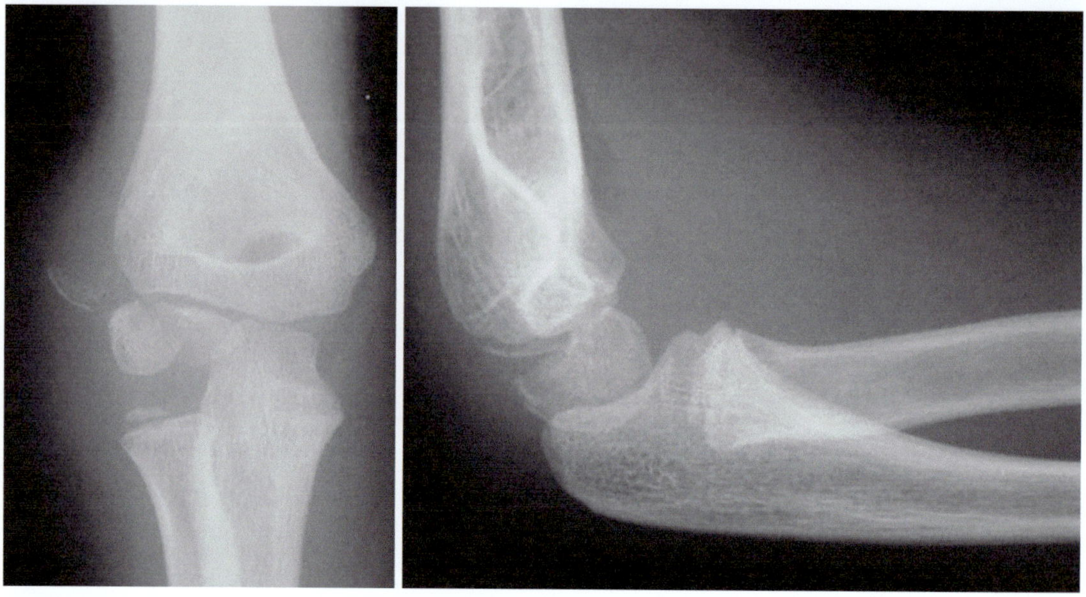

Fig. 22.12 If the trochlear ridge is intact, it is considered as a stable fracture

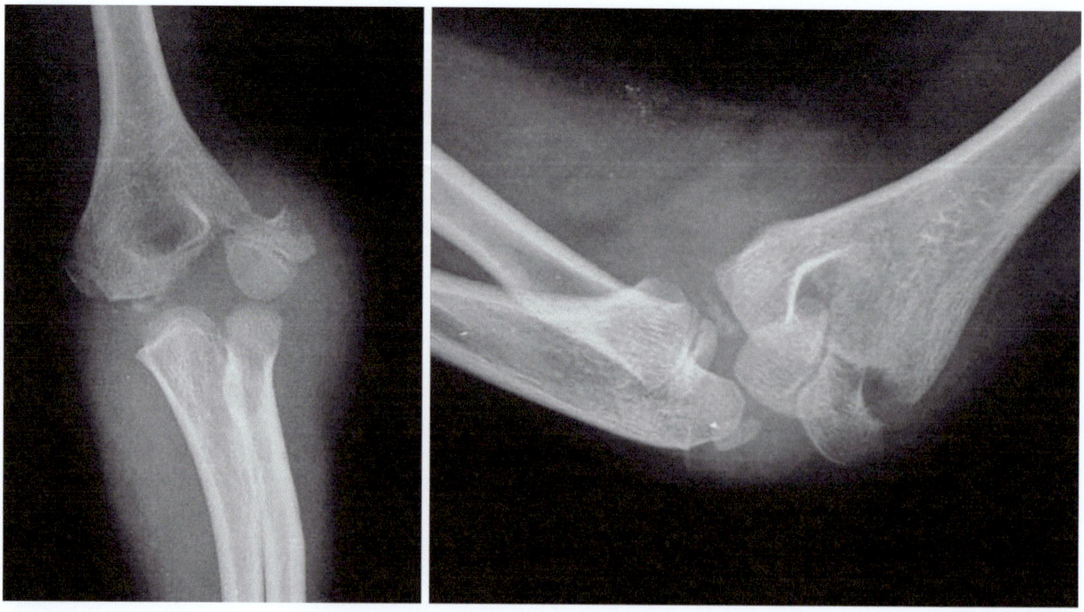

Fig. 22.13 If it extends to the capitellar-trochlear groove and involves the trochlea, it is considered as an unstable fracture

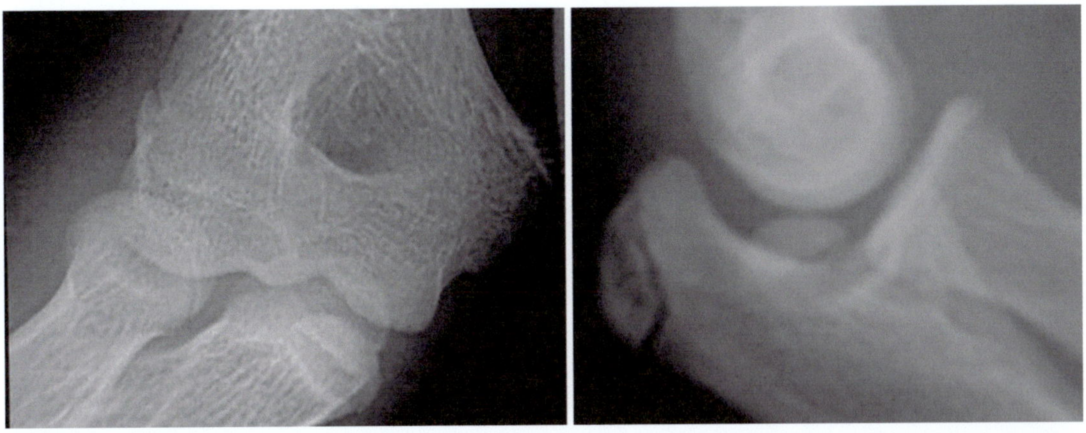

Fig. 22.14 Displacement of the avulsed fragment occurs by the flexor-pronator muscle mass, sometimes, the fragment may displace into the elbow joint

Athletes typically present acute medial elbow pain, and they are unable to continue their activities. Ulnar nerve paresthesia may develop following the injury. Clinically, patients present local tenderness on palpation of the medial epicondyle, frequently accompanied by edema and occasionally by ecchymosis. Pain typically limits full elbow extension, and assessment of elbow stability may be difficult (valgus stress test). Valgus instability is usually related to epicondylar avulsion (the physis of the medial epicondyle is the weak point in adolescent elbows). Acute ruptures of the ulnar collateral ligament are most commonly seen in skeletally mature individuals [15].

Careful assessment of the athletes' range of motion and stability is critical, as is the performance of thorough neurovascular examinations with particular attention to ulnar nerve function.

In case of suspected medial epicondyle avulsion fracture, X-rays assessment can be useful. Diagnosis may be facilitated by a comparison view of the contralateral elbow using the gravity stress test radiograph [16], or a manual stress view using dynamic fluoroscopy evaluations.

Displacement of the avulsed fragment *is consequence of the flexor-pronator muscle mass, sometimes, the fragment may displace into the elbow joint* (Fig. 22.14), *particularly when the medial epicondylar avulsion is related to an elbow dislocation* [14–16]. Woods and Tullos [14] presented a classification system related to patient age and fragment size:

Type I injuries occur in children under the age of 14 with fracture fragments typically involving the entire apophysis (Fig. 22.15). Displacement and rotation of the fragments are common.

Type II injuries occur in patients over the age of 15, and they consist in large fragments that may involve the anterior band of the ulnar collateral ligament (Fig. 22.16).

Type III injuries occur in patients above 15, and they consist in fragments smaller than Types I or II injuries.

Treatment depends on fragments displacement. Nondisplaced fragments are typically treated conservatively. Displaced fragment—with or without fragments incarcerated into the joint—are treated operatively. Presence of ulnar nerve dysfunction requires exploration with open reduction and internal fixation.

22.4.4 Proximal Radius Fracture

In children, fractures of the proximal radius involve the radial neck and/or the physis. Radial neck fractures occur when falling onto an out-

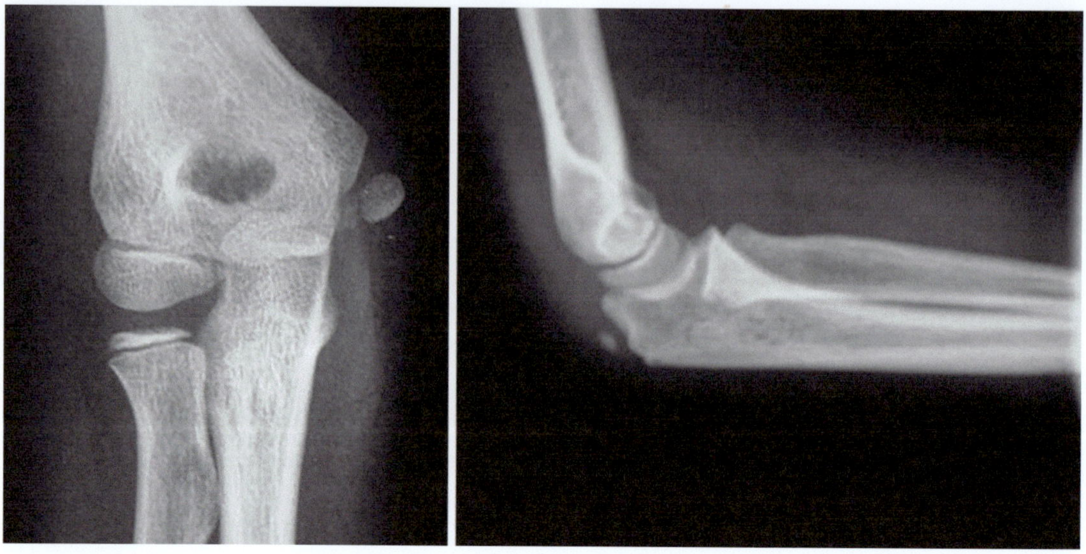

Fig. 22.15 In patients under the age of 14 with a fracture fragment typically involving the entire apophysis

stretched hand and forearm causing a Valgus force on the elbow. The proximal ulna impacts on the capitellum and fractures the radial neck [7]. Average age of injury ranges 9–10 years old [8]. Nondisplaced radial neck fractures are often very difficult to detect—they are frequently hidden in initial X-rays. Pain on the lateral compartment of the elbow joint and limited forearm rotation are frequently present. Intra-articular hematomas can be palped thought the soft spot and visualized on lateral X-rays (anterior fat pat).

In nondisplaced fractures, or those with less than 30 degrees angulation, treatment is conservative. If angulation is greater than 30 degrees, open or closed reduction with percutaneous pinning is the standard treatment (Fig. 22.17).

22.4.5 Olecranon Apophyseal Injury or Olecranon Fracture

Acceleration phase of throwing in overhead athlete subjects the olecranon to repetitive forceful contraction of the triceps, which may result in olecranon apophyseal injury [17–19]. These repetitive forceful triceps contractions place a distraction force across the epiphysis. Patients may report acute or—less frequently—chronic pain at the posterior aspect of the elbow, frequently accompanied by swelling and decreased range of motion. Physical examination findings include olecranon tenderness on palpation and pain on resisted elbow extension. Radiographs characteristically demonstrate widening or fragmentation of the olecranon physis. Treatment of olecranon apophyseal injuries depends on the

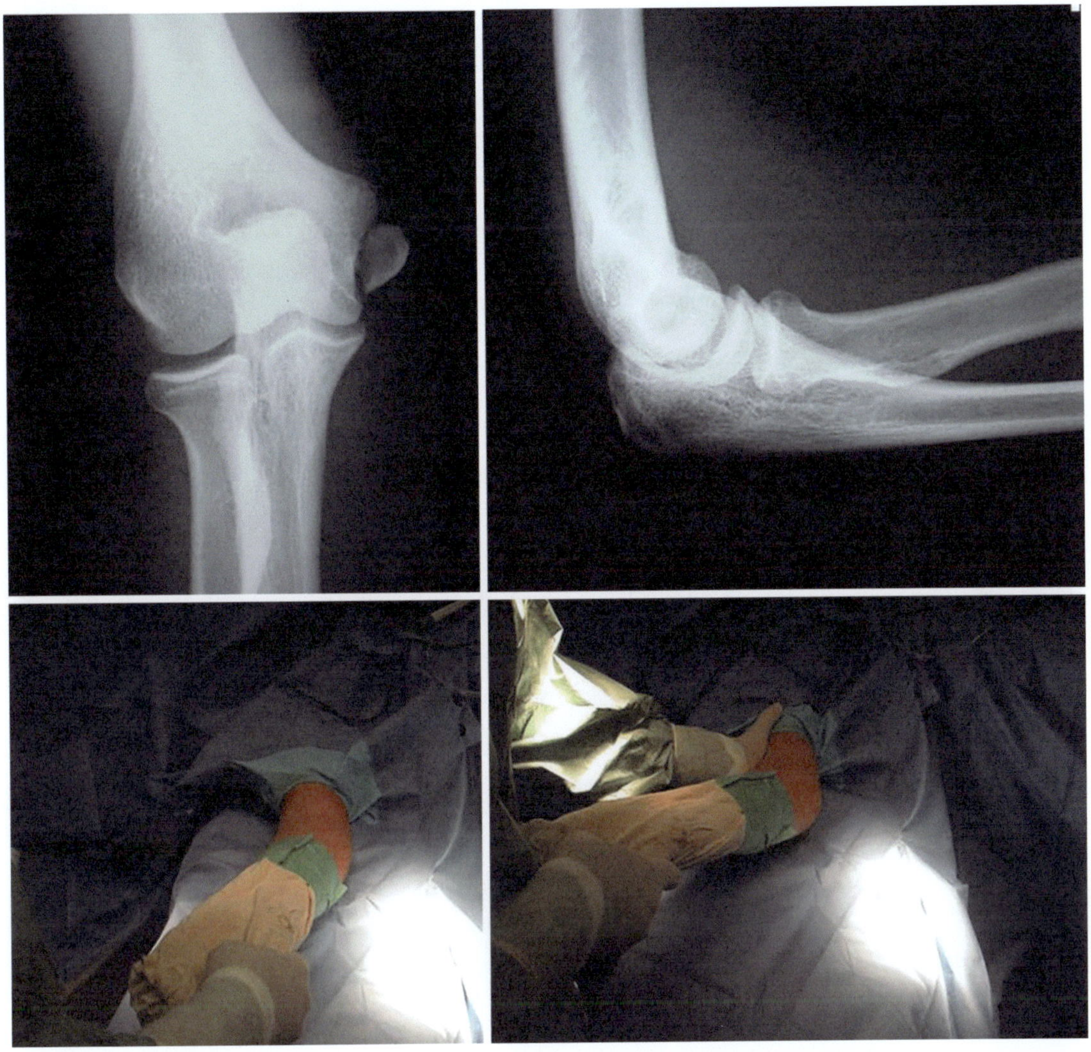

Fig. 22.16 In patients over the age of 15 with a large fragment that may potentially involve the anterior band of the ulnar collateral ligament

degree of fragment displacement, and on the athlete's symptoms. Initially, treatment consists of activity modification. Conservative management is normally *performed* within approximately 4 to 6 weeks, then surgical treatment may be required.

Olecranon fractures are relatively uncommon in children compared to adults. They are a consequence of a fall onto an outstretched hand or of a direct force being applied onto the olecranon.

Olecranon fractures are seen in association with radial head dislocations, radial neck fractures, or medial or lateral condylar fractures [20] (Fig. 22.18). Most olecranon fractures are non-displaced or minimally displaced (Fig. 22.19). Displaced olecranon fracture is managed operatively.

22.4.6 Elbow Dislocations, Fracture-Dislocations

Simple elbow dislocation is uncommon in juvenile and adolescent athletes (Fig. 22.20). Dislocations of the radial head most frequently occur as isolated injuries or in combination with

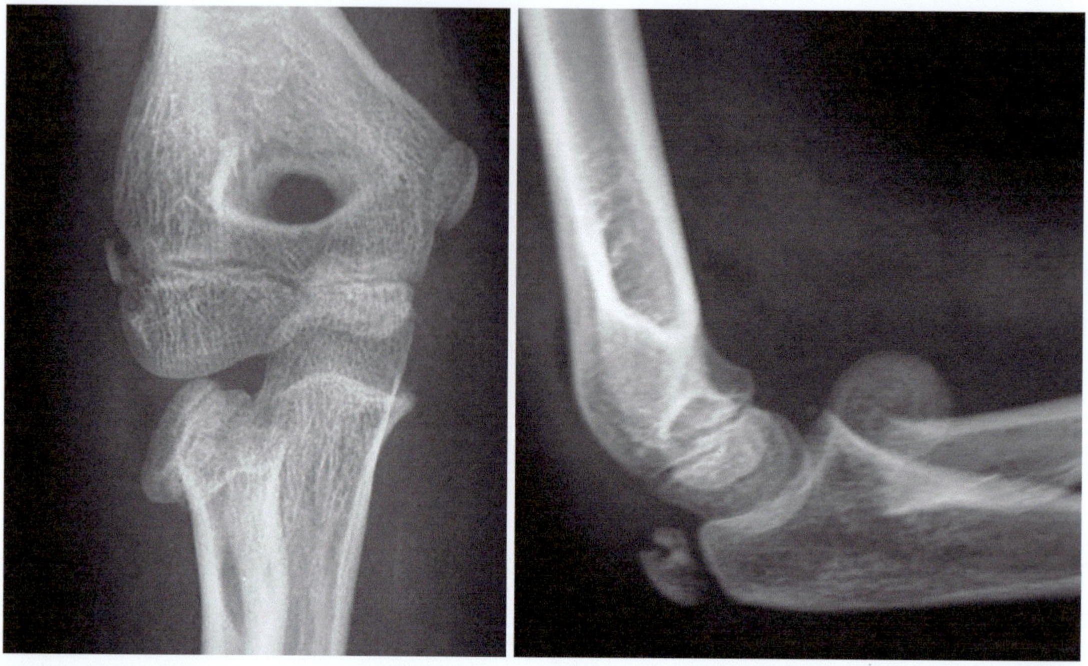

Fig. 22.17 If the radial head angulation is greater than 30 degrees, open or closed reduction with percutaneous pinning is the standard treatment

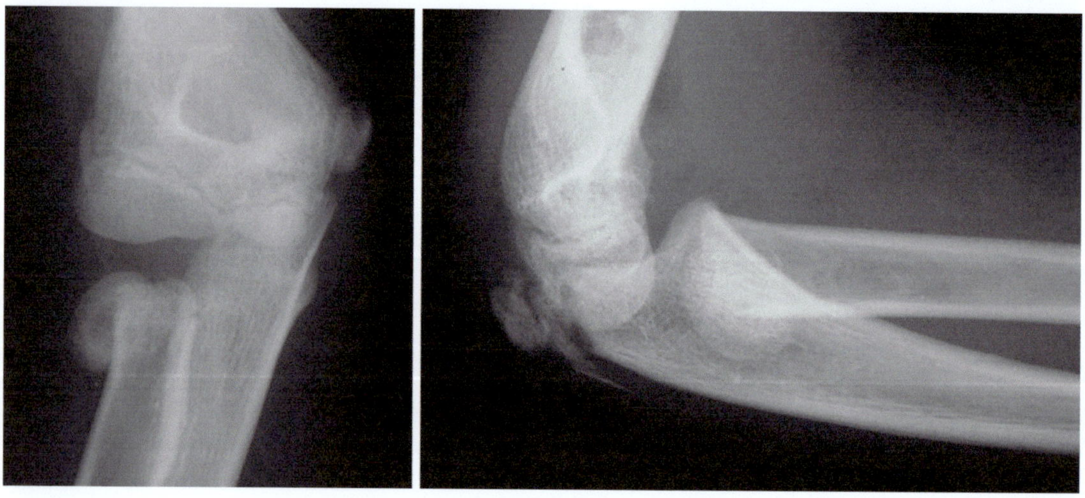

Fig. 22.18 Olecranon fractures are seen in association with radial head dislocation, radial neck fracture, or medial or lateral condylar fracture

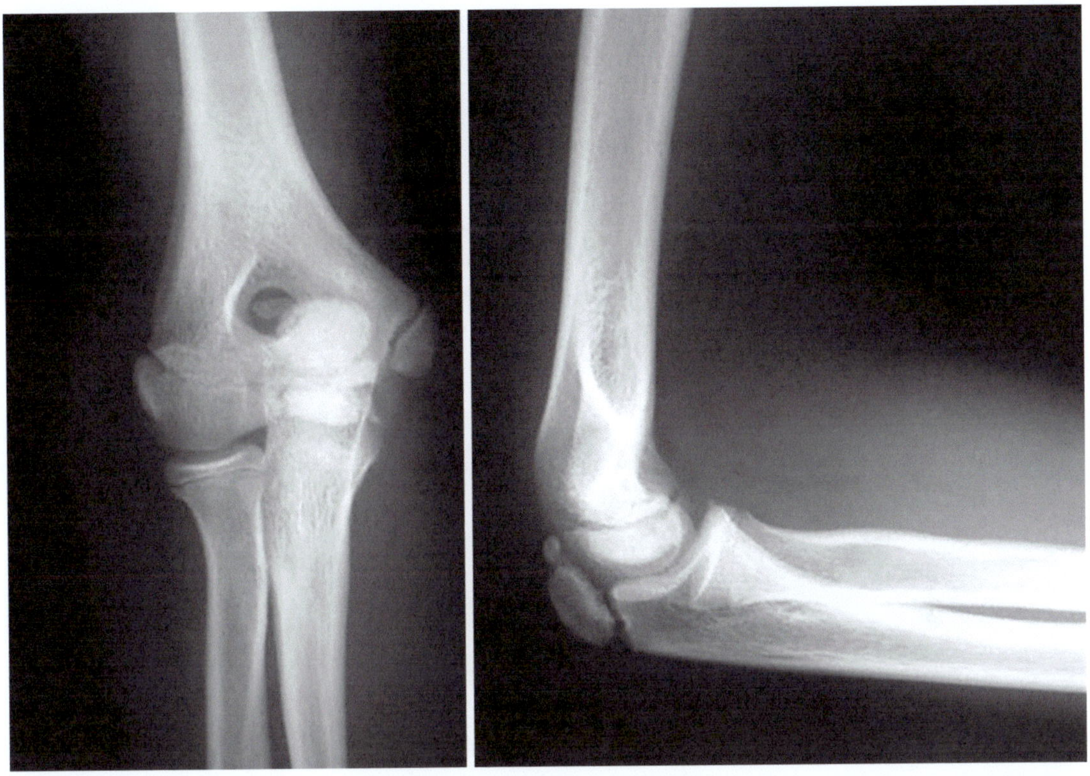

Fig. 22.19 Most olecranon fractures are nondisplaced or minimally displaced

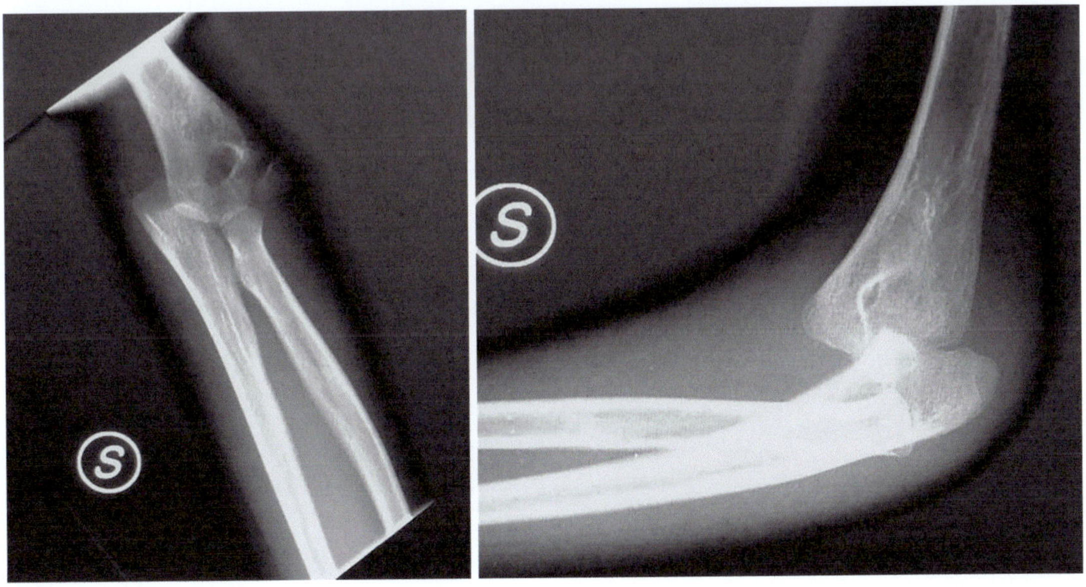

Fig. 22.20 Simple elbow dislocation are uncommon in youth and adolescence

forearm or humeral fractures (Fig. 22.21). Isolated radial head dislocations are identified on lateral X-ray views as loss of the radio-capitellar line. Proximal radioulnar joint dislocations should be assessed for the presence of combined ulnar fractures. Combination of proximal radioulnar dislocations and ulnar fractures are termed Monteggia fractures (Fig. 22.22).

The Bado classification [21] can be used to define Monteggia fracture-dislocations (Fig. 22.23).

Type I, anterior radial head dislocation
Type II, posterior radial head dislocation
Type III, lateral radial head dislocation
Type IV, anterior radial head dislocation in conjunction with proximal ulnar and radial fractures

Management of these injuries involves reduction of the radial head dislocation and fixation/reduction of the ulnar fracture.

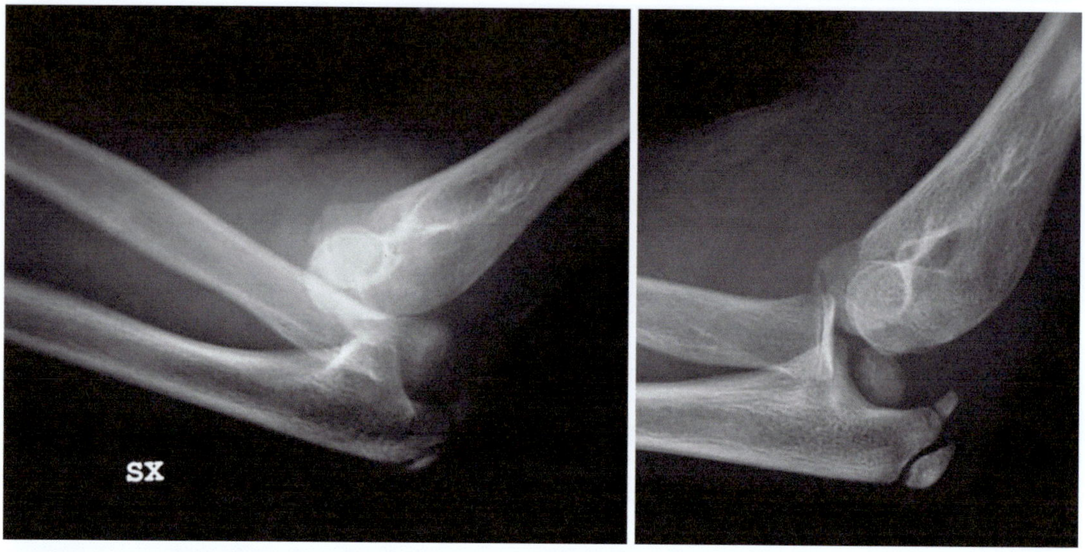

Fig. 22.21 Elbow dislocation most frequently occurs associated with radial head fracture or in combination with a forearm or humeral fracture

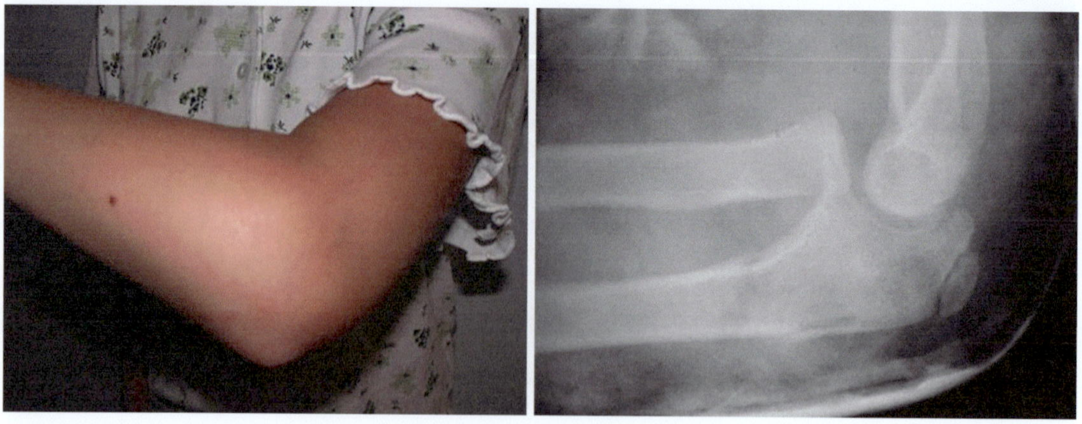

Fig. 22.22 The combination of proximal radioulnar dislocation and ulnar fracture is termed Monteggia lesion

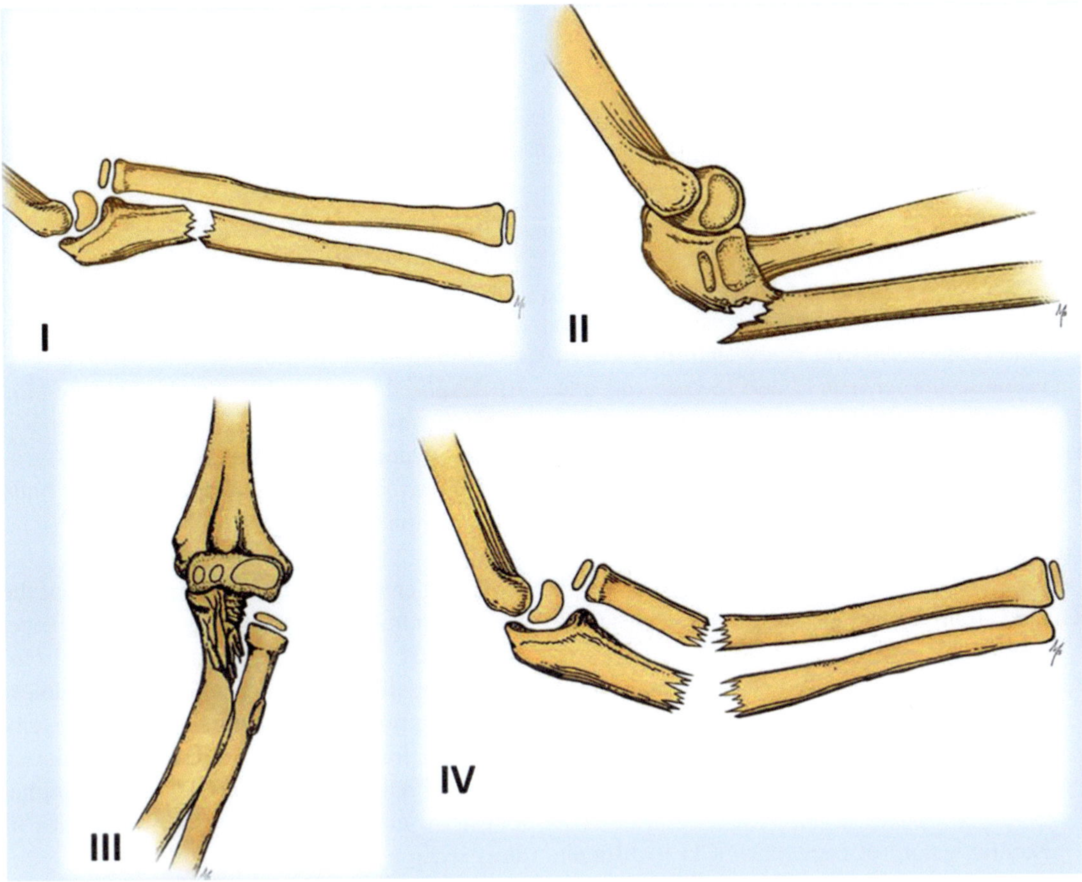

Fig. 22.23 Monteggia fracture-dislocation types

22.4.7 Panner's Disease

Osteochondrosis of the capitellum was identified in the context of its similarity to Legg-Calve-Perthes disease, and it was first described in 1927 [22].

Panner's disease is defined as a focal lesion of the capitellar subchondral bone and of its overlying articular cartilage. It is characterized by a disorder of the capitellar ossification center which begins as degeneration or necrosis and is followed by regeneration of the ossification nucleus [22]. Panner's disease is the most common cause of lateral elbow pain in young children, and it is most commonly seen in patients less than 10 years old. Etiology has not been determined yet. However, a proposed mechanism involves alteration in the vascularity of the developing capitellum. Its natural history involves a benign, self-limited process, with eventual restoration of the normal capitellar size contour. Collapse of the subchondral bone and persistent deformity are rare.

Children presenting Panner's disease typically report pain on the lateral elbow that is exacerbated with activity and relieved by resting. It is frequently accompanied by a loss of elbow motion and subjective stiffness. Physical examination may reveal lateral tenderness on palpation at the radio-capitellar joint, with elbow contracture. Imaging studies (MRI, US) typically reveal diffused involvement of the central aspect of the anterior capitellum, with an irregular appearance and a variable area of rarefaction and fragmentation.

22.4.8 Osteochondritis Dissecans (OCD)

First described in the knee by Konig in 1889, OCD was initially presented as a condition whose appearance was suggestive of a subchondral inflammatory process causing dissection of a fragment of the overlying articular cartilage [23]. However, this condition has never been shown to involve a true inflammatory process [24]. It is a common cause of lateral elbow pain in children and adolescents between 11 and 16 years old, and it is most frequently seen in athletes who sustain repetitive trauma through valgus stress and lateral compression across the elbow (Fig. 22.24).

Based on X-rays, it can be classified as follows:

Grade I, translucent shadow in central or lateral capitellum
Grade II, clear zone or split line between the subchondral bone and the lesion
Grade III, loose bodies identified. MRI can be useful to assess the lesion.

Natural history of capitellar OCD is difficult to predict, and no reliable criteria exists to predict which lesions will collapse with subsequent joint incongruity and which will go on to heal without further sequelae.

The lesions can be classified based on a combination of clinical examination, diagnostic imaging, and arthroscopy findings.

Baumgarten et al. proposed the following classification [24]:

I Smooth, soft, ballotable
II Fissuring or fibrillation
III Exposed bone with fixed osteochondral fragments
IV Loose, nondisplaced fragments
V Displaced fragments with loose body formation

Treatment is related to healing and size of the osteochondral bone. If healing does take place, this usually occurs by the time of physeal closure. If healing does not occur during adolescence, appropriate treatment is not only important for a potential return to pre-existing levels of activity, but imperative for optimal long-term elbow functioning with activities of daily living.

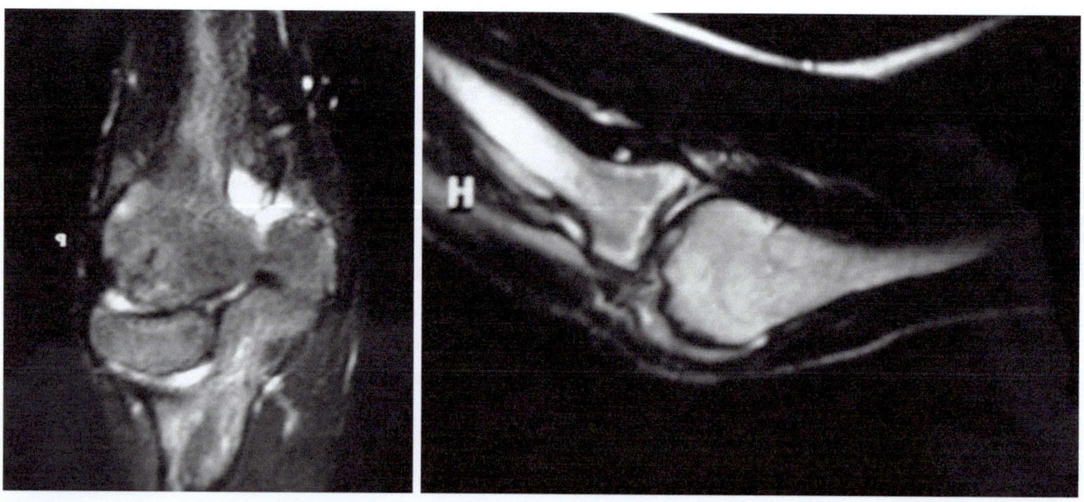

Fig. 22.24 OCD was initially presented as a condition with an appearance suggestive of a subchondral inflammatory process causing dissection of a fragment of overlying articular cartilage

References

1. Pappas AM. Elbow problems associated with baseball during childhood and adolescence. Clin Orthop. 1982;164:30–41.
2. Bradley JP. Upper extremity: elbow injuries in children and adolescents. In: Stanitski CL, DeLee JC, Drez D, editors. Pediatric and adolescent sports medicine. Orthopedic sports medicine: principles and practice, vol. 3. Philadelphia, PA: WB Saunders; 1994. p. 242–61.
3. DeFelice GS, Meunier MJ, Paletta GA. Elbow injury in the adolescent athlete. In: Altchek DW, Andrews JR, editors. The athlete's elbow. Philadelphia, PA: Lippincott Williams and Wilkins; 2001. p. 231–48.
4. Haraldsson S. On osteochondrosis deformans juvenilis capituli humeri including investigation of intra-osseous vasculature in distal humerus. Acta Orthop Scand. 1959;38(Suppl):1–23.
5. Gartland JJ. Management of supracondylar fractures of the humerus in children. Surg Gynecol Obstet. 1959;109(2):145–54.
6. Alburger PD, Weidner PL, Betz RR. Supracondylar fractures of the humerus in children. J Pediatr Orthop. 1992;12(1):16–9.
7. Ogden JA. Skeletal injury in the child. 3rd ed. New York, NY: Springer-Verlag; 2000. p. 456–567.
8. Rockwood CA, Wilkins KE, Beaty JH, editors. Fractures in children. Philadelphia, PA: Lippincott-Raven; 1996. p. 663–77.
9. Jacoby SM, Herman MJ, Morrison WB, Osterman AL. Pediatric elbow trauma: an orthopaedic perspective on the importance of radiographic interpretation. Semin Musculoskelet Radiol. 2007;11(1):48–56.
10. Milch H. Fractures of the external humeral condyle. J Am Med Assoc. 1956;160(8):641–6.
11. Gugenheim JJ Jr, Stanley RF, Woods GW, et al. Little League survey: the Houston study. Am J Sports Med. 1976;4:189–200.
12. Jobe FW, Nuber G. Throwing injuries of the elbow. Clin Sports Med. 1986;5:621–36.
13. Ogawa K, Ui M. Fracture-separation of the medial humeral epicondyle caused by arm wrestling. J Trauma. 1996;41:494.
14. Woods GW, Tullos HS. Elbow instability and medial epicondyle fractures. Am J Sports Med. 1977;5:23–30.
15. Adams IE. Injury to the throwing arm: a study of traumatic changes in the elbow joints of boy baseball players. Calif Med. 1965;102:127–32.
16. Rosendahl B. Displacement of the medial epicondyle into the elbow joint. Acta Orthop Scand. 1959;36:212–9.
17. Gill TJ, Micheli LJ. The immature athlete. Common injuries and overuse syndromes of the elbow and wrist. Clin Sports Med. 1996;15:401–23.
18. Ireland ML, Andrews JR. Shoulder and elbow injuries in the young athlete. Clin Sports Med. 1988;7:473–94.
19. DaSilva MF, Williams JS, Fadale PD, et al. Pediatric throwing injuries about the elbow. Am J Orthop. 1998;27:90–6.
20. Gokkus K, Kose O, Saylik M, Sagtas E, Turan AA. Pediatric olecranon fractures associated with radial neck fractures: review and report of two cases. Trauma Mon. 2016;21(3):e20686.
21. Bado JL. The Monteggia lesion. Clin Orthop Relat Res. 1967;50:71–86.
22. Panner HJ. A peculiar affection of the capitellum humeri, resembling Calve-Perthes disease of the hip. Acta Radiol. 1929;10:234–42.
23. Schenck RC Jr, Goodnight JM. Osteochondritis dissecans. J Bone Joint Surg Am. 1996;78(3):439–56.
24. Baumgarten TE, Andrews JR, Satterwhite YE. The arthroscopic classification and treatment of osteochondritis dissecans of the capitellum. Am J Sports Med. 1998;26(4):520–3.

Part III

Wrist and Hand Reviewer Dr Cage

Hand Anatomy

23

Christopher M. Stewart and Paul A. Ghareeb

23.1 Skin and Subcutaneous Tissue

The skin and subcutaneous tissues of the hand perform important functional tasks during use of the hand. The palmar (or volar) skin is highly specialized to perform manual tasks. The skin is found to be thicker, and the papillary ridges contain an increased number of sensory organs which make it ideal for grasping and manipulating small objects. The dorsal skin is more mobile and contains more lymphatic channels and veins.

Palmar creases are formed due to the motion of the surrounding joints and are found to be relatively constant. These creases can be used as important landmarks when identifying deeper structures. Just under the dermis is a layer of connective adipose tissue. Volarly, the adipose layer is thick and contains several small, fibrous bands that help anchor the dermis to the palmar fascia and deeper osseous structures. Dorsally, the hypodermis is a loose areolar layer just superficial to the dorsal aponeurosis.

The glabrous palmar skin is closely related to the deeper tissues and has limited mobility when compared with skin found in other areas of the body. This is due to the dense fibrous attachments between the skin and the palmar fascia (or palmar aponeurosis). The palmar fascia consists of longitudinal, transverse, and vertical bands of connective tissue that anchor the tissues of the hand to the osseous structures. The pretendinous bands are found within the retinaculum overlying the flexor tendons and continue distally as the central cord within the finger. The natatory ligaments run transversely just beneath the skin within the webspaces. The spiral cord branches from the pretendinous cord, wraps around the neurovascular bundle, and inserts laterally as the lateral digital sheet in the finger. The flexor tendons and neurovascular bundles are divided by the septa of Legueu and Juvara, which are thicker, vertical fibers that run from the skin to the underlying bony structures. Within the finger, the neurovascular structures are contained by the lateral digital sheet, Grayson's ligament volarly, and Cleland's ligament dorsally.

23.2 Nail Anatomy

The anatomy of the perionychium facilitates the compression of the volar soft tissue pad to the dorsal nail plate allowing for variable compression and thus variable sensory evaluation of objects. The germinal matrix of the nail begins at the base of the dorsal distal phalanx, just distal to the extensor tendinous insertion. As the germinal matrix cells die, they leave a keratinous material known as onchyn

C. M. Stewart · P. A. Ghareeb (✉)
Division of Upper Extremity Surgery, Emory University Division of Plastic Surgery, Atlanta, GA, USA
e-mail: chris.stewart@emory.edu; paul.ghareeb@emory.edu

J. G. Lane et al. (eds.), *The Art of the Musculoskeletal Physical Exam*,
https://doi.org/10.1007/978-3-031-24404-9_23

that provides mechanical and chemical production to the underlying nail bed. The germinal matrix begins at the phalanx periosteum and transfers the dead cells superficially. The distal extent of the germinal matrix can be seen grossly as the lunula, the white, moon-shaped proximal nail plate. The nail fold consists of a floor of germinal matrix and a roof of eponychium. The eponychium's soft tissue inserts onto the nail plate as the cuticle. The nail bed itself consists of the germinal matrix and the sterile matrix. The germinal matrix produces about 90% of the nail plate while the sterile matrix adds the remaining cells that contribute to nail adherence. The sterile matrix is intimately associated with the underlying distal phalanx, which provides rigid support to the nail bed. The nail plate is also supported by the paronychium, where the skin folds laterally to provide additional stability. Distally, the hyponychium is where the sterile matrix meets the finger pulp. It contains a keratinous plug, lymphocytes, and polymorphonuclear cells that protect the digit from infection. The digital nerves, located at the ulnar and radial aspect of the digit, send volar and dorsal branches while the digital arteries trifurcate distal to the joint. Glomus bodies are specialized neurovascular structures found within the nail bed itself and throughout the finger. These arteriovenous shunts are surrounded by a capsule of soft tissue and are involved in temperature regulation (Fig. 23.1).

23.3 Muscles and Tendons

There are numerous myotendinous units that intrinsically or extrinsically affect the hand. The flexor and extensors surface of the hand may be divided into zones that are convenient for describing an injury's location. The palmar aspect of the hand is divided into five zones. Zone I includes the flexor digitorum profundus tendon from its insertion on the distal phalanx to the level of the flexor digitorum sublimis tendon insertion on the middle phalanx. Zone II is found between the insertion of the FDS on the middle phalanx to the start of the flexor tendon sheath (A1 pulley at the distal palmar crease). Zone III is found between the beginning of the A1 pulley and the distal transverse carpal ligament. Zone IV is found within the retinaculum (carpal tunnel). Zone V runs from the proximal border of carpal tunnel to myotendinous junctions of the extrinsic flexors. There are eight extensor zones. Zone I found over the DIP, Zone II over the middle phalanx, Zone III over the PIP, Zone IV over the proximal phalanx, Zone V over the MCP joint, Zone VI over the dorsum of hand, Zone VII over the extensor retinaculum, and Zone VIII over the distal forearm.

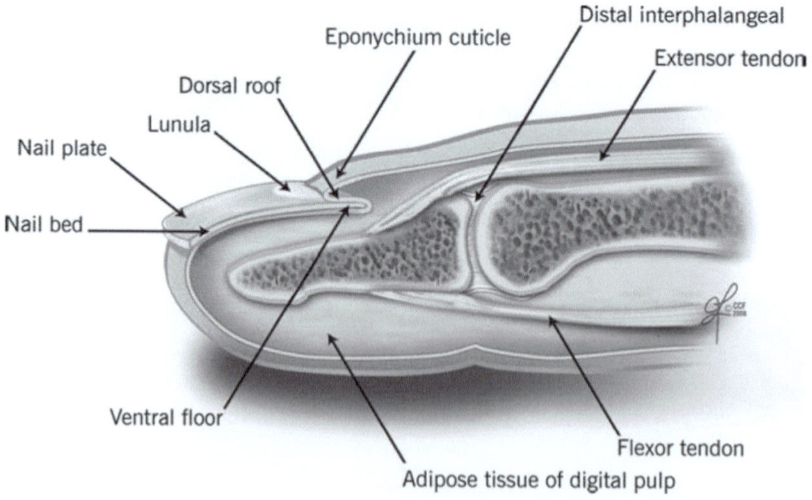

Fig. 23.1 Nail anatomy

23.3.1 Extrinsic Flexors

The majority of the myotendinous components of the hand originate outside of the hand. As they traverse the forearm and wrist crease, they transition from muscular bellies to fibrotendinous bands that insert at different points along the osseous structure. Such components are labeled extrinsic muscles. The superficial flexors share a common flexor-pronator origin at the medial epicondyle of the humerus just superficial to the flexor digitorum superficialis and pronator teres. The flexor carpi radialis (FCR) traverses on the radial aspect of the forearm and wrist and inserts at the base of the second metacarpal with accessory slips to the third metacarpal and trapezium. The palmaris longus (PL) fans distally to insert into the palmar fascia. The flexor carpi ulnaris (FCU) invests the pisiform bone as it passes distally to insert on to the base of the fifth metacarpal and may provide some slips to the hamate. The PL is the weakest of the three and can be absent in up to 20% of the population. The FCU is the strongest and is especially important in providing power grasp and strong ulnar deviation forces (Fig. 23.2).

The flexor retinaculum is created by the thick transverse carpal ligament (TCL), which is continuous with the antebrachial fascia proximally. The TCL forms the roof of the carpal tunnel and runs transversely from the scaphoid and trapezium to the pisotriquetrum and hook of hamate ulnarly. The transverse carpal ligament serves to prevent flexor tendon bowstringing during wrist flexion and contains the four FDS tendons, four FDP tendons, the FPL tendon, and the median nerve. The thenar muscles originate from the superficial radial surface of the TCL and ulnarly the TCL forms the floor of Guyon's canal. Guyon's canal is found between the pisiform and

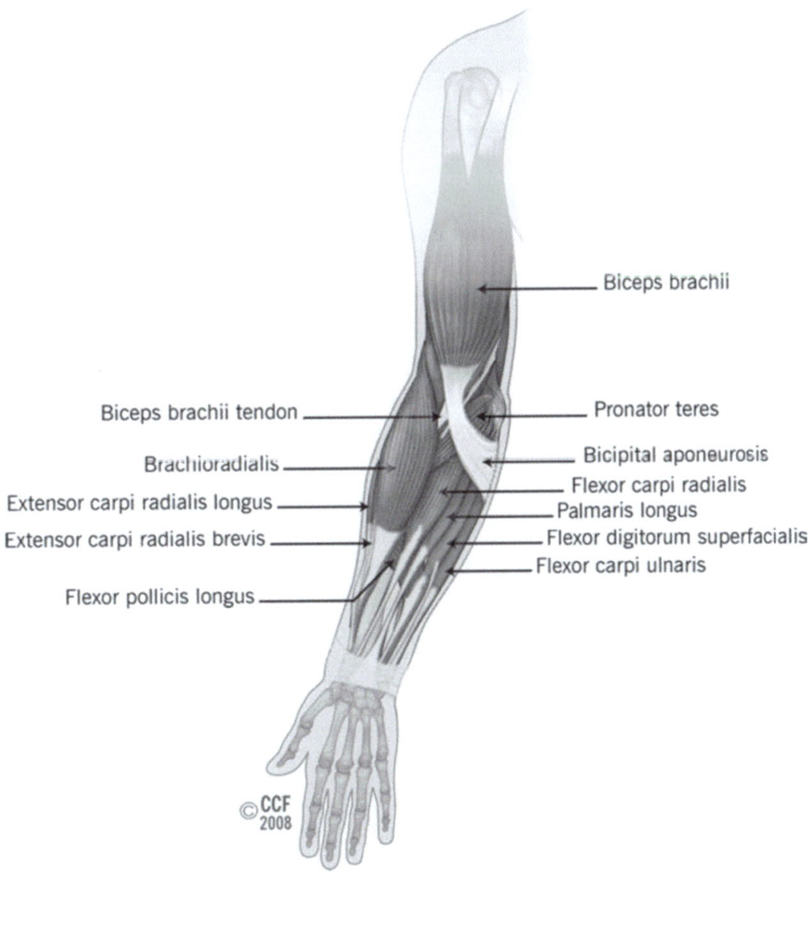

Fig. 23.2 Flexor anatomy

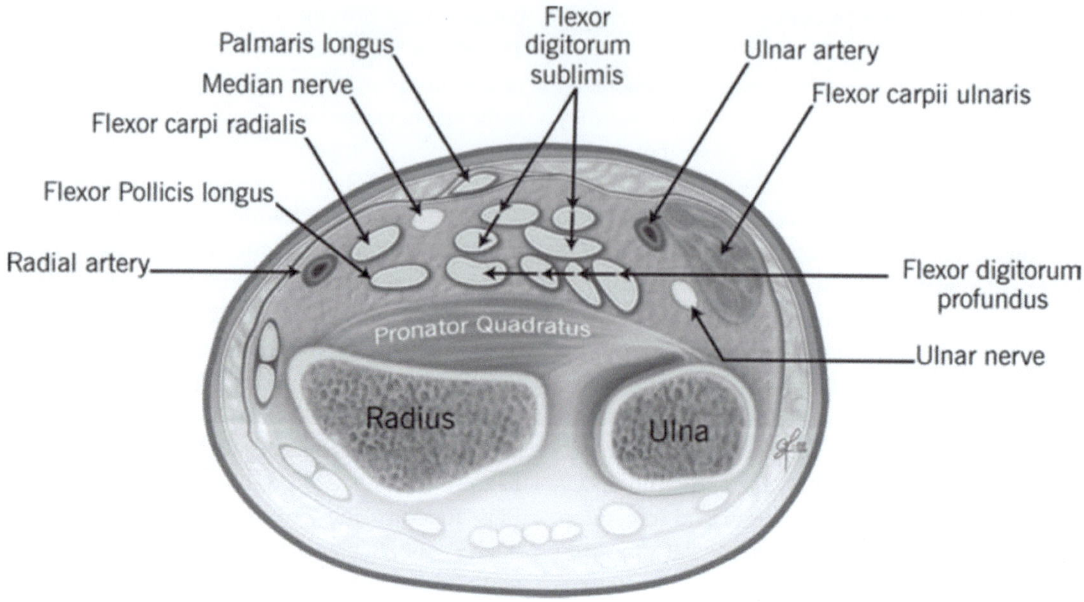

Fig. 23.3 Cross section of distal forearm flexor anatomy

the hamate. The floor is the TCL and the roof is the volar carpal ligament as it blends with the FCU insertion to the pisiform and the pisohamate ligament (Fig. 23.3).

The FDS originates just deep to the superficial flexors at the medial epicondyle and proximal ulna, enters the hand through the carpal tunnel, then passes deep to the superficial palmar arch and enters a fibro-osseous tunnel (flexor tendon sheath) that begins at the metacarpophalangeal joint. The FDS tendon then splits into two slips to create Campers Chiasm, and each slip undergoes a 180° turn as it inserts into the base of the middle phalanx of each digit. The FDS tendons to the middle and ring finger pass through the carpal tunnel superficially to the FDS tendons of the index and small finger. Each FDS tendon has its own muscle belly, with the occasional exception of the small finger. This allows for independent proximal interphalangeal joint (PIP) flexion.

Deep to these are the four FDP tendons. The FDP tendons originate from a common muscle belly more distal than the FDS origin in the mid-forearm and enter the carpal tunnel deep to the FDS tendons. They continue distally to enter the flexor tendon sheath just proximal to the MP joint and remain deep until they emerge between the two FDS slips inserting to the middle phalanx (Camper's chiasm). They continue this course to their insertion at the base of the distal phalanx. The FDP tendons allow for distal interphalangeal (DIP) joint flexion, but because the middle, ring, and small finger share a common muscle belly, they are prevented from independent DIP flexion. However, the index finger has its own muscle belly and thus the advantage of independent DIP flexion. The flexor pollicis longus (FPL) tendon originates from the radial half of the mid-forearm and is the radial-most structure as it passes through the carpal tunnel before inserting to the distal phalanx of the thumb. Occasionally, the FPL and FDP to the index finger share connections that results in simultaneous flexion (Linburg-Comstock variant).

23.3.2 Intrinsic Muscles

The muscles of the hand that originate and insert solely within the hand are referred to as the intrinsic musculature. These include the volar and dorsal interossei, lumbricals, hypothenar and thenar muscle groups. The thenar group involves the three muscles that originate from the transverse carpal ligament and trapezium: the abductor pollicis brevis (APB), the opponens pollicis, and the flexor pollicis brevis (FPB). The radial-most APB is median innervated and allows the thumb to oppose by inserting to the radial aspect of the proximal phalanx and MP joint. Ulnarly, the FBP has two muscle bellies separated by the coursing FPL tendon. The superficial head of the FPB, like the APB, is median innervated and inserts to the radial aspect of the proximal phalanx while the deep head (which is ulnar innervated) inserts to the ulnar aspect. Each head of the FPB inserts with a concomitant sesamoid bone. Slightly different from the other thenar muscles, the deep head of FPB originates from the thumb metacarpal as opposed to the transverse carpal ligament and trapezium. Lastly, the opponens pollicis is median innervated and lies deep to the APB and FPB and inserts to the radial aspect of the thumb's metacarpal. It is the only thenar muscle to do so, allowing for opposition and flexion at the trapezium-metacarpal joint.

Although not considered a thenar muscle, the adductor pollicis is an important muscle allowing for key pinch by inserting to the base of the thumb's proximal phalanx and ulnar sesamoid bone. It is ulnar innervated and has two origins, the transverse head from the middle finger metacarpal base and the oblique head from the index and middle metacarpals.

The hypothenar muscles are ulnar innervated and lie opposite to the thenar group and contain an analogous three muscles: the abductor digiti minimi (ADM), flexor digiti minimi, and opponens digiti minimi. Superficial to this trio is the palmaris brevis. It is a quadrangular muscle that originates from the palmar aponeurosis and transverse carpal ligament and inserts to hypothenar skin. Like the longus that tightens palmar skin, the brevis's role is to deepen the cup of the palm and increase surface area during grip. The hypothenar muscles lie deep to this muscle. Like the APB in the thumb, the ADM is the most superficial hypothenar muscle, originating from the pisiform and inserting on the proximal phalanx of the small finger. The flexor digiti minimi is deeper and more radial, originating from the hook of the hamate and inserting to the proximal phalanx which allows for flexion at the MP joint. The opponens digiti minimi, just like the opponens pollicis, is the deepest of the three, originating from the hamate body and inserting to its metacarpal. This allows for flexion of the small finger carpometacarpal (CMC) during power grip.

Between the thenar and hypothenar group sit the lumbricals. The lumbricals are unique in that they are the only muscles in the body that both originate and insert onto a tendon. They arise from the FDP tendons and insert onto the radial aspect of each extensor mechanism. Their muscle bellies begin on the FDP tendon at the level of the metacarpal base and fan distally with each inserting onto the radial lateral band of the extensor mechanism. The index and middle finger are median innervated and have unipennate muscle bellies that start from their respective FDP tendon, while the lumbricals to the ring and small finger ulnar innervated and are bipennate, arising from the FDP tendon on each flank. The lumbricals pass through a separate canal as they course volar to the metacarpophalangeal joint. Because the lumbricals pass volar to the MCP joint and then dorsal to the IP joints, they act to flex the MP joint and extend the IP joints.

Deep to the lumbricals are the ulnarly innervated unipennate palmar interossei (three) and bipennate dorsal interossei (four). The palmar interossei originate from the flanking metacarpals run superficial to the sagittal bands and insert to the lateral bands of the extensor hood. They adduct the digits towards the middle finger. Thus, the small and ring finger palmar interossei are radial and the index finger is ulnar. The middle finger, being central, does not have a palmar interossei. The dorsal interossei start the same as the volar group at the metacarpal but have two insertions: the proximal phalanx and the extensor mechanism. The first, second, and fourth dorsal interossei have two muscle bellies: a deep belly that runs superficial to the sagittal bands and inserts on the dorsal hood which allows for flexion at the MCP and extension at the PIP. The superficial belly runs deep to the sagittal bands and inserts on the tubercle of the proximal phalanx which abducts the digit. The third dorsal interosseous is unique and has a single belly that inserts to the dorsal hood, thus contributing little in terms of abduction. Contrast this to the first dorsal interosseus, where the superficial belly is predominant and allows for strong abduction forces to counteract the thumb during pinch. The small finger does not have a separate dorsal interossei, as it already has the ADM.

23.3.3 Extrinsic Extensors and the Extensor Retinaculum

The interplay of intrinsic and extrinsic hand flexors at the level of the digit allows for flexion at the wrist and MP joints with extension forces at the PIP and DIP joints. To extend the wrist and add additional extension forces at the digit, the extensor tendons assist. Understanding their orientation is easiest at the level of the distal radius and ulna, where the tendons are organized into six dorsal compartments.

These six extensor compartments are numbered radial to ulnar and allow for mechanical advantage by holding the tendons close to the bone. The six extensor compartments are a continuation of the palmar carpal ligament (thickening of antebrachial fascia). The first dorsal compartment is located on the volar-radial aspect of the radius. It contains the abductor pollicis longus (APL) tendon and the extensor pollicis brevis (EPB) tendon. The APL originates from the ulna and interosseous membrane in the mid forearm, passes through the first dorsal compartment and inserts onto the base of the thumb metacarpal. The APL allows for radial deviation of the wrist and abduction of the thumb. There are often multiple slips of the APL tendon at the level of the first dorsal compartment. Compared to the APL, the extensor pollicis brevis (EPB) originates more distal in the forearm from the radius and interosseous membrane. Although it originates more radially, the EPB joins the APL tendon ulnarly in the first dorsal compartment before inserting to the base of the proximal phalanx of the thumb. The EPB allows for MP joint extension. The EPB may be contained within a subsheath in up to 60% of the population, which predisposes it to tendinoses (De Quervain's syndrome). The APL and EPB tendons are often called the out-riggers due to their ulnar origin and radial insertion and can develop tendinoses as they pass over the radial wrist extensors (Intersection syndrome) (Fig. 23.4).

The second dorsal compartment is on the dorsal-radial aspect of the radius. It contains the extensor carpi radialis longus (ECRL) and extensor carpi radialis brevis (ECRB) tendons that insert onto the metacarpal base of the index and middle fingers, respectively. They originate proximal to the elbow at the lateral epicondyle of the humerus and with the brachioradialis form the mobile wad (of Henry). The ECRB shares a common origin with the extensor digitorum, extensor digiti minimi, and extensor carpi ulnaris. Analogous to the common flexor tendon which is also superficial, this common extensor tendon contains the superficial extensors. Together, the ECRL and ECRB tendons run deep to the APL and EPB muscle bellies before surfacing in the second compartment. Both the first and second dorsal compartments are radial to Lister's tubercle of the distal radius (Fig. 23.5).

The extensor pollicis longus (EPL) is the only tendon of the third compartment. This compartment is just ulnar to Lister's tubercle. Like the APL and EPB, the EPL originates in the mid forearm just distal to the APL. After passing Lister's tubercle at the level of the distal radius, it traverses radially over the ECRL and ECRB to insert at the base of the distal phalanx of the thumb. This muscle is unique as it is the only muscle that can extend the IP joint of the thumb and raise the thumb from a flattened, tabletop position (retropulsion). The tendons of the first and third compartments can be seen grossly as they form the volar and dorsal boundaries of the anatomical snuffbox.

The fourth compartment is found ulnar to the third, along the dorsal ulnar aspect of the distal radius. It contains the extensor digitorum communis (EDC) and the extensor indicis proprius (EIP). The EDC begins at the common extensor tendon at the lateral epicondyle and inserts on the extensor hood of all four fingers. The juncturae tendinae interconnect the EDC fibers. In some cases, the EDC tendon to the small finger is not

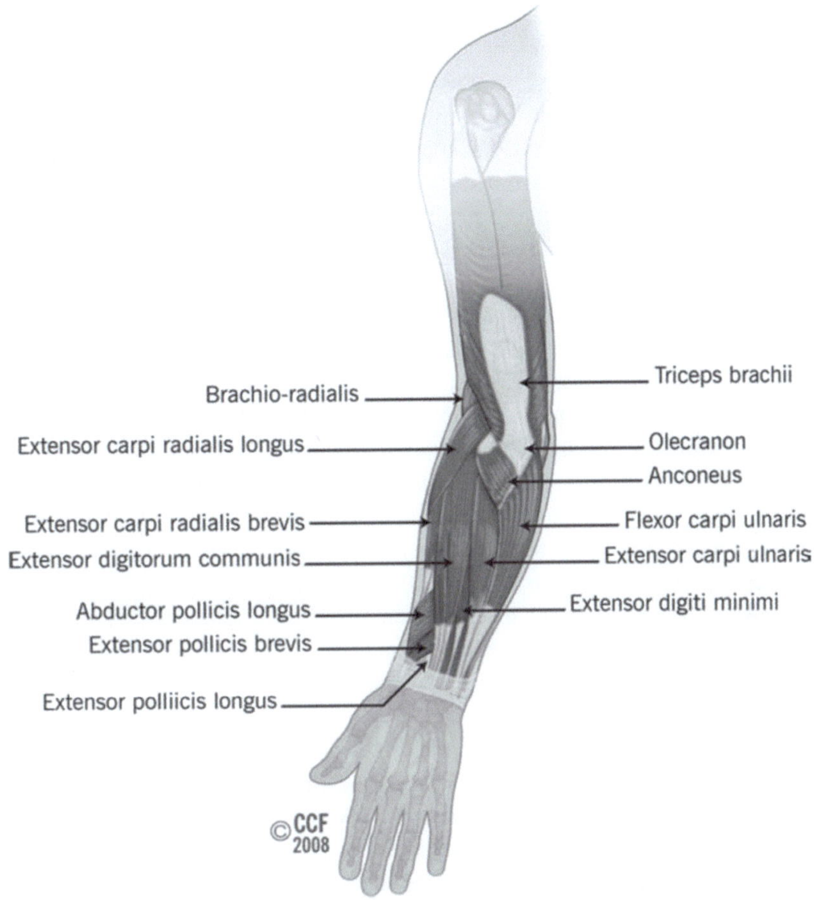

Fig. 23.4 Extensor musculature

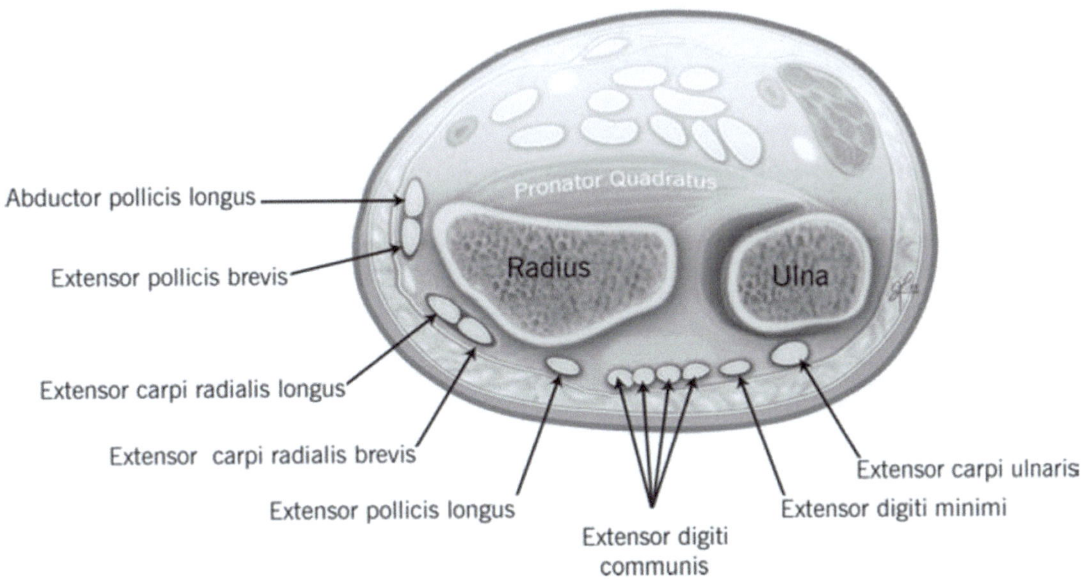

Fig. 23.5 Cross sectional anatomy of extensor compartments

present, and the juncturae between the ring and small fingers provides extensor function. Deep and ulnar to the EDC tendons, the EIP originates on the ulna in the distal third of the forearm and just distal to the EPL. It crosses the fourth compartment under the EDC and inserts ulnarly to the index EDC insertion on the extensor hood.

The fifth dorsal compartment is located on the ulna and contains the extensor digiti minimi (EDM). The EDM begins at the common extensor tendon, passes the fifth compartment, and inserts to the ulnar aspect of the dorsal hood, meeting with the EDC insertion. Often, the EDM tendon is bifid and contains two smaller tendons.

More ulnar and dorsal is the sixth compartment. This compartment contains the extensor carpi ulnaris (ECU), which acts as a powerful wrist extensor and ulnar deviator. The ECU begins with the common extensor tendon and inserts to the base of the fifth metacarpal.

Together, the extrinsic extensors counteract the flexion forces of the intrinsics. At the MP joint, the extrinsic extensors extend while the intrinsic muscles flex. The extrinsic extensors extend the MP joint through the extensor mechanism but do not actually insert on the proximal phalanx. Instead, they use the sagittal bands at the MP joint to raise the base of the proximal phalanx. Likewise, the extrinsic flexors also lack insertions to the proximal phalanx and flex the MP joint through FDS and FDP coupling to the extensor mechanism.

23.3.4 Digital Flexor Apparatus

The flexor tendons are held within a fibro-osseous sheath that keeps them in close proximity to the bone and offers increased mechanical advantage. The tunnel has thickenings known as pulleys that prevent bowstringing of the tendons. The odd numbered pulleys overlie the joints and the even pulleys cover the phalanges. A1 is over the MCP joint, A2 over the proximal phalanx, A3 covers the PIP joint, A4 covers the middle phalanx, and A5 over the DIP. A2 and A4 are the most important of these. Three cruciate pulleys supplement this system C1 and C2 flank the A3 pulley at the PIP joint and the C3 pulley is located just distal to the A4 pulley on the middle phalanx. Within this tunnel the FDS and FDP tendons each have two vinculae that are important for blood supply, the vinculum brevis and vinculum longus. Just like the flexor apparatus being tethered to the bone, the overlying skin is stabilized by anchor points at the PIP and DIP joints. Cleland's ligament runs dorsal to the neurovascular bundle and Grayson's ligament courses volar to the bundle (Fig. 23.6).

23.3.5 Digital Extensor Mechanism

The terminal extensors with the intrinsic and extrinsic flexor tendons form a hammock around the digit called the extensor mechanism. This complex is simplified by understanding the individual role of its components as they send contributions to and from each other. The components include the sagittal bands, central slip, the lateral bands, transverse retinacular ligament, oblique retinacular ligament, and triangular ligament. As the EDC tendons cross the MP joint, they are kept midline by the sagittal bands. The sagittal bands are the first component of the extensor hood and act independently from the MP joint capsule. They originate at the EDC tendon and wrap around the finger to insert on the MP volar plate at the metacarpal neck. This MP girdle assists in extension of the proximal phalanx as there is no extensor insertion onto the proximal phalanx. The EDC doesn't insert onto bone until the base of the middle phalanx as the central slip, which accomplishes PIP extension. Distal to the MP, the extensor hood trifurcates into a central slip and two lateral bands. The lateral bands run along the proximal phalanx dorsal to the central axis of rotation of the PIP and continue dorsally to the DIP before their insertion to the base of the distal phalanx. The lumbricals, EIP, palmar, and dorsal interossei insert onto the lateral bands as they traverse the proximal phalanx. Extension by these muscles occurs because the lateral bands run dorsal to the PIP and DIP axes of rotation. Terminally, the lateral bands are held together by an aponeurosis that begins at the middle phalanx called the triangular ligament. The triangular ligament prevents palmar subluxation of the lateral bands during PIP flexion. Disruption of this ligament leads to an alteration of the lateral band's

Fig. 23.6 Flexor pulley anatomy

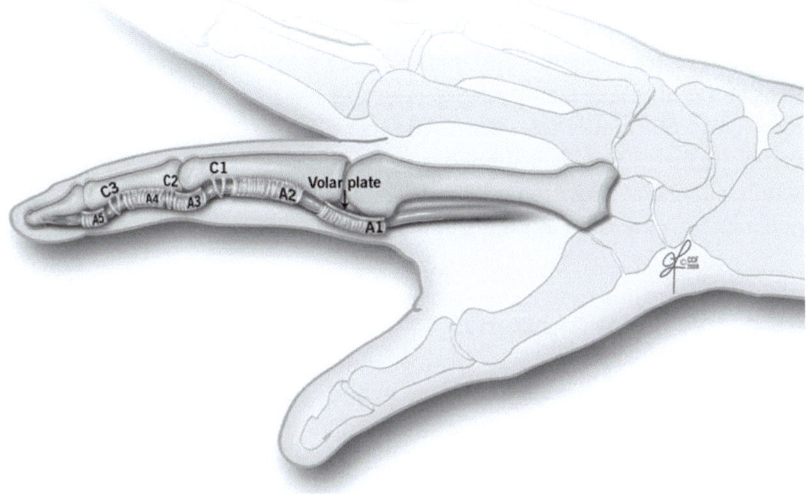

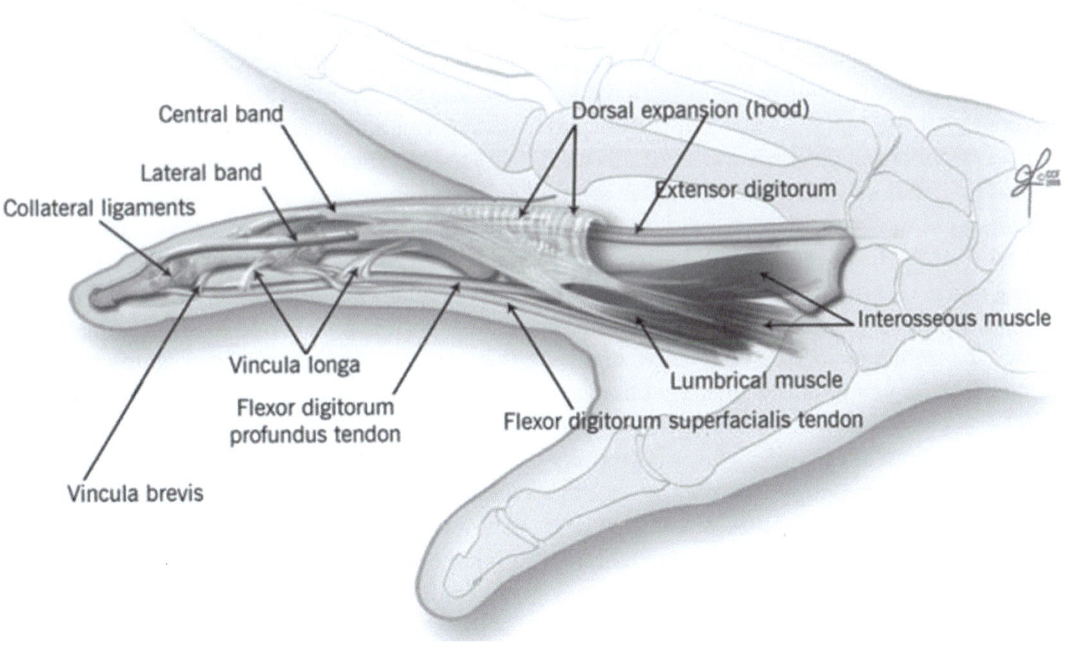

Fig. 23.7 Digital extensor mechanism

dorsal configuration at the PIP and DIP joints. The transverse retinacular ligament originates from the palmar flexor tendon sheath of the PIP joint and inserts onto the lateral bands preventing them from becoming too dorsal into subluxation. The oblique retinacular ligament (of Landsmeer) couples DIP and PIP joints motion. It originates deep to the transverse retinacular ligament on the volar proximal phalanx, volar to the PIP joint, before inserting to the dorsal extensor tendon dorsal to the DIP. In PIP flexion, the tendon is loose and permits DIP flexion. As the PIP extends, the ligament tightens and extends the DIP in conjunction (Fig. 23.7).

The lumbrical tendons are palmar to the interossei and separated by the deep transverse metacarpal ligament. As stated previously, the extensor hood trifurcates along the proximal phalanx into the central slip and lateral bands. The intrinsic tendons coalesce after the deep transverse metacarpal ligament and then immediately bifurcate sending a medial extension to join the central slip and a lateral extension to join the lateral bands.

23.4 Vascular Anatomy

With a thorough understanding of the underlying framework, the anatomy of the corresponding vascular is more easily understood. In the antecubital fossa, distal to the elbow joint, the brachial artery bifurcates into the radial and ulnar arteries which ultimately supply the hand and terminate as the digital arteries. This bifurcation occurs just superficial to the brachialis tendon, ulnar to the biceps brachii tendon and deep to the ulnarly coursing bicipital aponeurosis (lacertus fibrosus). The course of the radial artery is approximated by a line drawn from the ulnar aspect of this tendon to the radial styloid, while the ulnar artery parallels a line drawn to the pisiform. In the proximal third of the forearm, the radial artery gives off the radial recurrent artery and the common interosseous artery branches from the ulnar artery. The common interosseous soon divides into the anterior and posterior interosseous arteries. The anterior interosseous will rejoin the posterior interosseous artery when it dives posteriorly through the interosseous membrane at the metaphyseal flare of the radius. Together, these two will form the dorsal carpal arterial arch (Fig. 23.8).

The radial artery courses with the superficial branch of the radial nerve in the forearm deep to the brachioradialis (BR) muscle and reaches the wrist radial to the FCR tendon, between the FCR and BR tendons. At the radial styloid, it divides into superficial and deep branches along with smaller branches to the carpal bones. The deep branch travels deep to the anatomic snuffbox tendons of the first and third dorsal compartments at the level of the scaphotrapezial joint, rounds the thumb metacarpal and splits the bipennate first dorsal interosseous to meet up with a smaller, deep ulnar artery to form the deep palmar arch. The superficial branch of the radial artery crosses the wrist superficial to the flexor retinaculum (transverse carpal ligament) and enters the thenar muscles deep to the APB and superficial to the opponens pollicis. It is smaller than the deep branch and will sometimes incompletely form its arch with the superficial palmar arch from the ulnar artery.

Differing from the radial artery, the ulnar artery crosses the wrist before dividing into its superficial and deep branches. Like the radial artery, it gives off small carpal branches before entering superficial to the flexor retinaculum and radial to the FCU, pisiform, and ulnar nerve. There, it divides into deep and superficial branches which anastomose to their corresponding radial artery branches. The deep arch gives off perforators that run between the ring and middle fingers posteriorly, at the metacarpal base, and join the dorsal carpal arterial arch formed by the anterior and posterior interosseous arteries.

The superficial arch is the predominant blood supply to the digits. To approximate its location, Kaplan's cardinal line can be drawn from the first webspace to the pisiform bone. The superficial arch is found just deep to this line, and the deep palmar arch can be found 1 cm proximal to this line. The superficial arch is predominantly supplied by the ulnar artery and the deep arch by the radial. The superficial arch supplies the common digital arteries which bifurcate proximal to the webspace into the proper digital arteries on either side of the fingers. These vessels run with the digital nerves just beneath the palmar fascia and are separated from the tendons by the Septa of Legueu and Juvara. In the palm, the common digital arteries are volar to the common digital nerves; however, this arrangement changes at the arterial bifurcation where the proper digital arteries dive dorsal to the nerves.

Collateralization to the digits is provided by the dorsal carpal and deep palmar arches. The deep palmar arch sends branches called the palmar metacarpal arteries which join the common digital arteries before their division into proper

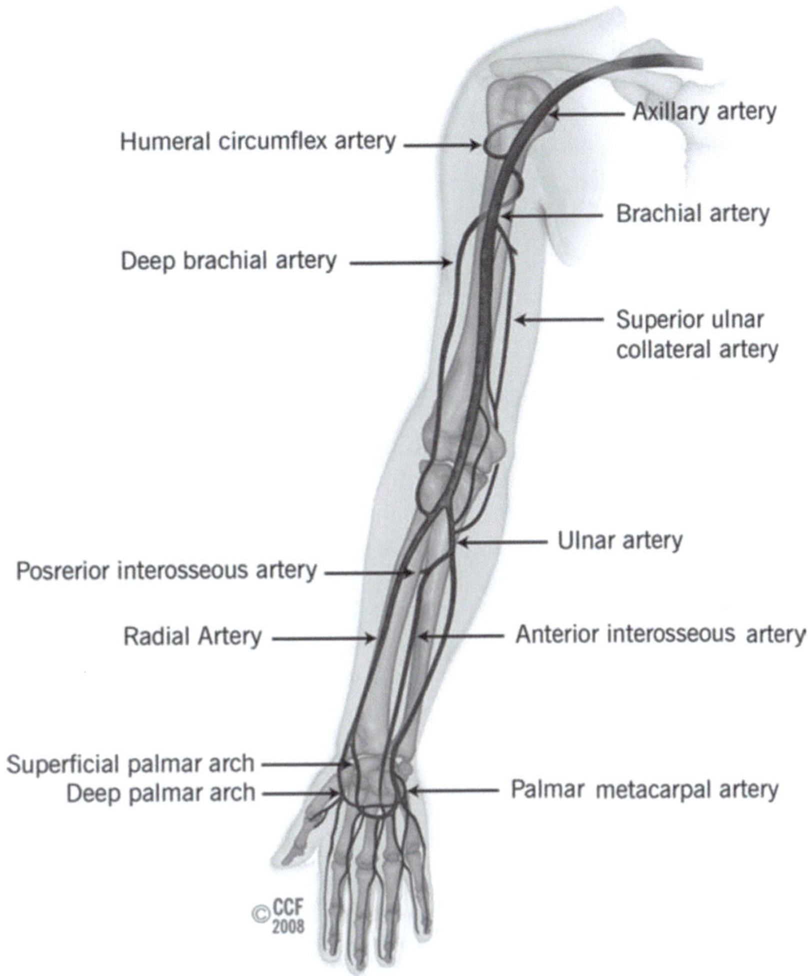

Fig. 23.8 Arterial anatomy of the upper extremity

digital arteries. The dorsal carpal arch sends dorsal metacarpal arteries to join the digital arteries at the PIP joints. The dorsal metacarpal arteries run in the subaponeurotic space of the dorsal hand. Each digit has a radial and ulnar digital artery. Just before reaching the PIP joint, the artery gives off a dorsal branch to supply the skin and at the DIP the arteries each trifurcate to supply the nail, pulp, and volar skin.

Like the arterial supply, the venous system has superficial and deep components. The superficial venous system is the dominant of the two and is located in the dorsal loose areolar tissue of the subcutaneous space just superficial to the dorsal aponeurosis. A few superficial veins also contribute and are located on the palmar aspect of the digits. Radially, these veins drain into the cephalic vein and ulnarly they drain into the basilic. The deep venous system are the veins that accompany the arterial network, appropriately named the venae comitantes. These vessels are important for free tissue transfer of the hand. The deep venous system connects with the superficial system via numerous perforating branches throughout the hand.

23.5 Bony and Ligamentous Anatomy

The bony anatomy of the hand and wrist truly begins at the distal forearm with the radius and ulna. Distal to these are eight carpal bones: scaphoid, lunate, triquetrum, pisiform, trapezium, trapezoid, capitate, and hamate. The carpus articulates with five metacarpals which in turn articulate with the phalanges. Each finger consists of a proximal (P1), middle (P2), and distal (P3) phalanx while the thumb has a proximal and distal phalanx. The bones are associated together by a complex ligamentous structure that allows for functional mobility while protecting stability of the hand.

23.5.1 Wrist

The bony framework of the hand is supported by the radius and ulna at the wrist. The distal radius has a metaphyseal flare with a terminal wrist articulation featuring two facets, one for the scaphoid and one for the lunate. The articular surface of the radius has relatively constant anatomical features, which includes a volar tilt of approximately 11° and a radial inclination of approximately 23°. The ulnar surface of the distal radius articulates with the ulna at the sigmoid notch, which defines the distal radioulnar joint and permits rotation of the radius around the ulna (pronation and supination). The radius articulates with the scaphoid and lunate of the proximal carpal row. The triquetrum articulates ulnar to the lunate, and the pisiform is found volar to the triquetrum at the pisotriquetral joint. The pisiform is essentially a sesamoid bone that is invested within the FCU tendon. In neutral deviation, the scaphoid and half of the lunate articulate with the radius, while the other half of the lunate and triquetrum articulate the triangular fibrocartilage complex (TFCC) disc between the ulnar head and carpus. The TFCC features dorsal and palmar radioulnar ligaments (ligamentum subcruentum), a central articular disc, a meniscus homolog, the ECU subsheath, and an ulnar collateral ligament (Fig. 23.9).

23.5.2 Extrinsic Ligaments

The proximal carpal row articulates with the distal carpal row which includes from radial to ulnar: the trapezium, trapezoid, capitate, and hamate bones. In this articulation, the scaphoid acts as the link between the proximal and distal carpal rows. The distal row then articulates with the metacarpals. The articulations of the wrist and hand are maintained by ligaments. The ligaments of the wrist can be divided into extrinsic and intrinsic in nature. Extrinsic ligaments originate outside of the carpus, while intrinsic ligaments originate and insert within the carpus. Extrinsic ligaments have a radius group, an ulnar

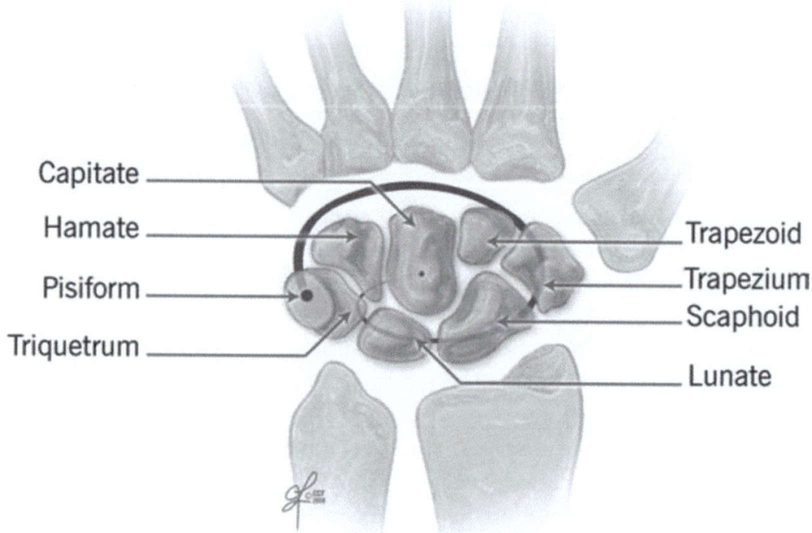

Fig. 23.9 Carpal anatomy

group, and a dorsal group. From the volar radius are the radioscaphoid, radioscaphoid-capatate (RSC), long radiolunate (LRL), and short radiolunate (SRL). Between the RSC and the LRL the fibers diverge and represent a weak spot known as the space of Poirier, a place where perilunate injuries often occur. Ulnarly are the ulnocapitate, ulnolunate, and ulnotriquetral ligaments. The ulnocapitate and the RSC form an inverted V known as the arcuate ligament. The space of Poirier is the central space within this arc. Dorsally are the dorsal radiocarpal (DRC) and the dorsal intercarpal (an intrinsic ligament) which converge on the triquetrum to form a lateral oriented V.

23.5.3 Intrinsic Ligaments

The proximal row is bound by the scapholunate interosseous ligament (SLIL) and the lunotriquetral interosseous ligament (LTIL). The scapholunate ligament is thickest in its dorsal aspect, while the lunotriquetral ligament is thickest in its volar component. The dorsal intercarpal (DIC) ligament also serves to stabilize the dorsal aspect of the carpus. The majority of wrist motion occurs at the radiocarpal and midcarpal joints. Likewise, there is a significant amount of motion that occurs between the bones of the proximal carpal row. In contrast, there is relatively little motion between the distal carpal row and the second and third metacarpals. This is known as the fixed unit of the hand, due to its relative immobility.

23.5.4 Metacarpals

The carpometacarpal joints involve the metacarpal bases with their corresponding carpal bones and are held tightly by volar and dorsal intermetacarpal ligaments. These ligaments are strong so injury is uncommon. The small finger CMC permits 15–20° of flexion and the ring finger CMC permits 5–10°. Injury of the CMC usually involves the more mobile border digits of the small finger or thumb. At the metacarpal head, there are the deep transverse metacarpal ligament and the superficial transverse metacarpal ligament (Natatory ligament). The neurovascular bundles run between these two. The deep transverse ligament lies 3–4 mm deep to the superficial ligament and connects the MP volar plates. It prevents abduction of the metacarpal heads and separates the lumbrical muscles running palmarly from the interossei running dorsally. The superficial transverse metacarpal ligament also resists abduction, runs just beneath the webspace skin, and is often involved in Dupuytren's disease (Fig. 23.10).

23.5.5 Fingers

The metacarpals bridge the carpus to the digits. Each finger has an MP joint, a PIP, and a DIP joint. The MP joints are relatively more mobile than the IP joints due to a few reasons. The MP joint is a condyloid joint with a rounded metacarpal head articulating to a shallow, concave base of the proximal phalanx. This allows for some radial and ulnar deviation during extension in contrast to the IP hinge joint that permits flexion and extension only. The IP joints also have slightly different osseous anatomy at the base of the middle and distal phalanx. Each has a prominent ridge at their midline that articulates with a groove in the head of the corresponding proximal and middle phalanx. The MP doesn't feature this tubercle-groove configuration and is more round in the dorso-volar view (Fig. 23.11).

These joints have thick radial and ulnar collateral ligaments that prevent deviation in these directions and a thick volar ligament, the palmar plate, preventing hyperextension. The palmar plate is most adherent laterally where the collateral ligaments insert and the central aspect of the palmar plate blends with the periosteum of the next phalanx. The palmar plate dovetails proximally into two periosteal extensions known as the checkrein ligaments. The collateral ligaments add ulnar and radial stability and include a proper collateral ligament and an accessory collateral ligament. The proper collateral ligament begins on the dorsal head of the metacarpal/phalanx and

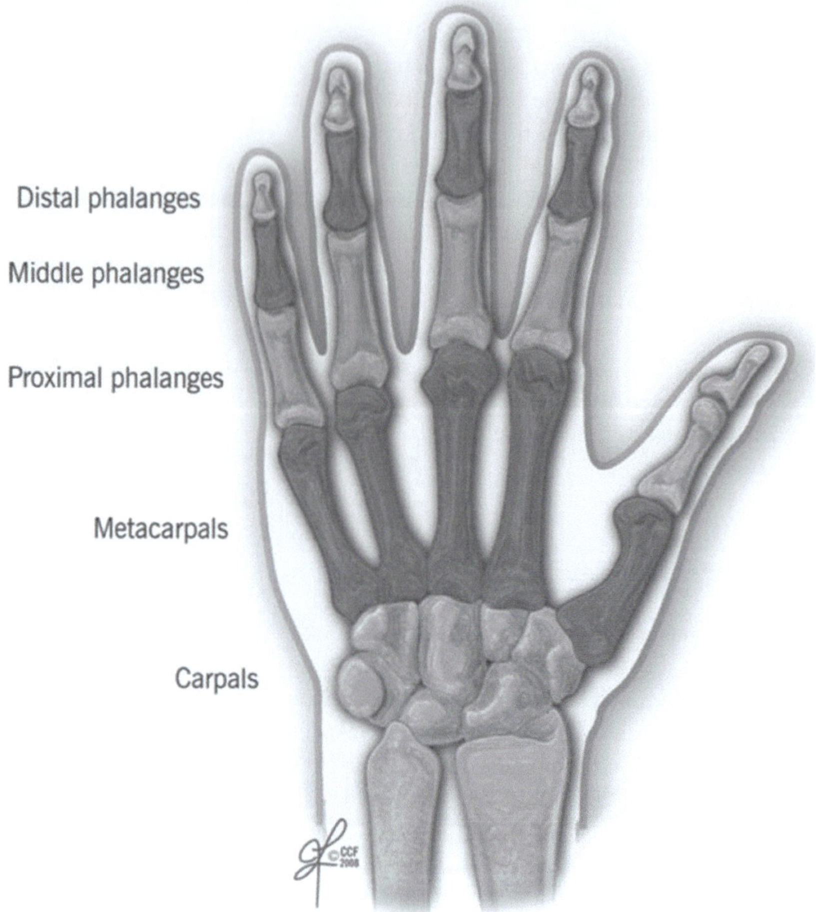

Fig. 23.10 Osseous anatomy of the hand

travels obliquely to the base of the next phalanx. The accessory collateral ligament starts similarly on the dorsal head but follows a more sagittal course and inserts directly to the palmar plate.

The importance of these collateral ligaments can be seen from the sagittal view. From this perspective, the metacarpal head is cam shaped. If one were to measure the radius from the center of rotation to the articular surface, the distance to the extension articular surface is shorter than the distance to the flexion articular surface. The proximal and middle phalanx heads are spherical in the sagittal plane and therefore the distance to the articular surface is the same in extension and flexion. Due to the cam of the metacarpal head the collateral ligaments are loose in extension and taut in flexion. Thus, there is more radial-ulnar stability when the hand MP is in flexion (during grasp or pinch). The spherical phalangeal heads of the PIP and DIP joints mean the collateral ligaments are roughly equally taut in flexion and extension, yet the volar plate is collapsed on flexion and expanded in extension. Together, the "intrinsic plus" position is the safest position to splint the hand because it allows maximum stretch of the MP collateral bands during MP flexion, and maximum expansion of the volar plates of the IP joints with PIP and DIP extension.

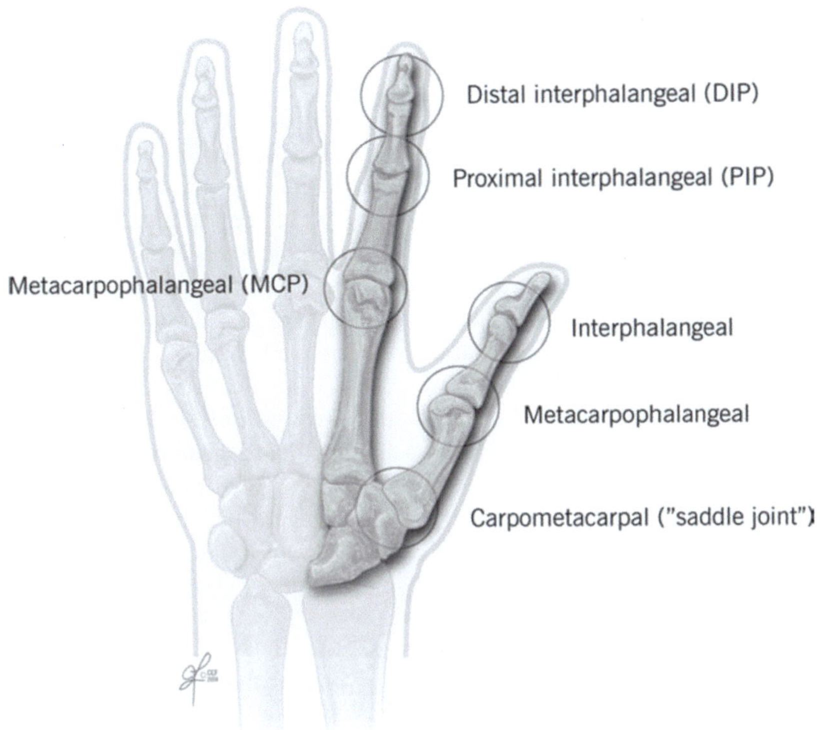

Fig. 23.11 Carpometacarpal, metacarpophalangeal, and interphalangeal joint anatomy

23.5.6 Thumb

The thumb is unique to the fingers as it only has a proximal and distal phalanx. Thus, it has an IP joint, an MP joint and the first CMC also known as the trapeziometacarpal (TM) joint. The TM joint is a biconcave saddle joint as each of the surfaces run perpendicular to one other allowing flexion, extension, abduction-adduction, and pronation-supination (opposition-retropulsion). The generous mobility of the osseous interface necessitates four strong ligaments for TM stability, most importantly the anterior oblique ligament or volar "beak" ligament. The anterior oblique ligament originates at the volar tubercle of trapezium and inserts to the ulnar metacarpal base. The anterior oblique ligament is supported by the strongest and thickest ligament of the thumb, the dorsoradial ligament which is a primary restraint to dorsal translation. These two are joined by the intermetacarpal ligament and the posterior oblique ligament to encircle the thumb. Stability provided by these ligaments is crucial as a pinch force at the distal phalanx of the thumb is amplified 12 times to the thumb CMC joint making it prone to arthritic degeneration. The MP joint has a concave-convex osseous interface like that of the fingers but is supported by much stronger radial and ulnar collateral ligaments that add stability during flexion, extension, and pinch. The two heads of the flexor pollicis brevis each have a sesamoid bone that add to the moment arm before insertion to the base of the proximal phalanx of the thumb. The IP joint, similar to the fingers, is a hinge joint.

23.6 Nerve Anatomy

23.6.1 Median Nerve

The median nerve is developed by fibers from the medial and lateral cords and enters the forearm deep to the lacertus fibrosus and medial to the brachial artery. It gives off no branches in the arm. The first branch is to the pronator teres as it passes between the ulnar and humeral heads. Here, it also supplies the FCR, PL, and FDS. After supplying the superficial flexors originating from the medial epicondyle, it gives off the Anterior Interosseous Nerve (AIN) branch which supplies motor innervation to the index and middle FDPs and FPL before coursing distally to supply the pronator quadratus muscle. The AIN branches from the radial aspect of the median nerve and courses volar to the FDP before diving deeper where it continues along the interosseous membrane. The median nerve proper continues distally between the flexor digitorum superficialis and flexor digitorum profundus. The palmar cutaneous branch of the median nerve exits approximately 5.5 cm proximal to the wrist crease, between the PL and FCR, and travels distally to provide sensation to the palm and thenar eminence (Fig. 23.12).

The median nerve then enters the carpal tunnel and is typically found radial and superficial at this level. The FDS and FDP tendons are often found ulnar to the nerve, and the FPL tendon can be found radial or deep to the nerve. After passing distal to the transverse carpal ligament, the nerve divides into six branches: the common digital nerve to the third webspace, the common

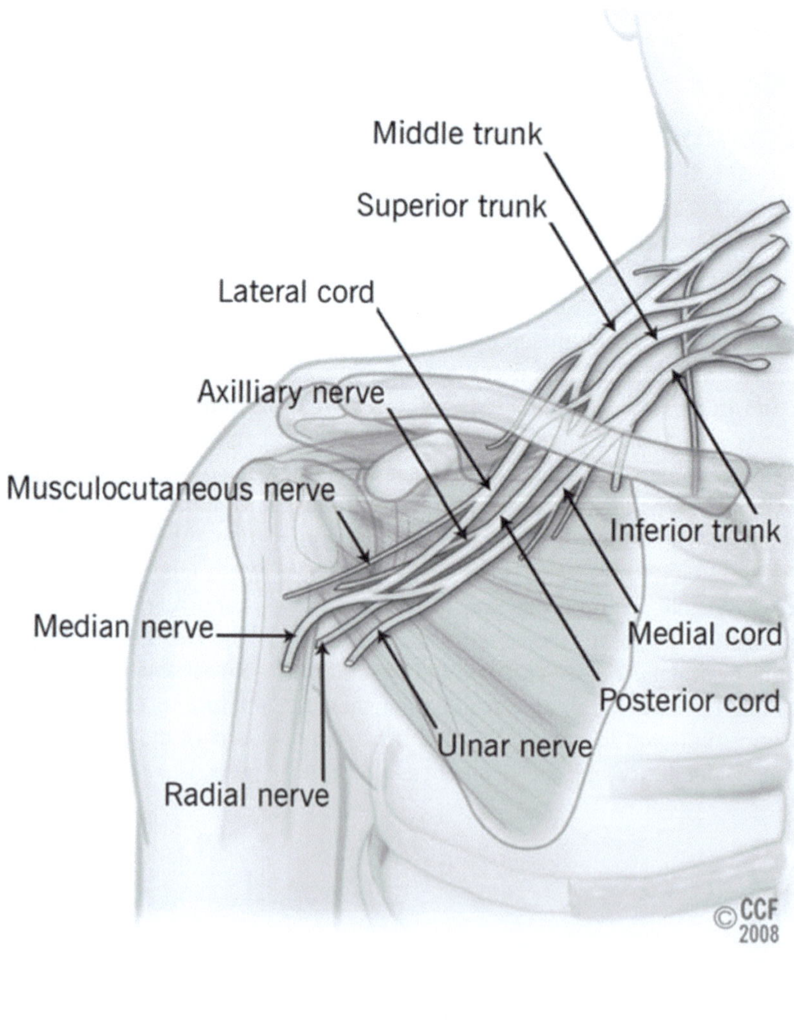

Fig. 23.12 Brachial plexus anatomy

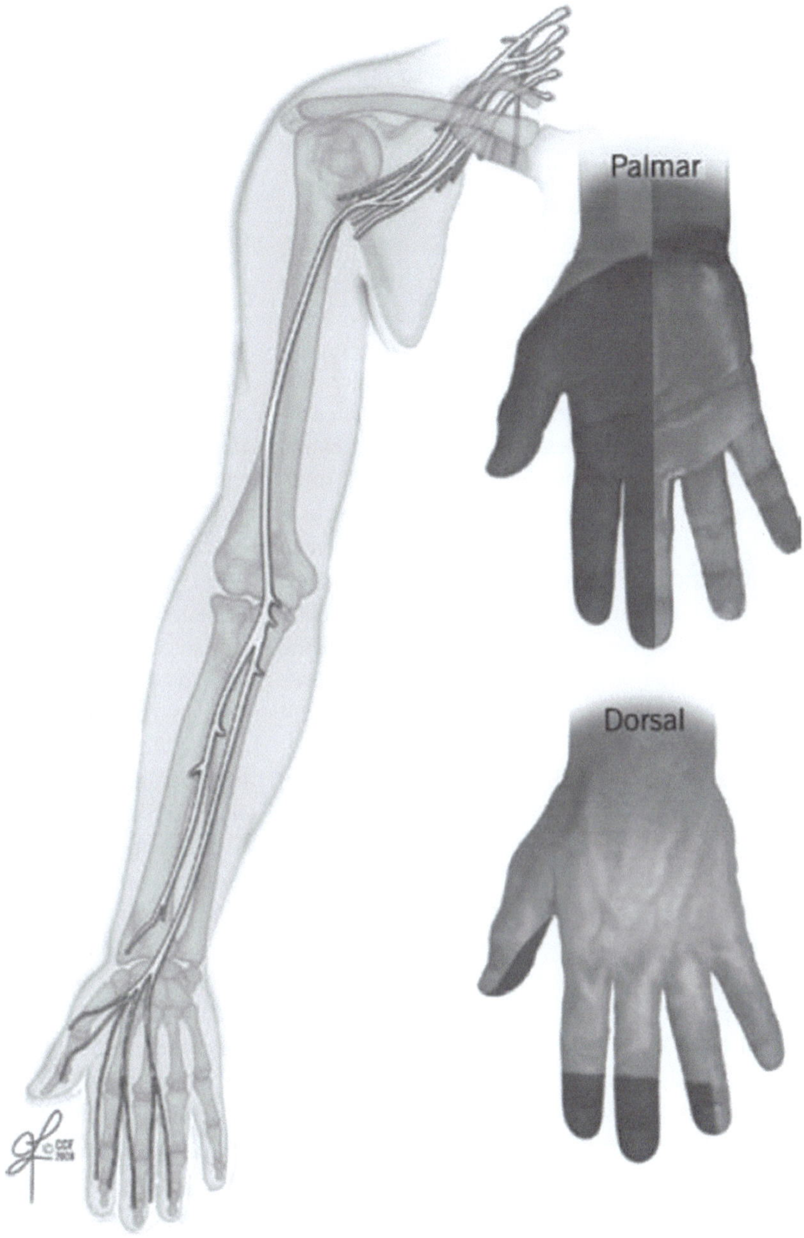

Fig. 23.13 Median nerve anatomy and sensory innervation

digital nerve to the second webspace, the radial digital nerve of the index finger, the ulnar digital nerve of the thumb, the radial digital nerve of the thumb, and the recurrent motor branch. The motor branch of the median nerve most commonly exits the nerve on the radial superficial border of the nerve and travels in recurrent fashion to supply the opponens pollicis, abductor pollicis brevis, and superficial head of the flexor pollicis brevis muscles (deep head is supplied by the ulnar nerve). The recurrent branch also has branches supplying the first and second lumbricals. The remainder of the nerve travels distally supplying sensation to the thumb through the radial half of the ring finger (Fig. 23.13).

23.6.2 Ulnar Nerve

The ulnar nerve is the continuation of the medial cord and enters the posterior compartment of the arm through the Arcade of Struthers approximately 8 cm proximal to the medial epicondyle. The ulnar nerve does not have any branches in the arm. The nerve courses just posterior to the medial intermuscular septum, and then wraps posterior to the medial epicondyle through the cubital tunnel. The cubital tunnel is bordered by the medial epicondyle anteriorly, the olecranon posteriorly, and the Ligament of Osborne superficially. The nerve then exits the cubital tunnel and courses between the two heads of the FCU muscle and sends off motor branches to the muscle at this level. The nerve then travels distally between the FCU and FDP muscles, giving motor branches to the FDP tendons to the ring and small fingers. In the mid forearm, the ulnar artery joins the nerve on its radial side and courses with the nerve across the wrist into the palm. Approximately 10 cm proximal to the wrist, the ulnar dorsal sensory branch exits the nerve on its ulnar most side. This branch moves from volar to dorsal at the wrist level and provides sensation to the dorsal ulnar aspect of the hand. The ulnar artery and nerve then pass superficial to the transverse carpal ligament, radial to the pisiform bone, and enter Guyon's canal. Guyon's canal is bordered by the pisiform ulnarly, the transverse carpal ligament on its deep and radial surfaces, and by the volar carpal ligament superficially. The ulnar motor branch exits the nerve on its deep side, and passes deep to the pisohamate ligament. It provides motor innervation to the palmaris brevis and hypothenar musculature (abductor digiti minimi, flexor digiti minimi, and opponens digiti minimi) and then courses radially to innervate all of the palmar and dorsal interossei, the third and fourth lumbricals, the adductor pollicis, and the deep head of the flexor pollicis brevis muscle. After the motor branch exits the canal, the ulnar nerve continues distally to provide the common and proper digital nerves to the small finger and the ulnar aspect of the ring finger (Fig. 23.14).

23.6.3 Radial Nerve

The radial nerve is the terminal branch of the posterior cord and courses along the radial groove of the humerus before traversing the lateral intermuscular septum into the anterior compartment of the arm. At the lateral epicondyle, it passes between the brachialis and brachioradialis. The nerve splits into a pure sensory nerve, the sensory branch of the radial nerve (SBRN), and the posterior interosseous nerve (PIN) at the radiocapitellar joint. The SBRN courses with the radial artery beneath the brachioradialis along until the distal third of the forearm when it becomes superficial between the BR and ECRL tendons to supply sensation to the dorsal thumb and hand. The ECRL is typically innervated by the radial nerve proper, with the remainder of the dorsal musculature being innervated by the PIN. The PIN innervates the supinator as it courses between its superficial and deep heads. The PIN then innervates the common extensor tendon group: extensor carpi radialis brevis, extensor digitorum communis, extensor digiti minimi and extensor carpi ulnaris, and the deep extensor group: abductor pollicis longus, extensor pollicis brevis, extensor pollicis longus, and finally the extensor indicis muscle. The PIN terminates on the floor of the fourth extensor compartment to provide sensation to the dorsal wrist capsule (Fig. 23.15).

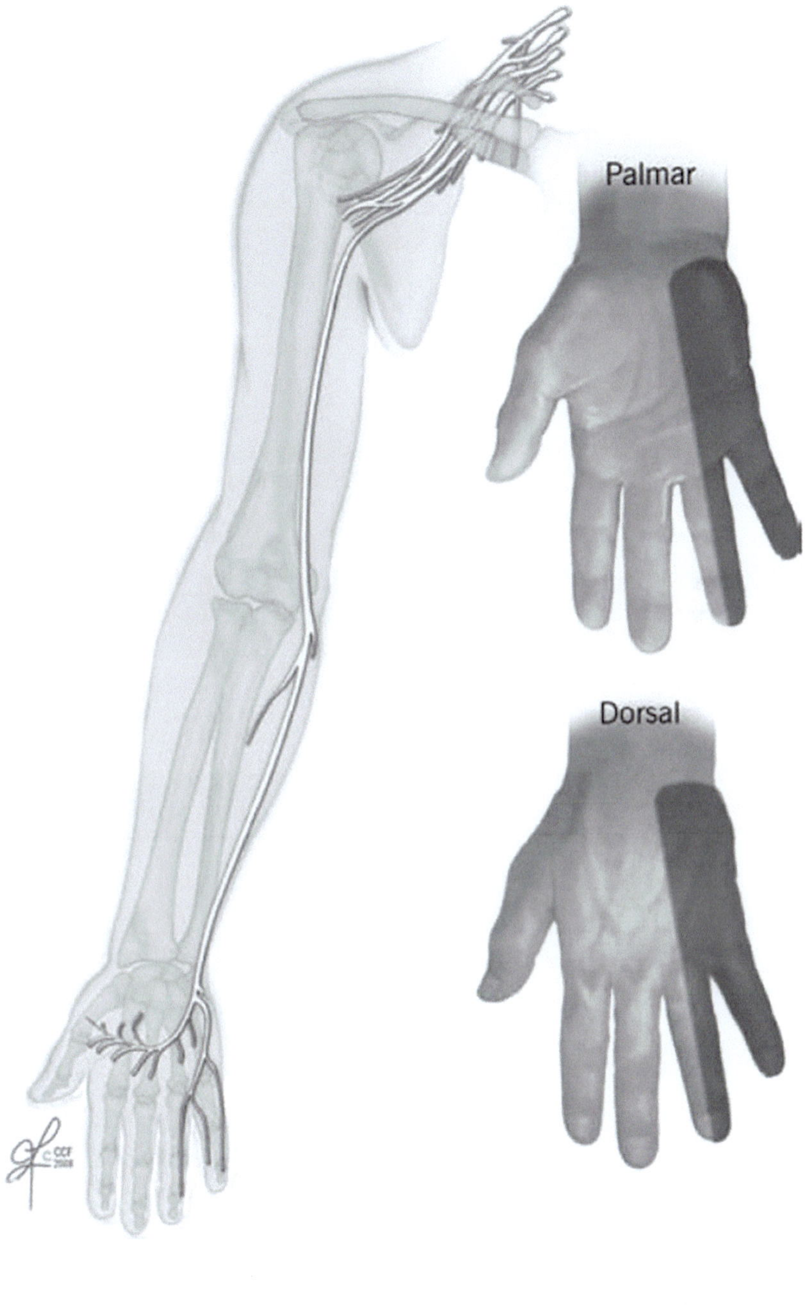

Fig. 23.14 Ulnar nerve anatomy and sensory innervation

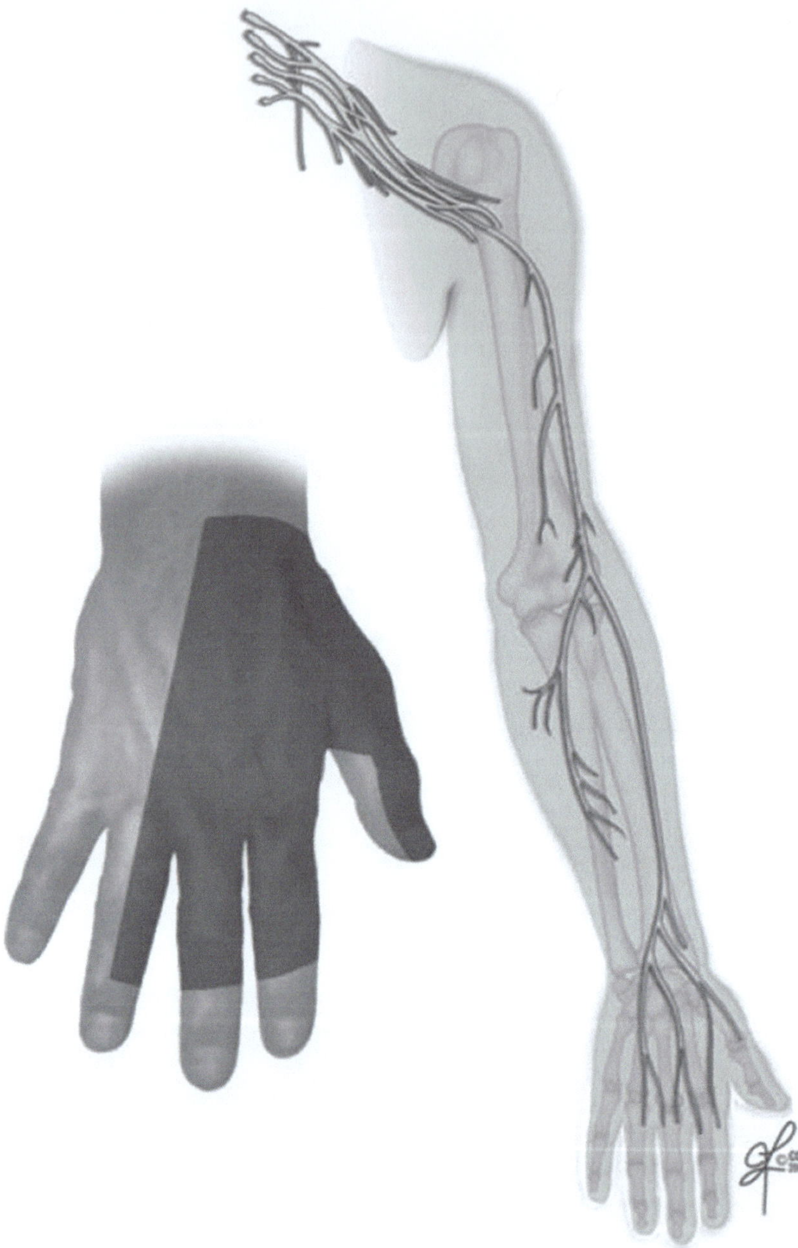

Fig. 23.15 Radial nerve anatomy and sensory innervation

References

1. Beasley RW. Beasley's surgery of the hand. New York: Thieme Medical; 2003.
2. Berger RA, Weiss APC, editors. Hand surgery, vol. 1. Philadelphia, PA: Lippincott Williams & Wilkins; 2004.
3. Bunnell S. Surgery of the hand. In Surgery of the hand. 1944. pp. xvii–734.
4. Hammert WC, Boyer MI, Bozentka DJ, Calfee RP. ASSH manual of hand surgery. Philadelphia, PA: Lippincott Williams & Wilkins; 2012.
5. Kaplan EB, Beasley RW. Functional and surgical anatomy of the hand. 1966

6. Rayan GM, Akelman E. The hand: anatomy, examination, and diagnosis. Philadelphia, PA: Lippincott Williams & Wilkins; 2012.
7. Schmidt HM, Lanz U. Surgical anatomy of the hand. Stuttgart: Thieme; 2011.
8. Trumble TE, Rayan GM, Baratz ME, Budoff JE, Slutsky DJ. Principles of hand surgery and therapy E-book. Washington, DC: Elsevier Health Sciences; 2016.
9. Wiesel SW. Operative techniques in orthopaedic surgery. Philadelphia, PA: Lippincott Williams & Wilkins; 2012.
10. Wolfe SW, Pederson WC, Hotchkiss RN, Kozin SH, Cohen MS. Green's operative hand surgery E-book. Philadelphia, PA: Elsevier Health Sciences; 2016.

24 Biomechanics of the Distal Forearm and Wrist

Toshiyasu Nakamura

24.1 Forearm Biomechanics

Main motion of the forearm is rotational movement of the radius around the ulna [1]. The radius and ulna contact at proximal end and distal end. In the proximal radioulnar joint (PRUJ), the radial head attaches to the sigmoid notch of the ulna and the sigmoid notch of the radius attaches to the ulnar head at the distal radioulnar joint (DRUJ). The ulna connects to the humerus with ligaments via humeroulnar joint. This joint is stable, which only arrows 6° rotation [1]. The radius rotates over the ulna approximately 135° with slight translation dorsally or palmary inside the DRUJ [1]. Especially in maximum pronation, radius moves palmary relative to the ulna by average 1.8 mm and ulnarly by 4.0° (Fig. 24.1). As additional 20–25° motion can be achieved inside the carpus and hand, the hand can rotate over approximately 180°, which we call pronation-supination. Rotation axis of forearm movement passes onto line between the center of the radial head and the fovea of the ulna with slight movement within 2 mm range (Fig. 24.2).

Main stabilizer of the PRUJ is the annular ligament. This ligament laps around the radial head to keep its position. When the annular ligament ruptures, the radial head dislocates either anteriorly or posteriorly. The main stabilizer of the DRUJ is the triangular fibrocartilage complex (TFCC) [2, 3], and the radioulnar ligament (RUL) is the proximal component of the TFCC which directly connects and stabilizes the radius to the ulna [3]. In the DRUJ, the ulnar head is round shape and the sigmoid notch of the radius has a concave shape, then the radius rotates around the ulna, while in the PRUJ, radial head indicates

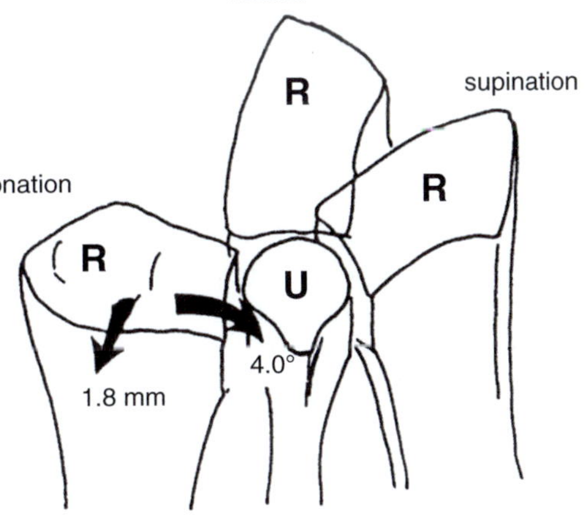

Fig. 24.1 Illustration of three-dimensional movement of the radius with ulna, indicating radius additionally translates 1.8 mm palmary and rotates 4.0° ulnary

T. Nakamura (✉)
Department of Orthopaedic Surgery, School of Medicine, International University of Health and Welfare, Tokyo, Japan
e-mail: toshiyasu@ae.em-net.ne.jp

J. G. Lane et al. (eds.), *The Art of the Musculoskeletal Physical Exam*,
https://doi.org/10.1007/978-3-031-24404-9_24

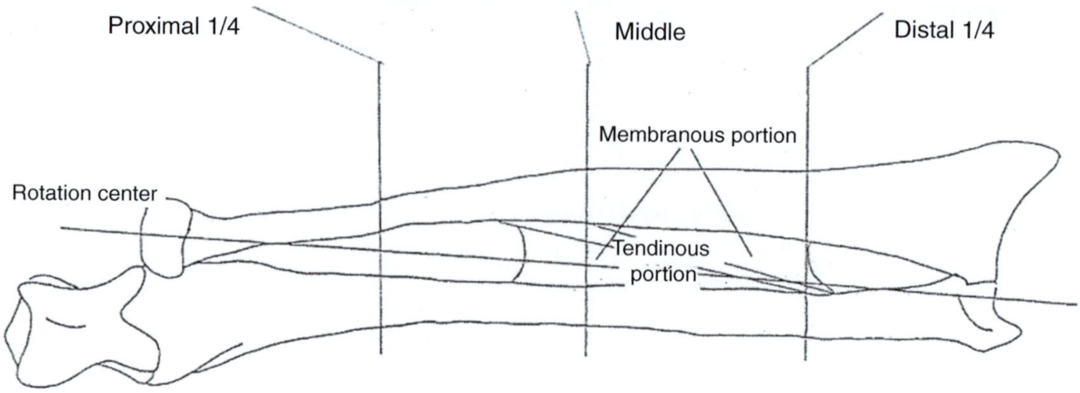

Fig. 24.2 Rotation axis of the forearm rotational movement passes onto the center of the radial head and central lesion of the ulnar fovea

round shape and ulnar sigmoid notch indicates concave shape.

Between the radius and ulna, there is the interosseous membrane [4]. There are several function of this structure; (1) intramuscular septum between the extensors and flexors; (2) origin of the flexors and extensors; (3) to keep the radius not away from the ulna; (4) load distribution from the radius to the ulna. The interosseous membrane consists of tendinous part (tendinous portion, interosseous ligament that runs from distal one-fourth of the ulnar to the middle part of the radius), membranous part (thin, membranous tissue), dorsal oblique cord (distal proximal cord that runs proximal one-fourth of the ulna to the middle part of the radius) [4]; and distal oblique band (40% existence rate; runs distal one-third of the radius to just proximal of the ulnar head) [5]. Especially, the tendinous part functions as the stabilizer between the radius and ulna and the load distributor from the radius to the ulna [4, 6] (Fig. 24.3). In case of Essex-Lopresti fracture which is a concomitant injury of the (comminuted) fracture of the radial head and dislocation of the DRUJ, the interosseous membrane ruptures with TFCC avulsion [6].

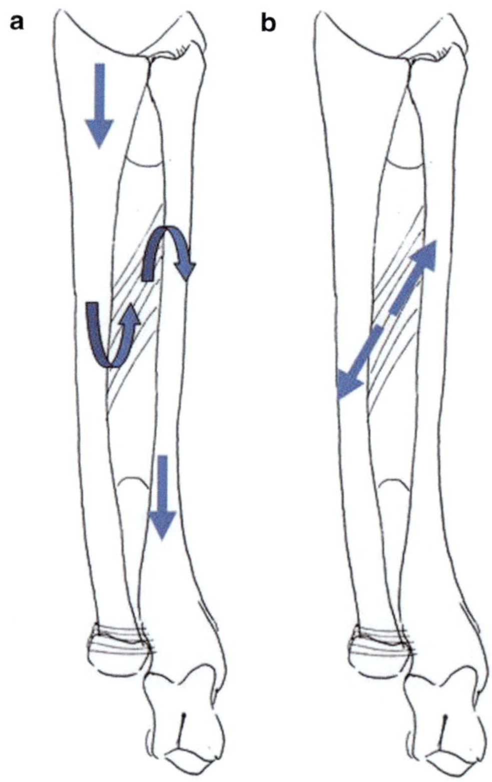

Fig. 24.3 Illustration of load distribution from the radius to the ulna through the tendinous part of the interosseous membrane (**a**). Stability of the radius and ulna can also be achieved through the tendinous part of the interosseous membrane (**b**)

24.2 Biomechanics of the DRUJ

The DRUJ consists of sigmoid notch of the radius, ulnar head, triangular fibrocartilage complex (TFCC), and dorsal and palmar joint capsules [7]. This joint can achieve pronosupination movement with the proximal radioulnar joint (PRUJ). In the DRUJ, the ulnar head indicates a round shape and the sigmoid notch of the radius has a concave shape, then the radius rotates around the ulna, while in the PRUJ, radial head shapes round and ulnar sigmoid notch indicates concave shape. The main stabilizer of the DRUJ is the RUL [7, 8] (Fig. 24.4).

During pronosupination, the radius rotates around the ulna with slight translation in the maximum pronation [4]. Rotation axis of the forearm rotation passes onto the center of the radial head and the central lesion of the ulnar fovea with slight positional changes inside [4] (Fig. 24.5).

The RUL is the proximal component of the TFCC and attaches to the ulnar fovea and base of the ulnar styloid [8] (Fig. 24.4). The RUL has three portions, dorsal, palmar, and central. The dorsal and palmar portions indicate strong, robust fiber arrangement from the fovea of the ulna to dorsal or palmar edge of the sigmoid notch [8]. Origin of the RUL at the fovea indicates nearly vertical fiber arrangement [9, 10]. Three-dimensional arrangement of the RUL achieves constant stability of the DRUJ during pronosupination. Centrally attaching fibers of the RUL to the fovea indicates relatively constant changes in length during forearm rotation. Eccentric fibers of the RUL show dorsal portion is longer in pronation and shorter in supination and vice versa [8].

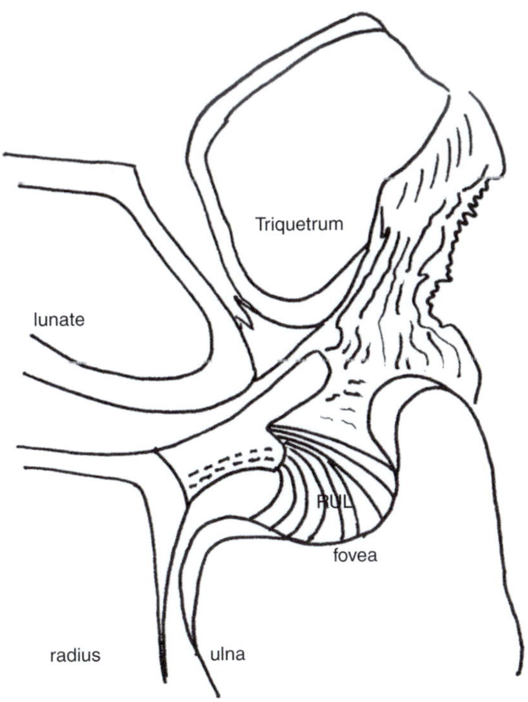

Fig. 24.4 Illustration of the DRUJ. The RUL, main stabilizer of the DRUJ, connects the radius to the ulnar fovea

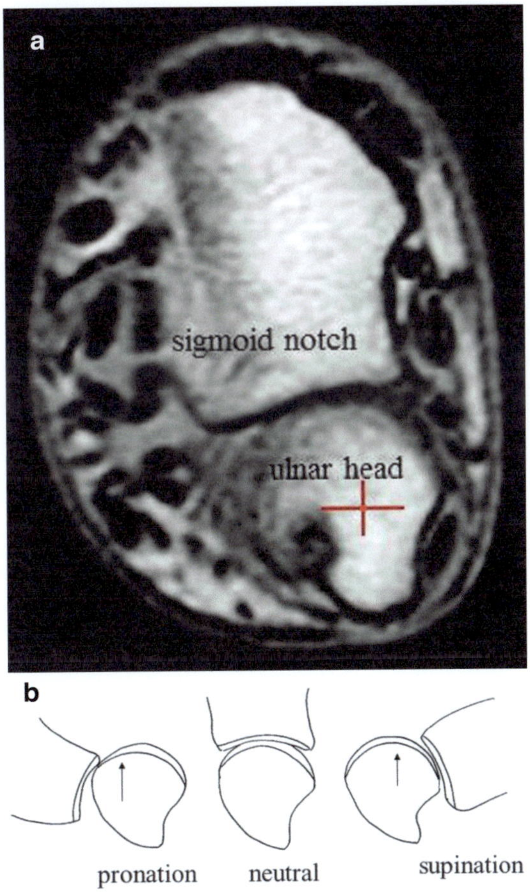

Fig. 24.5 (**a**) Axial MRI view of the DRUJ. Red cross indicates the position where the rotation axis passes inside the ulnar fovea. (**b**) Illustration of the movement of the radius relative to the ulna. See slight palmar translation is indicated in supination and slight dorsal translation in pronation

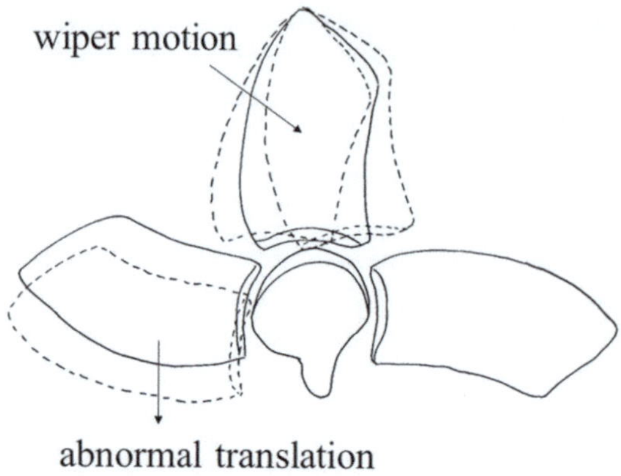

Fig. 24.6 Abnormal motion of the radius relative to the ulna in DRUJ instability patients. When the RUL is ruptured, the over-rotation of the radius (wiper action) and increased translation of the radius (arrow) in pronation can be occurred, showing protrusion of the ulnar head in pronation or slacking during pronosupination

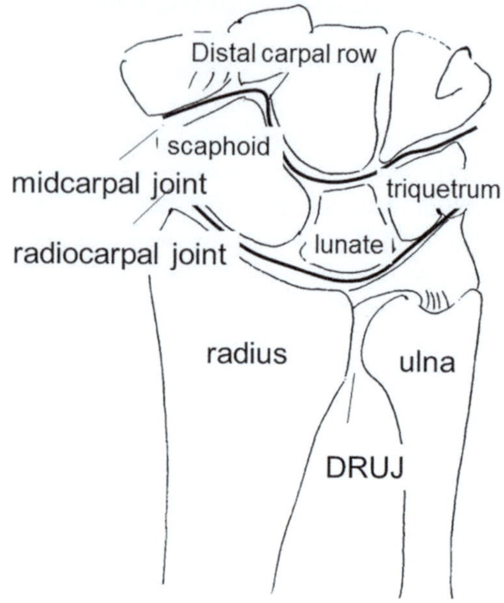

Fig. 24.7 Illustration of the wrist, consisting of midcarpal, radiocarpal, and DRU joints

When the patient indicates severe, no end-point instability of the distal radioulnar joint, the RUL should be ruptured or detached at the fovea where is the main inserting point of the RUL to the ulna. The DRUJ shows abnormal translation of the radius relative to the ulna or uncontrolled over-rotation of the radius is indicated with RUL avulsed patient (Fig. 24.6). The RUL should be reattached to the fovea in such cases [8, 11].

24.3 Biomechanics of the Wrist

The wrist consists of radiocarpal joint (RCJ), midcarpal joint (MCJ), and distal radioulnar joint (DRUJ) (Fig. 24.7). The radius, ulna, scaphoid-lunate-triquetrum (proximal row), and trapezium-trapezoid-capitate-hamate (distal row) also composed the wrist. In the RCJ and MCJ, wrist flexion-extension and radial-ulnar deviation movement can be achieved, while in the DRUJ, the pronosupination motion is achieved. The distal row is very stable and only slight motion is found between each bone, while the proximal row indicates relatively larger motion between scaphoid-lunate and lunate-triquetrum.

There are extrinsic ligaments between bones in the wrist [12]. In the dorsal side wrist, the radiocarpal ligament that runs from the radius to the triquetrum and the intercarpal ligament that runs triquetrum to the scaphoid ridge. Some fibers of the intercarpal ligament attaches to the dorsal surface of the lunate. The volar side is very complex. The radioscaphocapitate, long and short radiolunate ligament, ulno lunate and ulno triquetrum ligament is noted for RCJ stabilizing ligament. Scaphotrapeziotrapezoidal (STT) ligament, scapho capitate ligament, triquetrohamate ligament, capitohamate ligament, and ulnocapitate ligament are MCJ stabilizers (Fig. 24.8). In the proximal carpal row, two intrinsic ligaments are noted, i.e., the scapholunate (SL) ligament and lunotriquetral (LT) ligament [13]. These two intrinsic ligaments show horseshoe shape, in which dorsal and volar ligamentous parts and proximal membranous part. The ligamentous part shows true ligament and dorsal part of SL ligament and volar part of LT ligament are major

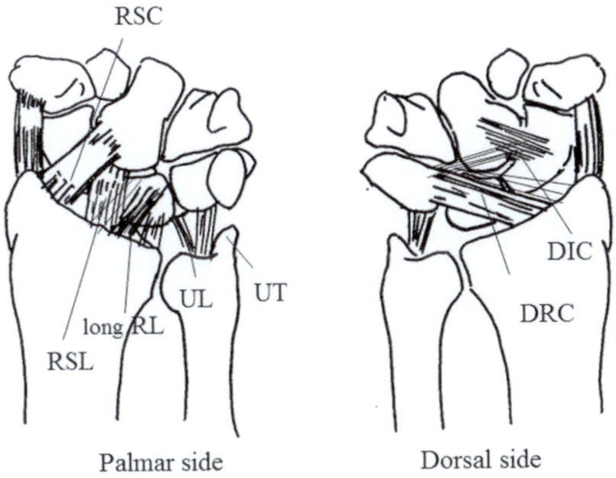

Fig. 24.8 Diagram of dorsal and palmar extrinsic ligaments of the wrist. *RSC* radioscaphocapitate ligament, *long RL* long radiolunate ligament, *UL* ulnolunate ligament, *UT* ulnotriquetral ligament, *DRC* dorsal radiocarpal ligament, *DIC* dorsal intercarpal ligament

stabilizers, while the proximal membranous part indicates fibrocartilaginous structure [13].

The scaphoid has a tendency to flex because of the axial load from the distal carpus and flexion force form the FCR tendon, while the triquetrum tends to extend [14]. If the links, such as the SL and LT ligaments, are intact, proximal carpal row behaves as nearly one unit in flexion-extension, whereas the SL ligament shows a slight twist motion during radial-ulnar deviation. When the SL ligament was ruptured, there are gap formation between the scaphoid and lunate which indicates scapholunate dissociation (Fig. 24.9). This gap formation also indicates carpal instability dissociative (CID). In this scenario, the scaphoid flexes and lunate and triquetrum extend. The lateral radiograph shows extension of the lunate called DISI (dorsal intercarpal segment instability) deformity (Fig. 24.10) [14]. When the LT ligament ruptures with DIC ligament avulsion, scaphoid, and lunate flex while the triquetrum extends, showing flexion of the lunate that is called VISI (volar intercarpal segment instability) deformity [15]. Same DISI deformity phenomenon is noted when the scaphoid waist fracture or nonunion occurs [16].

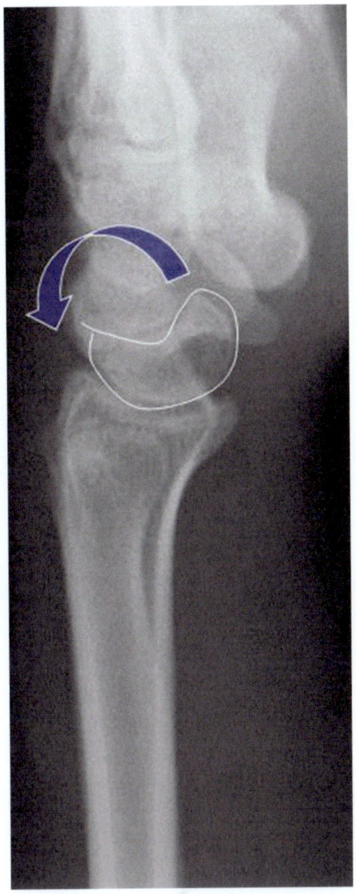

Fig. 24.9 Radiograph of posteroanterior view indicates gap formation between the scaphoid and lunate (white arrow)

There are other carpal instabilities. If there is no gap formation between the scaphoid and lunate and only the flexion of the scaphoid is indicated, we call it dynamic SL instability or carpal instability non-dissociative (CIND) both show DISI deformity [17].

When the patient shows DISI deformity with SL ligament rupture or scaphoid nonunion, DISI should be reduced with K-wire, then reconstruction of the SL ligament [18] or restoration of the scaphoid nonunion with bone grafting should be needed.

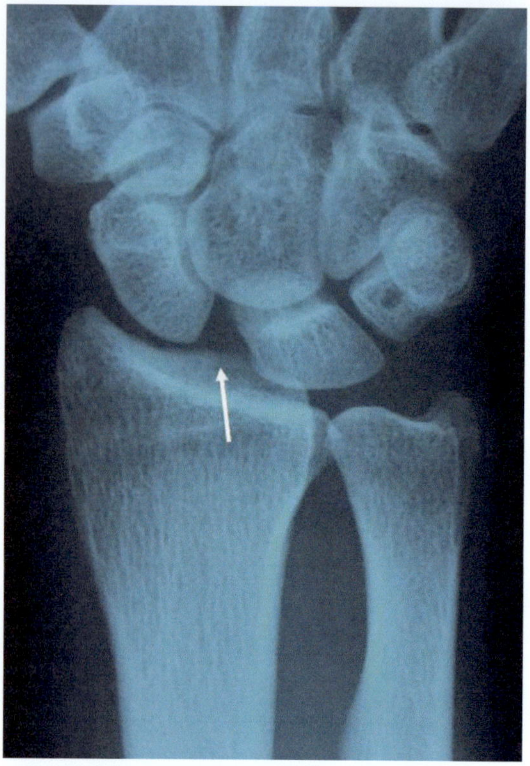

Fig. 24.10 Dorsal intercalated segment instability of the lateral radiograph (blue, round arrow), indicating dorsal rotation of the lunate (white line)

References

1. Nakamura T, Yabe Y, Horiuchi Y, Yamazaki N. In vivo motion analysis of forearm rotation utilising magnetic resonance imaging. Clin Biomech. 1999;14:315–20.
2. Nakamura T, Yabe Y, Horiuchi Y, Yamazaki N. Three-dimensional magnetic resonance imaging of the interosseous membrane of forearm – a new method using fuzzy reasoning. Magn Reson Imag. 1999;17:463–70.
3. Nakamura T, Yabe Y, Horiuchi Y. In vivo MR studies of dynamic changes in the interosseous membrane of the forearm during rotation. J Hand Surg. 1999;24B:245–8.
4. Nakamura T, Yabe Y, Horiuchi Y, Yamazaki N. Normal kinematics of the interosseous membrane of the forearm during rotation – a three-dimensional MRI study. Hand Surg. 2000;5:1–10.
5. Moritomo H. The distal interosseous membrane: current concept in wrist anatomy and biomechanics. J Hand Surg Am. 2012;37(7):1501–7.
6. Hotchkiss RN, An KN, Sowa DT, Basta S, Weiland AJ. An anatomic and mechanical study of the interosseous membrane of the forearm: Pathomechanics of proximal migration of the radius. J Hand Surg Am. 1989;14:256–61.
7. Nakamura T, Yabe Y, Horiuchi Y. Functional anatomy of the triangular fibrocartilage complex. J Hand Surg. 1996;21-B:581–6.
8. Nakamura T, Makita A. The proximal ligamentous component of the triangular fibrocartilage complex: functional anatomy and three-dimensional changes in the length of the radioulnar ligament during pronation and supination. J Hand Surg. 2000;25B:479–86.
9. Nakamura T, Yabe Y. Histological anatomy of the triangular fibrocartilage complex of the human wrist. Ann Anat. 2000;182:567–72.
10. Nakamura T, Yabe Y. Origins and insertions of the triangular fibrocartilage complex – a histological study. J Hand Surg. 2001;26B:446–54.
11. Nakamura T, Sato K, Okazaki M, Toyama Y, Ikegami H. Repair of the foveal detachment of the triangular fibrocartilage complex: open and arthroscopic transosseous techniques. Hand Clin. 2011;27:281–90.
12. Berger RA. The anatomy of the ligaments of the wrist and distal radioulnar joints. Clin Orthop. 2001;383:32–40.
13. Berger RA. The gross and histologic anatomy of the scapholunate interosseous ligament. J Hand Surg. 1996;21A:170–8.
14. Linscheid RL, Dobyns JH, Beabout JW, Bryan RS. Traumatic instability of the wrist: Diagnosis, classification and pathomechanics. J Bone Joint Surg Am. 1972;54:1612–32.
15. Horii E, Garcia-Elias M, An KN, Bishop AT, Cooney WP, Linscheid RL, Chao EY. A kinematic study of luno-triquetral dissociations. J Hand Surg Am. 1991;16:355–62.
16. Moritomo H, Murase T, Oka K, Tanaka H, Yoshikawa H, Sugamoto K. Relationship between the fracture location and the kinematic pattern in scaphoid nonunion. J Hand Surg Am. 2008;33:1459–68.
17. Wright TW, Dobyns JH, Linscheid RL, et al. Carpal instability on-dissociative. J Hand Surg Br. 1994;19:763–73.
18. Nakamura T, Abe K, Iwamoto T, Ochi K, Sato K. Reconstruction of the scapholunate ligament using capitohamate bone-ligament-bone. J Wrist Surg. 2015;4:264–8.

25 Evaluation of Range of Motion

Esther Ching San Chow

25.1 Introduction

Measurement of the hand and wrist range of motion is an essential part of a proper physical examination of the hand and wrist. It serves the purpose of evaluation, derives the proper diagnosis, and monitors the treatment progress and effectiveness.

25.2 Anatomy of the Joint

25.2.1 Wrist Joint

The wrist joint is a complex joint and is composed of several joints, these include the radiocarpal joint, the mid-carpal joint, and the distal radioulnar joint.

25.2.1.1 Radiocarpal Joint (RCJ)

The radiocarpal joint is the joint between the distal articular surface of the radius and the proximal row of the carpal bones (Scaphoid, Lunate, and triquetrum). This is a condyloid joint and synovial joint. Together with the mid-carpal joint, these two joints allow combined motions in multiple planes, including flexion, extension, radial deviation, and ulnar deviation.

25.2.1.2 Mid-Carpal Joint (MCJ)

The mid-carpal joint is the joint between the proximal row and distal row of carpal bones. Similar to the radiocarpal joint, the mid-carpal joint is a condyloid joint. It works together with the radiocarpal joint to provide motion of flexion, extension, radial and ulnar deviation, but each joint contributes a different proportional of movement. For radial and ulnar deviation motion, it occurs primarily in the mid-carpal joint. Previous study had shown that mid-carpal motion accounted for 60% of radial deviation and 86% of ulnar deviation [1]. Flexion occurs mainly at the mid-carpal joint (60 vs. 40% in radiocarpal joint). On the other hand, extension occurs mainly in the radiocarpal joint (66.5 vs. 33.5% in mid-carpal joint) [2].

25.2.1.3 The Distal Radio-Ulnar Joint (DRUJ)

The distal radioulnar joint is located anatomically at the wrist; however, it has a separate joint capsule from the radiocarpal joint and mid-carpal joint. This joint is formed by the ulna head (which is convex) and the sigmoid notch of the distal radius (which is flat or concave in shape). This is a pivot-type of synovial joint. This joint allows movement of rotation with pronation and supination, where the radius rotates around the rela-

E. C. S. Chow (✉)
Department of Orthopaedics and Traumatology, United Christian Hospital, Kowloon, Hong Kong

J. G. Lane et al. (eds.), *The Art of the Musculoskeletal Physical Exam*, https://doi.org/10.1007/978-3-031-24404-9_25

tively fixed ulna head. This joint also allows movement in translation, with the ulna located relatively palmar during supination and relatively dorsal during pronation [3].

25.2.2 Hand Joints

The hand joints include the carpometacarpal joints, metacarpophalangeal joints, proximal interphalangeal joints, and distal interphalangeal joints.

25.2.2.1 The Carpometacarpal Joint (CMCJ)

The carpometacarpal joints are the joints located between the distal carpal bones and the metacarpals. The first CMCJ is different from the other CMCJs, it is formed by the articulation by the first metacarpal and trapezium. It is known as a saddle-shape synovial joint. This saddle joint allows the thumb to move in multi-direction. The thumb CMCJ can move in abduction, adduction, extension, flexion, rotation, and opposition. Opposition is a combination of flexion, medial rotation, and abduction of the first CMCJ. However, not all of these motion at the thumb CMCJ can be measured easily. The measurement method of thumb CMCJ motion will be discussed later.

The carpometacarpal joints of the fingers are plane-type or gliding-type of synovial joints. However, the second and third CMCJ have nearly no motion, while the fifth CMCJ is the most mobile.

25.2.2.2 Metacarpophalangeal Joint (MCPJ)

The metacarpophalangeal joint (MCPJ) is classified as condyloid synovial joint. It is formed by the articulation of the convex head of the metacarpal with the concave base of the proximal phalanx of the corresponding digit. The MCPJs of the index, middle, ring, and little finger can move in flexion, extension, abduction, and adduction. The metacarpal head has a special shape, such that when the MCPJ is in flexion, the collateral ligaments are tightened. This is known as the "cam effect." Therefore, the abduction and adduction motion is limited when the joint is in flexed position. The motion of the thumb MCPJ is limited to flexion and extension only.

25.2.2.3 Interphalangeal Joint (IPJ)

The thumb has only a single interphalangeal joint. The index, middle, ring, and little finger has one proximal interphalangeal joint and one distal interphalangeal joint.

25.2.2.4 Proximal Interphalangeal Joint (PIPJ)

The proximal interphalangeal joint is the joint that articulates between the convex head of the proximal phalanx and the concave base of the middle phalanx. This is a hinge-type synovial joint, which allow motion of flexion and extension.

25.2.2.5 Distal Interphalangeal Joint (DIPJ)

The distal interphalangeal joint is the joint that articulates between the convex head of the middle phalanx with the concave base of the distal phalanx. Similar to the PIPJ, the DIPJ is a hinge-type of synovial joint. The movement in this joint included flexion and extension.

25.3 Measurement Techniques

25.3.1 Measurement Instruments

The most commonly used method for measuring hand and wrist motion is by using the protractor-type goniometers. There are several different types of goniometers. The most commonly used one is the universal goniometer with full-circle body (Fig. 25.1). This is mostly used for wrist motion measurements. The smaller size goniometers commonly used include the Devore pocket goniometer (Fig. 25.2), for measurement of finger joint motion. It can be placed on the dorsal, volar, or lateral aspect of the finger joint. Another commonly used hand goniometer is the Rolyan finger goniometer (Fig. 25.3), which is placed on the dorsal surface of the finger. It has a short arm (movable arm) that is only 2 cm in length, which is suit-

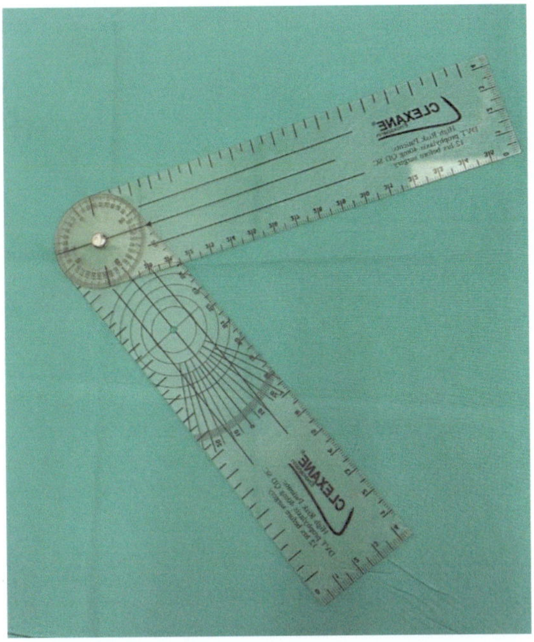

Fig. 25.1 Universal goniometer

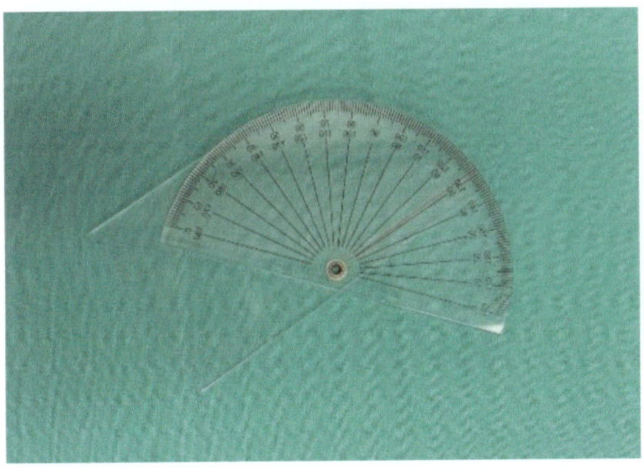

Fig. 25.2 Devore pocket goniometer

able for measurement of DIPJ motion. It can also allow measurement of hyperextension up to 30°. Another measurement instrument used is the inclinometer (Fig. 25.4). It is useful for measurement of forearm rotation.

25.3.1.1 Wrist Joint

The wrist joint motion that can be measured include: flexion, extension, radial deviation, ulnar deviation, pronation, and supination. The most commonly used method for measuring wrist motion is by using the protractor-type goniometer. Table 25.1 showed the normal wrist range of motion.

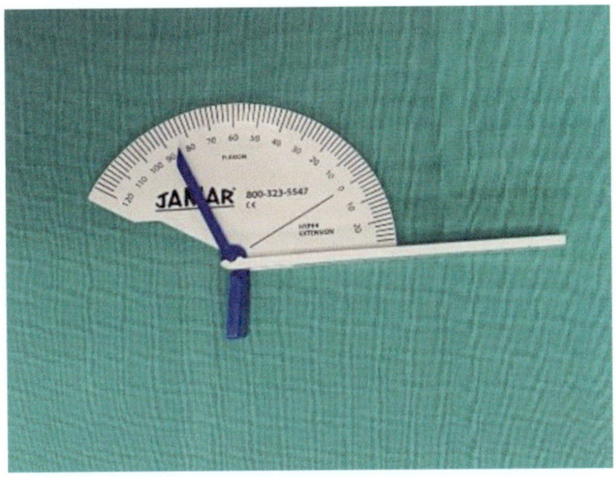

Fig. 25.3 Rolyan finger goniometer

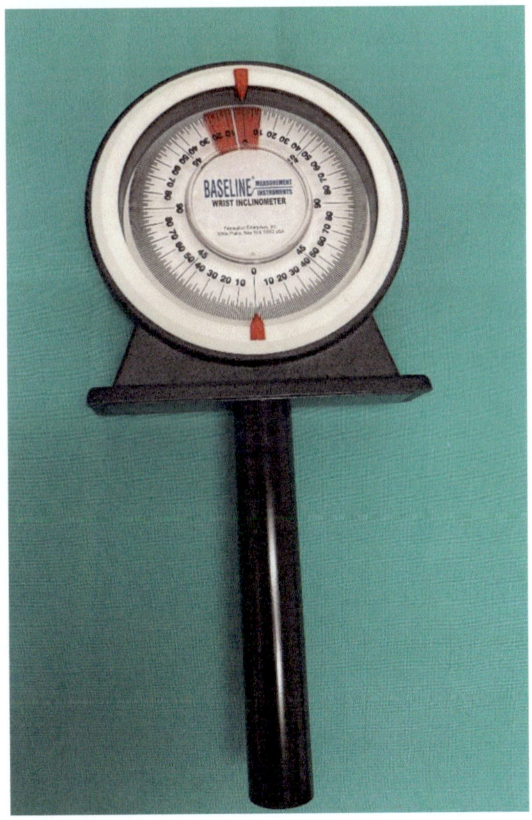

Fig. 25.4 Inclinometer

Table 25.1 Normal range of motion of the wrist (Boone and Azen [4])

Flexion	75 ± 6.6
Extension	74 ± 6.6
Radial deviation	21 ± 4.0
Ulnar deviation	35 ± 3.8
Pronation	75 ± 5.3
Supination	81 ± 4.0

25.3.2 Wrist Flexion-Extension

For wrist flexion and extension, there are three recommended techniques for measurement, these include the radial technique, the ulnar technique, and the dorsal-volar technique. Study by Lastayo showed that the dorsal-volar technique is the most reliable technique both within and between testers [5]. Therefore, it is recommended that the radial, ulnar, and dorsal-volar goniometric techniques should not be used interchangeably, as their results frequently will be inconsistent. Similarly, Carter et al. demonstrated that the dorsal-volar technique had the greatest inter-rater reliability [6]. During measurement of wrist in the plane of flexion and extension, the wrist should be in neutral position along the radial or ulnar deviation plane. For the positioning, the shoulder should be abducted 90°, elbow flexed 90°, with the forearm resting comfortably on the table. The three techniques are described as in Table 25.2 and illustrated in Fig. 25.5, 25.6,

Table 25.2 Measurement technique for wrist flexion-extension

Measurement technique	Position of goniometer	
	Stationary arm	Movable arm
Radial technique	Along the midline of radial shaft	Parallel the axis of the third metacarpal
Ulnar technique	Along midline of ulna shaft	Parallel the axis of the third metacarpal
Dorsal/volar technique	Flexion: Along dorsal surface of forearm Extension: Along volar surface of forearm	Along the axis of third metacarpal

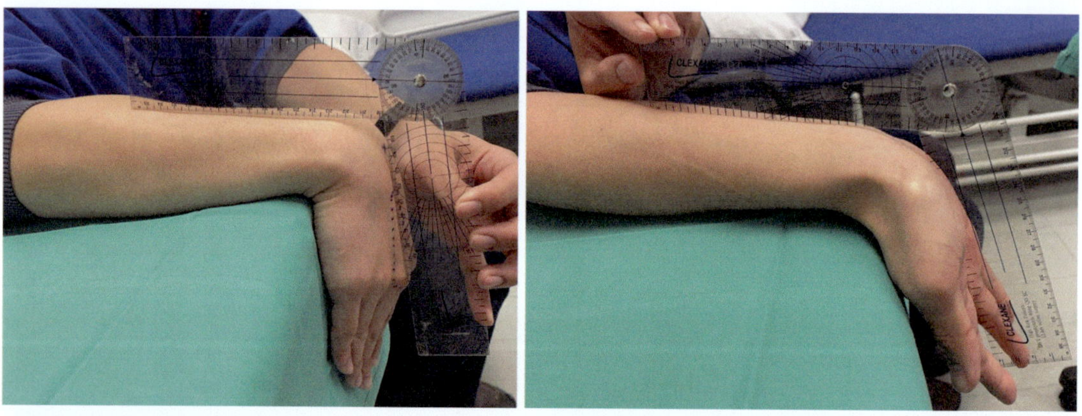

Fig. 25.5 Dorsal-Volar technique

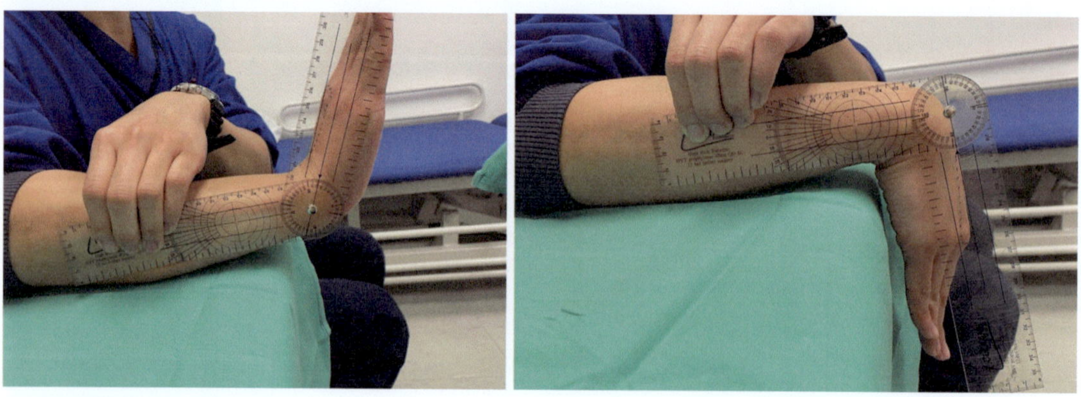

Fig. 25.6 Ulnar technique

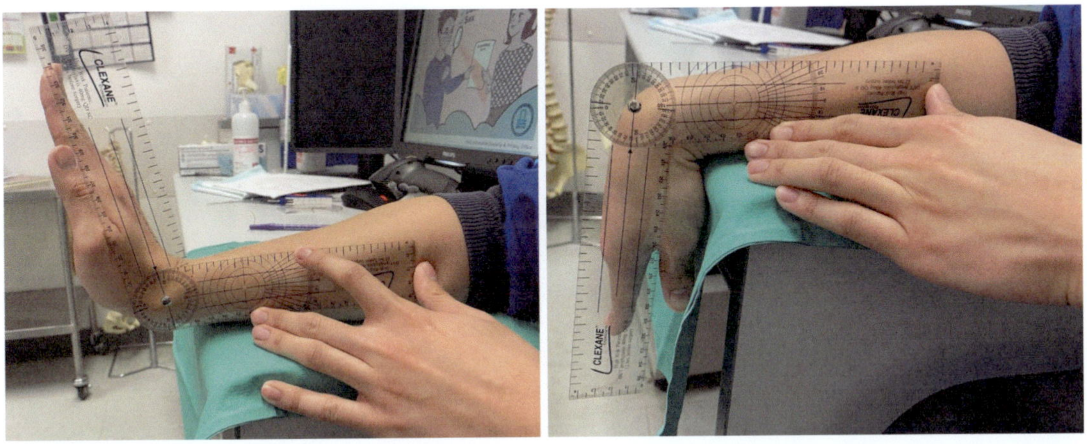

Fig. 25.7 Radial technique

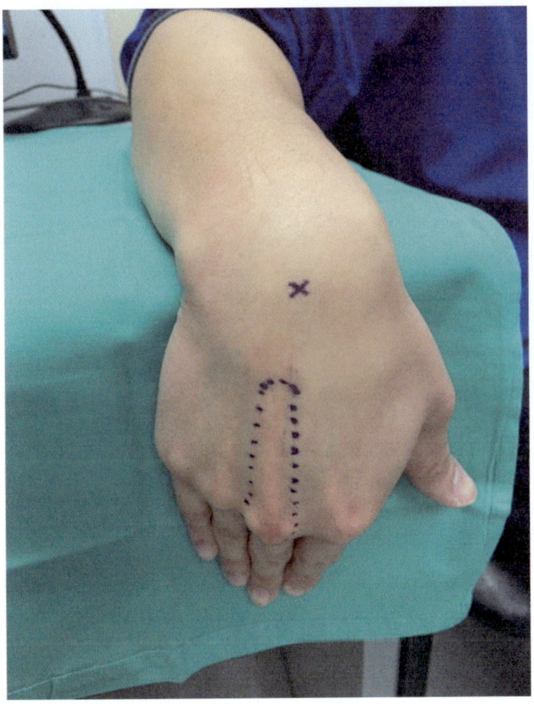

Fig. 25.8 Capitate as the center of rotation

and 25.7. The pivot point should be the center of motion of the wrist, which is the proximal capitate (Fig. 25.8).

25.3.3 Wrist Radial Deviation and Ulnar Deviation

Wrist abduction and adduction are measured using the standard technique. The position of the person is same for measuring wrist flexion/extension, with the shoulder abducted 90°, elbow flexed 90°, and the forearm resting comfortably on the table. The stationary arm of the goniometer is placed along the axis of the dorsal surface of the forearm. The pivot point should be at the center of motion of the wrist, which is the proximal capitate. The movable arm of the goniometer should be placed along the axis of the third metacarpal (Fig. 25.9). Similar to the measurement of wrist flexion/extension, while measuring the wrist along the plane of radial and ulnar deviation, the wrist should be in neutral position at the flexion/extension plane.

25.3.4 Pronation-Supination

The measurement of wrist motion cannot be completed without the measurement of pronation and supination. Strictly speaking, pronation and supination is a measurement of forearm rotation, and this motion is a composite motion occurring at the proximal and distal radioulnar joint. During measurement of forearm rotation, the shoulder should be adducted with the arm against the side of the chest wall, the elbow is flexed 90°. The pronation and supination can be measured with two methods: (1) Using the goniometer and (2) using the inclinometer. The goniometer technique was first described by Norkin and White [7]. For the measurement of supination, the stationary arm (Vertical arm) of the goniometer is placed just ulnar to the ulnar styloid, perpendicu-

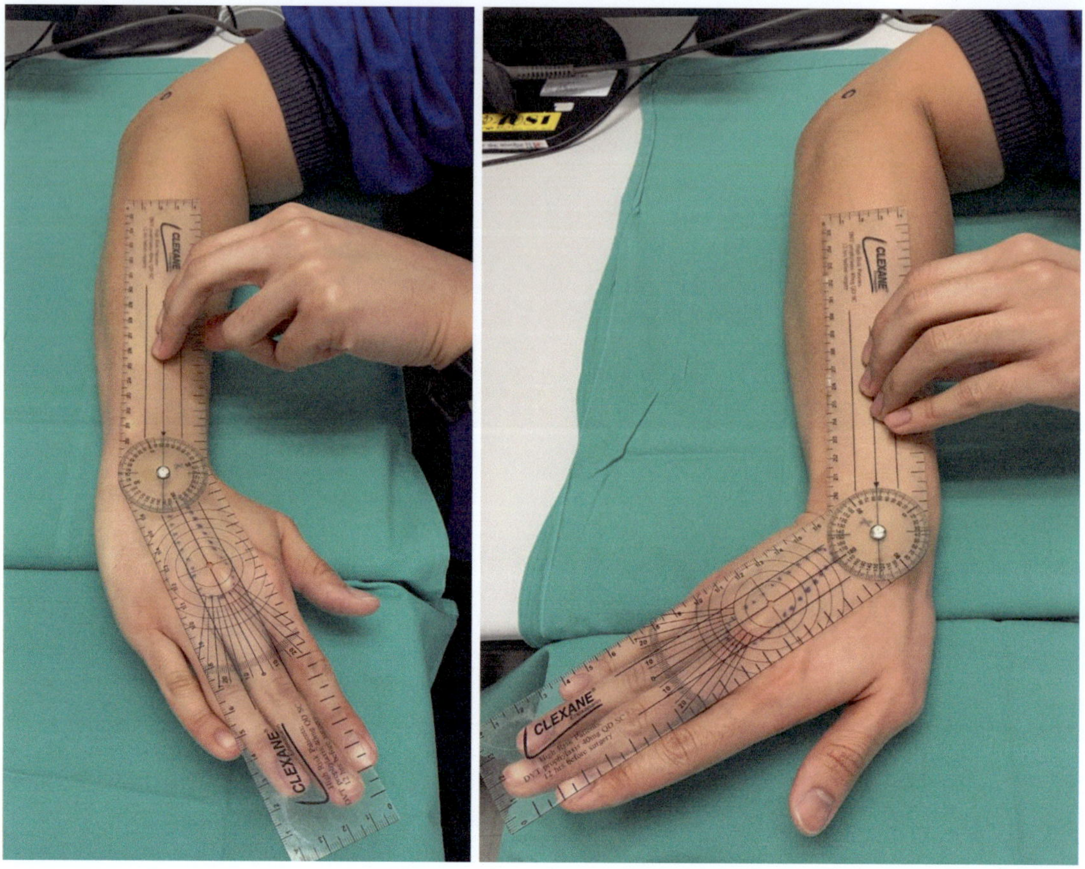

Fig. 25.9 Radial deviation and Ulnar deviation of the wrist

lar to the floor or aligned parallel to the humerus, the movable arm is placed on the volar aspect of the wrist. For measurement of pronation, the stationary arm (vertical arm) is placed just ulnar to the ulnar styloid, perpendicular to the floor or aligned parallel to the humerus, while the movable arm is placed over the dorsum of the wrist (Fig. 25.10). Another method is using the inclinometer, as shown in Fig. 25.11.

25.3.4.1 Hand Joints

Thumb Joints Measurement

The normal thumb range of motion is found in Table 25.3.

First Carpometacarpal Joint (CMCJ)

As mentioned previously, the first CMCJ is a saddle-shape joint which allows the thumb to move in multi-direction. The first CMCJ can move in abduction, adduction, extension, flexion, rotation, and opposition. The first CMCJ adduction, extension, and flexion are difficult to quantify and therefore rarely measured. We will discuss the measurement of the first CMCJ abduction and opposition in this chapter.

First CMCJ Abduction

Thumb CMCJ abduction can be measured in two directions: palmar abduction and radial abduction. Thumb palmar abduction is the movement in which the thumb metacarpal moves away from the index metacarpal, perpendicular to the plane of the palm (i.e., the sagittal plane). While thumb radial abduction is the angle between the thumb metacarpal and index metacarpal in the coronal plane. The measurement is done by a conventional goniometer, with the placement of the goniometer as shown in Figs. 25.12 and 25.13. The pivot point is located at the radial styloid,

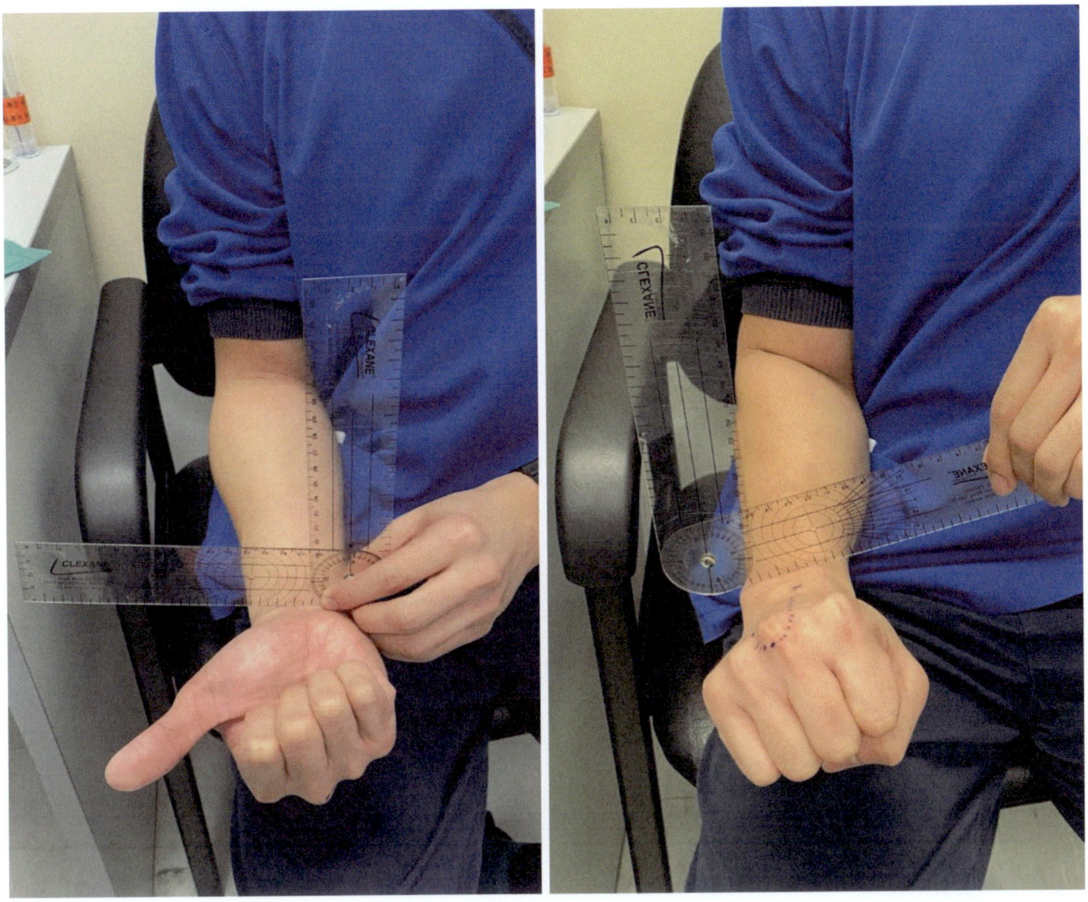

Fig. 25.10 Supination and pronation measurement with goniometer

while the stationary arm is aligned parallel to the second metacarpal shaft, the movable arm is aligned parallel to the first metacarpal shaft.

First CMCJ Opposition

The most reproducible and reliable measurement of the first CMCJ is by thumb opposition. Thumb opposition is a combination of flexion, medial rotation, and abduction of the first CMCJ. One commonly used method of thumb opposition measurement is by the Kapandji score [9]. Kapandji uses a simple method for measurement of thumb opposition, without the need for any measuring device. This method uses the patient's own hand as reference. A score from 0 to 10 is assigned according to the position of the thumb tip touching the position of the four fingers, which the highest score 10, where the thumb tip is able to touch the distal crease of the hand at the base of the little finger. The scoring system is illustrated in Fig. 25.14. Another method of measurement of the thumb opposition is by measuring the distance between the thumb IPJ crease to the distal palmar crease of the third metacarpal (Fig. 25.15).

25.3.5 Thumb Metacarpophalangeal Joint and Interphalangeal Joint

The standard technique for measuring thumb MCPJ and IPJ is by placing the goniometer over the dorsal surface of the thumb. It is important to keep the proximal joint in neutral position for accurate measurement of MCPJ and IPJ. Lateral positioning of the goniometer can be used if there is excessive swelling of the thumb for more accurate measurement.

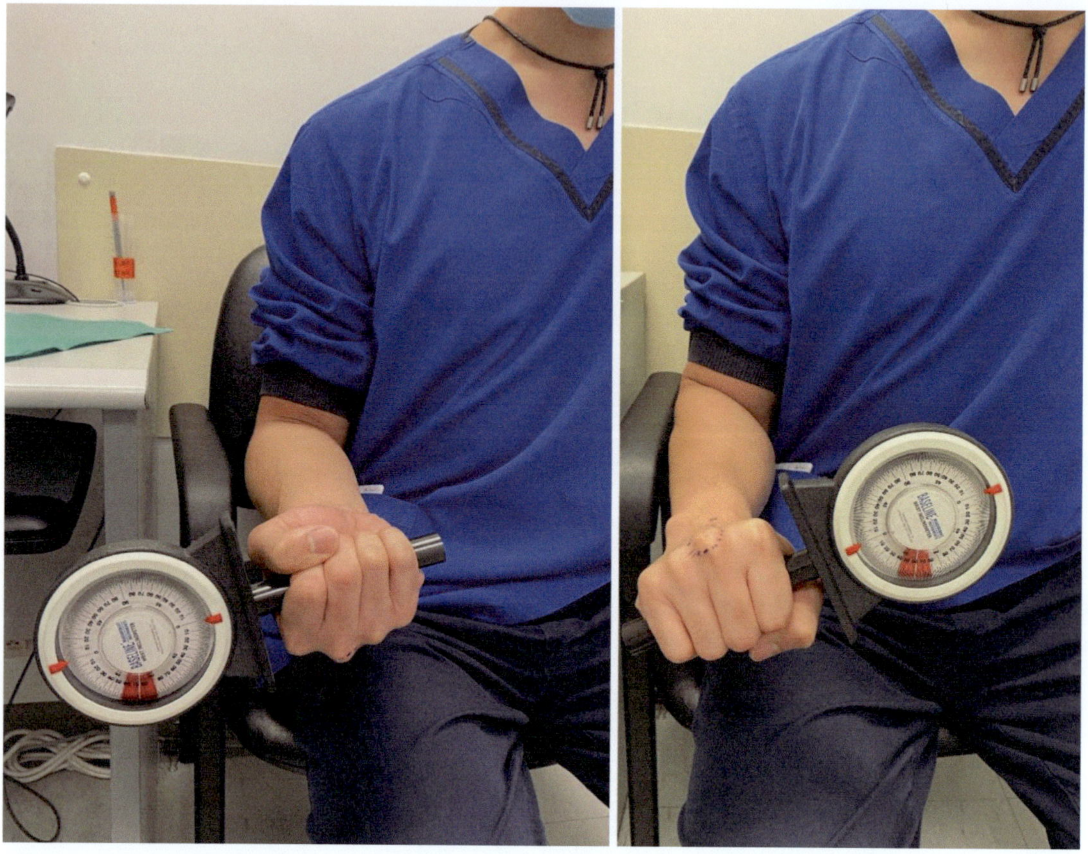

Fig. 25.11 Supination and pronation measurement with inclinometer

Table 25.3 Normal range of motion of thumb (Barakat et al [8])

CMCJ			MCPJ		IPJ	
Radial abduction	Palmar abduction	Kapandji score	Extension	Flexion	Extension	Flexion
61.2	62.9	9	8.1	60	12	88

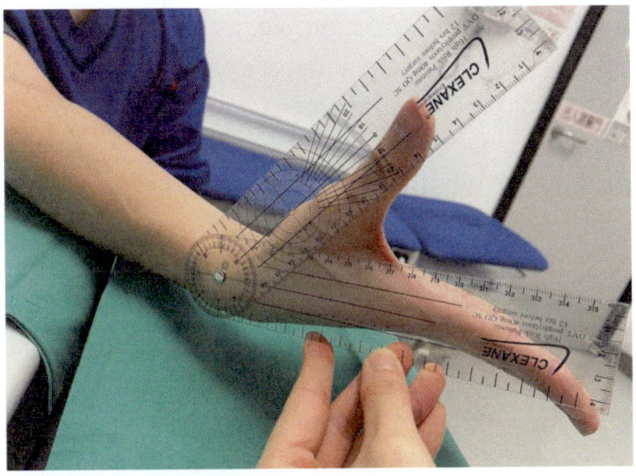

Fig. 25.12 Thumb palmar abduction

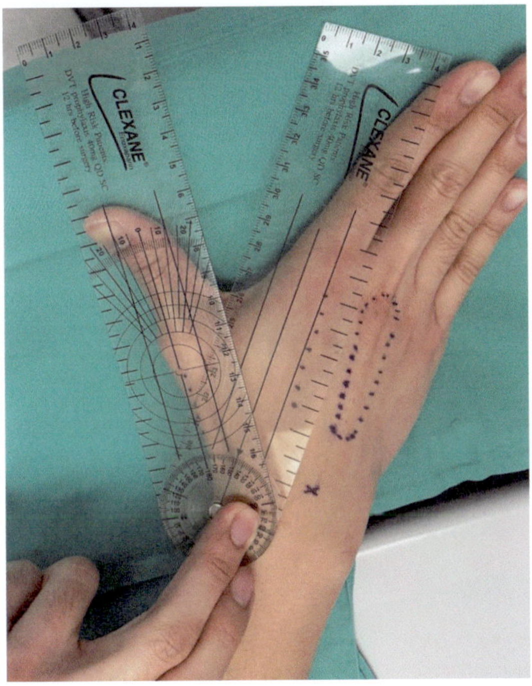

Fig. 25.13 Thumb radial abduction

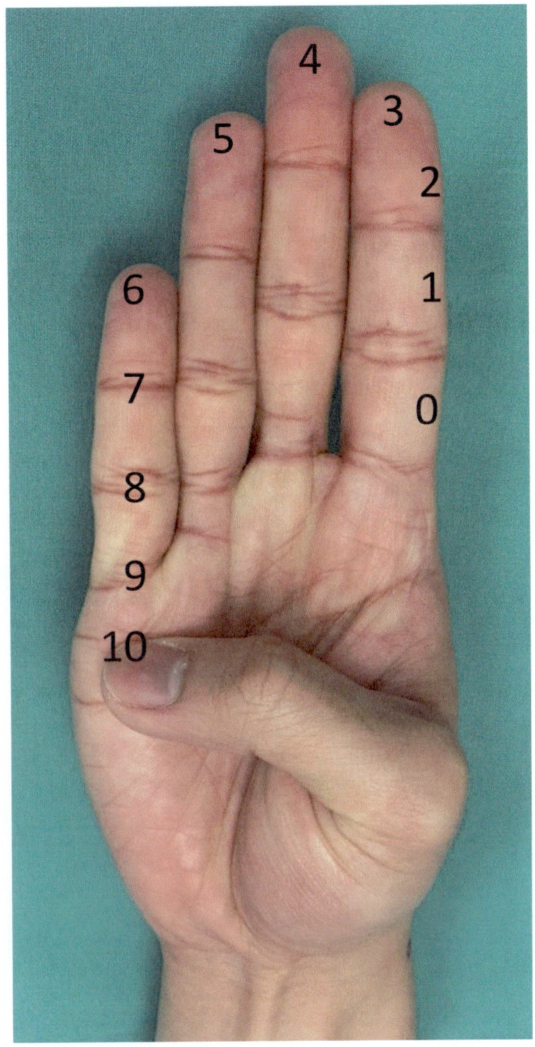

Fig. 25.14 Kapandji score

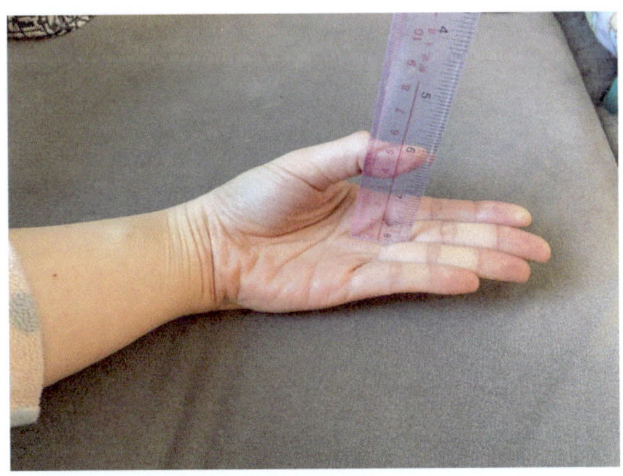

Fig. 25.15 Thumb opposition: measured as the distance between the thumb IPJ crease and the distal palmar crease at 3rd metacarpal

25.4 Finger Joints Measurement

The normal finger range of motion is shown in Table 25.4.

25.4.1 Metacarpophalangeal Joint (MCPJ)

The finger metacarpophalangeal joint has movement in flexion, extension, abduction, and adduction. When measuring the MCPJ motion, the wrist joint should be in neutral position. This is to prevent any tension over the extrinsic tendons which will affect the range of motion at the MCPJ. The goniometer placement can be placed in dorsal position or lateral position. In 1947, Hurt first described the dorsal technique for goniometric measurement of the hand [11]. In 1949, Moore suggested to apply the goniometer to the lateral side of a joint [12]. In some clinical conditions, for example, when there is significant edema, or masses on the dorsal surface of the finger, a lateral placement is preferred. On the other hand, there are conditions where lateral placement is difficult, for example, tightly adducted or overlapping fingers, enlarged joints and joints that are grossly immobile [13]. There is an increased flexion range of motion when progresses from the index to the little finger (Table 25.4). The MCPJ motion can be measured in isolated position (where the PIPJ and DIPJ is in extension) or composite position (where the PIPJ and DIPJ are in flexion). Functionally, the hand digits are used in a combined and synchronous manner to perform tasks. Therefore, it is more practical to measure the finger joint motions in composite position (fisted position) [14]. The measurement of MCPJ by dorsal placement is shown in Fig. 25.16.

25.4.2 Proximal Interphalangeal Joint (PIPJ)

The proximal interphalangeal joint has movement in flexion and extension. Similar to the MCPJ, there are two methods for placement of the goniometer: dorsal or lateral. The PIPJ motion can be measured in isolated position

Table 25.4 Normal finger range of motion (Mallon et al. [10])

Finger	MCPJ		PIPJ		DIPJ	
	Extension	Flexion	Extension	Flexion	Extension	Flexion
Index	22	86	7	102	8	72
Middle	18	91	7	105	8	71
Ring	23	99	6	108	8	63
Little	19	105	9	106	8	65

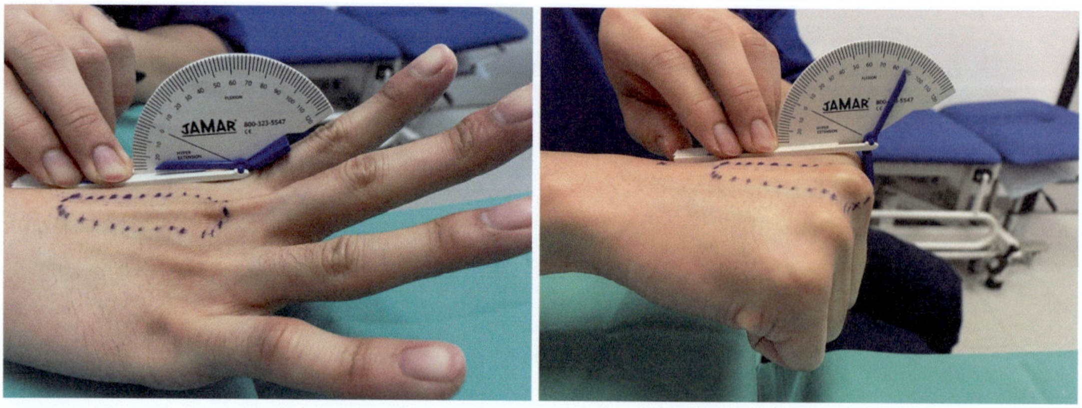

Fig. 25.16 Dorsal placement for measurement of MCPJ extension and flexion

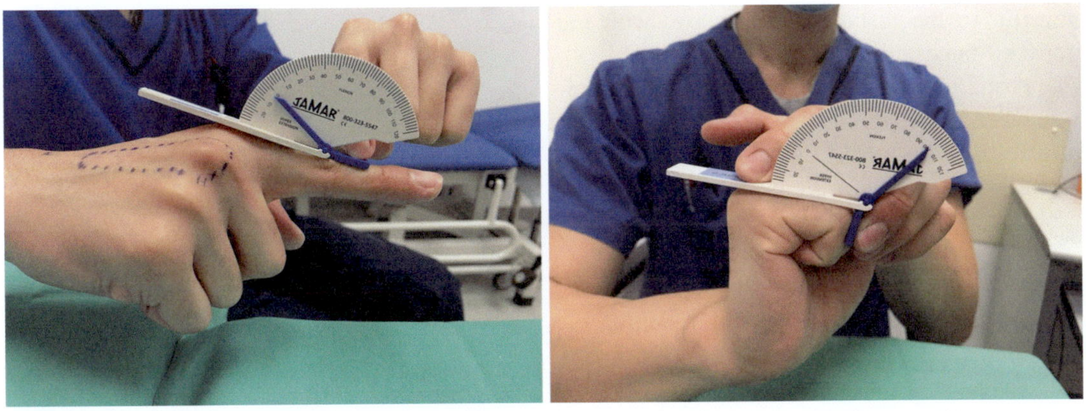

Fig. 25.17 Dorsal placement for measurement of PIPJ extension and flexion

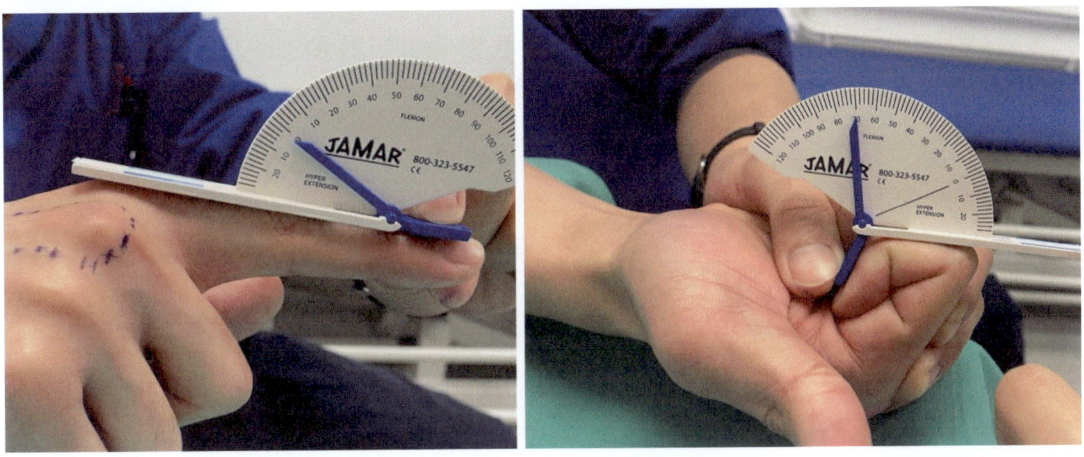

Fig. 25.18 Dorsal placement for measurement of DIPJ extension and flexion

(i.e., the MCPJ and DIPJ in extension, while the PIPJ in flexion) or in composite position (i.e., MCPJ, PIPJ, and DIPJ all in flexion or a fisted position. Similar to the MCPJ, it is recommended that the PIPJ should be measured in the composite position [14]. The measurement of PIPJ range of motion is shown in Fig. 25.17.

25.4.3 Distal Interphalangeal Joint (DIPJ)

The distal interphalangeal joint has motion in flexion and extension. Similar to the measurement of the PIPJ, the DIPJ can be measured with either the dorsal or lateral placement method. Similar to the MCPJ and PIPJ, it is recommended that the DIPJ should be measured in the composite position [14]. The measurement of DIPJ range of motion is shown in Fig. 25.18.

25.4.4 Composite Finger Flexion

In condition where there is significant impairment of finger motion, measuring the composite motion of flexion provides an informative measurement. This is the distance measured from the fingertip to the distal palmar crease, with the impaired finger in maximal flexion (Fig. 25.19). Composite finger flexion is also quicker to measure when multiple joints of multiple fingers are involved. It is a reliable measurement and is a useful alternative when goniometry is impracticable [15].

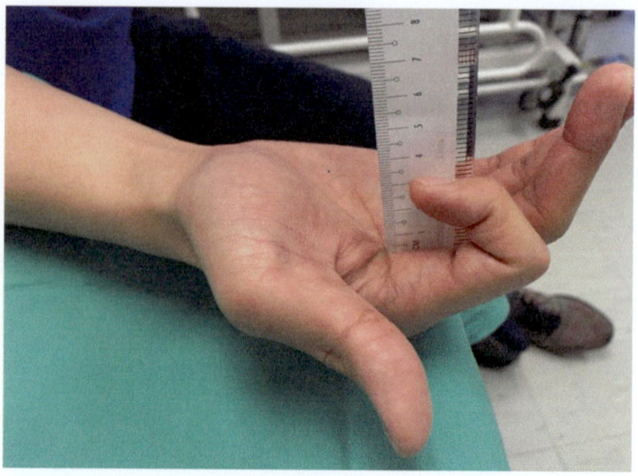

Fig. 25.19 Composite finger flexion

25.4.4.1 Total Active Motion (TAM)

The total active motion is the sum of the MCPJ, PIPJ, and DIPJ active flexion of the digit minus the total active extension deficit of the same digit's MCPJ, PIPJ, and DIPJ.

$$\text{TAM} = (\text{MCPJ} + \text{PIPJ} + \text{DIPJ active flexion}) - (\text{MCPJ} + \text{PIPJ} + \text{DIPJ active extension deficit})$$

25.4.4.2 Total Passive Motion (TPM)

The total passive motion is the sum of the MCPJ, PIPJ, and DIPJ passive flexion of the digit minus the total passive extension deficit of the same digit's MCPJ, PIPJ, and DIPJ.

$$\text{TPM} = (\text{MCPJ} + \text{PIPJ} + \text{DIPJ passive flexion}) - (\text{MCPI} + \text{PIPJ} + \text{DIPJ passive extension deficit})$$

25.5 Functional Range of Motion

Functional range of motion is the motion necessary for performing activities of daily living. This functional range of motion of the hand and wrist is commonly the treatment goal for surgeons and hand therapists.

Ryu et al. assessed the wrist functional range of motion and reported that 40° of wrist extension, 40° of wrist flexion, and a total of 40° of radial-ulnar deviation is needed to perform a majority of the activities of daily

living [16]. Gates et al. reported that the minimum wrist range of motion required for activities of daily living was 40° in flexion, 38° in extension, 28° in radial deviation, and 38° in ulnar deviation [17]. The range of motion needed for forearm rotation was 53° of supination and 13° of pronation [17].

The metacarpophalangeal joint (MCPJ) is important for many activities of daily living. According to Swanson, the finger MCPJ require a flexion range of 70° for functional activities [18]. However, Nalebuff suggested that 30–40° of MCPJ flexion is adequate for most hand function, provided that the proximal interphalangeal joint (PIPJ) has full range of motion [19]. Similarly, Beckenbaugh reported that 35° of motion at the MCPJ is adequate for most activities [20]. However, the studies by Swanson, Nalebuff, and Beckenbaugh were limited to patients with rheumatoid arthritis, where the functional demand may be lower than normal active people. Hume et al. reported data on the functional positions of small joints of the hand necessary for performance of 11 different activities of daily living [21]. The results showed that the averaged motion arc for MCPJ is 33 to 73°,

PIPJ is 36 to 86°, and DIPJ is 20 to 61°. However, this study measured the static posture. There are other studies that measure the dynamic functional posture. Hayashi and Shimizu found that the mean functional range of motion of finger MCP joints of the index, middle, ring, and little fingers were 10 to 60; 10 to 75; 10 to 80; and 10 to 85°, respectively, for performing 19 daily activities [22]. Bain et al. had performed similar study and found that the functional range of motion of the finger joints was 19° to 71°, 23° to 87°, and 10° to 64° at the MCPJ, PIPJ, and DIPJ, respectively [23]. Murai assessed the functional range of motion of the MCPJ of the thumb, as well as other fingers [24]. They found that the functional ROM values for MCPJ of the thumb, index, middle, ring, and little fingers were −7.5 to 35.3, 10.6 to 67.8, 4.0 to 79.9, 3.0 to 83.9, and 2.9 to 91.4°, respectively. A summary of the functional range of motion was found in Table 25.5.

Table 25.5 Functional range of motion of hand and wrist

Study	Joint	Function ROM
Ryu [16]	Wrist	Extension 40 Flexion 40 Radial/ulnar deviation arc 40
Gates [17]	Wrist	Extension 40 Flexion 38 Radial deviation 28 Ulnar deviation 38
Gates [17]	DRUJ	Pronation 13 Supination 53
Hume [21]	MCPJ	33–73
	PIPJ	36–86
	DIPJ	20–61
Hayashi [22]	MCPJ	Index 10–60 Middle 10–75 Ring 10–80 Little 10–85
Bain [23]	MCPJ	19–71
	PIPJ	23–87
	DIPJ	10–64
Murai [24]	MCPJ	Thumb −7.1 −35 Index 10.6–67.8 Middle 4–79.9 Ring 3.0–83.9 Little 2.9–91.4

25.6 Active and Passive Range of Motion

It is important to test both the active and passive range of motion of the hand and wrist joints. The discrepancies between active and passive range of motion might represent muscle or tendon pathology.

25.7 Fixed Flexion Contracture

A joint is defined as in fixed flexion contracture when it cannot be extended beyond the fixed degree of flexion passively. In contrary, an extension lag is that the joint cannot be extended actively but can be extended passively.

25.8 Factors Affecting Hand and Wrist Range of Motion: Gender, Ethnics, Age, Hand Dominance

Concerning the effect of gender on the hand and wrist range of motion, Mallon et al. reported that female had greater extension at the metacarpophalangeal joint in both active and passive motion and have a greater total active motion at all digits [10]. Female was also found to have greater motion at the wrist, as reported by Allander et al. in 1974 [25]. Doriot et al. reported that female subjects were observed to have greater motion in forearm pronation and wrist radial deviation [26]. However, there were also study that found no difference in wrist range of motion regarding the gender [27].

Different ethnic groups might show difference in range of motion. Kwok et al. compared the wrist range of motion in Chinese with Caucasians and found that Chinese has more wrist flexion while Caucasians has more wrist extension. The radial and ulnar deviation range of motion were similar in both ethnic groups [27].

Age might also be an important factor affecting the wrist and hand range of motion. Allander reported that increasing age had an effect on wrist joint motion, with increasing age resulted in lower joint mobility [25]. However, Doriot et al. found age had no effect on the wrist range of motion [26].

Several studies had found that the non-dominant hand had a greater range of motion in the hand and wrist. Cobe in 1928 reported that the left wrist had a greater range of movement than the right wrist [28]. Similarly, Allander et al. found that left hand showed better range of motion than the right hand at the wrist and thumb MCPJ [25]. Gunal et al. reported the non-dominant hand has a wider range of motion than the dominant hand, both actively and passively in forearm rotation, wrist extension and radial deviation [29]. However, Mallon reported that there was no difference in the finger range of motion regarding to the handedness [10]. Kwok et al. also showed handedness had no effect on wrist range of motion [27].

25.9 Other Measurement Techniques

The goniometer is the recommended instrument for measuring the hand and wrist range of motion [30]. There are also studies that used the video-based motion analysis [31]. New technology had been developed for multiple joint motion measurement. Gutiérrez-Martínez J reported the use of transducer systems to measure multiple hand joints motion with good accuracy [32]. Motion sensors for measuring dynamic motion of hand and wrist had also been used [33]. In recent years, new smart phone applications can be used to measure the wrist range of motion and studies had showed good reliability. These new smart phone applications are readily available, simple to use and can be used by the patients outside the clinical setting. [34–36]

References

1. Kaufmann R, Pfaeffle J, Blankenhorn B, Stabile K, Robertson D, Goitz R. Kinematics of the midcarpal and radiocarpal joints in radioulnar deviation: an in vitro study. J Hand Surg Am. 2005;30(5):937–42.
2. Sarrafian SK, Melamed JL, Goshgarian GM. Study of wrist motion in flexion and extension. Clin Orthop Relat Res. 1977;126:153–9.
3. King GJ, McMurtry RY, Rubenstein JD, Gertzbein SD. Kinematics of the distal radioulnar joint. J Hand Surg. 1986;11(6):798–804.
4. Boone DC, Azen SP. Normal range of motion of joints in male subjects. J Bone Joint Surg Am. 1979;61(5):756–9.
5. LaStayo PC, Wheeler DL. Reliability of passive wrist flexion and extension goniometric measurements: a multicenter study. Phys Ther. 1994;74(2):162–74; discussion 174–6
6. Carter TI, Pansy B, Wolff AL, Hillstrom HJ, Backus SI, Lenhoff M, Wolfe SW. Accuracy and reliability of three different techniques for manual goniometry for wrist motion: a cadaveric study. J Hand Surg Am. 2009;34(8):1422–8.
7. Norkin CC, White DJ. Measurement of joint motion: a guide to goniometry. Philadelphia, PA: FA Davis; 1985.
8. Barakat MJ, Field J, Taylor J. The range of movement of the thumb. Hand (N Y). 2013;8(2):179–82. https://doi.org/10.1007/s11552-013-9492-y.
9. Kapandji A. Cotation clinique de l'opposition et de la contre-opposition du pouce [clinical test of apposition and counter-apposition of the thumb]. Ann Chir Main. 1986;5(1):67–73. French
10. Mallon WJ, Brown HR, Nunley JA. Digital ranges of motion: normal values in young adults. J Hand Surg [Am]. 1991;16(5):882–7.
11. Hurt SP. Considerations in muscle function and their application to disability evaluation and treatment-joint measurement. Am J Occup Ther. 1947;1(4):209–14.
12. Moore ML. The measurement of joint motion, part II: the technic of goniometry. Phys Ther Rev. 1949;29(6):256–64.
13. Hamilton GF, Lachenbruch PA. Reliability of goniometers in assessing finger joint angle. Phys Ther. 1969;49(5):465–9.
14. Richard R, Parry IS, Santos A, Dewey WS. Burn hand or finger goniometric measurements: sum of the isolated parts and the composite whole. J Burn Care Res. 2017;38(6):e960–5.
15. Ellis B, Bruton A. A study to compare the reliability of composite finger flexion with goniometry for measurement of range of motion in the hand. Clin Rehabil. 2002;16(5):562–70.

16. Ryu JY, Cooney WP 3rd, Askew LJ, An KN, Chao EY. Functional ranges of motion of the wrist joint. J Hand Surg Am. 1991;16(3):409–19.
17. Gates DH, Walters LS, Cowley J, Wilken JM, Resnik L. Range of motion requirements for upper-limb activities of daily living. Am J Occup Ther. 2016;70(1):7001350010p1–7001350010p10.
18. Swanson AB. Flexible implant arthroplasty for arthritic finger joints: rationale, technique, and results of treatment. J Bone Joint Surg Am. 1972;54(3):435e455.
19. Nalebuff EA. The rheumatoid hand: reflections on metacarpophalangeal arthroplasty. Clin Orthop Relat Res. 1984;182:150e159.
20. Beckenbaugh RD, Dobyns JH, Linscheid RL, Bryan RS. Review and analysis of silicone-rubber metacarpophalangeal implants. J Bone Joint SurgAm. 1976;58(4):483–7.
21. Hume MC, Gellman H, McKellop H, Brumfield RH Jr. Functional range of motion of the hand. J Hand Surg Am. 1990;15(2):240–3.
22. Hayashi H, Shimizu H. Essential motion of metacarpophalangeal joints during activities of daily living. J Hand Ther. 2013;26(1):69e73.
23. Bain GI, Polites N, Higgs BG, Heptinstall RJ, McGrath AM. The Journal of hand surgery, European volume. J Hand Surg Eur Vol. 2015;40(4):406–11.
24. Murai T, Uchiyama S, Nakamura K, Ido Y, Hata Y, Kato H. Functional range of motion in the metacarpophalangeal joints of the hand measured by single axis electric goniometers. J Orthop Sci. 2018;23:504–10.
25. Allander E, Björnsson OJ, Olafsson O, Sigfússon N, Thorsteinsson J. Normal range of joint movements in shoulder, hip, wrist and thumb with special reference to side: a comparison between two populations. Int J Epidemiol. 1974;3(3):253–61.
26. Doriot N, Wang X. Effects of age and gender on maximum voluntary range of motion of the upper body joints. Ergonomics. 2006;49(3):269–81.
27. Kwok YY, Ho P-C, Feldman G, Lo E, Wells E, Wollstein R. Comparison of wrist motion and grip strength in a Normal Chinese and Caucasian population. J Hand Surg (Asian-Pac Vol). 2016;21:364–8.
28. Cobe HM. The range of active motion of the wrist of white adults. J Bone Joint Surg. 1928;10:763–74.
29. Gunal I, Kose N, Erdogan O, Gokturk E, Seber S. Normal range of motion of the joints of the upper extremity in male subjects, with special reference to side. J Bone Joint Surg Am. 1996;78(9):1401–4.
30. Adams I, Greene L, Topoozian E. Clinical assessment recommendations. Am Soc Hand Therap. 1999;2:56–70.
31. Chiu HY, Su FC, Wang ST, Hsu HY. The motion analysis system and goniometry of the finger joints. J Hand Surg. 1998;23B(6):788–91.
32. Gutiérrez-Martínez J, Ortiz-Espinosa A, Hernández-Rodríguez PR, Núñez-Gaona MA. System to measure the range of motion of the joints of the human hand. Rev Investig Clin. 2014;66(Suppl 1):S122–30.
33. Nizamis K, Rijken NHM, Mendes A, Janssen MMHP, Bergsma A, Koopman BFJM. A novel setup and protocol to measure the range of motion of the wrist and the hand. Sensors (Basel). 2018;18(10):3230.
34. Santos C, Pauchard N, Guilloteau A. Reliability assessment of measuring active wrist pronation and supination range of motion with a smartphone. Hand Surg Rehabil. 2017;36(5):338–45.
35. Kim T, Park D, Lee Y, Han D, Shim J, Lee Y, Kim P. A study on the measurement of wrist motion range using the iPhone 4 gyroscope application. Ann Plast Surg. 2013;73:215–8.
36. Pourahmadi MR, Ebrahimi Takamjani I, Sarrafzadeh J, Bahramian M, Mohseni-Bandpei MA, Rajabzadeh F, Taghipour M. Reliability and concurrent validity of a new iPhone® goniometric application for measuring active wrist range of motion: a cross-sectional study in asymptomatic subjects. J Anat. 2017;230(3):484–95.

26 Clinical Testing of the Wrist

Toshiyasu Nakamura

26.1 Popular Physical Test Commonly Used in Daily Practice

26.1.1 Tenderness Point

Careful examination of tenderness point is very useful to diagnose wrist disorders. When the patient indicates tenderness point on the radial side of the radius, especially at the first compartment, De Quervain tendinitis is the first candidate. When the tenderness point was found on the snuff box of the wrist, where the area between distal radius, extensor politis brevis tendon, extensor politis longs, and scaphotrapeziotrapezoidal joint, scaphoid fracture, or scaphoid nonunion is suspicious. When the area, we called 3–4 portal in arthroscopy, where the pit between the extensor politis longus tendon, second extensor digiti communis tendon, proximal carpal row and distal edge of the radius, was tenderness, scapholunate (SL) ligament injury, or SL dissociation is suspicious. In triangular fibrocartilage complex (TFCC) injury, the tenderness point are dorsal or palmar side of the ulnar styloid.

T. Nakamura (✉)
Department of Orthopaedic Surgery, School of Medicine, International University of Health and Welfare, Tokyo, Japan
e-mail: toshiyasu@ae.em-net.ne.jp

26.1.2 Watson Test

Watson test is a useful physical examination for SL injury or SL dissociation, indicating dorsal intercalated segment instability (DISI) deformity subsequently [1]. Scaphoid nonunion patient also indicates positive Watson test. In this test, the examiner's thumb is set on the scaphoid process and press the scaphoid, then the wrist is passively moved radial and ulnar deviation (Fig. 26.1). When the SL ligament is torn, sudden snap of the scaphoid process can be felt. Patient also claims pain during this maneuver.

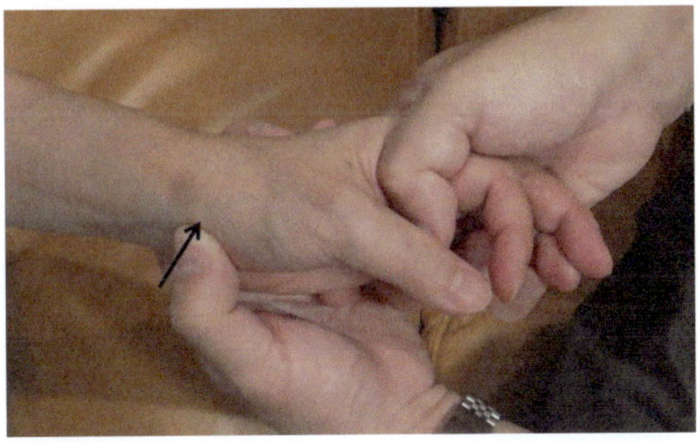

Fig. 26.1 Watson test. The examiner's thumb is set on the scaphoid process and press the scaphoid dorsally, then the wrist is passively moved radial and ulnar deviation

J. G. Lane et al. (eds.), *The Art of the Musculoskeletal Physical Exam*,
https://doi.org/10.1007/978-3-031-24404-9_26

26.1.3 Ulnocarpal Stress Test

Ulnocarpal stress test is an examination for ulnar side wrist pain, especially TFCC injury and ulnocarpal abutment syndrome. The examiner held patient's hand and kept forced ulnar deviated position [2]. In case of ulnar side wrist pain, patient claims pain. Also forced pronation-supination movement is applied with the forced ulnar deviated position, usually accelerates pain (Fig. 26.2). In my experience, forced supination with ulnar deviation is most sensitive in TFCC injury patient. Other problems including extensor carpi ulnaris (ECU) tendinitis, ECU dislocation, and dislocation of the distal radio-ulnar joint (DRUJ) are positive with this test.

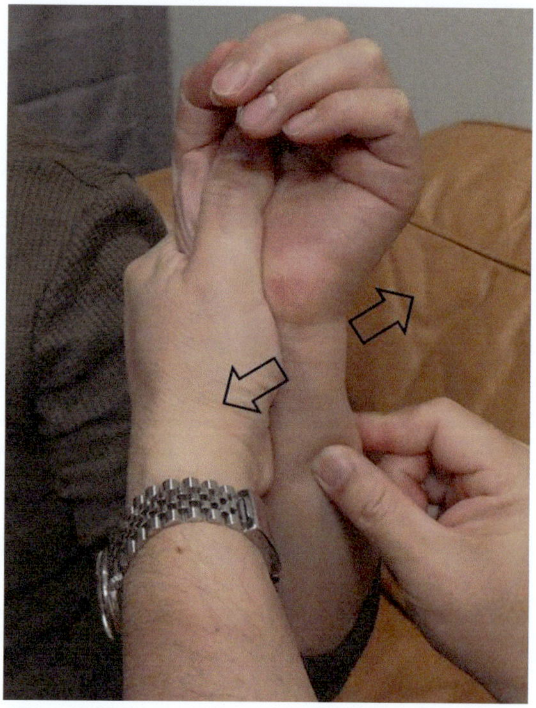

Fig. 26.3 Ballottement test. The examiner kept the radius and carpus in one unit with one hand, then moves the ulnar head dorsally or palmary by the other hand

26.1.4 Ballottement Test

AS the TFCC is the primary stabilizer, DRUJ ballottement test is very useful to diagnose TFCC injury, especially radioulnar ligament (RUL) rupture [3]. As the TFCC is also the stabilizer of ulnocarpal joint, the examiner kept the radius and carpus in one unit with one hand, then moves the ulnar head dorsally or palmary (Fig. 26.3). If the

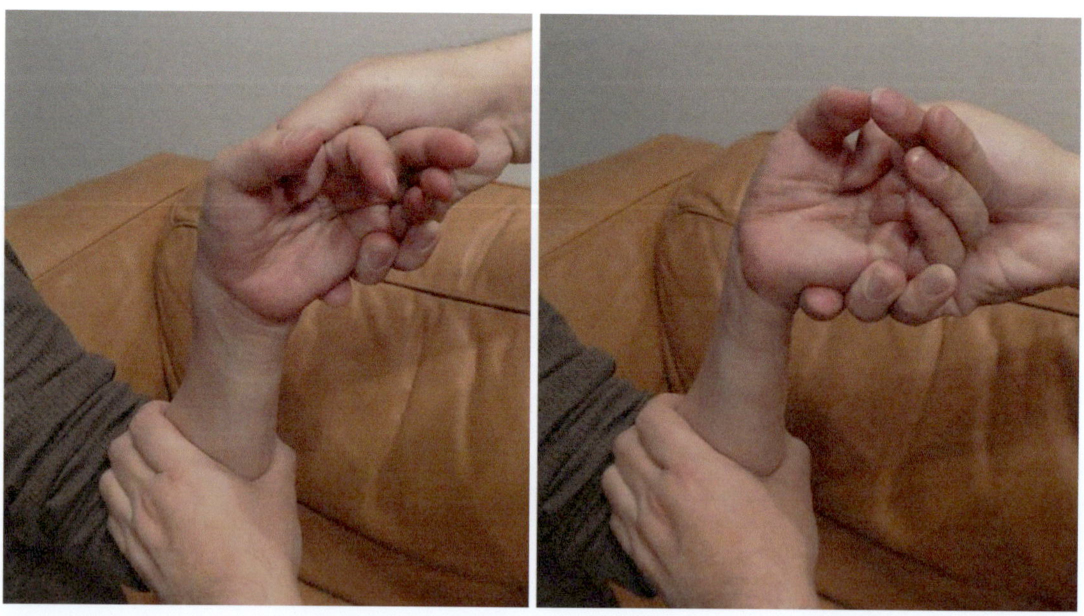

Fig. 26.2 Ulnocarpal stress test. The patient wrist is forced ulnar deviated with pronosupination

TFCC or RUL was intact, there should be end point with dorsopalmar movement. When the RUL was ruptured, there is no end point during this maneuver [4]. Especially in general anesthetic condition, this test is very sensitive for TFCC lesions because no muscle reinforcement can be applied by a patient.

26.1.5 Synergy Test

The synergy test is for ECU tendinitis or any other ECU disorders. The examiner holds the index, middle, ring, and little fingers in one hand, then asks the patient to abduct all fingers [5]. If the pain was felt in the ECU groove, the test is positive.

26.1.6 Fovea Sign

Tay et al. [6] reported the tenderness point on the palmar aspect of the ulnar styloid, as "Fovea sign," closely related with foveal avulsion of the TFCC and split tear of the ulnotriquetral ligament. The palmar aspect of the ulnar styloid is the origin area of palmar portion of the radioulnar ligament (RUL), this sign may relate with palmar or total avulsion of the RUL.

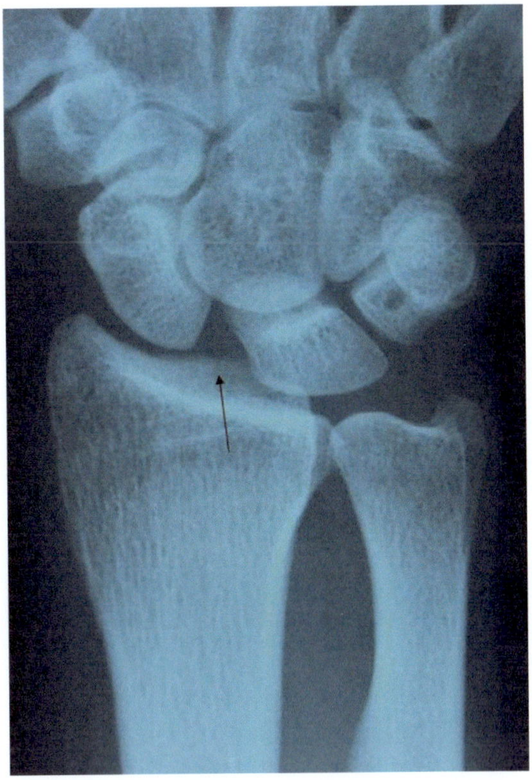

Fig. 26.4 Radiograph of the SL dissociation. Arrow indicates that wide gap formation between the scaphoid and lunate, showing SL dissociation

26.2 Popular Image Modalities for Wrist Disorders

26.2.1 X-Ray

X-ray shows fracture line(s) and malposition of carpus. In case of SL dissociation, there is a wide gap formation between the scaphoid and lunate (Fig. 26.4). The lateral view of the wrist shows dorsal rotation of the lunate, which calls dorsal intercalated segment instability (DISI) [7]. When the patient shows DISI deformity with SL ligament rupture or scaphoid nonunion, DISI should be reduced with K-wire, then reconstruction of the SL ligament or restoration of the scaphoid nonunion with bone grafting should be needed. In DRUJ dislocation, lateral X-ray shows displacement of the ulna with respect to the radius.

26.2.2 Computed Tomography (CT)

CT delineates details of the bone construction with multiple directions of sections, coronal, sagittal, and axial. It is very useful to indicate scaphoid fracture, intra-articular fracture of the radius, and SL dissociation. DRUJ dislocation can be easily diagnosed. 3-D reconstructed CT can be useful for preoperative planning of fracture fixation in intra-articular fracture of the distal radius.

26.2.3 MRI

MRI can delineate soft tissue because it can detect magnetic resonance signal from the body. Especially in TFCC injury, T2* weighted and fat-suppression water MRI can indicate abnormal signal(s) inside the TFCC on coronal sections [8,

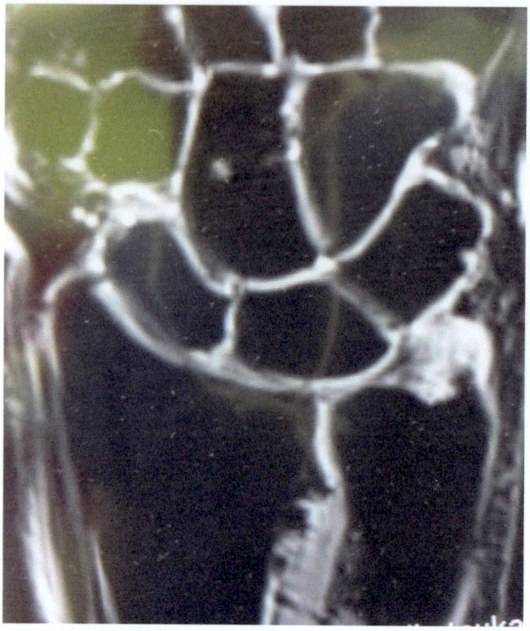

Fig. 26.5 T2* weighted MRI of the foveal avulsion type TFCC injury. The RUL is avulsed from the fovea of the ulna

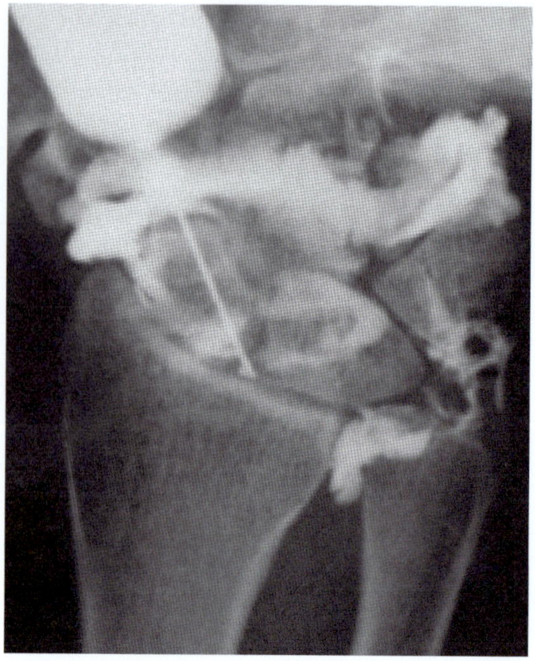

Fig. 26.6 Arthrogram shows peripheral injury of the TFCC. The radiographic dye is infiltrated into the ulnar side of the TFCC

9]. MRI is useful for the fovea avulsion of the radioulnar ligament (RUL) (Fig. 26.5), horizontal type tear of the TFCC and its degenerative condition. Peripheral slit tear can also be delineated. In case of SL ligament tear, wide gap formation can be detected in coronal MRI. Sometimes, MRI also can indicate ruptured SL ligament.

26.2.4 Arthrogram

In arthrogram, radiographic dye is injected into radiocarpal and DRU joints, which called double injection. When the SL ligament or lunotriquetral ligament was torn, the dye infiltrates into the midcarpal joint. Arthrogram is especially useful for TFCC injuries [10]. When the TFCC was ruptured, the dye goes into the DRUJ or to tear site in peripheral tear (Fig. 26.6). When patient wrist can be captured in neutral, radial or ulnar deviated position, abnormal widening of peripheral TFCC tear can be detected in arthrogram.

References

1. Watson HK, Ashmead D, Makhlouf MV. Examination of the scaphoid. J Hand Surg Am. 1988;13(5):657–60.
2. Nakamura R, Horii E, Imaeda T, Nakao E, Kato H, Watanabe K. The ulnocarpal stress test in the diagnosis of ulnar-sided wrist pain. J Hand Surg Br. 1997;22(6):719–23.
3. Moriya T, Aoki M, Iba K, Ozasa Y, Wada T, Yamashita T. Effect of triangular ligament tears on distal radioulnar joint instability and evaluation of three clinical tests: a biomechanical study. J Hand Surg Eur. 2011;34(2):219–23.
4. Nakamura T, Sato K, Okazaki M, Toyama Y, Ikegami H. Repair of the foveal detachment of the triangular fibrocartilage complex: open and arthroscopic transosseous techniques. Hand Clin. 2011;27:281–90.
5. Ruland RT, Hogan CJ. The ECU synergy test: an aid to diagnose ECU tendonitis. J Hand Surg Am. 2008;33(10):1777–82.
6. Tay SC, Tomita K, Berger RA. The "ulnar fovea sign" for defining ulnar wrist pain: an analysis of sensitivity and specificity. J Hand Surg Am. 2007;32(4):438–44.
7. Linscheid RL, et al. Traumatic instability of the wrist. J Bone Joint Surg Am. 1972;54:1612–32.
8. Nakamura T, Yabe Y, Horiuchi Y. Fat suppression magnetic resonance imaging of the triangular fibro-

cartilage complex -comparison with spin echo, gradient echo pulse sequence and histology. J Hand Surg Br. 1999;24(1):22–6.
9. Tanaka T, Yoshioka H, Ueno T, et al. Comparison between high-resolution MRI with a microscopy coil and arthroscopy in triangular fibrocartilage complex injury. J Hand Surg Am. 2006;31(8):1308–14.
10. Zinberg EM, Palmer AK, Coren AB, et al. The triple-injection wrist arthrogram. J Hand Surg Am. 1988;13(6):803–9.

27 Evaluation of the Triangular Fibrocartilage Complex

Zhixin Wang and Bo Liu

27.1 Introduction

The traumatic or degenerative triangular fibrocartilage complex (TFCC) injuries are the most common cause of ulnar-sided wrist pain. The TFCC is a fibrocartilage–ligament complex which is composed of seven distinct parts: the articular disc, volar and dorsal radioulnar ligaments (RULs), ulnolunate and ulnotriquetral ligaments, the extensor carpi ulnaris (ECU) subsheath, and the ulnocarpal meniscoid [1, 2]. The TFCC transmits axial load between the carpus and the ulna and stabilizes the ulnar aspect of the carpus. The volar and dorsal distal radioulnar ligaments are considered to be the most important stabilizer for distal radioulnar joint (DRUJ). These two ligaments arise from the sigmoid notch of distal radius and converge to form a strong foveal attachment (deep fibers) as well as a styloid attachment (superficial fibers) [3].

Before physical examination, the physician should always collect a detailed history about the location and characteristics of the wrist pain, any precipitating events, exacerbating and relieving factors, and associated symptoms. These could include clicking or snapping sensations, stiffness, or altered sensation. It is important to determine the chronicity of the problem, which is closely related to management. Establishing the patient's hand dominance, occupation and hobbies can help determine the etiology of the pain. In most cases, the patient is able to point out whether the pain is radial or ulnar to an imaginary line drown through the midline of the wrist. Besides, the information about when the pain usually arises and what movements relate to the symptom should also be recorded.

Due to the complex nature the TFCC, it is common that the diagnosis of TFCC pathology cannot be readily reached based on imaging alone. History taking and physical examination remain the vital elements for detecting the TFCC pathology. This chapter will review the common methods for evaluation of TFCC.

27.2 Inspection

When a wrist is examined, the physician and the patient sit on the opposite side of a hand table. The patient is asked to be relaxed without any resistance during examination. The exam and observations should be compared with the contralateral side because some findings may be symmetrical.

For acute TFCC injuries, skin abrasion, swelling, redness, and bony deformity may indicate accompanied injuries and may require further

Z. Wang · B. Liu (✉)
Department of Hand Surgery, Beijing Ji Shui Tan Hospital, The 4th Clinical College of Peking University, Beijing, China

J. G. Lane et al. (eds.), *The Art of the Musculoskeletal Physical Exam*,
https://doi.org/10.1007/978-3-031-24404-9_27

investigation. A dorsally subluxated ulnar head can be noticed if there is a distal radioulnar ligaments incompetence. The subluxation of ulnar head may be flexible or fixed and should be checked.

27.3 Palpation

Palpation is an important part in clinical examination for any wrist pathology. The goal of palpation is to locate the pathology and assess the characteristic of tenderness[4]. It is recommended to systemically follow a predetermined order of palpation and compare all findings with those from the contralateral normal side. Marc Garcia-Elias proposed a systemic way to palpate the wrist for ulnar-sided wrist pain, following the specific anatomical landmarks (Marc [5]). It is important to assess the whole wrist rather than "go to TFCC directly." Palpation can help the physician to locate the tenderness, find any mass, swelling or prominence around the joint or tendon, and abnormal flexibility between small bones.

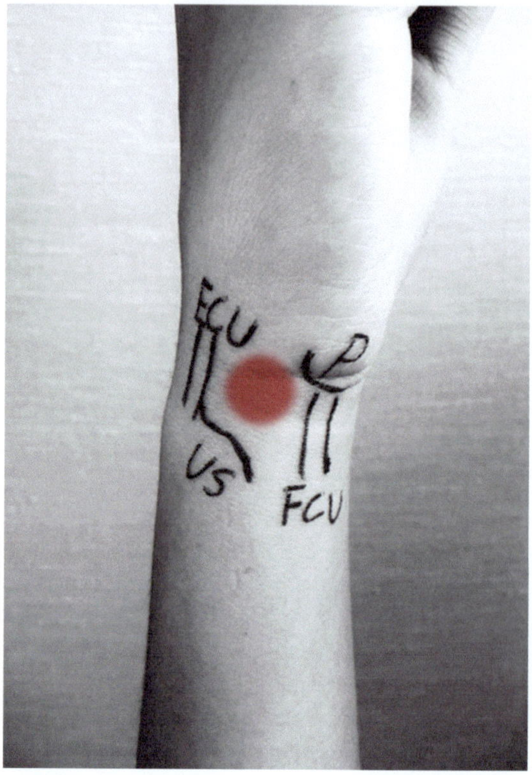

Fig. 27.1 Surface anatomy of the ulna fovea. *ECU* extensor carpi ulnaris, *P* pisiform, *FCU* flexor carpi ulnaris, *US* ulnar styloid

27.3.1 Fovea Sign

Eliciting tenderness in the region of ulnar fovea is a commonly done test for TFCC injury (Rosemary [6]). The ulnar fovea, the so-called the ulnar snuffbox, is a soft spot located between the ulnar styloid and the pisiform bone. Dorsally, it is bounded by the extensor carpi ulnaris (ECU) tendon and volarly by the flexor carpi ulnaris (FCU) tendon (Fig. 27.1). Shian Chao Tay reported that the sensitivity of the fovea sign in detecting foveal disruptions and/or UT ligament injuries was 95.2%. Its specificity was 86.5%. While in our study, the sensitivity and specificity of ulnar fovea sign was 64.8% and 74.2%, respectively ([7]).

During palpation, one should keep the potential differential diagnosis of ulnar-sided wrist pain in mind and look for particular signs that may help in diagnosis. The proximal ulnar corner of the lunate can be found distally between the fourth and fifth extensor compartment with the wrist slightly flexed. The tenderness in this area indicates a localized chondral defect, an ulnocarpal impaction syndrome, or a Kienböck's disease. The lunotriquetral interval is palpated dorsally between the fourth and fifth compartments 1 finger's breadth distal to the DRUJ and with the wrist in 30° of flexion. The tenosynovitis of ECU tendon is a common differential diagnosis of TFCC injury. The tenderness and swelling along ECU tendon can be assessed in forearm supination, the position when the ECU tendon is very prominent onto the top of the ulnar head.

27.4 Range of Motion (ROM)

The measurement of ROM of wrist usually includes six directions: flexion/extension, pronation/supination, and radial/ulnar deviation.

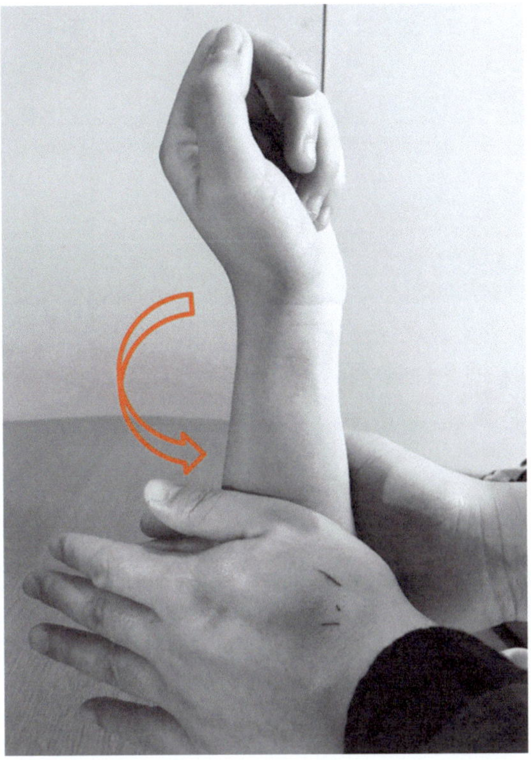

Fig. 27.2 The physician rotates the patient's forearm with both palms to elicit ulnar sided wrist pain, especially in extreme position

Patients will TFCC injury may loss some ROM of the wrist due to pain and swelling, particularly the supination and pronation. In addition to pain, it is also important to notice the presence of snaps or clunks and whether such during the wrist moves actively and passively.

DRUJ is a key component of rotation mechanism of forearm. Patients with TFCC pathology usually complains pain during rotational movements, such as twisting the towel or open the door knob. The physician should always check if there is pain or discomfort during forced passive rotation, especially in extreme position (Fig. 27.2).

27.5 Provocative Test

Several provocative tests have been reported for detecting TFCC injury. The most commonly used and recommended tests will be discussed below.

27.5.1 Ballottement Test

Ballottement test is designed for assessing the integrity of distal radioulnar ligaments (DRUL) [8]. Traumatic avulsion of the deep fibers of RULs from the ulnar fovea is a principle cause of DRUJ instability and has a greater effect on instability than disruption of the other insertion sites [9]. In vivo studies have demonstrated that the volar deep RUL tightens with forearm pronation and serves as restraint for dorsal translation of the ulnar head, while the dorsal deep RUL tightens with forearm supination and serves as a restraint for volar translation of the ulnar head [10]. In addition to pain, patients typically experience reduced grip strength, reduced forearm rotation and mechanical symptoms of "clicking" with pronosupination.

When conducting the ballottement test, the physician sits opposite to the patient, with the patient's elbow rests on the table, forearm in neutral rotation, wrist in neutral position, and fingers toward the ceiling. The physician uses one hand to hold distal radius (or distal radius and radial side of the wrist), the other hand holding the distal ulna. While stabilizing the radius, the physician mobilizing the distal ulna back and forth (Fig. 27.3). The degree of translation between distal radius and ulna and pain caused by this maneuver should always be compared with the contralateral side. Any excessive movement and pain induced by this provocative maneuver indicate laxity or tear of DRUL. In neutral position, up to 5 mm of translation may be normal. But in supination and pronation, there should be no translation [11].

To separately evaluate the integrity of the dorsal or volar deep fibers, the patient's forearm is passively rotated into extreme rotation positions. In full supination, dorsal deep fibers are under maximum tension. The physician uses one hand to hold the distal ulna with the other hand holding the distal radius. Then the physician simultaneously pushes the distal ulna toward the patient and pull the distal radius toward himself or herself. To examine the volar deep fibers of DRUL, the physician passively rotates the patient's fore-

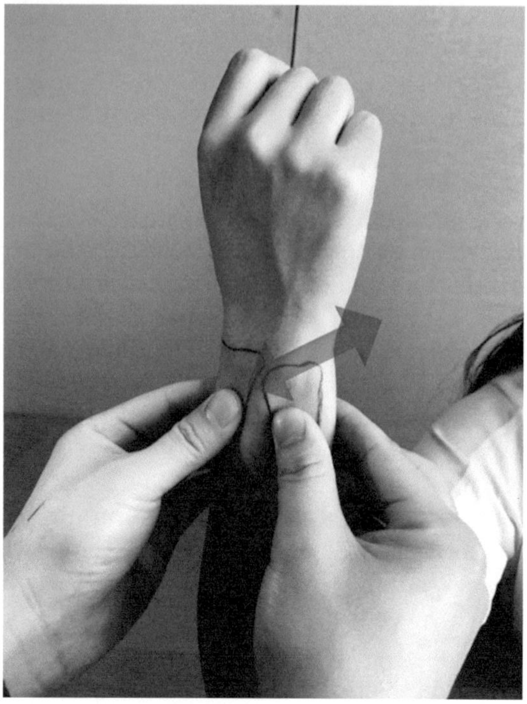

Fig. 27.3 Ballottement test

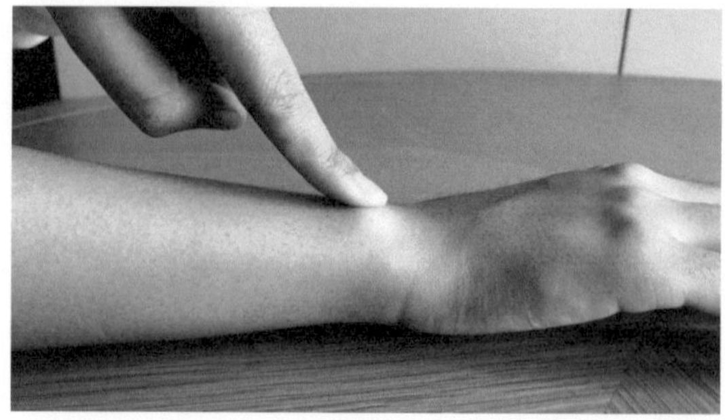

Fig. 27.4 The piano key test: the patient's hand is resting flat on the table in pronation, then the physician pushes onto the prominent ulnar head to reduce it. The ulnar head may spring back as soon as the pressure is released, which indicates a subluxated and unstable DRUJ

arm into extreme pronation to tighten the volar deep fibers. The ulna is then pushed away from the physician while the radius is simultaneously pulled toward the physician.

27.5.2 Piano Key Test

This test is performed with the patient's forearm pronated and rested on the table. Patients with TFCC injury and DRUJ instability will commonly present a dorsally subluxated ulna head. When the physician presses the ulna head from dorsal to volar, the dorsally subluxed ulnar head could be relocated. With the releasing of the depressing force, the ulna head will rebound to its subluxated position (Fig. 27.4). A positive result is characterized by painful laxity in the affected wrist compared with the contralateral wrist. This signify a significant tear or avulsion injury of the DRUL and resultant DRUJ instability [12].

27.5.3 TFC Grind Test /Ulnocarpal Stress Test

TFC grind test was designed to assess the ability of the TFC to sustain rotational shear stress. The physician sits opposite to the patient, with the patient's elbow rests on the table. The physician holds the patient's metacarpals with one hand and stabilize the distal forearm with the other hand. Place the patient's hand in maximum ulnar deviation and then apply an axial load to the hand. Then the forearm is passively rotated from full supination to pronation under consistent load (Fig. 27.5). A positive result includes a grinding sensation with pain or crepitus/click, which indicates TFC tear or ulnocarpal impaction syndrome [8].

If the positive results only appear during supination, the ulnar styloid triquetrum impaction syndrome (USTI) should be considered. The physician can passively extend the patient's wrist with ulnar deviation and axial load to reproduce the symptoms. This maneuver is also called the ulnar styloid triquetrum impaction test. Similarly, the ulnocarpal meniscal test (Waiter's test) consists of applying an axial load to the ulnar carpus, while the wrist is passively brought from extension to ulnar deviation and then to flexion. The original intention of these similar maneuvers is to target on specific structures, but clinically the positive results are not

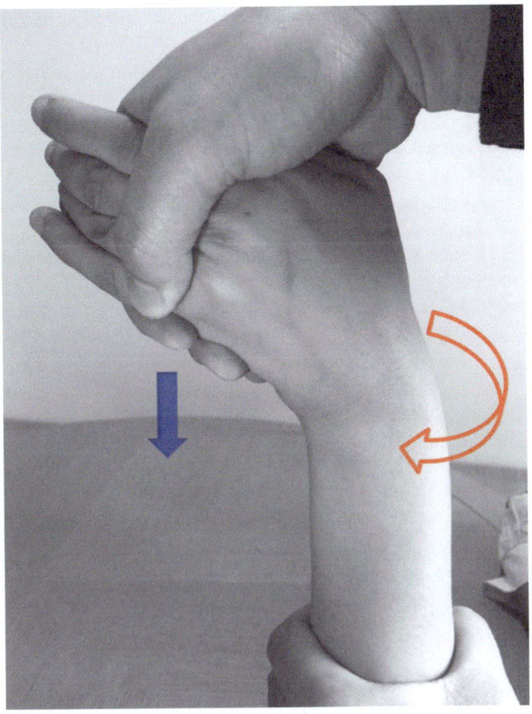

Fig. 27.5 TFC grind test: axial load is applied on the ulnar deviated wrist during passive rotation of the forearm

easy to be interpreted and the accuracy of these tests has not been well proved.

27.5.4 Press Test

A simple provocative diagnostic office test for detection TFCC tears is "press test" [13]. The seated patient is asked to push the body weight up off a chair using the affected wrist, creating an axial ulnar load. Similar to TFC grind test, the axial load may provoke ulnar wrist pain, and some patients may not be able to accomplish this task due to apprehension.

27.5.5 Provocative Tests for ECU Tendon Pathology

The subsheath of ECU tendon is considered as a component of TFCC. However, the pathology arise from ECU tendon or its sheath has to be differentiated from intracarpal problems. Burkhart described the maneuver to test ECU tendon subluxation. The patient's forearm is held in maximal supination and the examiner use one hand to palpate the ECU tendon. Then the patient is asked to deviate the wrist ulnarly against resistance. If the tendon subsheath is ruptured, the examiner can feel the subluxation of ECU tendon from its vertical groove located on the ulnar styloid process [14].

Patients with ECU tendonitis may present tenderness over the tendon or distributed vaguely over the ulnar aspect of wrist. Traditionally, the examiner asks the patient to extend and ulnar deviate the wrist with resistance. However, this maneuver may also elicit symptoms if intracarpal pathology exists. Ruland and Hogan described the ECU synergy test to avoid the false positive findings with traditional test [15]). The ECU synergy test is based on the synergistic activation of ECU with resisted thumb abduction. The patient is asked to place the affected forearm in supinated position and elbow in 90° flexion, and while the examiner holds the patients thumb and index finger with resistance, the patient is asked to separate the first web. This maneuver may provoke pain in ulnar aspect of wrist if there is tendonitis and avoid irritating the pathologic region directly.

References

1. Rein S, Semisch M, Garcia-Elias M, Lluch A, Zwipp H, Hagert E. Immunohistochemical mapping of sensory nerve endings in the human triangular fibrocartilage complex. Clin Orthop Relat Res. 2015;473:3245–53.
2. Semisch M, Hagert E, Garcia-Elias M, Lluch A, Rein S. Histological assessment of the triangular fibrocartilage complex. J Hand Surg Eur. 2016;41:527–33.
3. Kavi S. Ulnar-sided wrist pain: evaluation and treatment of triangular fibrocartilage complex tears, ulnocarpal impaction syndrome, and lunotriquetral ligament tear. J Hand Surg. 2012;37A:1489–500.
4. Young D, Papp S, Giachino A. Physical examination of the wrist. Orthop Clin North Am. 2007;38:149–65.
5. Garcia-Elias M. Clinical examination of the ulnar-sided painful wrist. In: del Piñal F, et al., editors. Arthroscopic management of ulnar pain. Berlin Heidelberg: Springer-Verlag; 2012. p. 25–44.

6. Prosser R, Harvey L, Lastayo P, Hargreaves I, Scougall P, Hebert RD. Provocative wrist tests and MRI are of limited diagnostic value for suspected wrist ligament injuries: a cross-sectional study. J Physiother. 2011;57:247–53.
7. Tay SC, Tomita K, Berger RA. The "ulnar fovea sign" for defining ulnar wrist pain: an analysis of sensitivity and specificity. J Hand Surg Am. 2007;32(4):438–44.
8. Nakamura R, Horii E, Imaeda T, Nakao E, Kato H, Watanabe K. The ulnocarpal stress test in the diagnosis of ulnar-sided wrist pain. J Hand Surg Br. 1997;22(6):719–23.
9. Haugstvedt J-R, Berger RA, Nakamura T, Neale P, Berglund L, An K-N. Relative contributions of the ulnar attachments of the triangular fibrocartilage complex to the dynamic stability of the distal radioulnar joint. J Hand Surg Am. 2006;31:445–51.
10. Kleinman WB. Stability of the distal radioulna joint: biomechanics, pathophysiology, physical diagnosis, and restoration of function what we have learned in 25 years. J Hand Surg Am. 2007;32:1086–106.
11. Rizzo C, Garret J, Guigal V, Gazarian A. Clinical examination of wrist ligaments. In: Overstraeten V, editor. Carpal ligament surgery. Springer-Verlag France; 2013. p. 39–66.
12. Manuel F, DaSilva Avi D, Goodman Joseph A, Akelman GE. Evaluation of ulnar-sided wrist pain. J Am Acad Orthop Surg. 2017;25(8):e150–6.
13. Lester B, Halbrecht J, Levy M, Gaudinez R. "Press test" for office diagnosis of triangular fibrocartilage complex tears of the wrist. Ann Plast Surg. 1995;35:41–5.
14. Burkhart SS, Wood MB, Linscheid RL. Posttraumatic recurrent subluxation of the extensor carpi ulnaris tendon. J Hand Surg Am. 1982;7:1–3.
15. Ruland RT, Hogan CJ. The ECU synergy test: an aid to diagnose ECU tendonitis. J Hand Surg Am. 2008;33:1777–82.

Compartment Syndrome of the Hand

28

Bernice Heng and Andrew Chin

28.1 Introduction

Acute Compartment Syndrome (ACS) is a limb-threatening condition in which increased interstitial pressure within a limited fibro-osseous space causes microvascular compromise. Compartment syndrome in the hand and digits are not well documented; however, a high index of suspicion should be maintained in patients exhibiting hallmark clinical signs and symptoms. It is a clinical diagnosis that requires urgent diagnosis and intervention.

28.2 Pathophysiology

Based on Matsen's "unified concept" of compartment syndrome [1], the initial pathophysiologic insult of increased intracompartmental pressures may gradually exceed capillary perfusion pressure. Increased intracompartmental pressures can be due to intrinsic injury (e.g., edema, bleeding, reperfusion) or extrinsic injury (e.g., compressive casts or dressings). An increase in local venous pressure leads to a narrowed arteriovenous perfusion gradient resulting in decreased tissue perfusion below the level necessary for cellular viability. This leads to local hypoxia, nerve dysfunction, and muscle necrosis.

A decrease in tissue perfusion subsequently results in an ongoing vicious cycle of further tissue insult, increased capillary leakage and resultant increase in intracompartmental pressures [2]. The extent of muscle injury depends on the duration of ischemia and the metabolic rate of the muscles involved. The extent of muscle injury depends on the duration of ischemia and the metabolic rate of the tissue. Prolonged ischemia can ultimately lead to necrosis of the affected muscle compartment.

28.3 Anatomy

Compartments are defined as enclosed myofascial spaces bound by connective tissues or bones [3].

The hand is made up of ten separate compartments. There are four dorsal interossei compartments, three volar interossei compartments, a thenar compartment, a hypothenar compartment,

B. Heng
Department of Hand and Reconstructive Microsurgery, Tan Tock Seng Hospital, Singapore, Singapore

A. Chin (✉)
Singapore General Hospital Academia, Singapore, Singapore

J. G. Lane et al. (eds.), *The Art of the Musculoskeletal Physical Exam*,
https://doi.org/10.1007/978-3-031-24404-9_28

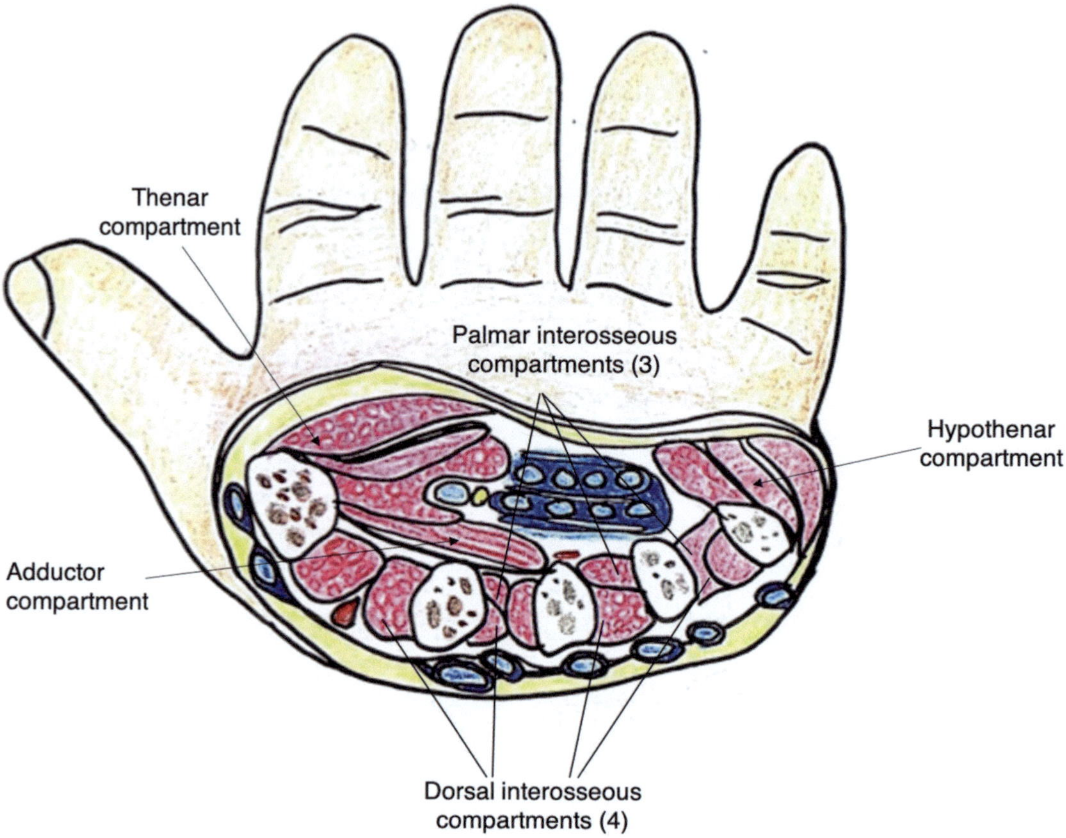

Fig. 28.1 Illustration demonstrating the 10 compartments in the hand at the level of the metacarpal bases

and an adductor compartment (Fig. 28.1) [3]. The carpal canal, although not a separate muscular compartment, is a defined space that with swelling or fracture can present with median nerve compression and acute carpal tunnel syndrome. The compartments of the hand are divided by the carpal bones, metacarpal bones, and individual investing fascial layers.

The compartments of the phalanges have not been well defined in the literature. The digital neurovascular bundles are confined by the Grayson's and Cleland's ligaments, volar and dorsally, respectively (Fig. 28.2) [4]. Therefore, any excessive digital swelling that occurs may constrict the neurovascular bundles in this space and compromise flow to the digit.

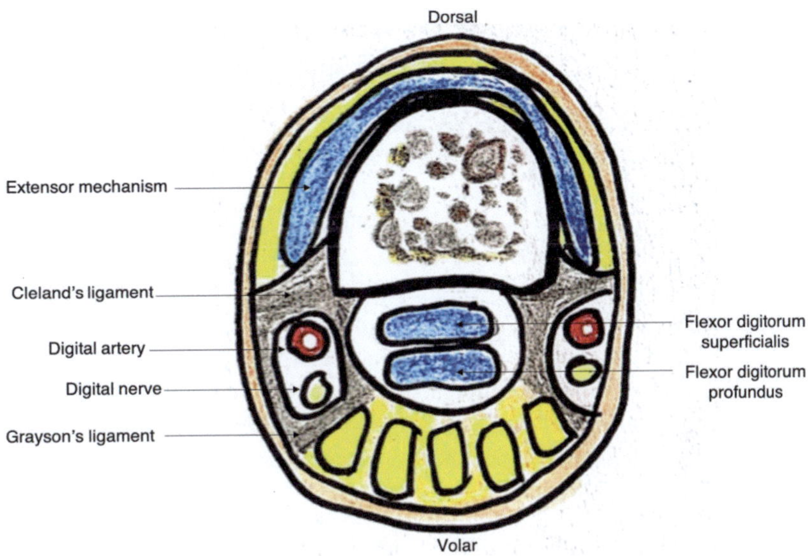

Fig. 28.2 Illustration demonstrating the cross section of a digit at the proximal phalanx level

28.4 Diagnosis

Acute compartment syndrome is a clinical diagnosis. Maintenance of a high index of suspicion, particularly in the setting of high risk injuries and conditions, aids in the prompt recognition and treatment of this condition.

Pain out of proportion to injury and pain with passive stretching of muscles in the involved compartment are the earliest indicators of ACS.

Although ACS commonly occurs in the setting of a closed fracture, other causes of increased compartment contents or decreased compartment volume should also be considered (Table 28.1). There are several conditions that can make the clinical diagnosis of ACS challenging. Neurologically obtunded patients secondary to an altered sensorium, patients with distracting pain from polytrauma, or patients with neurological deficits in the affected limb can mask the signs and symptoms of ACS.

Table 28.1 Intrinsic versus extrinsic causes of acute compartment

Increased compartment content	Decreased compartment size
Vascular injury, e.g., Ischemia-reperfusion injury Fractures or dislocations Crush injury [5] Burns causing soft tissue swelling, e.g., electrical burns Envenomation, e.g., pit viper	Compressive casts or dressings Circumferential burns Failure to remove a finger tourniquet [6] Prolonged limb positioning, e.g., in paralytics, during surgery

28.5 History

A thorough history of the patient's comorbidities including the use of antiplatelets or anticoagulants should be obtained. The history should be targeted towards the various causes of ACS. The hallmarks of ACS described as the six Ps include pain, pressure, paresthesia, pallor, paresis, and pulselessness, where pain is the most consistent.

Patients often complain of a deep, aching, often poorly localized pain that is disproportionate to the clinical signs. It continues to be persistent and progressive, and it is usually the earliest sign of ACS. Pain often peaks at about 2–6 h and may subside thereafter due to ongoing muscle ischemia. Therefore, patients with a late presentation may have more pain compared to those presenting early.

The classic findings of compartment syndrome in adults are not always detectable in children. Although the affected limb may appear tensely swollen, the child may not be able to comply with sensorimotor examination. Furthermore, the findings of pallor and absent pulses are late manifestations of the condition. In children, restlessness, agitation, and anxiety, with an increasing analgesic requirement, may be other indicators of an impending or established compartment syndrome [7].

28.6 Clinical Examination

Where the diagnosis of ACS may not be apparent in the initial setting, serial clinical examination of the limb should be performed to detect progression to a more severe clinical state and initiate timely intervention.

28.7 Inspection

Swelling, skin changes such as ecchymosis, blistering, and any bony deformities may be observed. The hand may be held in an intrinsic minus position (metacarpophalangeal joint extension; distal and proximal interphalangeal joint flexion) with severe swelling (Fig. 28.3).

28.8 Palpation

The affected compartments in compartment syndrome are firm and non-compressible. There is loss of normal skin creases due to edema. Palpation often elicits tenderness and this is further aggravated by passive stretch of the affected compartment. Pain with passive motion of the metacarpophalangeal joint corresponding to the affected intrinsic muscle has been described to be the most sensitive sign of ACS in the hand [8].

Vascularity of the hand should be determined by palpation of the distal pulses, capillary refill time and turgor. Sensory loss in the distribution of the nerve coursing through the involved compartment may be present and all pulses of the limb should be examined.

In the hand, the patient should be examined for acute carpal tunnel syndrome in the setting of high velocity trauma, for example, perilunate or lunate dislocations. Patients can present with paresthesia over the thumb, index, middle and radial half of the ring fingers.

28.9 Compartment Pressure Measurement

If the clinician is uncertain of the diagnosis based on clinical findings, intracompartmental pressures can be measured using a Stryker® Intracompartmental Pressure Monitoring System. Compartment pressures should be measured when ACS is suspected in a patient who is uncooperative, who has an altered

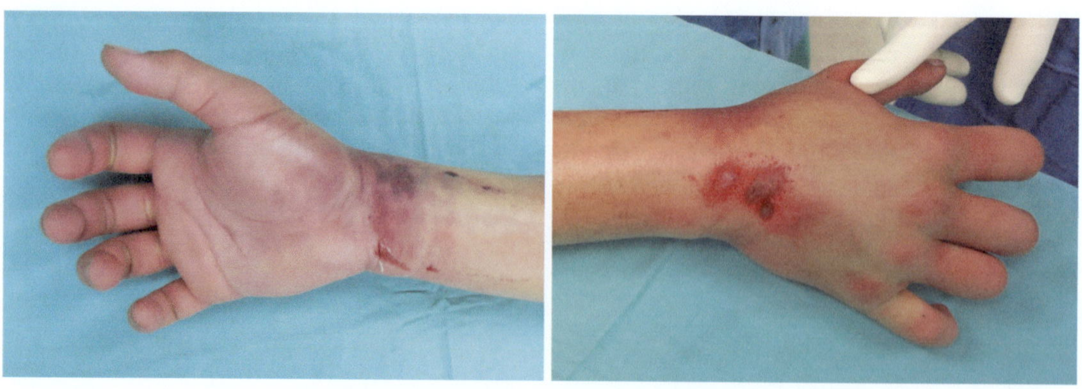

Fig. 28.3 Intraoperative photos demonstrating compartment syndrome of the right hand secondary to crush injury showing swelling, ecchymosis, and intrinsic minus posturing of the hand (Photos courtesy of Dr. Kang Yong Chiang, Singapore General Hospital)

mental status, or in the pediatric population where clinical examination may not be reliable.

Compartmental pressures are also measured in the operating theater to verify which compartment or compartments are affected and to ensure that they are fully decompressed following fasciotomy. Compartments that clinically appear to be involved are tested, and measurements are taken when the compartment appears to be the most tense and swollen.

Canine studies have demonstrated that the critical pressure threshold for ischemic muscle necrosis was ≤20 mm Hg less than the diastolic blood pressure [9] and the normal cellular metabolic state of healthy tissue to exist at 30 mmHg [10]. Hence, based on the current understanding of tissue metabolism and muscle ischemia, fasciotomy of the hand compartments is recommended if the pressures are within 30 mmHg of the patient's diastolic pressure [11].

If compartment syndrome is not diagnosed clinically or by pressure measurements, but remains a worry based on the patient's history or initial presentation, the patient should continue to be closely observed with serial examinations.

28.10 Management

Whilst the removal of extrinsic causes of ACS should be done emergently, emergency fasciotomy is the definitive treatment for hand compartment syndrome. Incisions should be extensile, and they should allow the surgeon to gain adequate access to all possible enclosed compartments in the hand (Figs. 28.4 and 28.5). It is important to leave a broad skin bridge between incisions to maintain adequate vascularity of the skin flaps.

Through longitudinal skin incisions, the fascia is released and necrotic tissue is excised.

The thenar and hypothenar compartments are released via longitudinal incisions are made at the junctions of the glabrous and non-glabrous skin over the radial border of the first metacarpal and the ulnar border of the fifth metacarpal bone to avoid the motor branches to the muscles of these compartments [3, 12]. The deep investing muscle fascia should also be released. The carpal tunnel can also be decompressed via a curvilinear incision ulnar and parallel to the thenar crease (Fig. 28.6a–c)

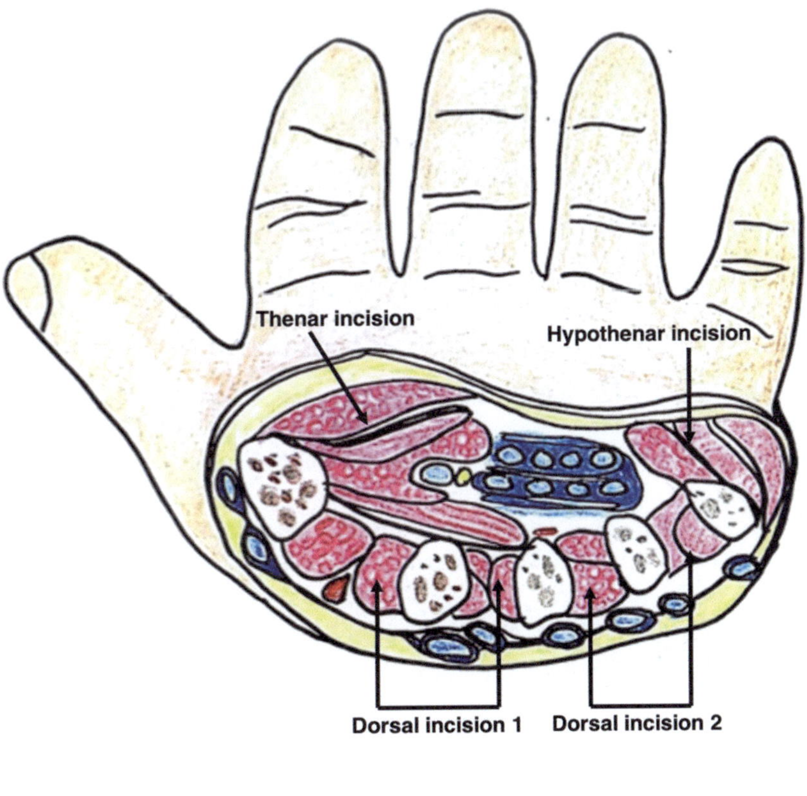

Fig. 28.4 Illustration demonstrating the 10 compartments of the hand which are released for each surgical incision

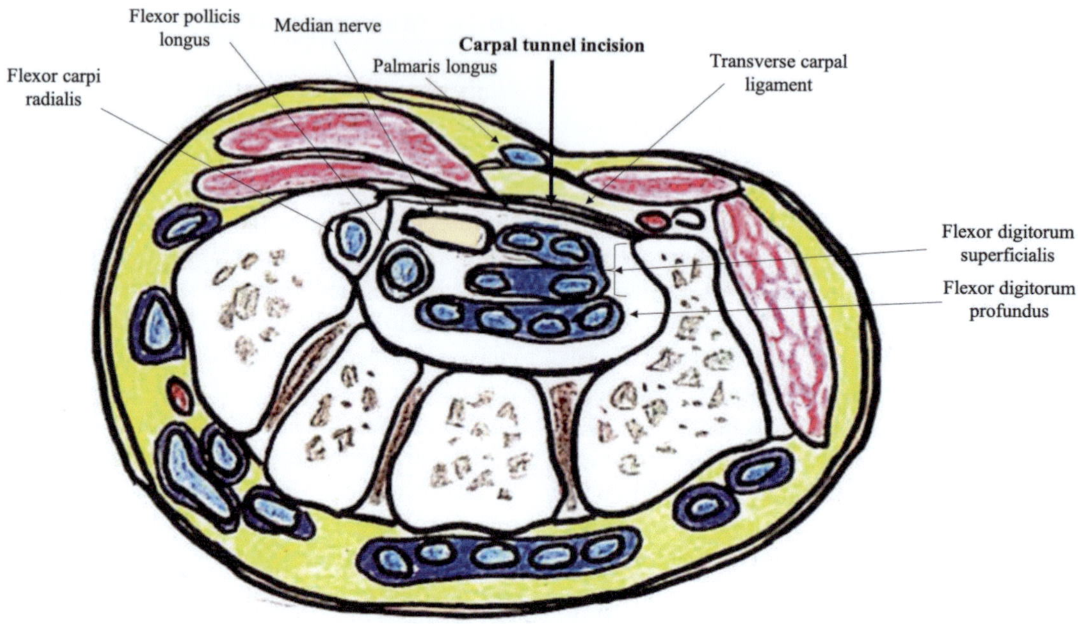

Fig. 28.5 Illustration demonstrating the carpal tunnel incision at the outlet of the carpal tunnel

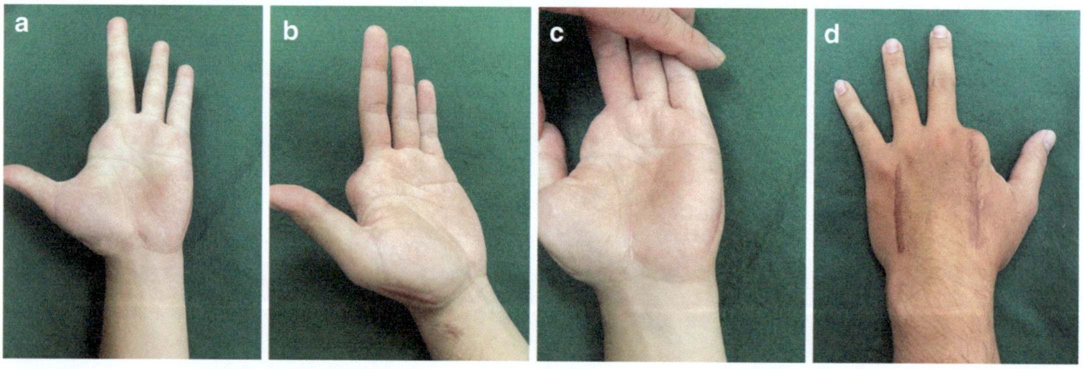

Fig. 28.6 Post-operative photos of a patient with a traumatic right index finger amputation with right hand compartment syndrome secondary to a road traffic accident. **a**. Healed volar incisions over the thenar eminence, carpal tunnel and hypothenar eminence **b**. Thenar eminence incision **c**. Hypothenar eminence **d**. Dorsal longitudinal incisions (Photos courtesy of A/Prof Andrew Chin, Singapore General Hospital)

The dorsal interosseous muscles (and volar interosseous muscles) are decompressed through dorsal incisions between the second and third metacarpals and fourth and fifth metacarpals (Figs. 28.6d and 28.7) The first dorsal interosseous muscle is decompressed through an incision placed in the first dorsal web space where the dorsal fascia of the adductor pollicis can also be released through this incision [12].

Digital fasciotomies are performed through midlateral incisions to prevent subsequent contractures (Fig. 28.8). They should be performed on the non-contact side of the digits. The incision is centered over the proximal interphalangeal joint (PIPJ) and may extend 2 cm proximal and 2 cm distal to the PIPJ.

Furthermore, the underlying cause of the ACS of the hand and digit should be treated adequately

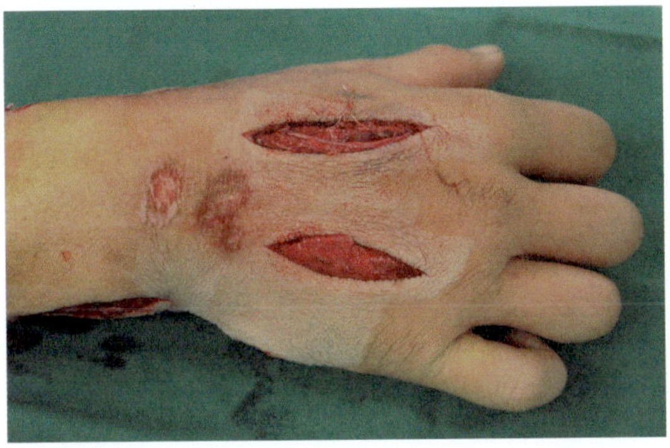

Fig. 28.7 Intraoperative photo demonstrating right hand dorsal incisions between the second and third metacarpals and fourth and fifth metacarpals and decompressed interosseous muscles (Photo courtesy of Dr. Kang Yong Chiang, Singapore General Hospital)

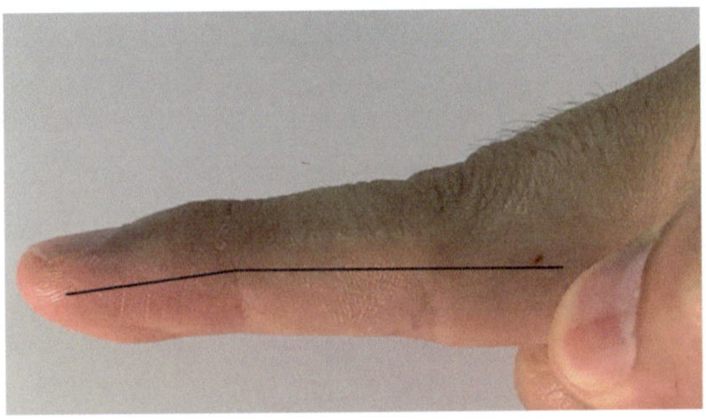

Fig. 28.8 Illustration demonstrating a midlateral fasciotomy incision of the digit

such as manipulation and reduction of carpal fracture-dislocations or fixation of closed fractures of the digit.

The wounds are usually left open and initially covered with sterile wet-to-dry dressings after the initial surgery. Vacuum-assisted closure (VAC) can also be considered to reduce the need for skin grafting, reduced hospital stay, and a lower risk of nosocomial infection. The patient can undergo relook irrigation and debridement after 48–72 h. However, if there is any suspicion of ongoing muscle necrosis, debridement should be performed even earlier. Eventually, the patient's wounds should be considered for secondary closure or skin grafting when the swelling has resolved.

28.11 Summary

ACS of the hand and fingers is a potentially limb-threatening condition that can lead to substantial loss of function if not recognized early and treated appropriately. A high index of suspicion should be maintained based on the patient's history and clinical findings and adjunct modalities such as compartmental pressure measurements can be performed to clinch the diagnosis. Following which, early surgical decompression is essential to reduce the morbidity from ACS.

References

1. Matsen F. Compartmental syndromes. Clin Orthop Relat Res. 1975;113:8–14.
2. Holden C. The pathology and prevention of Volkmann's ischaemic contracture. J Bone Joint Surg Br. 1979;61-B(3):296–300.
3. Chrysopoulo M, McGrouther D, Jeschke M, Kaufman B. Cleland's ligaments: an anatomic study. Plast Reconstr Surg. 2002;109(2):566–72.
4. Schwartz-Fernandes F, McDermott E, Culp J. Crush injury-induced finger compartment syndrome: a case report and literature review. Cures. 2020;12(4):e7509.
5. Avci G, Akan M, Yildirim S, Aköz T. Digital neurovascular compression due to a forgotten tourniquet. Hand Surg. 2003;08(01):133–6.
6. Bae D, Kadiyala R, Waters P. Acute compartment syndrome in children: contemporary diagnosis, treatment, and outcome. J Pediatr Orthop. 2001;21(5):680–8.
7. Halpern A, Greene R, Nichols T, Burton D. Compartment syndrome of the interosseous muscles. Clin Orthop Relat Res. 1979;140:23–5.
8. Matava M, Whitesides T, Seiler J, Hewan-Lowe K, Hutton W. Determination of the compartment pressure threshold of muscle ischemia in a canine model. J Trauma. 1994;37(1):50–8.
9. Heppenstall R, Sapega A, Scott R, Shenton D, Park Y, Maris J, Chance B. The compartment syndrome. Clin Orthop Relat Res. 1988;226:138–55.
10. Codding J, Vosbikian M, Ilyas A. Acute compartment syndrome of the hand. J Hand Surg Am. 2015;40(6):1213–6.
11. Oak N, Abrams R. Compartment syndrome of the hand. Orthop Clin N Am. 2016;47(3):609–16.
12. Prasarn M, Ouellette E. Acute compartment syndrome of the upper extremity. Am Acad Orthopaedic Surgeon. 2011;19(1):49–58.

Evaluation of the Neurological Conditions of the Elbow, Forearm and Hand

29

Margareta Arianni

29.1 Introduction

This chapter focuses on the assessment of neurological conditions caused by injuries or compression of major nerves (median, ulnar and radial nerves) and their branches in the elbow, forearm and hand. The discussion of each nerve begins with anatomy, followed by a section detailing the sites of compression in relevant entrapment neuropathies.

29.2 Median Nerve

29.2.1 Anatomy

The median nerve is composed of fibres from the lateral and medial cords of the brachial plexus, which receive contributions from all the anterior rami of C5 to T1. The sensory part of the median nerve is mainly constituted by the lateral cord, while the motoric part is mainly from the medial cord. At the level of the arm, the median nerve travels in the neurovascular compartment in the medial aspect of the arm, lateral to the brachial artery, lying between the biceps and brachialis. In a small percentage of the population, the supracondylar process is present at approximately 5–6 cm proximal to the medial epicondyle. It is a congenital bony protuberance that may measure up to 2 cm in length. The ligament of Struthers is another structure that is frequently related to the supracondylar process. Typically, this ligament arises from the medial epicondyle and inserts into the supracondylar process. However, this is not always the case. The ligament of Struthers may or may not be associated with a supracondylar process, and it may be present in the absence of the latter [1]. In relation to the median nerve, this fibrous band-like structure may impinge on the median nerve, causing compressive neuropathy, or simply travels along a path away from the median nerve, in which case it does not cause compression [1].

As it travels into the cubital fossa, the median nerve, now medial to the brachial artery, lies beneath the lacertus fibrosus and enters between the ulnar and humeral heads of the pronator teres. Here, it gives off a motor branch (or two branches) to the pronator teres. Slightly distal from these branches, the ulnar side of the median nerve gives motor branches to the flexor carpi radialis (FCR), palmaris longus (PL) and flexor digitorum superficialis (FDS). The median nerve, having passed through the pronator teres, goes behind the arch of the flexor digitorum superficialis (FDS) and anterior to the flexor digitorum profundus (FDP).

M. Arianni (✉)
Department of Orthopaedics (Hand Surgery Division), Ramsay Sime Darby Premier Bintaro Hospital, Tangerang, Indonesia

Department of Surgery, RSUD Pasar Minggu General Hospital, Jakarta, Indonesia

J. G. Lane et al. (eds.), *The Art of the Musculoskeletal Physical Exam*,
https://doi.org/10.1007/978-3-031-24404-9_29

The anterior interosseous nerve (AIN) emerges from the posterior or posterolateral side of the median nerve, just proximal to the arch of the flexor digitorum superficialis [2]. The AIN is a pure motor branch of the median nerve that innervates the FPL, FDP of the index and middle fingers and the pronator quadratus. The main trunk of the median nerve continues its course distally over the FDP muscle. At the distal part of the forearm, it becomes more superficial as it goes side by side the FDS and FCR tendons. Approximately 5 cm proximal to the wrist crease, the palmar cutaneous branch arises from the median nerve and penetrates the antebrachial fascia 2 cm proximal to the wrist flexion crease [3]. The median nerve itself enters the carpal tunnel at the wrist, at the distal end of which it gives rise to the recurrent motor branch that innervates the abductor pollicis brevis, opponens pollicis and superficial head of the flexor pollicis brevis.

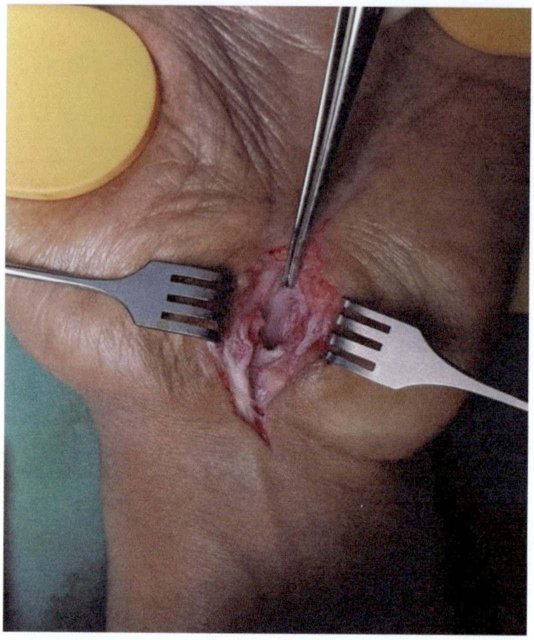

Fig. 29.1 Revision surgery for carpal tunnel syndrome. The forceps are pointing at the median nerve, which has been entrapped by scar tissue

29.2.2 Sites of Compression

Regarding compressive neuropathy, there are several sites along the path of the median nerve where it is susceptible to compression. Around the elbow joint, there are four possible sites of compression. At the distal humerus, the median nerve may be compressed by the supracondylar process and/or the ligament of Struthers. Just proximal the elbow, the median nerve may be compressed by thickened bicipital aponeurosis. At the elbow level, the pronator teres may cause compression of the median nerve, and further distally, the FDS arch may be thick enough to compress the median nerve. Compression of the median nerve by any of the aforementioned structures typically results in pronator syndrome. The AIN may also be compressed by the bicipital aponeurosis, pronator teres and FDS arch. But since the AIN is a deeply located branch, isolated compression by these structures is usually rare. Some possible AIN-compressing structures include anomalous muscles (Gantzer muscle), vascular malformation and tumour.

At the level of the wrist, carpal tunnel is the frequent site of compression of the median nerve (Fig. 29.1). Several sites of compression of the palmar cutaneous branch of the median nerve have also been described: fascia of the distal FDS [4], normal palmaris longus tendon [5], antebrachial fascia [6], carpal tunnel [7] and the tunnel of palmar cutaneous branch itself [8].

29.2.3 Evaluation

29.2.3.1 Observation or Inspection

Before embarking on a thorough sensory and motor examination, a simple observation is a good way to start the evaluation of any neurological condition. A patient's general health, sitting or standing posture, gait, upper limb posture or position can be observed during the first few minutes of the meeting. Look for any scars of previous surgeries or injuries, deformities and atrophy. Some conditions may provide signs that are immediately detectable on inspection. A patient with leprosy, for example, may appear with telltale signs, such as nodules, hypopigmented macules and claw hand.

Long-standing compression or injury of the median nerve frequently causes atrophy of the thenar muscles. The thenar muscles are innervated specifically by the recurrent motor branch of the median nerve, which is one of the most distal branches of the median nerve.

29.2.3.2 Sensory Function Evaluation

The median nerve sensory distribution (Fig. 29.2a and b) encompasses the palm (excluding the hypothenar area), palmar aspect of the thumb, index and middle fingers, radial half of the palmar aspect of the ring finger and the distal dorsal areas of the thumb, index, middle and ring fingers (radial half). It is important to note that the skin of the palm is innervated by the palmar cutaneous branch of the median nerve, while the palmar aspect of the digits is innervated by digital nerves. Carpal tunnel syndrome does not affect sensation on the palm as the compression is distal to the origin of the palmar cutaneous branch. Lesion of the median nerve proximal to the origin of the palmar cutaneous branch results in sensory changes in the palm. Lacertus syndrome, for example, frequently causes burning sensation and numbness in the distribution of the palmar cutaneous branch of median nerve.

Light touch sensation is assessed by applying light touch on the patient's right and left fingertips simultaneously with a wisp of cotton or the examiner's fingertips. The patient is asked whether he or she feels the stimuli, and also whether there is any difference of stimuli sensed between left and right fingers. Evaluation of light touch sensation is also performed on both radial and ulnar sides of the ring finger. With median nerve injury, patients report decreased sensation on the radial side of the ring finger, while the sensation on the ulnar side of the ring finger is maintained. Not infrequently, there is cross innervation between ulnar and median nerves. In such cases, sensation may not differ much between the radial and ulnar sides of the ring finger. Comparing sensations of the little finger pulp to index finger pulp in the same hand will usually reveal the difference since these areas receive exclusive innervations from ulnar and median nerves, respectively.

The two-point discrimination (2PD) test is an important examination to assess the loss of

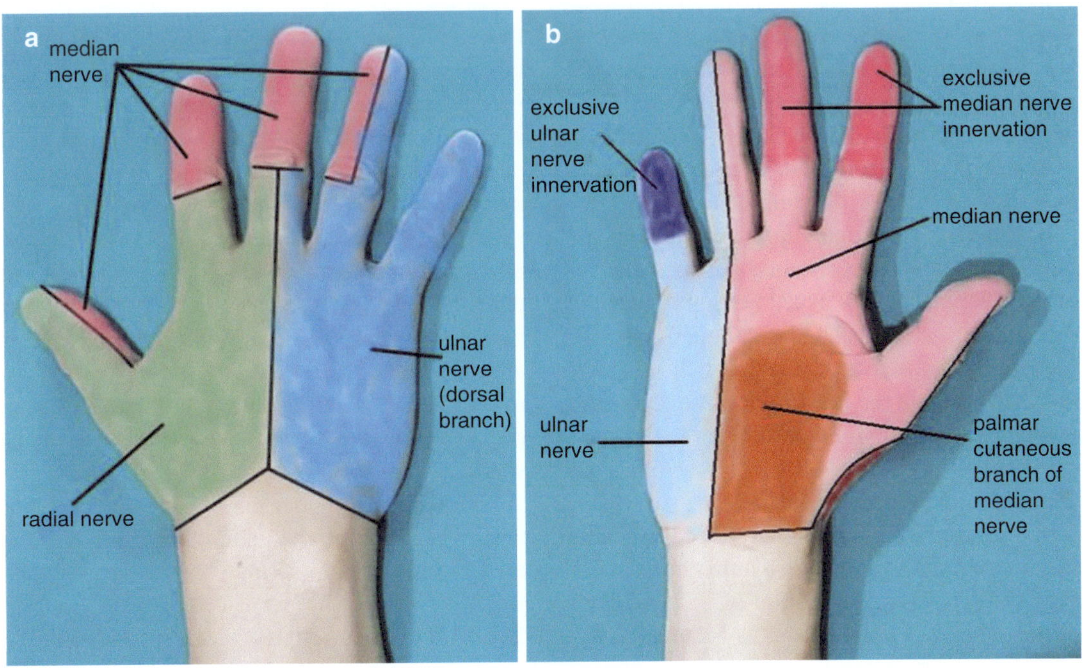

Fig. 29.2 Distribution of sensory innervation of the dorsal (**a**) and volar (**b**) aspects of the hand

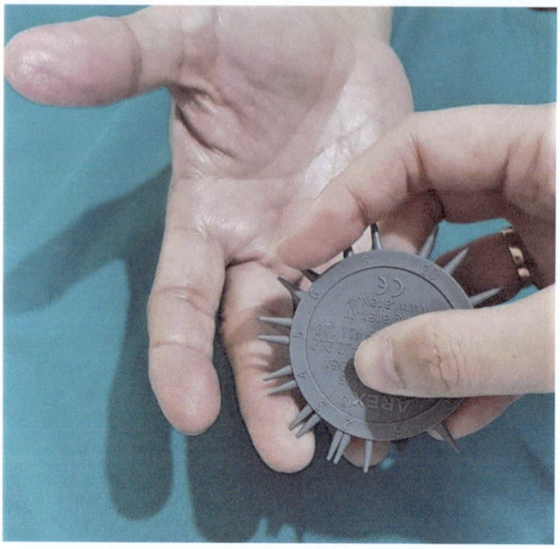

Fig. 29.3 Two-point discrimination testing

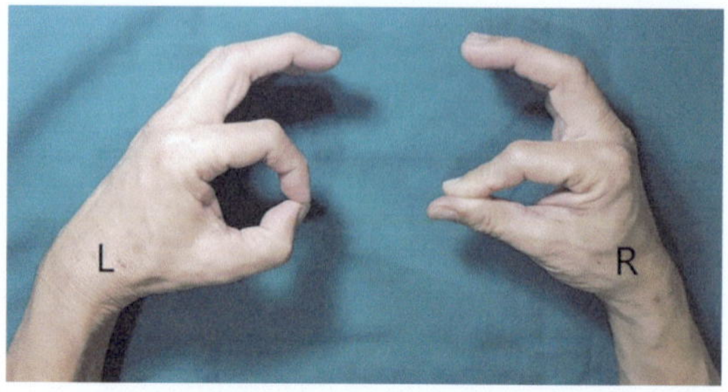

Fig. 29.4 The "OK" sign. Left hand, the normal hand, is able to make an OK sign. In the right hand with AIN palsy, the FPL and FDP of the index finger are paralysed, not able to flex the thumb IP joint and index DIP joint

discriminative sensation, which translates to the degree of denervated sensory receptors [9] [10]. The test describes the minimum distance between two points that the patient is able to feel as two contact points. Normal 2PD values vary depending on individuals, nerves, age and gender. The American Society for Surgery of the Hand recommends the following classification for 2PD values: 2PD < 6 mm = normal, 6–10 mm = fair, 11–15 mm = poor, and > 15 mm as protective [11]. The simplest tool needed is an unraveled paper clip. Aesthesiometer and disc-shaped two-point discrimination devices (Fig. 29.3) can also be used with more convenience and greater accuracy. The two-point discrimination test is first performed on the normal, uninjured hand, beginning with the widest distance between the two probes. The patient is asked whether she or he feels one or two probes being prodded onto their finger pulps. The test is repeated with a smaller distance between the two probes. The test is then performed on the injured or affected hand.

29.2.3.3 Motor Evaluation

In the setting of compressive neuropathy, motor dysfunction may not occur in mild compression. Frequently, motor dysfunction and atrophy only occur in severe compression. In traumatic nerve injury, however, loss of motor function is obvious soon after insult. Injury to the median nerve high at the elbow presents with weakness of muscles that are innervated by branches distal to the injury site, namely PT, FCR, PL, FDS, FPL, FDP and thenar muscles. When asked to make a fist, a patient with high median nerve injury is able to flex the ring and little fingers into full fist, but not the index and middle fingers. This is because the index and middle fingers have lost all of their flexors (FDP and FDS), but the ring and little fingers still retain the function of their FDPs that are innervated by the ulnar nerve. This posture is called the "Benediction Sign". The pronator syndrome and/or lacertus syndrome as well as AIN syndrome , present with weakness of the FPL and FDP of index and middle fingers. Significant motor weakness, while not common in the former, is a feature of the latter. Typically, the patient is unable to make an "OK" sign (Fig. 29.4). It must be remembered that AIN is a pure motor nerve, thus sensory symptoms are absent in AIN syndrome. Pronator/lacertus syndrome, on the other hand, presents with pain in the forearm and paresthesia/hypoesthesia similar to that in carpal tunnel syndrome, in addition to the motor signs aforementioned (Fig. 29.4). The only group of muscles that is affected by carpal tunnel syndrome is the thenar muscles, which are innervated by the recurrent branch of median nerve. Weakness of the thenar muscles results in weakness of pinch grip. The patient frequently reports difficulties in manipulating small objects.

29.2.3.4 Provocative Tests for Nerve Compression

Provocative tests to detect nerve compression are performed by either applying direct digital pressure on the affected nerve, or putting the elbow/forearm/hand in certain positions that impart indirect compression or create tension to the nerve. When provoked, compressed nerves will reproduce symptoms.

Median nerve within the pronator teres is most compressed when the forearm is fully supinated and elbow extended, or on resisted forearm pronation. When compression of median nerve is caused by lacertus fibrosus, the symptoms are exacerbated by resisted elbow flexion with the forearm supinated. Applying external pressure onto the lacertus fibrosus also elicits symptoms. Compression of the median nerve by FDS arch is triggered by resisted long finger flexion.

There are several tests to elicit the symptoms of carpal tunnel syndrome. The Durkan test is performed by pressing on the carpal tunnel, and the median nerve within, with the examiner's thumb for 30 s. Reproduction of pain and paresthesia along the median nerve distribution is a positive result. In the Phalen test, the wrist is flexed for 1 min (Fig. 29.5). With wrist flexion, more compression on the median nerve is generated, and symptoms will be reproduced in the appropriate nerve distribution.

Tinel sign is one of the provocative signs used to detect the site of a nerve irritation or lesion. A Tinel sign is considered positive when tapping on the nerve elicits symptoms in the nerve's distribution area. Tapping on the lacertus fibrosus or proximal part of the pronator muscle frequently produces a positive Tinel's sign. A positive Tinel sign in carpal tunnel syndrome is elicited by tapping on the median nerve in the wrist. Tinel sign is also useful to evaluate the progress of nerve regeneration. Tinel sign is positive at the site of injury. As the nerve regenerates, positive Tinel sign becomes elicited at more distal sites over time, hence advancing Tinel sign.

In the recent decade, more attention has been given to the scratch-collapse test, a provocative test developed by Susan Mackinnon [12]. The basis of this test is the cutaneous silent period, defined as brief interruption of voluntary muscle contraction that follows noxious stimulus to a cutaneous nerve. The patient and examiner are sitting face to face, with the former positioning his or her arm adducted and forearm in neutral rotation (Fig. 29.6). The patient is asked to maintain this position while the examiner is applying internal rotation force on both distal forearms. The skin overlying the carpal tunnel is scratched by the examiner's hand, after which the test is repeated. With carpal tunnel syndrome, the patient's affected arm will collapse

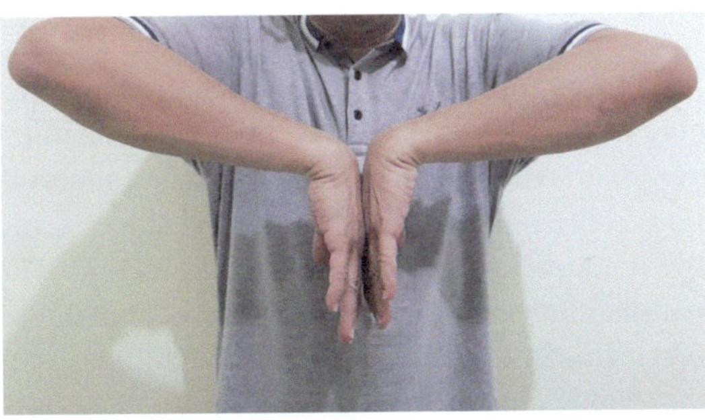

Fig. 29.5 Phalen test. The patient's wrists are put in flexion. Carpal tunnel syndrome will reproduce symptoms in the affected hand

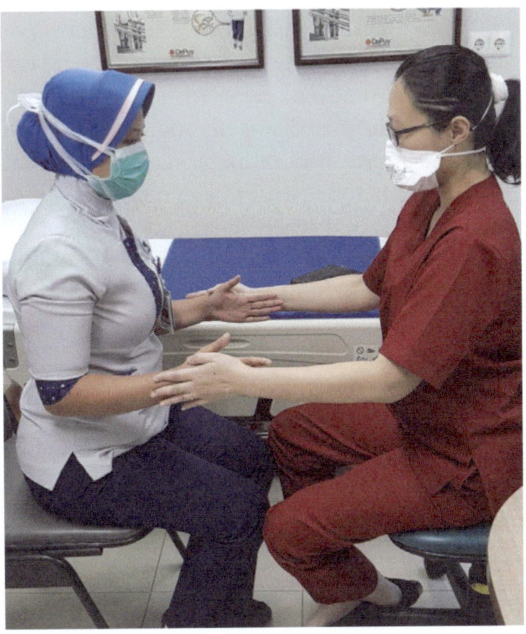

Fig. 29.6 The scratch collapse test. The examiner (right) is applying internal rotation force on the patient's distal forearms, while the patient is attempting to externally rotate her shoulders. The test is positive when, after scratching the skin over the carpal tunnel, the patient loses strength to resist the examiner's internal rotation force and folds the affected forearm toward the belly

in internal rotation. While various studies have come up with different results on the sensitivity and specificity of this test, the scratch-collapse test remains an interesting area of further studies.

29.2.3.5 Pitfalls

It is important to always remember the possibility of compression at more proximal sites. Compression of a nerve at a proximal site renders the distal part more sensitive to compression due to changes in axonal transport along the neuron. Vice versa, compression at a distal site may also cause inhibition of retrograde axonal transport and, eventually, changes in nerve cell bodies. These processes will lead to disturbance in the nerve segment proximal to the compression site (reverse double-crush phenomenon) [13]. Thus, Spurling test must be performed to rule out cervical radiculopathy in the presence of median nerve compression. The patient's head is slightly extended and tilted laterally, and the examiner applies axial compression on the patient's head. When radiating pain is elicited in the ipsilateral arm, the Spurling test is considered positive. Motor reflexes and muscle strength in the related myotome must also be checked.

Proximal median nerve entrapment at the level of lacertus fibrosus or pronator muscles needs to be excluded before carpal tunnel release. Caution must be exercised when thenar atrophy appears as an isolated finding (without sensory symptoms) as this can be a sign of motor neuron disease.

29.3 Ulnar Nerve

29.3.1 Anatomy

The ulnar nerve is one of the terminal branches of the brachial plexus. It receives fibres mainly from the C8 and T1 nerve roots. It branches off the medial cord in the axillary region and travels in the proximal half of the arm posteromedial to the brachial artery. In the middle third of the arm, the ulnar nerve, which has been in the anterior compartment, goes behind the medial intermuscular septum before its proximal border, and travels distally in the posterior compartment of the arm. In some cases, the ulnar nerve penetrates the septum to enter the posterior compartment [14]. In others, the ulnar nerve enters the medial intermuscular septum and travels within a fibrous tunnel in the septum before exiting into the posterior compartment [14].

The medial intermuscular septum is adjacent to the arcade of Struthers. The arcade of Struthers, which is distinct from the ligament of Struthers, is located at the medial aspect of the middle-lower third of the arm. It is a fibrous canal that is composed of the medial head of triceps muscle and its aponeurotic expansion, which extends to the medial intermuscular septum [15]. In relation to the ulnar nerve, the arcade can be posterior or anterior to the nerve. It can also encapsulate the ulnar nerve.

Approaching the elbow, the ulnar nerve goes posterior to the medial epicondyle and enters the cubital tunnel, which extends between the medial epicondyle of the humerus to the olecranon. The ceiling of the tunnel consists of the Osborne's ligament, which is a ligament extending from the medial epicondyle to the olecranon. The floor of the tunnel is composed of the medial collateral ligament (MCL) and the elbow joint capsule. The medial epicondyle and olecranon serve as the bony walls of both sides of the tunnel. At the elbow level, the ulnar nerve gives off an articular branch. The ulnar nerve enters the anterior compartment of the forearm between the two heads of the FCU, where it gives off the motor branches to the FCU and medial half of the FDP. In the forearm, the ulnar nerve travels alongside the ulnar artery, surrounded by the FCU, FDS and FDP at its ulnar, anterior and posterior aspects, respectively.

Approximately 9 cm proximal to the wrist crease, the ulnar nerve gives off the dorsal cutaneous branch, which provides sensory innervation to the dorsum of little and ring fingers, and the dorso-ulnar aspect of the hand [16, 17]. The palmar cutaneous branch of the ulnar nerve also originates in the forearm. This branch innervates the hypothenar skin near the wrist crease [18, 19]. At the wrist, the nerve enters the Guyon's canal, an anatomical fibro-osseous tunnel extending from the pisiform to the hook of hamate, together with the ulnar artery and vein. The roof of the canal is the volar carpal ligament, which is the extension of the antebrachial fascia, and the palmaris brevis muscle. The pisiform forms the ulnar or medial border of the canal, and the flexor retinaculum serves as the floor of the canal. The floor of the canal blends laterally with the ulnar portion of the roof of the carpal tunnel.

The Guyon's canal is classified into three zones, based on the branching of the ulnar nerve. Zone 1 extends from the proximal edge of the volar carpal ligament to the bifurcation of the ulnar nerve. Zone 2 extends from the bifurcation to the aponeurotic arch of the hypothenar muscles, containing the deep (motor) branch of the ulnar nerve. The deep motor branch winds around the hook of hamate in a radial direction. This motor branch gives rise to more branches that innervate the intrinsic muscles (hypothenars, palmaris brevis, adductor pollicis interosseous, third and fourth lumbricals). Zone 3, which contains the superficial (sensory) branch of the ulnar nerve, is distal to the bifurcation. This branch provides sensory innervation to the little finger and ulnar half of the ring finger. It also innervates the ulnar side of the palm, distal from the area innervated by the palmar cutaneous branch of the ulnar nerve [19].

29.3.2 Sites of Compression

Around the elbow, the most proximal site of compression of the ulnar nerve is the medial intermuscular septum (MIMS). Nakajima et al. [20] reported a case of ulnar nerve entrapment in the MIMS, and several authors [21, 22] mentioned the medial intermuscular septum as one of the culprits found in revision surgery for cubital tunnel syndrome. Two of the three types of MIMS that were described by Won et al. [14] put the ulnar nerve prone to compression. In type 2, the ulnar nerve pierced the MIMS from the anterior to posterior compartment of the arm, and in type 3, the ulnar nerve travels within a tunnel in the MIMS. Compression site distal to the MIMS is the arcade of Struthers. Cubital tunnel is the most common site of ulnar nerve entrapment. The volume of the cubital tunnel decreases when the elbow is flexed, compressing the nerve against the medial epicondyle. The Osborne ligament also becomes taut, decreasing the space for the nerve. The anomalous anconeus epitrochlearis muscle, when present within the cubital tunnel, may also cause compression on the ulnar nerve. Distal to the cubital tunnel, the ulnar nerve passes through the common flexor aponeurosis, which is also a possible site of compression.

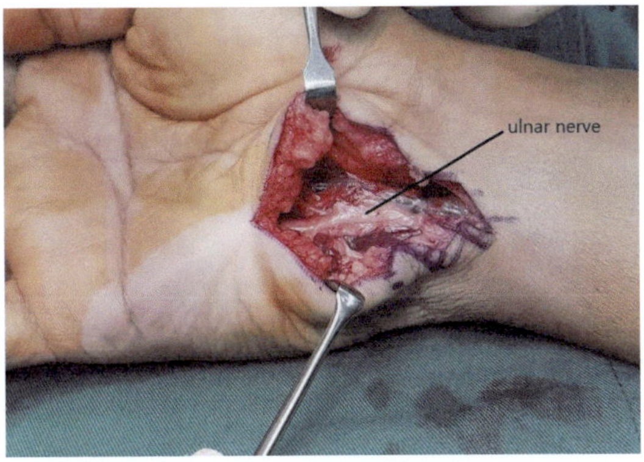

Fig. 29.7 Exploration of the entrapped ulnar nerve in the Guyon canal

In the wrist, the Guyon canal is a frequent site of compression (Fig. 29.7). The causes of compression include repetitive trauma, benign tumour, hook of hamate fracture, ulnar artery aneurysm, congenital bands, and palmaris brevis hypertrophy. Ulnar nerve compression in the wrist may give variable symptoms and signs, depending on the branch being entrapped. Shea and McCain classified the Guyon canal syndrome into three types [23]. Type I corresponds to lesion or entrapment proximal to the bifurcation of the nerve, resulting in both sensory and motor symptoms and signs. Type II is compression of the deep motor branch, and type III is compression of the sensory branch.

29.3.3 Evaluation

29.3.3.1 Observation

Ulnar claw hand is characterized by hyperextension of the MP joints and flexion of the interpha-

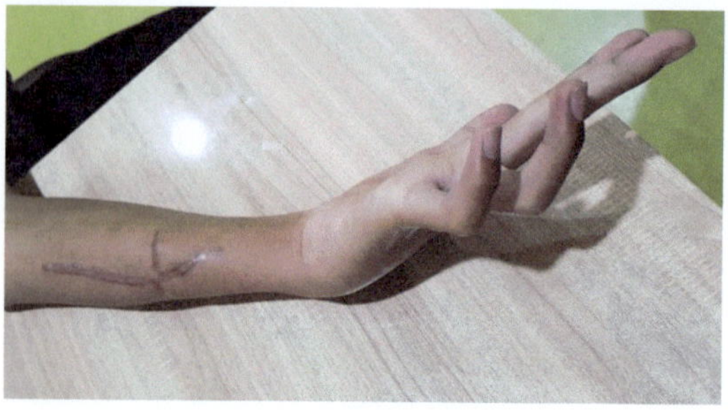

Fig. 29.8 Ulnar claw hand, resulting from transection of the ulnar nerve due to defensive injury

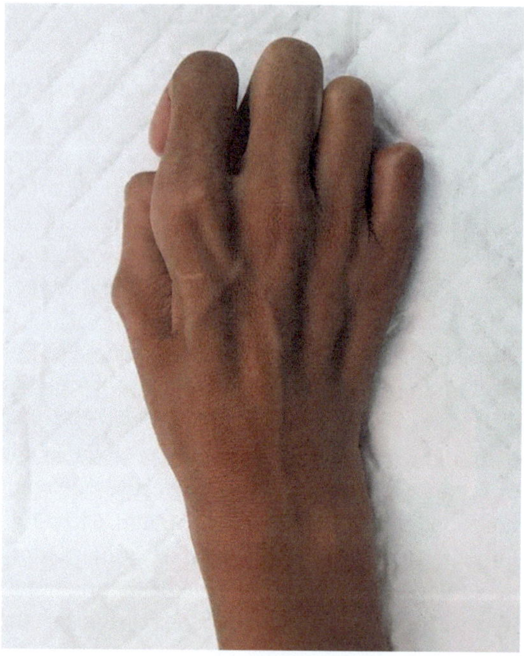

Fig. 29.9 Severe atrophy of all interosseous muscles

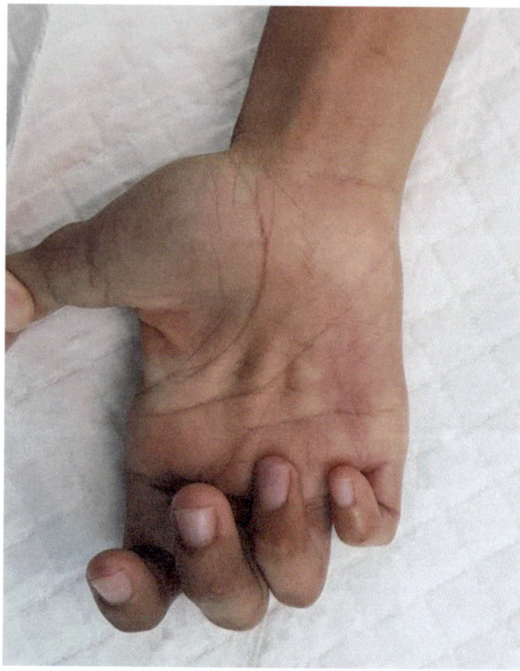

Fig. 29.10 The Masse's sign. The hypothenar muscles and metacarpal arch have flattened

langeal joints of the fourth and fifth fingers (Fig. 29.8). Loss of intrinsic functions on the fourth and fifth finger result in unopposed action of extrinsic extensors (EDC) and extrinsic flexors (FDS, FDP). When the lesion of the nerve is proximal to the branches to the ulnar half of the FDP, the paralysed FDP to the fourth and fifth digits no longer flexes the IP joint, giving the appearance of less claw and hence the term "ulnar paradox". Atrophy of the first dorsal interosseous muscle in the first dorsal webspace is frequently one of the first indicators of an ulnar nerve injury. In later stages of ulnar nerve palsy, all interosseous muscles will undergo atrophy (Fig. 29.9). The hypothenar muscles, also atrophied, no longer provide the bulkiness in the ulnar part of the hand. This results in a flat palm and metacarpal arch, known as the Masse's sign (Fig. 29.10).

29.3.3.2 Sensory Evaluation

Loss of sensation on the dorso-ulnar aspect of the hand and dorsal aspects of the fourth and fifth digits indicate that ulnar nerve lesion is proximal to the dorsal cutaneous branch of the ulnar nerve, which branches off the main trunk of the ulnar nerve in the distal third of the forearm. In Guyon canal syndrome, sensation provided by the dorsal cutaneous branch of the ulnar nerve is preserved. The ulnar palm (hypothenar skin) receives varying innervations from the following branches of the ulnar nerve: palmar cutaneous branch originating from the ulnar nerve in the forearm, and the multiple transverse branches originating from the ulnar nerve in the Guyon canal [19]. In Guyon canal syndrome, sensation of the hypothenar skin may be decreased in more distal area, but preserved in the more proximal area around the distal wrist crease [19]. The palmar cutaneous branch also exits the ulnar nerve in the forearm. It provides sensory innervation to the hypothenar area. This branch comes off the ulnar nerve slightly proximal to the dorsal cutaneous branch. The ring finger is the transitional area between the median and ulnar nerves. Classically, ulnar nerve injury results in diminished sensation on the ulnar side of the ring finger with preserved sensation on the radial side of the same finger. However, many patients with ulnar nerve injury do not have this particular textbook description of symptoms [24]. They may present with preserved sensation on the ulnar side of the ring finger even in the presence of ulnar nerve injury. This confusion may be accounted for by a communicating branch between the ulnar and median nerves in the palm, such as the Berrettini branch.

29.3.3.3 Motor Evaluation

Motor evaluation begins with assessment of muscles innervated by the ulnar nerve. Active flexion of the DIP joints of the little and ring fingers against resistance tests the function of ulnar half of the FDP. Abduction and adduction of the fingers test the function of dorsal and palmar interosseous muscles, respectively. Loss of palmar interosseous muscle function leads to abduction of the little finger, which is facilitated by the unopposed action of the extensor digiti minimi (Wartenberg's sign). As the function of intrinsic muscles are impaired, normal flexion of the MP joints and extension of the IP joints, which are parts of normal dynamics of finger flexion for gripping objects, are lost. When the fingers try to grasp an object, the object will be rolled away by the fingers since the IP joints flex before the MP joints do. Paralysis of the adductor pollicis, deep head of the FPB and the first dorsal interosseous muscles results in weakness of pinch. When the patient is asked to perform key -pinch, the FPL muscle works excessively to compensate for the loss of intrinsic functions, leading to the development of Froment's sign (Fig. 29.11). The adductor pollicis also acts as a flexor of the thumb MP joint. Weakness of this muscle, when coupled with MP joint laxity, leads to unopposed hyperextension of the MP joint of the thumb, the so-called Jeanne's sign.

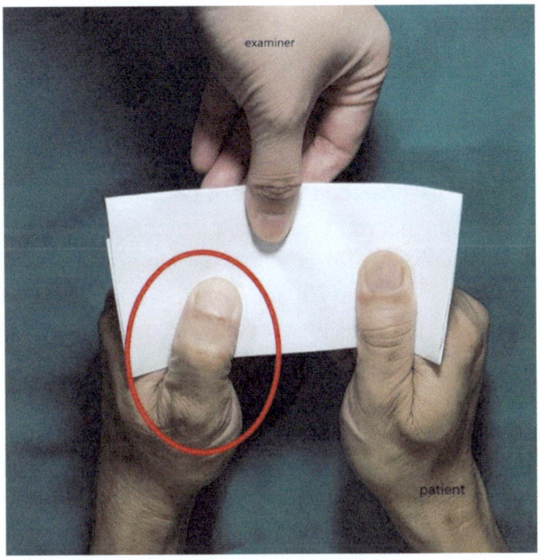

Fig. 29.11 Froment's sign. The left IP joint of the thumb flexes as it attempts to hold the paper against resistance

29.3.3.4 Provocative Tests

In the presence of cubital tunnel syndrome, maintaining elbow flexion (with forearm supinated and wrist extended) for 1–2 min will reproduce symptoms. Applying external pressure on the ulnar nerve in the cubital tunnel will also produce symptoms. Tinel sign is also frequently positive for cubital tunnel syndrome. Digital pressure on the ulnar nerve in the Guyon's canal will also reproduce symptoms in the presence of entrapment. Palpation on the Guyon's canal itself may help identify the cause of ulnar nerve entrapment. Tenderness on the hook of hamate, for example, indicates fracture, and palpable pulsating mass may indicate ulnar artery aneurysm as cause of ulnar nerve compression. Phalen test is also performed, as according to Grundberg [25], 92% patients with Guyon's canal syndrome present with positive Phalen test.

29.3.4 Pitfalls

Note that atypical findings may arise due to the presence of Riche-Cannieu or Martin-Gruber anastomosis [26]. These anastomoses may present as carpal tunnel syndrome-like symptoms or signs in ulnar nerve injury, or ulnar nerve injury-like symptoms or signs in median nerve lesion. More proximal compressions such as cervical radiculopathy and thoracic outlet syndrome, as well as motor neuron disease, must be excluded. Keep in mind that sensation of the volar medial forearm is provided by the medial antebrachial cutaneous nerve. This nerve emerges from the medial cord of the brachial plexus and receives contributions from C8 and T1. Sensory change in the distribution of this nerve is frequently a sign of neurogenic thoracic outlet syndrome. Atrophy of the APB in the midst of signs and symptoms suggestive of ulnar nerve impingement is also a sign of lower cervical radiculopathy, as APB is predominantly innervated by T1 nerve root.

29.4 Radial Nerve

29.4.1 Anatomy

Radial nerve, a terminal branch of the brachial plexus, receives nerve fibres from the C5, C6, C7, C8 and T1 nerve roots. Having emerged from the posterior cord, the radial nerve enters the posterior compartment of the arm. At the level of the axilla, the radial nerve gives off a motor branch to the long head of triceps muscle. As the radial nerve travels in the spiral groove, several branches emerge: posterior cutaneous nerve of arm, inferior lateral cutaneous nerve of arm, posterior cutaneous nerve of forearm, and motor branches to lateral head and medial head of triceps muscle. Approaching the distal third of the arm, the nerve penetrates the lateral intermuscular septum to enter the anterior compartment. In the anterior compartment of the arm, the radial nerve travels toward the lateral epicondyle between the brachioradialis and brachialis muscles. Just proximal to the lateral epicondyle, the nerve gives off branches to innervate the brachioradialis and ECRL. Then, in the cubital fossa, just proximal to the supinator muscle, the nerve divides into two main branches: the deep branch and the superficial sensory branch (Fig. 29.12).

The superficial branch travels beneath the brachioradialis muscle, accompanied by the radial artery at its medial side. Underneath this neurovascular bundle is the pronator teres tendon. At the distal third of the forearm, the nerve pierces the deep fascia of forearm and emerges at the interval between the ECRL and BR. The superficial branch of the radial nerve is a purely sensory nerve. The superficial branch of radial nerve gives rise to its first branch radially at approximately 4.5 cm proximal to the radial styloid [27]. This branch, along with the main trunk of radial sensory nerve, goes further distal to pass the extensor retinaculum of the first extensor compartment, and eventually innervates the radial part of the thumb. At the level of the radiocarpal joint, several branches come off the main trunk of radial sensory nerve to form an arborization at the level of radiocarpal joint, which cross the EPL tendon and travel in the ulnar direction toward the dorsum of the hand to innervate the radial three digits.

Meanwhile, the deep branch of the radial nerve heads enters the supinator and becomes the posterior interosseous nerve (PIN) (Fig. 29.12). The part of supinator muscle below which the PIN enters is called the arcade of Frohse. Just proximal to the arcade, the deep branch of the radial nerve gives off a motor branch to the supinator muscle, which receives fibres from mainly the C5 and C6 nerve roots. After emerging from the arcade of Frohse, PIN divides into six branches that innervate the muscles in the extensor compartment in the forearm (ECRB, EDC, EPL, EIP, EDM, ECU, APL, EPB). The fifth branch, which will eventually innervate the EPL muscle, divides into two subbranches (lateral and medial) [28]. The lateral subbranch passes deeply near the interosseous membrane. Eight centimetres proximal to the Lister's tubercle, this lateral subbranch gives off its last branch to the EPL. From that point, the subbranch becomes sensory and innervates the wrist capsule. At the wrist joint, this branch is found on the floor of the fourth extensor compartment. The PIN neurectomy, which is a pain-relieving procedure for chronic wrist pain, is performed specifically on this lateral subbranch.

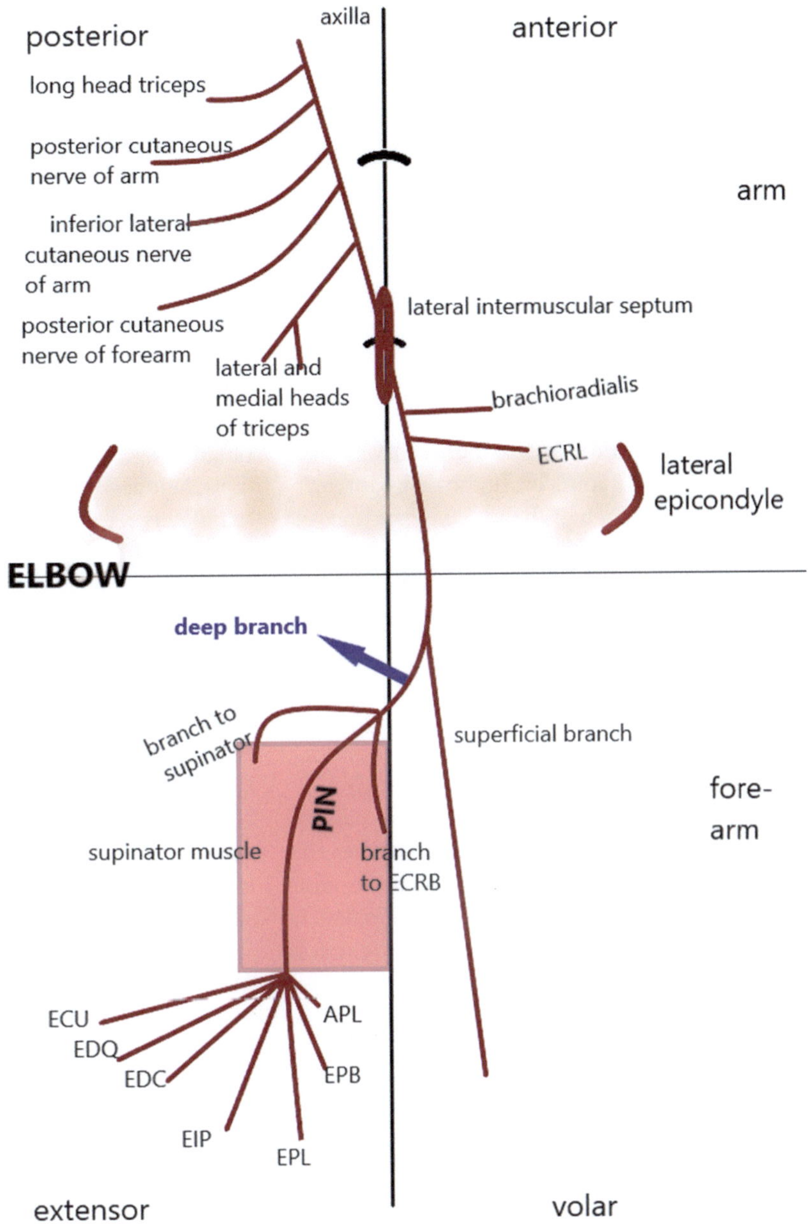

Fig. 29.12 Schematic diagram of the radial nerve

29.4.2 Sites of Injury and Compression

The radial nerve is the most commonly injured nerve in orthopaedic trauma [29]. The prevalence of radial nerve palsy in humeral shaft fracture has been reported to be as high as 12% [30]. Fracture of the middle third of the humerus and Holstein-Lewis fracture have higher association with radial nerve palsy [30]. In the former, the nerve's proximity to the bone (spiral groove) has been suggested to be the cause of injury. In Holstein-Lewis fracture, the radial nerve injury is more likely associated with fracture pattern rather than the nerve-bone proximity, particularly when the distal fragment is displaced proximally and the proximal fragment is displaced radially [31].

Injury to and procedures for the radial head or neck also put the PIN at risk for injury.

Although rare, the lateral intermuscular septum has been identified as a site of compression in several case reports [32, 33]. In one case report, the radial nerve was compressed between the brachioradialis and brachialis muscles as it just entered the anterior compartment [34]. The radial tunnel, extending from the radiocapitellar joint to the distal edge of the supinator, consists of several structures that are potential sites of compression for the PIN: fibrous bands anterior to the radiocapitellar joint, medial-proximal edge of the ECRB, arcade of Frohse, leash of Henry and the distal edge of the supinator. In the distal forearm, the superficial branch of the radial nerve may get compressed between the ECRL and the brachioradialis as they cross one another on forearm pronation (Wartenberg's syndrome).

29.4.3 Evaluation

29.4.3.1 Observation

The most noticeable sign of radial nerve palsy is drop hand, which shall be discussed in the following sections. Atrophy may not be perceptible as in median nerve palsy or ulnar nerve palsy. In the presence of high radial nerve palsy, which is proximal to the motor branches of brachioradialis and ECRL, wasting of brachioradialis may be noticeable, particularly in patients with bulky muscle mass, for example, body-builders [35].

29.4.3.2 Sensory Evaluation

Radial nerve sensory distribution in the upper extremity includes the lateral arm distal to the insertion of deltoid muscle (inferior lateral cutaneous nerve of the arm), posterior aspect of the arm (posterior cutaneous nerve of arm), mid-posterior aspect of the forearm (posterior cutaneous nerve of forearm) and the dorsal hand and dorsal radial fingers (superficial branch of radial nerve). The lower lateral cutaneous nerve of arm and posterior cutaneous nerve of forearm exit the radial nerve high in the arm. In such high radial nerve palsy, decreased sensation may be found in the areas supplied by these branches. In more distal radial nerve palsy, these areas are frequently spared.

The superficial branch of radial nerve provides sensory innervation to the dorso-radial surface of the hand and the dorsal surface of the radial 2 ½ digits [36], or the radial 3 digits [37]. Because of the variation of branching pattern, it is most reliable to assess sensibility in area that does not have overlapping innervations between the ulnar and radial nerves. The dorsal aspect of the thumb may receive contributions from both the lateral antebrachial cutaneous nerve and the radial nerve. The first dorsal webspace sensory innervation, however, is exclusively from the radial nerve.

It might be useful to test sensation using different Semmes Weinstein monofilament. Bertelli et al. found that as the radial nerve was recovering from injury, there is a difference between the perception of 2.0 and 0.05 monofilaments. Perception of 2.0 monofilament may be normal, but perception of 0.05 monofilament is impaired [38].

In Wartenberg's syndrome, the Tinel sign is elicited by tapping on approximately 9 cm proximal to the radial styloid, where the nerve becomes subcutaneous [39].

29.4.3.3 Motor Evaluation

Motor evaluation is important to determine the level and severity of injury, as well as to predict recovery. In high radial nerve palsy, the injury occurs proximal to the motor branches to the brachioradialis and ECRL. The wrist thus loses all extensors, resulting in wrist drop. When the injury is distal to the motor branch to ECRL but proximal to the origin of the PIN, ECRL function, preserved, will pull the wrist into radial deviation as the patient attempts to actively extend the wrist.

Loss of EDC function in radial nerve or PIN injury results in inability to actively extend the MP joint. Active extension of the IP joint is preserved since it is a function of the intrinsic muscles. It is therefore important to distinguish between MP extension and IP extension. To eval-

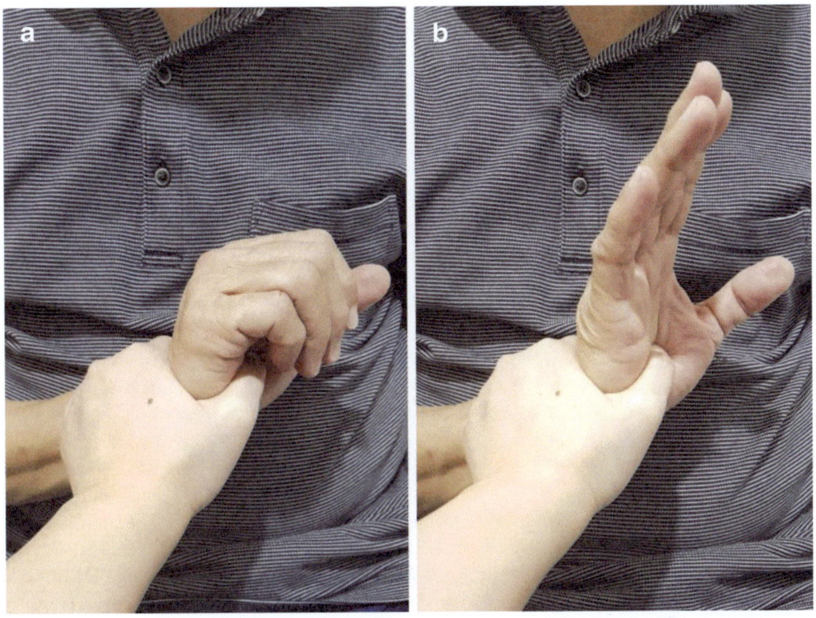

Fig. 29.13 To test for radial nerve or PIN injury, the wrist is held in extension by the examiner (**a**), and the patient is asked to extend the MP joints (**b**). These photographs show normal radial nerve function

uate the ability of active extension of the MP joint, the examiner holds the patient's wrist in extension and asks the patient to extend the fingers (Fig. 29.13a and b). By holding the patient's wrist in extension, the MP joints are flexed by tenodesis effect. Inability to extend the MP joint will therefore be more obvious. Should the patient's wrist be held in neutral or flexed position, tenodesis effect will pull the MP joint into some degree of extension and obscure the physical findings. Inability of the thumb to extend ("hitchhike" position) results from functional loss of thumb abductor and extensors. PIN syndrome may present with a variety of pattern of muscle weakness. Weakness of the ECU, EDM and EDC signifies involvement of the medial branch. Weakness of the APL, EPB, EPL and EIP ensues when the lateral branch is affected. When all of the muscles aforementioned are weak, the entrapment is proximal to the bifurcation.

29.4.3.4 Provocative Tests

Several provocative tests may be used to confirm the diagnosis of compression syndromes. In radial tunnel syndrome, wrist extension against resistance will elicit pain over the radial tunnel. Pain during resisted extension of middle finger is a typical sign of radial tunnel syndrome.

29.4.4 Pitfalls

It is important to distinguish between radial tunnel syndrome from PIN syndrome. The former does not present with motor dysfunction, while in the latter, motor dysfunction is the hallmark. The radial tunnel syndrome as a distinct clinical entity has been a subject of debate. Unlike other entrapment neuropathies, the pathogenesis of radial tunnel syndrome is unclear. Some argue that compression of the PIN is not severe enough to cause motor weakness, but enough to cause irritation that is perceived as pain [40]. It is also important to distinguish radial tunnel syndrome from lateral epicondylitis, as the two entities present with pain around the lateral epicondyle. The pain associated with radial tunnel syndrome, however, is 5 cm distal to the lateral epicondyle [41].

References

1. Caetano EB, Sabongi JJN, Vieira LA, Caetano MF, de Bona JE, Simonatto TM. Struthers' ligament and supracondylar humeral process: An anatomical study and clinical implications. Acta Ortop Bras. 2017;25(4):137–42.
2. Caetano EB, Vieira LA, Sabongi Neto JJ, Caetano MBF, Sabongi RG. Anterior interosseous nerve:

anatomical study and clinical implications. Rev Bras Ortop. 2018;53(5):575–81.
3. Davidge KM, Sammer DM. Median nerve entrapment and injury. In: Mackinnon SE, Yee A, editors. Nerve surgery. New York: Thieme Medical Publishers, Inc.; 2015. p. 207–50.
4. Shimizu K, Iwasaki R, Hoshikawa H, Yamamuro T. Entrapment neuropathy of the palmar cutaneous branch of the median nerve by the fascia of flexor digitorum superficialis. J Hand Surg Am. 1988;13(4):581–3.
5. Duncan GJ, Yospur G, Gomez-Garcia A, Lesavoy MA. Entrapment of the palmar cutaneous branch of the median nerve by a normal palmaris longus tendon. Ann Plast Surg. 1995;35(5):534–6.
6. Semer N, Crimmins C, Jones NF. Compression neuropathy of the palmar cutaneous branch of the median nerve by the antebrachial fascia. J Hand Surg Br. 1996;21(5):666–7.
7. Imai T, Wada T, Matsumoto H. Entrapment neuropathy of the palmar cutaneous branch of the median nerve in carpal tunnel syndrome. Clin Neurophysiol. 2004;115(11):2514–7.
8. al-Qattan MM, Robertson GA. Entrapment neuropathy of the palmar cutaneous nerve within its tunnel. J Hand Surg Br. 1993;18(4):465–6.
9. Novak CB. Evaluation of the patient with nerve injury or nerve compression. In: Mackinnon SE, Yee A, editors. Nerve surgery. New York: Thieme Medical Publishers, Inc.; 2015. p. 41–52.
10. Koo JP, Kim SH, An HJ, Moon OG, Choi JH, Yun YD, et al. Two-point discrimination of the upper extremities of healthy Koreans in their 20's. J Phys Ther Sci. 2016;28(3):870–4.
11. Boesch CE, Medved F, Held M, Bender D, Schaller HE, Fuchsberger T. Analysis of the two-point discrimination test in daily routine practice. Eur J Plast Surg. 2017;40(4):333–6.
12. Cheng CJ, Mackinnon-Patterson B, Beck JL, Mackinnon SE. Scratch collapse test for evaluation of carpal and cubital tunnel syndrome. J Hand Surg Am. 2008;33(9):1518–24. https://doi.org/10.1016/j.jhsa.2008.05.022.
13. Lundborg G, Dahlin LB. Anatomy, function, and pathophysiology of peripheral nerves and nerve compression. Hand Clin. 1996;12(2):185–93.
14. Won HS, Han SH, Oh CS, Chung IH, Kim SM, Lim SY. Topographic relationship between the medial intermuscular septum and the ulnar nerve in the upper arm. J Neurosurg. 2011;114(6):1534–7.
15. Caetano EB, Sabongi Neto JJ, Vieira LA, Caetano MF. The arcade of Struthers: an anatomical study and clinical implications. Rev Bras Ortop. 2017;52(3):331–6.
16. Andrews K, Rowland A, Pranjal A, Ebraheim N. Cubital tunnel syndrome: Anatomy, clinical presentation, and management. J Orthop. 2018;15(3):832–6.
17. Davidge KM, Boyd KU. Ulnar nerve entrapment and injury. In: Mackinnon SE, Yee A, editors. Nerve surgery. 1st ed. New York: Thieme Medical Publishers, Inc.; 2015. p. 251–88.
18. Tubbs RS, Rogers JM, Loukas M, Comert A, Shoja MM, Cohen-Gadol AA. Anatomy of the palmar branch of the ulnar nerve: application to ulnar and median nerve decompressive surgery. J Neurosurg. 2011;114(1):263–7.
19. Martin CH, Seiler JG 3rd, Lesesne JS. The cutaneous innervation of the palm: an anatomic study of the ulnar and median nerves. J Hand Surg Am. 1996;21(4):634–8.
20. Nakajima M, Ono N, Kojima T, Kusunose K. Ulnar entrapment neuropathy along the medial intermuscular septum in the midarm. Muscle Nerve. 2009;39(5):707–10.
21. Vogel RB, Nossaman BC, Rayan GM. Revision anterior submuscular transposition of the ulnar nerve for failed subcutaneous transposition. Br J Plast Surg. 2004;57(4):311–6.
22. Ruchelsman DE, Lee SK, Posner MA. Failed surgery for ulnar nerve compression at the elbow. Hand Clin. 2007;23(3):359–71. vi-vii
23. Coraci D, Loreti C, Piccinini G, Doneddu PE, Biscotti S, Padua L. Ulnar neuropathy at wrist: entrapment at a very "congested" site. Neurol Sci. 2018;39(8):1325–31.
24. Woo A, Bakri K, Moran SL. Management of ulnar nerve injuries. J Hand Surg Am. 2015;40(1): 173–81.
25. Grundberg AB. Ulnar tunnel syndrome. J Hand Surg Br. 1984;9(1):72–4.
26. Smith JL, Siddiqui SA, Ebraheim NA. Comprehensive summary of anastomoses between the median and ulnar nerves in the forearm and hand. J Hand Microsurg. 2019;11(1):1–5.
27. Folberg CR, Ulson H Jr, Scheidt RB. The superficial branch of the radial nerve: a morphologic study. Rev Bras Ortop. 2009;44(1):69–74.
28. Elgafy H, Ebraheim NA, Yeasting RA. The anatomy of the posterior interosseous nerve as a graft. J Hand Surg Am. 2000;25(5):930–5.
29. Chaudhry S, Ipaktchi KR, Ignatiuk A. Updates on and controversies related to Management of Radial Nerve Injuries. J Am Acad Orthop Surg. 2019;27(6):e280–e4.
30. Shao YC, Harwood P, Grotz MR, Limb D, Giannoudis PV. Radial nerve palsy associated with fractures of the shaft of the humerus: a systematic review. J Bone Joint Surg Br. 2005;87(12):1647–52.
31. Ljungquist KL, Martineau P, Allan C. Radial nerve injuries. J Hand Surg Am. 2015;40(1):166–72.
32. Bowman J, Curnutte B, Andrews K, Stirton J, Ebraheim N, Mustapha AA. Lateral intermuscular septum as cause of radial nerve compression: case report and review of the literature. J Surg Case Rep. 2018;2018(8):rjy226.

33. Chesser TJ, Leslie IJ. Radial nerve entrapment by the lateral intermuscular septum after trauma. J Orthop Trauma. 2000;14(1):65–6.
34. Cherchel A, Zirak C, De Mey A. The humeral origin of the brachioradialis muscle: an unusual site of high radial nerve compression. J Plast Reconstr Aesthet Surg. 2013;66(11):e325–7.
35. Ng AB, Borhan J, Ashton HR, Misra AN, Redfern DR. Radial nerve palsy in an elite bodybuilder. Br J Sports Med. 2003;37(2):185–6.
36. Sulaiman S, Soames R, Lamb C. The sensory distribution in the dorsum of the hand: anatomical study with clinical implications. Surg Radiol Anat. 2015;37(7):779–85.
37. Mok D, Nikolis A, Harris PG. The cutaneous innervation of the dorsal hand: detailed anatomy with clinical implications. J Hand Surg Am. 2006;31(4):565–74.
38. Bertelli JA, Cavalli E, Mendes Lehn VL, Ghizoni MF. Sensory deficits after a radial nerve injury. Microsurgery. 2018;38(2):151–6.
39. Spies CK, Unglaub F, Muller LP, Hahn P, Low S, Oppermann J. Endoscopically assisted release of the superficial radial nerve. Arch Orthop Trauma Surg. 2015;135(5):737–41.
40. Strohl AB, Zelouf DS. Ulnar tunnel syndrome, radial tunnel syndrome, anterior interosseous nerve syndrome, and pronator syndrome. Instr Course Lect. 2017;66:153–62.
41. Moradi A, Ebrahimzadeh MH, Jupiter JB. Radial tunnel syndrome, diagnostic and treatment dilemma. Arch Bone Jt Surg. 2015;3(3):156–62.

30 Evaluation of Tendinopathies/ Tendon Ruptures/Tendon Instability

Margaret Woon Man Fok

30.1 Introduction

Tendon pathologies are common in hand and wrist. They affect all ages and can be related to various causes, including trauma, overuse, inflammation and degeneration. They can be classified based on the chronicity of the presentation, of which they can present in an acute setting like tendon rupture or in a subacute/chronic manner like tendon subluxation and tendinopathy. Prior to the discussion on how to evaluate each individual condition, it is important to have a comprehensive understanding of the anatomy of the tendons around hand and wrist.

30.2 Anatomy

The main tendons which are involved in tendon pathologies in the hand and wrist are wrist extensors and finger extensors and flexors. Most of these tendons arise from the elbow (the origin of common flexor tendons at the medial epicondyle and the origin of common extensor tendons at the lateral epicondyle) or proximal ulnar/radius, before crossing the wrist joint and insert into the hand.

M. W. M. Fok (✉)
Department of Orthopaedics and Traumatology, Queen Mary Hospital, The University of Hong Kong, Hong Kong, China

At the wrist level, the two wrist flexors, flexor carpi radialis (FCR), and flexor carpi ulnaris (FCU) flank the sides of the nine finger flexors (flexor digitorum superficialis (FDS) and flexor digitorum profundus (FDP) of index finger, middle finger, ring finger, and little finger, and flexor pollicis longus (FPL)), of which all these nine tendons transverse through the carpal tunnel before inserting into the finger phalanges. In the cross section of the carpal tunnel, FDS of middle and ring finger lie in the most superficial layer, followed by FDS of index and little finger. The deeper layer lies FPL and FDP tendons. For the extensors, they are arranged in six compartments in the following order (from radial to ulnar):

- First compartment: abductor pollicis longus (APL) (the most radial tendon) and extensor pollicis brevis (EPB)
- Second compartment: extensor carpi radialis longus (ECRL) and extensor carpi radialis brevis (ECRB)
- Third compartment: extensor pollicis longus (EPL)
- Fourth compartment: extensor indicis (EI) and extensor digitorum communis (EDC) of index finger, middle finger, ring finger and little finger
- Fifth compartment: extensor digitorum minimis (EDM)
- Sixth compartment: extensor carpi ulnaris (ECU)

J. G. Lane et al. (eds.), *The Art of the Musculoskeletal Physical Exam*,
https://doi.org/10.1007/978-3-031-24404-9_30

At the finger level, the arrangement of the tendons is more complex. EI (ulnar aspect) and EDC of index finger join together at the MCP level to form the common extensor of index finger while EDM (ulnar aspect) and EDC of little finger join to form the common extensor of little finger in a similar fashion. Each common extensor of the finger splits into three bands at the proximal phalangeal (PP) level. While the central band continues distally to become the central slip, the two lateral bands, receive their contribution of the corresponding intrinsic muscles, travel along the dorsal aspect of the PIPJ and MP, and later join together to become the terminal extensor tendon at the base of DP. In addition, apart from the thumb, all common extensors of the fingers interlink with each other by the sagittal band at the metacarpal phalangeal joint (MCPJ) level. For the flexors, the FDS and FDP of each finger share the same tendon sheath. As the name implies, FDS is more superficial to FDP tendon till it divides into two slips at the PP, forming the Camper's chiasma, before inserting into the base of the middle phalanx (MP). Meanwhile the FDP continues its path and inserts itself into the base of distal phalanx (DP). It is important to recognize that the flexor and extensor tendons of each finger link with each other via the lumbricals. Thus, a rupture of any part of the tendon in the finger may tip the balance and cause compensatory deformity, e.g. swan neck deformity or boutonniere deformity.

30.3 Tendon Ruptures

30.3.1 General Examination

Unless there is a direct traumatic cut, closed tendon ruptures of the hands and wrist usually involve the finger extensor and flexors. They can occur at either the finger or wrist level.

The loss of finger cascade in the resting position gives the first hint of a rupture tendon (Fig. 30.1). Due to the unopposed force of action, the affected finger flexes more than the other fingers if its extensor tendon is ruptured. Similarly, it extends more if the flexor is not intact.

In addition, two simple manoeuvres can be performed for confirmation:

Tenodesis effect:

In normal circumstances, fingers are in extended position when the wrist is flexed and vice versa (Fig. 30.2). Thus, if there is a tendon rupture,

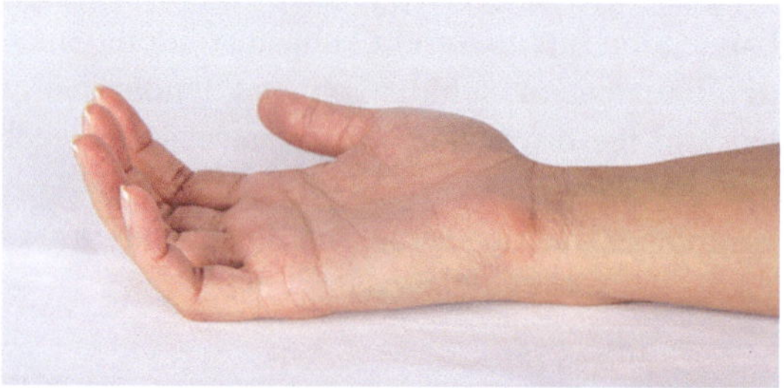

Fig. 30.1 Normal cascade of fingers in resting position (Copyright Dr. Margaret Fok)

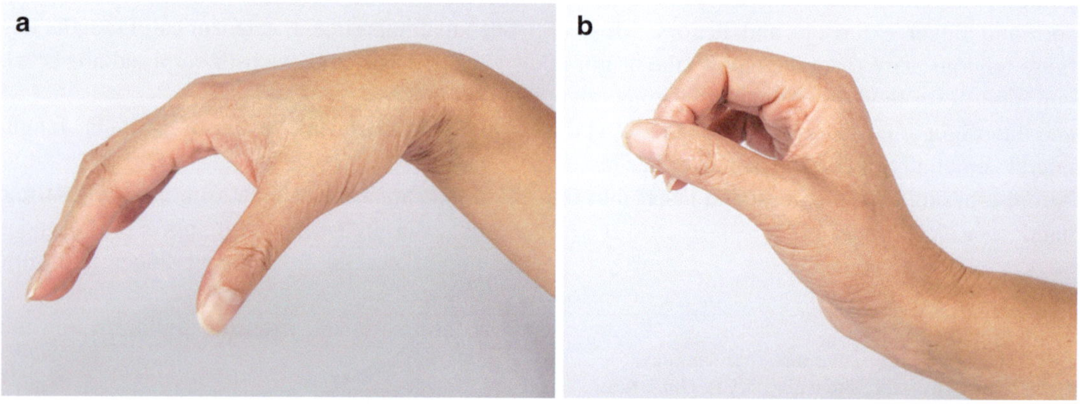

Fig. 30.2 Tenodesis effect (Copyright Dr. Margaret Fok). (**a**) Fingers are extended when the wrist is in flexion. (**b**) Fingers are flexed when the wrist is in extension

the affected finger will remain stationary during wrist movement (i.e. loss of tenodesis effect).

- in EPL rupture during a closed non-displaced distal radius fracture, the interphalangeal joint (IPJ) of the thumb remains in a flexed position on wrist movement.
- In EDC rupture of a particular finger, e.g. in Vaughan-Jackson syndrome i.e. rheumatoid wrist arthritis with distal radioulnar joint (DRUJ) instability, the affected finger fails to extend (especially the MCPJ)) when the wrist is moved from extended to flexed position.
- For flexor tendon rupture, the affected finger fails to flex its interphalangeal joints (IPJ) when the wrist is moved from flexed to extended position. Yet, as there are two flexor tendons for each finger, i.e. FDP and FDS, apart from the thumb, the failure to flex the finger may be less obvious, especially if FDP is intact.

The muscles of the finger flexors occupy the volar ulnar aspect of the proximal forearm and the muscles of the finger extensors occupy the dorsal radial aspect of the proximal forearm. In normal circumstances when all tendons are intact, fingers will flex when the volar forearm muscle bulk is being squeezed. Likewise, fingers will extend when the dorsal forearm muscle bulk is being gripped.

30.3.2 Finger Extensors [1]

Although the name EDC denotes finger extensors, extension of the distal interphalangeal joint (DIPJ) and proximal interphalangeal joint (PIPJ) of a finger can result from the action of the oblique retinacular ligament (flexing PIPJ while extending DIPJ) and intrinsic muscles of the hand (flexing MCPJ while extending PIPJ), respectively. Yet, the extension of MCPJ is purely acted by EDC tendon, pulling on the sagittal band. Thus, in situations in which the **EDC is ruptured**, DIPJ and PIPJ of a finger can still achieve extension. In order to evaluate whether EDC of a particular finger is ruptured, especially at the wrist level, the examiner must assess the extension of the MCPJ, but not DIPJ and PIPJ. The examiner can test the integrity of the common extensor of a finger by asking the patient to flex both DIPJ and PIPJ while trying to extend the MCPJ (Fig. 30.3).

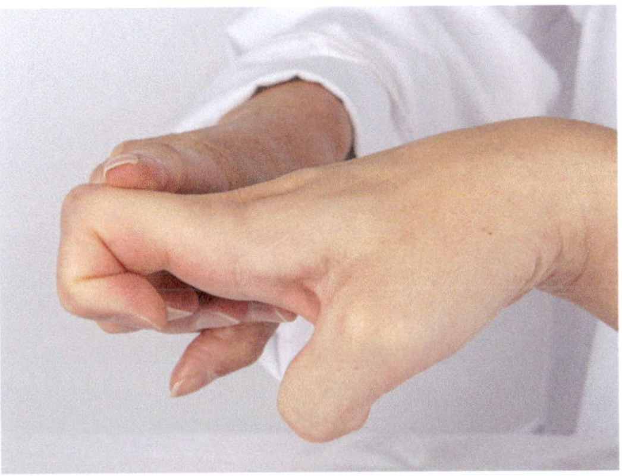

Fig. 30.3 Testing the integrity of the extensor digitorum communis (EDC): (Copyright Dr. Margaret Fok). With an intact EDC, the patient can resist downward force by keeping his metacarpal phalangeal joints in extension while flexing all his interphalangeal joints

Both the index and little fingers have two finger extensors, i.e. extensor indicis (EI) and EDC of index finger and extensor digitorum minimis (EDM) and EDC of little finger. EI and EDM enable the corresponding two fingers to extend individually (Fig. 30.4). If the **EDM is ruptured** only, e.g. with intact EDC of the little finger in a patient with Vaughan-Jackson Syndrome of rheumatoid wrist, the patient can still extend his little finger together with the other fingers but not by itself.

In addition to the index and little fingers, the thumb also has two extensors, namely extensor pollicis longus (EPL) and extensor pollicis brevis (EPB). EPB inserts into the base of proximal phalanx of the thumb and is responsible for the extension of the first MCPJ. Meanwhile EPL inserts into the base of DP of the thumb. It acts on the IPJ leading to IPJ extension. In situations in which the **EPL is ruptured** in a non-displaced distal radius fracture, [2] the examiner can examine the integrity of the EPL tendon by asking the patient to lift the thumb off a table (Fig. 30.5) or by actively extending the IPJ of the thumb.

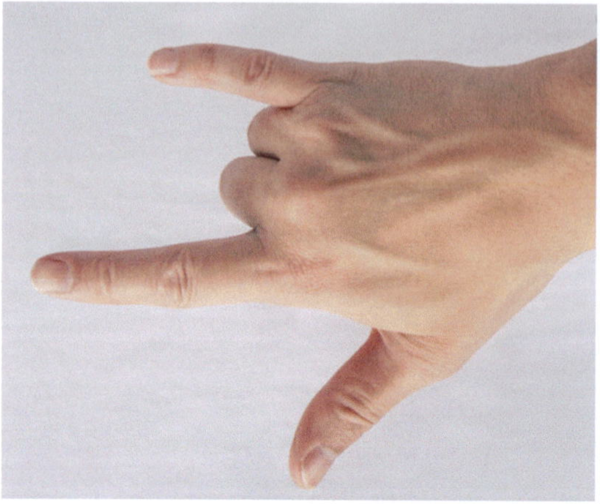

Fig. 30.4 Extensor indicis (EI) and Extensor digitorum minimi (EDM) in action: (Copyright Dr. Margaret Fok). Index finger and little finger can be extended individually due to the presence of EI and EDM

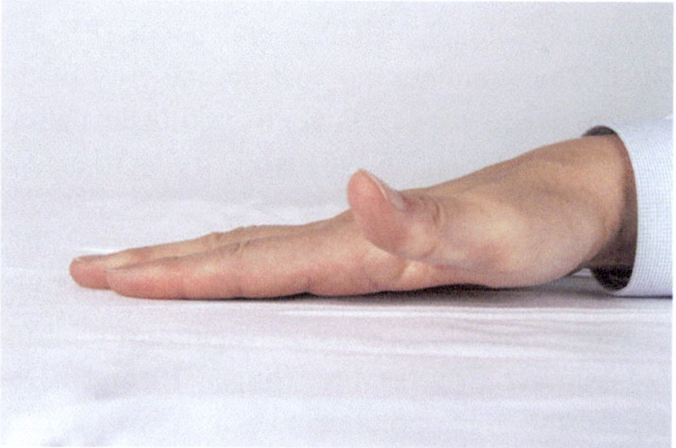

Fig. 30.5 Lift-off test for extensor pollicis longus (EPL) integrity: (Copyright Dr. Margaret Fok). The thumb can be lift off the table with an intact EPL

The extensor tendon apparatus of a finger is very complex. (refer to section: anatomy) **Central slip rupture** of a finger tips the balance of the extensor tendon apparatus. The action of the lumbricals becomes unopposed, pulling the lateral bands to migrate volarly. As a result, the finger may adopt a boutonniere deformity, i.e. flexion of the PIPJ and hyperextension of the DIPJ. **Elson test** [3] is used to confirm the rupture of a central slip rupture (Fig. 30.6). The patient is asked to bend the PIPJ of the questioned

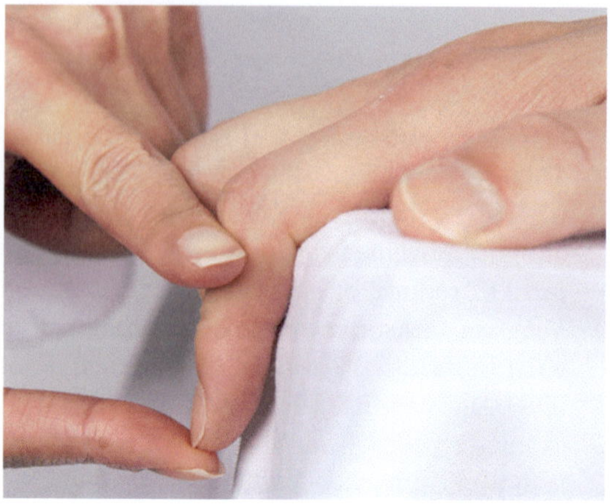

Fig. 30.6 Elson test to check the integrity of central slip (Copyright Dr. Margaret Fok). The proximal phalangeal joint of a questioned finger is flexed to 90 degrees at the edge of a table. The patient is asked by the examiner to extend the finger against resistance. If the central slip is intact, the distal interphalangeal joint (DIPJ) of the questioned finger is subtle. If there is a central slip rupture, the DIPJ of the questioned finger will be held in an extended position

finger 90^0 over the edge of a table. He is then asked to extend the finger against resistance. In the presence of central slip rupture, the PIPJ extension is weak and the DIPJ becomes rigid. In the absence of central slip injury, the lateral bands are not being activated and the DIPJ remains floppy.

Mallet finger [4] is a common sports injury, especially in ball games. It is due to the rupture of the terminal extensor tendon insertion or a corresponding avulsion fracture at the base of the DP (i.e. bony mallet). In a pure tendinous mallet finger, the patient complains of an inability to extend the DIPJ. If left untreated, a compensatory swan neck deformity (i.e. flexion of the DIPJ and hyperextension of the PIPJ) may be resulted.

30.3.3 Finger Flexors

Finger flexor injury is less common than extensor injury. The FDS is responsible for—flexion of

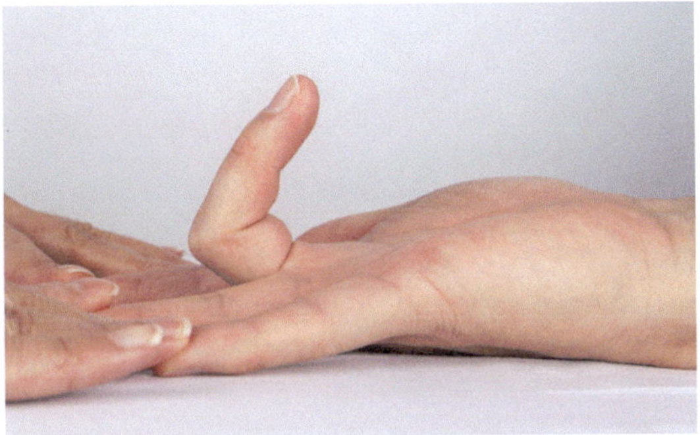

Fig. 30.7 Manoeuvre to check the integrity of flexor digitorum superficialis (FDS) tendon of a middle finger (Copyright Dr. Margaret Fok). All FDS have individual muscle belly while flexor digitorum profundus (FDP) of middle finger, ring finger and little finger share a common muscle belly. By holding all the other fingers in the extended position (eliminating the action of FDP), the ability of flexing the PIPJ of the middle finger denotes the integrity of the FDS of the middle finger

PIPJ while the FDP can act on both PIPJ and DIPJ. Each FDS tendon has a separate muscle belly while the FDP tendons of middle finger, right finger and little finger share a common muscle belly. As a result, to identify the presence of a rupture of the FDS tendon of a finger, the examiner needs to immobilize all the other fingers, in order to eliminate the mass action of FDP, before asking the patient to flex the questioned finger (Fig. 30.7) [1]. It is important to note that up to 42% of the population may not have an individual action for the FDS for their little fingers [5, 6]. The muscle of the FDS may be linked to the FDS of the adjacent ring finger (i.e. 21%) [6] or connected to the adjacent ring finger or be connected to the muscle belly of the index finger. Since this situation is usually symmetrical i.e. 74%) [6], in a case in which the FDS of the little finger is found to be absent, it is advisable to check the unaffected side to determine whether there is a rupture.

FDP tendon rupture is usually an avulsion at the insertion site, which is known as **jersey finger** [4, 7]. It results from the finger of a player being forcefully extended while in an actively flexed position, e.g. when the player is trying to grasp an opposition player's jersey. Similar to mallet finger, it can be an avulsion fracture instead of a pure tendon injury. Patient presents with a stiff swollen finger and has difficulty in flexing the DIPJ. Physical examination involves asking the patient to flex the DIP of the questioned finger while keeping the PIPJ in extension. Depending on the integrity of the vascular supply of the tendon during the injury (Fig. 30.8), the FDP stump may retract to the palm and be identified as a mass.

FPL tendon rupture is not common. The most known cause nowadays is attrition injury at the wrist level due to the prominence of the volar plate during the distal radius fracture fixation [8]. Apart from a loss of tenodesis effect, the patient is unable to flex the IPJ of the thumb.

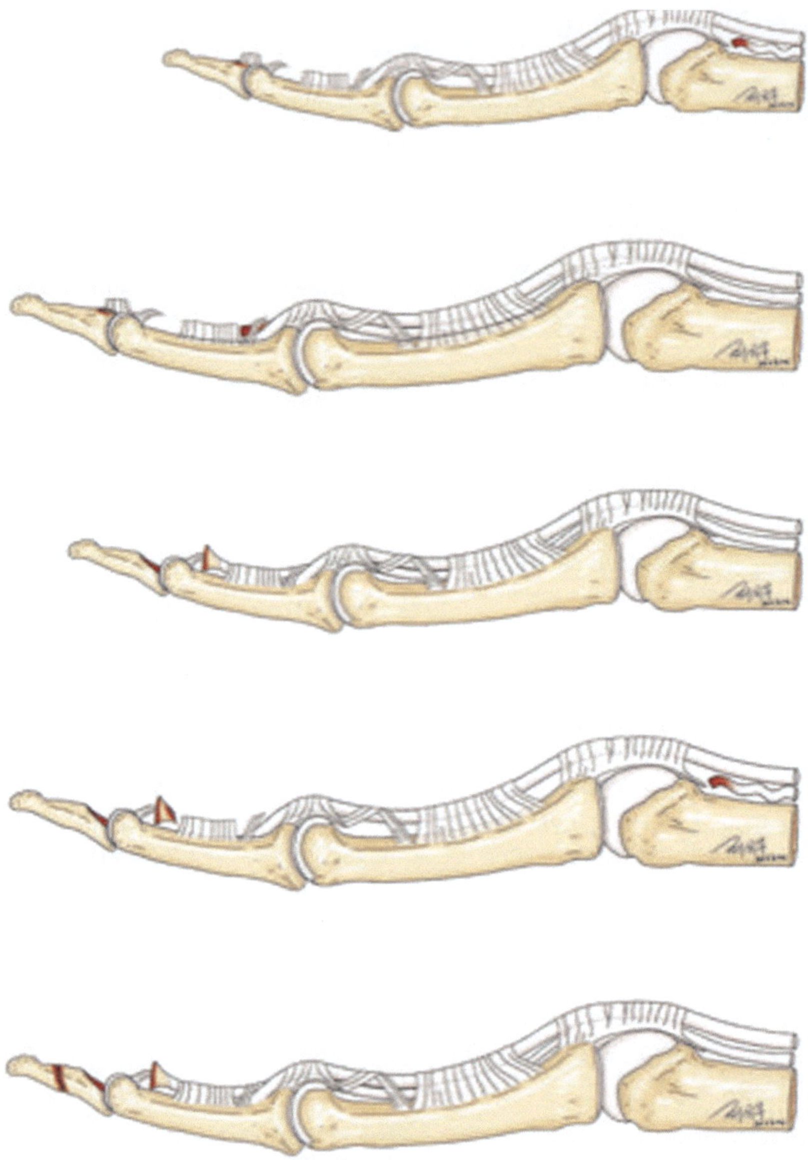

Fig. 30.8 Jersey Finger classification [7, 15, 16] (Copyright Dr. Margaret Fok). The retraction of the stump of the FDP is due to the point of failure and the integrity of the vincular system

30.4 Tendon Subluxation/ Dislocation

Apart from tendon rupture, a traumatic event may give rise to tendon subluxation. Unlike patients with tendon ruptures, they are often misdiagnosed as minor sprains and usually present late. The most classic conditions are ECU subluxation at the wrist level or extensor subluxation due to the rupture of the sagittal band at the MCPJ level.

30.4.1 ECU Tendon Subluxation/ Dislocation

ECU tendon occupies the sixth extensor compartment and is the most ulnar extensor tendon of the wrist. **ECU subluxation/dislocation** is a result of a traumatic event with forced supination, palmar flexion and ulnar deviation [4, 9]. Patient typically presents with dorsal ulnar wrist pain and complains of clicking and snapping in pronation and supination. Tenderness and swell-

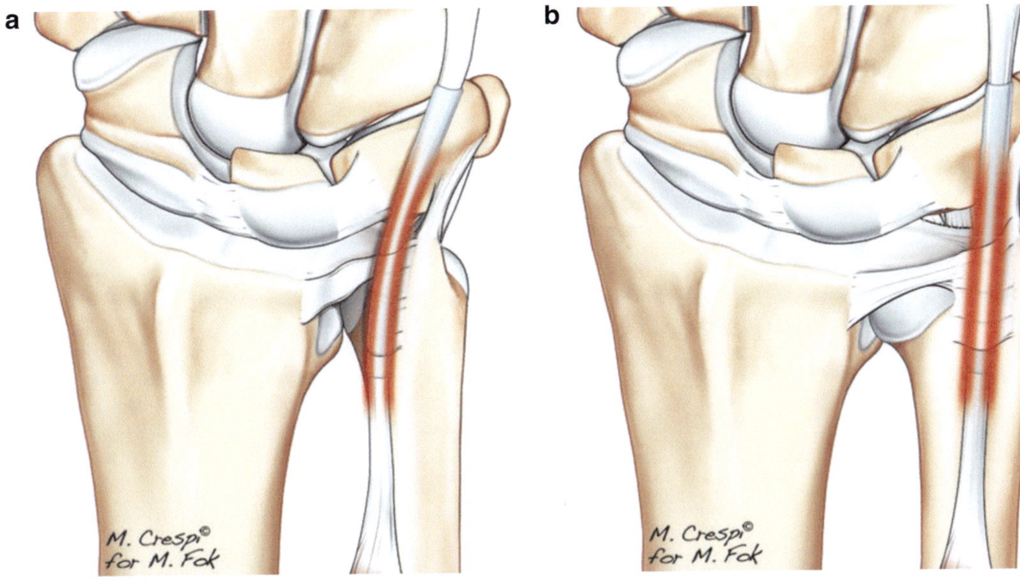

Fig. 30.9 Extensor carpi ulnaris (**a**) (ECU) subluxation during forearm supination (Copyright Dr. Margaret Fok). (**b**) ECU tendon is reduced when forearm is in pronation (Copyright Dr. Margaret Fok)

ing may be noted along the ECU tendon at the ulnar head. Subluxation/complete dislocation of the tendon can be reproduced by active forearm supination and ulnar deviation (Fig. 30.9). The tendon is reduced with the forearm in pronation. It is important to differentiate ECU subluxation/dislocation with ECU tendonitis (refer to section: ECU Tendonitis).

30.4.2 Extensor Tendon Subluxation

Subluxation of the extensor tendon at the MCPJ can be caused by direct trauma, forced flexion or resisted extension of the MCPJ. It is a result of the **rupture of the sagittal band**, the primary stabilizer of the extensor tendon at the MCPJ [10]. The patient often complains of a snapping sensation in the MCPJ of the affected finger during finger movement. At times, he may mistakenly regard himself as suffering from trigger finger due to the "clicking" sensation. (refer to section: Trigger Finger) In severe and chronic cases, the extensor tendon may not reduce to its natural position and the patient may complain of a loss of active extension of the affected finger. Subluxation of the tendon on active finger flexion is usually observed. In the passively extended finger position, the tendon can be maintained in its course. As the radial sagittal band is longer and thinner than the ulnar counterpart, the rupture usually occurs on the radial side, resulting in an ulnar subluxation of the tendon (Fig. 30.10).

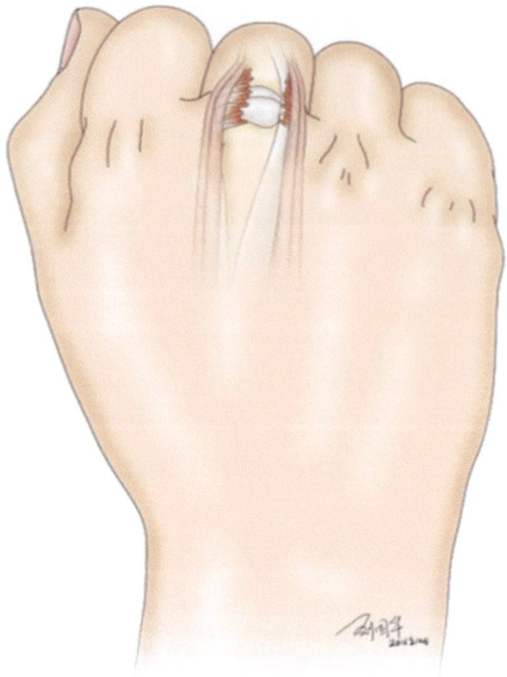

Fig. 30.10 Extensor tendon ulnar subluxation due to the rupture of radial sagittal band of middle finger (Copyright Dr. Margaret Fok)

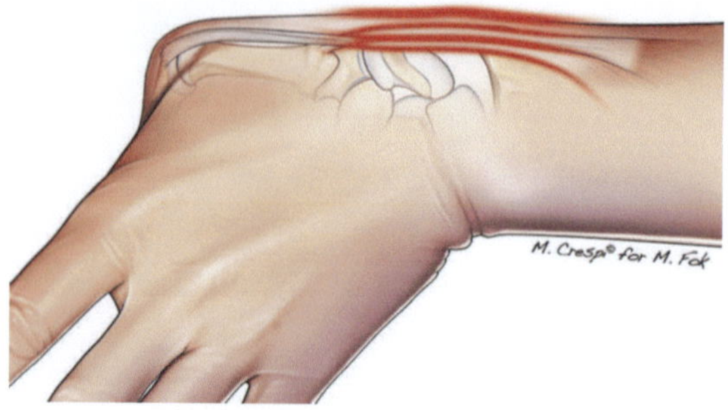

Fig. 30.11 Finkelstein test for DQV (Copyright Dr. Margaret Fok). The fisted hand (with thumb in palm) is taken from radial to ulnar direction. Pain can be felt at the first extensor compartment

30.5 Chronic Tendon Pathologies

The most common chronic tendon pathologies in the hand and wrist are stenosing tenosynovitis and tendonitis. While the causes for stenosing tenosynovitis and tendonitis are generally multifactorial, repetitive use of the hand and wrist usually plays a significant part in their pathogenesis . De Quervain's disease (DQV) and extensor carpi ulnaris (ECU) tendonitis occur in the wrist while trigger finger occurs in the finger.

30.5.1 de Quervain's Disease

This is stenosing tenosynovitis of the first extensor compartment of the wrist, i.e. APL and EPB tendons [11]. It is due to repetitive use of the thumb in activities which involve radioulnar deviation like hammering and climbing. The patient typically complains of pain over the dorsal radial aspect of the wrist. Tenderness and swelling along the first extensor compartment can be elicited. **Finkelstein test** is a specific test for DQV [12]. Pain is reproduced by stretching the affected tendons. The patient is asked to put his thumb in his palm before closing his fist. By taking the wrist from radial to ulnar deviation, the patient will experience pain (Fig. 30.11).

DQV is often been confused with **intersection syndrome** [13] and first carpal metacarpal joint (CMCJ) osteoarthritis. The former is synovitis due to inflammation at the crossing point of first (APL and EPB) and second extensor compartments (ECRL and ECRB). It is due to repetitive wrist extension and is common in rowers or weightlifters. The swelling and pain are typically more proximal than DQV, approximately 5 cm proximal to the wrist joint. Pain and crepitations can be elicited over the area with resisted wrist extension and thumb extension. Meanwhile first CMCJ osteoarthritis is a degenerative condition at the base of the thumb. There is localized pain and tenderness at the first CMCJ. **Finklestein test** may be positive in first CMC osteoarthritis. Pain can be elicited in first CMCJ osteoarthritis, but it is not usually present in DQV when the thumb is not flexed into the palm.-.

30.5.2 ECU Tendonitis

ECU tendonitis is a chronic inflammation of the ECU tendon [13]. It is a result from chronic loading of the tendon by performing repetitive wrist

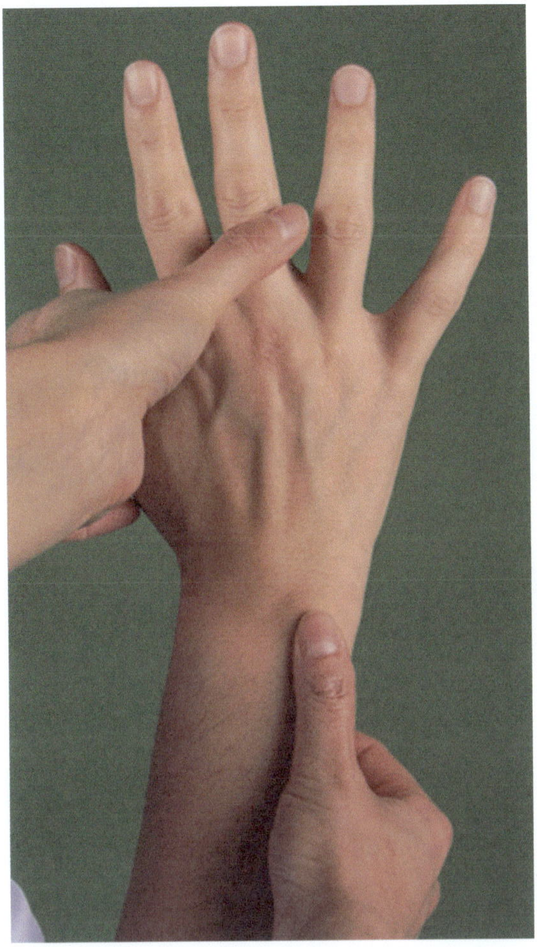

Fig. 30.12 ECU synergy test (Copyright Dr. Margaret Fok). Pain is noted at the ECU tendon when the examiner grasps the patient's extended thumb and middle finger while asking him to radially deviate his thumb against resistance

flexion and extension, particularly with forearm in supination. Patient typically complains of swelling and dull ache on the dorsa-ulnar aspect of the wrist. Tenderness can be felt along the ECU tendon and searing pain may be elicited along the tendon on resisted wrist extension. **ECU synergy test** is both a sensitive and specific test for ECU tendonitis [14]. It is performed with the wrist in supination. Patient experiences pain over the ECU tendon when the examiner grasps the extended thumb and middle finger of the affected hand while asking him to radially deviate his thumb against resistance (Fig. 30.12).

In cases of ECU tendonitis, it is important to assess the stability of the ECU tendon (refer to section: ECU Subluxation/Dislocation) and distal radioulnar joint (DRUJ). All three conditions may present with ulnar wrist pain in a similar fashion and can co-exist concomitantly.

30.5.3 Trigger Finger

Trigger finger is stenosing tenosynovitis of the finger flexors [13]. It is caused by the thickening of the A1 pulley (i.e. just volar to the MCPJ), resulting in triggering when flexor tendons try to glide within the narrowed and thickened tendon sheath. The patient typically complains of clicking, catching or locking of the finger during finger movement. Tenderness, with or without a nodule may be noted over the A1 pulley. Triggering can be felt when the patient attempts to extend the finger from a flexed position. In severe cases, PIPJ contracture may present as the patient is chronically reluctant to move the affected finger due to the triggering pain.

30.6 Conclusion

Tendinopathies are common in hand and wrist. Acute presentation is usually caused by trauma while chronic presentation is often partly due to overuse. With the complexity of the interrelationship of extensors and flexors, it can be misdiagnosed or missed. Thus, it is essential to understand the anatomy in order to make the correct diagnosis.

References

1. Trumble T, Budoff JE, Cornwall R. Hand, elbow & shoulder core knowledge in orthopaedics. 2006:1.
2. Roth KM, Blazar PE, Earp BE, Han R, Leung A. Incidence of extensor pollicis longus tendon rupture after nondisplaced distal radius fractures. J Hand Surg Am. 2012;37(5):942–7.
3. Elson RA. Rupture of the central slip of the extensor hood of the finger. A test for early diagnosis. J Bone Joint Surg. 1986;68-B(2):229–31.
4. Fok MWM, Liu B, White J, Bain G. Hand and wrist tendinopathies. In: Muscle and tendon injuries. 5th ed. Berlin, Heidelberg: Springer; 2017. p. 241–53.

5. Shrewsbury MM, Kuczynski K. Flexor digitorum superficialis tendon in the fingers of the human hand. SAGE Publications Sage CA: Thousand Oaks, CA; 1974;6(2):121–133.
6. Austin GJ, Leslie BM, Ruby LK. Variations of the flexor digitorum superficialis of the small finger. J Hand Surg Am. 1989;14(2 Pt 1):262–7.
7. Leddy JP, Packer JW. Avulsion of the profundus tendon insertion in athletes. J Hand Surg Am. 1977;2(1):66–9.
8. DeGeorge BR, Brogan DM, Shin AY. The relationship of volar plate position and flexor tendon rupture: should we question the validity of the Soong classification? Plast Reconstr Surg. 2020;146(3):581–8.
9. Campbell D, Campbell R, O'Connor P, Hawkes R. Sports-related extensor carpi ulnaris pathology: a review of functional anatomy, sports injury and management. Br J Sports Med. 2013;47(17):1105–11.
10. Rayan GM, Murray D. Classification and treatment of closed sagittal band injuries. J Hand Surg Am. 1994;19(4):590–4.
11. de Quervain F. On a form of chronic tendovaginitis by Dr. Fritz de Quervain in la Chaux-de-Fonds. 1895. Am J Orthop (Belle Mead NJ). 1997;26:4.
12. Finkelstein H. Stenosing tendovaginitis at the radial styloid process. JBJS. 1930;12(3):509.
13. Adams JE, Habbu R. Tendinopathies of the hand and wrist. J Am Acad Orthop Surg. 2015;23(12):741–50.
14. Ruland RT, Hogan CJ. The ECU synergy test: an aid to diagnose ECU tendonitis. J Hand Surg Am. 2008;33(10):1777–82.
15. Al-Qattan MM. Type 5 avulsion of the insertion of the flexor digitorum profundus tendon. J Hand Surg Br. 2001;26(5):427–31.
16. Smith JH. Avulsion of a profundus tendon with simultaneous intraarticular fracture of the distal phalanx--case report. J Hand Surg Am. 1981;6(6):600–1.

Evaluation of Hand Infections

31

Janus Siu Him Wong and Margaret Woon Man Fok

31.1 Introduction

A marvel of nature's engineering the human hand may be, both as a sensory organ and a motor effector channeling intellectual faculties to diverse actions requiring power and finesse, the human hand is not immune to infection. Highly specialised soft tissue and compartments give rise to distinct infective hand pathologies, where physical examination is key in differentiating different types of infection.

Although common, hand infections could prove challenging to manage and are potentially life threatening. Accurate and timely physical examination play an important role in avoiding delay in diagnosis and treatment, which would further lead to stiffness necessitating prolonged rehabilitation, delayed return to play and irreversible loss of function.

Systemic symptoms such as fever may be present but may be subdued in immunocompromised patients (such as diabetes mellitus, on steroid medication) incapable of mounting an adequate immune response; a higher degree of vigilance must therefore be maintained.

Meticulous physical examination is often rewarded by diagnostic clues pertaining to the site of primary pathology. *Dolor* (pain), *tumor* (swelling), *calor* (increased temperature) and *rubor* (erythema) are often conspicuous in superficial infections. Assessment of young children could nevertheless be challenging. Accurate diagnosis is often aided by sound knowledge of anatomy, accounting for commonly observed patterns of involvement and spread of deep infections.

Biochemical investigations such as elevated leucocyte count, C-reactive protein and erythrocyte sedimentation rate could be helpful but are not specific. Imaging could be helpful in the presence of radiographic foreign bodies; positioning of digit (semi-flexed position in tenosynovitis) may be a telltale sign. In most cases, however, the primary site of pathology remains elusive on plain radiographs, save for chronic infections with bony involvement. Exhaustive radiological investigations such as ultrasonography and radionucleotide imaging are therefore not universally performed, as they may lead to unnecessary delays to surgical treatment.

In this chapter, we shall focus on physical examination for evaluation and diagnosis of common hand infections.

J. S. H. Wong
Department of Orthopaedics and Traumatology, Li Ka Shing Faculty of Medicine, The University of Hong Kong, Hong Kong, China

M. W. M. Fok (✉)
Department of Orthopaedics and Traumatology, Queen Mary Hospital, The University of Hong Kong, Hong Kong, China

J. G. Lane et al. (eds.), *The Art of the Musculoskeletal Physical Exam*,
https://doi.org/10.1007/978-3-031-24404-9_31

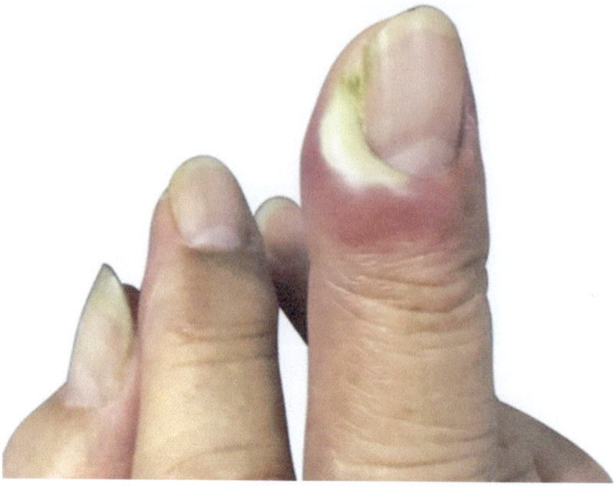

Fig. 31.1 Acute paronychia of the right middle finger over the radial nail fold (copyright by Dr. Janus Wong)

31.2 Paronychia

Acute paronychia describes infection of the soft tissue adjacent to the nail bed (i.e. paronychium) resulting from the breach of the nail fold through minor trauma [1]. Physical examination findings are apparent—patients clinically present with tenderness over the proximal and/or lateral fold, with an onset of a few days. Pus collection is often subcutaneous and accompanied by surrounding erythema and swelling, rendering diagnosis simple and straightforward in the majority of cases (Fig. 31.1). Involvement does not extend past the distal interphalangeal joint proximally. The dorsal, perionychial site of involvement distinguishes paronychia from other differential diagnoses such as felon (see Sect. 31.3).

31.3 Felon

Felon refers to the infection of the finger pulp of the distal phalanx. The term originates from the Old French "fel" referring to "bile" [2], referencing the severe throbbing pain reported by patients afflicted with this malady.

The finger pulp is made up of fat and eccrine sweat glands divided by multiple fibrous septa extending from the dermis to the distal phalanx periosteum. Under physiological circumstances,

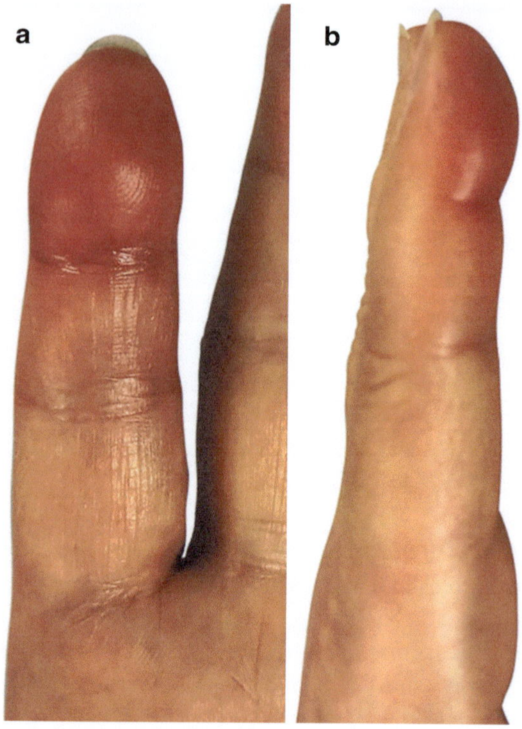

Fig. 31.2 Felon. Volar (Fig. 31.2a) and sagittal (Fig. 31.2b) views of felon involving the left index finger. Note the anatomical extent of involvement limited to the distal phalanx and not extending proximal to the distal interphalangeal joint (DIPJ), in contradistinction to tenosynovitis where the whole digit (flexor tendon sheath) is involved with swelling and erythema extending proximally past the DIPJ (copyright by Dr. Janus Wong)

these septa support and stabilise the pulp during grasping and pinching. Yet, during bacterial inoculation through either sweat glands or a penetrating wound, pus may build up and increase the intra-compartmental pressures. This in turn impedes blood supply and leads to tissue necrosis. Exquisite tenderness is commonly observed as finger pulps are densely innervated.

Tenderness, erythema and swelling do not usually extend proximal to the distal interphalangeal joint, due to the compartments defined by the septae (Fig. 31.2). Therefore, physical examination findings are focused and limited to the finger pulp at the distal phalanx. Ensuring adequate drainage of all septal compartments is key to eradicating infection.

31.4 Flexor Tenosynovitis

Flexor tendon sheaths are synovium-lined double-walled structures extending from the distal middle phalanx (just proximal to the distal interphalangeal joint) to the metacarpal neck, except for the thumb and little finger which continue proximally in communication with the radial bursa and ulnar bursa, respectively (Fig. 31.3). This is in contrast to extensor tendon sheaths, which stop distally at the carpus or metacarpal base.

Flexor tendon sheaths are lined by two layers—an outer parietal layer just beneath the finger pulleys and a visceral layer (essentially the epitenon) which is the invaginated reflected surface of the parietal layer. They contain synovial fluid and provide lubrication within the confines of the flexor pulley system.

Infection of the flexor tendon sheath is usually caused by penetrating trauma (e.g. animal bite) although there may be a lack of definite episode with an indolent presentation in immunocompromised patients. Involvement of the whole flexor tendon sheath translates clinically to a recognisable pattern of tenderness and swelling involving the whole digit on physical examination (Fig. 31.4). Allen Kanavel described three of his four eponymous cardinal signs in his seminal paper published in 1912 [3].

1. Exquisite tenderness over the course of the sheath, limited to the sheath (the most important sign according to Kanavel).
2. Flexion of the finger.
3. Exquisite pain on extending the finger, most marked at the proximal end.

The fourth sign - the whole of the involved finger is uniformly swollen (the most com-

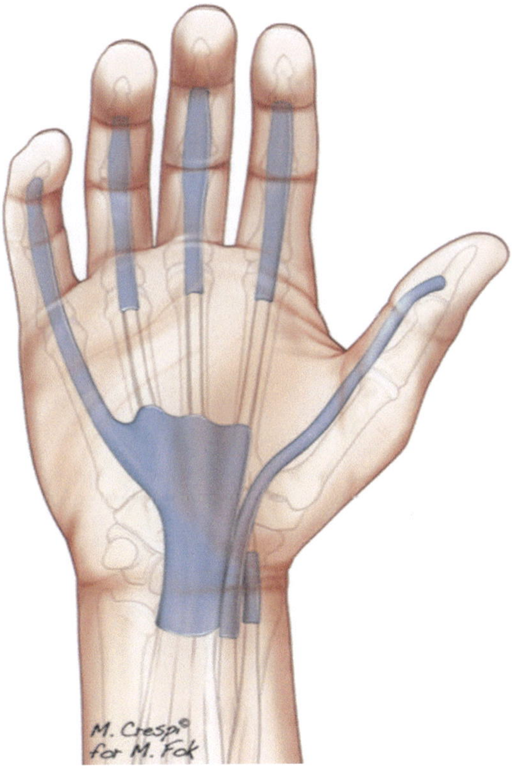

Fig. 31.3 Bursae anatomy (copyright by Dr. Margaret Fok)

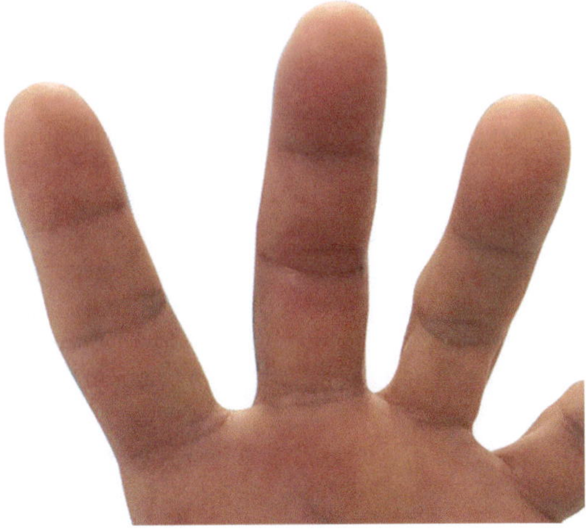

Fig. 31.4 Tenosynovitis of the left middle finger with fusiform swelling, held in semi-flexed position, tenderness along the flexor tendon sheath and pain on passive extension. Note the penetrating wound over the ulnar aspect of the proximal interphalangeal joint. (Copyright by Dr. Janus Wong)

mon sign) - was subsequently added and completes the quartet [5]. While widely taught to orthopaedic residents, the presence of all four signs is not *sine qua non* of tenosynovitis, and we caution against the exclusion of such diagnosis in the absence of any of the four signs [6].

Due to the continuation of the radial and ulnar bursae with the tendon sheaths of the thumb and little finger, respectively, tenosynovitis of the thumb and little finger may spread proximally. Clinical examination should therefore extend proximally to exclude involvement of the radial and ulnar bursae, which provide a path for the infection to spread to the other bursa in the case of horseshoe abscess, or even to other fingers depending on the bursal distribution. Scheldrup described wide variations in flexor tendon sheath and bursal configurations in his study of over 300 hand dissections [7]; as a result, in patients who present with tenosynovitis, surgeons need to examine all fingers for possible spread of infection. This is especially important in immunocompromised patients such as patients with diabetes mellitus. In case of proximal spread, one must also watch out for acute carpal tunnel syndrome.

31.5 Web Space Infection

Web spaces refer to interdigital palmar aponeurosis with overlying fat and skin, made up of loose areolar tissue around the intermetacarpal ligaments between metacarpal heads. Bacteria enter through breaches of the skin, which has dense attachments to the palmar aponeurosis. Infection therefore tends to spread towards areas of less resistance through the fascial hole to the dorsal web space [8]. Physical examination reveals tenderness, erythema, volar and dorsal swelling over the involved webspace resembling a dumbbell shape (the central restriction representing the space between metacarpal heads), hence the name "collar-button abscess" or "collar-stud abscess". Because of the swelling, the adjacent fingers are held in abduction in a V-shaped configuration—this differentiates it from a dorsal subcutaneous infection where, without webspace involvement, fingers are not abducted. Accordingly, volar and dorsal incisions are used to drain the abscess.

31.6 Deep Palmar Space and Bursae Infections

Three deep palmar spaces in the hand are potential loci for infection (Fig. 31.5), usually by direct inoculation through a penetrating wound.

Thenar space, bound by the adductor pollicis dorsally and volarly, flexor tendons of the index finger volarly, septum of Legueu and Juvara ulnarly, is the most commonly infected deep palmar space among the three. Physical examination reveals pain and swelling over thenar area and first webspace with painful thumb opposition and adduction. The thumb is held in wide abduction as a result.

The midpalmar space is bound by the second and third metacarpal and interossei dorsally, flexor tendons and lumbricals volarly, hypothenar muscles ulnarly and septum of Legueu and Juvara radially. Key physical examination find-

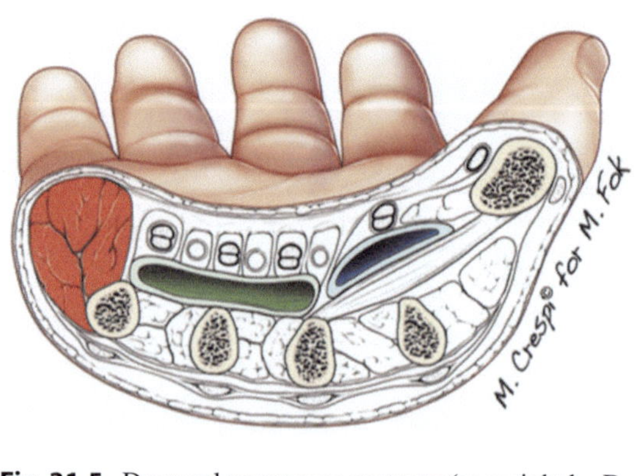

Fig. 31.5 Deep palmar spaces anatomy (copyright by Dr. Margaret Fok)

ings include loss of concavity of the volar palm (or even presenting with palmar convexity), tenderness and pain on passive motion over middle and ring fingers, at times with flexion deformity. Due to the presence of lumbricals within the space, infection may spread along the lumbricals (which insert to the extensor expansion in proximity to the metacarpophalangeal joint) into the dorsum of the hand. Dorsal hand swelling may therefore be present.

Hypothenar space infections, involving the area bound by septum laterally, fifth metacarpal bone dorsally, hypothenar muscle fascia medially and palmarly, are uncommon. There is an absence of digital signs on physical examination since flexor tendons are not involved.

The deep palmar spaces are in continuity proximally with Parona's space—thus deep palmar infections can spread to Parona's space and cause distal forearm infections (see Sect. 31.6).

The radial bursa begins distally at the level of the thumb metacarpophalangeal joint, contiguous of the sheath of the flexor pollicis longus tendon along its course through the carpal tunnel, ending 1–2 cm proximal to the transverse carpal ligament.

The ulnar bursa expands radially as it goes proximal from its beginning at the proximal end of the little finger flexor tendon sheath, overlapping the third and fourth metacarpal base.

Both bursae extend proximal to the transverse carpal ligament, sandwiched between the pronator quadratus (PQ) fascia dorsally and flexor digitorum profundus (FDP) tendons volarly (see Sect. 31.6 for proximal spread).

31.7 Parona's Space Infection

Parona's space refers to the plane between pronator quadratus fascia dorsally and flexor digitorum profundus tendons volarly, first described by the Italian surgeon anatomist Francesco Parona in 1876.

The radius and ulna bursae, proximal extensions of the flexor tendon sheaths of the thumb and little finger, respectively, extend proximal to the transverse carpal ligament, sandwiched between the PQ fascia and FDP tendons. Bursal rupture and spread of infection through Parona's space could occur, and as the pattern of bursae communication is not constant [7], involvement of any other digit(s) may occur. Similarly, infection from the deep palmar spaces can track into the Parona's space. Thus, it is important to check all potential spaces for infection, especially in immunocompromised patients.

31.8 Septic Arthritis of the Wrist

Patients with septic arthritis of the wrist present with swelling, pseudoparalysis with pain and reduced range of motion in all directions due to hyperaemia, build-up of inflammatory cytokines and intra-articular pressure. Bacterial seeding from bacteraemia is the usual cause. Universal swelling of the wrist—both palmar and dorsal—and global decreased range of motion for all directions differentiates this from other soft tissue infections. Atypical infections (e.g. due to *Mycobacteria* [9], Fig. 31.6) and inflammatory arthropathies (e.g. urate and calcium phosphate crystal arthropathy) could be difficult to differentiate from classical bacterial infection—further investigations such as arthrocentesis should be pursued [10], and a high level of suspicion maintained.

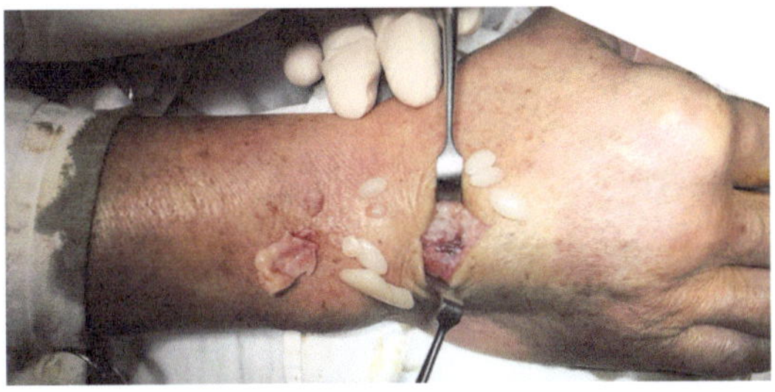

Fig. 31.6 Rice bodies in patient with *Mycobacterium* infection of the wrist (copyright by Dr. Janus Wong)

31.9 Conclusion

Physical examination is important in the evaluation of hand infections and key to diagnosis and subsequent management. Most hand infections are due to direct bacterial inoculations through penetrating trauma wounds, providing hints on the area involved. Yet, many potential spaces communicate with each other, therefore there is a need to assess the spread of infection by performing thorough physical examination. This is especially true for immunocompromised patients.

References

1. Wolfe SW. Green's operative hand surgery. Philadelphia, PA: Elsevier; 2011. p. 41–84.
2. Barger J, Garg R, Wang F, Chen N. Fingertip infections. Hand Clin. 2020;36(3):313–21.
3. Kanavel AB. The symptoms, signs, and diagnosis of tenosynovitis and fascial-space abscesses. Infections of the hand. Philadelphia, PA: Lea & Febiger; 1912. p. 201–26.
4. Pang HN, Teoh LC, Yam AK, Lee JY, Puhaindran ME, Tan AB. Factors affecting the prognosis of pyogenic flexor tenosynovitis. J Bone Joint Surg Am. 2007;89(8):1742–8.
5. Kanavel AB. Infections of the hand: a guide to the surgical treatment of acute and chronic suppurative processes in the fingers, hand and forearm. Philadelphia, PA: Lea & Febiger; 1939.
6. Dailiana ZH, Rigopoulos N, Varitimidis S, Hantes M, Bargiotas K, Malizos KN. Purulent flexor tenosynovitis: factors influencing the functional outcome. J Hand Surg Eur. 2008;33(3):280–5.
7. Scheldrup EW. Tendon sheath patterns in the hand; an anatomical study based on 367 hand dissections. Surg Gynecol Obstet. 1951;93(1):16–22.
8. Kalbfell E, Adams NS, Cullen WT. Collar button abscess. Eplasty. 2016;16:ic6.
9. Flevas DA, Syngouna S, Fandridis E, Tsiodras S, Mavrogenis AF. Infections of the hand: an overview. EFORT Open Rev. 2019;4(5):183–93.
10. Koshy JC, Bell B. Hand infections. J Hand Surg Am. 2019;44(1):46–54.

Diagnosis and Evaluation of Fractures of the Hand and Wrist

32

Lindsey S. Urband, Stephanie Wong, and Dori N. Cage

Fractures of the hand and wrist include 55% of all fractures of the upper extremities [1]. The diagnosis and evaluation of these fractures requires a general understanding of the mechanism of injury and physical exam characteristics that are often, though not always, unique to the injury. The information presented in this chapter is designed to provide basic, essential diagnostic tools for medical professionals treating a variety of orthopedic injuries. The chapter is organized in a distal to proximal pattern.

32.1 Fractures of the Phalanges

32.1.1 Fractures of the Distal Phalanx

Fractures involving the distal phalanx are relatively common injuries. They account for approximately half of all hand fractures [2]. The mechanism of injury is often an axial load [2] mechanism or "jamming" injury. The long digit is most often injured. There are several subtypes of distal phalanx fractures.

32.1.2 Dorsal Lip Avulsion Fracture

Dorsal lip avulsion fractures are commonly known as mallet injuries. The mechanism of injury is a direct blow to the tip of finger causing forceful flexion of the distal phalanx, or a dorsal laceration and occasionally, a crush injury [3]. The fracture is caused by an extensor tendon injury [3] causing avulsion fracture of the distal phalanx, and most commonly affects one of the medial three fingers of the dominant hand [4]. Physical examination findings will demonstrate pain over the dorsum of the distal interphalangeal (DIP) joint, swelling, ecchymosis, flexion deformity with the inability to actively extend the DIP joint fully, and flexed DIP joint at rest [3, 4]. When these fractures involve a substantial portion of the articular surface they can result in an unstable joint with subluxation or dislocation [5, 6].

32.1.3 Volar Lip Avulsion Fracture

Volar lip avulsion fractures are less common than dorsal injuries. These fractures are commonly referred to as jersey fingers. The mechanism of injury is forceful flexor digitorum profundus contraction during hyperextension of the distal

L. S. Urband · D. N. Cage (✉)
San Diego Hand Specialists, San Diego, CA, USA

S. Wong
California University of Science and Medicine, Colton, CA, USA

J. G. Lane et al. (eds.), *The Art of the Musculoskeletal Physical Exam*,
https://doi.org/10.1007/978-3-031-24404-9_32

interphalangeal joint. The force that is generated in these fractures is secondary to flexor digitorum profundus tendon rupture causing avulsion fracture of the distal phalanx at its insertion point [3]. On physical examination, swelling over the volar aspect of the DIP joint will be present. Additionally, pain along the course of the flexor tendon and inability to actively flex the DIP joint will also be noted [3]. As the flexor tendon involvement is the more critical issue with these fractures, urgent diagnosis and treatment is paramount to allow for restoration of function in the affected digit [3].

32.1.4 Seymour Fractures

Although Seymour fractures are technically only in the pediatric patient population, they can occur via a slightly different anatomic mechanism in adults and deserve mention [7, 8]. In the standard term, a Seymour fracture is a fracture through the physis of the pediatric distal phalanx which results in a disruption of not only the physis, but also the germinal matrix of the nail [8–11].

These are considered open fractures and warrant emergent operative reduction and irrigation. Delay in treatment may result in osteomyelitis of the distal phalanx [9, 11]. Although the physes are closed in adult patients, occasionally, these fractures can occur with a hyperflexion or crush injury which causes a transverse, extra-articular fracture at the metaphyseal-diaphyseal junction of the distal phalanx, resulting in volar displacement of the distal fragment. As the distal phalanx displaces volarly, the proximal aspect of the fracture fragment displaces through the germinal matrix and skin dorsally [9, 10]. As aforementioned, these fractures are often associated with nailbed [11] injury. Physical examination demonstrates a subungual hematoma. The nail plate is displaced dorsally and subsequently sits superficial to the eponychial fold. Fixed flexion is maintained at the distal interphalangeal joint [9, 11].

32.1.5 Distal Phalanx Tuft Fractures

The mechanism of injury for tuft fractures is often a crush injury [12]. A stellate pattern is present on radiographs in the AP view. Occasionally, there will be displacement through the diaphysis of the distal phalanx, most easily viewed on the lateral radiograph. If this displacement is greater than 50%, operative intervention may be required. However, in general, tuft fractures are treated without surgery in a splint for 3–4 weeks, allowing motion through the proximal interphalangeal joint [12, 13]. Physical examination demonstrates swelling of the distal phalanx with tenderness to [12, 13] palpation of the pulp of the digit. Subungual ecchymosis is often present. These fractures can result in radiographic nonunion which is usually not symptomatic [12].

32.2 Fractures of the Middle Phalanx

Fractures involving the middle phalanx can be further divided into intra-articular and extra-articular fracture patterns.

32.2.1 Intra-articular Middle Phalanx Fractures

Volar plate avulsion injuries are examples of intra-articular middle phalanx fractures involving the volar surface of the base of the middle phalanx. The mechanism of injury is a forced, sudden hyperextension of the proximal interphalangeal joint. Physical examination demonstrates pain is the greatest at the volar surface of the proximal interphalangeal joint. There is notable loss of stability in hyperextension. Diffuse swelling is present about the proximal interphalangeal joint, and there is usually significant tenderness to palpation at the proximal interphalangeal joint. This fracture pattern is sometimes accompanied by collateral ligament involvement resulting in joint instability [14].

Pilon fractures or fractures involving the dorsal and palmar articular surface of the PIP joint are unstable injuries. There is significant comminution at the articular surface and a central depression with a splay of the dorsal and palmar fracture fragments, causing a widened appearance of the base of the middle phalanx on lateral radiographs [15]. The mechanism of injury [14] for pilon fractures of the middle phalanx is an axial load to the extended digit. Physical examination findings may demonstrate dorsal or volar subluxation or dislocation of the middle phalanx. The PIP joint is unstable [15]. There is fusiform swelling about the PIP joint with limited active and passive range of motion.

Unicondylar or bicondylar fractures involve the head of the middle phalanx. The fracture line extends to the DIP joint and may involve one or both of the condyles. These are inherently unstable injuries and must be addressed with surgical correction to restore anatomy and functionality of the affected digit [14, 16]. The mechanism of injury is a torsional force applied to the digit. On physical examination, there is fusiform swelling at the middle phalanx and DIP joint [16]. There may also be associated rotational malalignment of the affected digit. Malrotation of the digits may be assessed using two simple physical examination findings. One is to have the patient actively extend their digits. All fingernails should be parallel to the floor. If angulation is present, the fingernail will also be malrotated. The second examination maneuver is to have the patient make a fist, to the best of their ability. All fingertips should point toward the scaphoid tubercle, at the base of the thenar eminence. Any scissoring, or overlap of one digit on another, is indicative of rotational malalignment.

32.2.2 Middle Phalanx Shaft Fractures

Extra-articular in nature, the fracture pattern can be transverse or spiral oblique [16]. Transverse fractures result from an axial load to an extended digit. On anteroposterior radiographs, the fracture line is parallel to the articular surface. Often, these fractures are stable and may be managed non-operatively if no major displacement is present [16]. However, a subset of these fractures, which may occur in the pediatric or adult populations, occur at the metaphyseal-diaphyseal junction and will result in flexion of the phalangeal head and extension of the shaft, resulting in an unstable, displaced pattern. These fractures will require operative intervention [16]. On physical examination, there will be fusiform swelling and ecchymosis, if the fracture is acute. Generally, no rotational malalignment of the digit is seen.

Spiral oblique fractures are secondary to a torsional force applied to the digit. On physical examination, there will be fusiform swelling, ecchymosis, and limitation in range of motion. Additionally, rotational deformity is also noted in the affected digit. These fractures are inherently unstable and most likely require operative intervention [16].

32.3 Proximal Phalanx Fractures

Similar to middle phalanx fractures, proximal phalanx fractures may be classified into intra-articular and extra-articular categories. Condylar fractures both unicondylar and bicondylar fracture patterns may occur involving the head of the proximal phalanx and warrant surgical intervention [16]. Transverse and spiral oblique shaft fractures also may occur. Physical examination findings associated with the aforementioned fracture patterns will present very much like fractures involving the middle phalanx. Ecchymosis, swelling, and limited range of motion of the affected digit is noted. The point of maximal tenderness to palpation is one method to differentiate the location of the fracture and may serve as a rudimentary diagnostic strategy before radiographic imaging is available. Fractures of the proximal phalanx need to be closely checked for rotational alignment. Fractures which heal with a palmar bony prominence proximal to the PIP joint may result in loss of PIP flexion.

Avulsion fractures at the base of the proximal phalanx are associated with an injury to either the radial or ulnar collateral ligament at the metacarpophalangeal joint [17]. These fractures are subtle and may be missed on initial radiographs. On physical examination, there will be point tenderness at the base of the proximal phalanx, especially in the affected webspace. There is associated laxity with varus or valgus stress of the metacarpophalangeal joint [17]. Swelling and ecchymosis may be more subtle than shaft fractures. The mechanism of injury for these fractures is hyperextension or hyperflexion of the digit with a radial or ulnar directed force on the digit or during a fall on an outstretched hand [18]. Depending on the percentage of articular surface which is [17] involved in collateral ligament avulsion fractures, surgical intervention is warranted.

32.4 Fractures of the Thumb

Fractures of the distal phalanx of the thumb are classified into the same patterns as those involving the distal phalanx of the digits. Physical examination demonstrates fusiform swelling, particularly at the tip of the thumb with ecchymosis and tenderness to palpation.

Collateral ligament injuries of the radial and ulnar collateral ligaments occur more frequently in the thumb than in the digits [18, 19]. The mechanism of injury is forced abduction and hyperextension [20] of the thumb at the MP joint. Injury to the ulnar collateral ligament of the thumb metacarpophalangeal joint is associated with the eponym "Gamekeeper's thumb" when occurring in a chronic setting and the eponym "Skier's thumb" when occurring in an acute setting [19, 20]. The injury to the ulnar collateral ligament may occur with or without an avulsion fracture of the base of the thumb proximal phalanx. On physical examination, there is instability at 30° of flexion of the thumb at the metacarpophalangeal joint with valgus stress [19]. Additionally, the region most tender to palpation is on the ulnar aspect of the thumb metacarpophalangeal joint. Less commonly, avulsion fracture of the thumb proximal phalanx can occur secondary to force from the radial collateral ligament attachment [20]. The physical examination finding in radial collateral ligament injury or avulsion fracture is maximal tenderness to palpation over the radial side of the metacarpophalangeal joint and instability at 0° of flexion with varus stress of the metacarpophalangeal joint [20]. Operative intervention is warranted with instability of the thumb metacarpophalangeal joint secondary to an avulsion fracture or a collateral ligament injury rupture [20].

Transverse fractures of the thumb proximal phalanx are secondary to an axial load on an extended thumb. The fracture pattern is most often stable and does not require surgical intervention. Physical examination demonstrates pain with palpation of the shaft of the proximal phalanx, swelling, and ecchymosis present about the thumb.

Spiral oblique fractures, which may or may not include intra-articular extension, are generally unstable fractures. Physical examination presentation is similar to transverse fractures of the proximal phalanx, but there may be an element of rotational malalignment present. The mechanism of injury is a torsional force applied to the extended thumb.

32.5 Fractures of the Metacarpals

Metacarpal fractures may be intra-articular or extra-articular. The majority of thumb fractures involve the metacarpal base; the most common pattern is extra-articular epi-basilar fracture. With regard to fractures of the first metacarpal, two intra-articular fracture patterns are most commonly seen and both require surgical intervention to restore the articular surface congruency.

32.5.1 Bennett Fracture Dislocation

In Bennett fracture-dislocations, there is dorsal-radial displacement at the base of the metacarpal while a small proximal fragment remains attached to the trapezium via the volar oblique ligament. The mechanism of injury is an axial force applied to the thumb in flexion or, less commonly, hyperabduction or hyperflexion of the thumb during a fall [21]. Physical examination demonstrates acute pain at the base of thumb, tenderness to palpation at the first carpometacarpal joint, pain with limited range of motion, and circumduction of the thumb. There may be a palpable step-off at the first carpometacarpal joint [21]. Additionally, swelling and ecchymosis are present at the carpometacarpal joint.

32.5.2 Rolando Fractures

Rolando fractures are comminuted fractures at the base of the thumb extending into the articular surface [21, 22]. The mechanism of injury and physical examination findings are identical to those of the Bennett fracture pattern. To restore the articular surface congruency of the first carpometacarpal joint, operative intervention is required [21, 22].

32.6 Second Through Fifth Metacarpal Fractures

Metacarpal fractures represent 18–44% of all fractures of the hand [22, 23]. Fracture types are divided into the anatomic location of the fracture as well as fracture pattern. The metacarpal is divided into head, neck, shaft, and base fractures. Fracture patterns include transverse, spiral oblique, and comminuted fracture patterns, most often involving the base or the head of the metacarpal.

32.6.1 Metacarpal Head Fractures

Metacarpal head fractures are intra-articular fractures that result from an axial load or crush injury to the hand. Physical examination demonstrates pain and swelling at distal aspect of the metacarpal. Additionally, there may be a visible loss of knuckle on the affected digit. Occasional rotational deformities may be present in the digit distal to the metacarpal involved; however, this is relatively rare. As these fractures are an intra-articular injury and likely involve significant step-off at the metacarpophalangeal joint articular surface, operative intervention is often required [22, 23].

32.6.2 Open Fractures at the Metacarpophalangeal Joint

These injuries are also called "fight bite" injuries. The mechanism of injury is a direct axial load to a clenched fist, most frequently impacted against the oral orifice of another person. Physical examination demonstrates a dorsal laceration over the metacarpophalangeal joint. As these injuries are open fractures, they must be treated with urgent irrigation and debridement of the wound to prevent pyogenic arthrosis. Definitive fixation of the fracture may be delayed depending on the status of the soft tissues [24, 25].

32.6.3 Metacarpal Neck Fractures

Metacarpal neck fractures are extra-articular in nature although some may involve intra-articular extension through the metacarpal head. The mechanism of injury is secondary to trauma to a clenched fist or blunt force trauma to the dorsum of the hand [22]. On physical examination, there is loss of visible knuckle of the affected metacarpophalangeal joint, tenderness to palpation at the neck of the metacarpal dorsally, fusiform swelling of the hand, and dorsal and palmar ecchymosis. Angulation of the fracture is generally well tolerated in the fourth and fifth metacarpal necks [22, 23]. However, significant angulation is not well tolerated in the second and third metacarpal necks and often must be treated with operative intervention [22, 23].

32.6.4 Metacarpal Shaft Fractures

Metacarpal shaft fractures are extra-articular fractures. The mechanism of injury may be one of several forces. An axial load to an extended digit may create a transverse shaft fracture.

Torsional force to a digit will cause a spiral, oblique, shear shaft fracture pattern; and a direct, crush injury to the hand may result in a comminuted shaft fracture [22]. Physical examination findings will demonstrate diffuse ecchymosis and swelling about the hand. If a torsional force was involved, rotational malalignment of the associated digit will be present. In spiral oblique, shear fractures, the digit may be shortened due to the unstable nature of the spiral oblique, shear fracture pattern. Tenderness to palpation will be present over the fracture site and loss of range of motion of the digit and wrist will be present.

32.6.5 Metacarpal Base Fractures and Fracture Dislocations of the Carpometacarpal Joint

Fractures involving the base of the metacarpals are most frequently caused by a direct, axial load to the clenched fist [22]. In the case of fourth and fifth carpometacarpal fracture dislocations, a longitudinally directed force along the fifth metacarpal results in proximal and dorsal subluxation of the metacarpal base [26]. Proximal displacement is secondary to the pull of extensor carpi ulnaris tendon. Additionally, these injuries may occur secondary to a fall onto outstretched upper extremity [22]. On physical examination, there is dorsal osseous prominence at the base of the affected metacarpal. Occasionally, rotational misalignment may be present. If significant force is endured as a component of this injury, there will be an associated dislocation of the carpometacarpal joint. Fracture dislocations of the carpometacarpal joints most commonly affect the fourth and fifth metacarpal bases [26]. Radiographs will demonstrate comminution of the base of the metacarpal and incongruency at the carpometacarpal joint. Lateral and oblique radiographs are important to assess as the fracture dislocations may be subtle. If there is any question regarding the diagnosis, advanced imaging, specifically CT scan imaging, is warranted.

32.7 Fractures of the Carpal Bones

32.7.1 Scaphoid

32.7.1.1 Acute Scaphoid Fractures

Acute fractures of the scaphoid account for 50–80% of all acute carpal bone fractures [27]. The mechanism of injury is a fall onto an outstretched hand with the wrist in dorsiflexion, ulnar deviation, and intercarpal supination [27, 28]. Physical examination findings demonstrate wrist pain.

These fractures may or may not have associated swelling and/or ecchymosis. Generally, there is point tenderness at the scaphoid tubercle and in the anatomic snuffbox of the affected wrist [28]. If the patient can tolerate, provocative testing may aid in the diagnosis. A positive scaphoid lift test is reproduction of pain with dorsal-volar shifting of the scaphoid. The scaphoid shift test is considered positive when painful dorsal scaphoid displacement occurs with radioulnar deviation and compression of the scaphoid tubercle. Scaphoid fractures are divided by the anatomic location of the fracture. Proximal pole fractures account for approximately 5–8% of all acute scaphoid fractures. Waist fractures account for approximately 80% of all scaphoid fractures; and distal pole fractures comprise 10–12% of all scaphoid fractures [28]. Anatomic location of the fracture becomes important when assessing and monitoring the healing potential and determining appropriate treatment.

32.7.2 Scaphoid Fracture Nonunion

Scaphoid nonunion is relatively common, occurring in 5–10% non-displaced fractures of the scaphoid [27]. The perfusion of the scaphoid is the primary reason some scaphoid fractures [27] progress to nonunion even in healthy, young patients. The blood supply to the scaphoid is from the dorsal and volar branches of the radial artery [28]. The volar branches, entering the scaphoid from the proximal pole region, supply 20–30% of the scaphoid with blood flow. The dorsal branches perfuse the scaphoid in a retrograde fashion, from distal to proximal, and supply 70–80% of the blood flow to the scaphoid. Therefore, fractures involving the scaphoid waist and proximal pole may not receive adequate perfusion to promote and allow for fracture healing [27, 28]. When monitoring even a nondisplaced scaphoid fracture, vigilanc and a threshold to proceed with operative intervention if there is evidence of delayed union should be considered. Sequelae of scaphoid nonunion is related to the anatomic location of the nonunion and the effect the nonunion has on the kinematics of the carpus. Scaphoid waist nonunion is associated with a Humpback deformity, essentially a flexed position through the fracture site which must be reduced during operative intervention [29]. Scaphoid nonunion may also progress to accelerated post-traumatic arthritis of the wrist, specifically a scaphoid nonunion advanced collapse, or SNAC, of the carpus [30]. Post-traumatic arthritides are out of the scope of this chapter. Secondary to poor perfusion of proximal pole fractures, avascular necrosis may occur in the proximal pole [27, 28]. Sequelae from proximal pole nonunion is particularly challenging to treat and requires a vascularized bone graft at the time of operative intervention.

32.7.3 Lunate Fractures

Fractures of the lunate are rare and comprise approximately 0.5–1% of all fractures of the carpal bones [31]. Often, these fractures go unrecognized until there is progression to osteonecrosis. Lunate body fractures are most frequently sustained secondary to a fall onto outstretched hand with the wrist in hyperextension or a strenuous push with the wrist maximally dorsiflexed [31]. On physical examination, there is tenderness to palpation over the volar wrist in the proximal row, midpoint. Often, associated swelling and ecchymosis of the wrist will be present. Pain with range of motion of the wrist, particularly with radial and ulnar deviation, will be present [31].

Lunate injuries are additionally associated with greater and lesser arc injuries. Greater arc injuries are associated with a perilunate fracture dislocation or a lunate dissociation. In a greater arc injury with perilunate fracture dislocation, the force begins with transmission through the scaphoid, capitate, and triquetrum and often results in trans scaphoid or trans scaphoid trans capitate fracture dislocations. Lesser arc injuries follow a path through the radial styloid or radioscaphoid joint and then transmit through the midcarpal joint and lunotriquetral space, resulting in perilunate and lunate dislocations. The mechanism of injury for perilunate dislocations and perilunate fracture dislocations is a load applied to the thenar eminence forcing the wrist into extension [32]. As aforementioned, the injury progresses through the carpus, either through scaphoid body, causing a fracture, or through scapholunate interval, causing scapholunate ligament dissociation; then force then moves ulnarly between lunate and capitate, after which it ultimately disrupts the lunotriquetral articulation [32].

On physical examination, there will be diffuse swelling about the wrist with ecchymosis. Tenderness to palpation will be present distal to Lister's tubercle. There variably may be dorsal prominence of the entire wrist.

Both greater and lesser arc injuries require immediate attempted reduction of the lunate, as lunate dislocation is the endpoint of these injuries. If the lunate is irreducible or if there is associated fracture of the carpal bones, operative intervention in an urgent setting is required. In these high energy injuries, there is a strong risk of acute median neuropathy, and carpal tunnel release should be considered simultaneously with open reduction and internal fixation of the osseous and ligamentous injuries [32].

32.7.4 Fractures of the Triquetrum

Triquetral fractures are the second most commonly fractured carpal bone after the scaphoid

[31]. Most frequently, the fracture is a dorsal avulsion fracture secondary to a fall onto an outstretched, dorsiflexed wrist. On physical examination, there is tenderness to palpation over the dorsal ulnar carpus and pain with radioulnar deviation and flexion and extension of the wrist. Most of these fractures are managed nonoperatively. It is common for patients to experience pain in the region for 3 months post-injury [31].

32.7.5 Pisiform Fractures

Pisiform fractures are rare injuries. The mechanism of injury is a direct blow to the volar wrist or a fall onto outstretched, dorsiflexed hand [31]. On physical examination, there is tenderness to palpation on the volar ulnar aspect of the wrist. Additionally, passive extension of the wrist is painful as the flexor carpi ulnaris is tensioned. In a non-displaced body fracture, non-operative management is warranted [31].

32.7.6 Trapezium Fractures

Fractures of the trapezium account for 1–5% of all carpal bone fractures [31]. Trapezial fractures may be classified into ridge avulsion fractures or vertical body fractures. The mechanism of injury in vertical body fractures involves an axial load to an adducted thumb which drives the first metacarpal into the articular surface of the trapezium. In trapezial avulsion fractures, the mechanism of injury is a forceful radial deviation with simultaneous longitudinal traction and rotation of the thumb [31]. On physical examination, there is ecchymosis and swelling with tenderness to palpation over the carpometacarpal joint as well as tenderness to palpation over the radial wrist [31]. There is a painful arc of active motion at the first carpometacarpal joint.

32.7.7 Trapezoid Fractures

Fractures of the trapezoid are rare. Fracture patterns include body fractures or dislocations of the second carpometacarpal joint. The mechanism of injury is an axial load transmitted through the second metacarpal [31]. On physical examination, there is maximal tenderness to palpation at the base of the second metacarpal. If the fracture is not displaced, nonoperative treatment is appropriate. Displaced fractures may require open reduction internal fixation.

32.7.8 Capitate Fractures

Fractures of the capitate occur uncommonly in isolation. More frequently, fractures of the capitate are involved in greater arc injuries, including trans scaphoid, trans capitate, and perilunate fracture-dislocations [31]. Scapho-capitate syndrome may also occur, which is defined as [31] simultaneous scaphoid and capitate fractures without associated dislocation of the lunate. The mechanism of injury of capitate fractures and greater arc injuries is generally a direct trauma to the dorsiflexed wrist or crush injury to the wrist. On physical examination, there is point tenderness to palpation over the dorsal aspect of the capitate and painful dorsiflexion of the wrist. Isolated capitate body fractures are generally managed non-operatively, but capitate fractures in association with greater arc injuries usually require operative intervention [31].

32.7.9 Hamate fractures

Fractures of the hamate may be subdivided into fractures of the hamate body, which are relatively rare, and hamulus fractures, or fractures of the "hook" of the hamate [31]. Importantly, a hamulus fracture should not be confused with an os hamulus proprium, which is a hamulus ossification center that failed to fuse.

Mechanism of injury for hamulus fractures is an athletic injury sustained when the palm of hand is struck by an object [31]. This occurs most commonly at the base of the hamulus although it may occur as an avulsion fracture. On physical examination, there is tenderness to palpation over the hamate [31]. It is important to consider that ulnar and median neuropathy can be seen, as well as [31] vascular injury to the ulnar artery. Therefore, a thorough neurovascular examination must be performed with a high index of suspicion for hamate fractures. Standard AP, lateral, and oblique views of the wrist may not exhibit well the hamate fracture. Thus, a carpal tunnel view may also be employed to further evaluate and diagnose the subtle hamulus fracture [31]. In the chronic setting, there is a risk of flexor tendon attritional rupture from ununited hamulus fractures.

Fractures of the body of the hamate may be trans-articular in nature and associated with a fifth metacarpal dislocation. The mechanism of injury is an axial load to a clenched fist which transmits the force through the base of the fifth metacarpal to the hamate [31].

32.8 Fractures of the Distal Radius

Fractures of the distal radius account for approximately 3% of all upper extremity injuries, with an incidence of greater 640,000 in the USA [33]. Distal radius fractures occur in a bi-modal age distribution in young males from high energy and athletic injuries and in elderly women associated with fragility fractures and a low energy fall, such as from standing [34]. Accepted radiographic parameters exist to aid in distinguishing between those fractures which require operative intervention and those that may be managed non-operatively. These parameters include radial inclination, in normal individuals averaging of 22°, radial height averaging of 11 mm, palmar tilt averaging 12°, and intra-articular step-off of greater than 2 mm [35]. There are many classification systems to describe distal radius fracture patterns. This text will discuss the eponyms associated with distal radius fractures, but the reader should familiarize themselves with the AO classification as well as various other classification schemes.

32.8.1 Colles Fractures

Colles fractures demonstrate radiographically dorsal angulation. Colles fractures may be intra- or extra-articular. They encompass 90% of distal radius fractures. The mechanism of injury is a fall onto outstretched hyperextended, radially deviated upper extremity with the forearm in pronation [36, 37]. Shear forces may also influence the injury pattern.

High energy injuries associated with intra-articular nature and increased comminution are generally more common in younger individuals [36, 37]. On physical examination, there is visible wrist deformity with swelling, ecchymosis, and displacement of the hand dorsally relative to the wrist [38]. Evaluation of the ipsilateral elbow and shoulder for concomitant injuries is important in any distal upper extremity fracture, but in particular with higher energy injuries about the carpus. In addition, neurovascular evaluation, including evaluation for compartment syndrome, is extremely important.

32.8.2 Smith Fractures

Smith fractures may be intra- or extra-articular in nature. They involve volar angulation of the distal radius [39]. The mechanism of injury is a fall onto the flexed wrist with the forearm in supination [40]. On physical examination, there is swelling and ecchymosis about the wrist with volar displacement of the hand relative to the wrist. Treatment is generally guided by the amount of displacement and how the fractures appear radiographically, with context of the accepted radiographic parameters considered.

32.8.3 Barton Fractures

Barton fractures of the distal radius are fracture-dislocations of the wrist when the dorsal or volar rim of the radius articular surface is displaced in relation to the hand and carpus [39, 41]. The mechanism of injury is a fall onto a dorsiflexed wrist with the forearm in pronation which results in a shear force [39]. Barton fractures are unstable fractures which most frequently require surgical intervention. On physical examination, there is swelling and deformity of the wrist with associated ecchymosis and tenderness to palpation diffusely about the wrist.

32.8.4 Chauffeur Fractures

Chauffeur fractures are radial styloid fractures [36, 41]. The fracture occurs as an avulsion fracture where extrinsic wrist ligaments remain attached to the radial styloid fragment. The mechanism of injury is compression of the scaphoid against the radial styloid with the wrist dorsiflexed and in ulnar deviation [36, 42]. These fractures often require operative intervention as well, dependent on the amount of displacement of the articular surface. On physical examination, there is swelling, ecchymosis, and tenderness to palpation about the distal radius.

32.9 Fractures of the Distal Ulna

Fractures of the distal ulna are most commonly associated with distal radius fractures and are usually ulnar styloid fractures. In ulnar styloid fractures, if the pattern is strictly an avulsion fracture, it is often left untreated unless there is associated distal radial-ulnar joint instability [43]. If the fracture pattern extends into the metaphysis or diaphysis of the ulnar shaft and is displaced, open reduction with internal fixation should be considered [43]. The mechanism of injury for distal ulna fractures is a fall onto the outstretched upper extremity with wrist dorsiflexed or volar flexed. On physical examination, there is ecchymosis, swelling, and tenderness to palpation about the ulnar side of the distal forearm.

32.10 Galeazzi Fractures

Galeazzi fractures are considered fracture dislocations. There is associated disruption of the distal radioulnar joint in addition to the comminuted radial shaft fracture [33, 44, 45]. Galeazzi fractures are considered true forearm axis injuries which compromise forearm stability. Mechanism of injury is usually a high energy trauma with an axial load to the affected distal upper extremity [44, 45]. Physical examination demonstrates diffuse swelling in the distal forearm, often with a piano key sign, or dorsal displacement of the distal ulna relative to the distal radius. Tenderness to palpation will be present both at the distal radioulnar joint as well as over the radius shaft fracture. As with distal radius fractures, neurovascular examination is paramount to evaluate for concomitant nerve or vessel injury, and compartment evaluation must be performed to examine for presence of compartment syndrome. Galeazzi fractures are notoriously unstable and warrant operative intervention [44, 45].

32.11 Essex-Lopresti Injuries

Essex-Lopresti injuries also involve disruption of the distal radioulnar joint and deserve mention. The radius and ulna articulate proximally and distally in the forearm. With a mechanism of injury in lower energy falls involving impact to an outstretched upper extremity, a radial head fracture may occur with force transmission through the interosseous membrane linking the radius and ulna together. This longitudinal force exits through the distal radioulnar joint (DRUJ) causing dislocation at the DRUJ [45, 46]. Higher energy traumas causing radial head fractures and even elbow dislocations may also cause a force to be transmitted longitudinally through the interosseous membrane and ultimately causing a DRUJ dislocation [45, 46]. A high index of suspicion for concomitant DRUJ dislocation is required when evaluating forearm and elbow injuries. Physical examination will indicate tenderness to palpation at the level of fracture, the radial head or elbow with concomitant pain and instability at the DRUJ. Ecchymosis and swelling may be present throughout the entire upper extremity. Surgical intervention of the principal fracture as well as stabilization of the DRUJ must be considered [45, 46].

32.12 Conclusion

In conclusion, fractures of the hand and wrist happen frequently and through a variety of injury mechanisms. It is important for the examining clinician to have a high index of suspicion when evaluating a patient with a trauma to the hand or wrist and utilize both clinical and radiographic findings to determine the correct diagnosis and treatment algorithm. Urgent referral to an orthopedic subspecialist is recommended particularly with complex injuries.

References

1. Chung KC, Spilson SV. The frequency and epidemiology of hand and forearm fractures in the United States. J Hand Surg Am. 2001;26(5):908.
2. Capo JT, Hastings H. Metacarpal and phalangeal fractures in athletes. Clin Sports Med. 1998;17(3):491–511.
3. Bachoura A, Ferikes AJ, Lubahn JD. A review of mallet finger and jersey finger injuries in the athlete. Curr Rev Musculoskelet Med. 2017;10(1):1–9.
4. Moradi A, Kachooei AR, Mudgal CS. Mallet fractures. J Hand Surg Am. 2014;39(10):2067–9.
5. Wehbe MA, Schneider LH. Mallet fractures. J Bone Joint Surg. 1984;66(5):658–69.
6. Husain SN, Dietz JF, Kalainov DM, Lautensclager EP. A biomechanical study of distal interphalangeal subluxation after mallet fracture injury. J Hand Surg. 2008;33(1):26–30.
7. Al-Qattan MM. Extra-articular transverse fractures of the base of the distal phalanx (Seymour's fracture) in children and adults. J Hand Surg Br. 2001;26(3):201–6.
8. Ugurlar M, Saka G, Saglam N, Milcan A, Kurtulmus T, Akpinar F. Distal phalanx fracture in adults: Seymour-type fracture. J Hand Surg Eur. 2014;39(3):237–41.
9. Abzug JM, Kozin SH. Seymour fractures. J Hand Surg Am. 2013;38(11):2267–70.
10. Krusche-Mandl I, Kottstorfer J, Thalhammer G, Aldrian S, Erhart J, Platzer P. Seymour fractures: retrospective analysis and therapeutic considerations. J Hand Surg Am. 2013;38(2):258–64.
11. Reyes BA, Ho CA. The high risk of infection with delayed treatment of pen Seymour fractures: Salter-Harris I/II or juxta-epiphyseal fractures of the distal phalanx with associated nailbed laceration. J Pediatr Orthop. 2017;37(4):247–53.
12. Abzug JM, Dua K, Bauer AS, et al. Pediatric phalanx fractures. J Am Acad Orthop Surg. 2016;24(11):e174–83.
13. Oetgen ME, Dodds SD. Non-operative treatment of common finger injuries. Curr Rev Musculoskelet Med. 2008;1(2):97–102.
14. Oak N, Lawton JN. Intra-articular fractures of the hand. Hand Clinics. 2013;29(4):535–49.
15. Stern PJ, Roman RJ, Kiefhaber TR, et al. Pilon fractures of the proximal interphalangeal joint. J Hand Surg Am. 1991;16(5):844–50.
16. Carpenter S, Rohde RS. Treatment of phalangeal fractures. Hand Clinics. 2013;29(4):519–34.
17. Kuhn KM, Dao KD, Shin AY. Volar A1 pulley approach for fixation of avulsion fractures of the base of the proximal phalanx. J Hand Surg Am. 2001;26(4):762–71.
18. Shewring DJ, Thomas RH. Avulsion fractures from the base of the proximal phalanges of the fingers. J Hand Surg Br. 2003;28(1):10–4.
19. Anderson D. Skier's thumb. Aust Fam Phys. 2010;39(8):575–7.
20. Schroeder NS, Goldfarb CA. Thumb ulnar collateral and radial collateral ligament injuries. Clin Sports Med. 2015;34(1):117–26.
21. Carlsen BT, Moran SL. Thumb trauma: bennet fractures, rolando fractures, and ulnar collateral ligament injuries. J Hand Surg Am. 2009;34(5):945–52.
22. Diaz-Garcia R, Waljee JF. Current management of metacarpal fractures. Hand Clinics. 2013;29(4):507–18.
23. Kollitz KM, Hammert WC, Vedder NB, et al. Metacarpal fractures: treatment and complications. Hand (NY). 2014;9(1):16–23.
24. Ruta D, Ozer K. Open fractures of the hand with soft tissue loss. Hand Clinics. 2013;29(4):551–67.
25. Okike K, Bhattacharyya T. Trends in the management of open fractures. A critical analysis. J Bone Joint Surg Am. 2006;88(12):2739.
26. Lilling M, Weinberg H. The mechanism of dorsal fracture dislocation of the fifth carpometacarpal joint. J Hand Surg. 1979;4(4):340–2.
27. Alshryda S, Shah A, Odak S, et al. Acute fractures of the scaphoid bone: systematic review and meta-analysis. Surgeon. 2012;10(4):218–29.
28. Mulligan J, Amblum J. Diagnosis and treatment of scaphoid fracture. Emerg Nurse. 2014;22(3):18–23.
29. Mathoulin CL, Arianni M. Treatment of the scaphoid humpback deformity – is correction of the dorsal intercalated segment instability deformity critical? J Hand Surg Eur Vol. 2018;43(1):13–23.
30. Penteado FT, dos Santos JBG, Caporrino FA, et al. Scaphoid nonunion advanced collapse classifications: a reliability study. J Hand Microsurg. 2012;4(1):12–5.
31. Suh N, Ek ET, Wolfe SW. Carpal fractures. J Hand Surg Am. 2014;39(4):785–91.
32. Apostolides JG, Lifchez SD, Christy MR. Complex and rare fracture patterns in perilunate dislocations. Hand (NY). 2011;6(3):287–94.
33. Wolfe SW. Distal radius fractures. In: Wolfe SW, editor. Green's operative hand surgery. Philadelphia, PA: Elsevier; 2011. p. 561–2.
34. Azad A, Kang HP, Alluri RK, et al. Epidemiological and treatment trends of distal radius fractures across multiple age groups. J Wrist Surg. 2019;8(4):305–11.
35. O'Malley MP, Rodner C, Ritting A, et al. Radiographic interpretation of distal radius fractures: visual estimations versus digital measuring techniques. Hand (NY). 2014;9(4):488–93.
36. Koval KJ, Zuckerman JD. Distal radius. Handbook of fractures. Philadelphia, PA: Lippincott Williams & Wilkins; 2002. p. 226–36.
37. Meena S, Sharma P, Sambharia AK, et al. Fractures of the distal radius: an overview. J Family Med Prim Care. 2014;3(4):325–32.

38. Porrino JA, Maloney E, Scherer K, et al. Fracture of the distal radius: epidemiology and premanagement radiographic characterization. Am J Roentgenol. 2014;203(3):551–9.
39. Caldwell RA, Shorten PL, Morrell NT. Common upper extremity fracture eponyms: a look into what they really mean. J Hand Surg Am. 2019;44(4):331–4.
40. Shah HM, Chung KC. Robert William Smith: his life and his contributions to medicine. J Hand Surg Am. 2008;33(6):948–51.
41. Mauck BM, Swigler CW. Evidence-based review of distal radius fractures. Orthop Clin North Am. 2018;49(2):211–22.
42. Andreotti M, Tonon F, Caruso G, et al. The "Chauffeur fracture": historical origins of an often-forgotten eponym. Hand (NY). 2020;15(2):252–4.
43. Logan AJ, Lindau TR. The management of distal ulnar fractures in adults: a review of the literature and recommendations for treatment. Strategies Trauma Limb Reconstr. 2008;3(2):49–56.
44. Atesok KI, Jupiter JB, Weiss AC. Galeazzi fracture. J Am Acad Orthop Surg. 2011;19(10):623–33.
45. Means KR, Graham TJ. Disorders of the forearm axis. In: Wolfe SW, editor. Green's operative hand surgery. Philadelphia, PA: Elsevier; 2011. p. 837–52.
46. Matson AP, Ruch DS. Management of the essex-lopresti injury. J Wrist Surg. 2016;5(3):172–8.

33 Evaluation of Instability and Joint Dislocations of the Hand

Hassan J. Azimi

33.1 Proximal Interphalangeal Joint

33.1.1 Introduction and Relevant Anatomy

Injuries to the PIP joint are common injuries [1]. When missed or managed incorrectly, they can be troublesome and devastating to the overall function of the finger. Even with proper management, the PIP joint is prone to stiffness and pain. Diligent and careful evaluation is particularly important when assessing the PIP joint.

The PIP joint is a bicondylar joint with slight asymmetry of the proximal phalanx condyles allowing for mild supination with PIP joint flexion [2]. Normal range of motion is 100–110° [3]. Stability of the joint comes from a combination of osseous anatomy, collateral ligaments, and volar plate. The hinge joint of the PIP allows for the collateral ligaments to be tight in all degrees of flexion. The collateral ligament originates on the lateral aspect of the proximal phalanx condyle, dorsal to the mid-axis, and inserts on the base of the middle phalanx slightly volar to the mid-axis. The collateral ligaments function to provide lateral stability to the PIP joint. The volar plate attaches firmly to the base of the middle phalanx and to the volar and lateral margins of the proximal phalanx, allowing for a central recess during flexion. The checkrein ligaments are a confluence of fascial structures that allows for a firm attachment of the volar plate at the lateral margins of the proximal phalanx. The volar plate and checkrein ligaments function to prevent hyperextension of the PIP joint [4].

33.1.2 History and Presentation

Injury to the PIP joint is typically an axial load force to a fully extended digit [5]. In athletics, this is typically associated in sports with larger balls such as basketball. The long lever arm from the force directed on the fingertip acts on the fixed hinge joint resulting in injury. Careful history should be obtained, as the chronicity and time from the injury may change management. Also, previous injuries to the digit and joint should be clarified, as a chronic injury with reaggravation may also change management. Previous treatments, including any reductions that may have been performed by the patient, bystander, or another provider should be noted. While the patient may present to the emergency department or urgent care with a deformity, the patient is unlikely to present to the orthopedic office with the joint dislocation [5]. Thus, the direction of the dislocation should be obtained if the patient can recall. Radiographs from the urgent care or

H. J. Azimi (✉)
San Diego Hand Specialists, San Diego, CA, USA

J. G. Lane et al. (eds.), *The Art of the Musculoskeletal Physical Exam*,
https://doi.org/10.1007/978-3-031-24404-9_33

emergency department if taken prior to a reduction will be particularly helpful. Classification and management of injuries to the PIP joint are in general described by the direction of the dislocation and will guide your examination to be more focused.

33.1.3 Physical Examination

Inspection of the injured digit will demonstrate edema and ecchymosis around the PIP joint. Dorsal PIP dislocations are most common and will often have circumferential edema. Volar dislocations are often associated with central slip injuries. Central slip injuries will have preferential edema to the dorsal aspect of the PIP joint (Fig. 33.1). If the examiner covers the dorsal aspect of a digit that has a central slip injury, the volar half of the digit will appear uninjured (Fig. 33.1b). Conversely, if the examiner covers the volar half of the digit, the dorsal half appears edematous and obviously injured (Fig. 33.1c). Resting posture of the injured digit should be noted. Injury to the PIP joint often results in the patient resting in a position of relative flexion. The distinction between a flexion contracture and extensor lag is important. An extensor lag is passively correctable by the examiner, who is able to fully extend the PIP joint, however, the patient is unable to actively maintain this extended posture. Extensor lags are most likely caused by an injury to the extensor mechanism, in the case of the PIP joint, a central slip injury. A flexion contracture is caused by contracture and adhesions of soft tissues about the joint and may be present if the patient is presenting delayed or after prolonged immobilization. Patients will often hold the PIP in relative flexion for comfort, avoiding range of motion. A digital block with local anesthetic should be performed in the office if the distinction between an extensor lag or flexion contracture cannot be obtained due to pain. Angular or rotational deformity may be associated with lateral dislocations or rotatory dislocations. Lateral or rotatory dislocations are associated with a

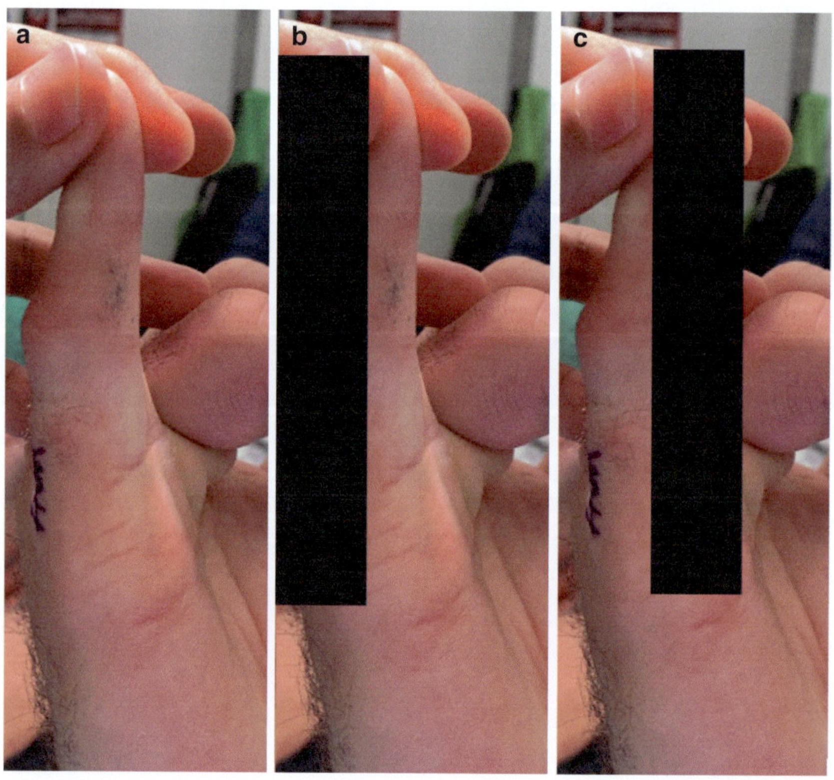

Fig. 33.1 a–c: (**a**) Patient with central slip injury associated with a volar dislocation. Note the edema is located primarily over the dorsal aspect of the PIP joint. (**b**) Covering the dorsal half of the digit demonstrates a normal appearing volar digit. (**c**) Covering the volar half of the digit demonstrates an obviously edematous PIP

block to reduction with the condyle button-holed between the central slip and lateral band.

Palpation of the PIP joint for an area of maximal point tenderness will help to determine the direction of the dislocation if the patient was unsure. Volar dislocations will be more painful dorsally, at the central slip injury. Dorsal dislocations may be associated with volar fractures of the middle phalanx base and volar plate injuries, thus maximal tenderness will be volar. Lateral or rotatory dislocations will have pain laterally along the collateral ligament origin or insertion.

Assessment of range of motion (ROM) of the distal interphalangeal (DIP) joint, PIP, and metacarpophalangeal (MCP) joint should be noted. Flexion contractures should be noted and are common during treatment of PIP injuries. In subacute or chronic central slip injuries, the patient may develop a boutonniere deformity (Fig. 33.2) and limited ROM. The boutonniere deformity is characterized by PIP joint flexion and DIP hyperextension. Dorsal dislocations causing volar plate insufficiency can lead to swan neck deformity. Swan neck deformity is characterized by hyperextension of the PIP joint and DIP joint flexion.

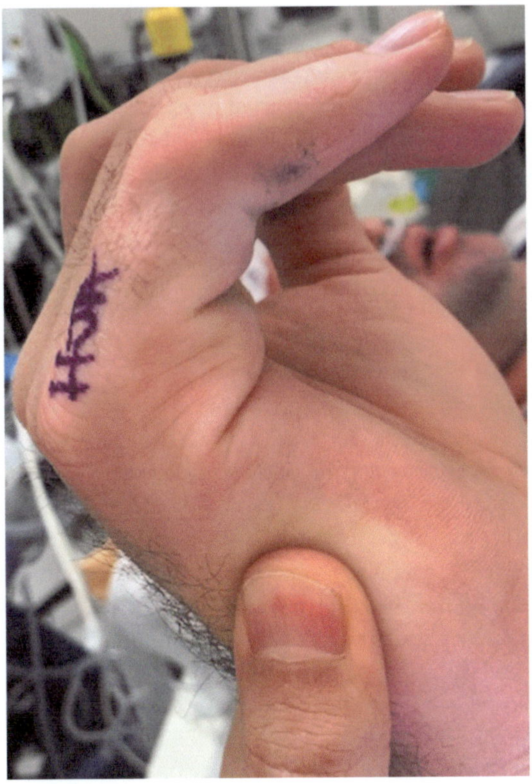

Fig. 33.2 Subacute central slip injury. Note the boutonniere posture with flexion at the PIP joint and hyperextension of the DIP

Provocative maneuvers for PIP joint instability and joint dislocations can be performed. Lateral stability of the PIP joint is assessed by stabilizing the proximal phalanx and placing a radial and then an ulnar deviated force to the middle phalanx. Lateral stress has been studied in cadaver models with serial sectioning [6]. These studies have shown that an intact collateral ligament allows for 5° of lateral deviation and 9° of supination throughout flexion and extension. Sectioning of the collateral ligament allows 20° of lateral deviation, which would be consistent with a grade 3 injury. When the accessory collateral ligament, volar plate, central slip and lateral band remain intact, or half of the proper collateral ligament remains intact with all other structures sectioned, then the PIP joint is not destabilized. Based on this data, 10° of lateral deviation in a fully extended position and 20° of lateral deviation in 30° of flexion is considered indicative of a complete tear of the proper collateral ligament. Greater lateral deviation in these positions would indicate tears of secondary stabilizers such as the volar plate and accessory collateral ligament. Stress examination should be always compared to adjacent or contralateral uninjured digits. Rotational deformity is assessed by observing the patient making a fist with both hands and looking for crossing of the injured digit with adjacent non-injured digits.

33.1.4 Radiographic Evaluation

Proper radiographic evaluation of the PIP is crucial in determining whether an injury is stable versus unstable. Plain radiographs should be obtained including with a minimum of a posteroanterior (PA) and lateral view of the affected digit. Radiographs of the digit often are easier to obtain optimal views than imaging of the entire hand. The lateral radiograph is most crucial. Dislocations are obvious; however, subluxation can be easily overlooked. Fluoroscopic imaging is often very helpful to obtain perfect lateral imaging when conventional radiographs are unsatisfactory. Dynamic fluoroscopy is particularly helpful when evaluating for joint stability. Dorsal dislocations are often associated with fracture of the middle phalanx base and are more stable in PIP flexion than in extension. Fluoroscopy of the PIP joint with varying ROM is helpful in determining stability. Concentric reductions with 30° or less of flexion and with less than 20% of the joint surface being fractured are considered stable. Injuries requiring greater than 30° of flexion or with 30–50% of joint involvement, are considered tenuous. Finally, greater than 50% joint involvement is considered unstable. The PIP joint should be assessed for proper gliding motion along the axis of the proximal phalanx condyles. Hinging of the fracture edge with flexion of the PIP joint should be noted and is a sign of instability. Static lateral radiographs should be analyzed for a V sign. The V sign is a radiolucent V that is created when the middle phalanx is dorsally subluxated; this is a sign of instability.

33.2 Metacarpophalangeal Joint

33.2.1 Introduction and Relevant Anatomy

Injuries to the MCP joint of the index, middle, ring, and small fingers are less common than the thumb. Injuries that do occur tend to be to the collateral ligaments with ulnar directed force and dorsal dislocations. The MCP joint is a condyloid joint, allowing for motion in three axes; flexion-extension, adduction-abduction, and circumduction. Normal range of motion is 15° of hyperextension to 100° of flexion, 10–20° of radial deviation, and 20–40° of ulnar deviation. The metacarpal head is trapezoid in shape, wider palmar compared to dorsal, allowing for more contact with the proximal phalanx in flexion and, thus, increased stability. The shape of the metacarpal head leads to a cam effect with the collateral ligaments being tight in flexion and loose in extension. The increased stability in flexion allows for more stability during grasping. The collateral ligament being loose in extension and tighter in flexion also explains why the MCP joint has a tendency to get stiff in extension, with a loss of joint flexion as the collateral ligaments shorten and scar. This stresses the importance of immobilizing the MCP joint in flexion after injury.

The MCP synovial capsule lines the joint circumferentially and is stabilized by thickening of connective tissue on all sides. On the volar side, the joint is reinforced by the volar plate, which is continuous with the deep transverse metacarpal ligament, stabilizing the MCP joints to each other. During dorsal dislocation, the volar plate tends to fail at its origin on the metacarpal neck. Laterally, the joint is stabilized to varus and valgus force primarily by the proper collateral ligaments. Secondary stabilizers include the accessory collateral ligament, dorsal capsule, and volar plate. The collateral ligament originates dorsal to axis of rotation on the metacarpal head, with the center being at about 40% from the dorsal aspect of the metacarpal head compared to its total height [7]. Its insertion onto the proximal phalanx base is slightly volar with the center being 46% from the volar border [7]. Dorsally, the capsule is reinforced by the extensor tendon and sagittal bands.

Dorsal dislocations of the MCP are classified as simple or complex. Simple dislocations are dislocations of the joint where the proximal phalanx continues to have contact with the metacarpal head while complex dislocations have loss of contact between the phalanx and metacarpal, frequently with interposed soft tissue. This distinction is important as complex dislocations are usually irreducible due to the pathoanatomy of the injury. As the proximal phalanx moves dorsal to the metacarpal head, the volar plate, which tears at its proximal origin, becomes interposed between the articular surfaces. With the volar plate interposed solely, the joint would likely reduce with traction. However, the surrounding volar tendons create a sling that prevents reduction with traction. In the index finger, the lumbrical and flexor tendons surround the metacarpal head and can create a noose preventing reduction. In the small finger, the lumbrical and flexor tendon radially and hypothenar muscles ulnarly create a similar noose.

33.2.1.1 History and Examination

Collateral ligament injuries to the MCP of digits not including the thumb, are uncommon and/or underreported. Injuries to the small finger were the most common, followed by the index then the middle fingers [8]. Patients will present after an injury in which they reported they "jammed" their finger when the finger remains painful. The mechanism of injury is likely to be an ulnar deviation force with the MCP in a flexed position. The radial collateral ligament (RCL) is the primary stabilizer of the index finger during pinch. The patient may complain of pain or weakness with pinch.

Examination of the MCP may demonstrate edema of the joint with thickening of the soft tissue. Tenderness to palpation at the origin or insertion will be present, depending on the location of the tear. The RCL is most likely to tear at its origin or insertion and less likely midsubstance [9]. The patient may have a deformity at rest with pronation of the index finger proximal phalanx and scissoring of the index under the middle finger [10]. Severe injuries involving the volar plate, RCL and dorsal capsule will result in pronation and ulnar drift as the proximal phalanx rotates around the intact UCL. Manipulation and stress testing will demonstrate pain with an ulnar-deviated force. Stress testing of the RCL should be performed in flexion. If the patient can tolerate MCP flexion, the MCP should be flexed to 60–90° and an ulnar force applied. It is vital that the examiner compare stability to the contralateral normal side to determine what is normal for each patient. Grade 1 injuries will demonstrate pain but no increased laxity. Grade 2 injuries will demonstrate laxity with an endpoint. Grade 3 injuries will demonstrate laxity and no endpoint. Biomechanical analysis of the MCP joint demonstrates about 15–20° of laxity in varus and valgus testing in the neutral position with a decrease in laxity to about 10° in 90° of flexion. While the small and ring fingers are more stable in flexion, the difference in laxity between 0 and 90° is less in the index and middle fingers [11].

Dorsal dislocations are usually caused by forced hyperextension and typically the MCP joint dislocates dorsally. Dislocations of the MCP joints are uncommon, occurring in the index finger most commonly, followed by the small finger. In a simple subluxation, the proximal phalanx does not completely dislocate and the volar plate does not become fully interposed. This prevents the volar structures from creating a noose around the metacarpal head. When a patient presents with the proximal phalanx subluxated, the MCP will be locked in hyperextension of 60–90°. This presentation and history and initial posture should cue the examining physician to avoid traction and hyperextension, as it will accentuate the deformity and potentially convert the subluxated joint into a complex complete dislocation. It may be the case that an attempted reduction has been performed prior to presentation, already converting a subluxated joint into a complete dislocation.

Complex complete dislocations (Fig. 33.3) will present with a patient's MCP locked in slight hyperextension. This is distinct from the more hyperextended position of the subluxated MCP. Inspection of the palm may demonstrate puckering of the skin with the dislocated digit deviated centrally. In complete dislocations, palpation will reveal a dorsally dislocated proximal phalanx.

33.2.1.2 Radiographic Evaluation

Standard hand radiographs are sufficient for evaluation for MCP injuries. Radiographs will confirm dorsal dislocations and subluxations. Subtle findings to confirm dislocation include an entrapped sesamoid, indicating the volar plate is entrapped. Occasionally, injuries will include concomitant fractures of the metacarpal head, these are best evaluated with a Brewerton view. This radiograph is an anteroposterior image with the fingers placed flat and the palm raised 60°. Severe injuries to the RCL will show volar subluxation of the proximal phalanx on the metacarpal head as well as ulnar drift and pronation [3].

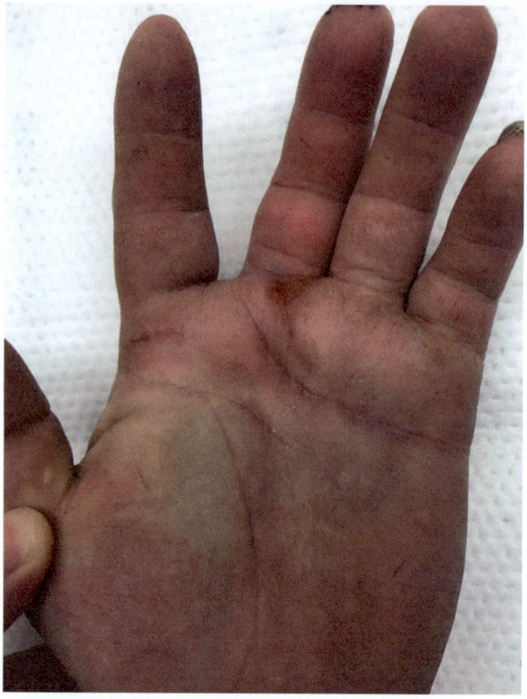

Fig. 33.3 Palmar view of complete dorsal dislocation of the middle finger MCP. Notice the deviation of the digit to the mid-line, puckering of the palmar skin at the level of the A1 pulley and distal palmar crease and the mild hyperextension of the MCP joint

33.3 Thumb Metacarpophalangeal Joint

33.3.1 Introduction and Relevant Anatomy

The thumb MCP joint is more commonly injured in comparison to the other digits with collateral ligament injuries being up to 86% of injuries to the thumb [12]. The ulnar collateral ligament (UCL) is more commonly injured than the RCL [13]. The metacarpal head is a single condyle and acts as a hinge joint with the primary motion being flexion and extension. The thumb MCP anatomy is variable, 10% of the population will have a flat metacarpal head, which may lead to increased radial and ulnar deviation.

The proper collateral ligaments provide the primary restraint to varus and valgus stress. The thumb proper collateral ligaments are more obliquely positioned in comparison to other digits with the origin being more dorsal and the insertion being more volar. Cadaver studies have shown the angle to be about 30° from proximal dorsal to distal volar [14]. Secondary stabilizers to the thumb MCP joint include the accessory collateral ligament, volar plate, and dorsal capsule. The accessory collateral ligament is volar to the proper collateral ligament and travels from proximal dorsal to distal volar at a 90 degree angle [14]. The proper collateral ligament is tight in MCP flexion and the accessory collateral ligament is tight in extension. The volar plate is similar to the other MCP joints in that it is thicker distally. The volar plate provides stability to varus and valgus stresses with the joint in full extension. Unique to the thumb MCP are the presence of radial and ulnar sesamoids which have insertional attachments for the flexor pollicis brevis and abductor pollicis brevis on the radial sesamoid and adductor pollicis on the ulnar sesamoid. These provide dynamic stabilization both to the volar joint and the lateral joint due to secondary insertions onto the extensor mechanism [12].

33.3.2 History and Examination

Acute injuries to the thumb UCL are termed skier's thumb. The mechanism of injury for an acute UCL injury is a radial abduction force and/or hyperextension. While the eponym alludes to skiing injuries, any athletic activities, fall or accident can lead to this injury. Acute injuries often present with edema, ecchymosis, and pain. Palpation should reveal point tenderness at the location of the tear, the distal insertion of the UCL is torn most often [15, 16]. While the entire MCP may be tender, maximal tenderness will often reveal the tear location. A Stener lesion may be present if the injury is severe enough, with displacement of the collateral ligament due to the overlying adductor aponeurosis. The examiner can carefully palpate for a mass that reveals a Stener lesion; however, ultrasound and MRI have been found to be more accurate [17].

Chronic thumb UCL injuries are termed gamekeeper's thumb, named after Scottish gamekeepers who would fracture the neck of rabbits between their thumb and index finger. The chronic attenuation of the UCL due to this type of pinch activity leads to instability of the MCP joint. Chronic UCL injuries will present with pain, weakness in pinch, and static deformity.

RCL injuries are less common than UCL injuries of the thumb. The mechanism of injury includes forced adduction of the thumb MCP. Presentation is similar with swelling, pain, and ecchymosis. There is not an overlying adductor aponeurosis so displaced tears are unlikely to be present, as such, palpation will not reveal a "Stener lesion." Rotational deformity of the MCP is more likely to be present with RCL tears, with palmar subluxation of the radial proximal phalanx. The examiner should observe this distinction from the UCL injury. Unlike the UCL, the RCL can tear with equal frequency at its origin and insertion. As such, palpation will reveal a point of maximal tenderness where the tear is located.

Stress of the radial and ulnar collateral ligaments should be performed and compared to the contralateral side. Examination is performed with the examiner stabilizing the thumb metacarpal with one hand and stressing the proximal phalanx with the other. As noted above, collateral ligament injuries may rest in a deviated position due to the forces placed across the joint from its dynamic stabilizers. Thus, it is important to ensure the thumb is resting in a neutral position. A goniometer is recommended to note the degree of deviation if present and to ensure the MCP is corrected to neutral. It is also important to ensure that the thumb is also in a rotationally neutral position. Evaluation of the thumb UCL with the MCP in pronation will give false negatives and examination in supination will give false positives [18]. The MCP is tested in both full extension to evaluate the volar plate and accessory collateral ligament and also in 30–40° of flexion to evaluate the proper collateral ligament [19]. Radial and ulnar deviated forces are applied to the proximal phalanx to test the UCL and RCL, respectively. The thumb can easily rotate and the examiner should be careful to not rotate the thumb metacarpal through the MCP during this maneuver. The grade of the injury is determined by the degree of laxity and the presence of a solid endpoint. Laxity is defined as greater than 30° of deviation or 15° greater than the contralateral side. Grade 1 injuries will demonstrate pain but no increased laxity. Grade 2 injuries will demonstrate laxity with an endpoint. Grade 3 injuries will demonstrate laxity and no endpoint [20]. Side-to-side variation can occur with one study of 200 thumbs showing that 34% of patients had greater than 10° of difference in full extension and 23% had greater than 10° of difference in 30° of flexion [20]. This data suggests that detecting the presence or lack of an endpoint is the most important component of this examination maneuver.

With both UCL and RCL injuries, if the injury is severe the dorsal capsule, accessory collateral ligament, and volar plate may also be torn. When this occurs, it may result in volar subluxation and rotatory malalignment of the MCP. When the RCL is torn and UCL remains intact the thumb pronates around the intact UCL and adductor aponeurosis. This will also cause a dorsal and radial prominence of the metacarpal head. The opposite is present in UCL injuries, with supination of the proximal phalanx on the metacarpal head [21]. While these deformities can be appreciated on physical examination, they are often subtle and should be confirmed by radiographs and compared to the contralateral side. PA, lateral, and oblique radiographs of the thumb should be ordered, a standard hand series is insufficient for evaluation of the thumb. Stress radiographs of the thumb MCP can be helpful but are not necessary [12].

References

1. Eaton RG, Littler JW. Joint injuries and their sequelae. Clin Plast Surg. 1976;3(1):85–98. http://www.ncbi.nlm.nih.gov/pubmed/1248198
2. Leibovic SJ, Bowers WH. Anatomy of the proximal interphalangeal joint. Hand Clin. 1994;10(2):169–78. http://www.ncbi.nlm.nih.gov/pubmed/8040195
3. Rozmaryn LM. The collateral ligament of the digits of the hand: anatomy, physiology, biomechanics, injury, and treatment. J Hand Surg Am. 2017;42(11):904–15.

4. Allison DM. Anatomy of the collateral ligaments of the proximal interphalangeal joint. J Hand Surg Am. 2005;30(5):1026–31. http://www.ncbi.nlm.nih.gov/pubmed/16182063
5. Saitta BH, Wolf JM. Treating Proximal interphalangeal joint dislocations. Hand Clinics. 2018;34(2):139–48.
6. Minamikawa Y, Horii E, Amadio PC, Cooney WP, Linscheid RL, An KN. Stability and constraint of the proximal interphalangeal joint. J Hand Surg Am. 1993;18(2):198–204. http://www.ncbi.nlm.nih.gov/pubmed/8463578
7. Dy CJ, Tucker SM, Kok PL, Hearns KA, Carlson MG. Anatomy of the radial collateral ligament of the index metacarpophalangeal joint. J Hand Surg Am. 2013;38(1):124–8.
8. Hayek SN, El Khatib A, Ibrahim A, Tamar Kabakian-Khasholian BA. Collateral ligament injury of the metacarpophalangeal joint: a meta-analysis of incidence, diagnosis, management and outcome. J Hand Surg Am. 2010;35(10):55–6.
9. Kang L, Rosen A, Potter HG, Weiland AJ. Rupture of the radial collateral ligament of the index metacarpophalangeal joint: diagnosis and surgical treatment. J Hand Surg Am. 2007;32(6):789–94. http://www.ncbi.nlm.nih.gov/pubmed/17606055
10. Gaston RG, Lourie GM. Radial collateral ligament injury of the index metacarpophalangeal joint: An underreported but important injury. J Hand Surg Am. 2006;31(8):1355–61.
11. Werner D, Kozin SH, Brozovich M, Porter ST, Junkin D, Seigler S. The biomechanical properties of the finger metacarpophalangeal joints to varus and valgus stress. J Hand Surg Am. 2003;28(6):1044–51.
12. Daley D, Geary M, Gaston RG. Thumb metacarpophalangeal ulnar and radial collateral ligament injuries. Clin Sports Med. 2020;39(2):443–55.
13. Moberg E, Stener B. Injuries to the ligaments of the thumb and fingers; diagnosis, treatment and prognosis. Acta Chir Scand. 1953;106(2–3):166–86. http://www.ncbi.nlm.nih.gov/pubmed/13137829
14. Bean CH, Tencer AF, Trumble TE. The effect of thumb metacarpophalangeal ulnar collateral ligament attachment site on joint range of motion: an in vitro study. J Hand Surg Am. 1999;24(2):283–7. http://www.ncbi.nlm.nih.gov/pubmed/10194011
15. Stener B. Skeletal injuries associated with rupture of the ulnar collateral ligament of the metacarpophalangeal joint of the thumb. A clinical and anatomical study. Acta Chir Scand. 1963;125:583–6. http://www.ncbi.nlm.nih.gov/pubmed/13983826
16. Tang P. Collateral ligament injuries of the thumb metacarpophalangeal joint. J Am Acad Orthop Surg. 2011;19(5):287–96. http://www.ncbi.nlm.nih.gov/pubmed/21536628
17. Hergan K, Mittler C, Oser W. Ulnar collateral ligament: differentiation of displaced and nondisplaced tears with US and MR imaging. Radiology. 1995;194(1):65–71. http://www.ncbi.nlm.nih.gov/pubmed/7997584
18. Mayer SW, Ruch DS, Leversedge FJ. The influence of thumb metacarpophalangeal joint rotation on the evaluation of ulnar collateral ligament injuries: a biomechanical study in a cadaver model. J Hand Surg Am. 2014;39(3):474–9. http://www.ncbi.nlm.nih.gov/pubmed/24495627
19. Gluck JS, Balutis EC, Glickel SZ. Thumb ligament injuries. J Hand Surg Am. 2015;40(4):835–42.
20. Malik AK, Morris T, Chou D, Sorene E, Taylor E. Clinical testing of ulnar collateral ligament injuries of the thumb. J Hand Surg Eur. 2009;34(3):363–6. http://www.ncbi.nlm.nih.gov/pubmed/19349298
21. Smith RJ. Post-traumatic instability of the metacarpophalangeal joint of the thumb. J Bone Joint Surg Am. 1977;59(1):14–21. http://www.ncbi.nlm.nih.gov/pubmed/833169

34 Rheumatoid and Other Arthritis of the Wrist and Hand

Gregory R. Mack and Dori J. Neill Cage

34.1 Introduction

Understanding the changes which occur in the hands and wrists with arthritis is essential for performing a comprehensive physical exam of these areas. Osteoarthritis, rheumatoid arthritis, crystalline arthropathy, psoriatic arthritis, and Systemic Lupus Erythematosus (SLE) have characteristic physical examination features. Hand and wrist changes may be the initial presenting abnormalities, and critical examination can frequently lead to a diagnosis.

Understanding the changes seen with inflammatory, crystalline, and other wrist and hand arthritis is important not only for physicians performing surgical treatment on these patients, but also for the physician evaluating the patient for possible infection, fracture, or dislocation. The changes seen in wrist and hand arthritis may mimic infection with warmth, swelling, and erythema. Bony erosions and structural deformities may mimic fractures and traumatic dislocations. Atraumatic dislocations may occur from rheumatoid arthritis and SLE, as well as advanced basilar thumb osteoarthritis. Treatment of a chronic arthritic dislocation involves correction of the underlying deformity and is addressed differently than an acute traumatic dislocation.

G. R. Mack (✉) · D. J. Neill Cage
San Diego Hand Specialists, San Diego, CA, USA

34.2 Common Features of Wrist and Hand Arthritis

Different types of arthritis affect the wrist and hand in unique ways. Osteoarthritis affects primarily the distal interphalangeal (DIP) joints and thumb carpal metacarpal (CMC) joint. Less often it involves the proximal interphalangeal (PIP) and wrist scaphotrapeziotrapezoidal (STT) joints. Joint enlargement occurs at the distal interphalangeal joints with loss of full DIP joint extension, and mucous cysts may be present with associated nail grooves. Osteoarthritis can also affect the proximal interphalangeal joints with limitation of range of motion, angular deformities, and pain. Arthritis of the base of the thumb has a higher prevalence in women but occurs frequently in both males and females. Basilar thumb arthritis physical examination features include deformity at the base of the thumb, enlargement of the thumb carpometacarpal joint, dorsal subluxation of the base of the metacarpal, and adduction contracture. Carpometacarpal grind testing axially loads the carpometacarpal joint and rotates the metacarpal, eliciting crepitus and pain. Osteoarthritis typically does not involve significant soft tissue swelling about the joints although variants known as inflammatory osteoarthritis do occur and have components that mimic inflammatory arthritis. Osteoarthritis generally does not involve the metacarpal phalangeal joints, whereas hemochromatosis and calcium

J. G. Lane et al. (eds.), *The Art of the Musculoskeletal Physical Exam*,
https://doi.org/10.1007/978-3-031-24404-9_34

pyrophosphate disease (CPPD) may result in osteoarthritis of the metacarpal phalangeal joints on the radial hand, mainly involving the index and long fingers.

Pseudogout occurs most frequently in the wrist with pain and swelling and mimics infection. As the treatment of pseudogout is nonoperative, it is important to be able to differentiate its inflammation from infection. Pseudogout generally occurs in older individuals and causes localized warmth and swelling without red streaking up the arm, fever, or systemic symptoms.

Gout is less common in the upper than the lower extremities. Like pseudogout, gout may present with wrist and finger swelling. Gouty tophi may erode soft tissue and give the appearance of a draining infection; they may also develop secondary infections. Gout may also cause thickening of the flexor tendon sheath and can result in flexor and extensor tendon ruptures. Physical examination of a finger with gout will demonstrate firm thickening which is generally irregular and can involve erosions with white granular crystals visible.

Psoriatic arthritis has many overlapping features with osteoarthritis, affecting primarily the distal interphalangeal joints. Nail pitting and patches of psoriasis are classic features of psoriatic arthritis. The so-called sausage finger swelling is a more advanced finding seen in psoriatic arthritis and can mimic infection.

Systemic lupus erythematosus has many of the same features as rheumatoid arthritis in terms of hand deformities by physical exam but tends to have less joint destruction on radiographs. Thinning of the fingertips, ulcerations of the distal digits, and tightening of the skin are all features seen in SLE.

34.3 Rheumatoid Arthritis

Rheumatoid arthritis is an inflammatory disease which presents with synovitis of the joints and tendons causing characteristic patterns of deformity. Examination of the hand with rheumatoid arthritis must not only address the patient's functional complaints but also be comprehensive in its scope. Patients with arthritis may present with specific complaints but are also at risk for further functional difficulties. Rheumatoid arthritis and its related conditions are progressive in nature, and understanding the basic pathomechanics is key to recognizing specific clinical patterns and identifying critical physical findings.

Patients with rheumatoid arthritis experience functional difficulties due to joint damage, soft tissue attenuation, ligament rupture, and/or tendon subluxation. These factors contribute to the imbalance of the dynamic forces which maintain the posture of the wrist, fingers, or thumb. As a result, pinch and grasp are affected, and activities of daily living are compromised. While multiple deformities may present in the same hand, there are predominant patterns which affect basic hand functions. A systematic approach to the patient will simplify the physical examination, bringing to light not only the basis for the patient's present complaints but also elements to be considered for future care.

34.4 General Considerations

Examination of the patient with arthritis is best preceded by a clear and relevant history including the duration of the disease, joints affected in all four limbs, spinal involvement, treatment to date, and the reason for the present visit.

For the hand and upper extremity, the initial examination is best done systematically. Proximal issues at the shoulder, elbow, or forearm may significantly limit upper extremity function and need addressing. Distal problems of the wrist or hand may adversely affect proximal function, for example, wrist or forearm stiffness which may lead to compensatory maladaptive postures at the shoulder.

The essential steps of evaluation include measurement of range of motion, identification of specific findings and deformities, functional assessment of current findings, and anticipation of further progression of hand impairment.

Findings in the rheumatoid hand are generally bilateral and frequently symmetrical. The patient presenting with problems on the one hand will invariably have findings on the other, even if asymptomatic. Similarly, deformities of the digit are rarely confined to a single finger. In every stage of the disease, rheumatoid arthritis effects will usually be generalized.

34.4.1 Range of Motion

It is important to measure joint range of motion at the initial evaluation for both a current assessment and to provide a baseline against which progression of disease and deformity may be identified at future examinations.

Measurements are recorded bilaterally and in a systematic fashion. Our practice is to record standard measurements at the shoulder, elbow, forearms, wrists, thumbs, and fingers in a proximal to distal order, entering values as right/left. Following a consistent order and using preprinted diagrams or electronic macros ensures accuracy and minimizes omission of potentially helpful values and documentation of deformities.

Joint range of motion will determine which specific surgical procedures will be predictable choices for a given deformity. Depending on the degree of contracture, flexion or extension contractures of the MP or PIP joints may be treated with capsulotomy. For example, PIP contractures of 30° or less may not be substantially improved by volar capsulotomy. Extension contracture of the PIP joint is more likely to benefit from contracture release when it involves 60° or less of flexion than for contractures which allow 60° or more of flexion. The functional gains of addressing an extension contracture which already allows 60 degrees of PIP flexion are not predictable. Similar criteria are helpful for decision-making at the MCP joint.

34.5 Identification of Findings and Deformities

In a single hand, a patient may demonstrate swan neck deformity, boutonniere deformity, joint crepitus, joint subluxation, joint instability, and tendon rupture. Each will need to be assessed and documented. Clear and accurate documentation of findings, deformities, and their severity provides the basis on which treatment decisions are based.

Assessment of joint laxity, instability, subluxation, and dislocation is essential to evaluation of the rheumatoid hand. Although any joint may be affected, the most common joint disruptions include dorsal displacement of the distal ulna, supination deformity of the carpus with respect to the forearm, volar-ulnar subluxation or dislocation of the MCP joints of the fingers, and joint abnormalities associated with boutonniere or swan neck deformities of the fingers or thumb.

34.6 Functional Assessment

The deformities alone do not constitute indications for surgical treatment unless the surgical procedures being considered are predictably expected to improve function or decrease pain that is unresponsive to conservative treatment. The broader goal of evaluation is to identify functional impairments and to direct treatment to preserve versus restore hand capabilities.

A quick functional assessment can be made by evaluating the patient's ability to pinch and grasp. The patient may be asked to make a circle by pinching the tip of the thumb against the index finger, holding it firmly as the examiner tries to "break" the circle. Similarly, the patient may be asked to demonstrate grasping a glass-size object. With these simple maneuvers, the examiner may observe the effectiveness of pinch and grasp for their mobility, strength, and prehension.

Pinch and grasp require movement of the thumb from the radial side of the hand to oppose and contact a finger or fingers from the ulnar side. This requires stability of the thumb and opposing fingers to generate strength and position to be effective. The strength with which the two sides oppose each other will be limited by the weaker of the two. Position is determined by the patient's voluntary ability to flex, close, and oppose the thumb to the finger or fingers, and his or her ability to voluntarily open the hand. In rheumatoid arthritis, this ability may be limited by joint damage or contracture, digital deformity, and/or weakness. Before a patient can grasp and lift a glass, he or she needs to expand the hand beyond the object's diameter. The elements of functional examination thus include assessment of voluntary opening, voluntary closing, effectiveness of opposition, and stability of the digits.

Some patients will be unable to voluntarily extend the fingers for grasp due do tendon ruptures or radial nerve palsy. The patient may then substitute extensor tenodesis function for voluntary finger extension by flexing the wrist.

A third functional capacity of the hand has been termed "hook," in which the fingers as a group can wrap around a handle such as would be found on door, suitcase, or handbag. This requires effective flexion of multiple interphalangeal joints at the same time. The hook function is most effective when full voluntary opening and full voluntary closing are present. For patients with rheumatoid arthritis and swan neck deformities, this function may be possible if those deformities are flexible, but if they are rigid, with PIP extension contractures, hook is impossible. Although this constitutes a significant impairment, loss of pinch and grasp is a far greater loss of functional capacity.

34.6.1 Inspecting the Wrist

Rheumatoid arthritis at the wrist may progress through four stages, including synovitis, instability, subluxation, and joint destruction. The areas which are most susceptible to attrition and secondary instability are the triangular fibrocartilage complex on the ulnar side of the wrist and the scapholunate ligament centrally. The earliest physical finding may be subtle swelling and tenderness of the ulnar fovea (overlying the sacciform recess) or the radiocarpal joint dorsally. Swelling in this area is to be differentiated from swelling due to extensor tenosynovitis. Either may demonstrate crepitus, but tenosynovium will move with the extensor tendons, whereas joint synovium will be less affected by tendon movement. Because both are commonly observed on the dorsum of the wrist, they need to be distinguished from wrist ganglia, which tend to be unilocular and may transilluminate.

Involvement of the TFCC results in increased translocation of the distal ulna in the dorsal-to-volar plane in any of the three standard testing positions. These positions include pronation, supination, and neutral rotation of the forearm.

As synovial disease weakens the critical anchors of the wrist, the consequence of ligament laxity is instability. It is not unusual for a rheumatoid wrist to demonstrate a dorsal intercalary segmental instability (DISI) pattern of deformity. More commonly, the distal ulna displaces and becomes hypermobile when the TFCC gives way and the ulna is disconnected from the ulnar side of the carpus.

As ligament attenuation progresses, the distal ulna may appear to subluxate dorsally. In fact, this abnormal position of the distal ulna is more commonly due to volar displacement of the ulnar side of the carpus, such that the hand and the carpus together as a unit tend to supinate with respect to the forearm. The result is exaggerated prominence of the distal ulna, which appears dislocated in pronation but may return to its proper location when the forearm is brought back to neutral or supination. In addition, the wrist may translocate ulnarly [1].

Unlike traumatic injury, scapholunate ligament laxity in rheumatoid arthritis tends not to present as wrist instability or clicking. Early in the disease, it may be a radiographic finding due to ligament attenuation and thereby a precursor of progressive synovial invasion and joint destruction. In later cases, the swelling and pain of aggressive wrist synovitis and altered radiocarpal relationships, including wrist dislocation and/or contracture, overshadow its involvement.

34.6.2 Fingers

The two classically described deformities of the fingers in rheumatoid arthritis are swan neck deformity and boutonniere deformity [2]. Both occur when the tightness of the intrinsic muscles overpower the stability of the interphalangeal joints. The joints may also develop instability due to the stress of normal use associated with invasive synovitis, and/or the anti-metabolic effect of arthritis medications, resulting in loss of capsular and ligament integrity. Assessing the finger includes first evaluation of soft tissue imbalances and then an assessment of the joints themselves. X-ray may be necessary.

34.7 Role of the Intrinsic Muscles

Because the intrinsic muscle-tendon units act to flex the MCP joints and extend the proximal and distal interphalangeal joints, they may cause the finger or fingers to assume that position, best described as the intrinsic plus position, at rest. When the tightness of the intrinsic overpower the resistance of the volar capsule of the PIP joint, hyperextension of that joint and a swan neck posture is likely. Rupture of the terminal tendon at the DIP joint may also precipitate or contribute to the development of this deformity. If intrinsic tightness increases over time, the deformity may become rigid.

34.8 Bunnell Test

Intrinsic tightness restricts passive PIP flexion when the MCP joint is concurrently extended. This phenomenon can be measured by first assessing the PIP joint's maximal flexion arc with the MCP flexed, i.e., in the composite "fisting" position. Passively extending the MCP with one of the examiner's hands tenses the intrinsics; normal intrinsic tension allows for a full or comparable degree of PIP flexion in the testing position, but a tight intrinsic will limit passive PIP flexion with the MCP held in extension. This test is especially significant when the test restricts PIP flexion to 60° or less, noting that "normal" PIP flexion is approximately 100°. This test can also be performed in each of three positions in the plane of radial versus ulnar deviation of the MCP joint. Those three positions are the neutral position, radial deviation, and ulnar deviation, allowing selective testing of the relative tension of the intrinsics on the radial versus ulnar side of the finger. Each intrinsic of each finger (radial versus ulnar) should be tested, especially for surgical planning.

34.9 Swan Neck Severity

When arthritis is more advanced, joint involvement and progressively increasing intrinsic tightness allow not only the proximal and distal interphalangeal joints, but also the metacarpal phalangeal joint to be overpowered. The MCP may sublux or dislocate, most commonly in a volar direction and/or ulnarly [3] causing the digits to assume the characteristic intrinsic plus posture, with PIPs extended, MPs flexed with the digits in a straight extended position. Ulnar drift is common with this deformity, and an attempt to do the Bunnell test for intrinsic tightness will usually demonstrate that tightness is greater on the ulnar side of the digit than the radial side. MCP involvement will commonly be addressed by arthroplasty, concurrent with which the ulnar-sided intrinsics are often released by tenotomy. Arthroplasty itself will be expected to partially relax the intrinsic tendon by shortening the "column of the finger" so to speak, which they are tethering.

34.10 Swan Neck Classification

Swan neck deformity develops with a progressive imbalance of forces which lead to increased functional impairment and additional considerations for medical management. Based on severity, there are four types or degrees of swan neck deformity [4].

Type I swan neck deformity. The finger at rest postures in the swan neck position, but PIP flexion is unrestricted in all positions.

Type II swan neck deformity. PIP motion is partially restricted only when the MCP is placed in extension, as demonstrated with the Bunnell test.

Type III swan neck deformity. PIP flexion is limited in all positions.

Type IV swan neck deformity. The PIP joint is not only fixed in extension with substantial restriction of motion, but radiographically there is evidence of joint damage or destruction.

These deformities are easily identified by measuring PIP motion in the "fist" position versus the Bunnell intrinsic testing position and assessing the integrity of the PIP joint. It is important to characterize the severity of this deformity in order to determine whether it will respond to soft tissue surgical procedures alone or will require arthroplasty or fusion of the PIP joint.

34.11 Boutonniere Deformity

Boutonniere deformity occurs when the central slip of the finger's extensor mechanism ruptures or attenuates, causing a loss of PIP extension. The lateral bands migrate volarly and accentuate the deformity by passing volarly to the PIP joint's axis of rotation. Intrinsic strength at the PIP is diminished or lost, causing hyperextension of the DIP joint with loss of DIP flexion.

Boutonniere deformity may be characterized as mild, moderate, or severe [5]. Mild and moderate deformities involve soft tissue alone and may respond to surgical rebalancing of the extensor mechanism. In a mild deformity voluntary loss of PIP extension is 30° or less, such that when the finger itself is extended, compensatory hyperextension of the MCP joint allows the tip of the finger to clear the plane of the palm. If the PIP flexion deformity is greater than 30°, MCP hyperextension may not be sufficient to allow the extended finger to clear the plane of the hand. A mild deformity may be well tolerated. A moderate deformity may interfere with placing a hand in a pocket or performing fine motor activities, and surgical correction is more often necessary. A moderate deformity where the PIP is supple but postures in 30 to 60 degrees of flexion, may respond to rebalancing of the extensor mechanism. In severe boutonniere deformity, loss of PIP extension and DIP hyperextension is not passively correctable.

With boutonniere deformity, it is also important to assess not only the PIP joint contracture but also the compensatory extension posture of the DIP joint. DIP joint hyperextension may preclude opposition of the fingertip to the tip of the thumb for pinch and fine motor activities. Over time, some patients develop a compensatory but abnormal thumb posture which preserves the ability to oppose the two abnormal digits (thumb and finger). Correcting one or more boutonniere deformities may facilitate voluntary opening and closing of the hand for grasp; however, any repositioning of the digit, even to a more "normal," i.e., anatomic position may alter that digit's contribution to pinch.

34.12 Rheumatoid Thumb

Nalebuff and Millender have described three patterns of thumb deformity seen in the rheumatoid hand [6]. Like the fingers, the thumb may develop an imbalance of flexion and extension forces among its three joints, resulting in deformities which are visually and functionally similar to the boutonniere or swan neck deformities of the fingers.

Type I thumb deformity. The type I deformity has the appearance of the boutonniere deformity with flexion of the MCP joint, dorsal hood attenuation, ulnar subluxation of the extensor pollicis longus, and hyperextension of the IP joint.

Type II thumb deformity. In this deformity, the MP is in extension and the IP joint is hyperextended. The CMC joint may be subluxed.

Tape III deformity. This is essentially a swan neck deformity with radial and dorsal subluxation of the CMC joint due to synovial disease. In this pattern, the metacarpal will be in adduction, the MP hyperextends, and the IP joint is drawn into flexion.

A Type IV deformity has also been described in which the ulnar collateral ligament of the MCP joint is attenuated resulting in a skier's thumb gamekeeper's thumb position. Metacarpal adduction aggravates and accentuates deformity.

34.13 Functional Assessment of the Thumb in Rheumatoid Arthritis

Regardless of the pattern of thumb deformity, a complete examination addresses the thumb's functional contribution to grasp and pinch. In the best of circumstances, the thumb provides a stable post for opposition to the finger for pinch and grasp, contributing to those functions by voluntarily extending its joints and flexing with sufficient power to hold and manipulate. When demonstrating opposition, both thumb rotation and the excursion of its joints are important. With soft tissue imbalance/or joint subluxation, the thumb may roll outward into supination, such that only lateral pinch is feasible and the center of its pulp fails to oppose the center of the tactile pad (distal segment) of the fingers.

34.14 Extensor Tendons

In the rheumatoid hand, extensor tendons are at risk for rupture and for subluxation [7]. Rupture of extensor tendons occurs most commonly at the wrist, and subluxation occurs most commonly at the level of the metacarpal phalangeal joints.

Extensor tendon rupture at the wrist may be immediately evident, or it may be a subtle finding obscured by subluxation and flexion contracture of the metacarpal phalangeal joints. Ruptures may be due to direct synovial invasion, mechanical irritation and abrasion, or medication side effects.

Extensor tendon ruptures commonly occur in association with dorsal subluxation of the distal ulna. Rupture of the TFCC due to synovitis may allow the carpus to rotate into a supinated position, which makes the distal ulna relatively more prominent. The extensor tendons are then subject to risk of attritional rupture as they glide over the prominent ulna.

Tendon rupture may present as the isolated extensor lag of a single digit, but even when only one tendon appears to be ruptured it is common for adjacent tendons to be ruptured a well. Extensor lag of the little finger may signal the rupture of extensor digiti minimi and also the extensor digitorum communis tendon to that finger. In that situation, rupture of the extensor digitorum communis tendon to the ring finger should be suspected as well. The adjacent finger may still extend even if its communis tendon is ruptured because the juncturae tendonae may allow it to do so.

34.14.1 Subluxation of Extensor Tendons

Subluxation of the extensor tendons to the fingers is most common at the metacarpal phalangeal joint level. Frequently, this will be associated with subluxation of the MCP joint. The synovitis which attenuates the joint capsule may attenuate the transverse retinacular fibers as well. This results in a flexion deformity of the MCP joint. Although it is logical to assume that lack of finger extension is due to subluxation of both the extensor tendon and the joint, the extensors should still be examined proximally to ensure that they have not ruptured at the level of the wrist.

34.15 Flexor Tendon Rupture

Flexor tendon rupture may occur at the wrist or in the digit, but rupture at the wrist is more common. Rupture is generally due to attrition and invasive synovial disease in the carpal tunnel but may also be associated with bony spurring formation in the wrist itself. Examination for tendon integrity isolates for testing each of the nine digital flexors—the four profundus tendons, the four sublimis tendons, and the flexor pollicis longus. Flexor rupture may also occur within the digit, but it is less common. A flexor ruptured at the wrist will not contribute to digital flexion at any joint level, but this is difficult to differentiate when the intrinsics or adjacent extrinsic flexors are still functional. Most commonly, the flexor pollicis longus and flexor digitorum profundus are affected [8].

34.15.1 Trigger Fingers and Carpal Tunnel Syndrome

Occasionally, a patient with rheumatoid arthritis will present with triggering of one or more fingers and or carpal tunnel syndrome. In the rheumatoid finger triggering may represent the catching of hypertrophic synovium, possibly with a nodular formation, at the A-1 pulley in the distal palm or at the sublimis decussation more distally within the fingers [9]. Similarly, carpal tunnel syndrome may be associated with aggressive flexor tenosynovitis, with limited active flexion in addition to the findings of carpal tunnel syndrome [10]. Hypertrophy of the flexor tenosynovium will be evident in the distal forearm just proximal to the carpal tunnel. Gliding of tenosynovium will move a mass with active flexion and extension of the fingers.

34.16 Summary

Examination of the rheumatoid hand deformity using a systematic approach which documents motion, identified findings, assesses the functional capacities of pinch and grasp, and anticipates progression of deformities will best contribute to the restoration and/or preservation of function. Correspondingly, the skills developed in examination and assessment of the rheumatoid hand will pay rich dividends in the physician's understanding and assessment of other hand problems.

Acknowledgment The authors wish to acknowledge the concepts of Edward Nalebuff and Lewis Millender, whose teaching and publications have identified timeless principles for evaluation of the rheumatoid hand.

References

1. Shapiro JS. The wrist in rheumatoid arthritis. Hand Clin. 1996;12(3):477–98.
2. Rizio L, Belsky MR. Finger deformities in rheumatoid arthritis. Hand Clin. 1996;12(3):531–40.
3. Stirrat CR. Metacarpophalangeal joints in rheumatoid arthritis of the hand. Hand Clin. 1996;12(3):515–29.
4. Nalebuff EA, Millender LH. Surgical treatment of the swan-neck deformity in rheumatoid arthritis. Orthop Clin North Am. 1973;6(3):733–52.
5. Nalebuff EA, Millender LH. Surgical treatment of the boutonniere deformity in rheumatoid arthritis. Orthop Clin North Am. 1973;6(3):753–63.
6. Stein AB, Terrono AL. The rheumatoid thumb. Hand Clin. 1996;12(3):541–50.
7. Wilson RL, DeVito MC. Extensor tendon problems in rheumatoid arthritis. Hand Clin. 1996;12(3):551–9.
8. Ferlic DC. Rheumatoid flexor tenosynovitis and rupture. Hand Clin. 1996;12(3):561–72.
9. Nalebuff EA. Surgical treatment of rheumatoid tenosynovitis in the hand. Surg Clin North Am. 1969;49(4):799–809.
10. Millender LH, Nalebuff EA. Evaluation and treatment of early rheumatoid hand involvement. Orthop Clin North Am. 1973;6(3):697–708.

Part IV

Hip/Pelvis Reviewer Dr's Rath & Hoelmich

35 Hip Anatomy

Domenico Potestio

The hip is a ball and socket joint and is a diarthrodial joint formed by the head and neck of the femur and the acetabulum which represents the joint cavity. The articular surfaces of the femoral head and of the acetabulum are both covered by cartilage. According to Li et al. [1], average cartilage thickness on the femoral side and on the acetabular side is, respectively, 0.94 mm and 1.1 mm.

The head of the femur is connected to the femur trough the femoral neck and precisely through the head–neck junction. The angle between the femur shaft and the neck is 125 ± 5°. The femoral neck also has a 15–20° anti-version from the femur shaft.

The angle formed by the femoral neck axis and a line that goes from the center of the femoral head to a point where the contour of the femoral head-neck junction exceeds the radius of the femoral head is called alfa angle and is of primary importance to define a CAM morphology. An angle >55° is considered to be pathological [2].

The acetabulum is formed by the innominate bone that is the union of three bones, the ilium, the ischium, and the pubis. These three bones represent, respectively, 40%, 40%, and 20% of the acetabular surface and before their fusion are separated by the triradiate cartilage. As described by Santori and Villar [3] in a small percentage of adults, a groove named ileopubic groove can be seen in the acetabular surface, and this might represent an incomplete fusion of the triradiate cartilage (Fig. 35.1). The entire acetabular surface is covered by cartilage except for the acetabula fossa which is a non-articular area in the medial part of the acetabulum. This fossa is filled with a fat pad covered by synovial tissue. In some patients, a small area of the acetabular articular surface might have cartilage which is less thick than normal. This area is named stellate crease [3].

In some patients, an additional fossa might be seen in high resolution MRI. This fossa is named supra-acetabular fossa and is located at the 12:00

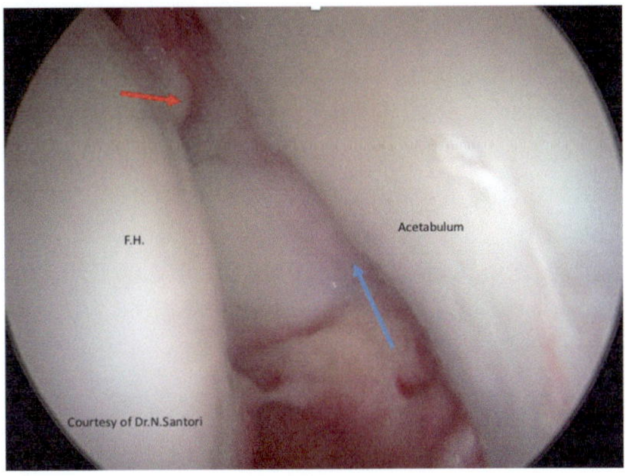

Fig. 35.1 FH: femoral head. Blue arrow: ileopubic groove. Red arrow: acetabular labrum

D. Potestio (✉)
ARS BIOMEDICA, Rome, Italy

J. G. Lane et al. (eds.), *The Art of the Musculoskeletal Physical Exam*,
https://doi.org/10.1007/978-3-031-24404-9_35

position. This fossa doesn't have a pathological meaning and must not be confused with a chondral defect [4].

35.1 Vascularization of the Hip

Vascularization of the hip is provided by the so-called ring circulation that provides a blood supply to the bone and to the soft tissue structures.

The major blood supply is granted by the superior and the inferior gluteal arteries originates from the internal iliac artery and supported by the medial and the lateral circumflex arteries and their branches. The medial circumflex arteries originate from the common femoral artery while the lateral circumflex arteries originates from the deep femoral artery. Both the common and the deep femoral arteries originate from the extern iliac artery. Grose et al. [5] demonstrated the anastomosis of these four vessels which occurs in the cruciate anastomosis. These connections may have a protective role towards vascular pathologies.

The first perforating branch of the deep femoral artery flows under the quadratus femoris muscle and gives a branch to this muscle, then creates an anastomosis with the medial femoral circumflex artery just behind the short rotators, and then moves deep, bringing blood supply to the capsule, the labrum, and the hip. After having created these anastomoses, it runs proximally and reaches the gluteal vessels completing the so-called cruciate anastomosis.

The superior gluteal artery moves under the gluteus minimus and the tensor fascia lata and reaches the lateral femoral circumflex artery anteriorly. These vessels form retinacular and perforating branches that give blood supply to the capsule, the labrum, and the acetabulum.

35.2 The Labrum

The rim of the of the acetabulum is covered by the acetabular labrum (Fig. 35.2). This fibrocartilaginous structure along with the transverse acetabular ligament, which is attached to

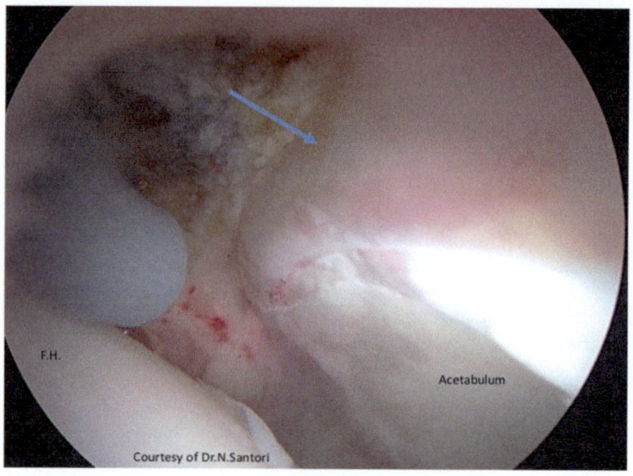

Fig. 35.2 Blue arrow shows an arthroscopic view of the acetabular labrum

the walls of the acetabular notch, surround the acetabulum. Together, these two structures form a basket that surrounds the femoral head over its equatorial line. Due to its conformation, the labrum seals the joint fluid between the cartilage layers of the femoral head and of the acetabulum [6].

The labrum has a triangular section, on the articular side it continues with the articular cartilage through the chondro-labral junction while on the capsular side it continues with the acetabular bone.

It has three differently arranged collagen layers, the first in which fibers have a random orientation, the second in which the fibers are arranged in bundles that intersect each other with variable angles and the third layer in which the fibers have a circular orientation. The collagen fibers also show a different orientation in the anterior and in the posterior labrum [7].

The collagen fibers of the anterior labrum are parallel to those of the acetabular bony edge, while on the posterior labrum the collagen fibers are perpendicular to those of the acetabular bony edge. This different fiber configuration is the reason why the anterior labrum is more frequently involved in lesions compared to the posterior labrum which is more resistant to share forces.

The labrum has a peculiar vascularization that was described by Kelly in 2005 [8]. Considering this vascularization, the labrum was divided by Kelly in two zones and in sub-zones. The half of

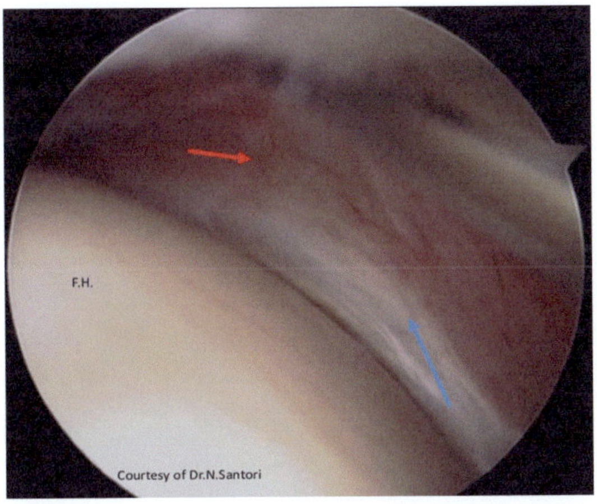

Fig. 35.3 Out in view of the hip. FH: femoral head. Blue arrow: acetabular labrum. Red arrow: peripheral vascularization of the labrum

the labrum on the capsular side is Zone 1 and the half on the articular side is Zone 2 (Fig. 35.3).

Zone 1 is divided into Zone 1A which is the peripheral part of the labrum on the capsular side and Zone 1B which is the part of the labrum on the capsular side closest to the acetabular bony rim. Zone 1B receives its blood supply from the bone which is the main source of labral circulation while Zone 1A is supplied by the capsular circulation.

Zone 2 is divided into Zone 2A which receives the same vascular circulation of Zone 1A and Zone 2B which is the part of the labrum which is closest and connected to the articular cartilage and is relatively an avascular zone.

35.3 The Ligamentus Teres

The ligamentus teres is a robust ligament and according to recent anatomic studies performed by Mikula et al. [9], it has six insertion zones on the acetabular side and a single one on the femoral side. The six insertion zones on the acetabulum are: the ischiatic insertion, the pubic insertion, the iliac insertion, the two margins of the acetabular notch and the transverse acetabular ligament. Its structure can be divided into three layers, an external one of synovial tissue, an intermediate one of loose connective tissue and a deeper one consisting of dense bundles of collagen tissue. Both in the intermediate and in the deepest layer, there are vessels and nerve endings, including the Pacini corpuscles. The dense presence of vessels in the foveal portion suggests a "nourishing" role of the round ligament also in adults. From a biomechanical point of view, its resistance is comparable to that of the anterior cruciate ligament of the knee. The round ligament has been shown to contribute to hip stability in flexion and extra-rotation, in extension and intra-rotation and in flexion and abduction. This occurs especially in cases of reduced capsule-ligament stability and reduced bone stability. In addition to the stabilizing role, Gray and Villar have proposed a role of the ligament in the distribution and circulation of the synovial fluid in the joint (windshield effect) [10].

35.4 The Capsule Ligaments

The hip joint capsule is reinforced by three longitudinal ligaments, the iliofemoral ligament, the ischiofemoral ligament, and the pubofemoral ligament. These three ligaments have a primary role in hip joint stability. The iliofemoral ligament has an inverted Y shape where the lateral and distal insertions are on the femoral intertrochanteric line while the proximal insertion is on the anterior inferior iliac spine. This ligament is also called the Bigelow ligament. The role of this ligament is to reinforce the capsule during external rotation and extension.

The ischiofemoral ligament has its insertion on the ischium, postero inferiorly to the acetabular rim and on the posterior intertrochanteric line. Its role is to reinforce the capsule during internal rotation in a neutral position and in a combination of flexion, adduction, and internal rotation.

The pubofemoral ligament has its insertion in the superior pubic ramus and converges on the femur with the medial portion of the iliofemoral ligament and the inferior part of the ischiofemoral ligament. Its role is to reinforce the capsule in abduction and in external rotation during extension.

A secondary role in capsular reinforcement and in increasing joint stability is played by the circular fibers of the zona orbicularis which forms a collar around the femoral neck. During hip extension, the zona orbicularis increases the stability of the femoral head anteriorly, while during flexion the zona orbicularis increases the stability of the femoral head posteriorly.

References

1. Li W, Abram F, Beaudoin G, Berthiaume MJ, Pelletier LP, Mrtel-Pelletier J. Human hip joint cartilage: MRI quantitative thickness and volume measurements discriminating acetabulum and femoral head. IEEE Trans Biomed Eng. 2008;55(12):2731–40.
2. Agnvall C, et al. Range of hip joint motion is correlated with MRI-verified cam deformity in adolescent elite skiers. Orthop J Sports Med. 2017;5(6):2325967117711890.
3. Santori N, Villar RN. The iliopubic groove: a possible consequence of incomplete triradiate fusion. Two case reports. J Anat. 1997;191(Pt 3):461–3.
4. Dietrich TJ, Suter A, Pfirrmann CW, Dora C, Fucentese SF, Zanetti M. Supraacetabular fossa (pseudo defect of acetabular cartilage): frequency at MR arthrography and comparison of findings at MR arthrography and arthroscopy. Radiology. 2012;263(2):484–91.
5. Grose AW, Gardner MJ, Sussmann PS, Helfet DL, Lorich DG. The surgical anatomy of the blood supply to the femoral head: description of the anastomosis between the medial femoral circumflex and inferior gluteal arteries at the hip. J Bone Joint Surg Br. 2008;90:1298–303.
6. Ferguson SJ, Bryant JT, Ganz R, Ito K. The acetabular labrum seal: a poroelastic finite element model. Clin Biomech. 2000;15(6):463–8.
7. Cashin M, Uhthoff H, O'Neill M, Beaulé PE. Embryology of the acetabular labral-chondral complex. J Bone Joint Surg Br. 2008;90(8):1019–24.
8. Kelly BT, Shapiro GS, Digiovanni CW, Buly RL, Potter HG, Hannafin JA. Vascularity of the hip labrum: a cadaveric investigation. Arthroscopy. 2005;21(1):3–11.
9. Mikula JD, Slette EL, et al. Quantitative anatomic analysis of the native ligamentum Teres. Orthop J Sport Med. 2017;5(2):2325967117691480.
10. Gray AJ, Villar RN. The ligamentum teres of the hip: an arthroscopic classification of its pathology. Arthroscopy. 1997;13(5):575–8.

Hip Biomechanics

36

Paolo Di Benedetto and Simona Cerulli

The human body is a structure designed in such a way that bone structures and soft tissues (muscles and ligaments) interact with each other both in static and dynamic situations in order to maintain balance and generate movement. Walking is a complex act resulting from the interactions between internal forces (muscles, joint constraints) and external forces (gravity, foot-ground reaction force) directed by an admirable postural and balance control system, which, through muscle contraction, exercises control over joint movements. This system allows man postural stability through the action of the antigravity muscles, both in a static position and in movement, adapting to continuous environmental changes. Anatomists and anthropologists over the years have explained how the anatomy and biomechanics of the pelvis have changed to adapt to the unique characteristics of bipedal walking [1]. They have shown us how the anterior rotation and the enlargement of the iliac wings, in addition to the structural change of the pelvis, are of fundamental importance for human bipedal gait because they modify the position and naturally the function of the hip abductor muscles (minimum and middle gluteus).

This anatomical configuration allows these muscles to make the greatest contribution to the lateral vector of the Ground Reaction Force (GFR) both during walking and running [2]. The GRF is the reaction force of the soil to the force discharged on it from the supporting limb, resulting from gravitational and inertial forces. It has the same direction and intensity as the latter, but in the opposite direction. While in an upright position, the GRF has vertical direction and intensity equal to the absolute gravity force of the body, while walking, the GRF varies continuously in intensity and direction, being the force discharged from the foot onto the ground the result of gravitational and inertial forces. Furthermore, the GRF can be broken down into a vertical component, an anteroposterior component and a medium-lateral component (the last two components correspond to the friction onto the ground). GRF generates external moments to the load-bearing joints, where it does not cross the articular axes. Moments are calculated by multiplying the GRF by the lever arm of this force with respect to the articular center of the joint taken into consideration. Without an adequate muscle contrasting action, this force could cause postural collapse while resting. That is confirmed by the gear change, known as Trendelenburg gait [3–5] in patients with hip abductor muscle deficit. When the hip abductors

P. Di Benedetto
Clinic of Orthopaedics, Friuli Centrale Healthcare and University Trust (ASUFC), Udine, Italy

DAME—University of Udine, Udine, Italy
e-mail: paolo.dibenedetto@uniud.it

S. Cerulli (✉)
University Policlinic Foundation, A. Gemelli. IRCCS, Rome, Italy

J. G. Lane et al. (eds.), *The Art of the Musculoskeletal Physical Exam*,
https://doi.org/10.1007/978-3-031-24404-9_36

are not working properly, the pelvis rotates away from the limb during the monopedal support phase, requiring a compensatory inclination of the trunk towards the support side.

For about three decades, gait analysis has established itself as the only movement analysis method widely used in the clinical setting. The use of gait analysis as an investigation tool is mainly due to two factors: first of all, this function is the one that best identifies the patient's autonomy in the context of daily life, and secondly, walking is an activity with a well-defined functional repetition, represented by the step cycle [6].

Static and dynamic biomechanical models have been used to understand the biomechanics of the hip.

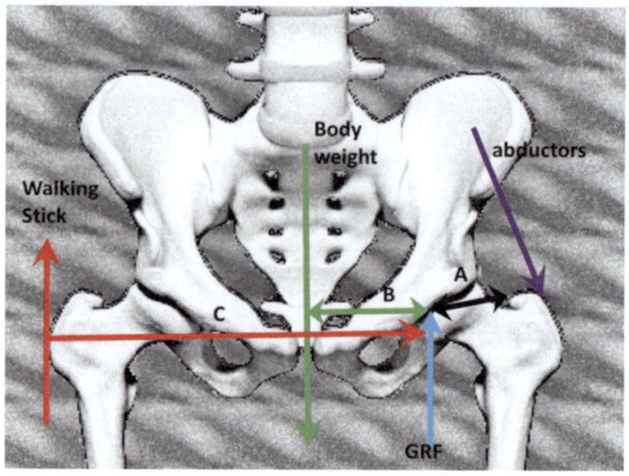

Fig. 36.1 Forces acting on the hip joint during single leg stance under conditions of equilibrium. Gravitational force (body weight, green arrow), abductor muscle force (violet arrow), hip joint reaction force (yellow arrow), and ground reaction force (blue arrow)

36.1 Static Analysis

During the bipedal station, the forces acting on any component must be zero. The same principle is applied to the rotational static equilibrium, in which the sum of the moments must be zero. The static model determines the muscle strength of the abductors as the product of the external force acting on the hip, measured as body weight minus the supporting limb weight.

The length of the hip abductor lever arm is defined as the perpendicular distance from the center of the hip joint rotation to the action line of the abductor muscles; that distance is comparable to the femur neck length which is often used as a reference size in skeletal samples. As shown in the study by Polkowski et al., under static conditions, the forces that act on the pelvis and hip joint to maintain the pelvis axis are as follows (Fig. 36.1): gravitational force (BW, green arrow), which is the body weight minus the weight of the contralateral lower limb; the abductor muscle strength that acts to maintain the pelvis axis (violet arrow); and the Ground Reaction Force (GFR, blue arrow), i.e., the reaction force of the soil to the force discharged on it from the supporting limb.

In standing or upright weight-bearing activities, at least half the weight of the HAT (gravitational force) passes down through the pelvis to the femoral head, whereas the ground reaction force (GRF) travels up the shaft. These two forces, nearly parallel and in opposite directions, create a force couple with a moment arm (MA) equal to the distance between the superimposed body weight on the femoral head and the GRF up the shaft (Fig. 36.2).

A problem with the static biomechanical model is that it does not take into account the moments generated by the ankle and knee during walking, which are however important for the body stability in standing position [7–9]. Furthermore, assuming that the external force vector remains close to the midline of the body during the monopedal phase of the gait cycle, the static model does not consider the lateral average component of the GRF [10].

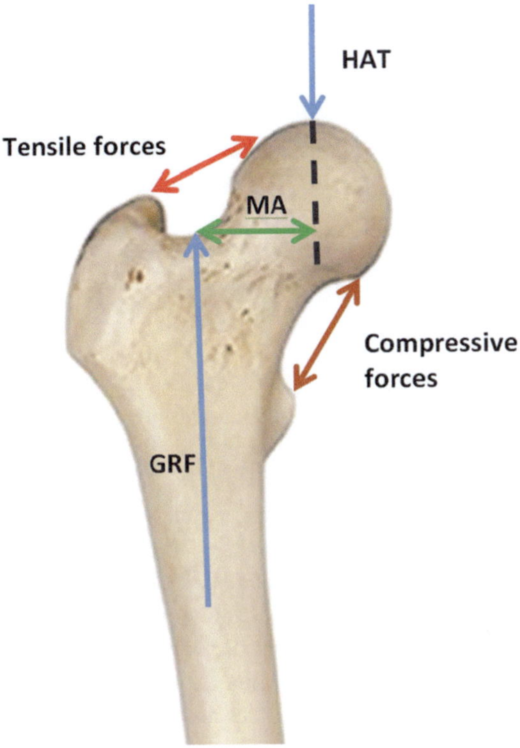

Fig. 36.2 The weight-bearing line of the head, arms, and trunks (HAT) loads the femoral head, whereas the ground reaction force (GRF) comes up the femoral shaft, resulting in a force couple that creates a bending moment, with a moment arm (MA) that is dependent on the length and angle of the femoral neck. The bending moment creates tensile stress on the superior aspect of the femoral neck and compressive stress on the inferior aspect

36.2 Dynamic Analysis

During the step, there are two distinctive phases: the "stance" or support phase, when the foot is in contact with the floor, and the "swing" or oscillation phase.

The stance phase is divided into three separate sub-phases: double initial support (both feet are in contact with the ground), single support (during which, one foot is in contact with the ground and the contralateral foot is oscillating), double terminal support (in which, both feet are again in contact with the ground). The swing instead consists of three phases that describe the trajectory traveled by the limb in flight during the progression. By convention, the gait cycle begins with the initial contact phase in which the reference foot hits the ground through the initial heel strike ("heel strike").

The role of the abductors during the gait cycle is to maintain balance on the loading side and to manage movement on the unloading limb while the leg moves forward.

Activation of the abductor muscles also during the unloading phase to control hip movement means that the joint is never completely unloaded even in the non-WB phase.

In a bipedal station, the body weight (BW) is supported by both hips and therefore, if the body is perfectly balanced, the action of the abductors is not necessary [11, 12]. In reality, it is unlikely that the body is ever perfectly balanced and therefore the GRF force vector varies from ½ BW to 3 BW of the bipedal station and the monopedal station, respectively.

Hip abductors therefore play a fundamental role in maintaining the balance and stability of the pelvis and their role becomes even more important if we consider biomechanics dynamically.

During the walk, the limb that starts the step leaves the ground to move forward; therefore, the body is temporarily loaded on one limb. In this instant, the body weight force (BW) acts downwards causing body inclination. However, this does not happen for the pelvis which is balanced by the action of the abductors. Thus, the hip behaves much like a lever with a load acting on both sides of a fulcrum (head of the femur).

The "dynamic problem" can be faced in two ways: the former called the direct dynamics problem according to which, given the motor actions and the external forces acting on a body (or on a system of bodies), its movement is predicted; and the latter called inverse dynamics problem, for which given the kinematics of a body (or of a system of bodies) and the forces exchanged with the environment, the forces and phenomena acting on it and causing a certain movement are predicted.

36.3 Inverse Dynamic

Inverse dynamics is a rooted approach in which the resolution of unknown forces acting on a body can be resolved by the known resultant force (GRF) and by the kinematic and inertial properties of the body itself [13, 14]. This data is now relatively easy to obtain, using force platforms and high-speed motion acquisition systems to monitor the movement of body segments during locomotion. The resulting net articular moments represent the total moment of a joint generated by all the anatomical structures that act on the adjacent body segments [15]. Contact forces can also be calculated using inverse dynamics.

Movement analysis traditionally takes place with the study of inverse dynamics. In this context, the human body is schematized as a chain of rigid segments. Given the relatively large number of segments, the complexity of the model becomes prohibitive. Hence, it is convenient to resort to a recursive procedure. The calculation is applied to each segment taken individually in sequence. The Newton–Euler mechanics allows us to calculate the forces and moments inside the joints of the various segments.

36.4 Forward Dynamic

Another powerful technique for understanding human movement is direct dynamics. It is a model-based approach in which the movement of a multi-segment body is driven by simulated forces. These models can be implemented through net joint moments or based on individual muscle contributions to segmental accelerations.

36.5 Muscle Action

The position and movements of the femur with respect to the hip joint are under the pelvis and spine muscle control.

Since these muscles are asymmetrically distributed and with different vector potential, if they enter tension excess they can initially make the joint muscle rigid and, if the tension excess persists over time, it causes joint misalignment.

Monoarticular muscles with a short line of force have favored vectors in joint stabilization; their tension excess is most manifested in the compaction of the femoral head within the acetabulum.

Polyarticular and monoarticular muscles with an important line of force (e.g., adductors) have better vectors for the femur movements and positioning (Table 36.1).

Flexion/extension and rotation dominances are expressed differently depending on whether the femur is a mobile point, in discharge, or a fixed point, in loading (Tables 36.2 and 36.3).

This concludes with a general overview of hip biomechanics. The demands of daily life often go beyond the relative basic description of mechanics and involve a complex interaction of the bone structures, muscles, capsule, and ligaments of this joint [16]. In the evaluation and treatment of anomalies in the soft tissues around the hip, bone anomalies must be taken into account, as they are

Table 36.1 Adduction–Abduction vector dominance

Adduction	Abduction	Vector dominance
• Adductors • Gracilis • Gluteus maximus (with insertion on the gluteal tuberosity) • Pectineus • Quadratus femoris • Obturator externus	• Tensor fasciae latae • Gluteus minimus and medius • Gluteus maximus (with insertion on the tensor fasciae latae) • Piriformis • Obturator internus	Abduction

Table 36.2 Flexion–Extension vector dominance

Flexion	Extension	Vector dominance
• Iliopsoas • Quadriceps • Sartorius • Adductors • Gracilis • Tensor fasciae latae • Pectineus	• Hamstrings • Gluteal muscles • Piriformis	• With the femur as a mobile point, vector dominance is in *flexion of the hip* • With the femur as a fixed point (under load) the dominance is in *anteversion of the pelvis and in lumbar hyperlordosis*

Table 36.3 Internal-external rotation vector dominance

Intrarotation	Extrarotation	Vector dominance
• Semitendinosus • Semimembranosus • Adductor magnus • Gracilis • Gluteus minimus and medius with its anterior fibers • Tensor fasciae latae	• Adductors • Biceps femoris • Sartorius • Gluteus maximus • Gluteus minimus and medius with its dorsal fibers • Iliopsoas • Quadratus femoris • Obturator internus • Piriformis	• Extrarotators when the femur is under load. Semitendinosus, semimembranosus (with adductor magnus and gracilis in co-agonism) in standing position and with the foot landed on *The resulting vector dominance is expressed in intrarotation*

recognized as contributing factors in the diagnosis and treatment of hip joint disorders.

References

1. Warrener AG. Hominin hip biomechanics: changing perspectives. Anat Rec (Hoboken). 2017;300(5):932–45.
2. Pandy MG, Andriacchi TP. Muscle and joint function in human locomotion. Annu Rev Biomed Eng. 2010;12(1):401–33.
3. Gottschalk F, Kourosh S, Leveau B. The functional anatomy of tensor fasciae latae and gluteus medius and minimus. J Anat. 1989;166:179–89.
4. Polkowski GG, Clohisy JC. Hip biomechanics. Sports Med Arthrosc Rev. 2010;18(2):56–62.
5. Nakagawa TH, Petersen RS. Relationship of hip and ankle range of motion, trunk muscle endurance with knee valgus and dynamic balance in males. Phys Ther Sport. 2018;34:174–9.
6. Rixrath E, Wendling-Mansuy S, Flecher X, Chabrand P, Argenson JN. Design parameters dependences on contact stress distribution in gait and jogging phases after total hip arthroplasty. J Biomech. 2008;41(5):1137–42.
7. Eschweiler J, Hawlitzky J, Quack V, Tingart M, Rath B. Biomechanical model based evaluation of total hip arthroplasty therapy outcome. J Orthop. 2017;14(4):582–8.
8. Pandy MG, Lin YC, Kim HJ. Muscle coordination of mediolateral balance in normal walking. J Biomech. 2010;43(11):2055–64.
9. Correa TA, Crossley KM, Kim HJ, Pandy MG. Contributions of individual muscles to hip joint contact force in normal walking. J Biomech. 2010;43:1618–22.
10. Simonsen EB. Contributions to the understanding of gait control. Dan Med J. 2014;61(4):B4823. Review
11. Zajac FE, Neptune RR, Kautz SA. Biomechanics and muscle coordination of human walking. Part I: introduction to concepts, power transfer, dynamics and simulations. Gait Posture. 2002;16(3):215–32.
12. Akrami M, Craig K, Dibaj M, Javadi AA, Benattayallah A. A three-dimensional finite element analysis of the human hip. J Med Eng Technol. 2019:1–7.
13. de Leva P. Adjustments to Zatsiorsky–Seluyanov's segment inertia parameters. J Biomech. 1996;29:1223–30.
14. Dempster WT. Space requirements of the seated operator. Geometrical, kinematic, and mechanical aspects of the body with special reference to the limbs, WADC Technical Report 55–159. Wright-Patterson Air Force Base, Ohio. Springfield, OH: Carpenter Litho & Printing Co.; 1955.
15. Jensen RH, Davy DT. An investigation of muscle lines of action about the hip: a centroid line approach vs the straight line approach. J Biomech. 1975;8(2):103–10.
16. Wilson BR, Robertson KE, Burnham JM, Yonz MC, Ireland ML, Noehren B. The relationship between hip strength and the Y balance test. J Sport Rehabil. 2018;27(5):445–50.

Evaluation of Dysplasia of the Hip (Children with DDH, Adolescents, and Adults)

37

Alessandro Aprato and Pietro Persiani

Developmental dysplasia of the hip (DDH) is a congenital anomaly, characterized by insufficient development of coxofemoral joints and concomitant laxity of the capsule-binding system, which may result in partial or total loss of coxofemoral joint relationships [1, 2].

It is the most common osteoarticular deformity, presents a predilection for female sex with a female/male ratio of 61. The incidence varies between 0.7% and 2.5%, and it is very common in the north Italy, France, Spain, and the Balkans where the incidence stands at 2–4%. Incidence is significantly lower in African and Asian patients [3, 4]. In 45% of cases, it is bilateral and associated with other congenital malformations such as clubfoot. DDH often leads to early degenerative hip arthritis if not treated [5].

37.1 Clinical Presentation and Diagnosis of DDH in Newborns

In newborns, clinical hip examination is essential for the diagnosis of DDH: semeiological maneuvers such as Ortolani and Barlow tests are still considered very useful [1]. Starting from the age of 8–12 weeks of life, these maneuvers are not easily assessed and not very indicative.

Currently, the instrumental investigation, which allows an accurate and early diagnosis, is based on ultrasound, which has been widely used in recent years and which has supplanted the use of conventional radiological examination (Fig. 37.1). The ultrasound has proven to be an accurate method of diagnosis in the first months of life, allowing to highlight all the anatomical structures of the baby's hip. Among the various ultrasound classifications, the Graf classification is certainly the most widespread in Europe and allows to classify the degree of severity of the pathology.

A. Aprato (✉)
Università degli Studi di Torino, Torino, Italy

P. Persiani
Università "La Sapienza"—P.le Aldo Moro, Rome, Italy
e-mail: pietro.persiani@uniroma1.it

J. G. Lane et al. (eds.), *The Art of the Musculoskeletal Physical Exam*, https://doi.org/10.1007/978-3-031-24404-9_37

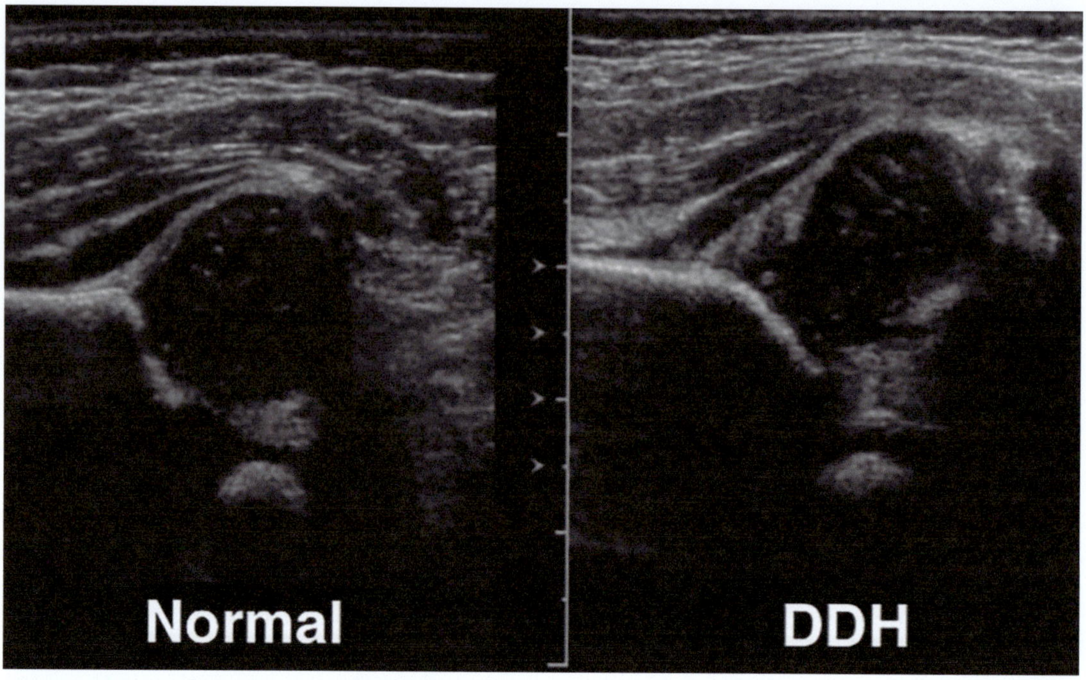

Fig. 37.1 Normal and DDH at ultrasound examination

37.2 Clinical Presentation and Diagnosis of DDH in Adolescents and Adults

The most common presenting symptoms of acetabular dysplasia in the skeletally mature individual are insidious onset of activity-related groin pain and/or lateral hip pain. Diagnosis is made through physical exam and radiographic findings [5]. Patients with hip dysplasia may initially present with mild peritrochanteric pain reflecting abductor fatigue, an almost universal finding in pre-arthritic hip dysplasia in the mature adult. Commonly, patients with dysplasia will have groin pain in the affected hip that worsens with activity. Pain worsening with weight-bearing and activity may indicate underlying joint pathology. Pain with flexion positions or prolonged periods of sitting may commonly be associated with femoroacetabular impingement (FAI). The sensation of locking or catching in the affected joint may indicate intra-articular mechanical problems, such as labral tear, chondral flap, or loose body. The physical examination of the patient with suspected hip dysplasia should also take into account general condition and body habitus, patient's sitting posture, and gait pattern. It is also important to assess for a leg-length discrepancy. Standing measurements will allow an assessment of pelvic balance and therefore are considered the best way to assess a leg-length discrepancy. Resting lower extremity rotation should be assessed with the patient lying supine on the exam table. Normal lower extremity rotation is between 10° and 30° of external rotation. Abnormal rotation may result from abnormal acetabular version, abnormal femoral version, or femoral head-neck abnormalities. Hip range of motion testing is essential for not only diagnosis but also pre-procedure planning. The examiner should steady the pelvis with one hand while performing range-of-motion testing, including hip flexion and internal and external rotation both in 90° of flexion and full extension.

In contrast to classic FAI, which may demonstrate restricted hip flexion and internal rotation, patients with classic dysplasia will generally demonstrate normal hip flexion and normal internal rotation.

Some special tests may help elucidate the nature of the hip pathology. The **anterior impingement test** is a special maneuver that is indicative of disorders of the anterior acetabular rim. It is very sensitive for a range of anterior hip lesions, including labral tears and rim fractures, but fairly nonspecific for intra-articular disease and joint irritability. The patient is placed supine, and the affected extremity is passively flexed, adducted, and internally rotated. The test is considered positive if this maneuver reproduces the patient's pain. The **apprehension test** is another special maneuver that is useful to assess for anterior hip instability, though it may be positive in patients with labral lesions as well. It is performed by having the patient lie supine on the table. The affected extremity is extended and rapidly externally rotated. The test is considered positive if it elicits apprehension or anterior hip pain. Finally, patients should be assessed for hip abductor weakness. The **Trendelenburg sign** is elicited if the patient leans away from the affected extremity during single-legged stance on the affected extremity. Alternatively, abductor strength can be tested by having the patient lie on his/her unaffected side and asking him/her to abduct the leg against resistance.

The evaluation of an adult patient with suspected hip dysplasia is confirmed with imaging [6, 7]. The radiographic features of adult hip dysplasia may range from subtle acetabular dysplasia to complete dislocation of the femoral head from the native acetabulum. The goals of imaging are to assess the structural anatomy of the hip, determine the congruency of the articulation, examine the integrity of the joint space, and assess the soft tissues.

The anterior-posterior (AP) view of the pelvis is the single most important view for defining acetabular coverage of the femoral head, femoral head sphericity, contour of the femoral head-neck junction, height of the greater trochanter, position of the joint center, joint space, and Shenton's line (Fig. 37.2).

An acetabular undercoverage is defined by a *lateral center edge angle* minor than 22° and/or an *acetabular index* >14° and/or a *femoral head extrusion index* >27%. Another feature of DDH is the internal rotation of the entire innominate bone. DDH presents also a decreased size of the lunate surface compared to normal hips and increased contact pressures (e.g., 23% increased pressure in the midstance phase of gait).

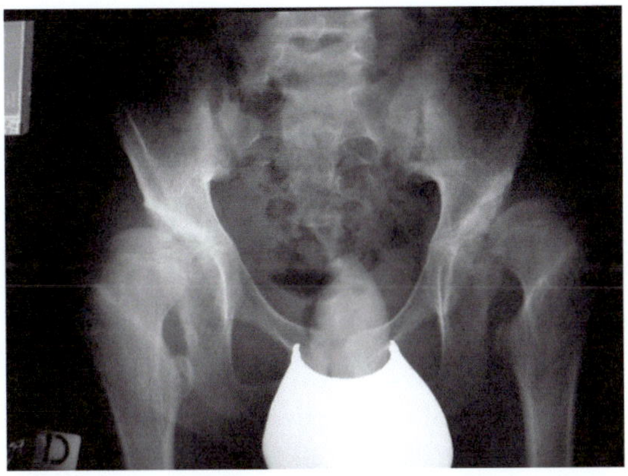

Fig. 37.2 A-P view of DDH

Pelvic AP X-rays allow evaluation of acetabular version: acetabular version refers to the orientation of the mouth of the acetabulum with respect to the sagittal plane. If the mouth faces forward, anteversion is present, but if the mouth faces posteriorly, retroversion is present.

In normal hips, the acetabulum usually is anteverted. Dysplastic hips are globally shallow, but the deficiency in coverage usually is more pronounced anterolaterally, thereby giving an impression of excessive anteversion. If the predominant deficiency occurs in the posterolateral wall of the acetabulum, the latter will seem retroverted. On an AP radiograph, it will be found that the posterior wall of the acetabulum meets the roof at a point medial to the junction of the anterior wall with the roof. This pattern is well recognized in certain types of dysplasias such as neuromuscular hip dysplasia, post-traumatic dysplasia, and proximal focal femoral deficiency.

Dysplastic hips are often anteverted, but DDH may also be associated with retroversion. In all cases of developmental hip dysplasia, the diagnosis of retroversion depends on the relationship of the anterior and posterior walls in the superior 1/3 of the acetabulum [8, 9].

Lateral radiographs allow better definition of the osseous anatomy of the proximal femur, anterior and posterior joint spaces, and acetabular rim. Lateral radiographs include the cross-table lateral and frog lateral. Also among the lateral views is the false profile view. This image is obtained by having the patient stand with their foot parallel to the radiographic plate and their pelvis rotated 65° relative to the film. This view is a true lateral view of the acetabulum. It allows measurement of the anterior coverage of the acetabulum and may also allow better detection of the degenerative changes that tend to begin at the anterior aspect of the joint.

Computer tomography (CT) is a complementary study in evaluating hip dysplasia when dysplastic signs have been recognized on X-rays and surgical correction is proposed. It allows reliable measurements of acetabular coverage, femoral neck anteversion, and appearance and position of the femoral head. It also allows better characterization of osseous impingement lesions.

MRI and MRI arthrogram are useful adjuvants to evaluate the labrum and should be obtained in patients with mechanical symptoms. A reactive labral hypertrophy may be detected on MRI or arthro-MRI resulting from an increased load (Fig. 37.3). The labrum often is torn along with a part of adjacent cartilage due to subluxation of the femoral head that tears the labrum from the acetabular rim together with a sleeve of cartilage.

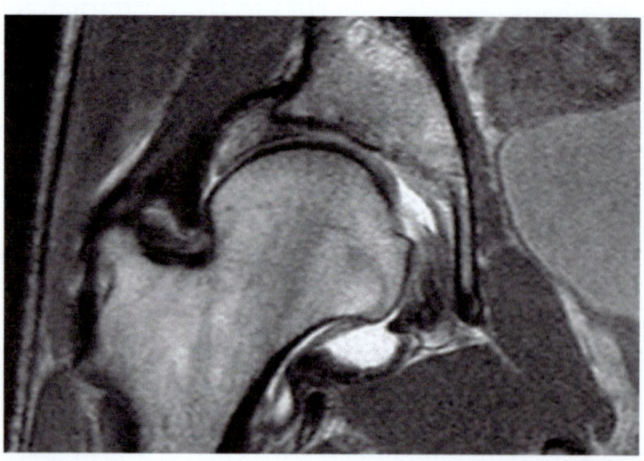

Fig. 37.3 MRI of hip dysplasia resulting in labrum hypertrophy

Furthermore, on MRI, a progressive thinning of the acetabular cartilage may appear in early adulthood and may result in full-thickness defects at the peripheral acetabular rim due to static overload. This chondrolabral damage is typically located superiorly.

37.3 Etiopathogenesis

It is multifactorial:

- Acetabular dysplasia with exiguous development of the cavity and that consequently escapes the acetabular roof
- Hyperlaxity of joint capsule structures
- Straight dysplasia: derived from muscle tensions with prevalence of the tone of flexor-adductor muscle with respect to the abductor projectors
- Response of the development tissue to possible hormonal tubes
- Congenital syndromes such as Larsen, Ehlers-Danlos, and Down or neuromuscular disorders such as myelodysplasia or spina bifida
- Mechanical causes including the position taken by hips during intrauterine life, type of delivery (breeding presentation), or oligohydramnios

At birth, the components of the hip are largely made by cartilage. As for the femur, the proximal portion, epiphysis, and part of the neck are cartilage for the first 3 months of life. The ossification nucleus appears from the third to the sixth month and is settled to the femoral metaphysis at an age ranging from 11 to 16 years.

In the newborn, the angle of inclination, the angle between the axis of the neck of the femur and axis of the diaphysis, is greater than that of the adult, with 130–135° in the newborn compared to 125 of the adult.

Also, the angle of declination, the angle that the axis of the neck of the femur forms with a frontal plan passing for the femoral condyles, appears wider: 35° of neonatal anteversion of the neck of the femur, 15/20° in the adult.

The first classification is established according to the time of diagnosis:

- From birth until 10–12 months (age of the walk) can have cases of preluxation or, rarer, embryo dislocation.
- From the beginning of the verticalization to the next development, the dysplastic hip may show itself to be dislocated or subluxed.
- From the end of the accretion, an inveterate dislocation is established.

37.4 Preluxation

At this early stage, acetabular cavity has an exiguous development, elusive acetabular roof, joint capsule laxity, and initial formation of a neo-limbus (semicircular saliency located in the upper region of acetabulum). Histologically, ossification will be slowed (hypoplasia and increased obliquity of the bone roof), and there will be acetabular cartilage dysplasia with irregular chondrocytes and laxity of extracellular matrix.

37.5 Subluxation

In this stage, the femoral head moves upward from socket, anteverted coxa valga, pulvinar hypertrophy, joint capsule distension, and increased muscle tone with retraction of pelvi-trochanteric muscle, adductors, and iliopsoas.

37.6 Dislocation

The femoral head exceeds the acetabular edge with neo-cotyle formation; the head is deformed and the neck becomes valgus and anteverted. A hypoplasia of the ossification nucleus is observed.

37.7 Inveterate Dislocation

Condition that occurs after the fourth–fifth years of life in the absence of treatment, with accentuation of the characteristics of dislocation and increased evidence of the neo-cotyle.

Diagnosis of DDH should be early and oriented by anamnestic news such as familiarity, ethnicity, and complications of gestation, but also by clinical finding of probability represented by the asymmetry of the skin, hypotrophy and slight shortening of the preluxated limb, tendency for extra rotation, slight flattening of the buttock, and limitation of the abduction.

In suspected case, clinical diagnosis is carried out with clinical maneuvers; just remember that a dysplastic hip is a non-luxated hip yet, but can be easily dislocated.

37.8 Ortolani Test (Reduction Test)

With the child in the supine position, the examiner puts his/her hands on the newborn's thighs and applies a slight pressure at the great trochanter while the thumb is on the thighs. It abducts and extra-rotates with the knees and hips in flexion, and a reduction of femoral head is warned with a click. The test is positive when the hip is subluxated.

37.9 Barlow Test (Dislocation Test)

With the child in the upper position and thighs flexed at the right angle from the pelvis, grab the knees with the palm of the hand and lay the thumbs on the middle legs and side at the great trochanters. By adjusting the thighs and bringing them to the midline, a slight pressure is applied on the knee, leading the force in the front-rear direction. As a forced maneuver, there is a risk of damage to structures that are already intrinsically unstable, so the Ortolani test is usually preferred.

37.10 Trelat Sign

Patient in prone position with the knees flexed at 90° shows major intra-rotation in the sliced hip.

37.11 Savariaud Sign

The hypometria of the limb affected during the transition from the supine position to the sitting with the knees extended.

Between the 10th and 12th months of life, the child starts walking and goes into a dislocation condition. When dysplasia has not been diagnosed properly on time, this represents the evolution of a preluxation or subluxation condition.

At this point, the deformity can be seen with extra rotation and shortening of the limb, major deficit of walking (Trendelenburg sign or gait anserine), muscle hypoplasia, and Galeazzi sign.

The Trendelenburg sign is shown in mono podalic stand caused by hypoplasia and deficiency of hip abductor (mainly medius gluteus). During walking, the Trendelenburg sign causes the so-called anserine gait.

The **Galeazzi sign** is the asymmetry of knees by observing the patient on the bed with the knees flexed at 90°. The knee is lower on the affected side due to the rear movement in the dysplastic hip.

37.12 The Instrumental Diagnosis (Ultrasound Scan and Radiography)

The ultrasound is certainly the elective test in the first 2–3 months of life for diagnosis and screening investigation since the proximal femoral epiphysis has not yet ossified. It is a noninvasive examination, and it is repeatable and extremely sensitive (so that it is possible to overestimate the pathology). It shows the soft parts and all components of the joint and allows the control of the evolution of the effects of treatment.

With the ultrasound, three lines are set:

1. A straight iliac line
2. The tip of the acetabular labrum
3. The transition from the os ilium to the triradiate cartilage

From these come two angles:

- **Angle α**: formed by the acetabular roof and the vertical cortex of the ilium in coronal plane. It is indicative of the depth bone portion of the acetabulum roof. Normal value is equal or greater than 60 degree.
- **Angle β**: formed by the vertical cortex of the ilium and the triangular labral fibrocartilage. It is an indication of the development of the cartilage part of acetabulum. A Graf β angle greater than 55° is abnormal.

Based on the ultrasound characteristics and values of angles, the proposed Graf classification divides the various stages of the disease into groups and subgroups, on which the treatment is based.

1. **Type I:** alpha angle >60° (normal) (*normal, fully mature hip; the acetabular rim is angular, and the acetabular cup is deep; the cartilaginous roof covers the femoral head*).
2. **Type II**:
    - **Type IIa**: alpha angle 50–59° (<3 months) (*physiologically immature at less than 3 months of age*).
    - **Type IIb**: alpha angle 50–59° (>3 months) (*similar to IIa but in infants older than 3 months; the joint is dysplastic and requires treatment to prevent further deterioration and dislocation*).
    - **Type IIc**: alpha angle 43–49°; beta angle 70–77° (*the hip socket is severely dysplastic and is close to decentering, but the cartilaginous roof still covers the femoral head*).
    - **Type II D**: alpha angle 43–49°; beta angle >77° (*similar to IIc but the hip is decentered; the cartilaginous roof bends cranially*).
3. **Type III**: alpha angle >43° (*dislocated femoral head with a shallow acetabulum*).
4. **Type IV**: alpha angle <43° (*dislocated femoral head with a severely shallow, dysplastic acetabulum with labrum interposed*).

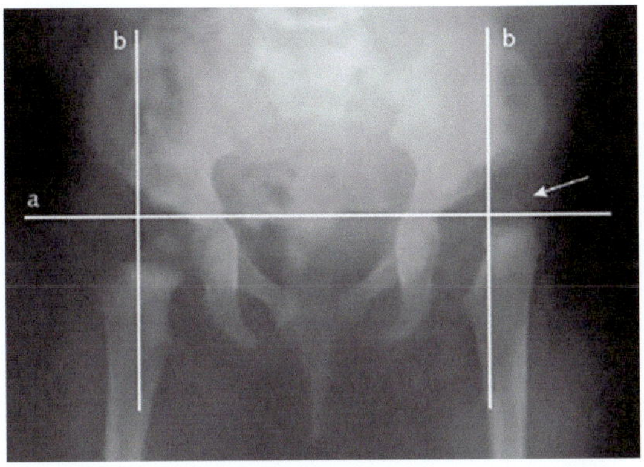

Fig. 37.4 A-P view with Hilgenreiner's and Perkin's line traced, showing the nucleus in the superior/external quadrant

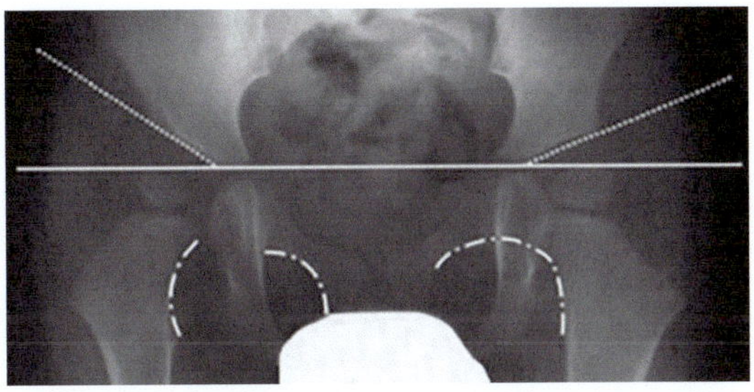

Fig. 37.5 A-P view showing an increased acetabular angle and interruption of Shenton ogive

The X-ray is performed after the first 4–5 months of life since the femoral head core is not ossified and therefore not visible. An A-P front projection of pelvis is performed.

We can obtain this by drawing three lines:

(a) **Hilgenreiner's horizontal line**: traced through the superior aspect of both triradiate cartilages of the pelvis
(b) **Two vertical lines of Perkin**: traced from the most lateral point of the acetabular perpendicular to the Hilgenreiner's line (Fig. 37.4)

The dysplastic hip has the femoral epiphysis in the superior/external quadrant rather than the inferior/medial quadrant, below the Hilgenreiner's line and medial to Perkin's line.

The whole radiographic signs, constituting the triad of putty, Fig. 37.1, enable DDH to be diagnosed: (1) Superolateral displacement of proximal femur; (2) Increase in acetabular angle; and (3) Small capital femoral epiphysis.

Distancing of the nucleus of the femoral epiphysis of the acetabular roof tends to lead to the superior/external quadrant, rather than being in the lower/internal quadrant of the Ombredanne diagrams, resulting in the interruption of the Shenton ogive (arc formed by the lower margin of the femoral metaphysis and the lower margin of the ileopubic branch).

Other X-ray alterations may be neo-cotyle formation; coxa valga (angle >130); and increased anteversion of the head of the femur (Fig. 37.5).

References

1. Vaquero-Picado A, González-Morán G, Garay EG, Moraleda L. Developmental dysplasia of the hip: update of management. EFORT Open Rev. 2019;4(9):548–56.
2. Shapira J, Chen JW, Bheem R, et al. Radiographic factors associated with hip osteoarthritis: a systematic review. J Hip Preserv Surg. 2020;7(1):4–13.
3. Westacott DJ, Perry DC. The treatment of neonatal hip dysplasia with splints in the United Kingdom: time for consensus? J Child Orthop. 2020;14(2):112–7.
4. Barrera CA, Cohen SA, Sankar WN, Ho-Fung VM, Sze RW, Nguyen JC. Imaging of developmental dysplasia of the hip: ultrasound, radiography and magnetic resonance imaging. Pediatr Radiol. 2019;49(12):1652–68.
5. Turchetto L, Massè A, Aprato A, Barbuio A, Ganz R. Developmental dysplasia of hip: joint preserving surgery in the adolescent and young adult. Minerva Ortop Traumatol. 2013;64:41–52.
6. Bixby SD, Millis MB. The borderline dysplastic hip: when and how is it abnormal? Pediatr Radiol. 2019;49(12):1669–77.
7. Breidel KE, Coobs BR. Evaluating and managing acetabular dysplasia in adolescents and young adults. JAAPA. 2019;32(8):32–7.
8. Jayasekera N, Aprato A, Villar RN. Hip arthroscopy in the presence of acetabular dysplasia. Open Orthop J. 2015;9:196–8.
9. Acuña AJ, Samuel LT, Mahmood B, Kamath AF. Systematic review of pre-operative planning modalities for correction of acetabular dysplasia. J Hip Preserv Surg. 2019;6(4):316–25.

Evaluation of Hip Osteoarthritis

38

Christian Carulli, Lorenzo Ius,
and Matteo Innocenti

Hip osteoarthritis (OA), also known as "wear-and-tear" arthritis, is the most common joint disease and represents the most frequent reason for an orthopedic consultation with knee OA [1–4]. It is a degenerative joint disorder that primarily affects the articular cartilage, then progressively involving all the surrounding tissues. OA is classified into primary and secondary types, depending on the cause producing these alterations. Primary OA is considered idiopathic and with multiple joint involvement, mainly related to aging. Secondary OA usually involves few joints and may be related to several causes: post-traumatic or rheumatic origin, evolution of extra-articular deformities, tumoral conditions, hemorrhagic diseases, and sequelae of joint infections [2–8].

Hip OA may be symptomatic (up to 5% of the whole population, mostly woman) or asymptomatic but evident at radiograms (up to 27% of the overall population, independently of sex). Whatever the presence of symptoms, hip OA results in limited range of motion (ROM) and physical impairment with loss of ability in daily life activities [3–10].

As for other joints, the evaluation of a patient starts from the observation of his static posture and gait, at the entrance of the outpatient office. After that, the medical history is important, starting from any personal or familiar issue (hip evaluation on the day of birth, US examination within the first 2 months of age, age of the standing ability and walking, history of developmental dysplasia in relatives) or risk factors (femoral-acetabular impingement, obesity, systemic diseases, history of trauma, avascular necrosis of the femoral head, leg deformities, ongoing medical therapies, occupation, or sports activity). Any previous instrumental analysis should be evaluated before proceeding with the visit [4, 9, 10].

With the patient on the medical couch, the inspection and palpation of the various structures, with specific attention to limb length, abdominal and thigh muscles, presence of any trigger point, or subcutaneous altered bone shape, are mandatory.

In OA, pain is generally located in the groin area, with irradiation to the trochanteric region, buttocks, or knee: depending on the stage of OA, pain is initially intermittent and prone to worsen over time. It is generally associated with stiffness (in the morning and evening, or all day long) and functional limitation, affecting weight-bearing activities and walking. Severe stages of hip OA are characterized by pain reduction as the patient is unable to move or walk, with pain onset at minimal movements, often associated with painful "clicking" or "crepitus."

C. Carulli (✉) · L. Ius · M. Innocenti
Orthopaedic Clinic, University of Florence,
Florence, Italy
e-mail: christian.carulli@unifi.it

J. G. Lane et al. (eds.), *The Art of the Musculoskeletal Physical Exam*,
https://doi.org/10.1007/978-3-031-24404-9_38

Then, an examination of hip active and passive movements with ROM measurements have to be evaluated: flexion, extension, abduction, adduction, internal rotation, and external rotation. Any ROM limitation or pain/clicking/snapping should be recorded. Both hips should be investigated.

Following this first part of the visit, specific maneuvers should be performed in order to assess the clinical suspect hip OA: however, the physician should keep in mind that no test is alone pathognomonic for this purpose.

The ***flexion-abduction-external rotation (FABER) or Patrick test*** of the hip is one of the most common maneuvers. With the patient lying supine, the hip is placed into the figure of "four" position with the hip flexed, abducted, and externally rotated with the contralateral pelvic being stabilized: a downward, external rotation force then is applied to the tested hip and the distance from the ipsilateral knee to the examination table and pain is measured, compared to the contralateral side (Fig. 38.1).

The ***flexion-adduction-internal rotation (FADIR) test***, also named "***anterior impingement test***," is useful and performed with the hip flexed, adducted, and internally rotated while an internal rotational force is applied too, compared to the other hip: reduced ROM, pain, and even clicking may be found (Fig. 38.2).

Thomas test consists of the investigation of a hip flexion contracture: the patient is asked to hold his knee onto the chest and try to extend the contralateral leg. If lumbar spine tract rises from the bed, a flexion contracture is present, and the test is positive.

Following this section of the visit, it is useful to evaluate the patient in standing position. It can reveal limb-length discrepancy, scoliosis, and any vicious posture due to pain. Asking the patient to walk may help to analyze any limping gait or the presence of the "*Trendelenburg's sign*": it consists of the finding of an abductor muscle weakness when the patient is on single leg stance standing, causing the fall of the pelvis from the contralateral healthy side.

Sitting position and raising from a chair are also very useful, to check both the loss of internal and external rotations, typically reduced in OA.

Crucial are the assessments of the spine (mostly lumbar and sacral tracts) and knees, specifically for a differential diagnosis of *groin pain syndrome* [11].

Finally, any positivity to clinical tests should be accompanied by an adequate instrumental imaging. A radiographic study in standing position is mandatory. OA may be classified by the use of dedicated X-ray evaluation according to two main classifications: the Kellgren and Lawrence's and the Tönnis' classifications [12, 13]. The latter is the most used and consists of the classification of radiographic aspect of the hip in four grades: starting from a normal hip

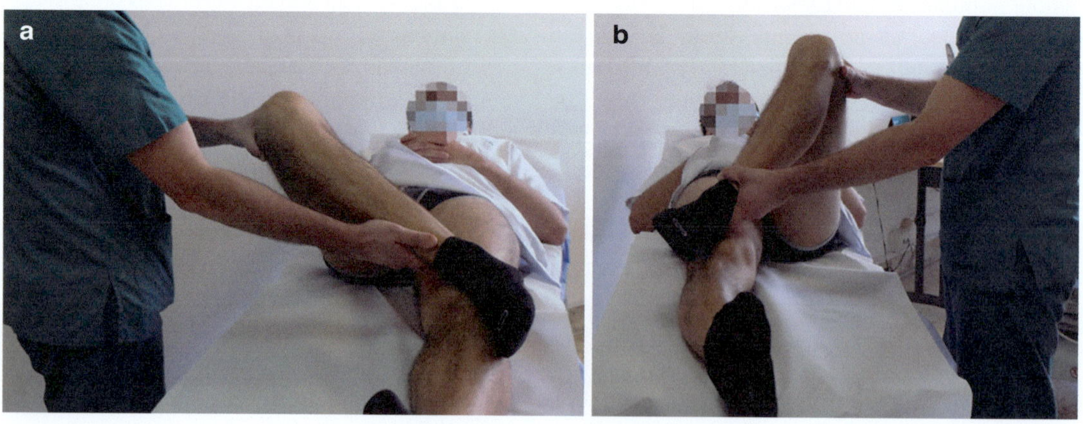

Fig. 38.1 FABER test. Its positivity in the hip joint may indicate FAI, labral tears, loose bodies or chondral lesions, and osteoarthritis. In this middle-aged and very active male patient, the left hip cannot be fully externally rotated due to groin and buttock pain and early osteophytes (**b**) compared to the almost normal right hip (**a**)

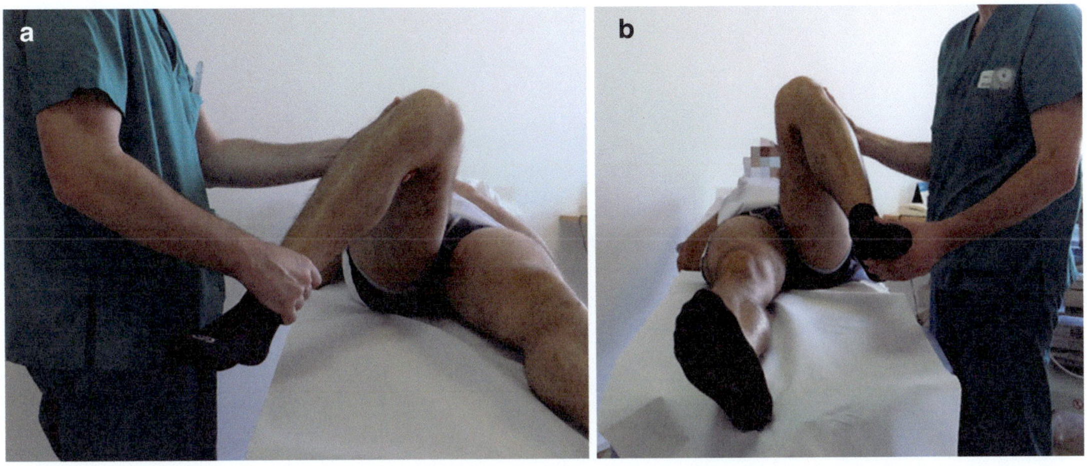

Fig. 38.2 FADIR test. Its positivity in the hip joint may indicate FAI, labral tears, loose bodies or chondral lesions, and osteoarthritis. In the same patient, the left hip cannot be fully internally rotated and adducted due to early osteophytes and labral degenerative changes (**b**) with respect to the contralateral one (**a**)

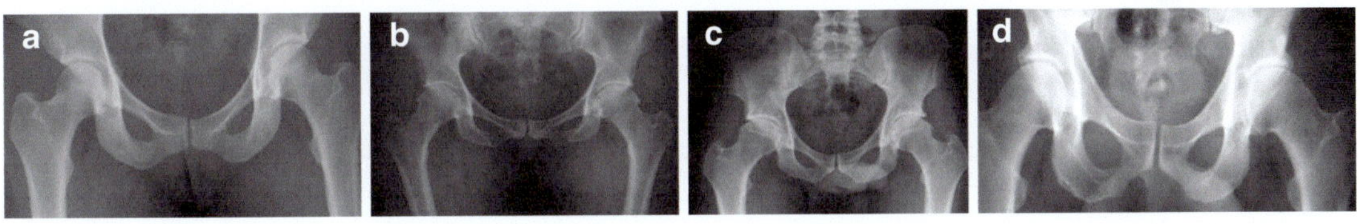

Fig. 38.3 Tönnis' classification: grades "**1**" to "**4**" corresponding to the radiologic aspect as in figures "**a**" to "**d**"

(grade 1), the presence of narrowing of joint space, osteophyte formation, subchondral sclerosis of femoral head and acetabulum, cysts, and finally loss of bone shapes and disappearance of joint space are progressively evaluated (grades 2–4) (Fig. 38.3).

References

1. Edwards J. In: Kuettner KE, Goldberg VM, editors. Osteoarthritic disorders. Rosemont, IL: American Academy of Orthopaedic Surgeons; 1995. p. 500.
2. Wenham CYJ, Conaghan PG. New horizons in osteoarthritis. Age Ageing. 2013;42(3):272–8.
3. Aronson J. Osteoarthritis of the young adult hip: etiology and treatment. Instr Course Lect. 1986;35: 119–28.
4. Hippisley-Cox J, Vinogradova Y. Trends in consultation rates in general practice 1995/1996 to 2008/2009: analysis of the QResearch database. London: QResearch Inf Cent; 2009.
5. Dagenais S, Garbedian S, Wai EK. Systematic review of the prevalence of radiographic primary hip osteoarthritis. Clin Orthop Relat Res. 2009;467(3):623–37.
6. Harris WH. Etiology of osteoarthritis of the hip. Clin Orthop Relat Res. 1986;213:20–33.
7. Ganz R, Parvizi J, Beck M, Leunig M, Nötzli H, Siebenrock K, a. Femoroacetabular impingement: a cause for osteoarthritis of the hip. Clin Orthop Relat Res. 2003;417:112–20.
8. Lievense AM, Bierma-Zeinstra SMA, Verhagen AP, van Baar ME, Verhaar JAN, Koes BW. Influence of obesity on the development of osteoarthritis of the hip: a systematic review. Rheumatology. 2002;41(10):1155–62.
9. Juhakoski R, Heliövaara M, Impivaara O, et al. Risk factors for the development of hip osteoarthritis: a population-based prospective study. Rheumatology. 2009;41(10):1155–62.
10. Harris EC, Coggon D. Hip osteoarthritis and work. Best Pract Res Clin Rheumatol. 2015;29(3):462–82.
11. Bisciotti GN, Auci A, Di Marzo F, Galli R, Pulici L, Carimati G, Quaglia A, Volpi P. Groin pain syndrome: an association of different pathologies and a case presentation. Muscles Ligaments Tendons J. 2015;5(3):214–22.
12. Kellgren JH, Lawrence JS. Radiological assessment of osteo-arthrosis. Ann Rheum Dis. 1957;16(4):494–502.
13. Tönnis D. Normal values of the hip joint for the evaluation of X-rays in children and adults. Clin Orthop Relat Res. 1976;119:39–47.

Evaluation of Snapping Hip and Extra-Articular Impingement

39

Manlio Panascì and Alberto Costantini

New forms of extra-articular hip impingement besides the snapping hip have recently been recognized as a cause of hip pain and limited function, especially in young active patients. These conditions include subspine impingement, avulsion, and ossification of the rectus femoris attachment, ischiofemoral impingement, and greater trochanteric-pelvic impingement.

39.1 Snapping Hip

The extra-articular causes of snapping hip are classified into two categories: external and internal. The conflict of iliotibial band (ITB) with the greater trochanter region determines the hip external snap. Intra-articular pathologies in differential diagnosis with internal snapping hip are due to articular loose bodies, labral injury, or articular instability [1, 2]. The underlying causes of a conflict or mechanisms that may determine the onset of an extra-articular snapping hip are shown in Table 39.1.

The most common cause of snapping hip syndrome is irritation of the greater trochanter by the ITB. The ITB is a large flat tendinous structure that originates on the anterior superior portion of the iliac crest, crosses over the greater trochanter of the femur, and inserts onto the lateral condyle of the tibia.

Table 39.1 Internal snapping hip

Tendon impingement at the level of ileum-pectineal eminence
Iliopsoas impingement with the acetabular component in a THA
Tendon snap at the level of the upper branch of the pubic bone
Conflict with the anterior acetabular margin
Conflict between two components of a bifid tendon
Impingement of the tendon at the level of the anterior inferior iliac spine
External snapping hip
Thickening of the posterior ITB or the gluteus maximus
Greater trochanteric-pelvic impingement
Big offsets in THA

When the hip is extended, the ITB is posterior to the greater trochanter. As the hip moves into flexion, the ITB moves anterior to the greater trochanter. Ordinarily, it glides smoothly over the greater trochanter with assistance from the underlying bursae [3]. When the posterior portion of the iliotibial tract or the anterior border of the gluteus maximus becomes thickened, however, this results in snapping of the tendon over the greater trochanter. The bursae can then become inflamed and further exacerbate the condition [4]. Coxa vara may predispose to a snapping hip. Other predisposing factors are hyperplasia of the

M. Panascì
San Carlo di Nancy Hospital, GVM Care and Research, Rome, Italy

A. Costantini (✉)
Concordia Hospital for Special Surgery, Rome, Italy

J. G. Lane et al. (eds.), *The Art of the Musculoskeletal Physical Exam*,
https://doi.org/10.1007/978-3-031-24404-9_39

trochanteric bursa, narrower bi-iliac width, prominent greater trochanters, and increased distance between the greater trochanters. A case of snapping hip secondary to fibrosis of the band/muscle related to multiple intramuscular injections has also been reported [5].

External snapping hip is seen in athletes who undergo repetitive knee flexion, such as runners, dancers, and cyclists. Athletes will have pain over the greater trochanter of the femur and lateral thigh or radiating pain down to the knee. Patients often report hip instability symptoms. If severe enough, the snapping sensation will occur during normal ambulation. Once this area becomes inflamed, running or rising from a seated position may hurt continuously. On physical examination, in addition to the evidence of the shot, which is often caused voluntarily by the patient, it can consist of tenderness upon insertion of IBD or the greater trochanter, with occasional painful side irradiation in the thigh. The most reliable method to assess the snapping hip is to request the patient to reproduce the mechanical phenomenon. Most legitimate snapping syndromes can be voluntarily reproduced in the medical office. Typically, the internal snapping hip originates from the iliopsoas tendon and is more audible than visible. On the other hand, the external snapping hip from the IT band or gluteus maximus is more visible than audible [6].

In external snapping hip, the snapping sensation occurs in the lateral upper thigh over the region of the greater trochanter. An audible snapping may occur during hip flexion and again when extending the hip from flexion. The majority of cases are palpable, and some cases may be visible under the skin. Snapping may also be found during internal and external rotation of the extended and adducted hip. Often, physical examination findings are best elicited with the patient in the lateral decubitus position with the IT band under tension. The amount of IT band tension is assessed side to side with the Ober test.

Physical examination of the internal snapping phenomenon is done with the patient supine by flexing the affected hip more than 90° and extending to neutral position. This maneuver will reproduce the snapping phenomenon at the front of the groin, which will be mentioned to the examiner by the patient when it occurs. The snapping phenomenon cannot be observed through the skin but frequently produces an audible snap. This may be accentuated with abduction and external rotation in flexion and adducting and internally rotating while extending. The snapping phenomenon may be palpated by placing the hand over the affected groin. When the snapping is symptomatic, there is always an apprehension response [6].

39.2 Subspine Impingement

Prominence of the anterior inferior iliac spine (AIIS) at the level of the acetabular rim is presented as another etiology of anterior hip impingement (Fig. 39.1). The rationale for this injury mechanism is that the morphology of the

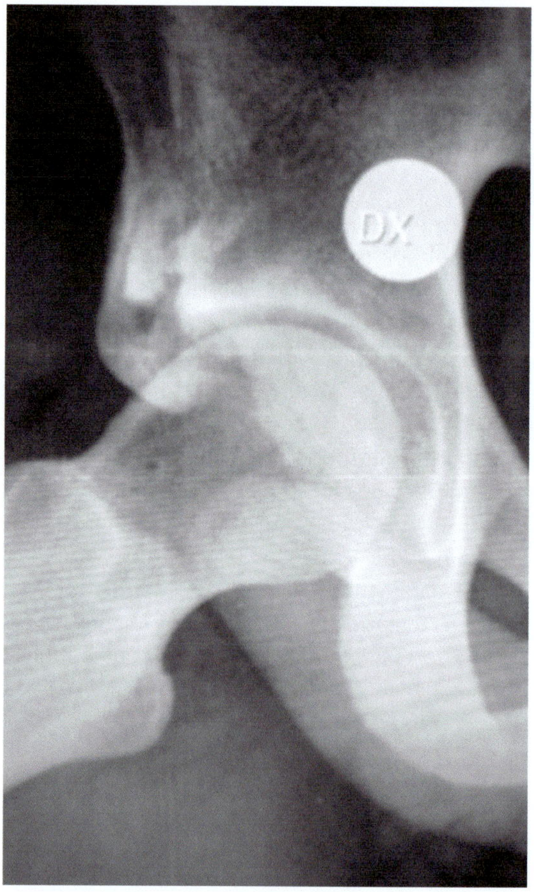

Fig. 39.1 Axial view of proximal femur showing a big prominence of AIIS

AIIS may result in decreased space available for soft tissue at the level of the acetabular rim during hip flexion. The result is impingement of anterior tissue (anterior capsule or iliocapsularis muscle origin) against the femoral head-neck junction. The clinical presentation in patients with subspine impingement includes tenderness to palpation over the AIIS that recreates typical pain and anterior hip or groin pain with straight or prolonged hip flexion [7]. On physical examination, hip flexion range of motion is limited. Partial pain relief and persistent hip flexion limitations after intra-articular anesthetic hip injection may be observed. This may be explained by the fact that the AIIS deformity may not have been the single cause for the preoperative symptoms since studies have shown the concomitant presence of abnormal cam morphology [8].

39.3 Avulsion and Ossification of the Rectus Femoris

Avulsions of the rectus femoris are common in the adolescent, skeletally immature population [9].

Several sports such as football, soccer, gymnastics, and dance are involved. Proximal avulsions of the rectus in the adult, skeletally mature population are a less common entity, while post-traumatic heterotopic ossification (PHO) of the rectus femoris can be the consequence of an avulsion or the result of a flexor strain. Ossification of the proximal rectus femoris tendon has been described following avulsion (with or without bony fragment) or tendon rupture (partial or complete) of the rectus femoris origin in young people. Flexion is often restricted, and a decrease in internal rotation in comparison with the opposite side is observed while abduction, adduction, extension, and external rotation are comparable to the other side. Tenderness and bruise in the region of the groin are commonly present [10].

Hip flexor strains commonly occur as a noncontact injury during either eccentric contraction of the muscle or hip flexion (kicking motion). Patients describe and immediate pain in the anterior hip. Ecchymosis and swelling often result. Clinical suspicion is confirmed by tenderness to palpation over the hip flexor origin, occasional step-off, and weakness and/or pain with resisted hip flexion [11].

39.4 Ischiofemoral Impingement

Ischiofemoral impingement is a cause of posterior hip pain recently being reported. It is described as a narrowing of the ischiofemoral space and an abnormal quadratus femoris muscle MR signal intensity. Both the ischiofemoral and quadratus femoris spaces have been found to be significantly narrower in affected subjects. Suggested cutoff values are ≤15 mm for the ischiofemoral space and ≤8 mm for the quadratus femoris space. Edema of the quadratus femoris muscle may be visible in patients with IFI, and some patients may present with fatty infiltration of the quadratus femoris muscle which is sometimes combined with muscle atrophy [12, 13].

Tenderness to palpation at the lateral surface of the ischial tuberosity is commonly present, which is distinguished from pain at the medial surface of the ischial tuberosity often associated with obturator internus pain and pudendal nerve entrapment.

Pain may be present with active or passive hip extension, external rotation, and adduction. Snapping may occur during hip flexion or extension during weight-bearing activities [14].

39.5 Greater Trochanteric-Pelvic Impingement

Impingement between an abnormally prominent or a high riding greater trochanter with a short femoral neck and the lateral pelvis when the hip is abducted may be a possible cause of greater trochanteric-pelvic impingement. Perthes-like hip is a common morphology example, in severe cases associated with important leg-length discrepancy and abductor muscle dysfunction. Pain and disability are present in the hip joint region, but it is not relieved by intra-articular injection or physical therapy [15].

Clinical examination often shows both lateral hip and groin pain, especially with active hip abduction and extension. Joint may be felt blocked at the end range of these combined motions. The examination may also reveal limited and painful active or passive hip abduction and extension, a shortening of the involved leg, and a positive Trendelenburg gait secondary to hip abductor weakness. The "gear-stick" sign is useful to differentiate other sources of hip pain. Patient is in a lateral position, and the symptomatic hip is passively abducted in extension with reproduction of the patient's symptoms representing a positive sign.

ROM abduction in flexion may be normal since the greater trochanter avoids contact with the ilium in this position. Since reliability of this test has not been proved yet, plain radiographs seem to be the "gold" standard for further diagnosing greater trochanteric-pelvic impingement.

References

1. Allen WC, Cope R. Coxa saltans: the snapping hip revisited. J Am Acad Orthop Surg. 1995;3:303–8.
2. Brignall CG, Brown RM, Stainsby GD. Fibrosis of the gluteus maximus as a cause of snapping hip. J Bone Joint Surg Am. 1993;75-A:909–10.
3. Zoltan DJ, Clancy WG, Keene JS. A new operative approach to snapping hip and refractory trochanteric bursitis in athletes. Am J Sports Med. 1986;14:201–4.
4. Faraj AA, Moulton A, Sirivastava VM. Snapping iliotibial band. Report of ten cases and review of the literature. Acta Orthop Belg. 2001;67:19–23.
5. Zini R, Munegato D, De Benedetto M, Carraro A, Bigoni M. Endoscopic iliotibial band release in snapping hip. Hip Int. 2013;23(2):225–32.
6. Winston P, Awan R, Cassidy JD, Bleakney RK. Clinical examination and ultrasound of self-reported snapping hip syndrome in elite ballet dancers. Am J Sports Med. 2007;35(1):118–26.
7. Blankenbaker DG, Tuite MJ. Non-femoroacetabular impingement. Semin Musculoskelet Radiol. 2013;17:279–85.
8. Hapa O, Bedi A, Gursan O, Akar MS, Güvencer M, Havitçioğlu H, Larson CM. Anatomic footprint of the direct head of the rectus femoris origin: cadaveric study and clinical series of hips after arthroscopic anterior inferior iliac spine/subspine decompression. Arthroscopy. 2013;29(12):1932–40.
9. Porr J, Lucaciu C, Birkett S. Avulsion fractures of the pelvis - a qualitative systematic review of the literature. J Can Chiropr Assoc. 2011;55(4):247–55.
10. Zini R, Panascì M. Post-traumatic ossifications of the rectus femoris: Arthroscopic treatment and clinical outcome after 2 years. Injury. 2018;49(Suppl 3):S100–4.
11. Gamradt SC, Brophy RH, Barnes R, Warren RF, Thomas Byrd JW, Kelly BT. Nonoperative treatment for proximal avulsion of the rectus femoris in professional American football. Am J Sports Med. 2009;37:1370–4.
12. Torriani M, Souto SC, Thomas BJ, Ouellette H, Bredella MA. Ischiofemoral impingement syndrome: an entity with hip pain and abnormalities of the quadratus femoris muscle. Am J Roentgenol. 2009;193:186–90.
13. Patti JW, Ouellette H, Bredella MA, Torriani M. Impingement of lesser trochanter on ischium as a potential cause for hip pain. Skelet Radiol. 2008;37:939–41.
14. Beckmann JT, Safran MR, Abrams GD. Extra-articular impingement: ischiofemoral impingement and trochanteric-pelvic. Oper Techn Sports Med. 2015;
15. de Sa D, Alradwan H, Cargnelli S, Thawer Z, Simunovic N, Cadet E, Bonin N, Larson C. Ayeni OR extra-articular hip impingement: a systematic review examining operative treatment of psoas, subspine, ischiofemoral, and greater trochanteric/pelvic impingement. Arthroscopy. 2014;30(8):1026–41.

40 Evaluation of Athletic Population with Hip/Hamstring/Quad Injuries

Paolo Di Benedetto, Giovanni Gorasso, Andrea Zangari, and Nunzio Lassandro

The hip represents the most complex anatomical district from the diagnostic point of view. History and physical examination of the athletic hip are the key for the evaluation of patients presenting with hip pain. The examiner should consider that independent pathologies could show concomitant and overlapping signs and symptoms. This subchapter describes a comprehensive and systematic approach to the athlete with hip, hamstring, and tight pain.

40.1 Introduction

The study of the hip represents a daily challenge for the orthopedic specialist. Defining precisely what the patient describes as "hip pain" is an effort often overlooked. The difficulties are further amplified in athletes, of all ages and of all sport levels. Recent studies show that groin pain is described as the main symptom in 10% of sports medicine visits [1, 2]; in addition, 5–9% of adolescent athletes suffer from hip and/or pelvic girdle disorders [3]. Sports which provide for frequent accelerations, decelerations, and changes of directions are increasingly popular (for example, football, tennis, American football), with a consequent increase in the risk of overstress of the coxofemoral joint and of lesions to the surrounding structures. The hip (and in general the pelvic girdle) represents perhaps the most complex anatomical district from the diagnostic point of view, so a precise knowledge of the anatomy and biomechanics of this district is necessary. To this complex picture, the wide range of differential diagnosis is added; independent pathologies could show concomitant and overlapping signs and symptoms, while related disorders could reveal a diversified symptomatology [4, 5]. It is therefore a compelling challenge. The orthopedic surgeon will have to take on the role of the investigator, recognizing fundamental signs and symptoms, isolating the essential details, and avoiding following incorrect diagnostic indications. As said, differential diagnosis is a fundamental and very complex context (Table 40.1).

A precise physical examination and an adequate radio-diagnostic study must always be preceded by a scrupulous clinical history of the patient. The history of a patient presenting with hip pain must necessarily include age, traumatic

P. Di Benedetto (✉)
Clinic of Orthopaedics, Friuli Centrale Healthcare and University Trust (ASUFC), Udine, Italy

DAME—University of Udine, Udine, Italy
e-mail: paolo.dibenedetto@uniud.it

G. Gorasso · N. Lassandro
Clinic of Orthopaedics, Friuli Centrale Healthcare and University Trust (ASUFC), Udine, Italy

A. Zangari
Clinic of Orthopaedics, Friuli Centrale Healthcare and University Trust (ASUFC), Udine, Italy

Clinic of Orthopaedics, Friuli Centrale Healthcare and University Trust (ASUFC), Udine, Italy

J. G. Lane et al. (eds.), *The Art of the Musculoskeletal Physical Exam*,
https://doi.org/10.1007/978-3-031-24404-9_40

Table 40.1 Differential diagnosis

Intra-articular	Extra-articular	Nonmusculoskeletal
Femoroacetabular impingement	Muscle strain	Sports hernia
Developmental hip dysplasia	Tendonitis/tendinopathy	Appendicitis
Acetabular labral tear	Ligament injury	Inflammatory bowel disease
Chondrosis	Bursitis	Diverticulitis
Legg-Calvè-Perthes disease	Pubic ramus fracture	Lymphadenitis
Osteochondritis dissecans	Apophyseal avulsion	Urinary tract infection
Avascular necrosis of femoral head	Greater trochanter pain syndrome	Prostatitis
Traumatic/stress facture	Snapping hip syndrome	Nephrolithiasis
Hip subluxation/dislocation	Lumbosacral radiculopathy	Pelvic inflammatory disease
Synovitis	Sacroiliac joint dysfunction	Ovarian cystitis
Tumor	Pelvic floor dysfunction	Ectopic pregnancy
Infection	Peripheral nerve entrapment	Urethritis

mechanism (if present and identifiable), quality, type, distribution, progression of the pain, exacerbating and alleviating factors, and associated signs and symptoms. All these data are essential to start the diagnostic investigation in the right direction, focusing on the most probable pathological situation. The location and irradiation of pain are useful information, although it is difficult to recognize and determine precisely. Anterior groin pain is often associated with intra-articular hip disorders (FAI, labral tear, intra-articular mobile bodies, osteoarthritis, etc.). Disorders of extra-articular structures can also manifest anterior pain (adductor muscles and iliopsoas). Lateral hip pain can occur in intra-articular pathological pictures but is more frequently associated with extra-articular disorders such as trochanteric bursitis and tendinopathies of the tensor fasciae latae or of the hip external rotators. Posterior pelvic pain represents the most nonspecific symptom, which best represents the difficulties to overcome towards a precise diagnosis; this clinical manifestation can be a consequence, in addition to the pathologies already mentioned (recent studies show the presence of posterior pain in 17% of patients with congenital hip dysplasia, in 38% of patients with isolated lesions of the acetabular lip, and in 29% of patients with femoroacetabular impingement), of sacroiliac joint disorders, hamstring muscle group, or spinal lumbosacral tract [6–8]. Also not to be forgotten in the clinical study are the characteristics of the pain described by the patient. Pain associated with active contractions or passive stretches of specific muscle groups is often suggestive of tendinopathies or muscle injuries. Symptoms exacerbated by coughing or sneezing can frame the diagnosis towards abdominal or intervertebral hernia disorders. Pops or blocks of movement can depend on extra-articular or intra-articular disorders; a sore and popping hip is often recognizable in sportsmen who perform frequent hip flexions and hyperextensions, such as footballers or dancers [9–11]. This symptomatology can be associated with impingement of the iliotibial band with the greater trochanter or of the iliopsoas tendon with the iliopectineal eminence [12]. Finally, burning pain is often associated with neuropathic symptoms. A hip examination must be conducted in a specific order to have a reproducibility between different observers, too. The order in which it is carried out must be simple so that the patient can easily follow the doctor's advice (Table 40.2). It starts with the standing position and then passes to sitting, prone, supine, and then lateral. In each position, specific tests can be performed (Table 40.3).

Table 40.2 The complete form for hip/hamstrings/quad injuries

Physical examination

Name ________ Surname ________ -	Date of birth ___/___/___
H:___ cm W:___ Kg BMI: ___	

Standing examination

Shoulder height: equal/not equal			
Iliac crest height: equal/not equal			
Active forward ________*			
Spine straight / scoliosis —> structural / non structural			
Hyperkyphosis +/-	Hyperlordosis +/-		
Iperlaxity test: thumb test +/-	elbows recurvatum +/-	Knees recurvatum +/-	
Gait			
Gait:	normal	antalgic	abductor deficient (Trendelemburg)
Pelvic wink	arm swing	short stride length	short stance phase
Foot progression angle:	external	neutral	hyperpronation
Single leg Stance Phase Test (Trendelenburg test)	Right	Left	None

Seated examination

Neurologic findings:			
Motor: ________			
Sensory: ________			
Deep tendon reflexes	Patella	Achilles	
Straight leg Phase:	Right	Left	
Seated R.O.M.	Right	Left	
Internal rotation:	External rotation:	Internal rotation:	External rotation
		Right +/-	Left +/-
Active slump test		☐	☐
Seated piriformis stretch test		☐	☐

Prone examination

	R +/-	L
Ely's test (rectus contracture)	☐	☐
Craig's test (femoral anteversion)	☐	☐
Palpation		
SI joint	☐	☐
Bursae Ischium	☐	☐
Spinous processes level________		
Muscle soreness	Right +/-	Left +/-
Hamstrings	☐	☐

(continued)

Table 40.2 (continued)

Supine Examination

Skin inspection: none/________________________________			
Lymphatic system: _ lymphedema	no lymphedema	pitting edema 1+ 2+	
Leg lenghts: R___cm	L___cm	equal/not equal	
R.O.M.	Right	Left	
Flexion	________	________	
Abduction	________	________	
Adduction	________	________	
		Right +/-	Left +/-
Thomas test (hip flexion contracture)		☐	☐
Modified Thomas test (quad injury)		☐	☐
FABER test		☐	☐
FADDIR test		☐	☐
DIRI test		☐	☐
DEXRIT test		☐	☐
Posterior rim impingement test		☐	☐
Log-roll test		☐	☐
Apprehension sign		☐	☐
Dial test		☐	☐
Stinchfield's test (straight leg raise against resistance)		☐	☐
Resisted knee extension test		☐	☐
Heel strike		☐	☐
Femoral nerve tinel		☐	☐
Abdomen	soreness	☐	☐
Adductor tubercle	soreness	☐	☐
Pubic symphysis	soreness	☐	☐
Muscle soreness		Right +/-	Left +/-
Iliopsoas		☐	☐
Quadriceps		☐	☐
Abductors		☐	☐
Neurologic findings		Right +/-	Left +/-
Lasegue		☐	☐
Dandy		☐	☐

Table 40.2 (continued)

Lateral examination

Palpation	tender	not tender
SI joint	☐	☐
Ischium	☐	☐
Great trochanter	☐	☐
ASIS	☐	☐
Piriformis	☐	☐
Sciatic nerve	☐	☐
Gluteus Max (insertion into ITB)	☐	☐
Glutes medius	☐	☐
Abductor strength: R L	R L	R L
straigth leg. ____ ____	Gluteus medius ____ ____	Gluteus max ____ ____
	R +/-	L +/-
FADDIR test	☐	☐
Lateral rim impingement	☐	☐
Gluteus Max. contracture test	☐	☐
Gluteus Med. contracture test	☐	☐
Tensor fascia lata contracture test	☐	☐
Passive adduction test	☐	☐
Active piriformi test	☐	☐
Muscle soreness	Right +/-	Left +/-
Gluteus med	☐	☐
Gluteus max	☐	☐
Tensor fascia lata	☐	☐
Piriformis	☐	☐

Specific tests

	R +/-	L
Philippon internal rotation test	☐	☐
McCarthy's sign	☐	☐
IFI test	☐	☐
Dynamic Trendelemburg	☐	☐
Scours	☐	☐
Foveal distraction	☐	☐
Bicycle	☐	☐
Pace sign	☐	☐
ABDEER	☐	☐
Supine abduction external rotation	☐	☐
Fulcrum	☐	☐
Long stride test	☐	☐

Table 40.3 Specific tests' description

Test	Patient position	Description	Positive test	Assessment
Log roll test	Patient in supine position with knee and hip extended.	The examiner passively internally and externally rotates the leg while stabilizing the knee and ankle	A positive test reproduces anterior or lateral hip pain	Intra-articular hip disease
Anterior hip impingement test	Patient in supine position.	The examiner passively flexes the hip and knee and internally rotates and adducts the hip.	A positive test reproduces anterior or lateral hip pain	Intra-articular hip disease, in particular impingement and/or an anterosuperior labral tear
Internal rotation over pressure test	Patient in supine position.	The examiner flexes passively the hip and the knee to 90 degree then internally rotates the hip to and range of motion and applies a posterior force.	A positive test reproduces anterior or lateral hip pain	Intra-articular hip disease
Hip scour test	Patient in supine position.	The examiner moves passively the hip in flexion and adduction end range of motion. A downward force is applied along the axis of the femur while moving the hip in a circular fashion.	A positive test reproduces groin, lateral hip or posterior pelvic pain	Intra-articular hip disease
FABER test / Patrick test	Patient in supine position.	The examiner rests the ankle of the leg on the opposite knee in a"figure 4" position; the opposite ASIS is stabilized with one band, and the other hand applies pressure lo the knee downward	A positive test for hip disease reproduces anterior or lateral hip pain. A positive test for SI joint or low back disease reproduces posterior pelvic pain	Hip disease or SI and low back disease
Stinchfiled resisted hip flexion test	Patient in supine position.	The patient flexes the hip to 30 degree against resistance with the knee extended.	A positive test reproduces anterior or lateral hip pain	Intra-articular hip disease Hip flexors disease
Thomas test	Patient in sitting position at the edge of the table	The patient flexes the contralateral knee to the chest and rolls onto the back while allowing the tested hip to remain extended and hanging off the edge of the table.	Lack of full hip extension indicates hip flexion contracture, while abduction of the leg indicates ITB tightness	Hip flexors contracture
Ober test	Patient in lateral position with the tested leg up. Hip and knee of the lower leg are flexed	The examiner passively extends the patient's upper leg with the knee flexed at 90 degree; while supporting the knee, the examiner slowly lowers the leg	If the ITB is shortened, the leg remains abducted	ITB tightness
Posterior hip impingement test	Patient in prone position with hip and knee extended	The examiner passively extends, adducts, and externally rotates the hip	a positive test reproduces anterior hip or posterior pelvic pain	Intra-articular hip disease in particular posterior wall or labral pathologies
Trendelemburg sign	Patient in standing position	the examiner places the hands on the top of both iliac crests while the patient stands on the tested leg in a single leg stance	the test is positive if the pelvis droops on the non stance side (hip abductor weakness on the side of the stance leg)	hip abductor weakness
Ely's test	Patient in prone position	The examiner passively flexes the knee.	The test is positive if the hip flexes as the knee flexes indicating a tight rectus femoris	Rectus femoris contracture
Piriformis test	Patient in lateral position with the tested leg up. Hip and knee of the upper leg are flexed	The examiner places a hand on the great trochanter with the thumb on the piriformis insertion. The patient actively external-rotates the hip	If the piriformis is tight pain is elicited. If the sciatic nerve is compressed the patient describes radicular-like symptoms	Contracture of the piriformis muscle or sciatic nerve compression

SI sacroiliac, *ASIS* anterior superior iliac spine, *ITB* iliotibial band

40.2 Standing Examination

The patient in this position indicates the point where he/she has pain with one finger; a characteristic sign that can be identified in this moment is the "C sign" [13]. The patient will hold their hand in the shape of a C and place it above the greater trochanter with the thumb positioned posterior and fingers extending to the groin. This finding can be misinterpreted as lateral soft-tissue pathology; however, the patient often describes deep interior hip pain [14]. The iliac crest heights and shoulder height are noted to evaluate leg-length discrepancies as the patient stands. Wooden tablets can be used as elevations to be placed under the feet to bring the shoulders and iliac spines to level and therefore determine any possible limb-length discrepancy. In this phase, it is also useful to evaluate the presence of eventual imbalances of the spinal column as scoliosis. At this stage of the physical examination, any abnormalities in the load are also assessed; this often helps to detect hip pathology owing to the transfer of dynamic and static load to the ligamentous and osseous structures [15]. A full gait of 6–8 stride lengths is observed from behind and the front of the patient. Particular attention should be paid to the rotation of the feet in all phases of the step. The foot progression angle will detect osseous or static rotatory malalignment. A shortened stance phase is very important to note because it can be indicative of neuromuscular abnormalities, trauma, or leg-length discrepancies. A step study can give many clues about the underlying pathology. An antalgic gait is characterized by a shortened stance phase on the painful side limiting the duration of weight-bearing [16]. A short-leg gait is evidenced by the drop of the shoulder in the direction of the short leg. Note also the alignment of the patellae, whose internal rotation or external rotation may indicate a patella-femoral malalignment. After observing the gait, we can conduct more detailed tests. The Trendelenburg test or single leg stance phase test (for description, see the table below) is conducted on both legs, with the non-affected leg examined first to establish a baseline reference for the patient's function. The abductor

Table 40.4 Normal hip motions and normal measurements for active range of motion [17]

Motion	Normal values	Comment
Flexion	110–120	Measure flexion with the knee flexed to prevent limitation caused by hamstring tightness
Extension	0–15	Increased lordosis of the lumbar spine may falsely compensate for decreased hip extension
Abduction	30–50	During active hip abduction, pelvic motion signifies the limit of abduction. Assess this motion by placing a hand on the patient's anterior-superior iliac spine
Adduction	30	Starting from neutral, the limit of adduction occurs when the pelvis moves
External rotation	40–60	Values may vary with increased femoral neck anteversion or retroversion. Test can be performed with the patient supine or prone
Internal rotation	30–40	Values may vary with increased femoral neck anteversion or retroversion. Tests can be performed with the patient supine or prone

deficient gait is an unbalanced stance phase attributed to proprioception disruption or abductor weakness. The abductor deficient gait may present in two ways: with a shift of the pelvis away from the body ("dropping out" of the hip on the affected side) or with a shift of the weight over the adducted leg (shift of the upper body "over the top" of the affected hip) [13]. A specific test during gait is the long stride test, for the ischiofemoral impingement, in which a long step causes pain (Table 40.4).

40.3 Seated Examination

During the physical examination in a sitting position, all possible vascular and nervous system problems must be analyzed; it is important to observe the state of the soft tissues, too. The strength in extension against knee resistance and the bilateral patella-femoral reflexes of the knee should be evaluated. In this position, it is essential to evaluate the passive intra- and extra rotation. In fact, the sitting position ensures that the ischium is square to the table, thus providing sufficient stability at 90 degree of hip flexion [13]. Passive and active intra-rotation should be examined gently and compared with the contralateral side. An increase in intra-rotation and a decrease in external rotation can be a manifestation of excessive femoral anteversion. A decrease in intra-rotation can indicate an intra-articular pathology. At the end of the seating position examination, iliopsoas contracture test is performed (Fig. 40.1).

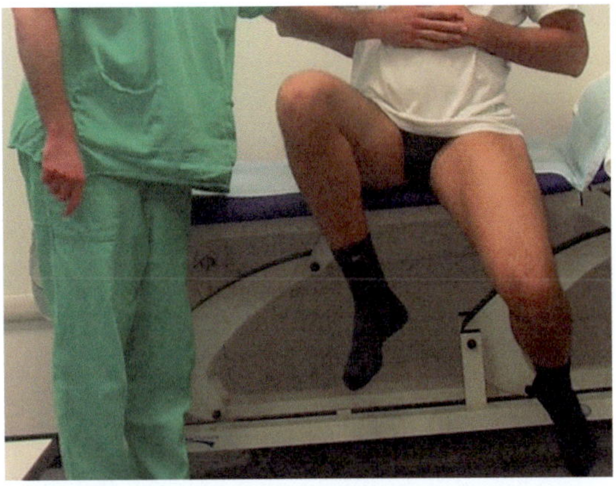

Fig. 40.1 Iliopsoas contracture test: reflecting the affected hip; note the internal rotation of the contralateral hip

40.4 Prone Examination

The prone examination involves palpation of four distinct areas: gluteus maximus origin, supra-sacroiliac joint, sacroiliac joint, and spine (facets). The femoral anteversion test (traditionally known as Craig test) will test the femoral anteversion or retroversion [19]. With the patient in the prone position, the knee is flexed to 90° and the examiner manually rotates the leg while palpating the greater trochanter so that it protrudes most laterally. Femoral version is assessed by noting the angle between the axis of the tibia and an imaginary vertical line, which normally is between 10 and 20°. This test will help to identify cases of retroversion. The rectus contracture test (also known as Ely test; see table below): restriction of hip flexion is indicative of a rectus femoris contracture [20].

40.5 Supine Examination

First of all, it is appropriate to measure the length of the lower limbs, comparing the spinomalleolar distance. The passive ROM in hip flexion is then examined, with the knee flexed. In addition to the amplitude of the articular excursion, it is useful to also observe the position of the pelvis since it could compensate for a hip flexion deficit by starting to rotate. The Thomas test is done by having the patient extend and relax the leg down towards the table, while holding the contralateral leg in full flexion. The inability of the thigh to reach the table shows a hip flexion contract [13]. Abduction and passive adduction are also tested. It is also useful to make a palpation of the abdomen to document the tension and any painful points; this is also to exclude any inguinal hernias [21]. It is useful to ask the patient to actively flex the back from the supine position and simultaneously palpate the abdomen to exclude any abdominal fascial hernias that can cause delays related to the pelvis. During this phase, articular sclerosis or snapping sensations can be felt; it is advisable to identify the movements with which this happens and reproduce them several times to identify if it is an internal or external snap. If it is an internal snap that has come from the iliopsoas tendon, many times this will be eliminated with a simultaneous abdominal contraction. A bicycle test (conducted in the lateral position) can help to distinguish the pop internally from the external pop of coxa with the externus one belonging to the subluxing iliotibial band over the greater trochanter. The FABER test (Patrick test) (for the description of the specific tests, see the table below) is helpful to distinguish hip versus lumbar complaints [22]. In the presence of hip pain, if the test is positive, it can be associated with musculotendinous or osseous posterior lateral acetabular incongruence or ligamentous injury. The DIRI test is also conducted in a supine position; positivity is determined if the same type of pain complained by the patient is re-created. Finally, the Tinel test for the femoral nerve is done. This test is found to be positive with hip flexion contractures of greater than 25°, as a result of the proximity of the psoas tendon and the femoral nerve. A heel strike is carried out by striking the heel abruptly, which if painful is indicative of some type of trauma or a stress fracture. After that, we can do the fulcrum test where the patient complains pain. The patient is seated on the examination table with his/her lower legs dangling. The examiner places one of his/her arms under the symptomatic thigh. The palm of the hand is facing up and touching the patient's leg. This arm will serve as a fulcrum. At one side of the fulcrum, the force is created by the patient's body weight. If the patient complains pain, a stress fracture or a traumatic incomplete fracture must be suspected [17].

The log roll test [14] involves passive intra- and extra-rotation of the femur, with the leg lying in an extended position. This test is conducted bilaterally, and any side-to-side differences of this maneuver can alert the examiner of the presence of laxity, effusion, or internal derangement. The straight leg raise against resistance test (also known as the Stinchfield test) is useful to evaluate the hip flexor/psoas strength [23]. A positive test is noted with recreation of the pain or weakness and is a sign of an intra-articular problem because of increasing compressive force across the hip joint or the psoas placing pressure on the labrum. The PRI [13] test or the lateral rim impingement test that is its variation is conducted with the patient at the edge of the examining table so that the examined leg hangs freely at the hip. The patient draws up both legs into the chest, eliminating lumbar lordosis. The affected leg is then extended off the table, allowing for full extension of the hip, abducted and externally rotated. This test causes a hip extension evaluating the congruence of the femoral neck and the posterior acetabular wall.

40.6 Lateral Examination

The objective examination in lateral decubitus begins with the evaluation of the healthy contralateral side and proceeds with the palpation of the area below and above the sacroiliac joint, of the abductor muscles, and in particular of the gluteus maximus which course must also be felt. Then we proceed with palpation of the ischium to exclude avulsion of the hamstring proximal tendons. Finally, we proceed with the palpation of the Valleix points along the course of the sciatic nerve, and therefore the piriform and gluteus medius muscles and the tensor fasciae latae must be evaluated. This must be done naturally on both sides. An active piriformis test (Fig. 40.2) is conducted by the patient pushing the heel down into the table, abducting and externally rotating the leg against resistance, while the examiner checks the piriformis [18, 24]. The Ober test (see in the table below) or passive adduction test is conducted with the leg in three positions: extension (tensor fasciae latae contracture test), neutral (gluteus medius contracture test), and flexion (gluteus maximus contracture test). Gluteus medius tension is assessed with knee flexion to relax the iliotibial band. The gluteus maximus is melted with the tensor fasciae latae anteriorly [25]. To distinguish the contribution of only the gluteus maximus when conducting the gluteus maximus contracture test, the hip is flexed and the knee is extended. If adduction cannot occur in this position, the gluteus maximus portion is contracted. The passive assessment of FADIR (this test can also be made in supine position) is done in a dynamic manner (see description in the table below). The difference from supine to lateral position is the position of the pelvis. The supine position eliminates lumbar lordosis, whereas the lateral tests the normal dynamic pelvic inclination. Pelvic inclination may affect testing, and both positions are helpful in evaluation [13, 21, 22]. The lateral rim impingement test is conducted with the hip externally rotated and passively abducted. The examiner cradles the patient's lower leg with one arm and monitors the hip joint with the opposing hand. The examiner passively brings the affected hip through a wide arc from flexion to extension in continuous abduction while externally rotating the hip [13]. The test is positive if the patient's pain is reproduced.

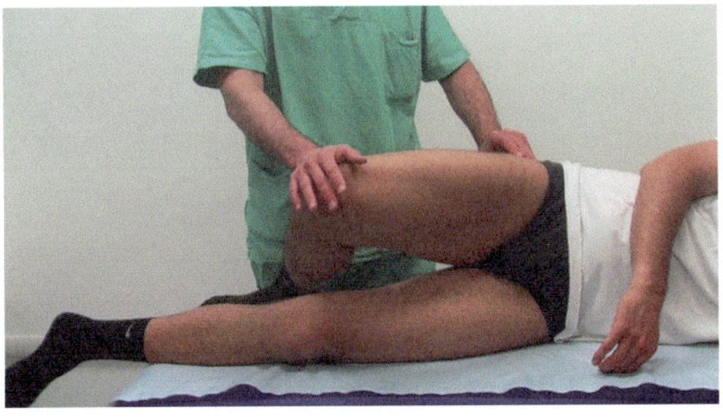

Fig. 40.2 Active piriformis test

40.7 Conclusions

A thorough knowledge of the anatomy and biomechanics of the hip is fundamental for a correct understanding of the pathology. A schematic and rigorous physical examination helps us to make a correct diagnosis.

References

1. Minnich JM, Hanks JB, Muschaweck U, et al. Sports hernia: diagnosis and treatment highlighting a minimal repair surgical technique. Am J Sports Med. 2011;39(6):1341–9.
2. Quinn A. Hip and groin pain: physiotherapy and rehabilitation issues. Open Sports Med J. 2010;4:93–107.
3. DeLee JC, Farney WC. Incidence of injury in Texas high school football. Am J Sports Med. 1992;20(5):575–80.
4. Lovell G. The diagnosis of chronic groin pain in athletes: a review of 189 cases. Aust J Sci Med Sport. 1995;27:76–9.
5. Offierski CM, MacNab I. Hip-spine syndrome. Spine (Phila Pa 1976). 1983, 8:316–21.
6. Burnett RS, Della Rocca GJ, Prather H, et al. Clinical presentation of patients with tears of the acetabular labrum. J Bone Joint Surg Am. 2006;88(7):1448–57.
7. Clohisy JC, Knaus ER, Hunt DM, et al. Clinical presentation of patients with symptomatic anterior hip impingement. Clin Orthop Relat Res. 2009;467(3):638–44.
8. Nunley RM, Prather H, Hunt D, et al. Clinical presentation of symptomatic acetabular dysplasia in

skeletally mature patients. J Bone Joint Surg Am. 2011;93(Suppl 2):17–21.
9. Anderson SA, Keene JS. Results of arthroscopic iliopsoas tendon release in competitive and recreational athletes. Am J Sports Med. 2008;36(12):2363–71.
10. Byrd JW. Evaluation and management of the snapping iliopsoas tendon. Instr Course Lect. 2006;55:347–55.
11. Sammarco GJ. The dancer's hip. Clin Sports Med. 1983;2(3):485–98.
12. Lee KS, Rosas HG, Phancao JP. Snapping hip: imaging and treatment. Semin Musculoskelet Radiol. 2013;17(3):286–94.
13. Martin HD, Shears SA, Palmer IJ. Evaluation of the hip. Sports Med Arthrosc Rev. 2010;18(2):63–75.
14. Byrd JWT. Physical examination. In: Byrd JWT, editor. Operative hip arthroscopy. 2nd ed. New York: Springer; 2005. p. 36–50.
15. Torry MR, Schenker ML, Martin HD, et al. Neuromuscular hip biomechanics and pathology in the athlete. Clin Sports Med. 2006;25(179–197):vii.
16. Perry J. Gait analysis Normal and pathological function. Thorofare, NJ: SLACK Inc.; 1992.
17. Scopp JM, Moorman CT. The assessment of athletic hip injury. Clin Sports Med. 2001;20(4): 647–59.
18. Pace JB, Nagle D. Piriform syndrome. West J Med. 1976;124:435–9.
19. Martin HD. Clinical examination of the hip. Oper Tech Orthop. 2005;15:177–81.
20. Marks MC, Alexander J, Sutherland DH, Chambers HG. Clinical utility of the Duncan-Ely test for rectus femoris dysfunction during the swing phase of gait. Dev Med Child Neurol. 2003;45:763–8.
21. Prather H, Cheng A. Diagnosis and treatment of hip girdle pain in the athlete. PM R. 2016;8(3 Suppl):S45–60. https://doi.org/10.1016/j.pmrj.2015.12.009.
22. Prather H, Colorado B, Hunt D. Managing hip pain in the athlete. Phys Med Rehabil Clin N Am. 2014;25(4):789–812. https://doi.org/10.1016/j.pmr.2014.06.012. Epub 2014 Aug 27
23. Reiman MP, Goode AP, Hegedus EJ, et al. Diagnostic accuracy of clinical tests of the hip: a systematic review with meta-analysis. Br J Sports Med. 2013;47(14):893–902.
24. Fishman L, Zybert P. Electrophysiologic evidence of piriformis syndrome. Arch Phys Med Rehabil. 1992;73:359–64.
25. Fearon AM, Scarvell JM, Neeman T, et al. Greater trochanteric pain syndrome: defining the clinical syndrome. Br J Sports Med. 2013;47(10):649–53.

41 Limping Child

Laura Ruzzini and Daniela Lamberti

Limping child is a challenging problem often faced by paediatricians, general practitioners and orthopaedic surgeons. The most common referred painful sites in the limping child are the hip (35%), the knee (20%) and the entire leg (18%) [1].

The first approach to establish the correct diagnosis is to divide aetiologies into traumatic and non-traumatic.

There are many non-traumatic causes of limping (Table 41.1), which are often age specific, ranging from surgical emergencies to cases who need observation only.

Begin with detailed history asking for the onset and duration of symptoms and past medical history including previous treatment and fever that may indicate the presence of an inflammatory or infectious condition, which are usually related with septic arthritis, transient synovitis and osteomyelitis [2].

General inspection should be done looking for joint swellings, muscle bulk, scoliosis, fixed flexion deformity of the hip or knee, rashes and bruising. It is also recommended to examine the range of movement in all lower limb joints to elicit any referred pain from the hip as hip pain refers often to the knee [3].

Table 41.1 Differential diagnoses for atraumatic limp by age

Age group	Likely diagnosis	Other possible diagnosis
<3 years	Septic arthritis, osteomyelitis	Developmental dysplasia of the hip, transient synovitis
3–10 years	Transient synovitis, septic arthritis, osteomyelitis, Perthes disease	
10–18 years	Slipped upper femoral epiphysis, osteomyelitis, septic arthritis, Perthes disease	

A full-body examination should be performed when assessing a limping child including arms and spine.

There are distinct patterns of deviation from normal gait, and their recognition is important to provide the right diagnosis. Antalgic and Trendelenburg gaits are the most common pathologic gait patterns [4]. An antalgic gait is caused by unilateral limb pain. It is typical of patients with infectious arthritis and fractures. A Trendelenburg gait is painless and due to muscle weakness or instability. It is characterized by a pelvic tilt away from the affected side caused by a relative weakness of the hip abductors [5]. This gait pattern is typically seen in developmental dysplasia of the hips (DDH). Mixed pathologic

L. Ruzzini (✉) · D. Lamberti
Ospedale Pediatrico 'Bambino Gesù', Rome, Italy
e-mail: laura.ruzzini@opbg.net; daniela.lamberti@opbg.net

J. G. Lane et al. (eds.), *The Art of the Musculoskeletal Physical Exam*,
https://doi.org/10.1007/978-3-031-24404-9_41

gait patterns can also be detected in slipped capital femoral epiphysis (SCFE) and Legg-Calvé-Perthes disease (LCPD).

When diagnosis is uncertain after examination, or when confirmation of clinical suspicion is needed, lab tests and radiographic or ultrasonographic studies may be useful.

A full blood count, C-reactive protein (CRP) and erythrocyte sedimentation rate (ESR) must be obtained when an infectious disease is suspected. CRP is a good negative predictor; if its value is <1.0 mg/dL, the probability that the patient does not have septic arthritis is 87% [6].

Plain radiographs of the hip, in anterior-posterior and frog leg lateral views, should be obtained in all those with hip pain and considered in those with knee pain. Ultrasound is 100% sensitive as well as specific for effusions of the hip joint in children, and it can also be used to obtain joint fluid by guided aspiration. Second-level imaging evaluation, such as magnetic resonance imaging (MRI) or computed tomography (CT), can help in difficult cases.

41.1 Children Under 3 Years of Age

In painful limping children without trauma history, the most important pathologies to diagnose are osteomyelitis and septic arthritis of the hip.

The resting position of the hip in slight flexion and external rotation should alert the examiner to a painful effusion of the hip joint, as seen in septic arthritis. This position is comfortable because the joint capsule is under the least tension.

Septic arthritis is an infection of the synovium and joint space, typically monoarticular and commonly caused by *S. aureus* [1].

Clinical manifestations are fever and a swollen tender joint that severely restricts joint movement. The definitive investigation is joint aspiration. The management consists of immediate administration of intravenous antibiotics and surgical drainage (open or arthroscopic) with washout of the affected joint.

When the infection is into the bone, it is called osteomyelitis. Infants presenting with fever, unexplained limp and reluctance to use the limb must be investigated for osteomyelitis. There may be joint tenderness, swelling and erythema. Neutrophilia and increased levels of CRP and ESR are detected at blood tests. MRI with contrast is the best test for diagnosing osteomyelitis and deep soft-tissue infections. Blood cultures are essential before starting targeted intravenous antibiotic therapy.

Pain-free limping children with a Trendelenburg gait should always do a pelvic X-ray to exclude an inveterate DDH.

41.2 Children Aged 3–10 Years

In this age group, the most common causes of limping are transient synovitis and Perthes disease. In any case, it is important to exclude traumatic and infectious diseases (septic arthritis and osteomyelitis).

Transient synovitis is a self-limiting inflammation of the hip synovium and generally preceded by a viral infection of the upper respiratory tract. Kocher's criteria can help to make a differential diagnosis between septic arthritis and transient synovitis (Table 41.2). The patient shows an antalgic gait with a limited range of motion of the affected hip, especially in abduction and internal rotation. X-ray is normal, and ultrasonography usually shows a fluid joint effusion. Treatment is rest and anti-inflammatory drugs. Complete symptom resolution is usually in 7–10 days.

Perthes disease is a condition that typically affects boys aged 6–8 and is characterized by avascular necrosis of the capital femoral epiphysis [7]. It is self-limiting, with revascularization and new bone formation occurring in 2–4 years.

Table 41.2 Differentiating septic arthritis of the hip from transient synovitis using Kocher's criteria

Temperature	>38.5 °C
White cell count	>12 × 109 L
Erythrocyte sedimentation rate	>40 mm/h
Weight bearing	Yes or no

One point is scored for each parameter. A score of 4 gives a 99% chance of the patient having septic arthritis, 3 gives a 93% chance of septic arthritis, 2 gives 43% chance of septic arthritis, and a score of 1 gives a 3% of septic arthritis

Outcome depends on the age of onset, disease stage upon presentation and resultant hip deformity after remodelling. The patient usually shows a sub-acute limp (Trendelenburg gait pattern); atrophic proximal thigh muscles, especially in long-lasting cases; and pain that is usually referred to the groin, thigh or knee. Blood tests are normal. X-rays are the gold standard in the diagnosis.

In the earliest stages, vascular occlusion causes hip synovitis, and it can be confused with transient synovitis. In these cases, X-rays may be normal and MRI should be considered if there are persistent symptoms or doubtful diagnosis. The subsequent phases of Perthes disease are fragmentation of the femoral epiphysis, during which the femoral head is vulnerable to deformity and reconstruction [8]. X-rays are useful for assessing the stage of disease and prognosis (Fig. 41.1).

Treatment aims to maintain the range of movement and to contain the femoral head in the acetabulum. Patients younger than 5–6 years and milder cases (with minimal lateral pillar collapse) are treated conservatively with analgesics and protected weight bearing. Corrective surgical osteotomy should be considered in more advanced cases.

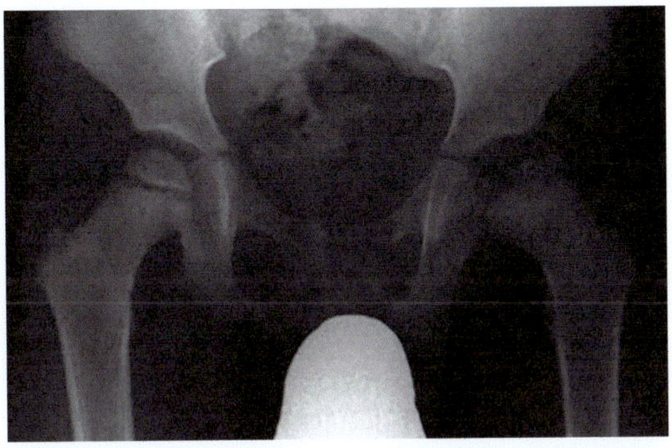

Fig. 41.1 Anteroposterior radiograph showing Perthes disease of the left hip. Note the fragmentation and flattening of the left capital femoral epiphysis

41.3 Children Aged 10–16 Years

Over the age of 10, it is important to consider the slipped capital femoral epiphysis (SCFE) that consists of displacement of the proximal femoral epiphysis relative to the metaphysis. It is more frequent in boys who are overweight or affected by endocrine disorders such as hypothyroidism [9].

Children show an acute limp with the lower leg externally rotated and progressive loss of hip abduction and internal rotation. SCFE is classified into stable or unstable, depending on the ability to weight bear. Diagnosis is performed with plain anteroposterior and frog leg lateral views and X-rays of the hip and pelvis (Fig. 41.2). Disease is bilateral in 25% of children. Most patients with unilateral onset will usually develop contralateral symptoms within 1.5 years.

If SCFE is suspected or diagnosed, patients should be referred urgently to orthopaedics.

Prompt diagnosis is crucial to avoid complications and long-term deformity of the hip.

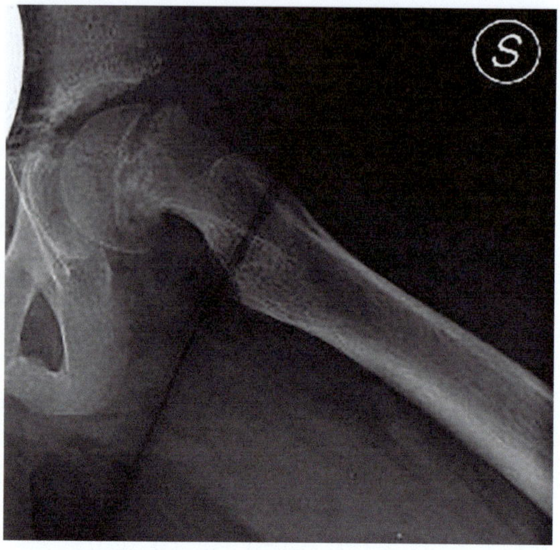

Fig. 41.2 Frog leg lateral views showing a left slipped upper femoral epiphysis (arrow)

Key Points

- A limping child can have a range of causes from mild to life threatening.
- Establishing the underlying pathology requires a detailed evaluation, full-body examination and imaging if required.
- Urgent referral to orthopaedics is required if septic arthritis, osteomyelitis or slipped upper femoral epiphysis is suspected.
- Atraumatic knee pain is highly suspicious for hip pathology. Not to consider the hip in a child with knee pain is a common pitfall.

References

1. Gunner KB, Scott AC. Evaluation of a child with a limp. J Pediatr Health Care. 2001;15(1):38–40. https://doi.org/10.1067/mph.2001.111950.
2. Fischer SU, Beattie TF. The limping child: epidemiology, assessment and outcome. J Bone Joint Surg Br. 1999;81(6):1029–34. https://doi.org/10.1302/0301-620x.81b6.9607.
3. Lawrence LL. The limping child. Emerg Med Clin North Am. 1998;16(4):911–29., viii. https://doi.org/10.1016/s0733-8627(05)70039-0.
4. Clark MC. The limping child: meeting the challenges of an accurate assessment and diagnosis. Ped Emerg Med Rep. 1997;2:123–34.
5. Myers T, Thompson G. Imaging the child with a limp. Pediatr Clin North Am. 1997;44:637–58.
6. Levine MJ, McGuire KJ, McGowan KL, Flynn JM. Assessment of the test characteristics of C-reactive protein for septic arthritis in children. J Pediatr Orthop. 2003;23(3):373–7.
7. Hall AJ, Barker DJ. The age distribution of Legg-Perthes disease. An analysis using Sartwell's incubation period model. Am J Epidemiol. 1984;120:531–6.
8. Kim HKW, Herring JA. Pathophysiology, classifications, and natural history of Perthes disease. Orthop Clin North Am. 2011;42:285–95.
9. Novais EN, Millis MB. Slipped capital femoral epiphysis: prevalence, pathogenesis, and natural history. Clin Orthop Relat Res. 2012;470(12):3432–8. https://doi.org/10.1007/s11999-012-2452-y.

Evaluation of Chronic Pelvic Pain (Athletic Pubalgia-Sports Hernia and Other Pain Conditions)

42

Bisciotti Gian Nicola

The groin pain syndrome (GPS) is an increasing problem in many sport activities requiring cutting, change of direction, and kicking such as soccer, football, ice hockey, handball, and rugby [1]. Indeed, its yearly incidence in some sport activities like soccer of 10–18% continues to increase due to many risk factors such as high loads and short recoveries [2]. A recent study reported a GPS incidence in soccer of up to 2.1/1000 h of total exposure [3]. Still in soccer, it is important to remember that the majority of injury surveillance studies in football are based on the so-called time loss concept [3]. In these type of studies, the injuries are recorded only if a player is unable to participate in soccer training and/or competition [4]. A recent study shows that the "time loss definition" is able to record only one-third of the GPS injuries in male soccer players [5]. Therefore, the "time loss injury" approach may be inappropriate to check the GPS incidence, and the recorded data may represent only the "tip of the iceberg" of a much more widespread problem [5]. Indeed, it is not uncommon for soccer players to continue training despite the pain due to GPS so as not registering any time loss injury. For this reason, the overuse is consolidated as an important element in most cases of GPS [3].

In agreement with the "Groin Pain Syndrome Italian Consensus Conference on terminology, clinical evaluation and imaging assessment in groin pain in athlete" (GPSICC) [6], the GPS can be defined as follows:

> *Any clinical symptom reported by the patient, located at the inguinal-pubic-adductor area, affecting sports activities and/or interfering with activities of daily living, and requiring medical attention.*

Always in agreement with the GPSICC [6], the GPS can be subdivided into three main categories:

1. GPS of traumatic origin, in which the onset of pain was due to any acute trauma, and this hypothesis is supported by medical history, clinical examination, and imaging.
2. GPS due to functional overload, characterized by insidious and progressive onset, without an acute trauma, or a situation to which the onset of pain symptoms can be attributed with certainty.
3. Long-standing GPS (LSGPS) or chronic GPS, in which the cohort of symptoms reported by the patient continues for a long period (over 12 weeks) and is recalcitrant to any conservative therapy.

B. G. Nicola (✉)
Qatar Orthopaedic and Sport Medicine Hospital, Doha, Qatar

Paris Saint Germain Football Club (F), Paris, France

J. G. Lane et al. (eds.), *The Art of the Musculoskeletal Physical Exam*,
https://doi.org/10.1007/978-3-031-24404-9_42

42.1 The Long-Standing Groin Pain Syndrome

In the case of LSGPS, in which several clinical situations overlap, often to make a correct diagnosis represents a challenge [1, 7, 8]. The LSGPS assessment requires clinical experience and a solid knowledge concerning the possible differential diagnoses [1].

The typical patient who complains of LSGP is a subject with a long history of groin pain (more than 3 months), who has already performed several clinical and imaging assessments and who, above all, has already performed unsuccessfully many types of conservative treatments. In this case, an inguinal disease must be strongly suspected [7, 9]. The term "inguinal pathologies" in accordance with GPSICC [6] includes a vast range of conditions, i.e.:

(i) Inguinal hernia
(ii) Posterior inguinal wall weakness
(iii) Conjoint tendon lesion
(iv) Inguinal ligament lesion
(v) Rectus abdominis distal insertion lesions
(vi) External obliquus, internal obliquus, and pyramidalis lesions
(vii) Rectus abdominis-adductor longus common aponeurosis lesions
(viii) Pre-aponeurotic capsule lesions

This series of clinical conditions may cause a situation of "groin disruption" [9]. In the most part of the cases, this situation of groin disruption is caused by the presence of a cam-FAI syndrome [7, 10]. Cam-FAI is an abnormal conformation of femoral head. In other words, cam-FAI is an osteochondral bump at femoral head-neck junction, leading to a diminution of the normal femoral head-neck offset [11]. Cam-FAI syndrome is identified by measuring the alpha angle on the Dunn-view X-ray, as showed in Fig. 42.1. An alpha angle measuring 55° or greater is considered radiographic evidence of cam-FAI [12]. Cam-FAI syndrome can cause both hip articular cartilage and labral lesion and a limitation of hip joint intra-rotation [13, 14]. Recently, cam-FAI

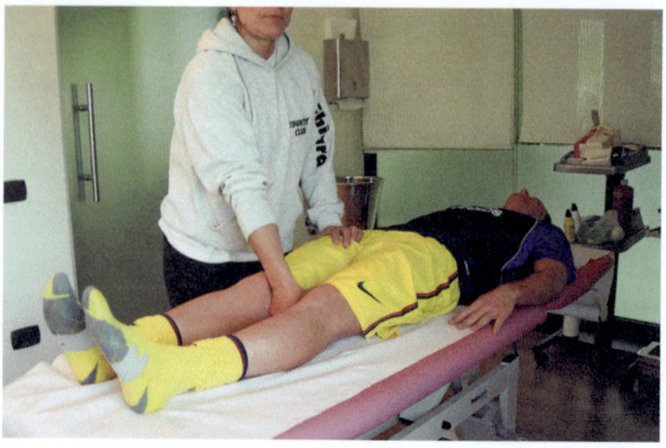

Fig. 42.1 In the squeeze test 1, the operator asks the patient to perform an isometric adduction with the resistance placed at the level of the patient's knees. This test shows a sensitivity and specificity equal to, respectively, 0.85 and 0.45. Since the likelihood ratio value is equal to 1.5, this test used alone results to be rarely useful

has been shown to be associated with GPS, especially in young high-level athletes [13, 15]. This association may be explained by the fact that the athlete suffering from cam-FAI generally shows a limitation of hip range of motion (especially in internal rotation) [13, 14]. This limitation of ROM may be compensated by a hypermobility of the symphysis joint. This hypermobility of the symphysis may stress the posterior inguinal wall causing the onset of inguinal pathology [6, 13, 14]. The inguinal pathology shows a low rate of positive outcome with conservative treatment [16]; in these cases, it is necessary to consider a surgical treatment. Nowadays, the most utilized surgical treatments in inguinal pathology are:

- Shouldice repair
- Open-suture repair technique
- Lichtenstein repair
- Transabdominal pre-peritoneal repair (TAPP)
- Total extraperitoneal repair (TEP)
- Trans-inguinal pre-peritoneal repair (TIPP)
- Minimal repair
- Inguinal ligament release procedure [17]

Sometimes, these procedures may be completed with a triple or selective neurectomy of ilioinguinal, iliohypogastric, and genital branch of femorogenital nerves [18]. Following surgical repair, independently of the surgical technique used, most of the series report that >90% of athletes return to full sport activity within 2–4 months after surgery [19]. Given the relationship between cam-FAI syndrome and inguinal pathologies, the problem arises if an athlete shows both pathologies, if the surgeon must limit the surgical intervention only to the inguinal pathology, or if he/she must also consider a second surgery intervention concerning the cam-FAI. This problematic had already been raised by other authors [13, 14].

42.2 Pubic Osteopathy and Adductor Tendinopathy

The symphysis instability due to the presence of a cam-FAI syndrome may often cause the association of other three different pathologies: osteitis pubis, longus adductor tendinopathy, and, as already mentioned, inguinal pathologies [6]. First of all, it should be noted that, as established during "The Italian Consensus Conference on FAI Syndrome in athletes" [20], pubic osteitis is not a correct term. Indeed, since this clinical condition is mainly degenerative in nature and not an inflammatory process, the term pubis osteopathy is to prefer. The radiological signs specific for pubic osteopathy are [20]:

(i) Bone marrow edema
(ii) Symphysis sclerosis
(iii) Symphysis irregularity
(iv) Subchondral cyst
(v) Central disc protrusion

In the presence of three, over five, of the abovementioned signs confirmed by clinical examination, it is possible to formulate the diagnosis of pubic osteopathy [20]. In cases of association of inguinal pathologies and chronic adductor tendinopathy unresponsive to conservative treatment, a double surgical intervention may be necessary, consisting, as already mentioned, of a mesh positioning or otherwise a surgical reinforcing inguinal canal posterior wall technique, coupled to adductor longus and partial or total tenotomy [19]. The adductor longus tenotomy presents the multiple vantage to relieve the mechanical stress at adductor and at rectus abdominis level and therefore may be the most optimal management in the case of chronic adductor tendinopathy [19]. Recent studies show that open and laparoscopic inguinal hernia repair, with or without mesh placement, coupled to adductor tenotomy demonstrates a return-to-full-activity rate of 95–100% in 3–4 months [19]. It is important to remember that some electromyographic studies demonstrate that adductor longus muscle shows a minimal activity during sprint [21] or cutting movements [22]. For this reason, some more recent studies demonstrate that the longus adductor tendon tenotomy does not compromise the high-level sports performance [23, 24].

42.3 The Profile of the Patient Affected by LSGPS

In the case of LSGPS, in which several clinical pictures overlap, often to make a correct diagnosis, it is necessary to adopt a clinical reasoning based on the association of the most frequent clinical situations [1, 25]. In one of our case series, in the process of publication, in which we considered 300 patients affected by LSGPS, the most frequent clinical situations and association were the following:

(a) 84% of the patients showed cam-FAI or mixed forms (pincer and cam).
(b) 68% of the patients showed inguinal pathologies.
(c) 46% of the patients had adductor longus tendinopathy (of which 87% showed a bilateral tendinopathy).
(d) 82% of the patients showed both cam-FAI and inguinal pathologies.
(e) 31% of the patients showed an association between inguinal pathologies, adductor tendinopathy, and cam-FAI (or mixed form).

42.4 The Clinical Evaluation

Given the complexity of the LSGPS and the frequent overlapping of several clinical situations, the clinical evaluation often appears as a real diagnostic challenge. Recommended clinical tests are shown in Tables 42.1, 42.2, 42.3, and 42.4.

The tests listed in Table 42.2 are shown in Figs. 42.1, 42.2, 42.3, 42.4, and 42.5, while the tests listed in Table 42.3 are shown in Figs. 42.6, 42.7, and 42.8.

Table 42.1 Palpation test and clinical exploration for clinical evaluation of inguinal diseases

Test	Sensitivity	Specificity	PPV	NPV	LR	Result
Pubic insertion of rectus abdominis	0.93	0.30	0.81	0.58	1.3	Rarely useful
Pubic tubercle	0.92	0.89	0.96	0.80	8.4	Moderately useful
Linea pectinea	0.91	0.88	0.96	0.76	7.6	Moderately useful
External inguinal ring	0.97	0.95	0.98	0.90	19.4	Conclusive

Legenda: *PPV* positive predictive value, *NPV* negative predictive value, *LR* likelihood ratio; the interpretations of the LR values are shown in Table 42.4

Table 42.2 Specific tests for adductor muscles

Test	Sensitivity	Specificity	PPV	NPV	LR	Result
Squeeze test 1	0.85	0.45	0.48	0.83	1.5	Rarely useful
Squeeze test 2	0.86	0.46	0.48	0.84	1.6	Rarely useful
Squeeze test 3	0.86	0.45	0.48	0.85	5.7	Sometimes useful
Squeeze test 4	0.65	0.46	0.45	0.65	1.2	Rarely useful
Adductor longus origin palpation	0.93	0.89	0.96	0.79	8.5	Moderately useful

Legenda: *PPV* positive predictive value, *NPV* negative predictive value, *LR* likelihood ratio; the interpretations of the LR values are shown in Table 42.4

Table 42.3 Specific test for the hip

Test	Sensitivity	Specificity	PPV	NPV	LR	Result
FABER test	0.90	0.93	0.98	0.72	12.9	Conclusive
FADIR test	0.93	0.42	0.86	0.61	1.6	Rarely useful
DIRI test	0.58	0.90	0.96	0.36	5.8	Moderately useful

Legenda: *PPV* positive predictive value, *NPV* negative predictive value, *LR* likelihood ratio; the interpretations of the LR values are shown in Table 42.4

Table 42.4 Interpretation of LR values

LR positive	LR negative	Change in the probability of disease	Result
>10	<0.1	Big	Conclusive
5–10	0.1–0.2	Moderate	Moderately useful
2–5	0.2–0.5	Small	Sometimes useful
<2	>0.5	Minimum	Rarely useful
1	1	Not any	Useless

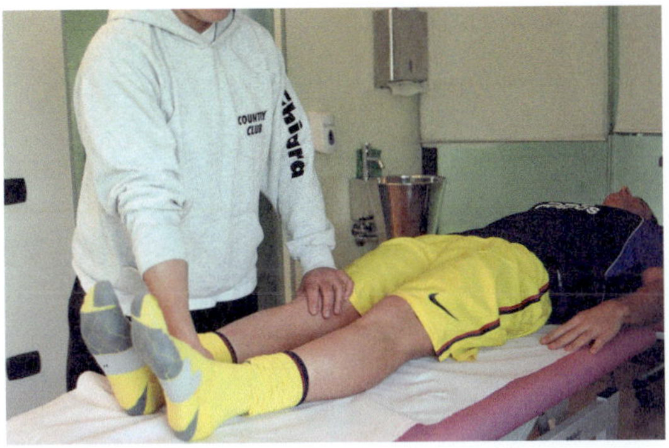

Fig. 42.2 In the squeeze test 2, the operator asks the patient to perform an isometric adduction with the resistance placed at the level of the patient's ankles. This test shows a sensitivity and specificity equal to, respectively, 0.86 and 0.46. Since the likelihood ratio value is equal to 1.6, this test used alone results to be rarely useful

Fig. 42.3 In the squeeze test 3, the operator asks the patient to perform an isometric adduction with the resistance placed at the level of the patient's ankles keeping his/her legs open. This test shows a sensitivity and specificity equal to, respectively, 0.86 and 0.45. Since the likelihood ratio value is equal to 5.7, this test used alone results to be sometimes useful

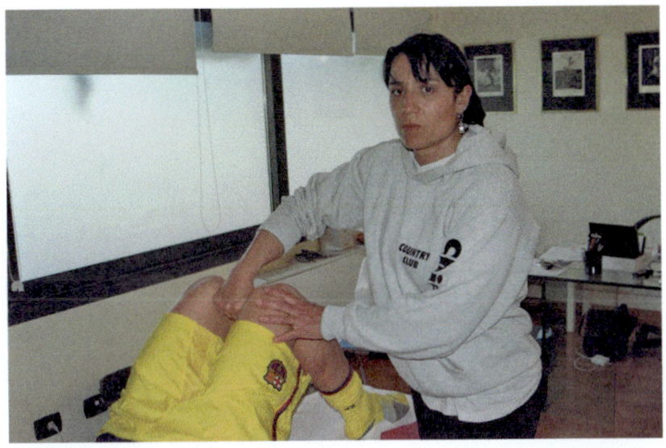

Fig. 42.4 In the squeeze test 4, the operator asks the patient to perform an isometric adduction with the resistance placed at the level of the patient's ankles keeping his/her legs flexed. This test shows a sensitivity and specificity equal to, respectively, 0.65 and 0.46. Since the likelihood ratio value is equal to 1.2, this test used alone results to be rarely useful

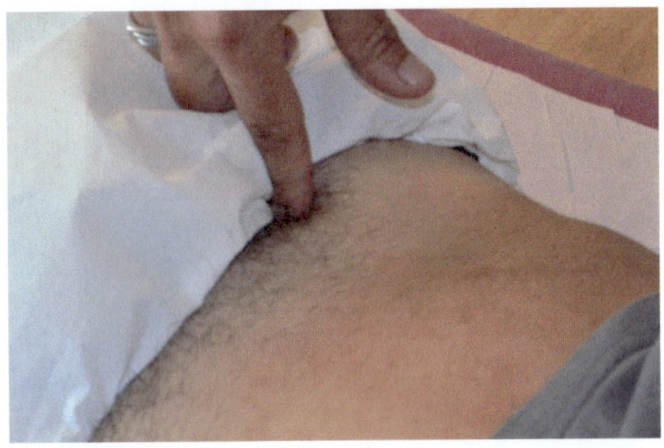

Fig. 42.5 The adductor longus origin palpation test shows a sensitivity and specificity equal to, respectively, 0.93 and 0.89. Since the likelihood ratio value is equal to 8.5, this test used alone results to be moderately useful

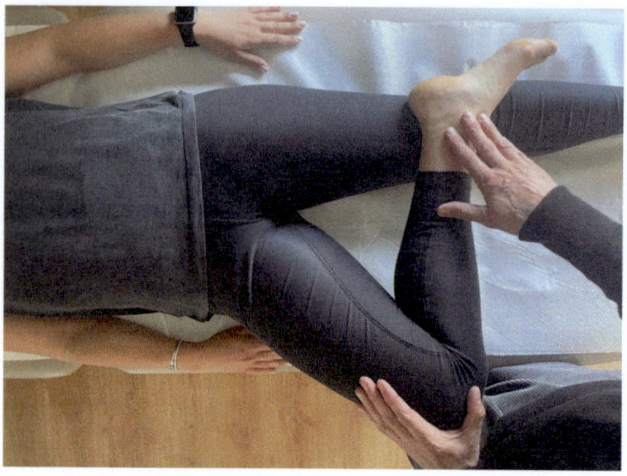

Fig. 42.6 Flexion-abduction-External Rotation test (FABER). The FABER test also known as the Patrick test, is used to assess labral damage, or FAI

Fig. 42.7 Flexion adduction internal rotation test (FADIR). During the FADIR test the hip and knee are passively flexed to 90° and the adducted with hip in internal rotation

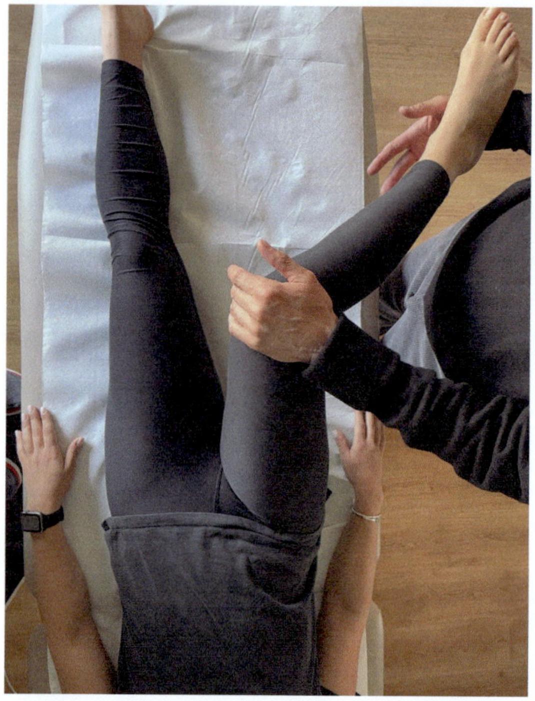

Fig. 42.8 The Dynamic Internal Rotation Impingement tests (DIRI). The DIRI is performed in supine position with the involved leg flexed to 90° at the hip and knee but the hip is passively taken through an arc of hip adduction and internal rotation

Adductor Clinic Assessment Pitfalls

Since all the squeeze tests (except squeeze test number 3) show a low likelihood ratio value, they must be interpreted in a global manner. In a clinical situation of adductor tendinopathies, the patient generally shows a pain value (on a VAS scale) increasing, respectively, at the squeeze tests 1, 2, and 3. The growing pain sensation is explained by the fact that the resistance arm is progressively increased, and in the squeeze test 3, in addition to the arm of greater resistance, a passive stretching of the adductors is added. The squeeze test 4 is more specific for gracilis muscle tendinopathy. In any case, the most suggestive test for adductor tendinopathy is the adductor longus origin palpation test

42.5 Conclusions

The LSGPS represents often a real diagnostic challenge made difficult by the frequent overlapping of different clinical situations. Moreover, these different clinical conditions are often linked to each other by a cause-effect relationship. One of the most distinctive aspects of LSGPS is the patient's poor responsiveness to any form of conservative treatment. It is very important for the clinician to be able to recognize the different clinical situations and their different role in the etiopathogenesis of pain reported by the patient. Furthermore, since the LSGP can have both a parietal and a musculoskeletal origin, for a correct diagnosis, a multidisciplinary approach would be strongly recommended.

References

1. Hölmich P. Long-standing groin pain in sportspeople falls into three primary patterns, a "clinical entity" approach: a prospective study of 207 patients. Br J Sports Med. 2007;41(4):247–52.
2. Mosler AB, Weir A, Eirale C, Farooq A, Thorborg K, Whiteley RJ, Hölmich P, Crossley KM. Epidemiology of time loss groin injuries in a men's professional football league: a 2-year prospective study of 17 clubs and 606 players. Br J Sports Med. 2018;52(5):292–7.
3. Werner J, Hägglund M, Ekstrand J, Waldén M. Hip and groin time-loss injuries decreased slightly but injury burden remained constant in men's professional football: the 15-year prospective UEFA Elite Club Injury Study. Br J Sports Med. 2019;53(9):539–46.
4. Noya Salces J, Gomez-Carmona PM, Gracia-Marco L, Moliner- Urdiales D, Sillero-Quintana M. Epidemiology of injuries in first division Spanish football. J Sports Sci. 2014;32(13):1263–70.
5. Harøy J, Andersen TE, Bahr R. Groin problems in male soccer players are more common than previously reported: response. Am J Sports Med. 2017;45(13):NP32–3.
6. Bisciotti GN, Volpi P, Zini R, Auci A, Aprato A, et al. Groin Pain Syndrome Italian Consensus Conference on terminology, clinical evaluation and imaging assessment in groin pain in athlete. BMJ Open Sport Exerc Med. 2016;2(1):e000142. https://doi.org/10.1136/bmjsem-2016-000142. eCollection 2016
7. Bisciotti GN, Auci A, Di Marzo F, Galli R, Pulici L, Carimati G, Quaglia A, Volpi P. Groin pain syndrome: an association of different pathologies and a case presentation. Muscles Ligaments Tendons J. 2015;5(3):214–22.
8. Bisciotti GN, Di Marzo F, Auci A, Parra F, Cassaghi G, Corsini A, Petrera M, Volpi P, Vuckovic Z, Panascì M, Zini R. Cam morphology and inguinal pathologies: is there a possible connection? J Orthop Traumatol. 2017;18(4):439–50.
9. Gilmore CJ, Diduch DR, Handley MV, Hanks JB. In: Diduch DR, Brunt LM, editors. Sports Hernia- History and physical examination: Making the diagnosis with confidence. New York: Springer; 2014.
10. Rambani R, Hackney R. Loss of range of motion of the hip joint: a hypothesis for etiology of sports hernia. Muscles Ligaments Tendons J. 2015;5(1):29–32.
11. Fairley J, Wang Y, Teichtahl AJ, Seneviwickrama M, Wluka AE, Brady SR, Hussain SM, Liew S, Cicuttini FM. Management options for femoroacetabular impingement: a systematic review of symptom and structural outcomes. Osteoarthr Cartil. 2016. pii: S1063-4584(16)30048-6
12. Bedi A, Chen N, Robertson W, Kelly BT. The management of labral tears and femoroacetabular impingement of the hip in the young, active patient. Arthroscopy. 2008;24:1135–45.
13. Hammoud S, Bedi A, Magennis E, Meyers WC, Kelly BT. High incidence of athletic pubalgia symptoms in professional athletes with symptomatic femoroacetabular impingement. Arthroscopy. 2012;28(10):1388–95.
14. Hammoud S, Bedi A, Voos JE, Mauro CS, Kelly BT. The recognition and evaluation of patterns of compensatory injury in patients with mechanical hip pain. Sports Health. 2014;6(2):108–18.
15. Larson CM, Sikka RS, Sardelli MC, Byrd JW, Kelly BT, Jain RK, Giveans MR. Increasing alpha angle is predictive of athletic-related "hip" and "groin" pain in collegiate National Football League prospects. Arthroscopy. 2013 Mar;29(3):405–10.
16. Omar IM, Zoga AC, Kavanagh EC, Koulouris G, Bergin D, Gopez AG, Morrison WB, Meyers WC. Athletic pubalgia and "sports hernia": optimal MR imaging technique and findings. Radiographics. 2008;28:1415–38.
17. Lloyd DM. Sportman's groin and the inguinal ligament release procedure. In: Jacob BP, Chen DC, Ramshaw B, Towfigh S, editors. The SAGES manual of groin pain. New York: Springer; 2016.
18. Santilli OL, Nardelli N, Santilli HA, Tripoloni DE. Sports hernias: experience in a sports medicine center. Hernia. 2016;20:77–84.
19. Rossidis G, Perry A, Abbas H, Motamarry I, Lux T, Farmer K, Moser M, Clugston J, Caban A, Ben-David K. Laparoscopic hernia repair with adductor tenotomy for athletic pubalgia: an established procedure for an obscure entity. Surg Endosc. 2015;29:381–6.
20. Zini R, Panascì M, Santori N, Potestio D, Bisciotti GN et al. The Italian consensus conference on FAI syndrome in athletes (Cotignola agreement). Joint. In press.
21. Mann RA, Moran GT, Dougherty SE. Comparative electromyography of the lower extremity in jog-

ging, running, and sprinting. Am J Sports Med. 1986;14(6):501–10.
22. Neptune RR, Wright IA, Van Den Bogert AJ. Muscle coordination and function during cutting movements. Med Sci Sports Exerc. 1999;31(2):294–302.
23. Schilders E, Dimitrakopoulou A, Cooke M, Bismil Q, Cooke C. Effectiveness of a selective partial adductor release for chronic adductor-related groin pain in professional athletes. Am J Sports Med. 2013;41(3):603–7.
24. De Queiroz RD, de Carvalho RT, De Queiroz Szeles PR, Janovsky C, Cohen M. Return to sport after surgical treatment for pubalgia among professional soccer players. Rev Bras Ortop. 2014;49(3):233–9.
25. Falvey ÉC, King E, Kinsella S, Franklyn-Miller A. Athletic groin pain (part 1): a prospective anatomical diagnosis of 382 patients—clinical findings, MRI findings and patient-reported outcome measures at baseline. Br J Sports Med. 2016;50(7):423–30.

43 Assessment of Outcome Scores of the Hip

Filippo Randelli, Gaia Santambrogio, Gennaro Fiorentino, Manuel Giovanni Mazzoleni, Alberto Fioruzzi, and Vittorio Calvisi

A comprehensive and comparable assessment of the outcomes following hip surgery, or any other treatment, is still a difficult task for orthopedic surgeons and all other physicians and physiotherapists involved. In the last 30 years, orthopedic surgeons, trying to overcome this barrier, have invited the patients to be much more interactive in their healing process. Standardized measurements of assessing the outcomes directly from the patients have been developed. They are called patient-reported outcome measurement (PROM). They are comprehensive and include symptoms (pain the most), physical functionality, mental health, social life, and general well-being of the patient.

The growth of these patient-reported outcome measurement (PROM) lies from the need of filling the gap in the clinical routine between the pure physician's point of view and the most important and real one, the one of the patient [1, 2].

Through PROMs, physicians may detect physical or psychological problems that otherwise might be missed, monitor the disease and its treatment, and recommend changes in therapeutic plans. They facilitate the communication between physician and patient and, of course, improve the quality and delivery of care [3].

Physicians are now starting to choose personalized treatments based on these new tools.

A question could arise: Why considering the point of view of the patient? There can be many answers, but above all, we have learned that only a patient himself/herself can assert if a symptom reduces, a disability improves, or the quality of life ameliorates. Furthermore, involving patients in their therapeutic plan can bring a benefit in their healing process focusing on the specific questions and details that really matter. With a standardized measurement, it is also much likely to avoid confounding bias by the observer [4]. Last but not least, PROMs transform the obtained results into numbers, which can therefore be easily compared.

For designating the perfect PROM for each patient and his/her clinical case, it is possible to consult databases that aim at offering the most exhaustive and updated collection of principal tools for patient-reported outcome measurement and indeed choose easily the most appropriate. For example, there is the French *MAPI Research Institute* that, in 2002, has started building a comprehensive database, nowadays named *Patient-Reported Outcome and Quality of Life Instruments Database* (PROQOLID) [5].

F. Randelli (✉) · G. Santambrogio · A. Fioruzzi
Hip Department (CAD) Gaetano Pini—CTO Orthopedic Institute University of Milan, Milan, Italy
e-mail: filippo.randelli71@gmail.com

G. Fiorentino
Ortopedia e Traumatologia, Humanitas Gavazzeni, Bergamo, Italy

M. G. Mazzoleni · V. Calvisi
Università degli Studi dell'Aquila, L'Aquila, Italy

J. G. Lane et al. (eds.), *The Art of the Musculoskeletal Physical Exam*,
https://doi.org/10.1007/978-3-031-24404-9_43

In the United States, there exists a national register independent from each hospital, which reunites all the different questionnaires and information from every state in the country. These questionnaires are part of the registration process within the personal information provided by entering the hospital. Firstly, they use the *Short Form-36* (*SF-36*) or its short version *SF-12* and then the *Hip disability and Osteoarthritis Outcome Score* (*HOOS*), and they, then, administer every step of the therapeutic process; in the western states of the country, the *UCLA activity score* is also used. In addition to these PROMs, other data are collected, such as information about adverse events or the prosthesis of each patient within their X-rays. This choice has been made for evaluating first a general PROM, such as the SF, for collecting information about the general health of the patient, and the second time a more specific PROM about the hip problem [6].

There are different types of questionnaires we can use today specifically for a hip problem.

Oxford Hip Score (OHS): The original version is from 1996 [7], and it was updated in 2007 [8] introducing a new scoring system. The aim of this score is mainly to evaluate the outcome of a total arthroplasty measuring the perception of the patient besides one of the physicians. The questionnaire is developed in 12 items, 6 grading the pain and 6 grading the functionality of the hip related to everyday activities, like walking, getting dressed, and sleeping. There are five possible answers for each question, and the original score divided them from 1 to 5 (from the best to the worst scenario) with a total possible score of 60, the worst scenario, and a minimum of 12, the best scenario. Today, the new scoring system, also supported by the original authors, is from 0 to 4 (the worst to the best) with a total score of 48, which is the best scenario and the minimum 0, the worst scenario. This score has also been validated for total hip joint revision operations [9]. Some cutoffs have been set to quickly explain the meaning of the results in terms of success of the procedure: >41 excellent, 34–41 good, 27–33 fair, and <27 bad. In an article published in 2005 by Kalairajah et al. [10], the OHS has been validated as a predictor for early revision intervention after a total hip joint replacement within 6 months. A fair or bad score has been associated with a risk of revision within the next 2 years of 7.6%, while a good or excellent score has a risk of just 0.7% [11]. This is one of the most used PROMs in the world, and this is why we can find different validated translations such as in Dutch [12], Japanese [13], German [14], and French [15], and we have its version also for total knee joint replacement [8] and shoulder surgery [16]. This is a useful measurement outcome score, prevalent and effective in total hip replacement. The reader can find the scoring system within the original work of Dawson et al. [7]. It is a straightforward questionnaire that takes just a few minutes to be completed. It is especially useful if repeated at different stages of the recovery.

There is, of course, some bias due to some confusion around a couple of questions, lack of information about hip flexion, and need for analgesics.

Hip Outcome Score (HOS): This score is dedicated to patients with hip disability more or less related to degenerative arthritis. There are two validated versions, slightly different from each other: LK1.1 and LK2.0 [17, 18]. The LK2.0 version includes the Western Ontario and McMaster Universities Osteoarthritis Index (WOMAC) LK3.0 [19] in its complete and original format, and from this, the WOMAC score can be obtained. In 2008, a short version with only five items was validated measuring the physical function, the HOOS-PS. The HOOS questionnaire consists of five sections: pain, other symptoms, functionality in daily life, functionality in sports and recreational activity, and quality of life related to the hip. It has 40 items in total divided into 10 items for pain, 5 items for other symptoms (3 movements, 2 stiffness), 17 for functionality in daily life, 4 for functionality in sports, and 4 for functionality in quality of life. There are five possible answers for each question with a score from 0 to 4. Scores are then added for each section and translated into a 0 to 100 score, in which 0 indicates the worst possible scenario and 100 the best one. This questionnaire is very commonly used around the world, and it has been translated and validated in Swedish [18], Dutch

[20], and French [21]. There are also some translated versions, but not validated, in Danish, English, German, Korean, and Lithuanian. The full items' questionnaire can be examined, by the reader, from the work of Nilsdotter et al. [18]. The strength of this score is an extension of the WOMAC score, which is also more suggested for young and active adults, so it is more effective and complete. It is proved to be very useful in the follow-up of patients affected by hip degenerative arthritis.

Harris Hip Score (HHS): It was developed for evaluating hip surgery outcomes. It was thought not only for total hip joint replacement but also for femur neck fracture and hip degenerative arthritis. The earliest version, available for consultation from the 1969 original William H. Harris' work [22], is still one of the most used questionnaires all over the world. It includes pain, functionality, absence of deformity, and range of motion. In this questionnaire, we only have ten items, with a maximum score of 100, which represents the best scenario. The items are pain, 1 item (maximum 44 points); functionality, 7 items (maximum 47 points); absence of deformity, 1 item (maximum 4 points); and mobility, 2 items (maximum 5 points). The total score is based on the addition of the four sections, and the higher the results, the better the condition; a score <70 is considered poor, between 70 and 80 fair, 80–90 good, and 90–100 excellent. Despite its popularity around the world, the validated translations are just a few: Portuguese [23], Italian [24], and Turkish [25]. The HHS is today one of the most used questionnaires around the hip (from total hip joint replacement to conservative therapies). Furthermore, HHS is the most used questionnaire also for femoroacetabular impingement (FAI). There is a shortened version of the HHS, the **modified Harris Hip Score (mHHS)**, designed within the original paper of Byrd et al. in 2000 [26]. The mHHS is an eight-item questionnaire that focuses mostly on the pain and functionality. The total score is from 0 to 100, from the worst scenario to the best possible. Although patients with FAI were not included in the creation of this short version of HHS, mHHS is the most used before and after hip arthroscopy.

The Copenhagen Hip and Groin Outcome Score (HAGOS): This is a score for young and middle-aged adults, physically active and with hip or groin pain [27]. It has been developed from the need to have a new tool considering the whole variety of different treatments for hip pain. It covers the anatomy structure and function of the hip joint, limitations related to activity, and limitations of social life. It evaluates 37 items, divided into six sections: pain (10 items), symptoms (7 items), daily activity (5 items), sports and recreational activity (8 items), participation in physical activity (2 items), and quality of life (5 items). The score goes from 0 to 100, where 100 means no problems at all and 0 means severe pain and disability. HAGOS has been the first PROM for young and active patients with impacting hip or groin pain. The reader can find the full questionnaire within the original Thorborg et al.'s work [27], which has been validated in several languages, such as Swedish [28], Dutch [29], and Chinese [30]. This questionnaire has also been considered very useful for young athlete patients, as football players [31].

International Hip Outcome Tool (iHOT-33 and iHOT-12): In 2012, this new questionnaire was developed evaluating the quality of life in young and active patients with hip pain [32]. The reader may find the full original questionnaire as appendix of the original work of Mohtadi et al. [32]. The aim was to create a specific tool for this selected and demanding subset of population. Indeed, in the inclusion criteria, we can find patients between 18 and 60 years with an activity level >4 in the Tegner activity scale [33]. The iHOT-33 is composed of 4 sections with a total of 33 items: symptoms (16 items), sports and recreational activities (6 items), job-related concern (4 items), and social, emotional, and lifestyle concerns (7 items). The maximum score possible is 100, which represents the best scenario, while 0 represents the worst scenario. In 2012, the same working group developed a shortened 12-item version, iHOT-12, available for consultation in the appendix of their original article [34]. The iHOT-12 has been demonstrated to be comparable to the iHOT-33 but much quicker. We already have different validated translations of the two scores: German [35], Chinese [36], Portuguese [37], and Spanish [38].

Victorian Institute of Sport Assessment-G (VISA-G): This is a score developed and validated in 2015 [39], aimed to evaluate the severity of disability of patients affected by a greater trochanteric pain syndrome (GTPS). Since then, no specific PROMs have been developed for this common condition, forcing the clinicians to use surrogate measuring tools designed for hip osteoarthritis or low back pain. Different VISA questionnaires, such as VISA-A [40], VISA-P [41], and VISA-H [42], have been widely adopted to assess, quantify, and monitor, respectively, patellar, Achilles, and hamstring tendinopathies. As the GTPS often involves a tendinopathy of the gluteus medius and minimus tendons, the VISA questionnaire structure may be a valid measurement to adopt for the lateral hip pain [39]. The VISA-G questionnaire, entirely reported as appendix of Fearon et al.'s original work [39], consists of 8 simple and quick queries, with a maximum score of 100 points. The first question concerns the pain related to the hip evaluated by a numeric scale rating. The following questions, from 2 to 7, concern pain and disability, caused by the hip, during the daily activity. The eighth and last question is instead divided into three sections, A, B, or C, of which the patient must answer only one of them, according to his/her ability to undertake weight-bearing activities related to the severity of hip pain. Since it has been developed for a few years, the VISA-G questionnaire is not yet widespread and has undergone only a validated translation in Danish [43].

Non-Arthritic Hip Score (NAHS): As the interest in hip pathologies of the young patients has grown, it has become more urgent for clinicians to develop clinical scores suitable for non-arthritic joint diseases. With this goal, in 2003 [44], a new system was designed to assess preoperative and postoperative hip pain and function in a younger and more athletic population. In order to maximize the patient compliance, this self-administered scoring system is symptom related only, avoiding also bias due to clinician interpretation or physical examination influences. This score has been created as a modified Western Ontario and McMaster Universities Osteoarthritis Index, including 20 concise unweighted multiple-choice questions, divided into four sections, concerning, in order, pain, mechanical symptoms, physical function, and level of activity. Each answer corresponds to a numeric value, of which the total sum must be multiplied by 1.25 to achieve the final score. The normal hip function is represented by the maximum score of 100 points. The NAHS is ultimately designed to assess and stratify the activity level in 20- to 40-year-old athletic patients who are limited in some way by their hip disease without a clear radiographic diagnosis. The reader may find the entire questionnaire in the appendix at the end of the original work of Christensen et al. [44]. This score has been translated and validated only in Brazilian Portuguese in 2013 [45].

References

1. Nelson EC, Eftimovska E, Lind C, Hager A, Wasson JH, Lindblad S. Patient reported outcome measures in practice. BMJ. 2015;350:1–3.
2. Mistry JB, Jauregui JJ, Lerner AL, Chughtai M, Elmallah RK, Mont MA. An assessment of the comprehensiveness of various hip outcome scores. Surg Technol Int. 2016;28:267–74.
3. Hahn EA, Cella D, Chassany O, Fairclough DL, Wong GY, Hays RD. Precision of health-related quality-of-life data compared with other clinical measures. Mayo Clin Proc. 2007;82(10):1244–54.
4. Prodinger B, Taylor P. Improving quality of care through patient-reported outcome measures (PROMs): Expert interviews using the NHS PROMs Programme and the Swedish quality registers for knee and hip arthroplasty as examples. BMC Health Serv Res. 2018;18(1):1–13.
5. Mapi Research Trust. ePROVIDE PROQOLID database. www.eprovide.mapi-trust.org
6. Franklin PD, Lewallen D, Bozic K, Hallstrom B, Jiranek W, Ayers DC. Implementation of patient-reported outcome measures in U.S. total joint replacement registries: Rationale, status, and plans. J Bone Joit Surg Am. 2014;96:104–9.
7. Dawson J, Fitzpatrick R, Carr A, Murray D. Questionnaire on the perceptions of patients about total hip replacement. J Bone Joint Surg Br. 2018;78-B(2):185–90.
8. Murray DW, Fitzpatrick R, Rogers K, Pandit H, Beard DJ, Carr AJ, Dawson J. The use of the Oxford hip and knee scores. J Bone Joint Surg Br. 2007;89-B(8):1010–4.

9. Dawson J, Fitzpatrick R, Frost S, Gundle R, Murray D. Instrument for assessment of outcome after revision hip replacement. JBJS Br. 2001;83-B:1125–9.
10. Kalairajah Y, Azurza K, Hulme C, Molloy S, Drabu KJ. Health outcome measures in the evaluation of total hip arthroplasties. A comparison between the Harris hip score and the Oxford hip score. J Arthroplast. 2005;20(8):1037–41.
11. Rothwell AG, Hooper GJ, Hobbs A, Frampton CM. An analysis of the Oxford hip and knee scores and their relationship to early joint revision in the New Zealand Joint Registry. J Bone Joint Surg Br. 2010;92-B(3):413–8.
12. Gosens T, Hoefnagels NH, de Vet RC, Dhert WJ, van Langelaan EJ, Bulstra SK, Geesink RG. The 'Oxford Heup Score': The translation and validation of a questionnaire into Dutch to evaluate the results of total hip arthroplasty. Acta Orthop. 2005;76(2):204–11.
13. Uesugi Y, Makimoto K, Fujita K, Nishii T, Sakai T, Sugano N. Validity and responsiveness of the Oxford hip score in a prospective study with Japanese total hip arthroplasty patients. J Orthop Sci. 2009;14(1):35–9.
14. Naal FD, Sieverding M, Impellizzeri FM, Von Knoch F, Mannion AF, Leunig M. Reliability and validity of the cross-culturally adapted German oxford hip score. Clin Orthop Relat Res. 2009;467(4):952–7.
15. Delaunay C, Epinette JA, Dawson J, Murray D, Jolles BM. Cross-cultural adaptations of the Oxford-12 HIP score to the French speaking population. Orthop Traumatol Surg Res. 2009;95(2):89–99.
16. Dawson J, Rogers K, Fitzpatrick R, Carr A. The Oxford shoulder score revisited. Arch Orthop Trauma Surg. 2009;129(1):119–23.
17. Klässbo M, Larsson E, Mannevik E. Hip disability and osteoarthritis outcome score: An extension of the Western Ontario and McMaster Universities Osteoarthritis Index. Scand J Rheumatol. 2003;32(1):46–51.
18. Nilsdotter AK, Stefan Lohmander L, Klässbo M, Roos EM, Lohmander LS. BMC Musculoskeletal Disorders Hip disability and osteoarthritis outcome score (HOOS) – validity and responsiveness in total hip replacement. 2003;8:1–8.
19. Bellamy N, Buchanan WW, Goldsmith CH, Campbell J, Stitt LW. Validation study of WOMAC: A health status instrument for measuring clinically important patient relevant outcomes to antirheumatic drug therapy in patients with osteoarthritis of the hip or knee. J Rheumatol. 1988;5(12):1833–40.
20. de Groot IB, et al. Validation of the Dutch version of the Hip disability and Osteoarthritis Outcome Score. Osteoarthr Cartil. 2007;15(1):104–9.
21. Ornetti P, et al. Cross-cultural adaptation and validation of the French version of the Hip disability and Osteoarthritis Outcome Score (HOOS) in hip osteoarthritis patients. Osteoarthr Cartil. 2010;18(4):522–9.
22. Harris WH. Traumatic arthritis of the hip after dislocation and acetabular fractures: treatment by mold arthroplasty. An end-result study using a new method of result evaluation. J Bone Joint Surg Am. 1969;51(4):737–55.
23. Guimarães R, Alves D, Silva G. Translation and cultural adaptation of the Harris Hip Score into Portuguese. Acta Ortop. 2010;18(3):142–7.
24. Dettoni F, et al. Validation and cross cultural adaptation of the Italian version of the Harris hip score. Hip Int. 2015;25(1):91–7.
25. Çelik D, Can C, Aslan Y, Ceylan HH, Bilsel K, Ozdincler AR. Translation, cross-cultural adaptation, and validation of the Turkish version of the Harris hip score. Hip Int. 2014;24(5):473–9.
26. Byrd JW, Jones KS. Prospective analysis of hip arthroscopy with 2-year follow-up. Arthroscopy. 2000;16(6):578–87.
27. Thorborg K, Hölmich P, Christensen R, Petersen J, Roos EM. The Copenhagen Hip and Groin Outcome Score (HAGOS): development and validation according to the COSMIN checklist [published correction appears in Br J Sports Med. 2011 Jul;45(9):742]. Br J Sports Med. 2011;45(6):478–91.
28. Thomeé R, et al. Cross-cultural adaptation to Swedish and validation of the Copenhagen Hip and Groin Outcome Score (HAGOS) for pain, symptoms and physical function in patients with hip and groin disability due to femoro-acetabular impingement. Knee Surg Sport Traumatol Arthrosc. 2014;22(4):835–42.
29. Brans E, De Graaf JS, Munzebrock AVE, Bessem B, Reininga IHF. Cross-cultural adaptation and validation of the Dutch version of the hip and groin outcome score (HAGOS-NL). PLoS One. 2016;11(1):1–12.
30. Cao S, Cao J, Li S, Wang W, Qian Q, Ding Y. Cross-cultural adaptation and validation of the Simplified Chinese version of Copenhagen Hip and Groin Outcome Score (HAGOS) for total hip arthroplasty. J Orthop Surg Res. 2018;13(1):1–8.
31. Thorborg K, Branci S, Stensbirk F, Jensen J, Hölmich P. Copenhagen hip and groin outcome score (HAGOS) in male soccer: Reference values for hip and groin injury-free players. Br J Sports Med. 2014;48(7):557–9.
32. Mohtadi NGH, et al. The development and validation of a self-administered quality-of-life outcome measure for young, active patients with symptomatic hip disease: The International Hip Outcome Tool (iHOT-33). Arthroscopy. 2012;28(5):595–610.e1.
33. Tegner Y, Lysholm J. Rating systems in the evaluation of knee ligament injuries. Clin Orthop Relat Res. 1985;198:43–9.
34. Griffin DR, Parsons N, Mohtadi NGH, Safran MR. A short version of the International Hip Outcome Tool (iHOT-12) for use in routine clinical practice. Arthroscopy. 2012;28(5):611–8.
35. Baumann F, et al. Validation of a German version of the International Hip Outcome Tool (G-iHOT33) according to the COSMIN checklist: how much improvement is clinically relevant? Arch Orthop Trauma Surg. 2016;136(1):83–91.
36. Li DH, et al. Development of a valid Simplified Chinese version of the International Hip Outcome

Tool (SC-iHOT-33) in young patients having total hip arthroplasty. Osteoarthr Cartil. 2017;25(1):94–8.
37. Polesello GC, Godoy GF, Trindade CA, Queiroz MC, Honda E, Ono NK. Translation and cross-cultural adaptation of the International Hip Outcome Tool (iHOT) into Portuguese. Acta Ortop Bras. 2012;20(2):88–92B.
38. Ruiz-Ibán MA, et al. The international Hip Outcome Tool-33 (iHOT-33): Multicenter validation and translation to Spanish. Health Qual Life Outcomes. 2015;13(1):1–7.
39. Fearon A, Ganderton C, Scarvell J, et al. Development and validation of a VISA tendinopathy questionnaire for greater trochanteric pain syndrome, the VISA-G. Man Ther. 2015;20(6):805–13.
40. Robinson JM, Cook JL, Purdam C, Visentini PJ, Ross J, Maffulli N, et al. The VISA-A questionnaire: a valid and reliable index of the clinical severity of Achilles tendinopathy. Br J Sports Med. 2001;35(5):335–41.
41. Visentini P, Khan K, Cook J, et al. The VISA score: an index of severity of symptoms in patients with jumper's knee (patellar tendinosis). Victorian Institute of Sport Tendon Study Group. J Sci Med Sport. 1998;1(1):22–8.
42. Cacchio A, De Paulis F, Maffulli N. Development and validation of a new visa questionnaire (VISA-H) for patients with proximal hamstring tendinopathy. Br J Sports Med. 2014;48(6):448–52.
43. Jorgensen J, Fearon A, Mølgaard C, et al. Translation, validation and test-retest reliability of the VISA-G patient-reported outcome tool into Danish (VISA-G. DK). PeerJ. 2020;8:e8724.
44. Christensen CP, Althausen PL, Mittleman MA, et al. The nonarthritic hip score: reliable and validated. Clin Orthop Relat Res. 2003;406:75–83.
45. Del Castillo L, Leporace G, Cardinot T, et al. Translation, cross-cultural adaptation and validation of the Brazilian version of the Nonarthritic Hip Score. Sao Paulo Med J. 2013;131(4):244–51.

Part V

Knee Reviewers Dr's Gobbi, Lane & Espregueira-Mendes

Anatomy of the Knee

44

Fabio Valerio Sciarretta and John G. Lane

The knee joint is the most complex and largest joint in the body. Although devoid of intrinsic stability, the knee is one of the most mobile joints of the body. It is a gliding hinge synovial joint and consists of two articulations—the tibiofemoral and the patellofemoral joint. The tibiofemoral joint is responsible for weight-bearing transmission, while the patellofemoral is responsible for the extension of the knee. We can further divide the tibiofemoral joint and identify the medial tibiofemoral compartment as the articulation between medial condyle of the femur and medial condyle or plateau of the tibia, and the lateral tibiofemoral compartment articulating with lateral condyle of the femur and lateral condyle or plateau of the tibia. The patellofemoral joint is the articulation between the patella and the trochlear surface of the femur and the proximal tibiofibular joint [1, 2]. The knee joins the lower leg and thigh bilaterally and is an essential component of efficient bipedal movements such as walking, running, and jumping. The main knee joint kinematics is based on rolling, gliding, and rotating movements through 6° of freedom of motion. The knee mainly allows for flexion and extension, but also allows a small degree of rotational movements (flexion–extension, internal–external rotation, and varus–valgus) and translational movements (anterior–posterior, medial–lateral, and compression and distraction direction of the knee joint). All these six freedoms of motion determine the complexity of the knee joint together with ligaments and tendons.

Femorotibial bone structures are eminently unstable with flat tibial condyles and convex femoral condyles. The medial tibial plateau is larger and nearly flat, while the lateral tibial plateau is concave. Both present a posterior slope of approximately 10° with respect to the tibial diaphysis. This partial congruency with instability is compensated for by the presence of the menisci, which enlarge the contact area and increase joint stability and congruency. Between the medial and lateral plateaus, the tibia presents with a slope and the tibial spines are followed posteriorly by the tibial tubercles separated by the intertubercle groove. This central area of the proximal articular tibia is occupied by the insertions of the anterior horn of the medial meniscus, the anterior cruciate ligament, the anterior horn of the lateral meniscus, and more posteriorly the posterior horns of medial and lateral menisci and finally the posterior cruciate ligament.

The stability of the knee depends on the action of the ligaments and the tendinous and muscular structures. At first glance, the stabilization of the knee must be achieved in both the frontal plane

F. V. Sciarretta (✉)
Nostra Signora della Mercede Clinic, Rome, Italy

J. G. Lane
Musculoskeletal and Joint Research Foundation, San Diego, CA, USA

J. G. Lane et al. (eds.), *The Art of the Musculoskeletal Physical Exam*,
https://doi.org/10.1007/978-3-031-24404-9_44

(valgus–varus) and the sagittal plane (anterior and posterior). In reality, stability is based on the fundamental principle of rotatory stabilization. Medial rotation of the tibia is associated with deviation in valgus of the tibia. The lateral tibial condyle rolls more than the medial tibial condyle, especially during the first 20° of knee flexion ("screw-home mechanism" of the knee, which locks the knee into extension) [3, 4]. The initial rotation of the femoral condyles in the first degrees of flexion follows a sliding phenomenon due to the tensioning of the central pivot composed by the anterior and posterior cruciate ligaments.

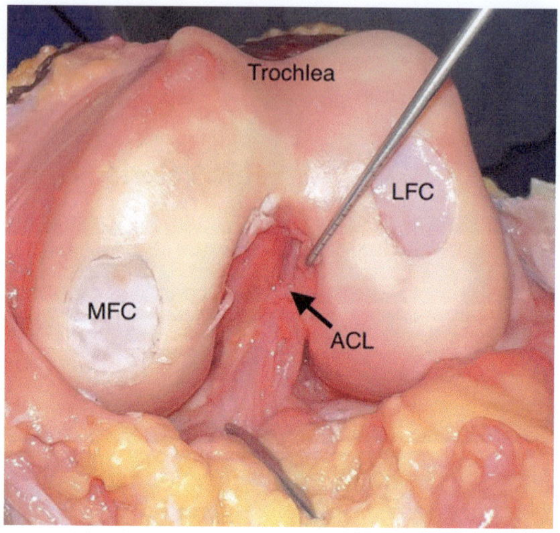

Fig. 44.1 Shows the course of the ACL, highlighting with arrows the femoral and tibial insertions. *MFC* medial femoral condyle, *LFC* lateral femoral condyle, *ACL* anterior cruciate ligament with tibial and femoral insertion sites noted by tips of probes

44.1 The Central Pivot

The anterior cruciate ligament (ACL) and posterior cruciate ligament (PCL) create the central pivot for the knee located within the joint capsule of the knee. These structures have been extensively studied, and their anatomy has been delineated in recent years, especially for the ACL, and it is more deeply appreciated and has been rediscovered, thus modifying the surgical strategies of their repair and reconstruction [1].

44.2 Anterior Cruciate Ligament

The ACL originates from the posterolateral surface of the intercondylar notch and courses distally and anteriorly to insert on the intercondylar eminence of the tibia (Fig. 44.1). It is composed of two bundles—an anteromedial and a posterolateral bundle—although also an intermediate bundle has been described; these structures have been named based on their tibial insertion [5]. The ACL presents with an irregular shape, which is especially variable depending on the degree of knee flexion.

As noted, the anterior cruciate ligament is located in the anterior intercondylar region, coursing from the anterior condylar area to the medial side of the lateral femoral condyle. On the tibia, its insertion surface is over a broad anterior area of the tibial plateau and is in an oval shape, between the intercondylar eminences of the tibia located on the anterior intercondylar area in front and in contact with the medial intercondylar tubercle extending to the jugal ligament and the lateral edge of the medial condyle. It is separated from the lateral tibial plateau by the horn of the lateral meniscus. The meniscus and ligament insertions seem to merge. The fibers go up, outside and back, towards the lateral femoral condyle, running therefore upwards and backwards from their origin. The insertion of the ACL is located on a semicircular area posteriorly on the femur, at the lateral wall of the intercondylar notch (Palastanga, 2002) [6]. The ACL fibers are divided into two bundles: one posterolateral and the other anteromedial. The intermediate component, described by Beasley et al. in 2005 [7], when present, although sharing fibers with the two previous groups, is closer to the anteromedial bundle. Finally, in 26% of cases, the ACL may show just a single bundle.

The tibial and femoral insertion planes of the anterior cruciate ligament are located along very different axes. The plane of insertion on the tibial side is horizontal and sagittal on the femoral condylar side.

The distance from the center of the ligament to the posterior superior edge of the femoral condyle corresponding to the "over the top" passage of the ligament femoral tunnel is 1.6 cm + 2 mm, and its distance from the anterior articular edge of the intercondylar notch is 2.3 cm + 4 mm.

The length of the anterior cruciate ligament is 3.3 cm on average (31–38 mm in range), with an average width of 11 mm (AM bundle 6–7 mm and PL bundle 5–6 mm). The average insertion surface is 0.250 cm^2 on the tibia (17–30 × 11 mm) and 0.215 cm^2 on the lateral condyle (16–24 × 11 mm).

The main function of the ACL is to resist anterior translation of tibia in respect to the femur, particularly at 20° of flexion (74% in full extension—85% at 30 and 90° of flexion). A secondary function of the ACL is to resist varus or valgus rotation of the tibia, especially in the absence of the collateral ligaments. The ACL also resists internal rotation of the tibia. In full extension, the anterior surface of the ACL lies against the intercondylar shelf assisting the ligament in preventing hyperextension. The ACL also has a very important proprioceptive action and allows normal helicon knee action, thus preventing meniscal damage.

The two ACL bundles have a different biomechanical action, with the anteromedial bundle being tighter in extension and the posterolateral bundle being tighter in flexion and in internal rotation due to the orientation of their tibial insertion and course to insertion on the lateral femoral condyle wall which changes during flexion. In extension, the origin of the AM bundle is located just above the origin of the PL bundle, but in flexion, the attachment site of the AM bundle moves behind the PL bundle origin [8].

The femoral insertion site of the ACL is semilunar. It shows an attachment length of 13–25 mm and a width of 6–13 mm. Others found a more circular or oval shape of the ACL attachment site [8].

In 2014, Smilgielski et al. in their anatomical study found that the ACL presented as a ribbon-like structure 2–3 mm from its bony femoral insertion, without a clear separation between AM and PL bundles, and the ribbon was in exact continuity with the posterior femoral cortex. The authors were able to calculate the mean cross-sectional area 2 mm from femoral insertion: 56.6 mm^2. More recent studies have concluded that the ACL presents a femoral insertion site that varies greatly individually and is eccentric, closer to the deep cartilage of the lateral femoral condyle [9].

The blood supply to the ACL arises through the anastomosis of the medial and lateral inferior geniculate arteries through the fat pad and the middle geniculate artery branching off the posterior capsule. This is the main reason why a tear of the ACL, especially if in the mid-substance, causes the disruption of the blood supply, thus limiting the ligament's own healing capacity and suggesting the need for reconstructive surgery.

44.3 Posterior Cruciate Ligament

The PCL has been less investigated than the ACL and is the strongest ligament of the knee. The PCL originates on the medial surface of the intercondylar notch extending distally to insert on the proximal tibia in the fovea centralis. In detail, it inserts on the posterior intercondylar area in its posterior lower portion, very distal from the intercondylar tubercules. Its insertion, unlike many schematic representations, is spread out on the posterior surface of the tibia in the axis of the posterior intercondylar area (retro-spinal surface) and is arranged in a frontal plane. It inserts superiorly and medially, towards the intercondylar fossa, and ends on the lateral face of the medial condyle.

The PCL has been described as composed of several bundles, but most authors suggest two bundles—a posteromedial and an anterolateral bundle. It is possible to identify four aspects of fibers: an anteromedial contingent, a posterior lateral, the meniscal femoral contingent of Wrisberg, and an inconsistent anterior portion or ligament of Humphry. The two anteromedial and posterolateral contingents are an integral part of the posterior cruciate ligament.

Both the lengths and the widths of the PCLs are larger than those of the ACL.

The posteromedial bundle is more taut in extension, and the anterolateral bundle is more taut in flexion.

The posterior cruciate ligament measures on average 3.6 cm in length with a diameter at its middle portion of 9.9 mm and a femoral insertion surface wider than the tibial insertion surface. In fact, the insertion surface is 0.46 cm^2 on the tibia and 0.50 cm^2 on the medial condyle.

The main function of the PCL is to allow femoral rollback in flexion and resist posterior translation of the tibia relative to the femur in extension (especially the anterolateral bundle from 30 to 90° of flexion) [10]. This ligament also increases the lever arm of the quadriceps mechanism and controls external rotation of the tibia with increasing knee flexion. Unlike the ACL, because of its association with the posterior capsule, blood supply is not permanently lost with a mid-substance tear of the PCL.

44.4 Passive and Active Peripheral Rotary Constraints

The ideal position of knee stability at 90° of flexion with a mechanical femoro-patello-tibial alignment is in everyday life difficult to maintain. The activities of walking and sport make the knee oscillate between two extreme positions where the rotational constraints are important, one of valgus-lateral rotation-flexion and the other of varus-medial rotation-flexion. To assume three-dimensional stability, the joint structures can be organized into passive ligament constraints and active musculo-tendon constraints. The knee joint is limited at its periphery by a capsule which, under the action of stresses, presents thickenings or passive peripheral formations reflecting the different medial, lateral, and condylar shells. To these ligament structures must be added the menisci, which participate in rotatory stability through all the ligament insertions.

44.5 The Medial Collateral Ligament

The medial collateral ligament is the wide and flat ligament found on the medial side of the joint, which takes its origin from the medial epicondyle and extends distally to insert on the tibia, deep to the sartorius and the pes anserinus, and it is divided into two bundles: one superficial, limited, is stretched obliquely below and in front of the internal condylar tubercle on the internal face of the tibia, while the other, deep bundle, is wider and protrudes in front and behind the superficial bundle. The medial collateral ligament inserts on the internal condylar tubercle and ends on the internal peripheric edge of the tibial plateau. The primary function of the medial collateral ligament is to restrain valgus rotation of the knee joint and secondly control the external rotation. The medial collateral ligament is tight in extension and lax in flexion [11].

44.6 The Lateral Collateral Ligament

Unlike the medial compartment, the lateral compartment is very mobile. The lateral collateral ligament, belonging to the category of extracapsular ligaments, is well individualized and runs from the femur to the fibula. It is thinner and rounder than the medial collateral ligament; it inserts on the lateral condylar tubercle and runs inferiorly and posteriorly, to insert on the head of the fibula, merging with the tendon insertion of the biceps femoris. The lateral collateral ligament restrains against varus rotation as well as resisting internal rotation.

44.7 Condylar Shells

This ligament structure is very important and plays a role in the passive limitation of extension. It consists of three parts: a medial and two lateral. The thickening of the middle portion or oblique popliteal ligament merges with the termination fibers of the semi-membranous muscle, delivering passage to the tendon of the popliteal muscle; the arched popliteal ligament, with its two bundles, completes the posterior median structures. The lateral portions are occupied by the medial and lateral condylar shells, which insert on the posterior surface of the condyles to terminate at the posterior aspect of the tibial condyles.

44.8 The Menisci

The two menisci, medial and lateral, are fibrocartilage structures that adapt the articular surface of the tibia to the rolling and gliding movements of the femoral condyles. They are fixed to the joint capsule at their periphery and to the tibia via meniscotibial ligaments, while their anterior and posterior roots directly insert into the tibial plateaus [12]. The menisci also play an important proprioceptive function.

The medial and lateral menisci have the shape of prismatic crescents of triangular cut with a capsular peripheral face, a tibial lower face, an upper femoral face, as well as an axial plateau edge. The meniscal horns are extended by means of fibrous elements, which ensure the fixation of the menisci onto the tibial surface, in addition to other fibro-tendinous structures.

The menisci constitute one of the knee barometers: their defailance is strictly correlated to a bony structure malfunctioning, and the joint derangement caused by meniscal damage causes the development of osteoarthritis in the long term.

The two menisci may appear similar. In reality, there is a very significant intrinsic and extrinsic morphological duality. The so-called para-meniscal peripheral area, including a quarter of the meniscal area, is vascularized from the capsuloligament plane, while the central area, extending up to three-quarters of the so-called articular meniscal area, is essentially fibrocartilaginous.

The upper, concave face of the menisci articulates with the femoral condyle. The relatively flat tibial surface is present on the periphery of each plateau cavity. The outer face, peripheral to the circumferential edge, thick, convex from front to back, is more or less adherent to the capsule. The inner, circumferential or central, concave, sharp edge circumscribes the center of the plateau cavity. The meniscal horns—one anterior and the other posterior—ensure the attachment of the menisci to the tibia.

The most evident difference between the two is that the lateral meniscus shape is closed nearly to form an O, but is interrupted medially at its axial slope by the thickness of the external tibial spine, while the medial meniscus is wider and is shaped like an open C with a continuous adhesion to the capsular peripheral face (Fig. 44.2). Another difference between the two menisci is mechanical, with a mobile lateral meniscus and a relatively fixed medial meniscus. The medial meniscus is, on average, 3.5 cm long, much larger posteriorly than anteriorly, and it possesses a solid attachment on the posterior intercondylar area of the tibia. Its anterior attachment is more variable and more lax. The lateral larger meniscus covers a wider portion of the articular surface than the medial meniscus. Its anterior horn is attached to the intercondylar area, laterally and posteriorly to the posterior cruciate ligament. The posterior horn inserts on the intercondylar area

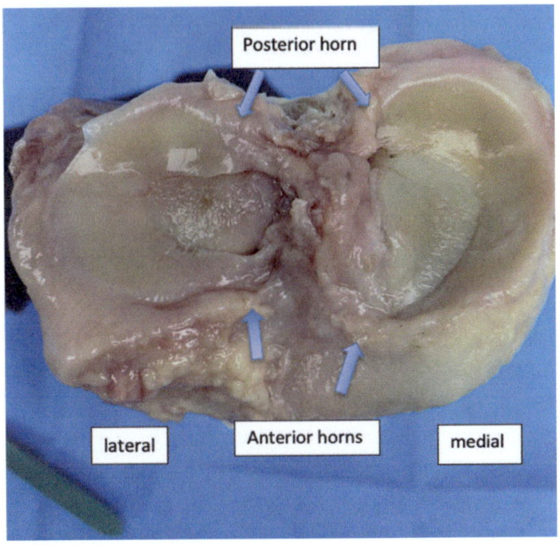

Fig. 44.2 Medial and lateral meniscus with insertions of anterior and posterior horns

anteriorly to medial meniscus posterior horn insertion site. Posterolateral to the lateral meniscus, the popliteal tendon runs along the posterior surface of the knee, sending some insertions into the superior peripheral meniscal margin.

The menisci, although variable, have several attachments: in fact, they are firmly anchored to the tibial condyles and to the joint capsule. The meniscal horns bind to the tibia via meniscal ligaments, four in number: the anteromedial meniscal ligament arises from the anterior aspect of the anterior horn of the medial meniscus and inserts on the medial side of the anterior intercondylar area in front and medial of the cruciate ligament insertion; the anterolateral meniscal ligament attaches over the entire extent of the lateral anterior horn and reaches the anterior intercondylar area in front of the lateral intercondylar tubercle, but lateral to the anterior cruciate ligament; the posteromedial meniscal ligament originates from the posterior horn of the medial meniscus and attaches to the posterior intercondylar area along a triangular surface behind the medial intercondylar tubercle and in front of the insertion of the posterior cruciate ligament; and the posterolateral meniscal ligament detaches from the posterior horn of the lateral meniscus and fixes itself in front of the posterior cruciate ligament insertions. The tibial insertions of the meniscal ligaments on the lateral tibial plateau are very close to each other, whereas the insertions on the lateral plateau are widely separated and are located in front of the insertions of each cruciate ligament. The meniscal ligaments of the medial meniscus are parallel to each other while those of the lateral meniscus converge towards the intercondylar tubercle. A fibrous strip called the jugal ligament or transverse meniscomeniscal ligament joins together the two anterior horns of the menisci. When the knee is being extended, the ligament prevents the anterior horns of the menisci from moving forward, and the condylar surfaces of the femur and tibia from exerting pressure on the menisci. It has a restricting effect on anterior-posterior excursion of the anterior horn of the medial meniscus at lower degrees of knee flexion.

Despite all these attachments, the menisci have great mobility and move and deform during all movements of the knee joint.

From a biomechanical point of view, the menisci have two functions: pressure transmission and dynamic stabilization. The internal structure is made up of collagen fibers oriented in the three planes of space with vertical, horizontal, and radial fibers. The menisci have a very important function of distributing the stresses between the femur and the tibia. Their shape and position adapt to the movements of the condyles, but their horns remain fixed. In extension, they are pulled forward. In contrast, in flexion, they move posteriorly with a displacement of 12 mm for the lateral meniscus and 6 mm for the medial meniscus.

During extension, the menisci are drawn forward by the meniscopatellar fins and the jugal ligament. During flexion, the medial meniscus is pulled back by the expansion of the semimembranous, its anterior horn being retracted by the fibers of the anterolateral bundle of the cruciate ligament which are inserted there. Likewise, the lateral meniscus is pulled posteriorly by the expansion of the popliteal muscle. During axial rotation, the menisci follow the movements of the condyles on the tibial fossae by deforming themselves. During the lateral rotation of the tibia on the femur, the lateral meniscus moves towards the anterior aspect of the lateral plateau while the medial moves posteriorly. During medial rotation, the opposite phenomenon occurs: the lateral meniscus recedes while the internal advances.

Once more, the loss of harmonic movement between the bone segments induces abnormal stresses and possible ruptures of the menisci.

The main problem of the knee is that of rotational stabilization. The central axis represented by the central pivot cannot oppose the considerable torsional stresses. The essential function of the cruciate ligaments is to maintain permanent femorotibial coaptation, thanks to their prestressed condition, to prevent anterior or posterior displacement of the tibial plateau, while rotatory stabilization is delegated to the peripheral structures, which act as tensile sensors of information making it possible to regulate the contraction of the periarticular muscles (by three types of mechanoreceptors present on anterior cruciate ligament—Ruffini mechanoreceptors, Golgi neurotendinous corpuscles, and Pacini corpuscles). All the ligamentous formations of the central pivot knee and peripheral structures work, in synergy and continuity, performing a screwing and unscrewing movement of the femur relative to the tibia.

Knee stability is finally completed by all of the symmetric arrangement of muscles inserting on both the medial and lateral sides of the joint and by the dynamic stability guaranteed by the ACL's isometry.

44.9 Patellofemoral Joint

The patellofemoral joint is also known as an extensor mechanism, composed of the patella that receives the direct insertion of the deep layer of the patellar tendon distally and of the vastus intermedius proximally.

The patella is the largest sesamoid bone in the body and functions as an attachment point for the quadriceps tendon and patellar ligament. The quadriceps tendon inserts into the proximal border of the patella, and the distal apex of the patella is connected to the anterior aspect of the proximal tibia at the tibial tuberosity by the patellar ligament. The patella tracks in the trochlear groove of the femur (or femoral sulcus or anterior articular surface of the distal femur) during knee flexion and extension. It also protects the anterior articular surface of the femoral portion of the knee.

The patella is described to possess seven facets; the medial and the lateral ones are divided vertically in three parts approximately of the same size, while the seventh facet, the odd facet, is located at the extreme medial margin of the bone. The medial facet is smaller and slightly convex while the lateral facet, that represents nearly the two-thirds of this bone, has a sagittal convexity with rest of its shape concave. There is a substantial mismatch between the size of the patella and the size of the anterior articulating surface of distal femur. The femoral trochlea has convex margins on both sides, with the lateral one larger than the medial one. The bony contact area between the femur and the patella varies depending on the exact knee position, with maximum contact occurring at 45° of flexion. At 90° of flexion, the patella contacts the femur on the upper portions of medial and lateral facets, while in complete extension, this contact occurs on the lower and distal parts of the facets.

Along the medial margin of the patella, the capsular reinforcements insert forming the medial retinaculum and its condensations. The medial patellar ligaments, that are responsible for maintaining the stability of the patellofemoral joint against lateral patella translation, are, proximally, the medial patellofemoral ligament (MPFL) and, distally, the medial patellotibial ligament (MPTL) and the medial patellomeniscal ligament (MPML). The MPFL is considered the primary restraint to lateral patellar dislocation, while the other two ligaments are considered secondary restraints. The medial patellofemoral ligament (MPFL—so named by Warren and Marshall in 1979) [13] recently has been of particular interest due to the various techniques developed for its reconstruction in the treatment of patellar dislocation; it courses from the adductor tubercle at Schottle's point (isometric point) to the vastus medialis obliquus insertion on the patella, thus connecting the patella to the medial femoral condyle and passively limiting the lateral translation of the patella. The medial patellotibial ligament (MPTL—identified by Terry, in 1989 [14], as a condensation of the medial retinaculum) extends from the inferomedial aspect of the patella to the anteromedial physeal scar of the proximal anteromedial tibia, 1.5 cm distally from the joint line. Its principal clinical significance is in patellar instability and distal extensor mechanism reconstruction in juvenile patients with excessive Q angles and an open proximal tibial physis. The medial patellomeniscal ligament (studied by Panagiotopoulos et al. 2006 and recently by Kaleka et al. 2017 and Hinckel et al. 2019) [15–17] lies deeper and originates in the inferomedial patella, just proximal to the MPTL, and inserts in the medial meniscus. MPTL and MPML exert their limiting action mostly in terminal extension, preventing lateral patellar subluxation, and are fundamental to restrain lateral patellar translation in flexion.

The stronger lateral retinaculum, described by Fulkerson and Gossling [18], courses from the proximal lateral epicondyle to the vastus lateralis tendon insertion on the patella and is composed of two principal components, including the oblique superficial retinaculum and the transverse deep retinaculum. The superficial one runs superficially from the iliotibial band to the patella, while the deep retinaculum is composed of three different components: the patella-epicondylar ligament, the deep transverse retinaculum, and the iliotibial band. Distally, the lateral patellotibial ligament courses from the inferolateral patella to the lateral tibial tubercle deep to the iliopatellar ligament.

The distal femoral condyles are asymmetric: the larger, medial one presenting a more symmetric curvature. The longitudinal axis of the lateral femoral condyle is slightly longer and located on the sagittal plane, while the medial one presents a 22-degree angle directed posteriorly.

The anterior muscles of the knee act primarily as knee extensors. The quadriceps femoris muscle is the principal muscle involved in knee extension. This muscle can be divided into four distinct parts: the rectus femoris, vastus medialis, vastus lateralis, and vastus intermedius. All four parts of this muscle come together to insert on the proximal edge of the patella, which then transfers their action, by way of the patellar tendon, to the tibia. The vastus intermedius originates on the anterior surface of the femoral shaft and inserts directly onto the superior pole of the patella. The rectus femoris parallels the vastus intermedius but runs in the retinacular layer, going from the anteroinferior iliac spine to the central tendon of the quadriceps. The rectus femoris, along with the vastus medialis obliquus, vastus medialis, vastus lateralis, and vastus lateralis obliquus, terminates in an aponeurosis that merges into the retinacular layer. The vastus lateralis represents a great part of the quadriceps muscle bulk. It originates from the anterolateral aspect of the femur and lateral intermuscular septum. The vastus medialis originates from the anteromedial aspect of the femur and medial intermuscular septum.

The patellar tendon extends from the inferior pole of the patella and inserts on the tibial tubercle, separated from the underlying tibia by the infrapatellar tendon bursa. The patellar tendon comprises a superficial layer, which is contiguous with the retinacular layer, and a deeper layer, which is again the deep layer of the extensor mechanism.

The primary muscles involved in knee flexion are the hamstring muscle group. This group is comprised of the biceps femoris, semitendinosus, and semimembranosus muscles. Their insertion occurs on the proximal tibia and head of the fibula. The biceps femoris muscle has an additional action of externally rotating the tibia while the semitendinosus and semimembranosus muscles also have an additional role of internally rotating the tibia. Other muscles participating in knee flexion and internal rotation are the sartorius and gracilis muscles. The popliteus muscle also serves to internally rotate the knee in a non-weight-bearing position. Additional muscles involved in isolated knee flexion include the gastrocnemius and plantaris muscles.

All joint surfaces of bones are covered with articular cartilage. Articular cartilage has a very low coefficient of friction and secretes a water lubricant under pressure. Its principal function is to provide a smooth, lubricated surface for articulation and to facilitate the transmission of loads. It works in strict conjunction with the underlying subchondral bone, forming what is defined as the subchondral unit. Cartilage integrity is of vital importance to protect bone, therefore preventing joint degeneration into osteoarthritis. Several studies have shown during knee arthroscopies that chondral lesions in the knee occur in greater than 60% of patients. Repair, reconstruction, resurfacing, and orthobiological treatments of cartilage defects are therefore mandatory, as well as reconstruction and preservation of meniscal tissue, in order to maintain healthy joint tissues and preserve the joint for a long-term survival.

The movement of the patellofemoral joint can be characterized as gliding and sliding. During flexion of the knee, the patella moves distally on the femur. This movement is governed by its attachments to the quadriceps tendon, ligamentum patellae, and anterior aspects of the femoral condyles. The muscles and ligaments of the patellofemoral joint are responsible for producing extension of the knee. The patella acts as a pulley in transmitting the force developed by the quadriceps muscles to the femur and the patellar ligament. It also increases the mechanical advantage of the quadriceps muscle relative to the instant center of rotation of the knee.

44.10 Synovial and Neurovascular Anatomy

The articular joint cavity of the knee is the largest articular space of all the body, located between the femoral condyles, the proximal tibia, and the posterior surface of the patella. The synovial membrane lines the joint capsule, reflects onto the bony segments of the joint, covers the cruciate ligaments except where the PCL is attached to the capsule, and creates several folds and recesses.

The knee also contains multiple bursas that facilitate structures gliding upon each other (suprapatellar bursa—between the quadriceps femoris and the femur; prepatellar bursa—between the apex of the patella and the skin; infrapatellar bursa—split into deep and superficial; the deep bursa lies between the tibia and the patella ligament; the superficial lies between the patella ligament and the skin; semimembranosus bursa—located posteriorly in the knee joint, between the semimembranosus muscle and the medial head of the gastrocnemius), reducing friction between the structures. Bursas are small sacs made up of synovial membranes and contain synovial fluid. Many of these structures mentioned above are part of the articular capsule, which serves to stabilize the knee further and contains synovial fluid. Synovial fluid is made of synovial membranes and serves to reduce friction between articular surfaces of the knee.

The structures of the knee receive much of their blood supply from a plexus of arteries with branches from the popliteal artery and femoral artery. The popliteal artery branches off the superficial femoral artery and runs posteriorly across the knee joint. The superior medial, inferior medial, superior lateral, and inferior lateral genicular arteries branch off the popliteal artery and travel anteriorly to anastomose with other parts of the plexus. Also, a descending genicular artery branches off the superficial femoral artery and anastomoses anteriorly with the other genicular arteries. The anterior and posterior tibial recurrent arteries travel laterally from the anterior tibial artery and also contribute to the plexus. The middle genicular artery travels directly into the joint. In addition, the sural arteries branch off the popliteal artery and travel inferiorly away from the midline. The flexor muscles are supplied mainly by the inferior gluteal, perforating, popliteal, deep femoral, and sural arteries. The femoral artery primarily supplies the extensor muscles.

A great part of the lymphatic drainage from the knee and lower leg drains to the popliteal lymph nodes, located in the popliteal fossa, which in turn drain into the deep inguinal and subinguinal nodes. The lymphatic system primarily follows the vasculature.

The structures of the knee and most of the flexor muscles receive innervation from branches of the femoral nerve (L1, L2, L3). The extensor muscles receive innervation from the sciatic nerve (L4, L5, S1, S2, S3), which branches into the tibial nerve and common peroneal nerve.

References

1. Hirschmann MT, Müller W. Complex function of the knee joint: the current understanding of the knee. Knee Surg Sports Traumatol Arthrosc. 2015;23:2780–8.
2. Flandry F, Hommel G. Normal anatomy and biomechanics of the knee. Sports Med Arthrosc Rev. 2011;19:82–92.
3. Blankevoort L, Huiskes HWJ, de Lange A. The envelope of passive knee joint motion. J Biomech. 1988;21(9):705–20.
4. Lafortune M, Cavanagh PR, Sommer H, Kalenak A. 3-dimensional kinematics of the human knee during walking. J Biomech. 1992;25(4):347–57.
5. Tsukada H, Ishibashi Y, Tsuda E, Fukuda A, Toh S. Anatomical analysis of the anterior cruciate

ligament femoral and tibial footprints. J Orthop Sci. 2008;13:122–9.
6. Palastanga N, Field D, Soames R. Anatomy and human movement, structure and function. 4th ed. Oxford: Butterworth and Heinemann.
7. Beasley LS, Weiland DE, Vidal AF, Chhabra A, Herzka AS, Feng MT, West RV. Anterior cruciate ligament reconstruction: A literature review of the anatomy, biomechanics, surgical considerations, and clinical outcomes. Operative Techniques Orthop. 2005;15(1):5–19.
8. Zantop T, Petersen W, Sekiya JK, Musahl V, Fu FH. Ante- rior cruciate ligament anatomy and function relating to ana- tomical reconstruction. Knee Surg Sports Traumatol Arthrosc. 2006;14:982–92.
9. Smigielski R, Zdanowicz U, Drwiega M, Ciszek B, Cisz-kowska-Lyson B, Siebold R. Ribbon like appearance of the midsubstance fibres of the anterior cruciate ligament close to its femoral insertion site: a cadaveric study including 111 knees. Knee Surg Sports Traumatol Arthrosc. 2014; https://doi.org/10.1007/s00167-014-3146-7.
10. Wijdicks CA, Kennedy NI, Goldsmith MT, Devitt BM, Michalski MP, Aroen A, Engebretsen L, LaPrade RF. Kinematic analysis of the posterior cruciate ligament, part 2: a comparison of anatomic single-versus double-bundle reconstruction. Am J Sports Med. 2013;41:2839–48.
11. Mueller W. The knee: form, function, and ligament reconstruction. New York: Springer; 1982.
12. Kapandji IA. Funktionelle Anatomie der Gelenke, vol. 2. Stuttgart: Hippokrates Verlag GmbH; 1999, Untere Extremität
13. Warren LF, Marshall JL. The supporting structures and layers on the medial side of the knee: an anatomical analysis. J Bone Joint Surg Am. 1979;61(1):56–62.
14. Terry GC. The anatomy of the extensor mechanism. Clin Sports Med. 1989;8(2):163–77.
15. Panagiotopoulos E, Strzelczyk P, Herrmann M, Scuderi G. Cadaveric study on static medial patellar stabilizers: the dynamizing role of the vastus medialis obliquus on medial patellofemoral ligament. Knee Surg Sports Traumatol Arthrosc. 2006;14(1):7–12.
16. Kaleka CC, Aihara LJ, Rodrigues A, de Medeiros SF, de Oliveira VM, de Paula Leite Cury R. Cadaveric study of the secondary medial patellar restraints: Patellotibial and patellomeniscal ligaments. Knee Surg Sports Traumatol Arthrosc. 2017;25(1):144–51.
17. Hinckel BB, Lipinski L, Arendt A. Concepts of the distal medial patellar restraints: medial patellotibial ligament and medial patellomeniscal ligament. Sports Med Arthrosc Rev. 2019;27(4):143–9.
18. Fulkerson JP, Gossling HR. Anatomy of the knee joint lateral retinaculum. Clin Orthop Relat Res. 1980;153:183–8.

Biomechanics of the Tibiofemoral and Tibiofibular Joints

45

Gwenllian Tawy, Alexander Jakubiec, and Leela Biant

45.1 Introduction

The knee is comprised of two joints: the tibiofemoral and patellofemoral joints. The tibiofemoral joint is defined as the articulation between the lateral and medial epicondyles of the femur and the tibial plateau. It has six degrees of freedom, making it a biomechanically complicated joint. The patellofemoral joint is defined as the articulation of the patellar notch on the anterior face of the femur with the posterior face of the patella. The combined roles of both tibiofemoral and patellofemoral joints in human movement are well described in the literature.

A joint which is less understood is the proximal tibiofibular joint. This joint is defined as the articulation between the fibular head and posterolateral surface of the tibia. Although not considered part of the knee joint proper, the proximal tibiofibular joint is understood to play an important role in the biomechanics and stability of the knee [1, 2].

This chapter outlines the biomechanics of the tibiofemoral and proximal tibiofibular joints to provide the reader with a basic understanding of their roles in knee motion. Joint kinematics and kinetics will be described in this chapter as per the following definitions:

(a) Translational kinematics are defined here as movements along a certain plane caused by a force whose action acts on the centre of gravity of the body. Translations are measured in millimetres (mm).
(b) Angular kinematics are defined here as movements along a certain plane caused by a force whose action does not pass through the centre of gravity of the body, resulting in rotation of the body. Rotations are measured in angles (°).
(c) Kinetics are defined as the forces which are involved in generating movement. Here, we refer to the moment of the force (torque), which is the application of a force at a perpendicular distance between the force's line of action and the point of rotation. Moments or torques are measured in Newton meters (Nm).

F. Tawy
Division of Cell Matrix Biology and Regenerative Medicine, University of Manchester, Manchester, UK
e-mail: gwenllian.tawy@manchester.ac.uk; Gwenllian.Tawy@mft.nhs.uk

A. Jakubiec · L. Biant (✉)
Manchester University NHS Foundation Trust, Manchester, UK
e-mail: alexander.jakubiec@postgrad.manchester.ac.uk; leela.biant@manchester.ac.uk

J. G. Lane et al. (eds.), *The Art of the Musculoskeletal Physical Exam*,
https://doi.org/10.1007/978-3-031-24404-9_45

45.2 Tibiofemoral Joint

Biomechanical research on the tibiofemoral joint (TFJ) shows that it is important for distributing external loading forces, providing stability during locomotion, and facilitating activities of daily living (ADL). Simplified, the TFJ can be described at a hinge joint of the sagittal plane; however, this oversimplification neglects the nuanced and complex mechanical phenomena that occur during flexion and extension of the knee. As such, a more accurate depiction is that the TFJ is capable of movement within 6 degrees of freedom [3]. The complexity of the TFJ is in part due to the simultaneous interaction between the structures that provide static (knee geometry/anatomy), active (muscle contractions) and passive stability (ligaments, menisci and retinacula) [4].

45.3 Angular Kinematics

45.3.1 Flexion-Extension (Sagittal Plane)

Whilst 0° indicates full extension of the TFJ, hyperextension of the knee by up to 5° is not uncommon [5]. Maximum knee flexion can vary between populations. On average, Caucasians exhibit approximately 140° of knee flexion, whereas individuals from Japan, India and the Middle East may exhibit up to 160° of knee flexion. This difference is likely due to the increased use of kneeling and squatting within Eastern cultures [6]. However, full range of motion in the sagittal plane is not needed for most ADL, with 100–110° of flexion being sufficient for walking, stair or slope ascent and descent, and standing up from a chair [7].

Freeman and Pinskerova divided knee flexion and extension into three sub-arcs: (1) the arc of terminal extension (full extension (−5° to 0°)–10°), indicating the limit of passive extension; (2) the arc of active function (10°–120°), in which ADL are readily achieved; and (3) the arc of passive flexion (120°–160°), whereby flexion is passive and is not supported by the neuromuscular system [6, 8]. It is useful to differentiate between these different arcs in order to understand when other kinematic phenomena, such as multiplanar rotations and translations, simultaneously occur during knee flexion and extension.

45.3.2 Valgus-Varus (Coronal/Frontal Plane)

Minimal coronal plane rotation occurs within the TFJ during loaded and unloaded conditions, with up to 2.8° of knee varus and 3.0° of knee valgus being reported in healthy knees during level gait [9]. Varus rotation of the knee occurs as a result of lateral femoral condyle lift-off due to the circular shape of the medial compartment in the coronal plane, and because of anatomical differences in the medial and lateral tibial plateau [6]. Varus rotation has been reported during initial swing phase and valgus during both stance phase and late-swing phase in normal level gait [9]. These findings suggest that varus and valgus rotations of the knee joint are a component of normal gait.

45.3.3 Internal-External (Transverse Plane)

As the medial compartment of the TFJ is more constrained and ball-in-socket-like than the lateral compartment, it can behave as an axial rotator from which the lateral femoral condyle is able to rotate around [6]. During the arc of terminal extension, sagittal plane rotation is heavily associated with the screw-home mechanism. This mechanism refers to coupled knee extension and external rotation of the tibia, whereby approximately 9° of external tibial rotation is reported [4]. The screw-home mechanism is thought to increase congruency of the TFJ and increase knee stability [10]. During the arc of active function, transverse and sagittal plane rotations can be independent from one another. Up to 30° of tibial transverse rotation has been shown to occur without simultaneous sagittal rotation, as a result of anterior-posterior (AP) translation of the lateral femoral condyle [6, 11].

45.4 Translational Kinematics

45.4.1 Anterior-Posterior (Sagittal Plane)

The concave anatomy of the medial tibial plateau and the gradient of the slope of the medial meniscus provide static and passive stability to the knee joint. These anatomical structures prevent the joint from translating excessively in the AP direction during loaded conditions [12, 13]. AP translations of up to 4 mm can be expected in the medial compartment of the TFJ [12, 13]. In contrast, the lateral tibial plateau is flat or convex, and the gradient of the slopes of its meniscal horns is less steep than the medial meniscus. Consequently, AP translations of 13–25 mm can be expected during loaded conditions in the lateral compartment of the TFJ [11, 14]. Due to the smaller articulating surface of the lateral tibial plateau, the anterior cruciate ligament (ACL) causes anterior translation of the lateral femoral condyle during the arc of active function in order to prevent subluxation of the femoral condyles off the tibial plateau during knee flexion [15]. This mechanism allows for greater posterior rollback of the femur, allowing for improved extensor mechanism function.

45.4.2 Medio-Lateral (Coronal/Frontal Plane)

Medio-lateral (ML) translation of the TFJ is seldom reported in the literature, implying that ML translation is less important to knee function than other translations. Lu and colleagues reported that the TFJ undergoes minimal medio-lateral (ML) translation [16]. When unloaded, the TFJ is reported to translate in the coronal plane by up to 5.0 ± 3.4 mm. This is increased to 5.8 ± 2.4 mm during loaded conditions [16].

45.4.3 Compression-Distraction (Transverse Plane)

Research suggests that translation of the TFJ in the transverse plane during active knee flexion is minimal. According to Shenoy et al. and Lafotune et al., the TFJ compresses and distracts by 2–5 mm when unloaded and by 3–7 mm during loaded conditions [5, 17]. The medial compartment of the TFJ has also been reported to distract by 2 mm during the arc of passive flexion. Caused by rollback of the medial femoral condyle, displacement of the medial femoral condyle onto the posterior horn of the medial meniscus distracts the joint, losing the cartilage-to-cartilage contact between the femoral condyles and the tibial plateau [6, 18]. Rollback of the lateral femoral condyle results in both the lateral femoral condyle and the posterior meniscal horn to translate 1.9 mm over the posterior border of the tibia [18].

45.5 Kinetics

45.5.1 Static Kinetics

Loading of the TFJ is split across the medial and lateral compartments of the joint. Joint loading is believed to be equally shared in a healthy joint, with the equilibrium being maintained by the mechanical alignment of the lower limb and the distribution of external forces by the meniscus [4, 19]. Given that mechanical alignment, measured by the hip-knee-ankle (HKA) angle, alters the moment arm, knees with greater HKA angles naturally yield higher moments and thus exhibit higher compartmental loading of the tibiofemoral joint [20]. Abnormal compartmental loading has been found to alter knee kinetics during dynamic movements and strongly correlates with the onset of osteoarthritis [20]. However, the clinical implication of increased compartmental loading is likely to be more severe for patients who exhibit knee varus rather than knee valgus [21].

45.5.2 Dynamic Kinetics

Due to the alignment and structure of the TFJ, total load is not distributed evenly across the medial and lateral compartments during dynamic activities. The TFJ exhibits two peaks of medial compartmental loading, 2.37 times body weight

(BW) of load at initial contact, and 1.80 times BW during the second half of stance phase during gait [9]. However, the lateral compartment exhibits only one peak, 1.30 times BW at initial contact. Loading through the lateral compartment then gradually decreases in the second half of stance phase, yielding maximum values of only 0.45 times BW [9]. Zhao and colleagues reported an approximated load split of 55% medial and 45% lateral to be reasonable for the approximation of TFJ joint loading during dynamic tasks [22].

45.6 Proximal Tibiofibular Joint

The proximal tibiofibular joint (PTFJ) dissipates torsional forces applied to the ankle, dissipates lateral bending moments of the tibia, and bears tensile forces under weight-bearing conditions [1, 23, 24]. However, the variable bony anatomy of the fibular head implies that the way in which the joint behaves *in vivo* to fulfil these biomechanical roles may vary considerably across a population [1, 2]. Its biomechanics is also heavily influenced by the kinematics of the tibiofemoral and ankle joints [25, 26].

45.7 Translational Kinematics

45.7.1 Anterior-Posterior (Sagittal Plane)

Recent cadaveric research has been used to model the effect of loading the tibiofemoral joint on the biomechanics of the PTFJ [1]. The model was designed to mimic loading conditions during stance phase of gait and stair climbing. According to the research, greatest anterior translation was observed when a force of 10 Nm was applied to the knee to externally rotate and adduct the joint (1.51 ± 0.66 mm). Maximal posterior translation of the PTFJ was observed under opposing conditions (1.63 ± 1.60 mm). These simultaneous translations and angular rotations are a result of the joint not being orientated in the true sagittal plane [2].

Translation of the PTFJ in the sagittal plane is also influenced by the angular kinematics of the knee joint [1, 24]. Extension of the knee joint causes the ligaments and tendons, which insert into the fibular head to become taut, resulting in posterior translation of the joint [26]. Flexion of the knee joint beyond 30° loosens these tissues, causing anterior translation of the PTFJ [2, 24]. However, the extent by which the joint translates is also dependent on distal movements at the ankle and foot. According to Soavi et al., the PTFJ translates anteriorly during dorsiflexion by 0.4 mm and posteriorly during plantarflexion by 0.6 mm [23].

Sagittal plane translations of up to 3 mm can therefore be expected in the PTFJ during common activities of daily living [1].

45.7.2 Medio-Lateral (Coronal/Frontal Plane)

Small medio-lateral translations of the PTFJ have been observed during dorsi- and plantar flexion [23]. The joint translates laterally during dorsiflexion (with simultaneous anterior translation) and medially during plantar flexion (with simultaneous posterior translation) [23]. Cadaveric research has found the PTFJ to translate both medially and laterally by 0.9 mm [23].

45.7.3 Compression-Distraction (Transverse Plane)

Scott and colleagues simulated transverse plane translation of the joint under conditions which would mimic walking (15° knee flexion) and stair climbing (45° knee flexion). 250 N was included in the model to account for compressive loading. The PTFJ compressed by up to 0.18 ± 0.16 mm when the knee was externally rotated and distracted by up to 0.31 ± 0.34 mm when the knee was internally rotated [1].

Small translations in the transverse plane have also been reported at the joint as a result of ankle movement. According to Soavi and colleagues, the joint distracts in plantar flexion and compresses in dorsiflexion (−0.1 mm to +0.3 mm) [23].

These consistent findings suggest that compression-distraction of the PTFJ is minimal, even under dynamic conditions.

45.8 Angular Kinematics

45.8.1 Flexion-Extension (Sagittal Plane) and Valgus-Varus (Coronal/Frontal Plane) of the PTFJ

Rotation of the PTFJ in the sagittal and coronal planes is reportedly inconsistent and negligible, although research to date has only studied this variable with respect to dorsi- and plantar flexion of the ankle (reported range of motion sagittal plane: −0.5 to +0.4°, range of motion coronal plane: −0.4 to +0.5°) [23].

45.8.2 Internal-External (Transverse Plane)

The PTFJ has been shown to externally rotate during dorsiflexion of the ankle [1]. This is necessary to allow for rotation of the talus and thus full range of motion at the ankle [2, 27]. Ranges of motion of −2.8 to +2.5° have been reported at the PTFJ [23]. However, the extent of rotation observed is dependent on the bony anatomy of the fibular head. This is clinically important as fibular heads with morphologies which limit rotation of the PTFJ may be at greater risk of dislocation or subluxation [2].

45.9 Kinetics

45.9.1 Static Kinetics

Static kinetics of the PTFJ remain poorly researched. According to Lambert, one-sixth of the static force generated by the fibula's articulation with the talus is transmitted to the PTFJ [28]. This suggests that the PTFJ plays an important role in dissipating forces when weight-bearing [28]. However, no further research has since been carried out to corroborate these findings. The PTFJ's role in weight-bearing thus remains unclear.

45.9.2 Dynamic Kinetics

Although contraction of the fibularis longus and fibularis brevis muscles is known to lead to movement at the PTFJ, our current understanding of the joint's kinetics relies mainly on research that has focused on the neighbouring joints [29]. When the tibia externally rotates, tensile force is transmitted to the fibular head via the fibular collateral ligament to anteriorly translate the fibula. During internal rotation of the tibia, the force vector is reversed, causing posterior translation of the fibula [1]. The load on the fibular collateral ligament during tibial rotation is reportedly highest when the knee is flexed by 0–30° [30].

Excess movement of the PTFJ is prevented by the joint's ligaments [24]. The anterior proximal tibiofibular ligament stabilises the joint during ankle plantar flexion and foot eversion [25]. Translations and rotations at the joint as a result of dorsiflexion, foot inversion and knee flexion are stabilised by the posterior proximal tibiofibular ligament [25]. The ligaments of the PTFJ have been reported to withstand a combined failure load of 517 ± 144 N anteriorly and 322 ± 160 N posteriorly [31]. This is important to consider when planning graft reconstructions of these structures [25].

45.10 Summary

Whilst the amount of research on the different planes can vary, biomechanical research on the tibiofemoral joint is extensive. The majority of the research focuses on sagittal and coronal plane rotations and translations of the TFJ; however, other studies have aimed to investigate the more nuanced movements that occur in the knee. The literature highlights the complexity of the TFJ and the different mechanisms that occur simultaneously. It is because of this that it may be easier to utilise Freeman and Pinskerova's subarcs to differentiate between different phases of flexion and extension [6]. The research also highlights the discrepancies between the medial and lateral TFJ compartments, and as such must be distinguished somewhat separately when discussing knee biomechanics.

Contrary to the TFJ, biomechanical research on the proximal tibiofibular joint is scarce and limited almost exclusively to in vitro cadaveric studies. Although movement at the ankle and knee has been reported to result in motion at the PTFJ, the degree of movement is small [23]. These ranges of motion may be greater in vivo; however, significant differences are unlikely due to the strong ligamentous support, which encapsulates the proximal head of the fibula. Further research is needed to better understand the function and biomechanics of this joint.

References

1. Scott J, et al. The effect of tibiofemoral loading on proximal tibiofibular joint motion. J Anat. 2007;211(5):647–53.
2. Semonian RH, Denlinger PM, Duggan RJ. Proximal tibiofibular subluxation relationship to lateral knee pain: a review of proximal tibiofibular joint pathologies. J Orthop Sports Phys Ther. 1995;21(5):248–57.
3. Wu G, Cavanagh PR. ISB recommendations for standardization in the reporting of kinematic data. J Biomech. 1995;28(10):1257–61.
4. Masouros SD, Bull AMJ, Amis AA. (i) Biomechanics of the knee joint. Orthopaedics and Trauma. 2010;24(2):84–91.
5. Shenoy R, Pastides PS, Nathwani D. (iii) Biomechanics of the knee and TKR. Orthopaedics and Trauma. 2013;27(6):364–71.
6. Freeman MAR, Pinskerova V. The movement of the normal tibio-femoral joint. J Biomech. 2005;38(2):197–208.
7. Rowe PJ, et al. Knee joint kinematics in gait and other functional activities measured using flexible electrogoniometry: how much knee motion is sufficient for normal daily life? Gait Posture. 2000;12(2):143–55.
8. Marchetti PH, et al. Muscle activation differs between three different knee joint-angle positions during a maximal isometric Back squat exercise. J Sports Med. 2016;2016:3846123.
9. Kumar D, Manal KT, Rudolph KS. Knee joint loading during gait in healthy controls and individuals with knee osteoarthritis. Osteoarthr Cartil. 2013;21(2):298–305.
10. Kim HY, et al. Screw-home movement of the tibiofemoral joint during Normal gait: three-dimensional analysis. Clin Orthop Surg. 2015;7(3):303–9.
11. Pinskerova V, et al. Does the femur roll-back with flexion? J Bone Joint Surg Br. 2004;86(6):925–31.
12. Hill PF, et al. Tibiofemoral movement 2: the loaded and unloaded living knee studied by MRI. J Bone Joint Surg Br. 2000;82(8):1196–8.
13. Reynolds RJ, Walker PS, Buza J. Mechanisms of anterior-posterior stability of the knee joint under load-bearing. J Biomech. 2017;57:39–45.
14. Dennis DA, et al. In vivo determination of normal and anterior cruciate ligament-deficient knee kinematics. J Biomech. 2005;38(2):241–53.
15. O'Keeffe M, et al. An evaluation of the use of opening-wedge osteotomy about the knee. Ir J Med Sci. 2011;180:S261.
16. Lu TW, et al. In vivo three-dimensional kinematics of the normal knee during active extension under unloaded and loaded conditions using single-plane fluoroscopy. Med Eng Phys. 2008;30(8):1004–12.
17. Lafortune MA, et al. Three-dimensional kinematics of the human knee during walking. J Biomech. 1992;25(4):347–57.
18. Johal P, et al. Tibio-femoral movement in the living knee. A study of weight bearing and non-weight bearing knee kinematics using 'interventional' MRI. J Biomech. 2005;38(2):269–76.
19. Johnson F, Leitl S, Waugh W. The distribution of load across the knee. A comparison of static and dynamic measurements. J Bone Joint Surg Br. 1980;62(3):346–9.
20. Nie Y, et al. The relationship between knee adduction moment and knee osteoarthritis symptoms according to static alignment and pelvic drop. Biomed Res Int. 2019;2019:7603249.
21. Fregly BJ, D'Lima DD, Colwell CW Jr. Effective gait patterns for offloading the medial compartment of the knee. J Orthop Res. 2009;27(8):1016–21.
22. Zhao D, et al. In vivo medial and lateral tibial loads during dynamic and high flexion activities. J Orthop Res. 2007;25(5):593–602.
23. Soavi R, et al. The mobility of the proximal tibiofibular joint. A roentgen Stereophotogrammetric analysis on six cadaver specimens. Foot Ankle Int. 2000;21(4):336–42.
24. Ogden JA. The anatomy and function of the proximal tibiofibular joint. Clin Orthop Relat Res. 1974;101:186–91.
25. Pessoa P, Alves-da-Silva T, Guerra-Pinto F. Repair with bicortical suspension device restores proximal tibiofibular joint motion. Knee Surg Sports Traumatol Arthrosc. 2019;27(2):412–8.
26. Alves-da-Silva T, et al. Kinematics of the proximal tibiofibular joint is influenced by ligament integrity, knee and ankle mobility: an exploratory cadaver study. Knee Surg Sports Traumatol Arthrosc. 2019;27(2):405–11.

27. Burkhart TA, et al. Anatomy of the proximal tibiofibular joint and interosseous membrane, and their contributions to joint kinematics in below-knee amputations. J Anat. 2015;226(2):143–9.
28. Lambert KL. The weight-bearing function of the fibula. A strain gauge study. J Bone Joint Surg Am. 1971;53(3):507–13.
29. Espregueira-Mendes JD, da Silva MV. Anatomy of the proximal tibiofibular joint. Knee Surg Sports Traumatol Arthrosc. 2006;14(3):241–9.
30. LaPrade RF, Tso A, Wentorf FA. Force measurements on the fibular collateral ligament, popliteofibular ligament, and popliteus tendon to applied loads. Am J Sports Med. 2004;32(7):1695–701.
31. Marchetti DC, et al. The proximal tibiofibular joint: a biomechanical analysis of the anterior and posterior ligamentous complexes. Am J Sports Med. 2017;45(8):1888–92.

46 Evaluation of Range of Motion of the Tibiofemoral Joint

Laura Ann Lambert and Mike McNicholas

46.1 Visual Estimation of Knee Range of Movement (KROM)

The knee can be described as a hinge joint with the main movement of flexion and extension, but we know this to be an oversimplification. The tibia unlocks at the beginning of flexion with external rotation of the joint. This is followed by lateral femoral condylar rollback and medial femoral condylar rotation [1–3]. If the tibiofemoral surfaces of the knee joint were of the same length and radii, evaluating the geometry of its motion would be very straightforward. However, both articulating surfaces have differing radii and lengths [4]. This results in both the sliding and rolling of the extended tibiofemoral joint culminating in external rotation of approximately 15 degrees at terminal extension [5, 6].

The normal flexion-extension range of movement for the knee is 0–140°. Most clinical tests assess this movement alone, as knee abduction–adduction and axial rotation movements are small and not appreciable with visual estimation [7]. The normal range for extension is 0° to −5° in the hypermobile female [8]. Hyperextension is difficult to appreciate in the supine position and is better measured when prone. Deep flexion is 130–140°. A typical examination of KROM consists of positioning the patient semi-recumbent on an examination table with the pelvis square and both legs extended. In a study by Peters et al. comparing visual estimation of knee range of motion to hand goniometry and radiographic measurements, the following clinical technique was used:

- To gauge full extension or hyperextension, the examiner places a hand above the knee. The contralateral hand cups the ipsilateral heel to lift it off the table until resistance is felt (Fig. 46.1a).
- The patient actively flexes their knee.
- With one hand, the examiner stabilises the thigh. The contralateral hand is placed to the front of the ankle with gentle pressure until an end point is reached. This determines maximal flexion (Fig. 46.1b).

This study noted that visual estimation had high intra-relater reliability (ICC 0.8), consistent with other literature on this method [9, 10] (a satisfactory ICC is generally accepted as >0.70, and excellent >0.90). However, there were some notable differences between visual estimation and other measurement methods. On average, goniometric measurement was 6° less than radio-

L. A. Lambert
The University of Manchester, Manchester, UK
e-mail: lauraann.lambert@postgrad.manchester.ac.uk

M. McNicholas (✉)
Division of Cell Matrix Biology & Regenerative Medicine, Medicine and Health, School of Biological Sciences, University of Manchester, Manchester, UK
e-mail: mike@mcnicholaskneeclinic.co.uk

J. G. Lane et al. (eds.), *The Art of the Musculoskeletal Physical Exam*,
https://doi.org/10.1007/978-3-031-24404-9_46

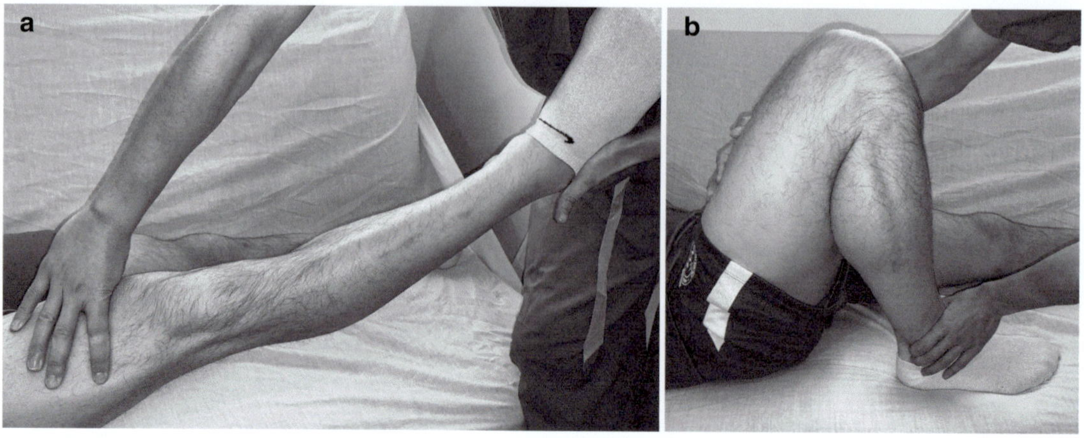

Fig. 46.1 (**a**) Positioning of the lower limb for assessment of knee extension (Courtesy of Dublin Hospitals Football Club). (**b**) Positioning of the lower limb for assessment of knee flexion (Courtesy of Dublin Hospitals Football Club)

graphic measurement and 8° less than visual estimation. They proposed 'is supine flexion of the knee synonymous with true full flexion?' as a potential explainer for some of this variation and advocated for not mixing methods [11]. Questions such as these are important as measurement of KROM has ramifications for gait and function. KROM is incorporated into orthopaedic knee scoring tools to assess disease severity and recovery after arthroplasty and other knee surgeries and is frequently used as a benchmark in physiotherapy to assess progress with rehabilitation. Surgeons will typically visually estimate KROM in clinic. However, patients generally see different doctors within most public healthcare settings, and this method is the least accurate between different observers [10].

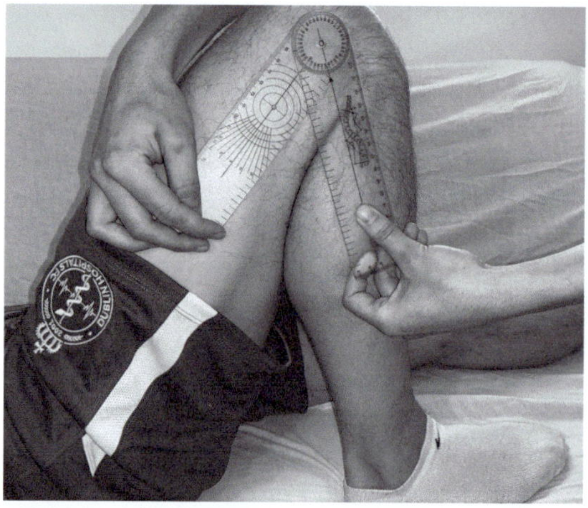

Fig. 46.2 Short-arm goniometer with proximal limb centred on the greater trochanter and distal limb towards the lateral malleolus (Courtesy of Dublin Hospitals Football Club)

46.2 Universal Goniometer (UG)

Goniometry is the measurement of the range of movement of a joint through the use of instruments. There are many instruments and techniques, the most common of which is the universal goniometer (UG). In its most basic form, the industry standard long-arm (50 cm) goniometer or short-arm (30 cm) goniometer gives a quick, gross measurement of static angles [12]. For assessment of the knee, the goniometer is placed with the proximal arm pointing towards the greater trochanter and the distal arm towards the lateral malleolus (Fig. 46.2) [13]. Measurement accuracy is contingent on the alignment of the device arms between bony landmarks [14]. In patients with a bigger soft-tissue envelope, finding the bony landmarks may be difficult and their position can change when cycling through flexion and extension [15]. While the UG does not provide information about dynamic movement, it is widely available, is simple to use and, in experienced hands, has good intra-observer reliability.

As time moves on, the digitised goniometer looks set to replace the manual goniometer at least in research fields, and perhaps, in the not-too-distant future, in clinical practice.

46.3 Electrical Digital Inclinometers (EDI)

An electrical digital inclinometer is a device that is affixed to bony landmarks and interprets the movement of the knee using the same technology that determines the position in mobile phones, car airbags and aircraft [16]. An accelerometer in the device monitors the effect of gravity on a tiny mass held within an elastic support structure. When the EDI tilts, the suspended mass moves slightly, causing a change in capacitance. The tilt angle is calculated from the measured capacitances. Several EDIs are available, including the Cybex EDI 320 (New York, USA), HALO Digital inclinometer (New South Wales, Australia) (Fig. 46.3) and the Limit Mini Digital Inclinometer (Alingsås, Sweden). When purchasing any equipment or adjunct to aid in clinical practice, the primary question is what does the device add and does it improve on the industry standard? The typical cost of a UG is several pounds (£) compared to several hundred for an EDI. Digital measurements have been reported as having similar validity and reliability as traditional goniometry measurements [17]. Hancock et al. report that an EDI has the smallest minimum significant difference, concluding that it is the most accurate compared to other standard measurements [10]. They did not compare to digital photographic goniometry, another modern modality of measurement in KROM.

Fig. 46.3 HALO Digital inclinometer (New South Wales, Australia). (Reproduced with permission from www.sportsphysio.ie)

46.4 Digital Photographic Goniometry (DPG)

Digital photographic goniometry has appeared in the literature on knee kinematic measurement as a viable means of measuring KROM over the last 10 years [15, 18, 19].

Recording and measuring knee joint motion using digital imaging were first described by Beverland et al. in 2009. High inter-observer (r > 0.948) and intra-observer repeatability (r > 0.906) was demonstrated in ten patients by two observers. The equipment needed was simply a digital camera and image analysis software (Rhinoceros, Seattle, USA).

The software available for interpreting KROM in DPG can account for variables such as camera lens quality and parallax errors [19]. The main benefit of DPG is that the digital images taken allow for further measurements by a different investigator at a later date and they can be rechecked for reproducibility. Even when the bony landmarks are not overtly identified in the image, the inter-rater reliability remains high [18]. The availability of such serial imaging may serve as a visual cue for the patient of their progress during rehabilitation and may even motivate them to engage in targeted improvements [20]. Use of a designated digital camera and separate software is cumbersome in an age where we strive for technology to work seamlessly across platforms. Smartphone applications look to fill this void by offering the ability to image and interpret the KROM on the device that most people carry in their pockets. In one such application, a virtual goniometer is placed on the image taken of the desired joint, with superimposed markers indicating the joint position and relationship to the floor. Their use has been validated across different joints, albeit only in healthy participants [21–23]. A variety of applications are available currently, including DrGoniometer (CDM S.r.L, Cagliari, Italy) (Fig. 46.4), Clinometer (Plaincode Software Solutions, Stephanskirchen, Germany) and ROM© goniometric application (Carci©, São

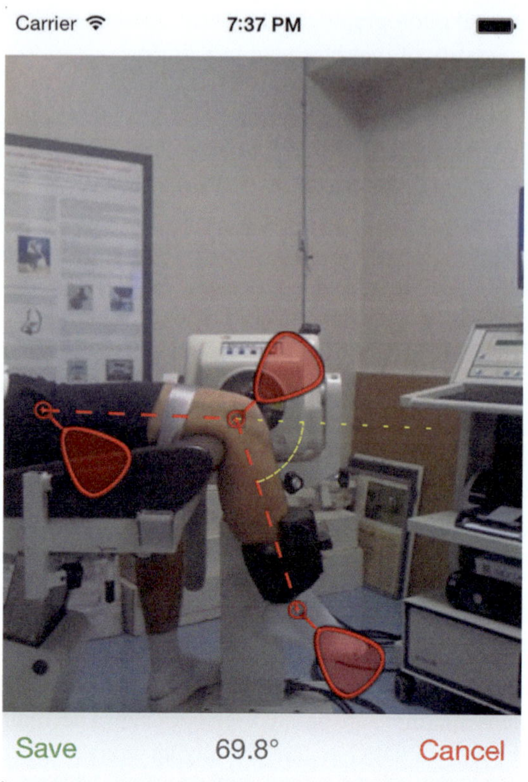

Fig. 46.4 DrGoniometer application interface. (Reproduced with permission from www.drgoniometer.com)

Paulo, Brazil), and it is likely that the market will become saturated with other iterations. In a systematic review by Milani et al., seven different applications were validated for use in lower limb measurements. DrGoniometer is an application that stood out over others for its ability to measure both static and dynamic angles, its potential to blind the rater to the measurement and its telemedicine integration [24].

It is not however available on Android devices for which a viable alternative is the ROM© application. This was used by Dos Santos et al. in healthy female population ($n = 34$) and demonstrated a high degree of correlation ($r \geq 0.90$; $p < 0.0001$) with the universal goniometer and importantly no significant difference in variation between the two methods in any analysis ($p > 0.05$) (Fig. 46.5) [25].

The following techniques in measuring KROM are mostly if not exclusively confined to use in research studies and include fluoroscopy and cross-sectional imaging, radiostereometric analysis and motion capture analysis. These methods lend themselves to an assessment of the nuance of knee movement beyond flexion and extension in the sagittal plane.

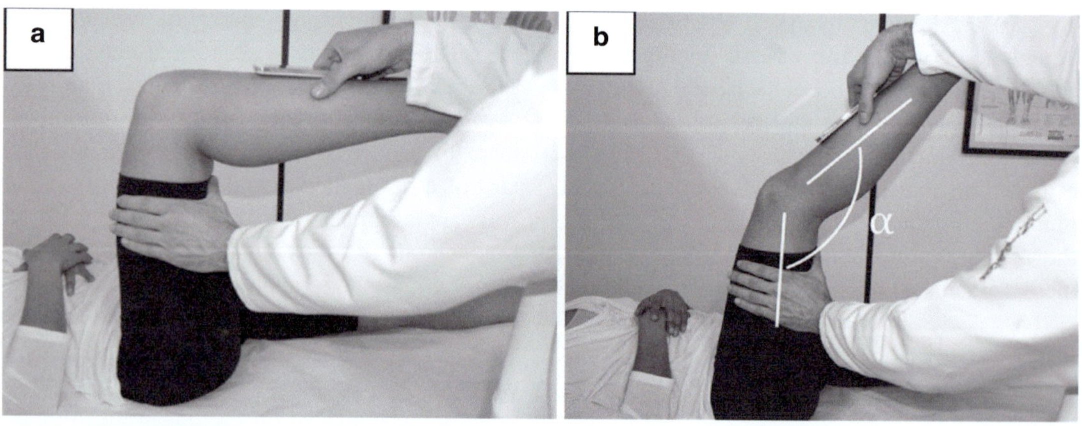

Fig. 46.5 Determination of knee angle with smartphone application; (**a**) initial position (0°); (**b**) final position; α = final joint angle of knee flexion. (Reprinted from the Journal of Bodywork and Movement Therapies, 21:3, Dos Santos et al., Evaluation of knee range of motion: Correlation between measurements using a universal goniometer and a smartphone goniometric application, p 699–703, Copyright (2017), with permission from Elsevier)

46.5 Fluoroscopy and Cross-Sectional Imaging

Single or biplane fluoroscopy measures in vivo joint kinematics using image intensifier(s) and provides in real time anatomic assessment during dynamic activities [26]. Fluoroscopy can be used to match 3-dimensional models from CT or computer-aided design (CAD) imaging of implants to 2-dimensionally acquired fluoroscopic images [27]. The set-up of the C-arm limits the ability of the imaging to only capture a small portion of the gait cycle, and some laboratories have made their systems mobile [28]. Biplane fluoroscopy has been traditionally used in combination with implanting markers into the bone; however, a number of studies have reported non-invasive model-based tracking techniques that provide submillimetre accuracy [29, 30].

Cross-sectional imaging techniques such as ultra-fast cine computed tomography or kinematic MRI studies can be used to measure KROM. Both modalities when first described in the late 1980s were focused on the evaluation of patellofemoral movement, mainly patellar tracking and subluxation [31, 32].

The first iterations of ultra-fast cine CT consisted of sequential static slices at different angles, and the first truly dynamic kinematic CT protocol was described by Elias et al. (2014) using a 256-multi-detector CT (MDCT) [33]. Compared to 64-MDCT, 256-MDCT is far superior in the acquisition of dynamic images [34]. Protocols are designed to minimise radiation exposure but again are focused on patellar tracking analysis rather than KROM. Kinematic MRI avoids exposure to ionising radiation and details the anatomy of bone and soft tissue in both static and dynamic positions. Dynamic MRI evolved from its initial cardiac applications of blood flow and valvular motion [35] to the measurement of joint movement. Conventional MR imaging is both non-weight bearing and static. Dynamic MRI can be divided into cine PC-MRI and real-time MRI. Cine PC-MRI has been used to measure tibiofemoral kinematics and to visualise cartilage contact during movement by Kaiser et al. In their study, external tibial rotation and anterior tibial translation of the knee were evident from extension to approximately 40° of flexion [36]. A significant rate-limiting step is that cine PC-MRI requires repeated repetitions of the movement cycle. Real-time MRI needs only one motion cycle and is preferable for patients who would be unable to participate in repeated movements due to pain or fatigue [37]. A lack of standardisation in musculoskeletal protocols including optimal acquisition time, field strength parameters, and patient and radiofrequency coil position limits the utility of this modality in standard clinical practice [38].

46.6 Radiostereometric Analysis (RSA)

Radiostereometric analysis (RSA) is a technique used to predict long-term prosthesis stability by studying its early behaviour. Traditionally, RSA involves the invasive implantation of tantalum beads into a joint at the time of arthroplasty, and subsequently the position of these beads can be evaluated by X-ray images. So how can this technique be applied to measure the ROM of the tibiofemoral joint? There are two obvious problems here: the invasiveness of implanting a foreign body, particularly in the setting where one wants to measure native rather than prosthetic knee joint, and the use of static imaging. Invasive insertion of any device to measure KROM, from tantalum beads to cortical pins, carries with them the inherent risk of infection [39]. In order to address this, a model-based RSA method was introduced in the early 2000s [40, 41]. This allows prosthesis or native knee tracking without the use of markers. This method has been validated by Stentz-Olesen et al., who measured the mean difference between the model method and the marker method for knee movement recorded by static and dynamic radiographs in a cadaveric study. They found that submillimetres of precision are lost compared to standard RSA; however (notwithstanding the added radiation of the CT image), the bone model has the potential to be developed as a clinical tool for measuring KROM [42].

46.7 Motion Capture Analysis

Modern non-invasive motion capture systems (MoCap) employ skin markers or virtual markers and video or optical sensors to capture trial data from an individual as they move during gait [43] or specific tasks such as squatting [44] or climbing stairs [45]. Indirect measures are taken from these systems to interpret the kinetics and kinematics of the upper or lower limbs. The main purpose of this approach is to determine the six degrees of freedom of different joint kinematics during activities of daily living and yield a more in-depth understanding of KROM than simple flexion and extension. Such systems use anatomical landmarks and functional models to resolve joint centres and axes, from which the range of motion of the joint can be ascertained [46]. The positions of the markers in space are determined using stereophotogrammetry, which requires a minimum of two cameras. In marker-based MoCap, there are issues with soft-tissue artefact, as the markers do not rigidly stay in place over bony landmarks but are mobile due to muscle and skin movement [47]. It is this artefact which renders MoCap less accurate than methods such as biplane fluoroscopy. The coordinate data from the markers is sent to either a commercial programme such as Vicon Plug In Gait (Oxford Metrics Limited, UK) or a bespoke programme to interpret the variables produced by standard coding software, e.g. MATLAB (MathWorks, USA). MoCap studies can determine flexion-extension angles, abduction-adduction angles and internal and external rotation angles. These studies have yielded some useful information, such as continued external rotation of the tibia during stair ascension [48, 49]. In other activities, e.g. squatting, there is no agreement in the literature with respect to abduction or adduction of the femur relative to the tibia [44, 50, 51]. Looking at these studies, the conditions under which MoCap is performed are variable, including some early post-operative patients [44] and stairs of different slopes and heights. A greater number of higher powered studies using standardised conditions and patient cohorts are necessary to make the findings generalisable to a normal population in the future.

46.8 Conclusion

At its simplest, tibiofemoral motion is measured in day-to-day clinical practice in the sagittal plane. There are devices and applications available, which improve on the universal goniometer and allow record-keeping for posterity and future treatment. Tibiofibular joint motion as well as other tibiofemoral movements (medial and lateral translational and knee abduction-adduction) are not routinely factored into consideration. Six degrees of freedom models consider knee movement in the sagittal, coronal and transverse planes. We have shown that this requires sophisticated and potentially costly equipment, such as fluoroscopy, cross-sectional imaging or motion capture analysis. These technologies are not routinely available and necessarily merited for everyday evaluation, but certainly have their place in a specialist gait laboratory for complex knee pathology.

References

1. Last RJ. The popliteus muscle and the lateral meniscus. J Bone Joint Surg. 1950;32-B(1):93–9.
2. Ahrens P, Kirchhoff C, Fischer F, Heinrich P, Eisenhart-Rothe R, Hinterwimmer S, et al. A novel tool for objective assessment of femorotibial rotation: a cadaver study. Int Orthop. 2011;35(11):1611–20.
3. Freeman MA, Pinskerova V. The movement of the knee studied by magnetic resonance imaging. Clin Orthop Relat Res. 2003;410:35–43.
4. Kim HY, Kim KJ, Yang DS, Jeung SW, Choi HG, Choy WS. Screw-home movement of the tibiofemoral joint during Normal gait: three-dimensional analysis. Clin Orthop Surg. 2015;7(3):303–9.
5. Kosel J, Giouroudi I, Scheffer C, Dillon E, Erasmus P. Anatomical study of the radius and center of curvature of the distal femoral condyle. J Biomech Eng. 2010;132(9):091002.
6. Bytyqi D, Shabani B, Lustig S, Cheze L, Karahoda Gjurgjeala N, Neyret P. Gait knee kinematic alterations in medial osteoarthritis: three dimensional assessment. Int Orthop. 2014;38(6):1191–8.
7. Desloovere K, Wong P, Swings L, Callewaert B, Vandenneucker H, Leardini A. Range of motion and repeatability of knee kinematics for 11 clinically relevant motor tasks. Gait Posture. 2010;32(4):597–602.
8. Norkin CC, White DJ, Sawant A. Measurement of joint motion: a guide to goniometry, Third Edition. Physiother Can. 2004;56:250.

9. Watkins MA, Riddle DL, Lamb RL, Personius WJ. Reliability of goniometric measurements and visual estimates of knee range of motion obtained in a clinical setting. Phys Ther. 1991;71(2):90–6. discussion 6–7.
10. Hancock GE, Hepworth T, Wembridge K. Accuracy and reliability of knee goniometry methods. J Exp Orthop. 2018;5(1):46.
11. Peters PG, Herbenick MA, Anloague PA, Markert RJ, Rubino LJ 3rd. Knee range of motion: reliability and agreement of 3 measurement methods. Am J Orthop (Belle Mead NJ). 2011;40(12):E249–52.
12. Richards J, Chohan A, Erande R. Biomechanics. In: Porter SB, editor. Tidy's physiotherapy. 15th ed. London: Churchill Livingstone; 2013. p. 331–68.
13. Jones A, Sealey R, Crowe M, Gordon S. Concurrent validity and reliability of the simple goniometer iPhone app compared with the universal goniometer. Physiother Theory Pract. 2014;30(7):512–6.
14. Edwards JZ, Greene KA, Davis RS, Kovacik MW, Noe DA, Askew MJ. Measuring flexion in knee arthroplasty patients. J Arthroplast. 2004;19(3):369–72.
15. Bennett D, Hanratty B, Thompson N, Beverland D. Measurement of knee joint motion using digital imaging. Int Orthop. 2009;33(6):1627–31.
16. Christ RD, Wernli RL. Navigational sensors. In: Christ RD, Wernli RL, editors. The ROV manual. 2nd ed. Oxford: Butterworth-Heinemann; 2014. p. 453–75.
17. Tveitå EK, Ekeberg OM, Juel NG, Bautz-Holter E. Range of shoulder motion in patients with adhesive capsulitis; intra-tester reproducibility is acceptable for group comparisons. BMC Musculoskelet Disord. 2008;9:49.
18. Naylor JM, Ko V, Adie S, Gaskin C, Walker R, Harris IA, et al. Validity and reliability of using photography for measuring knee range of motion: a methodological study. BMC Musculoskelet Disord. 2011;12:77.
19. Murphy M, Hides J, Russell T. A digital photographic technique for knee range of motion measurement: performance in a total knee arthroplasty clinical population. Open J Orthopedics. 2013;3:4–9.
20. Blonna D, Zarkadas PC, Fitzsimmons JS, O'Driscoll SW. Validation of a photography-based goniometry method for measuring joint range of motion. J Shoulder Elb Surg. 2012;21(1):29–35.
21. Ferriero G, Vercelli S, Sartorio F, Muñoz Lasa S, Ilieva E, Brigatti E, et al. Reliability of a smartphone-based goniometer for knee joint goniometry. Int J Rehabil Res. 2013;36(2):146–51.
22. Mitchell K, Gutierrez SB, Sutton S, Morton S, Morgenthaler A. Reliability and validity of goniometric iPhone applications for the assessment of active shoulder external rotation. Physiother Theory Pract. 2014;30(7):521–5.
23. Milani P, Coccetta CA, Rabini A, Sciarra T, Massazza G, Ferriero G. Mobile smartphone applications for body position measurement in rehabilitation: a review of goniometric tools. PM R. 2014;6(11):1038–43.
24. Vercelli S, Sartorio F, Bravini E, Ferriero G. DrGoniometer: a reliable smartphone app for joint angle measurement. Br J Sports Med. 2017;51(23):1703.
25. Dos Santos RA, Derhon V, Brandalize M, Brandalize D, Rossi LP. Evaluation of knee range of motion: correlation between measurements using a universal goniometer and a smartphone goniometric application. J Bodyw Mov Ther. 2017;21(3):699–703.
26. Barré A, Aminian K. Error performances of a model-based biplane fluoroscopic system for tracking knee prosthesis during treadmill gait task. Med Biol Eng Comput. 2018;56(2):307–16.
27. Bonanzinga T, Signorelli C, Bontempi M, Russo A, Zaffagnini S, Marcacci M, et al. Evaluation of RSA set-up from a clinical biplane fluoroscopy system for 3D joint kinematic analysis. Joints. 2016;4(2):121–5.
28. Guan S, Gray H, Schache A, Feller J, de Steiger R, Pandy M. In vivo knee kinematics and joint centre of rotation in total knee arthroplasty gait quantified by mobile biplane x-ray imaging. Orthop Procs. 2018;100-B(SUPP_6):18.
29. Giphart JE, Zirker CA, Myers CA, Pennington WW, LaPrade RF. Accuracy of a contour-based biplane fluoroscopy technique for tracking knee joint kinematics of different speeds. J Biomech. 2012;45(16):2935–8.
30. Anderst W, Zauel R, Bishop J, Demps E, Tashman S. Validation of three-dimensional model-based tibiofemoral tracking during running. Med Eng Phys. 2009;31(1):10–6.
31. Stanford W, Phelan J, Kathol MH, Rooholamini SA, El-Khoury GY, Palutsis GR, et al. Patellofemoral joint motion: evaluation by ultrafast computed tomography. Skelet Radiol. 1988;17(7):487–92.
32. Shellock FG, Mink JH, Deutsch AL, Foo TK. Kinematic MR imaging of the patellofemoral joint: comparison of passive positioning and active movement techniques. Radiology. 1992;184(2):574–7.
33. Elias JJ, Carrino JA, Saranathan A, Guseila LM, Tanaka MJ, Cosgarea AJ. Variations in kinematics and function following patellar stabilization including tibial tuberosity realignment. Knee Surg Sports Traumatol Arthrosc. 2014;22(10):2350–6.
34. Kalia V, Obray RW, Filice R, Fayad LM, Murphy K, Carrino JA. Functional joint imaging using 256-MDCT: technical feasibility. AJR Am J Roentgenol. 2009;192(6):W295–9.
35. Pettigrew RI. Dynamic cardiac MR imaging. Techniques and applications. Radiol Clin N Am. 1989;27(6):1183–203.
36. Kaiser J, Bradford R, Johnson K, Wieben O, Thelen DG. Measurement of tibiofemoral kinematics using highly accelerated 3D radial sampling. Magn Reson Med. 2013;69(5):1310–6.
37. Shapiro LM, Gold GE. MRI of weight bearing and movement. Osteoarthr Cartil. 2012;20(2):69–78.
38. Garetier M, Borotikar B, Makki K, Brochard S, Rousseau F, Ben SD. Dynamic MRI for articulating joint evaluation on 1.5T and 3.0T scanners: setup, protocols, and real-time sequences. Insights into. Imaging. 2020;11(1):66.

39. Tranberg R, Saari T, Zügner R, Kärrholm J. Simultaneous measurements of knee motion using an optical tracking system and radiostereometric analysis (RSA). Acta Orthop. 2011;82(2):171–6.
40. Kaptein BL, Valstar ER, Stoel BC, Rozing PM, Reiber JH. A new model-based RSA method validated using CAD models and models from reversed engineering. J Biomech. 2003;36(6):873–82.
41. Valstar ER, de Jong FW, Vrooman HA, Rozing PM, Reiber JH. Model-based roentgen stereophotogrammetry of orthopaedic implants. J Biomech. 2001;34(6):715–22.
42. Stentz-Olesen K, Nielsen ET, De Raedt S, Jørgensen PB, Sørensen OG, Kaptein BL, et al. Validation of static and dynamic radiostereometric analysis of the knee joint using bone models from CT data. Bone Joint Res. 2017;6(6):376–84.
43. Karatsidis A, Bellusci G, Schepers HM, de Zee M, Andersen MS, Veltink PH. Estimation of ground reaction forces and moments during gait using only inertial motion capture. Sensors (Basel). 2016;17(1):75.
44. Wang J-P, Wang SH, Zhao X, Hu H, Wang YQ, Liu JL, et al. A processing method for kinematics data of human knee joint obtained by motion-capture measurement 2020. https://doi.org/10.21203/rs.3.rs-37194/v1
45. Konrath JM, Karatsidis A, Schepers HM, Bellusci G, de Zee M, Andersen MS. Estimation of the knee adduction moment and joint contact force during daily living activities using inertial motion capture. Sensors (Basel). 2019;19(7):1681.
46. Klopfer-Kramer I, Brand A, Wackerle H, Mussig J, Kroger I, Augat P. Gait analysis–available platforms for outcome assessment. Injury. 2020;51(Suppl 2):S90–S6.
47. Camomilla V, Dumas R, Cappozzo A. Human movement analysis: the soft tissue artefact issue. J Biomech. 2017;62:1–4.
48. Moglo KE, Shirazi-Adl A. Cruciate coupling and screw-home mechanism in passive knee joint during extension–flexion. J Biomech. 2005;38(5):1075–83.
49. Kozanek M, Hosseini A, Liu F, Van de Velde SK, Gill TJ, Rubash HE, et al. Tibiofemoral kinematics and condylar motion during the stance phase of gait. J Biomech. 2009;42(12):1877–84.
50. Mizuno Y, Kumagai M, Mattessich SM, Elias JJ, Ramrattan N, Cosgarea AJ, et al. Q-angle influences tibiofemoral and patellofemoral kinematics. J Orthop Res. 2001;19(5):834–40.
51. Pianigiani S, Chevalier Y, Labey L, Pascale V, Innocenti B. Tibio-femoral kinematics in different total knee arthroplasty designs during a loaded squat: a numerical sensitivity study. J Biomech. 2012;45(13):2315–23.

Clinical Tests for Evaluation of Motor Function of the Knee

47

Gabriel Ohana Marques Azzini

47.1 Introduction

Injuries of the lower extremities are accompanied by a phase of reduced loading capacity of the knee structures and mostly with distinct losses of muscle mass and muscle strength [1]. Due to the lack of relevant information concerning the knee joint in functional loading situations, dynamic tests are highly recommended in recent sports medicine textbooks and articles [2]. The majority of the clinical tests of motor function give quantitative measures (e.g., time, height, distance) of abilities of the injured limb or important information about the rehabilitation process. The use of motor tests gives exclusive information about the knee, because in a usual clinical examination, correlations between the motor behavior in walking, running, jumping, and pivoting movements and results of the usual classic examination methods have not been found [3]. Motor function tests are frequently used in clinical practice because they do not require a lot of space or special equipment [4].

The clinical motor tests of the lower limbs are also very important to evaluate the effectiveness of the rehabilitation process after a complex surgery of the knee joint. Anterior cruciate ligament (ACL) rehabilitation programs have suffered a dramatic improvement in the last years. Most of the patients can now begin an active lifestyle after only 24 weeks of rehabilitation. To ensure that a patient has a satisfactory functional level, which will allow a safe return to a level of activity similar to that presented previously to an injury, we need concrete data to assess the motor capacity of the rehabilitated patient [5].

Nowadays, osteoarthritis is the most common knee disorder. Due to the stiffness, pain, and reduced muscle strength, it is a common cause of disability in older adults. With progression of the osteoarthritis of the knee, daily activities such as walking, getting in and out of a bath, and doing simple household chores become very difficult [6]. To evaluate the grade of disabilities caused by an osteoarthritic knee or the result of the applied treatment, we should use the clinical dynamic motor tests. In this chapter, we will detail the most used motor dynamic tests for evaluating orthopedic patients.

G. O. M. Azzini (✉)
O.A.S.I. Bioresearch Foundation Gobbi ONLUS, Milan, Italy

Instituto do Osso e da Cartilagem, Indaiatuba, Brazil

J. G. Lane et al. (eds.), *The Art of the Musculoskeletal Physical Exam*,
https://doi.org/10.1007/978-3-031-24404-9_47

47.2 Clinical History

A complete medical history must be performed prior to physical examination of the knee. During our academic training, we learned to characterize pain in terms of type, intensity, frequency, location, aggravating factors, and relief factors. Regarding the knee, there is extremely important information in the clinical history, such as the trauma mechanism, the course of the disease's development, and the patient's age. In addition to these, there are details that are often ignored by doctors who work in the treatment of pain, but which should be incorporated into the clinical evaluation of a patient with pain complaints, such as eating habits, stress level, and quality of sleep.

When investigating a complaint, we should always try to clarify.

- The trauma mechanism (if present)
  If the problem is due to direct trauma, torsional trauma, or excessive effort.
- The functional limitation caused by the health problem
  If the patient is unable to walk, he/she needs crutches, or is even unable to mobilize the knee.
- The problem development course
  We must investigate whether the problem had an acute development after trauma or another specific or insidious event. Or if it had an insidious course, evolving gradually.
- Previous treatments (medications, physiotherapy, immobilization, surgery, etc.)
- The patient's current level of physical activity
  We must have a clear notion of the patient's level of sporting and functional activity, as well as expectations after treatment.
- Presence of systemic inflammatory diseases (gout, seronegative arthritis, etc.)
- The patient's diet
  New studies indicate that an inflammatory diet can accelerate the progression of degenerative diseases such as osteoarthritis and lead to a higher prevalence of pain and functional limitations in our patients [7, 8].
- The presence of clinical signs of metabolic syndrome
  Signs such as hypertension, diabetes, increased abdominal circumference, and dyslipidemia are directly related to the quality of the subchondral bone of the knee [9]. These changes generate changes in the trabecular architecture of the subchondral bone, causing it to decrease its impact absorption capacity. Consequently, there is an overload of the knee joint cartilage, favoring the appearance of a cartilaginous lesion and pain in that joint [10, 11].
- Presence of depression or high stress levels
  The presence of depression or high levels of stress directly affect the threshold of sensitivity to pain and the result of the treatment employed. Patients with depressive symptoms show worse results in relation to pain relief and joint function after infiltration with hyaluronic acid in knees with osteoarthritis [12].
- Sleep quality
  Another point that we always evaluate in our patients is the quality of sleep. There is a direct relationship between the quality of sleep and the level of pain in patients with knee osteoarthritis [13].

47.3 Inspection

The first part of our patient's inspection begins even before the doctor's appointment. The doctor should assess the patient's gait looking for lameness and movement abnormalities and also assess the relationship of the knee to the adjacent joints and observe the use of some type of orthosis. After collecting the clinical history, with better exposure of the lower limbs, a new gait assessment should be performed (Fig. 47.1).

Fig. 47.1 Gait assessment being performed

Still with the patient in an orthostatic position, we assessed possible valgus or varus deformities of the knee, the position of the patella, and the presence of excessive pronation or supination of the foot. The skin is observed looking for edema, erythema, ecchymosis, or scars that indicate previous surgery or trauma. Muscle asymmetries can also be verified at this stage of the exam, where we also evaluate the popliteal region, in the search for masses and tumors.

47.4 Range of Motion

The evaluation of the knee range of motion begins with the patient in seated position. We start analyzing the active range of motion, and if there is any limitation, we do a passive evaluation. We search for a flexion range around 140°, and the comparison with the contralateral knee is always mandatory [14].

47.5 Palpation

The sequential palpation of the structures of the knee can find the correct location of the lesion in the majority of the cases. We also look for the presence of phlogistic signs. Heat, pain, and flushing may indicate exacerbated inflammatory activity, suggesting complementary laboratory tests [14].

47.6 Motor Tests of the Knee

47.6.1 Five-Time Sit-to-Stand Test (FTSST)

The patient is asked to sit in a chair with his/her arms folded across his/her chest. Then, the test taker is asked to do five repetitions of standing from seated position and return to sit as soon as possible without the use of the arms. We start measuring the time when the patient starts the movement and stop the count at the end of the fifth repetition (Figs. 47.2, 47.3, and 47.4).

The sit-to-stand movement is essential to assess not only the mobility of our patients but also their degree of independence. There is a high degree of dependence when this function is impaired. In addition, unsatisfactory results are related to an increased risk of falls and fractures in the elderly [4].

47.6.2 Five-Meter Walk Test (5mWT)

In this test, we mark the time it takes the patient to walk a distance of 5 m that must be somehow marked on the floor of the examination site. It is

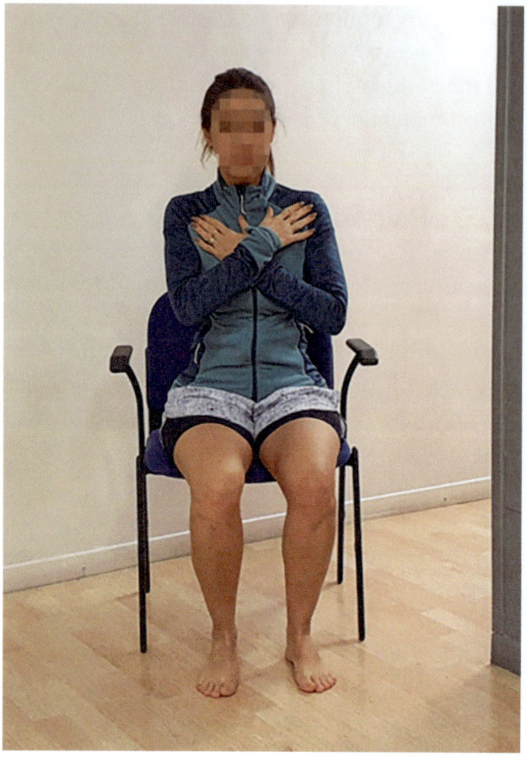

Fig. 47.2 Sequence demonstrating the sit-to-stand movement analysis

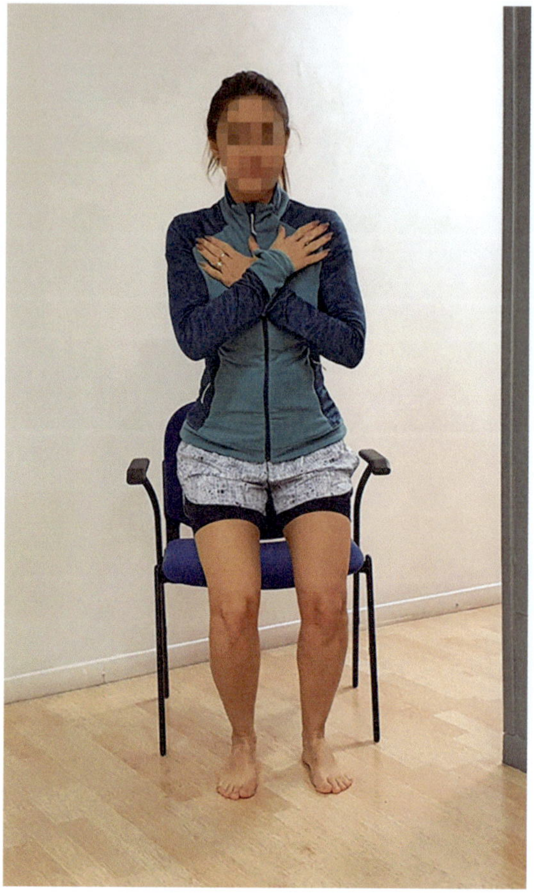

Fig. 47.3 Sequence demonstrating the sit-to-stand movement analysis

important to have an area before and after the 5 m that will be evaluated, so that the patient can accelerate and decelerate with tranquility and the time is measured at maximum speed. We can also calculate the patient's speed (m/s) by dividing the distance covered (5 m) by the time interval in which the patient completed the test [15].

47.6.3 Ascend/Descend Four Stairs

Having a safe and well-lit staircase, we can perform this test that gives us a lot of information regarding the patient's neuromotor function. We ask the test taker to go up four steps, turn around when he/she reaches the fourth step, and go down as quickly as possible. Time is recorded for future comparison [6] (Figs. 47.5, 47.6, and 47.7).

47.6.4 Maximal Hop for Distance

Hopping is the movement when we perform the body projection and the landing with the same lower limb. We ask the patient to maintain balance on one foot and, keeping the contralateral hip and knee flexed at approximately 90° and the hands on the hips, project his/her body as far as possible, then landing on the same foot. It is very important not to use the swing of the contralateral lower limb and arms to assist movement. After landing, the test taker can place both feet on the ground to maintain balance. The distance between the starting and ending positions of the foot is measured and saved for future comparisons [5] (Figs. 47.8 and 47.9).

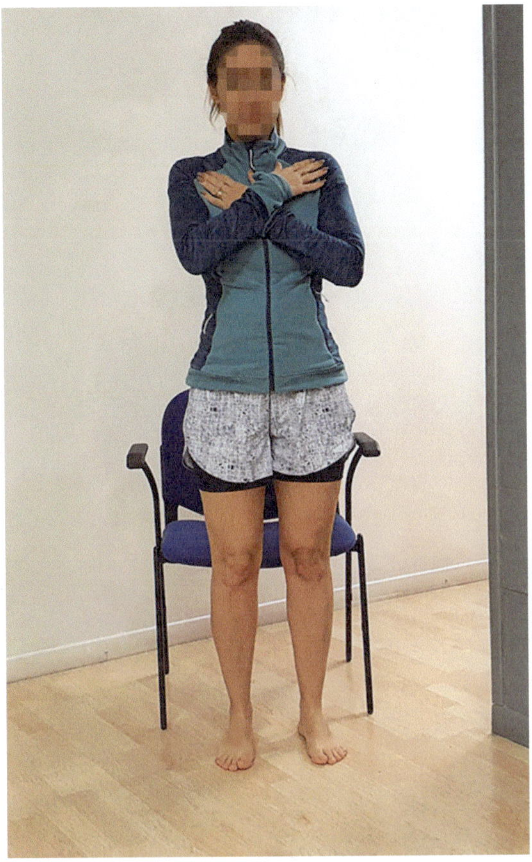

Fig. 47.4 Sequence demonstrating the sit-to-stand movement analysis

47.6.5 Maximal Controlled Leap

This test is very important, as this movement is present in the vast majority of sports. We ask patients to maintain balance with just one foot of support. He/she must maintain the contralateral knee and hip in approximately 90° of flexion and the hands on the hip. The patient projects the body forward with the force of the lower limb that is supported on the floor and lands with the other foot. It is very important that the test taker only extends the knee near the moment of landing and maintains total balance in approximately 1 s after the landing. The distance between the starting position of the foot and the final position of the landing foot is measured and saved for future analysis [5] (Figs. 47.10 and 47.11).

Fig. 47.5 Ascend/descend 4 stairs

47.6.6 Single-Legged Drop-Jump Landing Test

To perform the single-legged drop-jump landing test, we ask the patient to stand on a step of approximately 20 cm. We ask him/her to take a short takeoff with both feet and land with just one foot. After landing, it is important for the test

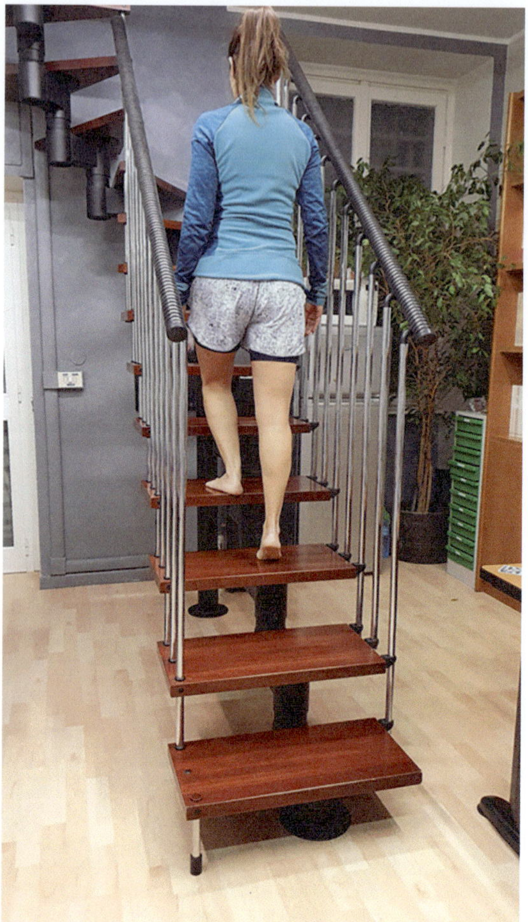

Fig. 47.6 Ascend/descend 4 stairs

Fig. 47.7 Ascend/descend 4 stairs

taker to stabilize the position by maintaining balance on just one foot and making as little movement as possible. The hands should be on the hips during the whole test, and the eyes should be fixing a point in the wall [16] (Figs. 47.12 and 47.13).

47.6.7 Y-Balance

The Y-balance test is used to assess muscle control, stability, and mobility of the lower limbs. The patient is asked to maintain balance in a single limb over the crossing of three lines marked on the floor, the first anteriorly, the second posterolateral, and the third posteromedial. Maintaining their hands on their hips, the test taker is asked to touch the tip of the foot that is out of contact with the ground as far as possible on all three lines marked on the ground. The examinee must maintain balance in only one member throughout the test. We then mark the distance reached on each line. Assessing the markings with the contralateral examination, we considered asymmetries equal to or greater than

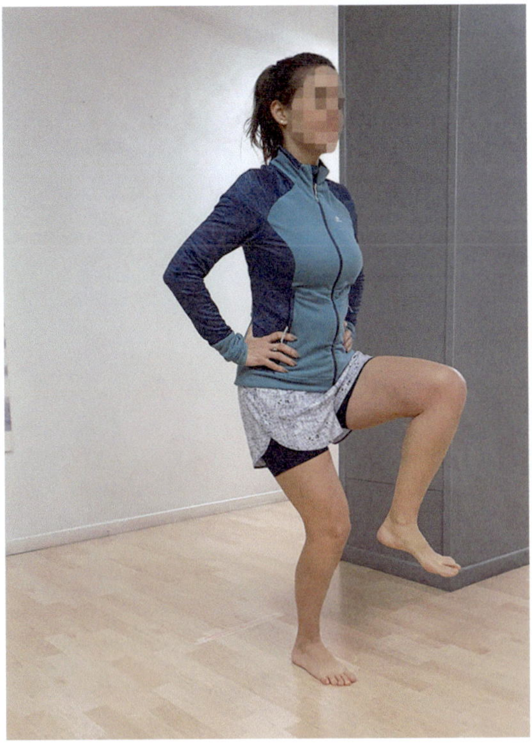

Fig. 47.8 Maximal hop for distance test. Start position

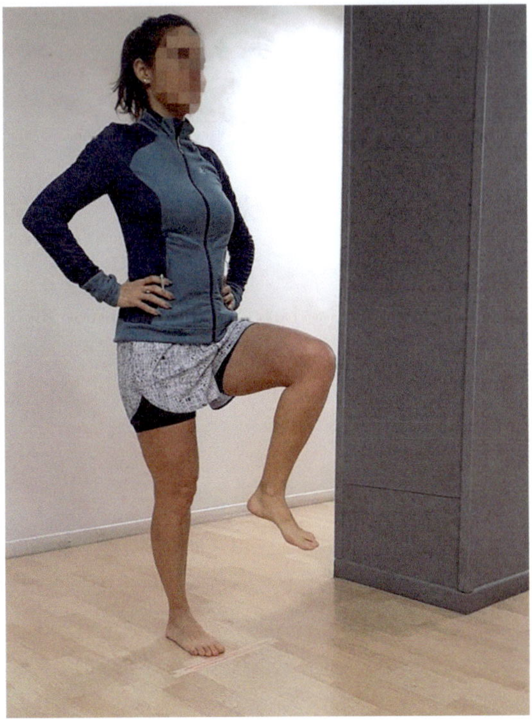

Fig. 47.9 Maximal hop for distance test. Final position

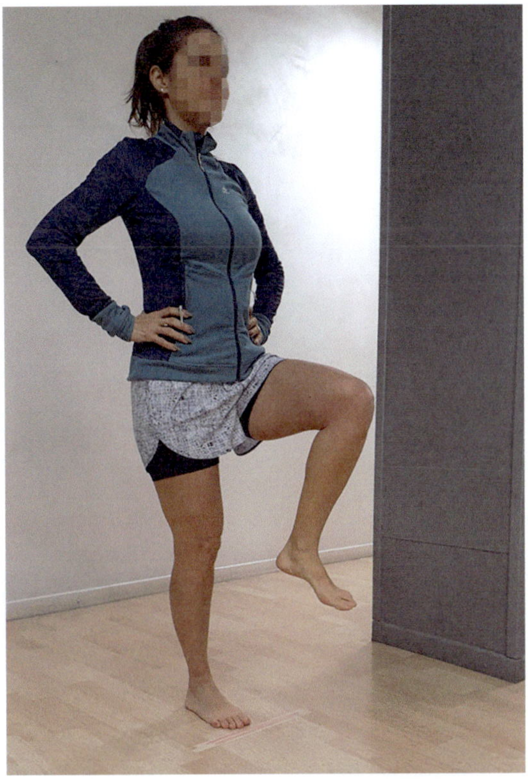

Fig. 47.10 Maximal controlled leap. Start position

Fig. 47.11 Maximal controlled leap. Final position

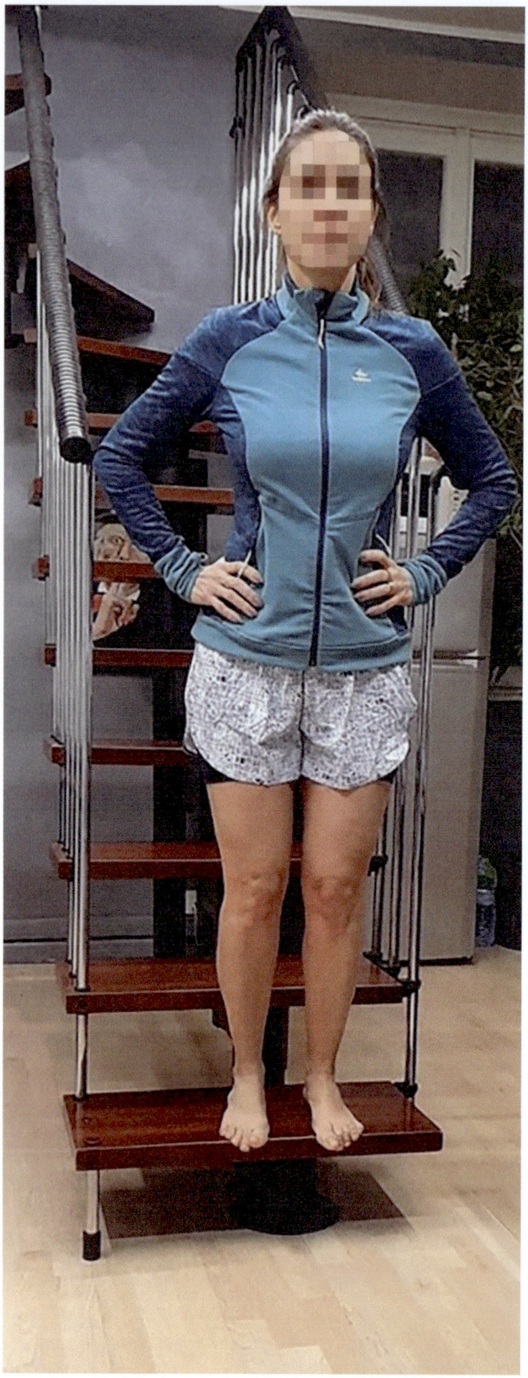

Fig. 47.12 Single-legged drop-jump landing test, initial position

Fig. 47.13 Single-legged drop-jump landing test, final position

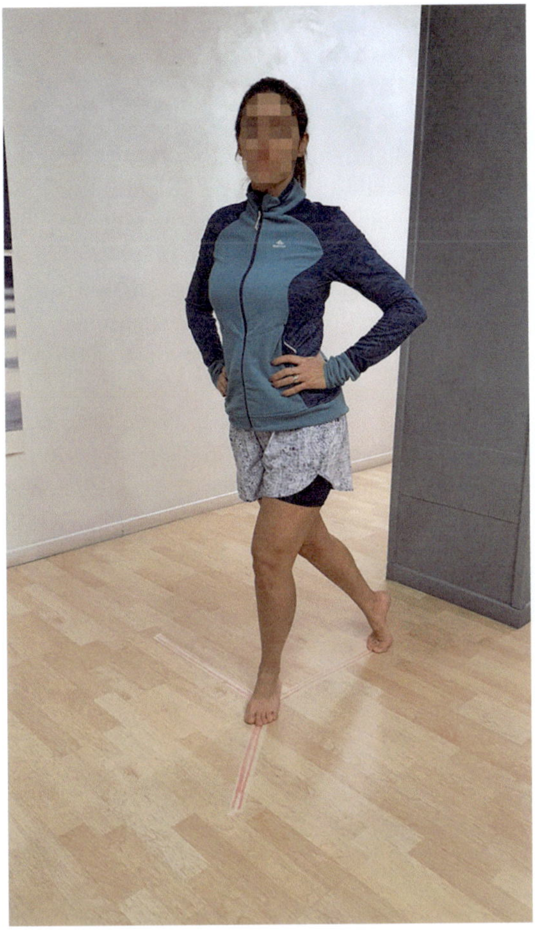

Fig. 47.14 Y-balance

Fig. 47.15 Y-balance

4 cm as a sign of neuromotor deficit [17] (Figs. 47.14, 47.15, and 47.16).

47.6.8 Modified T-Test

To perform this motor test, four cones (a, b, c, d) are used, arranged in a T shape with distances shown in Fig. 47.17. The test taker starts the examination on cone A and moves as fast as possible to the cone B, where he/she should touch the base of the cone with his/her right hand. Always looking forward and without crossing his/her legs, the patient moves laterally towards cone C on his/her left, touching the base of the cone with his/her left hand. In the next step, he/she must make a lateral displacement to cone D, touching the base of this cone with his/her right hand and returning as quickly as possible to cone B. After touching the base of cone B with his/her right hand, the patient must return to cone A always looking forward. If the test taker does not look forward or cross his/her legs during the activity, the test will not be considered valid and must be repeated [18] (Fig. 47.17).

Fig. 47.16 Y-balance

47.6.9 Ninety-Degree Medial Rotation Hop for Distance (MRH)

For the medial rotation hop for distance test, the patient is asked to stand in the tested lower limb with the medial side of the foot indicating the direction in which the body should be projected. The test taker is then asked to perform one single-leg jump, trying to reach the maximum distance. During the jump, the test taker must also rotate the trunk 90° in the medial direction and land with the foot facing forward. There may be a slight variation of up to 10° in the position of the foot with the indicated plane. If the variation is greater, the test should be repeated [19] (Figs. 47.18, 47.19, and 47.20).

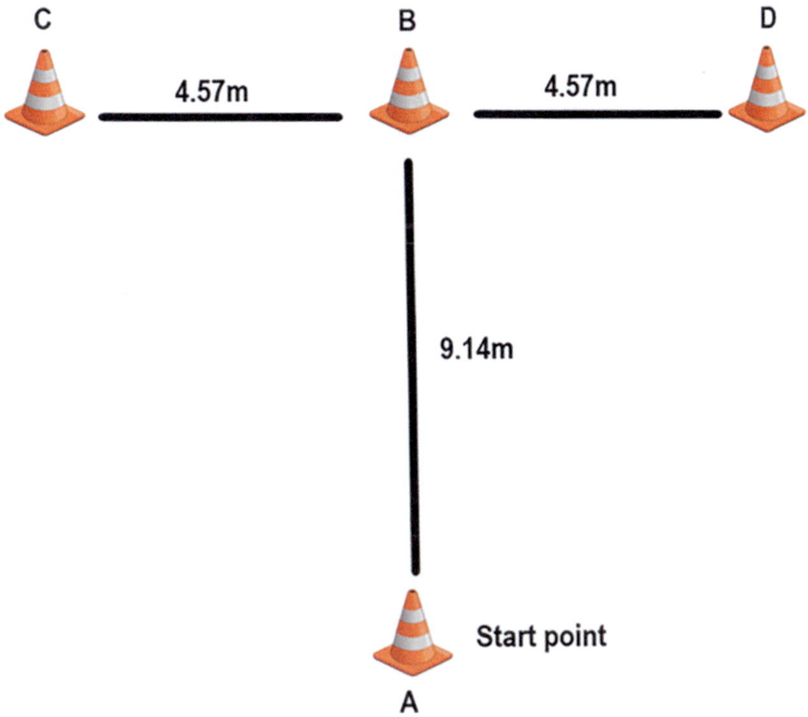

Fig. 47.17 Modified T-test cone arrangement

Fig. 47.18 90° medial rotation hop for distance (MRH). Sequence

Fig. 47.19 90° medial rotation hop for distance (MRH). Sequence

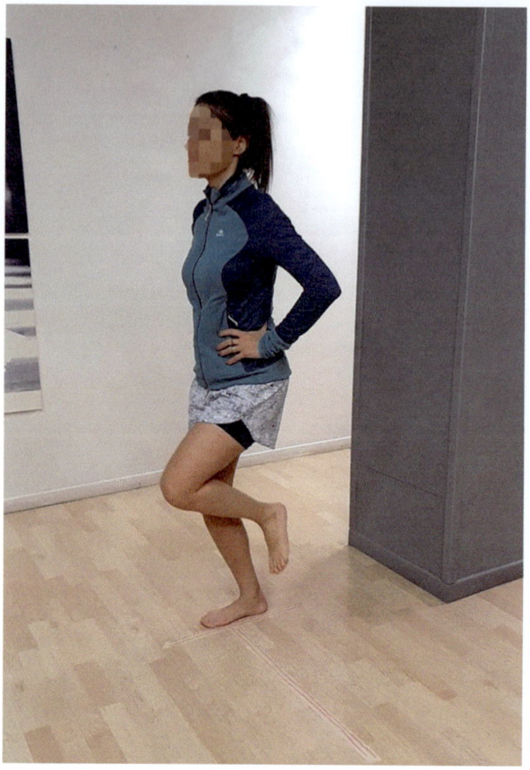

Fig. 47.20 90° medial rotation hop for distance (MRH). Sequence

47.7 Conclusion

The clinical tests for evaluation of motor function of the knee are complex and should be done by the examiner during the instructions prior to the test. The doctor should use the tests according to the patient's neuromotor ability and relating to the movements present during sports practice that is desired by the patient after complete recovery.

References

1. Kannus P, Jozsa L, Renstrom P, et al. The effects of training. immobilization and remobilization on musculoskeletal tissue-l. Training and immobilization. Scand J Med Sci Sports. 1992;2:100–8.
2. Harrelson GL. Knee Rehabilitation. In: Andrews JR, Harrelson GL, editors. Physical rehabilitation of the injured athlete. Philadelphia, PA: Saunders; 1991. p. 267–342.
3. Lephart SM, Perrin DH, Fu FH, Gieck JH, FC MC, Irrgang JJ. Relationship between selected physical characteristics and functional capacity in the anterior cruciate ligament-insufficient athlete. J Orthop Sports Phys Ther. 1992;16:174–81.
4. Amano T, Suzuki N. Minimal detectable change for motor function tests in patients with knee osteoarthritis. Prog Rehab Med. 2018;3:20180022. https://doi.org/10.2490/prm.20180022.
5. Juris PM, Phillips EM, Dalpe C, Edwards C, Gotlin RS, Kane DJ. A dynamic test of lower extremity function following anterior cruciate ligament reconstruction and rehabilitation. J Orthop Sports Phys Ther. 1997;26(4):184–91. https://doi.org/10.2519/jospt.1997.26.4.184. PMID: 9310909
6. Lin YC, Davey RC, Cochrane T. Tests for physical function of the elderly with knee and hip osteoarthritis. Scand J Med Sci Sports. 2001;11(5):280–6. https://doi.org/10.1034/j.1600-0838.2001.110505.x. PMID: 11696212
7. Liu Q, Hebert JR, Shivappa N, et al. Inflammatory potential of diet and risk of incident knee osteoarthritis: a prospective cohort study. Arthritis Res Ther. 2020;22:209. https://doi.org/10.1186/s13075-020-02302-z.
8. Thomas S, Browne H, Mobasheri A, Rayman MP. What is the evidence for a role for diet and nutrition in osteoarthritis? Rheumatology (Oxford). 2018;57(suppl_4):iv61–74. https://doi.org/10.1093/rheumatology/key011. https://www.ncbi.nlm.nih.gov/pmc/articles/PMC5905611/
9. Azzini GOM, Santos GS, Visoni SBC, Azzini VOM, dos Santos RG, Huber SC, Lana JF. Metabolic syndrome and subchondral bone alterations: the rise of osteoarthritis–a review. J Clin Orthop Trauma. 2020;11(S_5):S849–55. https://www.sciencedirect.com/science/article/abs/pii/S097656622030254X
10. Dickson BM, Roelofs AJ, Rochford JJ, et al. The burden of metabolic syndrome on osteoarthritic joints. Arthritis Res Ther. 2019;21:289. https://doi.org/10.1186/s13075-019-2081-x.
11. Wang H, Cheng Y, Shao D, et al. Metabolic syndrome increases the risk for knee osteoarthritis: a meta-analysis. Evid Based Complement Alternat Med. 2016;2016:7242478. https://doi.org/10.1155/2016/7242478. https://www.ncbi.nlm.nih.gov/pmc/articles/PMC5078652/
12. Chen YP, Huang YY, Wu Y, et al. Depression negatively affects patient-reported knee functional outcome after intraarticular hyaluronic acid injection among geriatric patients with knee osteoarthritis. J Orthop Surg Res. 2019;14:387. https://doi.org/10.1186/s13018-019-1419-z. https://link.springer.com/article/10.1186/s13018-019-1419-z#citeas
13. Parmelee PA, Tighe CA, Dautovich ND. Sleep disturbance in osteoarthritis: linkages with pain, disability, and depressive symptoms. Arthritis Care Res. 2015;67(3):358–65. https://doi.org/10.1002/

acr.22459. https://www.ncbi.nlm.nih.gov/pmc/articles/PMC4342277/
14. Kreder HJ, Hawker GA. The knee. In: Lawry GV, Kreder HJ, Hawker GA, Jerome D, editors. Fam's Musculoskeletal Examination and Joint Injection Techniques. 2nd ed. Maryland Heights: Mosby; 2010. p. 65–88.
15. Wilson CM, Kostsuca SR, Boura JA. Utilization of a 5-meter walk test in evaluating self-selected gait speed during preoperative screening of patients scheduled for cardiac surgery. Cardiopulm Phys Ther J. 2013;24(3):36–43.
16. Fransz DP, Huurnink A, Kingma I, de Boode VA, Heyligers IC, van Dieën JH. Performance on a Single-Legged Drop-Jump Landing test is related to increased risk of lateral ankle sprains among male elite soccer players: a 3-year prospective cohort study. Am J Sports Med. 2018;46(14):3454–62. https://doi.org/10.1177/0363546518808027.
17. Neves L. The Y Balance test–how and why to do it? Int Phys Med Rehab J. 2017;2:10.15406/ipmrj.2017.02.00058.
18. Radhouane HS, Dardouri W, Mohamed HY, Gmada N, Mahfoudhi M-E, Gharbi Z. Relative and absolute reliability of a modified agility t-test and its relationship with vertical jump and straight sprint. J Strength Cond Res. 2009;23:1644–51. https://doi.org/10.1519/JSC.0b013e3181b425d2.
19. Dingenen B, Truijen J, Bellemans J, Gokeler A. Test-retest reliability and discriminative ability of forward, medial and rotational single-leg hop tests. Knee. 2019;26(5):978–87. https://doi.org/10.1016/j.knee.2019.06.010. Epub 2019 Aug 17. PMID: 31431339.

48 The Stability and Function of the Patellofemoral Joint

Laura Ann Lambert
and Michael James McNicholas

48.1 Introduction

A systematic patellofemoral joint (PFJ) examination should yield specific answers about the cause of either instability or dysfunction of the joint. Both problems can present with anterior knee pain.

Anterior knee pain may arise from any of the bony or soft tissues that contribute to the knee, e.g. skin and subcutaneous tissue, investing synovium, subchondral bone of the femur, tibia or patella, neighbouring musculature or nerves traversing the joint [1]. The umbrella term often used for this anterior knee pain is the patellofemoral pain syndrome (PFPS) for which there are a myriad of underlying causes.

Patellofemoral instability (PFI) is a common cause of PFPS. It occurs in approximately 5.8 cases per 100,000 individuals with this increasing to 29 cases per 100,000 individuals amongst early adolescents [2].

The pathophysiology of PFI is poorly understood. It can be defined as abnormal patellar tracking in relation to the femoral trochlea as the knee extends/flexes [3].

The function and stability of the PFJ can be broken down into problems that arise in each of the following categories:

1. The inherent shape and position of the patella (patella alta or baja, Wiberg type C patella (hypoplastic medial facet) [4]
2. Local patellar attachments (traumatic rupture of stabilisers such as the medial patellofemoral ligament)
3. The extensor mechanism (increased Q-angle, muscular imbalance)
4. Lower limb alignment (increased femoral-tibial torsion, genu valgus, hindfoot position)
5. Soft-tissue laxity

Today's treatment algorithms seek to identify which of the above features are driving the pathology and whether or not PFI results in functional or mechanical instability. Discerning what is simply a painful patella versus true instability requires working through this algorithm.

The presentation of PFI is on a spectrum. At its most mild form, the patient may complain of anterior knee pain when the knee is flexed for prolonged periods to overt lateral patellar dislocation following dynamic activity.

L. A. Lambert
The University of Manchester, Manchester, UK

M. J. McNicholas (✉)
Division of Cell Matrix Biology & Regenerative Medicine, Faculty of Biology, Medicine and Health, School of Biological Sciences, University of Manchester, Manchester, UK

J. G. Lane et al. (eds.), *The Art of the Musculoskeletal Physical Exam*,
https://doi.org/10.1007/978-3-031-24404-9_48

48.2 Anatomy

We start at the beginning like all surgical conundrums, with the anatomy. The patella is a large heart-shaped sesamoid bone, the largest in the body. Invested in a sleeve of retinaculum, it is contiguous with the extensor mechanism of the knee made up of the proximal quadriceps (contributions from rectus femoris and vastus to form one common tendon) and the distal patellar tendon.

The patella is convex shaped on its superficial aspect.

On its deep articulating aspect, the patella is divided into facets, the lateral facet, the medial facet and the more medial odd facet. These facets are partitioned by a central vertical ridge. It can be further divided into two parts, an articular part superiorly and non-articular part inferiorly.

The corresponding femoral surface is made up of the medial and lateral femoral condyles. The lateral portion of the femoral sulcus is relatively broad and contains a higher lateral ridge than the medial portion [5]. Anatomic anomalies can be risk factors for PFPS, e.g. hypoplasia of the medial patellar facet, patella alta or trochlear dysplasia (rare).

48.3 History

A PFPS history should focus on the onset of pain, mechanism of injury (if any), location, character and severity of pain under different conditions and lastly any aggravating or alleviating factors. Eccentric contraction of the quadriceps when descending stairs or slopes may create pain due to the increased compressive loads placed upon the articular cartilage of the posterior patellar facets [6]. In a study of 745 knees, Stefanik et al. found that pain with climbing/descending stairs had the greatest sensitivity (90%) but poor specificity (15%) for identifying PFJ osteoarthritis. The combination of definite crepitus **with no pain** on walking had the greatest specificity (96%) and positive predictive value (53%) and likelihood ratio + (1.8), but poor sensitivity (7%) [7].

48.4 Examination

A comprehensive examination of the PFJ can be challenging. A clinical examination is considered to be the mainstay of diagnosing PFPS and pinpointing the factors at play that give rise to it. However, there is limited evidence available to support the diagnostic validity of different clinical tests described for patellofemoral pain [8].

Systematic examination of the knee should include both static and dynamic assessment with the patient in the following three positions:

1. Standing
2. Sitting
3. Supine

48.5 General Inspection

Lots of clues regarding the state of the patellofemoral joint can be deduced by simple observation.

An effusion may be appreciable and can be later palpated (when supine) and quantified as mild, moderate or large.

Assess for the hallmarks of previous surgical intervention such as telltale surgical scars, prominent metalwork or a bulky allograft.

Soft-tissue lesions may or may not be appreciable with general inspection. An infrapatellar fat pad with oedematous infiltration is called Hoffa's syndrome. In the final stages of this disease, there may be a palpable swelling adjacent to the patellar tendon due to the development of fibrocartilaginous tissue or secondary ossification [9]. The differential for such a swelling is plica syndrome, bursitis and cyclops syndrome (the arthroscopic or MRI appearance of an arthrofibrotic nodule post-ACL reconstruction) [10]. Plica syndrome refers to a pathological change in the medial patella plica resulting in anterior knee pain. It is frequently associated with a lateral patella tilt [11]. Clinically, it is difficult to differentiate from other intra-articular conditions. Palpation of the anteromedial joint line may identify a cord-like fold of tissue that clicks with knee extension [12].

48.6 Standing

48.6.1 Static Assessment while Standing

48.6.1.1 Frontal Assessment of the Lower Limbs

Lower limb alignment measurements can be divided into the anatomical axis, mechanical axis and patellar alignment. The anatomic axis is the angle yielded from lines that bisect the femur and the tibia, while the mechanical axis is a line drawn from the centre of the femoral head to the centre of the ankle joint. The anatomic axis of the femur has an approximate difference of 5–7° of inclination compared to mechanical axis of the femur [13]. Normal knee joint line alignment tends to be 2–3° of varus compared with the mechanical axis. Both anatomical and mechanical axes are most accurately determined using weight-bearing full-length radiographs [14].

However, in the standing position, frontal plane knee alignment can be approximated by observing the position of the lower limbs. Magee's method of visual assessment categorises lower limb alignment as varus, valgus or neutral. Adduction of the lower limbs results in the contact of bony landmarks in the following order: medial malleoli first; varus alignment, medial femur/proximal tibia first; valgus alignment and lastly knees and medial malleoli simultaneously; and neutral alignment [15]. This method is simple and easy to implement. It is not as accurate as using an inclinometer ($r = 0.8$) or callipers ($r = 0.76$), but the relationship of visual estimation ($r = 0.52$) correlates with radiographic measurement [14].

More formally, goniometric measurement may be undertaken, but there are challenges with consistent positioning of the goniometer on bony landmarks. In a study by Riddle and colleagues, they found that goniometric measurement error was large enough to skew interpretations of validity [16]. In obese patients, patients with lymphoedema or any condition where the soft-tissue envelope is increased, the alignment is likely to be falsely interpreted with goniometric measurements [13, 14].

Lastly, when assessing alignment, measure for a limb-length discrepancy. If a discrepancy is found, determine if there is tibial or femoral shortening. The examiner should perform Galeazzi's test and if required draw Bryant's triangle when examining the patient in the supine position to pinpoint the origin of shortening.

48.6.1.2 Q-Angle

The Q-angle was originally described by Scandinavian author Brattstrom, in his 1964 thesis, as "*the angle, with its apex at the patella, formed between ligamentum patellae and the extension of the quadriceps-resultant distally*". He described the normal position of the knee as lying between 8 and 10° of valgus. At terminal extension in the supine position, the angle of the knee is approximately 175°. The supplement angle to this valgus angle in the extensor mechanism is what he named the **Q-angle** [17]. It is well established that the "normal" Q-angle lies within the range of 12–20°. There are lots of factors that influence the Q-angle. The Q-angle is greater in women than men [18–20]. This is no great surprise as the generally wider pelvis of a woman means the distance from the ASIS to the patella is greater, producing a more obtuse angle. The range is greater amongst certain ethnic groups [18]. The angle is greater when standing than when sitting and is also influenced by foot position. Internal rotation and pronation of the foot increase the Q-angle [21–23]. There is better intra-observer reliability when assessing the Q-angle in 30° of flexion compared to a standing position [24]. Despite the abundance of literature written about the Q-angle, Fulkerson in his book on patellofemoral disorders asks if the Q-angle should even be measured and questions its clinical utility [25]. Whilst the Q-angle is commonly commented upon, it is not a particularly useful diagnostic tool. There ia a large variation in the value of the Q angle and increased Q angles do not correlate with patellofemoral pain. Increased Q-angles are not proven to have a definite relationship of patellofemoral subluxation [26].

48.6.1.3 Sagittal Assessment of the Knee

When assessing knee from the side, the examiner may appreciate a knee held in a flexed attitude, indicative of patellofemoral arthrosis. A hyperextended knee, genu recurvatum, is suggestive of hypermobility, a risk factor for PFJ instability [27]. The relative height of the patella is best observed from the side. A high-riding patella, patella alta, is associated with instability. A low-riding patella, patella baja, is associated with "*chondromalacia patellae*" [28]. PFJ literature interchanges the now outdated term "*chondromalacia patellae*" with anterior knee pain or patellofemoral pain syndrome. The senior author deems chondromalacia patellae a surgical and histopathological finding and not a diagnosis, as do others [29].

48.6.1.4 Foot Position

Similar to Q-angle, foot position and subtalar pronation impact the patellofemoral joint. In normal gait, the subtalar joint is supinated at the time of heel strike and pronated in the contact phase. When transitioning back to midstance, the joint should reverse to supination. This does not happen in patellofemoral pain, and the subtalar joint remains pronated. Consequently, this position impacts knee extension and tibial external rotation [30]. In a recent case-control study of patients with PFPS compared to healthy controls, those with PFPS had greater foot mobility and a more pronated posture than asymptomatic controls [31].

48.6.2 Dynamic Assessment while Standing

48.6.2.1 Gait

Gait is the term used to describe the manner in which we ambulate. It is a complex output of physiological, demographic and sociocultural inputs. Features of gait include walking speed, cadence (number of steps per unit of time), base width, step length and stride length [32]. All these features may be impaired by patellofemoral dysfunction. Arazpour et al. performed a systematic review of the literature on patellofemoral pain (n = 16 studies) and the resultant impact on kinematic, kinetic and spatio-temporal variables [33]. In their synthesis of the evidence, patellofemoral pain caused a significantly delayed peak rear foot eversion compared to healthy subjects. Mean walking speed and cadence are less in patients with patellofemoral pain. This lower speed is achieved through a shorter stride length. A short step length is also exhibited. All of these changes are likely to relate to pain avoidance techniques adopted in patellofemoral pain syndromes in a bid to reduce joint reaction forces and subsequent irritation of the PFJ itself.

48.6.2.2 Single-Leg Squat

The single-leg squat test was first described as a clinical test with nominal criteria by Livengood in 2004 [34]. Its original inception was an exercise technique, a progression from the double-leg squat in closed-chain knee rehabilitation, and was likened to a dynamic Trendelenburg test [35]. The patient stands on the limb being evaluated. The contralateral limb is lifted off the ground with the hip flexed to approximately 45° and the knee to approximately 90°. The shoulders are then forward flexed to 90° with the elbows in full extension and hands approximated together. Single-leg squat performance is then rated against Livengood's criteria for the squatting limb:

1. Hip flexion greater than 65°
2. Hip abduction/adduction greater than 100°
3. Knee valgus/varus greater than 10°

The outcome is graded as excellent if all three criteria are met, good if two are met, fair if just one criterion is met and poor if the patient can do none of the above, loses balance and falls over. Whilst these are objective outcome measures, there is currently no evidence to validate Livengood's criteria in PFPS.

There are various permutations of this test with differing protocols. They are also known by a multitude of names, e.g. the single-limb mini squat, unilateral squat, one-legged squat, single-legged squat, single-leg mini squat and single-leg small knee bend. The unifying factor for all test descriptions is that they visually assess balance, stability, knee control and quality of movement coordination. The single-leg squat test has been tested mostly in patients with patellofemoral pain. Those with patellofemoral pain experience a greater degree of knee valgus during unilateral limb loading than their contralateral asymptomatic limb or an asymptomatic control group [36]. Generally speaking, the clinical performance during the single-leg squat test when evaluated by experienced practitioners is reliable [37]. Both studies derive their outcome from frontal plane 2D analysis and subsequent digitisation of the images with markers on relevant bony landmarks.

This is echoed in a review and meta-analysis of the single-leg squat by Reesman et al. They found that the inter-rater reliability was 0.58 (95% CI 0.50–0.65) and intra-rater reliability was 0.68 (95% CI 0.60–0.74). The latter is indicative of substantial agreement. They concluded based on their findings that the single-leg squat is reliable for use in clinical practice [38].

48.7 Sitting

48.7.1 Static Assessment while Sitting

48.7.1.1 Patella Position

In the seated position, the patella may appear in the grasshopper position (upwards and outwards like the eyes of a grasshopper), i.e. lateral patella tilt in patella alta. This is much more appreciable if patella alta is present bilaterally [39].

48.7.1.2 Quadriceps Atrophy

Cadaveric studies consistently deny the existence of an observable, anatomically distinct muscle called the vastus medialis obliquus (VMO). Essentially, there is a functional region of the distal end of the vastus medialis, whose fibres insert obliquely into the patella [40, 41]. Contrary to the widely held perception that there is exclusive VMO wasting in PFJ pain, there is little evidence to support this theory. An ultrasound study assessing whole quadriceps thickness and wasting (including the rectus femoris, vastus lateralis, vastus medialis and vastus intermedius) observed that all portions of the quads are affected in patients with unilateral PFJ pain compared to an asymptomatic control group [42].

48.7.1.3 Tubercle Sulcus Angle

When the knee is flexed to 90° in a seated position, the patella is generally captured within the trochlea of the femur. The tubercle sulcus angle is approximated by observing the position of the tibial tubercle relative to the centre of the patella. While there is a paucity of literature on the consensus of what a normal angle is, lateral displacement of the tubercle with respect to the femoral sulcus is associated with patellofemoral pain [25].

48.7.2 Dynamic Assessment while Sitting

48.7.2.1 Passive Patellar Tracking

With the patient seated, they are encouraged to flex and extend their knee to look for the ubiquitous J sign and maltracking. Maltracking is an imbalance between the patella and the trochlea. As the knee is extended from 90° of flexion into extension, the patella deviates laterally at full extension. Smith et al. found that assessment of the J sign had moderate inter-observer reliability across five experienced examiners. Beckert et al. compared the J sign, assessed clinically through its arc of motion compared to dynamic MRI measurements and found that while imprecise, its presence can be accurately detected. However, both studies are underpowered and greater evidence is needed [24, 43].

48.8 Supine

48.8.1 Static Assessment while Supine

The medial patellar retinaculum should be palpated for tenderness. This finding demonstrates high inter-observer reliability in the evaluation of patients with patellofemoral instability compared to other more generic tests [24].

48.8.1.1 Apprehension Test

A clinical test producing a sense of apprehension in the patient when the clinician causes lateral displacement of the patella is called the apprehension test. A positive test produces "Fairbank's sign" after HA Fairbank, who originally described the luxating patella in 1936 [44]. Also widely known as Smillie's test, the apprehension of Fairbank's sign is mentioned in the book "Injuries of the Knee Joint" by Prof. Ian Smillie. He references Hughston et al.'s "Subluxation of the Patella", which gives the description of pressing on the medial side of the patella with the patient's knee flexed to 30° over the thigh of the examiner [45]. This recreates the position of greatest instability for the PFJ. Sallay et al. reported on clinical and arthroscopic findings associated with patellar dislocation and found that only 39% of patients with a history of dislocation were found to have a positive apprehension sign [46].

48.8.1.2 Medial Glide Test

During the medial glide test, the patella is grasped in the resting position and then translated medially using the examiner's index finger and thumb. This can be performed with the knee flexed over the examiner's knee [47] or over a pillow. The extent of displacement is described in relation to the width of the patella and measured in quadrants. Displacement of less than one quadrant medially indicates tightness of the lateral structures while more than three quadrants is considered hypermobile. This test can be performed in the opposite direction to assess the supporting medial structures [48].

48.8.1.3 Moving Patellar Apprehension Test

Detection of patellar apprehension can be inconsistent. In 1999, a modification of the patellar apprehension test was described in which the knee is flexed with lateral force concomitantly applied to the patella. A positive test is indicated by verbal anxiety or an involuntary quadriceps contraction [49]. This is because the patella is being forced into a position that recreates the dislocating event. This forms the first part of the moving patellar apprehension test (MPAT) later described by Ahmad et al. in 2009 and cited by Insall and Salvati in the sixth edition of Surgery of the Knee [50]. For the second part of the test, a medial force is applied to the patella as the knee is flexed. Apprehension is reduced because the symptoms of impending dislocation are removed as the patella now engages the trochlea in a normal fashion. Both manoeuvres must be completed, and to be positive, apprehension must be first provoked and then alleviated. The authors likened this to a pivot shift test, reporting a 100% positivity rate in patients who had patellar dislocation confirmed with examination under anaesthesia [51].

48.8.1.4 Femoral Rotation

With the patient supine, demonstrate femoral rotation with the hip and knee each flexed to 90°. Internally and externally rotate the femur, taking note of the maximal range. Femoral rotation can contribute to a varus or valgus angle at the knee [6]. The patient can alternatively be positioned prone to assess this.

48.8.2 Dynamic Assessment while Supine

Dynamic assessment of the PFJ is preferential in the weight-bearing position [52], so there is limited utility in the dynamic assessment of the PFJ while supine. Extensor mechanism dysfunction may be assessed, but an extensor lag is preferably checked from a seated position. Assessment of an extensor lag requires comparison to the contralateral "healthy" knee. The magnitude of asymmetry is difficult to gauge, and currently there are no absolute values of an extensor lag that correlate well with PFJ dysfunction [53].

48.9 Special Tests

One of the first and easiest identifiable signs of a PFJ abnormality is patellofemoral crepitus [7]. PFJ compression/glide test is also a useful assessment for PFJ OA as well as other pathologies [54].

48.10 Coarse Crepitus

Seat the patient on an adjustable examination couch with both knees flexed to 90° and feet plantigrade. The examiner places their hand over the PFJ. As the patient rises to a standing position, bearing weight evenly through both lower limbs, and sits back down again, the examiner can feel or palpate crepitus.

48.11 PFJ Glide/Compression Test [7]

Position the patient semi-recumbent on the examination couch with a pillow behind both knees. The examiner gently applies posteriorly directed pressure to the PFJ, ensuring no mediolateral or caudad/cephalad movement of the patella. This manoeuvre may elicit pain. If there is no pain, maintain downward pressure on the PFJ and glide the patella inferiorly as far as it will go. Downward compression is released once the patella returns to the resting position. The outcome is recorded as (i) no pain, (ii) pain with compression and (iii) pain with glide.

Clarke's test is where the patella is compressed against the trochlea manually while the examiner asks the patient to contract their quadriceps. It is a painful test, and the senior author believes that it should be performed (only once) at the end of the examination. A negative Clarke's test is useful in the formulation of the differential diagnosis. There is no available record of the original description of Clarke's test/sign, the technique amongst examiners is inconsistent, a positive test is ill-defined and it is often conflated with other testing techniques. The most common description is the one above with the knee extended and reflects active patellar contraction [55].

48.12 Patella Tilt Test

The patellar tilt test is a manual test of joint mobility during which the examiner applies a force to tilt the patella medially and at the same time prevents its lateral translation. It is performed to determine if the patient has a tight lateral restraint when considering a lateral release procedure. It has poor intra- and inter-relator reliability [56] and is highly subjective.

In a systematic review and meta-analysis by Decary et al., five tests of patellofemoral pain were evaluated from nine studies: (1) the active instability test, (2) pain during stair climbing, (3) Clarke's test/sign, (4) pain during prolonged sitting and (5) the patellar tilt test. Based on the evidence available, no individual tests could be recommended to diagnose or exclude patellofemoral pain [57].

48.13 Conclusion

Clinical examination of the patellofemoral joint has not been as widely assessed for intra- and inter-observer reliability as have tests of knee osteoarthritis and the status of the anterior cruciate ligament [58]. The standardisation of approach outlined in this chapter and use of correct examination techniques will allow the examiner to determine a working diagnosis. The subtle and subjective findings as described are of importance in achieving this working diagnosis and should ensure imaging is then used appropriately, to confirm that working diagnosis. Over-reliance on imaging is to be avoided. Better clinical examination will lead to improved management of PFJ pathology.

References

1. Fulkerson JP. Diagnosis and treatment of patients with patellofemoral pain. Am J Sports Med. 2002;30(3):447–56.
2. Osman NM, Ebrahim SMB. Patellofemoral instability: quantitative evaluation of predisposing factors by MRI. Egypt J Radiol Nucl Med. 2016;47(4):1529–38.
3. Rosa SB, Ewen PM, Doma K, Ferrer JFL, Grant A. Dynamic evaluation of patellofemoral instability: a clinical reality or just a research field? A literature review. Orthop Surg. 2019;11(6):932–42.
4. Panni AS, Cerciello S, Maffulli N, Di Cesare M, Servien E, Neyret P. Patellar shape can be a predisposing factor in patellar instability. Knee Surg Sports Traumatol Arthrosc. 2011;19(4):663–70.

5. Flandry F, Hommel G. Normal anatomy and biomechanics of the knee. Sports Med Arthrosc Rev. 2011;19(2):82–92.
6. Manske RC, Davies GJ. Examination of the patellofemoral joint. Int J Sports Phys Ther. 2016;11(6):831–53.
7. Stefanik JJ, Duncan R, Felson DT, Peat G. Diagnostic performance of clinical examination measures and pain presentation to identify patellofemoral joint osteoarthritis. Arthritis Care Res. 2018;70(1):157–61.
8. Décary S, Frémont P, Pelletier B, Fallaha M, Belzile S, Martel-Pelletier J, et al. Validity of combining history elements and physical examination tests to diagnose patellofemoral pain. Arch Phys Med Rehabil. 2018;99(4):607–14.e1.
9. Larbi A, Cyteval C, Hamoui M, Dallaudiere B, Zarqane H, Viala P, et al. Hoffa's disease: a report on 5 cases. Diagn Interv Imaging. 2014;95(11):1079–84.
10. Jackson DW, Schaefer RK. Cyclops syndrome: loss of extension following intra-articular anterior cruciate ligament reconstruction. Arthroscopy. 1990;6(3):171–8.
11. Dupont JY. Synovial plicae of the knee. Knee. 1994;1(1):5–19.
12. Lee PYF, Nixion A, Chandratreya A, Murray JM. Synovial plica syndrome of the knee: a commonly overlooked cause of anterior knee pain. Surg J (N Y). 2017;3(1):e9–e16.
13. Navali AM, Bahari LAS, Nazari B. A comparative assessment of alternatives to the full-leg radiograph for determining knee joint alignment. Sports Med Arthrosc Rehabil Ther Technol. 2012;4(1):40.
14. Hinman RS, May RL, Crossley KM. Is there an alternative to the full-leg radiograph for determining knee joint alignment in osteoarthritis? Arthritis Care Res. 2006;55(2):306–13.
15. Magee DJ. Orthopedic physical assessment-e-book. Amsterdam: Elsevier Health Sciences; 2014.
16. Riddle DL. Validity of clinical measures of frontal plane knee alignment: data from the osteoarthritis initiative. Man Ther. 2012;17(5):459–65.
17. Brattstroem H. Shape of the intercondylar groove normally and in recurrent dislocation of patella. A clinical and X-ray-anatomical investigation. Acta Orthop Scand Suppl. 1964;68(SUPPL 68):1–148.
18. Khasawneh RR, Allouh MZ, Abu-El-Rub E. Measurement of the quadriceps (Q) angle with respect to various body parameters in young Arab population. PLoS One. 2019;14(6):e0218387.
19. Abade PO, Okwoka BO, Ebeye OA. Influence of gender on quadriceps (Q) angle among adult urhobos in Nigeria population. J Exp Clin Anat. 2014;13(2):50.
20. Yablon IG, Heller FG, Shouse L. The key role of the lateral malleolus in displaced fractures of the ankle. J Bone Joint Surg Am. 1977;59(2):169–73.
21. Olerud C, Berg P. The variation of the Q angle with different positions of the foot. Clin Orthop Relat Res. 1984;191:162–5.
22. Kuhn DR, Yochum TR, Cherry AR, Rodgers SS. Immediate changes in the quadriceps femoris angle after insertion of an orthotic device. J Manip Physiol Ther. 2002;25(7):465–70.
23. Livingston LA, Spaulding SJ. OPTOTRAK measurement of the quadriceps angle using standardized foot positions. J Athl Train. 2002;37(3):252–5.
24. Smith TO, Clark A, Neda S, Arendt EA, Post WR, Grelsamer RP, et al. The intra- and inter-observer reliability of the physical examination methods used to assess patients with patellofemoral joint instability. Knee. 2012;19(4):404–10.
25. Fulkerson JP, Buuck DA. Disorders of the patellofemoral joint. Philadelphia: Lippincott; 2004.
26. Post WR. Clinical evaluation of patients with patellofemoral disorders. Arthroscopy. 1999;15(8):841–51.
27. Meininger A. Understanding the patellofemoral joint: from instability to arthroplasty; an issue of clinics in sports medicine. Amsterdam: Elsevier Health Sciences; 2014.
28. Outerbridge RE, Outerbridge HK. The etiology of chondromalacia patellae. Clin Orthop Relat Res. 2001;389:389.
29. Leblanc C, Houghton K. Chapter 51@ noninflammatory musculoskeletal pain. 2016.
30. Tiberio D. The effect of excessive subtalar joint pronation on patellofemoral mechanics: a theoretical model. J Orthop Sports Phys Ther. 1987;9(4):160–5.
31. Barton CJ, Bonanno D, Levinger P, Menz HB. Foot and ankle characteristics in patellofemoral pain syndrome: a case control and reliability study. J Orthop Sports Phys Ther. 2010;40(5):286–96.
32. Pirker W, Katzenschlager R. Gait disorders in adults and the elderly: a clinical guide. Wien Klin Wochenschr. 2017;129(3–4):81–95.
33. Arazpour M, Bahramian F, Abutorabi A, Nourbakhsh ST, Alidousti A, Aslani H. The effect of patellofemoral pain syndrome on gait parameters: a literature review. Arch Bone Jt Surg. 2016;4(4):298–306.
34. Carl GM, Ann LL, Mario AD, Tim LU. "Dynamic Trendelenburg": single-leg-squat test for gluteus medius strength. Int J Athl Ther Train. 2004;9(1):24–5.
35. Fitzgerald GK. Open versus closed kinetic chain exercise: issues in rehabilitation after anterior cruciate ligament reconstructive surgery. Phys Ther. 1997;77(12):1747–54.
36. Herrington L. Knee valgus angle during single leg squat and landing in patellofemoral pain patients and controls. Knee. 2014;21(2):514–7.
37. Crossley KM, Zhang WJ, Schache AG, Bryant A, Cowan SM. Performance on the single-leg squat task indicates hip abductor muscle function. Am J Sports Med. 2011;39(4):866–73.
38. Ressman J, Grooten WJA, Rasmussen BE. Visual assessment of movement quality in the single leg squat test: a review and meta-analysis of inter-rater and intrarater reliability. BMJ Open Sport Exerc Med. 2019;5(1):e000541.

39. Cibulka MT, Threlkeld-Watkins J. Patellofemoral pain and asymmetrical hip rotation. Phys Ther. 2005;85(11):1201–7.
40. Waligora AC, Johanson NA, Hirsch BE. Clinical anatomy of the quadriceps femoris and extensor apparatus of the knee. Clin Orthop Relat Res. 2009;467(12):3297–306.
41. Lefebvre R, Leroux A, Poumarat G, Galtier B, Guillot M, Vanneuville G, et al. Vastus medialis: anatomical and functional considerations and implications based upon human and cadaveric studies. J Manip Physiol Ther. 2006;29(2):139–44.
42. Giles LS, Webster KE, McClelland JA, Cook J. Atrophy of the quadriceps is not isolated to the vastus medialis oblique in individuals with patellofemoral pain. J Orthop Sports Phys Ther. 2015;45(8):613–9.
43. Beckert MW, Albright JC, Zavala J, Chang J, Albright JP. Clinical accuracy of J-sign measurement compared to magnetic resonance imaging. Iowa Orthop J. 2016;36:94–7.
44. Fairbank HA. Internal derangement of the knee in children and adolescents: (section of Orthopaedics). Proc R Soc Med. 1937;30(4):427–32.
45. Hughston JC. Subluxation of the patella. JBJS. 1968;50(5):1003–26.
46. Sallay PI, Poggi J, Speer KP, Garrett WE. Acute dislocation of the patella. A correlative pathoanatomic study. Am J Sports Med. 1996;24(1):52–60.
47. Hughston JC. Subluxation of the patella. J Bone Joint Surg Am. 1968;50(5):1003–26.
48. Kolowich PA, Paulos LE, Rosenberg TD, Farnsworth S. Lateral release of the patella: indications and contraindications. Am J Sports Med. 1990;18(4):359–65.
49. Reider B. The Orthopaedic physical examination. Philadelphia: WB Saunders; 1999.
50. Scott WN. Insall & Scott surgery of the knee E-book. Amsterdam: Elsevier Health Sciences; 2017.
51. Ahmad CS, McCarthy M, Gomez JA, Shubin Stein BE. The moving patellar apprehension test for lateral patellar instability. Am J Sports Med. 2009;37(4):791–6.
52. Muhle C, Brossmann J, Heller M. Kinematic CT and MR imaging of the patellofemoral joint. Eur Radiol. 1999;9(3):508–18.
53. Stillman BC. Physiological quadriceps lag: its nature and clinical significance. Aust J Physiother. 2004;50(4):237–41.
54. Crema MD, Guermazi A, Sayre EC, Roemer FW, Wong H, Thorne A, et al. The association of magnetic resonance imaging (MRI)-detected structural pathology of the knee with crepitus in a population-based cohort with knee pain: the MoDEKO study. Osteoarthr Cartil. 2011;19(12):1429–32.
55. Doberstein ST, Romeyn RL, Reineke DM. The diagnostic value of the Clarke sign in assessing chondromalacia patella. J Athl Train. 2008;43(2):190–6.
56. Watson CJ, Leddy HM, Dynjan TD, Parham JL. Reliability of the lateral pull test and tilt test to assess patellar alignment in subjects with symptomatic knees: student raters. J Orthop Sports Phys Ther. 2001;31(7):368–74.
57. Décary S, Ouellet P, Vendittoli PA, Roy JS, Desmeules F. Diagnostic validity of physical examination tests for common knee disorders: an overview of systematic reviews and meta-analysis. Phys Ther Sport. 2017;23:143–55.
58. Décary S, Ouellet P, Vendittoli PA, Desmeules F. Reliability of physical examination tests for the diagnosis of knee disorders: evidence from a systematic review. Man Ther. 2016;26:172–82.

Evaluation of the Stability and Function of the Tibiofemoral and Tibiofibular Joints

49

Felipe Galvão Abreu, Renato Andrade, Rogério Pereira, Ricardo Bastos, and João Espregueira-Mendes

49.1 Introduction

Despite improvements in advanced imaging techniques and manual instrumented devices, medical history and physical examination remain an essential step in evaluating the knee. Unlike meniscal and cartilage injuries which elicit painful stimuli and can often result in false-positive or false-negative results, ligament injury tests are much more sensitive. On the other hand, conclusions depend exclusively on the tactile sensitivity, experience, and training of the surgeon [1]. Evaluation of every patient should begin with a complete history of the symptoms and full description of the mechanism of injury. The clinical history often identifies the area of knee involvement. With this information, the clinician can perform a more direct and accurate diagnosis. When assessing the knee, the clinician needs to know the anatomy and fully understand the role of each ligament tested in the knee stability. A complete physical examination consists of inspection, palpation, and specific tests.

F. G. Abreu
Clínica Sportsmed, São José do Rio Preto, Brazil

Hospital Israelita Albert Einstein, São Paulo, Brazil

R. Andrade
Clínica Espregueira - FIFA Medical Centre of Excellence, Porto, Portugal

Dom Henrique Research Centre, Porto, Portugal

Porto Biomechanics Laboratory (LABIOMEP), Faculty of Sports, University of Porto, Porto, Portugal

R. Pereira
Clínica Espregueira - FIFA Medical Centre of Excellence, Porto, Portugal

Dom Henrique Research Centre, Porto, Portugal

Superior School of Health, University Fernando Pessoa, Porto, Portugal

R. Bastos
Clínica Espregueira - FIFA Medical Centre of Excellence, Porto, Portugal

Dom Henrique Research Centre, Porto, Portugal

J. Espregueira-Mendes (✉)
Clínica Espregueira - FIFA Medical Centre of Excellence, Porto, Portugal

Dom Henrique Research Centre, Porto, Portugal

ICVS/3B's–PT Government Associate Laboratory, Braga/Guimarães, Portugal

3B's Research Group–Biomaterials, Biodegradables and Biomimetics, University of Minho, Headquarters of the European Institute of Excellence on Tissue Engineering and Regenerative Medicine, Guimarães, Portugal

School of Medicine, University of Minho, Braga, Portugal
e-mail: espregueira@dhresearcentre.com

J. G. Lane et al. (eds.), *The Art of the Musculoskeletal Physical Exam*,
https://doi.org/10.1007/978-3-031-24404-9_49

49.2 Inspection

The examination should begin with observation. The observation of the gait pattern often provides important information on the patient's symptoms. The clinician should ask the patient to walk a few meters and assess the gait pattern and the stance position of the lower limb. It is important to note the patient's ability and speed of gait, if they are using any gait aids, and if there is any discomfort during the attempted gait test. A shortened stance phase of gait (antalgic gait) will confirm the side of involvement. Antalgic positions as flexed attitude of the knee are assumed to better accommodate the presence of joint effusion and relieve pain or could mean joint blocking. A short-leg gait requires confirmation of limb length [2].

Anatomical alignment of the lower limb is also a fundamental factor that needs to be measured. The clinician should note if there is varus or valgus alignment and/or medial or lateral thrust in the stance phase of gait. Clinical alignment is determined by measuring the femorotibial angle—which must be differentiated from the mechanical axis of the limb. The femorotibial angle is measured from the femoral head to the center of the knee joint (patella), and from there to the center of the ankle in a standing position. Using a goniometer to measure the femorotibial angle, the clinician can examine if there is clinical varus or valgus alignment. This measurement should be used along with the roentgenographic measurements [2]. In lateral knee evaluation, genu recurvatum should be evaluated. Increased knee hyperextension of 10° and hamstring flexibility are significantly associated with the risk of anterior cruciate ligament (ACL) lesion [3].

Clinical effusion is often visually identifiable. The clinician should measure the active range of motion (ROM) and if there are any limitations to full knee extension or flexion. Active ROM should also be evaluated with palpation and further compared with passive ROM. In a non-injured subject, it is expected that full extension is 0°. When the patient is unable to fully extend the knee, the clinician should suspect extension lag, a locked knee, or a flexion contracture. An inability to fully flex the knee may be associated with effusion, pain, or extension contracture.

Neurovascular evaluation should look for possible signs of numbness or tingling in the extremity or possible muscle weakness. A drop-foot sign can be indicative of a fibular nerve injury. The ability to contract the musculature must be evaluated and compared to the contralateral side. Quadriceps atrophy can sometimes be visually identified and will indicate the involved side. When there is gross atrophy, the clinician should measure the circumferential perimeter. Scars in the knee region often indicate previous surgeries, which are important to understand the patient's history and to plan any future surgical intervention.

49.3 Palpation

All bony landmarks should be palpated and identified. Any pain or deformity should be noticed and compared with the unaffected side. Effusions can be present and should be graded in size while compressing the suprapatellar pouch. The clinician should also note if there is any fluid.

Crepitation in and of itself may or may not represent evidence of a disorder. The location should be recorded for future reference. It can involve the medial or lateral side of the patellofemoral joint and/or the medial or lateral side of the tibiofemoral joint.

49.4 Special Tests: Anterior Cruciate Ligament (ACL)

49.4.1 Lachman-Noulis Test

Lachman test is considered the most sensitive test for ACL tear diagnosis, with high sensitivity and specificity [4, 5]. The sensitivity and specificity are both 81% [6]. The first description was made in 1875 by George K. Noulis in his doctoral thesis "Entorse du Genou." Indeed, his thesis is often reported as the first full description of this important test for ACL function that was attributed to John W. Lachman almost a century later [7, 8].

The test is performed with the patient in the supine position, with the knee in 30° of flexion. The tibia rests in a neutral position, maintaining secondary stabilizers relaxed. With one hand, the clinician holds the thigh to stabilize the distal femur, while with the other hand holds the proximal tibia firmly. The clinician then applies an anterior force to the proximal tibia while keeping the femur stabilized and notes if there is any feeling of anterior subluxation (Fig. 49.1) [9]. The test is considered positive if there is excessive anterior translation of the proximal tibia greater than the uninjured side and also a lack of a firm endpoint. Endpoints are graded from "hard" to "soft" and have been nominally classified as A (firm, hard endpoint) or B (absent, soft endpoint) [10]. The anterior translation result is graduated in *grade 1*–1–5 mm, *grade 2*–6–10 mm, and *grade 3*– > 10 mm.

49.4.2 Anterior Drawer Test

Although the anterior drawer test has been widely used in the diagnosis of ACL ruptures, its origin remains obscure [11]. Paul Segond described in 1879 an "abnormal anterior-posterior mobility" of the knee that is associated with ACL ruptures [12]. The anterior drawer test has a low sensitivity (38%), but relatively high specificity (81%) [6]. The patient is placed in supine position, with the knee in 90° of flexion. The clinician may sit on the foot to help stabilize the lower leg. Then, the clinician holds the proximal lower leg, just below the tibial plateau or tibiofemoral joint line, and applies a posteroanterior force, noting if there is any feeling of anterior tibial subluxation (Fig. 49.2) [13]. An abnormally increased anterior tibial displacement, when compared with the healthy contralateral knee, is most often indicative of an ACL tear [14]. The anterior drawer test must be performed in neutral, but also at 30° of internal and external rotation. In internal rotation,

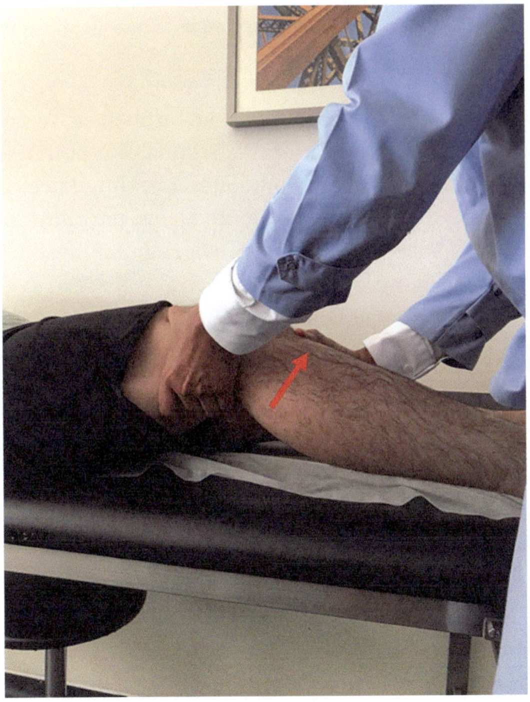

Fig. 49.1 The Lachman-Noulis test is performed with the patient supine and the knee flexed at 30°. The clinician applies a posteroanterior force on the proximal end of the tibia

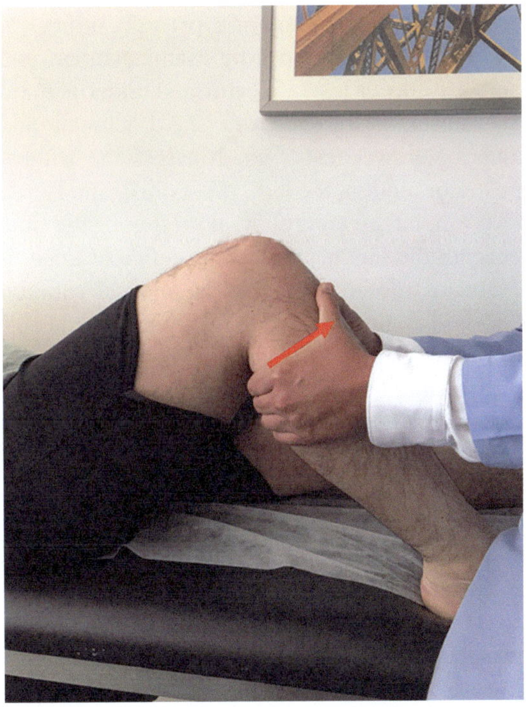

Fig. 49.2 The anterior drawer test is performed with the patient supine and the knee flexed at 90°. The clinician applies a posteroanterior force on the proximal end of the tibia

the posterior cruciate ligament (PCL) and the posterolateral corner are tightened, so a positive finding indicates a potential associated injury of these structures. Likewise, a positive finding during the external rotation—also known as the Slocum test—indicates potential injury of the posteromedial corner [15, 16]. In an acutely swollen knee, the clinician may perform the test but has to place the knee in a less flexed position (at 60–80°) to avoid excessive pain derived from the hemarthrosis. The menisci can mimic a hard stop during the anterior tibial displacement and thus lead to false-negative results. This is called as "doorstop" effect and is more common in the lateral meniscus [2].

49.4.3 Pivot-Shift Test

The pivot-shift test was first described by Galway and Macintosh in 1972 and has been considered a better predictor of clinical outcomes than any uniplanar maneuver and the gold standard for diagnosis of ACL injury [5, 17, 18]. It is more specific than Lachman and anterior drawer tests if performed under general anesthesia. Different from the Lachman and anterior drawer tests, which assess the anteroposterior motion, the jerk, pivot-shift, and Losee tests emphasize anterolateral motion of the tibia beneath the femur. In an ACL-deficient knee, it evaluates the combined tibial internal rotation and subluxation of the lateral tibial plateau followed by its reduction [19].

The clinician applies an internal rotation and valgus force to the extended knee. An anterolateral tibia subluxation is observed in case of an ACL-deficient knee. The clinician then applies a flexion and valgus force while flexing the knee (Fig. 49.3). The iliotibial band will change its role from being a knee extender to a flexor and, in case of a positive test, will visibly reduce the tibial subluxation with a palpable clunk. This signal is observed in the lateral compartment at about 20–30° of knee flexion. It is crucial to compare the results with the contralateral healthy knee. If there is a positive sign in both knees, the clinician should suspect and further evaluate the presence of underlying hyperlaxity.

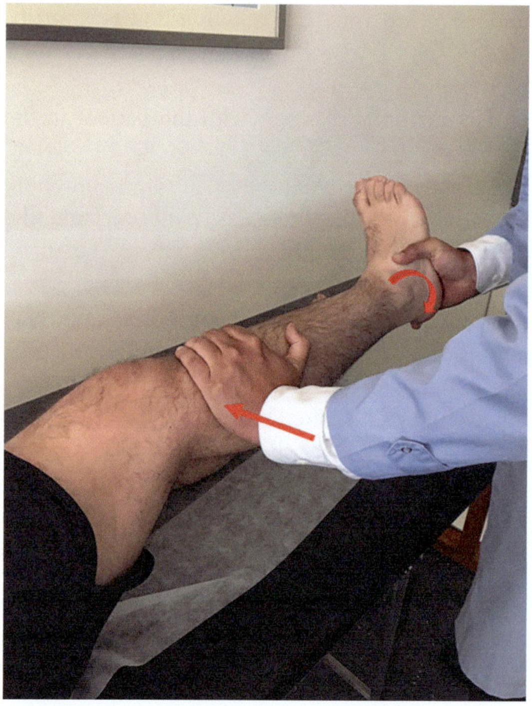

Fig. 49.3 The pivot-shift test starts with the patient supine and the knee in full extension. Then, the clinician applies internal rotation and valgus stress to the tibia while flexing the knee

This test can be limited in awake patients due to muscle guarding. This test can also be painful in patients that have a medial collateral ligament (MCL) injury. Sensibility and specificity will thus vary significantly depending on the patient's ability to abstain from muscle guarding. The reported sensitivity of pivot shift in ACL injuries is 73% with a specificity of 98% when the test is performed with the patient under anesthesia; in the alert patient, the sensibility and specificity are lower, 28% and 81%, respectively [6].

49.4.4 Jerk Test

The jerk test is initiated with the knee in flexion with associated internal tibial rotation, anterior pressure on the fibular head, and valgus stress. This combination results in anterior subluxation of the lateral tibial condyle. As the knee is brought into extension, in an ACL-deficient knee, the tibia reduces with a palpable clunk [2].

49.4.5 Losee Test

The Losee test is similar to the jerk test. It also begins with the knee in flexion and the tibia in external rotation, with the clinician applying valgus stress (Fig. 49.4a). Then, while maintaining the valgus stress, the knee is gradually extended and the clinician rotates the tibia internally (Fig. 49.4b). In a positive test, a clunk of reduction should be felt. The Losee test attempts to accentuate the tibial subluxation with external tibial rotation [20–22].

49.4.6 Lever Sign Test

To test the lever sign, the patient lies supine with the knees in full extension. The clinician places a closed fist under the proximal third of the calf, which will place the knee with a slight flexion. With the other hand, the clinician applies a moderate and downward force to the distal third of the anterior thigh (above the quadriceps tendon). The patient's leg will act as a lever over the clini-

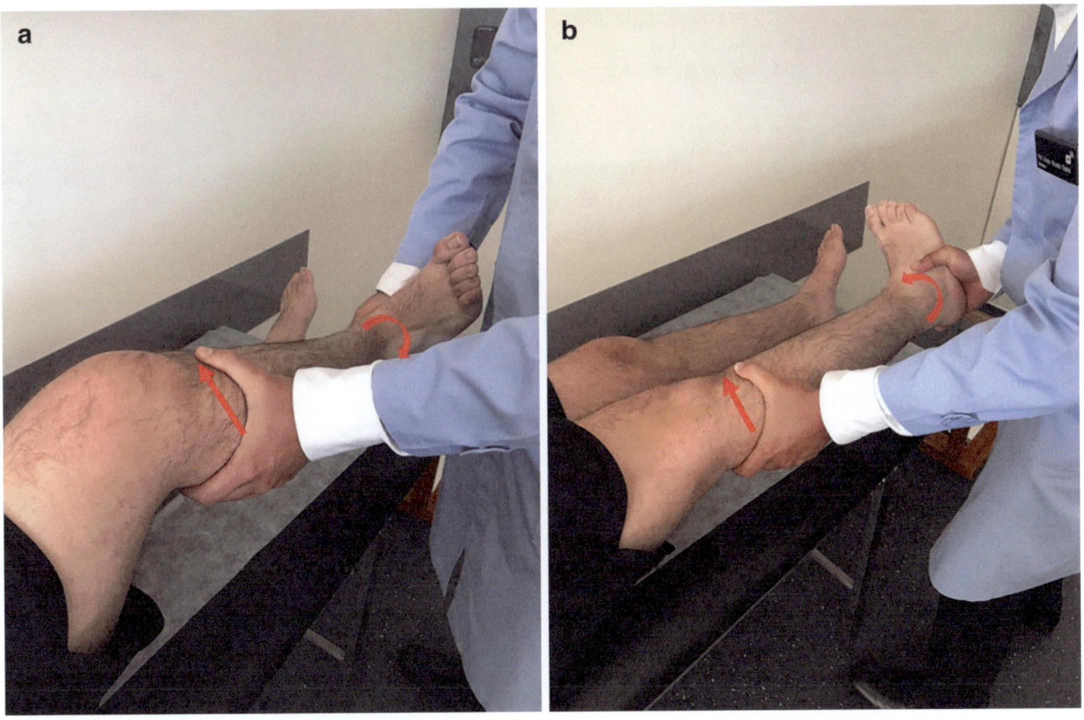

Fig. 49.4 The Losee test starts with the patient supine and the knee in 45° flexion and the tibia externally rotated. In this position, the clinician applies valgus stress (**a**) and then internal rotation to the tibia while extending the knee (**b**)

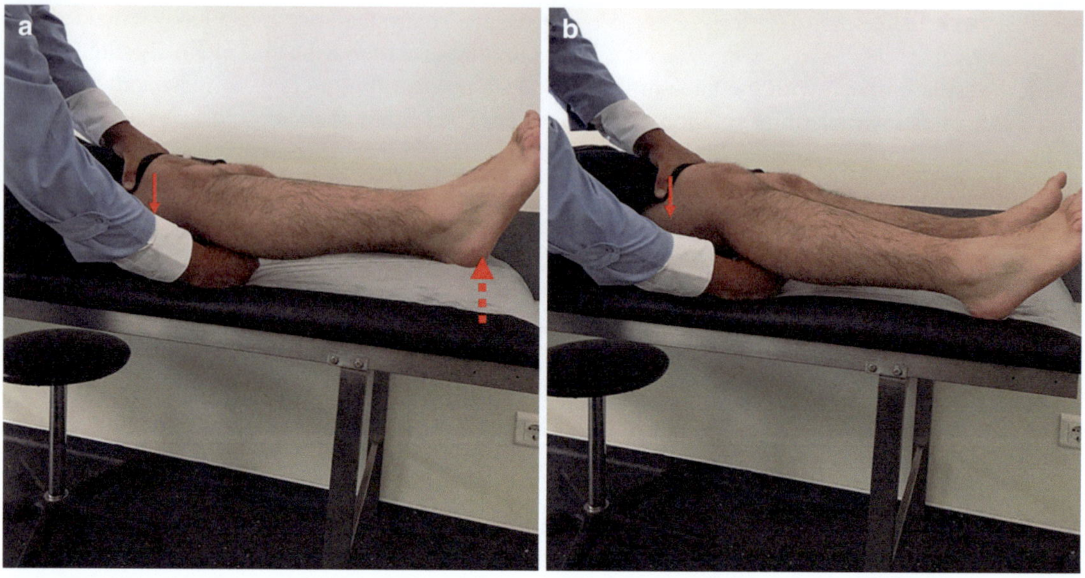

Fig. 49.5 The lever sign test starts with the patient lying supine. The clinician places a closed fist beneath the proximal third of the calf and with the other hand applies anteroposterior force at the distal end of the femur. In a patient with an intact and functional ACL, the patient's foot should rise (**a**); in an ACL-deficient knee, there is no lever effect and the foot will not rise (**b**)

cian's fist (which will serve as fulcrum). In an intact knee, the downward force on the quadriceps and the lever created by the ACL will outweigh the force of gravity and rotate the knee into full extension and the heel will rise up (Fig. 49.5a). In an ACL-deficient knee, the ability to outweigh the force of gravity of the lower leg is compromised and the tibial plateau will slide anteriorly without raising the patient's heel (Fig. 49.5b) [23].

49.5 Special Tests: Posterior Cruciate Ligament (PCL)

49.5.1 Posterior Sag Test

The posterior sag sign was first described by Mayo Robson in 1903, although it is unclear who coined the term *posterior sag sign* [24, 25]. The patient should lie supine with the hip flexed to 45° and the knee in 90° of flexion. The foot should be in neutral position and the clinician may sit on top of the foot (Fig. 49.6). In this position, if there is PCL tear, the tibia "rocks back" or sags back in relation to the femur. In healthy knees, the medial tibial plateau usually extends

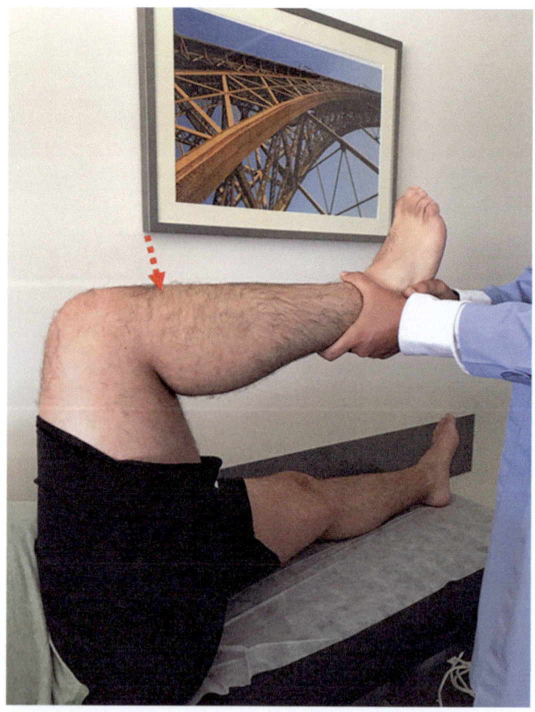

Fig. 49.6 The posterior sag test starts with the patient lying supine and relaxed, with the knee in 90° flexion and the foot resting in the examination table. The test is positive if the medial tibial plateau is posterior in relation to the medial femoral condyle. The dashed arrow indicates where the clinician should look for the drop of the tibia

1 cm anteriorly to the medial femoral condyle when the knee is flexed to 90°. If this step-off is lost, it is then considered positive for a PCL tear [26]. The posterior sag test is the most sensitive test for PCL injury [27].

49.5.2 Dial Test

The dial test—also called the posterolateral rotation test—should be performed with the knee flexed at 30° and 90° (Fig. 49.7). The patient is placed supine with the knee flexed off the edge of the table. The thigh is stabilized against the table and the foot externally rotated. An external rotation of the tibial tubercle greater than 15° indicates a posterolateral knee injury. When the test is repeated with the knee in 90° of flexion and there is also an increase in the external rotation, it indicates a combined PCL and PLC injury [28–30].

49.5.3 Posterior Lachman

The posterior Lachman test evaluates the PCL, but with less efficacy than anterior Lachman does the ACL. As in the anterior Lachman, the patient lies supine with the knee in 30° of flexion. The clinician applies posterior stress at the proximal tibia (Fig. 49.8). If an abnormal posterior movement is noted, the clinician should suspect a PCL tear [2].

49.5.4 Posterior Drawer Test

The posterior drawer test is easy to perform and considered the most accurate test for detecting a PCL injury [31], but requires some attention to avoid mistakes and for accurate interpretation. The test is performed at 90° of knee flexion and is positive when the application of posterior force to the proximal tibia promotes an abnor-

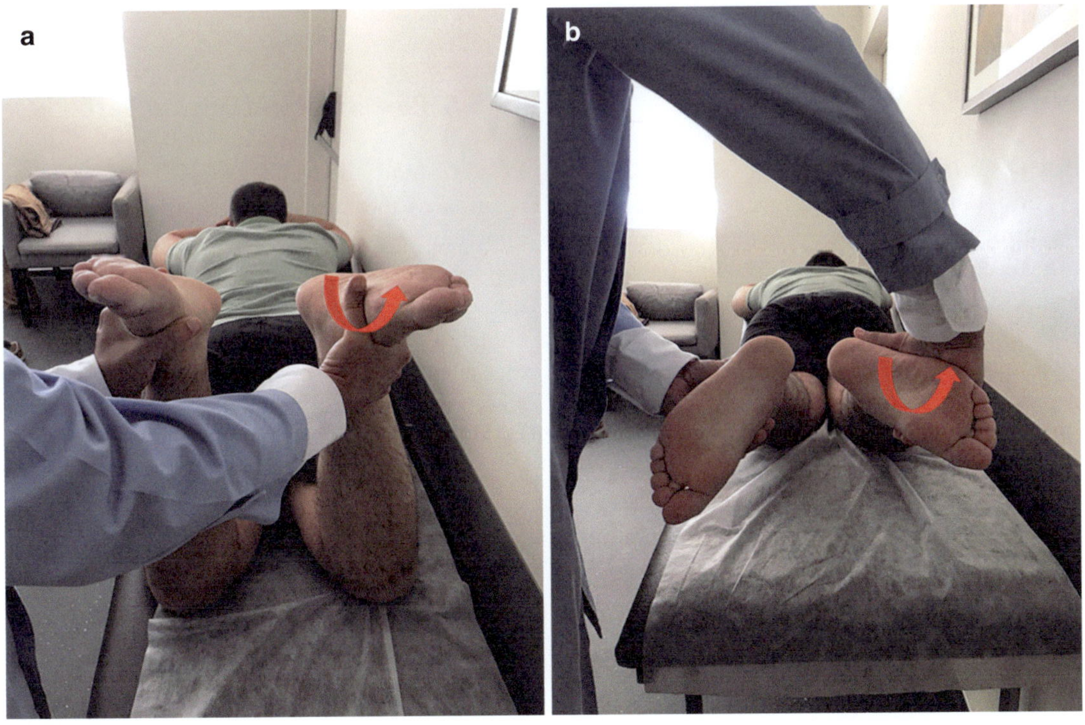

Fig. 49.7 The dial test may be performed supine or prone. The clinician applies an external rotation stress through the patient's feet. The test should be performed at 90° (**a**) and 30° (**b**) of knee flexion. The test should be performed bilaterally, and the clinician should look if there is an increased external rotation in the injured side. The patient thighs should be adducted and stabilized by an assistant (not shown in this figure)

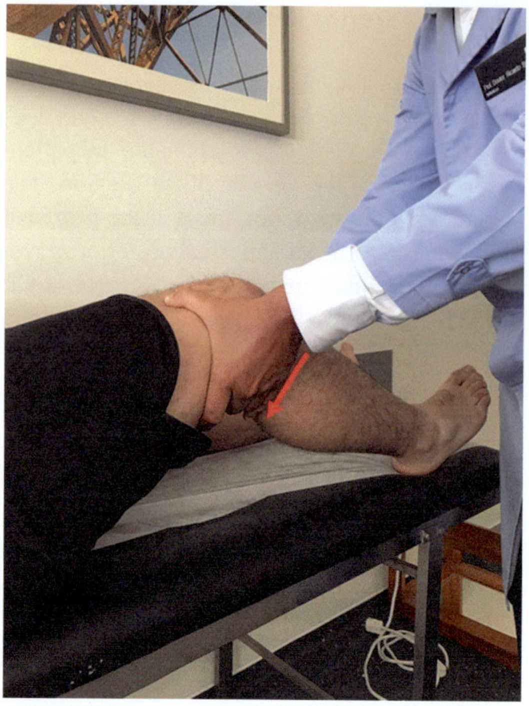

Fig. 49.8 The posterior Lachman test is performed with the patient supine and the knee flexed at 30°. The clinician applies an anteroposterior force on the proximal end of the tibia

mal posterior translation (Fig. 49.9) [32, 33]. Anterior and posterior drawer test should be performed simultaneously, and the clinician has to take care to rule out the amount of anterior and posterior tibial translation. When there is a PCL deficiency, the knee displays a posteriorized resting position and the reduction to the neutral position can mimic a positive anterior drawer test. The clinician should perform a careful evaluation to avoid this mistake. The clinician may use palpation to determine the correct starting point. In the neutral position, the tibial plateau and the medial condyle face one another, with a slight anterior step-off of the tibia (approximately 0.5–1 cm). This reference should be taken as the starting for both anterior and posterior drawer evaluation [2]. This test has a sensitivity and a specificity of 90% and 99%, respectively [31].

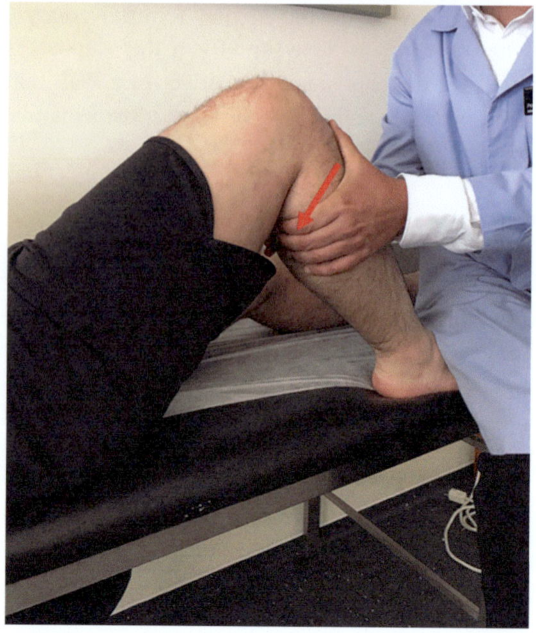

Fig. 49.9 The posterior drawer test is performed with the patient supine and the knee flexed at 90°. The clinician applies an anteroposterior force on the proximal end of the tibia

49.5.5 Posterolateral Drawer Test

Posterolateral drawer assesses posterolateral rotation of the knee at 90° of flexion, with the clinician forcing the tibia to external rotation (Fig. 49.10a). The increased posterolateral rotation is usually indicative of injury to the popliteus complex. This test must be differentiated from the posterior drawer test in neutral rotation, which tests the PCL. Posterolateral translation is increased when a combined posterior drawer and external rotation force is applied to the knee [34].

49.5.6 Posteromedial Drawer Test

Similar to the posteromedial drawer test, the posteromedial drawer test distinguishes between isolated PCL injury and combined injury of the PCL with superficial MCL and posterior oblique ligament (POL) injury [35]. The test is performed

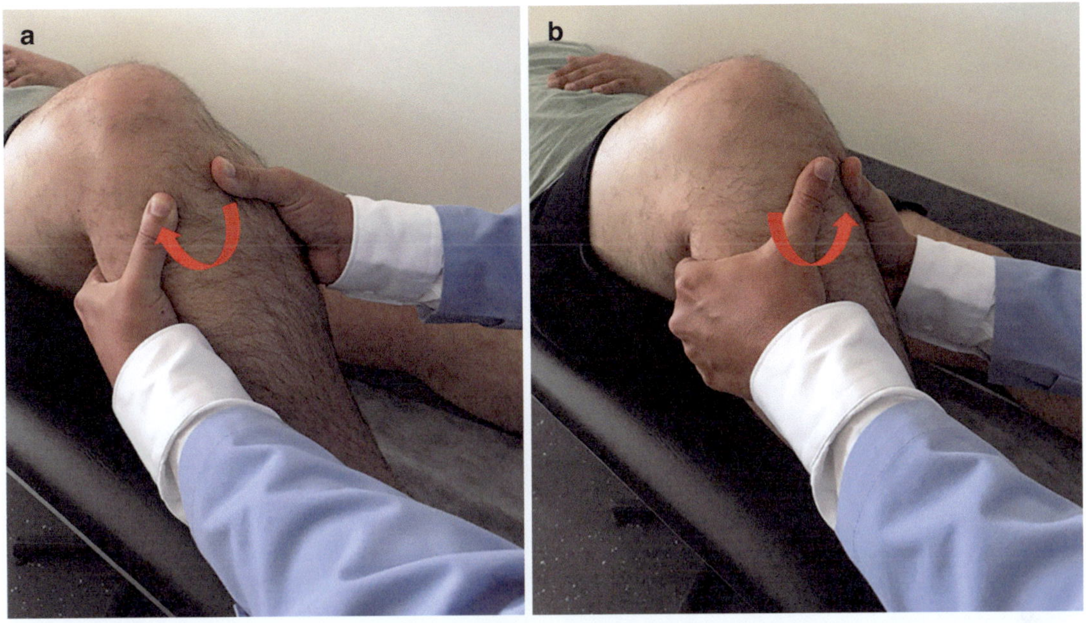

Fig. 49.10 The posterolateral and anterolateral drawer tests are performed with the patient supine and the knee flexed at 90°. The clinician applies an anteroposterior force on the proximal end of the tibia while rotating externally the tibia for the posterolateral drawer (**a**) or anteroposterior force while rotating internally the tibia for the posteromedial drawer test (**b**)

with the knee flexed at 90° with the clinician forcing the tibia to external rotation (Fig. 49.10b). It is positive when the application of anteroposterior force to the proximal tibia promotes an abnormal posterior translation.

49.5.7 Quadriceps Active Test

In *quadriceps active test,* the patient is placed in supine position with 90° of flexion and in neutral rotation (the clinician may sit on the top of the foot). The patient is then asked to contract the quadriceps muscle to extend the knee while the clinician applies counterpressure against the ankle. The quadriceps applies pressure through the patellar tendon to pull the tibial tubercle anteriorly, and thus any posterior translation of the tibia will be reduced back to a normal position [36]. The test is positive when the isometric quadriceps contraction dynamically reduces the tibia. This test is the most sensitive for PCL deficiency [27].

49.5.8 Reverse Pivot-Shift Sign

The *reverse pivot-shift sign* follows the same interpretation principles as the pivot shift, but in this case to assess PCL function. In a PCL-deficient knee, the lateral tibial plateau subluxes posteriorly when the tibia is stressed in external rotation and valgus, and that should reduce in knee extension (Fig. 49.11). The reverse pivot shift—which is very similar to a dynamic posterolateral drawer test—is not widely used because it can result in a large number of false positives as 35% of healthy knees can have a positive test when examined under anesthesia [37, 38].

49.5.9 External Rotation Recurvatum

The external rotation recurvatum test is performed with the patient in supine position. The clinician suspends the lower limb by holding the great toes and should observe if there is any

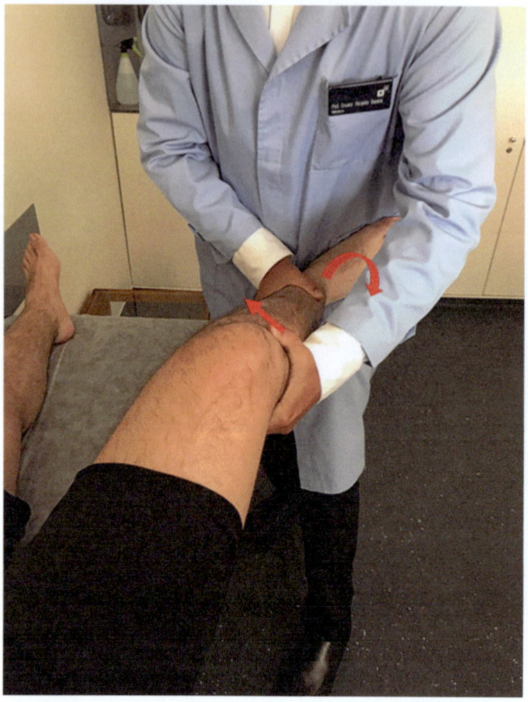

Fig. 49.11 The reverse pivot-shift test starts with the patient supine and the knee flexed about 60° and the tibia externally rotated. Then, the clinician extends the knee while maintaining the tibial external rotation and applying a valgus stress

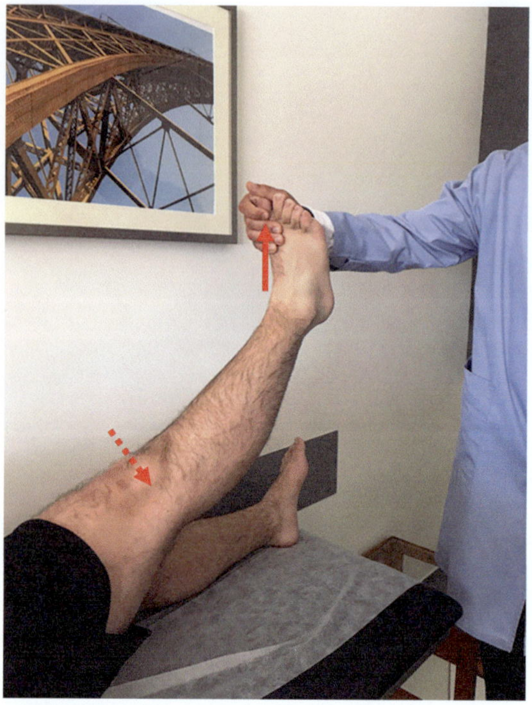

Fig. 49.12 The external rotation recurvatum test starts with the patient lying supine and relaxed. The clinician suspends the lower limb by grabbing the great toe and should look for any signs of knee hyperextension

abnormal hyperextension (Fig. 49.12). Historically, a positive test was described as relative hyperextension of the knee with relative external rotation of the tibial tubercle and varus deformity of the knee, indicating an injury to the ACL, PCL, and posterolateral corner (PLC) [39, 40]. More recently, after a rigorous evaluation of the external rotation recurvatum test, a positive external rotation recurvatum test is now defined as increased recurvatum combined with varus widening and anterior translation of the tibia, however displaying very low sensitivity (7.5%) [41]. If positive, it is indicative of a combined PLC and ACL injury.

49.6 Special Tests: Medial and Lateral Collateral Ligaments (MCL and LCL)

49.6.1 Valgus and Varus Stress Tests

Although the originator of the valgus and varus stress tests for detecting ligament laxity is unclear, Palmer, in 1938, described "abduction and adduction rocking" of the knee to determine the integrity of the collateral ligaments, which is an early reference to the valgus and varus stress tests used today [42]. Currently, varus and valgus testing is performed with the

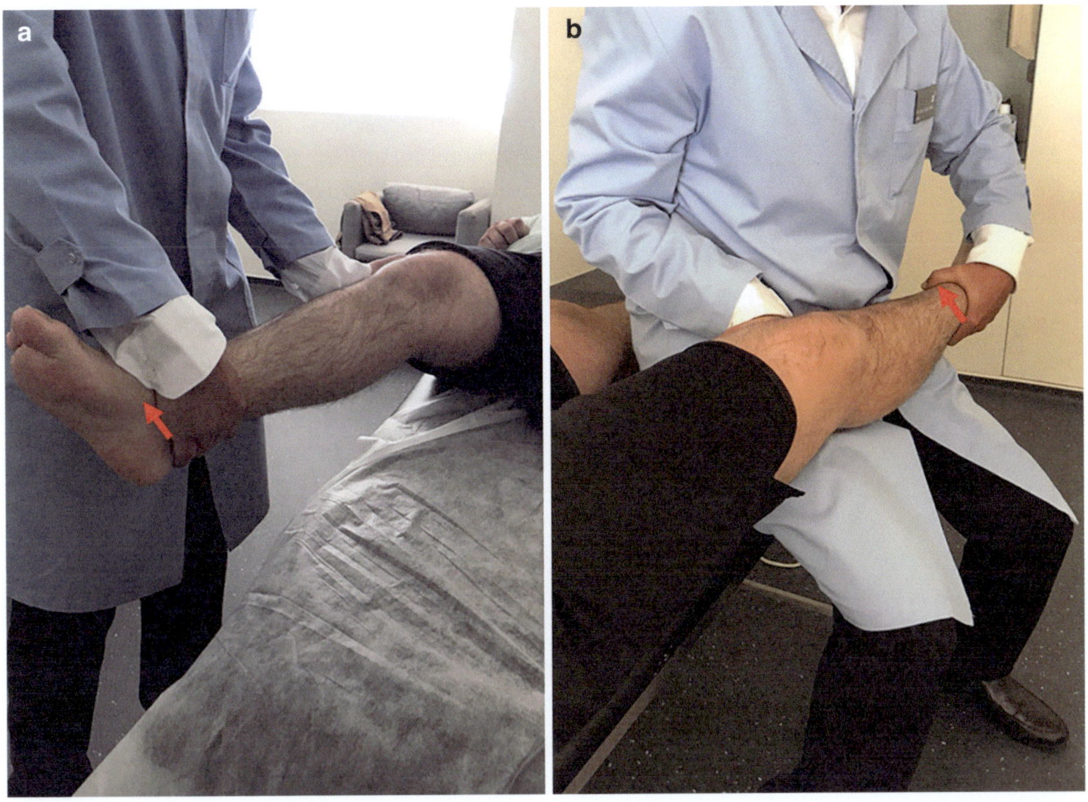

Fig. 49.13 The valgus and varus stress tests are performed with the patient supine and the knee in extension and then at 30° of knee flexion (figure shows 30° of flexion). For the valgus stress test, while palpating the medial joint line, the clinician stabilizes the distal femur externally and applies an abduction (lateral) stress at the medial ankle (**a**). For the valgus stress test, while palpating the lateral joint line, the clinician stabilizes the distal femur internally and applies an adduction (medial) stress at the lateral ankle (**b**)

patient placed in supine position, with slight abduction of the hip. Be cautious to not perform hip rotation. To perform the valgus stress test, the knee is flexed to 30° over the side of the table. Then, the clinician places one hand on the lateral aspect of the knee, holds the ankle with the other hand, and applies an abduction (valgus) stress to the knee (Fig. 49.13a). For the varus stress, the clinician must change hands' position, place one hand about the medial aspect of the knee with the other hand holding the ankle, and apply an adduction (varus) stress to the knee (Fig. 49.13b).

Both tests should be done with the knee in 30° of knee flexion and in extension. A valgus stress test positive at 30° and negative at 0° indicates a tear limited to the medial compartment ligaments (MCL with or without the posterior capsule), whereas a positive valgus stress test in extension implicates one or both cruciate ligaments in addition to the MCL and posterior capsule [43, 44]. On the lateral side, a positive varus stress test in flexion implicates the lateral collateral ligament (LCL), whereas a positive test in extension denotes a combined injury of the LCL, popliteus, and cruciate ligaments [44].

During stress testing, the clinician should record the degree of opening and the quality of the endpoint. It can be graded from I to III or by the number of millimeters that the joint opens as determined by the clinician [45]. Grade I corresponds to a stress examination that allows minimal to no opening with stress, but with the manipulation causing pain along the line of the collateral ligament. Grade II corresponds to a physical examination that shows some opening

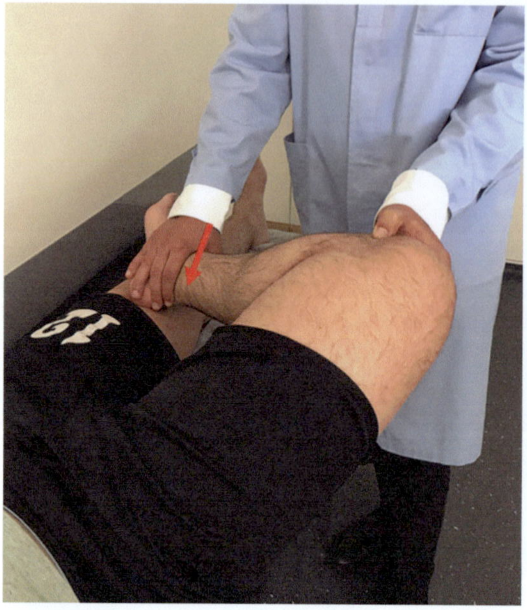

Fig. 49.14 The figure-of-four test starts with the patient lying supine. The knee is placed in a "figure-of-four" position, resting on the anterior thigh of the contralateral limb. The clinician applies a gentle varus stress while palpating the LCL

of the joint but with a distinct endpoint. Grade III shows no distinct endpoint to the stress evaluation [2].

49.6.2 Figure-of-Four

The figure-of-four is another stress test that evaluates the LCL. The knee is placed in a "figure-of-four" position, and a varus stress is applied to the knee joint (Fig. 49.14). When the LCL is intact, it can be distinctively palpated as a tight chord stretched from the fibular head to the lateral epicondyle. In an acute setting, this test is difficult to perform because it will elicit severe pain [46].

49.7 Clinical Evaluation: Proximal Tibiofibular Joint (PTFJ)

Injury of the proximal tibiofibular joint (PTFJ) is typically seen in athletes whose sports require violent twisting motions with a flexed knee, such as wrestling, parachute jumping, judo, jiujitsu, gymnastics, skiing, rugby, football, soccer, track, baseball, basketball, racquetball, and roller skating [47–52]. Although defined as a rare injury, some authors defend that PTFL injuries may be more common than previously thought. This is especially true when associated with significant trauma, making the acute injuries of the PTFJ go unnoticed and frequently missed [49, 53]. The precise diagnosis can be complex and may cause some confusion with other pathologies, such as posterolateral rotatory instability, hypermobile or torn meniscus, or biceps femoris tendinitis [47]. A better understanding of the anatomy of PTFJ, injury etiology, and most frequent symptoms can help the clinician in accurately identifying PTFJ injuries [54].

The PTFJ is a synovial membrane-lined joint covered by hyaline cartilage that is connected with the knee joint in 10–12% of people [55–57]. The joint capsule is thicker anteriorly than posteriorly. The anterior portion of the PTFJ is stabilized by three broad ligamentous bands that run obliquely upward attaching to the anterior aspect of the lateral tibial condyle. The posterior proximal tibiofibular ligament, composed of two thick ligamentous bands, runs obliquely from the fibular head to the posterior aspect of the lateral tibial condyle. The posterior band is covered and reinforced by the popliteus tendon [51, 55, 56]. The LCL contributes to securing the fibular head to the tibia that extends from the lateral aspect of the fibular head just anterior to the styloid and attaches to the posterior aspect of the lateral femoral condyle [51]. The LCL is tight from 0° to about 30° of flexion. With knee flexion, the proximal fibula moves anteriorly with relaxation of the LCL and biceps femoris tendon. With knee extension, the proximal fibula is pulled posteriorly because these same structures are tightened [57, 58]. As a result of the laxity in the joint capsule with flexion, injuries of the PTFJ usually occur with the knee in a flexed position [47].

During clinical presentation, patients with PTFJ injury most commonly report pain associated with swelling, or a painful mass in the lateral aspect of the knee. Pain is usually exacerbated by direct pressure over the fibular head or dorsiflexion and eversion of the foot, and as the flexed knee is extended [47]. During neurological examination, it is crucial to assess muscle contraction and active mobilization of the ankle to rule out fibular nerve injuries. During inspection, the clinician should observe any deformity, any presence of a mass on the fibular head, and the integrity of the local skin. The clinician should palpate the knee for tenderness. Joint laxity is assessed by translating the fibular head anteriorly and posteriorly while holding the fibular head between the thumb and index finger. It is helpful to ask the patient if this translation reproduces the symptoms or causes pain [59].

The Radulescu sign may be helpful to elicit the diagnosis. With the patient lying in prone position, the clinician stabilizes the thigh and the knee flexed to 90° with one hand and with the other hand rotates the leg internally in an attempt to subluxate the fibula anteriorly (Fig. 49.15). The PTFJ is usually stable with the knee in full extension; if it is not, injury of the LCL and posterolateral structures is likely [49, 60]. Physical examination in all patients with suspected PTFJ injuries should include an assessment of the integrity of the LCL and posterolateral structures of the knee, since they are frequently injured with a proximal tibiofibular dislocation [47].

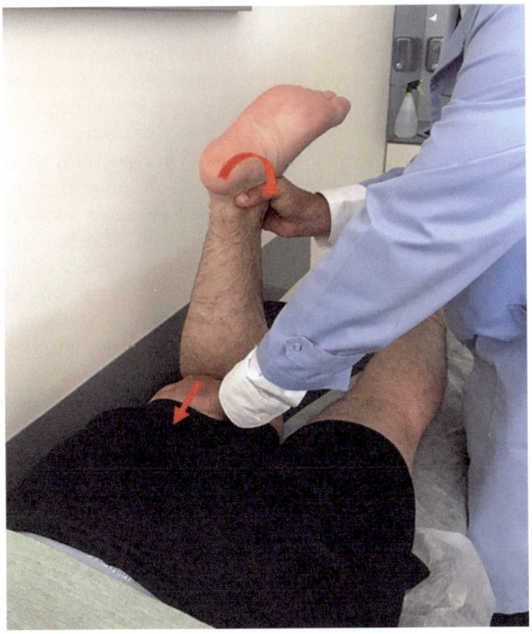

Fig. 49.15 The Radulescu test starts with the patient lying prone with the knee flexed at 90°. The clinician stabilizes the patient's thigh and rotates internally the tibia in an attempt to sublux the fibular head anteriorly

49.8 Conclusion

Musculoskeletal problems involving ligament injuries are among the most common reasons for patient visits to orthopedic clinicians, especially in the athletic population. The clinician needs to be well trained and ensure that they have a deep knowledge of anatomy and complete understanding of the biomechanical function of each ligament. In an unstable knee, a complete clinical history and well-performed physical examination clearly point out the injured structures. A thorough physical examination is essential for an accurate diagnosis, for minimizing unnecessary diagnostic testing, and for choosing the most appropriate treatment.

References

1. Branch TP, Mayr HO, Browne JE, Campbell JC, Stoehr A, Jacobs CA. Instrumented examination of anterior cruciate ligament injuries: minimizing flaws of the manual clinical examination. Arthroscopy. 2010;26(7):997–1004.
2. Scott N, Insall JN, Pedersen HB, Math KR, Vigorita VI, Cushner FD. Insall & Scott Surgery of the knee. Amsterdam: Elsevier; 2015. p. 146–51.
3. Posthumus M, Collins M, September AV, Schwellnus MP. The intrinsic risk factors for ACL ruptures: an evidence-based review. Phys Sportsmed. 2011;39(1):62–73.
4. Prins M. The Lachman test is the most sensitive and the pivot shift the most specific test for the diagnosis of ACL rupture. Aust J Physiother. 2006;52(1):66.
5. Katz JW, Fingeroth RJ. The diagnostic accuracy of ruptures of the anterior cruciate ligament comparing the Lachman test, the anterior drawer sign, and the pivot shift test in acute and chronic knee injuries. Am J Sports Med. 1986;14(1):88–91.
6. van Eck CF, van den Bekerom MP, Fu FH, Poolman RW, Kerkhoffs GM. Methods to diagnose acute anterior cruciate ligament rupture: a meta-analysis of physical examinations with and without anaesthesia. Knee Surg Sports Traumatol Arthrosc. 2013;21(8):1895–903.
7. Torg JS, Conrad W, Kalen V. Clinical diagnosis of anterior cruciate ligament instability in the athlete. Am J Sports Med. 1976;4(2):84–93.

8. Noulis G. Entorse du genou. These no. 142. Fac Med Paris. 1875; 1875: 1–53.
9. Jackson JL, O'Malley PG, Kroenke K. Evaluation of acute knee pain in primary care. Ann Intern Med. 2003;139(7):575–88.
10. Mulligan EP, McGuffie DQ, Coyner K, Khazzam M. The reliability and diagnostic accuracy of assessing the translation endpoint during the Lachman test. Int J Sports Phys Ther. 2015;10(1):52–61.
11. Malanga GA, Andrus S, Nadler SF, McLean J. Physical examination of the knee: a review of the original test description and scientific validity of common orthopedic tests. Arch Phys Med Rehabil. 2003;84(4):592–603.
12. Paessler HH, Michel D. How new is the Lachman test? Am J Sports Med. 1992;20(1):95–8.
13. Flynn TWCJ, Whitman JM. Users' guide to the musculoskeletal examination: fundamentals for the evidence-based clinician. Louisville, KY: Evidence in Motion; 2008.
14. Quillen WS. Special tests for orthopedic examination. J Athl Train. 1998;33(2):185.
15. Marshall JL, Wang JB, Furman W, Girgis FG, Warren R. The anterior drawer sign: what is it? J Sports Med. 1975;3(4):152–8.
16. Slocum DB, Larson RL. Rotatory instability of the knee. Its pathogenesis and a clinical test to demonstrate its presence. J Bone Joint Surg Am. 1968;50(2):211–25.
17. Citak M, Suero EM, Rozell JC, Bosscher MR, Kuestermeyer J, Pearle AD. A mechanized and standardized pivot shifter: technical description and first evaluation. Knee Surg Sports Traumatol Arthrosc. 2011;19(5):707–11.
18. Galway RBA, MacIntosh D. Pivot shift: a clinical sign of symptomatic anterior cruciate insufficiency. J Bone Joint Surg (Br). 1972;54(4):763–4.
19. Benjaminse A, Gokeler A, van der Schans CP. Clinical diagnosis of an anterior cruciate ligament rupture: a meta-analysis. J Orthop Sports Phys Ther. 2006;36(5):267–88.
20. Larson RL. Physical examination in the diagnosis of rotatory instability. Clin Orthop Relat Res. 1983;172:38–44.
21. Losee RE. Diagnosis of chronic injury to the anterior cruciate ligament. Orthop Clin North Am. 1985;16(1):83–97.
22. Losee RE, Johnson TR, Southwick WO. Anterior subluxation of the lateral tibial plateau. A diagnostic test and operative repair. J Bone Joint Surg Am. 1978;60(8):1015–30.
23. Lelli A, Di Turi RP, Spenciner DB, Dòmini M. The "lever sign": a new clinical test for the diagnosis of anterior cruciate ligament rupture. Knee Surg Sports Traumatol Arthrosc. 2016;24(9):2794–7.
24. Robson AW. VI. Ruptured crucial ligaments and their repair by operation. Ann Surg. 1903;37(5):716–8.
25. Barton TM, Torg JS, Das M. Posterior cruciate ligament insufficiency. A review of the literature. Sports Med. 1984;1(6):419–30.
26. Magee DJ. Orthopedic physical assessment. 3rd ed. Philadelphia: WB Saunders; 1997.
27. Kopkow C, Freiberg A, Kirschner S, Seidler A, Schmitt J. Physical examination tests for the diagnosis of posterior cruciate ligament rupture: a systematic review. J Orthop Sports Phys Ther. 2013;43(11):804–13.
28. Gollehon DL, Torzilli PA, Warren RF. The role of the posterolateral and cruciate ligaments in the stability of the human knee. A biomechanical study. J Bone Joint Surg Am. 1987;69(2):233–42.
29. Grood ES, Stowers SF, Noyes FR. Limits of movement in the human knee. Effect of sectioning the posterior cruciate ligament and posterolateral structures. J Bone Joint Surg Am. 1988;70(1):88–97.
30. Bleday RM, Fanelli GC, Giannotti BF, Edson CJ, Barrett TA. Instrumented measurement of the posterolateral corner. Arthroscopy. 1998;14(5):489–94.
31. Rubinstein RA Jr, Shelbourne KD, McCarroll JR, VanMeter CD, Rettig AC. The accuracy of the clinical examination in the setting of posterior cruciate ligament injuries. Am J Sports Med. 1994;22(4):550–7.
32. Allen CR, Kaplan LD, Fluhme DJ, Harner CD. Posterior cruciate ligament injuries. Curr Opin Rheumatol. 2002;14(2):142–9.
33. Covey CD, Sapega AA. Injuries of the posterior cruciate ligament. J Bone Joint Surg Am. 1993;75(9):1376–86.
34. Cooper JM, McAndrews PT, LaPrade RF. Posterolateral corner injuries of the knee: anatomy, diagnosis, and treatment. Sports Med Arthrosc Rev. 2006;14(4):213–20.
35. Ritchie JR, Bergfeld JA, Kambic H, Manning T. Isolated sectioning of the medial and posteromedial capsular ligaments in the posterior cruciate ligament-deficient knee. Influence on posterior tibial translation. Am J Sports Med. 1998;26(3):389–94.
36. Daniel DM, Stone ML, Barnett P, Sachs R. Use of the quadriceps active test to diagnose posterior cruciate-ligament disruption and measure posterior laxity of the knee. J Bone Joint Surg Am. 1988;70(3):386–91.
37. Cooper DE. Tests for posterolateral instability of the knee in normal subjects. Results of examination under anesthesia. J Bone Joint Surg Am. 1991;73(1):30–6.
38. Jakob RP, Hassler H, Staeubli HU. Observations on rotatory instability of the lateral compartment of the knee. Experimental studies on the functional anatomy and the pathomechanism of the true and the reversed pivot shift sign. Acta Orthop Scand Suppl. 1981;191:1–32.
39. Hughston JC, Norwood LA Jr. The posterolateral drawer test and external rotational recurvatum test for posterolateral rotatory instability of the knee. Clin Orthop Relat Res. 1980;147:82–7.
40. Levy BA, Boyd JL, Stuart MJ. Surgical treatment of acute and chronic anterior and posterior cruciate ligament and lateral side injuries of the knee. Sports Med Arthrosc Rev. 2011;19(2):110–9.
41. LaPrade RF, Ly TV, Griffith C. The external rotation recurvatum test revisited: reevaluation of the sagittal

plane tibiofemoral relationship. Am J Sports Med. 2008;36(4):709–12.

42. Palmer I. On the injuries to the ligaments of the knee joint: a clinical study. 1938. Clin Orthop Relat Res. 2007;454:17–22. discussion 14
43. Hughston JC, Andrews JR, Cross MJ, Moschi A. Classification of knee ligament instabilities. Part I. the medial compartment and cruciate ligaments. J Bone Joint Surg Am. 1976;58(2):159–72.
44. Marshall JL, Rubin RM. Knee ligament injuries–a diagnostic and therapeutic approach. Orthop Clin North Am. 1977;8(3):641–68.
45. Committee on the Medical Aspects of Sports, American Medical Association: Standard nomenclature of athletic injuries. Paper presented at the American Medical Association Chicago. 1968.
46. Rossi R, Dettoni F, Bruzzone M, Cottino U, D'Elicio DG, Bonasia DE. Clinical examination of the knee: know your tools for diagnosis of knee injuries. Sports Med Arthrosc Rehabil Ther Technol. 2011;3:25.
47. Sekiya JK, Kuhn JE. Instability of the proximal tibiofibular joint. J Am Acad Orthop Surg. 2003;11(2):120–8.
48. Turco VJ, Spinella AJ. Anterolateral dislocation of the head of the fibula in sports. Am J Sports Med. 1985;13(4):209–15.
49. Ogden JA. Subluxation and dislocation of the proximal tibiofibular joint. J Bone Joint Surg Am. 1974;56(1):145–54.
50. Thomason PA, Linson MA. Isolated dislocation of the proximal tibiofibular joint. J Trauma. 1986;26(2):192–5.
51. Ogden JA. Dislocation of the proximal fibula. Radiology. 1972;105(3):547–9. https://doi.org/10.1148/105.3.547.
52. Shapiro GS, Fanton GS, Dillingham MF. Reconstruction for recurrent dislocation of the proximal tibiofibular joint. A new technique. Orthop Rev. 1993;22(11):1229–32.
53. Semonian RH, Denlinger PM, Duggan RJ. Proximal tibiofibular subluxation relationship to lateral knee pain: a review of proximal tibiofibular joint pathologies. J Orthop Sports Phys Ther. 1995;21(5):248–57.
54. Espregueira-Mendes JD, da Silva MV. Anatomy of the proximal tibiofibular joint. Knee Surg Sports Traumatol Arthrosc. 2006;14(3):241–9.
55. Parkes JC 2nd, Zelko RR. Isolated acute dislocation of the proximal tibiofibular joint. Case report. J Bone Joint Surg Am. 1973;55(1):177–83.
56. Resnick D, Newell JD, Guerra J Jr, Danzig LA, Niwayama G, Goergen TG. Proximal tibiofibular joint: anatomic-pathologic-radiographic correlation. AJR Am J Roentgenol. 1978;131(1):133–8.
57. Ogden JA. The anatomy and function of the proximal tibiofibular joint. Clin Orthop Relat Res. 1974;101:186–91.
58. Andersen K. Dislocation of the superior tibiofibular joint. Injury. 1985;16(7):494–8.
59. Sijbrandij S. Instability of the proximal tibio-fibular joint. Acta Orthop Scand. 1978;49(6):621–6.
60. Baciu CC, Tudor A, Olaru I. Recurrent luxation of the superior tibio-fibular joint in the adult. Acta Orthop Scand. 1974;45(5):772–7.

Evaluation of the Menisci

50

Luís Duarte Silva, Philippe Tscholl, Ricardo Bastos,
Renato Andrade, and João Espregueira-Mendes

50.1 Introduction

The menisci play a key role in ensuring the stability of the knee, load transmission, shock absorption, and distribution of forces between the femur in the tibia, as well as lubrification and nutrition of the articular cartilage [1]. Despite the general belief that the meniscal injuries occur during the practice of sports, most occur during nonsporting activities. Hundreds of thousands of procedures addressing the meniscus are performed every year in the United States alone, and in some reports, the estimated frequency of injury is 0.33–8.27 per 1000 people every year [2, 3].

Traditionally, meniscal tears are divided into acute and chronic injuries. Acute tears usually occur after a traumatism to the knee and are associated to other knee injuries such as cartilage and

L. D. Silva
University Hospital Center of Algarve,
Faro, Portugal

Clinic of Physical Medicine and Rehabilitation of the Holy House of Mercy, Seia, Portugal

Clínica Espregueira - FIFA Medical Centre of Excellence, Porto, Portugal

Dom Henrique Research Centre, Porto, Portugal

P. Tscholl
Department of Orthopaedic Surgery and Traumatology, Geneva University Hospitals, Geneva, Switzerland

ReFORM (Réseau Francophone Olympique de la Recherche en Médecine du Sport) group, IOC Research Centre for Prevention of Injury and Protection of Athlete Health, University of Calgary, Calgary, AB, USA

R. Bastos
Clínica Espregueira - FIFA Medical Centre of Excellence, Porto, Portugal

Dom Henrique Research Centre, Porto, Portugal

R. Andrade
Clínica Espregueira - FIFA Medical Centre of Excellence, Porto, Portugal

Dom Henrique Research Centre, Porto, Portugal

Porto Biomechanics Laboratory (LABIOMEP), Faculty of Sports, University of Porto, Porto, Portugal

J. Espregueira-Mendes (✉)
Clínica Espregueira - FIFA Medical Centre of Excellence, Porto, Portugal

Dom Henrique Research Centre, Porto, Portugal

ICVS/3B's–PT Government Associate Laboratory, Braga/Guimarães, Portugal

3B's Research Group—Biomaterials, Biodegradables and Biomimetics, University of Minho, Headquarters of the European Institute of Excellence on Tissue Engineering and Regenerative Medicine, Barco, Guimarães, Portugal

School of Medicine, University of Minho, Braga, Portugal
e-mail: espregueira@dhresearchcentre.com

J. G. Lane et al. (eds.), *The Art of the Musculoskeletal Physical Exam*,
https://doi.org/10.1007/978-3-031-24404-9_50

ligament injuries that occur at the same time. Indeed, during an anterior cruciate ligament (ACL) tear, in half of the cases, there is also a meniscal injury [4]. Chronic meniscal injuries appear when some fragility of the meniscus is already present and a minor or repetitive trauma might even cause the lesion.

Physical examination is the cornerstone for the development of a musculoskeletal diagnosis, despite the developments of the complementary methods of diagnosis, such as magnetic resonance imaging (MRI) that shows in up to 50% of the patients of 40 years and older signs of degenerative meniscus lesions, which however are completely asymptomatic. The MRI has a high accuracy in detecting meniscal tears [5], but should only be used to confirm the suspicion of meniscal lesion and to assess the extent of the damage and the type and pattern of meniscal tear. On the other hand, the physical examination tests have poor diagnostic accuracy and low routine clinical value themselves [6–8] and should always be combined with a detailed history, about subjective and mechanical symptoms and injury mechanism. In this chapter, we will describe a step-by-step and systematic physical examination that should include history taking, inspection, palpation, evaluation of neurovascular status, range of motion, and special tests.

50.2 Clinical History

Meticulous history taking is mandatory, and several key points should be reviewed with the patient, including characterization of the type of pain (severity, localization, aggravating factors), circumstances of the beginning of the symptoms, swelling, presence or not of a "click," and knee blocking. Most of the meniscal injuries can be identified by the clinical history alone, and it is paramount to differentiate between traumatic and degenerative tears. A traumatic history is more common in people younger than 40 years old and a more insidious beginning in those older. The most common mechanisms are knee hyperflexion, pivot movement, and knee rotation with the foot fixed on the ground [9–11]. During the traumatic event, the patient often reports an audible "popping" in the knee during injury and there is a clicking when the torn part of the meniscus moves under the femoral condyle. Symptoms are often the result of joint instability due to a torn meniscal fragment or present as locking of the knee (in cases of bucket-handle tears [12]). Medial or lateral knee pain is another common symptom that is caused due to an increased tension at the joint capsule due to the meniscal injury. In a degenerative meniscal injury, the symptoms are similar, but there may also be problems with the patella and/or the articular cartilage. Symptoms are frequently aggravated while flexing the knee or load-bearing activities such as squatting and kneeling. Many patients might present diffuse medial knee pain after prolonged sitting, which might alleviate when stretching, or difficulties to perform the first few steps in the morning, which are no signs of a symptomatic meniscal injury and that might be seen especially in the degenerative knee.

50.3 Inspection

The clinician should evaluate the patient's walking and look for claudication, giving way, and/or instability. Muscle atrophies, scars, tumefactions, and alignment in the frontal (static or dynamic varus/valgus) and sagittal plane (flexum or recurvatum) should be registered.

50.4 Evaluation of the Vascular and Neurological Status

Vascular and neurological status needs to be evaluated, and the clinician should take special attention to the distal pulses as well as to the presence of venous insufficiency. Neurological evaluation of the obturator nerve must be performed as this nerve can cause knee pain.

50.5 Range of Motion and Palpation

Passive and active range of motion of the knee is normal mobility within 0–135° and tibial rotations within 0–15°. Insufficient range of motion especially in terms of side-to-side difference should be noted as abnormal. During the range-of-motion evaluation, the clinician may also identify the presence of knee locking.

Palpation must be performed in a systematic order and should include the bony landmarks, muscles, tendons, joint line, ligaments, and bursae. Palpation of the joint line is paramount as it can reproduce pain in case of meniscal injury or a meniscal cyst. The examiner needs however to bear in mind that many different structures cross the joint line, such as the medial collateral ligament, the hamstring tendons and sartorius fascia on the medial side, and the iliotibial tract and popliteus tendon on the lateral side. Palpation along these structures is necessary to interpret pain on palpation on the joint line as "meniscal" or associated soft-tissue pain.

50.6 Special Tests and Signs

The last part of the physical examination consists of special tests and signs. In the scientific literature, there are many tests and signs described, including Apley grinding test, McMurray test, joint-line palpation, Bragard test, Thessaly test, Ege's test, Payr sign, Steinman test, Bohler test, Merke test, Cabot test, Finochietto sign, Childress sign, Turner sign, Anderson medial and lateral compression test, Paessler rotational compression test, Tschaklin sign, and Wilson test [6, 13–17]. In the setting of an acute associated injury, for instance an ACL injury, these tests will most probably be positive due to the soft-tissue trauma/pain. Root tears and ramp lesions are found by increased instability and not by a painful McMurray or Apley test. In this chapter, we describe the tests that are specific to the meniscus and that are more commonly used.

50.6.1 Apley Grinding Test

With the patient in prone position, the knee is flexed at 90° with the hip stabilized, and the clinician should perform tibial rotation movements, either external (Fig. 50.1) or internal (Fig. 50.2). This should be performed while applying compression and traction to the knee. If pain is elicited with traction, it is more likely that a capsular ligament lesion is present. Pain during compres-

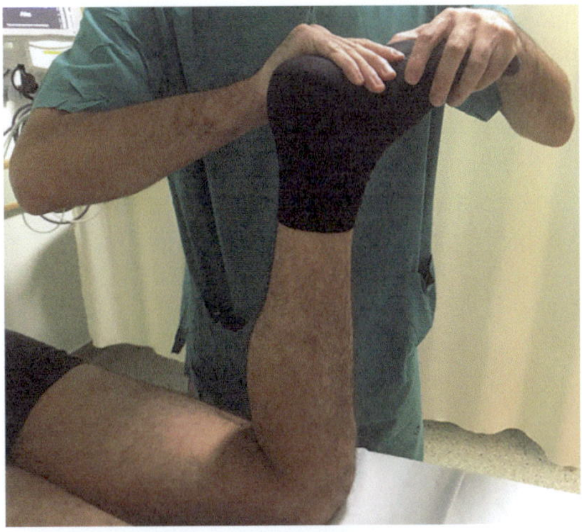

Fig. 50.1 Apley test, external rotation

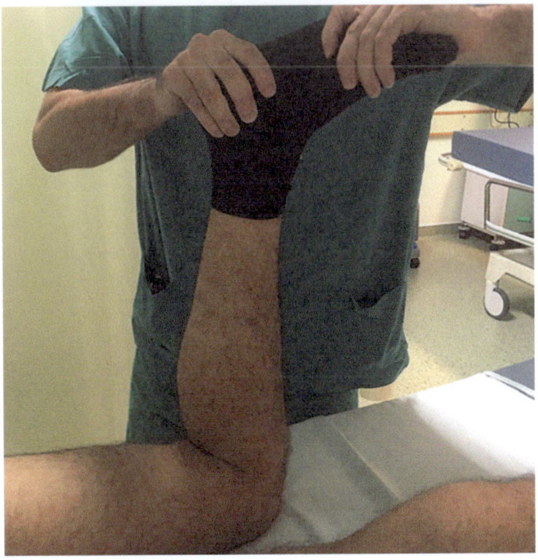

Fig. 50.2 Apley test, internal rotation

sion and rotation pinpoints a meniscal lesion (internal rotation points towards a lateral meniscus lesion and external rotation to a medial meniscus lesion). The Apley grinding test has 61% sensitivity and 70% specificity [6].

50.6.2 McMurray Test

The McMurray test is performed with the patient in the supine position with the knee and hip flexed at the maximum. The clinician holds the foot with one hand and the knee with the other. The knee is then extended up to 90° in external (Fig. 50.3) and internal rotation (Fig. 50.4). If pain occurs at the joint line during internal rotation, the clinician should suspect injury of the lateral meniscus. In turn, if pain at the joint line is elicited during external rotation, the injury is at the medial meniscus. The McMurray test has 71% sensitivity and 71% specificity [6].

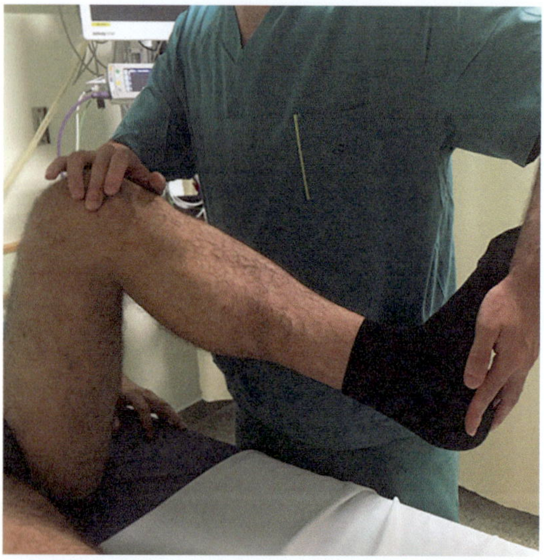

Fig. 50.3 McMurray test, external rotation

50.6.3 Joint-Line Palpation

The clinician should palpate the joint line and look for tenderness. The palpation should include the medial and lateral joint lines until the posterior part of the knee (Fig. 50.5). Tenderness while palpating the joint line has a sensibility of 67% sensitivity and 77% specificity [6]. If pain can be provoked 5 mm above or below the joint line, pain may be related to either a parameniscal cyst (in these cases, you may be able to palpate a bulging of the meniscus) or irritation of the meniscal fixation. Meniscal bulging can also be seen when there is a chronic root tear; the joint line that usually is concave in non-injured knees of young patients might become convex in older (if there is extrusion of the meniscus). If pain is more diffuse, injury to the surrounding soft tissues should be suspected.

50.6.4 Bragard Test

During the Bragard test, the examiner holds the foot with one hand and the knee with the other

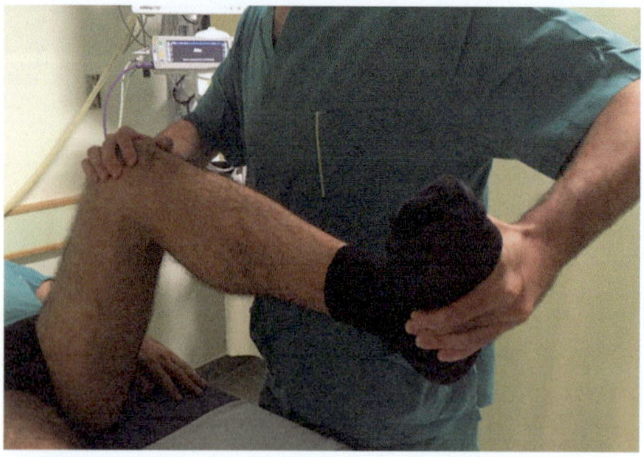

Fig. 50.4 McMurray test, internal rotation

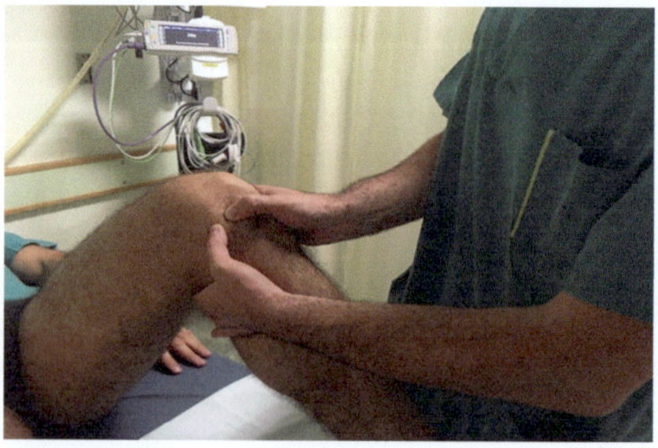

Fig. 50.5 Joint-line palpation

hand. The knee is fully extended in internal (Fig. 50.6) and external rotation (Fig. 50.7). If pain occurs during internal rotation, a lesion should be suspected at the lateral meniscus. In turn, if pain occurs during the external rotation, the lesion is expected to be on the medial meniscus.

50.6.5 Thessaly Test

The Thessaly test is performed with the patient standing on the symptomatic leg. The patient then rotates the lower limb internally and externally with a 5° of flexion, and then this test is repeated with 20° of knee flexion (Fig. 50.8). The clinician may assist the patient by holding his/her hands. The uninjured lower limb may be tested first to educate the patient on how to keep the knee in the correct flexed position. Performing this test with 20° of flexion appears to have better results in clinical evaluation [14]. It is considered positive if it elicits pain and/or a clunk (sense of locking/catching) at the joint line. This test has only moderate inter-examiner reliability [18]. The Thessaly test at 20° of knee flexion has 75% sensitivity and 87% specificity [7].

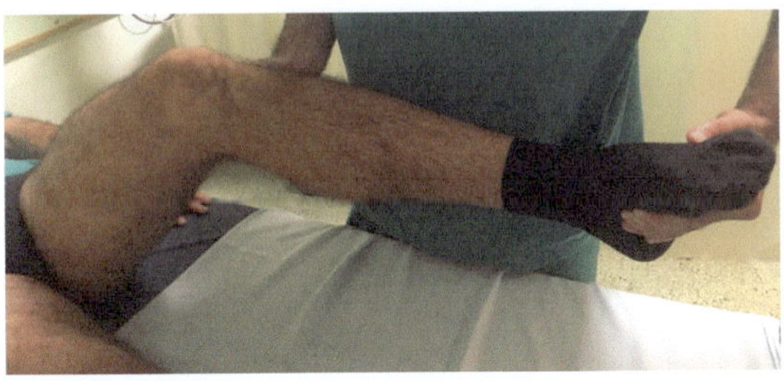

Fig. 50.6 Bragard test, internal rotation

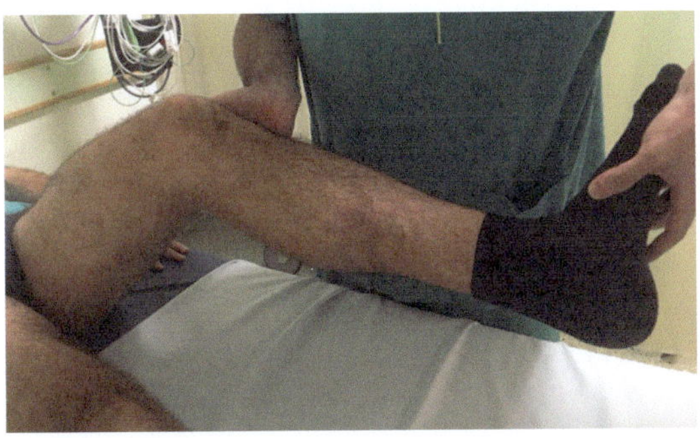

Fig. 50.7 Bragard test, external rotation

50.6.6 Ege's Test

The patient starts in the standing position with the knees fully extended and the feet placed 30–40 cm away from each other. The patient then squats with maximum external rotation (for medial meniscal injuries) or with maximum internal rotation (for lateral meniscal injuries). The internal and external tibial rotation will induce a weight-bearing valgus and varus, respectively. A full squat with tibial internal rotation is almost impossible even in healthy individuals, so expect a less than full squat. The test is positive if

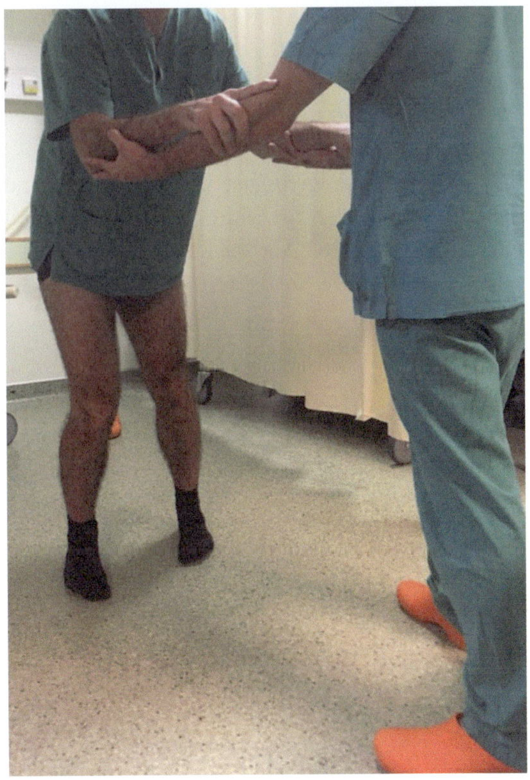

Fig. 50.8 Thessaly test. The left lower limb is being tested. This figure shows the initiation of the test, where the patient is internally rotating the left leg with 5° of knee flexion. Then, the patient will remain in a single-leg stance and rotate externally and internally three times. The test is then repeated with 20° of knee flexion

the patient reports pain and/or a clicking in the knee (that sometimes may be audible). As soon as the patient reports pain/clicking, further squatting is not needed; however, in some cases, the pain/clicking may not be felt even at deep squat, but may show as the patient comes out of the squat—which is also considered a positive sign for meniscal injury. Anterior meniscal tears are associated with symptoms during earlier knee flexion, whereas tears at posterior horn are associated with symptoms during deeper squats. This test has similar sensitivity rates (67% for medial and 64% for lateral meniscal injuries) as previously reported for McMurray's, Apley's, and joint-line palpation, but higher specificity rates (81% for medial and 90% for lateral meniscal tears). This test seemed to be prone to miss degenerative meniscal tears (missed 66% of all degenerative tears) [19]. These results require however replication in further studies.

50.6.7 Payr Sign

The patient is asked to seat "cross-legged," and the clinician applies pressure in an intermittent way to the affected leg (flexed and externally rotated). A positive test is defined by pain in the medial joint line that can represent a lesion mainly to the posterior horn.

50.6.8 Steinman Tests

The Steinman test is divided into two parts. In the first part of the test, the patient is seated with the knee at 90 degrees of flexion and the knees "hanging," and the clinician rotates the tibia internally and externally. If pain is elicited at the middle joint line while externally rotating the tibia, a medial meniscus lesion should be suspected. When pain is provoked at the lateral joint line while internally rotating the tibia, a lateral meniscus injury may be present. The test is repeated in various degrees of knee flexion. In the second part of the test—also known as Steinman tenderness displacement test—the clinician palpates the joint line looking for tenderness. The knee is then flexed while palpating the joint line. If the tenderness moves posteriorly with increasing knee flexion, the test is thus considered positive. The clinician then extends the knee while palpating the area of tenderness at the joint line. If the tenderness moves anteriorly during knee extension, the test is considered positive. Equally to the first part of this test, these maneuvers should be tested at various degrees of knee flexion. The second part of this test is used to differentiate meniscal injury from other knee injuries, because tenderness does not move during knee flexion and extension in cases of non-meniscal injuries. There is a scarcity of studies using this test, but one study has shown 87% specificity (76% for medial and 98% for lateral meniscus) and 96.5% specificity (98% for medial and 92% for lateral meniscus) [20].

50.6.9 Bohler Test

The clinician holds the patient leg (in full extension) with one hand in the femur and the other in the malleolar region and applies a valgus force to compress the lateral meniscus or a varus force to compress the medial meniscus. If pain is present in the medial joint line with a varus force, then a medial meniscus lesion should be suspected; in turn, with a valgus force and pain in the lateral joint line, a lateral meniscus lesion may be present.

50.7 Conclusion

Knee physical examination remains a difficult topic and no single test can be used isolated to perform the diagnosis of a meniscal injury, with the most important being the conjugation of clinical history, step-by-step physical examination, and imaging exams.

References

1. Makris EA, Hadidi P, Athanasiou KA. The knee meniscus: structure-function, pathophysiology, current repair techniques, and prospects for regeneration. Biomaterials. 2011;32:7411–31.

2. Lauder TD, Baker SP, Smith GS, Lincoln AE. Sports and physical training injury hospitalizations in the army. Am J Prev Med. 2000;18:118–28.
3. Jones JC, Burks R, Owens BD, Sturdivant RX, Svoboda SJ, Cameron KL. Incidence and risk factors associated with meniscal injuries among active-duty US military service members. J Athl Train. 2012;47:67–73.
4. Granan LP, Inacio MC, Maletis GB, Funahashi TT, Engebretsen L. Sport-specific injury pattern recorded during anterior cruciate ligament reconstruction. Am J Sports Med. 2013;41:2814–8.
5. Yan R, Wang H, Yang Z, Ji ZH, Guo YM. Predicted probability of meniscus tears: comparing history and physical examination with MRI. Swiss Med Wkly. 2011;141:w13314.
6. Hegedus EJ, Cook C, Hasselblad V, Goode A, McCrory DC. Physical examination tests for assessing a torn meniscus in the knee: a systematic review with meta-analysis. J Orthop Sports Phys Ther. 2007;37:541–50.
7. Smith BE, Thacker D, Crewesmith A, Hall M. Special tests for assessing meniscal tears within the knee: a systematic review and meta-analysis. Evid Based Med. 2015;20:88–97.
8. Blyth M, Anthony I, Francq B, et al. Diagnostic accuracy of the Thessaly test, standardised clinical history and other clinical examination tests (Apley's, McMurray's and joint line tenderness) for meniscal tears in comparison with magnetic resonance imaging diagnosis. Health Technol Assess. 2015;19:1–62.
9. Johnson LL, Johnson AL, Colquitt JA, Simmering MJ, Pittsley AW. Is it possible to make an accurate diagnosis based only on a medical history? A pilot study on women's knee joints. Arthroscopy. 1996;12:709–14.
10. Drosos GI, Pozo JL. The causes and mechanisms of meniscal injuries in the sporting and non-sporting environment in an unselected population. Knee. 2004;11:143–9.
11. Yeh PC, Starkey C, Lombardo S, Vitti G, Kharrazi FD. Epidemiology of isolated meniscal injury and its effect on performance in athletes from the National Basketball Association. Am J Sports Med. 2012;40:589–94.
12. Andrews JR, Norwood LA Jr, Cross MJ. The double bucket handle tear of the medial meniscus. J Sports Med. 1975;3:232–7.
13. Meserve BB, Cleland JA, Boucher TR. A meta-analysis examining clinical test utilities for assessing meniscal injury. Clin Rehabil. 2008;22:143–61.
14. Karachalios T, Hantes M, Zibis AH, Zachos V, Karantanas AH, Malizos KN. Diagnostic accuracy of a new clinical test (the Thessaly test) for early detection of meniscal tears. J Bone Joint Surg Am. 2005;87:955–62.
15. Harrison BK, Abell BE, Gibson TW. The Thessaly test for detection of meniscal tears: validation of a new physical examination technique for primary care medicine. Clin J Sport Med. 2009;19:9–12.
16. Rossi R, Dettoni F, Bruzzone M, Cottino U, D'Elicio DG, Bonasia DE. Clinical examination of the knee: know your tools for diagnosis of knee injuries. Sports Med Arthrosc Rehabil Ther Technol. 2011;3:25.
17. Gobbo Rda R, Rangel Vde O, Karam FC, Pires LA. Physical Examinations for Diagnosing Meniscal Injuries: correlation with surgical findings. Rev Bras Ortop. 2011;46:726–9.
18. Snoeker BA, Lindeboom R, Zwinderman AH, Vincken PW, Jansen JA, Lucas C. Detecting meniscal tears in primary care: reproducibility and accuracy of 2 weight-bearing tests and 1 non-weight-bearing test. J Orthop Sports Phys Ther. 2015;45:693–702.
19. Akseki D, Özcan Ö, Boya H, Pınar H. A new weight-bearing meniscal test and a comparison with McMurray's test and joint line tenderness. Arthroscopy. 2004;20:951–8.
20. Muellner T, Weinstabl R, Schabus R, Vecsei V, Kainberger F. The diagnosis of meniscal tears in athletes: a comparison of clinical and magnetic resonance imaging investigations. Am J Sports Med. 1997;25:7–12.

51 Evaluation of Muscle Injuries

Camila Cohen Kaleka, Pedro Henrique C. Andrade, Pedro Debieux, André Fukunishi Yamada, and Moisés Cohen

51.1 Introduction

Muscle injuries are among the most frequent injuries in sport, accounting for up to one-third of all sports-related injuries; they are responsible for 25% of days of absence away from training and competition [1]. Despite their considerable impact, there is still a lack of high-quality studies evaluating their specific control [2]. This entity is common, often occurring during sports practice, affecting the athlete's functional capacity and ability to continue [3].

Athlete's return to competition and injury prevention are the physician's main goal. The clinical evaluation and imaging information is crucial to dictate prognosis [3]. Since the 1980s, availability of ultrasound (US) and magnetic resonance imaging (MRI) has been permitting direct visualization of muscle tears, resulting in appreciable tools to confirm and address the extent of muscle injuries helping to lead the management [2].

Despite the recognition of many risk factors and prevention protocols, muscle injuries have been increasing over the last 12 years. Therefore, it is necessary to enhance our knowledge based on clinical experiences and supported by scientific evidence about this issue [4].

51.2 Epidemiology

Muscle injuries are frequent in high-demand sports, accounting for 10–55% of all acute sports injuries [5]. In view of all damaging consequences, this current scenario causes a necessity for better comprehension of muscle tears and specially their prevention has turned a considerable challenge to the sports medicine centers all over the world [4].

Regarding professional soccer players, muscle injuries represent 31% of all injuries and are responsible for 25% of days of absence from training and competition [5]. The majority of muscle injuries (92%) affect the lower extremity involving the four major muscle groups: hamstrings, adductors, quadriceps, and calves, with the hamstring injuries being the most common single type, representing 12–37% of all [2, 4]. In

C. C. Kaleka (✉) · P. Debieux
Department of Orthopedic, Israelita Albert Einstein Hospital, São Paulo, SP, Brazil

P. H. C. Andrade
Department of Orthopedic, Santa Casa de Belo Horizonte Hospital, Belo Horizonte, MG, Brazil

M. Cohen
Department of Orthopedic, Israelita Albert Einstein Hospital, São Paulo, SP, Brazil

Department of Orthopedic, Federal University of São Paulo, UNIFESP-EPM, São Paulo, SP, Brazil

A. F. Yamada
Department of Radiology and Diagnostic Imaging, Federal University of São Paulo, UNIFESP-EPM, São Paulo, SP, Brazil

J. G. Lane et al. (eds.), *The Art of the Musculoskeletal Physical Exam*,
https://doi.org/10.1007/978-3-031-24404-9_51

addition, the incidence risk of muscle tears in this population is six times higher during match play compared to training. Age represents another attribute involved with muscle injuries; the increase occurs in players aged over 30, emphasizing the sural triceps [4, 5].

In other modalities, the incidence of muscle injuries is variable: 16% in running sports, 11% in rugby, and 18% in basketball. In all these sports, the main affected muscles are the hamstrings, quadriceps, and adductors [5].

51.3 Etiology

The main function of skeletal muscle is to perform as the "motor" driving the muscle-tendon-bone unit, producing movement and locomotion. This unit is an extraordinarily organized complex of tissues consisting of the following: the muscle; the myotendinous junction, where muscle fibers interdigitate with collagen, allowing force to be transferred from the muscle to the tendon; the tendon, which may be at the ends of and/or within the muscle; the enthesis, where the tendon attaches to the bone; and the bone [3].

The injury therefore depends on the impact intensity, state of contraction of the muscle, traumatic moment, and muscle injured. Muscle injuries usually occur in the eccentric phase of the muscle contraction after an indirect insult, more common in noncontact sports, and after direct trauma, as in contact sports.

Trauma mechanisms can be divided into direct and indirect forms. The former indicates that the muscle is damaged directly by an externally applied force, such as contusion or penetration. In contrast, the indirect form refers to muscle trauma that occurs secondary to a nonphysiologically or abrupt elongation. Whereas the direct type of injury typically results in hemorrhage, hematoma, and/or laceration of the target muscle, the indirect type commonly results in muscle strains or tears. The direct injury mostly occurs where the maximal external force is exerted, whereas the indirect type frequently takes place across the myotendinous junction of the affected muscles [5].

Direct muscle injuries are a commonplace in sports that involve frequent body contact, such as rugby, basketball, and soccer. However, indirect muscle injuries more often occur during activities in which the contracted muscles undergo excessive traction [6]. Muscle strain is the most common presentation of an indirect-type injury resulting from extravagant stretching of a shortened muscle.

The most prevalent location for muscle strain is the myotendinous junction, where the distal tendons are interwoven with the muscle fibers to relay contraction forces. The myotendinous junction (MTJ) reveals less capacity for energy absorption than muscle and tendon. In animal models, the average tension leading to MTJ failure is only 20% greater than the maximum isometric tension normally produced during activity [3, 6].

The MTJ is situated at a variable distance from the bone, and the tendinous portion located between the bone and the MTJ is named as the "free tendon." Therefore, muscles with short free tendons, such as the deltoid and gluteus maximus, appear to insert virtually directly on the bone, whereas muscles such as the biceps brachii, rectus femoris, and plantaris have free tendons varying in length [3].

51.4 Physical Exam

Physical examination routine includes the inspection and palpation of the injured area and also function tests of the suspicious injured muscles both with and without resistance. The main objectives are to determine the precise location (region, muscle(s), tendon, or fascia involvement) and presume the severity of the injury. The diagnostic accuracy of the tests is quite variable; therefore, a bilateral comparison should always be performed to help the exam accuracy (Table 51.1) [4].

In the past, the earliest efforts to grade the severity of muscle injuries were based on the symptoms and signs present on physical exam and this evaluation composed the basis for grading a given injury as "mild," "moderate," or "severe" [7].

Table 51.1 Physical examination assessment for muscle injuries [4]

Inspection	Ecchymosis or deformities on the muscle belly profile (Fig. 51.1)	
	Important issues:	
	Swelling?	■ This usually determines the severity of the injury
	Hematoma?	■ This suggests a structural injury/tear
	Retraction of muscles?	■ This suggests a severe injury/tear
	Changes to contours of muscle? Hint: *Hematomas, muscle retraction, and changes to the contours of a muscle are only seen in structural injuries*	■ This suggests a severe injury/tear
Palpation	Identify the specific region injured through pain provocation and presence or absence of a palpable defect in the musculotendinous junction (Fig. 51.2)	
	Important issues:	
	Localized or larger area?	■ This helps to identify injury location: Localized pain is more likely in a structural injury. Larger pain area tends toward functional injury
	Muscle tone?	■ Helps to determine injury type
	Edema?	■ Helps to identify injury location and injury type
	Pain/relief on careful stretching?	■ Pain on stretching suggests a tear
	Palpable defect Note: Muscle tension/tightness can be assessed only by means of clinical examination	■ A defect is most likely to indicate a structural injury
Strength assessment	Manual resistance is applied distally to the injury site; multiple test positions are utilized to assess isometric strength and pain provocation; pain provocation is as relevant as noting weakness (Fig. 51.3)	
	Important issues: Injury location:	
	Muscle belly	This is typical of some functional injuries
	Muscle–tendon junction?	This is typical for structural injuries/tears
	Tendon–bone junction?	This is typical for tendinous avulsions
Range of motion	For hamstring injuries, passive straight leg raises (hip) and active and passive knee extension test (knee) are commonly used to estimate hamstring flexibility and maximum length	
Muscle length	Should be based on the onset of discomfort or stiffness reported by the patient. Acute injuries' tests are often limited by pain and may not provide a valid assessment of musculotendon extensibility	
Pain provocation maneuvers	Biceps injury can lead to more pain during stretching than contraction, while SM or ST injuries have more pain during contraction than stretching	

Many authors, since 1966, tried to better identify the severity of the injury considering diverse attributes such as the degree of pain, disability, swelling, ecchymosis, presence of a palpable defect or other features of the clinical history such as the type of trauma, ability to continue activity after the injury, and, not least of all, range-of-motion limitation and anatomic location of the tear in the muscle unit [1, 7].

Generally, a *grade I* or "mild" muscle injury has been considered to correspond to stretching or minimal disruption of muscle fibers and a clinical presentation marked by minimal, well-localized pain, contracture and hemorrhage, minor disability, a full pain-free range of motion (or <10° ROM deficit), and ability to continue the sporting activity immediately after the injury. *Grade II* or "moderate" injury has been considered as a tear of a considerable number of muscle fibers but no complete muscle rupture; a more severe clinical presentation occurs when compared with the previous grade, represented by moderate pain, disability, painful range of motion (or 10–25° ROM deficit), and inability to continue the sporting activity. *Grade III* or "severe" injury has been defined to be a complete muscle

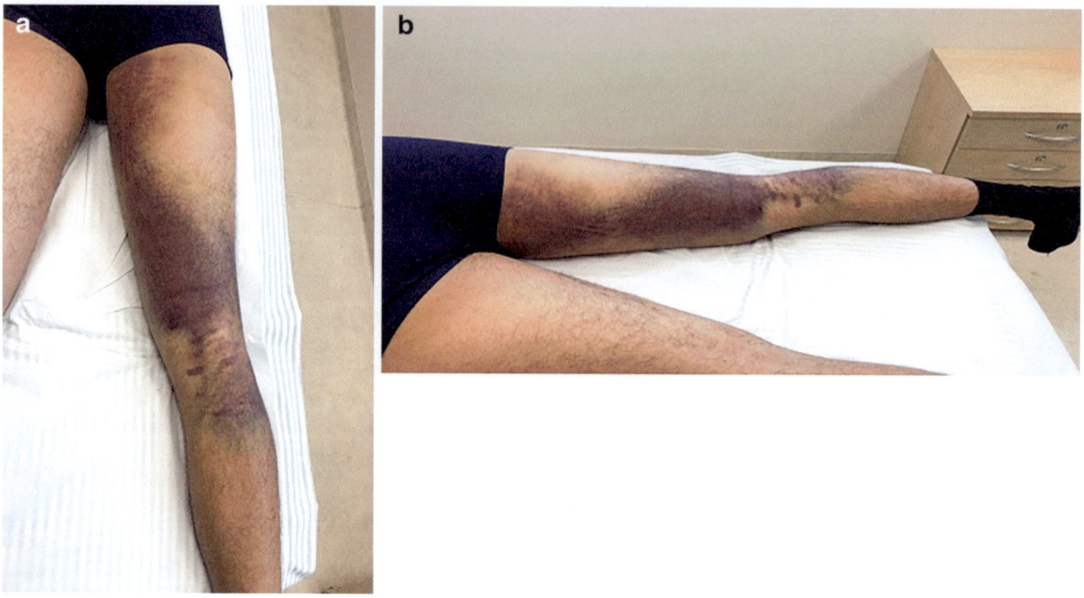

Fig. 51.1 A 45-year-old male presenting with enlarging areas of ecchymosis over inner and posterior aspects of the right thigh after a long head biceps femoris injury; (**a**) clinical inspection (posterior view) of a large bruising on the inner and posterior aspects of the right thigh; (**b**) lateral view

rupture, with the worst clinical scenario, recognized by the athlete's fall in pain immediately after the injury, more than 50% loss of motion (or <25° ROM), a rapid muscle circumference decrease of more than 12 mm when compared to the healthy contralateral muscle, diffuse pain, and hemorrhage (Figs. 51.1, 51.2, and 51.3) [7].

All these clinical presentations and physician's findings about the muscle injury described since the 1960s are estimated tools for an initial evaluation, mainly because of their reproducibility and simplicity, resulting in a quick approach. However, they were based only on specialist opinion and did not have a validated prognostic value [7].

Although the diagnosis is usually clinical, many physicians have an ultrasound in their office and this is especially important when the diagnosis is unclear, when recovery is taking longer than expected and when interventional or surgical management may be necessary [4]. It has advantages such as non-radiation, portability, excellent resolution of neuromuscular structures, and possibility for the examiner to perform dynamic examinations [6].

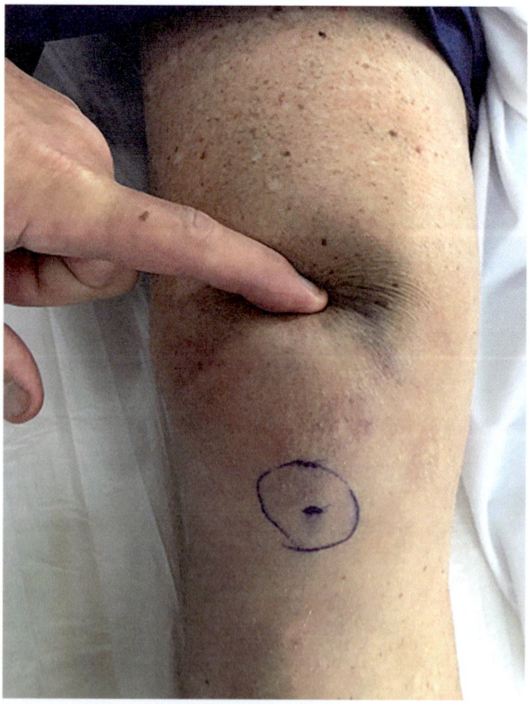

Fig. 51.2 Presence of a gap detected at physical examination in the place where the patient reports pain, featuring a more severe quadriceps muscle injury

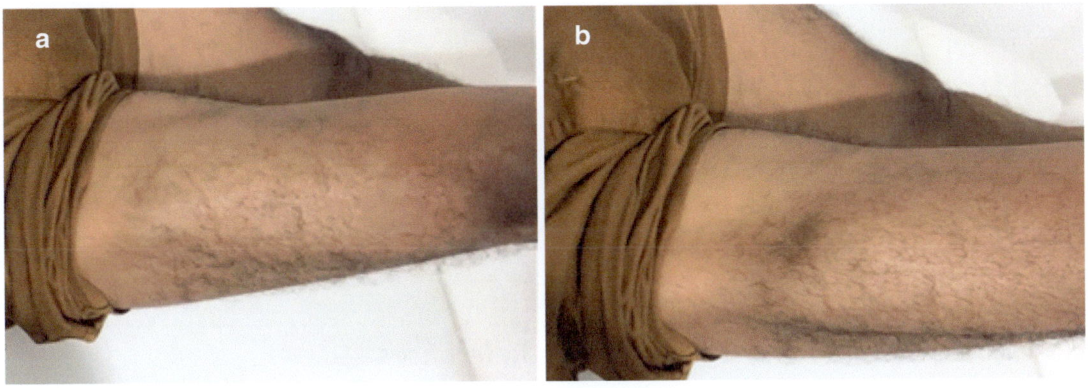

Fig. 51.3 A 32-year-old male soccer player presenting a distally retracted muscle belly after a subtotal muscle tear of the rectus femoris; (**a**) relax muscle; (**b**) active contraction for right-leg raises

Table 51.2 The Munich classification [7, 9]

Type of injury	Definition and symptoms	MRI
Direct	***Contusion:*** Blunt external force, muscle intact ***Laceration:*** Blunt external force, muscle rupture	Hematoma Hematoma
Indirect		
Functional	***Type 1: Overexertion-related muscle disorder***	
	1A: Fatigue-induced muscle disorder Muscle tightness	Negative
	1B: Delayed-onset muscle soreness Acute inflammatory pain	Negative or edema
	Type 2: Neuromuscular muscle disorder	
	2A: Spine-related neuromuscular muscle disorder. Increase of muscle tone due to a spinal disorder	Negative or edema
	2B: Muscle-related neuromuscular muscle disorder. Increase of muscle tone due to altered neuromuscular control	Negative or edema
Structural	***Type 3: Partial muscle tear***	
	3A: Minor partial muscle tear Tear with small maximum diameter	Fiber disruption
	3B: Moderate partial muscle tear Tear with increased maximum diameter	Retraction and hematoma
	Type 4: (Sub)total muscle tear avulsion	
	Complete muscle diameter involvement, defect	Complete discontinuity

51.5 Classification

Over the last decade, the "Munich Muscle Injury Classification" was established by an international group of experts in sports medicine, establishing an alliance with the International Olympic Committee (IOC) and the Union of European Football Associations (UEFA) (Table 51.2) [7–9]. Magnetic resonance imaging (MRI) is a crucial exam to assess the degree of muscle injury.

This classification is characterized by two main branches according to the nature of the muscle trauma: direct or indirect. Additionally, indirect muscle injuries are subdivided into four types according to MRI appearance where types 1 and 2 represent MRI-negative functional disease, and types 3 and 4 represent structural injuries that can be classified as minimal (type 3a), moderate (type 3b), or complete (type 4) [7–9].

Key Points

- Muscle injuries are among the most common injuries in sports and remain a notable issue since it causes competition time loss and a relatively high recurrence rate, despite all technological advances of science in this field.
- Indirect muscle tear is related to muscle strain/stretching; the most common area injured is the myotendinous junction.
- Direct muscle tear or muscle contusion is caused by a direct blow; the severity is determined by the force of impact, size of the striking body, and muscle's state of contraction.
- The examiner should start with a precise history of the circumstances surrounding the injury, before reviewing the symptoms and identifying any previous related injury problems.
- The next step in diagnosis is a careful clinical examination of the injured area. This should include an inspection, palpation, a comparison with the contralateral side, and functional testing of the muscles.
- The main purpose of the physical examination is to determine injury location and try to define severity. Acute injuries' evaluation is often limited by pain, and accuracy of the tests is quite variable.
- Diagnosis is based on history and meticulous clinical examination. Crucial imaging exams for evaluation are US and MRI. US is fast and cost effective, can offer a dynamic muscle assessment, and represents a good tool for serial evaluation to follow healing. MRI imaging is considered the gold standard method to assess the muscular architecture and to evaluate the extent of muscle injuries.
- Classifications and grading systems are currently going through a continuous progression. Additional studies may help validate new grading systems to develop a universally applicable terminology with more clinical relevance.

References

1. Guermazi A, Roemer FW, Robinson P, Tol JL, Regatte RR, Crema MD. Imaging of muscle injuries in sports medicine: sports imaging series. Radiology. 2017;282(3):646–63.
2. Hamilton B, Valle X, Rodas G, Til L, Grive RP, Rincon JAG, et al. Classification and grading of muscle injuries: a narrative review. Br J Sports Med. 2015;49(5):306.
3. Flores DV, Gómez CM, Estrada-Castrillón M, Smitaman E, Pathria MN. MR imaging of muscle trauma: anatomy, biomechanics, pathophysiology, and imaging appearance. Radiographics. 2018;38(1): 124–48.
4. Til L, Tol H, Bruce Hamilton E. Muscle injuries clinical guide 3.0. In: Muscle Injuries Clinical Guide 30; 2015. p. 1–60.
5. Buono D, Via G. Muscle injuries: a brief guide to classification and management. Transl Med UniSa. 2015;12(4):14–8.
6. Chang KV, Wu WT, Özçakar L. Ultrasound imaging and rehabilitation of muscle disorders: part 1. Traumatic injuries. Am J Phys Med Rehabil. 2019;98(12):1133–41.
7. Grassi A, Quaglia A, Canata G, Zaffagnini S. An update on the grading of muscle injuries. Joints. 2016;4(1):39–46.
8. Patel A, Chakraverty J, Pollock N, Chakraverty R, Suokas AK, James SL. British athletics muscle injury classification: a reliability study for a new grading system. Clin Radiol. 2015;70(12):1414–20. https://doi.org/10.1016/j.crad.2015.08.009.
9. Mueller-Wohlfahrt HW, Haensel L, Mithoefer K, Ekstrand J, English B, McNally S, et al. Terminology and classification of muscle injuries in sport: the Munich consensus statement. Br J Sports Med. 2013;47(6):342–50.

52 Evaluation of Neuropathies/Nerve Entrapment Around the Knee Joint

Dawid Szwedowski, Przemysław Pękala, and Radosław Grabowski

52.1 The Infrapatellar Branch of the Saphenous Nerve Injury

52.1.1 Anatomy

The saphenous nerve is a clinically important nerve, which may differ in its anatomical course not only among individuals but also between the left and right limbs of a person. It is the longest cutaneous branch of the femoral nerve and further divides into the sartorial and infrapatellar branches. The first branch, the sartorial, runs along the great saphenous vein to the medial side of the foot. The second one, the infrapatellar branch of the saphenous nerve (IPBSN), diverges from the main saphenous nerve at the level of the knee joint. It then courses around the posterior border of the sartorius muscle, and it pierces the fascia lata. It has three branches in the subcutaneous layer of the anteromedial part of the knee and supplies the skin over the medial and front sides of the knee, as well as the patellar ligament. There are many studies regarding the variants of the IPBSN. It may emerge anteriorly or posteriorly to the sartorius muscle or penetrate this muscle.

52.1.2 Causes

The potential for surgical trauma to the IPBSN has been known since 1945 [1]. It may often take a long time before a correct diagnosis of this syndrome is made, and it may be misdiagnosed or even overlooked by physicians. The infrapatellar pain syndrome may occur secondary to many factors, including post-traumatic and post-surgical ones. It may thus be caused by an everyday, direct fall on the knee, as well as plenty of surgical procedures including arthroscopy, total knee replacement, and bone-patellar tendon-bone or hamstring tendon harvesting for anterior cruciate ligament (ACL) procedures [1]. The injury mainly occurs during tendon harvesting (Figs. 52.1 and 52.2).

D. Szwedowski
Department of Orthopaedics and Trauma Surgery, Provincial Polyclinical Hospital, Torun, Poland

Orthopaedic Arthroscopic Surgery International (OASI) Bioresearch Foundation, Milan, Italy

P. Pękala
Orthopaedic Arthroscopic Surgery International (OASI) Bioresearch Foundation, Milan, Italy

Faculty of Medicine and Health Sciences, Andrzej Frycz Modrzewski Krakow University, Krakow, Poland

R. Grabowski (✉)
Orthopaedic Arthroscopic Surgery International (OASI) Bioresearch Foundation, Milan, Italy

Department of Orthopaedics and Traumatology, Medical University of Lodz, Lodz, Poland

J. G. Lane et al. (eds.), *The Art of the Musculoskeletal Physical Exam*, https://doi.org/10.1007/978-3-031-24404-9_52

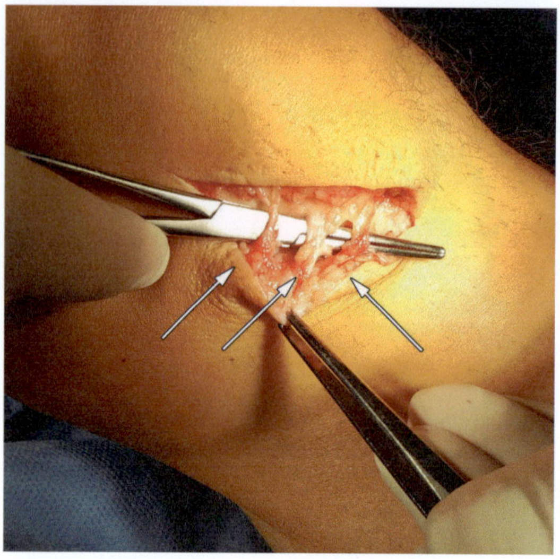

Fig. 52.1 Left knee of a 35-year-old male during revision surgery after ACL reconstruction. Infrapatellar nerve branches' entrapment in the scar tissue in the region of pes anserinus (arrow)

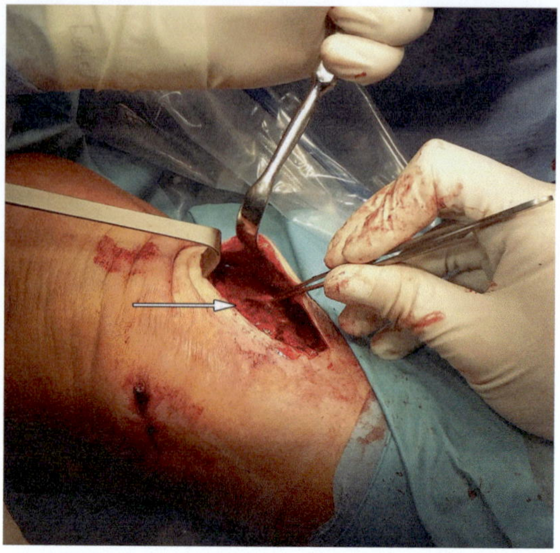

Fig. 52.2 Right knee of a 24-year-old male during revision surgery after ACL reconstruction. Histologically confirmed infrapatellar nerve neuroma formation in the area of the anteromedial portal

52.1.3 Examination

Figueroa et al. [2] described four types of disturbances including hypoesthesia, dysesthesia, painful neuroma, and reflex sympathetic dystrophy. The infrapatellar pain syndrome (IPS) has often been described as a pain similar to a toothache. This pain is present even at rest and increases during palpation of the skin at this region. Three trigger points were found and are depicted in Fig. 52.3.

Thinking about a neuroma due to injury of the IPBSN, it is crucial to bear in mind the other nerves located in this region: the lateral cutaneous nerve of the thigh, the intermediate cutaneous nerve of the thigh, and the anterior division of the medial cutaneous nerve of the thigh.

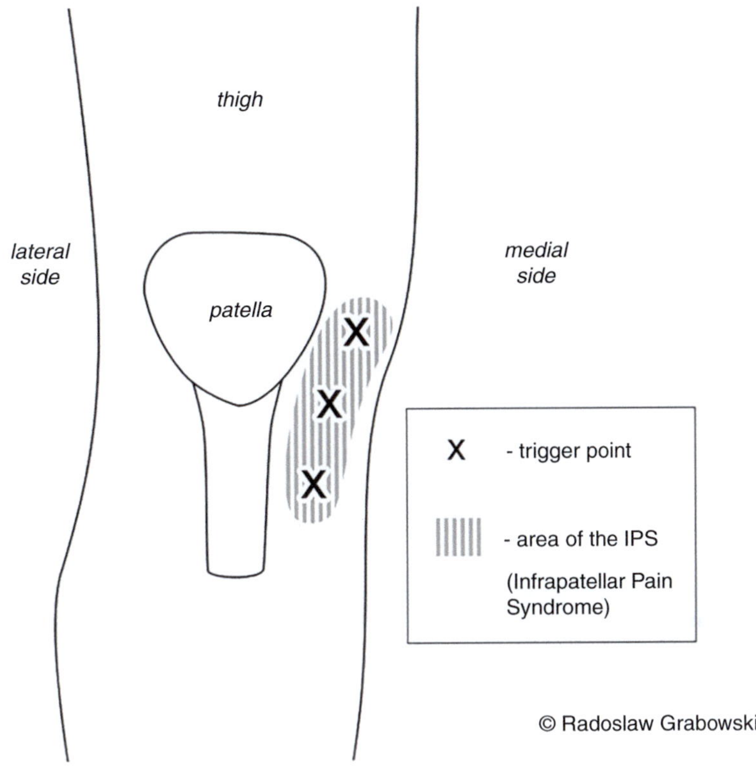

Fig. 52.3 Schematic presentation of the IPS trigger points

52.2 Peroneal Neuropathy

52.2.1 Anatomy

The common peroneal nerve (CPN) is a branch of the sciatic nerve that separates at the upper level of the popliteal space. In the distal thigh, it gives one motor branch to the short head of the biceps femoris muscle. This is the only motor branch proximal to the fibular head. The CPN travels superficially to the lateral head of the gastrocnemius and plantaris muscles as it provides the cutaneous innervation to the lateral aspect of the calf (by the lateral cutaneous nerve of the calf and a communicating branch that anastomoses with the sural nerve). At the fibular neck, the peroneal nerve passes through a fibro-osseous tunnel, the roof of which is formed by the origin of the peroneus longus and the intermuscular septum. It continues between the peroneus longus muscle and the fibula. At this point, it divides into two main branches: the deep peroneal and superficial peroneal nerves. The deep peroneal nerve innervates the anterior muscles of the leg by traveling deep to the peroneus longus. This nerve supplies the tibialis anterior, extensor digitorum longus, peroneus tertius, and extensor hallucis longus. These muscles control foot dorsiflexion and toe extension. Additionally, it provides a sensory branch to the first interdigital space. The superficial peroneal nerve supplies the peroneus longus and brevis muscles and then penetrates the lateral fascia at the junction of the mid and distal calf as the medial and lateral cutaneous branches to supply sensation to the anterolateral aspect of the distal calf and dorsum of the foot.

52.2.2 Causes

Causes of peroneal nerve palsy can be divided into two large categories. At the level of the knee, the CPN can be compressed by normal or pathological structures. Anatomical structures can compress the CPN secondary to repetitive activity or unusual position. The common peroneal nerve may be compressed by the tendinous origin of the peroneus longus as it winds around the fibular head and passes through the fibrosis tunnel [3]. Nerve entrapment at the level of the biceps femoris tendon in the popliteal fascia was also reported [4, 5]. Common peroneal nerve compression has been found secondary to kneeling, crossed-leg sitting, prolonged squatting, and lying on hard surfaces. Weight loss, malnutrition, and diabetes may predispose to this condition [4]. Peroneal neuropathy has also been reported after bariatric surgery [6]. The proposed mechanism is a loss of adipose tissue previously protecting the CPN. In addition, patients who require prolonged bed rest after surgery may be at risk for peroneal neuropathy due to the tendency of the lower limb to rest in external rotation of the hip. A preventative program should thus be instituted during hospitalization.

The second major cause of CPN neuropathy around the knee includes tumors. Intraneural ganglion cysts are the most common mass lesion in peroneal neuropathy. Other tumors, in order of decreasing frequency, include schwannoma, neurofibroma, osteochondroma, neurogenic sarcoma, focal hypertrophic neuropathy, desmoid tumor, and glomus tumor [7].

52.2.3 Examination

Pain can be the earliest symptom in peroneal neuropathy, usually preceding sensory changes in a similar distribution. Tapping over the area where the CPN winds around the head of the fibula may reproduce dysesthesia in the lateral calf or foot. A careful sensory examination can assist with localizing the lesion. The patient's leg should be examined for ecchymosis, edema, or ulcers. Evidence of trauma or vascular compromise may help to determine the cause of the lesion. All muscles of the lower limb should be examined for weakness and compared to the contralateral side. Weakness of the ankle dorsiflexors, toe extensors, and ankle evertors, commonly referred to as foot drop, is suggestive of peroneal neuropathy. Less frequently, lumbosacral plexopathy can result in weakness in a similar distribution. The tibialis posterior, innervated by the tibial nerve, contributes most to ankle inversion. If the ankle inversion is weak, the lesion involves more than the common peroneal nerve. In addition, weakness in peroneal neuropathy may lead to functional gait impairment.

The main manifestation of peroneal neuropathy is foot drop. Examining normal and pathologic reflexes can further narrow down the differential diagnosis. The presence of pathologic reflexes, such as a Babinski reflex, suggests foot drop of central origin.

Clinically, peroneal neuropathy appears most similarly to L5 radiculopathy. Concurrent low back pain or posterolateral thigh pain suggests spinal origin of symptoms. Absent or diminished patellar tendon (innervated by L2–L4) and Achilles tendon (innervated by S1) reflexes suggest a peripheral origin of foot drop. Sensation of the area between the first and second toes' plantar surface of the foot is disrupted in both cases.

References

1. Trescot A, Brown MN, Karl HW. Infrapatellar saphenous neuralgia–diagnosis and treatment. Pain Physician. 2013;16(3):E315.
2. Figueroa D, Calvo R, Vaisman A, Campero M, Moraga C. Injury to the infrapatellar branch of the saphenous nerve in ACL reconstruction with the hamstrings technique: clinical and electrophysiological study. Knee. 2008;15:360. https://doi.org/10.1016/j.knee.2008.05.002.
3. McCrory P, Bell S, Bradshaw C. Nerve entrapments of the lower leg, ankle and foot in sport. Sports Med. 2002;32:371.
4. Hunter RE. Peroneal nerve entrapment at the knee. Oper Tech Sports Med. 1996;4:46. https://doi.org/10.1016/S1060-1872(96)80010-4.
5. Dawson D, Wilbourn HM. Entrapment neuropathies 3rd ed. J Clin Neuromuscul Dis. 1999;1:54. https://doi.org/10.1097/00131402-199909000-00015.
6. Elias WJ, Pouratian N, Oskouian RJ, Schirmer B, Burns T. Peroneal neuropathy following successful bariatric surgery. J Neurosurg. 2008;105:631. https://doi.org/10.3171/jns.2006.105.4.631.
7. Kim JY, Ihn YK, Kim JS, Chun KA, Sung MS, Cho KH. Non-traumatic peroneal nerve palsy: MRI findings. Clin Radiol. 2007;62:58. https://doi.org/10.1016/j.crad.2006.07.013.

Evaluation of Malalignment of the Knee

53

Ignacio Dallo, John G. Lane,
Silvio Villascusa Marin, and Alberto Gobbi

53.1 Introduction

The clinical assessment of knee joint alignment is challenging for orthopedic surgeons because of the different pathologic conditions associated with various complexities causing the disorder. The origin of this entity is generally multifactorial. Therefore, the clinical examination is essential and should pay special attention to the history and physical exam and must include appropriate diagnostic tests to reach a precise diagnosis that allows treating the pathology effectively.

The clinical examination must include the evaluation of soft tissue and skin as well as the neurovascular status of the lower extremity. Systemic or local infection should be ruled out. Malunions, metabolic and neurological diseases, and congenital disorders can lead to complex deformities in the lower extremity. In patients with complex leg deformities, an accurate examination is mandatory to localize the level and the character of the deformity [1]. It should be remembered that a proper therapeutic approach can only be achieved with a thorough analysis of the patients' disorder.

I. Dallo
O.A.S.I. Bioresearch Foundation, Milan, Italy

J. G. Lane
Musculoskeletal and Joint Research Foundation, San Diego, CA, USA

S. V. Marin
Hospital Clínico Universitario Virgen de la Arrixaca, Murcia, Spain

A. Gobbi (✉)
O.A.S.I. Bioresearch Foundation Gobbi NPO, Milan, Italy
e-mail: gobbi@cartilagedoctor.it

53.2 Clinical Evaluation

53.2.1 History

An accurate and complete history focusing on the patient's symptoms, duration, and onset of the problem and the deformity's effects upon his/her daily life activities must be determined. The patient's age, occupation, degree of physical activity, and type of malalignment all play an essential role in the clinical diagnosis process and need for treatment. Besides, consideration of biological age should take priority over chronological age.

Identifying risk factors that can be modified, such as smoking, obesity, rheumatoid arthritis, and diabetes, will improve patient selection and decrease or avoid postoperative complications associated with realignment surgery [2, 3]. History of previous trauma or meniscectomy, ACL reconstruction, or cartilage procedures must be documented.

A careful history will point to the presence of knee pain with or without malalignment. While evaluating the history of pain, the clinician must ask for the exact location of pain, its timing, whether it is activity related, and an association

J. G. Lane et al. (eds.), *The Art of the Musculoskeletal Physical Exam*,
https://doi.org/10.1007/978-3-031-24404-9_53

of trauma. Any referred pain must also be evaluated, as sometimes a pain in the hip can radiate to the region above the knee.

53.2.2 Physical Examination

A thorough clinical examination of the knee in the standing posture is essential. Both knees should be examined together to check for symmetry. It could reveal effusion and coronal imbalance if any. Particular attention must be given to coronal and sagittal standing alignment and signs of varus/valgus or recurvatum thrusts during gait.

It is essential to understand that the frontal plane alignment analysis must involve the entire leg and consider the torsional alignment and the dynamic forces in standing and walking. Analysis of deformity should include the joint angles of femur, tibia, and ankle as well as the joint-line convergence angle. The normal ranges of these parameters must be considered (Fig. 53.1).

53.2.2.1 Gait Examination

A focused physical exam begins with a gait assessment. In a dynamic pattern, one needs to assess for the presence of a varus or valgus thrust as this suggests that there may be a combination of both bone and soft-tissue contribution to this abnormal finding. Varus or valgus alignment should be noted in the gait stance phase. A short leg gait is present when there is the asymmetry of gait length between the two legs, and it requires the assessment of limb length. This problem may be accompanied by a significant varus or valgus deformity at the knee or may represent an extra-articular deformity requiring further evaluation. During normal walking—the activity that imposes the largest number of loading cycles in daily life—the knee is subjected to two peak loads during the stance phase of every step. The first peak is caused by the large quadriceps tension necessary to arrest the body mass' descent when the weight transfers from the leg pushing off from the ground to the leg that accepts the load shortly after the heel strikes the ground (Fig. 53.2). This force curve includes a second peak, when the knee and hip are extended, the heel is raised from the floor, and the forefoot is pushing off, propelling the body forward.

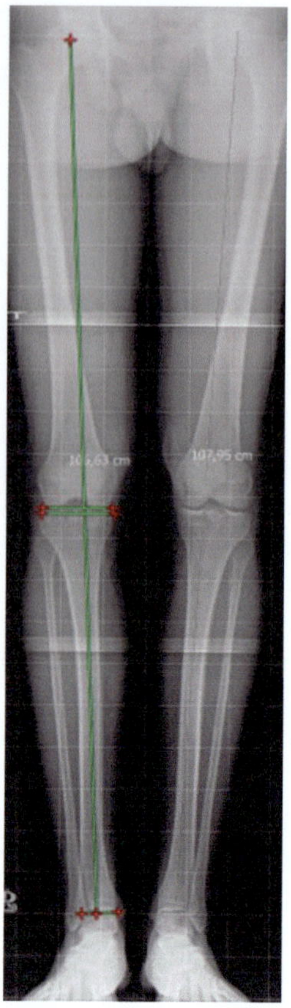

Fig. 53.1 Normal tibiofemoral coronal alignment on X-ray

53.2.2.2 Clinical Alignment

The patient should be asked to stand barefoot in bipodal position and face the examiner with both feet together and pointing ahead. From the front, the knee's alignment (varus or valgus deformity), orientation of the patellae (squinting or outfacing patellae), and patellar height are noted. Patients with a valgus angulation have a more predisposition to lateral subluxation; furthermore, patients with squinting patellae have a higher incidence of

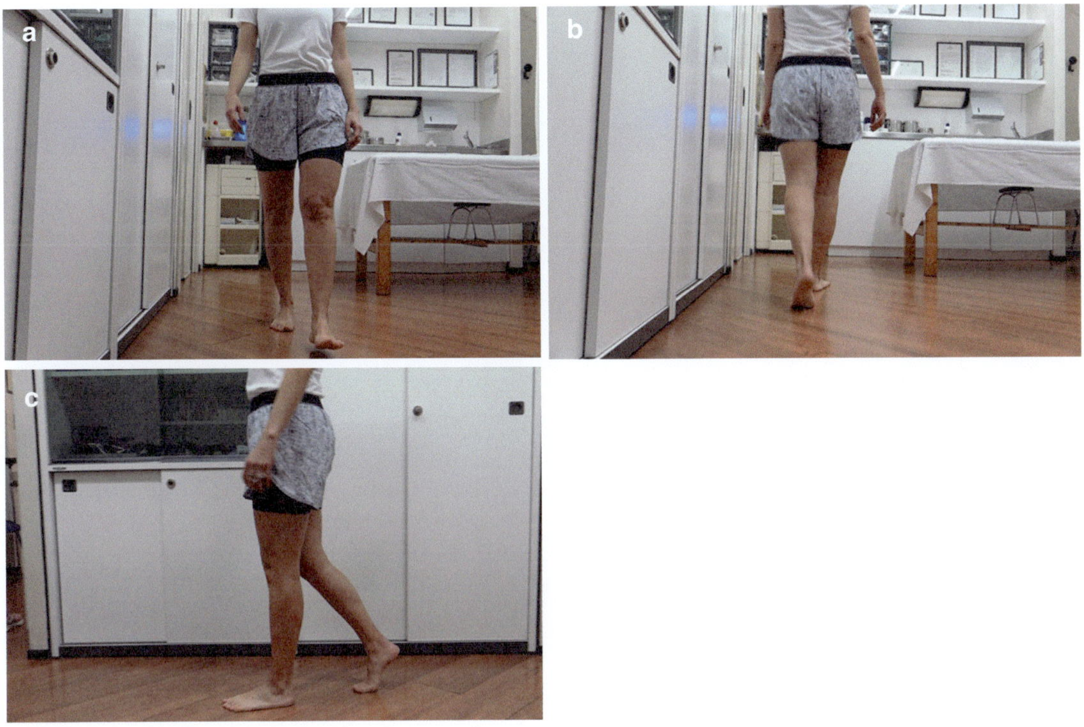

Fig. 53.2 (**a**) Contact phase (heel strike). (**b**) Foot-flat. (**c**) Mid-stance phase

patellofemoral pain, and patients with outfacing patellae are associated with habitual subluxation or dislocation of the knee cap. Inspection from the side includes knee flexion, foot arches, and toe deformity, whereas inspection from behind allows evaluation of the iliac crest alignment, gluteal muscle bulk, popliteal swelling, and hindfoot abnormalities (Fig. 53.3).

The complete evaluation of the patellofemoral joint is described in Chaps. 51 and 58.

The clinical alignment of the lower part of the leg (anatomical axis) measures the femorotibial angle (Fig. 53.4) and is different from the mechanical axis of the limb (Fig. 53.5), as measured from the femoral head through the knee to the center of the ankle on standing long-leg radiographs.

With a goniometer applied to the anterior aspect of the thigh as well as the lower part of the leg and on the patella, the examiner can report the clinical varus or valgus alignment. This measurement should be used along with the standing long-leg X-rays.

From the side, the inclination of the pelvis, spinal curvatures—particularly hyperlordosis—and the body's position concerning the pelvis can be assessed, and from the back, the presence or absence of a scoliotic curve pelvic tilt or foot rotation anomalies can be assessed.

53.2.2.3 Range of Motion (RoM)

A precise assessment will reveal genu recurvatum or any deformities. Range of motion should be compared to the contralateral knee, and any recurvatum or loss of flexion or extension should be noted. Limited mobility of the hip, especially its rotation, may influence both the gait pattern and dynamic load (Fig. 53.6). The RoM evaluation of the tibiofemoral joint was described in Chap. 49.

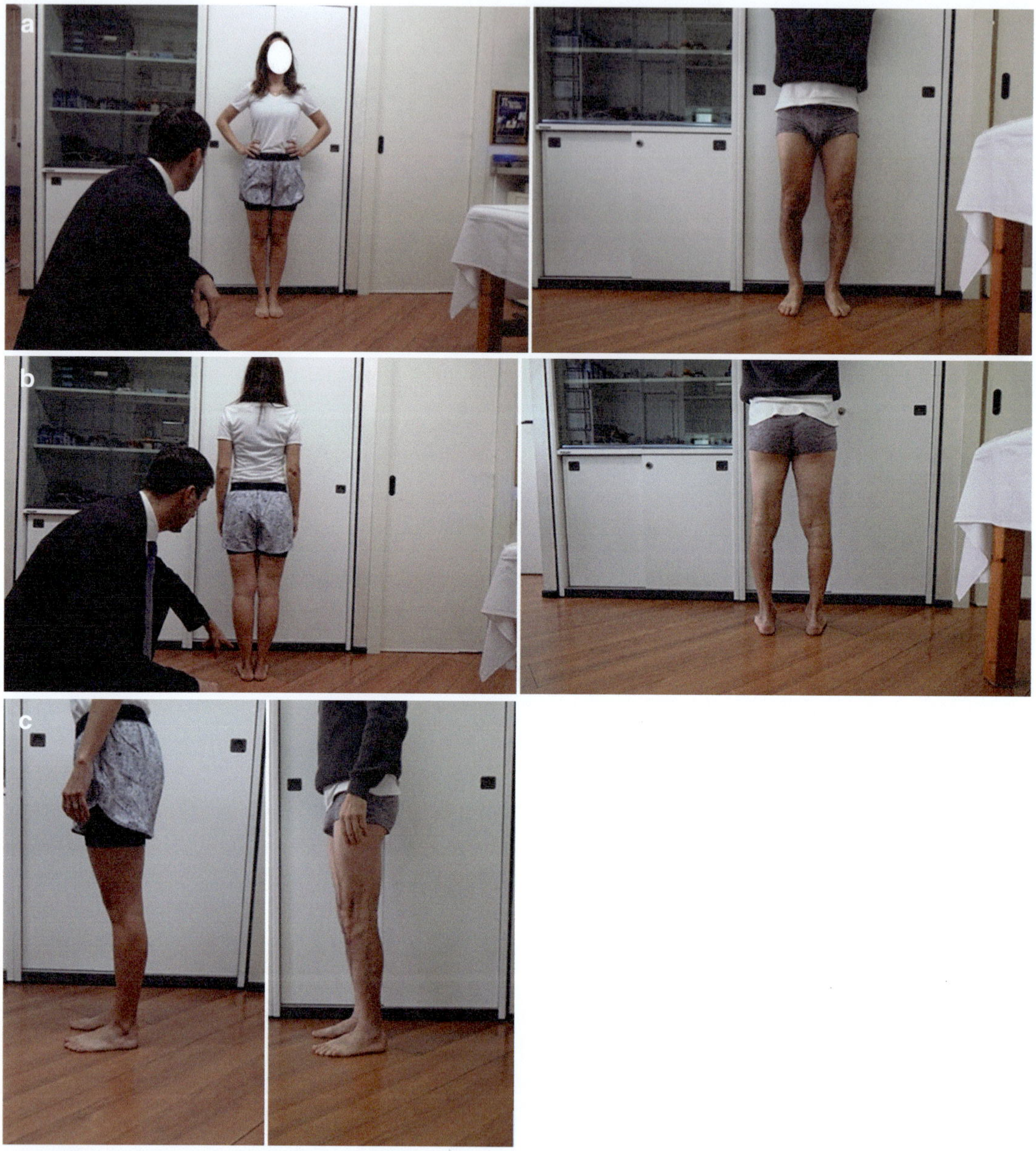

Fig. 53.3 (**a**) Barefoot stand in bipodal position, front evaluation: a) normal alignment, b) genu varus. (**b**) Barefoot stand in bipodal position, back evaluation: a) normal alignment, b) genu varus. (**c**) Barefoot stand in bipodal position, side evaluation: a) normal alignment, b) genu varus with a deficit of knee extension

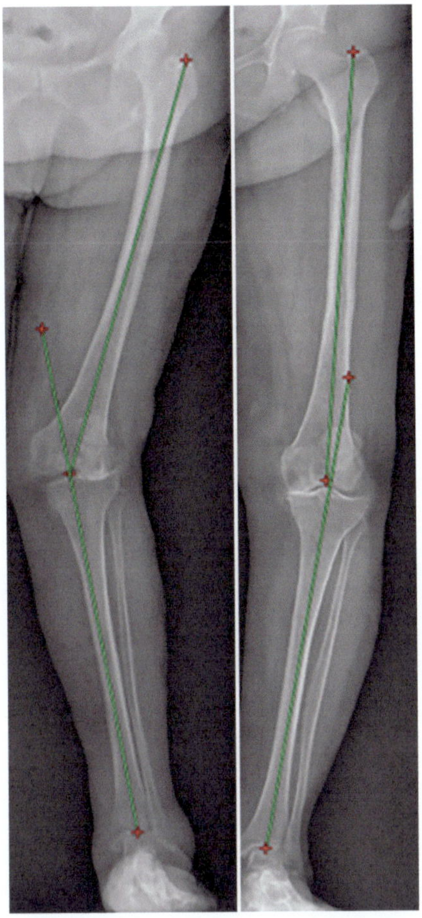

Fig. 53.4 The anatomical axis is measured by drawing lines parallel to the femur's long axis and the tibia and measuring the intercepting angle (from left: valgus and varus alignment)

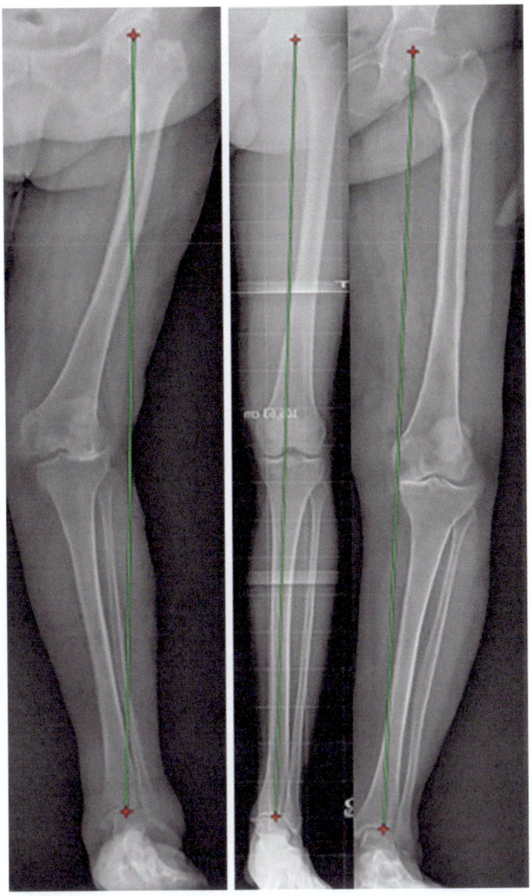

Fig. 53.5 The leg's mechanical axis is measured in the standing position with an imaginary "plumb line" dropped from the femoral head to the center of ankle joint. This angular measurement gives the best functional evaluation of lower extremity alignment (from left: valgus, normal, and varus alignment)

53.2.3 Special Tests

53.2.3.1 The Q Angle or Quadriceps Angle

This measurement reflects the valgus vector of the quadriceps pull acting on the patella between the iliac crest and the tibial tuberosity and gives an idea of tracking the patella (Fig. 53.7). It is usually measured in standing position and is done by drawing the first line between the center of the patella and the anterior-superior iliac spine and the second line between the center of the patella and the center of the tibial tubercle. It averages 15° in normal individuals: 14° in men and 17° in women. Anatomic variants produce either inwards-facing patellae or lateral displacement of the tibial tubercle and other external factors such as femoral anteversion and external torsion of the tibia. They can result in an increased Q angle, increasing the tendency for patellofemoral pain [4, 5]. The patient is then asked to walk while the examiner evaluates whether there is evidence of antalgic gait localized to the hip or knee joint. Moreover, it is possible to assess the symmetry of gait, stride length, patellae's orientation, varus or valgus alignment of the knee, and pelvic tilt.

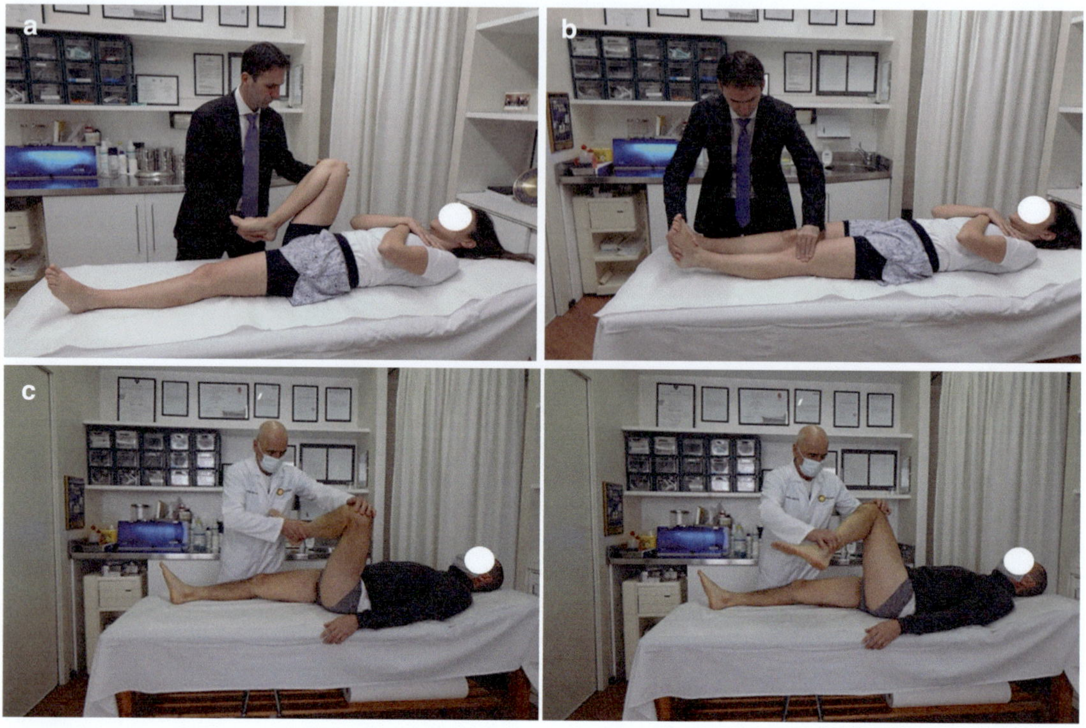

Fig. 53.6 (**a**) Knee flexion assessment. (**b**) Knee extension assessment. (**c**) Hip rotation assessment

53.2.3.2 The Single-Leg Knee-Bend Test

The single-leg knee-bend test is performed wherein the patient is asked to bend his/her leg by bringing the knee close to the thorax. This position helps assess the quadricep's support and evaluate the core stability at the hip and pronation of foot and ankle. Establishing the level of lower extremity support in any patient with knee pain is vital to guide physical therapy appropriately to improve the lower extremity's overall function [6].

53.2.3.3 Sitting Position with the Legs Hanging from the Table

On inspection, swelling of the tibial tuberosity or the patella's inferior pole might suggest osteochondrosis, insertional tendinitis, or a partial rupture of the patellar tendon. An overview of muscular atrophy, especially vastus medialis, is also evident. Besides, an assessment of patellar height can be made in this position (Fig. 53.8). In this position, the patella should face directly forward in a normal patient. In *patella alta*, the high-riding patella faces upward towards the ceiling, and in *patella infera/baja,* the patella is lower than the normal side and seems to be drawn into the sulcus between the femoral condyles. The patellar tracking is then evaluated by asking the patient to flex and extend the knee while sitting. In the presence of instability, the patella undergoes subluxation near full extension constituting the so-called *J-sign* [7].

Also, an assessment of the *tubercle-sulcus angle* can be made at 90-degree flexion of the knee and is a variation of the Q angle described by Kolowich [8] to eliminate the effect of femoral rotation and to detect the abnormal displacement of the tibial tubercle. The angle is measured by drawing the first line passing through the

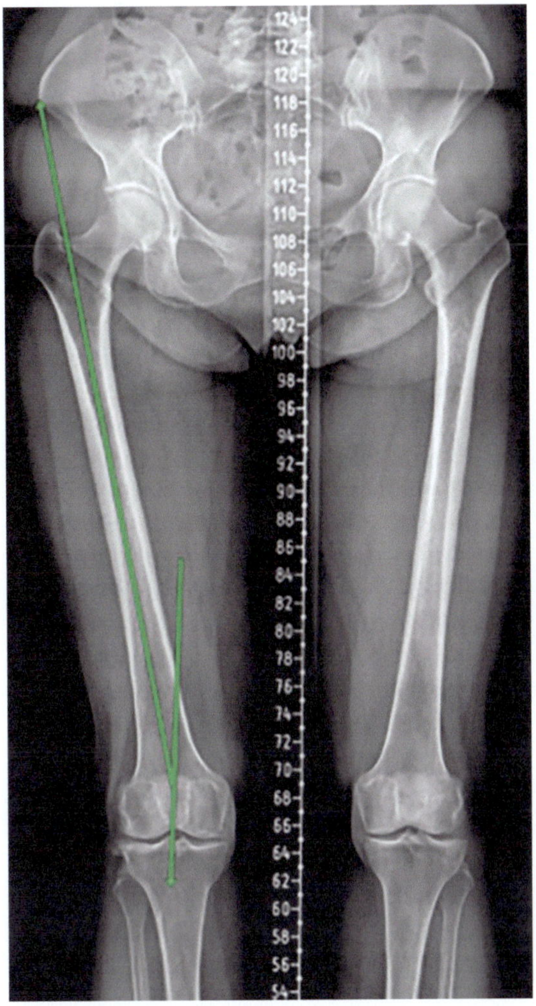

Fig. 53.7 Q angle formed by the intersection of lines from the anterior superior iliac spine and from tibial tuberosity through the midpoint of patella. Large Q angle predisposes to patellar subluxation

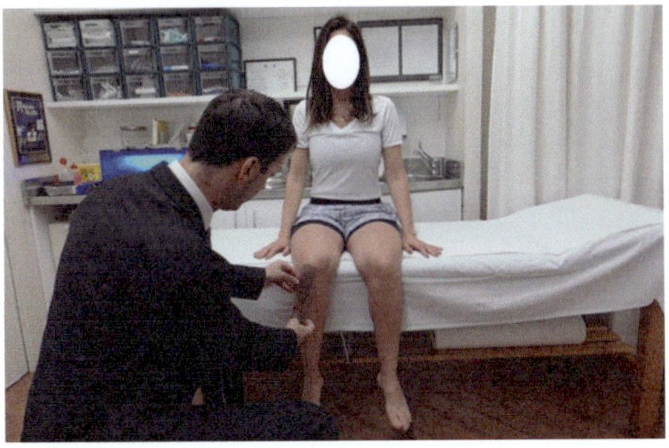

Fig. 53.8 Patellofemoral alignment assessment

patella center and the center of the tibial tubercle, and the second line is drawn perpendicular to the transepicondylar axis. This angle is usually less than 8° in women and less than 5° in men. An increase in this angle reflects lateral displacement of the tibial tubercle and is associated with patellofemoral pain and instability.

53.3 Unicompartmental Osteoarthritis

Many patients with malalignment have unicompartmental medial or lateral femorotibial osteoarthritis. Those patients have pain during weight-bearing in the affected joint compartment. If pain is not located exclusively either over the medial compartment or the lateral joint space, the indication for osteotomy/partial knee replacement (PKR) should be reconsidered. Patellofemoral pain with significant degenerative changes of the cartilage is considered as a relative contraindication for unicompartmental surgery. Particular attention should be addressed to the subjective pain level, such as the visual analogue scale (VAS). Patients with the worst pain usually have less relief and benefit from the osteotomy.

53.4 Knee Instability

We often see another group of patients in our office with varus malalignment and knee instability. Signs of cartilage damage in the medial joint are frequent in these patients. In contrast to patients with osteoarthritis, where the pain is the leading symptom, instability and "giving away" of the knee are prominent. Combined procedures with correction osteotomy and ligament reconstruction or two-plane osteotomy (valgization-flexion or valgization-extension osteotomy) are established treatment methods for these patients. The different exams for knee instability are described in Chap. 52.

Box 1 Tips for Knee Inspection

Skin:

- Discoloration, wounds, gross deformity, or previous scars

Soft tissues:

- Swelling, muscle atrophy, symmetry

Bone:

- Length: compare to the contralateral side
- Position: genu varum or valgus; flexion contractures
- Gross deformity or malalignment

Gait:

- Varus thrust can indicate LCL or PLC insufficiency or injury
- Antalgic (painful), shortened stance phase on the affected side
- Patellar tracking
- Flexed knee gait from tight Achilles tendon or hamstrings

Box 2 Tips for a Complete Knee Malalignment Clinical Examination

During the physical examination, the knee is examined to take particular note of:

- The severity of the deformity.
- The correctability of the deformity—how close to neutral alignment will the knee get with applied valgus force?
- The competence of the ligaments—does the lateral ligament have a firm endpoint? Does the medial ligament have a firm endpoint?
- Associated fixed flexion deformity.

53.5 Case 1

Active female 60 years old with knee varus malalignment.

Mechanical tibiofemoral axis 20°, JLCA: 10°, LDFA: 85.4°, mLDFA: 91.6°, MPTA: 82.4°, LDTA: 86°.

Treatment: Closing wedge high tibial osteotomy (Fig. 53.9).

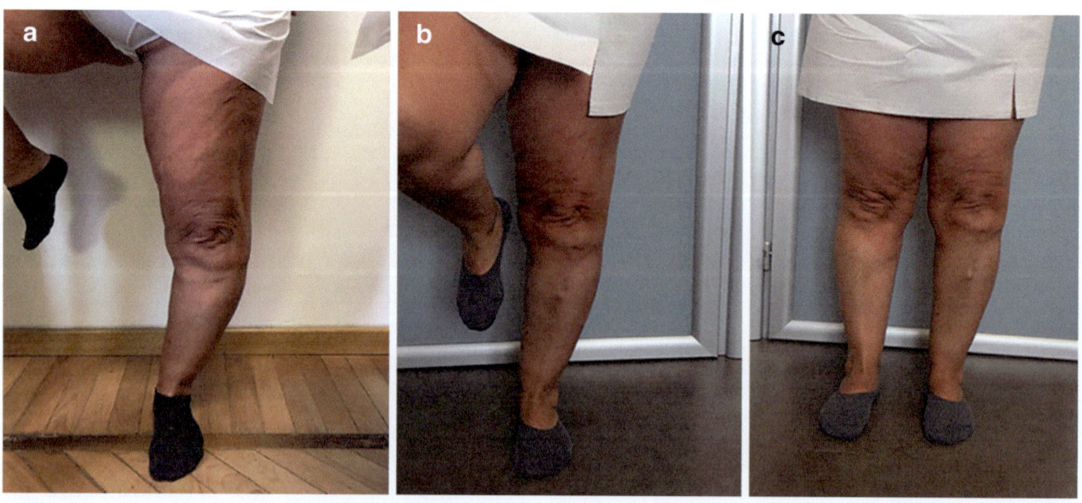

Fig. 53.9 (**a**) Pre-op monopodalic clinical evaluation of knee varus. (**b**, **c**) Four months post-op clinical evaluation after surgical correction. (**d**) Pre-op full-length AP weight-bearing radiograph. (**e**) Post-op full-length AP weight-bearing radiograph

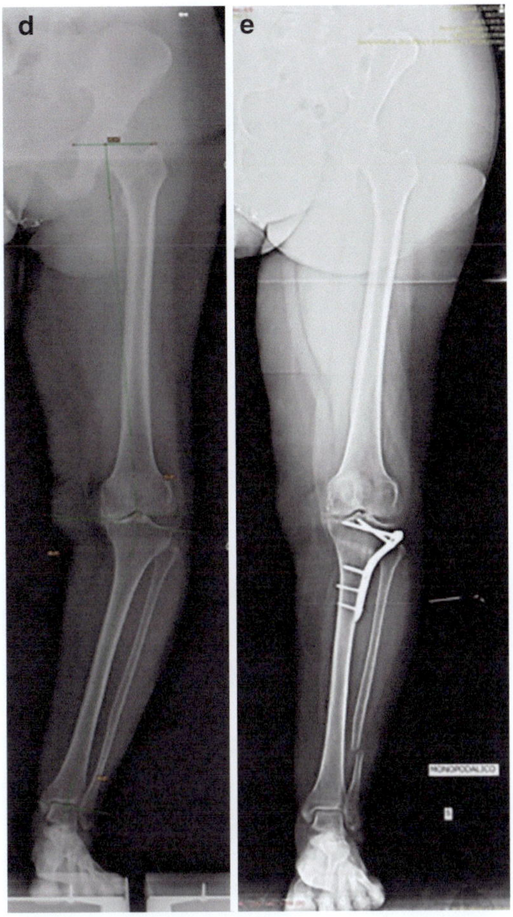

Fig. 53.9 (continued)

53.6 Case 2

Active male 58 years old, with right-knee varus malalignment.

Radiographs: Genu varum of 5° (mLDFA of 92° and mMPTA of 87°).

Varus knee correction with femoral closing wedge osteotomy with lateral subtraction (Fig. 53.10).

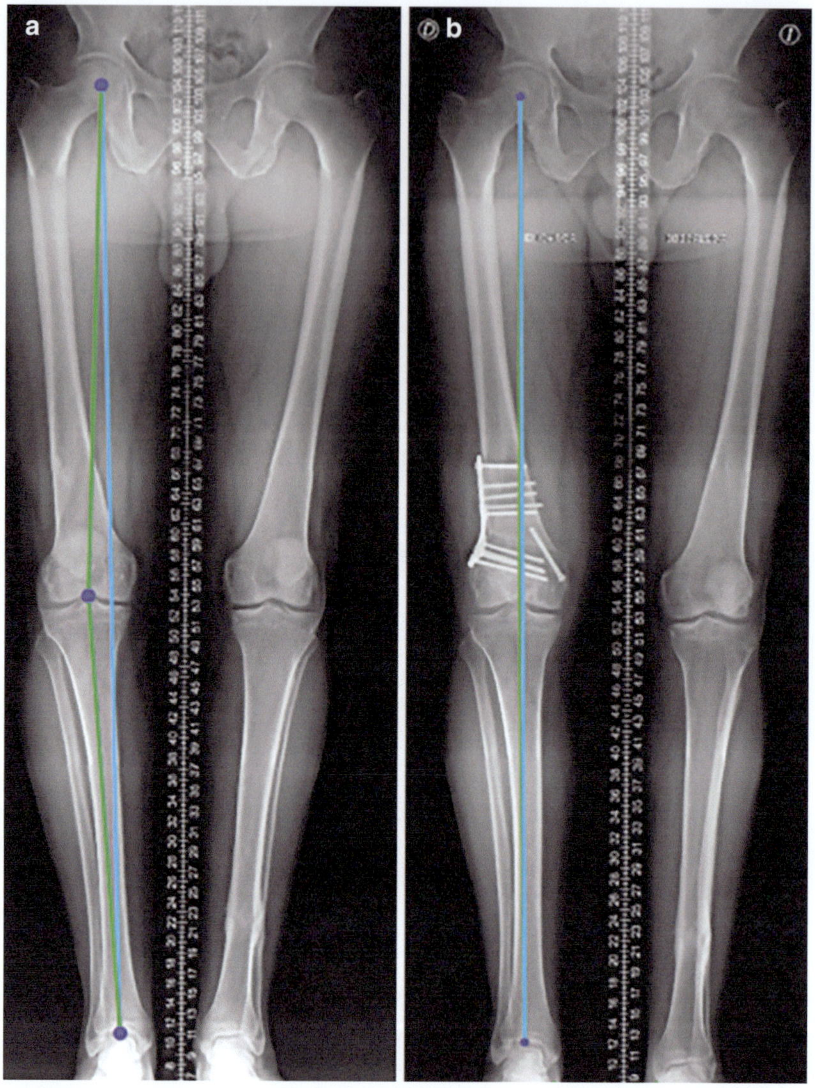

Fig. 53.10 (**a**) Pre-op full-length AP weight-bearing radiograph. (**b**) Post-op full-length AP weight-bearing radiograph

53.7 Conclusion

Clinical examination of the knee remains an essential step in the evaluation process, despite improvements in advanced imaging techniques. The complete assessment of knee malalignment should always begin with a meticulous clinical examination, including the patient's complete history of symptoms and a full description of the injury mechanism if it exists. Often, the history will direct the examiner to the knee's concerned area, but one must evaluate the patient as a whole. This will improve the physical examination, result in a more accurate diagnosis, and should allow us to identify the cause of misalignment. Finally, the clinical evaluation is mandatory to determine the additional investigations needed to complete the diagnosis and plan the strategy to treat the pathology efficiently.

References

1. Gao L, et al. Advances in modern osteotomies around the knee: report on the Association of Sports Traumatology, arthroscopy, Orthopaedic surgery, rehabilitation (ASTAOR) Moscow international osteotomy congress 2017. J Exp Orthop. 2019;6(1):9.

2. Herman BV, Giffin JR. High tibial osteotomy in the ACL-deficient knee with medial compartment osteoarthritis. J Orthop Traumatol. 2016;17(3):277–85.
3. Noyes FR, Barber-Westin SD. 26–Tibial and femoral osteotomy for varus and valgus knee syndromes: diagnosis, osteotomy techniques, and clinical outcomes. In: Noyes FR, Barber-Westin SD, editors. Noyes' knee disorders: surgery, rehabilitation, clinical outcomes. 2nd ed. Amsterdam: Elsevier; 2017. p. 773–847.
4. Kaya D, Doral MN. Is there any relationship between Q-angle and lower extremity malalignment? Acta Orthop Traumatol Turc. 2012;46(6):416–9.
5. Christian SR, et al. Imaging of anterior knee pain. Clin Sports Med. 2006;25(4):681–702.
6. Claudon B, et al. Knee kinetic pattern during gait and anterior knee pain before and after rehabilitation in patients with patellofemoral pain syndrome. Gait Posture. 2012;36(1):139–43.
7. Watson CJ, et al. Reliability of the lateral pull test and tilt test to assess patellar alignment in subjects with symptomatic knees: student raters. J Orthop Sports Phys Ther. 2001;31(7):368–74.
8. Kolowich PA, et al. Lateral release of the patella: indications and contraindications. Am J Sports Med. 1990;18(4):359–65.

54 Evaluation of Bursitis About the Knee

Katarzyna Herman, Przemysław Pękala, Dawid Szwedowski and Jerzy Cholewiński

54.1 Introduction

Bursae are synovium-lined bag-like structures that can be found throughout the human body and function as a buffer decreasing the friction between adjacent tissues [1]. Bursitis may occur due to many pathological conditions, both infectious and noninfectious. Typically, the inflamed bursa has a thickened synovial lining coexisting with an increased amount of fluid inside [2]. Diagnostic protocol includes usually ultrasound and magnetic resonance imaging (MRI); in some cases, a standard X-ray may also be helpful. Treatment algorithm depends on the etiology of the bursitis and varies from conservative therapy to surgery.

54.2 Anatomy and Pathophysiology

In the whole human body, there are over 140 bursae [3]. Depending on the anatomical location, structure, and surrounding tissues, bursae can be divided and described in different groups [1, 2, 4]. Based on the structure, two types can be distinguished, constant (native) and adventitial, and inflammation may occur in both. Constant bursae are formed during embryologic development and are lined with endothelial cells that produce fluid. Adventitial bursae form later in life as a reaction to repetitive trauma or friction; they lack endothelial lining, and the accumulated fluid is an effusion caused by increased local vascular permeability [2, 4].

K. Herman (✉)
Department of Orthopedics and Traumatology, Brothers Hospitallers Hospital, Katowice, Poland

Department of Medical Rehabilitation, Medical University of Silesia, Katowice, Poland

Orthopaedic Arthroscopic Surgery International (O.A.S.I.) Bioresearch Foundation, Gobbi N.P.O., Milan, Italy

P. Pękala
International Evidence-Based Anatomy Working Group, Department of Anatomy, Jagiellonian University Medical College, Krakow, Poland

Faculty of Medicine and Health Sciences, Andrzej Frycz Modrzewski Krakow University, Krakow, Poland

Lesser Poland Orthopedic and Rehabilitation Hospital, Krakow, Poland

D. Szwedowski
Orthopaedic Arthroscopic Surgery International (O.A.S.I.) Bioresearch Foundation, Gobbi N.P.O., Milan, Italy

Department of Orthopaedics and Trauma Surgery, Provincial Polyclinical Hospital, Torun, Poland

J. Cholewiński
Department of Orthopedics and Traumatology, Brothers Hospitallers Hospital, Katowice, Poland

Department of Medical Rehabilitation, Medical University of Silesia, Katowice, Poland

J. G. Lane et al. (eds.), *The Art of the Musculoskeletal Physical Exam*,
https://doi.org/10.1007/978-3-031-24404-9_54

Table 54.1 Location of the knee bursae

Aspect	Bursa	Location
Anterior	Suprapatellar	Between the quadriceps tendon and the anterior aspect of the femur (in 85% of adults communicates with the knee joint) [1, 10]
	Prepatellar	Subcutaneously over the patella [13]
	Prepatellar subcutaneous	Between skin and superficial fascia
	Prepatellar subfascial	Between superficial fascial and intermediate aponeurotic layer
	Prepatellar subaponeurotic	Between intermediate oblique aponeurotic layer and deep rectus fibers anterior to the patella
	Superficial infrapatellar	Between the tibial tubercle and the skin
	Deep infrapatellar	Between distal insertion of patellar tendon, anterior part of the tibia, and inferior border of the infrapatellar fat pad [4, 22]
Medial	Pes anserine	Upper medial aspect of the tibia, at the insertion of the conjoined tendon of the pes anserinus muscles [28]
	Medial collateral ligament	Between the deep and superficial layer of the MCL [37]
	Semimembranosus–medial collateral ligament	Between semimembranosus tendon and MCL with extension lying between semimembranosus tendon and medial tibial condyle [11]
Lateral	Lateral collateral ligament–biceps femoris	Between the distal insertion of the fibular collateral ligament (FCL) and the anterior arm of the long head of the biceps femoris tendon [46]
	Iliotibial	Between the iliotibial band insertion to the Gerdy's tubercle and the surface of underlying tibia [48]
Posterior	Gastrocnemius–semimembranosus (popliteal or baker's cyst)	Between the medial head of the gastrocnemius and the semimembranosus tendon [11]

Based on the location, they may be subcutaneous, subfascial, subtendinous, and submucosal [1]; also, deep and superficial may be distinguished. Bursae about the knee joint can be divided into four groups based on their location: anterior, posterior, lateral, or medial side of the joint (Table 54.1) [5].

Bursitis may be classified depending on the triggering factor: traumatic, inflammatory, infectious, and proliferative [4]. Traumatic etiology may be acute due to direct trauma to the area or chronic, being a result of repeated incidents of microtrauma or overuse. Inflammatory response of the synovial lining may be caused by systemic diseases such as rheumatoid arthritis, gout or pseudogout, systematic lupus erythematosus, idiopathic calcification, calcinosis, or uremia [6–8]. Less frequently, a proliferative process such as pigmented villonodular synovitis, primary synovial chondromatosis, lipoma arborescens, or amyloidosis may occur [4]. Some bursae adjacent to the joint may communicate with the joint cavity; in those situations, the bursal distention may be a representation of the intra-articular pathology [4, 9].

54.3 Anterior Aspect

54.3.1 Suprapatellar Bursa

A suprapatellar bursa is located between the quadriceps tendon and the anterior aspect of the femur. In 85% of adults, the bursa communicates with the joint. During embryonic development, the suprapatellar plica, separating the bursa from the joint cavity, is perforated and usually involutes; therefore, the bursa becomes a part of the knee joint cavity [1, 10]. The suprapatellar plica may be present only as a remnant, but in some patients, it can be a complete membrane separating the bursa from the joint [11, 12]. That is why in the absence of suprapatellar plica the bursa is a part of the joint cavity and therefore the distention is the result of an effusion of the joint and not a typical bursitis [1]. However, if the bursa is separated from the joint, and therefore is non-communicating, fluid collection may clinically appear as a soft-tissue tumor [11].

54.3.2 Prepatellar Bursa

Prepatellar bursa is located subcutaneously over the patella and is therefore prone to repetitive trauma. Typically, there is a trilaminar arrangement of the fibrous tissues overlying the patella, superficial transverse fascia, intermediate oblique aponeurotic layer, and deep rectus femoris tendinous fibers. Consequently, there are three possible locations of the bursa: subcutaneous, subfascial, and subaponeurotic [13].

Most commonly, patients with bursitis are males aged 40–60 years of age [7]. The pathomechanism varies depending on the etiology of the bursitis, and it is important to distinguish between septic and non-septic, since it determines both the diagnostic procedures and treatment.

Non-septic bursitis is most commonly caused by repetitive work-related or recreational trauma [7]. Usually, acute and chronic cases can be distinguished. Acute cases are a result of a direct trauma to prepatellar area followed by bleeding of the synovial lining of the bursa. Chronic cases are a result of repetitive microtrauma, especially in patients with professions that require frequent kneeling, like cleaners or tilers [14].

Septic bursitis occurs due to contamination of the bursa usually through the skin, directly from the injury, through injection, from aspiration, or from skin infection [3]. A hematogenous contamination is possible; however, it occurs rarely due to poor blood supply of the prepatellar bursa [15]. It is estimated that in about 80% of cases, infection is caused by *Staphylococcus aureus* [16]. Other bacteria such as Staphylococcus species, Streptococcus, Brucella, and Mycobacterium have also been reported [16]. It is important to assess risk factors such as trauma or overuse. However, immunocompromised patients, alcoholics, and patients with chronic diseases (diabetes mellitus and chronic kidney disease) are prone to septic bursitis [8, 14].

Patients usually report to emergency units or a general practitioner with pain over the patella, swelling, and warmth, sometimes with a local redness over the bursa; these symptoms however are not specific for infection. If the patient reports aforementioned symptoms with fever, a septic pathogenesis should be considered [7]. On examination, pain in palpation is present, range of motion is not usually affected, but pain exaggerates in high flexion due to compression of the bursa [8, 14]. Smith et al. reported that a difference in temperature between affected and contralateral side ≥2.2 °C was 100% sensitive and 94% specific for septic bursitis [17]. On the contrary, Baumbach et al. stated that the cause of bursitis cannot be established solely on clinical findings [7].

Differentiating between septic and aseptic pathogenesis of the prepatellar bursitis may be problematic. A standard radiograph of the knee joint should be considered, especially in patients with a history of trauma or when a loose body is suspected [7, 15]. The most common finding is the swelling of prepatellar tissues; in some cases, gas bubbles may be visible, suggesting a presence of a gas-producing microorganism [18]. Ultrasonography can be useful in visualizing the details and further characterizing the pathology of the inflamed bursa (Fig. 54.1). Usually, swelling of the surrounding tissues and thickening of the bursal walls are seen. Non-septic, chronic distention of the bursa with simple fluid collection would appear anechoic on ultrasound imaging [15, 18]. MRI is rarely used in prepatellar bursitis but may be useful to diagnose coexisting pathologies, i.e., evidence of osteitis [15].

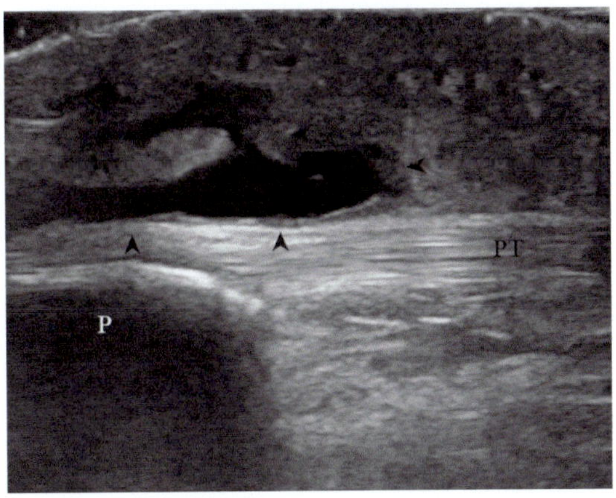

Fig. 54.1 Prepatellar bursitis. A long-axis ultrasound image over the patellar tendon shows an anechoic fluid collection (arrows) and coexisting synovial hypertrophy. *P* patella, *PT* patellar tendon

Generally, an aspiration is necessary not only to relieve pain but also to examine the fluid. The sole appearance of the aspirate may indicate the diagnosis. In non-septic bursitis, the fluid is usually yellow; in trauma or overuse-related cases, it may be blood-tinged, but a purulent and cloudy aspirate suggests infection [7]. It is crucial to obtain a bursal fluid culture to confirm the diagnosis but also to reassess the efficacy of the chosen antibiotic therapy in cases of septic bursitis. Blood sampling is useful in determining the etiology and severity of the process; elevated infection parameters such as C-reactive protein (CRP) and leukocyte level suggest septic bursitis [7].

54.3.3 Superficial Infrapatellar Bursa

Superficial infrapatellar bursa, also known as pretibial bursa, lies within soft tissues between the tibial tubercle and the skin [19]. The inferior margin of the bursa is located 0–16.7 mm from the inferior insertion of the patellar tendon [20]. Occasionally, it can communicate with the prepatellar bursa [5]. It is an adventitial bursa, so it does not have a synovial lining. In the anatomical study by Viegas et al., it was present in 55% of the specimens [20]. Chronic bursitis typically occurs in patients with occupational kneeling; that is why it is sometimes referred to as clergyman's knee. In acute cases though, it is generally a result of direct trauma [11].

Similarly, as in prepatellar bursitis, a superficial infrapatellar bursitis demonstrates as pain and local swelling, inflammation, and erythema anterior to the tibial tubercle [21]. Poorly defined focal fluid and inflammation of the surrounding tissues can be seen on MRI and on ultrasound [21].

54.3.4 Deep Infrapatellar Bursa

The deep infrapatellar bursa is a V-shaped structure located between distal insertion of patellar tendon, anterior part of the tibia, and inferior border of the infrapatellar fat pad [4, 19]. No connection between the bursa and the joint cavity was found in a cadaveric study [22]. Bursitis occurs usually due to overuse in knee extension mechanism, most commonly in runners and jumpers [19, 20]. It has also been reported in Osgood-Schlatter disease; in 71% of the cases, distended deep infrapatellar bursa was seen on MRI [23].

Clinically, patients complain of anterior knee pain around the insertion of patellar tendon, which usually worsens when palpated and therefore is often mistaken for patellar tendonitis [11, 20]. Both ultrasound and MRI show a fluid collection in the anatomical location of the bursa; however, it is considered normal to find a 2–3 mm of fluid in the deep infrapatellar bursa [11].

54.4 Medial Aspect

54.4.1 Pes Anserine Bursa

The sartorius, gracilis, and semitendinosus muscles attach to the medial side of tibia to form a shape reminiscent of a goose's foot. The pes anserinus, which is the Latin expression of the goose foot, is the conjoint tendinous insertion of these three muscles. The anserine bursa is located deep to these tendons. It serves to reduce friction between the superficially located pes anserinus tendons and the medial tibia and the medial collateral ligament that lies deep to it [12].

Although there have been anatomical studies investigating the shapes of the pes anserinus in various populations [24–27], the anatomical information of anserine bursa is limited. The anserine bursa is located at the upper medial aspect of the tibia, at the insertion of the conjoined tendon of the pes anserinus muscles. Lee et al. tried to identify the boundary of the anserine bursa to determine the recommended injection site and the shape of the insertion area of pes anserinus [28]. According to this anatomical study, the shape of anserine bursa was irregularly circular. Most of the anserine bursa in the specimens reached the proximal line of the tibia. The inferior border of anserine bursa was, on average, located at about 16 mm from the inferomedial point of tibial tuberosity. In the medial view of the tibia, the anserine bursa was located posteriorly and superiorly from the tibia's midline, and it followed the lines of the sartorius muscle. The study by LaPrade et al. confirmed that the sartorius tendon fascia is intimately attached to the superficial fascial layer [29]. The gracilis and semitendinosus tendons are located in the deep surface of the superficial fascial layer over the medial aspect of the tibia. These two muscles are also connected to the crural fascia [30].

The posteromedial part of the bursa contained the largest number of nerves and vessels, whereas the anteroinferior part, which is usually the injection site, contained the least [28]. The anserine bursitis can be the source of discomfort in patients experiencing knee pain and usually occurs in patients with the knee osteoarthritis (83.3% of anserine bursitis or tendinitis patients were reported to have radiographic evidence of knee osteoarthritis) [31, 32]. In contrast, this entity was clinically diagnosed in 46.8% of the patients with knee osteoarthritis [33].

Anserine bursitis should be suspected in patients with medial knee pain on activities such as going up and downstairs, rising from a chair, and lying in bed with one knee pressing the other. The patients with knee osteoarthritis present this entity with rapidly increasing knee pain on the medial side of the knee joint or with transition of the pain with activity into the pain at night [34]. Differential diagnoses include deep vein thrombosis and muscle bleeding after trauma or administration of anticoagulant treatment. The rupture of the pes anserinus bursa is a rare condition; however, it should also be considered as a differential diagnosis of acute unilateral lower leg swelling [35]. Imaging studies such as ultrasonography and MRI are an essential part of the assessment in order to make an appropriate diagnosis [5, 36].

54.4.2 Medial Collateral Ligament Bursa

The medial collateral ligament (MCL) bursa was described in 1943 by Brantigan and Voshell [37]. It is located between the deep and superficial layer of the MCL. Normally, there is no fluid in this space and therefore the MCL bursa is not visible in radiological imaging when not inflamed. However, existence of this potential anatomical space has been documented in most of the studied cadavers using dissection [38]. It may be located near the femoral and/or tibial cortex. It is hypothesized that the MCL bursa reduces friction between the MCL and bony surfaces during knee motion. Histologically, the cells inside the bursa are similar to those present in ganglion cysts [39].

The medial knee pain is one of the most prevalent clinical problems in the orthopedic practice. There are several causes of this condition such as medial meniscal tears, injuries of the medial ligamentous structures (the superficial MCL, the deep MCL, and the posterior oblique ligament), cartilage defects, bursitis, and many others [40]. The most common inflammation of bursa responsible for medial knee pain is pes anserinus bursitis. However, one should always differentiate it with less common, but important, MCL bursitis, which diagnosis requires certain knowledge and clinical experience.

The MCL bursitis is a rare clinical finding, and it can be observed in both adults (most commonly) and children (less commonly) [41]. To confirm this diagnosis, the radiological test should be performed such as MRI or ultrasound examination [42]. Noteworthy is the fact that the fluid in the MCL bursa can be observed in asymptomatic patients. In the comprehensive radiological study, De Maeseneer et al. emphasized that the superficial fluid collection, detected anteriorly or posteriorly to the superficial MCL, should not be considered as the MCL bursitis, and other causes should be considered [43]. Moreover, one should remember to differentiate this condition from the cyst of the medial meniscus (continuity with the meniscal tissue).

Noteworthy is the fact that the MCL bursitis can be caused by irritation of this region due to tibial and femoral osteophytes around the joint space, which are prevalent features of the osteoarthritis [40].

54.4.3 Semimembranosus–Medial Collateral Ligament Bursa

The semimembranosus–tibial collateral ligament (SMTCL) bursa has a U-shape with arms of the U forming two pockets: distal superficial and proximal deep pocket. The distal superficial pocket extends between the tendon of semimembranosus muscle and MCL, whereas the proximal deep pocket lies between the semimembranosus tendon and medial tibial condyle [44, 45]. Both join at the anterosuperior margin of the semimembranosus tendon [11]. The most proximal portion of the bursa lies beside the posterior horn of medial meniscus, and the most distal portion lies 1–2 cm lower than the joint line [11, 48]. During extension, external rotation, and valgus stress to the knee, the bursa protects the semimembranosus tendon which becomes trapped between the strained MCL and the medial tibial condyle. Chronic overuse or acute trauma results in inflammatory process within the bursa [11].

Patients usually report posteromedial pain and tenderness, localized superiorly and posteriorly to the pes anserine and at the level of the joint line or just below [19]. Symptoms can be misleading and be confused with medial meniscus lesion [21]. Conservative treatment and rest are sufficient, but symptoms may be recurring after returning to physical activity too suddenly [45].

54.5 Lateral

54.5.1 Lateral Collateral Ligament–Biceps Femoris Bursa

Lateral collateral ligament–biceps femoris bursa was described to be a constant bursa lying between the distal insertion of the fibular collateral ligament (FCL) and the anterior arm of the long head of the biceps femoris muscle tendon. It is described to be wrapping around along the lateral aspect of the FCL up to three-quarters of its circumference. Histologically, it was reported that this bursa is constant with synovial lining [46].

Clinically, the patient complains of lateral knee pain that aggravates on palpation of the distal aspect of FCL [12]. Intriguingly, a case of calcifying bursitis due to direct trauma was described in an 18-year-old man with a flexion contracture of 30° [47].

54.5.2 Iliotibial Bursa

The iliotibial bursa lies between the iliotibial band insertion to the Gerdy's tubercle and the surface of underlying tibia [48]. Normally, bursitis occurs due to repetitive microtrauma and varus stress to the knee, commonly in runners training on slopped terrain. Patients generally report anterolateral knee pain [11, 19]. Symptoms may be similar to those reported in iliotibial tendonitis and injury of the lateral meniscus or FCL [12].

To determine diagnosis, a postexercise ultrasound may be helpful, and the iliotibial tract will be seen as soft-tissue swelling rather than a fluid collection which is typical for bursitis [49].

54.6 Posterior

54.6.1 Gastrocnemius–Semimembranosus Bursa (Popliteal or Baker's Cyst)

Popliteal cyst or Baker's cyst is the most commonly encountered synovial cyst about the knee [19]. It is a distended fluid collection arising from posteromedial joint capsule into the gastrocnemius–semimembranosus bursa lying between the medial head of the gastrocnemius and the semimembranosus tendon [11]. The gastrocnemius–semimembranosus bursa has two parts: the subgastrocnemius bursa and semimembranosus bursa that form together a crescent-shaped structure [50]. They may be partially divided by a septum in the central part [51]. Though the location is typically between the gastrocnemius muscle and the semimembranosus tendon and generally the cyst expands between the fascia and medial head of the gastrocnemius, it may also expand in different directions: medial, lateral, or even superior [21, 48, 52].

Wolfe et al. suggested that the formation of the cyst depends on the communication between the joint and bursa with a coexisting effusion of the joint [53]. The bursa communicates with the joint cavity through an opening in a joint capsule, at the conjunction of the gastrocnemius tendon and the joint capsule [54]. Later, studies have confirmed the relationship between joint effusion and presence of the baker's cyst [55, 56]. In a more recent cadaveric study, a potential weak point in the posterior capsule was identified between the two extensions of the semimembranosus tendon [52]. It has been debated that possibly due to degenerative weakening of the joint capsule, the communication between joint cavity and bursa increases with age, being significantly higher over the age of 50 [57]. However, Labropoulos et al. questioned that theory and suggested that it rather forms due to the lack of supporting structures in the area [52]. In children, the popliteal cyst is a rare finding and is usually not associated with intra-articular pathology [12, 58]. The flow of the fluid is unidirectional, from joint to the cyst, due to valve-like mechanism found between them (Fig. 54.2) [58]. During intra-cystic and intra-articular pressure measurements, the pressure was found to be higher inside the cyst and it was not possible to lower it through various manipulations. Constant higher cyst pressure and failure to empty the cyst support the one-way valve mechanism theory [54]. The cyst usually develops in a presence of the joint effusion due to abnormalities inside the joint. Intra-articular pathologies include medial meniscal tear, anterior cruciate ligament lesions, chondral lesions, arthritic changes, systematic inflammatory diseases causing knee effusion, and both bacterial and fungal infection [58]. On MRI of the knee obtained for various reasons, baker's cyst was seen in 5–38% of patients [55–57]. Prevalence of this lesion depends on the studied population but is generally associated with an intra-articular pathology. Most commonly in over 60% of reported cases, tears of the posterior horn of the medial meniscus were associated with popliteal cyst [56]. Interestingly, Labropoulos et al. suggested that the posterior horn of the medial meniscus is the only reinforcing structure at the weak spot of the posterior capsule. That is why the degenerative lesions of the medial meniscus posterior horn coexist with distention popliteal bursa [52]. Fielding et al. reported a correlation of baker's cyst with tears of the lateral meniscus in 38%, bilateral meniscal tears in 27%, and anterior cruciate ligament (ACL) tears in 13% [56]. Interestingly, Han et al. found in their MRI study that internal derangement and joint effusion were not related to popliteal cyst presence. They stated that the presence of the cyst depends on specific relation and location of medial head of gastrocnemius tendon and semimembranosus tendon [60].

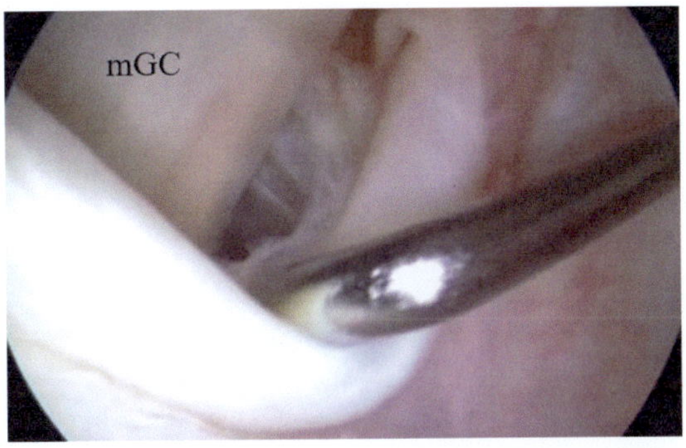

Fig. 54.2 An arthroscopic image of the valve mechanism; the retractor is pulling the synovial fold; seen behind it is the tendon of the medial head of the gastrocnemius muscle (mGC)

Clinically, patients may complain of a mass, swelling, tightness, or sometimes a vague pain in the popliteal region; symptoms generally exacerbate with physical activity. Because the popliteal bursa distention coexists with intra-articular pathology, the symptoms of the cyst may meld with those caused by the intra-articular problem [61]. Due to its anatomical location, the cyst can additionally compress adjacent structures and therefore the patient may present various symptoms. Compression of the popliteal vein will result in calf swelling and sometimes deep venous thrombosis, while a compressed popliteal artery will cause intermittent claudication [62, 63]. Tibial nerve entrapment is correlated with

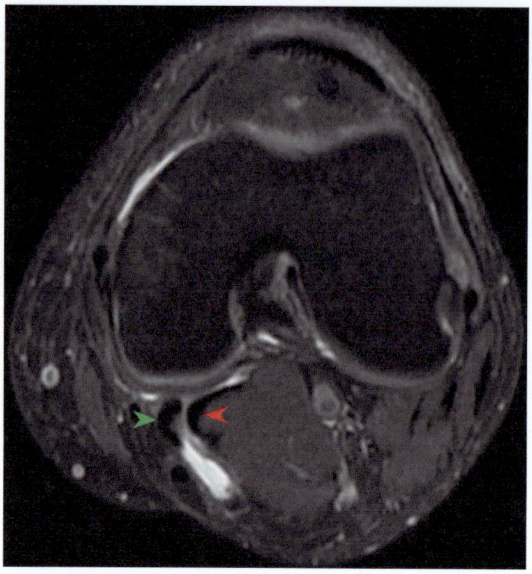

Fig. 54.3 MRI imaging of the baker's cyst—an axial MRI (proton density, fat-saturated) image showing a hyperintense fluid in the distended bursa communicating with the joint cavity between the medial head of the gastrocnemius (red arrow) and semimembranosus tendon (green arrow)

gastrocnemius muscle atrophy, paresthesia, and pain [62]. The cyst may rupture resulting in fluid leakage into the calf, between the muscles and fascia, causing pain, swelling, redness, and tenderness of the calf. This condition, also known as pseudothrombophlebitis, may be misdiagnosed as thrombophlebitis due to similarity of the symptoms.

For physical examination, the patient lies on the back and the examiner palpates the popliteal fossa and lifts the knee changing from full extension to 90° flexion. If a cyst is present, the examiner may feel the tender mass on extension that softens in flexion due to relaxation of the gastrocnemius and semitendinosus muscles; this is also known as Foucher's sign [58, 61, 64]. However, one should be aware that the test is subjective and relies mainly on examiner's experience and opinion [58]. If a popliteal cyst is suspected, ultrasound is the first diagnostic choice to confirm its presence. Typically, it is an anechoic distention of the bursa. If a hemorrhage, synovial hypertrophy, or other debris is present, the image is more heterogenous [4]. Identifying a cyst's opening between the medial head of the gastrocnemius and semimembranosus tendons suggests a popliteal bursa rather than other cysts or ganglia (Fig. 54.3) [50].

The gold standard for diagnosing popliteal cyst and coexisting intra-articular pathologies is MRI [65]. It is also vital to identify a possible coexisting intra-articular pathology, but also to differentiate from other possible lesions, i.e., a ruptured baker's cyst will be easily identifiable [66].

References

1. Hirji Z, Hunjun JS, Choudur HN. Imaging of the bursae. J Clin Imaging Sci. 2011;1:22.
2. Ivanoski S, Nikodinovska VV. Sonographic assessment of the anatomy and common pathologies of clinically important bursae. J Ultrason. 2019;19(78):212–21.
3. Zimmermann B, Mikolich DJ, Ho G. Septic bursitis. Semin Arthritis Rheum. 1995;24(6):391–410.
4. Ruangchaijatuporn T, Gaetke-Udager K, Jacobson JA, Yablon CM, Morag Y. Ultrasound evaluation of bursae: anatomy and pathological appearances. Skelet Radiol. 2017;46(4):445–62.
5. Draghi F, Corti R, Urciuoli L, Alessandrino F, Rotondo A. Knee bursitis: a sonographic evaluation. J Ultrasound. 2015;18(3):251–7.
6. McAfee JH, Smith DL. Olecranon and prepatellar bursitis. Diagnosis and treatment. West J Med. 1988;149(5):607–10.
7. Baumbach SF, Lobo CM, Badyine I, Mutschler W, Kanz K-G. Prepatellar and olecranon bursitis: literature review and development of a treatment algorithm. Arch Orthop Trauma Surg. 2014;134(3):359–70.
8. Aaron DL, Patel A, Kayiaros S, Calfee R. Four common types of bursitis: diagnosis and management. J Am Acad Orthop Surg. 2011;19(6):359–67.
9. Fang CSJ, McCarthy CL, McNally EG. Intramuscular dissection of Baker's cysts: report on three cases. Skelet Radiol. 2004;33(6):367–71.
10. Steiner E, Steinbach LS, Schnarkowski P, Tirman PF, Genant HK. Ganglia and cysts around joints. Radiol Clin N Am. 1996;34(2):395–425, xi–xii.
11. McCarthy CL, McNally EG. The MRI appearance of cystic lesions around the knee. Skelet Radiol. 2004;33(4):187–209.
12. Marra MD, Crema MD, Chung M, Roemer FW, Hunter DJ, Zaim S, et al. MRI features of cystic lesions around the knee. Knee. 2008;15(6):423–38.
13. Boya H, Pinar H, Ozcan O. Synovial osteochondromatosis of the suprapatellar bursa with an imperforate suprapatellar plica. Arthroscopy. 2002;18(4):E17.

14. Khodaee M. Common superficial bursitis. Am Fam Physician. 2017;95(4):224–31.
15. Lormeau C, Cormier G, Sigaux J, Arvieux C, Semerano L. Management of septic bursitis. Joint Bone Spine. 2019;86(5):583–8.
16. Cea-Pereiro JC, Garcia-Meijide J, Mera-Varela A, Gomez-Reino JJ. A comparison between septic bursitis caused by Staphylococcus aureus and those caused by other organisms. Clin Rheumatol. 2001;20(1):10–4.
17. Smith DL, McAfee JH, Lucas LM, Kumar KL, Romney DM. Septic and nonseptic olecranon bursitis. Utility of the surface temperature probe in the early differentiation of septic and nonseptic cases. Arch Intern Med. 1989;149(7):1581–5.
18. Simpfendorfer CS. Radiologic approach to musculoskeletal infections. Infect Dis Clin N Am. 2017;31(2):299–324.
19. Beaman FD, Peterson JJ. MR imaging of cysts, ganglia, and bursae about the knee. Radiol Clin N Am. 2007;45(6):969–82, vi.
20. Viegas FC, Aguiar ROC, Gasparetto E, Marchiori E, Trudell DJ, Haghighi P, et al. Deep and superficial infrapatellar bursae: cadaveric investigation of regional anatomy using magnetic resonance after ultrasound-guided bursography. Skelet Radiol. 2007;36(1):41–6.
21. Steinbach LS, Stevens KJ. Imaging of cysts and bursae about the knee. Radiol Clin N Am. 2013;51(3):433–54.
22. LaPrade RF. The anatomy of the deep infrapatellar bursa of the knee. Am J Sports Med. 1998;26(1):129–32.
23. Rosenberg ZS, Kawelblum M, Cheung YY, Beltran J, Lehman WB, Grant AD. Osgood-Schlatter lesion: fracture or tendinitis? Scintigraphic, CT, and MR imaging features. Radiology. 1992;185(3):853–8.
24. Tschirch FTC, Schmid MR, Pfirrmann CWA, Romero J, Hodler J, Zanetti M. Prevalence and size of meniscal cysts, ganglionic cysts, synovial cysts of the popliteal space, fluid-filled bursae, and other fluid collections in asymptomatic knees on MR imaging. AJR Am J Roentgenol. 2003;180(5):1431–6.
25. Zhong S, Wu B, Wang M, Wang X, Yan Q, Fan X, et al. The anatomical and imaging study of pes anserinus and its clinical application. Medicine. 2018;97(15):e0352.
26. Olewnik Ł, Gonera B, Podgórski M, Polguj M, Jezierski H, Topol M. A proposal for a new classification of pes anserinus morphology. Knee Surg Sports Traumatol Arthrosc. 2019;27(9):2984–93.
27. Pękala PA, Tomaszewski KA, Henry BM, Ramakrishnan PK, Roy J, Mizia E, et al. Risk of iatrogenic injury to the infrapatellar branch of the saphenous nerve during hamstring tendon harvesting: a meta-analysis. Muscle Nerve. 2017;56(5):930–7.
28. Lee JH, Kim KJ, Jeong YG, Lee NS, Han SY, Lee CG, et al. Pes anserinus and anserine bursa: anatomical study. Anat Cell Biol. 2014;47(2):127–31.
29. LaPrade RF, Engebretsen AH, Ly TV, Johansen S, Wentorf FA, Engebretsen L. The anatomy of the medial part of the knee. J Bone Joint Surg Am. 2007;89(9):2000–10.
30. Candal-Couto JJ, Deehan DJ. The accessory bands of Gracilis and semitendinosus: an anatomical study. Knee. 2003;10(4):325–8.
31. Rennie WJ, Saifuddin A. Pes anserine bursitis: incidence in symptomatic knees and clinical presentation. Skelet Radiol. 2005;34(7):395–8.
32. Gnanadesigan N, Smith RL. Knee pain: osteoarthritis or anserine bursitis? J Am Med Dir Assoc. 2003;4(3):164–6.
33. Yoon HS, Kim SE, Suh YR, Seo YI, Kim HA. Correlation between Ultrasonographic findings and the response to corticosteroid injection in pes Anserinus Tendinobursitis syndrome in knee osteoarthritis patients. J Korean Med Sci. 2005;20(1):109–12.
34. Helfenstein M, Kuromoto J. Anserine syndrome. Rev Bras Reumatol. 2010;50(3):313–27.
35. Yagi S, Sata M. Rupture of pes anserine bursa in a patient with pes anserine pain syndrome due to osteoarthritis. J Med Investig. 2019;66(1.2):211–2.
36. Forbes JR, Helms CA, Janzen DL. Acute pes anserine bursitis: MR imaging. Radiology. 1995;194(2):525–7.
37. Brantigan OC, Voshell AF. The tibial collateral ligament: its function, its bursae, and its relation to the medial meniscus. J Bone Joint Surg Am. 1943;25:121–31.
38. Warren LF, Marshall JL. The supporting structures and layers on the medial side of the knee: an anatomical analysis. J Bone Joint Surg Am. 1979;61(1):56–62.
39. de Haas WH, van Heerde P. Synovial nature of pathologic periarticular structures, including subcutaneous nodules. Descent from embryonic arthrogenic fibroblasts: a hypothesis. Z Rheumatol. 1979;38(9–10):318–29.
40. Nur H, Aytekin A, Gilgil E. Medial collateral ligament bursitis in a patient with knee osteoarthritis. J Back Musculoskelet Rehabil. 2018;31(4):589–91.
41. Kerlan RK, Glousman RE. Tibial collateral ligament bursitis. Am J Sports Med. 1988;16(4):344–6.
42. Tyson LL, Daughters TC, Ryu RK, Crues JV. MRI appearance of meniscal cysts. Skelet Radiol. 1995;24(6):421–4.
43. De Maeseneer M, Shahabpour M, Van Roy F, Goossens A, De Ridder F, Clarijs J, et al. MR imaging of the medial collateral ligament bursa: findings in patients and anatomic data derived from cadavers. AJR Am J Roentgenol. 2001;177(4):911–7.
44. Hennigan SP, Schneck CD, Mesgarzadeh M, Clancy M. The semimembranosus-tibial collateral ligament bursa. Anatomical study and magnetic resonance imaging. J Bone Joint Surg Am. 1994;76(9):1322–7.
45. Rothstein CP, Laorr A, Helms CA, Tirman PF. Semimembranosus-tibial collateral ligament bursitis: MR imaging findings. AJR Am J Roentgenol. 1996;166(4):875–7.
46. LaPrade RF, Hamilton CD. The fibular collateral ligament-biceps femoris bursa. An anatomic study. Am J Sports Med. 1997;25(4):439–43.

47. Keskin D. Fibular collateral ligament-biceps femoris calcific bursitis causing flexion contracture in the knee, external rotation in the leg, and equinus deformity in the ankle. J Manip Physiol Ther. 2008;31(3):247–50.
48. Janzen DL, Peterfy CG, Forbes JR, Tirman PF, Genant HK. Cystic lesions around the knee joint: MR imaging findings. AJR Am J Roentgenol. 1994;163(1):155–61.
49. Fairclough J, Hayashi K, Toumi H, Lyons K, Bydder G, Phillips N, et al. The functional anatomy of the iliotibial band during flexion and extension of the knee: implications for understanding iliotibial band syndrome. J Anat. 2006;208(3):309–16.
50. Ward EE, Jacobson JA, Fessell DP, Hayes CW, van Holsbeeck M. Sonographic detection of Baker's cysts: comparison with MR imaging. AJR Am J Roentgenol. 2001;176(2):373–80.
51. Lee K, Cox G, Neff JR, Arnett G, Murphey M. Cystic masses of the knee: arthrographic and CT evaluation. Am J Roentgenol. 1987;148(2):329–34.
52. Labropoulos N, Shifrin DA, Paxinos O. New insights into the development of popliteal cysts. Br J Surg. 2004;91(10):1313–8.
53. Wolfe RD, Colloff B. Popliteal cysts. An arthrographic study and review of the literature. J Bone Joint Surg Am. 1972;54(5):1057–63.
54. Jayson MI, Dixon AS. Valvular mechanisms in juxta-articular cysts. Ann Rheum Dis. 1970;29(4):415–20.
55. Miller TT, Staron RB, Koenigsberg T, Levin TL, Feldman F. MR imaging of baker cysts: association with internal derangement, effusion, and degenerative arthropathy. Radiology. 1996;201(1):247–50.
56. Fielding JR, Franklin PD, Kustan J. Popliteal cysts: a reassessment using magnetic resonance imaging. Skelet Radiol. 1991;20(6):433–5.
57. Lindgren PG, Willén R. Gastrocnemio-semimembranosus Bursa and its relation to the knee joint. Acta Radiol Diagn. 1977;18(5):497–512.
58. Herman AM, Marzo JM. Popliteal cysts: a current review. Orthopedics. 2014;37(8):e678–84.
59. Martí-Bonmatí L, Mollá E, Dosdá R, Casillas C, Ferrer P. MR imaging of Baker cysts – prevalence and relation to internal derangements of the knee. MAGMA. 2000;10(3):205–10.
60. Han DY, Ryu KN, Park JS, Jin W, Park SY, Yun SJ. The prevalence of baker cyst in relation to the arrangement pattern between the medial head of gastrocnemius tendon and the semimembranosus tendon. Eur Radiol. 2020;30(3):1544–53.
61. Handy JR. Popliteal cysts in adults: a review. Semin Arthritis Rheum. 2001;31(2):108–18.
62. Sanchez JE, Conkling N, Labropoulos N. Compression syndromes of the popliteal neurovascular bundle due to baker cyst. J Vasc Surg. 2011;54(6):1821–9.
63. Fujiyoshi K, Minami Y, Tojo T, Iwase D, Hirata M, Ako J. Lower limb ischemia due to popliteal artery compression by baker cyst. J Vasc Surg Cases Innov Tech. 2018;4(2):99–101.
64. Canoso JJ, Goldsmith MR, Gerzof SG, Wohlgethan JR. Foucher's sign of the Baker's cyst. Ann Rheum Dis. 1987;46(3):228–32.
65. Fritschy D, Fasel J, Imbert J-C, Bianchi S, Verdonk R, Wirth CJ. The popliteal cyst. Knee Surg Sports Traumatol Arthrosc. 2006;14(7):623–8.
66. Munk PL, Vellet AD, Levin MF. Leaking Baker's cyst detected by magnetic resonance imaging. Can Assoc Radiol J. 1993;44(2):125–8.

55 Evaluation of Patellofemoral Knee Pain

Fabio Valerio Sciarretta and John G. Lane

Patellofemoral pain syndrome (PFPS) represents one of the most common causes of knee pain affecting adolescents and adults, particularly in the physically active populations and especially in runners.

In a recent systematic review and meta-analysis [1] regarding the incidence and prevalence of patellofemoral pain, Smith [1] also found that approximately one in 10 military recruits and one in 14 adolescents were suffering from patellofemoral pain at any one time, and one in five of the general population had experienced pain within the last year. Reported annual prevalence in the general population was 22.7%, and in adolescents 28.9%. Elite athletes' reported prevalence varied from 16.7% to 29.3% (female university ballet dancers) and up to 35.7% in professional male cyclists. The pooled estimate for point prevalence in adolescents was 7.2%, and in female-only adolescent athletes was 22.7%.

PFPS has been reported to represent 16–25% of all injuries in runners [2], but Akodu and Nwakalor [3] noted that a 12-month prevalence of PFPS among amateur runners in Lagos state was 45.3% and the percentages of PFPS in males and females were 39.5% and 53.6%, respectively. PFPS was, therefore, 1.35-fold more prevalent in females than in males, slightly lower than 1.5 times the prevalence in females than males postulated by Taunton et al. [2] and lower than 2.23 times the incidence of PFPS reported at the United States Naval Academy (USNA) between females and males [4].

Anterior knee pain or PFPS remains, according to John Insall, an orthopedic "enigma" ("black hole of orthopedics"). At that time, in 1984 [5], John Insall devised a classification of patellofemoral disorders based on the presence of articular cartilage damage, which is present, variable, or absent. In the first group, Insall included chondromalacia (divided into four stages, the first three involving only the patella and the fourth stage involving the femoral trochlea), osteoarthritis (grade IV chondromalacia typical of older patients identified by roentgenographic findings of joint-space narrowing, sclerosis, and spurring and involving more frequently the lateral than the medial patellar facet associated with mirror femoral opposing surface involvement), osteochondral fractures, and osteochondritis dissecans. In the group with variable cartilage damage, he described malalignment syndromes, which include different levels of instability ranging from patella pain to patellar subluxation and dislocation as well as synovial plica. In the last group, Insall included peripatellar causes of pain including bursitis and tendinitis, overuse

F. V. Sciarretta (✉)
Chirurgo Ortopedico, Clinica Nostra Signora della Mercede, Rome, Italy

J. G. Lane
Musculoskeletal and Joint Research Foundation, San Diego, CA, USA
e-mail: jglane@san.rr.com

J. G. Lane et al. (eds.), *The Art of the Musculoskeletal Physical Exam*,
https://doi.org/10.1007/978-3-031-24404-9_55

syndromes, reflex sympathetic dystrophy, and patellar abnormalities.

In 2002, Fulkerson [6] described six major anatomic sources of patellofemoral pain: subchondral bone, synovium, retinaculum, skin, muscle, and nerve. These structures may be affected by many factors, including systemic disease, but in orthopedic sports medicine, the most common reasons for anterior knee pain are overuse, patellofemoral malalignment, and trauma.

For many years, many studies that have evaluated the origin of PFPS have focused on the biomechanics of the patellofemoral joint, expressing the idea that the ethology of this clinical entity was related to abnormal patellar alignment and movement.

Several possible causes of the laterally translated position of the patella (valgus knee Q-angle, anteversion of the hip) and of incongruence between the patella and the trochlea (patellar dysplasia, patella alta, patella baja) have been considered and studied as factors leading to lateral malalignment of the patellofemoral joint, contributing to patellar instability and dislocation. Among the advocates of this mechanical model, four clinical presentations are possible:

1. Objective patellar instability with a history of at least one patellar dislocation and objective radiological abnormalities
2. Potential patellar instability; no patellar dislocation but presence of objective radiological abnormalities
3. Patellofemoral pain without dislocation and without objective radiological abnormalities
4. Patellofemoral osteoarthritis

In the 1980s and 1990s, "maltracking" due to quadriceps atrophy related to altered activation pattern of the vastus medialis obliquus (VMO) medially and of the vastus lateralis (VL) part of the quadriceps muscle laterally was believed to be a major cause of the abnormalities of patellar tracking.

More recently, a biological model has been proposed that the tissue around the patella can be loaded physiologically within a pain-free zone. The disruption of tissue homeostasis, outside this "envelope of function," leads to an overload of the patellofemoral joint and peri-patellar pain.

PFPS is therefore a multifactorial syndrome where proximal factors intrinsic to the hip and distal factors intrinsic to the foot kinematics converge together with knee patellofemoral factors (patellar kinematics, patellar tracking, patellofemoral contact-stress forces, and possible altered muscular functions) in causing pain and altered knee function.

The diagnosis of PFPS is primarily a clinical finding. As always in the medical practice, the first step in the evaluation of patellofemoral knee pain begins with a detailed and complete history of the patient's symptoms and problems. The patient must precisely delineate the symptoms, which brought him/her for medical treatment.

Is it the pain? What causes the onset of symptoms and what is the duration? Is it an acute or chronic pain? Has it been precipitated by a trauma? Does it occur in one or both knees?

Obtaining detailed answers will enable the physician to determine if the patient's symptomatology is caused by an overuse or an overload of the knee joint and if the situation represents an acute exacerbation of previously experienced chronic problems. The exact location; character and ameliorating or aggravating factors, movements, or activities (running, prolonged sitting often defined as "theater sign," squatting, ascending, or descending steps); systematic and possible association with other symptoms as swelling; and range of motion limitation and stiffness need to be completely investigated.

If the patient is suffering from instability episodes, the definition of the duration, the frequency, and the specific circumstances that lead to giving way will help differentiating true instability from inhibitory reflexes due solely to pain.

Most commonly, the patients with PFPS report pain located "behind," "underneath," or "around" the patella often without a precise localization. Patients wanting to indicate their pain-trigger area may draw with their hands a circle around the patella—the "circle sign."

Even if PFPS and patellar instability represent two clearly distinct pathological entities, it may occur that the patient with PFPS may complain of giving way, in most cases due to transient quadriceps muscle inhibition because of pain or deconditioning.

55.1 Physical Examination

Once a detailed history has been obtained with particular attention to recent changes in the patient's activities and in the intensity, duration, and resistance of exercises and training, particularly squats and lunges, a complete physical examination of the knee, including careful assessment of the patellofemoral joint, may begin to highlight all possible static and dynamic factors predisposing to PFPS.

Observation, palpation, and specific patellofemoral testing are completed in a stepwise fashion following a standardized routine, including the hip, lumbar spine, foot and ankle, and a focused neurovascular examination.

55.2 Inspection

Initial observation of the patient should be performed as they progress from standing to seated and supine positioning.

PFPS findings are quite often subtle, and the examiners' experience and ability to facilitate relaxation of the patient may have a critical role in a successful assessment.

Collado and colleagues [7] found that torsional or angular malalignment of the lower extremity has a significant influence on patellofemoral joint mechanics. There is, however, difficulty in defining normal patellofemoral alignment with respect to malalignment. Normally, the patella seats in the trochlea and during flexion moves medially reaching maximum contact pressures at 90° of knee flexion.

Patellar height, trochlear dysplasia, and pathological tibial tubercle-trochlear groove (TT-TG) distance represent the principal factors to consider in a patient with potential or objective patellar instability associated with patellofemoral pain [8].

55.3 Standing Examination

A general standing examination should be evaluated in the three basic cardinal planes.

Each plane presents peculiar aspects that should be examined to enable the clinician to obtain a thorough knee postural and structural evaluation to evaluate possible abnormalities that may be a source of PFPS symptoms [9].

In the anterior sagittal plane, static alignment is assessed with consideration of knee biomechanics with potential varus or valgus malalignment as well as femoral anteversion and tibial external rotation. This may be evidenced by the presence of "squinting patellae," a condition in which the patella appears to be pointing inward or medially when the patient is standing, generally caused by excessive femoral anteversion, and this is associated with a chronically elongated medial retinaculum and a tight lateral knee retinaculum.

In the lateral plane, it is easy to observe either genu recurvatum or a flexion contracture, while in the transverse plane, femoral and tibial rotation is best understood and appreciated. Excessive internal tibial rotation will create increased medial rotation of the patella position, while excessive femoral anteversion as well as external tibial torsion will create increased lateral rotation of the patella position.

Abnormal soft-tissue length can, in different ways, affect patellofemoral biomechanics. Tightness of quadriceps, hamstrings, and gastrocnemius muscles can alter patellofemoral contact pressures, while tightness of iliotibial band can affect patellar excursion. The lateral retinaculum, medial patellofemoral ligament, and synovium may also be responsible for augmented stresses on patellofemoral joint and contribute to anterior knee pain.

55.4 The Q-Angle

Basic clinical measurement of the alignment of the lower extremity is the Q-angle.

The Q-angle is formed by a line connecting the anterosuperior iliac spine to the center of the patella and a line connecting the center of the patella to the center of the anterior tibial tuberosity. The Q-angle represents in this way the line of action of the quadriceps force and normally measures 10–22°. Lankhorst et al. [10] performed a meta-analysis to search for factors associated with PFPS and concluded that these patients demonstrated a larger Q-angle, sulcus angle, and patellar tilt angle (weighted mean differences (WMD) 2.08, and 1.66 and 4.34, respectively); less hip abduction strength; lower knee extension peak torque; and less hip external rotation strength. The authors concluded that Q-angle alone is not a risk factor for the development of patellofemoral pain.

55.5 Dynamic Evaluation

Dynamic malalignment may exist during movement as a consequence of poor muscular control of the lower extremity segments. Dynamic patellar tracking alignment can be tested by having the patient perform a "single-leg squat and stand test," where an imbalance between the medial and the lateral structure's tension can be appreciated by a sudden medial deviation of the patella as the patella engages the femoral trochlea in the early phase of flexion, known as the shape of hockey stick "J" sign. Nunes et al. [11] reported that this test represents the best available test for eliciting anterior knee pain. Patellofemoral pain is evident in 80% of people who are positive in this test. This dynamic hip and quadriceps strength test imposes higher mechanical demands than a bilateral squat, which may induce compensatory movements such as knee valgus, because it requires an increased dynamic in all planes. Patients with decreased strength of the hip abductors and external rotators show several compensatory kinematic movements as ipsilateral trunk lean, contralateral pelvic drop, hip adduction, and knee abduction [12].

This dynamic malalignment may also be detected inviting the patient to step slowly upon and down from a 20 cm stool. This "step-down test" can be performed to assess hip and leg strength and endurance and requires good control of the patellofemoral extensor mechanism as this applies significant stress on the anterior compartment of the knee. The patient stands facing the examiner and is asked to stand on a 20 cm stool or box with leg to be tested and the other leg on the floor. Patient should stand with arms folded across chest and be instructed to squat down on one lower extremity 5–10 times consecutively, in a slow and controlled manner until the heel touches the floor at a rate of approximately 1 squat per 2 s. The presence of any abnormal movement of the patella when it engages or disengages the trochlea, any rotation of the trunk, and eventual loss of hip control must be noted.

When asking the patients to perform a single-leg squat or step-down test, patients affected with PFPS may show excessive contralateral pelvic drop, hip adduction and internal rotation, knee abduction, tibial external rotation, and hyperpronation.

Alternatively, the "J" sign may be observed in the supine or seated position, inviting the patient to extend the knee from a 90° flexed position: the abnormal path traced by the patella will be evident in the terminal phase of extension as one can observe the lateral deviation of the patella as it exits the trochlear groove.

It is also critical in the diagnostic process to check muscle strength and tropism.

It is therefore important to evaluate the quadriceps muscle, especially the VMO, both by visual inspection and direct measurement of the muscle girth compared to the contralateral side, in order to evaluate for any muscle atrophy. The measurements obtained will represent the baseline referral for future follow-ups assessing the improvements obtained by the rehabilitation program prescribed.

Patellar tracking needs to be carefully observed to determine if the patella rides centrally into the trochlea during knee flexion, and if it does not, then evaluate if the patella's restraints may be too lax or too tight.

Tight lateral structures have been implicated in subjects presenting with PFPS. In fact, it has been proposed that a tight iliotibial band (ITB) through its attachment of the lateral retinaculum into the patella could cause lateral patella tracking, patella tilt, and compression.

The ITB is a continuation of the tensor fascia lata and the gluteus maximus at the lateral aspect of the thigh. As it approaches the knee, the ITB separates into two functional components, the ilio-patellar band and the iliotibial tract (ITT), which attach to Gerdy's tubercle of the lateral tibial condyle and the lateral patella retinaculum, respectively.

To evaluate contracture and stretch of the ITB or ITT, the Ober or modified Ober tests can be used. To perform the Ober test, the patient is positioned lying on the contralateral hip with the bottom knee flexed to flatten the lower back. Then the examiner stabilizes the hip on the top and passively flexes the hip and knee to 90°; the hip is then stretched passively through abduction and extension by the examiner, who finally allows the knee to slowly drop by gravity until reaching its final angle. If the ITB is normal, the leg will adduct with the thigh dropping down slightly below the horizontal and the patient will not experience any pain; in this case, the test is called negative. If the ITB is tight, the leg will remain in the abducted position and the patient will experience lateral knee pain; in this case, the test is called positive.

Kendall et al. [13] modified the Ober test by extending the knee to 0° before the stretch.

With the knee straight and the pelvis in a neutral position, the thigh drops about 10° below the horizontal. It is suggestive of normal length. If the leg fails to drop, it indicates a tight ITB and TFL. In these tests (Ober, modified Ober), a normal ITT or distal ITB is stretched and becomes taut when the hip adducts beyond 0° and 10°, respectively.

55.6 Palpation

Once one has completed the observation phase, the patient is placed supine on a table with the knee extended as the physician begins to palpate the joint. A joint effusion may be present, although it is not common and may be related to additional intra-articular pathologies. Quadriceps tone is assessed both at rest and in isometric contraction.

Palpation must be conducted very carefully, following a systematic routine around the patella in an attempt to clearly define the exact location of the pain, and should be extended to the entire anterior region of the knee, including distal quadriceps, quadriceps tendon, and patellar tendon assessing for possible further or alternative sources of pain. The ITB must be palpated as well. The patella must be moved in all directions and especially medially and laterally in order to verify if its displacement causes pain.

55.7 Range of Motion

Passive and active range of motion of both the knee and the hip must be assessed and compared to the contralateral joint. Patients with PFPS usually present with a complete range of motion in flexion and extension of the knee. An asymptomatic crepitus during flexion-extension is nonspecific, but a possible painful crepitus may indicate a cartilage defect or a degenerative patellofemoral joint disease, involving the patellar surface, the femoral trochlea surface, or both as a kissing lesion.

55.8 Special Tests

Critical for correct diagnosis is the execution of all the clinical tests for patellar mobility and position and provocative tests for pain.

55.8.1 Patellar Mobility Testing

55.8.1.1 Patella Glide

To simplify the measurement of patella position and mobility, the patella has been divided into four equal longitudinal segments or quadrants, each representing 25% of the width of the patella.

With the knee in a resting and relaxed position, flexed at 30° over a towel roll, the physician, on the lateral side of the patient's knee, first establishes visually the site of the midline of the trochlea, then grasps the patella, translates it passively medially and laterally (medial glide and lateral glide) using both thumbs, and takes note of the extent of displacement (Fig. 55.1a and b). The lateral glide is also known as the Fairbank's sign or apprehension test. Therefore, it should be performed after the medial glide testing. The extent of displacement is described in relation to the width of the patella and expressed in quadrants that can be converted in 25%, 50%, 75%, and 100% of displacement beyond the midline of the trochlea. The patella can normally be manually displaced both medially and laterally between 25% and 50% of width of the patella. The displacement of less than one quadrant medially or laterally indicates tightness of opposite retinacular structure. The displacement of more than three quadrants is defined as hypermobility of the patella [13], indicating loose patella restraints, a finding often observed in adolescent females. Two quadrants of passive mobility are considered normal.

Additional patellar glide testing can be performed to assess superior and inferior motion.

The examination test to assess for superior patellar glide is performed at 30° of knee flexion and assesses the limitation of mobility of the inferior patellar tendon and soft-tissue structures.

The examination test to assess for inferior patellar glide is also performed at 30° of knee flexion and assesses the limitation of mobility of the quadriceps tendon and quadriceps muscles [9].

It is quite important to compare all these glide tests to the other knee to obtain a more complete and reliable evaluation.

55.8.1.2 Patellar Tilt Test

This test assesses for tightness of the lateral structures, including the iliotibial band and the lateral patellofemoral and patello-tibial ligaments. Lateral deep retinacular tightness is a condition that is very common in patients with anterior knee pain.

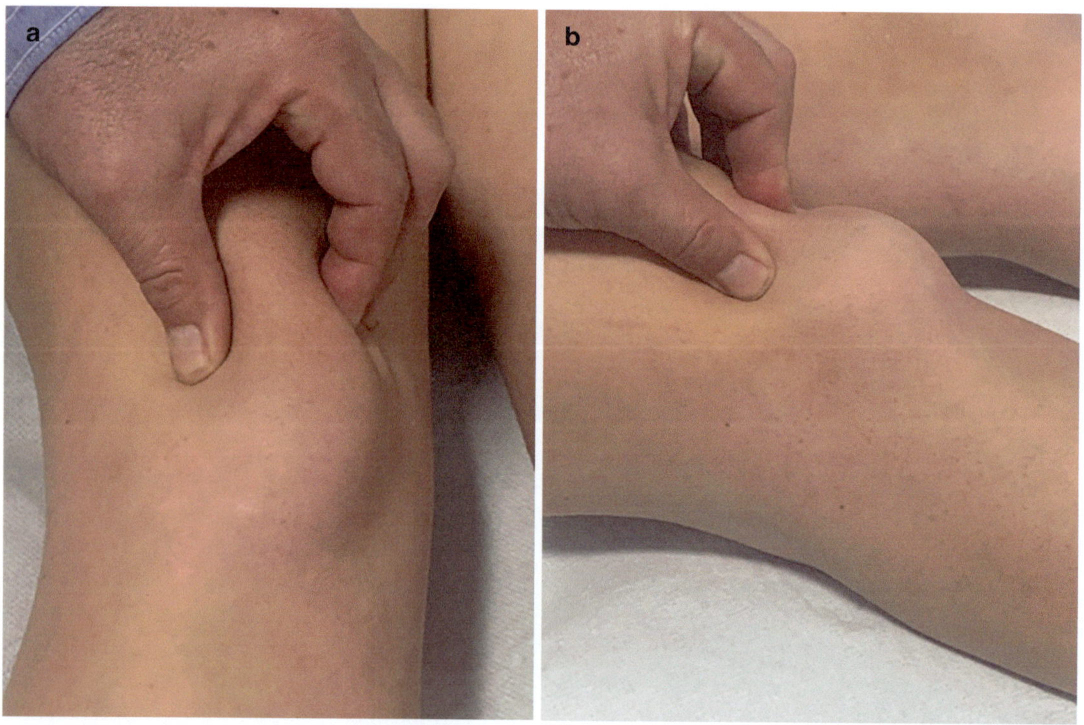

Fig. 55.1 (**a**, **b**) Patellar mobility testing. The patella is grasped between thumb and index finger in resting position (**a**) and then translated medially (**b**)

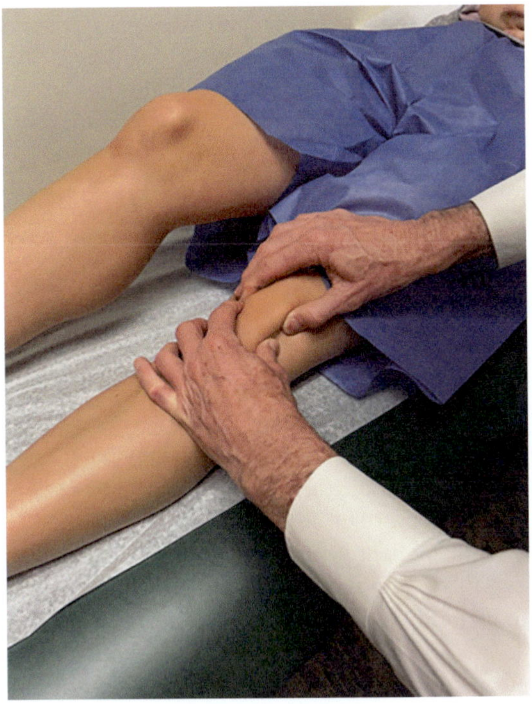

Fig. 55.2 Patellar tilt test. The patella is grasped between the thumbs and index fingers with the knee in extension, and then tilted anteriorly with the thumbs

The test is performed with the knee completely extended and the clinician standing on the lateral side of the knee to be examined. The patella is grasped between the thumb and forefinger. The medial aspect of the patella is then compressed posteriorly while the lateral aspect is elevated, trying to rotate the patella anteriorly (Fig. 55.2). If the lateral aspect of the patella is fixed and cannot be raised to at least the horizontal position (0°), the test is positive and indicates the presence of tight lateral retinaculum structures. Normally, the patella can be tilted upwards above the horizontal.

Executing comparisons with the opposite side is extremely important as wide ranges of variation of tightness are the norm.

The patella should tilt 15° with both medial and lateral tilts.

This situation can also be observed in patients with patellofemoral osteoarthritis.

55.8.1.3 Patellar Grind Test or Clarke's Patellofemoral Grind Test

While the patient is in the supine position with the knees extended, the examiner places one hand superior to the patella and displaces the patella inferiorly into the trochlear groove (Fig. 55.3). The patient is then asked to contract the quadriceps while the examiner continues to palpate the patella and provides resistance to superior movement of the patella. The test is positive if pain, crepitus, and apprehension of the patient as the irritated surfaces of the patella rub over the femur are produced, although comparison to the contralateral knee is needed to interpret the result.

55.8.2 Patellar Apprehension Test

The patient lies supine with the knee in full extension. The examiner grasps the limb at the ankle with one hand and allows the knee to be flexed over the side of the table. With the other hand, the

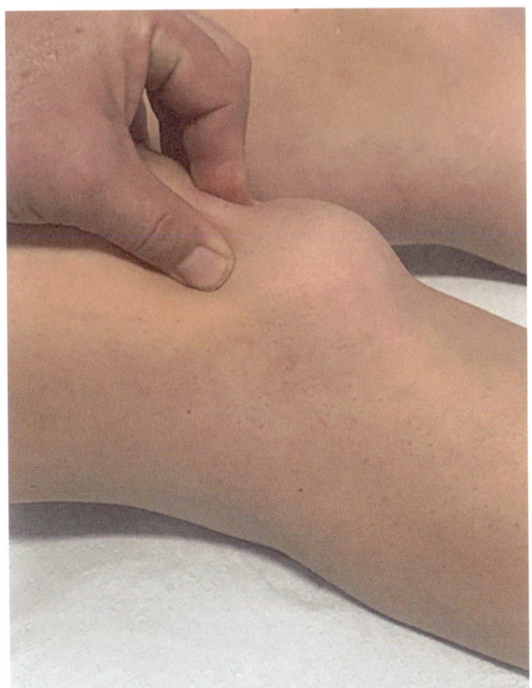

Fig. 55.3 Patellar grind test. The examiner's fingers displace the patella inferiorly into trochlear groove

examiner applies pressure from the medial side of the patella. As a result, the patella is forced as far laterally as possible with tension being maintained by the medial retinacular structures and medial patellofemoral ligament. The patient is asked to tighten the quadriceps muscle, and the examiner slowly flexes the knee with the other hand.

The test results positive if pain is produced.

Patient may refuse to do this test in anticipation of pain.

55.8.3 Moving Patellar Apprehension Test

The moving patellar apprehension test is performed with the patient supine with the thigh on the examining table and the clinician holding the leg in full extension off the table. The clinician then moves the patella laterally moving from an extended to a flexed position. The test is considered positive if the patient feels pain while performing this movement for lateral instability.

To confirm this test, the clinician translates the patella medially while performing the same maneuver. If the patient does not refer any symptoms on the second portion of this test, lateral instability is thought to be present.

55.9 Examination in Prone Position

The prone position enables the evaluation of the flexibility of the quadriceps.

Ely's test for rectus femoris flexibility is performed prone by placing the hip in neutral extension, a position that places the rectus femoris on stretch proximally at the hip and again distally as the knee is further flexed.

If the patient flexes the hip during this maneuver, this represents confirmation of rectus femur inflexibility or tightness.

In the prone position, it is also possible to evaluate the strength of hip extensors and hamstring muscles. Assessment of the femoral anteversion is detected by internal and external rotation of the femur with the hip in extension (Fig. 55.4a and b).

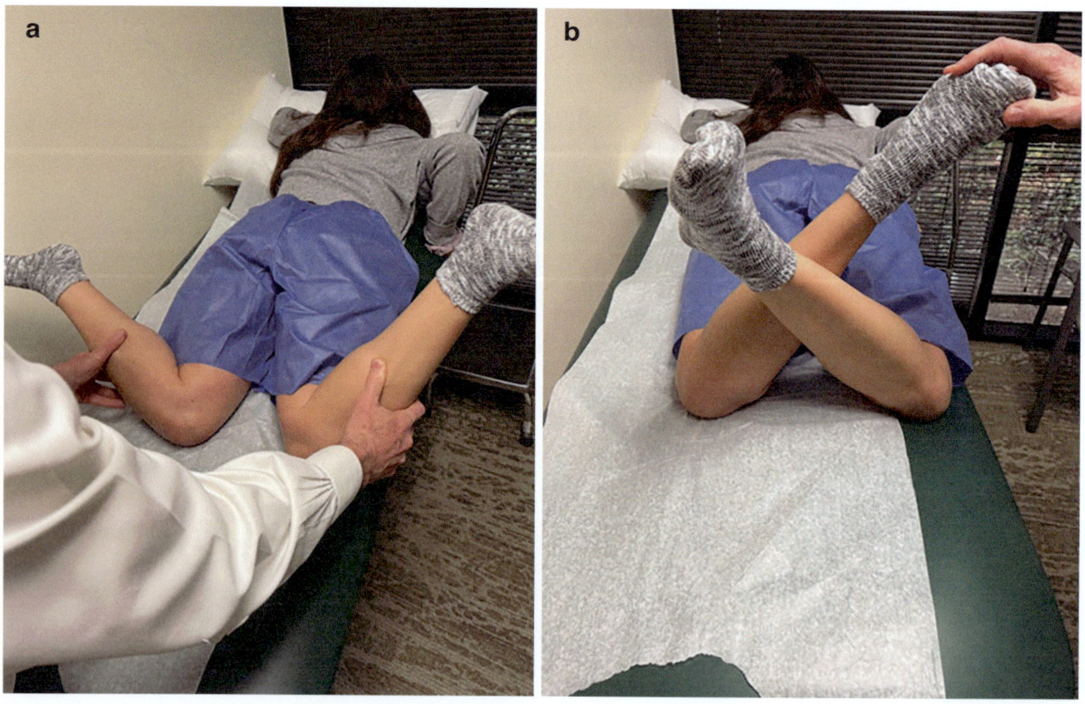

Fig. 55.4 (**a**) Assessment of femoral anteversion with external and internal rotation of the hip. (**b**) Internal rotation of the hip. (**c**) External rotation of the hip

55.10 Conclusions

The young population and the active sport population often complain of anterior knee pain due to PFPS.

Its definitive diagnosis and treatment are challenging due to the complex and multiple anatomical and functional variations, with the various bony, ligamentous, and tendinous restraints contributing to the condition.

PFPS evaluation needs a thorough history and a careful physical examination, including numerous special tests introduced over the years to define patellofemoral maltracking and instability. The clinician has therefore to be prepared and specifically informed and prepared.

Reaching the most appropriate diagnosis represents the milestone to be able to give the patient the most appropriate diagnosis and establish correct conservative or surgical treatment.

To obtain this paradigm, it is mandatory to complete the physical examination with both static and dynamic physical examination techniques. If a definitive diagnosis is not able to be determined, this should be augmented with conventional and cross-sectional imaging to assess for subtle soft tissue, bone abnormalities, or cartilage defects and lesions.

References

1. Smith BE, Selfe J, Thacker D, Hendrick P, Bateman M, Moffatt F, et al. Incidence and prevalence of patellofemoral pain: a systematic review and meta-analysis. PLoS One. 2018;13(1):e0190892. https://doi.org/10.1371/journal.ponc.0190892.
2. Taunton J, Ryan M, Clement D, McKenzie D, Lloyd-Smith D, Zumb B. A retrospective case-control analysis of 2002 running injuries. Br J Sports Med. 2002;36(2):95–101.
3. Akodu AK, Nwakalor NO. Patellofemoral pain syndrome: prevalence and coping strategies of amateur runners in Lagos state. Medicina Sportiva. 2018;XIV(2):3059–67.
4. Boling M, et al. Gender differences in the incidence and prevalence of patellofemoral pain syndrome. Scand J Med Sci Sports. 2010;20(5):725–30. https://doi.org/10.1111/j.1600-0838.2009.00996.
5. Insall JN. Disorders of the patella. In: Insall JN, editor. Surgery of the knee. New York: Churchill Livingstone; 1984. p. 191.
6. Fulkerson JP. Diagnosis and Treatment of patients with patellofemoral pain. Am J Sports Med. 2002;30(3):447–56.
7. Collado H, Fredericson M. Patellofemoral pain syndrome. Clin Sports Med. 2010;29:379–98.
8. Berruto M, Ferrua P, Carimati G, Uboldi F, Gala L. Patellofemoral instability: classification and imaging. Joints. 2013;1(2):7–14. Published 2013 Oct 24.
9. Manske RC, Davies GC. Examination of the patellofemoral joint. Int J Sports Phys Ther. 2016;11(6):831–53.
10. Lankhorst NE, Bierma-Zeinstra SMA, van Middelkoop M. Factors associated with patellofemoral pain syndrome: a systematic review. Br J Sports Med. 2013;47:193–206.
11. Nunes GS, Stapait EL, Kirsten MH, de Noronha M, Sanatos GM. Clinical test for diagnosis of patellofemoral pain syndrome: systematic review with meta-analysis. Phys Ther Sport. 2013;14(1):54–9.
12. Nakagawa TH, Moriya ETU, Maciel CD, Serrao FV. Trunk, pelvis, hip, and knee kinematics, hip strength, and gluteal muscle activation during a single-leg squat in males and females with and without patellofemoral pain syndrome. J Orthop Sports Phys Ther. 2012;42(6):491–501.
13. Kendall F, McCreary E, Provance P, Rodgers M, Romani W. Muscle testing and function with posture and pain 4th edn; Modified Ober test. Philadelphia: Lippincott; 1993. p. 56–9.

Part VI

Ankle Reviewer Dr Canata

Foot and Ankle Anatomy

56

Giovanna Stelitano, Vincenzo Candela, Calogero Di Naro, Carlo Casciaro, Giuseppi Longo, and Vincenzo Denaro

56.1 Introduction

The foot and the ankle work synchronically in a complex system made of 26 bones and joints. They help the feet to adapt to uneven surfaces while the other parts of the body rest standing. The ankle plays an essential role in the execution of plantar flexion and dorsiflexion movements. The foot can be subdivided into the hindfoot, the midfoot, and the forefoot. The hindfoot, which includes the subtalar, the talonavicular, and the calcaneocuboid bones, is the main responsible for foot inversion and eversion. The ankle and hindfoot have been considered as a global joint that makes possible the correct walking movement [1]. They, in fact, uniformly spread the body load. The potential loss of motion from one of the two articular complexes can bring to excessive load of the other, which tries to offset the lost movement. The midfoot is made of cuboid, navicular, and three cuneiform bones, and it links the hindfoot to the metatarsal heads. At this joint level, stability results to be more relevant than flexibility and, consequently, joint arthrodesis does not impair foot function. The forefoot is composed of five toes and the corresponding five metatarsi. The bones of the toes are named phalanges, and the first toe is made of two phalanges while the other four toes have three phalanges each [2]. The joints between the phalanges are called interphalangeal and those between the metatarsi and phalanges are called metatarsophalangeal (MTP). Toes, particularly the first, give the propulsion during the gait. Metatarsophalangeal (MTP) contributes to this function, while the interphalangeal role is not required. In a regular foot, the Achilles tendon contraction reduces the length between the calcaneal tuberosity and the rest of the ankle. In this way, the ankle performs a rotatory movement, transmitting the strength across the rigid foot to the metatarsal heads. The effect is the intensification of the Achilles strength for propulsion.

G. Stelitano · C. Casciaro · G. Longo (✉)
Research Unit of Orthopaedic and Trauma Surgery, Fondazione Policlinico Universitario Campus Bio-Medico, Rome, Italy

Research Unit of Orthopaedic and Trauma Surgery, Università Campus Bio-Medico di Roma, Rome, Italy
e-mail: g.longo@unicampus.it

V. Candela · C. Di Naro · V. Denaro
Research Unit of Orthopaedic and Trauma Surgery, Fondazione Policlinico Universitario Campus Bio-Medico, Rome, Italy

J. G. Lane et al. (eds.), *The Art of the Musculoskeletal Physical Exam*,
https://doi.org/10.1007/978-3-031-24404-9_56

56.2 The Ankle

The ankle joint works as a mortise, in which tibia and fibula create a tight socket for the talar dome. The talar dome results to be broader in its anterior part than in the posterior one, so that the dorsiflexion movement includes the tibia in the mortise and allows the lateral slip of the fibula [3]. At the ankle joint level, the weight-bearing strengths of the body are conformally spread thanks to the presence of highly conforming joint areas that provide the reduction of intra-articular pressure. If an alteration of these articular areas occurs (for example as the consequence of syndesmotic widening or intra-articular fracture), the intra-articular pressure could increase, causing the development of arthritis. In particular, the condition of persistent syndesmotic widening, after traumatic events, is marked by worse results [4]. The ankle articular cartilage, however, appears moderately compact. The fibula supports about one-sixth of weight-bearing strengths, while the remaining part moves towards the tibia. The distal tibial joint area (plafond) could present three degrees of valgus, and the mortise is characterized by an external rotation of 20–30° with respect to the knee.

56.3 Ankle Ligaments

Ankle ligaments, together with the conforming articular surfaces, play an essential role in joint stability during walking and running. The medial ankle compartment is characterized by the presence of two collateral ligaments: the superficial and deep deltoid ligaments [5, 6]. The superficial deltoid fibers cross the medial malleolus to arrive into the talus, navicular, and calcaneus bones. The deep deltoid, instead, links the medial malleolus to the medial body of the talus. The latter gives the main contribution to the maintenance of ankle stability [7, 8]. Moreover, the medial ligaments guarantee the bloody apport for the talus. The lateral ankle compartment stability is due to three collateral ligaments: the anterior talofibular ligament (ATFL), the calcaneofibular ligament (CFL), and the posterior talofibular ligament [9].

56.4 The Distal Tibiofibular Syndesmosis

The distal tibia articulates with the distal part of the fibula thanks to the presence of a notch in its posterolateral surface defining the tibiofibular syndesmosis. Several ligaments maintain in place the syndesmosis. They are the anterior-inferior tibiofibular, the posterior-inferior tibiofibular, the transverse tibiofibular ligaments, and the interosseous ligament and membrane.

56.5 The Talus

The talus identifies the central point of the ankle joint and, at the same time, represents the leading part of the hindfoot. It is characterized by the presence of ample articular cartilage, while muscular insertions to the osseous surface are not detectable. The talus presents a weak vascularization. The vessels implicated in its blood supply are the posterior tibial artery, the dorsalis pedis artery, and the peroneal artery [10]. The first of these gives a branch to the tarsal canal, which enters the talus crossing the medial side of the deltoid and the inferior area of the tarsal canal. The dorsalis pedis artery contributes to the vascularization of the talar dorsal neck, while the peroneal artery to the sinus tarsi. Any traumatic or surgical damage (such as a fracture or a dislocation) is able to interrupt this vascular network, causing the avascular necrosis. Even if this eventuality is unusual, the absence of consistent vascular support could provoke a bone nonunion. It is essential to pay attention to the talus during ankle arthroplasty: a talus surgical injury, in fact, could hinder the osseous ingrowth into the implant, causing aseptic loss of the talar segment.

56.6 The Calcaneus

The main characteristics of the calcaneus that distinguishes it from the talus are its copious vascular network and the presence of several soft-tissue insertions. The substantial calcaneal vascularization guarantees optimal healing also in case of multiple-fragment fractures, marked by a whole disruption of the vessels. Calcaneal fractures rarely develop avascular necrosis [11]. The calcaneus owns a wide, thick, medial prominence: the sustentaculum tali, from which the spring ligament origins and that includes the middle and anterior facets of the subtalar joint. The posterior tibial nerves and vessels cross the medial side of the calcaneus. For this reason, during a bone osteotomy of the lateral side, it is necessary not to work with the osteotome nearly to the medial surface.

56.7 Os Trigonum

The os trigonum is an accessorial osseous segment normally detectable in several subjects. Localized in the posterior side of the ankle, the os trigonum can appear as an ununited posterior process which articulates with the dorsal surface of the calcaneus [12]. The clinical relevance of this bone process is given by its ability to cause painful symptomatology that usually afflicts dancers. It is placed deeply and laterally to the flexor hallucis longus tendon. The excessive bone stimulation that occurs in dancers when they are in the extreme plantar flexion can cause os trigonum irritation, which could require a resection when persistent [13]. The resection should be performed safely using a posteromedial approach. The posterolateral one could bring damage to the sural nerve.

56.8 Hindfoot Joints and Stabilizers

The subtalar joint presents three different articular segments: the anterior and the middle facets, which usually are adjacent, and the posterior facet that results to be wider and isolated. For a long time, the subtalar joint was considered as a simple mechanical complex endowed of a single rotatory axis, crossing the joint with a posterolateral to anterodorsal direction [14]. The axis produces a 41° angulation with the terrain. More recently, the joint has been described as screw, with different axes of rotation. In fact, during the eversion movement, the calcaneus is also translated posteriorly, while during inversion, the anterior calcaneal translation occurs. Talonavicular and subtalar own movements strictly joined. Arthrodesis of the talonavicular joint completely removes any subtalar mobility. On the other hand, the fusion of the subtalar joint provides the loss of about 75% of normal talonavicular mobility. The calcaneocuboid joint covers a marginal role in hindfoot mobility [15]. The singular arthrodesis of the calcaneocuboid does not obstacle subtalar movements and does not disrupt talonavicular joint action. The calcaneus and navicular bones (together with the other elements of the foot) rotate around the talus. Usually, the mobility among the other tarsal bones is poorer and less significant than that among the talus, calcaneus, and navicular. Several interosseous ligaments guarantee the stabilization of the subtalar joint. Movements in varus are controlled by lateral ligaments, the CFL and the inferior extensor retinaculum, which provides a solid support. The cervical and the interosseous talocalcaneal ligaments contribute to subtalar stability in varus. Other ligaments are involved in longitudinal arch support. The spring ligament, which crosses the calcaneal sustentaculum tali to reach the navicular, is the main stabilizer. The long plantar ligament aids its action, running from the calcaneal to the cuboidal bone.

56.9 Midfoot Joints and Stabilizers

A little range of motion characterizes the midfoot joint [16]. The maximal degree of motion is detectable among the cuboid and fourth and fifth metatarsal bones: it is necessary to ease gait, making flexible the lateral column of the foot. For this reason, it would be more safe to avoid their fusion. The leading midfoot stabilizer is the Lisfranc ligament, which links the medial cuneiform to the second metatarsal base [17]. The plantar and dorsal ligaments are smaller and less relevant for joint stability.

56.10 Forefoot and Stabilizers

MTP joints and metatarsal heads cover an essential role during gait and in the body weight-bearing support. Metatarsal heads should be saved in any case where it is possible. In erect posture, the body weight is spread as follows: about 40% bears on the first metatarsal, while the remaining 60% is distributed between the minor metatarsal heads [18].

The interphalangeal joints do not result to be useful for walking. They contribute to grasping, even if this function is not widely required to satisfy human necessities [19, 20]. Consequently, interphalangeal fusion or resection does not give several problems. The sesamoid bones are placed between the tendons of the two flexor hallucis brevis. They contribute to keeping the MTP joint: the loss of these bones is not accompanied by the flexor hallucis brevis tendon repair and could lead to varus or valgus MTP deformity [21]. Usually, the resection of a single sesamoid results to be feasible without important consequences, but resection of both should not be executed [22].

56.11 Plantar Fascia

The plantar fascia is the thick aponeurosis that sustains the longitudinal arch on the plantar side of the foot. It originates from the calcaneal tuberosity and extends to the forefoot [23]. Its central band has two different insertions: the first occurs in the ball of the foot, while the second in the septa of the flexor tendons in the toes [24]. The entire section of the fascia alters longitudinal arch height, sometimes causing the total loss of its structure.

56.12 Plantar Fat Pad

The plantar fat pad is a specialized adipose tissue, organized in fibrous lamellae with an intricate spiral pattern, which supports the foot during gait. It cushions the impact of the foot with the ground [25]. Injury to the plantar subcutaneous tissue can happen subsequent to a high-energy trauma: the calcaneus fracture is the main risk factor.

56.13 The First Ray: Role and Pathological Conditions

The first ray is made up of the medial cuneiform and the first metatarsal. In humans, it results to be tightly packed with the other parts of the foot. Joint instability is the main harmful condition that can afflict the first ray. This state can represent the origin of several problems in the foot [26]. First of all, it is undeniable that the first ray instability causes the hallux valgus deformity, characterized by deformation at the metatarso-cuneiform and MTP joints. Ageing is not enough to explain the onset of the pathology, considering that the cuneiform and metatarsal do not alter their shape with the advancing age [27]. For this reason, the hallux valgus has to be considered as an acquired deformity in which instability has a dominant role [28]. Further, instability seems to be the leading risk factor for the elevation of the first metatarsal and to metatarsalgia [29]. The latter can be explained with the redistribution of weight-bearing strengths to the second metatarsal head in case of first ray inadequate support.

56.14 Arch Height

The medial longitudinal arch crosses the talonavicular, navicular-cuneiform, and metatarsal-cuneiform joints. Alteration, deformity, or instability of each of these elements causes the onset of pathological conditions like the fallen arch or the flatfoot [30].

The first metatarsal-cuneiform instability, for example, generates the hallux valgus deformity. Being a three-dimensional alteration, it is immediate to think that the patient affected from hallux valgus frequently develops a flatfoot [31].

The arch integrity, consequently, is essential. The weight-bearing lateral radiographs allow evaluating the talo-metatarsal angle, the leading indicator of natural arch conformation [32]. Talo-metatarsal angle is delimited by the axis of the talus and the axis of the first metatarsal. Physiologically, it is about ±4°. Arch height is subjected to variations that do not necessarily influence the normal angle range. When the angle assumes values outside its standard range, a pathologic mechanism is happening.

56.15 Foot Functional Model

The structure of the foot is described using a specific model, which compares it to a tripod. The tripod should be made of three "legs": the heel, the first metatarsal, and the fifth metatarsal head, strictly balanced with each other. This condition is vital to obtain the right mobility of the foot [33].

In the case of first metatarsal alteration, such as elevation or depression, the other parts of the foot could suffer a slip. Patients affected from flatfeet, subluxation, or sag at the first metatarsal-cuneiform joint develop almost securely the disruption of one of the tripod structures. This happens because, lacking medial support to equilibrate the foot, the hindfoot can collapse in valgus, developing a flatfoot. This acquired deformity is defined as forefoot-driven hindfoot valgus. Also, the failure at the navicular-cuneiform or talonavicular joints provokes the identical condition: hindfoot valgus [34].

Following the same pathogenetic mechanism, the plantar flexion of the first metatarsal could cause hindfoot in varus, defined as forefoot-driven hindfoot varus. The final outcome is a cavovarus foot.

References

1. Ray RG. Arthroscopic anatomy of the ankle joint. Clin Podiatr Med Surg. 2016;33:467–80.
2. Vazquez-Zorrilla D, Millan-Alanis JM, Alvarez-Villalobos NA, Elizondo-Omaña RE, Guzman-Lopez S, Vilchez-Cavazos JF, Fernandez-Rodarte BA, Quiroga-Garza A. Anatomy of foot compartments: a systematic review. Ann Anat. 2020;229:151465.
3. Feiwell LA, Frey C. Anatomic study of arthroscopic debridement of the ankle. Foot Ankle Int. 1994;15:614–21.
4. Lubberts B, Vopat BG, Wolf JC, Longo UG, DiGiovanni CW, Guss D. Arthroscopically measured syndesmotic stability after screw vs. suture button fixation in a cadaveric model. Injury. 2017;48(11)2433–7. https://doi.org/10.1016/j.injury.2017.08.066.
5. Panchani PN, Chappell TM, Moore GD, Tubbs RS, Shoja MM, Loukas M, Kozlowski PB, Khan KH, DiLandro AC, D'Antoni AV. Anatomic study of the deltoid ligament of the ankle. Foot Ankle Int. 2014;35:916–21.
6. Cromeens BP, Kirchhoff CA, Patterson RM, Motley T, Stewart D, Fisher C, Reeves RE. An attachment-based description of the medial collateral and spring ligament complexes. Foot Ankle Int. 2015;36:710–21.
7. Oh CS, Won HS, Hur MS, Chung IH, Kim S, Suh JS, Sung KS. Anatomic variations and MRI of the intermalleolar ligament. AJR Am J Roentgenol. 2006;186:943–7.
8. Longo UG, Loppini M, Fumo C, Lanotte A, Trovato U, Risi Ambrogioni L, Candela V, Forriol F, DiGiovanni CW, Denaro V. Deep deltoid ligament injury is related to rotational instability of the ankle joint: a biomechanical study. Knee Surgery Sports Traumatology Arthroscopy. 2021;29(5):1577–83. https://doi.org/10.1007/s00167-020-06308-7.
9. Uğurlu M, Bozkurt M, Demirkale I, Cömert A, Acar HI, Tekdemir I. Anatomy of the lateral complex of the ankle joint in relation to peroneal tendons, distal fibula and talus: a cadaveric study. Eklem Hastalik Cerrahisi. 2010;21:153–8.
10. Gorman D, Handy E, Wang S, Irwin AL. Morphomics of the talus. Stapp Car Crash J. 2016;60:287–300.
11. Kumar R, Matasar K, Stansberry S, Shirkhoda A, David R, Madewell JE, Swischuck LE. The calcaneus: Normal and abnormal. Radiographics. 1991;11:415–40.
12. Vosseller JT, Dennis ER, Bronner S. Ankle injuries in dancers. J Am Acad Orthop Surg. 2019;27:582–9.
13. Nault ML, Kocher MS, Micheli LJ. Os trigonum syndrome. J Am Acad Orthop Surg. 2014;22:545–53.
14. Xavier G, Oliva XM, Rotinen M, Monzo M. Talonavicular joint arthroscopic portals: a cadaveric study of feasibility and safety. Foot Ankle Surg. 2016;22:205–9.
15. Espinosa N. The subtalar joint. Foot Ankle Clin. 2018;23:xi–xii.
16. Ponkilainen VT, Laine HJ, Mäenpää HM, Mattila VM, Haapasalo HH. Incidence and characteristics of midfoot injuries. Foot Ankle Int. 2019;40:105–12.
17. Döring S, Provyn S, Marcelis S, Shahabpour M, Boulet C, de Mey J, De Smet A, De Maeseneer M. Ankle and midfoot ligaments: ultrasound with anatomical correlation: a review. Eur J Radiol. 2018;107:216–26.
18. Leiß F, Baier C, Schwarz T, Grifka J. Degenerative forefoot: diagnostics and treatment. Z Rheumatol. 2019;78:255–64.
19. Redfern D, Vernois J, Legré BP. Percutaneous surgery of the forefoot. Clin Podiatr Med Surg. 2015;32:291–332.

20. Veljkovic A, Lansang E, Lau J. Forefoot tendon transfers. Foot Ankle Clin. 2014;19:123–37.
21. Park CH, Chang MC. Forefoot disorders and conservative treatment. Yeungnam Univ J Med. 2019;36:92–8.
22. Longo UG, Marinozzi A, Petrillo S, Spiezia F, Maffulli N, Denaro V. Prevalence of accessory ossicles and sesamoid bones in hallux valgus. J Am Podiatr Med Assoc. 2013;103:208–12.
23. Stecco C, Corradin M, Macchi V, Morra A, Porzionato A, Biz C, De Caro R. Plantar fascia anatomy and its relationship with achilles tendon and paratenon. J Anat. 2013;223:665–76.
24. Snow SW, Bohne WH, DiCarlo E, Chang VK. Anatomy of the Achilles tendon and plantar fascia in relation to the calcaneus in various age groups. Foot Ankle Int. 1995;16:418–21.
25. Stecco C, Cappellari A, Macchi V, Porzionato A, Morra A, Berizzi A, De Caro R. The paratendinous tissues: an anatomical study of their role in the pathogenesis of tendinopathy. Surg Radiol Anat. 2014;36:561–72.
26. Roukis TS, Scherer PR, Anderson CF. Position of the first ray and motion of the first metatarsophalangeal joint. J Am Podiatr Med Assoc. 1996;86:538–46.
27. Shereff MJ, Bejjani FJ, Kummer FJ. Kinematics of the first metatarsophalangeal joint. J Bone Joint Surg Am. 1986;68:392–8.
28. Longo UG, Marinozzi A, Petrillo S, Spiezia F, Maffulli N, Denaro V. Prevalence of Accessory Ossicles and Sesamoid Bones in Hallux Valgus. Journal of the American Podiatric Medical Association. 2013;103(3) 208–12. https://doi.org/10.7547/1030208.
29. Camasta CA. Hallux limitus and hallux rigidus. Clinical examination, radiographic findings, and natural history. Clin Podiatr Med Surg. 1996;13:423–48.
30. Hill M, Naemi R, Branthwaite H, Chockalingam N. The relationship between arch height and foot length: implications for size grading. Appl Ergon. 2017;59:243–50.
31. López-López D, Vilar-Fernández JM, Barros-García G, Losa-Iglesias ME, Palomo-López P, Becerro-de-Bengoa-Vallejo R, Calvo-Lobo C. Foot arch height and quality of life in adults: a strobe observational study. Int J Environ Res Public Health. 2018;15:1555.
32. Tomassoni D, Traini E, Amenta F. Gender and age related differences in foot morphology. Maturitas. 2014;79:421–7.
33. Wang M, Gu Y, Baker JS. Analysis of foot kinematics wearing high heels using the oxford foot model. Technol Health Care. 2018;26:815–23.
34. Di Sipio E, Piccinini G, Pecchioli C, Germanotta M, Iacovelli C, Simbolotti C, Cruciani A, Padua L. Walking variations in healthy women wearing high-heeled shoes: shoe size and heel height effects. Gait Posture. 2018;63:195–201.

Biomechanics of the Ankle Syndesmosis

57

Kenneth J. Hunt

57.1 Introduction

A syndesmosis is defined as a fibrous joint in which two adjacent bones are linked by a strong membrane or ligaments allowing a small amount of motion. The primary function of the distal tibiofibular syndesmosis is to maintain the congruency of the ankle mortise during physiologic loading. Injuries to the ankle syndesmosis are increasingly common in impact sports. The nature of these injuries is that they often occur during impact/pileup and can be associated with variable or multiple injury mechanisms. Thus, these injuries can present with numerous and complex clinical manifestations, resulting in the need for a thorough ankle/lower leg physical examination. A complete and accurate examination of the ankle syndesmosis requires a detailed understanding of the anatomy and biomechanical function of the syndesmosis.

57.2 Anatomy of the Ankle Syndesmosis

Osseous structures. The ankle syndesmosis is made up of the confluence of the tibia, fibula, and talus on the lateral aspect of the ankle. The medial surface of the fibula is roughly convex and articulates with the lateral fibular notch (incisura) of the distal tibia to form a fibrous joint, which is linked by strong ligaments on all sides. The tibial part of the syndesmosis forms a concave triangle. The apex located around 6–8 cm above the level of the tibiotalar joint and the lateral ridge of the tibia bifurcates caudally [1]. The anterolateral margin forms the anterior tubercle (Chaput's tubercle), and the posterolateral margin becomes the posterior tubercle [2]. The anterior tubercle is larger, thus preventing forward slipping of the fibula. The posterior tubercle functions as a fulcrum during external rotation injury during which the fibula spins around its longitudinal axis.

At the base of the syndesmosis, there is a small area (called the contact zone) where the tibia and fibula articulate directly with one another and are in direct contact. Its facets are covered with a thin layer of hyaline cartilage. Its size varies by patient, and this articulation can serve as a guide for the alignment of the ankle mortise during ankle fracture fixation or syndesmosis repair [2].

Ligamentous structures: There are four ligaments in the lateral ankle, which form the syndesmosis ligament complex and stabilize the tibiofibular syndesmosis. These include the anterior inferior tibiofibular ligament (AITFL), posterior inferior tibiofibular ligament (PITFL), tibiofibular interosseous ligament (TFIL), and transverse tibiofibular ligament (TTFL). Proximal

K. J. Hunt (✉)
Department of Orthopaedics, University of Colorado, Denver, CO, USA
e-mail: Kenneth.j.hunt@cuanschutz.edu

J. G. Lane et al. (eds.), *The Art of the Musculoskeletal Physical Exam*,
https://doi.org/10.1007/978-3-031-24404-9_57

to this, the tibia and fibula are stabilized by an interosseous membrane. These ligaments work together with the bony restraint to maintain the integrity between the distal tibia and the fibula by resisting the axial, rotational, and translational forces. The deltoid ligament on the medial side of the ankle is an important stabilizer of the syndesmosis as well.

The AITFL runs from Chaput's tubercle of the distal tibia to the anterior tubercle of the distal fibular, obliquely from proximal-medially to distal-laterally. It is composed of three bundles forming a trapezoidal shape. In many patients, an accessory anteroinferior tibiofibular ligament, referred to as Bassett's ligament, runs inferior and parallel to the AITFL. This is a common cause of anterolateral ankle soft-tissue impingement related to tearing, synovitis, and/or scarring. Since 20% of the AITFL is intra-articular, the inferior fibers can be seen and addressed during ankle arthroscopy. AITFL is the weakest of the four syndesmotic ligaments and is the first ligament subjected to stress upon an external rotation force of the fibula around its longitudinal axis [3]. The PITFL extends from the posterior tibial malleolus to the posterior tubercle of the fibula and runs in a proximal-medial to distal-lateral manner.

Tear of the PITFL can occur in Lauge-Hansen supination eversion, pronation-eversion, or pronation-abduction injury. However, as the ligament itself is strong and thick, force often results in a posterior malleolus avulsion fracture rather than pure ligamentous tear [9].

The TTFL runs horizontally between the proximal margin of the fibular malleolar fossa and the dorso-distal rim of the tibia distal to the PITF; it may extend to as far as the dorsal aspect of the medial malleolus. It is a thick and roundish ligament that functions as a labrum analogue to deepen the posteroinferior rim of the tibia. Whether the transverse ligament and PITFL are distinct structures is still controversial, and one is rarely injured without the other. In an anatomical study, the relative importance of individual syndesmotic ligaments to the overall syndesmotic stability was found to be as follows: 35% AITFL, 33% TTFL, 22% TFIL, and 9% PITFL [4]. Progressive injury to these ligaments in this order has been found to significantly alter tibiotalar contact pressures during axial loading [3].

57.3 Biomechanics of the Syndesmosis

In 2003, Beumer et al. evaluated the normal kinematics of the syndesmosis during weight-bearing and external rotation stress using radiostereometry [5]. They discovered that only very small rotations and displacements were detected indicating that the fibula is closely attached to the tibia during loading and rotation. During the external rotation stress test with a 75 Nm force, it caused 2–5° external rotation, 0–2.5 mm medial translation, and 1.0–3.1 mm posterior displacement of the fibula. During simple movement of the ankle from plantar flexion to dorsiflexion, there is an estimated 1.25 mm lateral translation and 2° external rotation of the lateral malleolus [6]. In a high-speed stereoradiography study of healthy subjects, Hogg-Cornejo et al. noted that during dynamic sporting activities, the fibula rotates a total of 9°, translates up to 3 mm posteriorly and 2.5 mm supero-inferior, but only widened 2.3 mm from the tibia [7]. These biomechanical studies proved that distal tibiofibular syndesmosis is highly stable even under extreme athletic conditions.

Ramsey and Hamilton [8] discovered in a cadaveric model that when the talus moves laterally by as little as 1 mm, the contact area in the tibiotalar articulation is decreased by 42%. Hunt et al. [3] showed significant shifts in tibiotalar loading after even moderate syndesmosis injury. Furthermore, Burns et al. [9] demonstrated that a complete disruption of syndesmosis with a disruption of the deltoid ligament causes a 40% decrease in the tibiotalar contact area and a 36% increase in the tibiotalar contact pressures. Therefore, a stable and precise articulation is essential for normal motion of the ankle joint and to avoid articular injury during athletic activities, particularly in collision sports like football, hockey, and soccer.

57.4 Pathomechanics of Syndesmosis Injury

Injury to the ankle syndesmosis typically occurs during dorsiflexion and external rotation forces that exceed tissue failure thresholds. Injury to these structures occurs through rupture or bony avulsion of the syndesmotic ligament complex. It has been estimated that syndesmotic injury occurs in up to 24% of ankle sprains [10], with a higher incidence of 17–74% in certain type of sports activities, e.g., skiers and football and hockey players [11]. Syndesmotic disruption usually takes longer to heal than common lateral ligamentous injury of the ankle, while up to 40% of patients still have ankle instability symptoms 6 months after the injury. In injuries where the ligament complex fails to heal properly, prolonged disability can result [4].

During external rotation of the foot, the fibula is translated posteriorly and rotated externally, resulted in high tensile force on the AITFL. Syndesmotic injury is most commonly associated with ankle fracture, typically the Weber C variety. However, increased incidence of complete syndesmotic ligamentous disruptions is increasingly recognized. All of these injuries tend to involve a twisting load with traumatic supination injury. Due to the difficulty diagnosing these injuries on plain radiographs, and the static nature of MRI scans, these cases can be missed without a knowledge of injury mechanism and a high index of suspicion. Patients most commonly recall an injury and feel a pop or a series of pops and complain of pain during weight-bearing, difficulty weight-bearing, a feeling of instability, and weakness of the ankle.

References

1. Kelikian HKS. Disorders of the ankle. Philadelphia: W.B. Saunders; 1985.
2. Hermans JJ, Beumer A, de Jong TA, Kleinrensink GJ. Anatomy of the distal tibiofibular syndesmosis in adults: a pictorial essay with a multimodality approach. J Anat. 2010;217(6):633–45.
3. Hunt KJ, Goeb Y, et al. Ankle joint contact loads and displacement with progressive syndesmotic injury. Foot Ankle Int. 2015;36:1095–103.
4. Ogilvie-Harris DJ, Reed SC. Disruption of the ankle syndesmosis: diagnosis and treatment by arthroscopic surgery. Arthroscopy. 1994;10:561–8.
5. Beumer A, Valstar ER, Garling EH, et al. Kinematics of the distal tibiofibular syndesmosis: radiostereometry in 11 normal ankles. Acta Orthop Scand. 2003;74:337–43.
6. Peter RE, Harrington RM, Henley MB, Tencer AF. Biomechanical effects of internal fixation of the distal tibiofibular syndesmotic joint: comparison of two fixation techniques. J Orthop Trauma. 1994;8:215–9.
7. Hogg-Cornejo V, Hunt KJ, Bartolomei J, Rullkoetter PJ, Myers C, Shelburne KB. Normal kinematics of the syndesmosis and ankle mortise during dynamic movements. Foot Ankle Orthop. 2020;5(3):2473011420933007.
8. Ramsey PL, Hamilton W. Changes in tibiotalar area of contact caused by lateral talar shift. J Bone Joint Surg Am. 1976;58:356–7.
9. Burns WC II, Prakash K, Adelaar R, Beaudoin A, Krause W. Tibiotalar joint dynamics: indications for the syndesmotic screw–a cadaver study. Foot Ankle. 1993;14:153–8.
10. Hunt KJ, George E, Harris AH, Dragoo JL. Epidemiology of syndesmosis injuries in intercollegiate football: incidence and risk factors from National Collegiate Athletic Association surveillance system data from 2004-2005 to 2008-2009. Clin J Sport Med. 2013;23:278–82.
11. Boytim MJ, Fischer DA, Neumann L. Syndesmotic ankle sprains. Am J Sports Med. 1991;19:294–8.

58 Clinical Tests for Assessment of Instability of the Ankle and Syndesmosis

Flávio Cruz, Gustavo Vinagre, and Pieter D'Hooghe

58.1 Introduction

The ankle sprain incidence in sports has been reported between 0.324 and 9 per 1000 h of activity [1, 2], with a variability that is most likely due to different definitions of injury and population. In football for example, increased ankle sprain rates have been reported in older players, dominant leg, during competition, and at the end of each half of a game [3]. Historically, the ankle used to be the most common location of injury in professional football players (around 30% of total injuries). However, more recent studies suggest a lower ankle injury rate, accounting for 10–15% of all injuries [4–6]. Approximately 60% of ankle sprains occur as a result of player contact [7, 8] and the overall ankle sprain recurrence rate is between 4% and 29% [3, 7, 8].

A major problem accompanying ankle injury is the high rate of recurrence associated with chronic ankle instability [9–12]. With an inversion trauma mechanism, the anterolateral ankle ligaments are strained and even ruptured in more severe cases. The anterior talofibular ligament (ATFL) is the most commonly ruptured, followed by the calcaneofibular ligament (CFL) and more rarely the posterior talofibular ligament (PTFL) [13, 14]. Post-traumatic osteoarthritis has been shown to be the resultant of abnormal loading and changes in contact stress and distribution [15].

Injury to the syndesmotic ligaments (often referred to as a high ankle sprain) occurs in 1–18% of patients with an ankle sprain, with reports of injury in football players of 3–6% [9, 16]. It is likely that this is an underestimate, as 20% of athletes suffering from an acute ankle sprain show evidence of a syndesmotic injury on MRI [17]. Male gender, higher level of competition, and a planovalgus foot alignment are known risk factors for syndesmotic injury in athletes [18].

Chronic ankle instability presents through symptoms like recurrent sprains, pain, swelling, and avoidance of provoking activities and can be classified as functional instability or mechanical instability resulting in proprioceptive deficit [19].

Previous history taking and clinical physical examination are essential in the diagnosis of ankle instability. Differentiation between functional and anatomical ankle instability is also important to guide the proper treatment [14]. Functional instability depends on the patient-generated reports or complaints that could be accompanied by clinical laxity, while mechanical instability can be identified by physical examination [20, 21]. It is cur-

F. Cruz · G. Vinagre · P. D'Hooghe (✉)
Department of Orthopaedic Surgery,
Aspetar Orthopaedic and Sports, Medicine Hospital,
Doha, Qatar
e-mail: flavio.cruz@aspetar.com;
gustavovinagre@gustavovinagre.com;
pieter.dhooghe@aspetar.com

J. G. Lane et al. (eds.), *The Art of the Musculoskeletal Physical Exam*,
https://doi.org/10.1007/978-3-031-24404-9_58

rently believed that the mechanical instability may be due to laxity imparted by the trauma, whereas the functional instability is a result of neuromuscular insufficiencies, muscular weakness, and proprioceptive insults [11, 13, 22]. Lateral ankle sprains are most commonly involved in the development of CAI [19].

Associated injuries that may accompany chronic ankle instability are chronic regional pain syndrome, neuropraxia, sinus tarsi syndrome, tendon disorders such as peroneal tendinopathy, dislocation or subluxation, impingement syndromes, fractures such as anterior calcaneal process, fibula and lateral talar process, loose bodies, and osteochondral lesion of talar dome or distal tibia [19].

Fact Box

- Ankle inversion injury is one of the most common injuries in the general as well as in the athletic population.
- Differentiation between functional and anatomical ankle instability is essential to guide proper treatment.
- Associated injuries can develop after a failed treatment for ligament injury in the ankle. They regularly present as talar osteochondral injuries, loose bodies, and anterior and posterior impingement syndromes.

58.2 Anatomy

The ankle joint can be regarded as a fork, in which the tibia plafond and both malleoli form a mortise to receive the talus. As a hinge joint, there is a single axis of movement that allows for dorsiflexion (20°) and plantar flexion (50°). As the superior surface of the talus is narrower posteriorly, there is a looser fit within the fork during plantar flexion and most stability is then provided by the ankle ligaments alone. This reduced intra-articular stability can potentially explain why most ligamentous injuries are sustained in plantar flexion [23].

The distal part of the tibia and fibula also articulates at the inferior tibiofibular level, supported by the syndesmotic ligaments (Fig. 58.1), thereby forming the inferior or distal tibiofibular joint ("distal tibiofibular syndesmosis"). This syndesmotic joint allows the tibiofibular complex (as a whole) to adapt to the varying width of the upper articular surface of the talus.

In addition, inferiorly to the talocrural joint, the subtalar joint is formed between the inferior surface of the talus and the superior surface of the calcaneus. The subtalar joints provide 35° of inversion and 15° of eversion.

Overall stability over the ankle joint is provided by passive and dynamic factors. Passive stability depends on the form of the articular surfaces, articular capsule, surrounding ligamentous complexes (Fig. 58.2), and retinacula. The

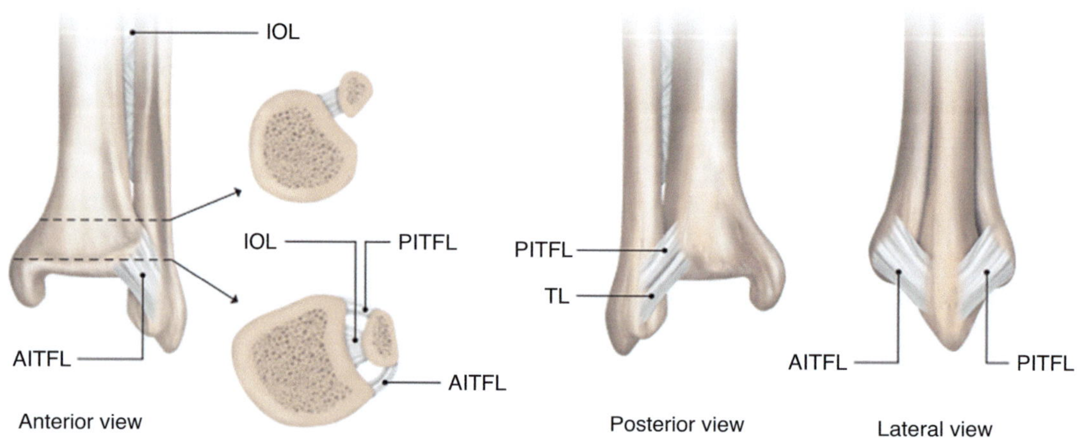

Fig. 58.1 The syndesmotic ankle ligaments presented. Copyright by Dr. Pieter D'Hooghe

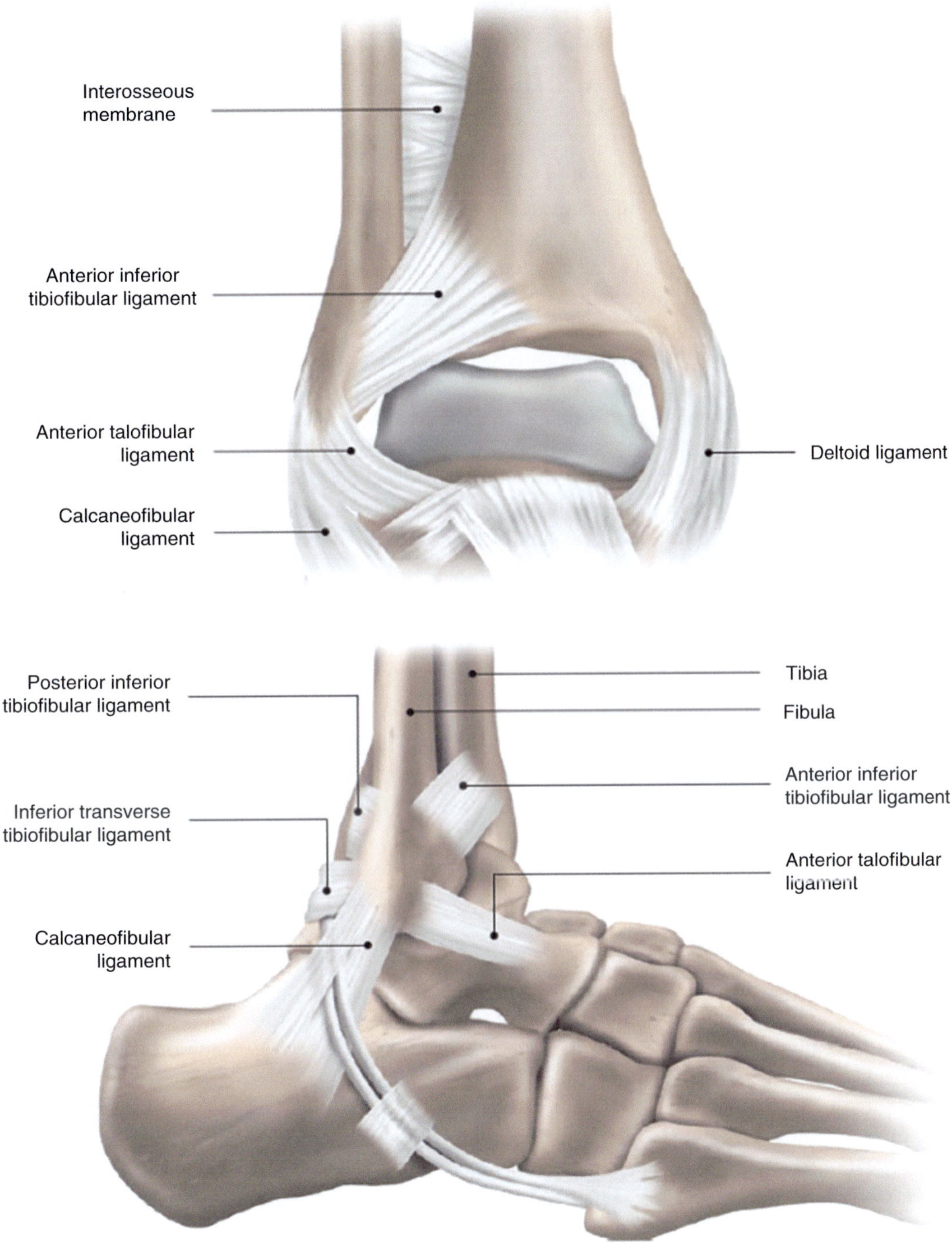

Fig. 58.2 Anteroposterior and lateral views to the lateral ankle joint ligaments. Copyright by Dr. Pieter D'Hooghe

dynamic stability is predominantly provided by muscle activation.

58.3 Mechanism of Injury

Lateral ankle sprains occur when the foot is in plantar flexion and inverted, causing damage to various structures most commonly involving the anterior talofibular ligament (ATFL) and calcaneofibular ligament (CFL). Between these two, damage to the ATFL is overwhelmingly more common, occurring in 66% of lateral ankle sprains with CFL sprains constituting about 20% of lateral ankle sprains [24]. Damage or disruption to the local soft tissue, nerves, and bony components of the lateral ankle, in addition to damaged ligamentous structures, can result in ankle pain [24]. Within the literature, chronic ankle instability (CAI) is most often described as chronic ankle pain (CAP) as a consequence of an acute or recurrent lateral ankle sprain. In a recent review, Miklovic et al. suggested that approximately 40% of individuals who sustain a lateral ankle sprain may end up in developing CAI and CAP [25].

Although medial ankle sprains are considered rare, recent studies have revealed that the deltoid ligament complex is injured more frequently than expected. This medial ankle injury sprain is thought to be due to a combination of eversion and external rotation of the hindfoot or a reverse inward body rotation on a fixed foot. With severe external rotation moments, the tibiofibular syndesmotic ligaments can join the mechanism of injury [26, 27]. Osteochondral lesions (OCLs) of the ankle typically present as a part of the acute medial ankle instability (AMAI) [28, 29].

The most accepted mechanism of injury for syndesmotic ankle sprains is a forceful external rotation of the foot and ankle with the ankle in dorsiflexion and the foot pronated [30]. This is most commonly experienced (1) when an athlete rapidly pivots internally off a foot planted in external rotation, (2) when contact with another player applies a valgus load to the leg while the foot is planted, or (3) when a direct blow to the lateral aspect of the heel forces the foot and ankle of a kneeling or fallen athlete into external rotation (toward the ground) [31]. As the talus rotates in the mortise, the fibula rotates externally and moves posteriorly and laterally, separating the distal tibia and fibula, sequentially tearing the anterior inferior talofibular ligament (AITFL) and deep medial ligament complex (MLC), or causing a malleolar fracture, involving the interosseous ligament (IOL) and finally the posterior inferior tibiofibular ligament (PITFL) [30, 32].

Severity of syndesmotic injury varies, ranging from a partially torn AITFL to a complete disruption of all ligaments with mortise widening. Combined deltoid and syndesmosis injury critically disrupts talar stability [33]. The amount of force and how long it is applied will determine how proximal the syndesmotic and interosseous injury extends [34], sometimes resulting in a Maisonneuve fracture. Another injury mechanism for syndesmotic ankle sprains is hyperdorsiflexion. Forced dorsiflexion of the ankle causes the wider anterior talus to act as a wedge that can cause injury to the syndesmotic ligaments.

Fact Box

- Lateral ankle sprains occur in plantar flexion—inversion and tearing the ATFL and CFL.
- Medially, the injury is a combination of eversion and external rotation of the hindfoot or a reverse inward body rotation on a fixed foot.
- Syndesmotic ankle sprains happen in a forceful external rotation of the foot and with the ankle in dorsiflexion and the foot pronated.

58.4 Clinical Presentation and Diagnosis

Patient history and video images are particularly useful in identifying the mechanism of injury, and it is important to distinguish a simple sprain from a ligament rupture, since adequate treatment is associated with a better prognosis [35,

36]. Ankle sprains may be accompanied by an audible snap or crack. All ligamentous and bony structures should be palpated for tenderness, including the whole length of the fibula and the base of the fifth metatarsal. Note that approximately 40% of the patients with a lateral ligament rupture have pain upon palpation over the medial malleolus, whereas 60% report tenderness over the anterior inferior tibiofibular ligament (AITFL)—without injury to the ligament, probably due to an anterior capsule tear [36, 37].

However, manual stress tests are less reliable in the acute phase, because of pain and swelling. Therefore, delayed physical examination (4–5 days) gives a better diagnostic timing and is considered the gold standard in the diagnosis of an acute lateral ligament injury. The sensitivity to diagnose an acute lateral ligament rupture during a delayed physical examination is 96%, with a specificity of 84% [36, 37].

In high ankle sprains, specific symptoms may be suggestive of syndesmotic injury; patients may have inability to bear weight, swelling, pain during the push-off phase of gait, and pain anteriorly between distal tibia and fibula, as well as postero-medially at the level of the ankle joint [38]. The presence of high ankle pain (proximally up the anterolateral leg) is suggestive of a more significant injury [17].

In acute injuries, inspection may reveal edema and ecchymosis around the lateral, medial, and/or anterior aspect of the ankle. In syndesmotic lesions, range of motion (ROM) is often limited with an empty or painful apprehension at terminal dorsiflexion [39]. There is a significant correlation between how far this anterolateral tenderness extends proximally in the leg and the injury severity. Local tenderness is, however, not specific in the acute setting, as 40% of the patients with an ATFL disruption report pain in the area of the AITFL, while arthroscopy shows no syndesmotic injury [36].

The Ottawa Ankle Rules (OAR) have been developed to help clinicians indicate X-ray imaging after acute ankle injury. It is an accurate instrument to rule out fractures over the ankle, with a sensitivity of almost 100% [40]. The OAR are recommended as a primary physical examination tool to rule out the likelihood of foot/ankle fractures by emergency physicians, general practitioners, or physiotherapists [41]. If a patient presents with negative findings following the OAR, there is less than a 2% chance of false negativity [42]. If grossly swollen ankle prevents proper palpation of bony structures, radiographs should be obtained [43, 44].

To apply the OAR, the physician might palpate the entire distal 6 cm of the fibula and the tibia/medial malleolus; palpate the entirety of the navicular bone (with special attention to the relatively avascular nickel-sized area at the central region of the proximal dorsal surface termed the "N" spot); palpate the base of fifth metatarsal bone; and observe the patient ambulate for at least four steps. Tenderness at these four locations is considered an indication for sending to radiography. Adequate imaging request should include AP, lateral, and ankle mortise views. Standard ankle radiographs (weight-bearing, if possible) are useful in acute cases to exclude widening of the medial ankle clear space, deltoid avulsion fragments, syndesmotic lesions, and ankle fractures. In chronic cases, the following weight-bearing radiographs are suggested: ankle mortise view, foot dorsoplantar view, foot lateral view, and hindfoot alignment (Saltzman) view [45].

If there is a clinical or radiographical suspicion of a Maisonneuve fracture (i.e., pain in the region of the proximal fibula, painful swelling on the medial side without a fracture, and isolated fractures of the medial malleolus or malleus tertius), radiographs with full-length views of the lower leg are required.

Several radiographic parameters have been developed to help identify ankle injuries. In cases of syndesmosis injury, the tibiofibular clear space (defined as the distance between the medial border of the fibula and the lateral border of the posterior tibia) provides the most reliable indicator of diastasis. This distance is measured at 1 cm proximal to the tibial plafond and should not exceed 6 mm in both the AP and mortise views.

Cross-sectional imaging such as magnetic resonance imaging (MRI) is fundamental to confirm clinical suspicion. Absence of tibiofibular diastasis no longer rules out the diagnosis. Computed tomography (CT) and comparative weight-bearing CT are helpful diagnostic tools, especially to evaluate potential avulsions or bony diastasis. Exhaustive osteo-ligamentous ankle assessment is necessary, as syndesmosis lesions may be just one component in more complex rotational instability [46].

Ultrasonography and MRI can be useful in diagnosing associated injury (bone, chondral, or tendon). Ultrasonography has been demonstrated to be an accurate investigation for ligamentous injury, but images may be difficult to interpret on retrospective review by other physicians. The sensitivity and specificity of ultrasonography for a lateral ligament rupture are 92% and 64%, respectively. However, if ultrasonography is performed after an inconclusive delayed physical examination, sensitivity increases to 100% and specificity to 72% [36]. A more recent study in ATFL injury comparing ultrasonography in the emergency room with MR images found no overall differences in diagnostic accuracy [47]. Ultrasonography can therefore be considered a valid alternative diagnostic method. It is less expensive and allows for a faster examination without radiation exposure, although it remains investigator dependent.

MRI is a reliable tool in the diagnosis of lateral ligamentous ruptures and other associated injuries such as tendinous and syndesmotic tears, occult fractures, and osteochondral lesions. The sensitivity and specificity of MRI for ATFL injuries are 92–100% and 100%, respectively [48, 49]. In comparison with arthroscopy, MR images can correctly locate the injured portion of the ATFL in 93%, whereas ultrasonography in 63% [49].

Ankle arthroscopy can also be helpful in the diagnosis of ankle instability. In AITFL injuries, the syndesmosis can be assessed by introducing an arthroscopic palpation hook into the tibiofibular joint line, also showing good correlation with MRI [46].

58.4.1 Physical Exam and Clinical Tests

Ankle instability can be defined as the inability to keep the normal relation between the bones in the ankle joint. The ankle that suffers from an acute high-grade sprain, when not properly treated, can be at risk for developing chronic instability (CAI) [50]. Some reports show that up to 20% of ankle sprains evolve to a CAI [19]. Therefore, physical examination of any ankle sprain should include clinical tests as well as inspection of the hindfoot (varus misalignment), and assessment of the ligamentous laxity and midfoot cavus alignment, which is considered a high arch foot condition [51], peroneal muscle strength, and hindfoot motion.

Two new concepts have recently been introduced (although there is no consensus currently between experts): micro-instability and rotational instability.

Micro-instability is a concept that can be presented as a lower grade of mechanical instability (due to a partial tear of the superior fascicle of the ATFL). This could relate to as why there is remaining pain and ankle discomfort after certain ankle sprains.

Rotational instability is defined as the increase of abnormal rotation of the talar bone into the mortise. This instability is explained as the result of chronic insufficiency over the lateral ligament complex, creating an excessive and continued internal rotation and anterior translation of the talus. Over the long term, it can lead to an injury of the most anterior component of the deltoid ligament, described as "open book tear" [50].

Ankle proprioception is commonly disturbed in patients with chronic ankle instability and therefore the Romberg's maneuver can be carried out [52]. Reduced ankle dorsiflexion and muscle strength should also be assessed since both are known to be directly related to chronic ankle instability [53].

Over the medial side, the clinical examination of medial ankle instability (MAI) starts with a careful bilateral inspection over the ankle in standing, walking, and sitting positions with a hanging leg over the examination table. It is essential to look for swelling, hematoma, malalignment, deformity, and potential scars. When a patient bears weight, asymmetrical planovalgus and abductus of the affected ankle and foot may indicate MAI. Palpation should include medial and lateral ligaments in addition to palpating the syndesmosis and posterior tibial, peroneal, and Achilles tendon. Typically, patients with MAI will present with tenderness, mostly over the medial gutter, over the deltoid ligament and the spring ligament. Some patients may have tenderness along the posterior tibial tendon, because of the common relation of MAI with posterior tibial tendon insufficiency (PTTI). In advanced cases, tenderness over the lateral ankle ligament (in rotational ankle instability) and subfibular area (due to hindfoot valgus) can be elicited [45].

In combined medial and syndesmotic injuries, the clinical presentation is very similar to the classic lateral ankle sprain one. Medial symptoms can occur if the deltoid ligament is injured in combination as well. Pain, swelling, ecchymosis, weight-bearing inability, and soreness around the ankle joint are some symptoms that can be noted. Numerous clinical tests are described to detect additional syndesmotic injury. The external rotation test and the squeeze test are the most commonly used, but the Cotton test, the fibular-translation test, and the cross-legged test have also been described.

58.4.1.1 Lateral Instability

Lateral ankle sprains typically result from supination and inversion over the foot in plantar flexion position usually during athletic activity or running on uneven surfaces. Initially, significant soft-tissue swelling and ecchymosis can be present following the acute injury. Sufficiency of the ATFL can be assessed by the anterior drawer test, while the integrity of the CFL can be assessed by the talar tilt test. During the physical examination, two hands of the examiner are used to manipulate the ankle—one hand stabilizing the distal tibia and the other hand measuring the displacement of the unfixed talus. Both the ADT and talar tilt test are based on the perception of the talar displacement in the joint, which is likely to be disturbed when compared to the normal non-injured side.

If a hematoma is present, accompanied by soreness at palpation or a positive stress test or both, it is most likely that a (partial) lateral ligamentous rupture exists [35, 36]. The integrity of the ATFL can be assessed directly via a manual anterior drawer test [54]. Similarly, the talar tilt test is performed to assess for integrity of the CFL [55]. The anterior drawer test has been reported to be a reliable test, with sensitivity and specificity as high as 96% and 86%, respectively, for the detection of a ligament injury [36].

Studies show that psychological factors of the examiners can influence the perception of a stimulus [56]; therefore, a new test is described to diagnose ATFL injuries, decreasing the sensation of relative hand movements. The new physical examination test, supplemented with more technical requirements, is named the "reverse anterolateral drawer test (RALDT)." This new clinical test can optimize the examination, avoid the contextual effects introduced by the clinician's hands, and provide a more sensitive and accurate measurement for diagnosing chronic ATFL injuries [57].

Anterior Drawer Test

With the ankle in 10–15° of plantar flexion and the examiner's hand stabilizing the distal tibia, the calcaneus is translated anteriorly, and the degree of laxity or subluxation is assessed. Anterior displacement of the talus creates a negative pressure, resulting in a "sulcus sign" whereby the skin dimples at the lateral side where the ATFL has ruptured [55] (Figs. 58.3, 58.4, and 58.5).

The test can be performed lying or in a seated position with the calf hanging over the edge of the examination bed. The examiner stabilizes the distal tibia of the participant with one hand and applies an anteriorly orientated force to the

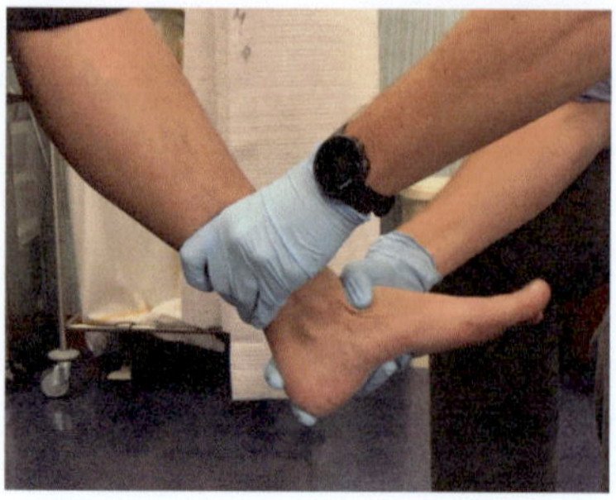

Fig. 58.3 Anterior drawer test: ankle in 10–15° of plantar flexion and the examiner's left hand stabilizing the distal tibia

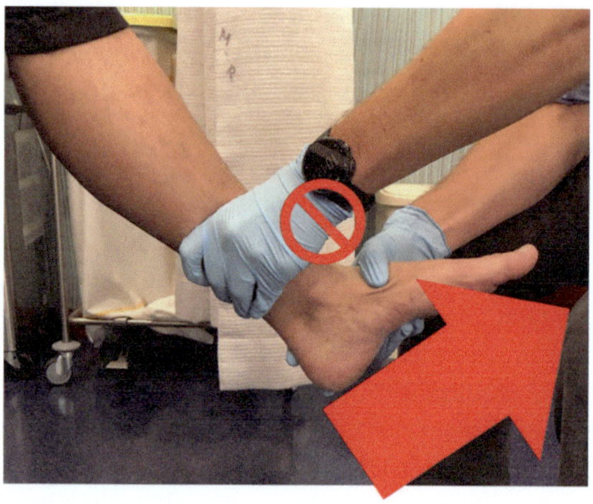

Fig. 58.4 The distal tibia is stabilized, the calcaneus is translated anteriorly, and the degree of laxity or subluxation is assessed

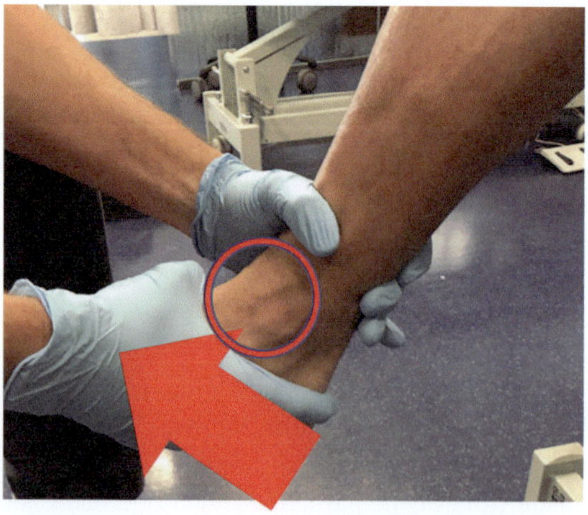

Fig. 58.5 Anterior displacement of the talus creates a negative pressure, resulting in a "sulcus sign" (circle) whereby the skin dimples at the lateral side where the ATFL has ruptured

calcaneus with the other hand as described by van Dijk et al. [36]. The examiner grades the amount of talus anterior-posterior displacement according to a 0–4 ordinal scale, with a grade of 0 as hypomobile, 1 as normal, 2 as mildly increased laxity, 3 as moderately increased laxity, and 4 as severely increased laxity. Grades 3 and above are considered "positive" for excessive laxity, whereas grades 0, 1, and 2 are considered "negative" or normal [58].

The sensitivity (84%) and specificity (96%) of a physical examination using the anterior drawer test are optimal if the clinical assessment is delayed for 4–5 days post-injury [35, 36, 59, 60]. In case of a suspected fracture, the OAR should be applied and sent to imaging where required [41]. In the anterior drawer test, when the ATFL is injured, the medial deltoid ligament can remain intact, leading to false-negative results. Lahde et al. reported that 28% of ATFL tears and 38% of combined ATFL and CFL (calcaneofibular ligament) tears were not detected by the anterior drawer test [61].

Anterolateral Drawer Test

The test is performed with one hand stabilizing the leg just above the ankle joint and the other hand providing a combination of an anterior orientated force, measuring the talus displacement, and control of ankle plantar flexion simultaneously with the other hand, as described by Phisitkul et al. More specifically, the index and middle fingers press firmly against the posterior aspect of the heel to provide the anteriorly directed force. The palm supports the sole of the foot to maintain a 10–15° plantar flexion and tighten the lateral ligaments. The thumb is placed along the relatively smooth plane of the lateral aspect of the anterior talar dome and the anterior aspect of the lateral malleolus 1 cm proximal to the tip. Anterior translation is applied at the posterior aspect of the heel, while the foot is allowed to rotate internally while a potential step-off is then palpable by the thumb [62].

The differences between the traditional ADT and ALDT are (1) 10–15° of controlled plantar flexion (which puts the ATFL in a tightened state) and (2) palpation at the smooth plane of the lateral aspect of the anterior talar dome and the anterior aspect of the lateral malleolus (which makes it easy to feel the displacement of the talus). Some techniques, such as unconstrained internal rotation of the forefoot (which puts the ATFL in a tightened state and relaxes the medial ligaments at the same time), can be applied in the diagnosis of ATFL injuries (to maximize the accuracy of the ALDT). This test has been modified and applied to diagnose ATFL injuries with or without CFL injuries with a higher accuracy [62].

Reverse Anterolateral Drawer Test

The exam is performed with the patient lying on the bed with the knee flexed and the angle of the knee adjusted to facilitate plantar flexion. The heel is completely pressed on the bed by the examiner with one hand after adjusting the ankle to a 10–15° of plantar flexion and unconstrained internal rotation. The index and middle fingers are placed along the relatively smooth plane of the lateral aspect of the anterior talus dome and the anterior aspect of the lateral malleolus (1 cm proximal to the tip). The other hand holds the distal tibia, and the base of the palm pushes against the tibia to induce a posteriorly oriented displacement of the tibia with a force parallel with that of the articular surface arch of the talus. The grading standard is similar to that of the ADT. Both sensitivity and specificity of the RALDT are higher than 85%. The RALDT (with both a relatively high specificity and sensitivity) shows a good ability to diagnose chronic ATFL ankle injuries [57].

Talar Tilt Test

The talar tilt test, or inversion stress maneuver, is performed with the patient supine or on the side, with the foot relaxed. The gastrocnemius muscle must also be relaxed by flexion of the knee. The talus is then tilted from side to side into adduction and abduction. The findings should be compared with the contralateral side. A sensitivity of 52% has been reported in one study for the talar tilt test [63]. If a firm end point cannot be felt when compared with the opposite ankle by abducting and inverting the heel, damage to the CFL can be apparent. (Note that the degree of tilt ranges from 0 to 23°.) Pain in the area of the ligament or a sensation of clunk would indicate a positive test. An outward translation in excess of 5° on the injured side compared to the uninjured side, or a spongy or indefinite end feel, may indicate a complete tear of CFL. In many cases, this test is difficult, if not impossible, to perform secondary to post-traumatic patient pain and swelling [64] (Fig. 58.6a, b).

58.4.1.2 Medial Instability

An acute injury to the deltoid ligament complex following a pronation (eversion) trauma usually presents with symptoms ranging from discomfort to a severely painful and swollen medial ankle with the inability to bear weight. The mechanism of trauma, level of daily activity, level of athletic activity, and any history of previous ankle sprain should be clarified. Chronically, a deltoid ligament injury presents as a medial giving way, especially when walking on uneven ground. The patient usually complains of anteromedial pain, and sometimes the pain emanates laterally due to the hindfoot valgus (subfibular impingement). Mechanical symptoms like catching or locking may be present with the development of intra-articular impingement (hypertrophic synovium, anteromedial osteophytes, OCLs). Any associated foot deformity or disease, history of ankle sprains, and the patient's activity level should also be evaluated [45].

Severe ankle sprains may affect the medial ligament complex, whose deterioration and suboptimal treatment can result in chronic rotational ankle instability (i.e., combination of medial and lateral ankle instability). In a prospective study, an arthroscopic exploration of the deltoid ligament done during the treatment of chronic lateral ankle instability revealed that 20% of patients had a concomitant lesion of the deltoid [26]. Therefore, small alterations in the deltoid ligament complex lead to secondary lateral ankle instability, respective rotational ankle instability, and increased risk for ankle osteoarthritis (OA) [45].

The integrity of the superficial deltoid ligament can be evaluated on a relaxed sitting patient with hanging legs using the "external rotation test," whereas the deep deltoid can be assessed by the "eversion stress test." A combined eversion and external rotation stress test addresses both

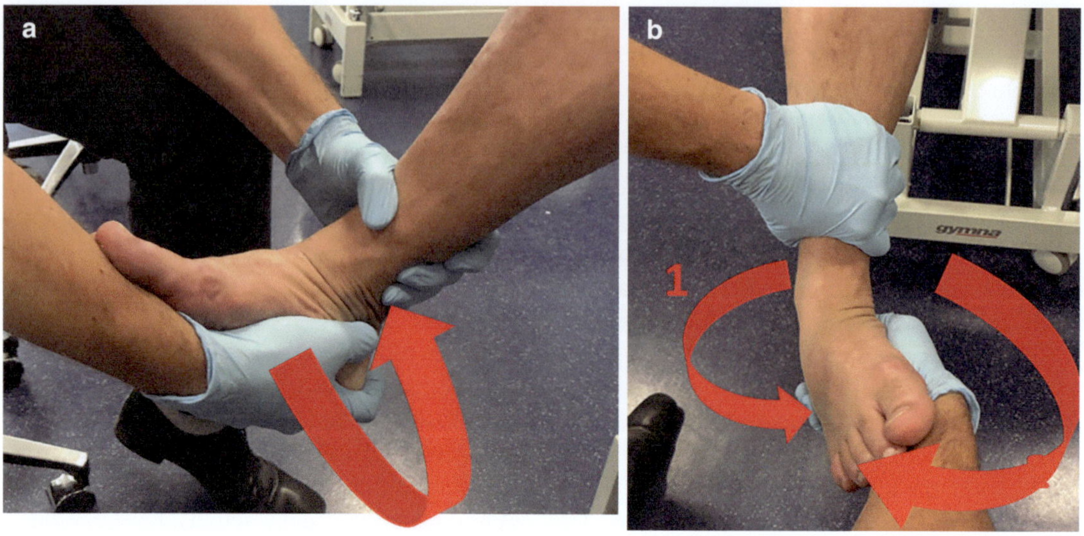

Fig. 58.6 (**a**) The talar tilt test, or inversion stress maneuver. (**b**) The talar tilt test. The talus is tilted from side to side into adduction (movement 1) and abduction (movement 2)

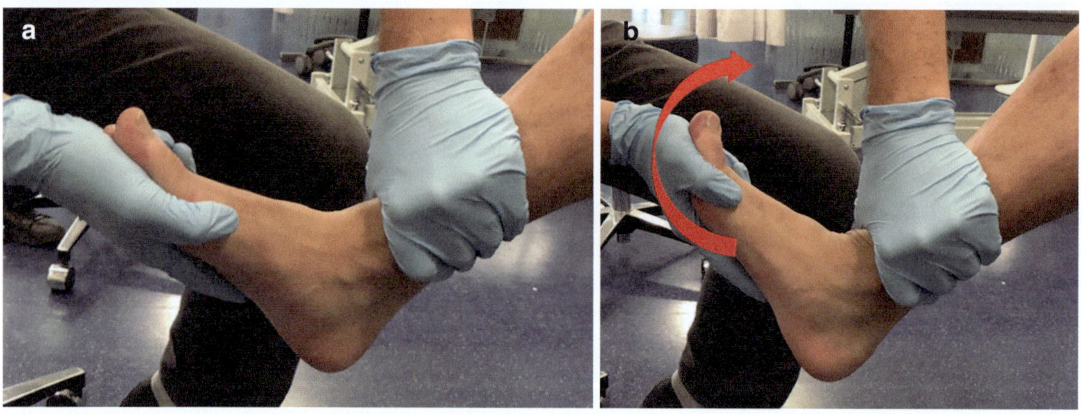

Fig. 58.7 (**a**) External rotation test: The foot is gently grasped and rotated laterally with the ankle locked in neutral. (**b**) External rotation test: The foot is gently grasped and rotated laterally with the ankle locked in neutral

the deep and the superficial deltoid ligament. The anterior drawer test is also a valuable tool for diagnosing anteromedial subluxation in deltoid ligament insufficiency [26, 27, 65].

External Rotation Stress Test

The external rotation test can demonstrate the integrity of the syndesmotic ligaments. The patient sits with the knee flexed to 90°. The foot is gently grasped and rotated laterally with the ankle locked in neutral. A positive test result occurs when the patient has pain over the syndesmosis. The external rotation test has a sensitivity of 20% and a specificity of 84.8% [66] (Fig. 58.7a, b).

Kleiger Test

This is also named the dorsiflexion external rotation stress test and seen as a variation of the external rotation test. The talus may displace from the medial malleolus, indicating a tear of the deltoid ligament. With the patient seated and the knee flexed approximately 90° and the ankle relaxed, use one hand to grasp and stabilize the leg from behind, making sure not to compress the fibula and tibia together. Use your other hand to fully dorsiflex the ankle and then externally rotate the foot (Fig. 58.8a–e). Pain over the location of the anterior inferior tibiofibular ligament is indicative of a syndesmosis sprain. A positive test

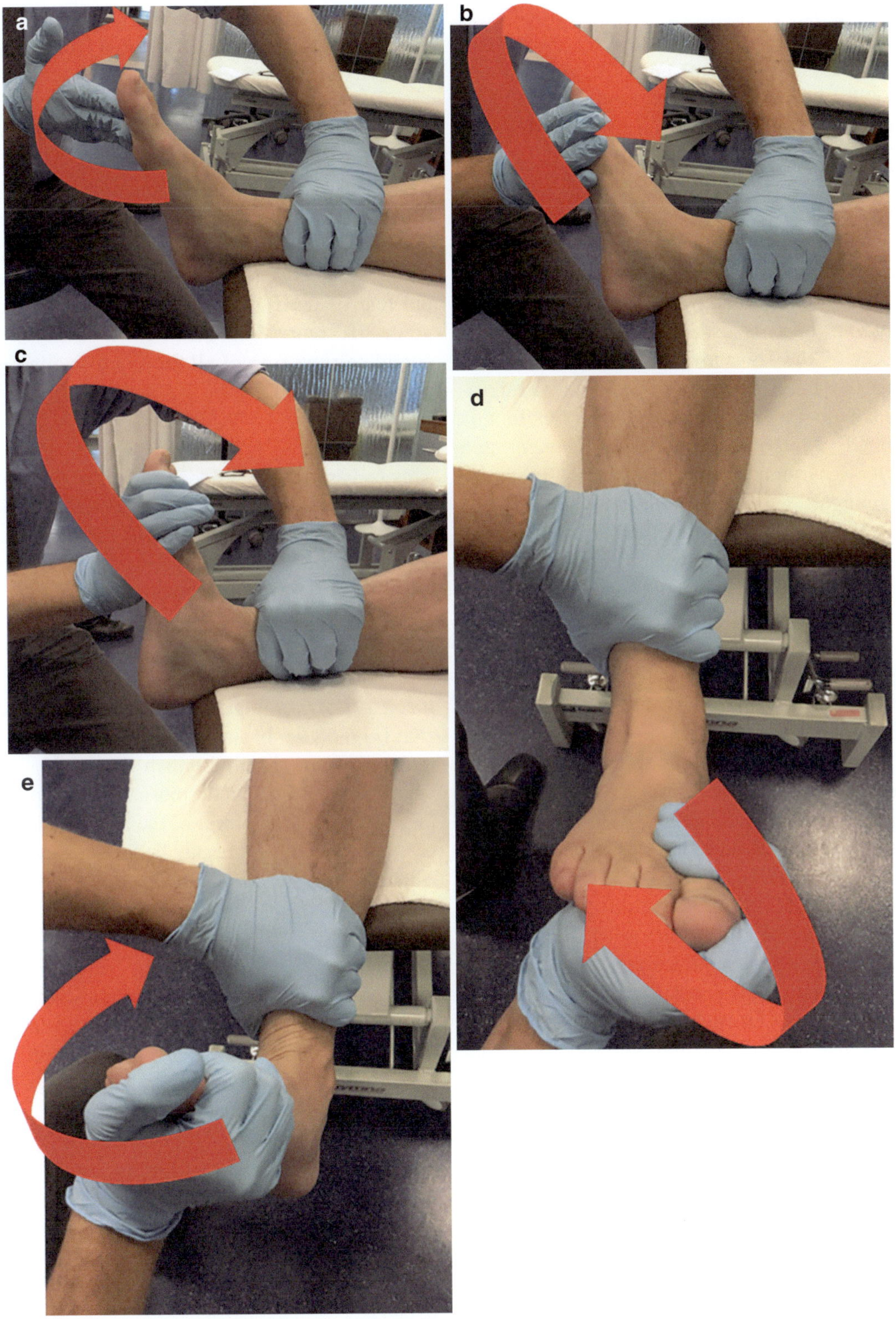

Fig. 58.8 (**a–c**) Kleiger test: one hand to grasp and stabilize the leg from behind, making sure not to compress the fibula and tibia together. Use your other hand to fully dorsiflex the ankle and then externally rotate the foot. (**d**) Upper view of the Kleiger test. External rotation. (**e**) Upper view of the Kleiger test. Hyperdorsiflexion

occurs when the patient has pain medially and laterally. Depending on the severity, the interosseous membrane may be involved, and pain can be radiated further up between the distal fibula and tibia. This test is also sensitive for fractures of the fibula [67].

Eversion Stress Test

The eversion stress test evaluates the integrity of the deltoid ligament. Patient is seated with knee flexed at approximately 90°, and ankle in neutral. One hand is used to stabilize the lateral aspect of the leg above the lateral malleolus. The other hand of the examiner is placed inferomedial on the calcaneus, and the hindfoot is everted. Pain over the deltoid ligament and increased eversion indicate possible injury to the midportion of the deltoid or a possible avulsion fracture of the medial malleolus. This test should be repeated while holding the ankle in full dorsiflexion to evaluate the posterior aspect of the deltoid and then repeated again in plantar flexion to evaluate the anterior aspect of the deltoid ligament [68].

58.4.1.3 Clinical Tests for Syndesmosis

Ruptures of the syndesmosis are rarely isolated injuries, but generally occur in association with other osteo-ligamentous injuries, especially fractures of either the fibula or the posterior and medial malleoli. It should be strongly suspected if there is an associated fracture of the proximal fibula (Maisonneuve fracture). Squeeze test, external rotation stress test, fibula translation test, Cotton test, and crossed-leg test are tests used to assess the integrity of the syndesmosis of the ankle. The external rotation test is specifically correlated with the presence of a syndesmosis sprain and is associated with a longer return to preinjury activities [69].

In a study comparing physical examination and MRI for lateral ankle sprain, sensitivity and specificity were 30% and 93.5% for the squeeze test, and 20% and 84.8% for the external rotation test, respectively. The sensitivity of the squeeze test and external rotation test was low, suggesting that physical examination often fails to diagnose syndesmotic injury. Conversely, the specificity was very high; nearly all patients with a positive test presented with a syndesmotic injury [66].

The classification of syndesmotic injury is divided into three grades: grade I represents a mild sprain to the AITFL without instability; grade II involves a tear of the AITFL and a partial tear of the IOL with some instability; and grade III represents definite instability with complete rupturing of all the syndesmotic ligaments [18].

Frick Test

This consists of forced dorsiflexion and external rotation of the ankle, with the knee in 90° flexion [16, 35, 70]. There is a variant of Frick test, performed under weight-bearing. Its sensitivity range is 30–71%, and it has a specificity of 85% [71–73].

Squeeze Test

It consists of proximal tibial and fibular compression (Fig. 58.9a, b), inducing a movement separating the two distally, which is possible, and painful, in case of syndesmosis ligament lesion [37, 40]. Pain at the level of the ankle joint indicates a positive test result. It has sensitivity of 30% and specificity ranging from 88 to 93.5% [36, 72, 74].

Cotton Test

The Cotton test is performed by translating the talus within the mortise from medial to lateral. Increased translation or pain may suggest syndesmosis involvement, as well as a deltoid (medial) ligament injury.

Single-Leg Jump Test

Exam consists of jumping with the injured leg; pain is alleviated when the same jumps are made with strapping at syndesmosis level, limiting abnormal motion of the distal tibiofibular mortise [47].

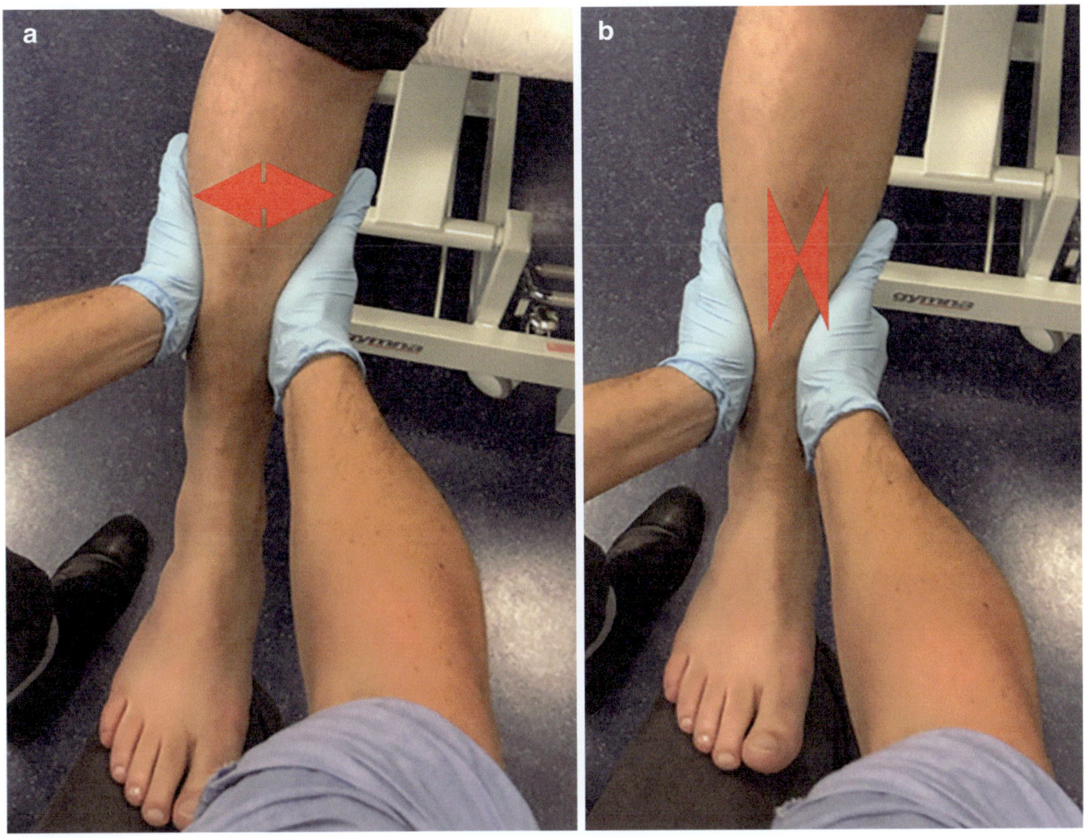

Fig. 58.9 (**a**, **b**) Squeeze test: proximal tibia and fibula compression

The External Rotation Stress Test

This is performed by placing the ankle in a dorsiflexion position and applying an external rotation force. Pain with this maneuver indicates a positive test result.

Fibula Translation Test

The examiner attempts to translate the fibula from anterior to posterior. In the normal ankle, there is a firm end point and little movement. Increased translation relative to the contralateral side and pain indicates a positive test result. The ligament tear causes abnormal motion revealing excessively anteroposterior fibular course compared to the tibia [70].

This test is performed by tightly applying several layers of 1.5 in. athletic tape just above the ankle joint to stabilize the distal syndesmosis. The patient is then asked to stand, walk, and perform a toe raise and jump. The test result is positive if these maneuvers are less painful after taping (Table 58.1).

Fact Box

- Syndesmotic injury generally occurs in high-grade ankle trauma.
- High number of syndesmotic injuries are missed during the initial clinical examination.
- Proper history taking and physical examination can lead to a fast diagnosis.

Table 58.1 Physical examination sequence on ankle instability and syndesmotic injury

Test	Description	Positive finding	Conditions associated with positive finding
Anterior drawer test	Patient supine, and the patient's foot in slight plantar flexion, brace the anterior shin with one hand and translate the heel anteriorly with the other hand. Comparison is made to the contralateral side	Laxity and absence of a firm end point laxity	ATFL rupture Possible CFL injury
Talar tilt test	With the patient seated or supine and the tibia and fibula stabilized, the ipsilateral foot is inverted and everted. Comparison is made to the contralateral side	Laxity and/or pain Pain	ATFL rupture CFL injury
Squeeze test	With the patient seated or supine, compress the patient's lower leg about midway up the test calf	Pain in the area of the distal tibiofibular and interosseous ligaments with proximal calf compression	Syndesmotic injury
External rotation test (syndesmotic stress test)	Patient seated on the examination table and the knee flexed over the edge of the table, the proximal lower leg is stabilized while the foot is grasped (plantar surface/heel) and dorsiflexed. The foot is then externally rotated	Pain in the area of the distal tibiofibular and interosseous ligaments	Syndesmotic injury
Kleiger test (dorsiflexion external rotation test)	With the patient seated and the knee flexed approximately 90° and the ankle relaxed, use one hand to grasp and stabilize the leg from behind, making sure not to compress the fibula and tibia together. Use your other hand to fully dorsiflex the ankle and then externally rotate the foot	Pain medially and laterally	Tear of the deltoid ligament Syndesmotic injury Interosseous membrane may be involved

58.5 Conclusion

Adequate history taking, clinical examination, and appropriate imaging are necessary to identify ankle ligament injuries, including potential associated pathology. Five to seven days of a delayed physical examination increases the sensitivity. Micro-instability and rotatory instability are new concepts that aim at fine-tuning a better understanding related to these injuries. Clinical examination of ankle ligament injuries is essential to avoid the ankle from developing instability and osteoarthrosis due to delayed diagnosis and rehabilitation.

References

1. Cloke DJ, Ansell P, Avery P, Deehan D. Ankle injuries in football academies: a three-Centre prospective study. Br J Sports Med. 2011;45(9):702–8.
2. Giza E, Fuller C, Junge A, Dvorak J. Mechanisms of foot and ankle injuries in soccer. Am J Sports Med. 2003;31(4):550–4.
3. Cloke DJ, Spencer S, Hodson A, Deehan D. The epidemiology of ankle injuries occurring in English football association academies. Br J Sports Med. 2009;43(14):1119–25.
4. Anderson RL, Engebretsen L, Kennedy N, LaPrade R, Wegner AM, Giza E. Epidemiology and mechanisms of ankle pathology in football. In: The ankle in football. Paris: Springer; 2014. p. 31–59.
5. Stubbe JH, van Beijsterveldt AM, van der Knaap S, Stege J, Verhagen EA, van Mechelen W, Backx FJ. Injuries in professional male soccer players in The Netherlands: a prospective cohort study. J Athl Train. 2015;50(2):211–6.
6. Ekstrand J, Hägglund M, Waldén M. Injury incidence and injury patterns in professional football: the UEFA injury study. Br J Sports Med. 2011;45(7):553–8.
7. Jain N, Murray D, Kemp S, Calder J. Frequency and trends in foot and ankle injuries within an English premier league football Club using a new impact factor of injury to identify a focus for injury prevention. Foot Ankle Surg. 2014;20(4):237–40.
8. Waldén M, Hägglund M, Ekstrand J. Time-trends and circumstances surrounding ankle injuries in men's professional football: an 11-year follow-up of the

UEFA champions league injury study. Br J Sports Med. 2013;47(12):748–53.
9. Woods C, Hawkins R, Hulse M, Hodson A. The football association medical research programme: an audit of injuries in professional football: an analysis of ankle sprains. Br J Sports Med. 2003;37(3):233–2388.
10. Waterman BR, Belmont PJ Jr, Cameron KL, De Berardino TM, Owens BD. Epidemiology of ankle sprain at the United States military academy. Am J Sports Med. 2010;38:797–803.
11. Al-Mohrej O, Al-Kenani NS. Acute ankle sprain: conservative or surgical approach? EFORT Open Rev. 2017;1:34–44.
12. Berson BL, Rolnick AM, Ramos CG, Thornton J. An epidemiologic study of squash injuries. Am J Sports Med. 1981;9:103–6.
13. Hertel J. Functional anatomy pathomechanics and pathophysiology of lateral instability. J Athl Train. 2002;37:364–75.
14. Hong C, Tan K. Concepts of ankle instability. OA Sports Med. 2014;2:3–11.
15. Blalock D, Miller A, Tilley M, Wang J. Joint instability and osteoarthritis. Clin Med Insights Arthritis Musculoskelet Disord. 2015;8:15–23.
16. Taylor DC, Englehardt DL, Bassett FH. Syndesmosis sprains of the ankle. The influence of heterotopic ossification. Am J Sports Med. 1992;20:146–50.
17. Roemer FW, Jomaah N, Niu J, Almusa E, Roger B, D'Hooghe P, Geertsema C, Tol JL, Khan K, Guermazi A. Ligamentous injuries and the risk of associated tissue damage in acute ankle sprains in athletes: a cross-sectional MRI study. Am J Sports Med. 2014;42(7):1549–57.
18. Williams GN, Jones MH, Amendola A. Syndesmotic ankle sprains in athletes. Am J Sports Med. 2007;35(7):1197–207.
19. Al-Mohrej O, Al-Kenani N. Chronic ankle instability: current perspectives. Avicenna J Med. 2016;6(4):103.
20. Ajis A, Maffulli N. Conservative management of chronic ankle instability. Foot Ankle Clin. 2006;11:531–7.
21. Rodriguez Merchan EC. Chronic ankle instability: diagnosis and treatment. Arch Orthop Trauma Surg. 2012;132(211):9.
22. Ortho Evidence. Approach to chronic ankle instability rehabilitation: does mobilization have a role? 2019. https://myorthoevidence.com/Blog/Show/28. This manuscript provides an approach to chronic ankle instability rehabilitation with a review of all the current literature
23. Giza E. Chronic footballer's ankle. In: Volpi P, editor. Football traumatology. Milan: Springer; 2006. p. 333–51.
24. Hamilton WG. Sprained ankles in ballet dancers. Foot Ankle. 1982;3(2):99–102.
25. Miklovic TM, Donovan L, Protzuk OA, Kang MS, Feger MA. Acute lateral ankle sprain to chronic ankle instability: a pathway of dysfunction. Phys Sportsmed. 2018;46:116–22.
26. Hintermann B, Boss A, Schäfer D. Arthroscopic findings in patients with chronic ankle instability. Am J Sports Med. 2002;30(3):402–9.
27. Hintermann B, Valderrabano V, Boss A, et al. Medial ankle instability an exploratory, prospective study of fifty-two cases. Am J Sports Med. 2004;32(1):183–90.
28. Valderrabano V, Hintermann B. Diagnostik und Therapie der medialen Sprunggelenkinstabilitat. Arthroskopie. 2005;18(2):112–8.
29. Taga I, Shino K, Inoue M, et al. Articular cartilage lesions in ankles with lateral ligament injury. An arthroscopic study. Am J Sports Med. 1993;21(1):120–6. [discussion: 126–7].
30. Calder JD, Bamford R, Petrie A, et al. Stable versus unstable grade II high ankle sprains: a prospective study predicting the need for surgical stabilization and time to return to sports. Arthroscopy. 2016;32(4):634–42.
31. Hunt KJ, Phisitkul P, Pirolo J, et al. High ankle sprains and syndesmotic injuries in athletes. J Am Acad Orthop Surg. 2015;23:661–73.
32. Taylor DC, Tenuta JJ, Uhorchak JM, et al. Aggressive surgical treatment and early return to sports in athletes with grade III syndesmosis sprains. Am J Sports Med. 2017;35:1833–8.
33. Van den Bekerom MP. Diagnosing syndesmotic instability in ankle fractures. World J Orthop. 2011;2:51–6.
34. Miller AN, Carroll EA, Parker RJ, Helfet DL, Lorich DG. Posterior malleolar stabilization of syndesmotic injuries is equivalent to screw fixation. Clin Orthop Relat Res. 2010;468(4):1129–35.
35. Van Dijk CN, Lim LS, Bossuyt PM, Marti RK. Physical examination is sufficient for the diagnosis of sprained ankles. J Bone Joint Surg (Br). 1996;78(6):958–62.
36. Van Dijk CN, Mol BW, Lim LS, Marti RK, Bossuyt PM. Diagnosis of ligament rupture of the ankle joint. Physical examination, arthrography, stress radiography and sonography compared in 160 patients after inversion trauma. Acta Orthop Scand. 1996;67(6):566–70.
37. van den Bekerom MP, Kerkhoffs GM, McCollum GA, Calder JD, van Dijk CN. Management of acute lateral ankle ligament injury in the athlete. Knee Surg Sports Traumatol Arthrosc. 2013;21(6):1390 5.
38. Verhage SM, Boot F, Schipper IB, Hoogendoorn JM. Open reduction and internal fixation of posterior malleolar fractures using the posterolateral approach. Bone Joint J. 2016;98(6):812–7.
39. Kent S, Yeo G, Marsland D, Randell M, Forster B, Lutz M, Okano S. Delayed stabilization of dynamically unstable syndesmotic injuries results in worse functional outcomes. Knee Surg Sports Traumatol Arthrosc. 2020;28(10):3347–53. https://doi.org/10.1007/s00167-020-05962-1).
40. Bachmann LM, Kolb E, Koller MT, Steurer J, ter Riet G. Accuracy of Ottawa ankle rules to exclude fractures of the ankle and mid-foot: systematic review. BMJ. 2003;326(7386):417.

41. Vuurberg G, Hoorntje A, Wink LM, Van Der Doelen BFW, Van Den Bekerom MP, Dekker R, et al. Diagnosis, treatment and prevention of ankle sprains: update of an evidence-based clinical guideline. Br J Sports Med. 2018;52(15):956.
42. Bachmann LM, Kolb E, Koller MT, et al. Accuracy of Ottawa ankle rules to exclude fractures of the ankle and mid-foot: systematic review. BMJ. 2003;326(7386):417.
43. Young CC, Niedfeldt MW, Morris GA, et al. Clinical examination of the foot and ankle. Prim Care. 2005;32(1):105–32.
44. Leddy JJ, Kesari A, Smolinski RJ. Implementation of the Ottawa ankle rule in a university sports medicine center. Med Sci Sports Exerc. 2002;34(1):57–62.
45. Alshalawi S, Galhoum AE, Alrashidi Y, Wiewiorski M, Herrera M, Barg A, et al. Medial ankle instability: the deltoid dilemma. Foot Ankle Clin. 2018;23(4):639–57.
46. Tourné Y, Molinier F, Andrieu M, Porta J, Barbier G. Diagnosis and treatment of tibiofibular syndesmosis lesions. Orthop Traumatol Surg Res. 2019;105(8):S275–86.
47. Gün C, Unlüer EE, Vandenberk N, Karagöz A, Sentürk GO, Oyar O. Bedside ultrasonography by emergency physicians for anterior talofibular ligament injury. J Emerg Trauma Shock. 2013;6(3):195–8.
48. Joshy S, Abdulkadir U, Chaganti S, Sullivan B, Hariharan K. Accuracy of MRI scan in the diagnosis of ligamentous and chondral pathology in the ankle. Foot Ankle Surg. 2010;16(2):78–80.
49. Oae K, Takao M, Uchio Y, Ochi M. Evaluation of anterior talofibular ligament injury with stress radiography, ultrasonography and MR imaging. Skelet Radiol. 2010;39(1):41–7.
50. Guerra-Pinto F, Corte-Real N, Vega J, Malagelada F, Guelfi M, Baduell A, Dalmau-Pastor M. Inestabilidad de tobillo: etiología, semiología y nuevos conceptos. Rev Esp Artrosc Cir Articul. 2020;27(1):52–60.
51. Chan KW, Ding BC, Mroczek KJ. Acute and chronic lateral ankle instability in the athlete. Bull NYU Hosp Jt Dis. 2011;69(17):26.
52. Nitz AJ, Dobner JJ, Kersey D. Nerve injury and grades II and III ankle sprains. Am J Sports Med. 1985;13(177):82.
53. Arnold BL, Linens SW, de la Motte SJ, Ross SE. Concentric evertor strength differences and functional ankle instability: a meta-analysis. J Athl Train. 2009;44:653–62.
54. Cho JH, Lee DH, Song HK, et al. Value of stress ultrasound for the diagnosis of chronic ankle instability compared to manual anterior drawer test, stress radiography, magnetic resonance imaging, and arthroscopy. Knee Surg Sports Traumatol Arthrosc. 2016;24:1022–8.
55. Gruskay JA, Brusalis CM, Heath MR, Fabricant PD. Pediatric and adolescent ankle instability: diagnosis and treatment options. Curr Opin Pediatr. 2019;31(1):69–78.
56. Meyers-Levy J, Zhu R, Jiang L. Context effects from bodily sensations: examining bodily sensations induced by flooring and the moderating role of product viewing distance. J Consum Res. 2010;37:1–14.
57. Li Q, Tu Y, Chen J, Shan J, Yung PSH, Ling SKK, et al. Reverse anterolateral drawer test is more sensitive and accurate for diagnosing chronic anterior talofibular ligament injury. Knee Surg Sports Traumatol Arthrosc. 2020;28(1):55–62.
58. Croy T, Koppenhaver S, Saliba S, Hertel J. Anterior talocrural joint laxity: diagnostic accuracy of the anterior drawer test of the ankle. J Orthop Sports Phys Ther. 2013;43:911–9.
59. Lin CW, Uegaki K, Coupé VM, et al. Economic evaluations of diagnostic tests, treatment and prevention for lateral ankle sprains: a systematic review. Br J Sports Med. 2013;47:1144–9.
60. Lin CY, Shau YW, Wang CL, et al. Quantitative evaluation of the viscoelastic properties of the ankle joint complex in patients suffering from ankle sprain by the anterior drawer test. Knee Surg Sports Traumatol Arthrosc. 2013;21:1396–403.
61. Lahde S, Putkonen M, Puranen J, Raatikainen T. Examination of the sprained ankle: anterior drawer test or arthrography? Eur J Radiol. 1988;8:255–7.
62. Phisitkul P, Chaichankul C, Sripongsai R, Prasitdamrong I, Tengtrakulcharoen P, Suarchawaratana S. Accuracy of anterolateral drawer test in lateral ankle instability: a cadaveric study. Foot Ankle Int. 2009;30:690–5.
63. Blanshard KS, Finaly DB, Scott DJ, Ley CC, Siggins D, Allen MJ. A radiological analysis of lateral ligament injuries of the ankle. Clin Radiol. 1986;37(3):247–51.
64. Lynch S. Assessment of the injured ankle in the athlete. J Athl Train. 2002;37(4):406–12.
65. Hintermann B, et al. Medial ankle instability. Foot Ankle Clin. 2003;8(4):723–38.
66. de César PC, Avila EM, de Abreu MR. Comparison of magnetic resonance imaging to physical examination for syndesmotic injury after lateral ankle sprain. Foot Ankle Int. 2011;32(12):111–4. (ISSN: 1071-1007).
67. https://www5.aaos.org/CustomTemplates/VideoGallery.aspx?id=4294972388&ssopc=1. Accessed 15 May 2020, 17h05.
68. https://www5.aaos.org/CustomTemplates/VideoGallery.aspx?id=4294972390&page=9&ssopc=1. Accessed 15 May 2020, 17h18].
69. Shore BJ, Kramer DE. Management of Syndesmotic ankle injuries in children and adolescents. J Pediatr Orthop. 2016;36(4):S11–4.
70. Khor YP, Tan KJ. The anatomic pattern of injuries in acute inversion ankle sprains: a magnetic resonance imaging study. Orthop J Sports Med. 2013;1(7):2325967113517078.
71. Andersen TE, Floerenes TW, Arnason A, Bahr R. Video analysis of the mechanisms for ankle injuries in football. Am J Sports Med. 2004;32(1 Suppl):69S–79S.

72. Krips R, de Vries J, van Dijk CN. Ankle instability. Foot Ankle Clin. 2006;11(2):311–29.
73. St Pierre RK, Rosen J, Whitesides TE, Szczukowski M, Fleming LL, Hutton WC. The tensile strength of the anterior talofibular ligament. Foot Ankle. 1983;4(2):83–5.
74. Hawkins RD, Hulse MA, Wilkinson C, Hodson A, Gibson M. The association football medical research programme: an audit of injuries in professional football. Br J Sports Med. 2001;35(1):43–7.

Evaluation of the Achilles Tendon

59

Niklas Nilsson, Annelie Brorsson, Katarina Nilsson Helander, Jón Karlsson, and Michael Carmont

59.1 Introduction and Epidemiology

The Achilles tendon is the largest tendon in the human body [1]. The tendon passes from the gastrocnemius, plantaris, and soleus muscle to the calcaneal bone. A functional and continuous Achilles tendon is essential for a normal gait pattern and the physical utility of the ankle and the lower leg. The Achilles tendon is situated subcutaneously in the posterior superficial compartment of the lower leg. Even though the Achilles tendon is renowned as the strongest tendon in the human body, it is often subject to acute traumatic injury and chronic overuse pathologies. These injuries and pathologies are most common among middle-aged athletes. The incidence of Achilles tendon rupture is approximately 23–47 per 100,000 person-years for males and 8–12 per 100,000 person-years for females [2, 3]. The correlative incidence for Achilles tendinopathy is more difficult to evaluate. However, the incidence has been reported to be 7–9% among active recreational runners and the prevalence is considered to be 2.35 per 1000 person-years [4, 5]. The incidence of these injuries and overuse pathologies has reportedly increased over the last decades [2, 6]. This may be a result of more people staying active in recreational sports at an older age. Although the Achilles tendon rupture and the Achilles tendinopathy are most known as injuries of the "middle-aged man," given the term the "weekend warrior," the injuries affect many professional athletes in sports such as long-distance running, football, track and field athletics, badminton, volleyball, racquetball, dancing, tennis, and basketball. An injured Achilles tendon can, and usually does, greatly impact the everyday life of a recreational athlete and an athlete's career in professional sports.

The Achilles tendon has a mean length of 15 cm (11–26 cm), the widest part averages 6.8 cm (4.5–8.6 cm) while the thinnest part averages 1.8 cm (1.2–2.6) [7]. The tendon is comprised of fascicles that rotate externally by 90°

N. Nilsson · K. N. Helander · J. Karlsson (✉)
Department of Orthopaedics, Institute of Clinical Sciences at Sahlgrenska Academy, University of Gothenburg, Gothenburg, Sweden

Department of Orthopaedic Surgery, Sahlgrenska University Hospital, Mölndal, Sweden
e-mail: jon.karlsson@telia.com

A. Brorsson
Department of Orthopaedics, Institute of Clinical Sciences at Sahlgrenska Academy, University of Gothenburg, Gothenburg, Sweden

IFK Kliniken Rehab, Gothenburg, Sweden

M. Carmont
Department of Orthopaedics, Institute of Clinical Sciences at Sahlgrenska Academy, University of Gothenburg, Gothenburg, Sweden

Department of Orthopaedic Surgery, Princess Royal Hospital, Shrewsbury and Telford Hospital NHS Trust, Shropshire, UK

J. G. Lane et al. (eds.), *The Art of the Musculoskeletal Physical Exam*,
https://doi.org/10.1007/978-3-031-24404-9_59

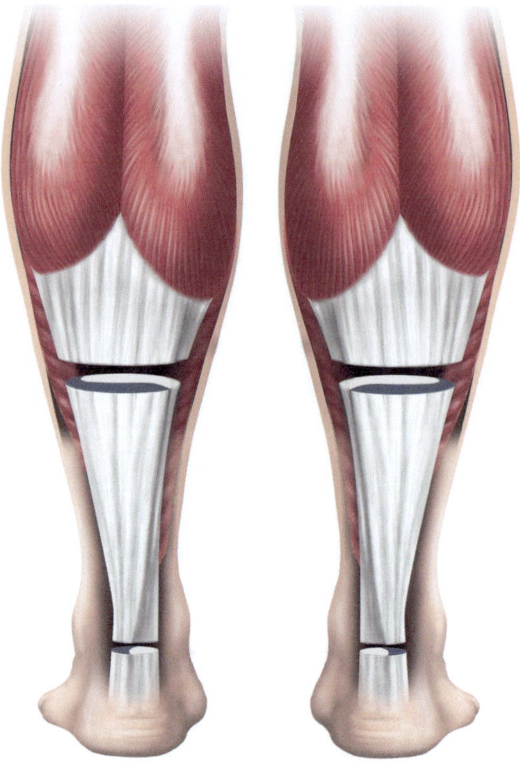

Fig. 59.1 **The anatomy of the Achilles tendon** depicting the 90° spiralling of the tendon from the gastrocnemius muscle to the calcaneal bone

from the muscle tendon junction to the insertion on the posterior aspect of the calcaneus bone (Fig. 59.1). The relative movement between these fascicles and sub-fascicles during loading, and unloading, means that the tendon is relatively strong compared to its thickness. During mobilization and exercise, the tendon twists and stretches up to 3%, during the gait cycle allowing energy to be stored and released. When the strain passes 4%, microscopic damage commences. When the strain exceeds 8%, macroscopic damage occurs and the tendon is at risk to deform and eventually rupture [8].

59.2 Trauma and Medical History

Acute injuries: The Achilles tendon rupture is a traumatic injury that in most cases occurs during sports or recreational activities [9]. The aetiology and trauma mechanism are multifactorial [10]. There is both a degenerative component and a traumatic component with increased tendon strains. The main traumas associated with Achilles tendon ruptures are concentric contraction of the gastrocnemius and soleus with the knee extended, eccentric contraction of the gastrocnemius and soleus with the ankle dorsiflexing and forced ankle dorsiflexion. Patients that are subject to Achilles tendon ruptures often describe the trauma as if they heard a loud "snap" followed by acute ankle pain. When a patient gives the history of a "pop" or "snap" to the Achilles tendon and ankle, they must be assumed to have a total Achilles tendon rupture, at least until the tendon is proven to be intact [11].

Achilles tendon ruptures are most common among "middle-aged men." Other intrinsic risk factors of Achilles tendon ruptures are hyperpronation of the foot, increased varus/valgus between forefoot and rearfoot, muscle weakness in the lower leg, being overweight, diabetes, chronic renal failure and being elderly [12]. The known extrinsic risk factors are excessive running, uneven surfaces and pharmaceuticals such as fluoroquinolone antibiotics, statins and corticosteroids [13–15]. However, the mechanism by which these variables predispose to rupture is still debated.

In addition to the details of the rupture, such as the date of the rupture, previous medical history should also be discussed. The main focus should be on periods of ankle pain, swelling and dysfunction relating to the tendon, previous ruptures, comorbidities and previous physical activity. The activity during which the injury occurs is usually a "new" activity for the patient, for example a charity or social football game for a recreationally competitive runner. Future aspirations also need to be considered. All of these are likely to affect management decisions. During the examination, it is important to consider the time that has elapsed since the point at which the rupture occurred. For example, patients with diabetes mellitus may present with bruising and dysfunction after a scorch from the Achilles tendon on top of a several months' history of intermittent swelling.

Overuse pathologies: Achilles tendon ruptures are acute injuries with a distinct traumatic episode. Overuse pathologies result after repetitive microtrauma over a longer period of time. As with acute Achilles tendon ruptures, the patient's history and clinical evaluation are sufficient to establish the diagnosis. The cardinal symptom is pain from the Achilles tendon and surrounding tissues during activities and rest [16]. A normal tendon has the ability to recover from these microtraumas. However, the pathological tendon is unable to respond acutely to the microtrauma becoming degenerate and altering the structure and thickness of the tendon. The most common histopathological sign of degenerative changes within the Achilles tendon is tendinosis with hypoxic, fibrinoid or calcific changes to the tendon [17]. Other common patterns are disorganization of the collagen fibres and separation of connective tissues. In some patients, even the surrounding paratenon is affected by the degeneration and shows signs of inflammation.

Achilles tendinopathy starts with an acute phase where the pain occurs mainly after heavy physical activities. As the disease progresses, symptoms occur during physical activities as well as after lighter activities in everyday life. This further progresses into a more chronic phase of injury where pain may occur even when resting. The time correlation between the phases is thought to vary between 1 and 6 months [18, 19]. In addition to pain, another common symptom is morning stiffness, which correlates with the severity of the pain experienced [20, 21].

59.3 Clinical Evaluation of the Achilles Tendon

Acute injuries: The classic clinical examination of the Achilles tendon consists of palpation of the tendon, the calf squeeze test (eponymously called either the Simmonds or the Thompson's test) and the resting position test (Matles' test). When observing or palpating the tendon, the examiner might be able to discern a defect at the rupture site 2–6 cm proximal to the calcaneal insertion (Fig. 59.2). More distal ruptures within

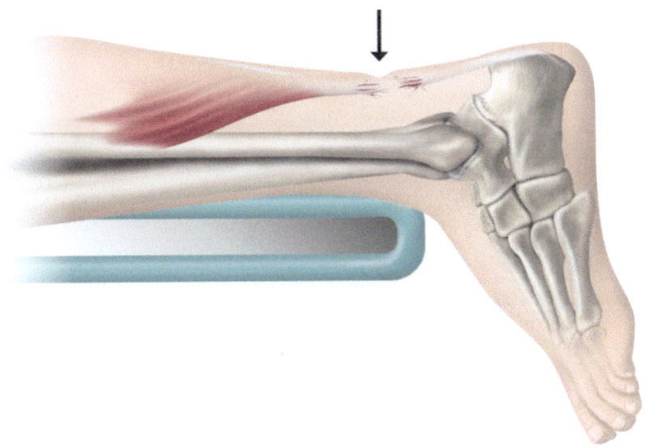

Fig. 59.2 **The tendon gap** that can be visualized and felt by palpation among patients with acute Achilles tendon ruptures

2 cm of the insertion occur much less commonly, and a small distal stump or even no remaining tendon may be palpable at the insertion. In midsubstance ruptures presenting late, the local swelling, haematoma and thickening of the paratenon may make the tendon defect harder to palpate. The ability to actively plantar flex the unloaded ankle may be preserved even though the tendon has been totally ruptured due to action of other preserved functional plantar flexors: tibialis posterior, flexor digitorum longus, flexor hallucis longus and peroneus longus and brevis. It is possible that the patient is able to mobilize on the injured side, but they will be unable to walk with a push-off or perform a single heel rise on the side with the rupture. Therefore, it is important to perform a full clinical examination of the Achilles tendon to avoid misdiagnosis. In recent studies, it has been shown that Achilles tendon ruptures have a delayed treatment, due to patient delay or misdiagnosis, in as many as 25% of cases [22, 23].

The most classic manoeuvre during the examination of the Achilles tendon is the "calf squeeze" test or Thompson's test. Thompson's test is performed with the patient in a prone position with both feet over the edge of the couch. The examiner then squeezes the calves of the patient at the level of the soleus muscle in turn. If the tendon is fully intact, the squeeze should induce a passive plantar flexion of the foot. If the plantar flexion is absent, or evidently weaker than

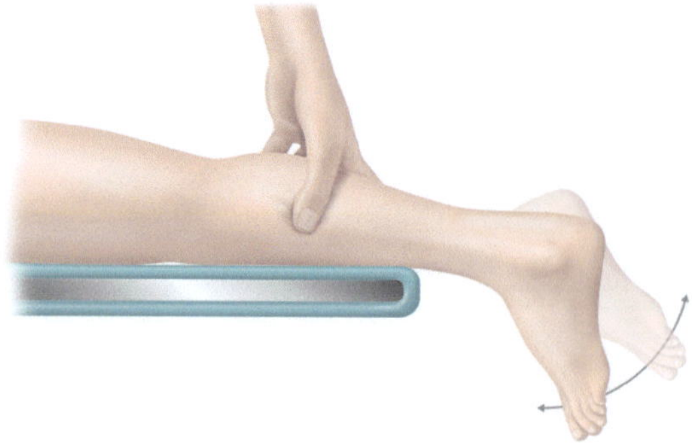

Fig. 59.3 **Calf squeeze test**: when a calf squeeze induces plantar flexion of the foot, as seen in the picture, the test is considered negative and the tendon is probably intact

Fig. 59.4 **Matles' test**: the test is considered positive when the injured side is more dorsiflexed than the uninjured side among patients with a suspected Achilles tendon rupture

the corresponding foot, it indicates that the tendon is injured (Fig. 59.3) [24]. Another test that is commonly performed is the "resting position test" or Matles' test. The patient starts in a prone position with both knees extended and with the feet over the edge of the stretcher. The patient is then asked to flex both knees 90° while the examiner observes the position of each foot. If one of the feet is more dorsiflexed than the other, it implicates that the tendon is discontinuous or ruptured (Fig. 59.4) [25]. For a more quantitative measurement of the tendon lengthening that occurs after an Achilles tendon rupture, the Achilles tendon resting angle (ATRA) might be used [26]. The measurement is executed with the patient in a prone position with the knee flexed to 90°. The angle is measured with a goniometer that is placed between the central fibula and the centre of the fifth metatarsal head (Fig. 59.5). The angle is presented in degrees.

Overuse injuries: While the clinical examination of a ruptured Achilles tendon mainly focuses on tendon continuation and length, the clinical examination of other Achilles tendinopathy focuses on tenderness and pain on palpation, oedema and erythema. The palpable pain is most likely located to the middle third of the tendon. On clinical examination, crepitation, focal thickening and tendon nodules are other signs for Achilles tendinopathy. Tendon thickening and nodules are most common in the more chronic phase of tendinopathy. The Achilles tendinopathy

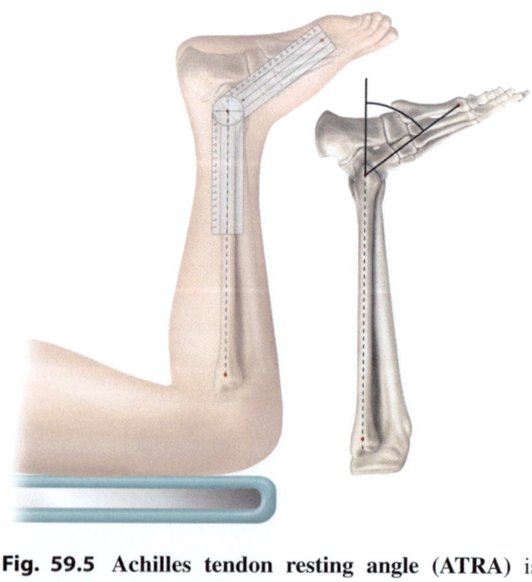

Fig. 59.5 **Achilles tendon resting angle (ATRA)** is measured as the angle between the long axis of the fibula and the fifth metatarsal head with the tip of the lateral malleolus as the centre. A larger angle indicates a longer or discontinuous tendon when compared to the uninjured foot

can, on clinical examination, be divided into a distal injury or a midportion injury depending on the location of the pain (Fig. 59.6). The midportion tendinopathy is most common (55–65%) and is located 2–6 cm from the calcaneal insertion [27]. The distal tendinopathy is located on the calcaneal insertion and is less common (20–25%) [28]. The distal tendinopathy is also called insertional Achilles tendinopathy and is commonly associated with retrocalcaneal bursitis and Haglund's deformity/Haglund's disease.

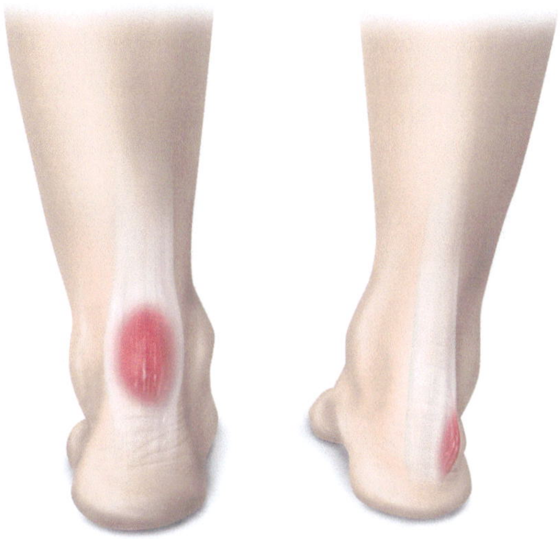

Fig. 59.6 Achilles tendinopathy with the sights of pain in midportion (left) and distal (right) tendinopathy

Aside from palpation of the tendon, the examining physician may practise some useful tests upon examining the Achilles tendon. The "painful arc sign" is a test where the examiner asks the patient to be placed in a prone position with legs straight and with both feet hanging over the edge of the examination bed. The patient is then asked to do a maximum plantar flexion followed by a maximum dorsiflexion of the foot. During these movements, the examiner inspects if the swelling moves along the tendon. The test is considered positive if the swelling is moving [29]. Another test is the Royal London Hospital test where the patient is placed in the same position. The examiner then locates the portion of the tendon with most palpable pain. After this, the patient is asked to do a maximum dorsiflexion of the foot. The examiner then palpates the same portion of the tendon as before, but in the dorsiflexed position. If the pain is decreased, the test is considered positive for Achilles tendinopathy (Fig. 59.7) [29]. The reason why the palpable pain of the Achilles tendon disappears upon dorsiflexion of the foot is still unknown. It is, however, specific for the diagnosis of Achilles tendinopathy.

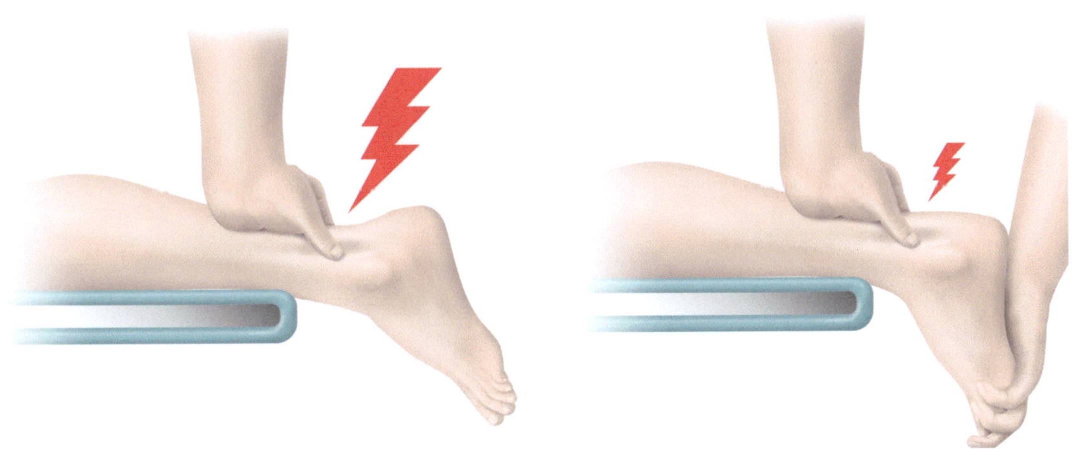

Fig. 59.7 Royal London Hospital test: if the palpable pain from the Achilles tendon decreases when the foot is dorsiflexed and if a possible intratendinous swelling moves with the tendon, the test is considered positive for Achilles tendinopathy

59.4 Imaging of the Achilles Tendon

Patient history and clinical examination are considered to be sufficient to establish the diagnosis of Achilles tendon ruptures and Achilles tendinopathy. This is largely due to the subcutaneous positioning of the tendon that makes it easy to inspect and palpate. When bone abnormalities (Haglund's deformity or lateral prominence) and fractures cannot be excluded, plain radiographs may be useful to specify the diagnosis (Figs. 59.8 and 59.9). Ultrasonography can be useful when the diagnosis of tendon rupture is uncertain. The imagining method is especially useful when the traumatic event and clinical examination do not match the classic symptomatology of an Achilles tendon rupture. Ultrasonography may also be of importance in mapping the tendon prior to surgical repair or in the selection of non-operative treatment (Fig. 59.10). This is operator dependent and has the advantage of allowing a dynamic assessment of the tendon. An alternative to ultrasonography is magnetic resonance imaging (MRI). The MRI is more sensitive to inflammation and other pathological conditions of the tendon than ultrasonography. Other advantages are the high resolution that may detect lesser anatomical abnormalities. The disadvantages of MRI are cost, availability and absence of dynamic examination.

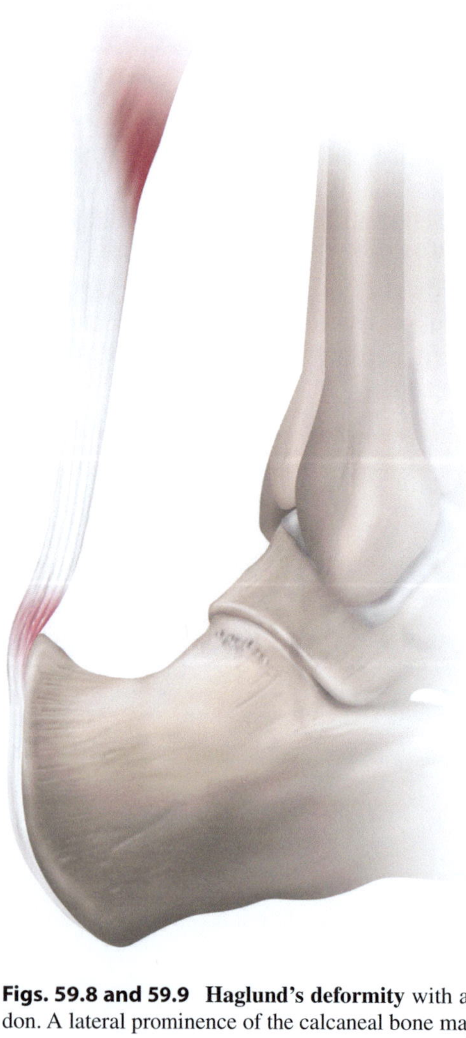

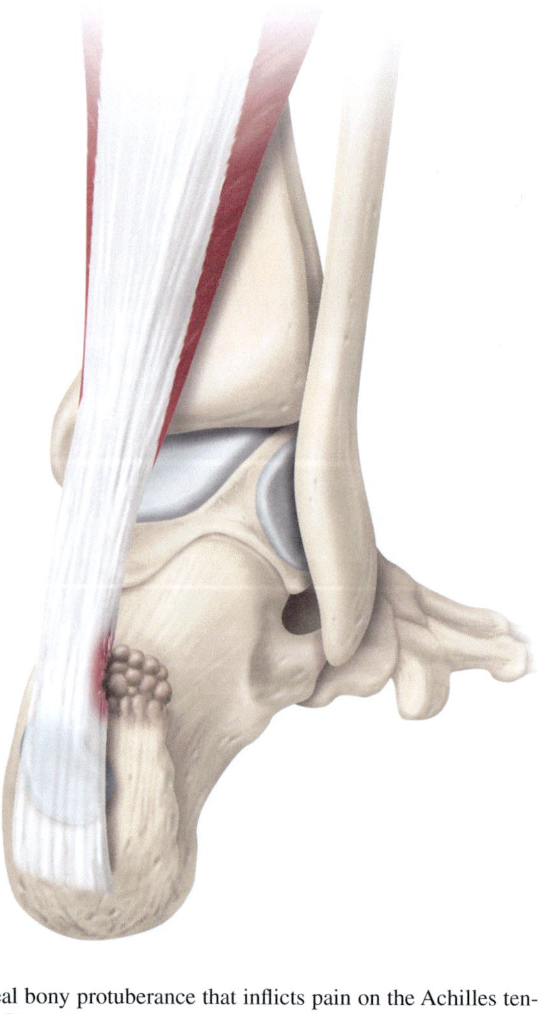

Figs. 59.8 and 59.9 Haglund's deformity with a calcaneal bony protuberance that inflicts pain on the Achilles tendon. A lateral prominence of the calcaneal bone may also inflict pain

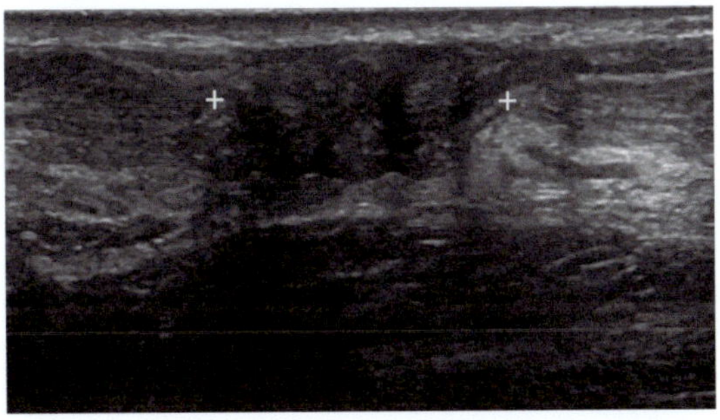

Fig. 59.10 **Ultrasonography** depicting the tendon gap after an acute Achilles tendon rupture

59.5 Post-treatment Evaluation

Regardless of whether the Achilles tendon rupture is treated operatively or non-operatively, it is essential that the patient follows a tailored rehabilitation programme at a physiotherapy department to optimize fully and to restore the pre-injury ankle function. To evaluate orthopaedic injuries and treatment, there are a wide variety of patient-reported outcome measures or PROMs. The Achilles tendon total rupture score (ATRS) is an Achilles tendon rupture-specific patient-reported outcome score developed by Nilsson-Helander et al. [30]. The instrument is well tested for validity and reliability. It is commonly used in Achilles tendon rupture follow-ups that evaluate the clinical treatment and rehabilitation of patients. The score involves ten inquiries with symptom scales ranging from 0 to 10, where 0 means the worst possible symptoms and 10 no affliction at all. The score 100 implicates a fully recovered patient with no limitation.

Regarding Achilles tendinopathy, outcome measures vary widely between institutions. The Victorian Institute of Sports Assessment therefore conducted an Achilles questionnaire named VISA-A [31]. This outcome measurement is a self-administered questionnaire that assesses symptoms of Achilles tendinopathy and their effect on physical activity. The questionnaire consists of eight questions regarding pain, function in daily living and sporting activity. The first seven questions have scales graded 1–10 and the eighth question a scale graded 0–30. The total score thereby ranges from 0 to 100. A total score of 0 out of 100 means lots of pain, and low activity, while 100 out of 100 implies no affliction at all. The outcome measurement is well tested for reliability and validity.

59.6 Summary

A careful insightful history and examination can identify and direct the majority of Achilles tendon pathologies. Imaging may be used to affirm or refute these findings but may have a role in directing individualized treatment of Achilles tendon rupture although the results of ongoing studies are awaited.

References

1. Doral MN, Alam M, Bozkurt M, Turhan E, Atay OA, Donmez G, et al. Functional anatomy of the Achilles tendon. Knee Surg Sports Traumatol Arthrosc. 2010;18(5):638–43.
2. Ganestam A, Kallemose T, Troelsen A, Barfod KW. Increasing incidence of acute Achilles tendon rupture and a noticeable decline in surgical treatment from 1994 to 2013. A nationwide registry study of 33,160 patients. Knee Surg Sports Traumatol Arthrosc. 2016;24(12):3730–7.
3. Huttunen TT, Kannus P, Rolf C, Fellander-Tsai L, Mattila VM. Acute achilles tendon ruptures: incidence of injury and surgery in Sweden between 2001 and 2012. Am J Sports Med. 2014;42(10):2419–23.
4. Nielsen RO, Rønnow L, Rasmussen S, Lind M. A prospective study on time to recovery in 254 injured novice runners. PLoS One. 2014;9(6):e99877.
5. Albers IS, Zwerver J, Diercks RL, Dekker JH, Van Den Akker-Scheek I. Incidence and prevalence of lower extremity tendinopathy in a Dutch general practice population: a cross sectional study. BMC Musculoskelet Disord. 2016;17(17):16.
6. Egger AC, Berkowitz MJ. Achilles tendon injuries. Curr Rev Musculoskelet Med. 2017;10(1):72–80.
7. Del Buono A, Chan O, Maffulli N. Achilles tendon: functional anatomy and novel emerging models of imaging classification. Int Orthop. 2013;37(4):715–21.
8. Maganaris CN, Narici MV, Maffulli N. Biomechanics of the Achilles tendon. Disabil Rehabil. 2008;30(20–22):1542–7.
9. Jozsa L, Kvist M, Balint BJ, Reffy A, Jarvinen M, Lehto M, et al. The role of recreational sport activity in Achilles tendon rupture. A clinical, pathoanatomical, and sociological study of 292 cases. Am J Sports Med. 1989;17(3):338–43.
10. Claessen FM, de Vos RJ, Reijman M, Meuffels DE. Predictors of primary Achilles tendon ruptures. Sports Med. 2014;44(9):1241–59.

11. Movin T, Ryberg A, McBride DJ, Maffulli N. Acute rupture of the Achilles tendon. Foot Ankle Clin. 2005;10(2):331–56.
12. Hess GW. Achilles tendon rupture: a review of etiology, population, anatomy, risk factors, and injury prevention. Foot Ankle Spec. 2010;3(1):29–32.
13. Seeger JD, West WA, Fife D, Noel GJ, Johnson LN, Walker AM. Achilles tendon rupture and its association with fluoroquinolone antibiotics and other potential risk factors in a managed care population. Pharmacoepidemiol Drug Saf. 2006;15(11): 784–92.
14. Spoendlin J, Layton JB, Mundkur M, Meier C, Jick SS, Meier CR. The risk of Achilles or biceps tendon rupture in new statin users: A propensity score-matched sequential cohort study. Drug Saf. 2016;39(12):1229–37.
15. van der Linden PD, Sturkenboom MC, Herings RM, Leufkens HM, Rowlands S, Stricker BH. Increased risk of Achilles tendon rupture with quinolone antibacterial use, especially in elderly patients taking oral corticosteroids. Arch Intern Med. 2003;163(15): 1801–7.
16. Alfredson H. Chronic midportion Achilles tendinopathy: an update on research and treatment. Clin Sports Med. 2003;22(4):727–41.
17. Józsa LKP. Human tendons. Anatomy, physiology and pathology. Champaign, IL: Human Kinetics; 1997.
18. el Hawary R, Stanish WD, Curwin SL. Rehabilitation of tendon injuries in sport. Sports Med. 1997;24(5):347–58.
19. Angermann P, Hovgaard D. Chronic Achilles tendinopathy in athletic individuals: results of nonsurgical treatment. Foot Ankle Int. 1999;20(5): 304–6.
20. Paavola M, Kannus P, Järvinen TA, Khan K, Józsa L, Järvinen M. Achilles tendinopathy. J Bone Joint Surg Am. 2002;84(11):2062–76.
21. Kader D, Saxena A, Movin T, Maffulli N. Achilles tendinopathy: some aspects of basic science and clinical management. Br J Sports Med. 2002;36(4):239–49.
22. Carden DG, Noble J, Chalmers J, Lunn P, Ellis J. Rupture of the calcaneal tendon. The early and late management. J Bone Joint Surg (Br). 1987;69(3):416–20.
23. Maffulli N, Via AG, Oliva F. Chronic Achilles tendon rupture. Open Orthop J. 2017;11:660–9.
24. Scott BW, al Chalabi A. How the Simmonds-Thompson test works. J Bone Joint Surg (Br). 1992;74(2):314–5.
25. Matles AL. Rupture of the tendo achilles: another diagnostic sign. Bull Hosp Joint Dis. 1975;36(1):48–51.
26. Carmont MR, Silbernagel KG, Mathy A, Mulji Y, Karlsson J, Maffulli N. Reliability of Achilles tendon resting angle and calf circumference measurement techniques. Foot Ankle Surg. 2013;19(4):245–9.
27. Järvinen TA, Kannus P, Maffulli N, Khan KM. Achilles tendon disorders: etiology and epidemiology. Foot Ankle Clin. 2005;10(2):255–66.
28. van Dijk CN, van Sterkenburg MN, Wiegerinck JI, Karlsson J, Maffulli N. Terminology for Achilles tendon related disorders. Knee Surg Sports Traumatol Arthrosc. 2011;19(5):835–41.
29. Maffulli N, Kenward MG, Testa V, Capasso G, Regine R, King JB. Clinical diagnosis of Achilles tendinopathy with tendinosis. Clin J Sport Med. 2003;13(1):11–5.
30. Nilsson-Helander K, Thomee R, Silbernagel KG, Thomee P, Faxen E, Eriksson BI, et al. The Achilles tendon Total rupture score (ATRS): development and validation. Am J Sports Med. 2007;35(3):421–6.
31. Robinson JM, Cook JL, Purdam C, Visentini PJ, Ross J, Maffulli N, et al. The VISA-A questionnaire: a valid and reliable index of the clinical severity of Achilles tendinopathy. Br J Sports Med. 2001;35(5):335.

Evaluation of Ankle Impingement 60

Nasef M. N. Abdelatif

60.1 Introduction

The term "impingement" was originally derived from the Latin verb *"impingere"* meaning "to force against"; thus, it describes the entrapment of an anatomic structure causing pain, bony and soft-tissue injury, decreased range of motion, and dysfunction [1–3]. Ankle impingement syndromes include a wide spectrum with varying etiologies, anatomic features, and presentations. These syndromes have been generally defined by the particular anatomical area involved, and any quadrant of the ankle joint could be affected. Specific anterior, anterolateral, anteromedial, posterior, posteromedial, posterolateral, and syndesmotic impingements have been described [4, 5]. Furthermore, some authors have described occurrence of these pathologies simultaneously and in combinations [6–9]. However, for the sake of simplicity, these pathologies are generally grouped into anterior and posterior impingement syndromes [10].

The exact etiology behind ankle impingement syndromes is not well understood. Anteriorly, it was hypothesized that spurs are enthesophytes caused by traction to the anterior capsule during repetitive plantar flexion, but that theory was anatomically rejected [11–14]. Another hypothesis was that this pathology occurs due to repetitive microtrauma and repetitive impaction injury to the anterior chondral margin from hyper-dorsiflexion or direct impact with an external object such as a soccer ball [15, 16].

Posteriorly, impingement specifically related to the os trigonum (OT) might occur acutely with a fracture or might also be a result of chronic injury due to repetitive microtrauma, or mechanical irritation of the surrounding soft tissues [17], or degenerative joint diseases' post-traumatic or malunion causes [18].

Whatever the type of affection, bony or soft tissue, and wherever the anatomical affection might be within the ankle joint, usually the end result is joint swelling, stiffness, pain, and eventual dysfunction.

Generally, impingement syndromes are readily appreciated by thorough history taking and adequate clinical examination. Various radiological investigations might be called upon, and ultimately a local infiltrative diagnostic injection might be very useful to confirm the diagnosis, but in some cases, it might also help to alleviate the presenting symptoms, becoming an adequate therapeutic tool.

N. M. N. Abdelatif (✉)
Department of Foot and Ankle Reconstructive Surgery, DrNasef OrthoClinic, Private Practice, Cairo, Egypt

J. G. Lane et al. (eds.), *The Art of the Musculoskeletal Physical Exam*,
https://doi.org/10.1007/978-3-031-24404-9_60

Fact Box

1. Impingement syndromes can occur individually in any of the anterior or posterior quadrants of the ankle joint but can also occur simultaneously both anteriorly and posteriorly in the ankle.
2. Thorough knowledge of anatomy and pathological causative agents is necessary to guide proper treatment.
3. Adequate physical examination and utilization of proper investigative measures are essential for correct diagnosis.

60.2 Relevant Anatomy

An adequate understanding of the ankle anatomy might be vital in understanding the various impingement syndromes. The talus articulates inferiorly with the calcaneal bone, and the talar axis is in line with the first web space of the foot, while the calcaneal bone axis is in line with the fourth web space (Fig. 60.1). Consequently, during dorsiflexion, bony impingement will occur anteromedially between the neck of the talus and the anterior lip of the tibia. However, with plantar flexion, bony impingement might occur posterolaterally between the os calcis and the posterior lip of the tibia [11, 12]. Therefore, anteromedial and posterolateral impingement will usually be bony in nature, while anterolateral and posteromedial impingement will usually be a consequence of a soft-tissue pathology [12]. Impingement conditions are most commonly seen in the anterior ankle; however, they can also occur in all quadrants of the ankle joint, including anterior, lateral, posterior, and medial.

Anteriorly, a structure of anatomical importance is the Bassett's ligament. Bassett's ligament is the distal fascicle of the anterior tibiofibular ligament (AITFL), which becomes thickened and hypertrophied. It can be separated from the AITFL by fibrofatty septum and appears as a separate ligament [19, 20]. This can consequently lead to impingement on the lateral shoulder of the talus when the ankle is plantarflexed (Fig. 60.2).

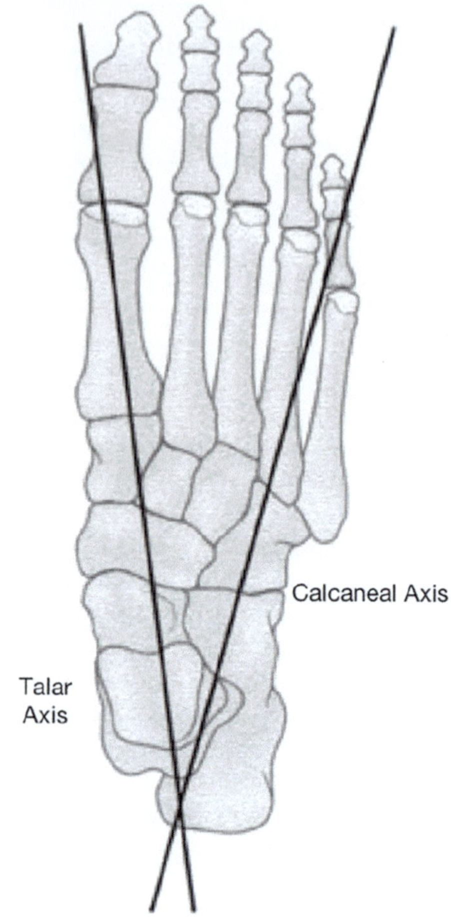

Fig. 60.1 Axis of talus and axis of os calcis

Similarly, knowledge of the posterior ankle anatomy is important in order to appropriately manage the impingement syndromes. The posterior aspect of the talus has two tubercles: the medial and the lateral tubercles (Fig. 60.3). The lateral tubercle is the origin of the posterior talofibular ligament (PTFL). An enlarged or elongated lateral talar tubercle is referred to as a posterior process of the talus or Stieda process.

In 7–11% of the population, this posterior process is separate from the talus and connected by a fibrous synostosis; in these cases, it is referred to as an os trigonum (OT). The OT is the second most common accessory bone in the foot, after the accessory navicular or os navicularum. When these anatomical variants occur in combination with a traumatic injury such as an ankle sprain or repetitive microtraumas, then collectively these can lead to posterior impingement syndrome [10, 17, 20].

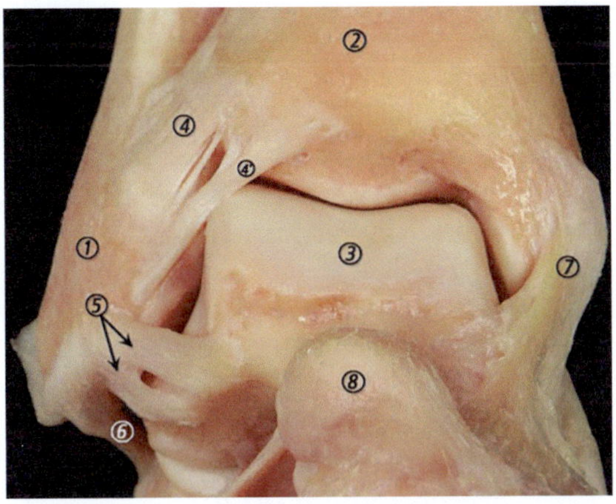

Fig. 60.2 Anatomic view of the anterior ligaments of the ankle. (*1*) Lateral malleolus, (*2*) tibia, (*3*) talus, (*4*) anterior tibiofibular ligament, (*4*′), distal fascicle of the anterior tibiofibular ligament, (*5*) anterior talofibular ligament, (*6*) calcaneofibular ligament, (*7*) superficial and deep layers of the MCL, (*8*) head of the talus. Reproduced with permission from [20]

In addition to bony pathologies, there are also important soft-tissue structures of the posterior ankle that can contribute to posterior ankle impingement such as the flexor hallucis longus (FHL) tendon, posterior ankle ligaments, joint capsule, and synovium. The FHL tendon passes in a groove on the posterior tibial plafond and through a fibro-osseous tunnel between the medial and lateral posterior tibial tubercles (Fig. 60.4). It runs from its proximal origin on the fibula to its insertion on the distal phalanx of the hallux. Tendinitis and synovitis within this tunnel can produce posterior medial pain. Generally, the most common sources of posterior ankle impingement pain are os trigonal impingement laterally and FHL tendinitis medially [20].

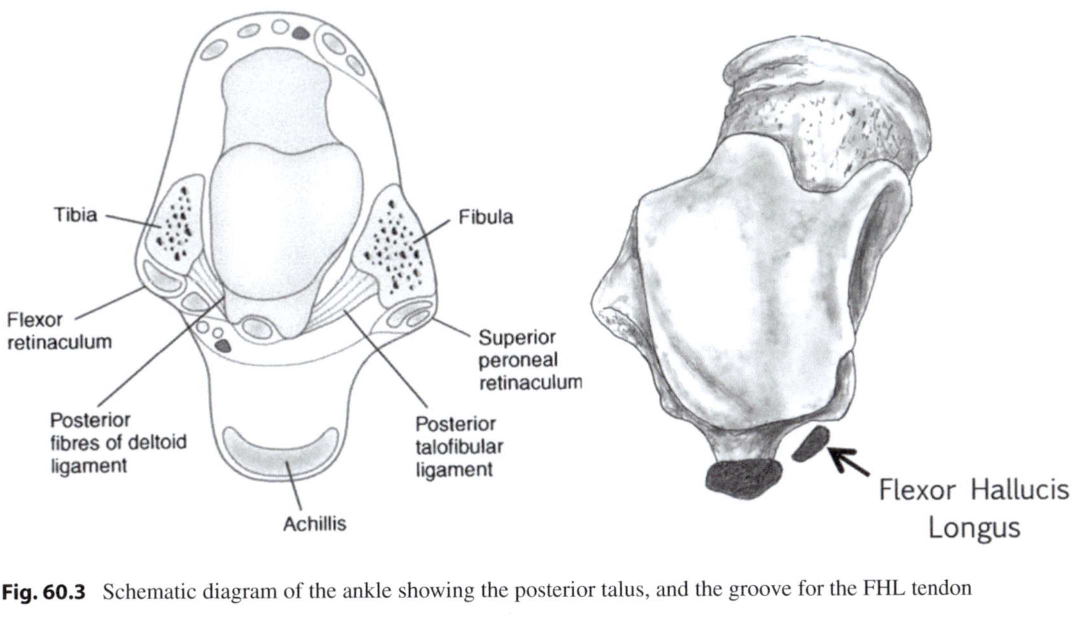

Fig. 60.3 Schematic diagram of the ankle showing the posterior talus, and the groove for the FHL tendon

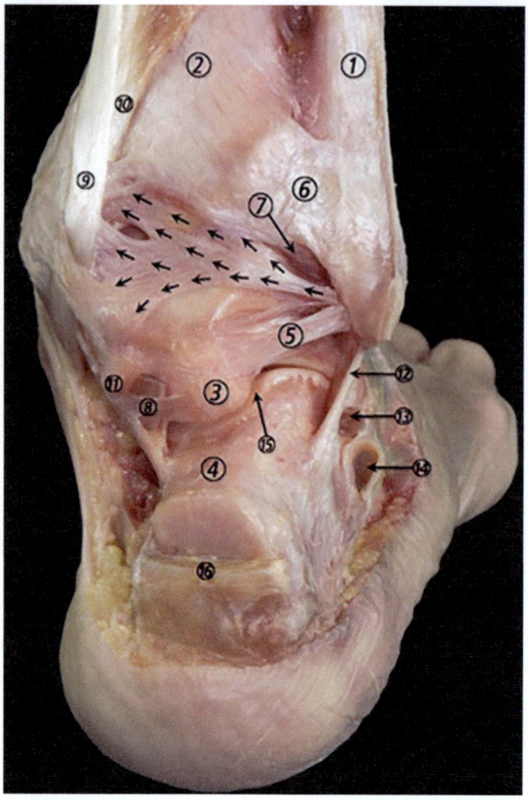

Fig. 60.4 Posterior view of anatomic dissection of the ankle ligaments. Posterior intermalleolar ligament and bundles indicated with arrows. (*1*) Fibula, (*2*) tibia, (*3*) lateral talar process, (*4*) dorsal surface of the calcaneus, (*5*) posterior talofibular ligament, (*6*) superficial component of the posterior tibiofibular ligament, (*7*) transverse ligament or deep component of the posterior tibiofibular ligament, (*8*) flexor hallucis longus tendon with its retinaculum, (*9*) flexor digitorum longus tendon, (*10*) tibialis posterior tendon, (*11*) medial talar process, (*12*) calcaneofibular ligament, (*13*) peroneus brevis tendon, (*14*) peroneus longus tendon, (*15*) subtalar joint, (*16*) calcaneal tendon (cut at the level of insertion). Reproduced with permission from [20]

60.3 Etiology

There might be a distinction between the causative agents in the different types of ankle impingement. Anteriorly, the impingement could be anteromedial ankle impingement (AMAI) or anterolateral ankle impingement (ALAI). In addition, syndesmotic and lateral impingement syndromes can also occur as separate entities.

60.3.1 Anteromedial Ankle Impingement (AMAI)

AMAI is usually bony and frequently results from recurrent acute ankle inversion injuries, which is usually combined with a rotational mechanism, leading to an abnormal soft-tissue growth in the joint recess [20]. During ankle inversion, the tibiotalar ligaments might become wedged between the medial malleolus and medial talar border, resulting in *"kissing"* contusions and ligament damage. Recurrence of this mechanism can stimulate bony spur growth in the anteromedial tibial plafond, anterior aspect of the medial malleolus, and talar neck [5, 21, 22].

60.3.2 Anterolateral Ankle Impingement (ALAI)

Anterolateral ankle impingement on the other hand is frequently of a soft-tissue nature. Synovitis or entrapment of either fibrotic tissues or ligaments at the anterolateral ankle joint is believed to be the cause of ALAI symptoms [23]. Inversion sprains result in microtrauma or microtears of the lateral ligament complex, and this is the primary cause of ALAI [24]. Most inversion ankle injuries will heal adequately with conservative treatment. However, an estimated 20% will progress to chronic ankle instability and repetitive trauma [25]. Excess anterior talar displacement that might occur with this chronic ankle instability might cause repeated contact of the talar dome with the distal fascicle of the anterior inferior tibiofibular ligament (AITFL). It has been estimated that a third of patients with chronic ankle instability will suffer from ankle impingement [26].

Hypertrophy of the inferior fascicle of the AITFL has been implicated as a cause of ALAI [20]. Several different terminologies were found to have been used to describe the same structure, "AITFL," "distal anterior tibiofibular ligament," "accessory anterior inferior tibiofibular ligament," "distal AITFL," and "Basset's ligament" [27, 28]. An anatomical study was performed that showed how the AITFL had a constant distal fascicle that touched the anterolateral corner of the talus in all examined ankles [28]. This contact was shown to increase with plantar flexion and should not be interpreted as pathological [28]. Similarly, the distal fascicle of the AITFL might be considered pathological when it impinges against the talar dome with the ankle in full dorsiflexion or when it is bent on top of the anterolateral margin of the talus [28]. Post-traumatic fibrous bands and synovial hypertrophy are other causes of soft-tissue impingement limiting ankle plantar flexion [29]. More rarely, congenital plica might also be another cause of soft-tissue impingement [30]. There are other causes of soft-tissue impingement such as tibiotalar ligament inflammation, capsular and synovial thickening, and syndesmotic synovial fringe impingement [31]. Anterolateral bony impingement is much less common, is usually traumatic in origin, and is mostly due to trauma. As previously stated, these might be enthesophytes caused by traction on the joint capsule or ligaments [12–14, 32].

60.3.3 Syndesmosis Impingement

Anterolateral ankle pain can also be caused by anterior syndesmosis pathology. This should not be regarded as a true impingement; it can be exacerbated by ankle dorsiflexion as the broader anterior aspect of the talus engages the malleoli and places tension on the anterior tibiofibular ligament. There are three types of anterolateral syndesmosis pathology: a sprain of the syndesmotic ligaments and interosseous membrane, also known as a high ankle sprain; the Tillaux fracture, which is an avulsion fracture of the insertion of the AITFL usually on the distal tibia; and the herniation of synovium into tears of the anterior tibiofibular ligament [33].

60.3.4 Lateral Ankle Impingement

Structures that form the lateral osteo-ligamentous complex of the ankle could be the source of impingement laterally to the ankle. Most of these will occur insidiously or following an insignificant or "simple" ankle inversion sprain. Some of these are the following:

Meniscoid lesion: Soft-tissue impingement between the lateral malleolus and lateral shoulder of talus that was originally described in soccer players [34].

Symptomatic os fibulare: A previously asymptomatic accessory ossicle that can loosen or fracture following an injury and become symptomatic.

Distal fibula avulsion fracture: Usually, an avulsion at the insertion site of the calcaneofibular ligament (CFL). Very rarely, a similar situation can occur at the insertion of the anterior talofibular ligament (ATFL).

Fracture of the lateral process of the talus: It is often misdiagnosed for an ankle sprain. It is also known as "snowboarder's fracture" and can be a source of impingement beneath the lateral malleolus [35].

Os calcis fractures: Malunited fractures of the os calcis can present with lateral ankle pain and sub-fibular impingement, which may be difficult to differentiate from subtalar joint pain and dysfunction [36].

Os calcis anterior process avulsion fracture: Occurring from the origin of the EDB and EHB and not considered a true impingement syndrome [37].

Accessory anterolateral talar facet: First described by Sewell in 1904, who found it to be present in 10.2% of cadaveric tali [38]. This can lead to talocalcaneal impingement and a symptomatic rigid flatfoot. It is more common in overweight adolescent boys who might present with lateral sided ankle and hindfoot pain and symptoms consistent with a rigid flatfoot deformity. In addition, dorsal talar beaking is also frequently present [39].

Peroneal tendon dysfunction: This can include peroneal tendon subluxation, longitudinal tears [40], and fractures of the os peroneum with retraction of the peroneal tendon [41]. Also, this is not considered true impingement but should be considered during the differential diagnosis.

60.3.5 Posterior Ankle Impingement (PAI)

Posterior ankle impingement results from compression of structures posterior to the tibiotalar and talocalcaneal articulations during last degrees of ankle plantar flexion. Like anterior ankle impingement, this can also be caused by multiple osseous and soft-tissue etiologies either in isolation or in combination. Likewise, it can also be anatomically divided into posterolateral ankle impingement and posteromedial ankle impingement [10].

60.3.6 Posterolateral Ankle Impingement (PLAI)

Perhaps bony impingement associated with the lateral (trigonal) process on the posterior talus is the most common cause of all posterior impingement pathologies and etiologies [10]. There are various anatomic variants of this structure: Stieda process refers to an elongated tubercle. An os trigonum may signify fusion failure of a secondary ossification center to the talar body. Impingement related to the trigonal process can result from acute fracture, chronic injury due to repetitive microtrauma, or mechanical irritation of the surrounding soft tissues [13].

60.3.7 Posteromedial Ankle Impingement (PMAI)

This is usually soft tissue in nature and more commonly related to tenosynovitis and tendinosis of the flexor hallucis longus (FHL) tendon, as it passes through its anatomical location between the medial and lateral posterior processes of the talus, to the level of the sustentaculum tali like a rope through a pulley. The tendinopathy may result from overuse or secondary to irritation from surrounding abnormal bony anatomy. As the FHL tendon passes through this pulley, it can be easily strained, and so it begins to bind, resulting in irritation and swelling. Repetitive chronic inflammation and hypertrophy of the musculotendinous unit within this tunnel can lead to a painful stenosing tenosynovitis, similar to de Quervain disease in the wrist. If a nodule or partial tear is present, triggering of the big toe may occur; this is known as hallux saltans. Sometimes in rare cases, the tendon may become completely encased in the sheath causing a pseudo-hallux rigidus.

Anatomic soft-tissue variants, such as the posterior intermalleolar ligament and anomalous muscles, have been described as other sources of impingement [17, 42–45]. Scarring and thickening of the posterior capsule, posterior inferior tibiofibular ligament, and posterior deltoid ligament might also occur due to repetitive or acute ankle hyperflexion causing posterior impingement also [46–49].

Nonspecifically and less commonly, posterior impingement symptoms may also result due to osteophyte impingement or reactive hypertrophic capsule and synovium as a consequence of ankle or subtalar degenerative joint diseases. It may also result as post-traumatic sequelae from fracture malunion of the posterior malleolus, talus, or calcaneus [18].

Fact Box

1. Anteromedial and posterolateral impingement syndromes are usually bony in nature.
2. Anterolateral and posteromedial impingement syndromes are usually of a soft-tissue origin.
3. Impingement syndromes might also occur in the syndesmosis, lateral gutter, or lateral ankle.
4. Anterior ankle impingement might more often occur following an inversion ankle sprain.
5. Posterior ankle impingement will more commonly occur due to repetitive ankle plantar flexion movements.

60.4 Clinical Presentation and Diagnosis

60.4.1 Anterior Impingement

Generally, anterior impingement syndrome will present as an anterior ankle pain during ankle range of motion especially at terminal dorsiflexion. Daily activities commonly as climbing stairs, running or walking up hills, ascending ladders, and deep squatting will usually exacerbate the condition. A classic association with competitive soccer players has long been recognized, but the exact reason for that association remains unclear [15, 16]. Later, this dorsiflexion may be limited secondary to mechanical block or pain, creating a cycle of progressive joint stiffness and loss of function. Anteriorly, the clinical picture may be differentiated into anteromedial and anterolateral impingements.

60.4.1.1 Anteromedial Impingement

Patients typically complain of anteromedial joint-line pain, which is aggravated by walking and running as well as planting and pivoting during sporting activities. In addition, medial clicking and limited, painful dorsiflexion may be the presenting symptom.

Upon examination, these patients often have anteromedial ankle swelling and pain over the anterior aspect of the deltoid ligament, and the spur can often be palpated.

60.4.1.2 Anterolateral Impingement

As discussed previously, the impingement in this location is usually soft tissue in origin, as the tibia and talus do not come in contact on the lateral side. Consequently, impingement in this location is usually a result of one of the two pathological conditions: Bassett's ligament and synovial impingement. Patients with both these pathologies present with almost similar complaints. There is usually an anterior ankle pain and a vague or remote history of a previous inversion ankle sprain. Physical exam findings include anterolateral talar dome and AITFL point tenderness exacerbated with ankle eversion and dorsiflexion.

On physical examination, one might encounter anterolateral talar dome and AITFL point tenderness exacerbated with ankle eversion and dorsiflexion [50]. The ankle is usually found to be stable on physical exam. Pain with palpation over the anterolateral gutter with the ankle plantarflexed typically reproduces the symptoms [50].

60.4.2 Posterior Impingement

Posterior impingement syndrome generally presents as a vague pain deep to the Achilles tendon and may often be confused with Achilles or peroneal tendon pathology. Symptoms may be worsened by activities involving plantar flexion and repetitive push-off maneuvers, such as descending stairs, high-heeled shoe wear, and downhill running and walking. Clinically, there seems to be a distinction between posterolateral and posteromedial impingement.

60.4.2.1 Posterolateral Impingement

Classically, this has been described in dancers, specifically those participating in classic ballet, presumably due to repetitive weight-bearing in the plantarflexed "en-pointe" and "demi-pointe" positions [51–55]. In a recent systematic review, 61% of patients undergoing surgery for posterior impingement were dancers [56]. Posterolateral impingement is thought to occur when the ankle is placed in full plantar flexion with weight-bearing, especially if an os trigonum (OT) is present. This position results in the fact that the OT (or posterior structures) is caught between the posterior lip of the tibia and the superior os calcis (Fig. 60.5).

On physical examination, patients will be found to show point tenderness posterior to the peroneal tendons behind the lateral malleolus, which might be exacerbated with forced passive plantar flexion of the ankle known as the "hyper-plantar flexion test" or "passive forced plantar flexion test" [56]. Due to the anatomical proximity of the peroneal tendons, one must look for and exclude peroneal tendinitis, which can also occur in association. Sometimes to confirm the diagno-

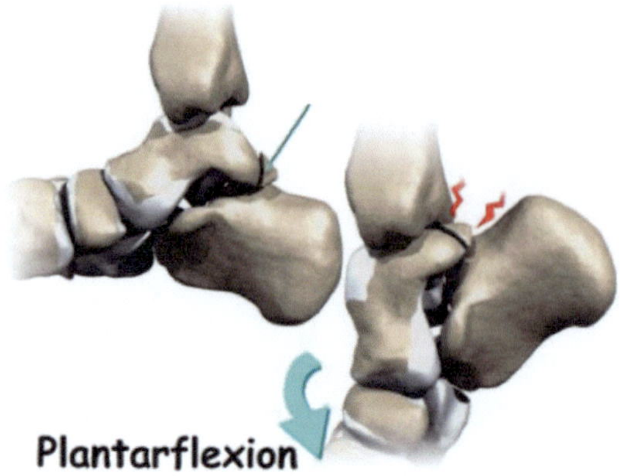

Fig. 60.5 Schematic illustration of posterior impingement with compression of os trigonum

sis, a 0.5 mL injection of a local anesthetic into the posterior soft tissues behind the peroneal tendons can be performed. One should always be wary that the presence of an os trigonum radiologically does not necessarily indicate a pathology.

60.4.2.2 Posteromedial Impingement

This is usually a soft-tissue impingement, and consequently the patient might complain of swelling and pain deep to the posteromedial aspect of the medial malleolus and deeper than the Achilles tendon. The pain is usually aggravated by plantar flexion ankle movement such as descending stairs and running or during playing sports as soccer. In some cases, patients might experience triggering of the big toe (hallux saltans) or more rarely a pseudo-hallux rigidus as previously described.

Physical examination findings include localized swelling and tenderness along the FHL sheath behind and lateral to the medial malleolus. Palpation of the sheath with active and passive motion of the hallux will simulate the patient's symptoms. Usually, there is no pain with forced plantar flexion of the ankle. FHL tendinitis can often be misdiagnosed as posterior tibial or Achilles tendonitis. In addition, one must be aware that fluid collection or effusion in the FHL tendon sheath might be a normal finding [57].

60.4.3 Physical Exam and Clinical Tests

A comprehensive physical examination of the foot and ankle should be performed when assessing for impingement syndromes. The ankle and foot are inspected for abnormal alignment, joint effusion, or soft-tissue edema.

In contrast to clinical tests for ankle instability and syndesmosis injuries, for example, the clinical tests described for various impingement disorders are yet to be validated or evaluated according to specificity and sensitivity.

Generally, these clinical tests could also be divided into those for the anterior impingement and tests for posterior impingement.

60.4.3.1 Anterior Impingement

The diagnosis of anterior impingement can be considered clinically based on physical examination.

Palpation of Different Anatomical Locations of Anterior Impingement

Recognizable local pain on palpation is present anteriorly, and the osteophytes may be palpable with the ankle joint in slight plantar flexion. A differentiation can be made between anteromedial and anterolateral impingement (Fig. 60.6).

Ankle Impingement Sign

Molloy and co-authors described an "ankle impingement sign" as a clinical test for synovial impingement in the anterolateral gutter [58]. During this test, the foot is initially held in a plantarflexed position and anterolateral pressure is applied with the thumb. This in itself may produce some pain, but will be neither sensitive nor specific. The foot is dorsiflexed after the thumb applying the pressure is removed. Thumb pressure is reapplied over the lateral gutter with the foot in plantar flexion, and with continued thumb pressure, the foot is moved from a plantarflexed position to a fully dorsiflexed position. This maneuver will produce pain or will intensify the pain already felt with pressure alone in the initial plantarflexed position (Fig. 60.7). The concept is that when hypertrophic synovium is

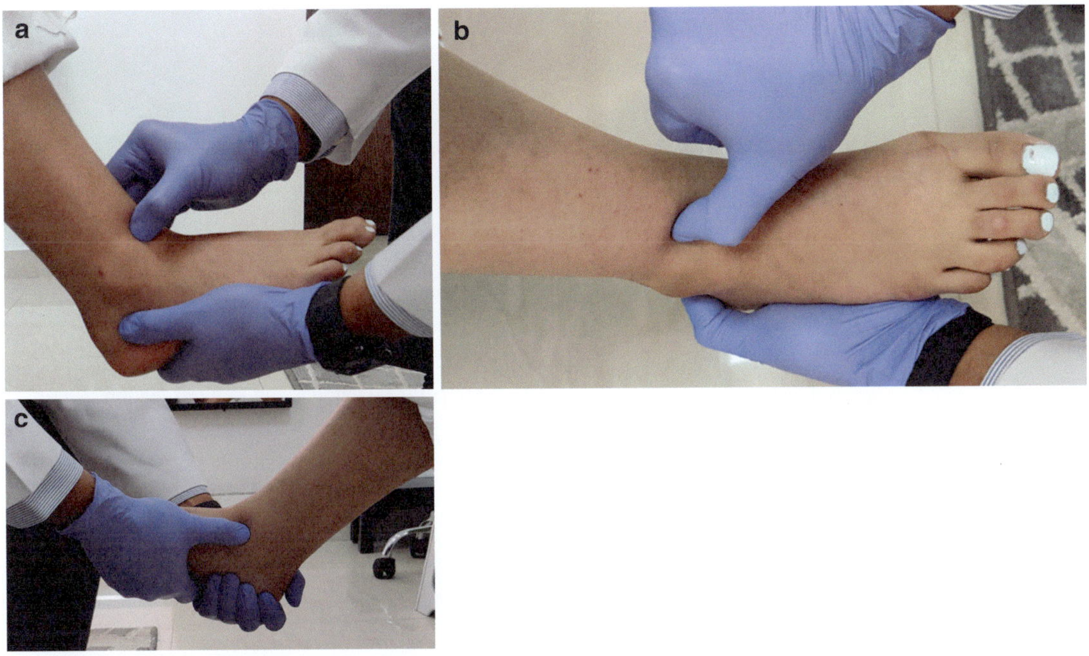

Fig. 60.6 Palpation and tenderness on the (**a**) anterolateral, (**b**) syndesmotic, and (**c**) anteromedial aspects of the ankle joint line

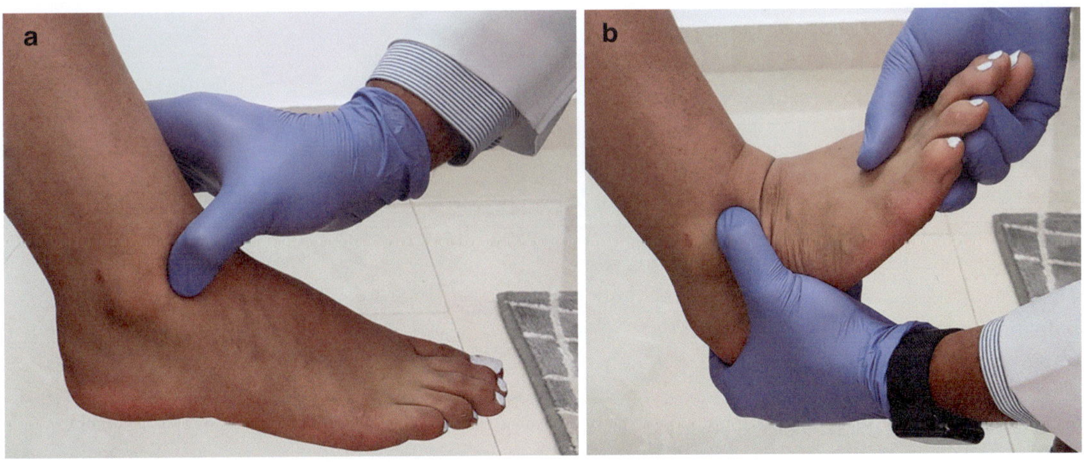

Fig. 60.7 Ankle impingement sign for synovial impingement of the anterolateral gutter [57]. (**a**) With knee flexed and ankle hanging in plantar flexion, tenderness is elicited by palpation of the anterolateral gutter. (**b**) Then palpation of the gutter is performed with passive dorsiflexion of the ankle joint

present, it will be forced into the joint by the examiner's thumb and will consequently be caught between the neck of the talus and the distal tibia while the ankle reaches full dorsiflexion. If the pain is not increased with ankle dorsiflexion, the sign is considered negative.

Single-Leg Squat Test

Examination consists of squatting only on the injured leg (Fig. 60.8); this also recreates a weight-bearing forced dorsiflexion of the affected ankle and might reproduce the patient's pain [59].

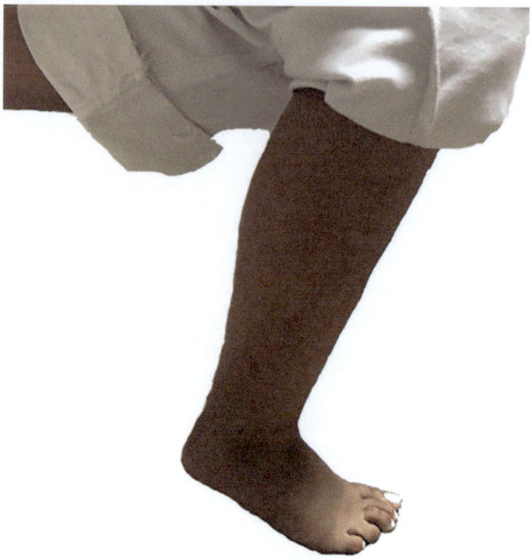

Fig. 60.8 Single-leg squat test

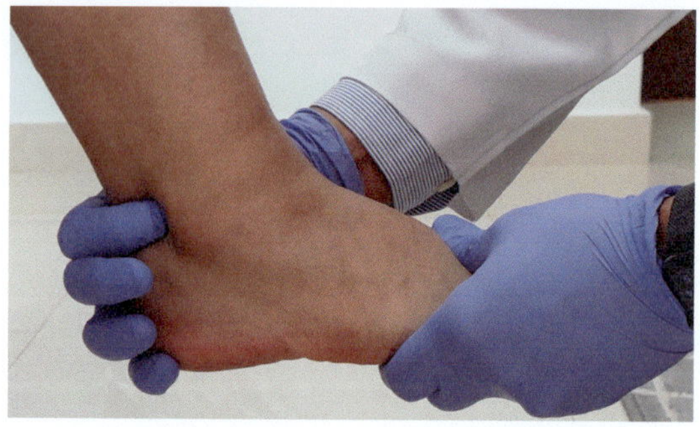

Fig. 60.9 Palpation of the posterolateral talar process between peroneal tendons and Achilles tendon

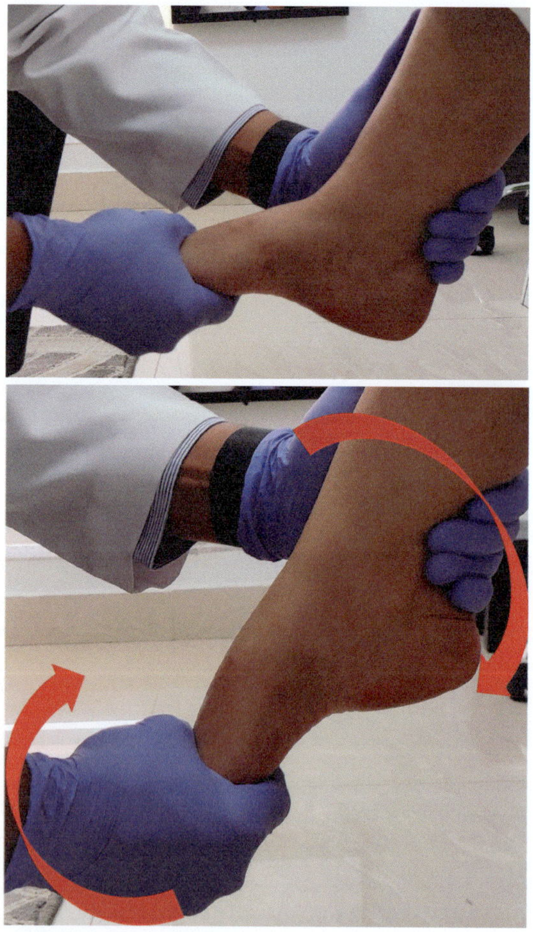

Fig. 60.10 Passive forced plantar flexion test [56]. With the knee at 90°, the ankle is taken repetitively from neutral to maximal hyperflexion. This can be done with slight degrees of foot internal or external rotation

60.4.3.2 Posterior Impingement

Palpation of Different Anatomical Locations of Posterior Impingement

On examination, there is pain on palpation of the posterior aspect of the talus. Posterolaterally, this may be of more significance when palpating the posterior talar process between the peroneal tendons and the Achilles tendon (Fig. 60.9). However, due to the presence of the neurovascular bundle and flexor tendons on the posteromedial aspect of the talus, posteromedial pain on palpation should not automatically imply impingement.

Passive Forced Plantar Flexion Test

For posterolateral impingement, a passive forced plantar flexion test has been described [56]. With the knee flexed to 90° and the foot and ankle in neutral, the affected ankle is supported by one of the examiner's hands at the heel, and the other hand holds the whole foot at the mid-foot or metatarsal heads. The examination is then performed with repetitive quick passive hyper-plantar flexion movements (Fig. 60.10). The test can be repeated in

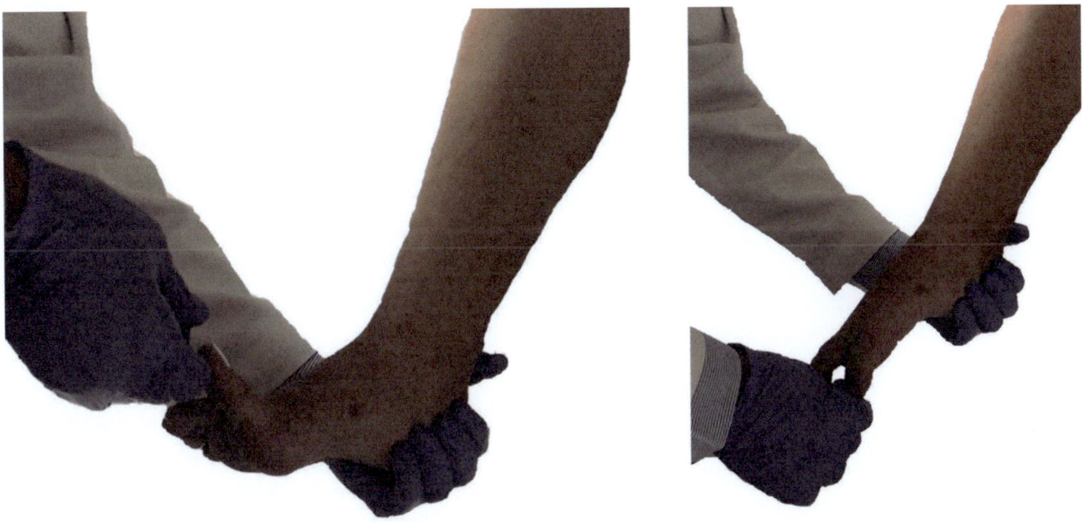

Fig. 60.11 Examination of FHL pain, tenderness, and possible crepitus with dorsiflexion and plantar flexion of big toe

slight external rotation or slight internal rotation of the foot on the tibia. A negative test might rule out a posterior impingement syndrome. On the other hand, a positive test, in combination with pain on posterolateral palpation, might best be followed up by a diagnostic infiltration of local anesthetic. If the presenting pain disappears after that injection, a diagnosis of posterolateral impingement can be confirmed.

Big Toe Dorsiflexion Plantar Flexion Motion Test

Also, palpation on the posteromedial aspect of the ankle with plantar flexion and dorsiflexion motion of the big toe while the knee is in 90° of flexion might be positive and elicit pain or crepitus [60]. This might be present especially in more severe cases of stenosing tenosynovitis (Fig. 60.11).

Resisted Dorsiflexion Test of Big Toe

Similarly, another test has been described for the assessment of posterior impingement by resisting dorsiflexion of the big toe [61]. Again, with the knee in 90° of flexion, and with palpation on the posteromedial aspect of the ankle, resisted dorsiflexion motion of the big toe is examined (Fig. 60.12).

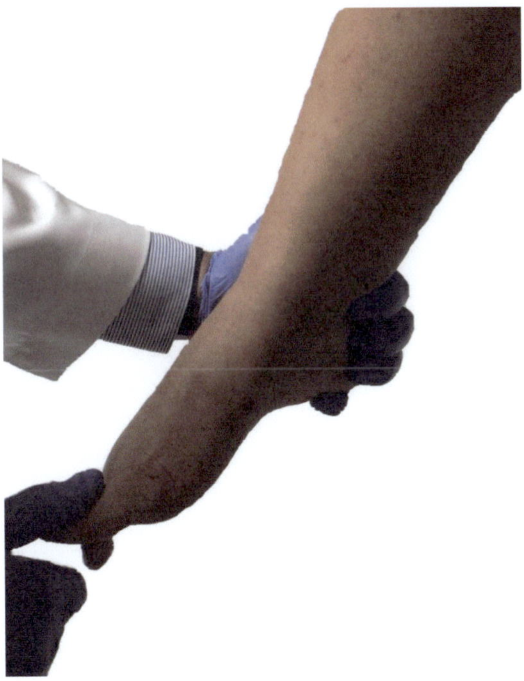

Fig. 60.12 Examination for resisted dorsiflexion test of the big toe with the knee flexed at 90°

Particularly posteriorly, it might be beneficial to differentiate between the two most common causes of posterior impingement, namely posterolateral bony impingement and posteromedial FHL impingement (Table 60.1).

Table 60.1 Differentiation between posterolateral bony impingement and posteromedial FHL tenosynovitis

	Posterior impingement	FHL tenosynovitis
Site	Posterolateral	Posteromedial
Location of tenderness	Tenderness behind lateral malleolus and fibula	Tenderness over FHL tendon course
Clinical symptom	Pain with plantarflexion of ankle	Pain and triggering with big toe movement
Clinical test	Passive forced plantar flexion test	Resisted dorsiflexion of big toe in plantar flexion of ankle
Main differential diagnosis	Peroneal tendinitis	Tibialis posterior tenosynovitis

Fact Box

1. Direct palpation of the anatomical ankle quadrants is an important aspect of physical examination for impingement syndromes.
2. Ankle impingement sign is useful to detect anterolateral synovial impingement.
3. Posterior passive forced hyperflexion test is used for detection of posterolateral impingement.
4. Examination for posteromedial impingement due to FHL might be aided by dorsiflexion and plantar flexion of big toe during ankle plantarflexion or by resisted dorsiflexion test.

60.5 Radiological Investigations

Imaging of an ankle suspected of impingement should begin with a standard plain X-ray series. Initial views should include weight-bearing AP, lateral, and mortise projections. Advanced imaging, such as MRI, may also be considered when the diagnosis remains inconclusive. MRI should be evaluated for soft-tissue lesions such as effusion, synovitis, and tenosynovitis, as well as bone edema and concomitant chondral injury. Computed tomography can be used in outlining the morphology and location of bony lesions for adequate planning of surgical resections [62]. Recently, ultrasound has also gained popularity as a reliable and inexpensive modality in evaluating impingement lesions and administering therapeutic injections [63, 64].

60.5.1 Anterior Impingement

In addition to the standard plain X-ray series, certain specific radiographs can be utilized. These standard views might show anterior osteophytes. In addition, the "plié view," which is a lateral weight-bearing view in maximal dorsiflexion, can be useful to demonstrate ***anterior impingement*** [33]. However, especially regarding anteromedial osteophytes, plain anteroposterior and lateral radiographs can often miss these osteophytes. Some studies have demonstrated that spurs over 7 mm in size can be missed in these locations due to radiographic projection issues and relative positions of the anterolateral tibia and talus. An oblique anteromedial impingement view (AMI) has been described to overcome this issue [15]. This radiograph is obtained by plantarflexing the foot and externally rotating the leg to 30° with the X-ray beam tilted 45° cranial.

With regard to ***synovial impingement***, the initial radiographs might be used solely for exclusion of other bony pathologies. Advanced imaging modalities such as CT scan and conventional magnetic resonance imaging (MRI) have shown moderate sensitivity and specificity but often depend on the interpreter's experience level [59, 65–68]. Magnetic resonance (MR) arthrogram, however, has been associated with a sensitivity of 96%, specificity of 100%, and accuracy of 100% in the assessment of anterolateral impingement when clinical signs are present [23].

In the radiographic workup for ***lateral impingement***, there are a few points worth mentioning. Plain radiographs might easily miss subtle fracture as fracture lateral process of talus and so CT would be considered the investigation of choice. In contrast, standard radiographs would pick up other pathologies as os fibulare, avulsion fractures of anterior process of os calcis, and malunions of calcaneal fractures. Likewise, for accessory anterolateral facets of the calcaneus, standard lateral radiographs will typically show a broadening of the anteroinferior lateral talar process, and perhaps dorsal talar beaking [38, 39]. In these patients, CT scans will both confirm the diagnosis and exclude tarsal coalitions. However, if performed, MRI will show bone marrow edema localized to the anterolateral talar facet and adjacent calcaneus [39].

60.5.2 Posterior Impingement

Standard plain X-ray series can show certain posterior pathologies particularly an os trigonum or an elongated Stieda process, which could be readily appreciable on standard lateral radiographs. A posterior impingement view was described in 2014 to better delineate the posterior talar processes and posterior talar facet [69]. This specialized view is obtained by placing the cassette directly against the medial side of the weight-bearing ankle and foot, and the beam is directed 25° in external rotation relative to the projection required for a direct conventional lateral weight-bearing view (Fig. 60.13). The authors reported an increase in sensitivity from 17% to 35% and an increased specificity from 3% to 15% [69].

Computed tomography images are not often utilized as much as they should be when examining cases for posterior impingement. These images are particularly useful in cases of subtle bony avulsions, bone surface irregularities, and other less obvious bony lesions [70].

On the other hand, MRI would be useful mainly in posterior impingement cases of soft-tissue origin. FHL tenosynovitis, low-lying bellies of FHL, or other anatomical aberrations of that musculotendinous structure, in addition to effusion, thickenings, or synovitis of the ankle or subtalar joint capsules and joints, would be

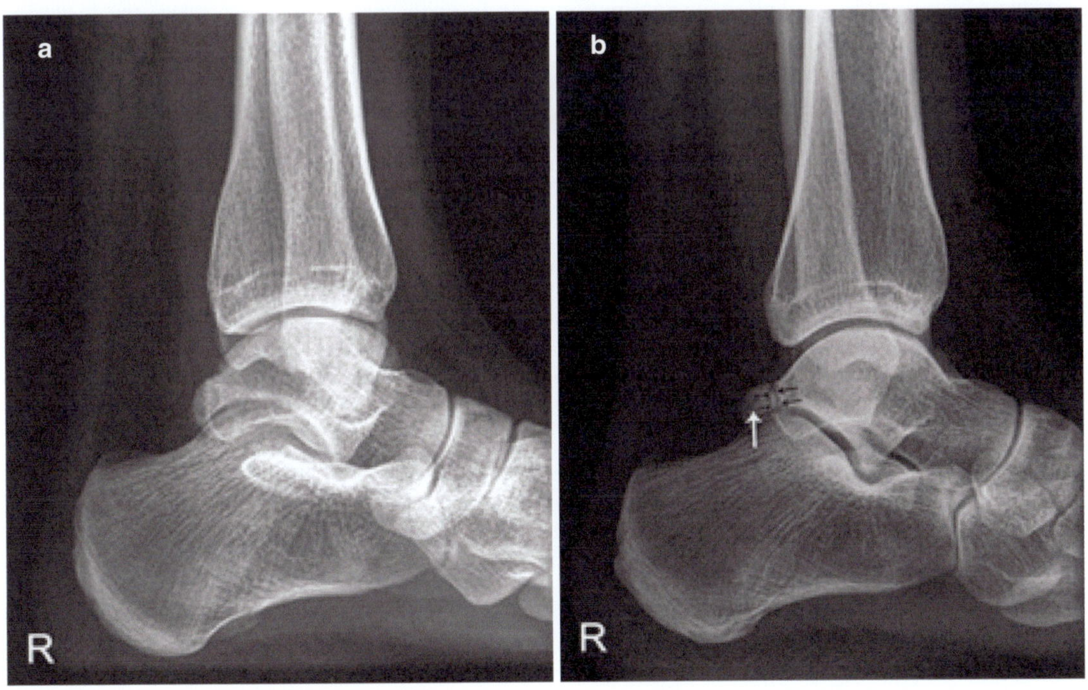

Fig. 60.13 Plain weight-bearing radiographs of the same patient. Image on the left shows the standard weight-bearing lateral radiograph, while image on the right shows the posterior impingement view, which elicits the posterior talar process more apparently (White arrow). Small black arrows show the synchondrosis between the talus and the posterior talar process [69]

clearly demarcated. Another aspect of value might be bone marrow edema of the posterior talar process or the os trigonum [71].

Fact Box

1. Standard ankle radiograph series will delineate most impingement syndromes, especially those that are bony in nature.
2. For anterior bony impingement, an anteromedial impingement view can be used.
3. Likewise, a posterior impingement view might be done for posterior impingement.
4. CT and MRI can be used according to the clinical condition suspected.

60.6 Conclusion

Adequate history taking and clinical examination are necessary to identify ankle impingement pathologies. These clinical examinations might be coupled with appropriate diagnostic modalities to ensure an accurate identification of the impingement syndrome.

References

1. Ogilvie-Harris DJ, Gilbart MK, Chorney K. Chronic pain following ankle sprains in athletes: the role of arthroscopic surgery. Arthroscopy. 1997;13(5):564–74.
2. Biedert R. Anterior ankle pain in sports medicine: aetiology and indications for arthroscopy. Arch Orthop Trauma Surg. 1991;110(6):293–7.
3. van den Bekerom MP, Raven EE. The distal fascicle of the anterior inferior tibiofibular ligament as a cause of tibiotalar impingement syndrome: a current concepts review. Knee Surg Sports Traumatol Arthrosc. 2007;15(4):465–71.
4. Sanders TG, Rathur SK. Impingement syndromes of the ankle. Magn Reson Imaging Clin N Am. 2008;16(1):29–38.
5. Dimmick S, Linklater J. Ankle impingement syndromes. Radiol Clin N Am. 2013;51(3):479–510.
6. Bojanic I, Bergovec M, Smoljanovic T. Combined anterior and posterior arthroscopic portals for loose body removal and synovectomy for synovial chondromatosis. Foot Ankle Int. 2009;30(11):1120–3.
7. Allegra F, Maffulli N. Double posteromedial portals for posterior ankle arthroscopy in supine position. Clin Orthop Relat Res. 2010;468(4):996–1001.
8. Kim HK, Jeon JY, Dong Q, Kim HN, Park YW. Ankle arthroscopy in a hanging position combined with hindfoot endoscopy for the treatment of concurrent anterior and posterior impingement syndrome of the ankle. J Foot Ankle Surg. 2013;52(6):704–9.
9. Abdelatif NMN. Combined arthroscopic management of concurrent posterior and anterior ankle pathologies. Knee Surg Sports Traumatol Arthrosc. 2014;22(11):2837–42.
10. Lavery KP, Mchale KJ, Rossy WH, Theodore G. Ankle Impingement. J Orthop Surg Res. 2016;11:97. https://doi.org/10.1186/s13018-016-0430-x.
11. O'Donoghue DH. Impingement exostoses of the talus and tibia. J Bone Joint Surg Am. 1957;39-A(4):835–52.
12. Tol JL, van Dijk CN. Etiology of the anterior ankle impingement syndrome: a descriptive anatomical study. Foot Ankle Int. 2004;25(6):382–6.
13. Hayeri MR, Trudell DJ, Resnick D. Anterior ankle impingement and Talar bony outgrowths: osteophyte or Enthesophyte? Paleopathologic and cadaveric study with imaging correlation. Am J Roentgenol. 2009;193(4):W334–8.
14. Berberian WS, Hecht PJ, Wapner KL, DiVerniero R. Morphology of tibiotalar osteophytes in anterior ankle impingement. Foot Ankle Int. 2001;22(4):313–7.
15. Tol JL, Slim E, van Soest AJ, van Dijk CN. The relationship of the kicking action in soccer and anterior ankle impingement syndrome. A biomechanical analysis. Am J Sports Med. 2002;30(1):45–50.
16. Massada JL. Ankle overuse injuries in soccer players. Morphological adaptation of the talus in the anterior impingement. J Sports Med Phys Fitness. 1991;31(3):447–51.
17. Mouhsine E, Crevoisier X, Leyvraz PF, Akiki A, Dutoit M, Garofalo R. Post-traumatic overload or acute syndrome of the os trigonum: a possible cause of posterior ankle impingement. Knee Surg Sports Traumatol Arthrosc. 2004;12(3):250–3.
18. Lui TH. Posterior ankle impingement syndrome caused by malunion of joint depressed type calcaneal fracture. Knee Surg Sports Traumatol Arthrosc. 2008;16(7):687–9.
19. Nikolopoulos CE, Tsirikos AI, Sourmelis S, Papachristou G. The accessory anteroinferior tibiofibular ligament as a cause of talar impingement: a cadaveric study. Am J Sports Med. 2004;32(2):389–95.
20. Golano P, Vega J, Perez-Carro L, Gotzens V. Ankle anatomy for the Arthroscopist. Part II: role of the ankle ligaments in soft tissue impingement. Foot Ankle Clin. 2006;11(2):275–96.
21. LiMarzi GM, Khan O, Shah Y, Yablon CM. Imaging manifestations of ankle impingement syndromes. Radiol Clin N Am. 2018;56(6):893–916.

22. Linklater J. MR imaging of ankle impingement lesions. Magn Reson Imaging Clin N Am. 2009;17(4):775–800.
23. Robinson P, White LM, Salonen D, Ogilvie-Harris D. Anteromedial impingement of the ankle: using MR arthrography to assess the anteromedial recess. AJR Am J Roentgenol. 2002;178(3):601–4.
24. Umans H, Cerezal L. Anterior Ankle Impingement Syndromes. Semin Musculoskelet Radiol. 2008;12(2):146–53.
25. Robinson P, White LM. Soft-tissue and osseous impingement syndromes of the ankle: role of imaging in diagnosis and management. Radiographics. 2002;22(6):1457–69.
26. de Vries JS, Krips R, Sierevelt IN, Blankevoort L, van Dijk CN. Interventions for treating chronic ankle instability. Cochrane Database Syst Rev. 2011;(8):CD004124.
27. Ross KA, Murawski CD, Smyth NA, Zwiers R, Wiegerinck JI, van Bergen CJA, Dijk CN, Kennedy JG. Current concepts review: arthroscopic treatment of anterior ankle impingement. Foot Ankle Surg. 2017;23(1):1–8.
28. van Dijk CN, van Bergen CJA. Advancements in ankle arthroscopy. J Am Acad Orthop Surg. 2008;16(11):635–46.
29. Dalmau-Pastor M, Malagelada F, Kerkhoffs GMMJ, Karlsson J, Manzanares MC, Vega J. The anterior tibiofibular ligament has a constant distal fascicle that contacts the anterolateral part of the talus. Knee Surg Sports Traumatol Arthrosc. 2020;28(1):48–54.
30. Berman Z, Tafur M, Ahmed SS, Huang BK, Chang EY. Ankle impingement syndromes: an imaging review. Br J Radiol. 2017;90(1070):20160735.
31. Rosenbaum AJ, Positano RG, Positano RCJ, Dines JS. Ankle impingement caused by an intra-articular plica: a report of 2 cases. Foot Ankle Spec. 2016;9(1):79–82.
32. Hermans JJ, Beumer A, de Jong TAW, Kleinrensink GJ. Anatomy of the distal tibiofibular syndesmosis in adults: a pictorial essay with a multimodality approach. J Anat. 2010;217(6):633–45.
33. Taylor M, Amendola N. Chapter 2: impingement syndromes of the ankle. In: Porter D, Schon L, editors. Baxter's the foot and ankle in sport. 3rd ed. Amsterdam: Elsevier; 2020. p. 8–21. ISBN: 9780323549424.
34. Mccarroll JR, Schrader JW, Shelbourne KD, Rettig AC, Bisesi MA. Meniscoid lesions of the ankle in soccer players. Am J Sports Med. 1987;15(3):255–7.
35. Valderrabano V, Perren T, Ryf C, Rillmann P, Hintermann B. Snowboarder's talus fracture: treatment outcome of 20 cases after 3.5 years. Am J Sports Med. 2005;33(6):871–80.
36. Hawkins LG. Fractures of the neck of the talus. J Bone Joint Surg Am. 1970;52(5):991–1002.
37. Harburn TE, Ross HE. Avulsion fracture of the anterior calcaneal process. Phys Sportsmed. 1987;15(4):73–80.
38. Sewell RB. A study of the astragalus. J Anat Physiol. 1904;38(Pt 3):233–47.
39. Martus JE, Femino JE, Caird MS, Kuhns LR, Craig CL, Farley FA. Accessory anterolateral talar facet as an etiology of painful talocalcaneal impingement in the rigid flatfoot: a new diagnosis. Iowa Orthop J. 2008;28:1–8.
40. Sammarco GJ, DiRaimondo CV. Chronic peroneus brevis tendon lesions. Foot Ankle. 1989;9(4):163–70.
41. Thompson FM, Patterson AH. Rupture of the peroneus longus tendon. Report of three cases. J Bone Joint Surg Am. 1989;71(2):293–5.
42. Fiorella D, Helms CA, Nunley JA 2nd. The MR imaging features of the posterior intermalleolar ligament in patients with posterior impingement syndrome of the ankle. Skelet Radiol. 1999;28(10):573–6.
43. Lohrer H, Arentz S. Posterior approach for arthroscopic treatment of posterolateral impingement syndrome of the ankle in a top-level field hockey player. Arthroscopy. 2004;20(4):e15–21.
44. Rosenberg ZS, Cheung YY, Beltran J, Sheskier S, Leong M, Jahss M. Posterior intermalleolar ligament of the ankle: normal anatomy and MR imaging features. AJR Am J Roentgenol. 1995;165(2):387–90.
45. Best A, Giza E, Linklater J, Sullivan M. Posterior impingement of the ankle caused by anomalous muscles. A report of four cases. J Bone Joint Surg Am. 2005;87(9):2075–9.
46. Seipel R, Linklater J, Pitsis G, Sullivan M. The peroneocalcaneus internus muscle: an unusual cause of posterior ankle impingement. Foot Ankle Int. 2005;26(10):890–3.
47. Koulouris G, Connell D, Schneider T, Edwards W. Posterior tibiotalar ligament injury resulting in posteromedial impingement. Foot Ankle Int. 2003;24(8):575–83.
48. Paterson RS, Brown JN. The posteromedial impingement lesion of the ankle. A series of six cases. Am J Sports Med. 2001;29(5):550–7.
49. Peace KA, Hillier JC, Hulme A, Healy JC. MRI features of posterior ankle impingement syndrome in ballet dancers: a review of 25 cases. Clin Radiol. 2004;59(11):1025–33.
50. Diniz P, Sousa D, Batista J, Abdelatif N, Pereira H. Diagnosis and treatment of anterior ankle impingement: state of the art. J ISAKOS. 2020;5:295–303. https://doi.org/10.1136/jisakos-2019-000282.
51. Moser BR. Posterior ankle impingement in the dancer. Curr Sports Med Rep. 2011;10(6):371–7.
52. Hamilton WG, Geppert MJ, Thompson FM. Pain in the posterior aspect of the ankle in dancers. Differential diagnosis and operative treatment. J Bone Joint Surg Am. 1996;78(10):1491–500.
53. Russell JA, Kruse DW, Koutedakis Y, McEwan IM, Wyon MA. Pathoanatomy of posterior ankle impingement in ballet dancers. Clin Anat. 2010;23(6):613–21.
54. Russell JA, Kruse DW, Koutedakis Y, Wyon MA. Pathoanatomy of anterior ankle impingement in dancers. J Dance Med Sci. 2012;16(3):101–8.

55. Ribbans WJ, Ribbans HA, Cruickshank JA, Wood EV. The management of posterior ankle impingement syndrome in sport: a review. Foot Ankle Surg. 2015;21(1):1–10.
56. van Dijk CN. Anterior and posterior ankle impingement. Foot Ankle Clin. 2006;11:663–83.
57. Na JB, Bergman AG, Oloff LM, Beaulieu CF. The flexor hallucis longus: tenographic technique and correlation of imaging findings with surgery in 39 ankles. Radiology. 2005;236(3):974–82.
58. Molloy S, Solan MC, Bendall SP. Synovial impingement in the ankle. A new physical sign. J Bone Joint Surg (Br). 2003;85(3):330–3.
59. Liu SH, Nuccion SL, Finerman G. Diagnosis of anterolateral ankle impingement. Comparison between magnetic resonance imaging and clinical examination. Am J Sports Med. 1997;25(3):389–93.
60. Scholten PE, Seirevelt IN, van Dijk CN. Hindfoot endoscopy for posterior ankle impingement. J Bone Joint Surg Am. 2008;90:2665–72.
61. Hedrick MR, McBryde AM. Posterior ankle impingement. Foot Ankle Int. 1994;15:2–8.
62. Takao M, Uchio Y, Naito K, Kono T, Oae K, Ochi M. Arthroscopic treatment for anterior impingement exostosis of the ankle: application of three-dimensional computed tomography. Foot Ankle Int. 2004;25(2):59–62.
63. McCarthy CL, Wilson DJ, Coltman TP. Anterolateral ankle impingement: findings and diagnostic accuracy with ultrasound imaging. Skelet Radiol. 2008;37(3):209–16.
64. Pesquer L, Guillo S, Meyer P, Hauer O. US in ankle impingement syndrome. J Ultrasound. 2014;17(2):89–97.
65. Hauger O, Moinard M, Lasalarie JC, Chauveaux D, Diard F. Anterolateral compartment of the ankle in the lateral impingement syndrome: appearance on CT arthrography. AJR Am J Roentgenol. 1999;173(3):685–90.
66. Farooki S, Yao L, Seeger LL. Anterolateral impingement of the ankle: effectiveness of MR imaging. Radiology. 1998;207(2):357–60.
67. Rubin DA, Tishkoff NW, Britton CA, Conti SF, Towers JD. An-terolateral soft-tissue impingement in the ankle: diagnosis using MR imaging. AJR Am J Roentgenol. 1997;169(3):829–35.
68. Duncan D, Mologne T, Hildebrand H, Stanley M, Schreckengaust R, Sitler D. The usefulness of magnetic resonance imaging in the diagnosis of anterolateral impingement of the ankle. J Foot Ankle Surg. 2006;45(5):304–7.
69. Wiegerinck JI, Vroemen JC, van Dongen TH, Sierevelt IN, Maas M, van Dijk CN. The posterior impingement view: an alternative conventional projection to detect bony posterior ankle impingement. Arthroscopy. 2014;30(10):1311–6.
70. Masciocchi C, Catalucci A, Barile A. Ankle impingement syndromes. Eur J Radiol. 1998;27(Suppl 1):S70–3.
71. Donovan A, Rosenberg ZS. MRI of ankle and lateral hindfoot impingement syndromes. AJR Am J Roentgenol. 2010;195:595–604.

Stress Syndromes Around the Ankle

61

Julie Amendola and Annunziato Amendola

61.1 Introduction

Stress fractures are described as fatigue injuries caused by repeated submaximal stress [1, 2]. Instead of one event delivering enough force to break the bone, a repetitive application of submaximal forces leads to bone insufficiency and eventual fracture [1, 2]. Stress fractures initially present with pain only during exercise or activity but can progress to pain at rest. The patient's history often reveals a recent increase in either training volume or intensity.

Stress fractures are more common in athletes compared to the general population because of the repetitive stress and high physical demands they often endure within their specific sport. With an increase in athletics participation in recent years, the recreational and collegiate athletes have become a closely studied group. It is estimated that the general incidence of stress fractures in athletes is 1%; however, in some sports such as cross-country and long-distance running, the incidence may be as high as 15% [3–5].

The tibia is the most common bone to suffer a stress fracture, comprising 25–46% of reported stress fracture, with the fibula comprising an additional 5–12% of injuries. Injuries involving the leg, therefore, comprise a third to half of all reported cases of stress fractures [1, 3–8].

In addition to bone-related stress syndromes, chronic tendon disorders make up a significant portion of all sport-related injuries. Studies report that chronic tendon problems represent 30% of all running-related injuries and have a prevalence of almost 40% in tennis players [9, 10]. Many chronic tendinopathies are a result of excessive mechanical loading. Around the ankle, the most commonly affected are the posterior tibial tendon, peroneal tendons, and Achilles tendon. In most cases of chronic tendon injuries, there is often a combination of intrinsic predisposing factors and superimposed extrinsic factors leading to injury.

61.2 Pathophysiology and Etiology

Tendon and soft tissue overload generally occurs from overuse (duration and intensity of activity) and can often be aggravated by other extrinsic factors such as poor technique, inappropriate shoe wear, and, depending on the sport, a lack of protective gear. It is important to identify these modifiable factors on the history and examination as they are essential to treatment. Anatomic characteristics of the athlete including malalign-

J. Amendola
East Carolina University Family Medicine, Greenville, NC, USA

A. Amendola (✉)
Orthopaedic Surgery, Sports Medicine at Duke University, Durham, NC, USA
e-mail: ned.amendola@duke.edu

J. G. Lane et al. (eds.), *The Art of the Musculoskeletal Physical Exam*, https://doi.org/10.1007/978-3-031-24404-9_61

ment, poor flexibility (i.e., ankle ROM), and muscle weakness or imbalance would be considered as intrinsic factors. In addition, age-related factors contributing to chronic tendinopathies include tendon degeneration, decreased healing response, increased tendon stiffness, and decreased vascularity. Systemic factors that also contribute include diabetes mellitus, obesity, smoking, and inflammatory enthesopathies [9, 11–13]. These features are important to gather on a complete physical examination.

Stress fractures and bone stress reactions are thought to occur by one of the two ways. In healthy metabolic bone, stress fractures occur from repetitive physiological overloading of the involved bone [1, 2]. The repetitive submaximal forces initially cause elastic deformation to the bone, which progresses to plastic deformity and eventually microfractures. Without sufficient time for the bone to undergo remodeling and repair before the next insult, the microdamage accumulates, predisposing the bone to complete fractures (Fig. 61.1). Although a single load is insufficient to cause an acute fracture, the constant and repetitive overloading without time for repair causes this chronic injury. This situation most likely occurs in athletes and military recruits [1, 2].

The other cause of stress fractures is seen with the female athlete triad or metabolic bone disease. In this case, you have normal forces and strain, but decreased bone remodeling due to a lack of energy availability, thereby predisposing the bone to a stress injury. Factors such as menstrual disturbances, caloric restriction, decreased bone density, muscle weakness, and leg-length differences are well-known risk factors for stress fractures [14, 15].

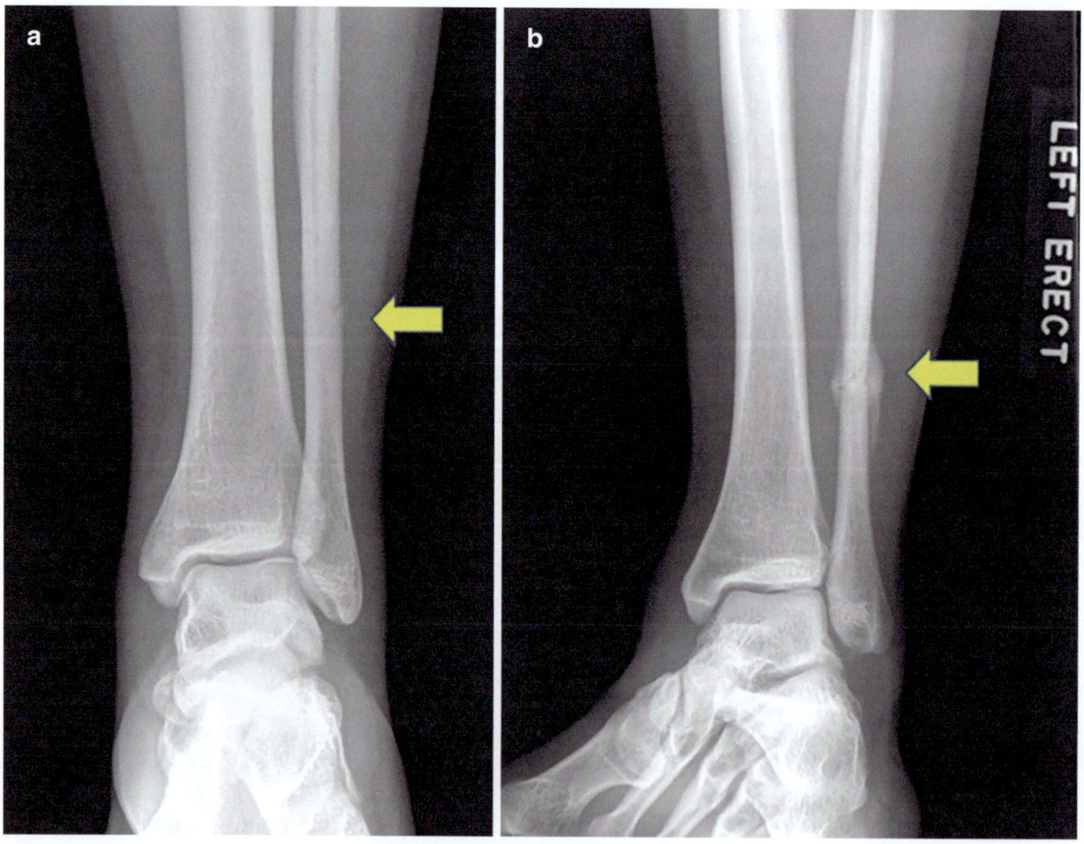

Fig. 61.1 (**a, b**) Initial fibular stress fracture in a 22-year-old female long-distance runner; (**b**) normal healing with reduced activity

In general, biomechanical alignment of the foot and ankle is a key section of the examination in these syndromes. Planovalgus alignment predisposes to Achilles and tibialis posterior tendon overload, medial tibial stress syndromes, fibular stress fractures, and talonavicular overload (Fig. 61.2). Cavovarus alignment predisposes to peroneal tendon tears, chronic lateral ankle instability, medial malleolar, tibial stress fractures, and navicular stress fractures (Fig. 61.3).

Range of motion of the ankle and subtalar joint is important. Decreased dorsiflexion of the ankle increases the risk of chronic ankle instability, as well as tendinopathy around the ankle. Silfverskiöld test is performed to examine the presence and severity of equinus [19]. Poor subtalar motion decreases the ability of the foot to dissipate shock during loading, which may predispose to overuse injuries and stress fractures.

61.3 Presentation and Physical Evaluation

As noted above, an organized and detailed history and exam are important to discern intrinsic and extrinsic factors that may be contributing to the overuse syndrome or stress reaction. A detailed history cannot be overemphasized in its importance.

An anatomically based physical examination will identify the structural location of the problem with localized tenderness, swelling, and dysfunction. This applies to the bony structures as well as the tendons around the ankle and joint-line tenderness [16].

Initially, a biomechanical exam is essential to identify any underlying malalignment, assessing the knee, leg, and foot and ankle for overall alignment, but mainly looking for a cavovarus or planovalgus hindfoot deformity (Figs. 61.2 and 61.3).

Patients should be examined standing and walking with both their lower extremities exposed and barefoot. Basic gait evaluation/progression in clinic is important to assess loading characteristics.

The normal hindfoot is aligned in neutral or a few degrees of varus. Assess both limbs for excess heel varus or valgus from behind. It may be difficult to discern subtle varus from the back, but the "peek-a-boo heel" sign is sensitive from the frontal view [17] (Fig. 61.4). The Coleman block test can be used to differentiate between forefoot- and hindfoot-driven varus or valgus [18].

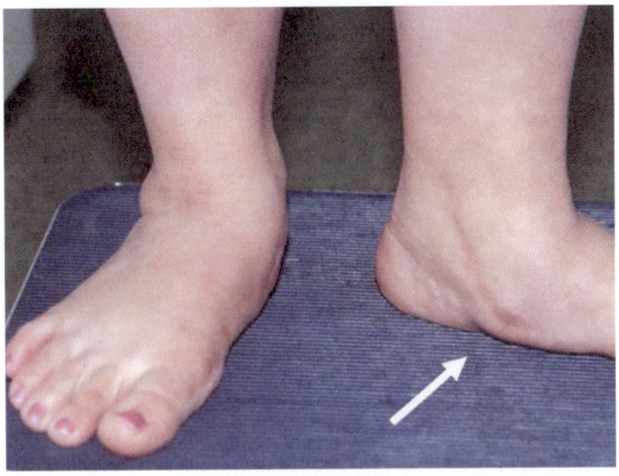

Fig. 61.2 Pes planus with medial swelling and tibialis posterior tendon inflammation

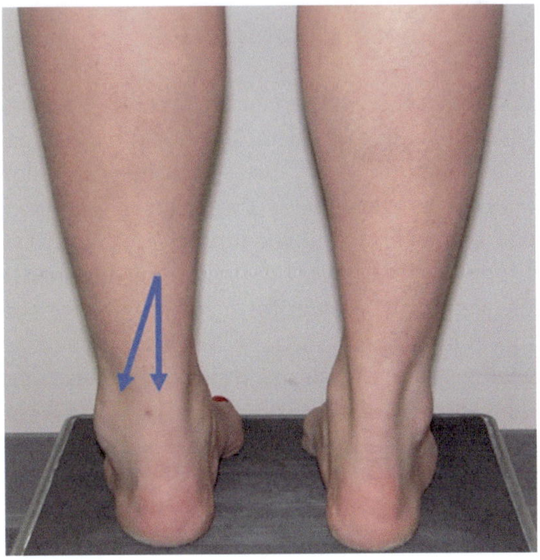

Fig. 61.3 Mild pes cavovarus with chronic lateral ankle instability

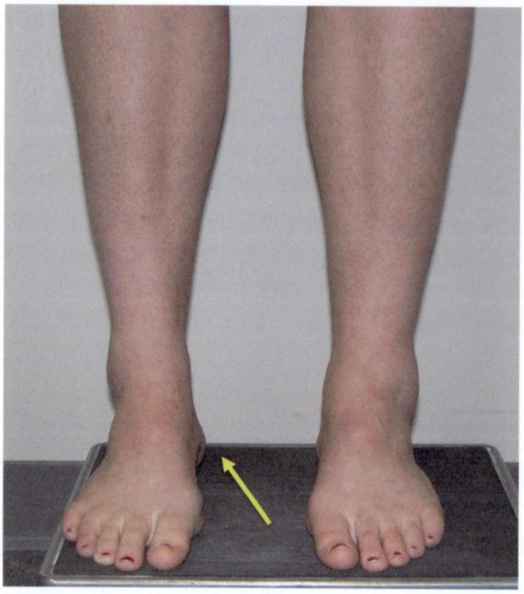

Fig. 61.4 Other exam features include assessment of arch height, which is typically evaluated with patient standing and comparing one side to the other

General examination, including a vascular exam and a full neurologic assessment, should be performed to discriminate between neuromuscular and idiopathic cavovarus deformity.

61.4 Specific Examination

Special tests are important for specific examination and diagnosis. Lateral ligamentous examination should be performed since chronic ankle laxity is associated with peroneal tendon injuries and other chronic stress injuries [20, 21]. The anterior drawer test is used to test the integrity of the anterior talofibular ligament. It is performed by holding the calcaneus on the posterior and the anterior of the tibia and trying to translate the talus anterior relative to the tibia. Increased laxity is seen compared to the contralateral side as a positive result. The lateral talar tilt test is used to test the integrity of the calcaneal fibular ligament. It is performed with one hand on the talus and calcaneus and the other hand on the tibia. Using a supination force on the hindfoot, a positive result is increased laxity compared to the normal side.

Tendinopathies around the ankle require anatomical inspection and palpation with and without resistance to assess integrity. Achilles tendinopathy generally presents as midsubstance 2–6 cm above the insertion or insertional as part of Haglund's syndrome [22]. Patients with posterior tibial tendinitis may complain of medial foot and ankle pain, and they are tender to palpation over the tendon around and distal to the medial malleolus [23]. Mechanical malalignment, excessive pronation, and muscular imbalances are other anatomic abnormalities that can be present and predispose patients to posterior tibial tendon overload. Peroneal tendon pathology is common and consists of tendinosis, tenosynovitis, tendon subluxation or dislocation, and commonly associated tears, all of which are clinically similar in presentation [24, 25]. Direct examination is done around the lateral malleolus for swelling, tenderness, and instability. The patient can often reproduce subluxation or snapping between the tendons, or dislocation of the peroneus brevis by actively plantarflexing, and everting against resistance (Fig. 61.5).

The physical examination of an athlete with shin splints should reveal tenderness to palpation over a wide region of the tibia and the tibialis muscle, whereas the pain from stress fractures tends to be localized to a specific area on the tibia.

Rare in the general population, navicular stress fractures are more common in elite-level athletes, accounting for 25% of all diagnosed stress fractures [12, 26] (Fig. 61.6). They are frequently observed in short-distance runners and basketball players due to high-demand jumping, explosive movement, and rapid cutting that leads to forceful and repetitive loading of the forefoot [12].

The navicular is an important structural link between the midfoot and hindfoot, allowing for forceful push-off during gait. Patients present with vague midfoot or medial arch pain that is aggravated by weight-bearing activities [26]. Initially, patients may present with pain just at the initiation of activities. They will have tenderness to palpation over the navicular bone, particularly the proximal dorsal area. They may have limited subtalar motion or dorsiflex-

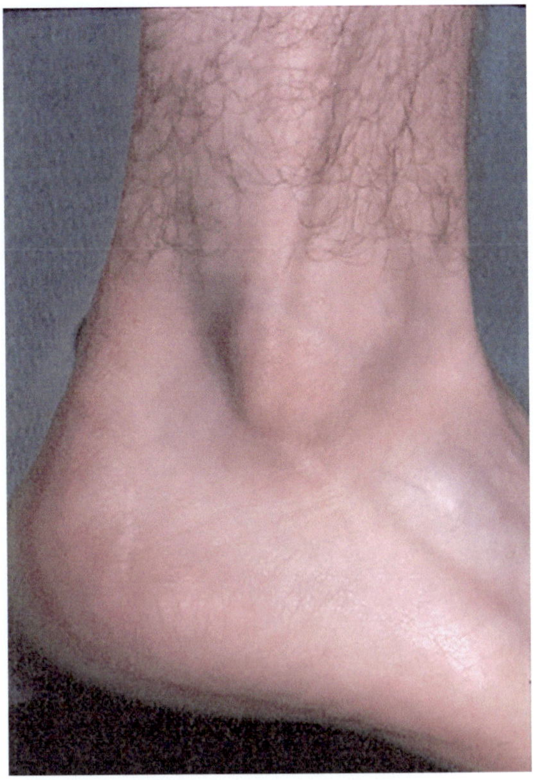

Fig. 61.5 Peroneal tendon instability with resisted eversion

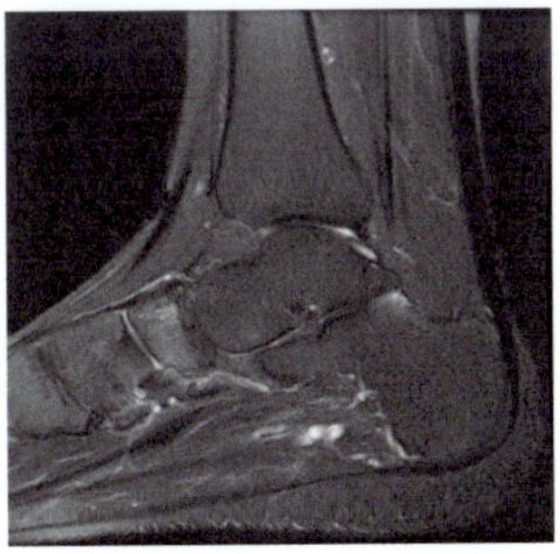

Fig. 61.6 Navicular stress reaction without fracture

ion of the ankle. They will often have pain standing on their tip toes and pain with a single-leg hop.

61.5 Summary

Stress syndromes in bone or tendons around the ankle are often multifactorial and the result of both intrinsic and extrinsic factors. Intrinsic factors include anatomic factors, age-related factors, and systemic factors, whereas extrinsic factors include mechanical overload and improper form and equipment. These comprise an important group of sports injuries seen in high-level and recreational athletes due to their physical demand. Diagnosis of these injuries can be difficult, but detailed clinical examination is key in order to avoid delay in treatment and for faster return to play.

References

1. Kaeding CC, Spindler KP, Amendola A. Management of troublesome stress fractures. Instr Course Lect. 2004;53:455–69. PMID: 15116634
2. Pepper M, Akuthota V, McCarty EC. The pathophysiology of stress fractures. Clin Sports Med. 2006;25(1):1–16. vii
3. Welck MJ, Hayes T, Pastides P, Khan W, Rudge B. Stress fractures of the foot and ankle. Injury. 2017;48(8):1722–6.
4. Fredericson M, Jennings F, Beaulieu C, Matheson GO. Stress fractures in athletes. Top Magn Reson Imaging. 2006;17(5):309–25.
5. Astur DC, Zanatta F, Arliani GG, Moraes ER, Pochini Ade C, Ejnisman B. Stress fractures: definition, diagnosis and treatment. Rev Bras Ortop. 2016;51(1):3–10.
6. Kaeding CC, Yu JR, Wright R, Amendola A, Spindler KP. Management and return to play of stress fractures. Clin J Sport Med. 2005;15(6):442–7. https://doi.org/10.1097/01.jsm.0000188207.62608.35. PMID: 16278549
7. Behrens SB, Deren ME, Matson A, Fadale PD, Monchik KO. Stress fractures of the pelvis and legs in athletes: a review. Sports Health. 2013;5(2):165–74.
8. Boden BP, Osbahr DC. High-risk stress fractures: evaluation and treatment. J Am Acad Orthop Surg. 2000;8(6):344–53.
9. Almekinders LC. Tendinitis and other chronic tendinopathies. J Am Acad Orthop Surg. 1998;6(3):157–64.
10. James SL, Bates BT, Osternig LR. Injuries to runners. Am J Sports Med. 1978;6(2):40–50.
11. Ackermann PW, Hart DA. General overview and summary of concepts regarding tendon disease topics addressed related to metabolic disorders. Adv Exp Med Biol. 2016;920:293–8.
12. Vopat B, Beaulieu-Jones BR, Waryasz G, et al. Epidemiology of navicular injury at the NFL combine and their impact on an Athlete's prospective NFL career. Orthop J Sports Med. 2017;5(8):2325967117723285.
13. McCarthy MM, Hannafin JA. The mature athlete: aging tendon and ligament. Sports Health. 2014;6(1):41–8.
14. Nattiv A, Armsey TD Jr. Stress injury to bone in the female athlete. Clin Sports Med. 1997;16(2):197–224.

15. Chen YT, Tenforde AS, Fredericson M. Update on stress fractures in female athletes: epidemiology, treatment, and prevention. Curr Rev Musculoskelet Med. 2013;6(2):173–81.
16. Korkola M, Amendola A. Exercise-induced leg pain: sifting through a broad differential. Phys Sportsmed. 2001;29(6):35–50. https://doi.org/10.3810/psm.2001.06.825. PMID: 20086577
17. Manoli A 2nd, Graham B. The subtle cavus foot, "the underpronator". Foot Ankle Int. 2005;26(3):256–63.
18. Coleman SS, Chesnut WJ. A simple test for hindfoot flexibility in the cavovarus foot. Clin Orthop Relat Res. 1977;123:60–2.
19. Singh D. Nils Silfverskiöld (1888-1957) and gastrocnemius contracture. Foot Ankle Surg. 2013;19(2):135–8.
20. Krause F, Seidel A. Malalignment and lateral ankle instability: causes of failure from the Varus tibia to the Cavovarus foot. Foot Ankle Clin. 2018;23(4):593–603.
21. Strauss JE, Forsberg JA, Lippert FG 3rd. Chronic lateral ankle instability and associated conditions: a rationale for treatment. Foot Ankle Int. 2007;28(10):1041–4.
22. de Jonge S, van den Berg C, de Vos RJ, et al. Incidence of midportion Achilles tendinopathy in the general population. Br J Sports Med. 2011;45(13):1026–8.
23. Gluck GS, Heckman DS, Parekh SG. Tendon disorders of the foot and ankle, part 3: the posterior tibial tendon. Am J Sports Med. 2010;38(10):2133–44.
24. DiGiovanni BF, Fraga CJ, Cohen BE, et al. Associated injuries found in chronic lateral ankle instability. Foot Ankle Int. 2000;21(10):809–15.
25. Roster B, Michelier P, Giza E. Peroneal tendon disorders. Clin Sports Med. 2015;34(4):625–41.
26. Jones MH, Amendola AS. Navicular stress fractures. Clin Sports Med. 2006;25(1):151–8. https://doi.org/10.1016/j.csm.2005.08.007. PMID: 16324981

Evaluation of Common Injuries of the Ankle and Calf Areas

62

Gian Luigi Canata, Giacomo Zanon, Valentina Casale, Alberto Castelli, and Alberto Polizzi

62.1 Introduction

Calf and ankle injuries are common in middle-aged and young patients, usually due to muscle fatigue or impaired coordination. The well-known term "tennis leg" is an example testifying the popularity of calf lesions. Ankle injuries are even more common with an incidence rate of 2.15 per 1000 person-years in the United States. Lateral sprains occur more frequently than syndesmotic and medial injuries. A thorough knowledge of their physical characteristics is mandatory.

Beyond these common injuries, there is a variety of less frequent but not less important pathologies requiring a careful physical examination. A wide knowledge of the different possible causes of symptoms and their clinical presentation is important to start a proper management.

G. L. Canata (✉) · V. Casale
Centre of Sports Traumatology, Koelliker Hospital, Torino, Italy
e-mail: studio@ortosport.it

G. Zanon · A. Castelli · A. Polizzi
Clinica Ortopedica e Traumatologica, IRCCS Policlinico San Matteo, Pavia, Italy
e-mail: a.castelli@smatteo.pv.it; alberto.polizzi01@universitadipavia.it

J. G. Lane et al. (eds.), *The Art of the Musculoskeletal Physical Exam*,
https://doi.org/10.1007/978-3-031-24404-9_62

62.2 Calf

62.2.1 Calf Anatomy

The lower leg divides into three fascial compartments: anterior, lateral, and posterior.

The posterior compartment of the leg (usually referred to as the "calf") is further divided into the superficial and deep compartments by the transverse intermuscular septum.

The larger, superficial posterior compartment includes the true "calf muscles," which are the gastrocnemius, soleus, and plantaris muscles. The gastrocnemius and the soleus are known together as the *triceps surae*. The *triceps surae* attaches to the Achilles tendon and is responsible for extending the foot and rising up the toes. It forcefully contracts whenever a takeoff or change of direction occurs.

The gastrocnemius and the plantaris are biarticular muscles, and both contribute to knee flexion and ankle plantar flexion (Fig. 62.1). The gastrocnemius consists of a mix of muscle fibers, mainly fast-twitch fibers (type 2), which are responsible for the explosive/powerful contractions of this muscle [1]. Conversely, the plantaris muscle is considered a vestigial muscle providing only about 0.7% of strength to plantar flexion [2].

The soleus is a uniarticular muscle responsible for plantar flexion. It plays an important role as a postural muscle through a high percentage of slow-twitch muscle fibers (type 1) [3].

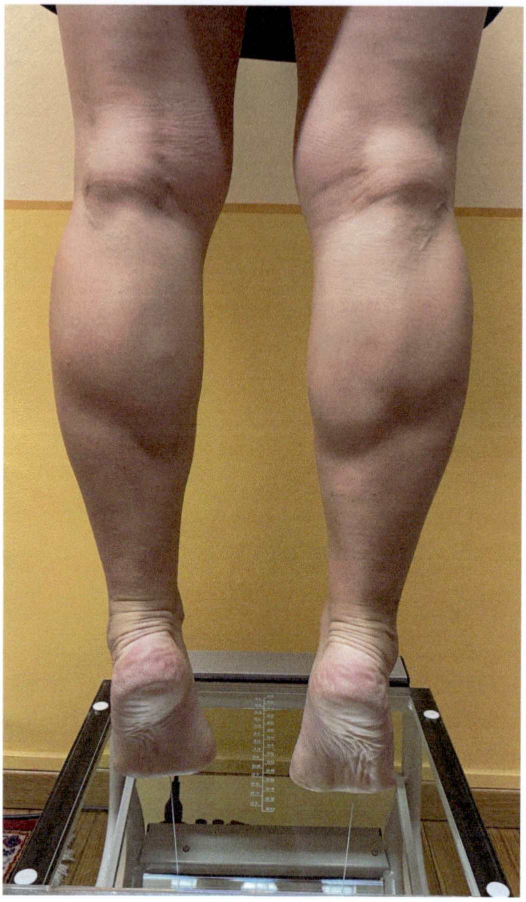

Fig. 62.1 Soleus muscles with normal contour

The contribution of fibers from the gastrocnemius and soleus to the Achilles tendon is variable: this receives fibers mainly from the soleus [4].

Considering these characteristics, it can be stated that the *triceps surae* musculotendon unit is composed of anatomically distinct parts, with a consequent nonuniform mechanical loading. Furthermore, the sites where mechanical stress increases the most are the medial/central sites of the enthesis and the tendon midportion [5].

The plantaris tendon runs between the gastrocnemius and soleus muscles and then inserts on the medial calcaneus or Achilles tendon [6].

The deep posterior compartment contains the popliteus, flexor digitorum longus, flexor hallucis longus, and tibialis posterior muscles. Injuries involving these structures typically present with pain in the shin or near the ankle [1] and usually cause a reduction of the sensation of plantar aspect of the foot, as well as a reduction of the great toe and lateral four toes' plantar flexion, a reduction of the foot inversion, and a reduction of the posterior tibial pulse [7].

Both the superficial and the deep calf muscles are in fact responsible for plantar flexion of the foot and the ankle. They are especially engaged in activities such as running and jumping.

The sural nerve, in particular the tibial nerve, provides innervations to the plantaris muscle through the L5–S1 nerve roots, while the gastrocnemius and the soleus muscle are innervated through the S1–S2 roots [8].

62.2.2 Calf Injuries

Calf pain affects many athletes, typically runners [9]. The etiology includes vascular, neurologic, skeletal, and musculotendinous issues [1].

Several affections should be considered, especially in case of bilateral calf pain. Some examples are as follows [10]:

- Musculoskeletal causes: shin splints, stress fractures of the tibia or fibula, gastrocnemius or soleus muscle strain or tear, tenosynovitis or tendinopathy, periostitis, myopathy, hip or knee osteoarthritis
- Vascular causes: vascular claudication, vasculitis, bilateral deep vein thrombosis, bilateral popliteal aneurysms, thrombophlebitis
- Neurological causes: lumbosacral spinal stenosis, lumbosacral disc herniation, peripheral neuropathy
- Other causes: infection (such as cellulitis), lower leg edema, neoplasia or metastatic disease, psychogenic affections, fibromyalgia

Calf pain in runners usually has a musculotendinous origin, especially a gastrocnemius or a soleus muscle injury [11]. Plantaris muscle injuries are the least common injuries involving the Achilles tendon and calf area but should be hypothesized when athletes report insidious medial mid-calf pain [11, 12].

No specific anatomical risk factors have been reported as predictors of calf injuries [13, 14], not even the cavus foot despite its association with Achilles tendon injuries [1].

Among runners, master athletes show a consistently higher risk for calf injuries, as well as men compared to women [9, 15]. Apparently, age and sex differences in fatigability, cellular metabolism, and tendon elongation/stiffness are the main causes of this disproportionate distribution [16, 17]; however, contributing factors to age and sex differences are unclear [1].

History and specific aspects of the assessment ease differential diagnosis. For example, if an athlete had fever, systemic symptoms, or a history of diabetes mellitus, the clinician should consider and screen for infections or myonecrotic causes of deep calf pain [1].

62.2.2.1 Muscle and Tendon Injuries of the Calf Area

A thorough history and physical examination are fundamental when assessing calf injuries. Firstly, the potential mechanism of injury (acute versus chronic) and other associated secondary trauma must be investigated.

Usually, an acute event mainly leads to gastrocnemius and popliteus injuries, frequently after sprinting or jumping. This mechanism of injury explains the definition of "tennis leg," commonly used for this type of trauma. Similarly, runners typically develop gastrocnemius and popliteus lesions during faster interval training, racing, or high-speed tempo runs [1].

Soleus injuries are usually caused by fatigue or overtraining instead. The athlete typically does not recall an acute event but reports a progressive development of calf pain, which eventually forces him to quit even light physical activity [1].

When a calf strain occurs, although epidemiology and clinical history may be helpful to individuate the muscle involved, it is the physical exam that really allows to assess the site and the severity of injury [18].

Palpation should be practiced along the entire length of the muscles and the aponeuroses. It is one of the most important phases of clinical assessment (Fig. 62.2).

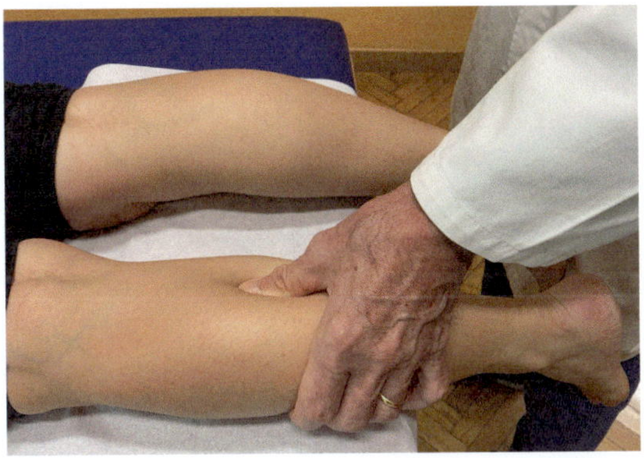

Fig. 62.2 Palpation of medial belly of the gastrocnemius muscle

Tenderness, swelling, thickness, defects, and masses must be investigated. Tenderness caused by gastrocnemius injuries is usually localized at the distal insertion of the medial or lateral head into the proximal Achilles fascia. In case of soleus muscle injuries, tenderness may be palpated deep and distal to the muscle bellies of the gastrocnemius on either medial or lateral leg. If the popliteus muscle is involved, tenderness is deep and usually at the intersection of the two gastrocnemius heads or along the medial proximal Achilles tendon [1]. If a mass is detectable, a more severe injury should be considered.

Strength testing is a further method of assessment and localization of injury. In fact, with the knee in maximal flexion, the soleus is the main force generator in plantar flexion. When the knee is fully extended, the gastrocnemius becomes the primary contributor instead [19]. This way, the clinician can effectively delineate which muscle has been injured.

Flexibility testing using passive ankle movements and stretching not only helps to determine the site of injury, but may also guide a rehabilitation program with stretching exercises [18] (Fig. 62.3).

The medial gastrocnemius is a pennate muscle which fascicles are arranged obliquely towards the longitudinal axis of the whole muscle. The pennation angle affects the amount of the fascicle length change with joint movement [20]. Longer and less pennate gastrocnemius muscle fascicles increase muscle shortening velocity and thus

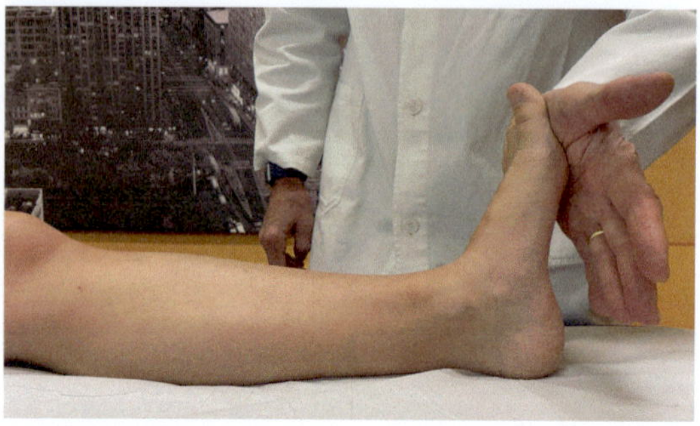

Fig. 62.3 Flexibility test of triceps surae showing lack of dorsiflexion

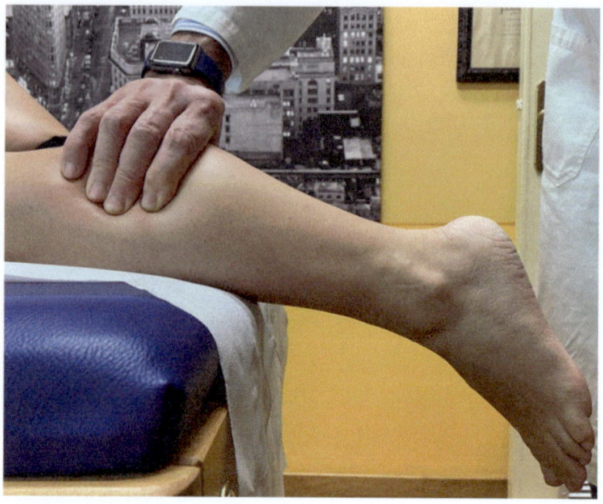

Fig. 62.4 Negative Thompson's test

power output [21], while pennation angle and muscle thickness are positively correlated with muscle force production [22, 23].

Several studies point out that gastrocnemius inflexibility can cause overuse injuries of the lower extremity, including patellofemoral syndrome [24], Achilles tendinopathy [25], plantar fasciitis [26], muscle strains [27], stress fractures, and shin splints [25, 28, 29]. Flexibility is usually evaluated by measuring maximal range of joint motion [30] or, alternatively, from the joint angle-passive torque relationship [31]. The most frequently used technique is performed with the knee in maximal extension, and then the ankle is passively dorsiflexed to cause a relatively isolated stretch of both the gastrocnemius and the soleus muscles [18].

Joint flexibility or passive joint torque is influenced by musculotendinous structures around the joint [30, 32–35]. In fact, it has been demonstrated that both the gastrocnemius muscle belly and Achilles tendon are elongated by passive stretching of the muscle-tendon unit performed with progressive ankle dorsiflexion [36].

To assess muscle groups' flexibility, several methods with different devices have been proposed [31, 37–39]. For example, the range of motion can be measured using goniometers or inclinometers, as well as with a (video) camera coupled with motion analysis software [40]. Movement can be generated either by the examiner (manually or mechanically aided) or actively by the patient himself or herself [38]. However, the final values are usually subjectively defined by the patient's tolerance to stretch [39]. Only few studies have measured the endpoint angle with the application of a standardized force [41] or torque [39], demonstrating to be more accurate than methods related to patient feedback [37]. It must be always considered that tissue compliance and tolerance to tissue stretch are necessary for a greater joint flexibility. One of the major limiting tissues of passive ROM is the muscle in fact. Considering that muscles atrophy with age, the effect of muscle stiffness on passive range of motion (ROM) might become weaker for older people compared with young. Nonmuscular structures have also been referred to play a role in limiting passive ROM [42]. Thus, it has been suggested that tissues limiting passive ankle dorsiflexion ROM are muscle and nerve for young and older people, respectively, whereas stretch tolerance influences passive ROM for both groups [43].

Thompson's test is usually performed to exclude the involvement of the Achilles tendon (Fig. 62.4). The patient lies in a prone position, with the foot off the examining table. If the tendon is intact, when the examiner compresses the calf area, the ankle plantar flexes. If the tendon is ruptured, the ankle remains static [44].

Clinical tests may also be used for the classification of calf injuries. A commonly used grading system for muscle strains includes clinical, pathologic, and radiology findings for a three-grade classification [45], as reported in Table 62.1.

Table 62.1 Classification of muscle strains [18]

Grade	Symptoms	Signs	Pathologic correlation	Radiology correlation
Grade 1 First degree mild	Acute pain at the time of injury or pain during activity. Usually able to continue activity	Mild pain and localized tenderness Mild spasm and swelling No or minimal loss of strength and ROM	<10% muscle fiber disruption	Bright signal on fluid-sensitive sequences. Feathery appearance <5% muscle fiber involvement
Grade 2 Second degree moderate	Unable to continue activity	Clear loss of strength and ROM Tenderness to palpation, pain with resisted muscle testing	>10–50% disruption of muscle fibers	Change in myotendinous junction Edema and hemorrhage
Grade 3 Third degree severe	Immediate severe pain, disability	Complete loss of function Palpable defect or mass Significant tenderness, discoloration, and swelling Possible positive Thompson's test	50–100% disruption of muscle fibers	Complete disruption of discontinuity of muscle Extensive edema and hemorrhage Wavy tendon morphology and retraction

ROM range of motion

If an athlete reports intense acute pain and marked swelling or if the calf injury has the characteristics of a complete rupture with a muscular defect, the evolution of the symptoms must be carefully monitored to avoid the development of a compartment syndrome or deep venous thrombosis (DVT) [1].

However, it must be underlined that both the acute and the chronic compartment syndrome of the lower leg rarely involve the calf area. The acute cases are frequently a consequence of tibial shaft fractures, thus affecting the anterior lower leg compartment [46]. The chronic compartment syndrome is often bilateral and usually involves the anterior and lateral compartments [10].

62.3 Ankle

62.3.1 Ankle Anatomy

The talocrural or ankle joint is formed by the articular surfaces of the tibial and fibular distal epiphyses and the talus in its superior, lateral, and medial aspects. The morphology of these surfaces forms a hinge-type synovial joint. Although the talocrural joint is considered a simple hinge joint, articular, ligamentous, and tendinous anatomy makes it much more complex.

The rotational axis is dynamic and changes in dorsal and plantar flexion, with a single axis of movement: the bimalleolar axis. This dynamic axis allows dorsiflexion (flexion) and plantar flexion (extension) of the ankle and foot in the sagittal plane. The normal range of motion is 10° of dorsiflexion and 30° of plantar flexion starting from the neutral position. Dorsiflexion moves the joint in a "locked" position with the extreme congruency of the articular surfaces and ligamentous tension. In this position, the malleoli lock the talus and never allow for lateral movements. The talar superior surface is larger in the anterior side than in the posterior one; therefore, the intermalleolar distance increases in dorsiflexion, due to a minimal stretch of the inferior tibiofibular syndesmosis.

The anterior capsular insertion in the tibia and talus occurs at a distance from the cartilaginous layer. The distance is approximately 6–8 mm in the tibia and 8–10 mm in the talus. The joint capsule is reinforced posteriorly by the transverse ligament, which is the deepest part of posteroinferior tibiofibular ligament (PITFL). The joint capsule is reinforced by the medial and lateral ligament complex. The ligaments around the ankle can be divided into three groups—the lateral ligaments (forming the collateral lateral ligament), the deltoid ligament (medial collateral ligament) on the medial side, and the ligaments of the tibiofibular syndesmosis which maintains the relationship between distal tibia and fibula [47].

The **lateral collateral ligament (LCL)** consists of three ligaments: the anterior talofibular, the calcaneofibular, and the posterior talofibular ligaments.

The **anterior talofibular ligament (ATFL)** is the most frequently injured ligament of the ankle observed in the emergency room [48]. This ligament plays an important role in limiting anterior displacement of the talus and plantar flexion of the ankle [49].

This flat, quadrilateral ligament blends with the ankle joint capsule and is typically composed of two separate bands. The anterior talofibular ligament originates at the anterior margin of the lateral malleolus. The center is on average 10 mm proximal to the tip of the fibula [50]. The width of the anterior tibiotalar ligament is about 6–10 mm. It runs from the anteroinferior margin of the fibula to the lateral margin of the talus inserting near the junction of the talar body and neck and seems to be horizontal to the ankle in the neutral position [50, 51].

The **calcaneofibular ligament (CFL)** is a cordlike ligament, thicker than the ATFL, and originates on the inferior margin of the fibula distal to the ATFL and runs underneath the peroneal tendons to insert on the lateral tubercle of the calcaneus. CFL has a cross section of about 6–8 mm; its length is about 20 mm. This ligament is separate from the ankle joint capsule but is intimately associated with the posteromedial part of the peroneal tendon sheath [47]. The calcaneofibular ligament is a biarticular ligament and bridges both talocrural and subtalar joints; therefore, it is implicated also in the subtalar movement.

The **posterior talofibular ligament (PTFL)** originates from the malleolar fossa, located on the medial surface of the lateral malleolus, coursing almost horizontally to insert in the lateral tubercle of the posterior process of the talus. In dorsiflexion, the ligament is stretched.

Considering the **posterior intermalleolar ligament**, its prevalent occurrence both in radiological and in anatomical studies may vary widely, ranging from 19 up to 100% [47]. In the study of Chang-Seok Oh et al., MRI was performed on all ankles (26 specimens) and anatomic variations of the intermalleolar ligament were observed [52]. These differences can probably be explained by its limited size. The posterior intermalleolar ligament is situated between the transverse ligament and the PTFL. It runs obliquely from lateral to medial and from downward to upward. The posterior intermalleolar ligament tenses during dorsiflexion and relaxes during plantar flexion. Therefore, trauma that causes forced dorsiflexion of the ankle can be assumed to produce rupture of this ligament, or osteochondral avulsion. Plantar flexion would cause relaxation and become susceptible to trapping between the tibia and the talus, leading to impingement.

The **medial collateral ligament (MCL)**, or deltoid ligament, is composed of two layers: the superficial and deep [47]. The MCL is a multifascicular ligament, originating from the medial malleolus to insert in the talus, calcaneus, and navicular bone. Six bands have been described for the MCL. Most of the posterior and middle part of the MCL is covered by the tendon sheath of the tibialis posterior muscle.

Since each band is continuous to one another, pointing out individual bands is very difficult [53].

The **syndesmotic articulation** allows the tibia and fibula to adapt to the varying width of the upper articular surface of the talus by slight ascending and medial rotation movements of the fibula during extreme dorsiflexion (maximum width) and by inverse movements during plantar flexion (minimum width) [47].

The **syndesmotic ligament complex** ensures the stability between the distal tibia and the fibula and resists the axial, rotational, and translational forces that attempt to separate the tibia and fibula. The three ligaments responsible are the anterior or anteroinferior tibiofibular ligament, the posterior or posteroinferior tibiofibular ligament, and the interosseous tibiofibular ligament. The inferior segment of the interosseous membrane also helps stabilize the tibiofibular syndesmosis. Distal to the insertion site of this interosseous ligament, the remaining anterior surface corresponds to the tibiofibular synovial recess of the ankle joint, and at the posterior surface, there is a small bundle of adipose tissue called the fatty synovial fringe. Syndesmosis injuries have been referred to as a cause of chronic pain following ankle sprain in the condition known as anterolateral soft tissue impingement or more specifically syndesmotic impingement [47].

The **anterior or anteroinferior tibiofibular ligament (ATFL or AITFL)** originates from the anterior tubercle of the tibia (5 mm in average above the articular surface), and its fibers extend in a distal and lateral direction to the insertion site in the anterior margin of the lateral malleolus, with increased length of the fibers distally. This ligament is perforated by small branches from the peroneal artery. The inferior fascicles of the AITFL are responsible for anterolateral soft tissue impingement. Moreover, removal of this distal fascicle frequently resolves the patient's complaints [47].

The **posterior or posteroinferior tibiofibular ligament (PTFL or PITFL)** is formed by two independent components, the superficial and deep component. The superficial component originates at the posterior edge of the lateral malleolus, directs proximally and medially to insert in the posterior tibial tubercle, and is homologous to the anterior tibiofibular ligament. The deep component is cone shaped and originates in the proximal area of the malleolar fossa to insert in the posterior edge of the tibia. Its insertion is immediately posterior to the cartilaginous covering of the inferior tibial articular surface, and its fibers may reach the medial malleolus. This component is also known as the transverse ligament, forming a true labrum to provide talocrural joint stability and to prevent posterior talar translation [47].

The **interosseous tibiofibular ligament** is a dense mass of short fibers, which, together with adipose tissue and small branching vessels from the peroneal artery, span the tibia to the fibula. It can be considered a distal continuation of the interosseous membrane at the level of the tibiofibular syndesmosis. The biomechanical role of this structure is still a matter of concern [47, 54].

62.3.2 Ankle Ligamentous Injuries

The incidence of ankle sprain is every 10,000 people each day. This kind of injury constitutes 7–10% of all admission diagnoses to emergency departments. Moreover, the inversion mechanism represents 25% of all injuries to the musculoskeletal system and about 50% are sport related [55].

The highest prevalence of ankle sprains occurs in people aged 15–19 years, and these account for up to 40% of all athletic injuries. The highest prevalence of these injuries occurs in basketball, soccer, running, and dancing activities. 53% of basket injuries and 29% of soccer injuries are due to ankle sprains. The injury rate in volleyball players was 0.9 per 1000 player-hours with 0.7% during practice and 2.6% during games. Females seem to be more prone to these injuries than males, and in 75% of cases, the lateral compartment is involved [55].

The ATFL is the first ligament to be damaged in 75% of all injuries, and very often, it is the only one involved. Combined lesions of ATFL and CFL occur in 20% of cases, and isolated rupture of CFL is rare. PTFL is usually not involved in inversion trauma because this ligament is lax when the ankle is plantar flexed and its maximal load to failure is three times that of the ATFL [55].

62.3.2.1 Classification

The term "ankle sprain" is used to describe a wide range of pathologies ranging from overstretching of the ligament to the complete rupture with ankle instability. The aim of classification is to ease prognosis and treatment. There are several grading systems for ankle ligament injuries based on anatomy, symptoms, and trauma mechanism. One of the most popular is the system introduced by Hamilton and Kaikkonen in 1982 that considers anatomy and symptoms. This system grades the injuries from I to III. In clinical practice, it is very important to recognize a simple sprain (grade I) and an injury that may lead to joint instability (grades II and III), which require additional treatment.

Konradsen et al. classified the ankle ligament injuries into **grade I (mild)**, including stretching of the ligament without rupture, with minor swelling and not increased laxity; **grade II (moderate)** including partial macroscopic ligament ruptures, with moderate swelling, tenderness, and pain, mild increment of laxity, and functional disability; and **grade III (severe)** including complete ligament and capsular rupture, with severe bruising, swelling, pain, major loss of function, and increased laxity [56].

The supination trauma is the most common ankle trauma and is associated with distraction forces on the lateral side of the ankle and compression forces on the medial side. In this movement, the foot rotates inward with a plantar flexion of the ankle in equinus position, with the narrowest part of the talus engaged in the mortise and the ankle being lax [57]. The medial malleolus acts as a fulcrum when the ankle moves into increased inversion increasing the strain on the lateral side. Rupture of the lateral ligament complex occurs in about 10–15% of all inversion traumas [58].

Isolated lesion of the ATFL occurs in 50–75% of supination trauma injuries with ligament rupture (grades II–III). Increasing the force, the calcaneofibular CFL is also damaged. The combined rupture of ATFL and CFL occurs in 15–25% of the trauma with ankle ligament rupture. Instead, isolated rupture of CFL accounts for about 1% of all cases and much more rare is the isolated rupture of PTFL [47].

Isolated deltoid ligament injuries are infrequent and account for about 2.5% of all ankle ligament injuries [57]. The mechanism of this trauma is an extreme eversion, outward rotation of the foot during simultaneous inward rotation of the tibia. Pain and swelling are present for a longer time than in the injured lateral side of the ankle. Increasing the supination and dorsiflexion force, the avulsion of a bone fragment from the talus or the avulsion of flexor retinaculum can occur.

The associated lesion of the syndesmosis is due to an external rotation of the foot or occurs after a forced dorsiflexion or with a combination of external rotation, axial compression, and forced dorsiflexion like in skiers, soccer players, and dancers [47]. These lesions are very painful and heal slowly but with a low residual instability [59]. The AITFL occurs in 1–5% of patients after severe inversion trauma, and the clinical examination does not discriminate between isolated lateral ankle ligament rupture and lesion accompanied by a partial syndesmotic ligament rupture.

Additional damages to cartilage of the anterodistal tibia, anterior part of medial malleolus, and medial talar facet can be observed in many patients who sustained an inversion-supination trauma.

62.3.2.2 Lateral Compartment Injuries

Patients undergoing a lateral ankle injury refer to almost the same history, and the following symptoms are widely referred to: feeling the ankle bent in supination plantar flexion movement, cracking sound at the time of the injury, intense pain with swelling, and loss of function with activity discontinuation. It is still very difficult to distinguish between a distortion and a lateral ligament rupture only with physical examination in an acute setting, because clinical tests are almost unreliable due to pain and swelling. However, if physical examination is repeated 4–5 days after the trauma, the specificity and sensitivity to detect presence or absence of ligament rupture increase up to 84 and 96%, and the inter-observer variation is good (kappa value of 0.7) [47].

First of all, when acute lateral ankle injury occurs, we must rule out a fracture or avulsion. Standard ankle X-rays are very useful to detect a fracture, but Brooks et al. (1981) showed that only 15% of patients with a painful swollen ankle visiting the emergency department of Western hospitals sustained an ankle fracture. Therefore, Ottawa ankle rules are routinely applied to reduce the number of ankle X-ray [60, 61]. In spite of this, X-rays are routinely made in the majority of professional athletes neglecting the Ottawa rules, to avoid delayed diagnosis and for defensive medicine reasons.

The delayed physical examination is the gold standard to detect lateral ankle acute ligament rupture. The more frequent findings are the following:

- Swelling: ligament rupture causes immediate and evident swelling.
- Pain on palpation: pain on palpation on ATFL is strongly associated with acute ligament rupture; pain over ATFL has a low specificity.
- Pain on palpation on ATFL associated to hematoma discoloration has 90% chance of acute lateral ligament rupture.
- A positive anterior drawer test has a sensitivity of 73% and a specificity of 97%. However, when this test is combined with pain on palpation on ATFL and hematoma discoloration, the sensitivity grows up to 100% and specificity to 77%.

Additional clinical findings in patients with lateral ankle injury could be as follows:

- Medial pain at the level of the medial malleolus due to the compression forces during supination sprain mechanism (60%), associated with cartilage damage on the anteromedial facet of the talus and distal anteromedial border of the tibia.
- Syndesmotic pain at the level of the AITFL with or without a rupture of the syndesmotic ligament (40%).
- Hematoma discoloration: Few days after trauma, it is easier to distinguish between edema and hematoma discoloration.

Clinical tests for acute lateral ankle instability are talar tilt test and anterior drawer test.

They can be considered the most important tests to detect an ankle ligament rupture, even if in acute settings, they can be less reliable if not performed immediately after the injury (van Dijk 1994) [62].

The talar tilt test is particularly useful in evaluating the CFL, while the anterior drawer test focuses on the ATFL (Lynch 2002) [63].

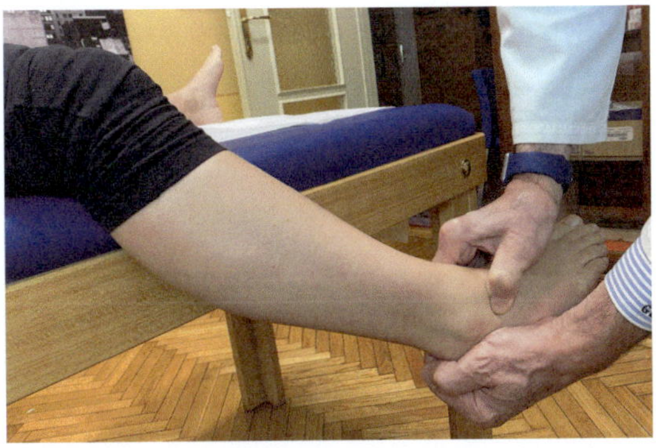

Fig. 62.5 Talar tilt test with knee flexed. Right thumb on the talus checking selective tibiotalar laxity

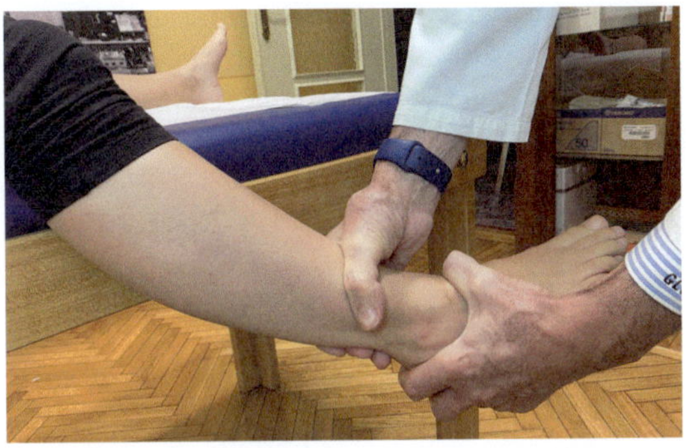

Fig. 62.6 Anterior drawer test checking an ATFL laxity

The first one is performed by applying an inversion-eversion torque or rotation of the foot and stressing the CFL (Hanlon 2016) [64] (Fig. 62.5). There are several ways to perform an anterior drawer test. It has been shown that the anterior drawer test is not a straightforward translation of the talus in relation to the tibia but is a rotatory movement. The intact medial malleolus and deltoid ligament act like a fulcrum and prevent the talus to move medially; therefore, the movement is a combination of anterior translation and internal rotation movement [65] (Fig. 62.6).

The best setting to perform a correct anterior drawer test is with the patient sitting at the end of the table, with the knee flexed at 90°. Starting from this position, the talus rotates out of the ankle mortise anteriorly with the intact deltoid ligament as a fulcrum. The examiner should guide the foot curving inward during the anterior translation of the talus. An anterior translation of about 1 cm is predictive of ATFL rupture, and sometimes the intra-articular negative pressure generated draws the skin inward at the site of ligament rupture.

62.3.2.3 Syndesmotic Injuries

The syndesmosis is an osseous and ligamentous complex which maintains the relationship between distal tibia and fibula. The anterolateral (Chaput) tubercle of the tibia, anterior (Wagstaffe) tubercle of the fibula, and incisura fibularis are the bone components, and the anterior inferior tibiofibular ligament (AITFL), posterior inferior tibiofibular ligament (PITFL), and interosseous membrane with interosseous ligament are the soft tissue components [66].

Despite the frequency of ankle sprain injuries, the diagnosis and treatment of a syndesmotic torn are still difficult [67]. In the diagnostic process, medical history and physical examination are the most important tools for proper radiological exams.

Syndesmotic lesions, partial or total, can be classified into acute and chronic injuries. The event leading to acute total injuries is the combination of foot external rotation and hyper-dorsiflexion of the ankle with an axial loading. On the other hand, acute partial injuries are usually caused by a supination trauma.

A chronic syndesmotic rupture should be suspected if the patient refers to a continued loss of function, which does not improve over time. Generally, the patient complains of persistent ankle pain after a history of injury [66].

The reliability of physical examination can be enhanced when the investigation is repeated a few days after the trauma. In this case, the specificity and sensitivity of physical examination for the presence or absence of a lateral ankle ligament rupture were 84% and 96%, respectively [68]. A few days after the trauma, the swelling and pain have diminished, the pain on palpation has become more localized, and the anterior drawer test can now be performed. Pain on palpation over the anterior talofibular ligament itself cannot distinguish between a rupture and a distortion. Considering the physical examination,

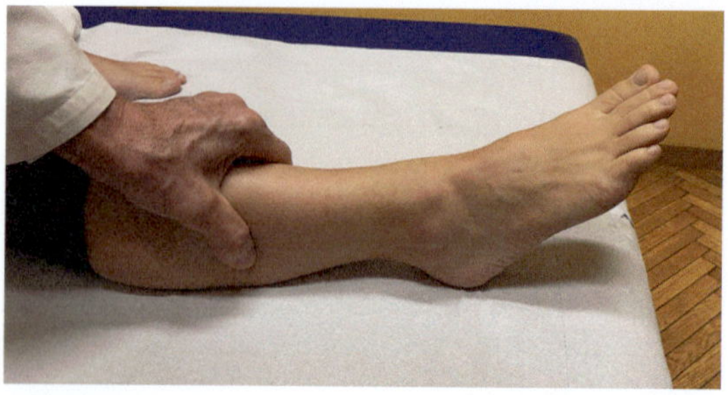

Fig. 62.7 Squeeze test for syndesmosis assessment

we look for swelling and hematoma discoloration. In addition, different physical tests can be selected [69]:

- External rotation stress test: This is almost impossible to be performed in acute setting (due to pain and swelling) and may lead to false-positive test. The clinician stabilizes the leg proximal to the ankle joint while grasping the plantar aspect of the foot and rotating the foot externally relative to the tibia. The test is positive if pain is elicited in the region of the anterior to the lateral malleolus and proximal to the ankle joint.
- Squeeze test: To perform this test, place the thumb on the tibia and the fingers on the fibula above the midpoint of the calf. Then squeeze the tibia and fibula together; pain down the length of the fibula indicates a positive test (Fig. 62.7).
- Dorsiflexion compression test: It is performed with the patient standing and actively dorsiflexing the ankle to simulate the normal loading of the syndesmosis ligaments. The patient dorsiflexes once independently and once with the therapist applying a manual compressive support around the malleoli. A significant increase in the ankle range of motion with added compression and/or a decrease in the end of range pain with added compression indicates a positive test.
- Crossed-leg test: It is the most accurate clinical test for diagnosing the syndesmosis sprain; it requires no manipulation, and it uses gravity and the force pair of gravity (action and reaction). The test mimics the mechanism of the

squeeze test, but avoids some of its disadvantages; the pivot point where the leg is squeezed may be inconsistent among different examiners, and the test is difficult and requires more power for pressure in large legs [67].

62.3.2.4 Medial Compartment Injuries

Deltoid ligament injuries account for 5–15% of ankle sprain cases. Isolated deltoid ligament injury is rare; up to 10% of deltoid ligament injuries in the ankles are associated with syndesmosis injuries. Acute injuries to the deltoid ligament can also occur in association with ankle fractures [70].

Single or repeated episodes of ankle sprain cause insufficiency of the deltoid ligament and may lead to pain, instability, or impingement syndrome of the ankle.

The event that leads to deltoid ligament injuries may occur either because of pronation or because of supination-external rotation injury. Pronation injuries apply a direct abduction force to the deep deltoid ligament and may result in medial malleolar fracture or deltoid ligament tear. The deltoid can be injured in isolation, but, in more severe trauma, it is associated with a fibular fracture or a tear of the syndesmosis [71]. The twisting component of the supination-external injury mechanism may also result in deltoid ligament injury. Either the superficial or the deep deltoid ligament can rupture as an avulsion from the medial malleolus (majority), as an avulsion from the distal insertion (minority), or as a midsubstance tear (least common) [72, 73].

A thorough examination including the patient history with possible injury mechanisms often provides valuable information on the diagnosis of injuries to the MCL or spring ligament complex [74].

Swelling, hematoma discoloration, and medial tenderness are physical examination findings that have been advocated for use in determining whether there is injury to the deltoid ligament in the setting of a lateral malleolar fracture with a normal medial clear space [72].

Chronic medial instability is fairly common. A significant number of patients with chronic lateral instability have concomitant medial instability, presenting as medial ankle pain, a feeling of giving way, and a pronation deformity of the hindfoot [75]. An MR imaging study found deltoid ligament injuries in 72% of patients with chronic ankle instability. When the detection of injuries to the superficial and deep deltoid ligament was considered separately, tears of the superficial deltoid were detected with a sensitivity of 83% and specificity of 94%; tears of the deep deltoid were detected with a sensitivity of 96% and specificity of 98% [76].

Physical examination for the presence of medial injuries is not reliable in the setting of acute lateral ankle injury; thus, stress radiographs are often used to diagnose deep deltoid injuries. However, it has been reported that stress radiographs and MR imaging in lateral malleolar fractures may be falsely positive in patients with a partial rupture of the deep deltoid detected on MR imaging criteria and who do not need operative treatment. MR imaging is uncommonly performed for acute ankle sprain but should be considered in severe cases [77].

62.4 Conclusions

Calf and ankle injuries are quite common. Anatomical and clinical knowledge is of utmost importance for a reliable assessment and proper management.

References

1. Fields KB, Rigby MD. Muscular calf injuries in runners. Curr Sports Med Rep. 2016;15:320–4.
2. Silver RL, de la Garza J, Rang M. The myth of muscle balance. A study of relative strengths and excursions of normal muscles about the foot and ankle. J Bone Joint Surg (Br). 1985;67:432–7.
3. Carr AJ, Norris SH. The blood supply of the calcaneal tendon. J Bone Joint Surg (Br). 1989;71:100–1.
4. Cummins EJ, Anson BJ. The structure of the calcaneal tendon (of Achilles) in relation to orthopedic surgery, with additional observations on the plantaris muscle. Surg Gynecol Obstet. 1946;83:107–16.
5. Toumi H, Larguech G, Cherief M, Batakis A, Hambli R, Jennane R, et al. Implications of the calf muscula-

ture and Achilles tendon architectures for understanding the site of injury. J Biomech. 2016;49:1180–5.
6. Spang C, Alfredson H, Docking SI, Masci L, Andersson G. The plantaris tendon: a narrative review focusing on anatomical features and clinical importance. Bone Jt J. 2016;98-B:1312–9.
7. Pechar J, Lyons MM. Acute compartment syndrome of the lower leg: a review. J Nurse Pract JNP. 2016;12:265–70.
8. Hansen JT. Netter's clinical anatomy E-book. Elsevier Health Sciences; 2014.
9. Marti B, Vader JP, Minder CE, Abelin T. On the epidemiology of running injuries. The 1984 Bern grand-prix study. Am J Sports Med. 1988;16:285–94.
10. Siau K, O'Rourke KP, Khanna A, Laversuch CJ. Bilateral calf chronic compartment syndrome in an elderly male: a case report. Cases J. 2009;2:8346.
11. Cushman DM, Shihab S. Ultrasound-guided corticosteroid injection for proximal plantaris tendinopathy: a case report. Radiol Case Rep. 2020;15:2175–7.
12. Delgado GJ, Chung CB, Lektrakul N, Azocar P, Botte MJ, Coria D, et al. Tennis leg: clinical US study of 141 patients and anatomic investigation of four cadavers with MR imaging and US. Radiology. 2002;224:112–9.
13. Campbell JT. Posterior calf injury. Foot Ankle Clin. 2009;14:761–71.
14. Gallo RA, Plakke M, Silvis ML. Common leg injuries of long-distance runners: anatomical and biomechanical approach. Sports Health. 2012;4:485–95.
15. Taunton JE, Ryan MB, Clement DB, McKenzie DC, Lloyd-Smith DR, Zumbo BD. A retrospective case-control analysis of 2002 running injuries. Br J Sports Med. 2002;36:95–101.
16. Kent-Braun JA, Ng AV, Doyle JW, Towse TF. Human skeletal muscle responses vary with age and gender during fatigue due to incremental isometric exercise. J Appl Physiol Bethesda Md. 1985;2002(93):1813–23.
17. Onambele GL, Narici MV, Maganaris CN. Calf muscle-tendon properties and postural balance in old age. J Appl Physiol Bethesda Md. 1985;2006(100):2048–56.
18. Bryan DJ. Gastrocnemius vs. soleus strain: how to differentiate and deal with calf muscle injuries. Curr Rev Musculoskelet Med. 2009;2:74–7.
19. Bojsen-Møller J, Hansen P, Aagaard P, Svantesson U, Kjaer M, Magnusson SP. Differential displacement of the human soleus and medial gastrocnemius aponeuroses during isometric plantar flexor contractions in vivo. J Appl Physiol Bethesda Md. 1985;2004(97):1908–14.
20. Gans C, Gaunt AS. Muscle architecture in relation to function. J Biomech. 1991;24(Suppl 1):53–65.
21. Baxter JR, Hullfish TJ, Chao W. Functional deficits may be explained by plantarflexor remodeling following Achilles tendon rupture repair: preliminary findings. J Biomech. 2018;79:238–42.
22. Randhawa A, Wakeling JM. Associations between muscle structure and contractile performance in seniors. Clin Biomech Bristol Avon. 2013;28:705–11.
23. Morse CI, Thom JM, Reeves ND, Birch KM, Narici MV. In vivo physiological cross-sectional area and specific force are reduced in the gastrocnemius of elderly men. J Appl Physiol Bethesda Md. 1985;2005(99):1050–5.
24. Lun V, Meeuwisse WH, Stergiou P, Stefanyshyn D. Relation between running injury and static lower limb alignment in recreational runners. Br J Sports Med. 2004;38:576–80.
25. Wilder RP, Sethi S. Overuse injuries: tendinopathies, stress fractures, compartment syndrome, and shin splints. Clin Sports Med. 2004;23:55–81. vi
26. Riddle DL, Pulisic M, Pidcoe P, Johnson RE. Risk factors for plantar fasciitis: a matched case-control study. J Bone Joint Surg Am. 2003;85:872–7.
27. Ekstrand J, Gillquist J. The frequency of muscle tightness and injuries in soccer players. Am J Sports Med. 1982;10:75–8.
28. Messier SP, Davis SE, Curl WW, Lowery RB, Pack RJ. Etiologic factors associated with patellofemoral pain in runners. Med Sci Sports Exerc. 1991;23:1008–15.
29. Neely FG. Biomechanical risk factors for exercise-related lower limb injuries. Sports Med Auckl NZ. 1998;26:395–413.
30. McHugh MP, Kremenic IJ, Fox MB, Gleim GW. The role of mechanical and neural restraints to joint range of motion during passive stretch. Med Sci Sports Exerc. 1998;30:928–32.
31. Magnusson SP, Aagaard P, Nielson JJ. Passive energy return after repeated stretches of the hamstring muscle-tendon unit. Med Sci Sports Exerc. 2000;32:1160–4.
32. Gajdosik RL, Vander Linden DW, Williams AK. Influence of age on length and passive elastic stiffness characteristics of the calf muscle-tendon unit of women. Phys Ther. 1999;79:827–38.
33. Strickler T, Malone T, Garrett WE. The effects of passive warming on muscle injury. Am J Sports Med. 1990;18:141–5.
34. Toft E, Espersen GT, Kålund S, Sinkjaer T, Hornemann BC. Passive tension of the ankle before and after stretching. Am J Sports Med. 1989;17:489–94.
35. Wilson GJ, Elliott BC, Wood GA. Stretch shorten cycle performance enhancement through flexibility training. Med Sci Sports Exerc. 1992;24:116–23.
36. Kawakami Y, Kanehisa H, Fukunaga T. The relationship between passive ankle plantar flexion joint torque and gastrocnemius muscle and Achilles tendon stiffness: implications for flexibility. J Orthop Sports Phys Ther. 2008;38:269–76.
37. Ben M, Harvey LA. Regular stretch does not increase muscle extensibility: a randomized controlled trial. Scand J Med Sci Sports. 2010;20:136–44.

38. Folpp H, Deall S, Harvey LA, Gwinn T. Can apparent increases in muscle extensibility with regular stretch be explained by changes in tolerance to stretch? Aust J Physiother. 2006;52:45–50.
39. Law RYW, Harvey LA, Nicholas MK, Tonkin L, De Sousa M, Finniss DG. Stretch exercises increase tolerance to stretch in patients with chronic musculoskeletal pain: a randomized controlled trial. Phys Ther. 2009;89:1016–26.
40. Fourchet F, Materne O, Horobeanu C, Hudacek T, Buchheit M. Reliability of a novel procedure to monitor the flexibility of lower limb muscle groups in highly-trained adolescent athletes. Phys Ther Sport Off J Assoc Chart Physiother Sports Med. 2013;14:28–34.
41. Fredriksen H, Dagfinrud H, Jacobsen V, Maehlum S. Passive knee extension test to measure hamstring muscle tightness. Scand J Med Sci Sports. 1997;7:279–82.
42. Nordez A, Gross R, Andrade R, Le Sant G, Freitas S, Ellis R, et al. Non-muscular structures can limit the maximal joint range of motion during stretching. Sports Med. 2017;47:1925–9.
43. Hirata K, Yamadera R, Akagi R. Associations between range of motion and tissue stiffness in Young and older people. Med Sci Sports Exerc. 2020;52:2179–88.
44. Thompson TC. A test for rupture of the tendo achillis. Acta Orthop Scand. 1962;32:461–5.
45. Järvinen TAH, Järvinen TLN, Kääriäinen M, Kalimo H, Järvinen M. Muscle injuries: biology and treatment. Am J Sports Med. 2005;33:745–64.
46. Mauser N, Gissel H, Henderson C, Hao J, Mauffrey C, Hak D. Acute lower-leg compartment syndrome. Orthopedics. 2013;36:619–24.
47. van Dijk CN. Ankle arthroscopy: techniques developed by the Amsterdam foot and ankle school. New York: Springer; 2014.
48. Boruta PM, Bishop JO, Braly WG, Tullos HS. Acute lateral ankle ligament injuries: a literature review. Foot Ankle. 1990;11:107–13.
49. van den Bekerom MPJ, Oostra RJ, Alvarez PG, van Dijk CN. The anatomy in relation to injury of the lateral collateral ligaments of the ankle: a current concepts review. Clin Anat. 2008;21:619–26.
50. Burks RT, Morgan J. Anatomy of the lateral ankle ligaments. Am J Sports Med. 1994;22:72–7.
51. Milner CE, Soames RW. Anatomical variations of the anterior talofibular ligament of the human ankle joint. J Anat. 1997;191(Pt 3):457–8.
52. Oh C-S, Won H-S, Hur M-S, Chung I-H, Kim S, Suh J-S, et al. Anatomic variations and MRI of the intermalleolar ligament. Am J Roentgenol. 2006;186:943–7.
53. Golanó P, Vega J, Pérez-Carro L, Götzens V. Ankle anatomy for the Arthroscopist. Part II: role of the ankle ligaments in soft tissue impingement. Foot Ankle Clin. 2006;11:275–96.
54. Hoefnagels EM, Waites MD, Wing ID, Belkoff SM, Swierstra BA. Biomechanical comparison of the interosseous tibiofibular ligament and the anterior tibiofibular ligament. Foot Ankle Int. 2007;28:602–4.
55. van den Bekerom MPJ, Kerkhoffs GMMJ, McCollum GA, Calder JDF, van Dijk CN. Management of acute lateral ankle ligament injury in the athlete. Knee Surg Sports Traumatol Arthrosc. 2013;21:1390–5.
56. Konradsen L, Hølmer P, Søndergaard L. Early mobilizing treatment for grade III ankle ligament injuries. Foot Ankle. 1991;12:69–73.
57. Fallat L, Grimm DJ, Saracco JA. Sprained ankle syndrome: prevalence and analysis of 639 acute injuries. J Foot Ankle Surg. 1998;37:280–5.
58. Bouysset M. Bone and joint disorders of the foot and ankle a Rheumatological approach. Berlin: Springer; 2013.
59. Xenos JS, Hopkinson WJ, Mulligan ME, Olson EJ, Popovic NA. The tibiofibular syndesmosis. Evaluation of the ligamentous structures, methods of fixation, and radiographic assessment. J Bone Jt Surg. 1995;77:847–56.
60. Shell IG. Decision rules for the use of radiography in acute ankle injuries: refinement and prospective validation. JAMA. 1993;269:1127.
61. Brooks SC, Potter BT, Rainey JB. Inversion injuries of the ankle: clinical assessment and radiographic review. BMJ. 1981;282:607–8.
62. van Dijk CN. On diagnostic strategies in patients with severe ankle sprain. Rodopi. 1994.
63. Lynch SA. Assessment of the injured ankle in the athlete. J Athl Train. 2002;37:406–12.
64. Hanlon S, Caccese J, Knight CA, Swanik CB, Kaminski TW. Examining ankle-joint laxity using 2 knee positions and with simulated muscle guarding. J Athl Train. 2016;51:111–7.
65. Kerkhoffs GMMJ, Blankevoort L, Sierevelt IN, Corvelein R, Janssen GHW, van Dijk CN. Two ankle joint laxity testers: reliability and validity. Knee Surg Sports Traumatol Arthrosc. 2005;13:699–705.
66. Swords M, Brilhault J, Sands A. Acute and chronic syndesmotic injury. Foot Ankle Clin. 2018;23:625–37.
67. Yu G-S, Lin Y-B, Xiong G-S, Xu H-B, Liu Y-Y. Diagnosis and treatment of ankle syndesmosis injuries with associated interosseous membrane injury: a current concept review. Int Orthop. 2019;43:2539–47.
68. van Dijk CN, Lim LS, Bossuyt PM, Marti RK. Physical examination is sufficient for the diagnosis of sprained ankles. J Bone Joint Surg (Br). 1996;78:958–62.
69. Alonso A, Khoury L, Adams R. Clinical tests for ankle syndesmosis injury: reliability and prediction of return to function. J Orthop Sports Phys Ther. 1998;27:276–84.

70. Jeong MS, Choi YS, Kim YJ, Kim JS, Young KW, Jung YY. Deltoid ligament in acute ankle injury: MR imaging analysis. Skelet Radiol. 2014;43:655–63.
71. Koval KJ, Egol KA, Cheung Y, Goodwin DW, Spratt KF. Does a positive ankle stress test indicate the need for operative treatment after lateral malleolus fracture? A preliminary report. J Orthop Trauma. 2007;21:449–55.
72. Crim JR, Beals TC, Nickisch F, Schannen A, Saltzman CL. Deltoid ligament abnormalities in chronic lateral ankle instability. Foot Ankle Int. 2011;32:873–8.
73. Lee S, Lin J, Hamid KS, Bohl DD. Deltoid ligament rupture in ankle fracture: diagnosis and management. J Am Acad Orthop Surg. 2019;27:e648–58.
74. Jordan M, Thomas M, Elser F, Fischer W. Verletzungen des medialen Kollateralband– und Spring-ligament-Komplexes. Unfallchirurg. 2018;121:704–14.
75. Knupp M, Lang TH, Zwicky L, Lötscher P, Hintermann B. Chronic ankle instability (medial and lateral). Clin Sports Med. 2015;34:679–88.
76. Crim J, Longenecker LG. MRI and surgical findings in deltoid ligament tears. AJR Am J Roentgenol. 2015;204:W63–9.
77. Crim J. Medial-sided ankle pain: deltoid ligament and beyond. Magn Reson Imaging Clin N Am. 2016;25:63–78.

63 Assessment of Outcome Scores of the Ankle

Cortez L. Brown, Stephen Canton, Lorraine Boakye, and MaCalus V. Hogan

Outcome scoring systems are common throughout orthopaedics. Hunt and Lakey define patient-reported outcome metrics as tools that capture the health status of a patient throughout an episode of care for treatment of an injury [1]. Outcome metrics entered the medical field as early as the nineteenth century with Florence Nightingale during the Crimean War, in which she implemented simple safety measures to reduce infection risk [2]. By reducing soldier mortality from 43% to 2% in 5 months, she is considered one of the founders of evidence-based medicine. From 1990 to 2001, Button and Pinney found that 49 outcome scoring systems had been mentioned in published literature [3]. Only 18 were mentioned more than once. From 2002 to 2011, Hunt and Hurwit found 139, and Safavi et al. found 74 between 2012 and 2017 [4, 5].

Of the many scoring systems, Scott and Huskisson are known for developing the well-known visual analog scale (VAS) [6]. This pain assessment system is used throughout medicine but was validated for orthopaedic pathology in 1980 [6]. According to Safavi et al., VAS is the second most common scoring system with roughly 21,000 patients evaluated and used in 308 articles between January 2012 and July 2017 [5]. VAS has been the second most common scoring system since 2001 [4]. Previous literature supports this scoring system's validity, reliability, and acceptability [7]. Although commonly used and validated for orthopaedics, VAS is known to have low specificity [8]. According to Safavi et al., 3000 more patients and 85 more articles used the American Orthopaedic Foot and Ankle Society Ankle-Hindfoot (AOFAS-AH) outcome scoring system compared to VAS between January 2012 and July 2017 [5].

An AOFAS subcommittee developed the AOFAS-AH outcome scoring system in 1994 [9]. To date, it is the most commonly used outcome scoring for ankle pathology [3, 5]. In addition to the ankle-hindfoot pathology, the subcommittee developed outcome scoring systems for the midfoot, hallux, and lesser toes [9]. The committee's goal was to design accessible systems which allowed for the comparison of results of different methods of treatment in patients with the same

C. L. Brown · L. Boakye
Department of Orthopaedic Surgery, University of Pittsburgh Medical Center, Pittsburgh, PA, USA

S. Canton · M. V. Hogan (✉)
Department of Orthopaedic Surgery, University of Pittsburgh Medical Center, Pittsburgh, PA, USA

University of Pittsburgh School of Medicine, Pittsburgh, PA, USA
e-mail: hoganmv@upmc.edu

J. G. Lane et al. (eds.), *The Art of the Musculoskeletal Physical Exam*,
https://doi.org/10.1007/978-3-031-24404-9_63

disorder. An additional goal is to enable the surgeon to follow the clinical progression of their patients before and after treatment [9]. The AOFAS scoring system has a max score of 100 points and focuses on function, pain, and alignment. It requires input from both the patient and clinician [9]. The maximum amount of points is given in the following scenarios: full range of sagittal and hindfoot motion, no ankle or hindfoot instability, ability to walk more than six blocks, ability to ambulate on any walking surface, no limitation of daily or recreational activities, no discernable limp, no assistive devices needed for ambulation, no pain, and good alignment. The maximum points for function, pain, and alignment are 50, 40, and 10, respectively. Although there were several scoring systems prior to the AOFAS metric [10–14], the AOFAS subcommittee's system has stood the test of time and its long-standing use has led to many clinicians and researchers investigating its usability. Button and Pinney found the AOFAS-AH scoring system to lack reliability, responsiveness, and both criterion and construct validity. Madeley et al. included that there is no literature available stating a clear minimal clinical important difference (MCID) for this scoring system [15].

The following scoring systems are disease-specific outcome systems for ankle pathology: Foot Function Index (FFI), Ankle Osteoarthritis Scale (AOS), Foot and Ankle Disability Index (FADI), Foot and Ankle Ability Measure (FAAM), Short Musculoskeletal Function Assessment (SMFA), and Foot and Ankle Outcome Score (FAOS). Recent publications have reported FFI use for rheumatoid arthritis and osteoarthritis of the ankle [3, 4, 16]. AOS is also known to report ankle arthritis outcomes [3, 17]. The athlete with chronic ankle instability would benefit from FADI [18]. Both FAAM and FAOS are also validated for use in patients with chronic ankle instability, while SMFA is validated generally for musculoskeletal disease [5]. Although it is not a disease-specific outcome scoring system, the general patient-reported Short Form-36 (SF-36) is frequently used throughout orthopaedics (e.g., ankle osteoarthritis, orthopaedic trauma, and knee osteoarthritis) [5].

There are a plethora of outcome scoring systems for foot and ankle pathology, and those which are validated are not the most frequently used. The accessibility and ease of the AOFAS-AH may be a primary factor for its high use rate while being an unvalidated scoring system. Even the AOFAS called for the termination of its use due to insufficient reliability and validity [19]. Along with supporting disease-specific outcome scoring systems, this leading society for foot and ankle surgeons supports the use of recently introduced patient-reported outcome instruments such as the Patient-Reported Outcomes Measurement Information System (PROMIS) [20, 21], Physical Function Computerized Adaptive Test (PF CAT) [22–24], and Lower Extremity Computerized Adaptive Test (LE CAT) [24]. These systems aim to standardize scoring and ameliorate the crowded field of outcome scoring systems. Fidai et al. published a systematic review that compared PROMIS to commonly used scores (e.g., FAAM, FFI, and SF-36) [25]. They found PROMIS to correlate strongly with older scoring systems in foot and ankle pathology. Importantly, Fidai et al. reported quicker administration and a broader patient population for PROMIS when compared to older scoring systems.

We believe that there is a clear need for standardized scoring throughout foot and ankle pathology, especially since healthcare reimbursement shifted from volume-based to value-based practice after the signing of the Patient Protection and Affordable Care Act in 2015 [26]. Additionally, an effective and nationally supported outcome scoring system has the potential to augment orthopaedic surgery practice and compensation. Having such a system provides an opportunity for current and future registries to improve health outcomes. Ferguson et al. emphasize the transformational potential of access to patient-reported outcome research and total costs of a complete cycle of care [27]. Understanding these two components could increase the quality of care for future patients. Clinicians developing foot and ankle outcome scoring systems ought to learn from the benefits and limitations of the many outcome scoring systems available to date.

Therefore, we support the position of the AOFAS in regard to moving forward with disease-specific and valid outcome scoring systems. Not doing so would be in opposition to providing quality care to our patients.

References

1. Hunt KJ, Lakey E. Patient-reported outcomes in foot and ankle surgery. Orthop Clin North Am. 2018;49(2):277–89. https://doi.org/10.1016/j.ocl.2017.11.014.
2. McDonald L. Florence nightingale and the early origins of evidence-based nursing. Evid Based Nurs. 2001;4(3):68–9. https://doi.org/10.1136/ebn.4.3.68.
3. Button G, Pinney S. A meta-analysis of outcome rating scales in foot and ankle surgery: is there a valid, reliable, and responsive system? Foot Ankle Int. 2004;25(8):521–5. https://doi.org/10.1177/107110070402500802.
4. Hunt KJ, Hurwit D. Use of patient-reported outcome measures in foot and ankle research. J Bone Joint Surg Am. 2013;95(16):e118(1–9). https://doi.org/10.2106/JBJS.L.01476.
5. Shazadeh Safavi P, Janney C, Jupiter D, Kunzler D, Bui R, Panchbhavi VK. A systematic review of the outcome evaluation tools for the foot and ankle. Foot Ankle Spec. 2019;12(5):461–70. https://doi.org/10.1177/1938640018803747.
6. Scott J, Huskisson EC. Graphic representation of pain. Pain. 1976;2(2):175–84.
7. Hawker GA, Mian S, Kendzerska T, French M. Measures of adult pain: Visual Analog Scale for Pain (VAS Pain), Numeric Rating Scale for Pain (NRS Pain), McGill Pain Questionnaire (MPQ), Short-Form McGill Pain Questionnaire (SF-MPQ), Chronic Pain Grade Scale (CPGS), Short Form-36 Bodily Pain Scale (SF-36 BPS), and Measure of Intermittent and Constant Osteoarthritis Pain (ICOAP). Arthritis Care Res (Hoboken). 2011;63(Suppl 11):S240–52. https://doi.org/10.1002/acr.20543.
8. Lee JS, Hobden E, Stiell IG, Wells GA. Clinically important change in the visual analog scale after adequate pain control. Acad Emerg Med. 2003;10(10):1128–30. https://doi.org/10.1111/j.1553-2712.2003.tb00586.x.
9. Kitaoka HB, Alexander IJ, Adelaar RS, Nunley JA, Myerson MS, Sanders M. Clinical rating systems for the ankle-hindfoot, midfoot, hallux, and lesser toes. Foot Ankle Int. 1994;15(7):349–53. https://doi.org/10.1177/107110079401500701.
10. Kitaoka HB, Anderson PJ, Morrey BF. Revision of ankle arthrodesis with external fixation for non-union. J Bone Joint Surg Am. 1992;74(8):1191–200.
11. Kitaoka HB. Salvage of nonunion following ankle arthrodesis for failed total ankle arthroplasty. Clin Orthop Relat Res. 1991;268:37–43.
12. Kitaoka HB, Romness DW. Arthrodesis for failed ankle arthroplasty. J Arthroplast. 1992;7(3):277–84. https://doi.org/10.1016/0883-5403(92)90049-v.
13. Mazur JM, Schwartz E, Simon SR. Ankle arthrodesis. Long-term follow-up with gait analysis. J Bone Joint Surg Am. 1979;61(7):964–75.
14. Buechel FF, Pappas MJ, Iorio LJ. New Jersey low contact stress total ankle replacement: biomechanical rationale and review of 23 cementless cases. Foot Ankle. 1988;8(6):279–90. https://doi.org/10.1177/107110078800800603.
15. Madeley NJ, Wing KJ, Topliss C, Penner MJ, Glazebrook MA, Younger AS. Responsiveness and validity of the SF-36, ankle osteoarthritis scale, AOFAS ankle Hindfoot score, and foot function index in end stage ankle arthritis. Foot Ankle Int. 2012;33(1):57–63. https://doi.org/10.3113/FAI.2012.0057.
16. Domsic RT, Saltzman CL. Ankle osteoarthritis scale. Foot Ankle Int. 1998;19(7):466–71. https://doi.org/10.1177/107110079801900708.
17. Naal FD, Impellizzeri FM, Rippstein PF. Which are the most frequently used outcome instruments in studies on total ankle arthroplasty? Clin Orthop Relat Res. 2010;468(3):815–26. https://doi.org/10.1007/s11999-009-1036-y.
18. Hale SA, Hertel J. Reliability and sensitivity of the foot and ankle disability index in subjects with chronic ankle instability. J Athl Train. 2005;40(1):35–40.
19. Pinsker E, Daniels TR. AOFAS position statement regarding the future of the AOFAS clinical rating systems. Foot Ankle Int. 2011;32(9):841–2. https://doi.org/10.3113/FAI.2011.0841.
20. Cella D, Yount S, Rothrock N, Gershon R, Cook K, Reeve B, et al. The patient-reported outcomes measurement information system (PROMIS): progress of an NIH roadmap cooperative group during its first two years. Med Care. 2007;45(5 Suppl 1):S3–S11. https://doi.org/10.1097/01.mlr.0000258615.42478.55.
21. Smith MV, Klein SE, Clohisy JC, Baca GR, Brophy RH, Wright RW. Lower extremity-specific measures of disability and outcomes in orthopaedic surgery. J Bone Joint Surg Am. 2012;94(5):468–77. https://doi.org/10.2106/JBJS.J.01822.
22. Hung M, Clegg DO, Greene T, Saltzman CL. Evaluation of the PROMIS physical function item bank in orthopaedic patients. J Orthop Res. 2011;29(6):947–53. https://doi.org/10.1002/jor.21308.
23. Hung M, Clegg DO, Greene T, Weir C, Saltzman CL. A lower extremity physical function computerized adaptive testing instrument for orthopaedic patients. Foot Ankle Int. 2012;33(4):326–35. https://doi.org/10.3113/FAI.2012.0326.
24. Hung M, Baumhauer JF, Latt LD, Saltzman CL, SooHoo NF, Hunt KJ, et al. Validation of PROMIS® physical function computerized adaptive tests for orthopaedic foot and ankle outcome research. Clin Orthop Relat Res. 2013;471(11):3466–74. https://doi.org/10.1007/s11999-013-3097-1.

25. Fidai MS, Saltzman BM, Meta F, Lizzio VA, Stephens JP, Bozic KJ, et al. Patient-reported outcomes measurement information system and legacy patient-reported outcome measures in the field of Orthopaedics: a systematic review. Arthroscopy. 2018;34(2):605–14. https://doi.org/10.1016/j.arthro.2017.07.030.
26. Squitieri L, Bozic KJ, Pusic AL. The role of patient-reported outcome measures in value-based payment reform. Value Health. 2017;20(6):834–6. https://doi.org/10.1016/j.jval.2017.02.003.
27. Ferguson CM, Rocha JL, Lalli T, Irrgang JJ, Hurwitz S, Hogan MV. Developing performance and assessment platforms in foot and ankle surgery. Foot Ankle Int. 2016;37(6):670–9. https://doi.org/10.1177/1071100716649169.

Part VII

Foot and Toes Dr Canata

64 Anatomy of the Foot

Ivan Saenz, Ignasi Manent, Anna Rubio, and Fernando Conejo

64.1 Osteology

The foot is composed of 26 bones, classified into two groups: the tarsi, formed of seven bones, and the metatarsi, formed of five bones. The bones of the toes are the remaining 14. At its turn, the tarsi can be divided into two parts: the hindfoot, that is the most posterior part of the foot, formed of the talus above and the calcaneus below, and the midfoot, anteriorly, formed of the navicular, the cuboid, and the three cuneiforms. The metatarsi are formed of the five metatarsal bones and, together with the toe phalanges, form the forefoot, which is the most anterior part of the foot [1].

The ankle is the joint that connects the leg and the foot. It is an articulation where the surfaces are highly conforming, and this minimizes the joint pressures. The tibia and fibula make a tight socket (mortise) for the talar dome. The dome narrows progressively from anterior to posterior, so the talus fits tighter in dorsiflexion position than in plantarflexion. In dorsiflexion, the fibula is slightly laterally displaced [2].

The hindfoot is composed of the talus and the calcaneus and works together with the ankle: as the ankle is the principal joint for plantarflexion and dorsiflexion, the hindfoot joints (the talocalcaneal—or subtalar—the talonavicular, and the calcaneocuboid) on the whole provide the movements of inversion and eversion [2]. The midfoot is composed of the cuboid, navicular, and cuneiform bones, and the forefoot is formed of the metatarsal bones and the phalanges.

64.1.1 Talus

The talus is a short and irregular saddle-shaped bone that has three main parts, head (anteroinferior part), body (posterosuperior part), and neck (connecting both), and it is the center of the ankle. It has no tendon attachments. The superior part of the body is the talar dome, a curved and smooth surface that is convex when it is seen from a lateral view and slightly concave seen from an anteroposterior view. The medial surface, in its top margin, articulates with the tibial malleoli. The lateral surface articulates with the fibular malleoli and inferiorly forms a bony projection (lateral process), which supports the lower portion of the lateral articular facet. The posterior aspect has a groove for the tendon of the flexor hallucis longus. The inferior surface has a

I. Saenz (✉)
Orthopaedic and Trauma Service of Fundación Hospital de l'Esperit Sant, Barcelona, Spain

Human Department (Campus Clinic), University of Barcelona, Barcelona, Spain

I. Manent · A. Rubio · F. Conejo
Orthopaedic and Trauma Service of Fundación Hospital de l'Esperit Sant, Barcelona, Spain

J. G. Lane et al. (eds.), *The Art of the Musculoskeletal Physical Exam*,
https://doi.org/10.1007/978-3-031-24404-9_64

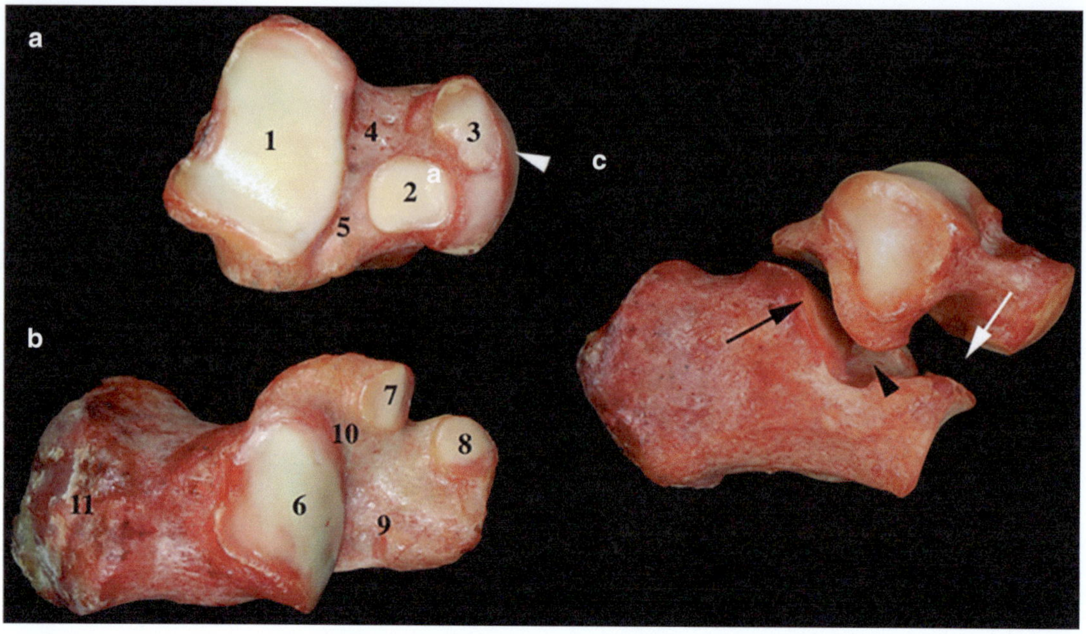

Fig. 64.1 (**a**) Caudal view of the right talus. (**b**) Superior view of calcaneal. (**c**) Articular joint formed by talus and calcaneal bones. 1 Posterior calcaneal articular surface; 2 middle calcaneal articular surface; 3 anterior calcaneal articular surface; 4 sinus tarsi; 5 tarsal canal; 6 posterior talar articular surface; white head, navicular articular surface of talus; 7 middle talar articular surface; 8 anterior talar articular surface; 9 sinus tarsi; 10 calcaneal canal; 11 calcaneal tuberosity. Black arrow, posterior subtalar joint; black head, sinus tarsi; white arrow, anterior subtalar joint

deep groove (talar sulcus) that separates the two calcaneal anteroinferior facets (anterior and middle) and the posteroinferior (posterior) one [3] (Fig. 64.1a).

It articulates superiorly with the tibia and the fibula (ankle joint); inferoposteriorly with the calcaneus, through its posterior large concave oblique facet (Fig. 64.1a1); anteroinferiorly also with the calcaneus through another two smaller facets, middle and anterior calcaneal articular surfaces (Fig. 64.1a2, 3); and anteriorly with the talar head joints with the navicular bone (talonavicular joint) (Fig. 64.1a).

64.1.2 Calcaneus

The calcaneus is the biggest tarsal bone, an irregular bone with a cuboid form, and the largest axis is anteroposterior. The lateral part has the fibular trochlea, which separates the fibularis brevis (passing above) and the fibularis longus (passing below) tendons. The posterior third of the superior surface is nonarticular and composed of the posterior tuberosity. The middle third of the superior surface supports the large posterior calcaneal articular surface (Fig. 64.1b6). It inclines anteriorly and creates a step contour. The angle of inclination in 50 calcanei was average 65.5°, maximum 75°, and minimum 55°. The surface of the anterior third is formed by the sinus tarsi, the tarsal canal, and the middle and anterior calcaneal articular surface (Fig. 64.1b7, 8).

On the medial surface, on its most superior part, protrudes the sustentaculum tali, which supports the head of the talus.

The calcaneal sulcus and the talar sulcus (tarsal canal) (Fig. 64.1) are the floor and the ceiling of a width tubular gap between the two bones, the tarsal sinus, which has its opening at the anterior end of the lateral aspect of the hindfoot, 1 cm anterior and inferior from the fibular malleoli. The anterior surface is convex and articular.

It articulates superiorly with the talus (talocalcaneal joint) and anteriorly with the cuboid (calcaneocuboid joint).

64.1.3 Navicular

The navicular is a boat-shaped bone, with its posterior concave surface that articulates with the talar head, its lateral surface that articulates with the cuboid, and its anterior surface, divided into three facets (medial, intermediate, and lateral). These articulate with the cuneiform bones (Figs. 64.2 and 64.3). The superior, inferior, and medial surfaces are irregular and narrow.

64.1.4 Cuboid

The cuboid bone is, as its name suggests, cubic, with six surfaces. It is located lateral to the navicular, playing a role in the lateral support of the foot. The posterior surface articulates with the calcaneus (calcaneocuboid joint), the medial surface with the navicular (cuboideonavicular joint) and with the lateral cuneiform in the most anterior part of the medial surface (cuneocuboid joint), and the anterior surface with the fourth and the fifth metatarsal bases (lateral tarsometatarsal joint) (Figs. 64.2 and 64.3).

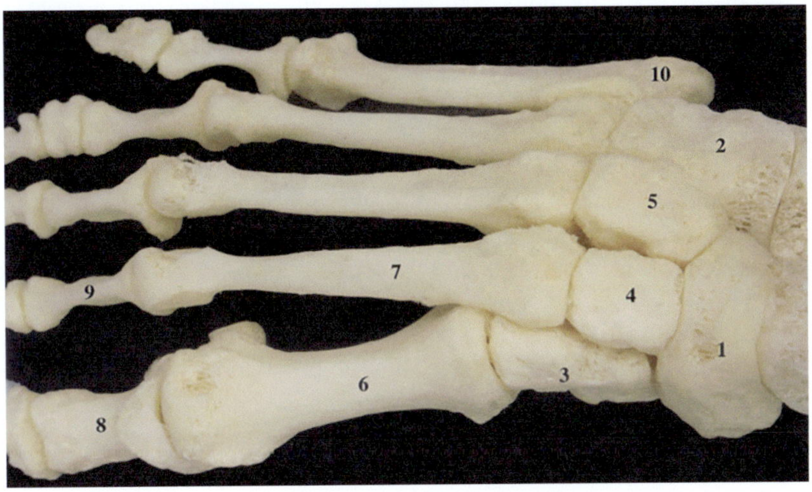

Fig. 64.2 Cranial view of skeletal foot. 1 Navicular; 2 cuboid; 3 medial cuneiform; 4 middle cuneiform; 5 lateral cuneiform; 6 first metatarsal; 7 second metatarsal; 8 fifth metatarsal tuberosity; 9 proximal phalanx of the big toe; 10 proximal phalanx of lesser toe

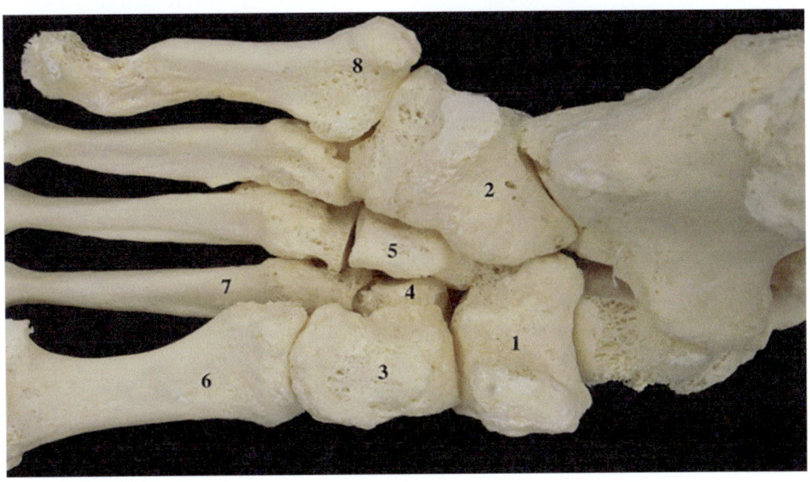

Fig. 64.3 Caudal view of skeletal foot. 1 navicular; 2 cuboid; 3 medial cuneiform; 4 middle cuneiform; 5 lateral cuneiform; 6 first metatarsal; 7 second metatarsal; 8 fifth metatarsal tuberosity

64.1.5 Cuneiforms

The cuneiforms are three wedge-shaped bones (medial, intermedial, and lateral), with the base located above and the vertex located below. The intermedial is shorter than the other two. The cuneiforms articulate with the navicular posteriorly and with the base of the three first metatarsals anteriorly (Figs. 64.2 and 64.3).

64.1.6 Metatarsal Bones

The metatarsals are five dorsally convex large bones and can be divided into three main parts: the base, which is proximal and articulates with the midfoot; the body, which has a triangular section of dorsal base; and distally, the head, which is convex and articulates with the proximal phalanx of the corresponding toe. They are slightly curved over their axis, and that is why their superior surface becomes the medial surface progressively from the base to the head. The metatarsals are numbered as first to fifth from medial to lateral (Figs. 64.2 and 64.3).

They have some differences: the first metatarsal is the shorter, but the wider. The fifth one is also short but the thinnest one. The second is the largest one, and the third and the fourth, that are similar in longitude and volume, are slightly displaced from each other (the third is anterior).

The first three metatarsals articulate with the three cuneiforms, respectively, but characteristically the base of the second metatarsal acts as a keystone in the three cuneiform bones, because the intermedial cuneiform is shorter than the other two. The fourth and the fifth metatarsals articulate with the cuboid.

The joints that are formed between the hindfoot and the midfoot (the talonavicular, the calcaneocuboid, and the talocalcaneal joint) on the whole are named as Chopart joint, and the others formed between the hindfoot and the forefoot are named as Lisfranc joint.

64.1.7 Phalanges

64.1.7.1 Big Toe

The big toe has two phalanges: proximal and distal. The proximal phalanx has a large base directed transversely and bears an oval, concave articular surface, the glenoid cavity, smaller than the corresponding articular surface of the metatarsal head. A transverse crest or bony prominence is present on the dorsum of the base. On the plantar aspect, two tubercles, lateral and medial, give insertion to the intrinsic muscles of the big toe. The shaft of the proximal phalanx is convex dorsally and flat on the plantar aspect, and the articular surface is trochlear and is larger and more concave on the plantar aspect. The distal phalanx has a large, transversely oriented base bearing the articular surface corresponding to the trochlear surface of the proximal phalangeal head. On the plantar aspect of the shaft, the flexor hallucis longus tendon inserts on a tuberosity directed toward the fibular side and strongly marked on the tibial side (Fig. 64.2).

64.1.7.2 Lesser Toes

The lesser toes have three phalanges: proximal, middle, and distal. The proximal phalanx is the longest of the three, being slightly longer than the other two phalanges combined. The base is large and transverse, with an oval articular surface for the metatarsal head. Two plantar tubercles give insertion to the interossei muscles. The medial and lateral tubercles of the base give insertion to interossei muscles, depending on the toe. The shaft is convex dorsally and flat inferiorly. The head is flat and supports a trochlear type of articular surface that extends more on the plantar aspect (Fig. 64.2).

The middle phalanx is very short. The base bears a transverse articular surface corresponding to the trochlear contour of the proximal phalangeal head. The distal articular head is also transversely oriented but presents a strong convexity in the dorsoplantar direction.

The distal phalanx is rudimentary. It is more or less triangular, with a crescent-shaped contour of the distal tuberositas unguicularis. The dorsum of the shaft and the crescent-shaped distal tuberosity support the ungual matrix.

64.2 Arthrology

We will resume the most important concepts, articular surfaces, and ligaments of the foot's joints. We have to understand that the ligaments and biomechanics of the hindfoot are indivisible from the ankle.

64.2.1 Foot Joints

64.2.1.1 Subtalar Joint

The subtalar joint is a complex structure, with great variability between individuals. Anatomically, the subtalar joint is made up of two joints with separate synovial sheaths. The talocalcaneal joint (TCJ) constitutes the posterior part, while in the anterior part is the talocalcaneonavicular joint (TCNJ) (Fig. 64.3). Both are separated by the interosseous calcaneal ligament and the sinus tarsi, a sulcus formed between the posterior and medial facets of the talus and calcaneus [4–7].

The talocalcaneal joint (TCJ) articulates with the posterior part of the talus and calcaneus. The talus has two large oval biconcave facets, while the calcaneus has a complementary biconvex facet [4, 7, 8].

The talocalcaneonavicular joint (TCNJ) articulates with the anterior and medial facets of the calcaneus and talus, the inferomedial facet at the talar head and the spring ligament complex, and the convex ball-shaped distal surface of the head of the talus articulates with the posterior articular surface of the navicular. This joint complex has been referred to in the literature as the acetabulum pedis, due to its similarity to the structure of the hip joint, where the head of the talus forms the femoral head, and the calcaneal, navicular, and spring ligament surfaces with which it articulates form the acetabulum [4–7]. This morphology would provide the TCNJ with a wide range of motion in three dimensions; nevertheless, its movement is coupled to the TCJ, with a more limited range of motion due to its concave-convex morphology [5].

The motion of the subtalar joint is triplanar, due to the complex morphology of articular surfaces and distribution of motion axes in the three dimensions. Clinically, this triplanar motion is classified as inversion (flexion, supination, and adduction) and eversion (dorsiflexion, pronation, and abduction, with a range of movement of 25–30° and 5–10°, respectively) [4, 5].

64.2.1.2 Midtarsal Joint (Chopart Joints)

The Chopart joint works closely linked to the subtalar joint, to a greater extent through the talonavicular joint, which is considered part of both joint complexes, thus forming a triple joint complex.

Midtarsal joint complex comprises the talocalcaneonavicular and calcaneocuboid joints. Nevertheless, there is also a functional unit between cuboid and navicular bones that slide on the anterior surface of the calcaneus and the head of the talus [9]. The calcaneocuboid joint is a condyloid joint formed by anterior calcaneal surface, having a convex transverse and concave vertical configuration, and articular surface of the cuboid, concave transversely and convex vertically, both in saddle-shaped configuration with a rotational mobility range of 25° during the inversion-eversion movement of the foot. All this joint complex provides flexibility and stiffness at different moments of the gait cycle, allowing full load of the body weight during planted foot and providing stability during the propulsion phase of the gait [9–11].

64.2.1.3 Cuneonavicular Joint

The cuneonavicular joint is a flat joint composed of four bones: navicular and the three cuneiform bones: medial, intermediate, and lateral. The distal surface of the navicular is entirely articular, with three facets that articulate with each of the three cuneiforms. Although it has three articular surfaces, the complex is a single joint surrounded by a dense capsule and ligaments that provide stability. The intermediate cuneiform is distally shortened with respect to the medial and lateral cuneiforms approximately 8 mm and 4 mm, respectively [12, 13]. The joint complex participates in the pronation-supination and flexion-extension movements of the foot with a mobility gradient that decreases from medial to lateral [9].

64.2.1.4 Tarsometatarsal Joint (Lisfranc Joint)

The Lisfranc joint complex is made up of intertarsal, intermetatarsal, and tarsometatarsal gliding joints. At the proximal level, it is composed of the three cuneiform and cuboid bones, and the distal level includes the five metatarsals. This complex forms three independent columns with their own joint capsules, forming part of the plantar arch: medial, central, and lateral [14]. This column system is crucial for a normal gait cycle, especially during weight transfer on uneven terrain [15, 16].

The medial column is made up of the first metatarsal and medial cuneiform. There is inter-individual morphological variability of the articular surfaces between these two bones, which gives a greater or lesser range of mobility and stability to the joint [17]. Its range of motion in the sagittal plane is 3.5 mm. The intermediate column is formed by the intermediate and lateral cuneiforms, and the second and third metatarsals. The base of the second metatarsal is lengthened proximally until it is embedded in a mortise formed by the medial and lateral cuneiforms. This configuration allows less sagittal mobility with a range of 0.6 mm; however, it provides great joint stability. In addition, another locking system is established between the distal portion of the lateral cuneiform and the bases of the second and fourth metatarsals that reinforce the stability of the articular complex of the midfoot. The lateral column is formed by the joints that the cuboid forms with the fourth and fifth metatarsals. It is the one with the greatest range of sagittal mobility, reaching 13 mm [6, 14–18].

64.2.1.5 Metatarsophalangeal Joints

The metatarsophalangeal joint of the hallux is a condyloid joint surrounded by a fibrous capsule, which articulates the proximal part of the first phalanx with the head of the first metatarsal. The articular surface of the phalanx is concave, while that of the first metatarsal is convex. Unlike the lesser metatarsals, the head of the first metatarsal has a central plantar ridge that separates two grooves in which the sesamoid bones are located. It can perform movements in flexion-extension, abduction-adduction, and longitudinal rotation, with flexion-extension being the one with the greatest range of motion, around 74° [6, 19–21].

The lesser metatarsophalangeal joints are also condyloid joints, with a bicondylar articular surface, in which the lateral condyle is greater. The heads of the lesser metatarsals, unlike the metatarsal head of the hallux, are larger vertically than transversely. The convex heads articulate with the concave articular surfaces of the first phalanges, which are small relative to the size of the metatarsal heads. This results in a more unstable joint, similar to the shoulder joint. They can perform the same movements as the hallux joint, with a decreasing flexion-extension range from medial to lateral, ranging from 105° of the second toe to 95° of the fifth toe [6, 21, 22].

64.2.1.6 Interphalangeal Joints

Interphalangeal joints are hinge joints, more stable than metatarsophalangeal joints but with less mobility. They can perform flexion-extension movements. There are differences between the hallux and the lesser toes. The hallux has a single interphalangeal joint, while the lesser toes have a proximal and a distal joint. The interphalangeal joint of the hallux allows for extension above 0° (flexion 46°/extension 31°), while the interphalangeal joints of the lesser toes have a variable range of flexion, with a maximum extension up to 0° [6, 23].

64.2.2 Ligaments

64.2.2.1 Ligaments of the Talocalcaneonavicular Joints

The calcaneonavicular complex is a functional unit moving around the talus.

Cervical Ligament

The cervical ligament (external talocalcaneal ligament, anterolateral talocalcaneal ligament, anterior talocalcaneal ligament) is the strongest ligament connecting the talus and the calcaneus [24–27]. It originates in the anteromedial segment of the sinus tarsi from the cervical tubercle located on the medial aspect of the bony eminence giving origin to the extensor digitorum brevis muscle (Fig. 64.4). It inserts on the inferior aspect of the talar neck, where a tubercle may be present. The cervical ligament is in the form of a rectangular. As mentioned by Smith, the cervical ligament is separate from the ligament of the canalis tarsi in the majority [27].

Ligament of the Tarsal Canal

This ligament (interosseous talocalcaneal ligament) of the tarsal canal is a flat, oblique band that originates from the sulcus calcanei of the tarsal canal close to the anterior capsule of the posterior talocalcaneal joint but independent of it [24, 26, 27]. It inserts on the sulcus tali medially. The ligament is crossed anteriorly in an X by the canalicular portion of the medial root of the inferior extensor retinaculum.

Lateral Talocalcaneal Ligament

The lateral talocalcaneal ligament is a flat, short, rectangular ligament, parallel to the calcaneofibular ligament. It originates from the anteroinferior aspect of the lateral talar process, extends downward and posteriorly, and inserts on the os calcis just lateral to the posterior articular surface. It is slightly anterior and medial to the calcaneofibular ligament. Milner and Soames observed this ligament in 55% [28].

Posterior Talocalcaneal Ligament

The posterior talocalcaneal ligament is a short, flat, quadrilateral ligament directed downward and laterally. It originates from the lateral surface and apex of the posterolateral talar tubercle and inserts on the superior and medial aspect of the os calcis [29] (Fig. 64.5).

Medial Talocalcaneal Ligament

The medial talocalcaneal ligament is a strong, short ligament. It originates from the talar posteromedial tubercle, courses anteriorly and inferiorly, and inserts on the posterior border of the sustentaculum tali (Fig. 64.5).

Talonavicular Ligament

The talonavicular ligament is located in the dorsal interval between the superomedial calcaneonavicular ligament and the lateral calcaneonavicular ligament. It has two components: superficial and deep [24]. The superficial component originates from the dorsum of the

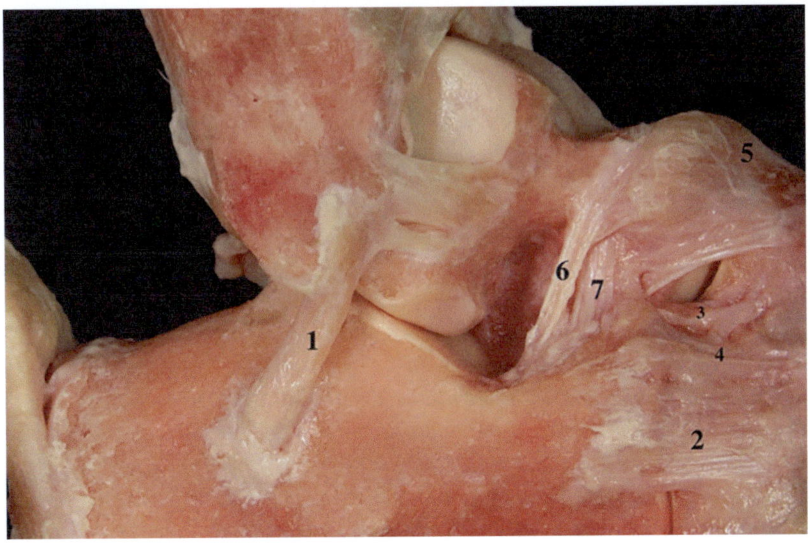

Fig. 64.4 Dorsolateral view of midtarsus and ankle ligaments. 1 Calcaneofibular ligament; 2 dorsolateral calcaneocuboid ligament; 3 lateral calcaneonavicular ligament; 4 medial calcaneocuboid ligament; 5 talonavicular ligament; 6 intermediate root of inferior extensor retinaculum; 7 cervical ligament

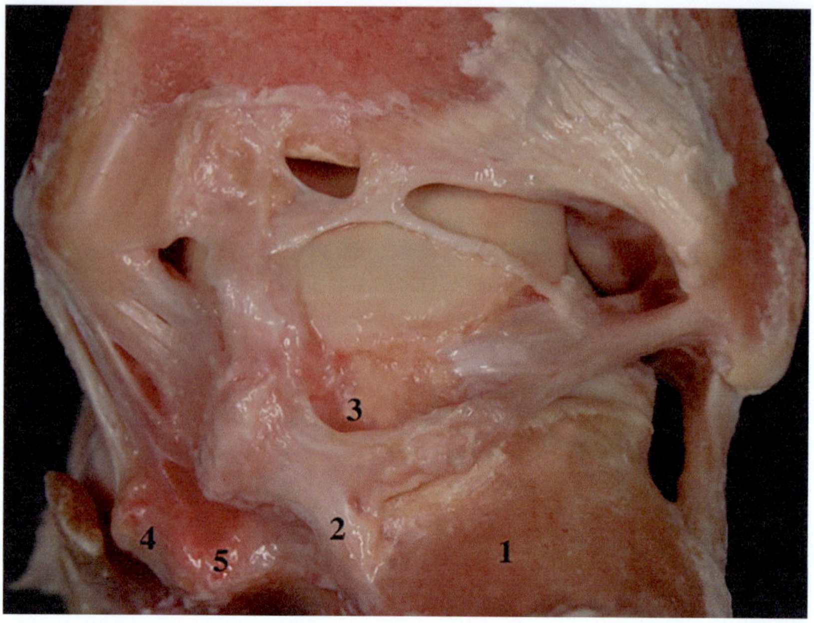

Fig. 64.5 Posterior view of ankle and subtalar joints. 1 Calcaneum; 2 posterior talocalcaneal ligament; 3 fibrous tunnel for flexor hallucis longus tendon; 4 sustentaculum tali; 5 medial talocalcaneal ligament

talar neck and the deep component originates from the superomedial aspect of the talar neck and inserts on the dorsum of the navicular (Fig. 64.4).

64.2.2.2 Ligaments of the Calcaneonavicular Joint and Acetabulum Pedis

The head of the talus is received into a deep socket or acetabulum pedis formed by the navicular, the anterior and middle calcaneal articulating surfaces, the calcaneonavicular component of the bifurcate ligament, the superomedial calcaneonavicular ligament, and the plantar calcaneonavicular ligament.

Spring Ligament Complex

The spring ligament is composed of the superomedial and inferior calcaneonavicular ligament.

Superomedial Calcaneonavicular Ligament

It was described as an individual ligament by Lane [30]. The superomedial calcaneonavicular ligament is quadrilateral, inseparable from the inferior calcaneonavicular ligament (Fig. 64.6). It originates from the medial and anterior borders of the sustentaculum tali and inserts on the superomedial aspect of the navicular and, to a lesser degree, on the lateral aspect of the tuberosity of the navicular. The tibionavicular and tibiocalcaneal components of the superficial deltoid ligament interlace with it. The superficial aspect of the ligament is further hidden by a thick fascial or fibrocartilaginous layer forming the floor of the tunnel crossing the tibialis posterior tendon (Fig. 64.7). It requires great care to dissect away this fascial stratum, and to remove the deposited cartilage, in order to expose the proper calcaneonavicular fibers [24].

Inferior Calcaneonavicular Ligament (Spring Ligament)

The inferior calcaneonavicular ligament (spring ligament) is trapezoidal and fasciculated and corresponds to the inferior segment of the talar head unsupported by the articular surfaces. It originates from the upper part of the calcaneus between the anterior border of the sustentaculum tali and the cuboidal articular surface. This ligament inserts into the plantar aspect of the navicular bone, laterally to the navicular beak, and just lateral to the insertion of the superomedial calcaneonavicular ligament [31]. This ligament has a fascicular morphology; it has two fascicles: the lateral fascicle, which inserts into the navicular beak, is the strongest (Fig. 64.6). Between the

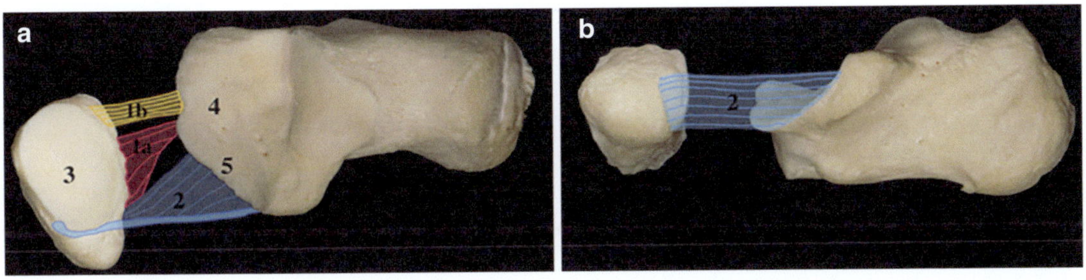

Fig. 64.6 Digital composition of the spring ligament complex. (**a**) Superior view of coxa pedis floor. (**b**). Medial view of calcaneus and navicular without talus. 1a, Medial fascicle; 1b lateral fascicle of inferior calcaneonavicular ligament; 2 superomedial calcaneonavicular ligament; 3 articular talar surface of navicular; 4 anterior talar articular surface; 5 middle talar articular surface

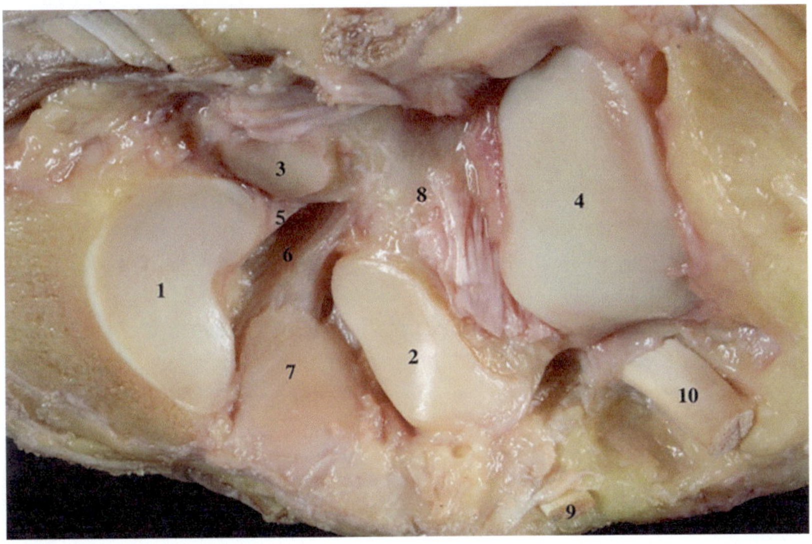

Fig. 64.7 Cranial view of coxa pedis. 1 Talar articular surface of navicular; 2 middle talar articular surface; 3 anterior talar articular surface; 4 posterior talar articular surface; 5 lateral fascicle of inferior calcaneonavicular ligament; 6 medial fascicle of inferior calcaneonavicular ligament; 7 superomedial calcaneonavicular ligament; 8 sinus tarsi; 9 flexor digitorum longus; 10 flexor hallucis longus

fiber bundles, there is a series of longitudinal intervals, with an adipose layer spread through them that covers the plantar aspect of this ligament.

Its dorsal surface is fibrocartilaginous for the support of the head of the talus and appears as a triangular morphology (Fig. 64.7). Medially, its plantar surface is in contact with the tendon of the posterior tibialis muscle, and, laterally, with the tendons of the flexor hallucis longus and flexor digitorum longus muscles [32].

Together with the anterior superficial fibers of the deltoid ligament, the long plantar ligament, and the plantar fascia, the inferior calcaneonavicular ligament is one of the main static stabilizers of the longitudinal arch of the foot [33].

Bifurcate Ligament

The bifurcate ligament (or Chopart ligament) is made up of the lateral calcaneonavicular ligament and the medial calcaneocuboid ligament, which are arranged in a "Y" or "V" with different origins at the os calcis.

The lateral calcaneonavicular ligament (Fig. 64.4) arises from the anteromedial angle of the sinus tarsi, just lateral to the anterior talar articular surface, and reaches the lateral aspect of the intermediary tubercle [34]. It extends anteriorly, dorsally, and medially to its insertion in the superior segment of the lateral end of the navicular.

The medial calcaneocuboid ligament (Fig. 64.4) arises from the anterior aspect of the intermediary tubercle, lateral to the origin of the lateral calcaneonavicular ligament. It runs anteriorly and slightly inferiorly and attaches into the dorsal aspect of the cuboid (approximately 1.5 cm anterior to the posterior margin of the cuboid).

They form an angle of approximately 30° in the transverse plane and 20° in the sagittal plane [33].

64.2.2.3 Ligaments of the Calcaneocuboid and Cubonavicular Joints

Medial Calcaneocuboid Ligament

The medial calcaneocuboid ligament has already been described as the outer component of the bifurcate ligament.

Dorsolateral Calcaneocuboid Ligament

The dorsolateral calcaneocuboid ligament (Fig. 64.4) originates from the dorsolateral corner of the anterior segment of the calcaneus close to the margin of the anterior articular surface of the calcaneus. It is directed, as a flat band, anteromedially and inserts on the dorsum of the cuboid at least 0.5 cm distal to the articular interline. Occasionally, a smaller, more lateral band is present.

Inferior Calcaneocuboid Ligament

The inferior calcaneocuboid ligament (long and short plantar ligaments) (Fig. 64.8) is a thick, powerful, longitudinally oriented ligament with two components: superficial or long and deep or short.

The superficial or long plantar ligament originates from the segment of the inferior surface of the os calcis extending from the anterior surface of the posterior tuberosities and their intertubercular segment to the anterior tuberosity. The strong longitudinal fiber passes over the calcaneocuboid joint and divides into two sets of fibers, deep and superficial. The deep fibers, representing the bulk of this component, insert on the oblique crest of the cuboid. The more superficial fibers form a thinner layer, cross the tunnel of the peroneus longus (contributing to its formation), and divide into four thinner slips inserting over the metatarsal bases 2–5.

Cubonavicular Ligaments

There are three cubonavicular ligaments: dorsal (Fig. 64.9), plantar, and interosseous.

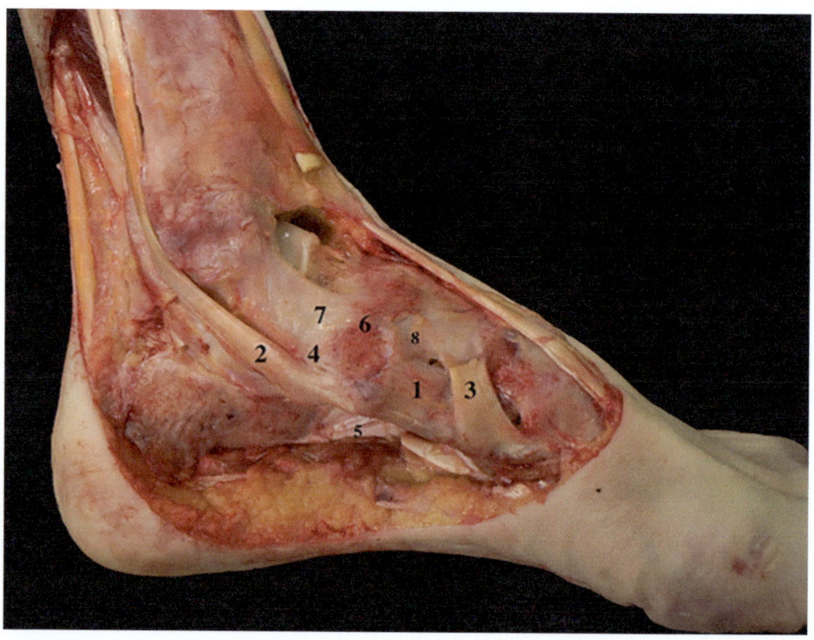

Fig. 64.8 Medial view of medial tendinous and ligamentous structures of the foot. 1 Medial cuneonavicular ligament; 2 tibialis posterior tendon; 3 tibialis anterior tendon; 4 superomedial calcaneonavicular ligament; 5 long plantar or long calcaneocuboid ligament; 6 navicular; 7 tibionavicular ligament

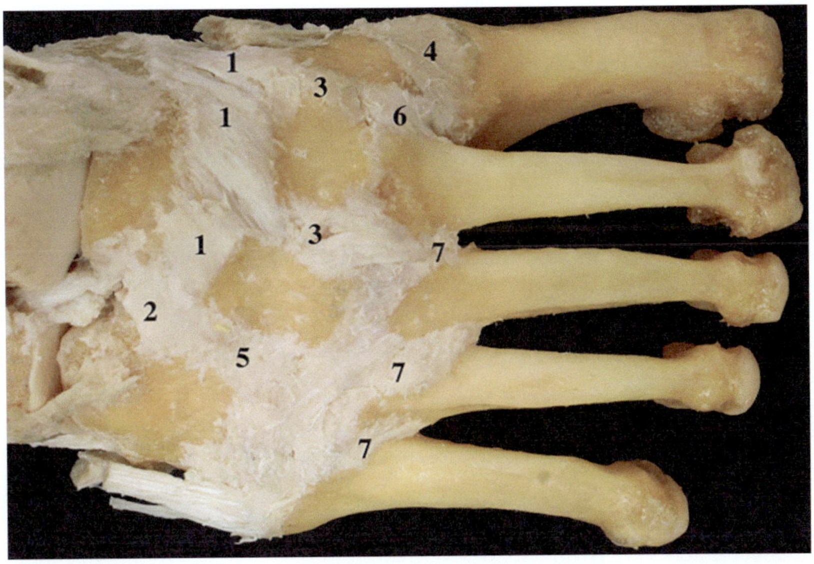

Fig. 64.9 Dorsal view of midtarsal ligaments. 1 Dorsal cuneonavicular ligaments; 2 dorsal cubonavicular ligament; 3 dorsal intercuneiform ligaments; 4 dorsal cuneo$_1$metatarsal$_1$ ligament; 5 dorsal cubocuneiform ligaments; 6 dorsal cuneo$_1$metatarsal$_2$; 7 intermetatarsal ligaments

64.2.2.4 Ligaments of the Cuneonavicular and Cuneocuboid Joints

Cuneonavicular Ligaments

Each cuneiform is united to the navicular by a dorsal and a plantar ligament. In addition, the first cuneiform is united to the navicular by the medial cuneonavicular ligament (Fig. 64.9).

The first cuneonavicular ligament extends straight anteriorly and inserts on the dorsum of the first cuneiform; it is the strongest of the three. They are partially covered by an expansion from the talonavicular ligament.

There are also three plantar cuneonavicular ligaments. The first arises from the anterior and plantar aspect of the tuberosity of the navicular; extends anteriorly as a thick, rectangular, flat cuff; and inserts on the plantar tuberosity of the first cuneiform. The second and third ligaments arise from the inferior surface of the navicular, between the tuberosity and the beak. They are thin and deep and are masked by the expansions of the tibialis posterior tendon. The third ligament is oriented anterolaterally and is the longest; the second ligament is the deepest.

The medial cuneonavicular ligament is a thick, strong ligament that extends from the medial aspect of the tuberosity of the navicular to the medial aspect of the first cuneiform and receives a few fibers from the tibialis posterior tendon.

Cuneocuboid Ligaments

There are three cuneocuboid ligaments: dorsal (Fig. 64.9), plantar, and interosseous.

Intercuneiform Ligaments

There are two dorsal (Fig. 64.9), two interosseous, and one plantar intercuneiform ligaments.

64.2.2.5 Ligaments of the Tarsometatarsal Joint (Lisfranc Joint)

The tarsometatarsal joint (Lisfranc joint), connecting the cuneocuboid block to the bases of the metatarsals, has a complex joint interline configuration. The line is oblique, directed laterally and posteriorly, and presents a dorsolateral convexity corresponding to the transverse cubocuneiform arch.

The ligaments connecting the cuboid and the cuneiforms to the metatarsal bases are the dorsal, plantar, and interosseous ligaments.

There are seven dorsal tarsometatarsal ligaments. The base of the first metatarsal is united by a large, thick ligament to the first cuneiform. This cuneo$_1$metatarsal$_1$ ligament is in a dorsomedial location and is the strongest.

The plantar cuneometatarsal ligaments are always present on the medial aspect of Lisfranc's joint, but they are very variable in number and disposition on the lateral side.

The cuneo$_1$metatarsal$_{2,3}$ ligament is a very strong ligament, considered by Sappey to be the key of the tarsometatarsal arch [35]. The ligament originates from the inferolateral surface of the cuneiform$_1$ and soon divides into two bands.

There is no ligament between cuneiform$_2$ and metatarsal$_2$ on the plantar aspect.

There are three sets of interosseous ligaments, corresponding to the first, second, and third cuneometatarsal spaces. There are none in the fourth interspace.

The first interosseous cuneo$_1$metatarsal$_2$ ligament (Fig. 64.9) (Lisfranc ligament, medial interosseous ligament) is the strongest ligament of the three. It arises from the lateral surface of the first cuneiform in front of the intercuneiform ligament and under the articular surface corresponding to the second metatarsal; it is directed obliquely and slightly downward and inserts on the lower half of the medial surface of the second metatarsal base. The ligament measures nearly 1 cm in height and approximately 0.5 cm in thickness.

The second interosseous cuneometatarsal ligament (middle interosseous ligament) has a complex ligamentous arrangement. It is located between the cuneiforms and metatarsals 2 and 3.

64.2.2.6 Intermetatarsal Ligaments

The intermetatarsal ligaments are dorsal, plantar, and interosseous. There is no ligament between metatarsal$_1$ and metatarsal$_2$ on the dorsal or plantar aspect (Fig. 64.9).

64.2.2.7 Ligaments of Metatarsophalangeal Joints and Proximal Phalangeal Apparatus

At the level of the metatarsophalangeal joint of the lesser toes, the proximal phalanx and the fibrocartilaginous plantar plate form an anatomic and functional unit. They are both suspended from the sides of the metatarsal head through the collateral and the suspensory glenoid ligaments. Furthermore, the plantar plate is connected on each side by the deep transverse intermetatarsal ligament. On the dorsal aspect, the plantar plate gives insertion to the accessory collateral ligament or the metatarsoglenoid suspensory ligament, the transverse lamina of the extensor aponeurosis, and the corresponding interossei muscles at the junction of the deep transverse intermetatarsal ligament with the plantar plate [36].

The anatomical unit formed by the proximal phalanx, the plantar plate, and their insertional connections is called the phalangeal apparatus.

Metatarsophalangeal Ligaments of the Lesser Toes

The lateral ligaments of the metatarsophalangeal joints are divided into metatarsophalangeal collateral ligaments and metatarsoglenoid suspensory ligaments. The lateral ligaments on the peroneal side are thicker and stronger than those on the tibial side. The metatarsophalangeal ligament originates on the lateral tubercle of the metatarsal heads and inserts on the lateral tubercle of the base of the proximal phalanx.

The metatarsoglenoid ligament or suspensory ligament originates from the posteroinferior aspect of the lateral metatarsal tubercle of the head. It is triangular (fan shaped), and the fiber descends vertically in the posterior part and obliquely in the anterior part and inserts on the lateral border of the plantar plate.

Interphalangeal Joint Ligaments of the Lesser Toes

The collateral ligaments of the interphalangeal joints extend from the lateral aspect of the head of the corresponding phalanx to the base of the distally located phalanx.

Proximal Phalangeal Apparatus of the Big Toe

The two sesamoids, embedded in the thick fibrous plantar plate and united to the proximal phalanx of the big toe, form an anatomic and functional unit called (by Gillette) the sesamophalangeal apparatus (Fig. 64.10b). In hallux valgus or in traumatic displacements, the sesamoids always follow the proximal phalanx and are displaced with the latter, not with the metatarsal head.

The two sesamoids are united by a thick intersesamoid ligament. Each sesamoid is united, by an ill-defined short sesamophalangeal ligament (Fig. 64.10a), to the base of the proximal phalanx, which makes the distal attachment to the plantar plate stronger on the sides and weaker centrally.

The proximal border of the plantar plate is complex. The central segment is in continuity proximally with synovial type tissue and anchors on the neck of the first metatarsal. This central segment provides attachment to the two vertical septa of the plantar aponeurosis of the first ray and also blends with the proximal segment of the fibrous tunnel of the flexor hallucis longus tendon. The lateral and medial segments of the proximal border of the plantar plate give partial insertion to the lateral and medial heads of the flexor hallucis brevis, respectively.

The plantar surface of the plantar plate is raised on each side by the medial and lateral sesamoids. Between the two, a groove for the flexor hallucis longus tendon is formed, converted into a fibrous tunnel by arcuate fibers. These fibers receive contributions from the two vertical septa of the plantar aponeurosis.

The proximal segment of each sesamoid gives insertion to the corresponding head of the flexor hallucis brevis. The deep transverse metatarsal ligament attaches longitudinally along the lateral sesamoid. Dorsal to this ligament, the lateral sesamoid gives insertion to the oblique and transverse components of the adductor hallucis muscle. Vertical fibrous bands extend from the sides of the sesamoids and the flexor hallucis longus tendon sheath and connect with the superficial band of the plantar aponeurosis.

Metatarsophalangeal Ligaments of the Big Toe

The metatarsal head is united to the proximal phalangeal apparatus by two sets of ligaments: the lateral collateral ligaments and the metatarsosesamoid suspensory ligaments (Fig. 64.10). These two sets originate from the lateral tubercle of the metatarsal head, the origin of the lateral collateral ligament slightly covering the origin of the suspensory ligament. The lateral collateral ligament is directed downward anteriorly and inserts on the tubercle at the base of the proximal phalanx. The thick metatarsosesamoid or suspensory ligament descends vertically and inserts on the lateral and medial borders of the plantar plate.

Interphalangeal Joint Ligaments of the Big Toe

The interphalangeal joint of the big toe may possess a sesamoid bone, median and transversely oriented, embedded in the plantar plate and located above the flexor hallucis longus tendon. The sesamoid is connected to the sides of the distal phalanx with two small ligaments [37]. Two collateral ligaments extend from the lateral aspect of the proximal phalangeal head to the tubercle at the base of the distal phalanx.

64.2.2.8 Plantar Aponeurosis

The plantar aponeurosis is the strong, fibrous, investing layer of the sole of the foot. It is subcutaneous and extends from the heel to the ball of the foot. Two longitudinally oriented intermuscular septa connect the plantar aponeurosis to the deep planta pedis. The plantar aponeurosis has three components: central, lateral, and medial.

The central or major component of the plantar aponeurosis is triangular, with a posterior apex and an anterior base. It originates from the plantar aspect of the posteromedial calcaneal tuberosity. It conforms to the convexity of the tuberosity and may receive contributions from the Achilles tendon and especially from the plantaris tendon.

The peroneal component (lateral) of the plantar aponeurosis is variable. It was present in 92% and absent in 7% in a study of 410 plantar aponeuroses [38]. This peroneal component may be of four types: complete and well developed, complete and thin, incomplete with only partial distal extension, or absent distal segment [38]. The well-developed peroneal component originates from the lateral margin of the medial calcaneal tubercle in close connection with the origin of the abductor digiti minimi muscle. It inserts on the base of the fifth metatarsal and forms the calca-

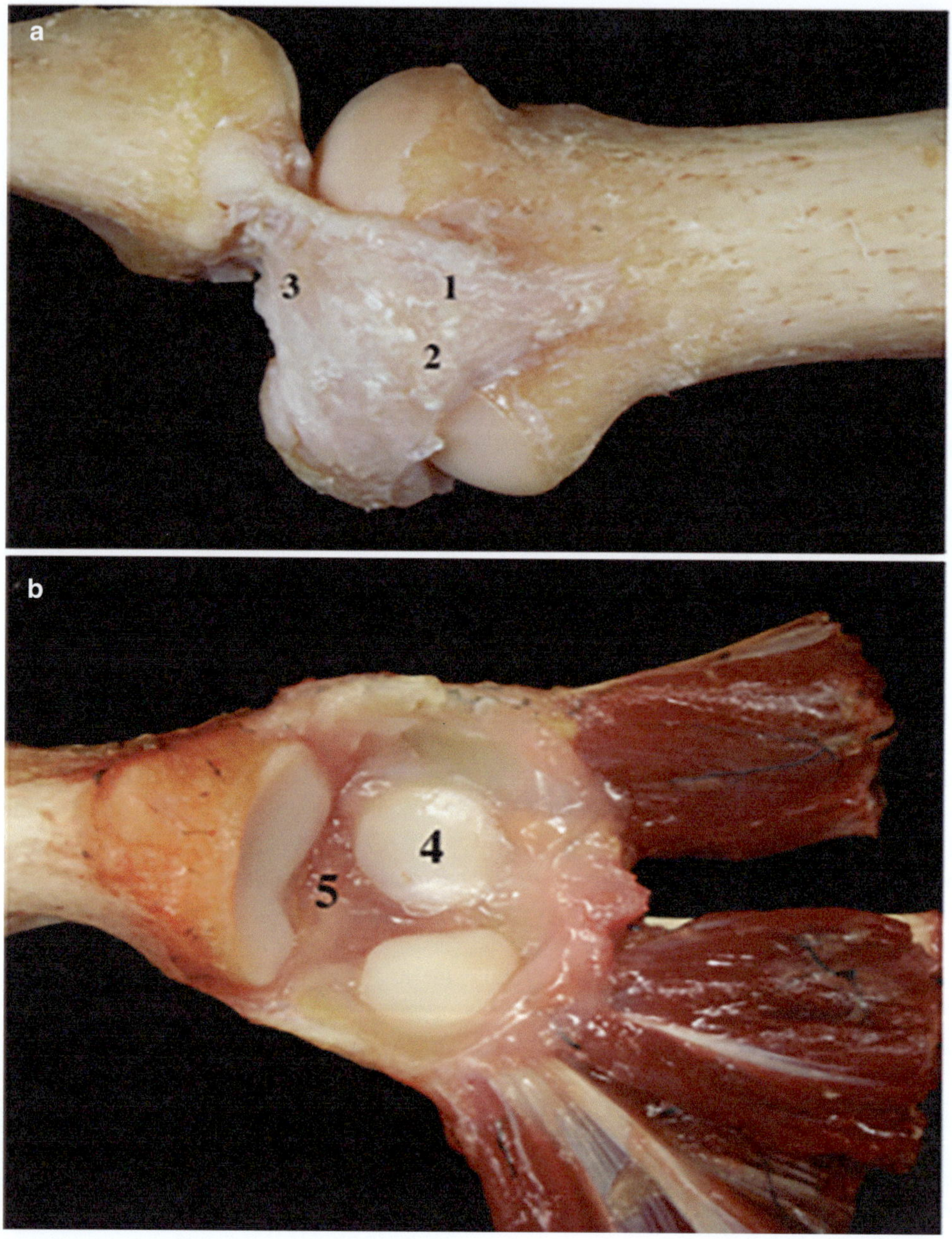

Fig. 64.10 Metatarsophalangeal joint of big toe. (**a**) Medial view of metatarsophalangeal joint with medial sesamoid. (**b**) Cranial view of sesamoid bones. 1 Medial collateral ligament; 2 medial metatarsoglenoid suspensory ligament; 3 sesamophalangeal ligament; 4 lateral sesamoid bone; 5 sesamophalangeal apparatus

neometatarsal ligament, and a medial band blends with the plantar plate of the fourth and, occasionally, the third metatarsophalangeal joints.

The medial component of the plantar aponeurosis is thin posteriorly and thicker anteriorly. It forms the covering fascia of the abductor hallucis muscle.

64.3 Myology

When talking about the muscles of the foot and ankle, we have got to differentiate between extrinsic muscles (those originated outside the region of the foot) and intrinsic muscles (those originated within the foot region).

64.3.1 Extrinsic Muscles

As per the extrinsic muscles, which arise in the leg but act on both the foot and ankle, functional muscle groups are organized in fascial compartments: the anterior compartment, in which the tibialis anterior (Fig. 64.8), extensor digitorum longus (EDL), extensor hallucis longus (EHL), and peroneus tertius muscles are contained; the lateral compartment, in which the peroneus longus and brevis muscles are contained; the superficial posterior compartment, in which the gastrocnemius, soleus, and plantaris muscles are contained; and finally the deep posterior compartment, in which the tibialis posterior, flexor hallucis longus (FHL), flexor digitorum longus (FDL), and popliteus muscles are contained. We will now proceed to develop a thorough description of their most important traits, such as origin, insertion, and main function [6, 39].

Within the anterior compartment, the tibialis anterior muscle arises from the lateral tibial surface, and it passes deep to the extensor retinaculum along the medial foot to insert in the medial cuneiform and plantar portion of the first metatarsal, its main function being that of dorsally flexing and inverting the foot. The EDL muscle originates from the lateral tibial condyle and anterior fibula, passing deep to the extensor retinaculum and dividing into four slips that insert on the dorsal aspects of the four lateral radiuses, being its main function that of dorsally flexing the toes. The EHL muscle originates from the midfibula, running deep to the tibialis anterior and EDL, becoming superficial in the lower leg and inserting into the distal phalanx of the hallux, its main function being that of extending the first radius. Finally, the peroneus tertius muscle originates from the distal anterior fibula and could be considered a portion of the EDL due to its intimate anatomic relation, having its insertion on the dorsal aspect of the base of the fifth metatarsal, varying in size and sometimes being absent [40].

The lateral compartment consists of the peroneus longus and brevis muscles, both arising from the superior lateral surface of the fibula, with the origin of the peroneus longus being more superior and running superficial to the peroneus brevis. Their tendons run together through a common tendon sheath and go posterior to the lateral malleolus, deep to the peroneal retinaculum, inserting on the base of the fifth metatarsal (peroneus brevis) and the base of the first metatarsal (peroneus longus) (Fig. 64.11d). Their

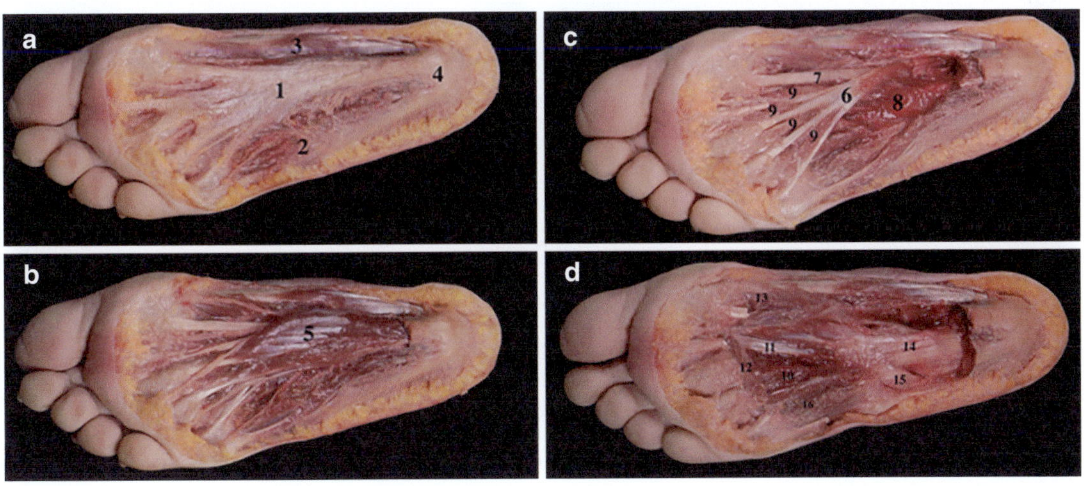

Fig. 64.11 (**a**) Most superficial layer of the plantar of the foot. (**b**) Fascia plantar rejected. (**c**) Second layer of plantar muscles of the foot. (**d**) The deepest layer of the intrinsic muscles. 1 Plantar aponeurosis; 2 abductor digiti minimi; 3 abductor hallucis; 4 calcaneal tuberosity; 5 flexor digitorum brevis; 6 flexor digitorum longus; 7 flexor hallucis longus; 8 quadratus plantae; 9 lumbrical muscles of the foot; 10 interosseus muscles; 11 adductor hallucis (oblique head); 12 adductor hallucis (transverse head); 13 flexor hallucis brevis; 14 long plantar ligament; 15 peroneus longus tendon; 16 flexor digiti minimi brevis

main function is that of everting the foot and assisting in plantar flexion [40, 41].

The superficial posterior compartment consists of three muscles. The gastrocnemius muscle, which two heads originate from the femoral condyles, is the most superficial muscle of the calf, transitioning into the flat tendon which, together with the distal portion of the soleus tendon (the next muscle in depth in this compartment), comprises the tendo calcaneus. These two muscles' main function is that of plantar flexing the foot. The plantaris is the third muscle of this compartment, and it originates from the lateral femoral condyle, being a small muscle that passes between the gastrocnemius and soleus muscles in an oblique direction, with a long tendon running to distally insert in the calcaneus, calcaneus tendon, or flexor retinaculum. Its main function is to minimally aid in knee flexion and plantar flexion of the foot.

Finally, the deep posterior compartment contains four different muscles, but we will only discuss the ones acting on the foot, leaving the popliteus muscle. The FHL originates from the lateral aspect of the middle third of the fibula, running from lateral to medial superficial to the tibialis posterior and deep to the flexor retinaculum (Fig. 64.7), inserting on the plantar aspect of the distal phalanx of the first radius, with flexing the hallux being its main function. The FDL originates from the posteromedial aspect of the tibia, running medially and superficial to the tibialis posterior and posterior to the medial malleolus, dividing into four slips which insert on the distal phalanges of the lateral four radius, with its main function being to flex the toes. Lastly, the deepest muscle in this compartment is the tibialis posterior muscle, which originates from the upper and central posterior aspect of the tibia and fibula, running deep between the FHL and FDL and passing behind the medial malleolus to insert on the navicular bone, tarsal bones, and bases of the second to fourth metatarsal as it spreads its tendon distally. Its main function is to aid in plantar flexion, adduction, and inversion of the foot.

It is of importance to note that these extrinsic muscles run, more specifically their tendons, through two important structures before they enter the foot region, which are the superior extensor retinaculum and the inferior extensor retinaculum in the anterior aspect of the ankle. As for the posterior aspect, there are also two anatomic structures of importance through where these tendons run, the posterior aspect of the tibia malleolus and the posterior aspect of the fibula malleolus [6].

64.3.2 Intrinsic Muscles

As per the intrinsic muscles of the foot, there is a distinction between those in the dorsal aspect of the foot and those in the plantar region of the foot. The extensor digitorum brevis (EDB) is the main dorsal muscle of the foot. On the other hand, the plantar region of the foot must be divided into layers, rather than compartments, given the fact that there are no clearly differentiated fascial structures in between them. Therefore, four different layers of muscles form the plantar region of the foot: the superficial layer of the plantar muscles includes the abductor hallucis, flexor digitorum brevis (FDB), and abductor digiti minimi; the second layer of the plantar muscles includes the quadratus plantae and lumbrical muscles, together with the tendons of the FHL and FDL; the third layer of the plantar muscles includes the flexor hallucis brevis (FHB), flexor digiti minimi brevis, and adductor hallucis muscles; finally, the deepest fourth layer of the plantar muscles is formed by the seven interosseous muscles (four dorsal and three plantar) [42, 43].

The main dorsal muscle of the foot is the EDB, arising from the superior calcaneus and the extensor retinaculum, with an oblique course toward the medial side and ending in four tendons which insert on the lateral aspect of the proximal phalanx of the hallux and the lateral EDL tendons of the second to fourth radiuses. Its main functions are to extend the hallux and second to fourth radiuses.

The most superficial layer (Fig. 64.11b) of the plantar muscles of the foot is formed by three muscles. The abductor hallucis originates from the medial process of the calcaneal tubercle and inserts on the medial side of the plantar surface of the proximal phalanx of the hallux, with its main function being to abduct the metatarsophalangeal joint of the hallux. The flexor digitorum brevis originates from the medial tubercular process of the calcaneus and plantar fascia, dividing into four tendons, each one with two distal slips inserting on the plantar aspect of the middle phalanges of the second to fifth lateral radiuses, with their main function being to flex the said lateral radius. The abductor digiti minimi arises from the lateral process of the calcaneal tubercle, running laterally and inserting on the lateral aspect of the base of the proximal phalanx of the fifth radius, serving to flex and abduct the said toe.

The second layer (Fig. 64.11c) of the plantar muscles of the foot contains two intrinsic muscles and tendons of the extrinsic muscles of the deep posterior compartment (FHL and FDL, which cross to connect in the midfoot location known as the knot of Henry) [44]. The quadratus plantae has two heads that originate from the medial and lateral plantar aspects of the calcaneal tuberosity, inserting distally in the posterolateral margin of the FDL just before its division into its four tendons, assisting the said muscle in flexing the four lateral radiuses. The four lumbrical muscles originate from the FDL tendon and insert distally into the medial aspect of the metatarsophalangeal joints of the four lateral radiuses, acting as flexors of the said joints.

The third layer (Fig. 64.11d) of the plantar muscles of the foot includes three intrinsic muscles. The FHB originates from the plantar aspect of the cuboid bone, with its tendons inserting on both sides of the base of the proximal phalanx of the hallux and sesamoid bones, serving as a flexor of the hallux. The adductor hallucis has two bellies: an oblique one originated from the second to fourth plantar metatarsal bases and a smaller transverse one originated from the plantar aspect of the third to fifth metatarsophalangeal joints. These two bellies join and insert on the lateral plantar aspect of the proximal phalanx of the hallux, acting as an adductor and flexor. The flexor digiti minimi brevis originates from the cuboid and plantar base of the fifth metatarsal, inserting on the lateral base or the proximal phalanx of the fifth radius and serving as a flexor.

The fourth layer of the plantar muscles of the foot contains the deepest muscles, which are the seven interosseous muscles (four dorsal and three plantar). The four dorsal interossei originate in two heads from both sides of the metatarsal bases and insert into the lateral side of the bases of the proximal phalanges, whereas the three plantar interossei originate from the bases of the third to fifth metatarsal bones and insert into the medial sides of the proximal phalanges of the said radius. Therefore, the dorsal interossei act as abductors and the plantar interossei as adductors [6, 39, 42, 43].

References

1. Viladot A, Viladot R. 20 lecciones sobre patología del pie. 1st ed. Barcelona: Ediciones Mayo; 2009.
2. Greisberg J, Vosseller T. Core knowledge in orthopaedics: foot and ankle. 2th ed. Columbia: Elsevier; 2018.
3. Testut L, Latarjet A. Anatomía humana: tomo 1. 7th ed. Barcelona: Salvat editores; 1943.
4. Sangeorzan A, Sangeorzan B. Subtalar joint biomechanics: from Normal to pathologic. Foot Ankle Clin. 2018;23(3):341–52.
5. Bartoníček J, Rammelt S, Naňka O. Anatomy of the subtalar joint. Foot Ankle Clin. 2018;23(3):315–40.
6. Kelikian AS. Sarrafian's anatomy of the foot and ankle: descriptive, topographic, functional. 3rd ed. Philadelphia, PA: Lippincott; 2011.
7. Krähenbühl N, Horn-Lang T, Hintermann B, Knupp M. The subtalar joint: a complex mechanism. EFORT Open Rev. 2017;2(7):309–16.
8. Maceira E, Monteagudo M. Subtalar anatomy and mechanics. Foot Ankle Clin. 2015;20(2):195–221.
9. Krähenbühl N, Lenz AL, Lisonbee RJ, Peterson AC, Atkins PR, Hintermann B, Saltzman CL, Anderson AE, Barg A. Morphologic analysis of the subtalar joint using statistical shape modeling. J Orthop Res. 2020;38(12):2625–33.
10. Walter WR, Hirschmann A, Alaia EF, Tafur M, Rosenberg ZS. Normal anatomy and traumatic injury of the Midtarsal (Chopart) joint complex: an imaging primer. Radiographics. 2019;39(1):136–52.
11. Walter WR, Hirschmann A, Tafur M, Rosenberg ZS. Imaging of Chopart (Midtarsal) joint complex: Normal anatomy and posttraumatic findings. AJR Am J Roentgenol. 2018;211(2):416–25.
12. Renner K, McAlister JE, Galli MM, Hyer CF. Anatomic description of the Naviculocuneiform articulation. J Foot Ankle Surg. 2017;56(1):19–21.

13. Ford LA, Hamilton GA. Naviculocuneiform arthrodesis. Clin Podiatr Med Surg. 2004;21(1):141–56.
14. DiDomenico LA, Cross D. Tarsometatarsal/Lisfranc joint. Clin Podiatr Med Surg. 2012;29(2):221–42. vii-viii
15. Mulcahy H. Lisfranc injury: current concepts. Radiol Clin N Am. 2018;56(6):859–76. https://doi.org/10.1016/j.rcl.2018.06.003. Epub 2018 Sep 17
16. Chen J, Sagoo N, Panchbhavi VK. The Lisfranc injury: a literature review of anatomy, Etiology, Evaluation, and Management. Foot Ankle Spec. 2020; 14:458.
17. Koury K, Staggers JR, Pinto MC, Godoy-Santos AL, Smyth NA, Shah AB, de Cesar NC. Radiographic assessment of first Tarsometatarsal joint shape and orientation. Foot Ankle Int. 2019;40(12):1438–46.
18. Arastu MH, Buckley RE. Tarsometatarsal joint complex and midtarsal injuries. Acta Chir Orthop Traumatol Cechoslov. 2012;79(1):21–30.
19. Rajczy RM, McDonald PR, Shapiro HS, Boc SF. First metatarsophalangeal joint arthrodesis. Clin Podiatr Med Surg. 2012;29(1):41–9.
20. Ashimolowo T, Dunham G, Sharp JW, Porrino J. Turf toe: an update and comprehensive. Radiol Clin North Am. 2018;56(6):847–58. https://doi.org/10.1016/j.rcl.2018.06.002.
21. Nery C, Baumfeld D, Umans H, Yamada AF. MR imaging of the plantar plate: Normal anatomy, turf toe, and other injuries. Magn Reson Imaging Clin N Am. 2017;25(1):127–44.
22. Finney FT, Cata E, Holmes JR, Talusan PG. Anatomy and physiology of the lesser metatarsophalangeal joints. Foot Ankle Clin. 2018;23(1):1–7.
23. Manganaro D, Dollinger B, Nezwek TA, Sadiq NM. Anatomy, Bony Pelvis and Lower Limb, Foot Joints. In: StatPearls [Internet]. Treasure Island (FL): StatPearls Publishing; 2020.
24. Barclay-Smith E. The astragalo-calcaneo-navicular joint. J Anat Physiol. 1896;30:390.
25. Jones WF. Structure and function as seen in the foot. 2nd ed. London: Baillière; 1949. p. 120.
26. Cahill DR. The anatomy and function of the contents of the human tarsal sinus and canal. Anat Rec. 1965;153:1.
27. Smith JW. The ligamentous structures in the canalis and sinus tarsi. J Anat. 1958;92:616.
28. Milner CE, Somas RW. Anatomy of the collateral ligaments of the human ankle joint. Foot Ankle Inter. 1998;19(11):757–60.
29. Sarrafian SK. Anatomy of the foot and ankle. Descriptive, topographic, functional. 2nd ed. Philadelphia: JLippincott; 1993. p. 163–223.
30. Lane AS. The causation, pathology and physiology of several of the deformities which develop during young life. Guys Hosp Rep. 1887;44:254.
31. Golanó P. The anatomy of the navicular and periarticular structures. Foot Ankle Clin N Am. 2004;9:1–23.
32. Borton DC, Saxby TS. Tear of the plantar calcaneonavicular (spring) ligament causing flatfoot: a case report. J Bone Joint Surg Br. 1997;79B:641–3.
33. Davis WH, Sobel M, DiCarlo EF, et al. Gross, histological, and microvascular anatomy and biomechanical testing of the spring ligament complex. Foot Ankle. 1996;17:95–102.
34. Kokturk C. Remarques sur le ligament de Chopart (lig. Bifurcatum) [Remarks on Chopart's ligament (lig.bifurcatum)]. Comptes-Rendus Assoc Anat. 1957;44:380. [in French]
35. Sappey PC. Traité d'Anatomie Descriptive. 4th ed. Paris, Delahaye, Lecrosnier; 1888. p. 712–4. 728
36. Meyer P. Contribution à l'étude de la region métatarsophalangienne. Comptes-Rendus Assoc Anat. 1958;44:500.
37. Gillette. Des os sesamoides. J Anat Physiol Normal Pathol. 1872;8:506.
38. Loth EM. Etude anthropologique Sur l'aponévrose plantaire. Bull Mem Soc Anthro Paris. 1913;4:601.
39. Berquist TH, editor. Imaging of the foot and ankle. Lippincott; 2010.
40. Wang XT, Rosenberg ZS, Mechlin MB, Schweitzer ME. Normal variants and diseases of the peroneal tendons and superior peroneal retinaculum: MR imaging features. Radiographics. 2005;25(3):587–602.
41. Bejjani FJ, Jahss MH. Le Double's study of muscle variations of the human body part II: muscle variations of the foot. Foot Ankle. 1986;6(4):157–76.
42. Goodwin DW, Salonen DC, Yu JS, Brossmann J, Trudell DJ, Resnick DL. Plantar compartments of the foot: MR appearance in cadavers and diabetic patients. Radiology. 1995;196(3):623–30.
43. Loeffler RD Jr, Ballard A. Plantar fascial spaces of the foot and a proposed surgical approach. Foot Ankle. 1980;1(1):11–4.
44. O'Sullivan E, Carare-Nnadi R, Greenslade J, Bowyer G. Clinical significance of variations in the interconnections between flexor digitorum longus and flexor hallucis longus in the region of the knot of Henry. Clin Anat. 2005;18(2):121–5.

The Art of the Musculoskeletal Physical Exam: Foot and Toes Biomechanics of the Foot

65

Masato Takao, Kosui Iwashita, and Yasuyuki Jujo

65.1 Terminology of the Motion

The terminology to describe the motion of the foot and ankle contains a number of ambiguities [1, 2]. The American Academy of Orthopaedic Surgeons (AAOS) [3], Kapandji [4], the American Orthopaedic Foot and Ankle Society (AOFAS) [5], and the International Society of Biomechanics (ISB) [6] use different terms to describe the same movement to use the three planes, transverse (horizontal), sagittal, and frontal (coronal), to express the joint motions (Table 65.1). About the terms of inversion/eversion and supination/pronation, there are two opinions. One is the concept defining inversion/eversion as coronal plane motion and supination/pronation as triplane motion, advocated by the AOFAS and the ISB. The other is the concept defining inversion/eversion as triplane motion and supination/pronation as coronal plane motion, advocated by Kapandji and the AAOS. In sagittal plane, it was defined as extension/flexion (dorsiflexion/plantar flexion) in the terminology of the AAOS; however, extension/flexion is defined as the reverse motion by Kapandji.

The Ad Hoc Committee of Terminology of the Japanese Society for Surgery of the Foot (JSSF) reviewed previous articles describing the transverse (horizontal) plane, sagittal plane, and frontal (coronal) plane motions of the hindfoot, midfoot, and forefoot (Fig. 65.1) [7]. According to the terminology utilized in a total of 160 literatures being explored by examining related academic papers retrieved through a search of the PubMed medical literature database (www.pubmed.com) up to 2008, the Ad Hoc Committee of Terminology of the JSSF proposed for the foot/ankle motion as follows (Table 65.2) [8]: abduction/adduction comprises motion of the ankle joint, subtalar joint, and Chopart/Lisfranc joint in the transverse (horizontal) plane (Fig. 65.2); dorsiflexion/plantar flexion comprises motion of the ankle joint, subtalar joint, and Chopart/Lisfranc joint in the sagittal plane (Fig. 65.3); and inversion/eversion comprises motion of the ankle joint, subtalar joint, and Chopart/Lisfranc joint in the frontal (coronal) plane (Fig. 65.4). Supination/pronation is a complex joint motion in the triplane. Pronation is the complex motion of dorsiflexion, abduction, and eversion comprised of motion of the ankle joint, subtalar joint, and Chopart/Lisfranc joint (Fig. 65.5, left). Supination is the complex motion of plantar flexion, adduction, and inversion comprised of motion of the ankle joint, subtalar joint, and Chopart/Lisfranc joint (Fig. 65.5, right). Proposed terminology for joint motion of toes is as follows: abduction/adduction is composed of the motion of MTP joint, IP joint of great toe, and PIP and DIP joints of toes in transverse (horizontal) plane; extension/

M. Takao (✉) · K. Iwashita · Y. Jujo
Clinical and Research Institute for Foot and Ankle Surgery, Jujo Hospital, Chiba, Japan
e-mail: m.takao@carifas.com

J. G. Lane et al. (eds.), *The Art of the Musculoskeletal Physical Exam*,
https://doi.org/10.1007/978-3-031-24404-9_65

Table 65.1 Definition of joint motion of foot and ankle by the AAOS, Kapandji, AOFAS, and ISB

Plane	AAOS [3]	Kapandji [4]	AOFAS [5]	ISB [6]
Transvers (horizontal)	Abduction Adduction	Abduction Adduction	Forefoot/midfoot Abduction Abduction Ankle/hindfoot External rotation Internal rotation	Forefoot/midfoot External rotation Internal rotation Ankle/hindfoot External rotation Internal rotation
Sagittal	Extension (dorsiflexion) Flexion (Planatarflexion)	Flexion Extension	Dorsiflexion Plantarflexion	Dorsiflexion Plantarflexion
Frontal (coronal)	Forefoot Pronation Supination Hindfoot Eversion Inversion	Forefoot Pronation Supination Hindfoot Eversion Inversion	Eversion Inversion	Eversion Inversion

AAOS the American Academy of Orthopaedic Surgeons, *AOFAS* the American Orthopaedic Foot and Ankle Society, *ISB* the International Society of Biomechanics

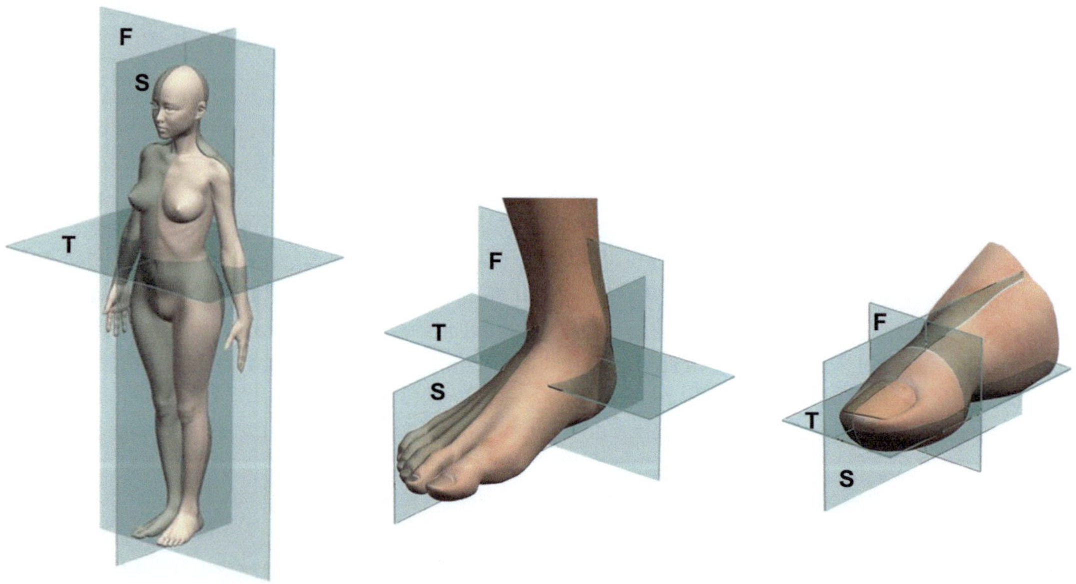

Fig. 65.1 Triplane of the whole body (left), hindfoot and midfoot (center), and forefoot (right). *T* transverse (horizontal) plane, *S* sagittal plane, *F* frontal (coronal). Reproduced from "Terminology of Foot & Ankle Surgery" with permission from the Glossary of the Japanese Society of Foot Surgery

flexion is composed of the motion of MTP joint, IP joint of great toe, and PIP and DIP joints of toes in sagittal plane; and supination/pronation is composed of the motion of MTP joint, IP joint of great toe, and PIP and DIP joints of toes in frontal (coronal) plane.

Although there still remains terminological chaos to describe the motion of the foot and ankle, it is considered that the proposed terminology described by the Ad Hoc Committee of Terminology of the JSSF reflects the opinion of majority.

Table 65.2 Definition of joint motion of foot and ankle proposed by the Ad Hoc Committee of Terminology of the Japanese Society for Surgery of the Foot

Plane	Foot and ankle	Toe (MTP/PIP/DIP)
Transverse (horizontal)	Abduction Adduction	Abduction Adduction
Sagittal	Dorsiflexion Plantarflexion	Extension Flexion
Frontal (coronal)	Eversion Inversion	Pronation Supination
Three planes (complex motion in triplane)	Pronation (Abduction in transvers plane Dorsiflexion in sagittal plane Eversion in frontal plane) Supination (Adduction in transvers plane plantarflexion in sagittal plane Inversion in frontal plane)	–

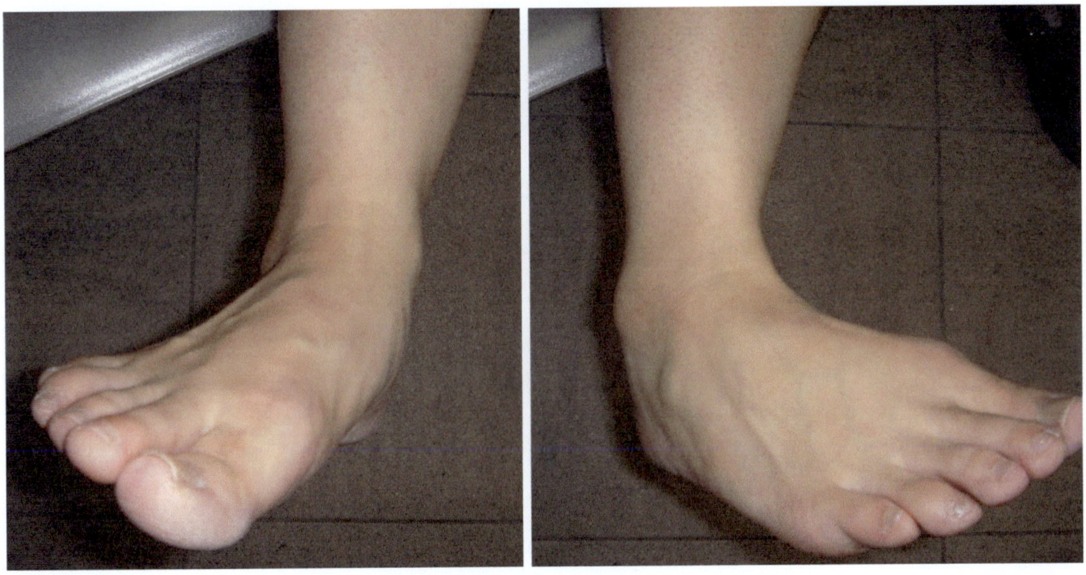

Fig. 65.2 Abduction (left) and adduction (right) of the hindfoot and midfoot in transverse (horizontal) plane

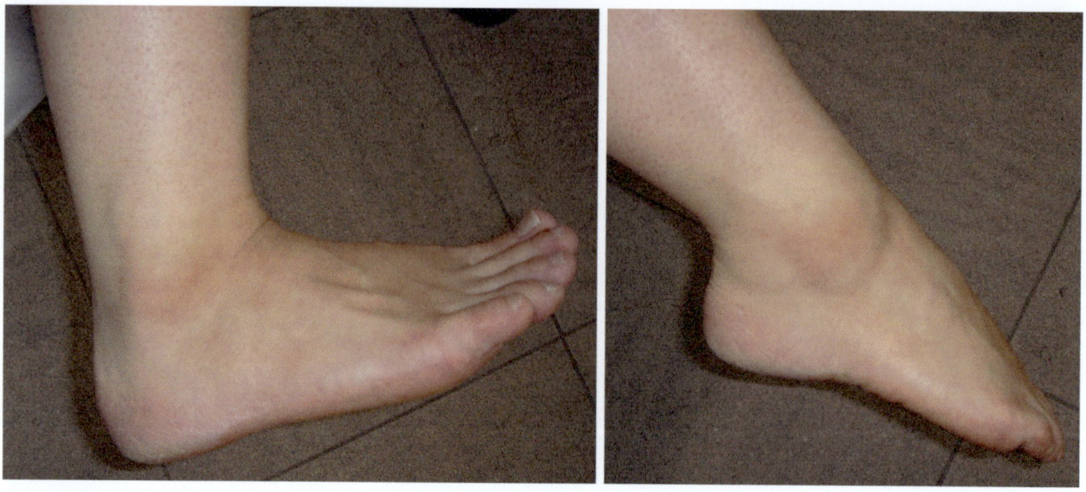

Fig. 65.3 Dorsiflexion (left) and plantar flexion (right) of the hindfoot and midfoot in sagittal plane

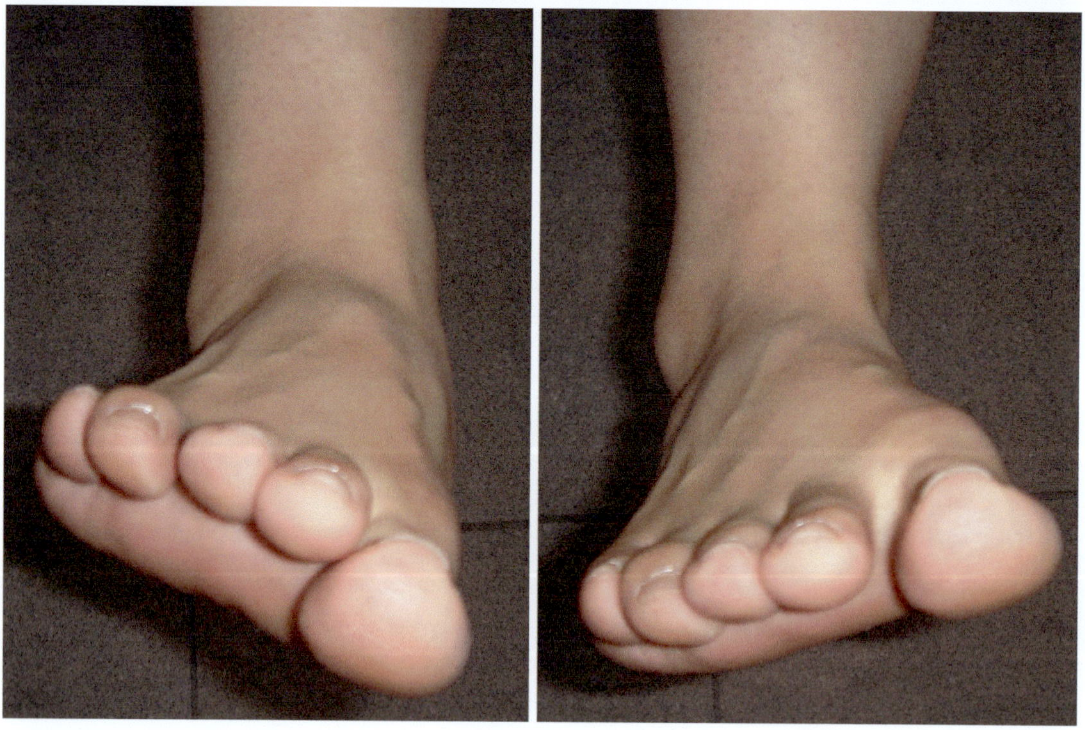

Fig. 65.4 Eversion (left) and inversion (right) of the hindfoot and midfoot in frontal (coronal) plane

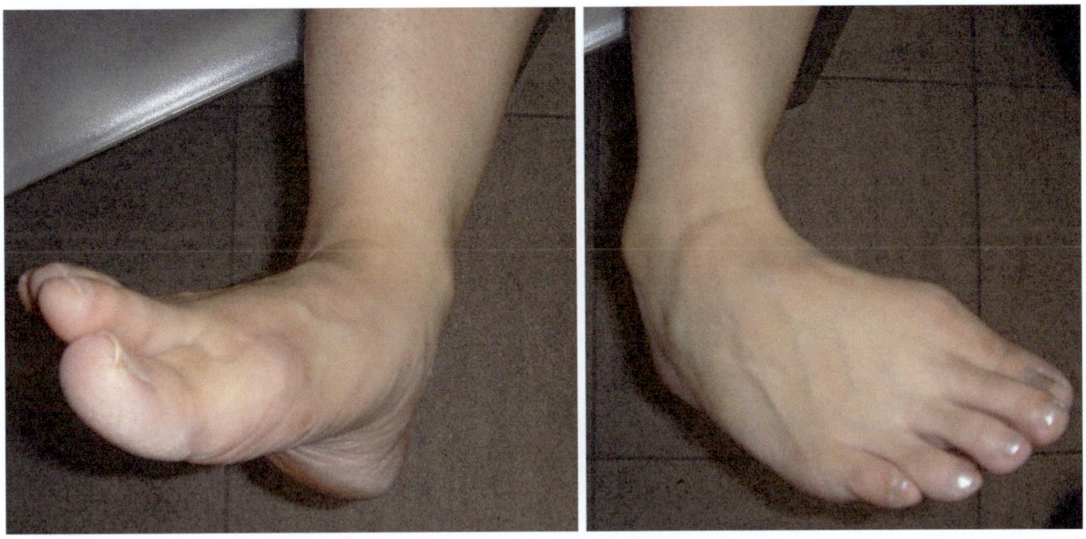

Fig. 65.5 Pronation (left) and supination (right) of the hindfoot and midfoot in triplane. Pronation: dorsiflexion in sagittal plane, abduction in transverse (horizontal) plane, eversion in frontal (coronal) plane. Supination: plantar flexion in sagittal plane, adduction in transverse (horizontal) plane, inversion in frontal (coronal) plane

65.2 Biomechanics of the Foot

65.2.1 Functional Anatomy of Subtalar Joint

Subtalar joint has three functional axes and moves three-dimensionally [9]. Although the direction of the functional axis of the subtalar joint and the normal range of motion vary depending on the literature, it basically moves in plantar flexion, dorsiflexion, inversion, eversion, supination, and pronation. The subtalar joint is connected by the interosseous talocalcaneal ligament and the cervical ligament. These are called the subtalar cruciate ligaments, and the medial fibers of the cervical ligament in the tarsal canal contribute to "swing motion" of the subtalar joint [10].

A characteristic motion of the subtalar joint is supination and pronation. Most of the movement of the foot in the supination/pronation depends on the subtalar joint with the range of motion as 30–50° (25–30° of supination, 5–20° of pronation). Therefore, in patients with talocalcaneal coalition (Fig. 65.6, left) or severe posterior ankle impingement syndrome, it is difficult to take a position of the supination of the foot due to limited movement of the subtalar joint (Fig. 65.6, right). During surgery, it is possible to know that a sufficient resection has been performed by confirming that the foot can take a normal supination position after excision of the lesion.

65.2.2 Functional Anatomy of Midfoot

The midfoot consists of the Chopart joint and the Lisfranc joint. Chopart joint consists of the calcaneocuboid joint and the talonavicular joint. Lisfranc joint consists of the metatarsocuneiform joint and the metatarsocuboid joint. The joint is moved by the bone being pulled by the tendon, and the movement is characterized by three factors: the distance between the axis of the motion and the tendon, the tendon's traction force, and the position where the tendon runs. Among them, the running position of the tendon is an important factor when considering the functional anatomy of the joint. The moment of tendon force with a short distance from the action axis is small, and the moment of tendon force with a long distance from the action axis is large. Figure 65.7 shows the motion axes of the midfoot. The tibialis pos-

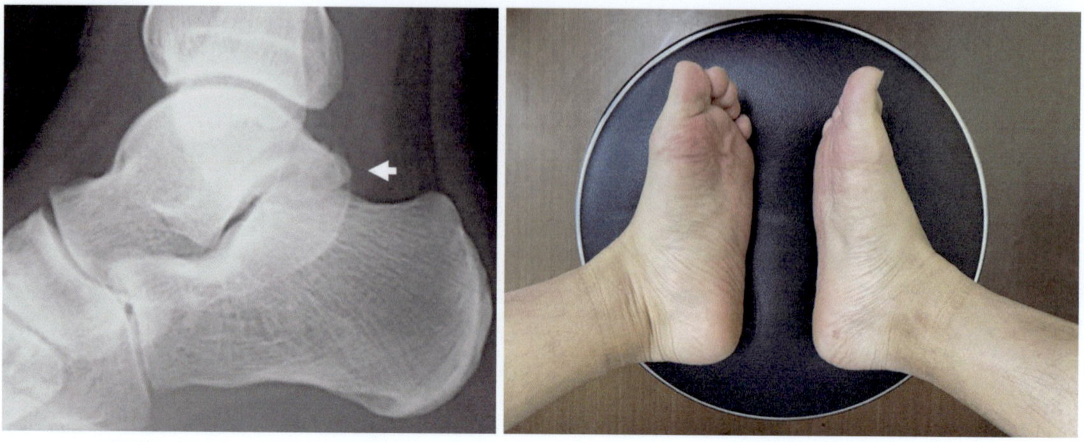

Fig. 65.6 Patient with talocalcaneal coalition. Left: Lateral X-ray showed a coalition of subtalar joint (arrow). Right: Supination in right hindfoot is restricted relative to left side

Fig. 65.7 Motion axes of the midfoot. *TA* tibialis anterior tendon, *EHL* extensor hallucis longus tendon, *EDL* extensor digitorum longus tendon, *PT* peroneus tertius tendon, *PL/PB* peroneus longus and brevis tendons, *TS* triceps surae (Achilles) tendon, *FHL* flexor hallucis longus tendon, *FDL* flexor digitorum longus tendon, *TP* tibialis posterior tendon

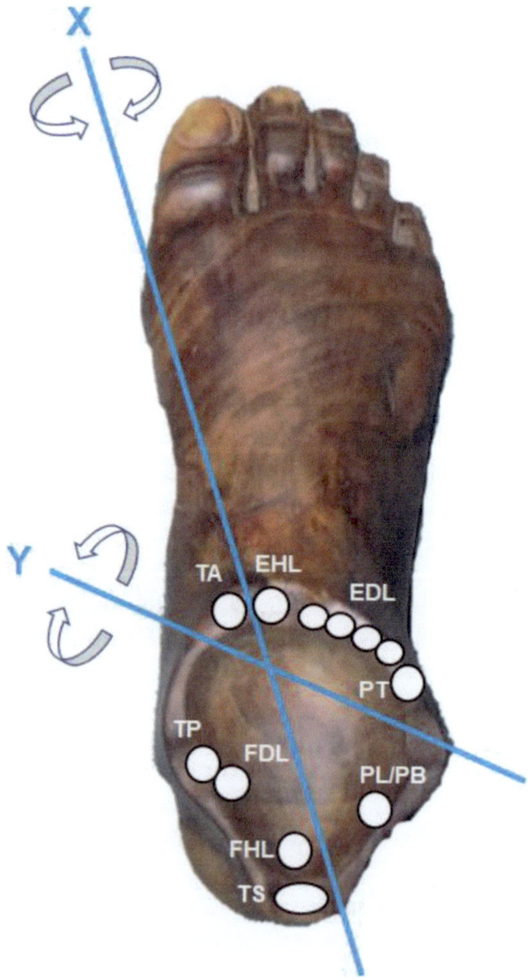

terior tendon and the flexor digitorum longus tendon running inward from X-axis contribute to the inversion, and the peroneus tendons located on the outside contribute to the eversion. At the same time, in standing position when the sole is in contact with the ground, the evertors rotate the lower leg internally, and the inverters rotate the lower leg externally. Dorsiflexion is performed by the traction force of the tibialis anterior tendon, the extensor hallucis longus tendon, and the extensor digitorum longus tendon. The plantar flexion is performed by the traction force of the Achilles tendon, the flexor hallucis longus tendon, and the flexor digitorum longus tendon.

The flexor hallucis longus and the flexor digitorum longus tendons not only are a source of toe movement in their anatomical attachment, but also contribute to the stabilization of each joint of the foot by coordinating with the plantar aponeurosis. In the toe-off phase, the toes stabilize each joint of the foot by tensioning the plantar aponeurosis while grasping the ground. At the same time, they act as a plantar flexor force of the ankle joint and the inversion force of the subtalar joint. On the other hand, the extensor hallucis longus and the extensor digitorum longus tendons act as the dorsiflexor force of the ankle joint and the eversion force of the subtalar joint during the period from the heel-strike to the foot-flat phase.

The Lisfranc ligament is the strongest ligament in the Lisfranc joint [11, 12] and plays an important role in the formation of the longitudinal arch of the foot and the stiffness of the foot during the toe-off phase [13].

65.2.3 Functional Anatomy of Forefoot

The long axis of each metatarsal is inclined with respect to the horizontal plane (Table 65.3), and disturbing this angle is considered to be the major cause of metatarsal pain [14]. The MTP joint has a larger range of motion in the dorsal direction and cooperates with the metatarsal and the IP joint [15, 16]. In the first stage of the toe-off phase in walking, the axis of motion of the forefoot passes between the second toes at an average angle of about 62° with respect to the line connecting the heads of the second and fifth metatarsals. In the next stage, the axis of motion moves to the first and second metatarsal heads [17].

Table 65.3 Angle of the inclination of each metatarsal with respect to horizontal plane

Metatarsal	Angle
1st	18–25°
2nd	15°
3rd	10°
4th	8°
5th	5°

65.2.4 Functional Anatomy of Toe

65.2.4.1 Hallux

Under non-weight-bearing, the IP joint of hallux exhibits a slightly dorsiflexion position due to the traction force of the extensor hallucis longus tendon. The extensor hallucis longus tendon runs through the center of the MTP and IP joints. In the hallux valgus, the extensor hallucis longus tendon is displaced laterally, so that it acts as an adductor for the hallux. As a result, the first metatarsal head is pushed medially, and the first metatarsal becomes inversion (Fig. 65.8).

The flexion of the MTP joint is carried out by the medial and lateral head of the flexor hallucis brevis, the adductor hallucis, and the abductor hallucis. The flexion force of the MTP joint is greater than the extension force due to the role of the sesamoid bone. Sesamoid bone has the function of increasing the bending moment of the intrinsic muscle by moving the plantar intrinsic muscle away from the motion axis of the MTP joint. Although resection of one side of sesamoid bone is usually not expected to cause significant problems, resection of both sides of sesamoid bones reduces the bending moment described above and reduces the flexion force required for MTP joints. In addition, since the sesamoid bone prevents the hyperextension of the IP joint in foot-flat phase, resection of both sides of sesamoid bones causes cock-up deformity of the IP joint.

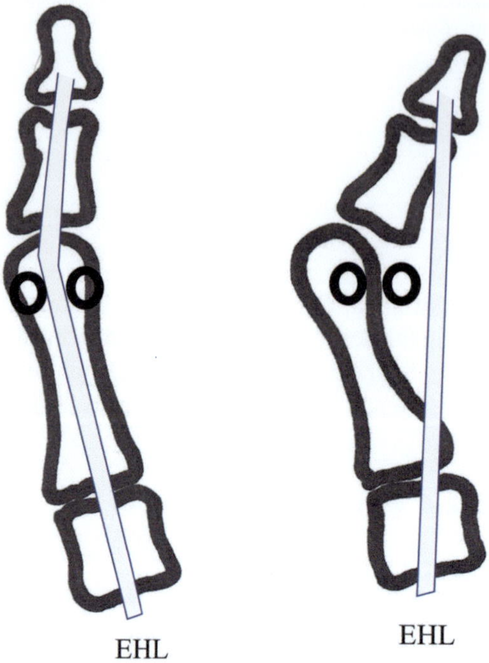

Fig. 65.8 Hallux valgus deformity and extensor hallucis longus tendon (EHL)
Right: normal. Left: hallux valgus

Table 65.4 The lesser toes' functional unit

Structures	Affected toes
The long extensors with the extensor digitorium brevis	2nd–4th toes
The flexor digitorum longus with the flexor digitorum brevis	2nd–5th toes
The planatar interosssei	3rd–5th toes
The dorsal interossei	2nd–4th toes
The lumbricals	2nd–5th toes
The abductor and the flexor digiti quinti	5th toe
The deep sagittal band attachment of the plantar aponeurosis	Proximal phalanx and the plantar plate of 2nd–5th toes

65.2.4.2 Lesser Toes

The lesser toes' functional units are divided into six (Table 65.4). All tendons passing to the axes of motion act as extensors, and all tendons passing plantar to the axes of motion act as flexors. At the MTP joint level, the dorsal and plantar interosseous tendon and the lumbrical tendon pass on the plantar side relative to the axis of motion and contribute to the flexion of the MTP joint. Distal to the MTP joint, lumbrical tendon and a part of interosseous tendon pass dorsally with respect to the axis of motion of the PIP joint, acting as extensor tendons.

The long extensors and the extensor digitorum brevis are strongly attached to the plantar aspect of the proximal phalanx through the transverse portion of the extensor aponeurotic lamina. They have only a loose connection with the dorsal capsule or the dorsum of the proximal phalanx. A pull applied on the long extensor tendon results in a hyperextension of the proximal phalanx at the MTP joint [18, 19]. The proximal phalanx is lifted into extension through the dorsal pull exerted by the proximal transverse segment of the dorsal extensor apparatus as the sling mechanism of extension. The proximal phalanx is extended, but the middle and distal phalanges remain slightly flexed.

The plantar aponeurosis, interosseous tendon, and lumbrical tendon are responsible for flexion of the MTP joint. When the plantar aponeurosis relaxes as in the case of the pronation foot, or when the interosseous muscle and the lumbrical muscle atrophy, the flexion force of the MTP joint decreases and the MTP joint is in the extended position. At that time, tension is applied to the long flexor tendon at the plantar of the MTP joint, so that the PIP/DIP joint will be in flexion, resulting in a hammertoe.

References

1. Reese NB, Bandy WD. Joint range of motion and muscle length testing. Philadelphia: Saunders; 2002.
2. Resch S. Functional anatomy and topography of the foot and ankle. In: Myerson MS, editor. Foot and ankle disorders. Philadelphia: Saunders; 1999. p. 25–9.
3. American Academy of Orthopaedic Surgeons. Joint motion: method of measuring and recording. Rosemont, IL: AAOS; 1965.
4. Kapandji IA. The physiology of the joints. 5th ed. London: Churchill Livingstone; 1987.
5. Saltzman C, Alexander I, Kitaoka H, Trevino S. Orthopaedic Foot and Ankle Society Ad Hoc Committee Report, January 1996. Foot Ankle Int. 1997;18:310–1.
6. Wu G, Siegler S, Allard P, Kirtley C, Leardini A, Rosenbaum D, et al. ISB recommendation on

defi nitions of joint coordinate system of various joints for the reporting of human joint motion. Part 1. Ankle, hip, and spine. J Biomech. 2002;35: 543–8.
7. Kelikian AS, et al. Terminology of foot and ankle surgery. 3rd ed. The Japanese Society for Surgery of the Foot; 2017.
8. Doya H, Haraguchi N, Niki H, Takao M, Yamamoto H. Ad Hoc Committee on Terminology of the Japanese Society for Surgery of the Foot. Proposed novel unified nomenclature for range of joint motion: method for measuring and recording for the ankles, feet, and toes. J Orthop Sci. 2010;15: 531–9.
9. Inman TV. The joints of the ankle. Baltimore: Williams & Wilkins; 1976.
10. Huson A. Ein Ontleedkundig-Functioneel Onderzoek van de Voetwortel. Leiden: Drukkerij; 1961.
11. Ambagtsheer JBT. The function of the muscles of the lower leg in relation to movement of the talus. An experimental study in human subjects. Acta Orthop Scand Suppl. 1978;172:1–196.
12. Van Langelaan EJ. A kinematical analysis of the tarsal joints. An X-ray photogrammetric study. Acta Orthop Scand Suppl. 1983;204:1–269.
13. Lundberg A, Svensson OK, Nemeth G, et al. The axis of the rotation of the ankle joint. J Bone Joint Surg Br. 1989;71:94–9.
14. Viladot PA. Patologia del Antepie Balcarona. Barcelona: Ediciones Toray, SA; 1974.
15. Joseph J. Range of movement of the great toe in men. J Bone Joint Surg Br. 1954;36:450–7.
16. Myerson MS, Shereff MJ. The pathological anatomy of claw and hammer toes. J Bone Joint Surg Am. 1989;71:45–9.
17. Re I, Inman VT. Anthropometric studies of the human foot and ankle. Technical report. Berkeley CA: University of California, Berkeley, Biomechanics Laboratory; 1968.
18. Sarrafian SK, Topouzian LK. Anatomy and physiology of the extensor apparatus of the toes. J Bone Joint Surg Am. 1969;51:669–79.
19. Kelikian AS, et al. Sarrafian's anatomy of the foot and ankle. 3rd ed. Philadelphia: Lippincott; 2011.

66 Ankle Joint Range of Motion Evaluation (ROM) Using Smartphone Calculators

Marco Quaranta, Francesco Oliva, and Nicola Maffulli

When planning the management of orthopedic patients following injuries, an important part of physical examination is the evaluation of joint function in terms of measuring range of motion (ROM) [1], classically defined as distance and direction of joint movement between flexion and extension.

ROM expresses quantitatively the motion of a joint, and its evaluation allows to ascertain the presence of impairments and evaluate the effectiveness of rehabilitation programs, guiding the choice of interventions during rehabilitation and defining the progress of treatment [2]. ROM is joint specific and changes according to the sex and age of the subject.

Recently, advanced smartphones, which integrate a series of functions that can be used in the medical field with different purposes, have become widespread. Modern smartphones are equipped with various movement sensors, including accelerometer, gyroscope, and magnetometer, all of which can perform various inclinometric functions.

The accelerometer allows to detect the linear acceleration (m/s) of the device along the three axes of space (Fig. 66.1), thus allowing to determine the linear displacements of the device.

A gyroscope measures the variations in angular speed (degrees/s) of a given device to provide information on rotational movements.

The magnetometer, based on the earth's magnetic field, allows to detect the orientation of the device in space.

These sensors and their dedicated software can be used in the musculoskeletal system to evaluate the ROM of various joints quickly and reliably

M. Quaranta · F. Oliva
Department of Musculoskeletal Disorders, Faculty of Medicine and Surgery, University of Salerno, Baronissi, Italy

Clinica Ortopedica, Ospedale San Giovanni di Dio e Ruggi D'Aragona, Salerno, Italy

N. Maffulli (✉)
Department of Musculoskeletal Disorders, Faculty of Medicine and Surgery, University of Salerno, Baronissi, Italy

Clinica Ortopedica, Ospedale San Giovanni di Dio e Ruggi D'Aragona, Salerno, Italy

Queen Mary University of London, Barts and the London School of Medicine and Dentistry, Centre for Sports and Exercise Medicine, Mile End Hospital, London, UK

Guy Hilton Research Centre, Faculty of Medicine, School of Pharmacy and Bioengineering, Keele University, Thornburrow Drive, Hartshill, Stoke-on-Trent ST4 7QB, England, UK
e-mail: n.maffulli@qmul.ac.uk

J. G. Lane et al. (eds.), *The Art of the Musculoskeletal Physical Exam*,
https://doi.org/10.1007/978-3-031-24404-9_66

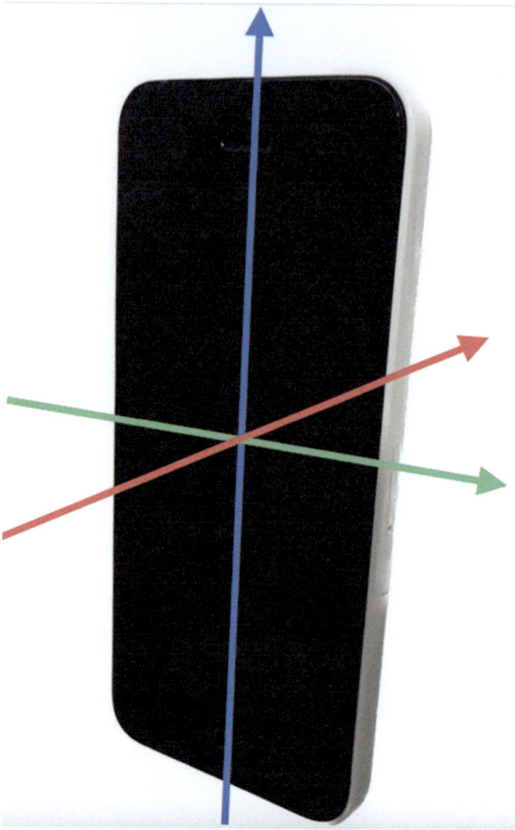

Fig. 66.1 Axes of space detectable with the accelerometer

66.1 Applications

Several applications are able to exploit these digital sensors, taking advantage of the potential of the smartphone for measuring the ROM of different joints [3].

The applications which use these sensors, therefore, in addition to being recreational, as with flight simulation apps, can measure and quantify the ROM of many joints [4, 5], with the advantage of not requiring specialized training [6].

Some of these applications allow to directly measure the angular excursion of the joint, while others require the acquisition of a photo of the joint in the initial and final phases of the movement and then calculate the angle of movement by means of a virtual protractor that overlaps the image.

The important advantage of using the smartphone as an ROM measurement tool is that, being commonly used, it can also be useful for rehabilitation purposes: patients themselves can obtain direct feedback with a self-assessment of the ROM; moreover, being equipped with a display, the smartphone can also be used to view the sequence of exercises that the patient must complete.

In this way, it is possible to avoid resorting to other more complex and expensive measuring instruments such as the three-dimensional motion capture system, which are difficult to access [7]. Consequently, more and more developers have produced dedicated medical applications to facilitate rehabilitation [8].

With the introduction of these new ROM measurement systems, the need arose to evaluate their reliability and validity. Therefore, numerous studies have focused on these smartphone applications designed to measure ROM, comparing the results obtained with those relating to instruments considered the gold standard [9].

66.2 Fields of Use

In several studies, the smartphone inclinometer has been tested and compared with the preexisting measurement systems considered gold standards for the evaluation of ROM, including the ankle (Fig. 66.2). Some of the apps validated for ankle ROM measurements are shown in Table 66.1.

The foot and ankle are frequently injured, both in the general population and in athletes [14]. Ankle sprains are the most common cause of attendance of an emergency department [15]. Although ankle sprains in most cases recover completely [16, 17], in some patients they may have adverse outcomes [18, 19].

Foot and ankle injuries are generally responsible for pain and decrease of ROM, with poor quality of life in affected patients [20, 21]. Therefore, a primary goal of management is to obtain the greatest functional recovery of the affected joint. For this purpose, the measurement

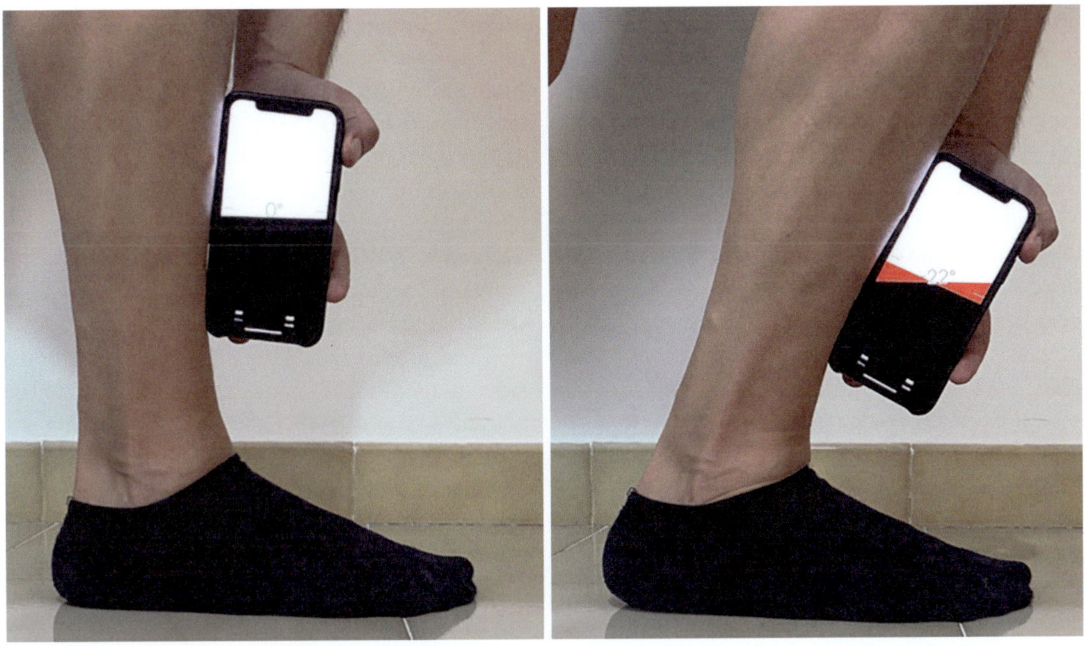

Fig. 66.2 Ankle ROM measurement with smartphone inclinometer

Table 66.1 Apps validated for ROM measurements

App	Seller	Price ($)	Year	Article
TiltMeter	Carlos Hernandez	1.99	2013	Williams et al. [10]
iHandy level	iHandy Ltd.	Free	2015	Vohralik et al. [11]
Clinometer	Peter Breitling	1.99	2018	Cox et al. [12]
My ROM	Carlos Balsalobre	9.99	2019	Balsalobre-Fernàndez et al. [13]

of ROM has an extremely important role both for the initial clinical evaluation and for the evaluation of rehabilitation [11].

Many ankle conditions may require the evaluation of the ROM, including sprains, dislocations, and fractures [22]. However, the evaluation of the ROM has an important role in degenerative conditions such as osteoarthritis as well [23].

Different methods can be used to evaluate ROM, including radiographic measurement [24, 25]. However, these are not practical in clinical practice, and the gold standard is the universal goniometer [26].

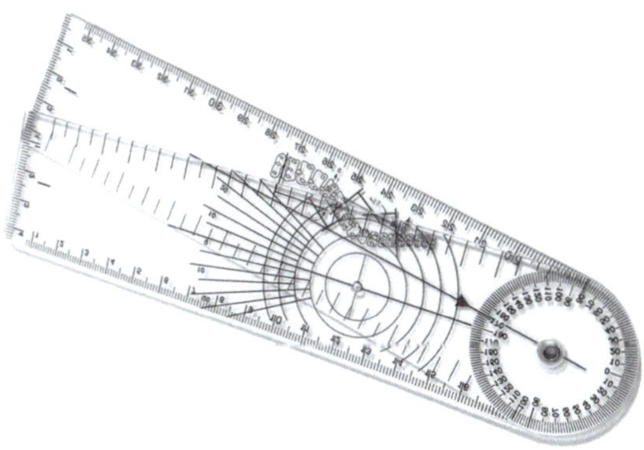

Fig. 66.3 Goniometer

66.3 Ankle

In the evaluation of patients with ankle injuries, the measurement of plantar flexion is commonly used and carried out using a goniometer [26] (Fig. 66.3), with a fulcrum, a fixed arm (stationary arm), and a movable arm.

The fulcrum of the goniometer must be placed at the center of the joint and in the case of the ankle, for the measurement of plantar flexion, the fulcrum must be at the lateral malleolus. The fixed arm is aligned with the fibula, and the mobile arm is in line with the shaft of the fifth metatarsal.

The alternative to the use of the goniometer is represented by the digital inclinometer, which, for the measurement of plantar flexion, is positioned in contact with the sole of the foot and aligned with it [26–28].

With the introduction of smartphone applications dedicated to ROM measurement, there has been a progressive increase in the use of inclinometry. Among the various applications available, one already considered valid for the measurement of shoulder ROM is the Clinometer app (Smartphone ApplicationTM produced by Plaincode App).

The measurement is carried out with the patient sitting keeping the knee extended and the foot outside the examination couch. With the smartphone aligned to the foot, the application is started and, after calibration, the patient is asked to perform an active plantar flexion to measure the ROM.

In 2018, the validity of this measurement was compared with the use of the goniometer [12]: the measurements were repeated three times per each limb in a sample of 100 individuals. The results were compared, with no statistically significant differences between the two procedures, reporting an average plantar flexion value of 62.79° (7.90°) with the goniometer and average values of 62.96° (7.85°) with the Clinometer app. Similar results were obtained by Alawna et al. [29] in 2019.

An important aspect is also the evaluation of the weight-bearing ankle dorsiflexion using the Dorsiflex iPhone, now renamed My ROM, application [13].

The latter allowed to analyze the validity and reliability of the measurement of ankle dorsiflexion with the iPhone inclinometer, comparing it with a professional digital inclinometer [13]. The study highlighted a perfect correlation between the measurements obtained with the two methods, demonstrating that a smartphone app can replace a professional tool to measure ankle dorsiflexion [13].

Many authors have proposed weight-bearing ankle dorsiflexion (WBDF) measurement as a relevant parameter in assessing the risk of injury to the lower limbs [30], such as anterior cruciate ligament injury [31], patellar tendinopathy [32], and plantar fasciitis [33]. During exercises such as the squat or landing after a jump, the joints of the lower limbs must absorb and shield an important force that acts on the sagittal plane: a decreased WBDF produces limitation of movements of the knee and trunk, with greater valgus strain at the knee [34].

To measure WBDF, the patient stands with the sole of the foot in contact with the floor. The patient is asked to flex the knee by moving the weight forward to dorsiflex the ankle. During the exercise, the digital inclinometer, or the smartphone, is applied to the anterior tibial surface immediately below the tibial tuberosity. The measurement is carried out bilaterally to establish the possible asymmetry between the two limbs, which represents an additional risk factor for injury [35]. The use of the smartphone as an inclinometer can provide valid and reliable measurements of the WBDF angle of the ankle, not different to those of the goniometer or professional digital inclinometer.

Other applications for the measurement of ankle ROM, whose results have been validated by comparison with gold standard instruments, are the iHandy Level app for iPhone to evaluate ankle dorsiflexion and the TiltMeter to evaluate dorsiflexion under load, for example during the weight-bearing lunge test [10, 11].

66.4 Results

The use of the inclinometer in the smartphone is a useful and easy-to-use tool to measure the ROM of the ankle, with reliable results comparable with those obtained with the use of instruments considered to be gold standards such as the goniometer and the digital inclinometer. The technological evolution therefore provides important advantages in the medical field, from the point of view of accessibility, costs, and practicality, allowing to improve the physician's clinical evaluation skills. In addition, another advantage lies in the possibility for patients themselves to use the smartphone to self-evaluate their motor skills during the rehabilitation phases, so as to have immediate feedback during the exercises.

However, for a correct use of this instrument in the clinical setting, a familiarity with the application is required, and adherence to developer's instructions is necessary.

References

1. Verrall GM, Slavotinek JP, Barnes PG, Esterman A, Oakeshott RD, Spriggins AJ. Hip joint range of motion restriction precedes athletic chronic groin injury. J Sci Med Sport. 2007;10(6):463–6.
2. American Physical Therapy Association. Guide to Physical Therapist Practice. Phys Ther. 2001;81(1):9–746.
3. Braden HJ. Goniometry apps: do they measure up? Exploring the accuracy of Mobile device apps. GGS. 2019;5(2):GGS.000609.2019. https://crimsonpublishers.com/ggs/fulltext/GGS.000610.php
4. Peters FM, Greeff R, Goldstein N, Frey CT. Improving acetabular cup orientation in Total hip arthroplasty by using smartphone technology. J Arthroplast. 2012;27(7):1324–30.
5. Shin SH, Ro DH, Lee O-S, Oh JH, Kim SH. Within-day reliability of shoulder range of motion measurement with a smartphone. Man Ther. 2012;17(4):298–304.
6. Cunha AB, Babik I, Harbourne R, Cochran NJ, Stankus J, Szucs K, et al. Assessing the validity and reliability of a new video goniometer app for measuring joint angles in adults and children. Arch Phys Med Rehabil. 2020;101(2):275–82.
7. Charlton PC, Mentiplay BF, Pua Y-H, Clark RA. Reliability and concurrent validity of a smartphone, bubble inclinometer and motion analysis system for measurement of hip joint range of motion. J Sci Med Sport. 2015;18(3):262–7.
8. Hafeez K, Kaim Khani GM, Jawaid M, Bux M. Is smartphone a necessity or luxury among orthopedic specialty? J Pak Med Assoc. 2014;64(12 Suppl 2):S27–9.
9. Rothstein J, Echternach J. Primer on measurement: an introductory guide to measurement issues. Alexandria, VA: American Physical Therapy Association; 1993.
10. Williams CM, Caserta AJ, Haines TP. The TiltMeter app is a novel and accurate measurement tool for the weight bearing lunge test. J Sci Med Sport. 2013;16(5):392–5.
11. Vohralik SL, Bowen AR, Burns J, Hiller CE, Nightingale EJ. Reliability and Validity of a Smartphone App to Measure Joint Range. Am J Phys Med Rehabil. 2015;94(4):325–30.
12. Cox RW, Martinez RE, Baker RT, Warren L. Validity of a smartphone application for measuring ankle plantar flexion. J Sport Rehabil 2018;27(3).
13. Balsalobre-Fernández C, Romero-Franco N, Jiménez-Reyes P. Concurrent validity and reliability of an iPhone app for the measurement of ankle dorsiflexion and inter-limb asymmetries. J Sports Sci. 2019;37(3):249–53.
14. Hunt KJ, Hurwit D, Robell K, Gatewood C, Botser IB, Matheson G. Incidence and epidemiology of foot and ankle injuries in elite collegiate athletes. Am J Sports Med. 2017;45(2):426–33.
15. Cooke MW, Lamb SE, Marsh J, Dale J. A survey of current consultant practice of treatment of severe ankle sprains in emergency departments in the United Kingdom. Emerg Med J. 2003;20(6):505–7.
16. Birrer RB, Fani-Salek MH, Totten VY, Herman LM, Politi V. Managing ankle injuries in the emergency department. J Emerg Med. 1999;17(4):651–60.
17. Wilkerson GB, Horn-Kingery HM. Treatment of the inversion ankle sprain: comparison of different modes of compression and cryotherapy. J Orthop Sports Phys Ther. 1993;17(5):240–6.
18. Braun BL. Effects of ankle sprain in a general clinic population 6 to 18 months after medical evaluation. Arch Fam Med. 1999;8(2):143–8.
19. Kerkhoffs GMMJ, Rowe BH, Assendelft WJJ, Kelly K, Struijs PA, van Dijk CN. Immobilisation and functional treatment for acute lateral ankle ligament injuries in adults. Cochrane Database Syst Rev 2002;(3):CD003762.
20. Ferran NA, Oliva F, Maffulli N. Ankle instability. Sports Med Arthrosc Rev. 2009;17(2):139–45.
21. Fereydounnia S, Shadmehr A, Talebian Moghadam S, Olyaei G, Jalaie S, Shiravi Z, et al. The comparison of dynamic postural control and muscle activity in time domain in athletes with and without chronic ankle instability. Muscle Ligaments Tendons J. 2019;08(04):552.
22. Dehghan N, McKee MD, Jenkinson RJ, Schemitsch EH, Stas V, Nauth A, et al. Early Weightbearing and range of motion versus non-Weightbearing and immobilization after open reduction and internal fixation of unstable ankle fractures: A randomized controlled trial. J Orthop Trauma. 2016;30(7):345–52.
23. Khlopas H, Khlopas A, Samuel LT, Ohliger E, Sultan AA, Chughtai M, et al. Current concepts in osteoarthritis of the ankle: review. Surg Technol Int. 2019;10(35):280–94.
24. Coetzee JC, Castro MD. Accurate measurement of ankle range of motion after total ankle arthroplasty. Clin Orthop Relat Res. 2004;424:27–31.
25. Dayton P, Feilmeier M, Parker K, Otti R, Reimer R, Kauwe M, et al. Experimental comparison of the clinical measurement of ankle joint dorsiflexion and radiographic Tibiotalar position. J Foot Ankle Surg. 2017;56(5):1036–40.
26. Norkin CC, White DJ. Measurement of joint motion: a guide to goniometry. 4th ed. Philadelphia: F.A. Davis; 2009. p. 450.
27. Cosby NL, Hertel J. Relationships between measures of posterior Talar glide and ankle dorsiflex-

ion range of motion. Athl Train Sports Health Care. 2011;3(2):76–85.
28. Denegar CR, Hertel J, Fonseca J. The effect of lateral ankle sprain on dorsiflexion range of motion, posterior Talar glide, and joint laxity. J Orthop Sports Phys Ther. 2002;32(4):166–73.
29. Alawna MA, Unver BH, Yuksel EO. The reliability of a smartphone goniometer application compared with a traditional goniometer for measuring ankle joint range of motion. J Am Podiatr Med Assoc. 2019;109(1):22–9.
30. Malloy P, Morgan A, Meinerz C, Geiser C, Kipp K. The association of dorsiflexion flexibility on knee kinematics and kinetics during a drop vertical jump in healthy female athletes. Knee Surg Sports Traumatol Arthrosc. 2015;23(12):3550–5.
31. Amraee D, Alizadeh MH, Minoonejhad H, Razi M, Amraee GH. Predictor factors for lower extremity malalignment and non-contact anterior cruciate ligament injuries in male athletes. Knee Surg Sports Traumatol Arthrosc. 2017;25(5):1625–31.
32. Malliaras P, Cook JL, Kent P. Reduced ankle dorsiflexion range may increase the risk of patellar tendon injury among volleyball players. J Sci Med Sport. 2006;9(4):304–9.
33. Petraglia F, Ramazzina I, Costantino C. Plantar fasciitis in athletes: diagnostic and treatment strategies. A systematic review. Muscles Ligaments Tendons J. 2017;7(1):107–18.
34. Dill KE, Begalle RL, Frank BS, Zinder SM, Padua DA. Altered knee and ankle kinematics during squatting in those with limited weight-bearing–lunge ankle-dorsiflexion range of motion. J Athl Train. 2014;49(6):723–32.
35. Bishop C, Turner A, Read P. Effects of inter-limb asymmetries on physical and sports performance: a systematic review. J Sports Sci. 2018;36(10):1135–44.

Assessment of Instability of the Calcaneus and Lisfranc

67

Silvampatti Ramasamy Sundararajan, Rajagopalakrishnan Ramakanth, Harsh Jalan, and Shanmuganathan Rajasekaran

67.1 Introduction

The term calcaneus instability refers to subtalar instability caused by ligamentous damage to the subtalar joint (STJ). The subtalar instability was attributed to the stretched calcaneofibular ligament [1]. Additional ligaments situated between the talus and calcaneum contribute to the STJ's stability. Injuries to these ligaments are prevalent in athletics, and first examinations of the foot almost invariably miss subtle injuries. Up to 80% of patients presenting with chronic ankle instability have subtalar instability, however it is rarely recognised in the diagnosis or therapy [2].

The instability of the Lisfranc joint is caused by a combination bone and ligamentous injury to the tarsometatarsal joint, accompanied with metatarsal base displacement. The injury may be solely ligamentous or accompanied with metatarsal, cuneiform, scaphoid, or cuboid bone fractures. It can be observed in vehicle accidents and sports injuries. This name is credited to a French surgeon and gynaecologist, Jacques Lisfranc de St. Martin (1790–1847), surgeon in Napoleon's army on the Russian front in 1815, who was the first to describe the damage and an amputation at that level [3]. Twenty percent of Lisfranc fracture-dislocations are undiagnosed or manifest late, particularly low-energy injuries or injury that are solely ligamentous. Late complications of this injury, such as post-traumatic osteoarthritis and foot deformities, might result in severe disabilities for the patients [4]. A prompt and precise diagnosis of these injuries is essential for their proper treatment and prevention of long-term complications.

S. R. Sundararajan (✉) · R. Ramakanth · H. Jalan
Athroscopy & Foot and Ankle Department, Ganga Hospital, Coimbatore, India

S. Rajasekaran
Orthopaedics, Ganga Hospital, Coimbatore, India

67.2 Patho-anatomy for Subtalar Instability and Midfoot/Lisfranc's Instability

Between the inferior surface of the talus and the superior surface of the calcaneus, the subtalar joint (STJ) contains three articular facets, namely the posterior facet, middle facet, and anterior facet. STJs are held together by the calcaneofibular ligament (CFL), the interosseous talocalcaneal ligament (ITCL), the cervical ligament, and the lateral root of the inferior extensor retinaculum [5] (Fig. 67.1). The CFL is perpendicular to the posterior facet of the STJ, aids in maintaining congruence between the articular surfaces [6], and is the primary stabiliser of the STJ complex. A CFL injury causes a considerable increase in joint motion and instability [1, 7]. The ITCL is a

J. G. Lane et al. (eds.), *The Art of the Musculoskeletal Physical Exam*,
https://doi.org/10.1007/978-3-031-24404-9_67

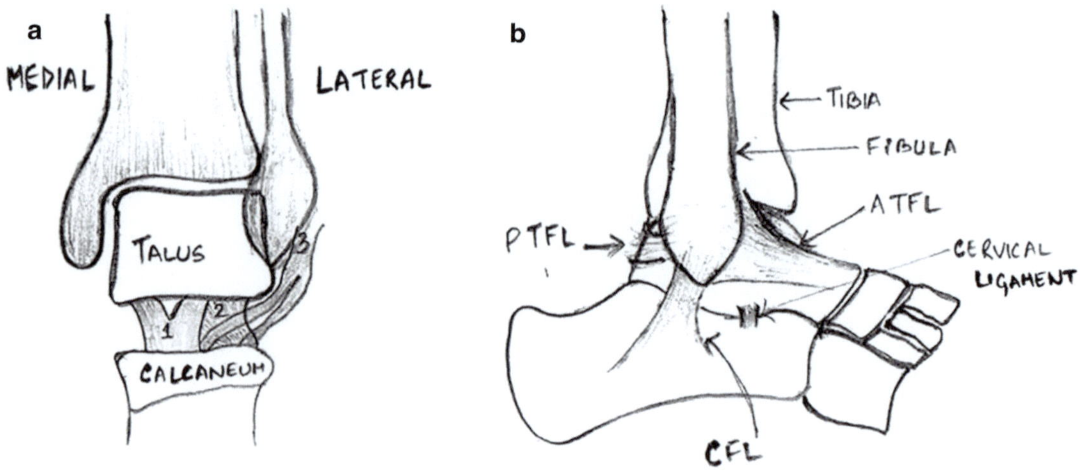

Fig. 67.1 This figure shows (1) Interosseous talocalcaneal ligament (ITCL), (2) cervical ligament, (3) lateral root of the inferior extensor retinaculum. *PTFL* posterior tibiofibular ligament, *ATFL* anterior talofibular ligament, *CFL* calcaneofibular ligament

fundiform ligament (fascia thickening) that lies in the sinus tarsi and tarsal canal and is essential for subtalar stability [7, 8]. The cervical ligament borders the anterolateral region of the STJ; it is the strongest ligament connecting the talus and calcaneus [9].

Subtalar instability is caused by the transmission of an inversion force to the foot architecture while the ankle is in a dorsiflexed position. This force acts directly on the CFL, resulting in ligament rupture and instability [10]. Other STJ ligaments, such as the ITCL and cervical ligament, can also be ruptured throughout the injury phase if the deforming stresses increase [8]. In basketball and volleyball players, a whiplash-like injury to the hindfoot might disrupt the ITCL [11]. The ITCL is typically ruptured when a player abruptly stops while the talus continues to slide anteriorly on the planted calcaneus due to inertia. Due to the fact that both lateral ankle instability and subtalar instability are inversion injuries, they frequently co-occur. 10% to 75% of patients with ankle instability also have STJ instability [12–14].

Two crucial anatomical features of the midfoot provide osseous support to the structure: the metatarsal's Roman arch arrangement and its recessed second metatarsal base. Middle cuneiform is 8 mm proximal to medial cuneiform and 4 mm proximal to lateral cuneiform [15, 16]. Thus, the second metatarsal base is extremely vital to the final stability of the midfoot joint complex, and an apparently trivial fracture here, if left untreated, can cause significant instability. Lisfranc ligamentous complex provides joint stability, and plantar ligaments are stronger and larger than dorsal ligaments [16].

The mechanism of injury for Lisfranc's and midfoot instability is typically an axial load applied to the heel of a plantarflexed ankle (with the toes extended), resulting in a tensile tension to the convexity of the tarsometatarsal articulations [17]. Conversely, a twisting or bending force over the midfoot, with the foot in a stationary posture and the body rolling over the midfoot, can cause disruption of the supportive ligaments constituting the Lisfranc complex. Development of Lisfranc's instability is dependent on the involved pathoanatomical structure. Understanding the ligamentous components that stabilise the Lisfranc joint is essential. There is no inter-metatarsal ligament between the first and second metatarsals, which explains why dorsal, inter-metatarsal ligaments that connect the second to the fifth metatarsals are frequently displaced. The plantar interosseous ligament from the lateral zone of the first cuneiform bone to the medial zone of the second metatarsal, also known as the Lisfranc ligament (large, oblique plantar Y-shaped ligament), is the thickest and strongest, which is why it is essential to maintain the second metatarsal and, consequently, the midfoot arch.

The insertions of the tibialis anterior and peroneus longus tendons dynamically stabilise the tarsometatarsal joint. In certain injuries, the tibialis anterior tendon becomes stuck between the medial and middle cuneiform, making reduction impossible. The dorsalis pedis artery and deep peroneal nerve constantly cross the TMT joint complex and are positioned laterally to the extensor hallucis brevis tendon. The plantar arch is formed by the Dorsalis pedis artery diving between the first and second metatarsal bases.

67.3 Clinical Assessment of Subtalar Instability

Subtalar joint dislocation and subsequent instability are commonly associated with talocrural joint instability and recurrent ankle instability [18]. During clinical examination, the signs and symptoms of subtalar joint instability are less evident than those of lateral ligamentous instability. Nonetheless, monitoring the subtalar joint during the evaluation of a suspected recurrent ankle instability provides crucial insight into the persistent progression of chronic ankle instability. Athletes with chronic ankle and/or STJ instabilities have a history of recurrent lateral ankle sprains and may present with an acute-on-chronic injury [19]. The assessment of subtalar joint injury in association with ankle instability has been generally ignored because it is overlooked in ankle joint instability. At the acute stage, lateral swelling, ecchymosis, and discomfort may be palpable, but are nonspecific. After severe trauma, the patient's discomfort and muscular resistance limit the clinical stress testing. In general, patients presenting with STJ instability do not do so after a single traumatic event [20]. Several patients report a sensation of giving way or tumbling, especially when walking on uneven terrain, which increases with greater physical exertion [21]. Several patients report experiencing pain in the region of the sinus tarsi and the lateral hindfoot [20]. Patients with STI prefer to wear high-top shoes or stabilizing orthotics [22], on contrary patients with CAI, prefer flat-top shoes [21].

Patients' both ankles have to be examined and compared; the three clinical examination tests for combined ankle and subtalar instability are talar tilt (TT), anterior drawer (AD), and medial subtalar glide (MSTG). The TT test is used to evaluate for severe talar inversion within the mortise (Fig. 67.2a, b). The AD test (Fig. 67.2d) evaluates anterior talar displacement within the mortise. The MSTG test evaluates for excessive calcaneus medial translation on the talus in the transverse plane. Examiner performs the MSTG test by holding the talus in subtalar neutral with one hand while gliding the calcaneus medially on the fixed talus with the other (Fig. 67.3c, d). This test

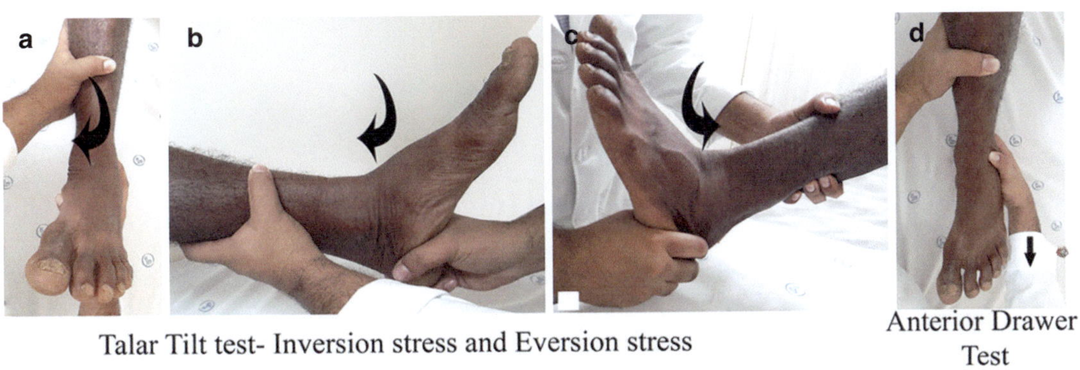

Fig. 67.2 Clinical examination of talar tilt test (**a**, **b**) Inversion test, (**c**)-Eversion test, and (**d**) anterior drawer test

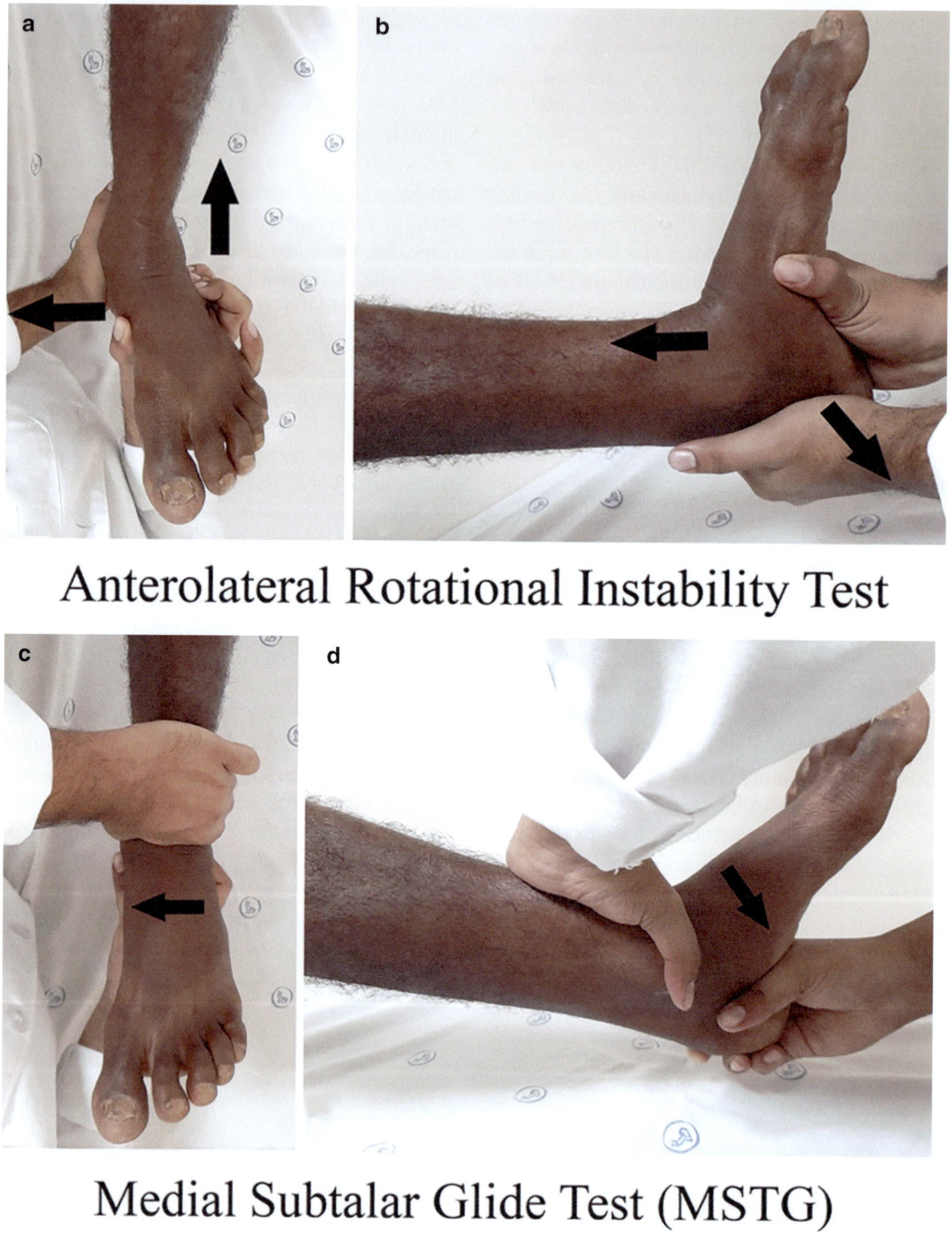

Fig. 67.3 Clinical examination of (**a**, **b**) anterolateral rotatory instability test and (**c**, **d**) medial subtalar glide test

is a modification of a joint mobilization technique described by Loudon and Bell [23].

Specialized test for chronic STI is the anterolateral drawer test [24], in which a combined inversion, internal rotation, and adduction stress is applied to the forefoot while the examiner holds the hindfoot in maximal extension (dorsiflexion) to prevent accidental motion at the ankle joint [21, 25] (Fig. 67.3a, b). A positive test demonstrates an enhanced anterior and medial displacement in addition to a varus tilt of the calcaneus beneath the talus. Relevance and validity of this test were confirmed in a cadaver model utilising selective ligament transection and direct assessment of enhanced instability diagnostic; a 4 mm threshold provided 100 % sensitivity and 67 % specificity with excellent intra-observer reliability [26].

67.4 Clinical Assessment of Lisfranc's Instability

These are caused by direct trauma, such as that which occurs in motor vehicle accidents, sports injuries, or low-velocity falls, jumping, etc. Individuals with a Lisfranc injury frequently appear with excruciating pain and an inability to bear weight. Crush injuries frequently manifest as a swollen foot, and these injuries can be linked to compartment syndrome. In these patients, the pain will be disproportionate to the clinical findings, and passive flexion and extension will be painful. It is easier to diagnose high-energy foot injuries, which can be closed or open wounds. They are characterised by painful foot abnormalities, edoema, and an inability to walk.

Occasionally, Lisfranc injuries might be subtle and manifest with less severe symptoms. These low-energy injuries can be difficult to diagnose and need a high level of suspicion from the examiner. Typically, patients present with pain and swelling without deformity, and they are frequently able to walk. Individuals with minor Lisfranc subluxation experience or hear a "snap" in the midfoot, followed by acute pain with weight-bearing and gradually diminishing pain. Midfoot sprains are the second most common foot injury among athletes, following metatarsophalangeal (MTF) joint injuries. The following are highly suggestive indicators of a Lisfranc lesion: (1) plantar ecchymosis at the midfoot level; (2) TMT joint discomfort upon palpation or manipulation [27] (Fig. 67.4): Stress testing with passive abduction and pronation of the forefoot can be useful while holding the hindfoot still; (3) altered sensitivity at the back of the first inter-metatarsal region can aid in an early diagnosis. Lisfranc lesions are associated with post-traumatic neuropathy of the medial terminal branch of the deep peroneal nerve and a decrease in two-point discrimination; (4) the "piano key test," which involves moving the head of the affected metatarsal while firmly holding the midfoot and hindfoot, is useful in identifying and isolating the affected TMT joint; and (5) an increase in the distance between the hallux and the second

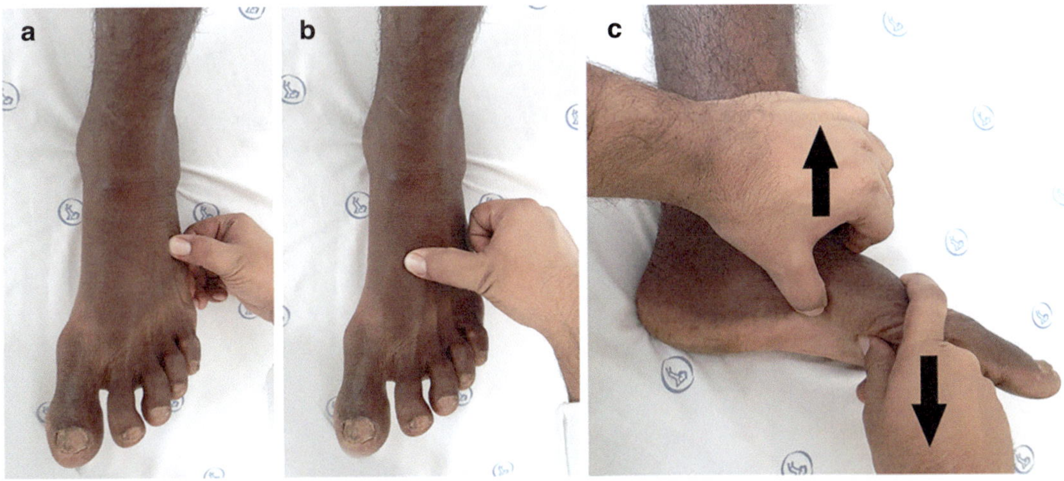

Fig. 67.4 Clinical examination of tender points in suspected Lisfranc's instability (**a, b**); assessment of medial column instability at tarsometatarsal joint (TMT) (**c**)

finger, also known as a "positive gap," which correlates with intercuneiform instability [4].

67.5 Radiological Assessment of Subtalar Instability

Conventional anteroposterior and lateral weight-bearing radiographs of the ankle, foot, and hindfoot may reveal bony avulsions at the subtalar joint caused by acute or chronic ligament injuries. Along Hellpap's [28] fracture line of supination, which extends from the ankle to the tuberosity of the fifth metatarsal, avulsions or rounded ossicles are frequently observed [28]. The utility of stress radiography to detect and quantify STJ instability has been the subject of a long-standing and contentious debate [20, 21, 29–31]. The Telos device (METAX Inc., Hungen-Obbornhofen, Germany) can be used for stress testing for subtalar joint [21] (Fig. 67.5), and this radiograph resembles Broden's views of the subtalar joint. With a varus force of 150 N applied to the hindfoot, a talocalcaneal angle of less than 10, a talocalcaneal tilt of less than 5, and a medial offset of the talus of less than 5 are considered normal [21].

A difference in talocalcaneal tilt of greater than 5 speaks in favor of STI [32, 33]. Radiographic modifications such as digital tomosynthesis can help to improve the image quality and to better read out the quantitative values of instability. The value of CT imaging has a limited role because the ligaments cannot be visualized directly, and stress MRI has been reported to better reflect the 3D motion at the subtalar joint [32]. MRI scan helps to identify a ligamentous lesion at the medial malleolus,

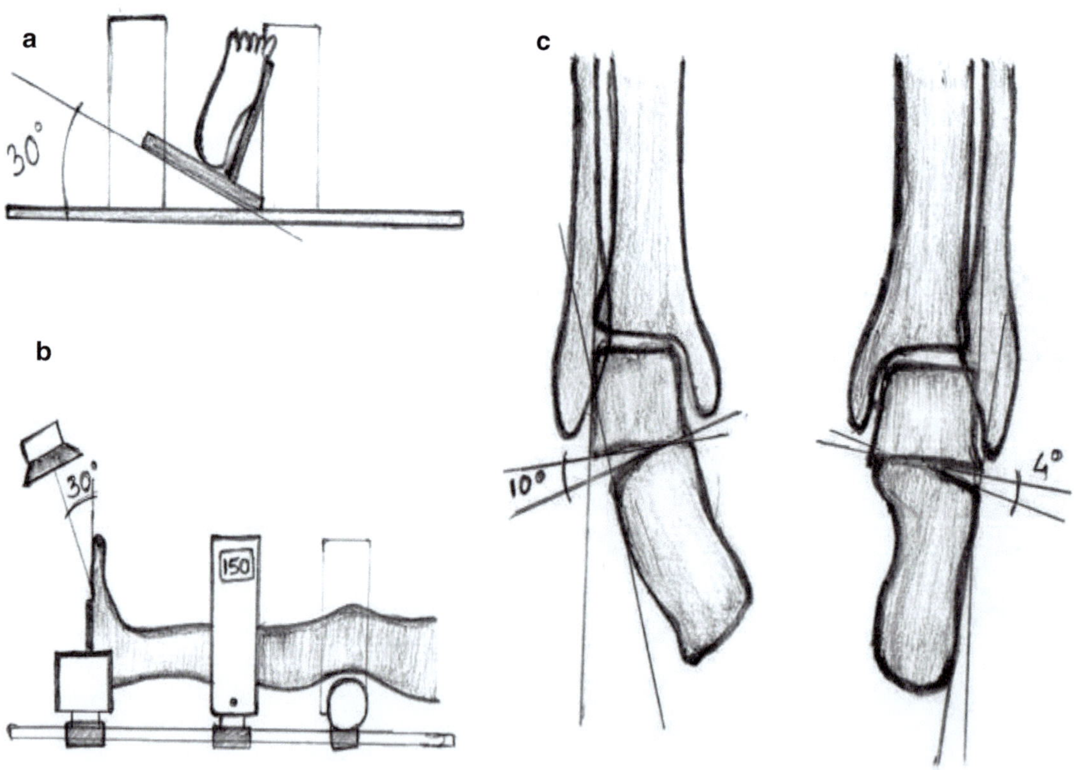

Fig. 67.5 (**a, b**) Position of leg, X-ray beam, and stress device; (**c**) pictorial representation of stress radiograph showing the joint opening

osteochondral lesions, or a disorder of the spring ligament and the tibialis posterior tendon [34, 35]. With the use of 3D isotropic high-resolution MRI with a dedicated coil for the ankle joint, the dimensions of the subtalar ligament components (ATCL, OTCL, ITCL, CFL, AFTL) can be measured directly [36]. Apart from all these noninvasive test approaches, subtalar joint arthroscopy has been advocated as a valuable diagnostic tool, in particular in cases in which the working diagnosis of a sinus tarsi syndrome was inconclusive for a further differentiation and assignment to a distinct diagnosis [30, 37, 38]. In summary, STI can be diagnosed clinically when four of the five following criteria are fulfilled [36, 39]:

1. Recurrent ankle/hindfoot sprains
2. Sinus tarsi syndrome and tenderness
3. Hindfoot laxity or giving way
4. Hindfoot instability during clinical examination
5. Hindfoot instability in stree radiographs

67.6 Radiological Assessment of Lisfranc's Instability

Radiographic weight-bearing study based on three views: (1) Anteroposterior (AP): the alignment between the medial edge of the second metatarsal and the medial edge of the second cuneiform bone should be checked. The distance between the bases of the first and second metatarsals should not exceed 2 mm. The "fleck sign" (Fig. 67.7b), a small bone fragment in the first inter-metatarsal space, indicates the avulsion of the Lisfranc ligament. (2) Internal oblique: the alignment between the medial border of the cuboid bone and the medial border of the fourth metatarsal should be checked (Fig. 67.6a). (3) Lateral view: the dorsal/plantar displacement of the metatarsals should be assessed (Fig. 67.7a). Weight-bearing radiographs are superior to the radiographs performed under stress maneuvers, in which the applied force is limited compared with that of the entire weight of the body. It has recently been shown that an angle of 28.9° in the AP radiograph of the foot improves visualization of the Lisfranc joint in

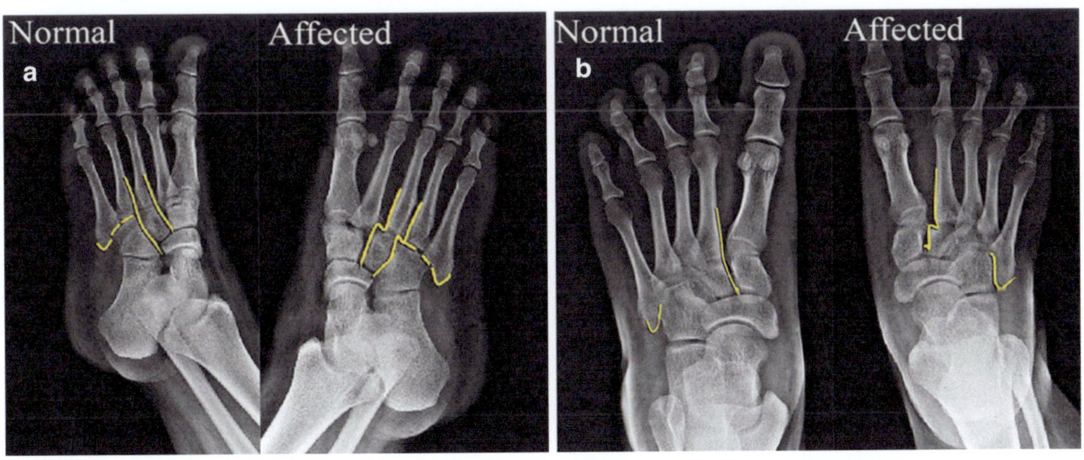

Fig. 67.6 (**a**) Internal oblique view of foot. Left side (normal foot) showing line drawn on the medial cortex of third and fourth metatarsal bone that continues/aligns along the medial border of lateral cuneiform bone and medial border of cuboid bone, respectively, whereas in right side (affected foot) incongruency/step-off in line is seen; (**b**) AP radiograph of foot—left side (normal foot) showing line along the medial border of second metatarsal bone that continues along the medial border of middle cuneiform, whereas right side (affected foot) shows incongruency/step-off in line

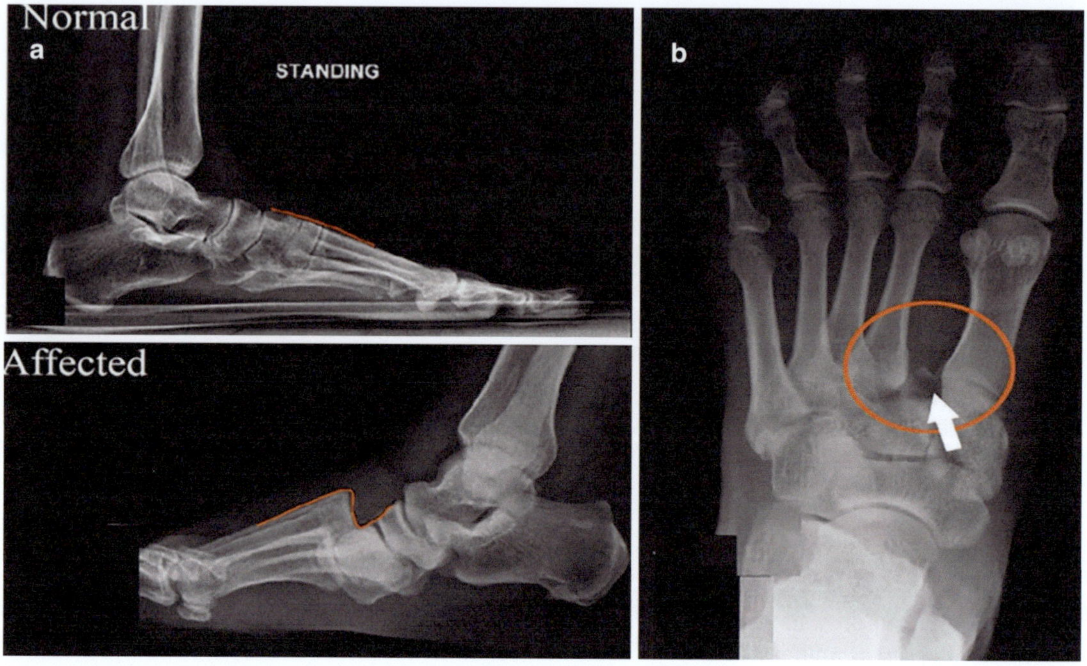

Fig. 67.7 (**a**) Lateral weight-bearing radiograph showing normal foot with continuous smooth line along the dorsal surface of metatarsal, tarsometatarsal (TMT), and cuneiform bones, whereas right foot shows break in this line due to plantar displacement at TMT joint; (**b**) "fleck sign" with widened first inter-metatarsal space is seen

most patients [40, 41]. A computed tomography (CT) scan allows a more accurate assessment of the Lisfranc joint. It allows the diagnosis of more subtle fractures and subluxations that are not observed in simple radiographs. The CT scan is generally used for surgical planning. Magnetic resonance imaging (MRI) is very useful for detecting soft-tissue injuries and ligamentous injuries. It has a sensitivity and predictive value of 94% in determining Lisfranc's instability and can therefore be useful for the diagnosis of subtle injuries [42, 43].

The classification of Lisfranc lesions was originally described by Quenu and Kuss in 1909 based on the concept of the three columns. The lesions were classified as homolateral, isolated, and divergent [44]. The classification was modified by Hardcastle et al. in 1982: in type A, all metatarsals are displaced in one direction and there is a total incongruence; in type B, there is partial incongruence with one or more displaced metatarsals; and type C has a divergent pattern of incongruence. Myerson proposed a modification of the classification that showed greater complexity of the lesion [43]. Recently, the introduction of a fourth category to the Myerson-modified Hardcastle classification (Table 67.1) has been proposed, which corresponds to subtle injury of the Lisfranc joint or lesion type D [45]. It is subdivided into D1, distance from the first cuneiform bone (c1) to the second metatarsal (m2) [(c1-m2)] $\leq$2 mm, which does not require surgical fixation, and D2, distance between c1 and m2 >2 mm, which requires surgical fixation. Type D2 can be purely ligamentous (D2L) or with bone avulsion (D2B).

Table 67.1 Meyerson-modified Hardcastle classification for describing Lisfranc injury

Types	Description	Subtype
A	Homolateral (all metatarsals displaced in one direction, total incongruity)	
B	Isolated Partial incongruity (displacement affects either first ray or lateral four metatarsals)	**B1**—partial medial incongruity
		B2—partial lateral incongruity
C	Divergent	**C1**—first metatarsal medially and lateral four in other concomitant pattern with partial incongruity
		C2—divergent pattern with total incongruity
D	Subtle Lisfranc injury	**D1**—distance between medial cuneiform and second metatarsal ≤2 mm
		D2—distance between medial cuneiform and second metatarsal >2 mm **D2B**—if bony avulsion **D2L**—if ligamentous disruption

67.7 Conclusion

Clinicians must recognize the occurrence of subtalar instability as a clinical entity associated with persistent ankle instability. The integrity of the talocrural (ankle) and subtalar joints must be frequently evaluated in patients with recurrent ankle instability. The clinical and radiographic examination of subtalar instability is highly essential for restoring hindfoot stability, particularly in athletes. A Lisfranc injury disrupts the midfoot's keystone and leads to midfoot instability may result from a soft tissue injury or a bone injury to the base of the second metatarsal. Patients frequently present with complaints of extreme discomfort and difficulty to bear weight. Lisfranc injuries can have terrible long-term repercussions if left untreated. In order to avoid these injuries, appropriate static and dynamic images are required, as well as additional images for surgical planning, such as computed tomography or magnetic resonance imaging scans will be required.

References

1. Kamada K, Watanabe S, Yamamoto H. Chronic subtalar instability due to insufficiency of the calcaneofibular ligament: a case report. Foot Ankle Int. 2002;23(12):1135–7.
2. Choisne J, Hoch MC, Alexander I, Ringleb SI. Effect of direct ligament repair and Tenodesis reconstruction on simulated subtalar joint Instability. Foot Ankle Int. 2017;38(3):324–30.
3. Cassebaum WH. Lisfranc fracture-dislocations. Clin Orthop. 1963;30:116–29.
4. Moracia-Ochagavía I, Rodríguez-Merchán EC. Lisfranc fracture-dislocations: current management. EFORT Open Rev. 2019;4(7):430–44. https://www.ncbi.nlm.nih.gov/pmc/articles/PMC6667981/
5. Budny A. Subtalar joint instability: current clinical concepts. Clin Podiatr Med Surg. 2004;21(3):449–60. viii
6. Heilman AE, Braly WG, Bishop JO, Noble PC, Tullos HS. An anatomic study of subtalar instability. Foot Ankle. 1990;10(4):224–8.
7. Weindel S, Schmidt R, Rammelt S, Claes L, Campe AV, Rein S. Subtalar instability: a biomechanical cadaver study. Arch Orthop Trauma Surg. 2010;130(3):313–9.
8. Zwipp H, Rammelt S, Grass R. Ligamentous injuries about the ankle and subtalar joints. Clin Podiatr Med Surg. 2002;19(2):195–229. v
9. Sarrafian SK. Anatomy of the foot and ankle: descriptive, topographic, functional. Philadelphia: Lippincott; 1993.
10. Meyer J-M, Garcia J, Hoffmeyer P, Fritschy D. The subtalar sprain: a roentgenographic study. Clin Orthop Relat Res. 1988;226:169–73. https://journals.lww.com/clinorthop/Abstract/1988/01000/The_Subtalar_Sprain__A_Roentgenographic_Study.23.aspx
11. Pisani G. Chronic laxity of the subtalar joint. Orthopedics. 1996;19(5):431–7.
12. Hertel J, Denegar CR, Monroe MM, Stokes WL. Talocrural and subtalar joint instability after lateral ankle sprain. Med Sci Sports Exerc. 1999;31(11):1501–8.
13. Chrisman OD, Snook GA. Reconstruction of lateral ligament tears of the ankle. An experimental study and clinical evaluation of seven patients treated by a new modification of the Elmslie procedure. J Bone Joint Surg Am. 1969;51(5):904–12.
14. Larsen E. Tendon transfer for lateral ankle and subtalar joint instability. Acta Orthop Scand. 1988;59(2):168–72.
15. Weatherford BM, Anderson JG, Bohay DR. Management of Tarsometatarsal Joint Injuries. J Am Acad Orthop Surg. 2017;25(7):469–79.
16. Pearce CJ, Calder JD. Surgical anatomy of the midfoot. Knee Surg Sports Traumatol Arthrosc. 2010;18(5):581–6.
17. Preidler KW, Wang YC, Brossmann J, Trudell D, Daenen B, Resnick D. Tarsometatarsal joint: anatomic details on MR images. Radiology. 1996;199(3):733–6.

18. Gribble PA. Evaluating and Differentiating Ankle Instability. J Athl Train. 2019;54(6):617–27. https://doi.org/10.4085/1062-6050-484-17.
19. Lee MS, Hofbauer MH. Evaluation and management of lateral ankle injuries. Clin Podiatr Med Surg. 1999;16(4):659–78.
20. Keefe DT, Haddad SL. Subtalar instability: etiology, diagnosis, and management. Foot Ankle Clin. 2002;7(3):577–609. http://www.sciencedirect.com/science/article/pii/S1083751502000475
21. Zwipp H, Rammelt S. Ligamente: Rupturen und Luxationen. In: Zwipp H, Rammelt S, Tscherne Unfallchirurgie: Fuß [Internet]. Berlin, Heidelberg: Springer; 2014. p. 215–270. doi:https://doi.org/10.1007/978-3-540-68883-9_3.
22. Barg A, Tochigi Y, Amendola A, Phisitkul P, Hintermann B, Saltzman CL. Subtalar instability: diagnosis and treatment. Foot Ankle Int. 2012;33(2):151–60.
23. Loudon JK, Bell SL. The foot and ankle: an overview of arthrokinematics and selected joint techniques. J Athl Train. 1996;31(2):173–8.
24. Mittlmeier T, Rammelt S. Update on subtalar joint Instability. Foot Ankle Clin. 2018;23(3):397–413.
25. Thermann H, Zwipp H, Tscherne H. Treatment algorithm of chronic ankle and subtalar instability. Foot Ankle Int. 2017;18(3):163–9. https://journals.sagepub.com/doi/10.1177/107110079701800309
26. Vaseenon T, Gao Y, Phisitkul P. Comparison of two manual tests for ankle laxity due to rupture of the lateral ankle ligaments. Iowa Orthop J. 2012;32:9–16.
27. Aronow MS. Treatment of the missed Lisfranc injury. Foot Ankle Clin. 2006;11(1):127–42. ix
28. Hellpap W. The neglected lower ankle. The "fracture" line of supination. Arch Orthop Unfallchir. 1963;55:289–300.
29. Kato T. The diagnosis and treatment of instability of the subtalar joint. J Bone Joint Surg Br. 1995;77(3):400–6.
30. Lee KB, Bai LB, Song EK, Jung ST, Kong IK. Subtalar arthroscopy for sinus tarsi syndrome: arthroscopic findings and clinical outcomes of 33 consecutive cases. Arthrosc. 2008;24(10):1130–4.
31. Ishii T, Miyagawa S, Fukubayashi T, Hayashi K. Subtalar stress radiography using forced dorsiflexion and supination. J Bone Joint Surg Br. 1996;78(1):56–60.
32. Zwipp H, Krettek C. Diagnosis and therapy of acute and chronic ligament instability of the lower ankle joint. Orthopade. 1986;15(6):472–8.
33. Jung H-G, Kim T-H. Subtalar instability reconstruction with an allograft: technical note. Foot Ankle Int. 2012;33(8):682–5.
34. Lötscher P, Lang TH, Zwicky L, Hintermann B, Knupp M. Osteoligamentous injuries of the medial ankle joint. Eur J Trauma Emerg Surg. 2015;41(6):615–21.
35. Steginsky B, Vora A. What to do with the spring ligament. Foot Ankle Clin. 2017;22(3):515–27.
36. Kim TH, Moon SG, Jung H-G, Kim NR. Subtalar instability: imaging features of subtalar ligaments on 3D isotropic ankle MRI. BMC Musculoskelet Disord. 2017;18(1):475.
37. Lee BH, Choi K-H, Seo DY, Choi SM, Kim GL. Diagnostic validity of alternative manual stress radiographic technique detecting subtalar instability with concomitant ankle instability. Knee Surg Sports Traumatol Arthrosc. 2016;24(4):1029–39.
38. Frey C, Feder KS, DiGiovanni C. Arthroscopic evaluation of the subtalar joint: does sinus tarsi syndrome exist? Foot Ankle Int. 1999;20(3):185–91.
39. Jung H-G, Park J-T, Shin M-H, Lee S-H, Eom J-S, Lee D-O. Outcome of subtalar instability reconstruction using the semitendinosus allograft tendon and biotenodesis screws. Knee Surg Sports Traumatol Arthrosc. 2015;23(8):2376–83.
40. Pourcho AM, Liu YH, Milshteyn MA. Electrodiagnostically confirmed posttraumatic neuropathy and associated clinical exam findings with lisfranc injury. Foot Ankle Int. 2013;34(8):1068–73.
41. Rankine JJ, Nicholas CM, Wells G, Barron DA. The diagnostic accuracy of radiographs in Lisfranc injury and the potential value of a craniocaudal projection. AJR Am J Roentgenol. 2012;198(4):W365–9.
42. Nunley JA, Vertullo CJ. Classification, investigation, and management of midfoot sprains: Lisfranc injuries in the athlete. Am J Sports Med. 2002;30(6):871–8.
43. Myerson MS, Fisher RT, Burgess AR, Kenzora JE. Fracture dislocations of the tarsometatarsal joints: end results correlated with pathology and treatment. Foot Ankle. 1986;6(5):225–42.
44. Stavlas P, Roberts CS, Xypnitos FN, Giannoudis PV. The role of reduction and internal fixation of Lisfranc fracture-dislocations: a systematic review of the literature. Int Orthop. 2010;34(8):1083–91.
45. Sivakumar BS, An VVG, Oitment C, Myerson M. Subtle Lisfranc injuries: A topical review and modification of the classification system. Orthopedics. 2018;41(2):e168–75.

68 Evaluation of Hindfoot Varus and Valgus Conditions

Hamed Mazoochy

68.1 Inspection

68.1.1 Shoes

It starts by looking at the patients' shoes. The wear pattern of the soles of patients' shoes reflects how their weight is distributed when they walk and the pattern of foot contact with the ground. Lateral or medial heel wear is suggestive of varus or valgus hindfoot, respectively. It is also beneficial to see if patients are already using medical insoles and whether it has improved their problem or not.

68.1.2 Whole Foot

There are some clues on other parts of the foot, which are related to the hindfoot conditions [1, 2].

Ask the patient to lie on their abdomen (prone position) and bend their knees. Look at the plantar surface carefully, and look for any hyperkeratosis which shows the overload area on foot; thus, its pattern is important. Hyperkeratosis on the lateral or medial side of the sole will usually suggest varus or valgus hindfoot alignment, respectively (Fig. 68.1); however, hyperkeratosis may be secondary to an antalgic posture, as well. Look for forefoot adduction or abduction, when assessing forefoot alignment. Although forefoot adduction or abduction is frequently accompanied by hindfoot varus or valgus, respectively, they can also be seen alone without any hindfoot pathologies. A severe metatarsal

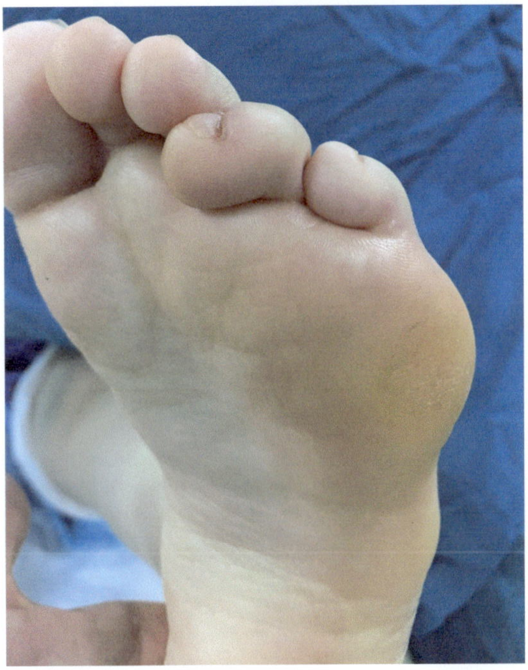

Fig. 68.1 Lateral plantar hyperkeratosis indicating excessive pressure area in a patient with cavovarus deformity

H. Mazoochy (✉)
Foot, Ankle and Limb Reconstruction at Milad Hospital, Tehran, Iran

J. G. Lane et al. (eds.), *The Art of the Musculoskeletal Physical Exam*,
https://doi.org/10.1007/978-3-031-24404-9_68

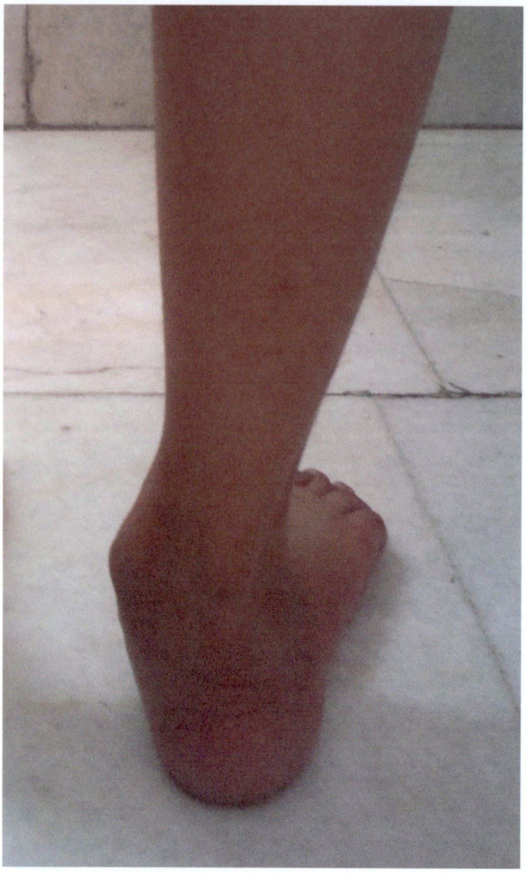

Fig. 68.2 "Too-many-toes" sign in a patient with severe hindfoot valgus and forefoot abduction as parts of pes planovalgus deformity. Four toes can be seen. While normally only the fifth toe and half of the fourth toe are seen

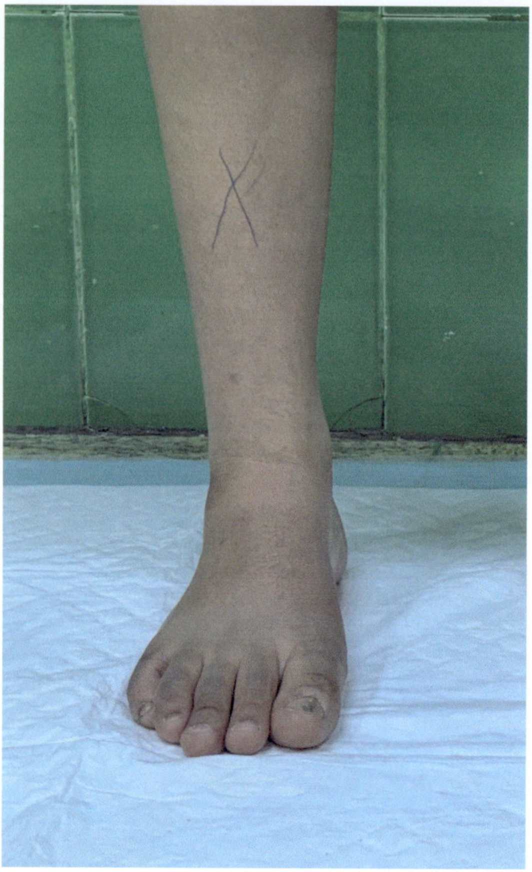

Fig. 68.3 "Peek-a-boo" sign in a patient with Charcot-Marie-Tooth and cavovarus deformity. Looking from the front in the standing position shows heel peeking around the ankle

adductus or abductus may mimic hindfoot varus or valgus malalignment on standing physical examination.

In the weight-bearing position, first, assess the patient's feet from behind and count how many toes are visualized on the lateral side of the leg. Normally, the fourth and fifth toes are exposed. In the patients with forefoot abduction associated with hindfoot valgus, more toes are visualized, and the number of toes that are seen can be used to classify the severity of the deformity and is called "too-many-toes" sign [3] (Fig. 68.2).

Still, in the standing position, look at the patient's feet from the front, while the patient's feet are pointing straight ahead. The medial border of the heel is normally not seen in this view. However, in patients with hindfoot varus deformity, it becomes exposed, which is referred to as the "peek-a-boo" sign [4] (Fig. 68.3). Also, this sign can be found in patients with a very large heel pad or significant metatarsus adductus deformity. In the same position, assess the medial longitudinal foot arc. A sagging arc (pes planus) or a high arc (pes cavus) is commonly associated with a hindfoot valgus or varus, respectively.

68.1.3 Hindfoot

The key test of the hindfoot alignment is assessing the patient in the standing position from behind with their feet about shoulder-width apart. The vertical heel axis is a line over the middle of the distal part of Achilles tendon at the level of the

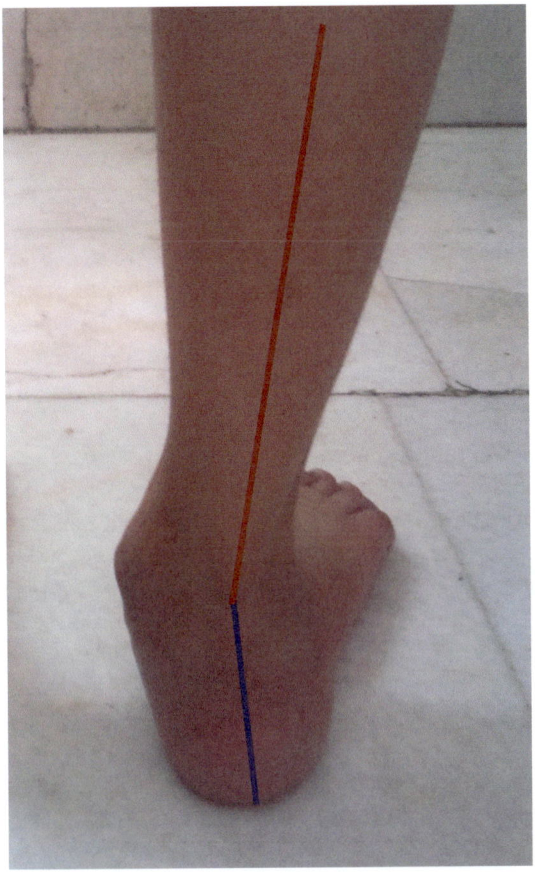

Fig. 68.4 Hindfoot coronal alignment. The vertical heel axis (blue line) and the vertical leg axis (red line) to assess hindfoot coronal alignment

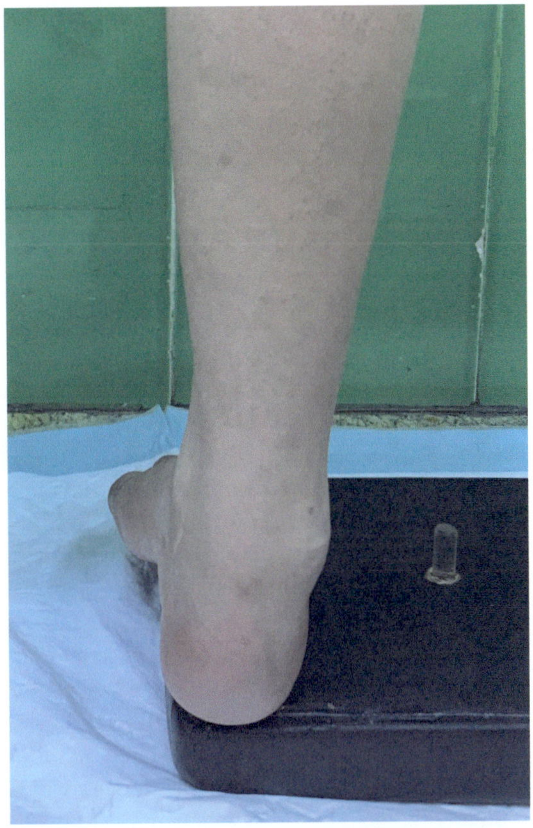

Fig. 68.5 "Coleman block test" in a patient with Charcot-Marie-Tooth and cavovarus deformity. Hindfoot varus is partially corrected, which shows degrees of flexibility

ankle joint to the middle of its attachment to the calcaneal tuberosity. The vertical leg axis is a line from the middle of the distal third of the calf to the middle of Achilles tendon at the level of the ankle joint. Assessing the hindfoot coronal alignment is by measuring the angle between the vertical heel axis and vertical leg axis (Fig. 68.4). It is measured by a goniometer and is referred to as a valgus hindfoot if the heel axis deviated away from the mid-axis of the body, or a varus hindfoot if the heel axis deviated toward the mid-axis of the body.

There are two specific tests for assessing a hindfoot coronal deformity. First, in a patient with hindfoot valgus deformity, after measuring the valgus angle from behind, ask the patient to perform a toe raise test. Normally, the heel goes to the varus with the toe raise test due to normal function of the posterior tibialis tendon. This test shows the amount of flexibility of the hindfoot valgus deformity along with the posterior tibialis tendon function. The other test is called the "Coleman block test" and is used to evaluate a varus hindfoot deformity [5]. An abnormal plantarflex first ray pronates forefoot and elevates the longitudinal foot arc (cavus deformity). As a result, medial forefoot strikes the ground first, which supinates the subtalar joint to bring the lateral forefoot on the ground and maintain three-point contact resulting in hindfoot varus. While this compensatory hindfoot varus is initially flexible, it can become rigid over time. Coleman block test is performed by asking the patient to stand on a block while the block is underneath the lateral part of the foot and the first ray is unloaded by hanging over the edge of the block (Fig. 68.5). This test eliminates the first ray as the possible pathology of the hindfoot varus deformity. If the hindfoot varus deformity is corrected into valgus, it is called a forefoot-driven hindfoot varus deformity. If the hindfoot deformity remains in varus, it may be a rigid deformity, or it means that the etiology is not from the forefoot and it is due to an abnormal pathology of the hindfoot itself.

68.1.4 Gait

Looking at the initial part of the stance phase, where the foot is contacting the ground, it typically starts with a heel strike and extends from the heel to the toes. Patients with hindfoot varus deformity frequently have reverse abnormal ground contact starting from toes and extending to the heel, which is due to associated Achilles tightness and equinus deformity. Patients with hindfoot valgus deformity may have abnormal foot-flat phase of gait, and they may also have toes to the heel ground contact, similar to patients with hindfoot varus deformity, depending on the severity of possible associated Achilles tightness and equinus deformity. The associated Achilles tendon tightness causes early heel rise, which normally happens at 35% of the gait cycle.

Gait analysis studies show that patients with cavovarus deformity have an unstable base of support, increased double-limb stance, and decreased single-limb stance.

68.2 Palpation

Ask the patient for permission to palpate their heel. Any mass in this area such as a large soft-tissue bursa or a massive Haglund's deformity may falsely impact the hindfoot alignment assessment.

68.3 Range of Motion

Hindfoot inversion and eversion occur primarily at the talocalcaneal (subtalar) joint. At extremes of motion, the tibiotalar joint can also contribute to inversion and eversion. Talocalcaneal motion is evaluated while the patient is sitting, by holding the patient's heel in one hand and the patient's forefoot in the other. Bring the subtalar joint into inversion and eversion. Inversion is usually twice as much as eversion. Talocalcaneal range of motion is measured while the patient's ankle is held in neutral dorsiflexion to maximally eliminate talus motion in the ankle mortise. Studies show variability in subtalar motion due to variation in subtalar axis, anatomy, and kinematic coupling with the ankle joint.

Achilles tightness examination is an essential exam in any foot and ankle pathology. Achilles tightness is more common in hindfoot varus or valgus deformity than other foot and ankle pathologies. Thus, it is always important to address this issue during the surgical procedures for hindfoot varus or valgus correction. The "Silfverskiold test" that is used to isolate the gastrocnemius tightness from the gasterosteus tightness is discussed in the related chapter [6]. It is essential to passively correct hindfoot varus or valgus deformity to the neutral when assessing Achilles tightness to isolate tibiotalar joint (Fig. 68.6a and b).

If the hindfoot problem is unilateral, comparing its range of motion to the healthy side is helpful. In the case of stiffness, the position in which the stiffness is present should be documented. If the deformity is not stiff, then it is essential to know how much it is reducible. Reducibility of the deformity is a key factor in making the surgical plan by choosing between soft-tissue release, tendon transfer, corrective osteotomy, and joint fusion or the combination of these procedures.

Any cavovarus deformity especially when it is unilateral needs a thorough neurologic workup since neurologic pathologies are among the most common etiologies of such deformities [7]. In contrast, most hindfoot valgus deformities are idiopathic or secondary to a posterior tibialis tendon insufficiency [8].

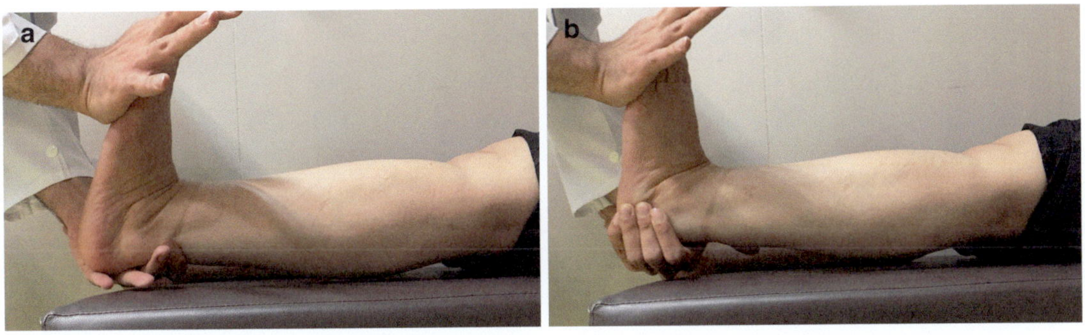

Fig. 68.6 (**a**) Incorrect demonstration of the "Silfverskiold test." Hindfoot and mid-foot joints are unlocked. Dorsiflexion is gained through the ankle, hindfoot, mid-foot, and forefoot, rather than isolating the tibiotalar joint. (**b**) Correct demonstration of the "Silfverskiold test." The subtalar and mid-foot joints are locked. Dorsiflexion is gained through the ankle joint

References

1. Krause F, Seidel A. Malalignment and lateral ankle instability. Foot Ankle Clin. 2018;23(4):593–603.
2. Bluth B, Eagan M, Otsuka N. Stress fractures of the lateral rays in the Cavovarus foot: indication for surgical intervention. Orthopedics. 2011;34(10):e696–9.
3. Johnson KA. Tibialis posterior tendon rupture. Clin Orthop Relat Res. 1983;177:140–7.
4. Manoli A 2nd, Smith DG, Hansen ST Jr. Scarred muscle excision for the treatment of established ischemic contracture of the lower extremity. Clin Orthop Relat Res. 1993;292:309–14.
5. Coleman SS, Chesnut WJ. A simple test for hindfoot flexibility in the cavovarus foot. Clin Orthop Relat Res. 1977;123:60–2.
6. Singh D. Nils Silfverskiöld (1888-1957) and gastrocnemius contracture. Foot Ankle Surg. 2013;19(2):135–8.
7. Neumann JA, Nickisch F. Neurologic disorders and Cavovarus deformity. Foot Ankle Clin. 2019;24(2):195–203.
8. Pomeroy G, Pike R, Beals T, Manoli A. Acquired flatfoot in adults due to dysfunction of the posterior Tibial tendon. J Bone J Surg. 1999;81(8):1173–82.

Hindfoot Tendinopathies

69

Pim A. D. van Dijk

69.1 Introduction

Tendinopathies are common hindfoot injuries that can result in high morbidity for the patient. While the exact incidence in the hindfoot remains unknown, around 30% of all musculoskeletal consultations in the general practice are associated with this disabling condition [1]. In the acute phase, hindfoot tendinopathies are often missed or mistaken for an ankle sprain [2]. On the other hand, overuse or chronic injuries are difficult to diagnose due to the slow onset of symptoms. As tendinopathies can have a strong impact on the patient's quality of life, timely and accurate diagnosis is important for effective management and to avoid socioeconomic burden [3].

In recent times, our understanding of tendon injuries has evolved significantly and today tendinopathy is considered a chronic, degenerative pathology. The cause of tendinopathies is mostly multifactorial and can be an expression of systemic or general musculoskeletal problems. A combined thorough patient history and physical examination are the cornerstone for diagnosing the tendinopathy itself as well as to find possible contributing factors or underlying diseases. Additional diagnostic modalities such as ultrasound and magnetic resonance imaging can be used to confirm the clinical suspicion of tendinopathy.

This chapter provides an overview of the pathophysiology, patient history, risk factors, and physical examination of tendinopathy with special emphasis on the hindfoot tendons.

69.2 Nomenclature

Historically, many confusing terminologies have been used to define tendon injuries in the clinical setting. In 1998, the umbrella term "tendinopathy" was introduced to describe the combination of pain, swelling, and impaired function of the tendon in the clinical setting [4]. It was found that only after histologic examination of the damaged tissue, one can properly distinguish between subtypes of tendinopathy such as tendinitis, tenosynovitis, and tendinosis. In the context of physical examination of hindfoot tendinopathies, as in this chapter, the general term tendinopathy is therefore most appropriate to use.

P. A. D. van Dijk (✉)
Department of Orthopaedic Surgery, Amsterdam University Hospital–Academic Medical Centre, University of Amsterdam, Amsterdam, The Netherlands

Academic Centre for Evidence Based Sports Medicine, Amsterdam, The Netherlands

Amsterdam Collaboration on Health and Safety in Sports, Amsterdam, The Netherlands
e-mail: p.a.d.vandijk@gmail.co; p.a.vandijk@amc.uva.nl

J. G. Lane et al. (eds.), *The Art of the Musculoskeletal Physical Exam*,
https://doi.org/10.1007/978-3-031-24404-9_69

69.3 Pathophysiology

Healthy tendons are highly organized structures to allow the transmission of high-energy forces between bones and muscles. Regulation of this strict organization is essential to maintain the tendon's integrity and mechanical properties and to prevent injuries [5, 6]. Normal response of tendon tissue to damage consists of inflammation, followed by proliferation or deposition of collagen matrix, and remodeling [7, 8]. In case of an inadequate inflammatory response, the tendon's structure becomes hypervascularized and disorganized by an increase of fibrin depositions, collagen breakdown, and a reduction of macrophages and neutrophils: the first step in the development of tendinopathy.

Both magnitude and distribution of the tendon's load play a major part in the development of tendinopathy [9, 10]. During repeated motion and loading, tendons become susceptible for tendinopathy because the tendon's repair mechanism has less time to heal possible microtraumas [9, 11]. Moreover, muscle activation and force misdistribution within the tendon can result in nonuniform loading of the tendons, which in turn may cause microtraumas [9]. In case of ongoing mechanical forces in spite of these microtraumas, further tendon deterioration occurs which eventually leads to tendinopathy.

No consensus exists as to the exact etiology of tendinopathy. Several theories have been proposed, most of them being related to either the inadequate vascularization of the tendon or the tendon's inflammatory response to damage [12–17]. Ischemia might be induced by maximal loading of the tendon, followed by relaxation and subsequent generation of free radicals. These radicals potentially contribute to the development of degenerative changes in the tendon [17]. Moreover, local hypoxia may result in inadequate levels of ATP, which may also lead to tendon degeneration [12]. Others describe that tenocyte apoptosis is induced by high strain application [14]. In animal and human studies, tendon stress was related to the release of harmful inflammatory mediators such as interleukin-6 and prostaglandin E_2, which potentially induce histological degenerative changes [15, 16].

While tendons have some regenerative capacity, it is not likely that the tendon's tissue will return to the pre-injury level of mechanical properties and highly structured organization [18].

69.3.1 What Causes Pain in Tendinopathy?

While pain is the main reported symptom in tendinopathy, the cause remains an enigma. As per definition, pain serves to protect the tendon and surrounding tissues. Tendon pain is closely linked to extensive loading and activities, while it is barely experienced during rest or low-load activities. Moreover, "warming-up" of the tendons reduces pain [19, 20].

Many theories have been proposed in literature, as the molecular biology of tendons covers many potential contributors to the development of pain [21]. Some authors suggest that degeneration and disruption of the tendon tissue, including deposit of fibrin and mechanical breakdown of collagen, contribute to the development of pain as this disruption may stimulate nociceptors [22, 23].

Others hypothesize that the cause of tendon pain has to do with the tendon's complex network of neurons and neuronal mediators' innervation. The latter is also involved in the tendon's inflammatory response with levels increasing after injury [24–26]. This might explain a synchronized mechanism of tissue repair and stimulation of the nerves at the same time [25]. In 2003, Kahn et al. raised "the biochemical hypothesis" of pain in tendinopathy, relating to the release of biochemical agents in tendon inflammation. These agents might irritate nociceptors in the surrounding tissue or the tendon itself and in this way cause pain [27].

More recently, a theory was published suggesting that tendons itself cannot be the cause of pain due to their aneural and avascular nature. The fact that up to 32% of asymptomatic, unpainful Achilles tendons show histopathological changes contributes to this theory [28, 29]. Instead, it is more likely that the pain is originated from outside the tendon. Damaged tendons generate cytokines, which induce neovascularization and neoinnervation from the paratenon or vincula into the tendon, as seen in chronic painful tendons. It is these neonerves that are responsible for the pain [30].

69.4 Patient History and Risk Factors

In general, up to 83% of all patients provide the diagnosis in their clinical history [31]. In patients with a hindfoot tendinopathy, there is little to suggest that this is any different. The cardinal complaint in patients with tendinopathy is dose-dependent pain during loading of the tendon, or in other words "too soon, too much, and/or too often" [32]. In early stages, pain is typically provoked after longer periods of loading while proper "warming-up" decreases pain complaints. In later stages of disease, pain can even be present in rest with worsening during activities. Seldomly, patients report an insidious onset of pain. Other complaints in tendinopathy include overuse, swelling, and functional impairment.

Tendinopathies are most common in the active population, as physical activities and/or occupation often determine the amount of force and stress on the tendons. Any tendon can be affected, depending on the type of sports or occupation performed. Runners and dancers, for example, show higher prevalence of flexor hallucis longus or Achilles tendinopathy. In fact, 52% of male elite runners are affected by Achilles tendinopathy at one point in their life [33].

Besides activity and overuse, several extrinsic and intrinsic factors are associated with the development of tendinopathy. Extrinsic factors include training errors such as disbalanced intensity, hill work, uneven grounds, fatigue, wrong technique, and excessive distances. Especially changes in load, technique, and/or equipment are associated with the onset of tendinopathy [32]. Also, environmental changes and outdoor training in cold weather are considered risk factors [34].

Intrinsic factors include biomechanical abnormalities such as malalignment and difference in leg length, age, gender, and health status. Adults are more prone to tendinopathy in the lower extremity in comparison to adolescents [35]. In contrast to other anatomical regions of the body, in hindfoot tendinopathies, there is no difference in incidence between male and female patients [36].

Diseases such as diabetes, obesity, and autoimmune disorders may affect the biology and mechanical function of tendons, making the patient more prone to the development of tendinopathies. Moreover, tendinopathy may be the first clinical presentation of various metabolic disorders such as gout or hypercholesterolemia. Several pharmacological subtracts are associated with the development of tendinopathy. Among them are quinolone antibiotics, corticosteroids, statins, and aromatase inhibitors [37].

69.5 General Physical Examination

69.5.1 Inspection

Every orthopedic physical examination starts with a general inspection [38]. Since this is thoroughly described elsewhere in this book, it is not further discussed in this chapter. In hindfoot tendinopathies, special attention should be paid to the patient's habitus, walking pattern, (mal) alignment, and possible leg length difference.

More specific for hindfoot tendinopathies, inspection should focus on the foot and ankle. Check the shoes for signs of abnormal or asymmetrical wear, which can indicate malalignment. With the patient on bare feet, check the alignment of the foot and ankle from behind and check the arch height (Fig. 69.1). A normal hindfoot has a valgus of approximately 5°. Depending on the type of malalignment, load of specific tendons is increased, which puts them at higher risk for the development of tendinopathy (see paragraph F). In case of a flatfoot, the Coleman block test can be used to evaluate the flexibility of the malalignment. A block is placed under the lateral side of the foot in order to eliminate contribution of the first ray and pronation of the forefoot. In case of a flexible flatfoot, the alignment of the foot is corrected to neutral or valgus. A rigid hindfoot will not correct during the Coleman block test.

Check the lower extremity for scars, swelling, muscle atrophy, change of color, and hair growth. Swelling along the course of a specific tendon may indicate tendinopathy. Poor blood supply, an intrinsic risk factor for tendinopathy, may be indicated by decreased hair growth, hypertrophic nails, and a thinner skin.

69.5.2 Palpation

Palpation of the hindfoot is best performed with the patient sitting in a chair and the examiner's chair opposed. It should be done in a systemic manner and always compared to the other ankle. Due to the foot's superficial contours, the surface anatomy can be used as a guide for identification of possible pathology (Fig. 69.2). Check for areas of tenderness and bony or soft-tissue thickening: recognizable tenderness on palpation of the affected tendon is one of the most specific findings for tendinopathy during physical examination. Proper knowledge of the hindfoot's anatomy

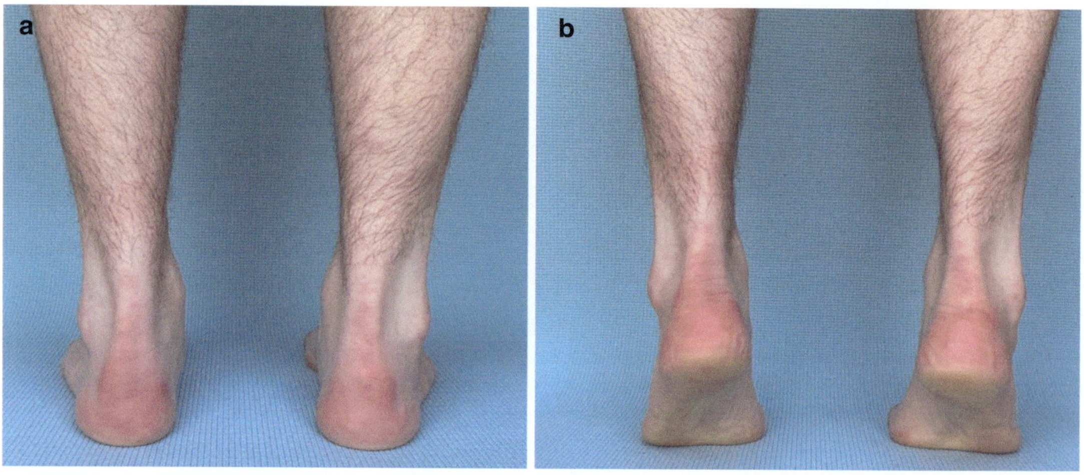

Fig. 69.1 Hindfoot evaluation in (**a**) neutral and (**b**) forefoot loaded position

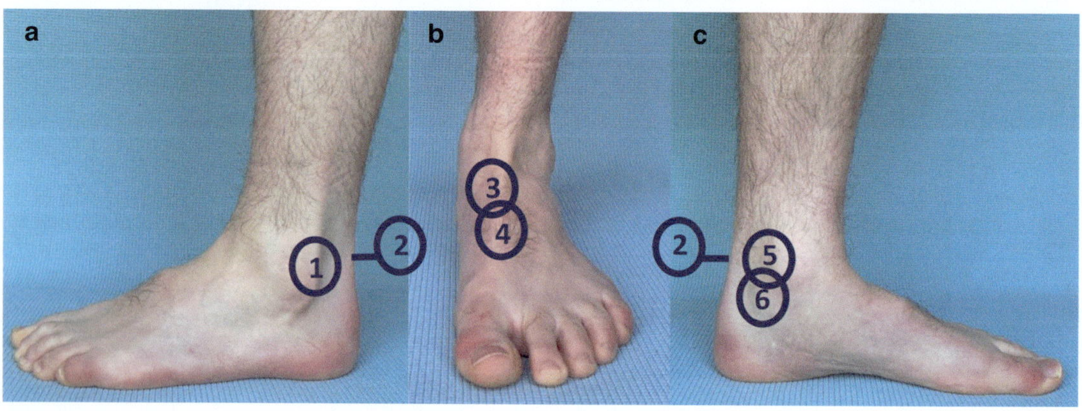

Fig. 69.2 Evaluation of tendons in the ankle via surface anatomy

Table 69.1 Hindfoot tendon function

Hindfoot tendon	Function
Flexors and plantiflexors	
Posterior tibial tendon	Plantarflexion, inversion and adduction of the ankle
Peroneus brevis tendon	Plantarflexion, abduction, eversion of the ankle
Peroneus longus tendon	Plantarflexion of the hallux, abduction and eversion of the ankle
Flexor hallucis longus tendon	Flexion of the hallux (distal phalanx)
Flexor digitorum longus tendon	Flexion of digiti 2-5
Achilles tendon *Discussed in chapter 5 of the Ankle section*	Plantarflexion of the ankle
Extensors and dorsiflexors	
Anterior tibial tendon	Dorsiflexion, inversion and adduction of the ankle
Peroneus tertius*	Dorsiflexion and eversion of the ankle
Extensor hallucis longus	Extension of the hallux (distal phalanx)
Extensor digitorum longus	Extension of digiti 2-5

is therefore essential. Once areas of tenderness are found, define the limits of tenderness and possible changes by loading and motion of the ankle.

69.5.3 Range of Motion and Muscle Strength

Range of motion of the ankle should be performed systematically. It is important to test both active and passive range of motion and compare to the contralateral side. Please refer to Table 69.1 for an overview of the function of specific hindfoot tendons. Deficits in range of motion or reduction of muscle strength may be explained by injury of specific tendons. It should be kept in mind that the culprit may be in a different anatomic region. For example, lack of active hallux extension may indicate pathology of the extensor hallucis longus but can also be related to a nervus peroneus injury of L5 radiculopathy.

69.5.4 Pain Provocation Test

As tendon pain is loading dose-dependent, recognizable pain may be provoked during extensive loading or active motion against resistance. In Table 69.1, and as further discussed in paragraph F, the specific function to test for each tendon is summarized.

69.6 Physical Examination of Hindfoot Tendinopathy

69.6.1 Posterior Tibial Tendon

The posterior tibial tendon acts as the primary invertor of the ankle and dynamic stabilizer of the medial arch and plays an important role in the medial ankle stability during walking. Dysfunction of the tendon is the most common cause of acquired flatfoot deformity. While changes in the tendon's structure become obvious in later stages of tendinopathy, the degenerative process begins far before clinical disease is apparent. Adequate clinical history and physical examination lead to earlier diagnosis, potentially preventing the patient from further deterioration and invasive treatments.

Posterior tibial tendinopathy is often seen after overuse or extensive loading in high-impact sports such as tennis, soccer, and basketball, as the tendon may suffer repetitive microtraumas due to extensive loading. Woman and patients older than 40 years are also at

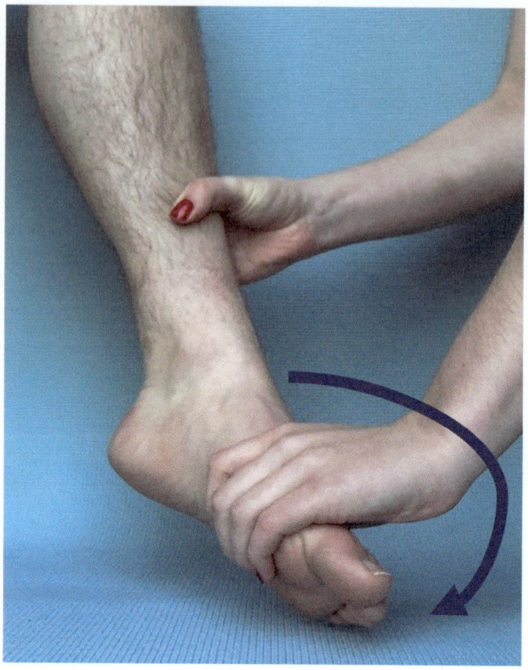

Fig. 69.3 Assessment of Posterior Tibial tendon function

greater risk. Patients typically demonstrate pain at the medial side of the ankle during activities such as running and walking. In later stages of disease, pain can also be experienced in front of the foot or even on the lateral side of the ankle.

During inspection, a pes planus or "too-many-toes sign" may be observed. A Coleman block test often reveals a flexible flatfoot in early stages of the disease, and a more rigid flatfoot in severe tendinopathy. Swelling may be presented over the course of the tendon, and pain is often provoked during palpation of the tendon (Fig. 69.2). Active plantar flexion and inversion against resistance may reveal pain and loss of muscle strength (Fig. 69.3). Single-leg heel raise is a useful clinical tool to differentiate the severity of tendinopathy: in an early stage, patients are able to stand tiptoes on the affected foot without pain. In later stages, standing on the affected foot and coming up on tiptoes will provoke pain or the rigid deformity may prevent the patient from completing the test at all [39].

69.6.2 Anterior Tibial Tendon

The anterior tibial tendon functions as a dorsiflexor and invertor of the ankle. Moreover, it stabilizes the ankle during walking as the foot hits the ground and pulls it clear of the ground during push-off. Tendinopathy of the anterior tibial tendon may result from overuse, with specific risk in activities such as running up and down hill. During uphill running, the tendon must lift the foot higher than normal. In downhill running, the tendon is lengthening and contracting at the same time, which puts even higher loads on the tendon. Moreover, the tendinopathy may be caused by tight shoelaces or strapping over the front of the ankle, causing higher compression on the tendon.

Patients typically present with pain complaints in front of the ankle that worsens with activity. Other symptoms include weakness during dorsiflexion or "slapping" of the foot on the ground when walking.

During physical examination, pain and swelling can be found over the course of the tendon, especially in front of the ankle (Fig. 69.3). Active dorsiflexion against resistance often exacerbates pain and may reveal loss of muscle strength (Fig. 69.4).

69.6.3 Peroneal Tendons

Together, the peroneal tendons act as the primary evertors and abductors of the foot. In this way, they play an important role in the active lateral ankle stability, the foot's eversion strength, and stabilization of the lateral column of the foot during stance. Tendinopathy of the tendons is often described after a gross ankle sprain in the patient's history or in case of chronic lateral ankle instability.

Typically, patients present with lateral ankle pain or pain along the course of the tendons that worsens with activity. Other symptoms include tenderness, giving way, and lateral ankle instability. On inspection, a cavovarus foot may be observed just as swelling over the course of the peroneal tendons. Palpation of the tendons often generates recognizable tenderness, pain, and/or crepitus (Fig. 69.2). Passive inversion and plantar flexion or active eversion and dorsiflexion against resistance typically exacerbate pain (Fig. 69.5). Moreover, muscle strength can be weaker when compared to the contralateral side.

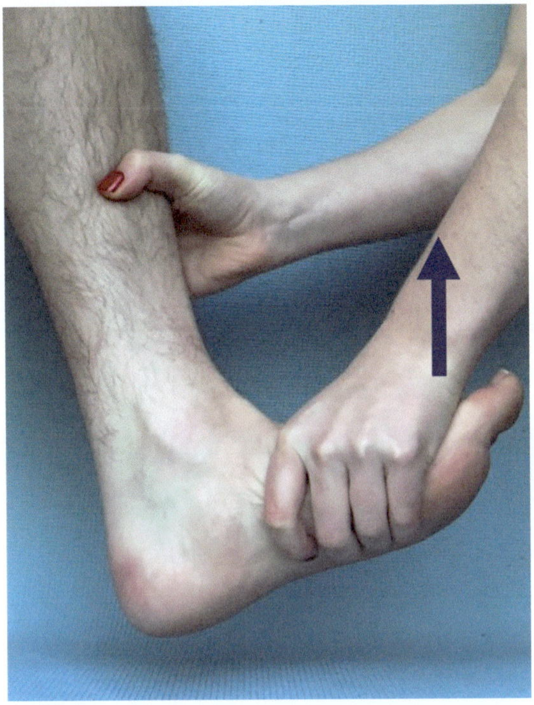

Fig. 69.4 Evaluation of Anterior Tibial tendon

69.6.4 Flexor Hallucis Longus

The flexor hallucis longus tendon is the primary flexor of the hallux by plantarflexing the talocrural, metatarsophalangeal, and interphalangeal joints of the hallux while also aiding in plantar flexion and inversion of the ankle. Tendinopathy of the flexor hallucis longus, also known as "dancer's tendinopathy," may result from high tendon loads in extreme plantar flexion of the ankle and flexion of the hallux. It is most often found in classical ballet dancers but is also related to athletes performing frequent push-off maneuvers such as gymnasts, skaters, long-distance runners, and swimmers.

Pain typically starts around the posteromedial side of the ankle or on the medial side of the subtalar joint, below the tip of the medial malleolus. It may be provoked by plantar flexion such as push-off in running or relevé in ballet. Moreover, it is related with other pathologies such as an os

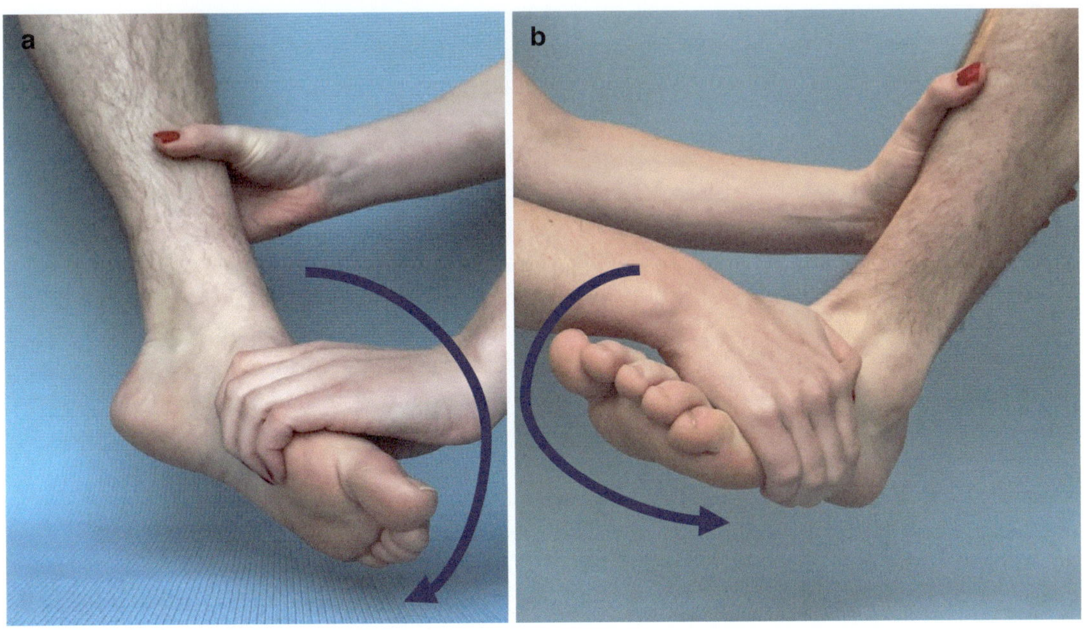

Fig. 69.5 Evaluation of Peroneal tendon function

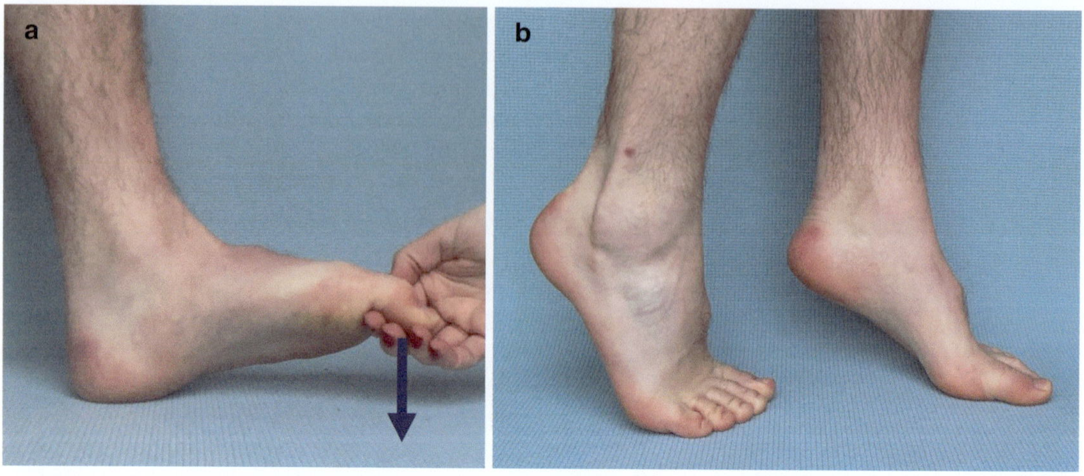

Fig. 69.6 Evaluation of Flexor Hallucis Longus tendon function

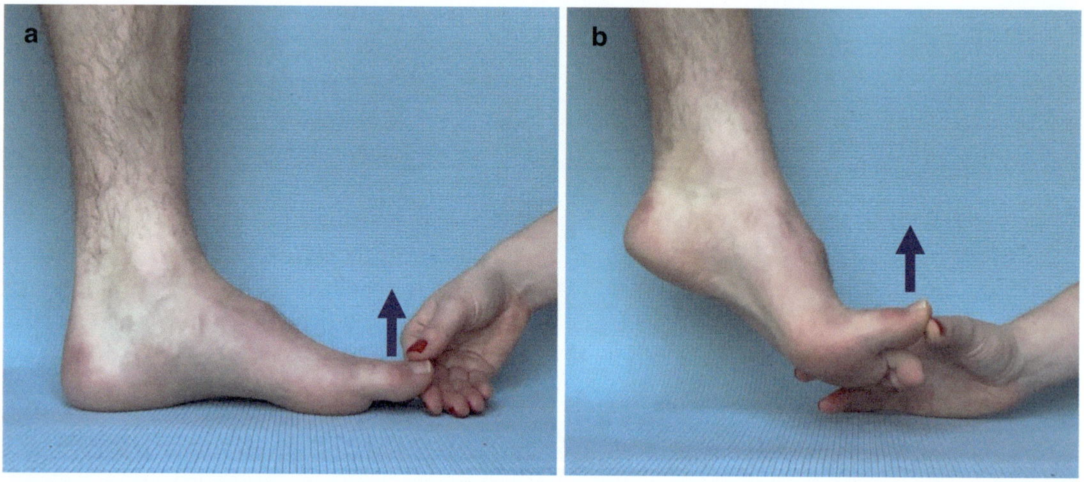

Fig. 69.7 Evaluation of Extensor Hallucis Longus tendon function

trigonum, bony avulsions, ankle and subtalar arthrosis, and loose bodies [40–43].

Tenderness on palpation of the flexor hallucis longus usually reveals flexor hallucis longus tendinopathy (Fig. 69.2). The tendon can be palpated behind the medial malleolus in its sliding canal by asking the patient to flex the toes repetitively with the ankle in a position of slight plantar flexion (10–20°). Flexion of the hallux against resistance or walking on tiptoes may provoke recognizable pain (Fig. 69.6). Moreover, passive extension of the hallux with the ankle in both neutral and plantarflexed positions can be compared: in case of limited to no extension in neutral position that disappears in plantar flexion, the flexor hallucis longus was likely entrapped which might indicate tendinopathy (Fig. 69.7). Passive extension of the hallux with the ankle in plantar flexion can also provoke pain complaints (Fig. 69.7b).

69.6.5 Extensor Hallucis Longus

The extensor hallucis longus tendon acts as the primary extensor of the hallux and contributes to

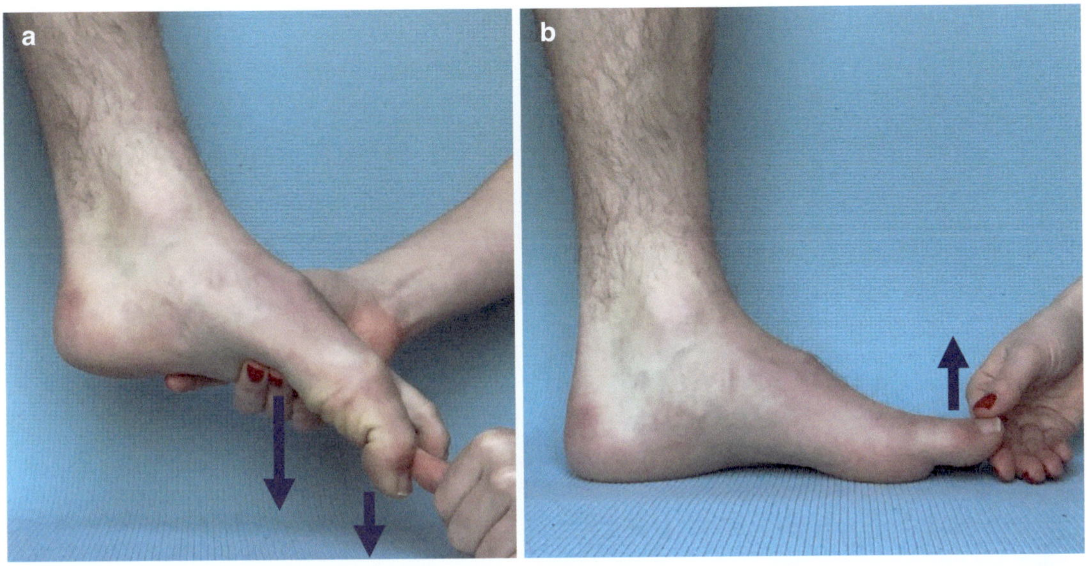

Fig. 69.8 Evaluation of Extensor Hallucis tendon strength and function

dorsiflexion of the ankle. Moreover, it dorsiflexes the ankle and inverts the foot. In this manner, it allows clearance off the ground during walking. Isolated tendinopathy of the extensor hallucis longus is uncommon [44]. It may be caused by eccentric overload with repeated friction of the tendon passing under the extensor retinaculum at the front of the ankle, also known as "ultramarathoner's ankle" [45]. Other causative factors include excessive pronation, eccentric overload, muscle imbalance, and tight shoelaces or strapping over the front of the ankle [46–48].

Patients typically indicate pain over the course of the tendon, especially in front of the ankle (Fig. 69.2). Other symptoms include weakness during extension of the hallux, "slapping" of the foot on the ground when walking, and a characteristic ultramarathoner's gait.

During physical examination, pain and swelling can be found over the course of the tendon, especially in front of the ankle (Fig. 69.2). Forced passive flexion of the hallux with the ankle in plantar flexion and active extensor of the hallux against resistance often exacerbate pain (Fig. 69.8). The latter may also reveal loss of muscle strength.

69.7 Conclusion

Tendinopathies are common injuries in the hindfoot and can result in high morbidity for the patient. While significant deterioration of the tendon's structure becomes obvious in later stages of tendinopathy, the degenerative process begins far before clinical disease is apparent. Adequate clinical history and physical examination are key to early diagnosis, potentially preventing the patient from further deterioration and invasive treatments.

Typically, tendinopathy occurs in the active population and results from overuse or "too soon, too much, and/or too often." In case of symptoms such as loading dose-dependent pain, swelling, tenderness along the course of a specific tendon, and provocation of pain during active motion against resistance, always consider the possibility of a tendinopathy.

69.8 Pearls and Pitfalls

- Adequate clinical history and physical examination are key to early diagnosis of tendinopathy, potentially preventing the patient from further deterioration and invasive treatments.
- The cardinal complaint in patients with tendinopathy is dose-dependent pain during loading of the tendon, or in other words "too soon, too much, and/or too often."
- Tenderness on palpation of the affected tendon is one of the most specific findings for tendinopathy during physical examination.
- Proper knowledge on the anatomy of the hindfoot is mandatory for diagnosing hindfoot tendinopathies.

References

1. Kaux JF, Forthomme B, Goff CL, Crielaard JM, Croisier JL. Current opinions on tendinopathy. J Sports Sci Med. 2011;10(2):238–53.
2. Gorter K, de Poel S, de Melker R, Kuyvenhoven M. Variation in diagnosis and management of common foot problems by GPs. Fam Pract. 2001;18(6):569–73.
3. Yelin E, Weinstein S, King T. An update on the burden of musculoskeletal diseases in the U.S. Semin Arthritis Rheum. 2019;49(1):1–2.
4. Maffulli N, Khan KM, Puddu G. Overuse tendon conditions: time to change a confusing terminology. Arthroscopy. 1998;14(8):840–3.
5. Bi Y, Ehirchiou D, Kilts TM, Inkson CA, Embree MC, Sonoyama W, et al. Identification of tendon stem/progenitor cells and the role of the extracellular matrix in their niche. Nat Med. 2007;13(10):1219–27.
6. Lin TW, Cardenas L, Soslowsky LJ. Biomechanics of tendon injury and repair. J Biomech. 2004;37(6):865–77.
7. Fu SC, Rolf C, Cheuk YC, Lui PP, Chan KM. Deciphering the pathogenesis of tendinopathy: a three-stages process. Sports Med Arthrosc Rehabil Ther Technol. 2010;2:30.
8. Andarawis-Puri N, Flatow EL, Soslowsky LJ. Tendon basic science: development, repair, regeneration, and healing. J Orthop Res. 2015;33(6):780–4.
9. Nakama LH, King KB, Abrahamsson S, Rempel DM. Effect of repetition rate on the formation of microtears in tendon in an in vivo cyclical loading model. J Orthop Res. 2007;25(9):1176–84.
10. Nakama LH, King KB, Abrahamsson S, Rempel DM. Evidence of tendon microtears due to cyclical loading in an in vivo tendinopathy model. J Orthop Res. 2005;23(5):1199–205.
11. Arndt AN, Komi PV, Bruggemann GP, Lukkariniemi J. Individual muscle contributions to the in vivo Achilles tendon force. Clin Biomech (Bristol, Avon). 1998;13(7):532–41.
12. Birch HL, Rutter GA, Goodship AE. Oxidative energy metabolism in equine tendon cells. Res Vet Sci. 1997;62(2):93–7.
13. Goodship AE, Birch HL, Wilson AM. The pathobiology and repair of tendon and ligament injury. Vet Clin North Am Equine Pract. 1994;10(2):323–49.
14. Arnoczky SP, Tian T, Lavagnino M, Gardner K, Schuler P, Morse P. Activation of stress-activated protein kinases (SAPK) in tendon cells following cyclic strain: the effects of strain frequency, strain magnitude, and cytosolic calcium. J Orthop Res. 2002;20(5):947–52.
15. Skutek M, van Griensven M, Zeichen J, Brauer N, Bosch U. Cyclic mechanical stretching enhances secretion of interleukin 6 in human tendon fibroblasts. Knee Surg Sports Traumatol Arthrosc. 2001;9(5):322–6.
16. Stone D, Green C, Rao U, Aizawa H, Yamaji T, Niyibizi C, et al. Cytokine-induced tendinitis: a preliminary study in rabbits. J Orthop Res. 1999;17(2):168–77.
17. Wang MX, Wei A, Yuan J, Clippe A, Bernard A, Knoops B, et al. Antioxidant enzyme peroxiredoxin 5 is upregulated in degenerative human tendon. Biochem Biophys Res Commun. 2001;284(3):667–73.
18. Frank C, McDonald D, Shrive N. Collagen fibril diameters in the rabbit medial collateral ligament scar: a longer term assessment. Connect Tissue Res. 1997;36(3):261–9.
19. Bagge J, Gaida JE, Danielson P, Alfredson H, Forsgren S. Physical activity level in Achilles tendinosis is associated with blood levels of pain-related factors: a pilot study. Scand J Med Sci Sports. 2011;21(6):e430–8.
20. Ferretti A. Epidemiology of jumper's knee. Sports Med. 1986;3(4):289–95.
21. Rio E, Moseley L, Purdam C, Samiric T, Kidgell D, Pearce AJ, et al. The pain of tendinopathy: physiological or pathophysiological? Sports Med. 2014;44(1):9–23.
22. Snedeker JG, Foolen J. Tendon injury and repair–a perspective on the basic mechanisms of tendon disease and future clinical therapy. Acta Biomater. 2017;63:18–36.
23. Gotoh M, Hamada K, Yamakawa H, Inoue A, Fukuda H. Increased substance P in subacromial bursa and shoulder pain in rotator cuff diseases. J Orthop Res. 1998;16(5):618.
24. Danielson P, Alfredson H, Forsgren S. Distribution of general (PGP 9.5) and sensory (substance P/CGRP) innervations in the human patellar tendon. Knee Surg Sports Traumatol Arthrosc. 2006;14(2):125–32.
25. Ackermann PW, Ahmed M, Kreicbergs A. Early nerve regeneration after Achilles tendon rupture--a prerequisite for healing? A study in the rat. J Orthop Res. 2002;20(4):849.

26. Alfredson H, Thorsen K, Lorentzon R. In situ microdialysis in tendon tissue: high levels of glutamate, but not prostaglandin E2 in chronic Achilles tendon pain. Knee Surg Sports Traumatol Arthrosc. 1999;7(6):378–81.
27. Khan KM, Cook JL, Maffulli N, Kannus P. Where is the pain coming from in tendinopathy? It may be biochemical, not only structural, in origin. Br J Sports Med. 2000;34(2):81–3.
28. Haims AH, Schweitzer ME, Patel RS, Hecht P, Wapner KL. MR imaging of the Achilles tendon: overlap of findings in symptomatic and asymptomatic individuals. Skelet Radiol. 2000;29(11):640–5.
29. Khan KM, Forster BB, Robinson J, Cheong Y, Louis L, Maclean L, et al. Are ultrasound and magnetic resonance imaging of value in assessment of Achilles tendon disorders? A two year prospective study. Br J Sports Med. 2003;37(2):149–53.
30. van Sterkenburg MN, van Dijk CN. Mid-portion Achilles tendinopathy: why painful? An evidence-based philosophy. Knee Surg Sports Traumatol Arthrosc. 2011;19(8):1367–75.
31. Hampton JR, Harrison MJ, Mitchell JR, Prichard JS, Seymour C. Relative contributions of history-taking, physical examination, and laboratory investigation to diagnosis and management of medical outpatients. Br Med J. 1975;2(5969):486–9.
32. Renström PA, Woo SL. Tendinopathy: a major medical problem in sport. In: Woo SL, Renström PA, Arnoczky S, editors. Tendinopathy in athletes. London: Wiley-Blackwell; 2008.
33. Kujala UM, Sarna S, Kaprio J. Cumulative incidence of Achilles tendon rupture and tendinopathy in male former elite athletes. Clin J Sport Med. 2005;15(3):133–5.
34. Milgrom C, Finestone A, Zin D, Mandel D, Novack V. Cold weather training: a risk factor for Achilles paratendinitis among recruits. Foot Ankle Int. 2003;24(5):398–401.
35. Albers IS, Zwerver J, Diercks RL, Dekker JH, Van den Akker-Scheek I. Incidence and prevalence of lower extremity tendinopathy in a Dutch general practice population: a cross sectional study. BMC Musculoskelet Disord. 2016;17:16.
36. Morton S, Williams S, Valle X, Diaz-Cueli D, Malliaras P, Morrissey D. Patellar tendinopathy and potential risk factors: an international database of cases and controls. Clin J Sport Med. 2017;27(5):468–74.
37. Knobloch K. Drug-induced tendon disorders. Adv Exp Med Biol. 2016;920:229–38.
38. van Dijk PA, van Dijk CN. Orthopedisch onderzoek. In: Wiersinga WJ, Levi M, Schimmer B, Savelkoel J, van der Zande JMJ, editors. Handboek voor de coassistent. Houten: Springer; 2021.
39. Ellis SJ, Deyer T, Williams BR, Yu JC, Lehto S, Maderazo A, et al. Assessment of lateral hindfoot pain in acquired flatfoot deformity using weightbearing multiplanar imaging. Foot Ankle Int. 2010;31(5):361–71.
40. Hamilton WG. Tendonitis about the ankle joint in classical ballet dancers. Am J Sports Med. 1977;5(2):84–8.
41. Sammarco GJ, Cooper PS. Flexor hallucis longus tendon injury in dancers and nondancers. Foot Ankle Int. 1998;19(6):356–62.
42. Michelson J, Dunn L. Tenosynovitis of the flexor hallucis longus: a clinical study of the spectrum of presentation and treatment. Foot Ankle Int. 2005;26(4):291–303.
43. Solomon R, Brown T, Gerbino PG, Micheli LJ. The young dancer. Clin Sports Med. 2000;19(4):717–39.
44. Perlman MD, Leveille D. Extensor digitorum longus stenosing tenosynovitis. A case report. J Am Podiatr Med Assoc. 1988;78(4):198–9.
45. Fallon KE. Musculoskeletal injuries in the ultramarathon: the 1990 Westfield Sydney to Melbourne run. Br J Sports Med. 1996;30(4):319–23.
46. Bishop GW, Fallon KE. Musculoskeletal injuries in a six-day track race: ultramarathoner's ankle. Clin J Sport Med. 1999;9(4):216–20.
47. So CH, Siu TO, Chan KM, Chin MK, Li CT. Isokinetic profile of dorsiflexors and plantar flexors of the ankle–a comparative study of elite versus untrained subjects. Br J Sports Med. 1994;28(1):25–30.
48. Stanish WD, Curwin S, Rubinovich M. Tendinitis: the analysis and treatment for running. Clin Sports Med. 1985;4(4):593–609.

70 Examination of Common Heel and Forefoot Conditions

Kenneth J. Hunt

70.1 Introduction

Before proceeding with examination, it is vital to obtain a thorough history as this enables examiners to narrow the differential diagnosis. There are key subjective issues within the history that tend to raise clinical suspicion for a particular diagnosis such as increased physical activity and metabolic disorders in stress fractures, and relief of pain when walking barefoot in Morton's neuroma. Foot examination follows the general orthopedic sequence of look, feel, move, and special tests. It is helpful to develop a routine systematic approach to foot examination, which can be either proximal (hindfoot) to distal (forefoot) or medial to lateral to avoid overlooking critical parts of the examination. The first step should include having the patient remove both shoes and socks and lift up pant legs if necessary. Inspection of the foot needs to be done in sitting and standing positions if the patient is able to bear weight. Inspecting the plantar aspect of the foot is equally important. For example, plantar ecchymosis can be the only sign of a subtle Lisfranc injury. Callosities follow areas of increased plantar pressure and provide a clue to altered mechanics such as lateral column overload or failed medial column and transfer metatarsalgia. Palpation follows inspection and should be done firmly to develop a true sense of the level of pain or edema that a patient may be experiencing. Subsequently, range of motion should be observed both passively and actively, again comparing both sides. Additionally, the foot examination is not complete without gait assessment, examination of shoe wear and orthotics if any, and Silfverskiöld test. This chapter provides the reader with knowledge and tips on how to approach the examination of common causes of heel and forefoot pain.

70.2 Plantar Heel Pain

Plantar heel pain is among the most common conditions seen by the orthopedic and sports medicine providers. The heel is an anatomic location that experiences a great deal of force during athletic activities and therefore is surrounded by tissues uniquely designed to withstand greater forces. The plantar heel consists of a thickened skin covering a dense fat pad that overlies the calcaneus bone. At the center of this is the origin of the plantar fascia, a structure which contributes to the windlass mechanism during gait. The branches of the tibial nerve run through the "tarsal tunnel" along the medial aspect of the calcaneus to reach the foot. While many conditions

K. J. Hunt (✉)
Foot and Ankle Surgery, Quality, Patient Safety and Outcomes, Department of Orthopaedics, University of Colorado, Boulder, CO, USA
e-mail: KENNETH.J.HUNT@cuanschutz.edu; kenneth.j.hunt@ucdenver.edu

J. G. Lane et al. (eds.), *The Art of the Musculoskeletal Physical Exam*,
https://doi.org/10.1007/978-3-031-24404-9_70

can present with heel pain, plantar fasciitis, calcaneal stress fractures, and tarsal tunnel syndrome should be high on the list of differential diagnoses. Since neurogenic pain is covered elsewhere, this section will focus on plantar fasciitis and calcaneal stress fractures.

70.2.1 Plantar Fasciitis

Plantar fasciitis is the most common cause of heel pain in adults, usually between ages 40–60, but may occur in younger ages in athletes with repetitive heel stress, such as runners [1]. Despite its previous characterization as an inflammatory condition, "fasciitis" is now believed to be a misnomer, with the more appropriate term being "plantar fasciosis" to describe the degenerative changes caused by repetitive microtrauma [2, 3].

Due to the nature of the medial longitudinal arch, the insertion of the medial chord is the most common location of degeneration. Patients typically complain of sharp nonradiating plantar heel pain that is commonly but not exclusively medial. They may also report medial arch pain and start-up pain after getting out of bed or after periods of rest that improves with physical activity and stretching [4].

On physical examination, focal tenderness to palpation is usually found over the insertion of the plantar fascia onto the medial calcaneal tuberosity. Patients may have tenderness throughout the plantar fascia. Commonly, a tight Achilles or gastrocnemius may contribute to the pathology, restricting ankle dorsiflexion and exacerbating the trauma to the plantar fascia [4]. In this case, it is critical to perform a Silfverskiöld test to determine which part of the musculotendinous complex is tight. Further, it is important to palpate the length of the plantar fascia to its insertion on the long flexors of the toes, feeling for thickened areas of fibromatosis, or particularly tight regions of the structure. It is key to differentiate between the medial chord of the plantar fascia and the FHL as these structures run close and parallel with one another.

70.2.2 Calcaneal Bone Stress Injuries

Despite the calcaneus bone's important role in weight-bearing, it is largely a cancellous structure which makes it prone to stress fractures, especially after repeated activity, when activity level is rapidly increased or there is a change to footwear or surfaces. It is important to consider biologic/metabolic risk factors as well. Generally, stress fractures are caused by high loads (athletes and military recruits) on normal bones, normal stresses on abnormally weak bones (osteopenia), or high loads on weak bones (female athlete triad). The stress fracture is typically located posteroinferior to the posterior facet of the calcaneus [3]. Patients often complain of pain about the medial and lateral aspects of the tuberosity, diffuse in nature and worse with activity. History frequently includes an identifiable increase in exercise or training regimen leading up to the onset of pain.

On physical examination, patients can often ambulate with an antalgic gait due to pain at heel strike. They may elicit tenderness to palpation medially or laterally along the calcaneal tuberosity. A positive "calcaneal squeeze test," with pain reproduced with compression of medial and lateral calcaneus, is pathognomonic for calcaneal stress fracture [5]. The tenderness is often accompanied by swelling medially, laterally, or both.

70.2.3 Heel Pad Atrophy

The heel pad has a unique structure that allows it to withstand a great degree of repetitive force and loading, enabling the calcaneus to serve as a primary weight-bearing area. The adipose tissue is highly structured and septate, creating an attachment between the heel pad and the periosteum of the calcaneus, preventing the heel pad from shifting or displacing in relation to the skeleton. The most at-risk groups are diabetic patients, and elderly patients that are more prone to loss of water content, collagen structure, and elastic properties of the fat pad [6].

On physical examination, patients typically have focal, nonradiating pain over the plantar, weight-bearing aspect of the calcaneal tuberosity. Unlike other pathologies of the heel, it is not reproduced by palpation along medial and lateral aspects of the calcaneus, calcaneal squeeze test, or passive motion of the ankle or toes [4]. The pain is often not focal as is the case with plantar fasciitis and is often posterior to the plantar fascia insertion.

70.3 Forefoot Pain

70.3.1 Hallux Valgus

Bunions or hallux valgus (HV) is a common condition of the hallux that affects women 15 times as frequent as men [7]. HV describes a deformity of the hallux metatarsophalangeal joint where the first metatarsal is in varus and the proximal phalanx assumes a compensatory valgus position at the MTP joint—Fig. 70.1. There is commonly a pronation deformity due to malrotation of the proximal phalanx caused by adductor hallucis tendon. Patients with hallux valgus can present with various complaints such as difficulty with shoe wear, plantar or dorsal pain due to joint capsule stretching, nerve pain and paresthesia due to stretching of the medial dorsal cutaneous nerve, crossover toes and metatarsalgia of the lesser toes, or undesirable cosmesis. It is crucial to determine the location of pain and underlying cause for successful management. History should include history of inflammatory arthropathy and neuromuscular disorders as well as family history for bunions.

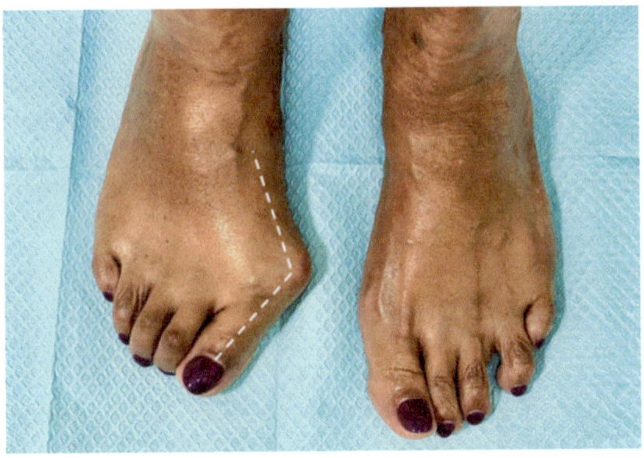

Fig. 70.1 Right hallux valgus. Note the medial deviation of the first metatarsal and lateral (valgus) deviation of the toe at the MTP joint (dashed line). The lesser toes show claw toe deformity with dorsal callosities, more pronounced on the second toe

Physical examination of hallux valgus includes assessment of the components of the deformity, its magnitude, unilateral vs. bilateral, and whether it is passively correctable. Assess for medial eminence protrusion and dorsal osteophytes. Assess the pronation using the toenail as a guide, which should be parallel to the ground while standing. It is extremely important to check for arthritis in the MTP joint. Assess and document the active and passive ranges of motion, especially dorsiflexion. Normal ROM ranges between 70–90° dorsiflexion and 20–30° plantar flexion. MTP grind test, palpation of the sesamoids, and sesamoid grind test provide crucial information, in addition to radiographs, to guide motion-sparing vs. fusion decisions.

Medial column stability assessment should be done in all HV cases. One hand applies dorsiflexing pressure to the lesser metatarsals while stabilizing the lesser TMT joints. Then the other hand attempts to dorsiflex and plantarflex the first ray at the TMT1. The ROM is compared to the other side. Any translation of more than 1 cm is indicative of hypermobility—Fig. 70.2. Generalized ligamentous laxity should also be ruled out. Metatarsalgia and lesser MTP stability are discussed later but should be included in HV evaluation in addition to any forefoot abduction, medial arch collapse, hindfoot deformity, or ankle equinus.

70.3.2 Metatarsalgia

Metatarsalgia is a general term used to describe forefoot pain rather than a specific pathology. Common causes of metatarsalgia include lesser metatarsophalangeal joint disorders such as hammertoes, instability, inflammatory arthropathies, and synovitis. They also include Morton's neuroma, ankle equinus, stress fractures, Freiberg's disease (AVN of second metatarsal head), and

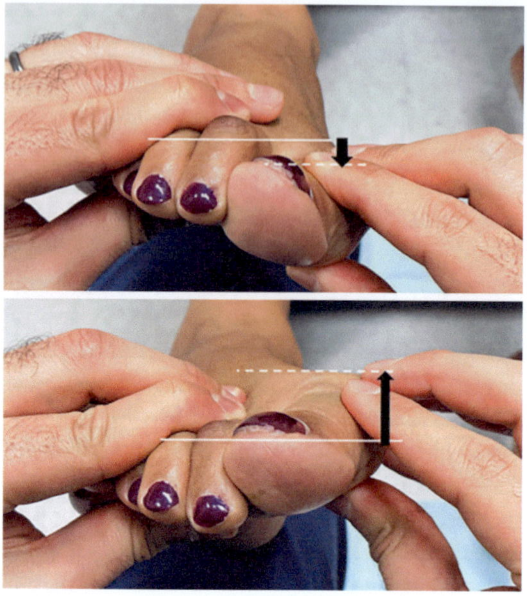

Fig. 70.2 Assessment of medial column stability. Note the plantar (top) and dorsal (bottom) translation of the first metatarsal head (dashed line) relative to the lesser metatarsal heads (solid line)

transfer metatarsalgia. It is important to point out that these conditions are not mutually exclusive.

Detailed history and thorough examination can help to narrow down the list to the most likely etiology and guide successful management. For example, walking barefoot typically causes relief of symptoms in patients with Morton's neuroma but exacerbates metatarsalgia in those with MTPJ pathologies and ankle equinus. Using a cushioned insole or a metatarsal pad offers no to minimal relief to those with Freiberg's disease and stress fractures. The most important clue is the location of the pain whether plantar or dorsal and whether isolated to one metatarsal head or not. Tingling and paresthesia affecting third and fourth toes are typical of Morton's neuroma.

In addition to the examination of the hallux, physical examination includes inspection for lesser toe deformities such as hammertoes, hallux valgus, and scars of previous surgeries. Plantar callosities under the second and occasionally the third metatarsal head often point towards transfer metatarsalgia. Next is palpation of each metatarsal head and MTPJ for tenderness both dorsally and plantarly. Assess the stability of each MTPJ by performing the vertical drawer test by translating the proximal phalanx dorsally while stabilizing the metatarsal. This test has been reported to be 99.8% specific and should be done if any lesser toe deformity is observed [8]. Assess the range of motion of the forefoot joints and the ankle. Toe ROM is typically normal except in lesser toe deformities, which can be partially correctable. In Freiberg's disease, MTPJ degeneration and dorsal osteophytes limit dorsiflexion. Silfverskiöld test should be performed in all forefoot pain patients to rule out gastrocnemius tightness as a causative or contributing pathology. Additionally, neurological and vascular examination should be done. Provocative tests such as Mulder's click, while not highly sensitive, can point out to the presence of a Morton's neuroma. The test is performed by applying alternating plantar and dorsal pressure to the affected web space while compressing the foot at the level of the metatarsal heads. Pain radiating into the affected toes and a click are both required for a positive test. Finally, deep imprints in the shoe insoles point to excessive pressure under the metatarsals in transfer metatarsalgia.

References

1. Buchbinder R, et al. Ultrasound-guided extracorporeal shock wave therapy for plantar fasciitis: a randomized controlled trial. JAMA. 2002;288:1364–72.
2. Lemont H, et al. Plantar fasciitis: a degenerative process (fasciosis) without inflammation. J Am Podiatr Med Assoc. 2003;93:234–7.
3. Thomas JL, et al. The diagnosis and treatment of heel pain: a clinical practice guideline. J Foot Ankle Surg. 2010;49:S1–S19.
4. Lareau CR, et al. Plantar and medial heel pain: diagnosis and management. J Am Academy Orthopaed Surg. 2014;22:372–80.
5. Fredericson MJ, et al. Stress fractures in athletes. Top Magn Reson Imaging. 2006;17:309–25.
6. Coughlin MJ. Mann's surgery of the foot and ankle. Philadelphia: Mosby; 2013.
7. Nery C, Coughlin MJ, Baumfeld D, et al. Hallux valgus in males—part 1: demographics, etiology, and comparative radiology. Foot Ankle Int. 2011;32:917–22.
8. Klein EE, Weil L Jr, Weil LS Sr, Coughlin MJ, Knight J. Clinical examination of plantar plate abnormality: a diagnostic perspective. Foot Ankle Int. 2013;34(6):800–4.

Evaluation of Stress Fractures

71

Gustavo Vinagre, Flávio Cruz,
and Pieter D'Hooghe

71.1 Introduction

A stress fracture is a partial or complete disruption of bone continuity, representing the inability of the skeleton to withstand repetitive submaximal loading, which results in structural fatigue, and resultant signs and symptoms of localized pain and tenderness [1].

Although stress fractures are rare, they can present as common overuse injuries, leading to fatigue fractures (gradual onset—overload injury; high recurrence) in physically active individuals like athletes and military recruits [2]. These injuries are associated with a prolonged absence from sports [3, 4] and can potentially be career-ending [5]. In addition, stress fractures have a high rate of recurrence (22%) and often present as a competition season-ending injury (20%) [6].

This chapter aims at outlining the most common stress fractures of the ankle and foot with focus on the initial management by means of clinical examination. It also provides diagnostic, therapeutical, and return-to-play guidelines for these types of injuries.

G. Vinagre · F. Cruz · P. D'Hooghe (✉)
Department of Orthopaedic Surgery, Aspetar Orthopaedic and Sports, Medicine Hospital, Doha, Qatar
e-mail: gustavovinagre@gustavovinagre.com; flavio.cruz@aspetar.com; pieter.dhooghe@aspetar.com

71.2 Epidemiology

Stress fractures are rare, but they most commonly occur in athletes who are usually exposed to highly repetitive weight-bearing activities and they can lead to long absences from sports.

There is a lack of evidence-based epidemiologic data among studies regarding the rate of stress fractures in recreational or competitive athletes, ranging from 0.5% to 4.8% of all injuries in multiple sports [7]. The incidence of stress fractures in military recruits is higher when comparing to athletes and varies from 5% to 30% per year, being higher in female recruits [8–11].

According to Ekstrand et al. [12], in elite male football (soccer) players, stress fractures represent only 0.5% of all injuries, giving an injury incidence of 0.04 injuries/1000 h (a team of 25 players can expect one stress fracture every third season), but the median absence from sports is high—80 days (3 months).

It seems that the injury rate among females is higher when compared to males due to the specific physiopathogenesis of the female athlete triad [13]. According to Rizzone et al.'s study [6] in collegiate student-athletes, the ratio of female/male was 2.5–3 to 1. There are some sports in which stress fractures are more common such as running, gymnastics, ballet, football (soccer), and figure skating [14].

It is also known that younger players are more likely to develop stress fractures in comparison to

J. G. Lane et al. (eds.), *The Art of the Musculoskeletal Physical Exam*,
https://doi.org/10.1007/978-3-031-24404-9_71

older players and they also have a higher reinjury rate (29%) [12].

Anatomically, all bones of the human body can be subjected to stress fractures, but there is a clear predominance over the lower limbs, especially over the weight-bearing bones [15].

The most common locations for stress fractures are the metatarsals (fifth and second), tibia (anterior cortical bone), femoral neck (superolateral), ribs, medial and lateral malleolus, talus, calcaneus, navicular, cuboid, patella, sesamoids (hallux), pelvis (ischiopubic rami), sacrum, and pars interarticularis. According to Matheson et al. [16] in a study of 320 athletes, the tibia (49.1%), tarsals (25.3%), and metatarsals (8.8%) were the most frequently involved bones affected by a stress fracture (Fig. 71.1). Certain anatomical sites are specific of certain sports [17].

71.3 Etiopathogenesis and Mechanism of Injury

The exact pathophysiology of stress fractures is unknown. Bone is a dynamic tissue, and its turnover cycle, remodeling, and mineralization phases generally require a period of 3–4 months.

The applied bony load during sports activities or weight-bearing activities results in external forces (strain) and internal forces (stress) [18], both of which are vital for the maintenance of the normal bony strength. Stress fractures occur when there is an imbalance between the bone that is formed and the bone that is remodeling. This will consequently result in a progressive discontinuity of the bone, and eventually, the bone becomes unable to resist repetitive (cyclic) loads. It can also be explained by an imbalance between the strength of the bone itself (bone resistance)

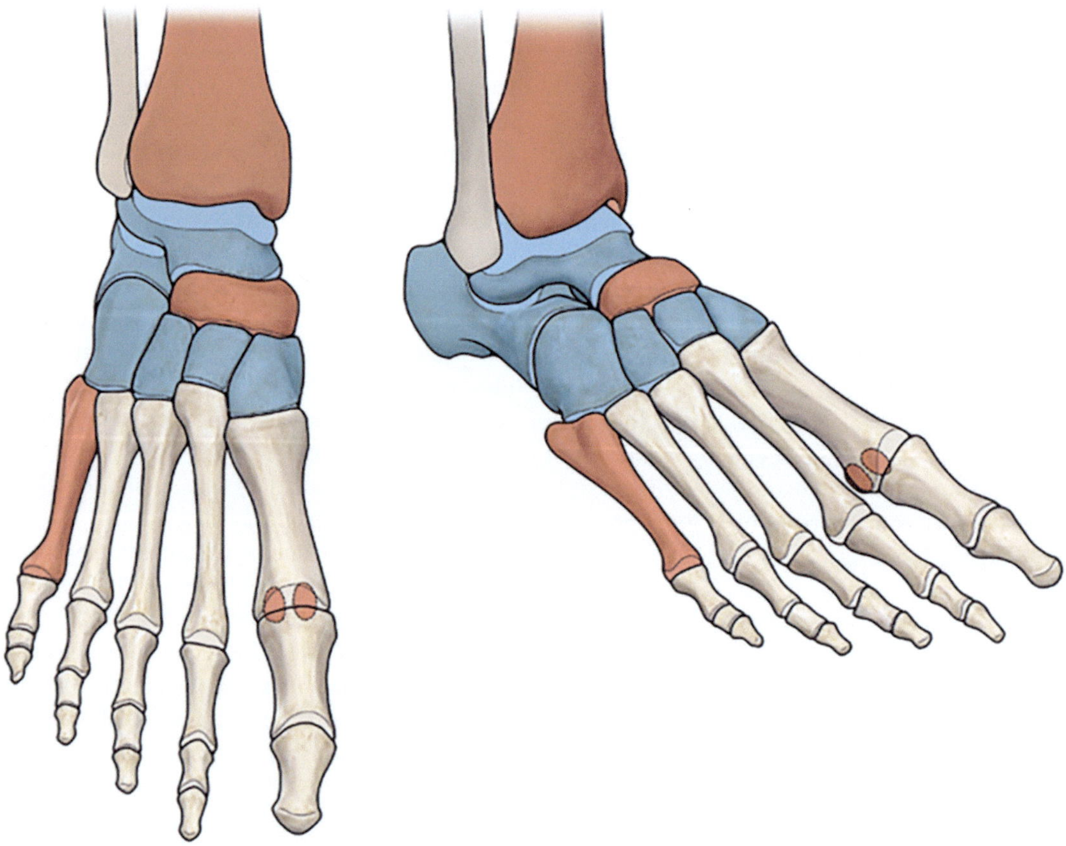

Fig. 71.1 "High-risk (brown)" and "low-risk (blue)" foot and ankle stress fractures' anatomical sites

Table 71.1 Intrinsic and extrinsic risk factors for stress fractures [19]

Intrinsic factors	Extrinsic factors
• Female gender (hypoestrogenic state/menstrual dysfunction/female athlete triad) • Younger age: In males, incidence decreases after 17 years of age; in females, it increases after menarche • Low body mass index (BMI <19) • Decrease muscle strength • Misalignment • Low physical activity • Bone geometry and bone turnover (osteopenia/osteoporosis) • Genetic predisposition • Hormonal factors • Nutritional aspects and eating disorders • Cavus feet • Forefoot varus • Leg-length discrepancies • Forefoot varus • Tarsal coalitions	• Training schedule changes • Inadequate/changing footwear (low-impact absorption) • High-volume and -intensity training (external load) • New training regimen/technical staff • Preseason training • Improper technique • Hard/rigid field surfaces • Long-distance runners • Alcohol • Smoking • Low vitamin D • Sleep deprivation

and chronic mechanical overload (forces of tension and compression/impact) on the bone that makes it unable for the bone to deform and absorb these forces (repetitive loading), exceeding the range of normal bone elasticity.

Stress fractures can be divided (according to the bone quality and load) into fatigue fractures (in those that result from an increase load/cyclic forces on a normal bone like a high volume of exercises in a short period of time) and insufficiency fractures, produced by normal load within weakened bones (i.e., osteoporosis/osteomalacia).

However, not only the bone pathophysiology and biomechanical factors are involved, but also hormonal, nutritional, and genetic factors. Other intrinsic and extrinsic risk factors are detailed in Table 71.1.

It is important to identify these risk factors and target these groups in order to prevent and anticipate the occurrence of stress fractures.

71.4 Diagnosis

71.4.1 Clinical Assessment and Physical Examination

A detailed clinical assessment and a focused physical examination with a presumptive clinical suspicion are key for a correct non-delayed diagnosis of a stress fracture.

Most of the patients have an insidious and progressive onset of mild pain with no history of trauma. The pain is aggravated with repeated weight-bearing activities and is relieved by rest. However, many athletes with stress fractures can present asymptomatic. It is critical to have a high index of suspicion on the presence of predisposing risk factors since many athletes with complaints may present with normal radiographs. A complete history should include detailed questions on diet, nutrition, medication, daily activities, training schedule, equipment/footwear, training surface, menstrual cycle, and external load. Ask if something has changed, regarding training load, coach and/or staff, footwear, or surface.

In females, it is mandatory to evaluate gender-specific risk factors such as menstrual disorders, eating disturbances, and recent weight loss.

The physical examination should assess limb alignment, length discrepancies, gait, muscular asymmetry, active and passive range of motion, abnormalities of the plantar arch, callosities (repetitive stress in a focal area), fixed or flexible deformities, and laxity or stiffness.

Usually, tenderness on palpation or percussion of the involved bone is found, and sometimes it is possible to palpate a periosteal thickening (sign of inadequate callus formation), especially in chronic stress fractures with delayed union. Swelling, erythema, ecchymosis, and warmth may also be apparent. If the pain increases with passive inversion stretch, you should look for a fifth metatarsal stress fracture. In the case of a calcaneus stress fracture, the pain should increase with a calcaneal squeeze test (Fig. 71.2).

The hop test (Fig. 71.3) should be performed to rule out foot and ankle stress fractures. This test consists of hop up and down on the affected limb several times barefoot. A large amount of pain in a localized area of the lower extremity is a positive test and may signify a fracture.

Cavovarus foot alignment and pain upon tiptoe standing can also be suspicious.

Fact Box 1
Always suspect a stress fracture in an athlete who presents with dull foot pain, localized to the bone, that increases on activity and is relieved with rest.

71.4.2 Imaging

Radiographic imaging is considered critical for the diagnosis, prognosis, and follow-up of stress fractures.

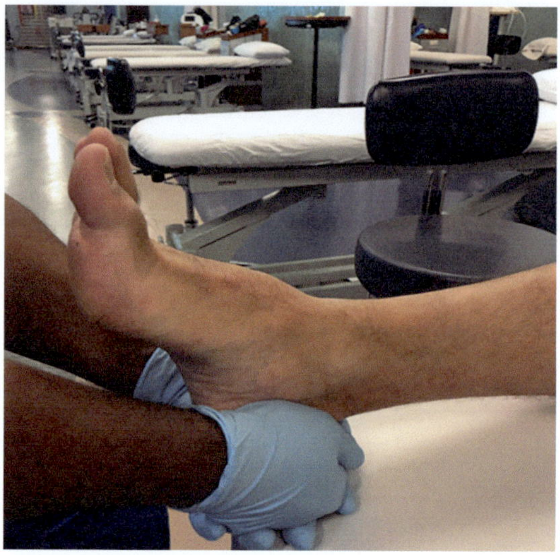

Fig. 71.2 Calcaneal squeeze test. The test is positive if the patient feels pain with the medial-lateral compression of the calcaneus

Fig. 71.3 Hop test. The patient should be asked to jump (**a**) and land in one foot. (**b**) The test is positive if the patient feels pain in the localized area of the stress fracture when landing (**b**)

71.4.2.1 Radiography

Radiographs are considered the initial imaging modality in stress fractures due to their easy access and relatively low cost. Plain weight-bearing anteroposterior, lateral, and oblique radiographic views are the current standard.

Although radiographs have a high specificity, they are not always reliable and can be missed (up to 87% false-negative rate), particularly during the initial 2–3 weeks after the onset of symptoms [20, 21]. Most fractures are incomplete, which makes them invisible on plain radiographs until bony osteoclastic resorption has taken place [22].

The earliest stress fracture sign on radiographs is a localized periosteal thickening or a subtle radiolucent area. Especially at later stages in the stress fracture development, periosteal reaction and callus formation will appear (Fig. 71.4). In the absence of positive radiographic findings and a high index of stress fracture suspicion, further imaging should be considered by means of magnetic resonance imaging (MRI), computed tomography (CT), or bone scan.

> **Fact Box 2**
> Do not rely on plain radiographs only for the diagnosis of foot stress fractures. MRI, CT scan, and/or bone scan are often required for diagnostic confirmation.

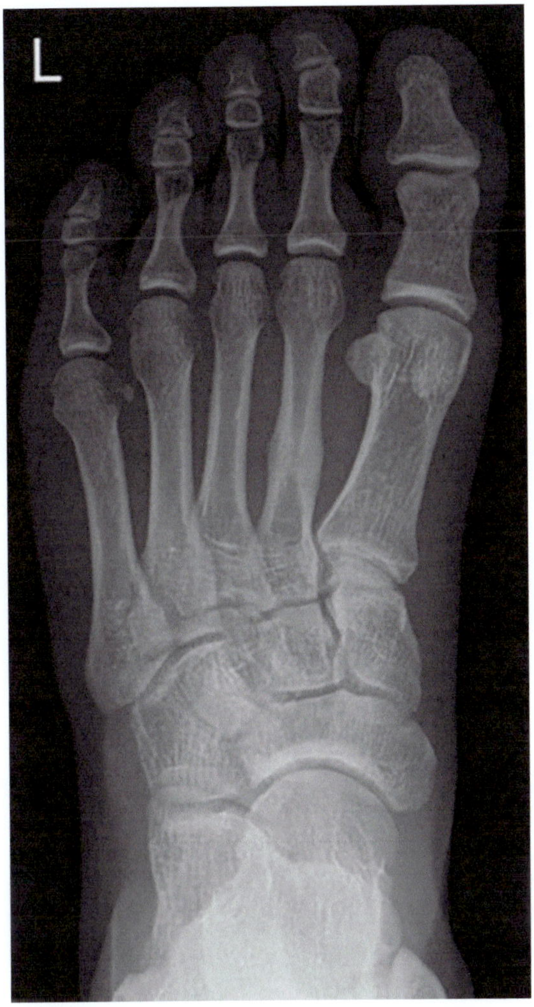

Fig. 71.4 Anteroposterior plain foot radiographic view with a periosteal reaction on the second metatarsal shaft (stress fracture) in a national team football player

71.4.2.2 Ultrasonography

Ultrasound is a sensitive diagnostic tool and easily accessible and can identify early bony stress reaction. In case of a normal radiograph, it is advised to ask for an ultrasound to identify periosteal reaction as a primary sign of the bony stress reaction. Although previously thought to have a poor sensitivity and specificity, a recent trial found increased pain to have a positive predictive value of 99% (sensitivity, 81.9%; specificity, 66.6.%) in the application of therapeutic ultrasound at the site of a bony stress injury [23].

71.4.2.3 CT Scan

Computed tomography (CT) scan can be helpful to highlight a small fracture line, can confirm the diagnosis of a complete stress fracture, and can also be used to monitor healing (Fig. 71.5). This modality can also be useful when there is a contraindication for the use of magnetic resonance imaging (MRI) or in chronic cases (where it is shown to be more useful than MRI or bone scan).

71.4.2.4 MRI

Magnetic resonance imaging (MRI) is the most sensitive and specific imaging modality for the

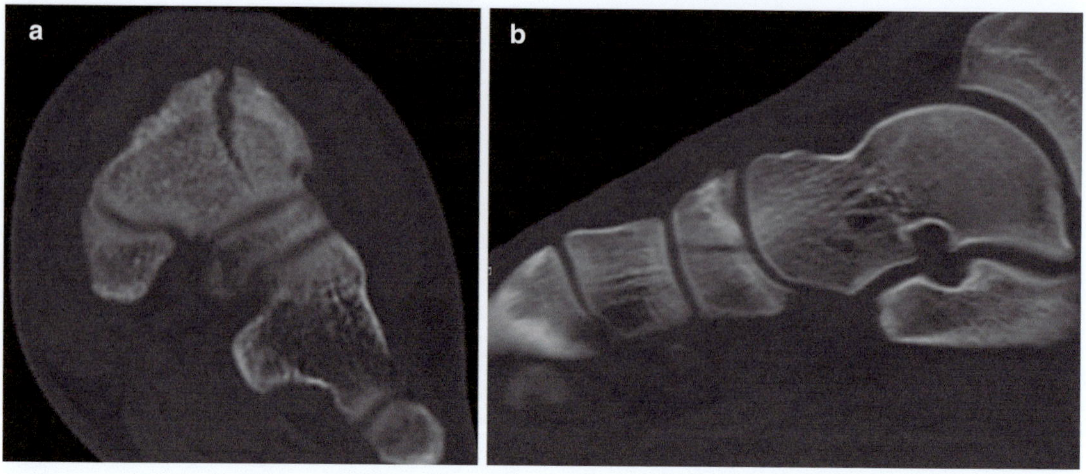

Fig. 71.5 CT scan of an elite football player with a navicular stress fracture. (**a**) Axial CT view, (**b**) sagittal CT view

Table 71.2 Generic MRI grading system of stress fractures. From Arendt EA, Griffiths HJ [29]

Low grade	Grade 1	STIR signal changes
	Grade 2	STIR and T2 changes
High grade	Grade 3	STIR, T1, and T2 changes No fracture line present
	Grade 4	STIR, T1, and T2 changes Fracture line present

diagnosis of stress fractures, and it is currently considered the golden standard [24].

The abnormalities caused by the fracture can usually be identified 1–2 days after the start of symptoms, with an early detection of bony edema [25–27].

MRI can differentiate medullary damage from the cortical, endosteal, and periosteal bone, allowing it to formulate a severity and prognosis staging (Table 71.2). MRI can also accurately delineate the exact anatomic location and the extent of the stress injury (by detecting bone edema and changes in cortical density), but it remains important to correlate the radiological images with the patient's clinical findings. In a study involving 21 asymptomatic runners [28], 43% presented with bone edema on MRI suggesting a potential stress fracture because athletes are subjected to repetitive microtrauma during endurance running.

It is important to highlight MRI as the imaging modality of choice in the detection of pre-fracture stages with bone edema (Fig. 71.6) and stress reac-

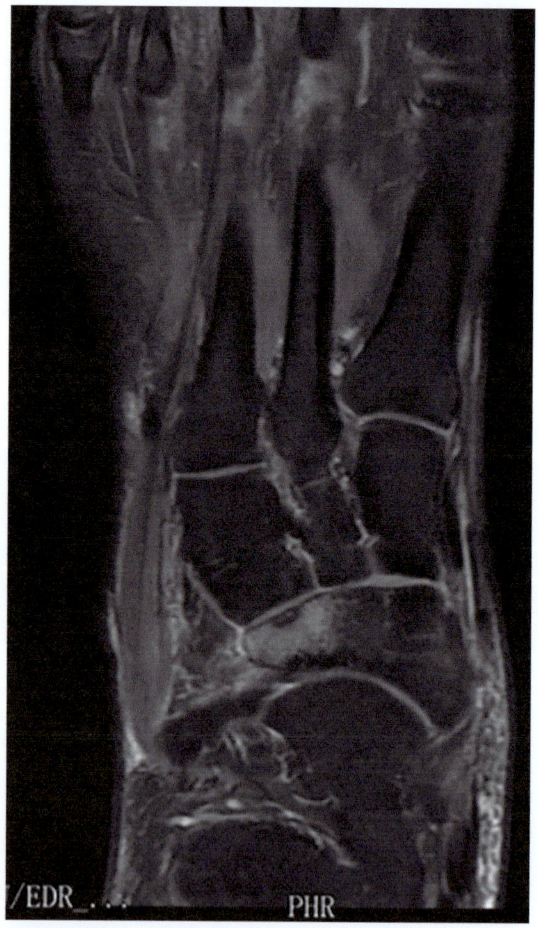

Fig. 71.6 Axial T2 MRI view of a navicular stress reaction over the typical "nutcracker effect" area in a young football player

tion, but MRI is shown to be inferior to CT scan in the identification of the stress fracture itself.

When a stress fracture is clinically suspected, the initial imaging modality should be CT scan in order to visualize the fracture line. If the CT scan does not demonstrate any signs of fracture, an MRI is the next appropriate imaging modality to determine whether a stress reaction is occurring.

71.4.2.5 Bone Scan (Scintigraphy)

A bone scan has a high sensitivity but low specificity and should not be used as a first-line imaging modality in bony stress fractures. However, a bone scan can be helpful when CT and/or MRI fail to demonstrate a clear diagnosis. In case of a stress fracture, the bone scan can present with positive signs in all three phases, in contrast to soft-tissue injuries where it can only hypercaptate during the first injury phase.

71.5 General Treatment Concepts

Although there are no evidence-based guidelines supported by literature regarding best treatment, below is a summary of the current treatment options.

71.5.1 Conservative Treatment

Most nonoperative treatment strategies should include decrease in physical activity and training load, avoidance of pain-related activities and relative weight-bearing restriction, or immobilization. Pain control with oral analgesic medications (NSAIDs should be avoided), cold therapy, proper rehabilitation, and a personalized conditioning alternative physical program is recommended (minimal impact aerobic activities to maintain flexibility and strength; consider using antigravity treadmill). Other modalities of nonoperative treatment can be described below:

- *Biomechanical therapy*

  The major components in the treatment of any stress fracture are "rest" and "activity modification." Especially during in-season, a team approach (medical staff, coach, agent, director) is key in the management of setting up treatment goals and return-to-play criteria. Although this sometimes means to temporarily unload and immobilize a player/athlete, cycling, rowing, swimming, and even specific running (to maintain cardiovascular fitness and lower limb strength) can be continued. The focus should be on altering the nature of the training, individualizing the program, and type of surface and shoe. Hydrotherapy and antigravity treadmill are useful in this stage to keep the player's fitness while allowing the stress fracture to heal. Careful evaluation of malalignment, muscular imbalance, and abnormal loading patterns need to be assessed and corrected meanwhile.
- *Biological therapies*

  Several adjuvant treatment strategies, such as bone morphogenic protein (BMP), bisphosphonates, teriparatide, and hyperbaric oxygen, have been recommended in literature to enhance stress fracture healing. Controversy remains due to the heterogeneity of patient populations, the interventions offered, as well as the different outcome measures considered. In addition, most of the published reports are case series. We did not find any evidence regarding platelet-rich plasma (PRP) injections for stress fracture treatment.
- *Extracxorporal Shock-wave therapy (ESWT)*

  ESWT can have a role in the upregulation of local angiogenesis and concentration of growth factors. Although quality research is lacking, ESWT has shown improved healing potential and earlier return to play in chronic and nonunion cases [30, 31].
- *Low-intensity pulsed ultrasound therapy (LIPUS)*

  LIPUS has been introduced as a promising alternative to treat nonunion or delayed union. A recent systematic review however concluded that LIPUS does not reduce the time to return to activity for conservatively treated stress fractures [32].
- *Electrical stimulation (ES)*

  There is laboratory evidence that endochondral bone formation and growth factor expression may be appropriately stimulated by the application of an electromagnetic field. Electrical stimulation (ES) therapy can be provided via several modes: direct current, capacitative coupling, inductive coupling, and pulsed electromagnetic field. Although in football for example the pressure can be high to try out these "new tools," there is little evidence for the use of ES in the management of stress fractures [33].

- *Bone morphogenic proteins (BMPs)*

  BMPs belong to "transforming growth factor b superfamily proteins" and act as osteoinductive agents to enhance fracture healing. Several BMPs have been isolated like BMP-2 and BMP-7 that are subjected to clinical trials. In stress fractures, there is no evidence to support the use of BMPs [33].

- *Teriparatide*

  Teriparatide is a bone anabolic agent (recombinant human parathyroid hormone analogue) used in the treatment of osteoporosis by stimulating osteoblasts. A recent RCT showed that it can shorten the time to fracture healing compared to placebo, but no credible data in stress fractures is available yet [34].

- *Vitamin D*

  In football, there is a growing awareness of the importance of the role of vitamin D levels, particularly in Northern Europe (even the elite players), with several studies showing hypovitaminosis (some among the lowest levels), especially in wintertime [35].

71.5.2 Surgical Treatment

Stress fractures can be classified as "low risk" or "high risk," and this can influence the proper decision-making (Table 71.3). High-risk fractures are prone to displacement, delayed union, and nonunion. In case of nonoperative treatment, they require a prolonged period of healing and can potentially become career-ending.

In patients with "high-risk" stress fracture sites or displaced fractures (especially in athletes), early surgical fixation is preferred due to a high rate of biological failure, extended healing time, nonunion, and possibility of refracture [36–38].

Table 71.3 Foot and ankle "high-risk" and "low-risk" stress fractures

"High-risk" stress fractures	"Low-risk" stress fractures
Medial malleolar fractures	Calcaneus
Talar neck fractures	Cuboid
Central-dorsal navicular fractures	Cuneiform
Hallux sesamoid fractures	Lateral malleolus
Fifth metatarsal fractures	

71.6 Site-Specific Stress Fractures

According to a stress fracture study in 2379 elite male football players [12], all fractures (51) were noted in the lower extremities: the most common site was the fifth metatarsal (78%); 12% were in tibia and 6% in the pelvis. Twenty-nine percent of those stress fractures were reinjuries.

Additionally, it is important to be aware of "high-risk" stress fracture sites such as fifth metatarsal (zone 3—metaphysio-diaphyseal junction); talar neck; femoral neck; patella; anterior tibial diaphysis; medial malleolus; navicular; sesamoids of the hallux; and neck of the second to fourth metatarsals. These site-specific stress fractures are prone to complications such as displacement, delayed and nonunion, or prolonged specific treatment management requirements. "Low-risk" stress fracture sites are the posteromedial tibial shaft, metatarsal shaft, distal fibula, medial femoral neck, femoral shaft, and calcaneus and often heal with a proper diagnosis and treatment (Table 71.3).

71.6.1 Metatarsal Stress Fractures

The metatarsal area is a common site for developing stress fractures [5]. Stress fractures of the fifth and the second metatarsals (MT) are more common than the rest. Because of its unique anatomy and function, it is conventional to group MT fractures into fractures of the medial column (first MT), central column (second–third MT), and lateral column (fourth–fifth MT).

> **Fact Box 3**
> Always enquire about changes in the training schedule and playing surface since these are important risk factors in football particularly.

71.6.1.1 Fifth Metatarsal Stress Fractures

Fractures of the base of the fifth metatarsal have been classified into three zones according to Lawrence and Botte [39], with each zone representing a different mechanism of injury, treatment strategy, and prognosis (Fig. 71.7). Zone 1 (tuberosity area) is associated with avulsion fractures; zone 2 (tuberosity—metaphyseal area) is the so-called Jones fracture; and zone 3 (metaphyseal-diaphyseal area) represents true stress fractures and involves the junction of the metaphysis and diaphysis of the fifth metatarsal. Zones 2 and 3 are predisposed to delayed healing due to a vascular watershed zone between the insertion of peroneus brevis and the diaphyseal blood supply.

Surgical strategies include percutaneous, mini-open, and open reduction and internal fixation techniques with intramedullary screw fixation, bone grafting, and tension band wiring (by means of a metal cerclage or a novel high-resistance suturing method) (Fig. 71.8) [40]. In

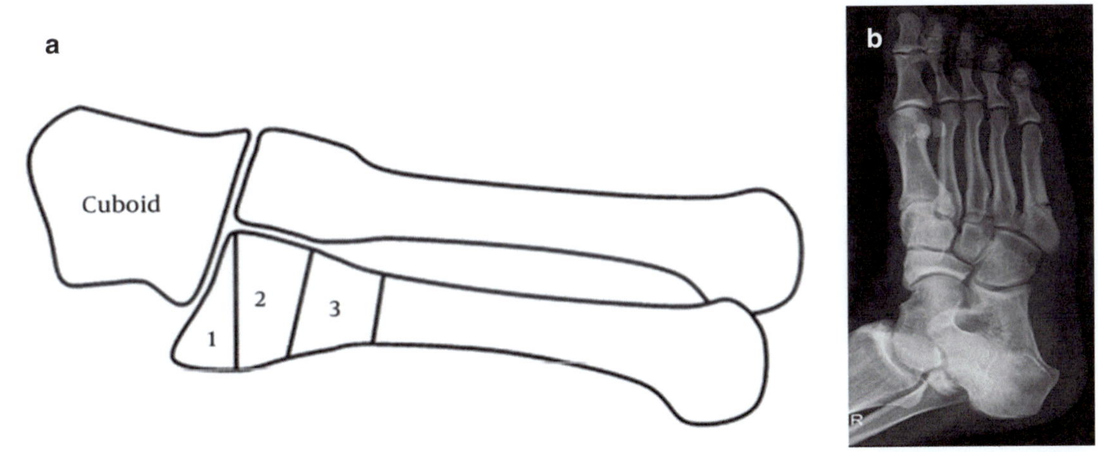

Fig. 71.7 (**a**) Lawrence and Botte's classification of proximal fifth metatarsal fractures: zone 1 (avulsion fractures), zone 2 (Jones fracture), and zone 3 (stress fractures). (**b**) Typical fifth metatarsal stress fracture (zone 3)

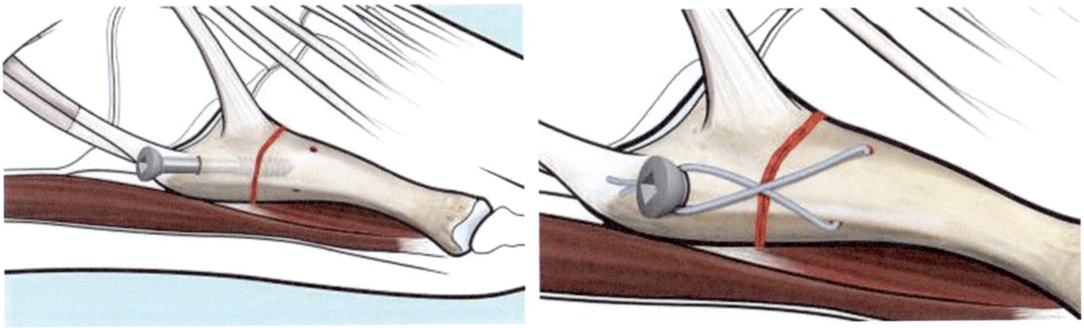

Fig. 71.8 High-resistance suture cerclage technique [40]

addition, some authors have advocated adding bone marrow aspirate concentrate (BMAC) to improve the biological environment and healing potential of the fracture.

A recent systematic review from Mallee et al. [41] showed proven benefit of surgical management, in terms of return to sport. Conservative management varies from limitations of activities to non-weight-bearing cast immobilization and remains a valid option for the recreational athlete.

Especially for fifth metatarsal Torg type II and III fractures (Table 71.4) and for stress fractures displaced more than 2 mm, surgery is recommended [42, 43]. A 5.5 mm partially threaded cannulated screw is the current golden standard for fixation although good results are reported with headless compression screws also [44]. It is recommended to use the largest diameter screw that fits the width of the intramedullary canal with a minimum diameter of 4 mm, and with sufficient screw length, covering the fracture (Fig. 71.9) with threads jutting beyond the fracture site in order to generate maximum compression [45, 46].

Table 71.4 Torg classification of fifth metatarsal stress fracture

Torg 1	Torg 2	Torg 3
• Acute fracture • No periosteal reaction • No intra-medullary sclerosis	• Delayed union • Periosteal reaction • Intramedullary sclerosis • Widened fracture line	• Nonunion Periosteal reaction Intramedullary narrowing • Widened fracture line

Refractures after the surgical treatment of fifth metatarsal stress fractures can be as high as 10–30% in some series. They are found to be associated with increased body weight and zone 2–3 Torg-type fractures in the hypo-vascularized metatarsal stress fracture zone.

Postoperatively, a non-weight-bearing short leg cast or plaster splint for 1–2 weeks is started, followed by a controlled ankle movement walker boot for 2 more weeks [47]. After 6–8 weeks postoperatively, full weight-bearing is allowed and normal activities can be resumed.

71.6.1.2 Second Metatarsal Stress Fractures

Among the metatarsal bones, the second is known to be the most commonly injured, followed by the third, base of the fifth, fourth, and first metatarsal [3, 48]. Stress fractures of the proximal second metatarsal are most commonly observed in female ballet dancers, due to the "en pointe" position. They are also frequently encountered in runners, due to the high-force vector that is transmitted through the meta-diaphyseal region, leading to a

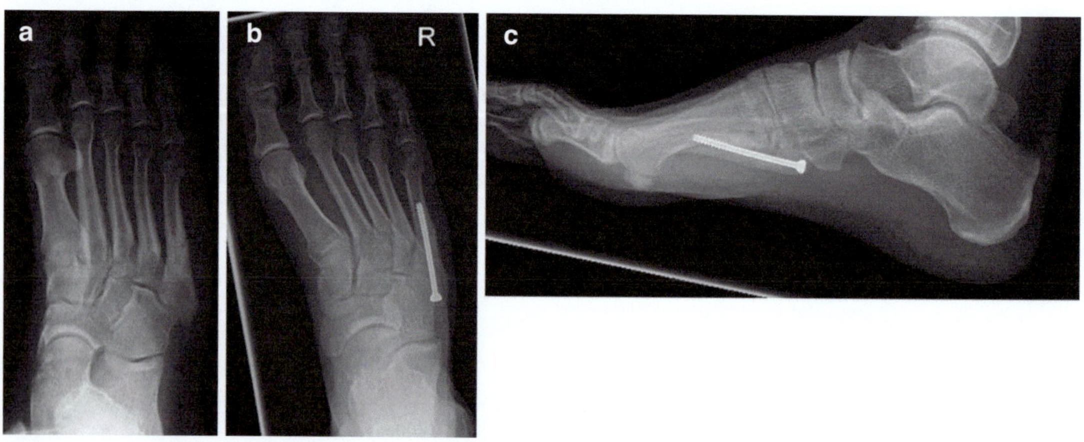

Fig. 71.9 Anteroposterior and lateral plain radiograph view of fifth MT stress fractures in a national team football treated with a mini-open fixation technique. (**a**) Double-stress fracture line, (**b**, **c**) 2.5 months postoperative

high metatarsal bending stress. Although no direct link with a specific forefoot morphology has been reported, a shorter and hypermobile first metatarsal or a longer second metatarsal is hypothesized to be a risk factor for these stress fracture types [3, 49, 50]. Clinically, these second metatarsal base fractures can be very difficult to differentiate from Lisfranc joint synovitis. They are also at high risk for delayed union and may significantly affect the athlete's ability to return to elite level.

Distal fractures have a good prognosis and a relatively fast recovery through nonoperative management. In these patients, relative rest and partial to full weight-bearing in a CAM boot are recommended.

71.6.2 Navicular Stress Fractures

Stress fractures of the tarsal navicular account for up to 35% of all foot and ankle stress fractures [51]. Navicular fractures are typically seen in track and field athletes, especially sprinting athletes, professional tennis players, jumpers, and athletes engaging in explosive push-off activities [52].

Navicular stress fractures are considered at high risk for delayed union or nonunion, due to the relatively poor vascular supply in the central third of the bone [53]. However, nonoperative treatment in a non-weight-bearing plaster cast or boot for 5 weeks followed by 4–6 weeks of rehabilitation is still recommended for the general population [54–58]. In case of displaced fractures, delayed union, or nonunion, surgery can be indicated [24, 59].

Currently, there is no evidence-based consensus on best therapeutic strategies in the athletic population.

According to Saxena's classification [58], nonoperative treatment should be advised in type 1 fractures (dorsal cortex-only involvement) and surgery in type 2 (propagation of the fracture into the navicular body) and type 3 (bicortical disruption) fractures.

Usually, surgical internal fixation involves the use of two partial-threaded cannulated screws placed lateral to medial in a parallel or cross configuration.

71.6.3 Medial Malleolus Stress Fractures

Medial malleolus stress fractures are relatively uncommon. They account for 0.6–4.1% of all stress fractures [51, 60]. They typically occur in high-level runners and jumpers. Stress fractures of this anatomical site seem to be due to repetitive impingement of the talus on the medial aspect of the distal tibia during forced dorsiflexion of the ankle [48].

The treatment of medial malleolus stress fractures varies depending on fracture propagation, displacement, athletic level, and seasonal timing. In case of a clear fracture line or displacement (especially in elite and "in-season" athletes), surgical internal fixation is recommended [61–65].

71.6.4 Other Stress Fractures of the Foot

71.6.4.1 Calcaneus

Calcaneal stress fractures are rare, and most studies report on the occurrence of these fractures in army recruits rather than in athletes. The diagnosis is often delayed due to similarity of symptoms with plantar fasciitis, retrocalcaneal bursitis, Achilles tendinitis, Baxter's nerve disorder, and Sever's disease. The typical presentation is localized tenderness at the heel and/or with positive calcaneal compression test, which increases with activity and is relieved with rest/immobilization. At radiographic evaluation, a thin radiolucent or sclerotic line can become apparent, 2–3 weeks after the onset of symptoms. MRI can be a useful tool to identify early bone edema and fracture lines. In most cases, calcaneal stress fractures can be managed with nonoperative treatment and activity modification.

71.6.4.2 Talus

A talar stress fracture is a relatively rare injury but may present in athletes due to repetitive cycles of axial loading activities. The possibility of secondary displacement should be considered. MRI is often required as conventional radiographs are often unable to reveal talar stress fractures (Fig. 71.10).

Despite the lack of a general consensus, undisplaced fractures are often managed nonopera-

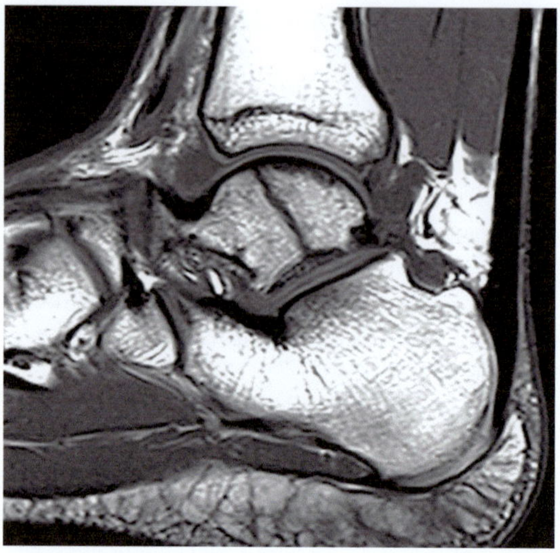

Fig. 71.10 Sagittal T1 MRI view of a talar body stress fracture in a young football player

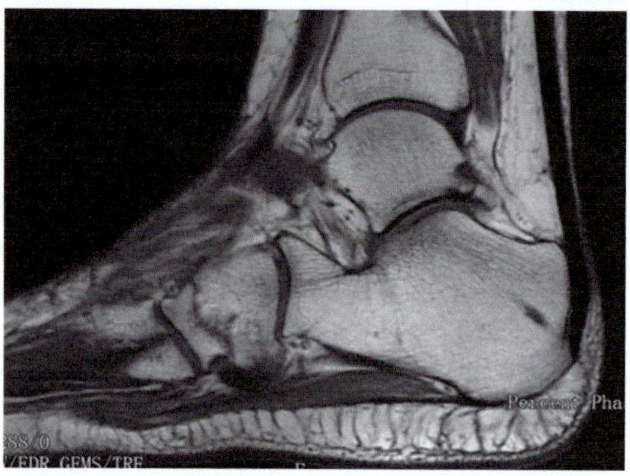

Fig. 71.11 Sagittal T1 MRI view of a cuboid stress fracture in a national league female football player

tively by means of 6-week non-weight-bearing cast or boot immobilization. Surgical fixation is indicated in case of secondary displacement, in order to reduce the risk of avascular necrosis [66].

71.6.4.3 Cuboid

Cuboid stress fractures (Fig. 71.11) are usually treated conservatively with non-weight-bearing with crutches for the initial 2 weeks. Then, when the patient is pain-free, move into protected weight-bearing and then progress into a CAM boot and staged rehabilitation.

The role of podiatry and tailored orthotics is key in treatment and prevention.

71.6.4.4 Sesamoid

Typically, the medial sesamoid is the most affected by stress fractures. Plain radiographs can be difficult to interpret, and the use of a CT scan is advised in case of clinical suspicion. A bipartite sesamoid may be confused with a fracture but is frequently bilateral (up to 75% of cases) and has smooth edges [66]. Normally, conservative management with rest, boot, orthotics, and progressive loading is standard. In the case where surgery is required, a partial sesamoidectomy is the treatment of choice.

71.7 Return to Play

The return-to-play decision depends on multiple factors such as the stress fracture localization and pattern, sports activity, severity of the injury, and risk factor modification/control.

Stress fractures can lead to prolonged absences from sports, and the return-to-play process is key to reincorporate the athlete as soon as possible, respecting the biology and healing process.

In case of fifth metatarsal stress fractures, according to Ekstrand and Torstveit [12], the mean absence from football was 3 months and the average time to return to sports is reported to be 24 weeks if conservative treatment is done, and 12 weeks (average time) in case of surgery [67].

For navicular stress fractures, according to Saxena et al., in general population, patients treated conservatively with non-weight-bearing cast last 5.6 months to return to activity while patients treated surgically last 3.8 months to return to activity [58].

In case of medial malleolus stress fractures, 7.6 weeks' mean time to return to sport has been reported [68], although resolution of symptoms may take 4–5 months [48].

The criteria for allowing an athlete to return to his/her practice can include:

- Absence of pain 10–14 days before starting to do sports activities, during sports movements, and on physical exam
- Formation of bony callus and obliteration of the fracture line on simple radiographs or CT scan imaging support if necessary

In summary, it is important to have clinical and imaging evidence of healing before the athlete fully returns to play.

71.8 Prevention

Prevention of stress fractures is essential. Gradual training load and monitoring through different training principles/methods, as well as injury prevention programs and risk factor identification and adaptation, are crucial to prevent these injuries. For example, injury prevention programs such as the FIFA 11+ are very important, specially in low-skill-level youth teams [69, 70].

There is limited evidence at present for biological treatment of stress fractures, but biological agents may be useful adjuncts.

Susceptible individuals and risk factors should be identified, modified, and addressed to avoid unnecessary stress fractures. In addition, realistic return-to-play strategies should be implemented from the start.

References

1. Warden SJ, Creaby MW, Bryant AL, Crossley KM. Stress fracture risk factors in female football players and their clinical implications. Br J Sports Med. 2007;41:i38.
2. Wright AA, Hegedus EJ, Lenchik L, Kuhn KJ, Santiago L, Smoliga JM. Diagnostic accuracy of various imaging modalities for suspected lower extremity stress fractures. Am J Sports Med. 2016;44(1):255–63.
3. Mayer SW, Joyner PW, Almekinders LC, et al. Stress fractures of the foot and ankle in athletes. Sports Health. 2014;6(6):481–91
4. Miller TL, Jamieson M, Everson S, et al. Expected time to return to athletic participation after stress fracture in division I collegiate athletes. Sports. Health. 2018;10(4):340–4.
5. Ekstrand J, Van Dijk CN. Fifth metatarsal fractures among male professional footballers: A potential career-ending disease. Br J Sports Med. 2013;47(12):754–8.
6. Rizzone KH, Ackerman KE, Roos KG, et al. The epidemiology of stress fractures in collegiate student-athletes, 2004-2005 through 2013-2014 academic years. J Athl Train. 2017;52(10):966–75.
7. Snyder RA, Koester MC, Dunn WR. Epidemiology of stress fractures. Clin Sports Med. 2006;25(1):37–52.
8. Välimäki VV, Alfthan H, Lehmuskallio E, et al. Risk factors for clinical stress fractures in male military recruits: a prospective cohort study. Bone. 2005;37:267–73.
9. Milgrom C, Giladi M, Stein M, et al. Stress fractures in military recruits. A prospective study showing an unusually high incidence. J Bone Joint Surg (Br). 1985;67:732–5.
10. Doral MN, Mann G, Constantini N, et al. Stress fractures: overview. In: Doral MN, Tandoğan RN, Mann G, Verdonk R, editors. Sports injuries. Berlin Heidelberg: Springer; 2012. p. 787–813.
11. Miller TL, Kaeding CC, Wasserstein D, et al. Pathophysiology and epidemiology of stress fractures. In: Miller TL, Timothy L, Kaeding C, editors. Stress fractures in athletes. New York: Springer; 2014. p. 3–11.
12. Ekstrand J, Torstveit MK. Stress fractures in elite male football players. Scand J Med Sci Sports. 2012;22(3):341–6.
13. Torstveit MK, Sundgot-Borgen J. The female athlete triad: are elite athletes at increased risk? Med Sci Sports Exerc. 2005;37(2):184–93.
14. Astur DC, Zanatta F, Arliani GG, Moraes ER, Pochini Ade C, Ejnisman B. Stress fractures: definition, diagnosis and treatment. Rev Bras Ortop. 2016;51(1):3–10.
15. Schneiders AG, Sullivan SJ, Hendrick PA, Hones BDGM, Mcmaster AR, Sugden BA, et al. The ability of clinical tests to diagnose stress fractures: a systematic review and meta-analysis. J Orthop Sports Phys Ther. 2012;42(9):760–71.
16. Matheson GO, Clement DB, McKenzie DC, Taunton JE, Lloyd-Smith DR, Macintyre JG. Stress fractures in athletes: a study of 320 cases. Am J Sports Med. 1987;15:46–58.
17. Brukner P, Bradshaw C, Khan KM, White S, Crossley K. Stress fractures: a review of 180 cases. Clin J Sport Med. 1996;6:85–9.
18. Bennell K, Matheson G, Meeuwisse W, et al. Risk factors for stress fractures. Sports Med. 1999;28(2):91–122.
19. Dhillon MS, Ekstrand J, Mann G, Sharma S. Stress fractures in football. J ISAKOS. 2016;1(4):229–38.
20. Kaeding CC, Yu JR, Wright R, Amendola A, Spindler KP. Management and return to play of stress fractures. Clin J Sport Med. 2005;15(6):442–7.
21. Murray SR, Reeder M, Ward T, et al. Navicular stress fractures in identical twin runners: high-risk fractures require structured treatment. Phys Sportsmed. 2005;33(1):28–33.
22. Mann JA, Pedowitz DI. Evaluation and treatment of navicular stress fractures, including nonunions, revision surgery, and persistent pain after treatment. Foot Ankle Clin. 2009;14(2):187–204.
23. Papalada A, Malliaropoulos N, Tsitas K, et al. Ultrasound as a primary evaluation tool of bone stress injuries in elite track and field athletes. Am J Sports Med. 2012;40:915–9.
24. Murray SR, Reeder MT, Udermann BE, et al. High-risk stress fractures pathogenesis, evaluation, and treatment. Comprehen Ther. 2006;32(1):20–5.

25. Bennell KL, Malcolm SA, Thomas SA, Wark JD, Brukner PD. The incidence and distribution of stress fractures in competitive track and field athletes. A twelve-month prospective study. Am J Sports Med. 1996;24(2):211–7.
26. Fredericson M, Bergman AG, Hoffman KL, Dillingham MS. Tibial stress reaction in runners. Correlation of clinical symptoms and scintigraphy with a new magnetic resonance imaging grading system. Am J Sports Med. 1995;23(4):472–81.
27. Sofka CM. Imaging of stress fractures. Clin Sports Med. 2006;25(1):53–62.
28. Bergman AG, Fredericson M, Ho C, Matheson GO. Asymptomatic tibial stress reactions: MRI detection and clinical follow-up in distance runners. AJR Am J Roentgenol. 2004;183(3):635–8.
29. Arendt EA, Griffiths HJ. The use of MR imaging in the assessment and clinical management of stress reactions of bone in high-performance athletes. Clin Sports Med. 1997;16(2):291–306.
30. Moretti B, Notarnicola A, Garofalo R, et al. Shock waves in the treatment of stress fractures. Ultrasound Med Biol. 2009;35(6):1042–9.
31. Furia JP, Rompe JD, Cacchio A, et al. Shock wave therapy as a treatment of non-unions, avascular necrosis, and delayed healing of stress fractures. Foot Ankle Clin. 2010;15(4):651–62.
32. Busse JW, Kaur J, Mollon B, et al. Low intensity pulsed ultrasonography for fractures: systematic review of randomised controlled trials. BMJ. 2009;338:b351.
33. Beck BR, Matheson GO, Bergman G, et al. Do capacitively coupled electric fields accelerate tibial stress fracture healing? A randomized controlled trial. Am J Sports Med. 2008;36(3):545–53.
34. Raghavan P, Christofides E. Role of teriparatide in accelerating metatarsal stress fracture healing: a case series and review of literature. Clin Med Insights Endocrinol Diabetes. 2012;5(5):39–45.
35. Clutton J, Perera A. Insufficiency and deficiency of vitamin D in patients with fractures of the fifth metatarsal. Foot (Edinb). 2016;27:50.
36. Volpi P, Ezequiel P, Eran K, et al. Stress fractures of the foot in footballers. In: Football traumatology. New York: Springer; 2005. p. 371–83.
37. Roche AJ, Calder JD. Treatment and return to sport following a Jones fracture of the fifth metatarsal: a systematic review. Knee Surg Sports Traumatol Arthrosc. 2013;21:1307–15.
38. Polzer H, Polzer S, Mutschler W, et al. Acute fractures to the proximal fifth metatarsal bone: development of classification and treatment recommendations based on the current evidence. Injury. 2012;43:1626–32.
39. Lawrence SJ, Botte MJ. Jones' fractures and related fractures of the proximal fifth metatarsal. Foot Ankle. 1993;14(6):358–65.
40. D'Hooghe P, Caravelli S, Massimi S, Calder J, Dzendrowskyj P, Zaffagnini S. A novel method for internal fixation of basal fifth metatarsal fracture in athletes: a cadaveric study of the F.E.R.I. technique (fifth metatarsal, extra-portal, rigid, innovative). J Exp Orthop. 2019;6(1):45.
41. Mallee WH, Weel H, van Dijk CN, van Tulder MW, Kerkhoffs GM, Lin CW. Surgical versus conservative treatment for high-risk stress fractures of the lower leg (anterior tibial cortex, navicular and fifth metatarsal base): a systematic review. Br J Sports Med. 2015;49:370–6.
42. Torg JS, Balduini FC, Zelko RR, et al. Fractures of the base of the fifth metatarsal distal to the tuberosity. Classification and guidelines for non-surgical and surgical management. J Bone Joint Surg Am. 1984;66(2):209–14.
43. Logan AJ, Dabke H, Finlay D, et al. Fifth metatarsal base fractures: a simple classification. Foot Ankle Surg. 2007;13(1):30–4.
44. Nagao M, Saita Y, Kameda S, et al. Headless compression screw fixation of jones fractures: an outcomes study in Japanese athletes. Am J Sports Med. 2012;40(11):2578–82.
45. Tsukada S, Ikeda H, Seki Y, et al. Intramedullary screw fixation with bone autografting to treat proximal fifth metatarsal metaphyseal-diaphyseal fracture in athletes: a case series. Sports Med Arthrosc Rehabil Ther Technol. 2012;4(1):25.
46. Ochenjele G, Ho B, Switaj PJ, et al. Radiographic study of the fifth metatarsal for optimal intramedullary screw fixation of Jones fracture. Foot Ankle Int. 2015;36(3):293–301.
47. Hunt KJ, Goeb Y, Esparza R, et al. Site-specific loading at the fifth metatarsal base in rehabilitative devices: implications for Jones fracture treatment. PM R. 2014;6(11):1022–9.
48. Brockwell J, Yeung Y, Griffith JF. Stress fractures of the foot and ankle. Sports Med Arthrosc. 2009;17:149–59.
49. Maitra RS, Johnson DL. stress fractures clinical history and physical examination. Clin Sports Med. 1997;16:260–74.
50. Gross TS, Bunch RP. A mechanical model of metatarsal stress fracture during distance running. Am J Sports Med. 1989;17:669–74.
51. Brukner P, Bradshaw C, Khan KM, et al. Stress fractures: a review of 180 cases. Clin J Sport Med. 1996;6:85–9.
52. Mandell JC, Khurana B, Smith SE. Stress fractures of the foot and ankle, part 2: site-specific etiology, imaging, and treatment, and differential diagnosis. Skelet Radiol. 2017;46:1165–86.
53. Kaeding CC, Yu JR, Wright R, et al. Management and return to play of stress fractures. Clin J Sport Med. 2005;15(6):442–7.
54. Khan K, Fuller P, Brukner P, et al. Outcome of conservative and surgical management of navicular stress fracture in athletes: 86 cases proven with computerized tomography. Am J Sports Med. 1992;20:657–66.
55. Khan K, Brukner P, Kearney C, et al. Tarsal navicular stress fracture in athletes. Sports Med. 1994;17:65–7.

56. Fowler JR, Gaughan JP, Boden BP, et al. The non-surgical and surgical treatment of tarsal navicular stress fractures. Sports Med. 2011;41(8):613–9.
57. Torg JS, Moyer J, Gaughan JP, et al. Management of tarsal navicular stress fractures: conservative versus surgical treatment: a meta-analysis. Am J Sports Med. 2010;38(5):1048–53.
58. Saxena A, Fullem B, Hannaford D. Results of treatment of 22 navicular stress fractures and a new proposed radiographic classification system. J Foot Ankle Surg. 2000;39(2):96–103.
59. Guermazi A, Hayashi D, Jarraya M. Sports injuries at the Rio de Janeiro 2016 summer Olympics: use of diagnostic imaging services. Radiology. 2018;287(3):922–32.
60. Iwamoto J, Takeda T. Stress fractures in athletes: review of 196 cases. J Orthop Sci. 2003;8(3):273–8.
61. Shelbourne KD, Fisher DA, Rettig AC, McCarroll JR. Stress fractures of the medial malleolus. Am J Sports Med. 1988;16:60–3.
62. Sherbondy PS, Sebastianelli WJ. Stress fractures of the medial malleolus and distal fibula. Clin Sports Med. 2006;25:129–37.
63. Kor A, Saltzman AT, Wempe PD. Medial malleolar stress fractures: literature review, diagnosis, and treatment. J Am Podiatr Med Assoc. 2003;93(4):292–7.
64. Shabat S, Sampson KB, Mann G, et al. Stress fractures of the medial malleolus: review of the literature and report of a 15-year-old elite gymnast. Foot Ankle Int. 2002;23(7):647–50.
65. Orava S, Karpakka J, Taimela S, et al. Stress fracture of the medial malleolus. J Bone Joint Surg Am. 1995;77(3):362–5.
66. D'Hooghe P, Wiegerinck JI, Tol JL, et al. 22 year old professional soccer player with atraumatic ankle pain. Br J Sports Med. 2015;49(24):1589–90.
67. Kerkhoffs GM, Versteegh VE, Sierevelt IN, Kloen P, van Dijk CN. Treatment of proximal metatarsal V fractures in athletes and non-athletes. Br J Sports Med. 2012;46(9):644–8.
68. McInnis KC, Ramey LD. High-risk stress fractures: diagnosis and management. PM R. 2016;8:S113–24.
69. Junge A, Rösch D, Peterson L, Graf-Baumann T, Dvorak J. Prevention of soccer injuries: a prospective intervention study in youth amateur players. Am J Sports Med. 2002;30(5):652–9.
70. Soligard T, Myklebust G, Steffen K, Holme I, Silvers H, Bizzini M, Junge A, Dvorak J, Bahr R, Andersen TE. Comprehensive warm-up programme to prevent injuries in young female footballers: cluster randomised controlled trial. BMJ. 2008;337:a2469.

72 Clinical Examination: Evaluation of Neurologic Conditions of the Foot (Interdigital Neuromas, Charcot-Marie-Tooth Disease)

Giovanna Stelitano, Calogero Di Naro, Vincenzo Candela, Casciaro Carlo, Laura Risi Ambrogioni, Giuseppi Longo, and Vincenzo Denaro

72.1 Introduction

The foot is an intricate "machine" made up of several bones and joints held in perfect equilibrium by extrinsic (leg) and intrinsic (foot) muscles. Any kind of alteration on muscle balance can entail the onset of foot deformity. If the disruption occurs before skeletal maturity, the bone growth will be irregular, while joints will result dysplastic. If the imbalance occurs when skeletal maturity is complete, the bones will be mostly regular in shape, and the joints will bear any deformation. Over time, the joints can decay as a consequence of deformity, causing secondary bone dysplasia. At the level of the ankle and hindfoot, different pairs of muscles support each other. The anterior tibialis dorsiflexes the first ray, while the peroneus longus flexes it plantarly. The posterior tibial muscle reverses the foot, and the peroneal brevis everts it. The triceps of the sura is the most potent muscle in legs and feet. The anterior tibialis, the extensor of the fingers, and the long extensor of the big toe work together, opposing to the triceps of the sura. In the forefoot, intrinsic and extrinsic flexor and extensor balance makes the fingers aligned with the metatarsophalangeal (MTP) and interphalangeal joints. Although theoretically muscle imbalance can cause numerous kinds of deformity of the foot, neuromuscular deformities generally manifest themselves with a cavus foot. The neuromuscular cavus foot is usually characterized by a high arch, varying quantities of varus of the hindfoot, plantar flexion of the first ray, and claw toes. Even if numerous neuromuscular feet present the same general shape as the cavus foot, particular elements of cavus differ on the base of the imbalance mechanism. Weakness is the main sign in neuromuscular disease (NMD); it usually occurs in infancy and negatively affects gait and functional walking. Muscular stiffness, contracture, musculoskeletal malformations, poor standing balance, and diminished resistance also affect practical outpatient activities such as standing and walking. For the majority of NMD of childhood, the pathology is progressive and leads to growing weakness and inability throughout life. Gait alteration has significative impacts on the function in daily life. In this chapter, we

G. Stelitano · C. Carlo · G. Longo (✉)
Department of Orthopaedic and Trauma Surgery, Campus Bio-Medico University, Rome, Italy

Università Campus Bio-Medico Roma, Rome, Italy
e-mail: g.longo@policlinicocampus.it

C. Di Naro · V. Candela · L. R. Ambrogioni · V. Denaro
Department of Orthopaedic and Trauma Surgery, Campus Bio-Medico University, Rome, Italy

J. G. Lane et al. (eds.), *The Art of the Musculoskeletal Physical Exam*,
https://doi.org/10.1007/978-3-031-24404-9_72

focus our attention on the most important and diffused neurological foot conditions: Charcot-Marie-Tooth disease and interdigital neuroma (Morton's neuroma).

72.2 Charcot-Marie-Tooth Disease

Charcot-Marie-Tooth (or CMT) is one of the most diffuse genetic neurological diseases, characterized by heterogeneous groups of hereditary motor and sensory neuropathies with an estimated prevalence of 0.5–1 in 2500 [1]. CMT was described for the first time by Charcot and Marie in France in the nineteenth century, and then it was independently conducted by Tooth in the United Kingdom during 1886. In particular, Tooth had referred to atrophy of the gastrocnemius muscle, which had caused atrophy of weak muscles. Afterwards, Dyck and Lambert had focused their attention on the main electrophysiologic characteristics of the disease, basing on them their first classification [2]. They identified two different groups: CMT1 and CMT2. The first is characterized by slow nerve conduction velocities (NCVs) and evident signs of hypertrophic demyelination; the second is marked by physiological or moderately decreased NCVs along with clear evidence of axonopathy [3]. As specified above, CMT disease is a whole of hereditary neuropathies, which affect both sensory and motor peripheral nerves. Over 80 genes have been recognized as responsible for CMT onset [4]. The genetic transmission of the disease can have autosomal dominant, recessive, and X-linked inheritance patterns. Currently, on the basis of the genetic defect involved, it is divided into different classes: type 1A, type 1B, type 2, and type X. Nowadays, 17 genetic variants of the disease are known, and many more are probably still to be discovered. Among these, Charcot-Marie-Tooth type 1 (CMT1), the demyelinating type with autosomal dominant inheritance pattern, represents the most widespread with 70% of all inherited neuropathies which fall into this class.

CMT2 groups, as well as the different dominant forms of axonal neuropathies, often have overlapping clinical features of CMT1 form, except the preservation of myotatic reflexes and NCVs. CMT is slow but has a certain degree of progression, and usually does not cause death. Many patients have a usual standard of living, full and productive. The limbs may be involved, presenting weakness during the distal movements. Tendon reflexes are decreased as is typical of peripheral neuropathies. CMT usually also affects the feet, strongly.

About 94% of patients with this condition will develop foot deformities with the characteristic weakness of the tibialis anterior and peroneus brevis. When the tibialis anterior muscle is weak, the peroneus longus is aligned straight, causing plantar flexion of the first ray. This condition will lead to a high arch (cavus). In some cases, the imbalance among the tibialis anterior and the Achilles tendon can lead to equinus deformity onset, even before other deformities' development. With the imbalance of the muscles in the toes and the excessive activity of the long toe extensors, which substitute for the anterior tibialis, clawing of the toes may occur. Loss of sensation is not as severe as loss of mobility, but patients may show two points of discrimination and failure of vibratory sense.

CMT is defined as a length-dependent neuropathy as the longest nerves of the body are affected earliest and most severely than the shorter ones. Patients affected by CMT present a broad range of symptoms and signs; among these, the distal limb weakness and muscle atrophy represent the most relevant since the beginning of the disease. Distal limb weakness causes considerable difficulty on the voluntary toe walking, except when a heel cord contracture coexists. The lower limbs ordinarily are involved earlier than the upper ones during the progression of the pathology. Toddlers and older children usually have a different clinical presentation of the disease. In the former, retarded motor development, toe walking, and tripping or falling represent the earliest and not specific signs. Several pediatric pathological conditions, such as cerebral palsy or myopathies, can occur in the same way. On the other hand, older children predominantly show a slow ability to running, ankle damages, and inability in sports practice. Numerous typical foot abnormalities also mark CMT. Among these, the most frequent are the pes cavus and the hammertoes (claw feet). Less commonly, flat feet can appear. Muscular weakness plays a key role in the clinical presentation of CMT. Patients affected by the deficit of the tibialis anterior can show the fall of the foot, feeling awkward or tripping on their feet also during walking. They usually walk with a big step. Muscular weakness plays a key role in the clinical presentation of CMT. Patients affected by the deficit of the tibialis anterior can show the fall of the foot, feeling awkward or tripping on their feet also during walking. They usually walk with a big step. On the other hand, the weakness of the peroneal brevis could lead to the lack of eversion of the hindfoot, causing ankle ligament injuries and fractures. Alterations in the foot structure and the muscular weakness lead to irregular weight distribution and stress on the plantar region of the foot. Thanks to the use of the motion analysis, it is possible to identify three different patterns of the foot and ankle function: flail foot, cavovarus foot, and toe walking [5]. Gait disorders can also be classified considering the sagittal ankle plane kinematic and kinetic patterns. The foot dorsiflexion range in terminal stance could result in smaller than the physiological control one. In most patients (90% about), the ankle plantar flexor deficiency is the main kinematic finding and, often, it could be the unique gait irregularity existing [6]. At the same time, many patients refer to plantar flexor weakness responsible for walking inability and/or difficulties in sports activities, which presuppose running. Foot drop and equinus deformity represent two other common clinical signs which occur in about 60% of patients, causing trips and falls. Patients affected by CMT can show specific neurological signs. First of all, the myotatic reflexes can often result to be decreased, more frequently in ankles and knees rather than in biceps and triceps, even if they could remain in axonal forms of the disease. Sensorineural hearing loss and vocal cord or phrenic nerve paralysis could occur. Clinical evaluation is essential to make the diagnosis of CMT, but it requires accompanying specific diagnostic tests to confirm neuropathies. Electrodiagnostic test (EMG/NCV) is the most important test to perform in case of suspected CMT. It is essential to distinguish between demyelinating and axonal forms of the disease [5]; it helps to exclude other neuromuscular pathologies (motor neuron diseases, neuromuscular defects, myopathic disorders). Although the importance of the biopsy has decreased, pathological characteristics of nerve and muscles can be helpful to define specific kinds of neuropathies and to distinguish them from different diseases (vasculitic neuropathies). Features of peripheral neuropathy can be shown through nerve conduction tests. Neural biopsies may be useful in particular in CMT type 1, which often shows the characteristic swelling. Specific CMT subtypes can now be detected with a blood test. Finally, it is important to make a careful investigation of family history. Single genetic tests, such as the research of the connexin 32 gene (GJB1) and the MFN2 gene, are, respectively, indicative of demyelinating CMT and dominant axonal CMT, especially in a context of maternal transmission evidence [7].

72.3 Interdigital (Morton) Neuroma

Interdigital neuroma, also known as Morton's neuroma, is the compressive neuropathy of the digital plantar nerve. It is one of the most common causes of metatarsalgia, typically located in correspondence of the third intermetatarsal space. Patients refer to a burning and tingling sensation at the forefoot/toes level, occasionally spread to the legs. The affected toes may feel numb. Pain predominantly occurs at night and can be improved by massage and taking off shoes. Civinini was the first to analyze this condition, while later, Durlacher focused on the description of the clinical complex of symptoms [8]. Some years later, Thomas George Morton defined the disease as "a painful affection of the fourth metatarsophalangeal articulation," attributing the pain to the fourth metatarsophalangeal joint [9].

Morton's neuroma mainly affects women, with a female:male ratio of 4:1 and an average age of 50 years [10]. It occurs bilaterally in 21% of patients and affects third space in 66% of cases, the second in 32%, and the fourth in 2%. Multiple presentations are not common.

The etiology of Morton's neuroma is still unclear. Nerve irritation, described as interdigital neuritis, would seem to be provoked by the compression of the interdigital nerve, proximally to the distal tip of the transverse intermetatarsal ligament. Ischemic alterations in the perineural tissue and repeated neurovascular trauma have also been identified as possibly responsible for neuroma onset. Each process which reduces the space available for the nerves can lead to interdigital neuritis. For this reason, synovitis, ganglion cyst, or trauma swelling could increase the risk of disease. Excessive stress caused by incorrect wear of the shoe (high heels) is related to the development of interdigital toe neuritis. Excessive hyperactivity (walking, running, or crouching) on the metatarsophalangeal joint could also represent the cause of this problem. A reduction in shock absorption, due to atrophy of the fat pad of the forefoot, can produce an increase in the irritative pain of the nerves. The neuroma has a preference for the third intermetatarsal space as a double contribution of the digital nerve coming from both the medial and lateral plantar nerves, which creates greater traction. The second interspace is the next most frequent position. Concurrent injuries are not common, so neuralgia rather than interdigital neuritis should be suspected when the patient has symptoms at more than one space. Clinical evaluation is essential to make the diagnosis of Morton's neuroma [11]. The pain must be localized accurately. Patients affected generally do not refer to painful symptom on the metatarsal heads. As currently reported in the literature, it is important to investigate any forefoot deformities, as well as the presence of instability or arthritis of the metatarsophalangeal joint and exclude Frieberg's disease existence. Tenderness and dorsal bulging can occur. Enlargement of the interdigital space may also be present. The pain is elicited through the application of an axial pressure in correspondence of the involved intermetatarsal space. Mulder's sign is the specific clinical sign which allows the surgeon to make the diagnosis. It is performed exerting pressure while tightening the metatarsals with the other hand: in this way, a painful sensation and palpable clicking perception appear [12]. This clinical examination has a 94–98% of sensitivity. The other applicable test is the so-defined pressure test which implicates the application of a plantar pressure eliciting tenderness and paresthesia among the metatarsal heads and right next to the heads. Also, with this examination, the patient's pain is reproduced. The audible, painful click during simultaneous lateral, dorsal, and plantar compression of the metatarsal heads in the hit area is named Mulder's click. This sign definitively supports the diagnosis [12]. Radiological evaluation in patients affected by Morton's neuroma is considered not necessary. The resonance magnetic imaging currently has shown a sensitivity of 93% and specificity of 68%. On the other hand, ultrasound sensitivity has been estimated at 90% and specificity 88%. Ultrasound, also in qualified hands, shows an elevated rate of accidental findings of asymptomatic nerve enlargement between the fingers, leading to a wrong definition of interdigital neuroma. At the same time, although little changes are challenging to be diagnosed through imaging, they can cause symptoms as more massive changes. The clinical trial remains the gold standard in the process of diagnosis. For this reason, the absolute request of imaging for patients with a high clinical suspicion is not necessary. Imaging evaluation should be performed only in two specific cases: patients with doubtful clinical evaluation and patients with multiple spaces affected. Ultrasound examination should be the first examination. On the other hand, MRI examination for Morton's neuroma does not presume symptoms. The exact relationship between clinical and MR imaging results is necessary before Morton's neuroma appears clinically relevant [13]. X-rays can make a useful contribution only for those patients in which it is essential to exclude other bone disorders. Local anesthetics to relieve pain is helpful for diagnosis. Specifically, the use of local infiltration with

2 mL of lidocaine below the intermetatarsal ligament has a compelling diagnostic power, driving the surgeon to Morton's neuroma diagnosis when it causes momentary pain alleviation. Nerve conduction investigations could show clear lateness in the feet and aid diagnosis. Subjects with interphalangeal neuritis could have increased forefoot pressure due to gastrocnemius muscle tension. Hypermobility of the first metatarsus may occur, causing an overload of the second and third metatarsal heads. Pain in the second or third intermediate space could be caused by other several different conditions such as neuritis or metatarsal synovitis. The interdigital neuroma is a presumably overdiagnosed disease. It is essential not to forget all factors which can provoke metatarsal pain, limiting the diagnosis of Morton's neuroma only to those patients with tenderness in a single space and without any other concomitant diagnosis.

References

1. Jani-Acsadi A, Ounpuu S, Pierz K, Acsadi G. Pediatric Charcot-Marie-tooth disease. Pediatr Clin N Am. 2015;62(3):767–86. PubMed PMID: 26022174. Epub 2015/04/15. eng
2. Dyck PJ, Lambert EH. Lower motor and primary sensory neuron diseases with peroneal muscular atrophy. I. Neurologic, genetic, and electrophysiologic findings in hereditary polyneuropathies. Arch Neurol. 1968;18(6):603–18. PubMed PMID: 4297451. eng
3. Martyn CN, Hughes RA. Epidemiology of peripheral neuropathy. J Neurol Neurosurg Psychiatry. 1997;62(4):310–8. PubMed PMID: 9120441. Pubmed Central PMCID: PMC1074084. eng
4. Timmerman V, Strickland AV, Züchner S. Genetics of Charcot-Marie-Tooth (CMT) disease within the frame of the human genome project success. Genes (Basel). 2014;5(1):13–32. PubMed PMID: 24705285. Pubmed Central PMCID: PMC3978509. Epub 2014/01/22. eng
5. Saporta MA, Shy ME. Inherited peripheral neuropathies. Neurol Clin. 2013;31(2):597–619. PubMed PMID: 23642725. PubMed Central PMCID: PMC3646296. Epub 2013/03/05. eng
6. Yagerman SE, Cross MB, Green DW, Scher DM. Pediatric orthopedic conditions in Charcot-Marie-tooth disease: a literature review. Curr Opin Pediatr. 2012;24(1):50–6. PubMed PMID: 22189393. eng
7. Lawson V, Gharibshahi S. Alphabet soup: making sense of genetic testing in CMT. Semin Neurol. 2010;30(4):373–86. PubMed PMID: 20941670. Epub 2010/10/12. eng
8. Pasero G, Marson P. Filippo Civinini (1805–1844) and the discovery of plantar neuroma. Reumatismo. 2006;58(4):319–22. PubMed PMID: 17216022. ita
9. Thomas G, Morton, M.D. The Classic. A peculiar and painful affection of the fourth metatarso-phalangeal articulation. Clin Orthop Relat Res. 1979;(142):4–9. PubMed PMID: 387325. eng.
10. Kasparek M, Schneider W. Surgical treatment of Morton's neuroma: clinical results after open excision. Int Orthop. 2013;37(9):1857–61. PubMed PMID: 23851648. Pubmed Central PMCID: PMC3764278. Epub 2013/07/13. eng
11. Giannini S, Bacchini P, Ceccarelli F, Vannini F. Interdigital neuroma: clinical examination and histopathologic results in 63 cases treated with excision. Foot Ankle Int. 2004;25(2):79–84. PubMed PMID: 14992707. eng
12. JD M. The causative mechanism in Morton's metatarsalgia. J Bone Joint Surg (Br). 1951;33 B(1):94–5. PubMed PMID: 14814167. eng
13. Bencardino J, Rosenberg ZS, Beltran J, Liu X, Marty-Delfaut E. Morton's neuroma: is it always symptomatic? AJR Am J Roentgenol. 2000;175(3):649–53. PubMed PMID: 10954445. eng

MIX
Papier aus verantwortungsvollen Quellen
Paper from responsible sources
FSC® C105338

If you have any concerns about our products,
you can contact us on
ProductSafety@springernature.com

In case Publisher is established outside the EU,
the EU authorized representative is:
Springer Nature Customer Service Center GmbH
Europaplatz 3, 69115 Heidelberg, Germany

Printed by Libri Plureos GmbH
in Hamburg, Germany